17th DAAAM International Symposium on Intelligent Manufacturing and Automation (DAAAM 2006)

Focus on Mechatronics and Robotics

Vienna, Austria
8 – 11 November 2006

Editor:

Branco Katalinic

ISBN: 978-1-5108-3999-1

AA004857

Printed from e-media with permission by:

Curran Associates, Inc.
57 Morehouse Lane
Red Hook, NY 12571

Some format issues inherent in the e-media version may also appear in this print version.

Copyright© (2006) by Danube Adria Association for Automation and Manufacturing (DAAAM)
All rights reserved.

Printed by Curran Associates, Inc. (2017)

For permission requests, please contact DAAAM International Vienna
at the address below.

DAAAM International Vienna
Karlsplatz 13/311, A-1040
Wien, Austria, European Union

Phone: 0043-1-58801-31121
Fax: 0043-1-58801-31199

secretary@daaam.info

Additional copies of this publication are available from:

Curran Associates, Inc.
57 Morehouse Lane
Red Hook, NY 12571 USA
Phone: 845-758-0400
Fax: 845-758-2633
Email: curran@proceedings.com
Web: www.proceedings.com

ISSN 1726-9679

DANUBE ADRIA ASSOCIATION FOR AUTOMATION & MANUFACTURING

DAAAM INTERNATIONAL VIENNA

www.daaam.com

ANNALS OF DAAAM FOR 2006 & PROCEEDINGS

OF THE 17TH INTERNATIONAL DAAAM SYMPOSIUM
"INTELLIGENT MANUFACTURING & AUTOMATION:
FOCUS ON MECHATRONICS AND ROBOTICS"
8-11TH NOVEMBER 2006, VIENNA, AUSTRIA

ORGANIZED BY:
DAAAM INTERNATIONAL VIENNA
VIENNA UNIVERSITY OF TECHNOLOGY
UNIVERSITY OF APPLIED SCIENCES TECHNIKUM, AND
AUSTRIAN SOCIETY OF ENGINEERS AND ARCHITECTS - ÖIAV 1848

UNDER THE AUSPICES OF:
THE DANUBE RECTORS' CONFERENCE &
RECTORS' HONOR COMMITTEE OF DAAAM INTERNATIONAL

EDITOR: B. KATALINIC

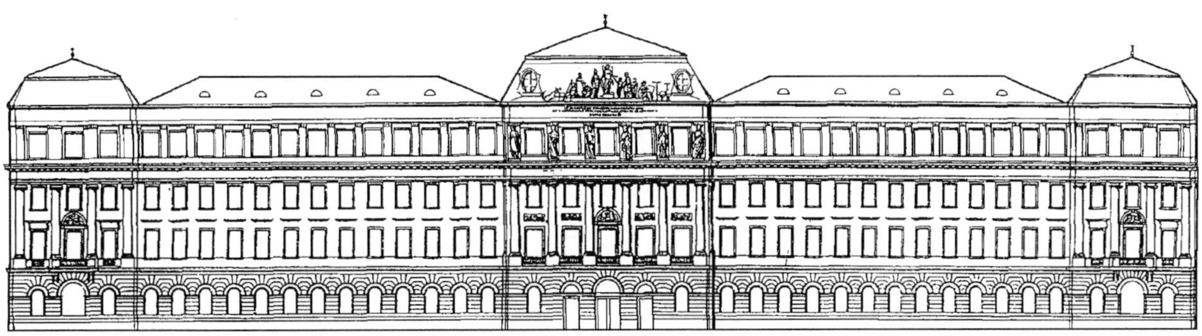

DAAAM INTERNATIONAL TABULA GRATULATORIA

Happy Birthday to:

Dr Adam Slota (35 Years) / Professor Bogomir Mursec (35 Years) / Hendrik Jähn (35 Years) / Professor Tobias Teich (40 Years) / Dr Marcin Andrzej Królikowski (40 Years) / Professor Bojan Acko (45 Years) / Professor Toma Udiljak (50 Years) / Professor Joze Balic (55 Years) / Professor Josip Brnic (55 Years) / Dr Tatjana Karaulova (55 Years) / Professor Karol Velíšek (55 Years) / Professor Soemon Takakuwa (55 Years) / Professor Zvonko Dragcevic (60 Years) / Professor Kanji Ueda (60 Years) / Professor Bozo Vranjes (65 Years) / Professor Frantisek Hruska (65 Years) / Professor Emil Wessely (70 Years) / Professor Jaroslav Balate (75 Years) / Professor Rudolf Julius Beer (80 Years)

Jena University of Applied Sciences (15 Years) / Faculty of Material Sciences and Technology Trnava - Slovak Technical University Bratislava (20 Years), University of Maribor (30 Years) / Faculty of Mechanical Engineering, Tallinn University of Technology Tallinn Technical College (70 Years) / Technical Faculty University of Zagreb (80 Years)

Congratulations to:
Professor Dzemo Tufekcic - Election for Rector Magnificus University of Tuzla
Professor Jozef Gawlik - Election for Rector Magnificus Cracow University of Technology
Professor Csaba Gyenge - Doctor Honoris Causa of University of Miskolc

Welcome to the 3358th DAAAM Author:
During the last fiveteen years **3358** authors and co-authors published their scientific papers in the Annals & Proceedings of DAAAM International Symposium.

Special thanks for long time mobility support to
Central European Exchange Program for University Studies - CEEPUS
Secretary General Mag. Elisabeth Sorantin - Central CEEPUS Office
Dr. Lydia Skarits - National CEEPUS office Austria

With love and appreciations
World-wide DAAAM International Community

Who is in the field, and not in the DAAAM he/she does not exist.
(Jyri Papstel 1996)

EDITOR'S NOTE This publication was reproduced by the photo process, using the manuscripts and soft copies supplied by their authors. The layout, the figures and tables of some papers did not conform exactly with the standard requirements. In some cases was the layout of the manuscripts rebuild. All mistakes in manuscripts there could not been changed, nor could the English be checked completely. The readers are therefore asked to excuse any deficiencies in this publication which may have arisen from the above causes. The editor and the DAAAM International Vienna are not responsible either for the statements made or for the opinion expressed in this publication.

Copyright © 2006 by the DAAAM-2006 editor Abstracting and nonprofit use of the material is permitted with credit to the source. Libraries are permitted to photocopy for private use of patrons. Instructors are permitted to photocopy isolated articles for noncommercial classroom use without fee. After this work has been published by the DAAAM International Vienna, the authors have the right to republish it, in whole or part, in any publication of which they are an author or editor, and to make other personal use of the work. Any republication, referencing, or personal use of the work must explicitly identify prior publication in the *Annals of DAAAM for 2006 & Proceedings of the 17th International DAAAM Symposium, ISSN 1726-9679, Editor B.[ranko] Katalinic, Published by DAAAM International, Vienna, Austria 2006,* including page numbers.

DAAAM International is a member of CrossRef. Contributions in DAAAM Scientific Book 2006 and DAAAM International Journal of Simulation Modelling will be crossreferenced and available on Internet. DAAAM International publications are cited and referenced in the: Inspec Database and Cambridge Scientific Abstracts

ANNALS OF DAAAM FOR 2006 & PROCEEDINGS OF THE 17TH INTERNATIONAL DAAAM SYMPOSIUM, VOL 17, NO1,
Abbreviated key title: ***Ann. DAAAM proc. int. DAAAM Symp.***
ISSN 1726-9679, Editor: B.[ranko] Katalinic
Published Annually by DAAAM International Vienna, Vienna 2006

Layout & Design: B.[ranko] Katalinic / Technical Editors: Lazinica A.[leksandar] Stuja K.[emajl] / DAAAM Home Page Administrator: Malisa V.[iktorio] / Cover graphics: Vienna University of Technology - Main Building / Graphics used in this Publication: M.C. Escher; T. Krauth and F.S. Meyer

Publisher: DAAAM International Vienna, TU Wien, Karlsplatz 13/311, A-1040 Vienna, Austria
E-mails: president@daaam.com katalinic@mail.ift.tuwien.ac.at / DAAAM homepage: http://www.daaam.com

The 17th International DAAAM Symposium

"INTELLIGENT MANUFACTURING & AUTOMATION: FOCUS ON MECHTRONICS AND ROBOTICS"

8-11th November 2006, Vienna, Austria

Organized by:

VIENNA UNIVERSITY OF TECHNOLOGY	DAAAM INTERNATIONAL VIENNA	UNIVERSITY OF APPLIED SCIENCES TECHNIKUM	AUSTRIAN SOCIETY OF ENGINEERS AND ARCHITECTS

Under the auspices of
Danube Rectors' Conference &
Rectors' Honor Committee of DAAAM International for 2006

Vladimír Báleš	Rector, Slovak University of Technology Bratislava
Gabriele Beibst	Rector, Jena University of Applied Science
Lajos Besenyei	Rector, University of Miskolc
Vladimir Bobal	Vice-Rector, Tomas Bata University University Zlin
Ion Bostan	Rector, Technical University of Moldova
Antonio Caştorani	Rector, Polytechnic of Bari
Tomáš Čermak	Rector, Technical University of Ostrava
Giovanni Del Tin	Rector, Politecnico di Torino
Jerzy Dembczyński	Rector, Poznan University of Technology
Mikhail I. Demchuk	Rector, Byelorussian State Polytechnic Academy
Akos Detreköi	Rector, Budapest Univ. of Technology and Economics
Karol Florian	Vice-Rector, Technical University of Kosice
Jozef Gawlik	Rector, Cracow University of Technology
Boris V. Gusev	President, Russian Academy of Engineering, Moscow
Said Irandoust	President, Asian Institute of Technology
Andres Keevallik	Vice Rector, Tallinn Technical University
Zejnel Kelmendi	Rector, University of Prishtina
Michael L. Kheifetz	Vice-Rector, Polotsk State University
Christian Kollmitzer	Vice-Rector, Univ. of Applied Sciences Technikum, Vienna
Gordana Kralik	Rector, Strossmayer University of Osijek
Frano Ljubic	Rector, University of Mostar
Damir Magas	Rector, University of Zadar
Salvatore Marzano	Rector, Polytechnic of Bari
Jasna Helena Mencer	Rector, University of Zagreb
Radu Munteanu	Rector, Technical University of Cluj-Napoca
Hasan Muratović	Rector, University of Sarajevo
Tomoyuki Nogami	President, Kobe University
Ivan Pavic	Rector, University of Split
Karel Rais	Rector, Brno University of Technology
Ivan Rozman	Rector, University of Maribor
Daniel Rukavina	Rector, University of Rijeka
Petr Saha	Rector, Tomas Bata University University Zlin
Takeshi Sasaki	President, University of Tokyo
Fritz Schmöllebeck	Rector, Univ. of Applied Sciences Technikum, Vienna
Juraj Sinay	Rector, Technical University of Kosice
Peter Skalicky	Rector, Vienna University of Technology
Yurij M. Solomenzev	Rector, Moscow State University of Technology
Peep Sürje	Rector, Tallinn Technical University
Mario T. Tabucanon	Provost, Asian Institute of Technology
Wieslaw Trampczynski	Rector, Technical University of Kielce
Dzemo Tufekcic	Rector, University of Tuzla
Ante Uglesic	Vice-Rector, University of Zadar
Vojo Visekruna	Vice-Rector, University of Mostar
Wojciech Zieliński	Rector, Silesian University of Technology, Gliwice

Annals of DAAAM for 2006 & Proceedings of the 17th International DAAAM Symposium, ISSN 1726-9679
Editor B. Katalinic, Published by DAAAM International, Vienna, Austria 2006
150th Birthday of Nikola Tesla (1856-1943) Inventor of Mobile Robot Teleoperation

Deans' and Academic Directors' Honor Committee for 2006

Badida, Miroslav
Bender, Klaus
Benic, Djuro
Berce, Petru
Bliedtner, Jens
Bolanca, Stanislav
Botezatu, Cornelia P.
Boyadjiev, Ilia
Budic, Ivan
Capek, Jan
Catovic, Fuad
Cevik, Kemal
Chlebus, Edward
Clausen, Uwe
Colak, Ivo
Cosic, Ilija
Curko, Tonko
Döbröczöni, Ádám
Domazet, Zeljko
Dulcic, Zelimir
Filaretov, Vladimir.F.
Galantucci, Luigi Maria
Gecevska, Valentina
Gordienko, Anatolii I.
Grösel, Bruno
Heusch, Christian
Hodolic, Janko
Jarvis, Christine W.
Jörgl, Hanns Peter
Kals, H.J.J.
Kavran, Zvonko
Kenter, I. Muhlis
Kovalenko, Volodymyr S.
Kozak, Drazan
Kozar, Ivica
Kraftova, Ivana
Krasniqi, Fejzullah
Louda, Petr
Majstorovic, Vlado
Marcic, Milan
Marquardt, Hans-Georg
Meghdari, Ali
Michailov, Y.K.
Michalowski, Stanislaw
Mikac, Tonci
Milcic, Diana
Moravcik, Oliver
Munitić, Ante
Nordmann, Rainer
Penninger, Antal
Petrescu, Valentin
Phien, Huynh Ngoc
Pivonka, Petr
Popp, Karl
Przybylski, W.
Rogale, Dubravko
Ružbarský, Juraj
Seda, Milos
Sljivic, Milan
Spiryagin, Maksym
Szadkowski, Jan
Tipuric, Darko
Vackar, Josef
Vasek, Vladimir
Veza, Ivica
Vitecek, Antonin
Weigang, Wu
Wiendahl, Hans-Peter

Scientific DAAAM International Committee for 2006

Alander, Jarmo Tapani (FIN)
Arai, Tamio (JPN)
Arndt, Günter (AUS)
Bachynski, Morrel (CAN)
Balate, Jaroslav (CZE)
Balic, Joze (SLO)
Beer, Rudolf Julius (AUT)
Belajev, Gennady J. (BLR)
Bilberg, Arne (DEN)
Bliedtner, Jens (DEU)
Bohez, Erik L. J. (THA)
Bouzakis, Konstantin (GRC)
Boyadjiev, Ilia (BGR)
Bright, Glen (RSA)
Browne, J. (IRL)
Celar, Stipe (HRV)
Covic, Dragan (BIH)
Cruz, Fernando M. M. (POR)
Cselenyi, Jozsef (HUN)
Cus, Franci (SLO)
Cyklis, Jerzy (POL)
Dar-El, E. (ISR)
De Filippi, A. (ITA)
Dikic, Mahmud (CAN)
Dincmen, M. (TUR)
Dong, Jianhua (CHN)
Dornfeld, David (USA)
Dove, Rick (USA)
Dudas, Illes (HUN)
Egorov, Sergey B. (RUS)
Emura, Takashi (JAP)
Endter, R. (DEU)
Eversheim, W. (DEU)
Filaretov, Vladimir F. (RUS)
Gawlik, Jozef (POL)
Giacaglia, G. E. (BRA)
Gibson, I. (AUS)
Grabenweger, Johann (BRA)
Gusev, Boris V. (RUS)
Gyenge, Csaba (ROM)
Han, Qing Yao (CHN)
Hesselbach, J. (DEU)
Iltschev, Velko (BGR)
Irfan, Mohammad A. (PAK)
Janac, Alexander (SVK)
Kals, H. J. J. (NDL)
Katalinic, Branko (AUT)
Kimura, Fumihiko (JPN)
Kinnander, Anders (SWE)
Knosala, Ryczard (POL)
Koch, Jan (POL)
Kondratenko, Yurij (UKR)
Kopylov, Yurii M. (RUS)
Krolikowski, M.A. (POL)
Kulenovic, Malik (BIH)
Kuljanic, Elso (ITA)
Kuosmanen, Petri (FIN)
Kusiak, Andrew (USA)
Kyttner, Rein (EST)
Lazea, Gheorghe (ROM)
Lee, Young H. (KOR)
Ljubic, Frano (BIH)
Llorente, J. I. (ESP)
Lombardi, F. (ITA)
Low, Kin Huat (MAL)
Lu, S. S. (TWN)
Majstorovic, Vlado (BIH)
Meghdari, Ali (IRI)
Meissner, Klaus (DEU)
Munteanu Radu (ROM)
Mertins, Kai (DEU)
Michelini, R. C. (ITA)
Monostori, Laszlo (HUN)
Nanasi, Jozef (SVK)
Neumann, H. J. (DEU)
Nicoletto, G. (ITA)
Novakovic, Branko (HRV)
Oczos, Kazimierz E. (POL)
Otto, Tauno (EST)
Pandey, P. C. (IND)
Park, Hong-Seok (KOR)
Pegden, Denis (USA)
Peklenik, Janez (SLO)
Poduraev, Yuriy V. (RUS)
Polajnar, Andrej (SLO)
Porter, Brian (HKG)
Pritschow, Günter (GER)
Przybylski, W. (POL)
Radermacher, F. J. (DEU)
Radhakrishnan, V. (IND)
Ruiz Herrera, Gilberto (MEX)
Russell, Stuart
Sam Salem (AUS)
Solomenzev, Juri. M. (RUS)
Tabucanon, Mario T. (THA)
Takakuwa, Soemon (JAP)
Takala, J. (FIN)
Tikka, H. (FIN)
Timoschenko, Victor (MOL)
Torres-Leza, F. (ESP)
Torvinen, S. (FIN)
Tsai, B. S. Y. (TWN)
Udiljak, Toma (HRV)
Ueda, Kanji (JPN)
Uyarr, Erol (TUR)
Van Brussel, Hendrik (BEL)
Van Oudheusden, Dirk L. (BEL)
Varga, Gyula (HUN)
Vavrin, Petr (CZE)
Velisek, Karol (SVK)
Venkatesh, V. C. (SGP)
Vitecek, Antonin (CZE)
Vranjes, Bozo (HRV)
Waldir, P. O. (BRA)
Wertheim, Rafi (ISR)
Weseslindtner, Helmar (AUT)
Wessely, Emil (SVK)
Westkämper, Engelbert (DEU)
Wiendahl, Hans-Peter (DEU)
Williams, D. J. (GBR)
Winston, Patrick H. (USA)
Zalzala, Ali M. S. (GBR)
Zitek, Pavel (CZE)
NB Country A3 codes ISO 3166

Annals of DAAAM for 2006 & Proceedings of the 17th International DAAAM Symposium, ISSN 1726-9679
Editor B. Katalinic, Published by DAAAM International, Vienna, Austria 2006
150th Birthday of Nikola Tesla (1856-1943) Inventor of Mobile Robot Teleoperation

INTERNATIONAL PROGRAM COMMITTEE FOR 2006

ACKO, BOJAN
ALBERS, ALBERT
ANISIC, ZORAN
AFZULPURKAR, NITIN
AURITE, TRAIAN
BADIDA, MIROSLAV
BALC, NICOLAE
BARISIC, BRANIMIR
BENIC, DJURO
BENKÖ, IMRE
BIBU, MARIUS
BLEY, HELMUT
BOBAL, VLADIMIR
BRANDASU, D.
BREZOCNIK, MIRAN
BRNIC, JOSIP
BUDIC, IVAN
BUNJAKU, AVDYL
BUTALA, PETER
CALA, IVO
CALVET, JOAN VIVANCOS
CAMPBELL, R. IAN
CAR, ZLATAN
CHAJDA, J.
CHOI, HOO-GON
COHODAR, MAIDA
COLEMAN, J. J.
COSIC, ILIJA
COSIC, PEJO
COTET, COSTEL EMIL
CUKOR, GORAN
DASIC, PREDRAG
DAVIS, STEVE
DEHELEAN, D.
DENKENA, BEREND
DOBI, DJORDJE
DRAGOI, GEORGE STELICA
DRAGCEVIC, ZVONKO
DRSTVENSEK, IGOR
DUFLOU, JOOST
DUHOVNIK, JOZE
DUNKIN, ANN
DÜRR, HOLGER
EGOROVA, TATYANA
FOLONOV, I.
FORRESTER, JILL
FRANK, WILLY
FUJII, NORIO
GALETTO, FAUSTO
GARIBALDO, F.
GERSAK, JELKA
GHEONEA, A.
GIURGIUMAN, HORIA T.
GONZÁLEZ, CRISTINA
GOLABCZAK, ANDRZEJ
GRIZELJ, BRANKO
GRONALT, MANFRED
GRZESIK, WIT
HATONO, ITSUO
HESSE, R.
HLAVA, JAROSLAV
HODOLIC, JANKO
HOFREITER, MILAN
HONCZARENKO, JERZY
IWATA, SHUICHI
JANACOVA, DAGMAR
JANECKI, DARIUSZ
JARVIS, CHRISTINE W.
JERBIC, BOJAN
JURAGA, IVAN
KATALINIC, BRANKO
KARAULOVA, TATYANA
KAWALEC, ANDRZEJ
KILIC, S. E.
KIRK, CHRIS. S.
KLICEK, BOZIDAR
KLIMENKO, S. A.
KLJAJIN, MILAN
KOHLMAIER, MARTIN
KOLIBAL, ZDENEK
KOLOMAZNIK, KAREL
KONDO, I.
KORDIC, VEDRAN
KOSMOL, JAN
KOZAK, DRAZAN
KRASNIQI, Z.
KRUCINSKA, IZABELLA
KUBALCIK, M.
KULENOVIC, MALIK
KUNDRAK, JANOS
KUZMAN, KARL
LAMPINEN, JUONI
LEE, KYO IL
LEGUTKO, STANISLAW
LEINONEN, T.
LI, GUIHUA
LJOLJIC, BERISLAV
LO VALVO, ERNESTO
LOPEZ DE LACALLE, LUIS N.
LOW, KIN HUAT
MAJDANDZIC, NIKO
MAJETIC, DUBRAVKO
MALAKOV, IVO
MALISA, VIKTORIO
MALOPOLSKI, WALDEMAR
MARJANOVIC, DORIAN
MARN, FERDINAND
MARSLAND, DAVID
MEHTA, NARINDER KUMAR
MEIER, HORST
MENZ, PETER
MICIETA, BRANISLAV
MIKAC, TONCI
MOK, H.
MORIWAKI, T.
MRSA, ZORAN
NAKAMOTO, NOBUYUKI
NOH, SANG DO
NURUL AMIN A.K.M.
OKAMOTO, J.
OSMANI, HYSNI
OSMERA, PAVEL
OTTOSSON, S.
PASCUAL, R.COMPAMYS
PAŠIĆ, MUGDIM
PAVIC, IVAN
PERIC, NEDJELJKO
PETERS, JACQUES
PETKOVIC, DARKO
PRLJACA, NASER
PRUTEANU, O.
PYTKO, STANISLAW
QUINN, ROGER D.
RADIC, JURE
RADOVANOVIC, MIROSLAV
RAY, T. S.
ROGALE, DUBRAVKO
ROSADO, V.O.G.
RUDAS, IMRE J.
SCHAFRIK, R. E.
SHIBASAKA, T.
SHIH, MING CHANG
SIEMIATKOWSKI, M.
SIHN, WILFRIED
SINAY, JURAJ
SKOCZYNSKI, WACLAW
SKOLUD, BOZENA
SLUGA, ALOJZ
STARCEVIC, GORDAN
STOPPER, MARKUS
STRINA, G.
STUJA, KEMAJL
SYSALA, TOMAS
TADAMURA, KATSUMI
TAKALA, JOSU
TOMASIC, ZELJKO
TAN, KOK KIONG
TEICH, TOBIAS
TEISCHINGER, ALFRED
TRUNSTEIN, I.
TSUZUKI, M. S. G.
ULRICH, GÜNTHER
VAARIO, JARI
VAMOS, T.
VASEK, VLADIMIR
VEZA, IVICA
VISEKRUNA, VOJO
VOSNIAKOS, GEORGE C.
WEIGANG, WU
WEIGL, K. HEINZ
WEISS, ZENOBIA
WINKLER, T.
XUE, D.
YAMAZAKI, K.
YOSHIDA, SUSUMU
YU, Y.-C.
ZAMFIRA, C. SORIN
ZAJAC, JERZY
ZHANG, GUANGMING
ZHIRABOK, ALEXEY
ZICHIL, VALENTIN
ZJAKIC, IGOR

ORGANIZING COMMITTEE FOR 2006

BUCHMEISTER, BORUT
DRAGCEVIC, ZVONKO
EGOROV, SERGEY
HÖLLER, LIANE
KATALINIC, BRANKO (CHAIR)
LAZINICA, ALEKSANDAR
MALISA, VIKTORIO
SAGMEISTER, NICOLE
STOPPER, MARKUS
STUJA, KEMAJL
VIKTORIK, HANS
VIKTORIK, SILVIA

Annals of DAAAM for 2006 & Proceedings of the 17th International DAAAM Symposium, ISSN 1726-9679
Editor B. Katalinic, Published by DAAAM International, Vienna, Austria 2006
150th Birthday of Nikola Tesla (1856-1943) Inventor of Mobile Robot Teleoperation

Reviewing Committee for 2006

Bilberg, Arne
Bostan, Ion
Botezatu Cezar
Botezatu, Cornelia P.
Bright, Glen
Buchmeister, Borut
Car, Zlatan
Celar, Stipe
Constantin, George
Cozminca, Irina
Duflou, Joost
Dulgheru, Valeriu
Egorov, Sergey
Filaretov, Vladimir F.
Galetic, Lovorka
Grabenweger, Johann
Gronalt, Manfred
Hook A. K.
Katalinic, Branko
Knego, Nikola
Kolar, Dusan
Kordic, Vedran
Ludovico, Antonio D.
Malisa, Viktorio
Meissner, Klaus
Mrsa, Zoran
Nanasi, Jozef
Osmera, Pavel
Park, Hong-Seok
Pedgen, Denis
Rametsteiner, Werner
Ruiz Herrera, Gilberto
Stopper, Markus
Stuja, Kemajl
Udiljak, Toma
Ueda, Kanji
Ungureanu, Catalin
Visekruna, Vojo
Weigl, Karl Heinz
Wessely, Emil

Young Researchers´ and Scientists´ International Committee for 2006

The main aim of this International Committee is international promotion of talented young scientists and researchers. Target group are: young engineers, master students, young master, doctoral students, young doctors, talented graduate students and other tolerant and talented young persons from one of the fields of interests of DAAAM International. The young winners of Best Paper Award of DAAAM International are also the member of this committee. The candidate has to be under age 35 years and has to be author or co-author of paper published in one of DAAAM publications. The candidate has to be nominated by the supervisor/professor/team leader. The nomination (short CV, research: status, field, course, targets, results) has to be send to the President of DAAAM International.

Afshari, Ali
Amaitik, S.M.
Ambrosch, Roland
Anisic, Zoran
Bernin, Fredrik
Blecha, Petr
Bravo, Unai
Brezak, Danko
Brol, Sebastian
Brzoska, Witold
Bucur, Carmen Cristiana
Camacho, Ana M.
Carutasu, N. Luminita
Celaya, Ainhoa
Chiciudean, Teodor Gelu
Ciobanu Lidia Florentina
Di Mascolo, Maria
Dobránsky, Jozef
Dzierwa, Andrzej Piotr
Ficko, Mirko
Fritz, Sebastian
Galantucci, Luigi M.
Georges, M. R. Rosa
Giannoccaro, Rosanna
Gogouvitis, Xenofon V.
Gonzalez, Ander
Hebbal, Shivanand
Hilmersson, Daniel
Ilie Zudor, Elisabeta
Indelli, Alessandro
Jacobsen, Jörn
Jaehn, Hendrik
Jalmuzna, Irena
Kavaz, Adis
Kim, Kwang-Soo
Kirola, Marijo
Knorr, Koldo Garrido
Kodriková, Klára
Korajda, Bartosz
Kovacic, Miha
Kulin, Admir
Kumagai, Masaaki
Lamikiz, Aitzol
Lazinica, Aleksandar
Lee, Gyu-Bong
Leffin, Tanja
Lengyel, Attila
Lika, Elma
Longeri, Tiziana
Macchi, Marco
Maidagan, Paul
Marić, Ivana
Masuda, Daisuke
Mathur, Akshay
Matusu, Radek
Michtchenko, Anna M.
Mihoci, Kristijan
Milfelner, Matjaz
Munar, Aldrin Cantor
Nepevný, Petr
Nica, Gabriela Beatrice
Obrovac, Karlo
Ortega, Naiara
Pai, Kei-Ren
Palcic, Iztok
Parasuraman, S.
Parpala Radu Constantin
Patrascu, Gabriela
Phornprapha, M.
Pološki, Nina
Pospisil, Richard
Prunic, Igor
Radaslic, Kemal
Rahman, Minhazur
Raich, Prince Ashif
Rasti, Javad
Remezkov, Andrey V.
Rivero, Asun
Romanov, Dmytro O.
Rovnak, Martin
Saaristo, Hannu PJ
Salgano, Miguel Angel
Salopek, Ivana
Samuhavinyoo, Kanchana
Sansomboonsuk, Siriluk
Schmidt, Michal
Shilpi, Kalra
Singh, Maharaj
Singh, Pradeep Kumar
Stojkic, Zeljko
Tache, Florin
Teich, Cornelia
Toma, Adina Cristina
Topcic, Alan
Umut, Durak
Wang, Teng-Yen
Vidučić, Vinko
Wong, H. Ken
Vos, Jeroen
Vuković-Obrovac, Jadranka
Yoshino, T.
Zuperl, Uros

Annals of DAAAM for 2006 & Proceedings of the 17th International DAAAM Symposium, ISSN 1726-9679
Editor B. Katalinic, Published by DAAAM International, Vienna, Austria 2006
150th Birthday of Nikola Tesla (1856-1943) Inventor of Mobile Robot Teleoperation

DAAAM International Industry, Business and Administration Committee for 2006

Agarwal, Beni Prasad
Akhonin, Andrey A.
Andreic, Andrej
Arsin, M.
Bader, Udo
Beer, Herbert
Beer, Michael
Birgersson, M.
Blaschko, Christian
Brajinovic, Marijan
Bruck, Peter
Buljan, Jakov
Burgjelez, Ivana
Carder, M.W.
Carlbom, Terri
Celar, Stipe
Cosic, Pejo
Coskun, Deniz
Crafer, Roger C.
Crnjac, Miljenko Mico
Culjak, Vlado
Dadic, Mihovil Braco
Darko, Csaba
Dicak, Josip
Dikic, Masha
Divic, Antun
Foltin, A.
Freiberger, Dietrich
Gloser, Manfred
Goja, Zeljko
Grabenweger, Johann
Haas, B.
Haas, Peter
Hanslik-Kupsa, R.
Haslauer, H.
Hebemägi, A.
Holjevac, Nikola
Ivanisevic, Darko
Jukic, Ivan
Kalinyaprak, Volkan
Katalinic, Anica Aca
Katalinic, Mihovil Milan
Keiner, Wolfgang
Knezevic, Sime
Kohlmaier, Martin
Kremnjak, Zvonko
Krezic, Davor
Kunstmann, Ruediger
Lahoda, Thomas
Lengauer, Markus
Lerch, Zoran
Levstek, Anton
Limbeck, Matthias
Ljuhar, Davul
Loktev, Alexander A.
Lörincz, S.
Magyar, Gabor
Marasovic, Tomislav
Marincic, Valentina
Martak, Lothar V.
Mayer, A.
Mazuran, Zeljko
Milovac, Ante
Moisseev, Iouri
Moll, Wolfgang
Morinaga, H.
Mortin, M.
Moskovenko, Igor B.
Muresan, Vasile
Murinde, Victor
Nader, Hans
Nakamoto, Nabuyuki
Nanasi, Jozef
Neuper, Peter
Niwinski, Janusz
Novels, Mike
Pfliegel, Reinhard
Pieslinger, Kurt
Poettinger, Klaus
Rametsteiner, Werner
Raming, Otto
Redtenbacher, Guenter
Reichel, Peter
Richler, Hannes
Santek, Mladen
Sobolewski, Andrzej
Szava, Tibor
Tsydenov, Dorzhi T.
Widtmann, Georg
Worener, K.
Yoshida, H.
Yoshida, Susumu

The Members of Permanent Committee of the University Network of DAAAM International for 2006 are:

Member organisation	***Delegate***
DAAAM International Vienna	*President Univ. Prof. Dipl.-Ing. Dr. techn. Dr.mult. h.c. Branko Katalinic*
Danube Rectors' Conference	*Professor Dr. sc. Franci Cus*
Budapest University of Technology and Economics, Budapest	*The members of Permanent Committee have to be nominated by Budapest University of Technology and Economics*
Josip Juraj Strossmayer University of Osijek	*Vice-Rector Professor Dr. sc. Dragomir Krumes*
Kobe University, Kobe	*Director, Chair and Professor Kanji Ueda*
Moscow State University of Technology "Stankin", Moscow	*Rector Professor Dr. Sci. (tech.) Member of Russian Academy of Science Juri M. Solomentsev, O. Univ. Prof. Dipl.-Ing. Dr. techn. Dr. h.c. Sergey Egorov*
Slovak University of Technology Bratislava	*Professor Dr. Marian Vesely Professor Ladislav Javorcik*
Tallinn Technical University	*Professor Rein Kyttner*
Technical University of Brno, Brno	*Senator Prof. Ing. Jaroslav Balate, DrSc. Doc. Ing. Vladimir Vasek, CSc.*
Technical University of Cluj-Napoca, Cluj-Napoca	*Dean Prof. Dr.Eng. Petru Berce, Prof.dr.Eng. Csaba Gyenge*
Technical University of Kosice, Kosice,	*Vice-rector Prof. Ing. Karol Florian, DrSc., Senator Prof. Dr. Emil Wessely*
University of Maribor, Maribor	*Professor Dr. sc. Joze Balic*
University of Miskolc, Miskolc	*Dean Professor Dr.-Ing. Dr.h.c. mult. Jozsef Cselenyi, Professor Dr Gabor Terstyanszky*
University of Mostar, Mostar, Bosnia and Herzegovina	*Rector Prof. Dr. Frano Ljubic, Vice-Rector Prof. Dr. sc. Vojo Visekruna*
University of Prishtina, Prishtina, Kosovo	*Vice-Rector Prof. Dr. Ahmet Geca, Prof.-Dr. Bajrush Bytyci*
University of Rijeka, Rijeka	*Professor Josip Brnic, Ph.D., Professor Zoran Mrsa, Ph.D.*
University of Split, Split	*Rector Professor Dr. Ivan Pavic*
University of Zagreb, Zagreb	*Rector Professor Dr. Jasna Helena Mencer*

Annals of DAAAM for 2006 & Proceedings of the 17th International DAAAM Symposium, ISSN 1726-9679
Editor B. Katalinic, Published by DAAAM International, Vienna, Austria 2006
150th Birthday of Nikola Tesla (1856-1943) Inventor of Mobile Robot Teleoperation

DAAAM INTERNATIONAL VIENNA:
BACKGROUND, HISTORY, AIMS, STRUCTURE, ACTIVITIES & FUTURE

Branko KATALINIC
President of DAAAM International
www.daaam.com

0.0 Contents:

1.0 BACKGROUND

The DAAAM International was founded and based on the idea and initiative of Professor B. Katalinic, on November 5th, 1990 at the Vienna Symposium, in order to mark the 175th anniversary of the Vienna University of Technology.

The purpose of DAAAM International is to provide a world forum for engineers, scientists and industrial managers to present and discuss the current status and impact of advanced manufacturing and automation.

During the last twelve years, the activities of DAAAM International have been supported by scientists and experts from over fifty countries all over the world. In the course of the last five years, DAAAM International has focused activities on the regional and global long-term co-operation between (technical) universities. The result of this initiative is the founding of the University Network of DAAAM International Vienna and DAAAM International Network for Advanced Technologies which was founded on 2002-10-25 in Vienna.

2.0 AIMS

The main aims of DAAAM are:

- To organise an annual DAAAM International Symposium each year.
- To found and organise the University Network of DAAAM International Vienna as a strategic alliance and permanent international open platform for long-term co-operation and partnership in all fields of technology, research and education.
- To organise long-term projects, initially known as Inter-university Doctoral Studies
- To organise a pilot project, initially known as the Virtual University.
- To enable and promote the exchange of people, scientific, and technical information.
- To enable and promote international co-operation among scientists and engineers.
- To enable and promote R & D, education and industrial application.
- To co-operate with other national and international organisations in matters of common interest.

3.0 ORGANISATION OF DAAAM INTERNATIONAL VIENNA

DAAAM International Vienna
President: Branko **Katalinic**

Vice presidents for 2006:
Joze **Balic /** Klaus **Meissner** / Malik **Kulenovic** / Andrew **Kusiak** / Toma **Udiljak** / Kanji **Ueda**

Honour Vice presidents for 2006:
Ferenc **Alpek**, Jaroslav **Balate**, P. C. **Pandey,** Janez **Peklenik,** Emil **Wessely**

Secretary
Aleksandar Lazinica, *secretary general* / Kemail Stuja, *designer*
Team: Borut Buchmeister / Liane Höller / Aleksandar Lazinica / Viktorio Malisa / Nicole Sagmeister

Presidents of National DAAAM Organisations for 2006

Jaroslav Balate	*Czech Republic*
Arne Bilberg	*Denmark*
Jens Bliedtner	*Germany*
Glen Bright	*Republic of South Africa*
Fernando M. M. Cruz	*Portugal*
Franci Cus	*Slovenia*
Mahmud Dikic	*Canada*
Jianhua Dong	*P.R. China*
Joost Duflou	*Belgium*
Sergey Egorov	*Russia*
Horia Traian Giurgiuman	*Switzerland*
Jan Goossenaerts	*Holland*
Johann Grabenweger	*Brazil*
Csaba Gyenge	*Romania*
Qing-Yao Han	*P.R. China*
Peter Holmstedt	*Sweden*
Velko Iltschev	*Bulgaria*
Mohammad A. Irfan	*Pakistan*
Yurij Kondratenko	*Ukraina*
Petri Kuosmanen	*Finland*
Andrew Kusiak	*USA*
Rein Kyttner	*Estonia*
Marcin Krolikowski	*Poland*
Luis Norberto Lopez de Lacalle	*Spain*
Vlado Majstorovic	*Bosnia and Herzegovina*
Ali Meghdari	*Iran*
P. C. Pandey	*India*
Zoran Pandilov	*Macedonia*
Hong-Seok Park	*Korea*
Gilberto Hererra Ruiz	*Mexico*
Kemajl Stuja	*Kosovo*
Dmitry Svirsky	*Belarus*
Kok Kiong Tan	*Singapore*
Mario T. Tabucanon	*Thailand*
Victor Timoschenko	*Moldavia*
Kanji Ueda	*Japan*
Toma Udiljak	*Croatia*
Erol Uyarr	*Turkey*
Ernesto Lo Valvo	*Italy*
Gyula Varga	*Hungary*
George Christopher Vosniakos	*Greece*

DAAAM International Coordinators for 2006

Miroslav **Badida**	*Technical Environmentalistics*
Joze **Balic**	*Intelligent Manufacturing*
Ivo **Cala**	*Maintenance*
Mahmud **Dikic**	*Assembly Automation*
Zvonko **Dragcevic**	*Textile and Garment*
Sergey **Egorov**	*Securities & Financial Management*
Vladimir F. **Filaretov**	*Teleoperators*
Johann **Grabenweger**	*Automotive Industry*
Manfred **Gronalt**	*Logistics*
Branko **Katalinic**	*Intelligent Manufacturing Systems*
Viktorio **Malisa**	*Mechatronics / Robotics*
Klaus **Meissner**	*Precision Engineering*
Zoran **Mrsa**	*Optimal Shape Design*
Jozef **Nanasi**	*Automotive Industry*
Markus **Stopper**	*Production Information Technology*
Kanji **Ueda**	*Biological Manufacturing Systems*
Heinz **Weigl**	*Simulation*
Igor **Zjakic**	*Graphics Technologies*

NB: The further International Coordinators for different topics/fields are needed and welcome, if you want to be one of them please contact the president of DAAAM International

4.0 DAAAM Symposiums Clubs

Since founding of DAAAM International Symposium in 1990 **3358** authors and co-authors published their scientific papers in the Annals & Proceedings of DAAAM International Symposium. The members of the club are the persons which presented and/or published the papers at all / not les than 15 / not les than 10 / not less than 5 DAAAM International Symposiums:

All DAAAM Symposiums Authors´ Club

Balic, J. / Cus, F. / Katalinic, B. / Polajnar, A. / Veza, I. / Wessely, E.

Members of 15+ DAAAM Symposiums Authors´ Club

Balate, J. / Buchmeister, B. / Udiljak, T.

2007 Candidates for 15+ DAAAM Symposiums Authors´ Club

Following DAAAM authors are candidates and have a chance to join the 15+ Club in 2007
Alpek, Ferenc / Boyadjiev, Iliya K / Budic, Ivan / Gyenge, Csaba / Novakovic, Branko

Members of 10+ DAAAM Symposiums Authors´ Club

Adamczak, Stanislaw / Alpek, Ferenc / Badida, Miroslav / Bobal, Vladimir / Boyadjiev, Iliya / Brezocnik, Miran / Buble, Marin / Budic, Ivan / Drstvensek, Igor / Dulcic, Zelimir / Grizelj, Branko / Gyenge, Csaba / Hodolic, Janko / Janac, Alexander / Janacova, Dagmar / Jerbic, Bojan / Kljajin, Milan / Kolomaznik, Karel / Krumes, Dragomir / Kubalcik, Marek / Kulenovic, Malik / Kundrak, Janos / Kunica, Zoran / Kuric, Ivan / Leber, Marjan / Majetic, Dubravko / Malakov, Ivo / Math , Miljenko / Matysek, Miroslav / Meissner, Klaus / Micieta, Branislav / Mrsa, Zoran / Mursec, Bogomir / Nanasi, Jozef / Nikolic, Gojko / Novakovic, Branko / Pahole, Ivo / Papstel, Jyri / Pavic, Ivan / Rogale, Dubravko / Siemiatkowski, Mieczyslaw / Szalay, Tibor / Vasek, Vladimir / Velisek, Karol / Visekruna, Vojo / Vranjes, Bozo

2007 Candidates for 10+ DAAAM Symposiums Authors´ Club

Following DAAAM authors are candidates and have a chance to join the 10+ Club in 2007
Acko, Bojan / Ancau, Mircea / Anna, Vojtech / Brnic, Josip / Canadija, Marko / Capek, Jan / Dragcevic, Zvonko / Dukovski, V / Kheifetz, M / Kostal, Peter / Koukhta, S / Mahovic, Sanja / Majstorovic, Vlado / Markos, Sandor / Milfelner, Matjaz / Neumann, P / Pandilov, Zoran / Park, Hong-Seok / Ueda, Kanji / Weiss, Zenobia

Members of 5+ DAAAM Symposiums Authors´ Club

Abhary, K / Abram-Zver, Marta / Acko, Bojan / Adamek, M / Agic, D / Ancau, Mircea / Anisic, Zoran / Anna, Vojtech / Apostol, B / Aracic, Stjepan / Arz, Gusztav / Badjgholi, Farsad / Balc, Nicolae / Baumann, Christian / Bekes, Jan / Belansky, Jan / Belohoubek, Pavel / Belous, V / Bleicher, F / Bliedtner, Jens / Bolanca, I / Bolanca, Zdenka / Bolcu, D / Brezak, Danko / Brnic, Josip / Cala, Ivo / Canadija, Marko / Capek, Jan / Car, Zlatan / Carija, Zoran / Casalino, G / Celar, Stipe / Chramcov, B / Cieloszyk, Janusz / Ciglar, Damir / Cohodar, Majda / Constantin, George / Cosic, Ilija / Cosic, Predrag / Costea, A / Cotet, Costel Emil / Covic, Dragan / Cristea, L / Crnekovic, Mladen / Croitoru, C / Cselenyi, Jozsef / Cubonova, Nadezda / Cyklis, Jerzy / Dabacan, M / De Filippis, L A C / Deacu, L / Dimitrovski, Krste / Dolga, Lia / Dostal, P / Dragcevic, Zvonko / Dudas, Illes / Dukovski, V / Dumitras, C / Durakbasa, M Numan / Dvorak, Jiri / Egorov, Sergey / Emert, R / Farcas, Cristian / Feld, Mieczyslaw / Ficko, Mirko / Filaretov, Vladimir F / First Rogale, Snjezana / Gatalo, Ratko / Gavlas, Josef / Gawlik, Jozef / Gecevska, Valentina / Ghionea, Adrian / Gojo, Miroslav / Grabenweger, Johannes / Gronalt, Manfred / Gudac, G / Guenther, U / Guran, M / Gyani, Karoly / Hajduk, Mikulas / Halbritter, Erno / Hernandez, O / Hesselbach, Juergen / Hlavenka, Bohumil / Hofreiter, Milan / Honczarenko, Jerzy / Hradocky, Ladislav / Hrman, M / Hrubina, Kamil / Hruska, F / Ivakovic, C / Ivancic, Ivan / Jablonski, W / Janecki, Darius / Jardzioch, Andrzej / Kachanak, Anton / Karaulova, T / Kasac, J / Kerekes, Laszlo / Kheifetz, M / Kittl, Burkhard / Knego, Nikola / Knezevic, P / Knosala, Ryczard / Kolakovic, M / Kopac, Janez / Kostal, Peter / Koukhta, S / Kralikova, Ruzena / Krolikowski, Marcin A. / Krys, N / Kubinger, Wilfried / Kurtovic, M / Lange, Dirk / Lazea, Gheorghe / Lee, Gyu -Bong / Lerch, Zoran / Ljubetic, Juraj / Lopez de Lacalle, Luis N. / Ludovico, A D / Luong, L H S / Maglic, Leon / Mahovic, Sanja / Majdandzic, Niko / Majernik, Milan / Majstorovic, Vlado / Malopolski, Waldemar / Mandic, L / Marchelek, K / Markos, Sandor / Marusic, Vlatko / Matjasic-Fris, Mateja / Mehta, Narinder Kumar / Meszaros, Imre / Mihoci, Kristijan / Mikac, Tonci / Milfelner, Matjaz / Monostori, Laszlo / Morar, L / Moskovenko, I B / Motika, Roman / Mrocheck, J / Muransky, Juraj / Nagy, Zoltan / Navratil, P / Navratil, Z / Nedelcu, Dumitru / Neumann, P / Nowicki, B / Obsieger, Boris / Olaru, Adrian / Olteanu, C / Olteanu, F / Oluic, C / Oravec, Ladislav / Osanna, Peter Herbert / Pajor, M / Palcic, Iztok / Pandilov, Zoran / Pandza, K / Park, Hong-Seok / Patrascu, Gabriela / Pavlic, M / Perinic, M / Peshkoun, S V / Peterka, Jozef / Petkovic, Darko / Petreus, D / Pierzchala, W / Pinteric, Andreja / Pisla, Arian / Pitica, D / Pivonka, Peter / Pokorny, Pavel / Poljak, Krunoslav / Pomenic, Loreta / Popescu, Diana / Popescu, Sorin / Poppeova, Viera / Powalka, B / Prokop, R / Prostrednik, Daniel / Pruteanu, Octavian / Przybylski, Wlodzimierz / Rachieru, N / Rados, Jerko / Radu, M / Rezic, Snezana / Rizescu, S / Ros, O / Rovan, Mario / Ryniewicz, A / Safran, M / Saks, A / Samardzic, Ivan / Sanchez, J A / Saric, Tomislav / Sebastijanovic, S / Seda, Milos / Seghedin, Neculai / Seminsky, Jaroslav / Seucek, Ivan / Sihn, Wilfried / Simunic, Juraj / Sinay, Juraj / Skenderi, Z / Slota, Adam / Sluga, Alojz / Sokansky, Karel / Somodji, Z / Sopta, Luka / Sostar, Adolf / Srdjak, Miroslav / Stefanic, N / Stoic, Antun / Stopper, Markus / Stuja, Kemajl / Sugar, Peter / Sysala, Tomas / Sysel, M / Szabo, Otto / Szadkowski, J / Szelig, Karoly / Tabucanon, Mario T / Tkalic, M / Tomas, A / Turkalj, Goran / Ueda, Kanji / Ujevic, Darko / Uricek, Juraj / Varga, Gyula / Vavrik, I / Viducic, Ljiljana / Vincze, Markus / Virkus, U / Vitecek, Antonin / Vitez, Ivan / Voloder, A / Vrljicak, Z / Wantuch, Edward / Weigl, Karl Heinz / Weiss, Zenobia / Zajac, Jozef / Zamfira, C. Sorin / Zeljkovic, Milan / Zhirabok, Alexey / Ziberna-Sujica, Milena / Zichil, Valentin / Zmijewski, W / Zorc, Davor / Zuperl, Uros / Zvirinsky, Vladimir

2007 Candidates for 5+ DAAAM Symposiums Authors´ Club

Following DAAAM authors are candidates and have a chance to join the 5+ Club in 2007

Adamczyk, Zbigniew / Andrlik, V / Babic, Darko / Bachvarov, Angel / Baksa, S / Bana, V / Banabic, Dorel / Banyai, T / Barisic, Branimir / Benes, P / Benic, Djuro / Berce, Petru / Bohez, Erik L J / Bolanca, S / Borza, Alexandru Ioan / Brandes, A / Budin, R / Bulgaru, Marius / Carutasu, G / Catrina, D / Cekic, A / Chodzko, M / Ciucescu, D / Comsa, Dan -Sorin / Cuesta, Eduardo Gonzales / Cukor, Goran / Dobras, B / Dolga, Valer / Dvorackova, M / Dvorakova, Jana / Egorova, Tatyana / Ekinovic, Sabahudin / Erdeljan, A / Evin, Emil / Gabrijelcic, Helena / Gallo, Mikulas / Gallova, Stefania / Gersak, Jelka / Giannoccaro, Rosanna / Gil, Stanislaw / Greguss, Pal / Grizelj, D / Grubisic, D / Haertig, Guenter / Hait, A L / Hajrizi, Edmond / Hebbal, Shivanand / Hlava, Jaroslav / Hruskova, Erika / Hrzic, Dragana / Iltschev, Velko / Jadlovska, A / Jaehn, Hendrik / Jain, Pramod K. / Johanyak, Z C / Jovic, Franjo / Jurecic, Denis / Jurum Kipke, J / Keran, Zdenka / Kirchfeld, Maria / Kiss, Akos / Kodacsy, J. / Kolibal, Zdenek / Kosmol, Jan / Kovacic, Miha / Kowalski, Marek / Kozak, Drazan / Krajci, V / Krolikowski, Marcin / Krzyzanowski, Jozef / Kuenanz, K / Kuzman, Karl / Lalic, Bojan / Lamikiz, Aitzol / Lazibat, T / Lipa, Z / Ljubic, Franjo / Ljuhar, Davul / Manriquez, J / Marincic, Valentina / Markic, B / Marunic, Gordana / Matusu, Radek / Mera, M / Mihelic-Bogdanic, Alka / Mlinaric, Tomislav J. / Moeller, K / Mudronja, Vedran / Nagla, Kuldeep Singh / Nardin, B / Neage, Irina / Nedeff, V / Nekrassov, G / Neslusan, M / Oberlaender, Rudolf / Osmani, Hysni / Osmera, Pavel / Palaghita, N / Parac-Osterman, D / Paralic, Jan / Pavletic, D / Pavlickova, Jirina / Pawlukowicz, P / Perez, J M / Peric, Nedjeljko / Perutka, Karel / Piculin, S / Pidany, Jan / Pisla, Doina / Pizmoth, Petja / Plancak, Miroslav / Ploj, Tone / Popescu, I / Radic, Jure / Rajkovic, B / Ravas, R / Repanovici, A / Robotin, R / Rogic, K / Rusu, L / Sabol, Tomas / Samanovic, Josip / Samek, David / Sariguil, Erdal / Sebastijanovic, N / Sebo, Dusan / Senchilo, I / Siminiati, Dubravka / Sisek, B / Situm, Z / Skoczynski, Waclaw / Skoko, M / Skolud, Bozena / Sladek, Jerzy / Sliwka, J / Smutny, Lubomir / Sogorovic, D / Sokhan, S / Solc, Frantisek / Staicu, S / Stan, Gheorghe / Starcevic, Gordan / Stefanek, Michal / Stjepanovic, Zoran / Sutlovic, I / Svirsky, D / Takakuwa, Soemon / Teich, Tobias / Tomic, D / Toth, Andras / Tudjarov, B / Uhlik, Alexander / Vaclavek, Pavel / Vukovic, Ivan / Weiland, Kristian / Weinert, Klaus / Zebala, Wojciech / Zizka, Jan / Zjakic, Igor / Zurek, W

Authors of DAAAM 2006 - NewComer Club

Afshari, A / Aghaallaei, N / Ahmed, S S / Akhavan, A N / Akhter, S / Albut, Aurelian / Antonescu, O / Arora, J S / Athanasiadis, G / Baek, J Y / Bajoghli, B / Barbu, M C / Barisic, M / Baser, O / Basher, T / Begovic, E / Belyi, A / Bilberg, A / Bob, D / Bob, M / Bockal, Z / Bogunovic, B / Borbely, S / Bostan, I / Botezatu, C / Botezatu, C P / Brautovic, H / Brnada, S / Brudniok, S / Brus, M / Cakolli, H / Cetin, A / Cetin, L / Coole, T / Cozminca, I / Cruz, F / Curkovic, P / Dan, V / Danut Tocut, P / Deac, C / Deac, V / Delic, V / Dewald, O / Dobransky, J / Dobre, T / Domek, S / Domingo, R / Dragicevic, M / Drga, T / Dulgheru, V / Dumitrascu, D / Dvorak, Z / Dzepina, I / Dzierwa, A / Fahimnia, B / Fan, K / Foldyna, J / Fulop, I / Galan, V / Garajayewa, G / Gheorghiu, H / Gjeldum, N / Gogouvitis, Xenofon Vasileios / Gombar, M / Gonzalez, C / Gordienko, A / Gupta, P / Haapalainen, P / Hakola, T / Hamidullah, H / Hamrlicek, J / Haq, S M / Haramina, H / Harbarchuk, V / Hassan, K / Heczko, M / Hoffman, M / Hricova, B / Ibrahimi, N / Irfan, M A / Isic, S / Islam, N / Itu, C / Jackson, M / Jamshidi, K / Jankovic, Z / Judele, A / Kacani, J / Kaloudis, A / Kalra, A S / Kandelbauer, A / Karic, E / Katyal, A / Kessler, R W / Kim, B H / Kleinova, J / Klipakov, M / Kolar, D / Kolar, V / Kondratenko, G / Kondratenko, Y / Konihodzic, A / Kovacevic, S / Kovacevic, Z / Kramarenko, S / Krsulja, M / Kuba, J / Kumar, R P / Kycyku, A / Lange, F / Lazarevic, M / Le, V S / Lee, H C / Lepadatescu, B / Licari, R / Lika, E / Lo Valvo, E / Louckova, P / Luburic, G / Lungu, G / Macan, M / Mach, V / Maekiranta, A / Maksumic, I / Manas, D / Markovic, M / Matsas, E / Mechler, G / Medved, T / Meghdari, A / Mihai, M / Millas, V / Mircea, D / Misak, S / Mladenova, C / Mohammad Esmaeil, M / Moore, P / Movahhedinia, N / Mueller, P / Munar, A C / Munitic, A / Nepevny, P / Nica, G B / Nidamarthi, S / Oflaz, H / Okulik, T / Olevnik, A / Oltean, G / Oral, O / Pap, K / Parida, S K / Park, S E / Parkin, R / Parus, A / Penker, A / Piacentini, M / Pietrusewicz, K / Pinos, R / Pogarcic, I / Polash, S A / Popescu, D P / Pospisil, R / Prester, J / Pupaza, C / Putnik, G / Rabinovitch, A / Radaslic, K / Radovanovic, M / Radvanska, A / Rahman, A / Rahman, I / Rahman, M / Rahmanovic, S / Raich, P A / Ramaj, V / Rasti, J / Reizer, R / Remenar, V / Rinnerthaler, F / Roman, S M / Romanov, D / Rosu, L / Rosu, S M / Rubesa, J / Ruzbarsky, J / Saaristo, H / Sandu, C / Sandu, Co / Schlickenrieder, K / Schmidt, C / Schwarz, I / Secujski, M / Sedlacik, R / Sedlar, D / Shemiakin, A / Shevchenko, E / Sibilia, A / Simon, A E / Singh, M / Skorokhodov, V / Slavu, B / Soporan, V / Spiryagin, M / Spiryagin, V / Srivastava, R K / Stanciu, V / Stanek, M / Stoica, A / Storch, B / Sulzbachner, C / Surcin, L / Sykorova, L / Takala, J / Tekic, Z / Theiner, A / Tirth, V / Todea, M / Topalova, I / Tos, Z / Tripe Vidican, A / Tripe Vidican, C / Uddin, M / Umesh-Babu, S / Uyar, E / Valentan, B / Valicek, J / Vanoli, R / Vidakovic, I / Viducic, V / Vivdenko, Y / Vosniakos, George Christopher / Vukovic-Obrovac, J / Wang, J / Watson, R / Werner, J / Wullschleger, C / Wuzella, G / Yoon, J / Zaeh, M F / Zander, J / Zavec Pavlinic, D / Zawada-Tomkiewicz, A / Zeqiri, H / Ziljak-Vujic, J / Zvolensky, R

5.0 IN MEMORIAM OF NICE DAAAM COLLEAGUES AND DEAR FRIENDS

Professor Dr Jyri Papstel (1940- 2004), Tallinn Technical University, Tallinn Estonia, Vice President of DAAAM International, Founder and President of DAAAM Estonia, Founding member of the pilot projects Interuniversity Doctoral Studies and Virtual University. Member of: International Organizing / Program / Scientific / Review Committees. President of Award Committee. Author. Honored with Medal of DAAAM International Vienna.

M. Sc. Dipl.-Ing. Ivan Ivancic (1952-2004), Ericsson - Tesla, Zagreb, Croatia, International Coordinator for the field Maintenance, Member of International Organizing / Program / Review Committees.

Professor Dr Zdenko Kordic (1949-2004), University of Mostar and University of Split, Founder and First Rector of the University of Mostar, Member of Rectors´ Honor Committee, Founding member of the pilot projects Interuniversity Doctoral Studies and Virtual University. Member of International Program and Scientific Committees. Author. Honored with Medal of DAAAM International Vienna.

6.0 DAAAM INTERNATIONAL MEDAL

On the occasion of 10th DAAAM International Symposium, DAAAM International introduced the Medal of DAAAM International "for excellence in the science, technology, and international scientific co-operation in the framework of DAAAM International Vienna" as a highest recognition for individual persons and institutions. The medal is designed by B. Katalinic, produced by Moscow State Mint Institute and sponsored by Moscow State University of Technology "Stankin" and Cassandra Trust – Ireland.

7.0 UNIVERSITY CHARTER OF DUBROVNIK

The text of the University Charter of Dubrovnik: We, the representatives of different Universities and Institutions, after two years preparation at the meeting initiated and organized by The Danube Adria Association for Automation & Manufacturing, DAAAM International under the auspices of The Danube Rectors' Conference on the occasion of the Eighth International DAAAM Symposium "Intelligent Manufacturing & Automation" held at the International Center of Croatian Universities (University of Zagreb) in Dubrovnik, Croatia, decided to found the University Network of DAAAM International Vienna as a strategic alliance and permanent open platform for long-term co-operation and partnership in all fields of technology, research and education.

The co-operation has to be continuously organized and is based on the partnership, friendship, tolerance, appreciation and positive experience of international scientific co-operation within the framework of DAAAM International during the last eight years. The co-operation can use different flexible forms, structures and aspects. The practical co-operation will be project-based organized. The highest priority is given to the organization of two long - term projects, initially known as: Interuniversitary Doctoral Studies and a pilot project Virtual University. Signed by the founding members in Dubrovnik, Croatia, on the twenty-third of October 1997

8.0 DAAAM INTERNATIONAL NETWORK FOR ADVANCED TECHNOLOGIES

The text of the Network founding document: We, the representatives of different Universities, Institutions, and Firms at the meeting initiated and organized by The Danube Adria Association for Automation & Manufacturing, DAAAM International Vienna under the auspices of The Danube Rectors' Conference on the occasion of the Thirteenth International DAAAM Symposium "Intelligent Manufacturing & Automation: Learning from Nature" held at the Vienna University of Technology and Austrian Society of Engineers and Architects in Vienna, Austria, decided to found the DAAAM International Network for Advanced Technologies as a strategic alliance and permanent open platform within European Research Area for long-term co-operation and partnership in all fields and aspects of research, developing, transfer, education and use of advanced technologies.

The co-operation has to be continuously organized and had to be based on the partnership, friendship, tolerance, appreciation and positive experience of international scientific co-operation within the framework of DAAAM International during the last fourteen years. The co-operation can use individual flexible project oriented forms, structures and aspects and support the young researchers' mobility as well. The practical co-operation will be project-based organized. The highest priority is given to the organization of long - term international cooperation projects focused on advanced technologies for balanced development and harmony between men, technology and nature. The network has to be platform for dissemination of relevant information and ideas, supporting international partnership and encouraging of cooperation initiatives. Signed by the founding members in Vienna, Austria, European Union, on the twenty-fifth of October 2002

Activity No. 1: The conference bags and flags for DAAAM 2003 in Sarajevo are sponsored by the Faculty of Textile Technology, University of Zagreb, and organised by Professor Z. Dragcevic as the first activity in the frame of DAAAM International Network for Advanced Technologies.

9.0 DAAAM INTERNATIONAL NETWORK FOR MECHATRONICS AND ROBOTICS

The text of the Network founding document: We, the representatives of different Universities, Institutions, and Firms at the meeting initiated and organized by The Danube Adria Association for Automation & Manufacturing, DAAAM International Vienna under the auspices of The Danube Rectors' Conference on the occasion of the Fifteenth International DAAAM Symposium "Intelligent Manufacturing & Automation: Globalisation - Technology - Men - Nature" held at the Vienna University of Technology and Austrian Society of Engineers and Architects in Vienna, Austria, decided to found the DAAAM International Network for Mechatronics and Robotics as a strategic alliance and permanent open platform within European Research Area for long-term co-operation and partnership in all fields and aspects of research, developing, transfer, education and use of advanced technologies.

The co-operation has to be continuously organized and had to be based on the partnership, friendship, tolerance, appreciation and positive experience of international scientific co-operation within the framework of DAAAM International during the last fiveteen years. The co-operation can use individual flexible project oriented forms, structures and aspects and support the young researchers' mobility as well. The practical co-operation will be project-based organized. The highest priority is given to the organization of long - term international cooperation projects focused on advanced technologies for balanced development and harmony between men, technology and nature. The network has to be platform for dissemination of relevant information and ideas, supporting international partnership and encouraging of cooperation initiatives. Signed by the founding members in Vienna, Austria, European Union, on the fifth of November 2004

10.0 DAAAM INTERNATIONAL SYMPOSIUMS (1990-2006)

One important activity is organising the DAAAM International Symposium. The main objective of the Symposium is to provide a world forum, which takes place annually in Central Europe to exchange knowledge, experience, results and information related to various aspects of the advanced manufacturing and modern automation.

The scope of the Symposium covers scientific, technological and practical concepts concerning research, development and realisation of advanced manufacturing and automation concepts, offering a unique opportunity for experts to meet and exchange ideas. The Symposium is concerned with the current status of automation and manufacturing, methods and results of research and development, as well as application problems and publishes proceedings accordingly.

2006	**17th International DAAAM Symposium "Intelligent Manufacturing & Automation: Focus on Mechtronics and Robotics"**
On the occasion of	150th Birthday of Nikola Tesla (1856-1943) Inventor of Mobile Robot Teleoperation
Date & Place	8-11th November 2006, Vienna, Austria
Organized by	DAAAM International Vienna, Vienna University of Technology - Department of Production Engineering / Intelligent Manufacturing Systems (IFT-IMS), ÖIAV 1848 Austrian Society of Engineers and Architects , University of Applied Sciences - Technikum
Under the auspices of	Danube Rectors' Conference & Rectors' Honor Committee of DAAAM International
In cooperation with	Vienna Convention Bureau
Organizer	Buchmeister, Borut / Dragcevic, Zvonko / Egorov, Sergey / Höller, Liane / Katalinic, Branko (Chair) / Lazinica, Aleksandar / Malisa, Viktorio / Sagmeister, Nicole / Stuja, Kemajl / Viktorik, Hans / Viktorik, Silvia
Publication	Annals of DAAAM for 2006 and Proceeding of the 17th International DAAAM Symposium, ISSN 1726-9679, ISBN 3-901509-57-7
Editor	Katalinic B.
2005	**16th International DAAAM Symposium "Intelligent Manufacturing & Automation: Focus on Young Scientists and Researchers"**
On the occasion of	15 Years of DAAAM International Vienna, 1990-2005
Date & Place	19-22nd October 2005, Opatija, Croatia
Organized by	DAAAM International Vienna, Vienna University of Technology - Department of Production Engineering / Intelligent Manufacturing Systems (IFT-IMS), ÖIAV 1848 Austrian Society of Engineers and Architects
Under the auspices of	Danube Rectors' Conference & Rectors' Honor Committee of DAAAM International
In cooperation with	Vienna Convention Bureau
Organizer	Anisic, Bojan / Barisic, B. / Blazevic, Dalibor / Buchmeister, Borut / Car, Zlatan / Dragcevic, Zvonko / Egorov, Sergey / Grabenweger, Johannes / Gronalt, Manfred / Höller, Liane / Huber, Ingrid / Jurkovic, Z. / Katalinic, Branko (Chair) / Lazinica, Aleksandar / Malisa, Viktorio / Merdan, Munir / Mikac, Tonci (Co-Chair) / Mirjanic, Valentina / Nanasi, Jozef / Schaumberger, Werner / Stopper, Markus / Stuja, Kemajl / Udiljak, Toma / Widtmann, Georg
Publication	Annals of DAAAM for 2005 and Proceeding of the 16th International DAAAM Symposium, ISSN 1726-9679, ISBN 3-901509-46-1
Editor	Katalinic B.
2004	**15th International DAAAM Symposium "Intelligent Manufacturing & Automation: Globalisation - Technology - Men - Nature"**
On the occasion of	155 Years of Austrian Society of Engineers and Architects ÖIAV 1848
Date & Place	3-6th November 2004, Vienna, Austria, European Union
Organized by	DAAAM International Vienna, Vienna University of Technology - Department of Production Engineering / Intelligent Manufacturing Systems (IFT-IMS), ÖIAV 1848 Austrian Society of Engineers and Architects
Under the auspices of	Danube Rectors' Conference & Rectors' Honor Committee of DAAAM International
In cooperation with	Vienna Convention Bureau
Organizer	Bartak, Harald; Buchmeister, Borut; Dragcevic, Zvonko; Egorov, Sergei; Frank, Willy; Grabenweger, Johann; Gronalt, Manfred; Höller, Liane; Hrelja, Igor; Katalinic, Branko (Chair); Kettler, Roman; Kordic, Vedran; Lazinica, Aleksandar; Löschner, Adam; Magyar, Teresa; Malisa, Viktorio; Merdan, Munir; Müller, Simon; Nanasi, Jozef; Piller, Johannes; Reinisch, Roland; Roth, Michael; Saliger, Florian; Schaumberger, Werner; Schmidt, Thomas; Stuja, Kemajl; Trauzettel, Ulrich; Udiljak, Toma; Widtmann, Georg; Wirths, Thomas.
Publication	Annals of DAAAM for 2004 and Proceeding of the 15th International DAAAM Symposium, ISSN 1726-9679, ISBN 3-901509-42-9
Editor	Katalinic B.

2003	**14th International DAAAM Symposium "Intelligent Manufacturing & Automation: Focus on Reconstruction and Development"**
On the occasion of	155 Years of Austrian Society of Engineers and Architects ÖIAV 1848 / 45 Years of the Faculty of Mechanical Engineering University of Sarajevo
Date & Place	22-25th October 2003, Sarajevo, Bosnia and Herzegovina
Organized by	DAAAM International Vienna, University of Sarajevo - Faculty of Mechanical Engineering, Vienna University of Technology - Department of Production Engineering / Intelligent Manufacturing Systems (IFT-IMS), ÖIAV 1848 Austrian Society of Engineers and Architects
Under the auspices of	Danube Rectors' Conference & Rectors' Honor Committee of DAAAM International
In cooperation with	Federal Ministry of Education, Sciences and Culture - Vienna, ÖIAV 1848 Austrian Society of Engineers and Architects
Organizer	Katalinic, Branko (Chair), Kulenovic, Malik (Co-chair), Widtmann, G., et al.
Publication	Annals of DAAAM for 2003 and Proceeding of the 14th International DAAAM Symposium, ISSN 1726-9679, ISBN 3-901509-34-8
Editor	Katalinic B.
Information	Contact: DAAAM International Vienna (Katalinic B.) president@daaam.com www.daaam.com
2002	**13th International DAAAM Symposium "Intelligent Manufacturing & Automation: Learning from Nature"**
On the occasion of	150 Years of Austrian Society of Engineers and Architects ÖIAV 1848
Date & Place	23-26th October 2002, Vienna, Austria
Organized by	DAAAM International Vienna, Vienna University of Technology - Department of Production Engineering / Intelligent Manufacturing Systems (IFT-IMS), ÖIAV 1848 Austrian Society of Engineers and Architects
Under the auspices of	Danube Rectors' Conference & Rectors' Honor Committee of DAAAM International
In cooperation with	Federal Ministry of Education, Sciences and Culture - Vienna, ÖIAV 1848 Austrian Society of Engineers and Architects
Organizer	Katalinic B. (Chair), Buchmeister B., Doganac A., Dragcevic Z., Gersak J., Grabenweger J., Kordic V., Ljoljic B., Nanasi J., Stopper M., Stuja K., Widtmann G., et al.
Publication	Annals of DAAAM for 2002 and Proceeding of 13th International DAAAM Symposium, ISBN 3-901509-29-1
Editor	Katalinic B.
2001	**12th International DAAAM Symposium "Intelligent Manufacturing & Automation: Focus on Precision Engineering"**
On the occasion of	10 Years of Jena University of Applied Science / 150 Years of ÖIAV 1848 Austrian Society of Engineers and Architects
Date & Place	24-27th October 2001, Jena, Germany
Organized by	DAAAM International Vienna, Jena University of Applied Science, Vienna University of Technology - Department of Production Engineering / Intelligent Manufacturing Systems (IFT-IMS), Austrian Society of Engineers and Architects ÖIAV 1848
Under the auspices of	Danube Rectors' Conference & Rectors' Honor Committee of DAAAM International
In cooperation with	Thuringian Ministry of Sciences, Research and Culture - Erfurt,
Organizer	Katalinic B. (Chair), Meissner K. (Co-chair), Beez S., Buchmeister B., Doganac A., Dragcevic Z., Fidi L., Gersak J., Grabenweger J., Jacob M., Kordic V., Ljoljic B., Nanasi J., Pachmann A., Schoele H., Selke K., Stopper M., Stuja K., Walther J., Widtmann G.
Publication	Annals of DAAAM for 2001 and Proceeding of 12th International DAAAM Symposium, ISBN 3-901509-19-4
Editor	Katalinic B.
2000	**11th International DAAAM Symposium "Intelligent Manufacturing & Automation: Men – Machine - Nature"**
On the occasion of	150 Years of ÖIAV 1848 Austrian Society of Engineers and Architects / 40 Years of Faculty of Engineering – University of Rijeka / 10 Years of DAAAM International
Date & Place	19-21st October 2000, Opatija, Croatia
Organized by	DAAAM International Vienna, Vienna University of Technology - Department of Production Engineering / Intelligent Manufacturing Systems (IFT-IMS)
Under the auspices of	Danube Rectors' Conference & Rectors' Honor Committee of DAAAM International
In cooperation with	Federal Ministry of Education, Sciences and Culture - Vienna, ÖIAV 1848 Austrian Society of Engineers and Architects, Union of Croatian Universities
Organizer	Katalinic B. et al.
Publication	Annals of DAAAM for 2000 and Proceeding of 10th International DAAAM Symposium, ISBN 3-901509-13-5
Editor	Katalinic B.

1999	**10th International DAAAM Symposium "Intelligent Manufacturing & Automation: Past – Present - Future"**
On the occasion of	10th International DAAAM Symposium, Last International DAAAM Symposium in this Millennium
Date & Place	21-23rd October 1999, Vienna University of Technology - Department of Production Engineering / Intelligent Manufacturing Systems (IFT-IMS), Austrian Society of Engineers and Architects ÖIAV 1848, Vienna, Austria
Organized by	DAAAM International Vienna, Vienna University of Technology, Department of Production Engineering / Intelligent Manufacturing Systems (IFT-IMS)
Under the auspices of	Danube Rectors' Conference & Rectors' Honor Committee of DAAAM International
In cooperation with	Federal Ministry of Sciences and Traffic - Vienna, ÖIAV 1848 Austrian Society of Engineers and Architects
Organizer	Katalinic B. (Chair), Adamczak S., Alpek F., Balate J., Balic J., Beer M., Buchmeister B., Celar S., Cus F., Dinnyes C., Dragcevic. Z., Duflou J. R., Egorov S., Fidi L., Grabenweger J., Gyenge C., Hajrizi E., Kerekes L., Kordic V., Kordic Z., Ljoljic B., Llorente J. I., Meissner K., Nanasi J., Papstel J., Park H.-S., Stuja K., Tabucanon M. T., Udiljak T., Ueda K., Vranjes B., Wessely E.
Publication	Annals of DAAAM for 1999 and Proceeding of 10th International DAAAM Symposium, ISBN 3-901509-10-0
Editor	Katalinic B.

1998	**9th International DAAAM Symposium "Intelligent Manufacturing, Automation & Networking"**
On the occasion of	50 Years of the Technical University of Cluj - Napoca 150 Years of ÖIAV 1848 Austrian Society of Engineers and Architects Founding of Permanent Committee of University Network of DAAAM International Vienna
Date & Place	22-24th October 1998, Technical University of Cluj - Napoca, Cluj - Napoca, Romania
Organized by	DAAAM International Vienna, Technical University of Cluj - Napoca, Vienna University of Technology - Department of Production Engineering / Intelligent Manufacturing Systems (IFT-IMS)
Under the auspices of	Danube Rectors' Conference & Rectors' Honor Committee of DAAAM International
In cooperation with	Ministry of Research and Technology - Bucuresti, Federal Ministry of Sciences and Traffic - Vienna, Journal of Intelligent Manufacturing, ÖIAV 1848 Austrian Society of Engineers and Architects
Organizer	Arsin M., Beer M., Bejan M., Berce P., Borzan M., Buchmeister B., Calinescu O., Car Z., Celar S., Dusa P., Fidi L., Gheorghe I., Gramescu T., Gyenge C., Hajrizi E., Ivan N.V., Katalinic B. (Chair), Kerekes L., Lazea G., Ljoljic B., Lörincz S., Nanasi J., Nemes S., Popescu S., Pusca M., Szava T., Udiljak T., Vranjes B.
Proceedings	ISBN 3-901509-08-9
Editor	Katalinic B.

1997	**8th International DAAAM Symposium "Intelligent Manufacturing & Automation"**
On the occasion of	Founding of the University Network of DAAAM International Vienna
Date & Place	23-25th October 1997, ICCU - International Centre of Croatian Universities (University of Zagreb), Dubrovnik, Croatia
Organized by	DAAAM International Vienna, University of Zagreb, Vienna University of Technology, Department of Production Engineering / Intelligent Manufacturing Systems (IFT-IMS)
Under the auspices of	Danube Rectors' Conference & Rectors' Honor Committee of DAAAM International
In cooperation with	Ministry of Science and Technology - Zagreb, Federal Ministry of Sciences and Traffic - Vienna, Journal of Intelligent Manufacturing, ÖIAV 1848 Austrian Society of Engineers and Architects, Union of Croatian Universities, Croatian Society - KoREMA - Zagreb.
Organizer	Benic D., Brunsko V., Budic I., Burgjelez I., Celar S., Gojo M., Hajrizi E., Ivancic I., Jerbic B., Katalinic B. (Chair), Kovacic B., Majstorovic V., Mencer H.J., Nanasi J., Pavic I., Polic-Bobic M., Udiljak T. (Co-Chair), Veza I., Visekruna V.
Proceedings	ISBN 3-901509-04-6
Editor	Katalinic B.

1996	**7th International DAAAM Symposium "Product & Manufacturing: Flexibility, Integration, Intelligence"**
On the occasion of	1,000 Years of Austria
Date & Place	17-19th October 1996, Vienna, Austria
Organized by	DAAAM International Vienna, Vienna University of Technology, Department of Production Engineering (IFT-IMS)
Under the auspices of	Danube Rectors' Conference
Organizer	Katalinic B., Celar S., Fabian E., Nanasi J.
Proceedings	ISBN 3-901509-02-X
Editor	Katalinic B.
1995	**6th International DAAAM Symposium "Intelligent Manufacturing Systems"**
On the occasion of	50 Years of Technical University of Krakow
Date & Place	26 - 28th October 1995, Poland
Organized by	DAAAM International Vienna; Technical University of Krakow Technical University of Kielce; Vienna University of Technology - Department of Production Engineering / Intelligent Manufacturing Systems (IFT-IMS)
In cooperation with	IMECO; Ministry of Education, Warsaw; Ministry of Sciencees and Technology, Warsaw; Federal Ministry of Sciences, Technology and Art, Vienna; BIT, Bureau for International Research and Technology Cooperation, Vienna
Organizer	Adamczak S., Duch A., Gawlik J., Katalinic B., Osanna H. P., Sladek J., Wantuch E., Rusiniak K., Zebala W.
Proceedings	ISBN 3-901509-01-1
Editor	Katalinic B.
Co-Editors	Osanna H.P., Gawlik J.
1994	**5th International DAAAM Symposium "Automation and Metrology: Challenge and Chance"**
On the occasion of	700 Years of City of Maribor, 35 Years of Faculty of Technical Sciences
Date & Place	27-28th October 1994, Maribor, Slovenia
Organized by	DAAAM International Vienna, University of Maribor, Faculty of Technical Sciences
In cooperation with	IMECO; Ministry of Sciencees and Technologies, LjubljanaFederal Ministry of Sciences, Technology and Art, Vienna
Organizer	Balic J., Cus F., Katalinic B., Novak D., Osanna H. P., Sujica M. Z.
Proceedings	ISBN 3-901509-00-3, ISBN 86-435-0084-4
Editors	Katalinic B., Osanna H.P., Balic J., Cus F.
1993	**4th International DAAAM Symposium "Automation and Metrology: Man, Technology, Environment"**
On the occasion of	750 Years of City of Brno
Date & Place	16-18th September 1993, Brno, Czechia
Organized by	DAAAM International Vienna; Technical University of Brno, Faculty of Mechanical Engineering
In cooperation with	IMECO, IUTAM
Organizer	Balate J., Dvorak J., Kadrnozka J., Katalinic B., Lacko B., Osanna H. P., Skopal M., Vavrin P.
Editors	Katalinic B., Osanna H.P:
1992	**3rd International DAAAM Symposium "Flexible and Intelligent Automation"**
Date & Place	4-6th November 1992, Budapest, Hungary
Organized by	DAAAM International Vienna; Technische Universität Budapest, Department of Production Engineering
In cooperation with	IMECO
Organizer	Alpek F., Arz G., Katalinic B., Osanna H. P., Szelig T., Jerzsabek P.
Editor	Katalinic B.
1991	**2nd International DAAAM Symposium "Flexible Automation"**
Date & Place	9-13th December 1991, Strbske Pleso, Slovakia, Czechoslovakia
Organized by	DAAAM International Vienna; House of Technology, Bratislava; Technical University of Kosice, Faculty of Mechanical Engineering, Department of Automated Production Systems
Organizer	Sebo D., Wessely E., Kovacik J., Katalinic B.
Proceedings	ISBN 80-233-0225-6

1990	1st International DAAAM Symposium „Automatization and Measurement Technique“
On the occasion of	175 Years of Vienna University of Technology
Date & Place	5th November 1990, Vienna, Austria
Organized by	DAAAM International Vienna, Vienna University of Technology - Department of Production Engineering / Intelligent Manufacturing Systems (IFT-IMS), ÖIAV, Sections: Measurement and Automation, ASESA, Robotics
Organizer	Beer R. J., Osanna H. P., Katalinic B.
Editor	Katalinic B.

11.0 BEST PAPER / PRESENTATION DAAAM AWARD

Since 1999 the DAAAM International Vienna recognizes excellence of contributions in a number of ways. There are three categories for papers:

- Best Paper Award of DAAAM International for the best paper presented at Symposium,
- DAAAM International Award for Best Oral Presentation of Paper,
- DAAAM International Award for Best Visual Presentation of Paper.

The Awards are made at the Closing Section and Best Paper Award Ceremony of the Symposium takes part on Friday at 17:00 hours. The Award Committee of DAAAM International makes the evaluation of presented papers during the Symposium. The final decisions are made by the President of the Award Committee of DAAAM International. Former Presidents of the Paper Award Committee of DAAAM International are: 2005 Professor Ueda, 2004 Professor Franc Cus, 2003 Professor Jyri Papstel, 2002 Professor Kanji Ueda, 2001 Professor Toma Udiljak, 2000 Professor Jyri Papstel, and 1999 Professor Kanji Ueda). The President and the final list of the members of the Committee will be announced at Opening Ceremony of the Symposium. All (co)authors of awarded papers are getting the certificate, the main author additionally getting Best Paper Medal. The layout of the Best Paper Medal you can see at the cover of this publication. The papers which authors are member of the Award committee and the papers which are not presented at Symposium are excluded from competition and can not get the Award.

12.0 BEST DIGITAL PHOTO DAAAM AWARD

Importance of digital photography for research, publishing and social relations is rapidly increased during last time. For that reason we are going to establish DAAAM International Photo Gallery with photos which are related to the symposium and our activities (place, people, situation, art, special effects). The gallery will be free accessed and searchable. For this project we are looking for amateur photographs among the participants to be official photographs of DAAAM International for 2006 and member of DAAAM International Photo Team. If you want to join this team please take your digital camera with you to Vienna and make as many as possible nice photos. DAAAM International will recognise three best digital photos related to the Symposium with the DAAAM International Best Digital Photo Award. If you want to join "DAAAM International Photo Team 2006" please send us: Your name and photo, type of camera and type of memory card.

13.0 BEST TASTE AND SMELL OF DAAAM AWARD

"Best Taste and Smell of DAAAM International": Many of colleagues in DAAAM International network are excellent hobby cooks, liquor maker or winemaker. This year we want to make one small tasting of their specialities under the motto "Best taste and smell of DAAAM International". We will be very happy if you are one of us. In this case please be free to take some your specialities to Opatija. The best will be honoured with "Best Taste and Smell DAAAM International Award".

14.0 DAAAM INTERNATIONAL CONFERENCES "ADVANCED TECHNOLOGIES FOR DEVELOPING COUNTRIES"

2006	5th DAAAM International Conference on Advanced Technologies for Developing Countries, ATDC-2006
Founder and Chair	Katalinic B.
Date & Place	28-30 June 2006, Rijeka, Croatia
Organized by	University of Rijeka, Faculty of Enineering, DAAAM International Vienna
Organizers	Mikac T.[onci] (Chair), Car Z.[latan] (Co-chair), Barisic B.[ranimir], Cukor G.[oran], Ikonic M.[ilan], Pavletic D.[usko], Perinic M.[laden], Blazevic D.[alibor], Jurkovic Z.[oran]
Proceedings	ISBN 3-901509-54-2 (Book), ISBN3-901509-55-0 (CD), ISBN 953-6326-45-0
Editors	Mikac T.[onci] & Katalinic B.[ranko]

Published by	University of Rijeka, Faculty of Engineering, DAAAM International Vienna
2005	**4th DAAAM International Conference on Advanced Technologies for Developing Countries, ATDC-2005**
Founder and Chair	Katalinic B.
Date & Place	September, 21-24th, 2005, Slavonski Brod, Croatia
Organized by	Faculty of Mechanical Engineering, Josip Juraj Strossmayer University of Osijek; DAAAM International Vienna
On the occasion of	30th Anniversary of Josip Juraj Strossmayer University of Osijek
Organizers	Budic, Ivan (Chair); Kljain Milan (Co-Chair); Kozak, Drazen (Secretary); Katalinic, Branko (Chair of Program Committee), Raos, Pero; Saric, Tomislav; Stojic, Antun; Simunovic, Goran; Novoselovic, Daniel; Karakasic, Mirko
Proceedings	ISBN 3-901509-49-6 ISBN 953-6048-29-9
Editors	Kljain M.[ilan], Katalinic B.[ranko] & Budic I.[van]
Published by	DAAAM International Vienna; Faculty of Mechanical Engineering Slavonski Brod Josip Juraj Strossmayer University of Osijek
2004	**3rd DAAAM International Conference on Advanced Technologies for Developing Countries, ATDC-2004**
Founder and Chair	Katalinic B.
Date & Place	June 23-26, 2004, Split, Croatia
Organized by	Faculty of Electrical Engineering, Mechanical Engineering and Naval Architecture, University of Split; DAAAM International Vienna
Organizers	Ivica Veža (Chairman); Željko Domazet (Co-Chairman); Branko Katalinic (Co-Chairman); Tonći Grubić (Secretary), et al.
Proceedings	ISBN 3-901509-41-0
Editors	Katalinic B.[ranko] & Veza, I.[vica]
Published by	DAAAM International Vienna; Faculty of Electrical Engineering, Mechanical Engineering and Naval Architecture, University of Split;
2003	**2nd DAAAM International Conference on Advanced Technologies for Developing Countries, ATDC-2003**
Founder and Chair	Katalinic B.
Date & Place	25-28th June 2003, Tuzla, Bosnia and Herzegovina
Organized by	Faculty of Mechanical Engineering, University of Tuzla; DAAAM International Vienna
On the occasion of	40th Anniversary of Mechanical Engineering Study in Tuzla
Organizers	Tufekcic, Dzemo (Chair); Katalinic, Branko (Co-Chairman); Widtmann, Georg (Co-Chairman); Topcic, Alen (Secretary); Kapetanovic, Izrudin; Alagic, Sandira; Alic, Izet; Buljubasic, Indira; Hadzic, Seniha; Mehmedovic, Muhamed; Mujic, Elvedin; Nezirevic, Salahudin; Saric, Emir; Kapetanovic, Irena
Proceedings	ISBN 3-901509-35-6
Editors	Katalinic B.[ranko] & Tufekcic, Dz.[emo]
Published by	DAAAM International Vienna; University of Tuzla - Faculty of Mechanical Engineering
2002	**1st DAAAM International Conference on Advanced Technologies for Developing Countries, ATDC-2002**
Founder and Chair	Katalinic B.
Date & Place	12-14the September 2002, Slavonski Brod, Croatia
Organized by	Faculty of Mechanical Engineering, Strossmayer University of Osijek; DAAAM International Vienna
On the occasion of	40th Anniversary of Mechanical Engineering Study in Slavonski Brod
Organizers	Majdandzic, Niko (Chair), Kljajin M. (Co-Chair), Kozak D. (Secretary), Cicak M., Galeta T., Hnatko E., Ivandic Z., Lujic R., Maglic L., Maticevic G., Raos P., Stojic A., Saric T., Simunovic K., Simunovic G., Vujcic D.
Proceedings	ISBN 3-901509-32-1; ISBN 953-7611-64-X
Published by	DAAAM International Vienna; Strossmayer University of Osijek - Faculty of Mechanical Engineering; Grafika Osijek
Editors	Katalinic B.[ranko] & Kljajin M.[ilan]

15.0 DAAAM INTERNATIONAL WORKSHOPS "COMPUTER AIDED SYSTEMS AND TECHNOLOGIES"

This workshop is based on the bilateral co-operation between two similar departments of the Faculties of Mechanical Engineering of the University of Zilina, Slovakia and Cracow University of Technology - CUT, Cracow, Poland. Those activities are supported by DAAAM National Organisations of Poland and Slovakia, and DAAAM International. In order to install this workshop as stabile long-term regional platform for the *CA Systems and Technologies* fields it was decided to found:

Permanent Committee of Workshop:

Permanent Members:

I. Kuric *(University of Zilina, Zilina, Slovakia),*
B. Katalinic *(Vienna University of Technology and DAAAM International, Vienna, Austria),*
J. Duda *(CUT, Cracow, Poland).*

Academic Year	Members	Institutions
2006/2007	B. Barisic	*University of Rijeka, Croatia*
	R. Cotetiu	*University of Baia Mare, Romania*
	L. Goldhahn	*FH Mittweida, Germany*
	J. Marcinčin	*TU Kosice, Slovakia*
	H. Sjolander	*University of Kristianstad, Sweden*
	P. Šugár	*TU Zvolen, Slovakia*
	N. Ungureanu	*University of Baia Mare, Romania*
2005/2006	R. Cotetiu	*University of Baia Mare, Romania*
	J. Marcinčin	*TU Kosice, Slovakia*
	J. Matuszek	*University of Bielsko Biala, Poland*
	S. Legutko	*PUT Poznan, Poland*
	P. Šugár	*TU Zvolen, Slovakia*
	N. Ungureanu	*University of Baia Mare, Romania*
2004/2005	L. Goldhahn	*FH Mittweida, Germany*
	J. Madl	*CVUT Prague, Czech Republic*
	J. Marcincin	*TU Kosice, Slovakia*
	E. Pay	*University of Baia Mare, Romania*
	D. Plinta	*University of Bielsko Biala, Poland*
	G. Varga	*University of Miskolc, Hungary*
2003/2004	D. Georgiev	*TU Varna, Bulgaria*
	L. Goldhahn	*FH Mittweida, Germany*
	J. Manlig	*TU Liberec, Czech Republic*
	J. Marcincin	*TU Kosice, Slovakia*
	G. Varga	*University of Miskolc, Hungary*
2002/2003	D. Kumicakova	*University of Zilina, Slovakia*
	S. Legutko	*PUT Poznan, Poland*
	J. Marcincin	*TU Kosice, Slovakia*
	I. Neaga	*University of Loughborough, U.K.*
	A. Pendrey	*TU Birmingham, U.K.*
	J. Peterka	*STU Trnava, Slovakia*
	L. Soos	*STU Bratislava, Slovakia*
2001/2002	N. Cubonova	*University of Zilina, Slovakia*
	S. Legutko	*PUT Poznan, Poland*
	J. Marcincin	*TU Kosice, Slovakia*
	I. Neaga	*TU Iasi, Romania*
	J. Peterka	*STU Trnava, Slovakia*
	L. Soos	*STU Bratislava, Slovakia*
2000/2001	N. Cubonova	*University of Zilina, Slovakia*
	S. Legutko	*PUT Poznan, Poland*
	J. Marcincin	*TU Kosice, Slovakia*
	I. Neaga	*TU Iasi, Romania*
	J. Peterka	*STU Trnava, Slovakia*
	L. Soos	*STU Bratislava, Slovakia*
1999/2000	S. Legutko	*PUT Poznan, Poland*
	J. Marcincin	*TU Kosice, Slovakia*
	I. Neaga	*TU Iasi, Romania*

The main aim of the Permanent Committee is to prepare the DAAAM International workshops oriented on CA systems and technologies and propagate all accompanying activities during academic year (projects, WEB journal, common publications, PhD seminars, exchange of students and lectures, etc.). The list of workshops organised during the past.

2006	**11th DAAAM International Workshop "CA Systems and Technologies"**
Date & Place	2 – 4th October 2006, Cracow, Poland
Organized by	CUT Cracow, University of Zilina, DAAAM International Vienna
Organisers	Duda J., Kuric I., Katalinic B.
Proceedings	ISBN 3-901509-56-9
Editors	Duda J., Kuric I., Katalinic B.
2005	**10th DAAAM International Workshop "CA Systems and Technologies"**
Date & Place	12 – 14th December 2005, Strecno - Zilina, Slovak Republic
Organized by	University of Zilina, CUT Cracow, DAAAM International Vienna
Organisers	Kuric, I., Duda J., Katalinic B.
Proceedings	ISBN 3-901509-51-8
Editors	Kuric I., Katalinic B.
2004	**9th DAAAM International Workshop "CA Systems and Technologies"**
Date & Place	5 - 6th October 2004, Cracow, Poland
Organized by	CUT Cracow, University of Zilina, DAAAM International Vienna
Organisers	Duda J., Kuric I., Katalinic B.
Proceedings	ISBN 83-7242-320-2
Editors	Duda J., Katalinic B.
2003	**8th DAAAM International Workshop "CA Systems and Technologies"**
Date & Place	29th September - 1st October 2003, Strecno - Zilina, Slovak Republic
Organized by	University of Zilina, CUT Cracow, DAAAM International Vienna
Organisers	Duda J., Kuric I., Katalinic B.
Proceedings	ISBN 3-901509-39-9
Editors	Kuric I., Katalinic B.
2002	**7th DAAAM International Workshop "CA Systems and Technologies"**
Date & Place	8-9th October 2002, CUT Cracow, Poland
Organized by	CUT Cracow, University of Zilina, DAAAM International Vienna
Organisers	Duda J., Kuric I., Katalinic B.
Proceedings	ISBN 3-901509-33-X
Editors	Duda J., Katalinic B.
2001	**6th DAAAM International Workshop "CA Systems and Technologies"**
Date & Place	25-26th September 2001, Sulov, Slovak Republic
Organized by	University of Zilina, CUT Cracow, DAAAM International Vienna
Organisers	Duda J., Kuric I., Katalinic B.
Proceedings	ISBN 3-901509-27-5
Editors	Kuric I., Katalinic B.
2000	**5th DAAAM International Workshop "CA Systems and Technologies"**
Date & Place	28-29th September 2000, Cracow, Poland
Organized by	CUT Cracow, University of Zilina, DAAAM International Vienna
Organisers	Duda J., Kuric I., Katalinic B.
Proceedings	ISBN 3-901509-16-X
Editors	Duda J., Kuric I., Katalinic B.
1999	**4th DAAAM International Workshop "CA Systems and Technologies: Advanced Methodology of Process Planning"**
Date & Place	12-13th October 1999, Zilina, Slovak Republic
Organized by	University of Zilina, VUT Vienna, CUT Cracow, DAAAM International Vienna, DAAAM Slovakia, DAAAM Poland
Organisers	Kuric I., Katalinic B., Duda J., Poppeova V., Marcincin J.
Proceedings	ISBN 3-901509-12-7
Editors	Kuric I., Katalinic B.
1998	**3rd DAAAM International Workshop "CA Systems and Technologies"**
Date & Place	29-30th September 1998, Cracow, Poland
Organized by	CUT Cracow, University of Zilina, DAAAM Poland, DAAAM Slovakia, DAAAM International Vienna
Organisers	Kuric I., Katalinic B., Duda J., Poppeova V., Marcincin J.
Editor	Kuric I.

1997	**2nd International DAAAM Workshop "CA Systems and Technologies"**
Date & Place	19th March 1997, Zilina, Slovakia
Organized by	University of Zilina, Faculty of Mechanical Engineering, DAAAM International Vienna; DAAAM Slovakia
Organizer	Cubanova N., Janac A., Kuric I., Micieta B., Wessely E.
Editor	Kuric I.

16.0 DAAAM INTERNATIONAL WORKSHOPS "INTELLIGENT MANUFACTURING SYSTEMS"

Founder and chair of DAAAM International workshop on Intelligent Manufacturing Systems is Professor Emil Wessely, Technical University of Kosice.

2002	**4th DAAAM Workshop "Human Factor and Environmentalist"**
Date & Place	5th December 2002, Kosice, Slovakia
Organized by	Department of Environmental Studies and Operations Management of Mechnaical Engineering Faculty of Technical University of Kosice, DAAAM International Vienna
Organized in	Slovak Technical Museum of Kosice
cooperation with	Oesterreichische Sued-Ost Institut - Vienna
Organisers	Badida M., Wessely E., Bosak M., Majernik M., Lumnitzer E., Kralikova R., Evin E., Rovnak M., Hricova B., Karaszova I., Liberkova L., Verebova H.
Supported by	Action Austria - Slovakia
Proceedings	ISBN 3-901509-37-2
Editors	Katalinic B., Wessely E.
2001	**3rd DAAAM Workshop "Intelligent Manufacturing Systems"**
Date & Place	29 November 2001, Kosice, Slovakia
Organized by	Technical University of Kosice, DAAAM International Vienna, DAAAM Slovakia
Organized in cooperation with	Action Austria-Slovakia, Faculty of Mechanical Engineering - Technical University of Kosice, Slovak Community for System Integration, Slovak Mechanical Engineering Association
Organisers	Wessely E.(Chair), Badida M., Seminsky J., Evin E., Ljoljic B., Kordic V., Pecha C., Rovnak M., Taticek F.
Supported by	Action Austria - Slovakia
Proceedings	ISBN 3-901509-31-3
Editors	Katalinic B., Wessely E.
2000	**2nd DAAAM Workshop "Intelligent Manufacturing Systems"**
Date & Place	30 November 2000, Kosice, Slovakia
Organized by	Technical University of Kosice, DAAAM International Vienna, DAAAM Slovakia
Organized in cooperation with	Action Austria-Slovakia, Faculty of Mechanical Engineering - Technical University of Kosice, Slovak Committee for Management - ZVTS Bratislava, Slovak Mechanical Engineering Association
Organizer	Wessely E.(Chair), Seminsky J., Badida M., Evin E., Paralic J., Knoflicek R., Hradocky L., Bosak M., Kozak L., Vargova J.
Supported by	Action Austria - Slovakia
Proceedings	ISBN 3-901509-18-6
Editors	Katalinic B., Wessely E.
1997	**1st DAAAM Workshop "Intelligent Manufacturing Systems"**
Date & Place	28th November 1997, Kosice, Slovakia
Organized by	Technical University of Kosice, DAAAM International Vienna, DAAAM Slovakia
Organized in cooperation with	Action Austria-Slovakia, Faculty of Mechanical Engineering - Technical University of Kosice, Slovak Committee for Management - ZVTS Bratislava, Slovak Mechanical Engineering Association
Organisers	Wessely E. (Chair), Seminsky J., Evin E., Kovacik J.
Proceedings	ISBN 3-901509-09-7
Editor	Seminsky J.[aroslav]

17.0 DAAAM INTERNATIONAL WORKSHOPS

As an extension of the Symposiums, International Workshops focussing on key issues are arranged. A number of these are detailed below.

1999	**1st International Regional DAAAM-CEEPUS Workshop on Intelligent Machines and Technologies in the 21st Century**
Date & Place	27-29th May 1999, Miskolc, Hungary
Organized by	Faculty of Mechanical Engineering, University of Miskolc in co-operation with DAAAM International Vienna
Proceedings	ISBN 3-901509-14-3
Organizer	Cselenyi J., Alpek F., Katalinic B., Kalmar L., Terstyanszky G., Kerekes I., Siposs I., Peter J.
1996	**4th International DAAAM Workshop "CAD & Intelligent Manufacturing Systems"**
Date & Place	22nd February 1996, Kosice, Slovakia
Organized by	DAAAM International Vienna; DAAAM Slovakia; Technische Universität Kosice, House of Technology, Informatech
Proceedings	ISBN 80-967223-8-7
Organizer	Katalinic B., Seminsky J., Wessely E.
1995	**3rd International DAAAM Workshop "Intelligent Manufacturing Systems"**
Date & Place	31st May 1995, Kosice, Slovakia
Organized by	DAAAM International Vienna; DAAAM Slovakia; Technische Universität Kosice, KAVS; Slovak-Austrian Society Kosice
Organizer	Katalinic B., Seminsky J., Wessely E.
1995	**2nd International DAAAM Workshop "Integration of Applications"**
Date & Place	15th May 1995, IBM Austria, Vienna, Austria
Organized by	DAAAM International Vienna; IBM Österreich; IBM Central Europe and Russia; AKS
Organizer	Beer M., Hansjakob K., Katalinic B., Kaufmann A., Lerch Z., Steinbichler J., Zucker R.
1994	**1st International DAAAM Workshop "Automatic Identification in Theory and Praxis",**
Date & Place	28th April 1994, Vienna University of Technology, Vienna, Austria
Organized by	DAAAM International Vienna; DATALOGIC; Vienna University of Technology, Department of Production Engineering, IFT-IMS
Organizer	Bruck P., Katalinic B., Lerch Z.
1991	**1st International DAAAM Seminar "Concurrent Engineering"**
Date & Place	17-18th April 1991, Kosice, Czechoslovakia
Organized by	Schlumberger Technologies, CAD/CAM Division – Wien; Department for Automatic Production Systems, Faculty of Mechanical Engineering, Technical University of Kosice
Organizer	Wessely E., Katalinic B., Ljuhar D.

18.0 INTERNATIONAL CONFERENCES ORGANIZED IN COOPERATION WITH DAAAM INTERNATIONAL AND/OR DAAAM NATIONAL

2006	**3rd International Textile Clothing & Design Conference, ITC&DC 2006**
Date & Place	8-11th October 2006, Dubrovnik, Croatia
Organized by	Faculty of Textile Technology, University of Zagreb
Under the auspices of	Croatian Ministry of Science, Education and Sport; University of Zagreb; Croatian Ministry of Economy, Labour and Enterpreneurship; Croatian Chamber of Economy
In cooperation with	DAAAM International Vienna; Association of Universities for Textile, Ghent, Belgium; The Textile Institute, Manchester, Great Britain; The International Textile Academia New Delhi; Croatian Academy of Engineering; Croatian Societs of Chemical Engineers; AATCC, United States of America; IFKT, Reutlingen, Germany; INA Oil Industry, Zagreb, Croatia; City of Dubrovnik, Croatia
Organizers	Dragcevic Z. (Chair), Vuljanic N. (Co-Chair), Bobovcan M., Bogovic S., Geratovic I., Hursa A., Jandric Dj., Martek M., Matusic K., Mauser B., Petrunic I., Vujasinovic E.
Proceedings	953-7105-12-1
Editors	Dragcevic, Z.[vonko]
Published by	Faculty of Textile Technology, University of Zagreb
2006	**11th International FMTÜ Conference for Young Researchers: Focus on IMS and Environmentally friendly technologies**
Date & Place	24-25 March 2006, Cluj -Napoca, Romania
Organized by	Technical Section of Transylvanian Museum Society (EME)
Under the auspices of	DAAAM International Viena
In cooperation with	Hungarian Academy of Sciences

Organizers	Gyenge Csaba - president of Scientific Committee Bitay Eniko - president of Organizing Committee
Proceedings	ISBN 973-82331-50-7
Editors	Gyenge C.[saba] & Katalinic B.[ranko]
Published by	Transylvanian Museum Society (EME) - external section of Hungarian Academy of Science
2006	**5th International Conference of DAAAM Baltic "Industrial Engineering – Adding Innovation Capacity of Labour Froce and Entrepreneur"**
Date & Place	20-22nd April 2006, Tallinn Technical University, Estonia
Organized by	Tallinn University of Technology, Faculty of Mechanical Engineering, DAAAM Estonia
Under Auspices of	DAAAM International Vienna
In cooperation with	DAAAM International Vienna , BALTECH Consortium, Estonian Academy of Science, Federation of Estonian Engineering Industries, Associations of Estanian Mechanical Engineers
Proceedings	ISBN 9985-894-92-8
Referred/indexed in	ISI Proceedings
Editor	Kyttner R.
2004	**3rd International Conference Business Systems Management UPS 2004**
Date & Place	23-24th September 2004, Mostar, Bosna and Herzegovina
Organized by	University of Mostar Faculty of Mechanical Engineering and Computing, DAAAM International Vienna, Austrian Society of Engineers and Architects OEIAV-1848, Scientific and Technical Forum of Mostar
Under Auspices of	Presidency of Bosnia and Herzegovina
Organizers	Majstorovic V. (Chair), Katalinic B. (Chair), Covic D., Visekruna, V.
Proceedings	ISBN 3-901509-24-0
Editors	Majstorovic V., Katalinic B., Covic D., Visekruna, V.
2004	**The 5th International Workshop on Emergent Synthesis**
Date & Place	May 24-25, 2004, Budapest, Hungary
Organized by	Computer and Automation Research Institute, Hungarian Academy of Sciences
Chair	L. Monostori, B. Katalinic, K, Ueda
2004	**4th International DAAAM Conference "Industrial Engineering – Innovations as Competitive Edge for "**
Date & Place	29-30th April 2004, Tallinn Technical University, Estonia
Organized by	Tallinn Technical University - Department of Machinery, DAAAM Estonia, DAAAM International Vienna
In cooperation with	Baltic Rector's Council; Association of Estonian; Mechanical Engineers, Federation of Estonian Engineering Industries
Proceedings	ISBN 9985-894-59-6
Editors	Papstel J., Katalinic, B.
2003	**The 1st European-Pacific Congress Globalization: Interaction of Economies, Cultures, Technologies, and Nature,**
Date & Place	23-26th September, 2003, Vladivostok, Russia
Organised by	Association "International Assembly of Capitals and Cities"; DAAAM International Vienna; Far Eastern State Technical University; Far Eastern National University; Maritime State University; International Engineering Academy Primorsk Chamber of Commerce & Industry; Far Eastern Branch of the Russian Academy of Sciences; Russian PEN Center; Vladivostok City Administration; Vladivostok Program Committee; Vladivostok State University of Economics
Organizers	Yu.M. Kopylov (Chairman); (Co-Chairman); G.S. Vishnyakova (Co-Chairman); I.V. Abrosjkin (Co-Chairman); Branko Katalinic (Co-Chairman); M.M. Sokolov, (Co-Chairman); A.A. Akhonin (Coordinator); R.G. Kazmiruk (Secretary)
Proceedings	in Print
2003	**The 8th Scientific Conference of Young Technical Researches, FMTU-2003**
Date & Place	21-22nd March 2003, Cluj-Napoca
Organised by	Technical Section of The Transilvanian Museum Association
Organizers	President of Scientific Committee: Professor Dr Csaba Gyenge
Proceedings	953-96408-8-1
Editors	Gyenge C.[saba]
2002	**1st International Textile Clothing & Design Conference, ITC&DC**
Date & Place	6-9th October 2002, Dubrovnik, Croatia
Organized by	Faculty of Textile Technology, University of Zagreb
Under the auspices of	DAAAM International Vienna; Association of Universities for Textile, Ghent, Belgium; The Textile Institute, Manchester, Great Britan; International Textile Academia; University of Zagreb; Croatian Academy of Engineering

On the occasion of	40th Anniversary of University Education and Scientific Research in the Field of Textiles and Textile Technology in Croatia
Organizers	Dragcevic Z. (Chair), Vuljanic N. (Co-Chair), Grundler D., Hursa A., Krmpotic N., Rolich T., Rogale D., Vujasinovic E.
Proceedings	953-96408-8-1
Editors	Dragcevic, Z.[vonko]
Published by	Faculty of Textile Technology, University of Zagreb
2002	**8th International Conference on Soft Computing: MENDEL 2002**
Date & Place	5-7th June 2001, Brno, Czech Republic
Organized by	Brno University of Technology, Faculty of Mechanical Engineering, Institute of Automation and Computer Science; DAAAM International Vienna, Austria; Foundation for Advanced Information Technology GENOPS, Ltd
Organizers	Ošmera P., Šeda M., Dvořák J., Roupec J., Matousek R.
2001	**5th International Conference Modern Technologies in Manufacturing MTeM 2001:** Assembly and disassembly, CIM, Cold Metal forming technologies, Concurrent Engineering, Ecological Design, Flexible Manufacturing Systems,Industrial Automation,, Nonconventional technologies, Rapid Prototyping, Quality Assurance, Robotics
Date & Place	4-6th October 2001, Cluj-Napoca, Romania
Organized by	Technical University of Cluj Napoca, Faculty of Machine Building, DAAAM International Vienna, S.C. Fortpres Cluj-Napoca, S.C. Plastor Oradea, AGIR- General Association of Engineers from Romania
Organizers	Gyenge C. (Chair) et all.
Proceedings	ISBN 973-85354-1-7
Editors	Gyenge C.
2001	**7th International Conference Maintenance 2001**
Date & Place	25-26th June 2001, Opatija, Croatia
Organized by	Croatian Maintenance Society
In cooperation with	DAAAM International Vienna, Union of Croatian Engineers
Organizers	Ivančić I., Cala I., Frković D., Sacevic T., Podgorčić D., Madžarević P., Kristovic D.,Poljak K., Kudelić, R., Ugarković Ž., Susanj B., Pavlovic A.
2001	**7th International Conference on Soft Computing: MENDEL 2001:** Evolutionary Computation, Genetic Programming, Fuzzy Logic, Rough Sets, Neural Networks, Fractals, Bayesian Methods
Date & Place	6--8th June 2001, Brno, Czech Republic
Organized by	Brno University of Technology, Faculty of Mechanical Engineering, Institute of Automation and Computer Science; DAAAM International Vienna, Austria; Foundation for Advanced Information Technology GENOPS, Ltd
Organizers	Ošmera P., Šeda M., Dvořák J., Roupec J., Matousek R.
2001	**2nd International Conference Business Systems Management UPS 2001**
Date & Place	31st May - 2nd June 2001, Mostar, Bosna and Herzegovina
Organized by	University of Mostar Faculty of Mechanical Engineering, DAAAM International Vienna, Austrian Society of Engineers and Architects OEIAV-1848
Organizers	Majstorovic V. (Chair), Katalinic B. (Chair), Covic D., Visekruna, V.
Proceedings	ISBN 3-901509-26-7
Editors	Majstorovic V., Katalinic B., Covic D., Visekruna, V.
2000	**6th International Conference Maintenance 2000**
Date & Place	19-20th June 2000, Opatija, Croatia
Organized by	Croatian Maintenance Society
In cooperation with	EFNMS - European Federation of National Maintenance Societies, DAAAM International Vienna, Union of Croatian Engineers
Organizers	Ivančić I., Cala I., et al.
2000	**5th International Conference on Soft Computing: MENDEL 2000:** Evolutionary Computation, Genetic Programming, Fuzzy Logic, Rough Sets, Neural Networks, Fractals, Bayesian Methods
Date & Place	7-9th June 2000, Brno, Czech Republic
Organized by	Brno University of Technology, Faculty of Mechanical Engineering, Institute of Automation and Computer Science; DAAAM International, Vienna, Austria; Foundation for Advanced Information Technology GENOPS, Ltd
Organizers	Ošmera P., Šeda M., Dvořák J.
Proceedings	ISBN 80-214-1609-2
2000	**2nd International Conference on Industrial Engineering – Activities Targeted to SME**
Date & Place	27 - 29th April 2000, Tallinn Technical University, Estonia
Organized by	Tallinn Technical University - Department of Machinery, DAAAM Estonia, DAAAM

	International Vienna
Under the auspices of	DAAAM International Vienna
In cooperation with	Baltic Rector's Council; Association of Estonian; Mechanical Engineers, Federation of Estonian Engineering Industries
Organizers	Kyttner R. (chair) Katalinic B., Kulu P., Papstel J., Reedik V., Riives J.
Proceedings	ISBN 9985-894-21-9
Editors	J. Papstel / Katalinic B.
1999	**District Heating Control ´99 – Control of Extensive Heating Systems**
Date & Place	26-27th October 1999, Zlin, Czech Republic
Organized by	Brno University of Technology, Faculty of Technology in Zlin; Czech Power Plants – Ltd.; Association for District Heating of Czech Republic; EUROHEAT & Power, Unichal – Brussels; DAAAM International, Vienna, Austria; DAAAM - Czech Republic
Organizers	Balate J., Vasek V., Vachutova I., Sysala T.
Proceedings	ISBN 80-214-1461-8
1999	**5th International Conference on Soft Computing: MENDEL ´99:** Evolutionary Computation, Genetic Programming, Fuzzy Logic, Rough Sets, Neural Networks, Fractals
Date & Place	9-12th June 1999, Brno, Czech Republic
Organized by	Technical University Brno, Faculty of Mechanical Engineering, Institute of Automation and Computer Science; DAAAM International, Vienna, Austria; Foundation for Advanced Information Technology GENOPS, Ltd
Organizers	Ošmera P., Šeda M., Dvořák J.
Proceedings	ISBN 80-214-1131-7
1999	**5th International Conference Maintenance 99**
Date & Place	7- 8th June 1999, Opatija, Croatia
Organized by	Croatian Maintenance Society
In cooperation with	DAAAM International Vienna, EFNMS - European Federation of National Maintenance Societies
Organizers	Ivančić I., Poljak K., Ugarković Ž., Frković D., Podgorčić D., Madžarević P., Kuzmanić A. Kudelić, R.
Proceedings	ISBN 953-97615-1-4
Editors	Ivančić I., Čala I.
1999	**XXIVth Seminary ASR 2000: Instruments and Control**
Date & Place	28th April 1999, Ostrava, Czech Republic
Organized by	VŠB - Technical University Ostrava, Faculty of Mechanical Engineering, Institute of Automation Equipment and Control; DAAAM National - Czech Republic, Zlín, Czech Republic; IMECO - Czech National Committee; Committee of Applied Cybernetics and Information Technology Ostrava
Organizers	Farana R., Smutny J., Szlachta D., Fojtik D., Piechota H.
Proceedings	ISBN 80-7078-774-0
Editors	Farana R., Smutny J., Szlachta D., Fojtik D., Piechota H.
1999	**XXIIth Seminary ASR ´99 : Computers on Measurements, Diagnostics and Control**
Date & Place	22nd May 1999, Ostrava, Czech Republic
Organized by	VŠB - Technical University Ostrava, Faculty of Mechanical Engineering, Institute of Automation Equipment and Control; DAAAM National - Czech Republic, Zlín, Czech Republic; IMECO - Czech National Committee
Organizers	Smutný L., Farana R., Vítecek A., Baláte J.
Proceedings	ISBN 80-7078-666-3
1998	**3rd International Conference „ New Trends in Automation of Energetic Processes ´98"**
Date & Place	19 - 20th May 1998, Zlín, Czech Republic
Organized by	Technical University Brno, Faculty of Technology in Zlín; Czech Power Company, Ltd.- CEZ, a.s.; Association for District Heating, Czech Republic; STU Bratislava, Faculty of Electrical Engineering and Information Technology; Slovak Power Plants, Ltd. - SE, a.s., Bratislava; DAAAM International Vienna, Austria; DAAAM National - Czech Republic
Organizer	Balate J., Vasek V., Sysala T. & Co.
Proceedings	ISBN 80-214-1094-9
Editor	Balate J., Sysala T.
1998	**4th International Mendel Conference: MENDEL ´98** on Genetic Algorithms, Optimization Problems, Fuzzy Logic, Neural Networks, Rough Sets
Date & Place	24 - 26th June 1998, Brno, Czech Republic
Organized by	Technical University Brno, Faculty of Mechanical Engineering, Institute of Automation and Information Technology; DAAAM International, Vienna, Austria; Foundation for Advanced Information Technology GENOPS, Ltd.
Organizer	Osmera P., Seda M., Simoník I.

Proceedings	ISBN 80 – 214 - 1199 - 6
1998	**XXIth Seminary ASR ´98 : Computers on Measurements, Diagnostics and Control**
Date & Place	22nd May 1998, Ostrava, Czech Republic
Organized by	VSB - Technical University Ostrava, Faculty of Mechanical Engineering, Institute of Automation Equipment and Control; DAAAM National - Czech Republic; IMECO - Czech National Committee
Organizer	Smutny L., Farana R., Vítecek A., Balate J.
Proceedings	ISBN 80-7078-559-4
1998	**9th International Conference on Information and Intelligent Systems IIS´98**
Date & Place	23-25th September 1998, University of Zagreb, Faculty of Organization and Informatics Varazdin, Croatia
Organized by	University of Zagreb, Faculty of Organization and Informatics Varazdin & DAAAM International Vienna
Organizers	Aurer B., Katalinic B., Kero K., et. al.
Proceedings	UDC: 007:681.3:651(082)
Editors	Aurer B., Logozar, R.
1998	**3rd National DAAAM Conference "Science ´98"**
On the occasion of	80 Years of Tallinn Technical University
Date & Place	September 1998, Tallinn Technical University, Estonia
Organized by	DAAAM Estonia, Tallinn Technical University, Department of Machinery
Organizers	Papstel J.
Proceedings	ISBN 9985-894-02-2
Editor	Papstel J.
1997	**4th International Conference Maintenance 97**
Date & Place	16-17th October 1997, Opatija, Croatia
Organized by	Croatian Maintenance Society
In cooperation with	DAAAM International Vienna, Croatian Academy of Science and Art, Zagreb EFNMS - European Federation of National Maintenance Societies
Organizers	Ivančić I., Čala I., Vidić J., Udiljak T., Ugarković Ž., Širola D., Podgorčić D., Keser A.
Proceedings	ISSN-1331-4114
Editors	I. Ivančić
1997	**1st International Conference "Industrial Engineering - Actual Activities"**
Date & Place	25-27th September 1997, Tallinn Technical University, Tallinn, Estonia
Organized by	Tallinn Technical University - Institute of Machinery, DAAAM Estonia Tallinn, DAAAM International Vienna
Under the auspices of	DAAAM International Vienna
In cooperation with	Baltic Rectors´ Council, Association of Baltic Material Societies, Federation of Estonian Engineering Industries, Association of Estonian Mechanical Engineers
Organizer	Kulu P. (vice-chair), Kyttner R. (general chair), Papstel J.(vice-chair), Katalinic B., Kübarsepp J., Reedik V., Riives J.
Proceedings	ISBN 3-901509-07-0
Editors	Papstel J., Katalinic B.
1997	**2nd International Conference "Enterprise in Transition"**
On the occasion of	1700 Years of the City of Split
Date & Place	22-24th May 1997, Split - Brela, Croatia
Organized by	University of Split, Faculty of Economics, Split, Croatia; DAAAM International Vienna
Under the auspices of	The Government of the Republic of Croatia, The Croatian Chamber of Economy, Bank of Zagreb
Organizer	Alfirevic N., Buble M., Derado D., Goic S., Katalinic B., Mijacika D., Pavic I. (Chair), Reic Z. (Secretary)
Proceedings	ISBN 953-6024-14-4
Publisher & Editor	Faculty of Economics, University of Split, Split
Co-Publisher	DAAAM International Vienna
1997	**3rd International Mendel Conference on Genetic Algorithms MENDEL '97**
Date & Place:	25-27th June 1997, Brno, Czech Republic
Organisation:	Technical University of Brno, Faculty of Mechanical Engineering, Institute of Automation and Computer Science; DAAAM International Vienna; Foundation Advanced Information Technology; GENOPS, Ltd.
Organizer:	Osmera P., Seda M., Simonik I.
1996	**The Second National DAAAM Conference SCIENCE '96**
Date & Place:	1996, Tallinn, Estonia
Organisation:	Tallinn Technical University - Estonia, DAAAM International Vienna

Organizer:	Papstel J. & Co
Proceedings:	ISBN 9985-59-041-4
Editor:	Papstel J.
1996	**2nd International Mendel Conference on Genetic Algorithms MENDEL '96**
Date & Place:	26-28th June 1996, Brno, Czech Republic
Organisation:	Technical University of Brno, Faculty of Mechanical Engineering, Institute of Automation and Computer Science; DAAAM Czech Republic & DAAAM International Vienna; Foundation Advanced Information Technology; AutoCont Brno Ltd. Computer Systems
Organizer:	Osmera P. & Co.
Proceedings:	ISBN 80-214-0769-7
1996	**13th Conference BIAM '96**
Date & Place:	18-20th June 1996, Zagreb, Croatia
Organisation:	KoREMA, Croatian Society for Communications, Computing, Electronics, Measurement and Control, DAAAM International Vienna
Organizer:	Surina T., Vranjes B., Katalinic B., Jerbic B., Cosic P., Husic S., Kovacic Z., Kunica Z., Majetic D., Momcilovic D., Schauperl Z., Stambuk-Borsic N., Stefanic N., Udiljak T.
Proceedings:	ISBN 953-6037-12-2
1996	**Nanotechnologie - Nanometrologie**
Date & Place:	26th April 1996, Brno, Czech Republic
Organisation:	Technical University of Brno, Faculty of Mechanical Engineering, Vienna University of Technology; DAAAM Czech Republic
Organizer:	P. H.Osanna, J. Svejcar, J. Vackar, M. Skopal
1996	**Rechenmaschinen in Messungen, Diagnostik und Steuerung**
Date & Place:	18th April 1996, Ostrava, Czech Republic
Organisation:	Technical University of Ostrava TU-VSB, KAKI Ostrava, DAAAM Czech Republic, IMECO Czech Republic
Organizer:	Smutny L., Farana R., Vitecek A.
1995	**„New Trends in the Automation of Energetical Processes"**
Date & Place:	17-18th May 1995, Zlin, Czech Republic
Organisation:	TU-Brno / FT-Zlin, KART, DAAAM Czech Republic & DAAAM International Vienna, and others
Organizer:	Balate J., Janacova J., Kubalcik M., Sysala T., Neumann P., Vasek V.
Proceedings:	ISBN 80-214-0654-2

19.0 DAAAM & CEEPUS JEP

A-104 CEEPUS Joint European Project "Intelligent Manufacturing and Automation"
Network Co-ordinator: *Branko Katalinic - Vienna University of Technology*

Project Code:	***A-104***
Project Duration:	***1999-2004***
Project Coordinator:	***Branko Katalinic - Vienna University of Technology***
2003/2004 Name:	**Intelligent Manufacturing and Automation**
Project Partner 03/04	Institute of Production Engineering, Vienna University of Technology, Vienna, Austria / Department of Manufacturing Engineering, Budapest University of Technology and Economics, Budapest, Hungary / Department of Manufacturing Engineering, Technical University of Cluj-Napoca, Cluj-Napoca, Romania / Department of Technology, Faculty of Mechanical Engineering and Naval Architecture, University of Zagreb, Zagreb, Croatia / Department of Technological Devices and Systems, Faculty of Material Science and Technology Trnava, Slovak University of Technology Bratislava, Trnava, Slovakia / Institute of Production Engineering, University of Maribor, Maribor, Slovenia / Department of Environmental Studies and Process Control, Technical University of Kosice, Kosice, Slovakia / Department of Machine Technology, The Silesian University of Technology, Gliwice, Poland / Department of Production Engineering, Faculty of Engineering, University of Rijeka, Rijeka, Croatia / Institute of Mechanical Technology, Poznan University of Technology, Poznan, Poland
Organizer 03/04	**Katalinic B.[ranko],** Alpek F.[erenc], Balic J.[oze], Gyenge C.[saba], Kosmol J.[an], Legutko S.[tanislaw], Mikac T.[onci], Velisek K.[arol], Wessely E.[mil], Udiljak T.[oma]
2002/2003 Name:	**Intelligent Manufacturing and Automation**
Project Partner 02/03	Institute of Production Engineering, Vienna University of Technology, Vienna, Austria / Department of Manufacturing Engineering, Budapest University of Technology and Economics, Budapest, Hungary / Department of Manufacturing Engineering, Technical

	University of Cluj-Napoca, Cluj-Napoca, Romania / Department of Technology, Faculty of Mechanical Engineering and Naval Architecture, University of Zagreb, Zagreb, Croatia / Department of Technological Devices and Systems, Faculty of Material Science and Technology Trnava, Slovak University of Technology Bratislava, Trnava, Slovakia / Institute of Production Engineering, University of Maribor, Maribor, Slovenia / Department of Environmental Studies and Process Control, Technical University of Kosice, Kosice, Slovakia / Department of Machine Technology, The Silesian University of Technology, Gliwice, Poland / Department of Production Engineering, Faculty of Engineering, University of Rijeka, Rijeka, Croatia / Department of Industrial Robotics, University of Applied Sciences - Fachhochschule Technikum - Wien, Vienna, Austria / Institute of Mechanical Technology, Poznan University of Technology, Poznan, Poland
Organizer 02/03	**Katalinic B.[ranko],** Alpek F.[erenc], Balic J.[oze], Gyenge C.[saba], Kosmol J.[an], Legutko S.[tanislaw], Mikac T.[onci], Stopper M.[arkus], Velisek K.[arol], Wessely E.[mil], Udiljak T.[oma]
2001/2002 Name:	**Assembly Automation in Manufacturing Engineering**
Project Partner 01/02	Vienna University of Technology, Budapest University of Technology and Economics, Silesian University of Technology, Slovak Technical University Bratislava / Trnava, Technical University Cluj-Napoca, Technical University of Kosice, University of Maribor, University of Zagreb
Organizer 01/02	**Katalinic B.,** Alpek F., Kosmol J., Velisek K., Gyenge C., Wessely E., Balic, J., Jerbic B.,
2000/2001 Name:	**Assembly Automation in Manufacturing Engineering**
Project Partner 00/01	Vienna University of Technology, Slovak Technical University Bratislava / Trnava, Technical University Cluj-Napoca, Technical University of Budapest, Technical University of Kosice, University of Maribor, University of Zagreb
Organizer 00/01	Katalinic B., Velisek K., Gyenge C., Alpek F., Wessely E., Balic J., Jerbic B.
1999/2000 Name:	**Assembly Automation in Manufacturing Engineering**
Project Partner 99/00	Vienna University of Technology, Slovak Technical University Bratislava / Trnava, Technical University Cluj-Napoca, Technical University of Budapest, University of Maribor, University of Zagreb
Organizer 99/00	**Katalinic B.,** Velisek K., Gyenge C., Alpek F., Trzan F., Jerbic B.

DAAAM – CEEPUS Joint European Annual Projects which finished

Project Code:	***RO-01***
Project Duration:	***2002/2003***
Project Coordinator:	***Csaba Gyenge, Technical University of Cluj-Napoca***
2002/2003 Name:	**Computer-Aided Systems for Manufacture and Measurement of Machine Elements**
Project Partner	Department of Manufacturing Engineering / Technical University of Kosice, Department of Manufacturing Systems / University of Maribor, Faculty of Mechanical Engineering, Institute for Production Engineering / Poznan University of Technology, Faculty of Mechanical Engineering, Institute of Mechanical Technology / Kecskemet College, Faculty of Technology

Project Code:	***PL-1***
Project Duration:	***1999-2002***
Project Coordinator:	***Stanislaw Adamczak - Technical University of Kielce***
2001/2002 Name:	**Computer-Aided Systems for Manufacture and Measurement of Machine Elements**
Project Partner	Technical University of Kielce, Vienna University of Technology, University of Zilina, Faculty of Mechanical Engineering of The Technical University of Kosice, Faculty of Production Technologies of The Technical University of Kosice in Presov, University of Maribor, Technical University of Ostrava, University of Zagreb, Technical University of Bratislava, Technical University of Prague, Technical University of Brno
Organizer	**Adamczak S.,** Osanna P. H., Obmascik M., Wessely E., Vasilko K., Cus F., Vrba V., Mahovic S., Janac J., Dvorak R., Vackar J.
2000/2001 Name:	**Geometrical Surface Structure of Machine Parts, Constitution, Methods for Measurement and Assessment, Influence on the Operational Capacity of Mating Machine Parts**
Project Partner	Technical University of Kielce, Vienna University of Technology, University of Zilina, Technical University of Kosice, University of Maribor, Technical University of Ostrava, University of Zagreb, Technical University of Bratislava, Technical University of Prague, Technical University of Brno
Organizer	**Adamczak S.,** Osanna P. H., Obmascik M., Wessely E., Cus F., Vrba V., Mahovic S., Janac J., Dvorak R., Vackar J.

1999/2000 Name:	**Geometrical Surface Structure of Machine Parts, Constitution, Methods for Measurement and Assessment, Influence on the Operational Capacity of Mating Machine Parts**
Project Partner	Technical University of Kielce, Vienna University of Technology, University of Zilina, Technical University of Kosice, University of Maribor, Technical University of Ostrava, University of Zagreb, Technical University of Bratislava, Technical University of Prague, Technical University of Brno
Organizer	**Adamczak S.,** Osanna P. H., Obmascik M., Wessely E., Cus F., Vrba V., Mahovic S., Janac J., Dvorak R., Vackar J.

Project Code:	***RO-01***
Project Duration:	***1998/1999***
Project Coordinator:	***Csaba Gyenge, Technical University of Cluj-Napoca***
1998/1999 Name:	**Development and Application of Environment-Friendly Technologies in Manufacturing Engineering**
Project Partner	Technical University of Cluj-Napoca, Technical University of Budapest, University of Maribor, Technical University of Kosice, Technical University of Bucharest
Organizer	**Gyenge C.,** Alpek F., Balic J., Anna V., Iatan F.

Project Code:	***PL-1***
Project Duration:	***1998/1999***
Project Coordinator:	***Stanislaw Adamczak - Technical University of Kielce***
1998/1999 Name:	**Metrology in Quality Assurance Systems**
Project Partner	Technical University of Kielce, Vienna University of Technology, University of Zilina, Technical University of Kosice, University of Maribor, Technical University of Ostrava, University of Zagreb
Organizer	**Adamczak S.,** Osanna P. H., Obmascik M., Wessely E., Cus F., Vrba V., Mahovic S.

Project Code:	***H-78***
Project Duration:	***1998/1999***
Project Coordinator:	***Alpek,F.***
1998/1999 Name:	**Process Integrated Measurement Techniques and Methods for Quality Management**
Project Partner	Technical University of Budapest, Vienna University of Technology, Cracow Institute of Technology, Slovak Technical University Bratislava, University of Maribor
Organizer	**Alpek F.,** Osanna H. P., Sostar A., Gawlik J., Kralik,M.

20.0 DAAAM INTERNATIONAL BOOKS

The DAAAM International initiated publishing of books in the field of advanced manufacturing and automation. There are three series: "Manufacturing technology", "DAAAM International University Book" and “Management Sciences”. Till now the following books are published:

DAAAM Publishing series - Manufacturing technology:

1. **Joze Balic:** ***"Contribution to Integrated Manufacturing"***, DAAAM Publishing series - Manufacturing technology, Editor: B Katalinic, ISBN 3-901509-03-8, DAAAM International Vienna, Vienna 1999 (Promotion at DAAAM 1999 - Vienna)
2. **Borut Buchmeister:** ***"English-Slovene Glossary of Terms used in Production Engineering and Production Management"***, DAAAM Publishing series - Manufacturing technology, Editor: B Katalinic, ISBN 3-901509-17-8, DAAAM International Vienna, Vienna 2000 (Promotion at DAAAM 2000 - Opatija)
3. **Joze Balic:** ***"FMS: Development Structure, Operation, Handling, Tooling"***, DAAAM Publishing series - Manufacturing technology, Editor: B Katalinic, ISBN 3-901509-20-8, DAAAM International Vienna, Vienna 1999 (Promotion at DAAAM 2001- Jena)

DAAAM Publishing series - University Book:

4. **Vlado Majstorovic:** ***"Upravljanje proizvodnjom i projektima – Production and Project Management"***, (in Croatian) DAAAM International University Book, Editor: B Katalinic, ISBN 3-901509-23-2, DAAAM International Vienna and University of Mostar, Mostar – Vienna, 2001 (Promotion at International Conference in Mostar, May 31st, 2001)

5. **Dragan Covic, Vlado Majstorovic, Vojo Visekruna:** ***"Poslovni sustavi – Business Systems"***, (in Croatian) DAAAM International University Book, Editor: B Katalinic, ISBN 3-901509-22-4, DAAAM International Vienna and University of Mostar, Mostar – Vienna, 2001 (Promotion at International Conference in Mostar, May 31st, 2001)

DAAAM Publishing Series – Management Science:

6. **Dragan Covic:** ***"Inventivnost Tima – Vodic za menedzere"***, (in Croatian) DAAAM International University Book, Editor: B Katalinic, ISBN 3-901509-25-4, DAAAM International Vienna and University of Mostar, Mostar – Vienna, 2003. (Promotion October 2003, Sarajevo – Mostar)

7. **Zvonko Kremljak:** ***" Decision making under risk"***, DAAAM International University Book, Editor: B Katalinic, ISBN 3-901509-44-5, DAAAM International Vienna, Vienna 2004 (Promotion at DAAAM 2004 – Vienna)
8. **Zvonko Kremnjak & Borut Buchmeister:** ***"Uncertainty and Development of Capabilities"***, DAAAM Publishing Series – Management Science Editor: Branko KATALINIC, ISBN 3-901509-55-0, DAAAM International Vienna, Vienna, 2004 (Promotion at DAAAM 2006 – Vienna)

DAAAM International Scientific Book

9. ***DAAAM International Scientific Book 2002***, ISBN 3-901509-30-5, hard cover, 713 pages, Publisher DAAAM International Vienna, Editor: B.[ranko] Katalinic, Vienna, 2002. (Promotion of the book at DAAAM 2002 Vienna, 2002-10-23/26)
10. ***DAAAM International Scientific Book 2003***, ISSN 1726-9687, ISBN 3-901509-36-4, 654 pages, hard cover, Publisher DAAAM International Vienna, Editor: B.[ranko] Katalinic, Vienna 2003. *Abbreviated key title: DAAAM int. sci. book* (Promotion of the book at DAAAM 2003 Sarajevo 2003-10-22/25)
11. ***DAAAM International Scientific Book 2004***, ISSN 1726-9687, ISBN 3-901509-38-0, 730 pages, hard cover, Publisher DAAAM International Vienna, Editor: B.[ranko] Katalinic, Vienna 2004. *Abbreviated key title: DAAAM int. sci. book* (Promotion of the book at DAAAM 2004 Vienna 2004-11-03/06)
12. ***DAAAM International Scientific Book 2005***, ISSN 1726-9687, ISBN 3-901509-43-7, hard cover, Publisher DAAAM International Vienna, Editor: B.[ranko] Katalinic, Vienna 2005. *Abbreviated key title: DAAAM int. sci. book*
13. ***DAAAM International Scientific Book 2006***, ISSN 1726-9687, ISBN 3-901509-47-X, hard cover, Publisher DAAAM International Vienna, Editor: B.[ranko] Katalinic, Vienna 2006. *Abbreviated key title: DAAAM int. sci. book*

Book published under the auspices of DAAAM International

14. **Balc, N.; Gyenge, Cs. & Berce, P:** ***"Design for Competitive Manufacturing"***, Published under the auspices of DAAAM International, ISBN 973-7898-31-1, Alma mater Publishing, Cluj-Napoca, 2006.

Several other book projects are in process, if you want to join one of them or propose to write a new book please contact the Editor.

21.0 INTERNATIONAL JOURNAL OF SIMULATION MODELLING

Title	**International Journal of Simulation Modelling**
Abbreviated key title	**IJSIMM Int. j. simul. model.**
ISSN	ISSN 1726-4529
Editors	B.[orut] Buchmeister (Editor-In-Chief); B.[ranko] Katalinic (Editor)
Founded	2002
Published	Quarterly by DAAAM International Vienna
Citation and reference	Article published in Journal are crossreferenced and online available on Internet. Additionali they are cited and referenced in the: Inspec Database and Cambridge Scientific Abstracts
Info and Link	http://www.ijsimm.com, http://www.daaam.com,

22.0 20 YEARS DAAAM INTERNATIONAL 1990-2010 PROJECT

Aims: The main aims for future development of our organisation are to increase meaning, importance, quality, relevance and prestige of DAAAM International Network and members in international academic and scientific community. To achieve those aims it is necessary to improve quality and quantity of existing activities and to organize new creative future oriented activities using enormous potential of knowledge, abilities, prestige, relevance and creativity of our members.

Who: Each person in DAAAM International community is welcome to give his/her own contribution to achieving defined aims and can be responsible for one initiative or activity on the level of DAAAM International.

Focus: Focused on upper aims and the fact that we are going soon to celebrate first 20 years of our organisation we are starting an open platform under the name "20 Years of DAAAM International Project". The project is open platform the main parts and topics of the project are:

- To complete and publish Cumulative Bibliography of DAAAM International including all publications: proceedings of DAAAM Symposiums; proceedings of DAAAM Workshops; proceedings of ATDC conferences; DAAAM International Scientific Books; International Journal on Simulation Modelling; Books.
- Founding and organisation of web DAAAM International Archive
- Founding and organisation of web DAAAM International Photo Gallery
- Completing and publishing exclusive mongraphy "20 Years of DAAAM International"
- Founding, completing and publishing "Who is who in DAAAM International" (Founding members, members of committees since 1990, authors, honours, awards, personal contributions, …)
- improve quality and quantity of existing activities of DAAAM
- to motivate old and animate new people to take a part in our activities
- better use of human resources
- including of younger researchers in our activities
- sponsoring and partnerships
- to increase dissemination of information about our activities
- use of synergies of DAAAM Network in science, research, education and international scientific and academic cooperation
- awards and recognition for DAAAM International on the occasion of celebrating of anniversary of your institution
- to increase quality of our publications
- increase citing and indexing of our publications
- publishing of journals, monographs and student books
- international scientific and academic cooperation with joint application for co-operation of institutions and mobility of teachers and students
- cooperation in industrial and research projects
- DAAAM International certifications of Labs
- Marketing
- DAAAM International Experts data bank
- Cyber space and DAAAM
- Spectrum of Activities to celebrate 20 Years of DAAAM International in 2009
- and many of your ideas

Call for cooperation: Please think about your wishes, possibilities, and field of interest and propose me one to three initiatives you or your team want to start, organize, manage and be responsible in the frame of DAAAM International. I will pick out of your proposal one initiative which is at the moment the most important from the point of view of DAAAM International. I will give you additional information and suggestions to support your initiative and make it relevant and usable for DAAAM International and synchronised with other activities.

Contact: I am looking forward to see soon your ideas, creativity and initiatives. Please be free to contact me at president@daaam.com I will be happy to hear from you.

23.0 DAAAM - INTERNET & CONTACT ADDRESS

Additional information about DAAAM activities your can obtain from DAAAM homepage: http://www.daaam.com
If you are interested to take a part in DAAAM activities please contact:

⊠ PROF. DR. BRANKO KATALINIC, DAAAM INTERNATIONAL PRESIDENT,
TU-WIEN, KARLSPLATZ 13/311, A-1040 WIEN, AUSTRIA, EUROPE
Phone: 0043 1 58801 31121, *Fax#* 0043 1 58801 31199,
E-mails: president@daaam.com, katalinic@mail.ift.tuwien.ac.at

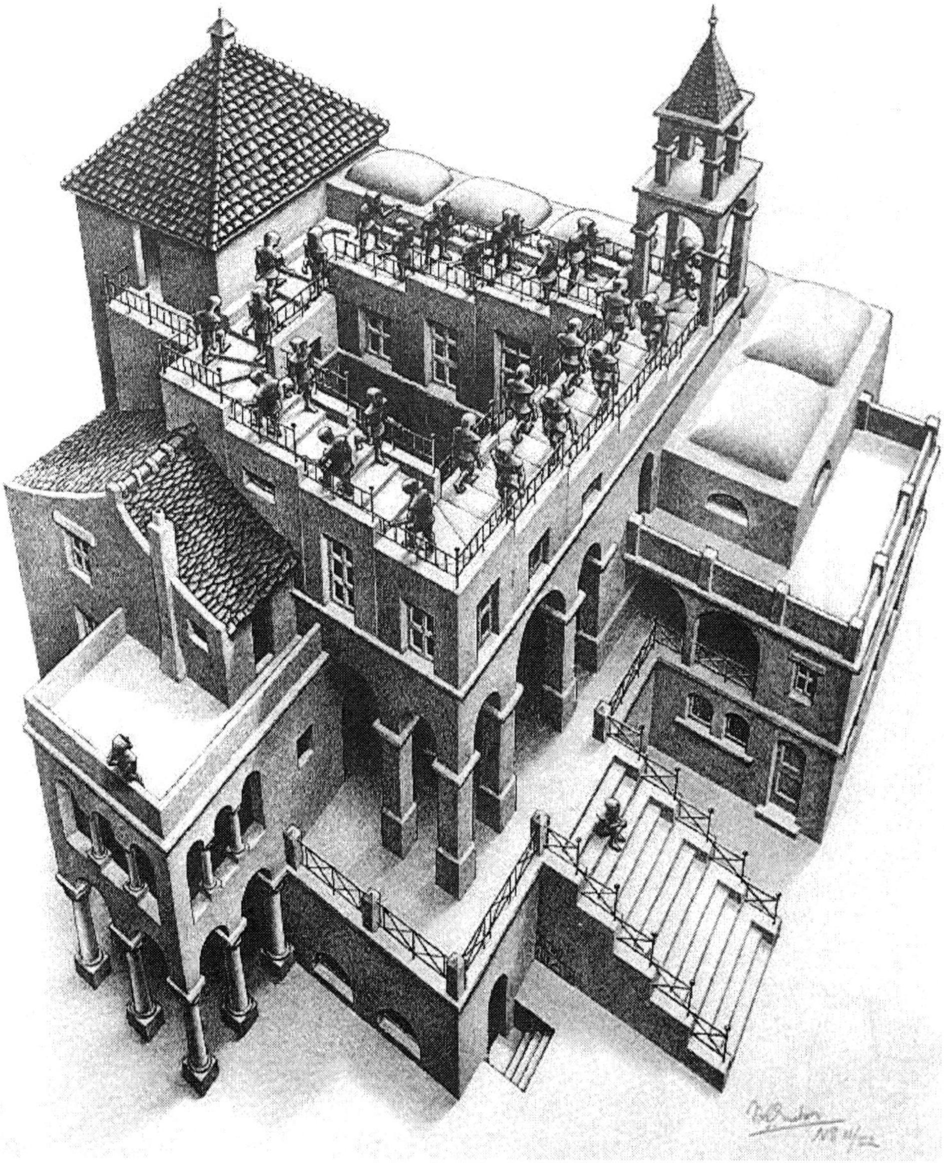

Annals of DAAAM for 2006 & Proceedings of the 17th International DAAAM Symposium, ISSN 1726-9679
Editor B. Katalinic, Published by DAAAM International, Vienna, Austria 2006
150th Birthday of Nikola Tesla (1856-1943) Inventor of Mobile Robot Teleoperation

CONTENTS

D

F

G

H

M

N

O

P

Q

R

S

T

U

V

W

Y

Z

WORK, FINISH, PUBLISH

MICHAEL FARADAY (1791-1867)

Annals of DAAAM for 2006 & Proceedings of the 17th International DAAAM Symposium, ISSN 1726-9679
Editor B. Katalinic, Published by DAAAM International, Vienna, Austria 2006
150th Birthday of Nikola Tesla (1856-1943) Inventor of Mobile Robot Teleoperation

MODELLING OF MODIFIED MICRO-FLOW SENSOR

ADAMEK, M.; MATYSEK, M. & MACKU, L.

Abstract: *In the presented investigation, the problem of accurate determination of micro mass flow was studied. The measurement of mentioned flow range is becoming more and more important for a lot of applications in the life science, analysis, biotechnologies, synthesis (of e.g. pharmaceuticals) and nanotechnology markets. In this paper a flowmeter design applicable for measurement of low gas flow amounts is presented. To this end, the numeric model of designed flowmeter was made and the flowmeter was numerically simulated using compressible Navier – Stokes equations in two dimensions. The computations were carried out for several types of agitated gases. The results were compared with experimental data.*
Key words: *flow, modelling, measurement, sensor words.*

1. INTRODUCTION

The modelling of microsystems requires appropriate compact or macro models for microdevices. In general, it is difficult or almost impossible to obtain closed form analytical models for microdevices. Thus, models are obtained by a simplification of the full physical model. The compact models allow for fast system level simulation of the microsystem. However, the accuracy of such a model can be questionable because of the simplifications made during the model development phase. Model accuracy and simplicity are important for the design of complex systems. Microflow sensor modelling has been examined by several authors (Lammerink, 1993). These approaches have used a solution of the partial differential equations (PDEs) for the coupled fluid/thermal problem. The work of (Kolomaznik, 2000) employs equivalent circuit descriptions that are solved in SPICE whereas (Lammerink, 1995) solve the PDEs using the finite-element and the finite-difference methods, respectively. However, none of these methods provide a simple and accurate macromodel for the flow sensor.
In this paper is presented modelling approach for microflow sensor. This paper is organized as follows: Section 2 contains description of the principle of the time – of – flight sensor; Section 3 presents a numerical model of the sensor (basic configuration, modelling approach, modelling results); Section 4 concludes the paper.

2. DESCRIPTION OF PRINCIPLES OF FLOW SENSOR

This work is focused on the modelling of a thermal method for mass flow determination of flowing media. This flowmeter type is called as time – of – flight sensor. The time – of – flight sensor consists of a heater and one or more temperature sensors downstream. The heater is activated by current pulses. The transport of the generated heat is a combination of diffusion and forced convection. The resulting temperature field can be detected by temperature sensors located downstream. The sensor output is the time difference between the starting point of the generated heat pulse and the point in time at which a maximum temperature at the downstream sensor is reached.
The time – of – flight sensors have the same limitations as the intrusive type of calorimetric sensors: corrosion, erosion and leakage. Since the signal processing needs a short while to measure the time difference, this sensor type is not suitable for dynamic measurement.
The transport of the heat generated in line source through a fluid is governed by the energy equation

$$\frac{\partial T}{\partial t} + v\nabla T = \left(\frac{\lambda}{\rho c}\right)\nabla^2 T + \frac{q^{'}}{\rho c} \qquad (1)$$

where T is the temperature, c is the specific heat at the constant pressure, ρ is the density, v is the kinematic viscosity of the fluid, λ is the thermal conductivity and q' (in $W\ m^{-3}$) is the amount of heat per unit of volume and time. The analytical solution of this differential equation for a pulse signal with input strength q'_o ($W\ m^{-1}$) is given in (Webster, 1999) as:

$$T(x,y,t) = \left(\frac{q_o^{'}}{4\pi\lambda t}\right)\exp\left\{-\frac{\left[(x-vt)^2\right]}{4at}\right\} \qquad (2)$$

where a denotes the thermal diffusivity.
The designed flowmeter is modified version of the time – of – flight sensor. In this case a temperature field is measured by semiconductor film located in top part of sensor tube. The temperature profile in semiconductor film depends on gas velocity. The heater is activated periodically; the period of the heat pulses generation can be different for different velocity size.

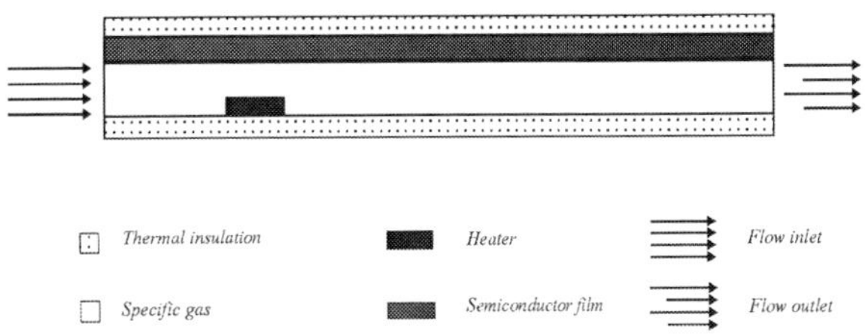

Fig. 1. Modified version of the time – of – Flight sensor

3. NUMERICAL MODEL

3.1 Basic equations for *T(x,y)*

The simulated time – of – flight sensor is a multiphysics model, meaning that it involves more than one kind of physics. In this case, there are Navier-Stokes equations from fluid dynamics together with a heat transfer equation that is essentially a convection-diffusion equation. There are four unknown field variables: the velocity field components u and v, the pressure p and the temperature T. They all are interrelated through bidirectional multiphysics couplings.

The pure Navier-Stokes equations consist of a momentum balance (a vector equation) and a mass conservation. The equations are

$$\rho\frac{\partial u}{\partial t}+\rho(u.\nabla)u=-\nabla p+\eta\nabla^2 u+F \qquad (3)$$

$$\nabla.u=0 \qquad (4)$$

where F is a volume force, ρ is the fluid density and v is the dynamic viscosity.
The heat equation is an energy conservation equation that only says that the change in energy is equal to the heat source minus the divergence of the diffusive heat flux

$$\rho c_p\frac{\partial T}{\partial t}+\nabla.(-k\nabla T+\rho c_p Tu)=Q \qquad (5)$$

where c_p is the heat capacity of the fluid and ρ is fluid density as before. The expression within the brackets is the heat flux vector and Q represents a source term. The heat flux vector contains a diffusive and a convective term, where the latter is proportional to the velocity field u.

3.2 Modelling approach

The two dimensional configuration of the designed flowmeter in the present work can be seen in Figure 2.

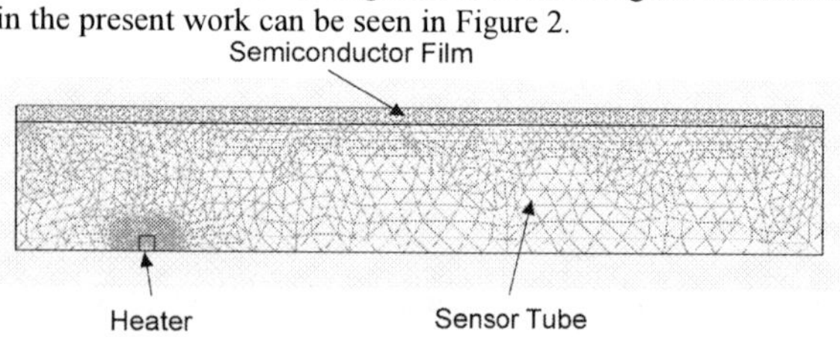

Fig. 2. Partial view of the computational grid for the basic configuration

The properties of the flow sensor were investigated at the relatively very low flow rates. In Table 1 are shown inlet speeds v of gas used in the sensor model and corresponding flow rate Q.

Q[ml.h^{-1}]	10	30	50	70	90
v [m.s^{-1}]	0.0309	0.0926	0.1543	0.2160	0.2778

Table 1. Gas flow volumes and velocities according to the pipe height 0.3 mm

The heater generates heat in the time interval <0,1> seconds. In the models the temperature change of the heater in time is shown in Figure 3.

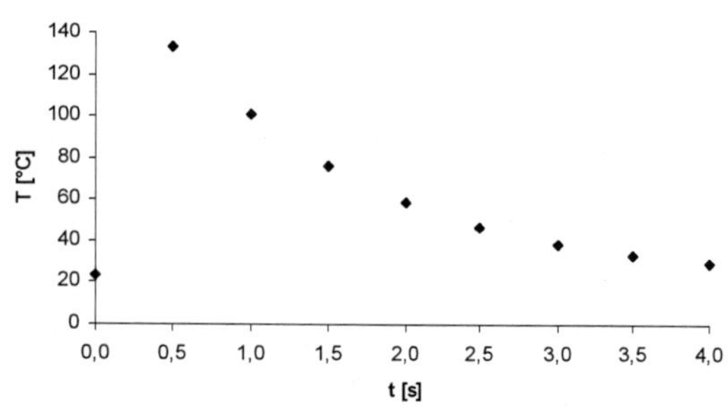

Fig. 3. Temperature change oh the heater in time

3.3 Modelling results

The properties of the modified time – of – flight sensor were simulated and verified for different flowing gases.The dependence of T_{MAX} on flow velocity v can be found in Figure 4. 5; the temperature T_{MAX} was measured in semiconductor film.

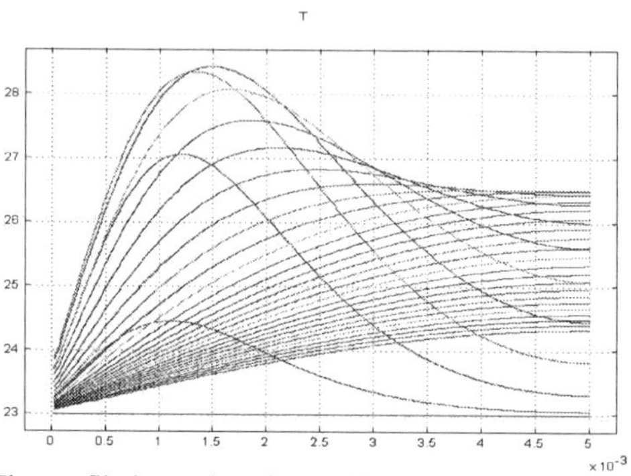

Fig. 4. The profile in semiconductor film

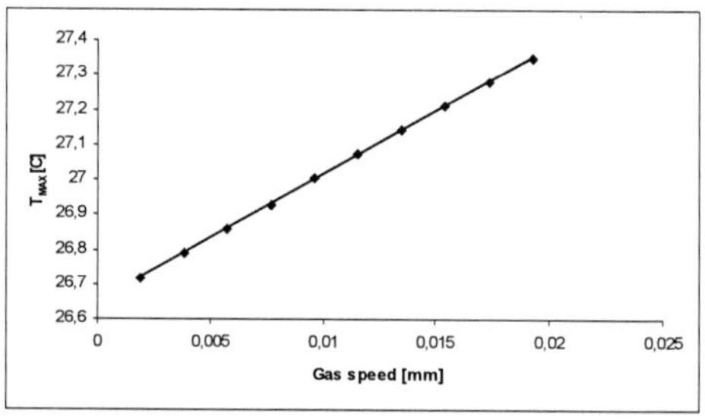

Fig. 5. The dependence of T_{MAX} in semiconductor film on flow velocity v, gas – carbon dioxide, the height of the fluid channel $d = 0{,}8mm$

4. CONCLUSION

This article describes properties of the modified time – of – flight sensor. The validation of the finite-element/analytical model of the designed flowmeter was directly compared with experimental data. The peristaltic pump was used for the calibration of the modified time – of – flight sensor. The flow sensor was calibrated for air and CO_2.The designed flowmeter has been used in a biochemical laboratory for study of reaction kinetic of a decomposition of sediments in waste water.

Acknowledgements

The work has been supported by the Ministry of Education of the Czech Republic under grant MSM 708835102. This support is very gratefully acknowledged.

5. REFERENCES

Hardy, J.; Hylton, O. 1999. *Flow measurement methods and applications*, USA, Viley –Interscience, 1999.

Kolomazník, K.; Taylor, M.; Kupec, J. 2000. The valuable products from wastes. In.: *5th World Congress R'2000, Recovery, Recycling, Re-integrat*, Canada. 2000, p.5.

Lammerink, T.; Niels, R.; Elwenspoek, M. 1993. Micro – liquid flow sensor, *Sensors and actuators A*, 1993, Vol. 37/38, 45 – 50.

Lammerink, T.; Dijkstra, F.; Houkes, Z.; Kuijk, J. 1995. Intelligent gas – mixture flow sensor, *Sensors and actuators A*, 1995, Vol. 46/47, 380 – 384.

Rasmussen, A.; Zaghloul, M. 1999. The design and fabrication of microfluidic flow sensors, In.: *ISCAS-99*, 1999, 136-139.

Webster, J. 1999. *Measurement, instrumentation and sensors,* USA, IEEE Press, 1999.

Annals of DAAAM for 2006 & Proceedings of the 17th International DAAAM Symposium, ISSN 1726-9679
Editor B. Katalinic, Published by DAAAM International, Vienna, Austria 2006
150th Birthday of Nikola Tesla (1856-1943) Inventor of Mobile Robot Teleoperation

NEW FORM OF JACOBIAN MATRIX AND EQUATIONS OF MOTION FOR A 6 D.O.F CABLE-DRIVEN PARALLEL ROBOT USING CONSTRAINED VARIABLES

AFSHARI, A. & MEGHDARI, A.

Abstract: *Cable-driven robots are a special type of parallel robots in which rigid links are replaced by cables. In this study, we introduce new motion variables to study kinematics and dynamics of a 6 d.o.f cable-driven parallel robot and then, we obtain equations of motion by Lagrange method. Because of using Lagrange multipliers, we encounter additional degrees of freedom and a new form of Jacobian matrix. We will show that the new method is computationally more efficient than the conventional method in workspace determination and solving equations of motion.*
Key words: *cable suspended robot, workspace, equations of motion*

1. INTRODUCTION

Cable-driven robots are a new type of robots which have recently attracted researchers. Cable-driven robots or cable robots are a special form of parallel robots in which rigid links are replaced by cables. Several cable robots have been studied in the past. An early robot is the Robocrane developed by NIST (Albus et al., 1993). Another cable robot is Charlotte, developed by McDonnell-Douglas (Campbell et al., 1995). The other cable robots, which have been built and tested, are: the Texas (Lindemann & Tesar, 1989), and the FALCON (Kawamura et al., 1995). Workspace is as an important issue in cable robots and has been studied by many researchers. Fattah and Agrawal (2005) as well as Barrette and Gosselin (2005) determined the workspace of planar cable robots. Alp and Agrawal (2002) determined the statically reachable workspace for a 6 d.o.f spatial cable robot. Pusey et al. (2004) studied the design and workspace of a 6-6 cable robot.
In this paper, using the concept of kinematic constraints in Lagrange method, a new Jacobian matrix and equations of motion for a 6 d.o.f cable robot are introduced and some of characteristics of this method are explained.

2. KINEMATIC AND DYNAMIC ANALYSES

In this study, the well-known NIST Robocrane as a 6 d.o.f cable robot was considered. Its movable platform and fixed support are two equilateral triangles often referred to as "lower triangle" and "upper triangle", respectively. Two frames are used for kinematic and dynamic analysis: The reference frame (XYZ), and the body frame (xyz) as shown in Fig. 1.
In this section, kinematical and dynamical relations of the robot are studied by use of two different sets of motion variables; the first set of variables which has been frequently used by other researchers and we call it the "conventional set", and the second set which can be called the "new set".

2.1 Analysis by the Conventional Set

Since the robot has six degrees of freedom, at least six independent motion variables are needed to study its kinematics and dynamics. Here, we have the vector X_C which represents the conventional set of variables as follows;

$$\{X_C\}=[x \quad y \quad z \quad \alpha \quad \beta \quad \gamma]^T \tag{1}$$

Where, x, y, and z are the Cartesian coordinate of center of mass of the moving platform and α, β, and γ are Euler angles. Now, using Newton-Euler equations we have:

$$[M_M]_{6\times6}\{\ddot{X}_C\}_{6\times1}+\{C\}_{6\times1}=\frac{1}{r_0}([R_M]^{-1}_{6\times6}[J_C]^T_{6\times6})\{\tau\}_{6\times1}-\{G\}_{6\times1}+\{L\}_{6\times1} \tag{2}$$

Here, $[M_M]$ is the modified inertia matrix, $[R_M]$ is the modified transformation matrix, $[J_C]$ is the conventional Jacobian matrix, $\{C\}$ is the vector of velocity terms, $\{\tau\}$ is the vector of winch torques, $\{G\}$ is the gravity vector, $\{L\}$ is the external load vector and r_0 is radius of the winch drums.

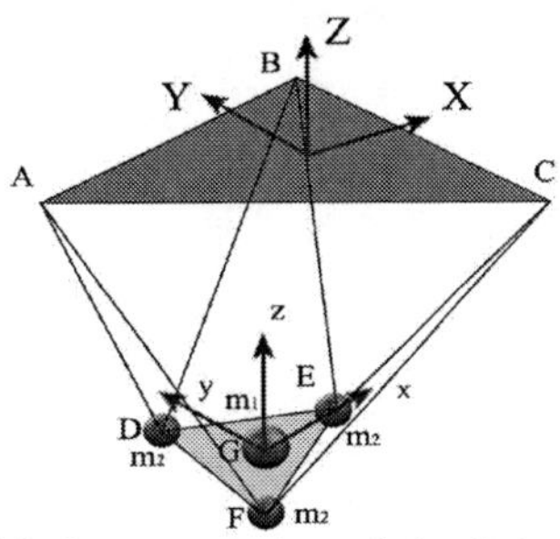

Fig. 1. Graphical representation of the Robocrane and the point masses used in this study

2.2 Analysis by the New Set

Here, we introduce $\{X_N\}$ as follows;

$$\{X_N\}=[X_D \quad Y_D \quad Z_D \quad X_E \quad Y_E \quad Z_E \quad X_F \quad Y_F \quad Z_F]^T \tag{3}$$

As it is observed, the elements of $\{X_N\}$ are the Cartesian coordinates of vertices of the moving platform written in the reference frame. Since the robot has six degrees of freedom, three constraint equations are also needed.

In order to express kinetic energy of the robot in terms of these motion variables, it is convenient to use point masses instead of distributed mass, otherwise we encounter complicated terms in the rotational part of kinetic energy. As it is shown in Fig. 1, four point masses are positioned in the moving platform; three identical masses (m_2) at the vertices and a single point mass (m_1) at the center of mass of the lower triangle. Now, we can determine m_1 and m_2 so that the masses and mass moments of inertia of these systems are identical;

$$m_1=9m_2=\frac{3}{4}m_{\text{moving platform}} \tag{4}$$

Now, kinetic and potential energies of the system are easily expressed in terms of the generalized coordinates. Then, using Lagrange equations, the equations of motion can be derived;

$$[M_N]_{9\times9}\{\ddot{X}_N\}_{9\times1}=\frac{1}{r_0}([P]^T_{9\times6})\{\tau\}_{6\times1}-\{G\}_{9\times1}+\{L\}_{9\times1}+[a]^T_{9\times3}\{\lambda\}_{3\times1} \tag{5}$$

In equation (5), $[a]$ is Jacobian constraint matrix, $\{\lambda\}$ is Lagrange multiplier vector, and $[P]$ is a 6x9 matrix that relates the vertex velocities to the joint speeds. In the preceding equation, $\{L\}$ is the generalized force vector which is obtained by calculation of virtual work done by external loads. In the case where we have external moments, because of difficulty in expression of virtual rotation of the moving platform in terms of the generalized coordinates, we replaced each of three external moment components (in xyz frame) with forces in the point masses so that the sum of them vanishes but they produce the moment in the corresponding direction.
Now, if we write equation (5) in the following form;

$$[M_N]_{9\times9}\{\ddot{X}_N\}_{9\times1}=\frac{1}{r_0}\left(\left[[P]^T_{9\times6} \;\vdots\; [a]^T_{9\times3}\right]\right)\left\{\begin{array}{c}\{\tau\}_{6\times1}\\ \hdashline \{\lambda\}_{3\times1}\end{array}\right\}-\{G\}_{9\times1}+\{L\}_{9\times1} \quad (6)$$

Then, we can define the new Jacobian matrix as;

$$[J_N]_{9\times9}=\left[\begin{array}{c}[P]_{6\times9}\\ \hdashline [a]_{3\times9}\end{array}\right] \quad (7)$$

Therefore, we have;

$$\left\{\begin{array}{c}\{\dot{l}\}_{6\times1}\\ \hdashline \{0\}_{3\times1}\end{array}\right\}=[J_N]_{9\times9}\{\dot{X}_N\}_{9\times1} \quad (8)$$

Or;

$$\left\{\begin{array}{c}\dot{l}_1\\ \dot{l}_2\\ \dot{l}_3\\ \dot{l}_4\\ \dot{l}_5\\ \dot{l}_6\\ \hline 0\\ 0\\ 0\end{array}\right\}=\left[\begin{array}{ccccccccc}
[u^x_{A/D} & u^y_{A/D} & u^z_{A/D}] & 0 & 0 & 0 & 0 & 0 & 0\\
[u^x_{B/D} & u^y_{B/D} & u^z_{B/D}] & 0 & 0 & 0 & 0 & 0 & 0\\
0 & 0 & 0 & [u^x_{B/E} & u^y_{B/E} & u^z_{B/E}] & 0 & 0 & 0\\
0 & 0 & 0 & [u^x_{C/E} & u^y_{C/E} & u^z_{C/E}] & 0 & 0 & 0\\
0 & 0 & 0 & 0 & 0 & 0 & [u^x_{C/F} & u^y_{C/F} & u^z_{C/F}]\\
0 & 0 & 0 & 0 & 0 & 0 & [u^x_{A/F} & u^y_{A/F} & u^z_{A/F}]\\
\hdashline
0 & 0 & 0 & [u^x_{E/F} & u^y_{E/F} & u^z_{E/F}] & -[u^x_{E/F} & u^y_{E/F} & u^z_{E/F}]\\
-[u^x_{D/F} & u^y_{D/F} & u^z_{D/F}] & 0 & 0 & 0 & [u^x_{D/F} & u^y_{D/F} & u^z_{D/F}]\\
[u^x_{D/E} & u^y_{D/E} & u^z_{D/E}] & -[u^x_{D/E} & u^y_{D/E} & u^z_{D/E}] & 0 & 0 & 0
\end{array}\right]_{9\times9}\left\{\begin{array}{c}\dot{x}_D\\ \dot{y}_D\\ \dot{z}_D\\ \dot{x}_E\\ \dot{y}_E\\ \dot{z}_E\\ \dot{x}_F\\ \dot{y}_F\\ \dot{z}_F\end{array}\right\} \quad (9)$$

Where, the elements of the Jacobian matrix are components of unit vectors in specified directions. The previous equation completely expresses all of the kinematical relations of the system. Since we used Lagrange multipliers, three additional degrees of freedom appeared in equation (9) and it is clear that their corresponding joint speeds are zero (i.e. $\dot{l}_7=\dot{l}_8=\dot{l}_9=0$).

3. CHARACTERISTICS OF THE NEW METHOD

3.1 Definition of the Jacobian Matrix Using the Concept of Constraint

In definition of the new Jacobian matrix, we inserted the velocity constraints in equation (9) to achieve a complete expression of the robot kinematics. In other words, we can consider the added degrees of freedom as three "virtual degrees of freedom" which appeared in the Jacobian matrix as a result of using constraints. This new approach employed to define Jacobian matrix has not been used before by other researchers and distinguishes this study from previous researches.

3.2 Efficiency in Workspace Determination

Checking both geometrical and equilibrium conditions is needed to determine the workspace of cable robots. For comparison between these methods, two programs were created in MATLAB to determine static constant-orientation workspace of the Robocrane for a special orientation. Therefore, a three dimensional grid of points was considered and the condition of falling into workspace in each point was investigated. We also considered the gravity as the only external load exerted to the robot. The workspaces obtained by these programs were identical, but the difference between them was in the time it took to obtain the workspace as shown in Table 1.

Computation time for the conventional method	*23.6 minutes*
Computation time for the new method	*8.55 minutes*

Table 1. Comparison between the conventional and the new methods in workspace determination

3.3 Efficiency in Solving Equations of Motion

Here, we assumed that it is desired to obtain the required motor torques (as functions of time) for a movement of the lower triangle in a special path within a specified time by both the methods. To do that, two programs were created in MATLAB and then the required time for computation in these programs were obtained and compared as shown in Table 2. We also observed that the motor torques in these programs were identical.

Computation time for the conventional method	*4.09 sec*
Computation time for the new method	*2.24 sec*

Table 2. Comparison between the conventional and the new methods in solving equations of motion

4. CONCLUSIONS

In this study, we introduced new motion variables to obtain equations of motion for NIST Robocrane and Lagrange method was used to derive equations of motion. In the next step, the Jacobian matrix associated with these new variables was introduced in which three virtual degrees of freedom appeared as a result of using Lagrange multipliers. The approach used to define this Jacobian matrix distinguishes it from the conventional one, which is usually used to study 6 d.o.f cable robots. Then, it was demonstrated that the new method is computationally more efficient than the conventional one in workspace determination and solving equations of motion.

5. REFERENCES

Albus, J.; Bostelman, R. & Dagalakis, N. (1993). The NIST Robocrane, *Journal of Robotic Systems,* (709-724)

Alp, A. & Agrawal, S. (2002). Cable suspended robots: design, planning, and control, *Proceedings of IEEE International Conference on Robotics and Automation*, pp. 4275–4280, May 2002, Washington

Barrette, G. & Gosselin, C. (2005). Determination of the Dynamic Workspace of Cable-Driven Planar Parallel Mechanisms, *Transactions of the ASME,* Vol. 127, (242-248)

Bostelman, R.; Jacoff, A.; Dagalakis, N. & Albus, J. (1996). RCS-Based RoboCrane Integration, *Proceedings of the International Conference on Intelligent Systems A Semiotic Perspective*, pp. 20-23, October 1996, Gaithersburg

Campbell, P.; Swaim, P. & Thompson, C. (1995). Charlotte Robot Technology for Space and Terrestrial Applications, *25th International Conference on Environmental Systems*, SAE Article 951520, San Diego

Fattah, A. & Agrawal, S. (2005). On the Design of Cable-Suspended Planar Parallel Robots, *Journal of Mechanical Design*, Vol. 127, (1021-1028)

Kawamura, S.; Choe, W.; Tanaka, S. & Pandian, S. (1995). Development of an ultra high speed robot FALCON using wire drive system, *Proceedings of the 1993 IEEE/ICRA International Conference on Robotics and Automation*, pp. 215-220, May 1995, Nagoya

Lindemann, R. & Tesar, D. (1989). Construction and Demonstration of a 9-String 6-DOF Force Reflecting Joystick for Telerobotics, *NASA International Conference on Space Telerobotics*, pp. 55-63

Pusey, J.; Fattah, A. & Agrawal, S. (2004). Design and workspace analysis of a 6–6 cable-suspended parallel robot, *Mechanism and Macdhine Theory*, (761–778)

Annals of DAAAM for 2006 & Proceedings of the 17th International DAAAM Symposium, ISSN 1726-9679
Editor B. Katalinic, Published by DAAAM International, Vienna, Austria 2006
150th Birthday of Nikola Tesla (1856-1943) Inventor of Mobile Robot Teleoperation

AN AUTONOMOUS VEHICLE WITH PERCEPTION AND VISION SYSTEM

AHMED, S.S.; HAQ, S.M.; RAICH, P.A. & POLASH, S.A.

Abstract: At present fully autonomous unmanned vehicle (UAV) is an important area of robotics research and development. But it is still in its infancy in Bangladesh in comparable with other countries. Some projects with semi autonomous capabilities and remote controlled vehicles using remote video links have been developed with varying degrees of success. The objective of this project is to build an autonomous vehicle with vision for object detection and tracking along with collision avoidance mechanism. The UAV built under this project can detect and chase a simple object without any human intervention. The UAV can also search for a target which has gone out of its field of vision. The project involves research and development in the fields of robotic and, machine vision
Key words: UAV, Machine Vision, Ultrasonic Ping Sensors, Sobel Edge Detector, Matlab, NeatVision

1. INTRODUCTION

An important application for robotic automation in everyday life is the development of intelligent or autonomous vehicles. To build a viable and working UAV we had to provide our vehicle with perception of its environment. Our project uses vision based physical sensor as its primary sensor. The primary sensor is augmented with ultrasonic ping sensor which provides collision avoidance information. A photo sensor is used to sense the illumination level in the environment. Navigation algorithms help the vehicle chase a specific target and seek a target when it is out of the field of vision of the camera. The control circuitry and logic of the vehicle issues and executes radio linked command signals to the vehicle to move it in any one of 8 directions. To keep the UAV design simple we had to assume that the target and the environment were of stark contrast. The illumination level would have to be maintained and the vehicle speed had to be optimized to allow for ample time for image processing and target acquisition.

2. ARCHITECTURE

We used a vision based perception system to 'see' and 'follow' the target. We had to ensure that our vehicle did not bump into walls or wander around in darkness. Our UAV can cease its normal mode of operation if the illumination level drops below a certain acceptable limit. The collision avoidance and photometer computations are performed onboard the vehicle using a microcontroller. The vehicle captures real time video using a wireless webcam and transmits the image to a PC. The PC then issues maneuvering commands to the vehicle using a remote control connected to the PC via LPT1. The collision avoidance subsystem is situated onboard. Finally the photometer is used to stop the vehicles motion when the room illumination drops below a predetermined safe level.

2.1 Collision Avoidance Subsystem

This subsystem is situated onboard the vehicle and comprises of a Parallax Basic Stamp-2 microcontroller module and a parallax ultrasonic range finder-SRF04. The ultrasonic range finder transmits and receives a 40 KHz signal and can compute the distance of a target within the range of 3-300cm. This subsystem takes decision independent of the remote processing node. Even if the vehicle is stopped because of an incoming wall, the main processing node will continue to analyze and compute incoming images and transmit command signals back to the vehicle.

2.2 Photo Meter Safety Subsystem

To overcome this problem we had used a simple light sensor circuit using a photo resistor along with the microcontroller. The application logic calculates the relative time required to drain the capacitor charge and transforms that value into a threshold limit to determine whether the vehicle is operating in a safe environment.

2.3 The Vehicle

Fig. 1. The vehicle

An ordinary remote controlled vehicle is used to house the microcontroller unit, ultrasonic sensor and photo sensor. The digital potentiometer circuitry is connected to the vehicles power system to control the vehicles velocity. We used an ordinary Creative webcam is a CMOS camera, approximately has a 60 degree field of vision, provides continuous video stream and provides 640x480 pixels image at 30 frames per second.

2.4 Wireless Processing Interface

2.4.1 Remote Control Interface

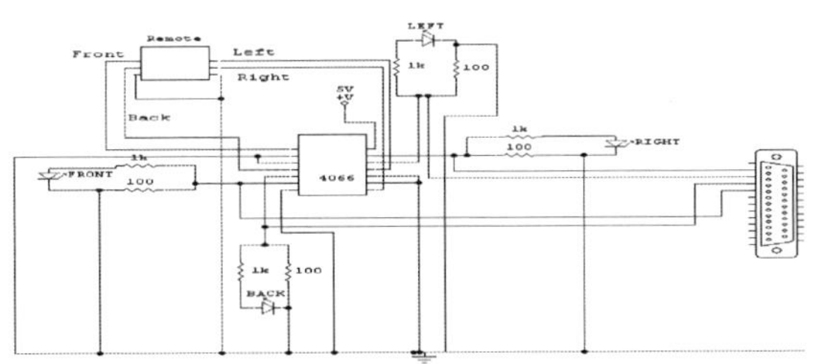

Fig. 2. Basic circuit layout for remote control interface

The standard handheld remote controller of the vehicle was interfaced with the PC parallel port. Parallel communication requires as many wires as the number of bits in a word for its transmission. The wire is connected from data port to a 4066IC. The 4066 is a quad bilateral switch intended for the transmission or multiplexing of analog or digital signals. If a data port is set to 1 then it generates +5V. The remote device then transmits signal and the car receives the signal. 4066IC basically performs switching task. It makes high to the remote pin with corresponding data pin of computer's data port. To make circuit complete, connection is made with the ground port of the LPT1 port.

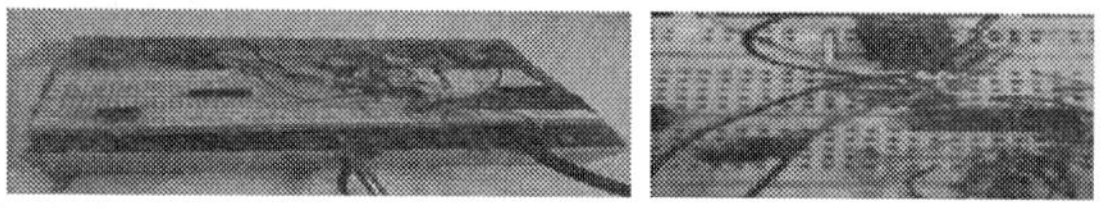
Fig. 3. Car remote controller connected to LPT1

2.4.2 Vision Subsystem

The system has a built-in safeguard against poor illumination condition. The field of vision (FOV) of the camera is an important factor to consider. Our webcam provides a field of vision of approximately 60 degrees. We have experienced problem with high relative speed of the vehicle as the narrow FOV tends to loose the target with little movement of the vehicle or target. So, we limit the forward motion command signal shortly.

2.5 Machine Vision Software

MATLAB integrates computation, visualization, and programming in an easy-to-use environment where problems and solutions are expressed in familiar mathematical notation. Our interest in Matlab software was due to its extensive image processing library, easy to use interface, and effortless interface with Java programming language.

2.6 Image Acquisition Module

Image acquisition involves the capture of digital/analogue image signal. We had to process approximately >10 frames per second. Without this rate the vehicle movement would appear to have a jerky motion. We had a small java program for image acquisition. The program can take images at 20 frames per second. We then convert the frames into a buffered image and save them in a pre-determined folder for processing with Matlab-7 software.

2.7 Object Tracking Module

Reflection of light on the ball causes errors while filtering out the circle. This problem is reduced by prudent use of threshold values and controlling the intensity of light to an acceptable value. Using Matlab we load a frame as a RGB24 image into the program and resize it into 320 x 240 pixel sizes. We apply the Matlab7 function `imgfilter(A,H)` to filter the image of unwanted noise. The image is then converted to grey scale to further reduce the amount of data to process. We than apply a brightness threshold value to the image.

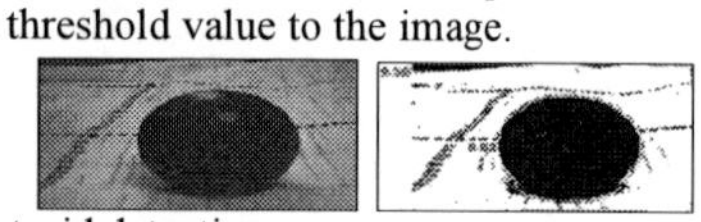
Fig. 4. Centroid detection

We use Sobel edge detector to detect the circle from the image. We set the threshold value ti 0.70 to mark objects above the threshold with a black circle if metric value is greater than threshold value. The circles that failed to meet the given threshold value are filled with black so that no more irrelevant circle exists. Then we can find the circle's centre point from relevant method of MATLAB. We compare the coordinate with the centre axis of the web camera and we can decide where the ball is. We use co-ordinate comparison in this case.

2.8 Decision Maker

The objective of the decision making part of the program is to keep the ball in the centre of the field of view of the camera. There can be two situations: The ball is in the field of view or the ball is NOT in the field of view.

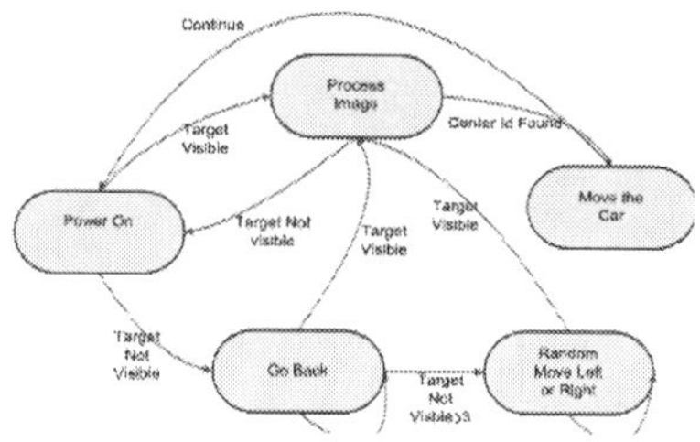

Fig. 5. State diagram for decision making

2.8.1 The ball is in the field of view

We had implemented a safety system to check whether successive frames are changing over each 30 seconds window. Whenever a collision situation is detected, the digital pot shuts down the vehicles propulsion mechanism. So the view of each frame remains unchanged. If within the 30 second window the frames do not change then the 'seek' part of the program is initiated.

2.8.2 The ball is NOT in the field of view

If the vehicle loses the ball from its FOV, it ceases its normal navigation mode and switches to 'seek' mode. In the 'seek' mode the vehicle attempts to increase it's FOV by going backward. The UAV will repeat this step for 3 times and if still the ball is not within it's FOV it will randomly choose a side to move. This algorithm is repeated until the UAV catches the ball within its FOV.

3. TEST RESULTS AND SUGGESTIONS

The test results for our Unmanned Autonomous vehicle was satisfactory.

2.9 Collision Avoidance Subsystem

Collision Avoidance Subsystem performed extremely well and was capable of detecting obstacles (walls) almost every time and could stop the vehicles motion before a collision occurred.

2.10 Photo Meter Safety Subsystem

Photo Meter Safety Subsystem was very handy in determining the suitable illumination level for testing the UAV. But there were situations when localized illumination on the target had affected the sub-systems ability to provide a suitable warning.

2.11 Vision Subsystem

We had chosen the environment to be as suitable as possible for detecting the target from the background. We were able to achieve a satisfactory motion with processing 10-15 frames per second and the vehicle moving at approximately 0.05 meter per second.

3. CONCLUSION

Now, we are working on the second phase of the UAV. This version would be able to detect a target object out of any kind of background. We shall also allow the system to be trained to identify objects of any pattern. The onboard microcontroller shall have a feedback system with the remote processing unit. Providing stereo vision capabilities to the system can eliminate the need to use physical range finding sensors. This project has given us insight into many core aspects of this fascinating area of computer science. We hope that our future researches will help to enrich the knowledgebase of UAV development.

4. REFERENCES

Tomi Engdahl, Get *Power out of PC RS-232 port*, Helsinki University of Technology,http://www.tkk.fi/Misc/Electronics/circuits/rspower.html, accessed on 7/15/2006.

Whelan, P.F., Molloy, D, *Machine Vision Algorithms in Java: Techniques and Implementations*, 2nd Edition, Springer, London (2000).

Alasdair, McAndrew, *An Introduction to Digital Image Processing With Matlab*, Thomson Course Technology, Boston (2004).

Quantitative Imaging Group, "Derivative-Based Operations",http://www.ph.tn.tudelft.nl/Courses/FIP/noframes/fip-Derivati.html, accessed on 4/6/2006.

Parallax Inc., http://www.parallax.com/, accessed on 5/6/2006.

Annals of DAAAM for 2006 & Proceedings of the 17th International DAAAM Symposium, ISSN 1726-9679
Editor B. Katalinic, Published by DAAAM International, Vienna, Austria 2006
150th Birthday of Nikola Tesla (1856-1943) Inventor of Mobile Robot Teleoperation

IMPORTANT ASPECTS OF TRAINING FOR LABOR-BASED ROAD CONSTRUCTION AND MAINTENANCE PERSONNEL IN DEVELOPING COUNTRIES

AKHAVAN, A.N.

Abstract:*This Paper discusses important aspects of training for labor based road construction and maintenance personnel in developing countries. It does not discuss training principles in general but hopes to also give some ideas for training in fields other than road work.*
Key words: *Training- Construction- Maintenance- Developing Countries- Personnel*

1. WHY LABOR-BASED METHODS FOR ROAD WORK?

The present poor state of the exiting road networks will most governments of developing countries to abandon new construction programmes in favor of rehabilitation and maintenance work. The governments, through their transport ministries, are responsible for determining the priorities of where and what type of road have to be built, rehabilitated and maintained. The most important links, which are generally trunk or main roads, have to guarantee the supply of vital commodities to and from the ports, the country's capital and its other main production centers. Historically, most of these roads were already built during colonial times and served primarily to allow military control and economic exploitation. Feeding into this main network are other roads of mainly regional character. Lower classes of roads, which are principally of local interest, are usually called district roads, feeder roads and access or rural roads. This secondary road network has been developed to some extent over the last twenty years and is meant to serve mainly the rural population.

2. WHY TRAINING FOR LABOR-BASED ROAD WORK?

The most important criticism towards labor-based methods is the production of "poor quality outputs". However, projects carried out in many countries over the last 20 years have convincingly demonstrated that labor-based projects can provide excellent quality results provided there is effective control and provided that quality standard are adhered to.
In general the competitiveness and quality of labor-based projects is determined:

- the introduction of appropriate planning, programming and control systems;
- the used of incentive schemes;
- the availability of adequate hand tools and light equipment; and, last but not least
- the availability of suitable trained management and supervisory staff.

If these aspects are not neglected, labor-based projects can produce good quality work at competitive costs in most developing countries.

The need for qualified manpower at all levels is very high in the road sector. Governments are making great efforts to improve and enlarge technical training centers, and universities. But in most countries will take many years until all vacancies in the construction industry and the related government services can be filled with their own graduates. Many training institutions have also adopted training curricula which were originally developed for a capital-based environment and which in many cases do not consider the actual local situation. Labor-based programmes require an intensive degree of supervision and deal with the application of appropriate technologies.

3. EDUCATIONAL BACKGROUND

Training for labor-based programmes involves of personnel basically "nobody" has a formal education background in labor-based methods.

4. MANAGERS AND ENGINEERS

The civil engineering faculties of most universities in developing countries have not yet managed to include special courses in labor-based road work, although some universities deal with appropriate technology aspects to a limited extent. Graduate engineers and managers working with a labor-based programme have very little knowledge about labor-based methods and management and can only learn through experience. However, there are still a large number of high level programme personnel who have little or on knowledge of labor-based techniques and so there is still a significant resistance towards appropriate work methods.

5. MEDIUM LEVEL SUPERVISORS

These are the people who actually work in the field and implement the work. They are in most cases technicians who have attended "traditional" civil engineering courses at a university or a Technical College. Apart from perhaps a few theoretical lessons in labor-based methods their knowledge and management capacity is very limited. The also lack the practical job experience as most of them are young technicians who finished their education recently or who have spent most of their professional life as administrators. Special training courses have to be developed for technicians and number of countries, through their labor-based programmes, have established such specialized courses.

6. SITE LEVEL SUPERVISORS

Site supervisors are seen as the "back-bone" of most of the labor-based programmes. The success or failure of a programme depends a great deal on the performance of the site supervisors. In most countries the technical back-ground of these people is rudimentary. Most ministries offer a few weeks of introduction training before the site supervisors are deployed on site. Generally no professional proficiency at this level can be expected and also the educational school background is more often than not just about elementary. Appropriate training courses for this very important cadre have therefore to be developed for labor-based programmes.

7. GANG LEADERS, HEADPERSONS

These people are from the lowest level of supervision. In most cases they must be able to write, read and they receive informal on-the-job training. However, in reality many of these supervisors are illiterate and it is very difficult and time consuming to achieve satisfactory training results.

8. SUPPORT STAFF

Support staff consist of administrators, store people, mechanics and the like. Many of the programmes are still implemented through the traditional civil service administration. Labor-based programmes have their specific support needs and certain knowledge about the programme, labor-based methods and management must be expected from the support staff if the project is to succeed.

9. TRAINERS

Qualified trainers for labor-based methods are very and most programmes have expatriates employed to set-up and run training courses. To be an effective trainer for labor-based methods it is essential to have pedagogical knowledge and a wide experience in labor-based techniques.

10. TRAINING NEEDS

Normally training is believed to be useful and it dose necessarily have to be justified in terms of financial return. The objective of a training programme for labor-baser road work is improved performance in terms of time, quality and cost. There are some problems which cannot be solved by training alone, for example, lack of fund, fluctuating material supply, shortage of highly experienced labor, and so on.

11. OBJECTIVES

Before the training needs can be assessed it is absolutely essential to decide what is to be achieved through training. The objectives are the basis for the training programme because they determine the input of resources and the way in which the programme is to be organized. For labor-based road work, setting objectives is for many trainers not an easy undertaking as there are few trainers who have the necessary experience. It is therefore recommended that trainers have discussions together with the managers of the labor-based programme and spend some time identifying the actual training objectives. This may seem to be a waste of time and energy to some managers and trainers as they might think they know what they are doing. But what they have in the back of their minds is not the some as objectives which describe explicitly the effects that the training can be expected to produce..

12. TRAINING NEEDS ANALYSIS

Training needs analysis is often not at all or only vaguely carried out. For a successful and objective oriented training programme the training needs analysis is a critical activity at the beginning of a training programme. An accurate analysis reduces the possibility of errors at a later stage.

The objective of an analysis is to collect and evaluate information on order to find out:

1- What is being done now?
2- What should be done in future?

The gap between the two defines the training intervention and allows a decision on:

- the content and method on training;
- the target population;
- the time frame for the training process;
- how to evaluate the training outcomes.

There is no straightforward recipe for training needs analysis and for each training programme another procedure may be appropriate. Through experience the following procedure seems to be advisable for labor-based road work programmes:

- identify the current programme stage (often the analysis takes off after the programme and training have already standard);
- review the vital characteristics and data of the programme (organization structure, responsibilities, job descriptions, technical and administrative procedures, production data, costs, planning, reporting and monitoring procedures, and so on);
- interview the management (to identify the clients' training expectations);
- interview the future trainees (to find out what their current level of understanding and knowledge is, through discussion and questionnaire);
- observe the product (visit sites, rate the actual performance and compare with the desired performance);
- Problem identification (organize a problem identification workshop with personnel from all levels of potential trainees plus the programme management).

13. TRAINING PROGRAMME DEVELOPMENT AND TRAINING IMPLEMENTATION

Although the need for training in the field of road construction and maintenance programmes seems to be indisputable, the design of training programmes is often unprofessionally carried out. Usually training programmes are developed and carried out by engineers with little or no principal pedagogical knowledge. This often leads to an inflexible training approach which does not sufficiently consider the specific learning and job situation of the trainees and dose not involve modern teaching methods

14. DIDACTIC PRINCIPLES

Infrastructure development projects are managed by adult professional staff and in most cases the training set-up for it is an in-service programme. The needs, the techniques, the general conduct and the way of measuring the effectiveness of training of adult professionals differ from learning in formal school systems:

	School Training	Training of Adult Professionals
Needs	Initial acquisition of knowledge	Difference felt between desired and actually realized job-function.
Techniques	The teacher transfer their superior knowledge to the students.	The trainers moderate the exchange of experience among the participants.
Conduct	Generally fixed formal curriculum controlled by supervising authority.	Flexible curriculum linked to situations and engagement of the participants in course-steering.
Evaluation	Success at the final tests: basis of knowledge sufficient for wide use.	Increased job-performance: higher capability to solve various problems.

Since, generally speaking, participants and trainers have much more experience in school training than in adult education, it is important to pay attention to this difference as it has consequence for the didactic approach.

Since the effectiveness of the training of adult professionals can only be measured reasonably by its impact on job performance, the responsibility of the training institution dose not end with formal tests. Course guides and instructors have to give special attention to the fact that job performance consists of a patchwork of difference elements. Even if many of them, such as organization of work, allocation of resources, pressure band reward systems, and promotion schemes, cannot be influenced by training, the training should have an impact on job performance at the following three levels:

Task-centered knowledge	Resources, technology, institutional framework
Operating skills	Technical handing, planning, organization, directing, monitoring, controlling
Human-centered attitudes	Responsibility, supervision, leadership and teamwork

Job performance is the result of the interrelation of different elements in a specific situation. Only under very specific circumstances will the isolated evolution of one element bring satisfactory results.

15. TRAINING PROGRAMMING

In general, operational staff cannot be absent from their job for a long time and their training has to provide them with immediate results for their day-to-day work situation. To fulfill these two conditions, the training events have to be:

- as short and concentrated as possible in terms of time;
- strongly linked to the work experience of the participants; and
- well prepared by a course instructor.

Operational staff normally have to solve specific practical problems, for example lack of resources, bad performance in the operation of machines and tools, deficiency in planning, weak management of people.

Furthermore, it is impossible to deal with all the problems in one single course. Courses should be planned according to the most pressing needs:

- At the beginning of a new project it is usually necessary to organize induction courses where the basics of the project or work process are taught. This allows the participants to start work and to produce an acceptable result. y.
- As the project proceeds, short problem oriented courses should be organized to improve the job performance where it is most needed. Such courses have to be practical and team oriented..

16. TEACHING METHODS

For infrastructure projects the training needs are, in most cases, of a very practical nature, although a certain amount of theory has to from the basis to ensure the necessary level of technical understanding. Very often training institutions concentrate far too much on the theoretical part. There are probably three main reasons for this:

- the trainers themselves have not the required background and competence to carry out practical training.
- there is still a belief that practical work is inferior to theory academic knowledge and hence is not given the necessary attention; and
- to carry out training in a well-established classroom with all the training aids in place requires much less organization and flexibility than to carry out practical training in the field where situations are frequently changing (for example, work site, weather, living conditions).

When developing a training programme these facts have to be considered is the necessary precautions should be taken. Often on-the-job training as pert of the training programme has proved to be the best solution to overcome the problem of too much theory and too little practice and then become a powerful tool for learning be the best solution to overcome the problem of too much theory and too little practice and then become a powerful tool for learning because of the following advantages:

- Immediate application of learning to the job.
- A strong emphasis on skill development.
- A trainer to trainer ratio of 1:1 (usually). This gives the trainer the time to give close attention to, and individualized feedback on, the trainee's performance. The trainer also has greater flexibility when choosing the content and the pace training.
- A continuing daily relationship between trainer and trainee. This helps the trainer to monitor performance and to offer appropriate rewards over an extended period of time.

For the theoretical training it is necessary to clarify the following two questions first:

1- Is the training means as part of the overall staff development of the particular authority (for example, for promotion, achieving professional or trade certificates)?
2- Is the training only meant to improve the job performance of the particular project staff?

Should the training fall within the framework of question 1, then the theoretical training part has to be co-ordinate within the existing requirements or curricula.

Should the training emphasize question 2 then the programmeing and organization can be independently planned and organized from other requirements. This usually gives the trainer more flexibility and the training can be specifically designed to meet the identified training needs of the project only.

17. TRAINERS

In most projects trainers are technical personnel who for one or the other reason become a 'trainer by coincidence'. More often than not such trainers have no, or very little pedagogical background and they tend to try to teach as they were once trained themselves.

The training of trainers therefore becomes an essential of the training programme. The development of trainers has out training before a sufficient trainer capacity is achieved should be avoided.

The training of trainers for road construction and maintenance work lasts approximately three years until the trainer is fully capable of carrying out training independently. This training period does not include the initial technical training, on-the-job training, training material design, course development and training evaluation or follow-up..

18. TRAINING MATERIAL

Design of training material is often seen as the most important aspect in the development of a training of a training programme. Another phenomenon is the fact that almost every training institution develops its training material from scratch. On the other hand, available training experience and material from similar projects are not being consulted and an enormous amount of energy is used to develop new sets of training manuals.

Invited for a dialogue and to exchange experiences.

Training material is often used like a cooking recipe. The trainers follow the pattern or modules of the material blindly and are than not able to react to the actual and present job situation. Trainers should therefore be coached to use training material rather as reference material which can be built into the training process.

19. DOES TRAINING ACHIEVE POSITIVE CHANGES?

This is the paramount question which trainer and client always have in mind. However, they should also be very well aware training alone cannot bring about a change in a project if a number of other actions are initiated:

- management support must be guaranteed at all levels;
- relevant technical and administrative manuals have to be developed;
- job descriptions with the relevant responsibilities, duties and authorities have to be identified;
- planning, reporting and monitoring systems have to be in place, plans of operation prepared;
- Follow-up of activities and work progress must be ensured (chasing both quality and quantity).

Only the interaction of all these aspects, together with training, can ensure a positive change in a project.

A small case study illustrates this: In a road project in country 'x' an improved road maintenance system was designed and the local training institution was asked by the project management to introduce this system through a short seminar for the engineers in charge of the implementation units. The seminar was successful-but no change took place in the field. The training institution was then asked to design and conduct for all medium level supervisors. The courses were successful-but no change took place in the field.

The management decided to engage a consultant to assist in introducing the system. The consultant designed an integrated programme in combination with the required management support. The training took place in the respective districts where all personnel involved, from engineer down to casual headmen, were invited. The courses were designed with implementation as the main objective, rather than understanding of the technicalities and theory of the maintenance system. Modern training techniques, including problem solving approaches and group work with strong visualization were used. The district operational units were given the group task of formulating their own implementation plans the units, then putting these plans into practice. The management if the project given guidelines for a standard follow-up of the field activities and monthly progress reports were sent to them for monitoring purposes. The management also followed up the results of these monitoring reports and discussed any problems with the engineer concerned.

20. REFERENCES

Beusch, A. and J. J. de Veen, 1991, *International Course for Engineers and Managers of Labor-Based Road Construction and Maintenance Programmes* (and pedagogical extracts from KEK consult) (Geneva: International Labour Office).

Hernes, Tor, 1999, *Training Contractirs for Results: A Guide for Trainers and Training Managers* (Geneva: International Labour office).

Imboden, N., 1997, *A Management Approach to project Appraisal Evaluation* (Paris: Development Centre of the Organisation for Economic Cooperation and Development).

Smith, B. J. and B. L. Delahaye. 1997, *How to be an Effective Trainer* (2nd ed.) (London: John Wiley & Sons).

Annals of DAAAM for 2006 & Proceedings of the 17th International DAAAM Symposium, ISSN 1726-9679
Editor B. Katalinic, Published by DAAAM International, Vienna, Austria 2006
150th Birthday of Nikola Tesla (1856-1943) Inventor of Mobile Robot Teleoperation

DEVELOPMENT OF THE HIGHLY INTEGRATED MECHATRONICAL SYSTEMS OF HUMANOID ROBOTS

ALBERS, A. & BRUDNIOK, S.

Abstract: *The development of complex mechatronical systems is a long iterative process. Therefore, there is a demand for specific development processes for systems that have similar basic properties and are able to simplify the development of such systems (Möhringer, 2004). This paper presents a procedure model with which humanoid robots can be developed according to specific requirements. The procedure model is based on the experience from the design of a humanoid robot for the Collaborative Research Center 588 (SFB588, 2006).*
Key words: *Humanoid robot, mechatronics, highly integrated systems, development process*

1. INTRODUCTION

Mechatronical systems feature multi-technological interactions, which are displayed by the existent development processes (e.g. VDI 2206) in a rather general and therefore abstract way. Examples for this are humanoid robots: highly integrated systems with spatial and functional connections of the components and assembly groups.
A humanoid robot is a multi-body system in which mechanical, electronic, and information-technological components are integrated into a very small design space and designed to interact with each other.

2. PREPARATIONS

2.1 Requirements on the robot

The demands result from the actions that the humanoid robot is to perform. The robot designed in the SFB 588 in Karlsruhe will interact with humans in their homes, especially in the kitchen. It will take on activities of the humans, such as the loading of a dish washer. For this task it is not necessary that the robot can walk on two legs, but it has to feature kinematics, especially in the arms, that enable the reaching for objects in the human surrounding. In addition, the robot needs the ability to move and to hold objects in its hand (Schulz, 2003).

2.2 Subdivision of the total system

The development of complex systems requires a subdivision of the total system into manageable partial systems and modules (Fig. 1).
The segmentation of the total system of the humanoid robot is oriented on the interactions present in a system. The total system can be divided into several subsystems. The relations inside the subsystems are stronger compared to the interactions between these subsystems. One part system of the humanoid robot is e.g. the upper body with the subsystem arm. Those systems that cannot be further divided into subsystems are here referred to as modules. In the humanoid robot's arm, this applies to the modules hand-, elbow-, and shoulder joint. With regard to the design, these modules can be exchanged with other modules that fulfill the same function. The modules again consist of function units, as e.g. the actuators for one of the module's joints. The function units themselves consist of components, here regarded as the smallest elements. In the entire drive, these components are the actuator providing the drive power and the components in the drive train connected in a series, e.g. a gear, drive belt, or worm gear transferring the drive power to the joint.

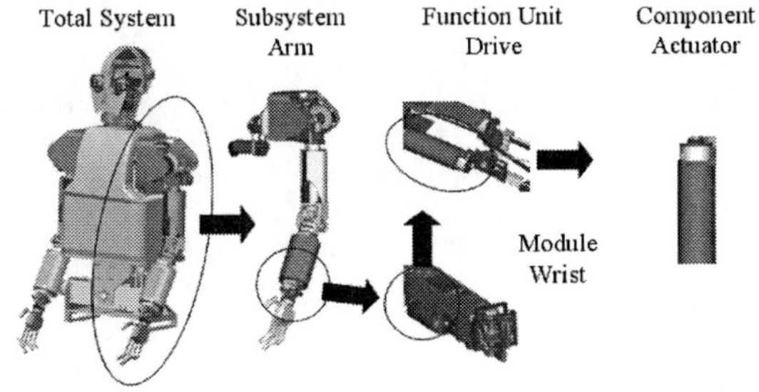

Fig. 1. Subdivision of the total system.

2.3 Selection and data base

The components of highly integrated systems are known and do not have to be newly developed. However, a humanoid robot consists of a large number of components, and each of them exists a variety of technical solutions. This fact results in an overwhelming number of combination possibilities, which complicates a target-oriented development. Therefore it is helpful to file the components of the joints, actuators and sensors as objects in a object-oriented classification. It enables a requirement-specific access to the objects and delivers information about the combination possibilities of the components.

2.4 Development sequence

The development sequence conforms with the order in which a component or information has to be provided for the further procedure. In general, the aim is a possibly small number of iterations in the procedure because they slow down the development significantly.
The development process can be roughly divided into two sectors. The first sector determines the basic requirements for the total system, which have to be known before the design process. This phase includes primarily two iterations: In the first iteration, the kinematics of the robot is specified according to the acting space of the robot and the kinematics again has to be describable in order to be regulable. In the second iteration, the control concept that operates the robot and the general possibilities for operating the joints are adjusted to the requirements for the desired dynamics of the robots. The second sector is the real design process. The sequence in which the modules are worked on is determined by their position in the serial kinematics of the robot. This means that e.g. in the arm, first the wrist, then the elbow joint and then finally the shoulder joint are designed.

Since all modules have in general a similar design, they can be designed according to the same procedure. The sequence in this procedure model is determined by the interactions existing between the function units and between the components. The relation between the components and the behavior of their interaction in case of a change of the development order can be displayed graphically in a design structure matrix (Browning, 2001). Iterations, which always occur in the development of complex systems, can be limited by considering early the properties of the components that are integrated at the end of the development process. One example is the torque measurement in the drive train. In the aforementioned data base, specifications of the components are given like the possibility for a component of the drive train to include some kind of torque measurement. It ensures that after the assembling of a drive train, a power measurement can be integrated.

3. DEVELOPMENT

All further considerations of this paper ably now only to the development of the robot's shoulder joint.

3.1 Design of the kinematics

For the tasks that are required from the robot, it is sufficient to maneuver the arm in front of its body. These movements can be performed by means of a ball joint in the shoulder without an additional pectoral girdle. In the available construction space, a ball joint can be modeled with the required performance of the actuators and sensors as a serial connection of three single joints. The axes of rotation of these joints intersect at one point. Here, a replacement joint is used that consists of a roll joint, a pitch joint, and then again another roll joint. The description of the kinematics can only be clarified together with the entire arm, which requires limiting measures, especially if redundant degrees of freedom exist (Asfour, 2003).

3.2 Design

Information about the mass of the arm and its distribution are requirements for the design of the shoulder joint module. In addition, information about the connection of elbow and shoulder has to be available. This includes the components that are led from the elbow to or through the shoulder, as e.g. cables or drive trains of lower joints. The development of the entire mechatronical system is displayed by the object-oriented means of SysML (System Modeling Language) (SysML, 2005) diagrams, with which it is possible to perform a very fast system test with regard to compatibility and operational reliability. It enables the representation of complex systems at different abstraction levels. Components that are combined in this way can be accessed in the aforementioned classification, which facilitates a quick selection of the components that can be used for the system. In addition, it makes a function design model possible at every point of the development.

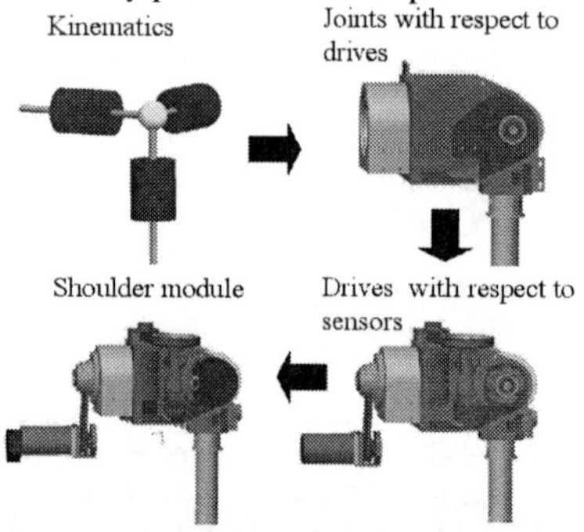

Fig. 2. Design of the shoulder module.

In the development of the shoulder module (Fig. 2), at first the function units of the joints for the three rotating axes are selected according to the kinematics. Then, the function unit drive, including the actors and the drive trains, are integrated. Hereafter, the sensors are selected and integrated. In order to prevent long-lasting iterations in the development, the components of the total system, integrated at a later stage, are already considered from the start with regard to their general ability to be integrated. Examples for this are the sensors, which can be assembled without problems since it is made sure that the already designed system offers the possibility to integrate them. During the next step the neighboring module is designed. Therefore information about the required interface of the shoulder and the mass of the arm and its distribution are given to the torso module.

4. EXAMPLE OF A TOTAL SYSTEM

The robot Armar III (Fig. 3), which consists of a variety of different modules, was built in the SFB588 project with a significant involvement of the Institute of Product Development.

Fig. 3. The upper body of the humanoid robot Armar III.

6. ACKNOWLEDGMENTS

This work is part of the Collaborative Research Center (Sonderforschungsbereich, SFB) No. 588 "Humanoide Roboter – Learning and Cooperating Multimodal robots" at the University of Karlsruhe. The SFB is supported by the Deutsche Forschungsgemeinschaft (DFG).

7. REFERENCES

Asfour, T. (2003). Sensomotorische Bewegungskoordination zur Handlungsausführung eines humanoiden Roboters, Dissertation, Fakultät für Informatik der Universität Karlsruhe (TH), ISBN 3-89863-146-X

Browning, T. R. (2001). Applying the Design Structure Matrix to System Decomposition and Integration Problems: A Review and New Directions, IEEE Transaction on Engineering Management, Vol. 48, No. 3

Möhringer, S. (2004). Entwicklungsmethodik für mechatronische Systeme, Habilitation, Heinz Nixdorf Institut Paderborn, HNI-Verlagesschriftenreihe, Band 156

Schulz, S. (2003). Eine neue Adaptiv-Hand-Prothese auf der Basis flexibler Fluidaktoren, Dissertation, Fakultät für Maschinenbau, Universität Karlsruhe (TH)

SFB588 (2006). Collaborative Research Center, SFB 588 Humanoid Robots – Learning and Cooperating Multimodal Robots, http://www.sfb588.uni-karlsruhe.de/

SysML Partners (2005). Systems Modeling Language (SysML), Specificationversion 1.0 alpha, 14. www.sysml.org

VDI2206. (2204). Verein Deutscher Ingenieure - Gesellschaft Entwicklung Konstruktion Vertrieb: Entwicklungsmethodik für mechatronische Systeme; VDI-Richtlinie.

Annals of DAAAM for 2006 & Proceedings of the 17th International DAAAM Symposium, ISSN 1726-9679
Editor B. Katalinic, Published by DAAAM International, Vienna, Austria 2006
150th Birthday of Nikola Tesla (1856-1943) Inventor of Mobile Robot Teleoperation

INVESTIGATION BY F.E.M. OF THE FRICTION COEFICIENT INFLUENCE ON SPRINGBACK EFFECT OF A U-SHAPED PART MADE BY TAILOR WELDED STRIPES

ALBUT, A.

Abstract: *This paper deals with some experimental and numerical tests related to tailor welded stripes forming and springback. Final shape of the parts manufactured by TWB is seriously affected by springback effect. This paper work is trying to prove the important role that friction coefficient has on the springback reduction. The simulations have been done using ABAQUS software. The F.E.M. analysis have been carried out using different friction coefficient and maintaining constant all other parameters.*
Key words: *tailor welded stripes, springback, friction coefficient, finite element method.*

1. INTRODUCTION

Recently, the automobile industries have been trying to develop various types of model and high-quality low-cost cars to meet the customer's requirements and to find new ways of establishing this goal effectively. For the purpose of achieving the above presented objectives, different methods using various welding processes (such as laser-welding, mash seam-welding processes, etc.) were developed. A tailor welded blank consists of two or more sheets that have been welded together in a single plane prior to forming. The sheets can be identical, or can have different thickness, mechanical properties or surface coatings.

Friction between sheets and tool plays an important role in sheet metal forming since fracture, wrinkling and weak spots are strongly influenced by frictional behaviour. In finite element model simulation of sheet metal forming, the Coulomb friction model, in which a constant model is assumed, is generally used. However, the frictional behaviour depends upon local contact conditions, such as local surface roughness, pressure and relative speed of contact surfaces.

In this study, the tailor welded stripes (joined together without taking in consideration the welding line) with two types of material having the same thickness, are used to investigate springback characteristics in U-shape bending.

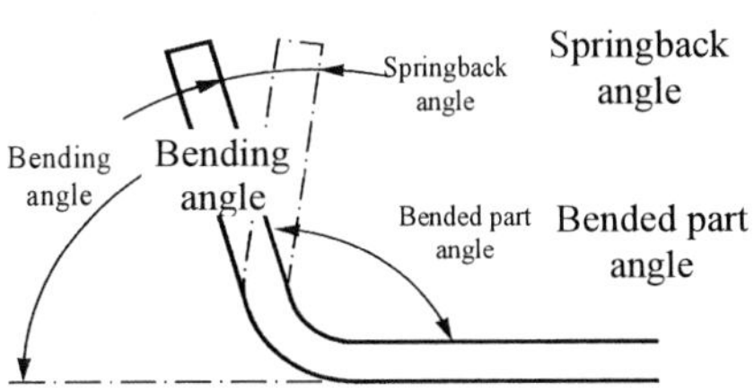

Fig. 1. Scheme of the bending process and springback

Springback (fig. 1)is mainly influenced by the punch and die profile radii, initial clearance between punch and die, friction conditions, rolling direction of the materials, blankholder force, material properties (elastic modulus, Poisson's coefficient, constitutive behaviour in plastic field) etc. The purpose of this study was to investigate the friction coefficient influence for minimizing the springback effect of the tailor-welded stripes. To achieve this goal, simulation tests were carried with different friction conditions.

2. ANALYSIS BY FINITE ELEMENT METHOD OF THE INFLUENCE OF FRICTION CONDITIONS

2.1 Simulation layout

The simulation of U-shape part forming was made using finite element method. The objective is to create a model that allows an accurate prediction of springback intensity, stress and strain state at the end of the forming process. The analyzed geometrical parameters are sidewall radius ρ and springback angles θ_1 and θ_2. In order to validate the model, the results of the FEM analysis were compared with the experimental results. The tailor welded stripes used in the experiments were made from FEPO and E220 steel. Springback parameters that were observed during the tests are presented in Fig. 2:

- θ_1 – sidewall angle between real profile and theoretical profile;
- θ_2 – flange angle between real profile and theoretical profile;
- ρ – curvature radius of the sidewall.

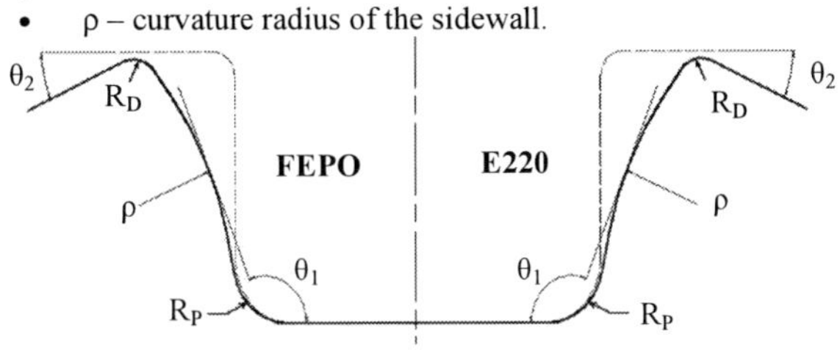

Fig. 2. Springback parameters

2.2 Simulation methodology

The material was modelled as elastic-plastic, where elasticity is considered isotropic and plasticity is modelled as anisotropic using Hill quadratic anisotropic yield criterion. In this simulation, fine mesh technology was applied to interested regions such as the punch profile radius and the tailor-welded strip model to improve the accuracy of the analysis for the springback. The geometrical model is presented in Fig. 3. The initial dimensions of the sheet were 350 mm length, 30 mm width and 0.7 mm thick. The sheet was modelled as deformable body with 400 shell elements (S4R) on one row with 5 integration points through the thickness. The tools (punch, die and blankholder) were modelled as analytical rigid because they have the advantage of reduced calculus efforts and a good contact behaviour. Rigid body movements are controlled by reference points. The boundary conditions imposed to the tools were intended to describe the experimental conditions as accurate as possible. For contact conditions a modified Coulomb friction law combined with penalty method was used.

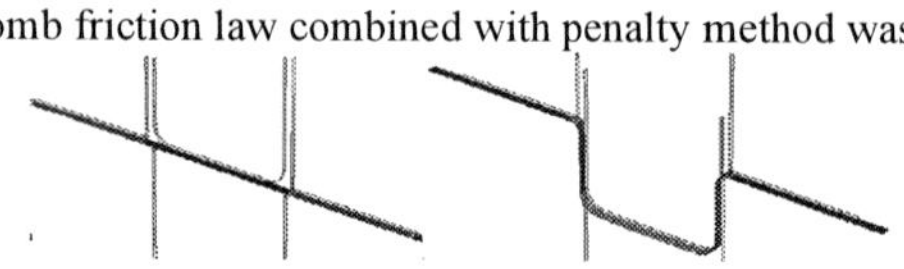

Fig. 3. Geometrical model

2.3 Simulation results concerning the influence of friction conditions

For determination of the influence of friction conditions between specimen and tools on the parameters of springback, the simulations have been done under the following conditions: the sample model is loaded with the material characteristics corresponding to 0° rolling direction; blank holder force F = 10 kN; the simulations were carried under dry friction (friction coefficient is 0.177) and with lubrication (friction coefficient 0.1) of the specimen and tools active surfaces.

The variations of springback parameters (θ_1, θ_2, ρ) as function of friction coefficient are presented in table 1.

	FEPO					
Friction coefficient	**Angle θ_1 [grd]**		**Angle θ_2 [grd]**		**Sidewall radius [mm]**	
	Theoretic value	Measured value	Theoretic value	Measured value	Theoretic value	Measured value
0.1	90	98.8	0	12.2	:	254.46
0.177	90	99.2	0	13.8	:	203.93
	E220					
Friction coefficient	**Angle θ_1 [grd]**		**Angle θ_2 [grd]**		**Sidewall radius [mm]**	
	Theoretic value	Measured value	Theoretic value	Measured value	Theoretic value	Measured value
0.1	90	99.2	0	17.1	:	162.51
0.177	90	101.4	0	18.4	:	103.34

Table 1. Springback parameters

The influence of the friction conditions on the parameters of springback is illustrated in figures 4, 15, 6.

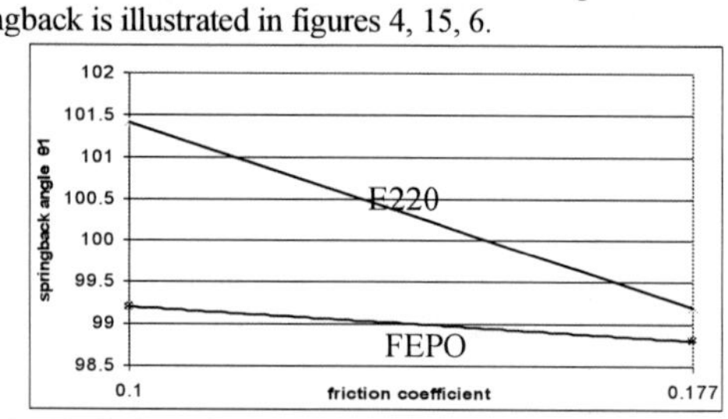

Fig. 4. Variation of angle θ_1

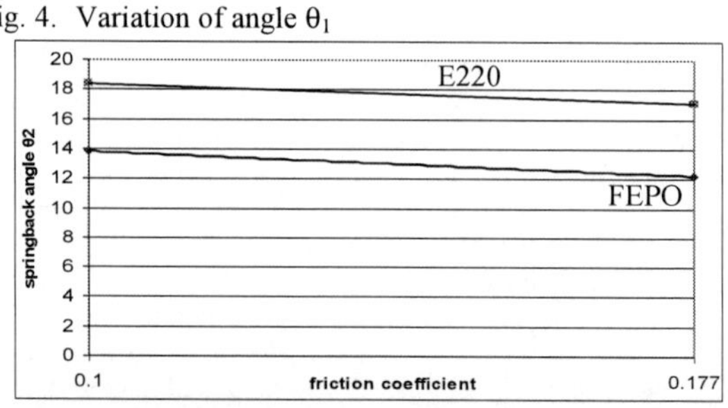

Fig. 5. Variation of angle θ_2

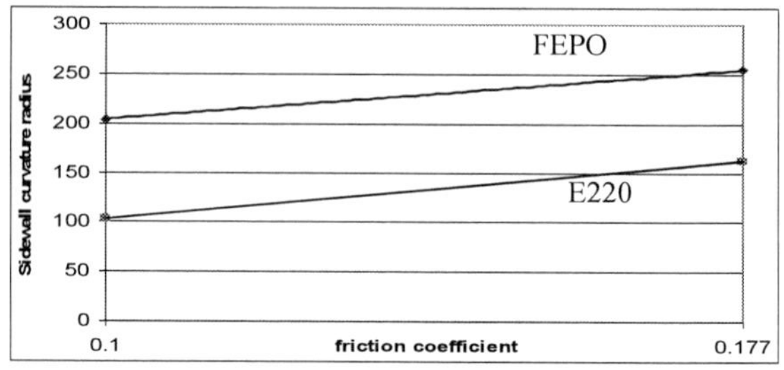

Fig. 6. Variation of sidewall curvature radius ρ

From the above presented results the following aspects can be remarked:

- the modification of the friction coefficient affects the springback parameters of the U-shaped part;
- the values of θ_1 and θ_2 angles attains higher values when the part made from TWBs is formed with lower friction coefficient;
- the sidewall radii ρ are smaller when the TWBs is formed with higher friction coefficient;
- the part area made from FEPO is not so much affected by the springback phenomenon in comparison with E220 steel area for both friction coefficients.

3. CONCLUSION

The following aspects stand out from the experimental researches of the influence of the rolling direction on springback parameters:

- the lubrication of specimens and tools active surfaces produced a small increase of springback intensity for both materials from TWB part;
- this phenomenon can be explained by the fact that lubrication lowers the friction coefficient allowing the material an easier flow into the die cavity.

It can be considered that the results generated by the analysis of springback phenomenon using finite element method are sufficiently accurate and can be considered valid. When properly used, simulation by finite element method can be considered a valuable tool in the study of the influencing factors of the springback phenomenon able to offer accurate data even from the design stage.

4. REFERENCES

Li G.Y., Tan M.J., Liew K.M. – *Springback analysis for sheet forming processes by explicit finite element method in conjunction with the orthogonal regression analysis*, Int. Journ. of Solids and Structures, 36, 1999

Papeleux L., Ponthot J.P. – *Finite element simulation of springback in sheet metal forming*, J. of Mat. Proc. Tech., 125-126, 2002

Yamamura N., a.o. – *Springback simulation by the static explicit FEM code, using a new algorithm for canceling the non-equilibrated forces*, Simulations of Materials Processing: Theory, Methods and Applications, Swets & Zeitlinger, Lisse, 2001

Pourboghrat F., Chu E. – *Prediction of spring-back and side-wall curl in 2-D draw bending*, J. of Mat. Proc. Tech., 50, 1995

Pourboghrat F., Chu E. – *Springback in plane strain stretch/draw sheet forming*, Int. J. Mech. Sci., vol. 36, no. 3, 1995

Lee J. K., Chun B. K., Kim H. Y. – *Numerical investigation of tailor welded blanks forming and springback*, Simulation of Material Processing, p 729-734, 2001

Annals of DAAAM for 2006 & Proceedings of the 17th International DAAAM Symposium, ISSN 1726-9679
Editor B. Katalinic, Published by DAAAM International, Vienna, Austria 2006
150th Birthday of Nikola Tesla (1856-1943) Inventor of Mobile Robot Teleoperation

BUILDING AN AUTONOMOUS, MOBILE DEMONSTRATION PLATFORM FOR EMBEDDED VISION APPLICATIONS

AMBROSCH, R.; RINNERTHALER, F. & KUBINGER, W.

Abstract: *This platform is a model of a real car to test different embedded vision algorithms and mobile robot algorithms. The test results of the platform can be used as a reference for real cars. To be able to evaluate different sorts of experiments, modularity is the basis of the platform. The built-in image processing unit is small in size to be close to automotive applications. Typical requirements of the automotive industry, like real-time behaviors, are considered. The developed software framework is highlighted to show the modularity of the platform.*
Key words: *embedded systems, embedded vision, mobile robots*

1. INTRODUCTION

This paper includes the results of the work on an autonomous test and demonstration platform for the evaluation of different algorithms for embedded vision and mobile robot applications. This platform was built for easy testing without real cars that cost much more and do not excuse errors. Mobile robots that are available are not as modular as this platform and mostly do not have the car typical non-holonomic actuation.
The main task was the sensor evaluation and the reproduction of a car's internal layout. To test the modularity of this platform, one application was implemented – platooning. Platooning is the process where one car autonomously follows the other one. The test bed setup and realization of the demonstrator are presented. Future works on this platform include the implementation of additional applications and the setup of a simulation framework.

2. REALIZATION OF THE PLATFORM AND EXPERIMENTS

2.1 Concept

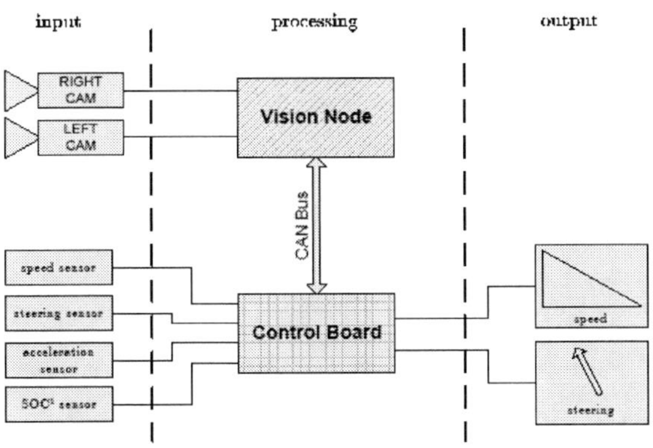

Fig. 1. – Input and Output of the platform

The platform is equipped with two cameras, speed and steering sensors, and a powerful image processing hardware board. Figure 1 shows a block diagram of the computer vision test platform. On the left side are the sensors of the system. In the middle the processing of the sensor data takes place. The Vision Node receives images from the cameras and gets information about the environment out of the images by applying a computer vision algorithm. The Control Board is responsible for controlling the truck with the help of the available information received by the sensors and the Vision Node.

2.2 Hardware

The base of the platform is an RC truck with differential gear and a maximum load of 10 kg. The steering mechanism is realized with a servo and the driving with a direct current motor. The rotational movement of the motor is transformed with a ratio of 1:90. To allow easy attaching of additional circuits, a stack design was used. On the lowest layer is the board that interferes with the platform's sensors and actuators. It has a CAN (Controller Area Network) module to be able to transfer data with high rates.

2.3 Electronics/Sensors

The sensing of the actual velocity is realized with an inductive sensor that detects a rotating plate on the car's axles.
The model car's maximum speed is 20 km/h. The minimum speed that can be measured is limited by the capture/compare module of the micro controller. The maximum period that can be measured has a period time t_{max} of six seconds which results in a minimum speed of 0,00785 m/s. The steering sensor is a potentiometer with a hollow shaft and tight tolerances. It is mounted around the moving part and is consistent against vibration. The resulting voltage of the built in voltage divider is relative to the position of the wheels. Two acceleration sensors are used to detect crashes with the environment. The chosen sensors are Dual-Axis Accelerometers with Duty-Cycle sensors. The sensors from Analog Devices produce PWM (Pulse Width Modulation) signals with a varying pulse width, depending on the actual acceleration A(g). For most accurate results, two sensors are used: one sensor has a range of $\pm$ 2 g and the second measures in the range of $\pm$ 10 g.

2.4 Software

The Software architecture is split in two parts. One part is the Firmware, which interacts with sensors, actuators and communication interfaces. This first layer sets PWM values to drive servos, converts voltages to determine the position of the steering potentiometer, communicates with other devices via CAN, configures the state-of-charge ICs (Integrated Circuits) and uses onboard hardware for the visualization of the current status. Above this layer there is the possibility to set up different controllers for varying applications. This layer could contain different controlling algorithms and for example communication functions. The concept of the software interface is shown in Figure 1.

The first layer (device access layer) sets up a framework to interact with the hardware of the car. The functions provided by this layer allow the user to initiate different mechanical actions.

The second layer (application layer) uses the first layer to interact with the car and has the possibility to easily set up a communication to external hardware and controlling algorithms. It is the application layer of the platform. Without knowing anything about the platform's hardware, it is possible to program the platforms behaviour on different situations.

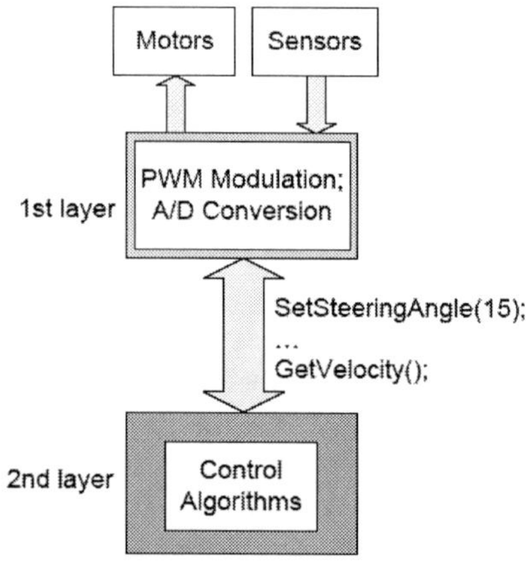

Fig. 2 Software Framework

3. EXPERIMENTS

To test the modularity of the platform, platooning was realized. The leading car is an RC car from TAMIYA (Mini Cooper Rallye) and the platform follows its path.

Run 1 - In the first part the actuation was tested by an acceleration and break sequence. This test shows whether the controller parameters are set correctly or not. The sequence is shown in Figure 3a. Black means acceleration and grey deceleration. After fine-tuning of the proportional parameters, the platform followed the leading car exactly at a distance of 0.35m. The sensors act like they should without reaching the limits of the sensing system.

Run 2 - Figure 3b shows the second part of the test run that evaluates the delay of the different mechanical and electronical parts. Therefore the leading car drives a sinusoidal path. The result showed that there is no overshooting and the platform follows with a tolerance of ± 0.05m.

Run 3 - The third and last part tested the behavior of the platform when the leading car drives in curves around corners. This sequence is shown in Figure 3c. This test run showed that there is an optimal distance of 0.35m between the two cars and a limit in the steering angle when driving around curves.

4. DISCUSSION

The evaluation of the velocity sensors showed that optical sensors are severely limited for dirt and light influences, incremental encoders with hollow shafts are too big and not economic for this vehicle. The chosen speed measurement sensors are small, have a high switching frequency, are insensible against dirt and are not to expensive.

The different methods and applications of measuring the front wheel positions mainly vary on the mounting place and on the monitored parts. A possible application is measuring the distance between the wheel and an optical sensor with reflectors. The problems are the same as they are with speed

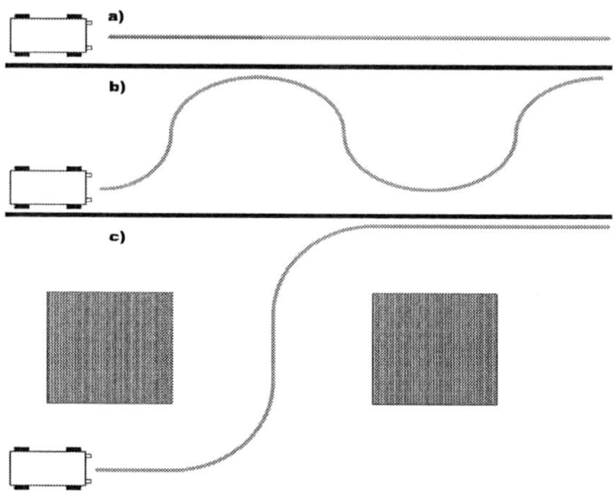

Fig. 3. Layout of Test Runs; a) = Run1; b) = Run2; c) = Run3

sensors. Dirt and light would influence the sensors and would not give useful values. Another problem is, that the optical sensor would not measure the distance at the same radius and probably the speed of the moving reflectors would influence the measured distance.

The solution to measure the steering position used in the previous platform had a problem with the additional degrees of freedom of the steering mechanism. The lightweight construction and its resulting flexibility did not deliver correct values. The new solution indirectly measures the position of the wheels via the rotating part of the steering mechanism and bypasses the mechanism failure.

5. CONCLUSION

This paper deals with the creation of an autonomous demonstration platform for embedded systems. The setup is viewed from the hardware and software view and is described in detail. The mechatronical solutions for the sensing include answers to mechanical problems with little space and problems with outdoor environment. Different sorts of sensors are thought-through and their advantages and disadvantages listed. The results represent the right sensor choice because there was no space left after the mounting of the sensors and their cables. The used potentiometer allowed a precise measurement of the current wheel positions although the steering mechanism is bendable.

The test runs with the platooning application showed, that the platform works and that the sensor-limited ranges are well designed.

6. REFERENCES

Altenburg, J. & Altenburg, U. (2001). *Mobile Roboter*, Hanser Verlag, ISBN 3446215107

Baehring, H. (2004). *Mikrorechner Technik*, Springer-Verlag, 3. Auflage, ISBN 3540436936

Borenstein, J., Everett, H.R. & Feng L. (1996). *Sensors and Methods for Mobile Robot Positioning*, University of Michigan.

Kubinger, W.; Hemetsberger, H. & Kogler, J. (2005). *Platooning Platform for analyzing embedded control algorithms*, 16th DAAAM International Symposium

The MathWorks; *Real-Time Workshop Embedded Coder User's Guide*; Version 4.1, October 2004.

Annals of DAAAM for 2006 & Proceedings of the 17th International DAAAM Symposium, ISSN 1726-9679
Editor B. Katalinic, Published by DAAAM International, Vienna, Austria 2006
150th Birthday of Nikola Tesla (1856-1943) Inventor of Mobile Robot Teleoperation

DESIGN OF TELESCOPIC UNIT WITH SYNCHRONIZED DOUBLE-SIDED FULL-STROKE

ANDRLIK, V. & JURDA, P.

Abstract: *The invention applies to the synchronized double-sided telescopic unit. This unit uses an unique mechanical structure to provide synchronisation of moving parts. Due to this mechanical kinematic structure it is possible to reliably control velocity and position of all moving parts during a working cycle and prevent collisions in working area .*
Key words: *Telescopic full-stroke, Synchronisation, Synchronized telescopic manipulator, Double-sided full-stroke, Telescopic manipulator*

1. TARGETS

Existing applications solving problems of full-strokes are based on using non-synchronized telescopic guides working with mechanical, hydraulic or pneumatic motion drives. In these cases the full-stroke in general is not possible to position and control reliably (Andrlík et al., 2002; Matička & Talácko, 1995).
One of the disadvantage of non-synchronized full-strokes using especially double-sided full-stroke is uncontrollable movement of extending parts, which can stagnate in various positions during the working cycle.
Creating of a synchronized full-stroke using mechanical kinematic structure between particular components of the telescopic unit can solve these problems mentioned above.

2. DESIGN OF SYNCHRONIZED TELESCOPIC UNIT

2.1 Basic idea

The basic idea of this invention is creating a mechanical kinematic structure using cogged belts with driving and tightening pulleys to achieve a synchronized movement of extending parts (fig. 1).

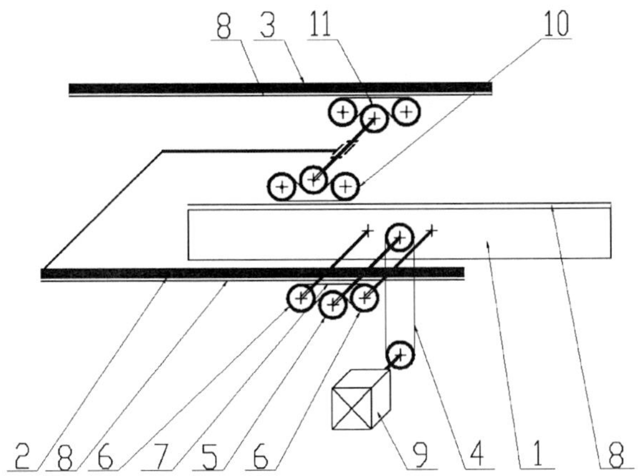

Fig. 1. Basic kinematic structure

Telescopic unit consists of the base body with driving pulley of the primary drive (pos. 1), extending sliders (pos. 2 and 3), parts for synchronisation (pos. 5, 10 and 11) and the primary drive cogged belt with possibility of electromotor adaptation (pos. 4 and 9).
Double-sided full-stroke synchronisation is provided by a triple mechanical "OMEGA" structure. One of this structure consists of driving and tightening pulleys and cogged belt (pos. 5, 6 and 7) organized into shape of "OMEGA", which creates the secondary mechanical drive for extending sliders (fig. 1).
Alternatively it is possible to use a triple driving pulleys and cogged belts without tightening pulleys (fig. 2) or system with one "OMEGA" cogged belt as shown on the figure 3. But the first constructions does not provide large contact especially the wrapping angle between driving pulley and fixed cogged belt to carry enough performance. In low-performance devices is this construction possible. And on the other hand the construction on figure 3 has enough wrapping angle, but enlarges the whole build-up area of the manipulator to provide the same ejection as in the case on figure 2. But the construction on the figure 3 absorbs due to the "OMEGA" structure any elastic strain in bend and torsion caused by external forces and torques strains.

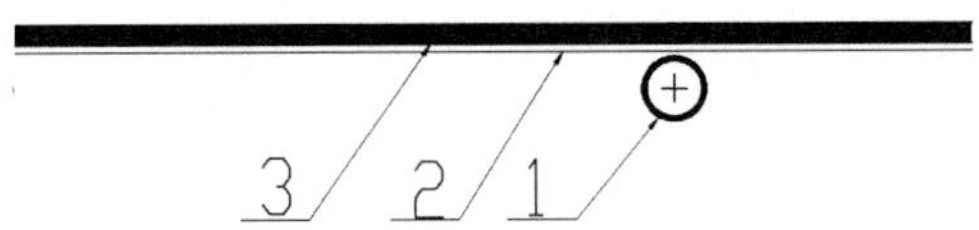

Fig. 2. Driving pulley and cogged belt
(1 – driving pulley, 2 – cogged belt, 3 – moving part)

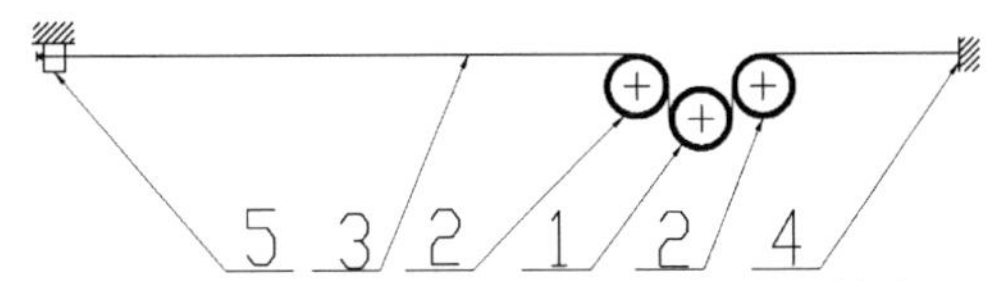

Fig. 3. Driving and tightening pulleys with cogged belt
(1 – driving pulley, 2 – tightening pulleys, 3 – cogged belt, 4 – fixed part, 5 – cogged belt tightening)

2.2 Principle of designed synchronisation (fig. 1)

Primary motion drive is mounted on the base body (pos. 1). The electromotor (pos. 9) rotates the primary driving pulley with cogged belt (pos. 4) to rotate the first mechanical "OMEGA" subsystem consists of the driving pulley (pos. 5), two tightening pulleys (pos. 6) and the cogged belt (pos. 7). Moving of this subsystem on the cogged belt (pos. 8) fixed on the extending slider (pos. 2) gets it actuating. The extending slider (pos. 2) carries remaining two "OMEGA" subsystems (pos. 10 and 11) to provide synchronisation of the second extending slider (pos. 3). The idea of actuating the second extending slider is the same with using pulleys nad cogged belts as shown on figure 1. The second "OMEGA" subsystem (pos. 10) moves with the extending slider (pos. 3) and at the same time does rotary motion caused by contact with the cogged belt (pos. 8) fixed on the base body (pos.

1). This rotary move actuates the third "OMEGA" (pos. 11) and moves the second extending slider (pos. 3).

2.3. Design of synchronized telescopic unit
Moving parts are jointed by telescopic rolling guides ROLLON, which provide very high rigidity and accuracy of this telescopic unit (Jurda, 2003; ŽIAČIK & LAHUČKÝ, 2005).
Secondary "OMEGA" drive´s structure is shown on the figure 4. For all "OMEGA" drives were used ball bearings instead of tightening pulleys and the driving pulley with side plates to prevent sliding of the cogged belt.

Fig. 4. 3D model of secondary "OMEGA" drive

General view on designed synchronized telescopic unit in full-closed state is shown on the figure 5. Primary drive´s cogged belt with tightening screw, hollow shaft and possibility of electromotor adaptation is evident from the axonometric view. Structural composition of "OMEGA" drives and semitelescopic rolling guides is evident from the front view.

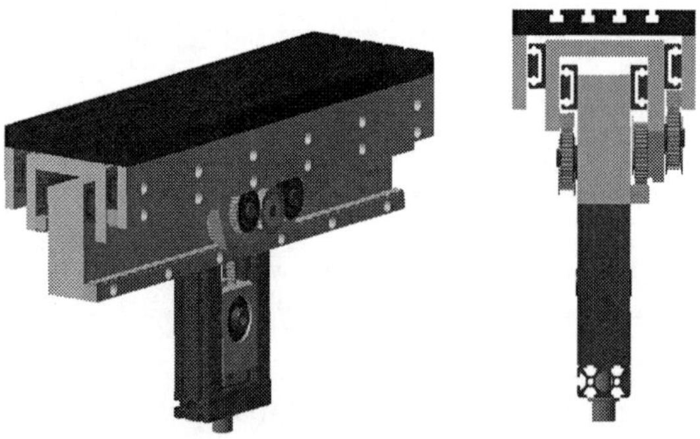

Fig. 5. 3D model of synchronized telescopic unit (axonometric and front view)

Designed telescopic unit in single-sided fully ejected state increases its length twice and this state is shown on the figure 6.

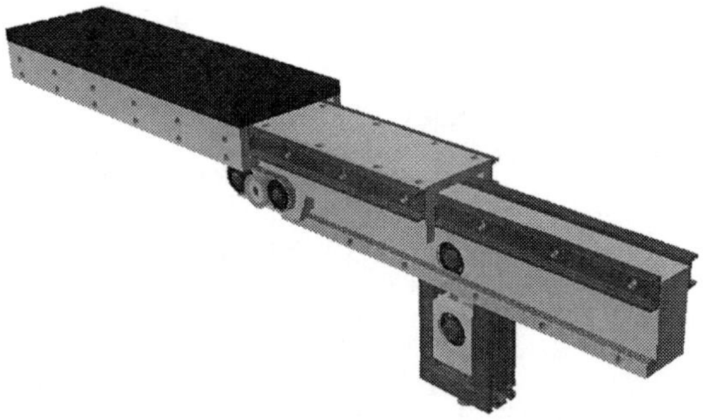

Fig. 6. Fully ejected synchronized telescopic unit

2.4. Design of a real synchronized telescopic unit
According to described design concept the prototype of this double-sided full-stroked telescopic unit was made. Three phase motor with worm-gear unit as a driving elektromotor was chosen and mounted on the the base body to the primary cogged belt drive. The whole telescopic unit was fit on a support construction to test its functionality during programmable working cycle.

Build-up area		Working parameters	
length	450 mm	load capacity	120 kg
width	136 mm	max. velocity	0,3 m/s
high	225 mm	extension	464 mm

Tab. 1. Technical parameters of designed unit

3. THE MAIN ADVANTAGES OF SYNCHRONIZED TELESCOPIC UNIT

The main advantage of the designed telescopic unit is in its full-stroke synchronisation, which can be reliably controlled by any motorized driving unit. The another advantage is in a modular and compact conception to provide using this unit alone or with another parallel units. These units can be connected to synchronous shaft and be driving with only one electromotor. Enough high carrying-capacity provides this unit due to a parallel structure of the telescopic rolling guides. In this way a high rigidity in bend and torsion is provided. Structural principle of this synchronized telescopic unit can be used even for multistage telescopes having severalfold larger working area than its build-up area in the closed stage is.

4. INDUSTRIAL USAGE

Using of a telescopic unit with synchronized stroke is possible in all fields of industry where working in limited area is needed, also in full-stroke applications to take control of velocity and position of extending parts. Mounting position of synchronized telescopic unit is not restricted due to mechanical kinematic structure, which allows the telescopic unit to be installed in various kinds of positions. Industrial manipulators can be also created with combination of this telescopic unit with other rotary units. The main usage of this unit seems to be a pallet rack handler, picking or laying down manipulator stations, stations for production and assembly lines and etc.

5. KEY CONCLUSIONS

This report solves issue of synchronized double-sided telescopic unit. According to described design concept the prototype of this double-sided full-stroked telescopic unit was made and patented as a utility patent of Vladimír Trhoň with registration number PUV 2005-16701 on 6th June 2005. Design concept and technical drawings were made in cooperation with Czech Technical University in Prague, Faculty of Mechanical Engineering. Designed synchronized double-sided telescopic unit solves reliably positioning and controlling of full-strokes and provides high rigidity in bend and torsion. This unit can be parallel connected to synchronous shaft and be driving with only one motorized driving unit and seems to have a great usage in various fields of all kinds of industry.

6. REFERENCES

Andrlík, V.; Pospíchal, J.; Talácko, J. & Vedral, K. (2002). Relevant Problems of Manipulators with Telescopic Arm, *Proceeding of Robtep 2002,* pp. 29-36, ISBN 80-7099-826-1, Tu in Kosice, 5/2002, TU Košice

Jurda, P. (2003). *Diploma thesis no.1431*, ČVUT, Department of Production Machines and Mechanisms, Prague

Matička, R. & Talácko, J. (1995). *Construction of industrial robots and manipulators,* ČVUT, ISBN 80-01-01291-3, Prague

Žiacik, A. & Lahucky, D. (2005). *Opotrebenie klzných a valivých uložení I. Strojárstvo 2*, MEDIA/ST, ISSN 1335-2938, Žilina

Annals of DAAAM for 2006 & Proceedings of the 17th International DAAAM Symposium, ISSN 1726-9679
Editor B. Katalinic, Published by DAAAM International, Vienna, Austria 2006
150th Birthday of Nikola Tesla (1856-1943) Inventor of Mobile Robot Teleoperation

THE INFLUENCE OF THE WAVE SIZE ON THE STRENGTH OF THE VERTICAL AND HORIZONTAL SURFACE OF CORRUGATED BOARD

BABIC, D.; JURECIC, D. & LAJIC, B.

Abstract: *This article discusses the testing of the resistance of corrugated board to pressure, defined as a force which deforms the material plastically, squeezes it, or destroys its structure. The strength of corrugated board equals the weakest force that can cause continuous deformities of the material (N/mm^2 surface of the material cross-section). The resistance to pressure depends solely on the properties of corrugated board, in this case on varying wave sizes in its layers, while the shape of the packed object has no influence whatsoever. The standard test tubes of defined shapes and dimensions were used for testing the strength of corrugated board. They were tested on a crush medium tester which measures the resistance of corrugated board to pressure. The results will show the influence of the wave size on the strength of corrugated board, i.e. on its horizontal and vertical surface.*
Key words: *corrugated board, strength, pressure*

1. INTRODUCTION

The mechanical stability of corrugated board is defined by its resistance to deformities or damages caused by outer influences or mechanical forces. Similar static and dynamic forces act on the packaging made of corrugated board. They cause strain of the material, which is met by its mechanical properties. If those properties of corrugated board are strong enough to resist the mechanical forces, the material has a great stability. Mechanical forces can break corrugated board, crush it, bend it or perforate it. Depending on the direction of the force and the type of loading which corrugated board resists, we distinguish different types of strength (resistance): resistance to deformation, pressure, bending, twisting and perforating. Corrugated board is a packaging material which consists of two or more layers of paper of different qualities glued together, at least one of which has waves. The basic raw material for the production of corrugated board are papers of different grammage and glues which connect the flat and the wavy layers. The choice of paper used for producing corrugated board depends on its mechanical properties in relation to its grammage and price. Soda paper can be used for the flat layers of corrugated board because it has a great strength to deformations. Paper with the grammage between 125 and 250 g/m^2 is usually used in the production of corrugated board, but board with the grammage of 400 g/m^2 can also be used. Those papers have a smooth side, which is very important in the production and processing of the packaging. The outer layer is coated and, as such, smooth and suitable for printing, while the inner one is coarse and suitable for gluing the wavy layer on it. If we want to produce corrugated board of a lesser strength, weaker paper such as common wrapping paper made of sorted old paper can be used for the flat layers. In the production of corrugated board resistant to water, impregnating paper is used for making the flat layer, most frequently soda-paper made of unbleached cellulose. For the production of the wavy layers, we use cheap paper made of semi-cellulose and common wrapping paper, as well as paper made of straw. The quality of corrugated board is influenced by the choice of glue used. The strength of the glued place must be somewhat greater than the strength of the paper itself, so that the glued parts cannot be detached without tearing the paper. Glues must neither penetrate deep into the paper nor pour out over the paper because of the low viscosity. The speed of adhesiveness must be high and match the working speed of the machine for the production of corrugated board. Today different kinds of synthetic adhesives are used. They are not toxic, dry very quickly, have great strength of bonding and are resistant to water. Starch glues and water glass are also used today. If the waves of the corrugated layer of the board are smaller, more glue will be used and vice versa and inner impregnation of corrugated board with smaller waves will be better.

2. EXPERIMENT

According to the number of layers, there are two-layered, three-layered, five-layered and seven-layered corrugated board (fig. 1).

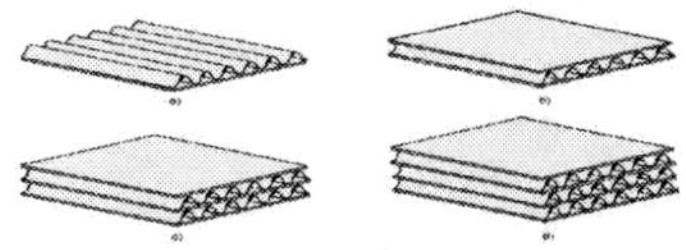

Fig. 1. Types of corrugated board {a) two-layered, b) three-layered, c) five-layered, d) seven-layered}.

The numbers denote the total number of the flat and wavy layers. Two-layered board has one flat and one wavy layer; three-layered board has two flat and one wavy layer; five-layered board has two wavy and three flat layers and seven-layered board has three wavy and four flat layers. The waves usually have the form of sine waves (fig. 2). That wave type is easy to make and ensures good mechanical properties of corrugated board. The peaks of the waves are glued to the flat layer and a wave has a similar function as an arch of a bridge.

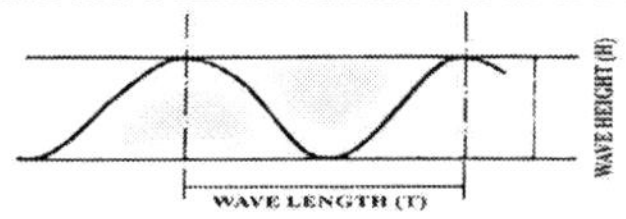

Fig. 2. Wave form

Waves give elasticity to corrugated board because of their form, which is extremely important for the packaging made of that material. The waves are functional only if they are glued well to the flat layers of paper. Depending on their height, waves are given marks. The standard marks are from A-wave (the greatest one) to F-wave and even smaller ones (microwaves) (table 1).

Wave name	Wave mark	Wave height (h) in corrugated board (mm)	Wave length (t) – step (mm)	Number of waves per 1m/length of corrugated board	Theoretical thickness (mm)	Glue quantity (g/m^2)
Great (coarse)	A	4,80	8,00-9,50	110	5,63	4,50-5,00
Small (fine)	B	2,40	5,50-6,50	150	3,65	5,50-6,00
Medium	C	3,60	8,80-8,90	130	4,72	5,00-6,50
Tiny	E	1,20	3,00-3,50	290	-	6,00-6,50
Microwaves	F,G,N	0,50-0,80	2,30-2,40	400-550	-	9,10-11,00

Table 1. Wave dimensions and marks

Each of these wave types has some characteristic mechanical properties which distinguish them and which are transferred to corrugated board. In general, the greater the wave height, the better the protection of goods packed in corrugated board packaging against blows and other mechanical damage. The testing was carried out by Lorentzen&Wettre Code 248, a device which was designed for testing deformations of corrugated board and its components. The device is constructed in such a way that any risk of folding and grooving is eliminated. Different samples of three-layered corrugated borad were used for the testing. The standard grammages and paper marks most frequently used in layers of corrugated board are the following: krafliner 125-200 g/m^2; testliner 150-250 g/m^2; fluting 105-150 g/m^2; common wrapping paper 150-175 g/m^2. Each producer of corrugated board in the Republic of Croatia has different marks. The company *Belišće d.d.* produces corrugated board of the mark 2T-b val, which appears on the market with marks 2K-b, 1919B, 2K-3, 2PK-b, T21 B, TRT/B/414 etc., when produced by other manufacturers. The mark of corrugated board TRT/B/414, used by the company *Valtrgovina d.o.o. Zagreb*, describes exactly the structure of the material (T= testliner; R= recycled fluting; B= b-wavel; 4=140 g/m^2; 1=112 g/m^2), so we used those marks in this paper. The used samples had the following marks:

1. TRT/A/414   4. TFT/A/414   7. TFŠ/A/414
2. TRT/B/414   5. TFT/B/414   8. TFŠ/B/414
3. TRT/C/414   6. TFT/C/414   9. TFŠ/C/414

Ten test-tubes were made from each sample of corrugated board and subjected to testing on an ECT (Edge Crush Test) and an FCT (Flat Crush Test) device. The resistance of the edge of corrugated board to pressure (fig. 4) and the resistance of its horizontal surface to pressure were tested (fig. 5).

Fig. 4. A schematic presentation of the tested corrugated board and of the procedure of testing the resistance of the vertical surface of corrugated board to pressure

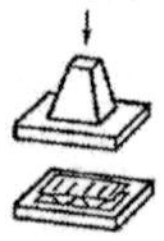

Fig. 5. A schematic presentation of the procedure of testing the resistance of a corrugated board edge to pressure and of testing the resistance of the horizontal surface of corrugated board to pressure

The testing was carried out in the following way: a sample of corrugated board of a defined size was inserted between the two plates, one of which was driven by an electrical engine and the other one, which resisted to pressure, was placed on a flexible plate with a digital control plate near it. The flexible plate is designed so that the determined pressure drives a mechanism on the control plate on which plate sagging is read. When the sample of corrugated board with waves that have to be crushed is submitted to pressure and when the crushing occurs, the plate stops and records the maximal pressure which the waves can bear before crushing. The greater the crushing pressure, the higher the quality of the paper used for producing corrugated board because it can resist greater loading. The pressure necessary to deform the wave of corrugated board is, in fact, the strength factor of the wave.

3. RESULTS AND DISCUSSION

The research results are presented in table 2 and diagram 1. For each sample the medium value is recorded. The tested waves A, B and C have different resistance to pressure. Samples 2, 5 and 8, which have B-waves, have more waves per meter length than samples 1, 3, 4, 6, 7 and 9, which have waves A and C. Therefore, they have a greater resistance to vertical pressure than waves A and C. If corrugated board is loaded parallelly to the wave direction, corrugated board with wave A can bear greater pressure than corrugated board with waves B and C.

Sample number	Sample mark	Kind of testing	
		ECT	FCT
Sample 1	TRT/A/414	4,80	1,90
Sample 2	TRT/B/414	6,60	3,80
Sample 3	TRT/C/414	5,40	2,90
Sample 4	TFT/A/414	4,40	1,80
Sample 5	TFT/B/414	6,30	3,60
Sample 6	TFT/C/414	5,20	2,70
Sample 7	TFŠ/A/414	4,50	1,60
Sample 8	TFŠ/B/414	6,20	3,40
Sample 9	TFŠ/C/414	5,20	2,50

Table 2. Results of the ECT and FCT testing of corrugated board samples

When a sample of corrugated board is pressed so that the pressure is vertical to the waves and not to their peaks, corrugated board with waves B can bear a higher pressure than corrugated board with waves A and C.

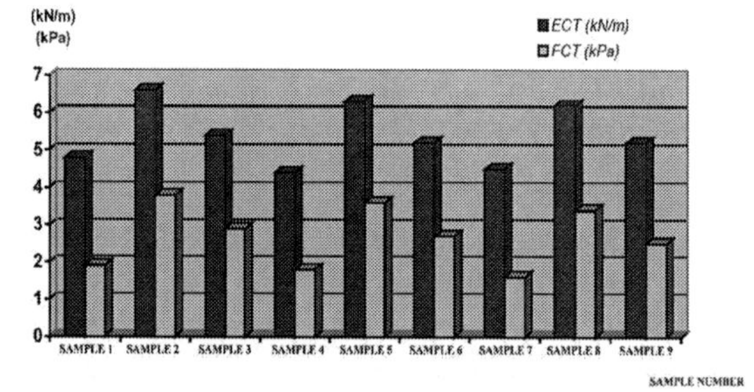

Diagram 1. A comparison of the resistance of the edge (ECT) and the horizontal surface (FCT) of corrugated board to pressure

4. CONCLUSION

The strengths of wave arches differ considerably, depending on the wave type used, the speed of waves per meter length and their form. It can be concluded that corrugated board with waves B and C can be used for the production of packaging for fruit and vegetables. As far as the above mentioned properties are concerned, corrugated board with wave C is in the middle between corrugated boards with waves B and A. Because of those specific characteristics, corrugated boards with waves A, B and C have a special role in the production of packaging, especially in the production of packaging with inner protection of goods. Because of that, packaging made of corrugated board must be produced so that the waves, regardless of their type, are set vertically to the bottom of the packaging. Only such a packaging can take the greatest pressure. It can further be concluded that boxes can be strongly pressed on the edges and less in the middle, which depends on the length of the box side. The right loading of the packaging is very important for practical usage, especially for storing. The packaging must be piled in storehouses in such a way that edges lie on edges. The boxes must be sorted according to their size and must not be put so that edges press the middle of box sides.

5. REFERENCES

Stern, W.: *The packaging engineering handbook*, Board Products Publishing Co., Chicago, 1991.

Stricevic, N.: *Suvremena ambalaza 1*, Skolska knjiga, Zagreb, 1981.

Stricevic, N.: *Suvremena ambalaza 2*, Skolska knjiga, Zagreb, 1983.

Tolliver, H.: *Packing trends*, Am. Ink Maker, no3. 2003.

Wybenga, G. L.; Roth, L.: *The Packaging Design Book of Patterns*, New York, 2002.

Annals of DAAAM for 2006 & Proceedings of the 17th International DAAAM Symposium, ISSN 1726-9679
Editor B. Katalinic, Published by DAAAM International, Vienna, Austria 2006
150th Birthday of Nikola Tesla (1856-1943) Inventor of Mobile Robot Teleoperation

THE WAYS OF THE IMPROVEMENT OF ENVIRONMENTAL PERFORMANCE OF INDUSTRIAL PRODUCTS

BADIDA, M.; VARGOVA, J.; HRICOVA, B. & WESSELY, E.

Abstract: *The authors described the concept of balancing both economic development and environmental preservation. In line with this concept, environmental policy has shifted from the end-of-pipe approach to the pollution prevention approach. Industrial products are a major source of today's environmental concerns. Sustainable development entails that manufacturers of industrial products take into account the environmental impacts throughout the entire life cycle of a product, not just focusing on environmental pollutants during the manufacturing of the product.*

Key words: *environmental performance, industrial product, sustainable development.*

1. INTRODUCTION

Permanently changing more and more demanding requirements for the tasks related to the management and preservation of environment, assurance of prosperity of companies as well as increasing pressure from competitors evoke constantly more urgent need of implementation of progressive methods and technologies which should assist to the companies by satisfaction their customers´ needs. The organisations have certain environmental tools to support and quide the positive changes towards better environmetal performance of their products.

2. SELECTED VOLUNTARY TOOLS TO ACHIEVE BETTER PRODUCT PERFORMANCE

Nowadays there are many tools to help to improve the environmental issues concerning the production and products. Among many tools which are known to be used within the Environmental Management Systems following will be focused only to two of them, which are voluntary tools, i.e. not ordered by certain laws:

- Ecodesign,
- LCA,

Both methods can influent a product and production significantly but in various ways.

3. ECODESIGN TOOLS

Ecodesign tools are a user-friendly eco-product development tool designed to be used by engineers as an integral part of the product launch activities. It follows both the ISO14062 and ISO 9001 processes for product development programs. The training segments of the tools are designed to train the user in eco-product development in easy stages, using the ACORN eco-management model. This allows the launch teams to learn to the depth that is appropriate for them. The tools analyses are modular so the companies can use just the parts relevant to them.

The new product development process has various stages as summarized below. The ecodesign team core led by the ecodesign champion, will overview and drive the project at all stages. Other functional participants may be involved at each stage depending on the company and product. Until now, the emphasis in business has been on minimizing the effects of own manufacturing processes or operations; the pressures for ecodesign require additional „life cycle" thinking. The main life cycle stages are described in Fig 1. (Ticha, 1998).

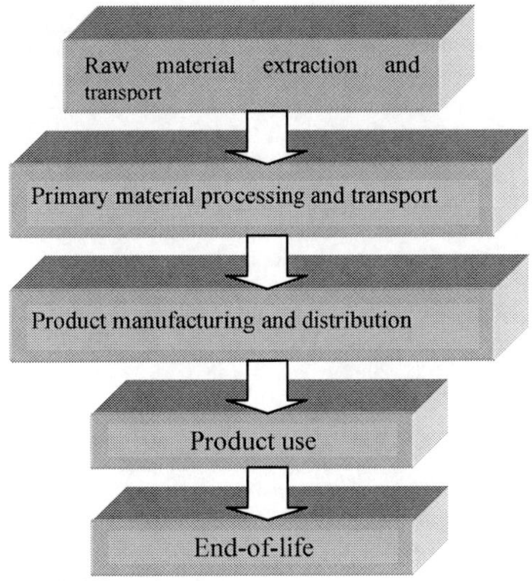

Fig. 1. The main life cycle stages

3.1 Ecodesign stages

***Planning*:** At the planning stage for a proposed new product the first stage is to know what environmental requirements from legislation, internal, customers etc need to be taken into account during the product launch.

Contract Review: Team needs the address whether the product targets set are realistic, and are compatible with other requirements for the product.

Concept Design: At this stage of the product launch different design solutions should be investigated, in order to find the best compromise between ecodesign, costs, quality and other factors. It is often possible to reduce costs whilst improving environmental product performance.

Concept Review: The various concepts should review to decide on the best option. The team needs to ensure the chosen design will meet the committed ecodesign targets.

Detail Design: At this stage of the product launch the supply chain often becomes a major factor in the eco-performance. It impacts the material content, transportation and packaging. It cans also sometimes effects recyclables and other ecodesign performance metrics.

Design Review: Once the final design has been completed, design review(s) should be held to determine whether all the programme objects have been met. Corrective actions should be put in place to address any issue.

Metrics: Most companies require metrics, so that they can have visibility on how the company is performing and for reporting externally to customers etc.

Requirements: This database holds environmental requirements of the users, companies and legislation for the different industry sectors. User companies can decide either to make their requirement visible to others, or hide them.

Material Properties: This database holds not only environmental data, but also general data on composition, mechanical, thermal and electrical properties, application, abbreviations, specifications and trade names.

Part and Assembly Declaration: This database holds the materials composition data for components; materials such as adhesives, solders etc; sub assemblies; and products. Each part is entered under the manufactures part number.

Ecodesign is likely to be most effective if considered and carried out, not as a separate exercise, or as technical activity alone, but as part of an environmental management approach integrated with other business processes and covering the company as a whole. The starting point should be an environmental review, which should identify and evaluate ecodesign and supply chain issues alongside other aspects of environmental performance, and the scope for improvement (Muransky & Badida, 2005).

4. PRODUCT LIFE CYCLE

The term product life cycle (LCA – Life Cycle Assessment) means all stages which each product goes through, i.e. the stage of raw material exploitation and material treatment for manufacturing of a particular product, production of the product itself, utilisation of a product connected eventually with the maintenance and product disposal as waste, Fig.1 (Badida et al., 2002).

4.1 The framework of the life cycle assessment

The life cycle assessment must contain this three phases:

I. phase – goal and scope definition

definition of the goals and the boundaries and the content of system which should be assessed. As well as the measures for comparison,

II. phase – inventory analysis

presents identifications and quantifications of all primary material and energy inputs and all the outputs. (wanted – unwanted),

III. phase – impact assessment

it is a quantitative and qualitative evaluation of data obtained in the inventory phase according to the size and the range of an impact.

4.2 The role of life cycle assessment in the environmental performance

A number of customers request their suppliers directly the material supply with environmental labelling and with the guarantee that at their production was no harm made to the environment. Similarly it is developed the relation to the transport, packaging etc.

The life cycle assessment techniques used to control the environmental aspects of products may have different applications. Private companies may use LCA-techniques internally for their development of new products, or optimisation of existing products, to reduce the environmental impact of the products in their whole life cycle.

Externally, the companies may use the assessments to document improvements for the consumer or environmental authorities, or to compare the environmental qualities of their product with those of competing products. The information provided by LCA can also be useful in applying an Eco-label licence (Frankl & Rubik, 2000).

5. CONCLUSION

In present time of global markets must the companies constantly innovate their products, processes and informational systems to remain on the market. They must collect, store and utilise their intellectual capital. The need of efficient management of the product life cycle is given by competitive pressure and constantly growing demanding ness of customer requirements. Adaptation of existing products may provide relatively little scope for the application of ecodesign. New product development offers opportunities for innovation, environmental improvement and contribution to business success. It is recognized that whiles both environmental product legislation and the environmental requirements of many multi-national companies on their suppliers is increasing dramatically, many companies just do not have the required in-house skills and experience to meet those requirements. Even many large companies suffer from their ecodesign skills existing only at corporate level, with Businesses launching products often not incorporating ecodesign into their products.

The consideration of LCA within decision –making depends on the institutionalisation stage, good justification and especially important of course, economic aspects. The economic aspects consist of different elements, for example the economic situation of the company,net benefits/burdens due to LCA – results. The further development of the tool LCA is one of the answers to the challenge of environmental pollution and deterioration. LCA seems the more complex and sophisticated instrument in the context of environmental improvements of products. (Frankl & Rubik, 2000).

The paper was elaborated in connection with the projects VEGA: "Modelling of factors of working environment and their optimisation in specific conditions of the engineering companies" (50%) and Theoretical-methodological aspects of disassembly of appliances after ending their life cycle for the needs of recycling"(50%), solved at the Department of Environmental Studies and Process Control Faculty of Mechanical Engineering at the Technical University in Košice.

6. REFERENCES

Badida,M., Králiková, R. Pauliková, A. (2001): *Environmental impact assessment in mechanical engineering*, Acta Mechanica Slovaca, Vienala, No. 3, Vol. 5, 1/2001, ISSN 1335-2393 pp 57-66

Badida,M., Hricová, B., Liberková ,L. (2002): *Complex method for valuation of the products by Life Cycle Assessment*. Acta Mechanica Slovaca, 4/2002, TU Košice, , pp. 17-24

Frankl;P., Rubik,F.(2000): *Life cycle Assessment in Industry and Business*, Springer, Berlin.

Muranský, J.- Badida, M.(2005): *Ekodizajn v strojárstve. Základy metodiky*. Technická univerzita v Košiciach, Vydavateľstvo Michala Vaška, Prešov, 2005, 304 s., ISBN 80-8073-119-5

TICHÁ,M.: *Design for environment*. Ad-hoc Commitee ASIP Club, Paris, 1998.

Annals of DAAAM for 2006 & Proceedings of the 17th International DAAAM Symposium, ISSN 1726-9679
Editor B. Katalinic, Published by DAAAM International, Vienna, Austria 2006
150th Birthday of Nikola Tesla (1856-1943) Inventor of Mobile Robot Teleoperation

A MONITORING SYSTEM FOR ACTIVATION OF GENES BY HEAT TREATMENT

BADJGHOLI, F.; AGHAALLAEI, N. & BAJOGHLI, B.

Abstract: *During the last decades, genetics have come to realize that fish (zebrafish and Medaka) are an excellent alternative model organism for studying human diseases and development. We are interested in the genes that are common between humans and fish and have important functions during embryonic development. Using a recently generated method we were able to activate our gene-of-interest during embryonic development by heat treatment most effectively. For this purpose, one problem was managing the time and dosage of the heat treatment. Manual calculations for each experiment revealed few errors. To overcome this problem, we developed - in cooperation between University of Veterinary Medicine and FH-Wien - a monitoring system (hardware and software) to control the normal- and "heat shock"- temperature during fish embryogenesis. Thus we are able to control the gene activation by regulating the temperature.*

Key words: *Monitoring system, Developmental biology, Genetics*

1. INTRODUCTION

During the embryonic development of all vertebrates, complex networks and interactions between several genes are important for the development of different organs and tissues as well as body shape. Each gene itself is regulated by "promoter"-structure that is responsible for turning gene activity on or off. Most of the native promoters are different and highly complex. To understand the function of each gene in these complex networks several methods are available. One method is the activation of individual genes by using an artificial promoter. Recently we have generated an artificial heat-inducible promoter (HSE promoter). This artificial promoter is switched off during normal temperature and is switched on at a higher temperature (Bajoghli et al., 2004).

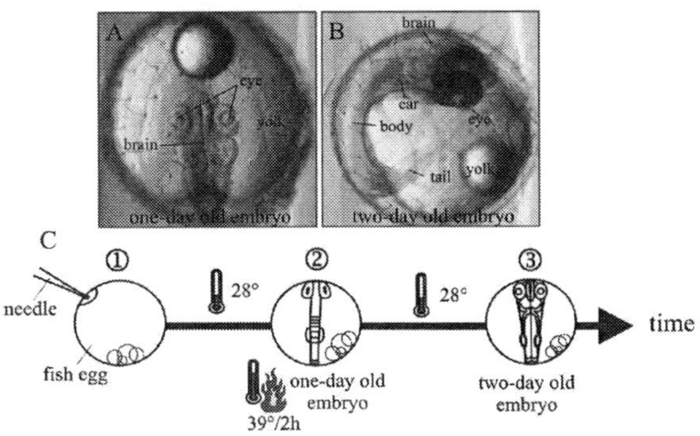

Fig. 1. An overview of biological experiment. (A,B) Medaka fish eggs are transparent. (A) one-day old medaka fish embryo. (B) two-day old embryo. (C) Experimental procedure: 1, micro-injection of DNA in the fish egg. 2, heat shock at desired stage. 3, observation of phenotypes under microscope.

To understand the function of genes during embryonic development we use Japanese Medaka fish as model organism. Medaka fish eggs are transparent and, therefore, it is easy to follow embryonic development (Fig.1A,B). The normal temperature for this species is 28°C.

At this temperature, our HSE promoter is inactive. After heat treatment at 39°C for 2h, the artificial promoter is switched on and is activated 1000-fold (Fig.1C; Bajoghli et al., 2004). So far, we have managed the temperature in our experiments manually; for each experiment, it took a long time and a lot of calculation to manage the "heat shock" time to the desired stage.

Therefore, in a cooperative project between the Institute of Animal Breeding and Genetics (University of Veterinary Medicine) and FH-Wien, we developed a monitoring system for controlling the embryonic temperature.

2. FRAMEWORKS

In planning the hardware system, we considered the following terms:

- Application for two different types of microscopes.
- Fit to four types of Petri dishes (6-well, 24-well, 4cm and 9cm- Petri-dishes)
- Waterproof
- Produce a constant temperature between 25°C and 45°C.
- Enrichment of each temperature without damage to embryos or to the hardware itself.
- No application of strong illuminant.
- Accuracy of measuring the 0.1°C temperature.
- No constriction for the objective lens of the microscopes.
- No constriction of UV-light observation.

3. THE MONITORING SYSTEM

3.1 Experiment field

The living embryos were positioned in different Petri dishes. To get an optimal temperature condition, the producing temperature had to be constantly distributed in the Petri dish.

3.2 The thermal origin

Due to the shortage of space in both microscopes, we decided to use an electric lamp as the thermal origin. A simple 100W lamp is not able to distribute the temperature equally in a Petri dish. Furthermore, it would interfere with the live-imaging camera. Therefore, we decided to use ceramic infrared heat emitters. Such a non-light emitting heat source is used for all types of reptile, amphibian, and plant terrariums. It can increase the inside air temperature of terrarium with considerable efficiency.

3.3 The Sensor

To control the experiment process, the medium temperature was measured every 5 minutes by using a waterproof sensor unit with PC port (DKT200 sensor with analogue exit from 0-1 VDC).

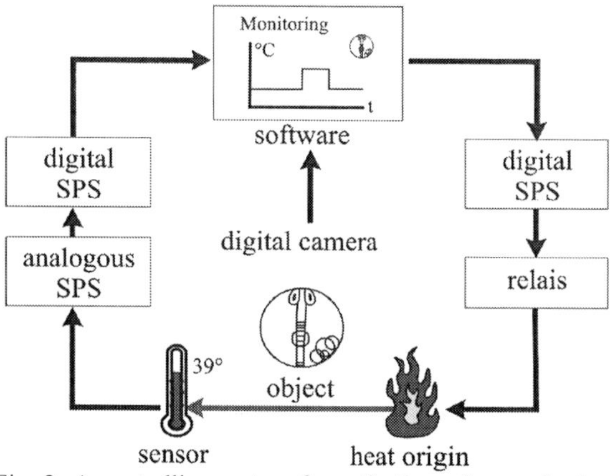

Fig. 2. A controlling system for activation of genes by heat treatment

3.4 The control system

A memory-programmable control unit (Simatic S7 200 CPU222 + EM235) should convert the analogous tension-changes measured by the temperature sensor into a digital signal and, at the same time, switch a relay to control the heat origin (Fig.2). In the SPS itself no programming was necessary, because the conceived software undertook the supervision and the exit.

The software had to perform a task to be user-friendly for the experiment programming. Furthermore we developed a databank in which the values of each experiment and new experience could be saved (Fig.3).

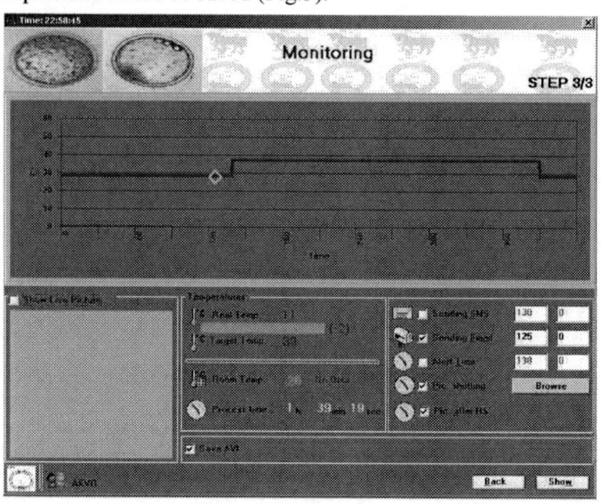

Fig. 3. Monitoring software control and manage the experimental procedures.

3.5 Internet connection

To reduce the complexity of the time-consuming trails (in part longer than 6 hours) and to enable experiments to be conducted independent of place, a web connection was already planned in the first concept. A suitably easy interface command provided for the representation of the topical test data (pictures and temperature) and likewise allowed the remote control from the homepage (Fig.4). Hence, the biologist is flexible and not bound by location. However, he can recall the ongoing data any time with a direct web connection and change the process. An additional alarm system provides for the timely notification using different possibilities such as sending an email or a text message (SMS) or an acoustic signal.

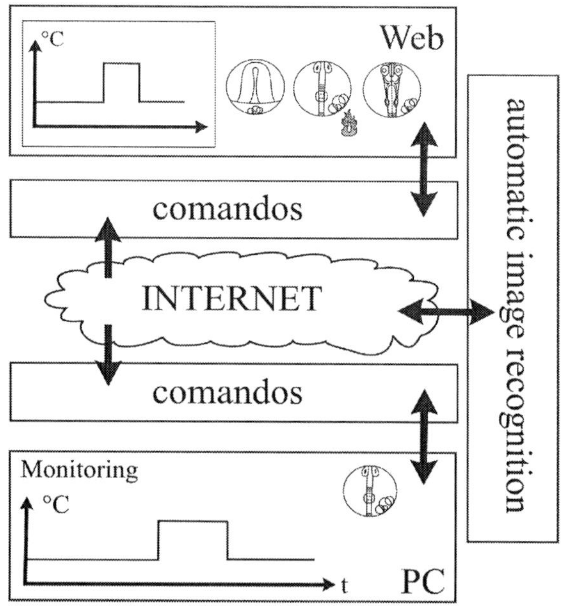

Fig. 4. The monitoring system by using internet connection makes experiments flexible and local-unbound

3.6 Automatic image recognition

An extension this project to the Vienna University of Technology (Institute of Computer Aided Automation) is planned. The first purpose this cooperation is the development of an optical recognition system of Medaka embryonic stages. This automatic image recognition software should be linked to the hardware. Thus, the software is able to recognize the right stage of the heat shock time and manage the experimental procedure automatically.

4. CONCLUSION

In this project, a new tool was developed allowing new possibilities in Life-Science research. The ability to control the normal growth temperature that is needed for gene activation, facilitates shorten experimental planning for biologist. With the extension of this project by automatic image recognition software there will be other even new possibilities in developmental biology research.

5. REFERENCE

Bajoghli B, Aghaallaei N, Heimbucher T, Czerny T. (2004). An Artificial Promoter Construct for Heat Inducible Misexpression during Fish Embryogenesis. *Developmental Biology*, Jul 15; 271 (2), 416-430; ISSN: 0012-1606

Annals of DAAAM for 2006 & Proceedings of the 17th International DAAAM Symposium, ISSN 1726-9679
Editor B. Katalinic, Published by DAAAM International, Vienna, Austria 2006
150th Birthday of Nikola Tesla (1856-1943) Inventor of Mobile Robot Teleoperation

DESIGN OF THE SUPPORT SYSTEM FOR THE PRODUCTION CONTROL OF INJECTION FORMING PROCESS USING DFSS METHODOLOGY

BAEK, J.Y.; KIM, B.H.; LEE, G.B.; PARK, S.E. & LEE, H.C.

Abstract: *The urethane injection forming process which mixes polyol and isocyanate material and makes a specific shape is the manufacturing technology which requires the chemical knowledge, the control know-how and the experience of manufacturing facilities. To analyze and control the manufacturing process of the injection forming systematically, this study introduces some DFSS (design for six sigma) tools such as process map, C&E (cause and effect) matrix, FMEA (failure mode effect analysis) and control plan. In other words, the manufacturing process is analyzed according to the following three steps: 1) making the process-map for analyzing the sequence of the manufacturing process and defining the input/output data of each process, 2) extracting the major factors affecting the product quality using C&E matrix, and 3) analyzing FEMA (failure mode effect analysis) of each process. Using the result of the analysis, this study makes the control plan and designs the detail control algorithms of the support system. Furthermore, the architecture of the support system is designed considering users' convenience and system portability.*

Key words: *Urethane injection forming process, Design for six sigma, Process map, Cause and effect matrix, FMEA, Control plan, Support system.*

1. INTRODUCTION

The urethane injection forming process which mixes polyol and isocyanate material and makes a specific shape is used to manufacture various products such as steering wheels, car interior parts, shoe soles, etc. To control this process requires not only the chemical knowledge of the resin but also the control know-how and the experience of the manufacturing facilities.

VOC list	Customer needs	Functional requirements
Check the production quantity	Check the production & defect quantity	The number of defect types collected by the support system
Check the defect quantity		
Record the history of injection conditions	Record the history of injection conditions	Storage size
Analyze the history of injection conditions		
Search for the past injection conditions	Setup the injection condition automatically	The number of try-out Run times
Suggest the new injection condition		
Setup the injection condition automatically		
Integrate the legacy system	Integrate the legacy system	Compatibility
Perform the statistical analysis		

Table 1. Customer needs and functional requirement

The purpose of this study is to develop the support system for controlling the injection forming process efficiently. To develop the user-centric support system of the production control systematically, this study introduces DFSS (design for six sigma) tools such as VOC (voice of customers) analysis, process map, C&E (cause and effect) matrix, FMEA (failure mode effect analysis) and control plan. The functional requirements of the support system, which can be measured quantitatively, are extracted from VOC list collected from managers and workers of injection forming shops[Table 1]. Before performing the detail design of the support system from the analysis of VOC, this study analyzes the current manufacturing process and prepares the stabilization tactics of the manufacturing process according to the following three steps. The first step is to analyze the current manufacturing process using the process-map which illustrates the sequence of the process and the input/output variables of each process. And the second step is to extract the major factors affecting the product quality using C&E matrix. The third step is to make FMEA (failure mode effect analysis) of each process(Huang et al., 2000).

This study finally makes the control plan for controlling the failure mode effectively and designs the detail algorithms of the support system using the functional requirements. To enhance the operation of the system, their existing work flow should not be interrupted, and additional works should not be required. Thus, this study designs the architecture of the support system using mobile technology(Ong & Sun, 2003; Gavalas et al., 2001). In other words, the production data can be collected from PDA in production line, and the information processing can be executed on mobile equipment using wireless network technology(Feridun & Krause, 2001; Raibulet & Demartini 2000). Fig. 1 shows the overall development procedure of the support system described above.

Extracting the functional requirements of the support system
- Collecting and analyzing VOC list from workers and production managers
- Define the functional requirements of the system

⬇

Analyzing the current manufacturing process and preparing stabilization tactics
1) making the process-map for analyzing the process in detail
2) extracting the major factors affecting the product quality
3) analyzing FEMA of each process

⬇

Detail design of the support system
- Make the control plan and design the control algorithms
- Design the system architecture under mobile operation environment

Fig. 1. Development of production management support system

2. ANALYSIS OF THE INJECTINON FORMING PROCESS

As shown in Fig. 2, the process is defined as a series of activities to generate the output data in which value added. The manufacturing process is complex because it consists of a lot of process linked in serial and parallel and is related to various factors such as input, output, noise factor, and outside variation. Thus, this study employs the divide-and-conquer paradigm, which divides the manufacturing process into each process and identifies input and output variables of each process to be measured quantitatively, to analyze the manufacturing process in detail. Using the process-map, this study defines production control data to be collected and extracts critical control variables from input/output data by the criteria of the functional requirements.

Fig. 2. Concept of process-map

3. DEFECT TYPES AND CAUSE & EFFECT ANALYSIS

To establish the guideline for solving defects of the injection forming process, this study defines defect types to be occurred in each process, analyzes the cause related to input data in the specific process, and proposes the solution based on process control variable. Table 2 summarizes the major defect type and solution of each process. The correlation of functional requirements is analyzed for evaluating input and output data of each process using C&E matrix. This study also performs FMEA which defines the failure mode to be occurred by the input data of each process and arranges its cause, solution, and frequency systematically.

Process	Defect type	Solution	Control parameter
Resin mixing	Raw material	Check lots of raw material	Raw material
	Polymerization	Check the mixing temp, time	Mixing temp., and time
Spraying	Spray	Preventive maintenance of the spray gun	Maintenance of spray gun
	Pigment	Check composition of pigment	Composition of pigment
	Gloss	Operation rule of spraying	Spray pattern
Mold closing	Flash	Increase the cramping pressure	Cramping pressure
	Hardness	Replace the bushing	Repair period of bushing
	Vent	Decrease the cramping pressure	Cramping pressure
Injection forming	Short shot	Increase the shot size	Shot size
	Flash	Decrease the shot size	Shot size
	Pigment	Check the composition of pigment	Composition of pigment
	Hardness	Adjust the composition of resin	Composition of resin
	Vent	Check the air leakage of valve	Maintenance of valve
	Polymerization	Adjust mold temp	Mold temp.
Cooling	Contour	Decrease cooling temp	Cooling temp.
	Solidity	Increase cooling temp	Cooling temp.
Ejecting	Surface contamination	Clean work's hands	Work's hands

Table 2. Solution of the defect type classified by process

4. DESIGN OF THE SUPPORT SYSTEM FOR THE INJECTION FORMING PROCESS

The final goal of the support system for the production control helps users maintain the non-failure mode state and recover the failure mode easily if happened. In the view of continuous monitoring and control, this study makes the control plan of process and considers it in the algorithm design of the support system.

This study designs the support system which can collect production data using PDA to not interrupt the existing work flow of workers and managers and to handle it easily.

The collected data is systematically stored in the control server using wireless interface. The stored data is provided with people requiring production data such as production managers, workers, product designers, and quality controllers after the data filtering or processing step. Fig. 3 shows the conceptual architecture and the application environment of the support system using mobile technology.

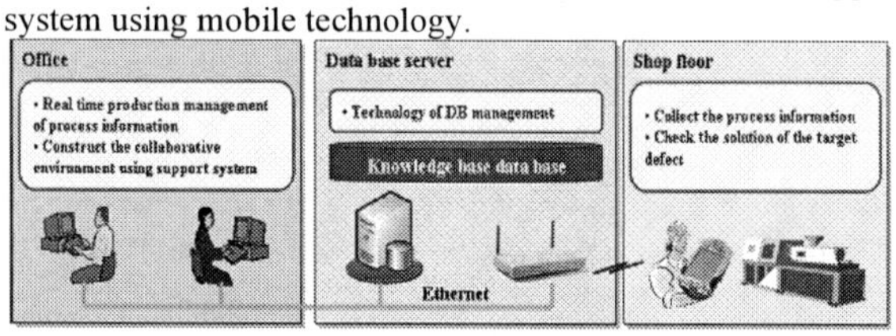

Fig. 3. Conceptual architecture and application environment of the support system

5. CONCLUSION

This study designs the support system of the production control in an injection forming shop floor using some DFSS tools. Researches performed in this study are summarized into the following four categories:

- The control parameters of each process are clearly extracted using the process-map which presents the sequence of manufacturing process and identifies the input & output variables of each process.
- Using C&E matrix, the major control parameters are identified to remove the cause of the major defect types.
- The control plan based on process FMEA is established and is applied to the algorithm of the support system.
- To consider users convenience and system portability, the system architecture is designed using mobile technology.

6. REFERENCES

Feridun, M. & Krause, J. (2001). A framework for distributed management with mobile component, *Computer networks*, 35(1), pp. 25-38.

Gavalas, D. Greenwood, D. Ghanbari, M. O'Mahony, M. (2001). Mobile software agents for decentralized network and systems management, *Microprocessors and microsystems*, 25, pp. 101-109.

Huang, G. Q. Shi, J. & Mak, K. L. (2000). Failure mode and effect analysis(FMEA) over the WWW, *International journal of advanced manufacturing technology*, 16(8), pp. 603-608.

Ong, S. K. & Sun, W. W. (2003). Application of mobile agents in a web-based real-time monitoring system, *International journal of advanced manufacturing technology* , 22(1/2), pp. 33-40.

Raibulet, C. & Demartini, C. (2000). Mobile agent technology for the management of distributed system a case study, *Computer networks*, 34(6), pp. 823-830.

Annals of DAAAM for 2006 & Proceedings of the 17th International DAAAM Symposium, ISSN 1726-9679
Editor B. Katalinic, Published by DAAAM International, Vienna, Austria 2006
150th Birthday of Nikola Tesla (1856-1943) Inventor of Mobile Robot Teleoperation

NON-LINEAR MODEL OF OPTIMISATION OF OPERATION OF THE HEAT SOURCE WITH COMBINED PRODUCTION OF HEAT AND POWER

BALATE, J.; CHRAMCOV, B. & NAVRATIL, P.

Abstract: *This paper deals with the access to solving optimisation of operation of heat source with combined production of heat and power by methods of non-linear programming.*
Key words: *heat, CPM – combined production of heat and power, non-linear, optimisation*

1. INTRODUCTION

It will be shown the access to solving optimisation of operation of an energy source with combined production of heat and power by methods of linear programming and non-linear programming. In generating plants of heat and electric energy, which have larger number of co-operating production units, there is a problem of economical distribution of load between these production units. The task of economical distribution of load is one of the basic tasks of optimum control. Its basic principle is usually the equality of the relative increase of cots (consumption of fuel) on the delivered heat output. The basic presumption of economical production is knowledge of energy economical characteristics of separate production equipments. In our specific case this applies to consumption characteristics of boilers, which generally have non-linear course. From the mathematical point of view, the aim of optimum control will be to achieve extreme value of the objective function. In our case i.e. at minimising the production by costs decreasing the costs of fuel, we shall look for the minimum of the objective.

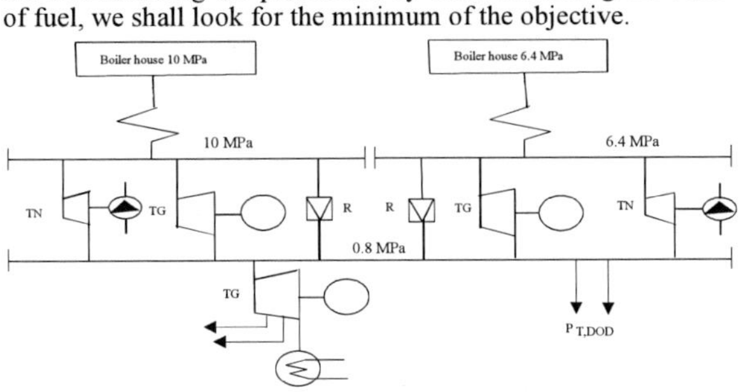

Fig. 1. A principle diagram of a heat and power plant.
Legend: TFP – Turbo-Feed-Pump, BPTG – Back-Pressure-Turbo-Generator, CTG – Condensing-Turbo-Generator, RS – Reducing Station, $P_{T,DOD}$ – Heat output supply

The total immediate electric output P and the heat output supplied to the heat network $P_{T,DOD}$ can be determined. There are two methods as the solutions to form mathematical models: linear and non-linear model.

2. MATHEMATICAL LINEAR MODEL

This model was formulated in these papers (Balátě, 1969), (Balátě et al., 2006), (Phan, 1996). In the main, the linear model (balance equations of steam pipe-lines) was created, the limiting non-linear conditions of dependent variables and objective function are linear. Independent variable is supplied heat output in the heat network $P_{T,DOD}$. In order to form the mathematical linear model, it is necessary to perform the following requirements:
- the consumption parameters of production units must have a convex monotonic continuance.
- it is necessary to approximate the convex curves of the consumption parameters by linear sections.

The way of linearization of the consumption parameters characteristics is illustrated on the Fig. 2a. In concrete case, it is able to approximate the consumption parameters of boiler by means of two linear sections. Then, according to the Fig.2a, we can describe the heat output of the boiler:

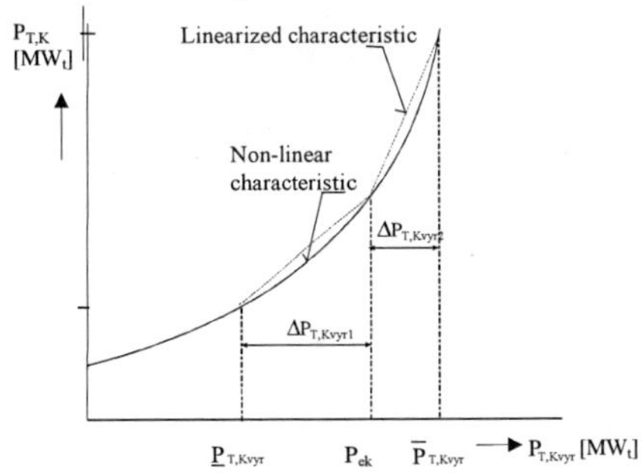

Fig. 2a Economic characteristic of steam boiler and its linearization.

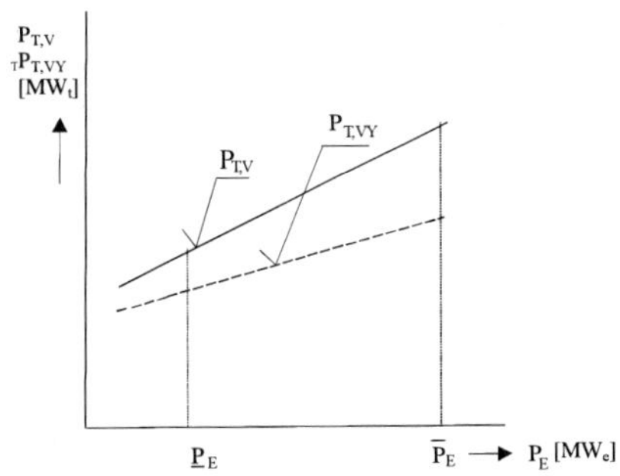

Fig. 2b Economic characteristic of back pressure turbine.

$$P_{T,K} = \underline{P}_{T,K} + \sum_{s=1}^{2} \Delta P_{T,K,s} \quad (1)$$

and the boiler heat consumption in fuel:

$$P_{PAL} = \underline{P}_{PAL} + \sum_{s=1}^{2} b_{K,s} \Delta P_{T,K,s} \quad (2)$$

Where $b_{K,s}$ is the consumption relative increment of boiler:

$$b_{K,s} = \frac{\Delta P_{PAL}}{\Delta P_{T,K,s}} \; [\text{ - }] \quad (3)$$

(In this case within linear programming $b_{K,s}$ is constant).
The further variables sign: $\underline{P}_{T,K}$ - the minimal heat output boilers, $\Delta P_{T,K,s}$ - the increments of output of boilers, s - the number of linearized section.
Consumption characteristics for back-pressure turbines have the linear course (Fig. 2b.) and therefore it is possible to approach it by one linear section: $P_{T,V} = \underline{P}_{T,V} + b_{T,V}.\Delta P$ (4)
that is the heat output on the input of turbine.
We can describe by analogy way the heat output on the exit of turbine: $P_{T,VY} = \underline{P}_{T,VY} + b_{T,VY}.\Delta P$ (5)
And than applies

$b_{T,V} = \frac{\Delta P_{T,V}}{\Delta P} = const$, $b_{T,VY} = \frac{\Delta P_{T,VY}}{\Delta P} = const$ in the range $\overline{P} - \underline{P}$,(6) (7)

where $b_{T,V}$ is relative increment of heat consumption on the input of turbine, $b_{T,VY}$ is relative increment of heat consumption on the exit of turbine and in mathematic model is used to definition of heat balance equation of steam pipe-line 0,8 MPa (eq. 10).

The mathematic model of determine scheme (see Fig. 1.):

It is described by heat balance equations for steam pipe-lines marked 10 MPa, 6,4 MPa, 0,8 MPa, by sum total equation of produce power energy and by definition of objective function E.

The following balance equations were created:

- for steam pipe-line 10MPa we can apply:

$$\sum_{i=1}^{n}\left(\underline{P}_{T,K}^{10,i}+\sum_{s=1}^{k}\Delta P_{T,K,s}^{10,i}\right)=\left[\sum_{j=1}^{m}\left(\underline{P}_{T,V}^{10,j}+b_{T,V}^{10,j}.\Delta P^{10,j}\right)+P_{T,N,V}^{10}+P_{T,R}^{10/0.8}\right].k_{q}^{10} \quad (8)$$

- for steam pipe-line 6.4MPa we can apply:

$$\sum_{i=1}^{n}\left(\underline{P}_{T,K}^{6,4,i}+\sum_{s=1}^{k}\Delta P_{T,K}^{6.4,i}\right)=\left[\sum_{j=1}^{m}\left(\underline{P}_{T,V}^{6.4,j}+b_{T,V}^{6.4,j}.\Delta P^{6.4,j}\right)+P_{T,N,V}^{6.4}+P_{T,R}^{6.4/0.8}\right].k_{q}^{6.4} \quad (9)$$

and also for steam pipe-line 0.8 MPa we can apply :

$$\sum_{j=1}^{m}\left(\underline{P}_{T,T,VY}^{10,j}+b_{T,VY}^{10,j}.\Delta P^{10,j}\right)+\sum_{j=1}^{m}\left(\underline{P}_{T,T,VY}^{6.4,j}+b_{T,VY}^{6.4,j}.\Delta P^{6.4,j}\right)+P_{T,N,VY}^{10}+P_{T,N,VY}^{6.4} \quad (10)$$
$$+P_{T,R}^{10/0.8}+P_{T,R}^{6.4/0.8}=\left[\left(\underline{P}_{T,T,V}^{0.8,j}+b_{T,el,s}^{0.8,j}.\Delta P^{0.8,j}\right)+P_{T,DOD}\right]k_{q}^{0.8}$$

The total output of power would be determined by the following relation :

$$\sum P=\sum_{j=1}^{m,10}\left(\underline{P}^{10,j}+\Delta P^{10,j}\right)+\sum_{j=1}^{m,6.4}\left(\underline{P}^{6.4,j}+\Delta P^{6.4,j}\right)+\sum_{j=1}^{m,0.8}\left(\underline{P}^{0.8,j}+\Delta P^{0.8,j}\right) \quad (11)$$

The solutions should respond the following inequalities or conditions:

$$0\le\Delta P^{6.4}\le\Delta\overline{P}^{6.4} \qquad 0\le\Delta P_{T,K,s}^{10,i}\le\Delta\overline{P}_{T,K,s}^{10,i} \qquad 0\le\Delta P^{10}\le\Delta\overline{P}^{10}$$
$$0\le\Delta P_{T,K,s}^{6.4,i}\le\Delta\overline{P}_{T,K,s}^{6.4,i} \qquad 0\le\Delta P_{T,R}^{6.4/0.8}\le\Delta\overline{P}_{T,R}^{6.4/0.8} \qquad 0\le\Delta P_{T,R}^{10/0.8}\le\Delta\overline{P}_{T,R}^{10/0.8} \quad (12)$$

$P_{T,el}$ is a constant in determined schema. It presents the own heat consumption of each turbine. Commonly it is possible to use a constant in determined schema.

The variables in mathematical model, contain not only the increments of boilers output $\Delta P_{T,K}$, but also the increments of electric output of individual turbines ΔP. The sum total of these variables, where each variable is multiplied with the appropriate price coefficient c_j (in case of $\Delta P^{i}_{T,K}$, the coefficients are expressed by the heat consumption relative increments $b^{i}_{K,s}$), defines the objective function E.

The objective function form:

$$E=\sum_{i=1}^{n,10}\sum_{s=1}^{k}b_{K,s}^{10,i}.\Delta P_{T,KVyr,s}^{10,i}+\sum_{i=1}^{n,6.4}\sum_{s=1}^{k}b_{K,s}^{6.4,i}.\Delta P_{T,KVyr,s}^{6.4,i} \quad (13)$$

Meaning of the variables in the equations:

The symbol $\underline{P}$ signs the minimum value and in opposite, the symbol $\overline{P}$ signs the maximum one of the appropriate quantity

$P_{T,Kvyr}=\dot{m}(i_{pK}-i_{nv})$ [MW_t] the boilers output

$P_{T,T,V}^{10}$, $P_{T,T,V}^{6.4}$ [MW_t] the heat outputs of turbine input

$P_{T,T,VY}^{10}$, $P_{T,T,VY}^{6.4}$ [MW_t] the heat outputs of turbine exit

$P_{T,N,V}^{10}$, $P_{T,N,V}^{6.4}$ [MW_t] the heat outputs of input of the turbo-feed pump

$P_{T,N,VY}^{10}$, $P_{T,N,VY}^{6.4}$ [MW_t] the heat outputs of exit of the turbo-feed pump

$P_{T,R}^{10/0.8}$, $P_{T,R}^{6.4/0.8}$ [MW_t] heat outputs of the reduction stations

$P_{T,DOD}$ [MW_t] the heat outputs supplied to the heat network

k_q^{10}, $k_q^{6.4}$, $k_q^{0.8}$ [-] the coefficients of heat consumption or losses in each phase of technology

$b_{T,el}$ [MW_t/MW_e] coefficient of own heat consumption of power production of each condensing turbine

The mathematical linear model includes the equations No. (8), (9), (10), (11), (12), (13).

3. MATHEMATICAL NON-LINEAR MODEL

Mathematical non-linear model is the main item of this article. All of signs presented for linear model are applicable for the non-linear model too.

a.The derivation of consumption characteristic:

Generally, steam-boiler consumption characteristics have non-linear continuances (Fig. 2a.). These continuances can be replaced by non-linear approximation, for example by the index form:

$$y=a.e^{b.x} \quad (14)$$

Their heat consumption proportionate increment are expressed by derivation:

$$y'=a.b.e^{b.x} \quad (15)$$

b. Creation of a non-linear mathematical model

b1. balance equations:

- for steam pipe-line 10MPa we can apply:

$$\sum_{i=1}^{n}\left(\underline{P}_{T,K}^{10,i}+\int_{\underline{P}_{T,K}^{10,i}}^{\overline{P}_{T,K}^{10,i}}dP_{T,K}^{10,i}\right)=\left[\sum_{j=1}^{m}\left(\underline{P}_{T,T,V}^{10,j}+\int_{\underline{P}_{T,T,VY}^{10,j}}^{\overline{P}_{T,T,VY}^{10,j}}b_{T,V}^{10,j}\left(P^{10,j}\right)dP^{10,j}\right)+P_{T,N,V}^{10}+P_{T,R}^{10/0.8}\right]k_q^{10} \quad (16)$$

- for steam pipe-line 6,4 MPa we can apply:

$$\sum_{i=1}^{n}\left(\underline{P}_{T,K}^{6.4,i}+\int_{\underline{P}_{T,K}^{6.4,i}}^{\overline{P}_{T,K}^{6.4,i}}dP_{T,K}^{6.4,i}\right)=\left[\sum_{j=1}^{m}\left(\underline{P}_{T,T,V}^{6.4,j}+\int_{\underline{P}_{T,T,VY}^{6.4,j}}^{\overline{P}_{T,T,VY}^{6.4,j}}b_{T,V}^{6.4,j}\left(P^{6.4,j}\right)dP^{6.4,j}\right)+P_{T,N,V}^{6.4}+P_{T,R}^{6.4/0.8}\right]k_q^{6.4} \quad (17)$$

and also for steam pipe-line 0.8 MPa we can apply:

$$\sum_{j=1}^{m}\left(\underline{P}_{T,T,V}^{10,j}+\int_{\underline{P}_{T,T,VY}^{10,j}}^{\overline{P}_{T,T,VY}^{10,j}}b_{T,V}^{10,j}\left(P^{10,j}\right)dP^{10,j}\right)+\sum_{j=1}^{m}\left(\underline{P}_{T,T,V}^{6.4,j}+\int_{\underline{P}_{T,T,VY}^{6.4,j}}^{\overline{P}_{T,T,VY}^{6.4,j}}b_{T,V}^{6.4,j}\left(P^{6.4,j}\right)dP^{6.4,j}\right)+ \quad (18)$$
$$+P_{T,N,VY}^{10}+P_{T,N,VY}^{6.4}+P_{T,R}^{10/0.8}+P_{T,R}^{6.4/0.8}=\left[P_{T,T,V}^{0.8}+\int_{\underline{P}_{T,T,V}^{0.8,j}}^{\overline{P}_{T,T,VY}^{0.8,j}}b_{T,el}^{0.8,j}\left(P^{0.8}\right)dP^{0.8}+P_{T,el}+P_{T,DOD}\right]k_q^{0.8}$$

The total electric output would be determined by the following relation:

$$\sum P=\sum_{j=1}^{m,10}\left(\underline{P}^{10,j}+\int dP^{10,j}\right)+\sum_{j=1}^{m,6.4}\left(\underline{P}^{6.4,j}+\int dP^{6.4,j}\right)+\sum_{j=1}^{m,0.8}\left(\underline{P}^{0.8,j}+\int dP^{0.8,j}\right) \quad (19)$$

b2. objective function:

The variables in mathematical model, contain not only the increments of boiler output $\Delta P^{i}_{T,K}$, but also the increments of power output of individual turbines ΔP^{i}. The total of these variables, where each variable is multiplied with the appropriate price coefficient c_j (in case of $\Delta P^{i}_{T,K}$, the coefficients are expressed by the heat consumption proportionate increments $b_{K,s}$), defines the objective function E. The objective function form:

$$E=\sum_{i=1}^{n,10}b_k^{10,i}.\int dP_{T,Kvyr}^{10,i}+\sum_{i=1}^{n,6.4}b_k^{6.4,i}.\int dP_{T,Kvyr}^{6.4,i} \quad (20)$$

In this case, b_K will be the non-linear functions (see equation 15),

$$\text{thus } b_k=f(P_{T,Kvyr})\approx\frac{dP_{T,K}}{dP_{T,Kvyr}} \quad (21)$$

that means the consumption relative increment of boiler.

Limiting non-negative conditions are following:

$$0\le dP^{6.4}\le\Delta\overline{P}^{6.4} \qquad 0\le dP_{T,K}^{10,i}\le\Delta\overline{P}_{T,K}^{10,i} \qquad 0\le dP^{10}\le\Delta\overline{P}^{10}$$
$$0\le dP_{T,K}^{6.4,i}\le\Delta\overline{P}_{T,K}^{6.4,i} \qquad 0\le dP_{T,R}^{6.4/0.8}\le\Delta\overline{P}_{T,R}^{6.4/0.8} \qquad 0\le dP_{T,R}^{10/0.8}\le\Delta\overline{P}_{T,R}^{10/0.8} \quad (22)$$

4. CONCLUSION

This paper has explained two methods to form mathematical model of a heat and power plant which produce heat and electric energy in order to optimisation its operation. Certainly, the differences between both models consist in reduction of the base variables number, the consumption characteristic approximation of non-linear equations and forming of the other objective function. In the case of the linear model, the objective function is linear (the consumption proportionate increments of boiler b_K are constant) and it is non-linear in the second one (the consumption proportionate increments of boiler b_K depend on index function). In order we would gain the values for the general coefficients in the above mentioned mathematical models, it is necessary to process the whole rank of bases and consumption parameters of the production units. The obtained results would be used as background for operative planning and own operation control of production of heat and power energy in real time too.

5. REFERENCES

Balátě, J. (1969). *Design of optimal control of heat output of district heating systems sources in Brno locality*, Brno (in Czech)

Balátě, J; Chramcov, B; Navrátil, P. & Phan,T. D. (2006). Optimisation of operation of the heat source with combined production of heat and power. In: *10th International Research/Expert Conference TMT 2006*, Spain, Barcelona-Lloret de Mar.

MathWorks. (1992). *Optimization Toolbox*, For Use with MATLAB.

Phan, T.,D. (1996). Diploma work, Zlín 1996 (in Czech)

Acknowledgement: This work was supported in part by the Grant agency of Czech Republic under grant No: 101/06/0920 and in part by the Ministry of Education of the Czech Republic under grant No. MSM 7088352102.

Annals of DAAAM for 2006 & Proceedings of the 17th International DAAAM Symposium, ISSN 1726-9679
Editor B. Katalinic, Published by DAAAM International, Vienna, Austria 2006
150th Birthday of Nikola Tesla (1856-1943) Inventor of Mobile Robot Teleoperation

PRELIMINARY GA BASED OPTIMIZATION IN PROCESS OF JIG DESIGN FOR MILLING PROCESSES

BARISIC, B.; KRSULJA, M. & CAR, Z.

Abstract: *In this paper we have presented a new method for preliminary (initial) optimization of the jig design for milling processes. Main reason for proposing this approach is shortening the time needed for preparation of the production process. In this method we have used GA in combination with TSP and classical 3-2-1 design method for obtaining the optimal solution of the process of design.*
Key words: *jig, milling processes, GA, optimization.*

1. INTRODUCTION

The fixtures and jigs are essential elements of mass production processes as they are required in most of the automated manufacturing, inspection, and assembly operations. A fixture is a device for locating, holding and supporting a workpiece during a manufacturing operation. In process planning, the correct selection of fixtures, jig and work holding devices represents an important specification in addition to cutting tool requirements (Jha & Hornik, 1995). To obtain an appropriate fixture design traditionally consists of several phases: careful study of a product specification, retrieval of any existing similar designs, and manual selection of fixture elements based on catalogues and items available in the shop. This procedure depends largely on the knowledge and experience of a tool designer and is both time-consuming and expensive.
Fixtures are normally designed for a definite operation to process a specific workpiece and are designed and manufactured individually. Jigs are similar to fixtures, but they not only locate and hold the part but also guide the cutting tools in drilling and boring operations. These workholding devices are collectively known as jigs and fixtures. Fig. 1 shows an example of a fixture commonly used on a horizontal CNC milling machine.

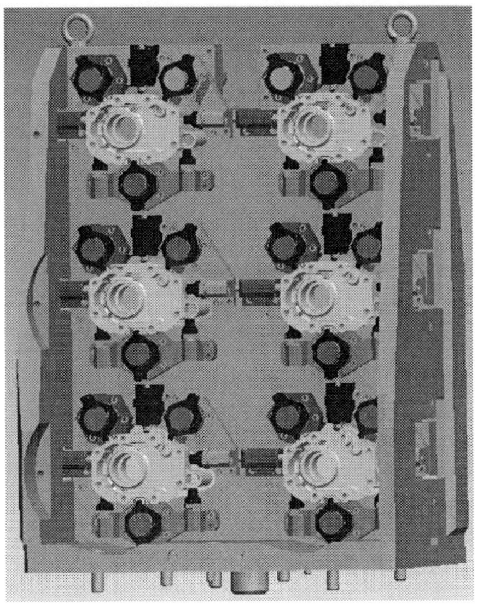

Fig 1. The fixture commonly used on horizontal CNC milling machine

Today design of fixtures and jugs is one of the most complex and time consuming processes, and in commonly in this process there is no or the automation is very low (Krishnakumur & Melkote, 2000). This process has been recognized as on of the bottlenecks in preparation for production in small and middle size of the production systems. The most common problem is how to utilize the knowledge obtained from previous design of jig or fixture and how to do that in short time period (Milfelner et al, 2005).

2. DEFINITION OF PRELIMINARY OPTIMIZATION

To be able to simplify the problem of the preliminary optimization of the jig design we have model the jig surface, or better defined the workpiece and the movement of the tool during machining process by mirroring it in the traveling salesman problem (TSP). By definition TSP problem is to find shortest or optimal path between the n-cities. We have use this method to model the workpiece from which we have defined the best (optimal) positions of the pins and clamps in design of the jig for milling process.
The method for locating the workpiece we have used the traditional 3-2-1 method (SME, 1998). With this method we have lock the workpiece and restrict the movement in nine directions (Fig.2). However, to be able to design jig with requirements which we are searching as our task we needed to add two more fixating clamps.

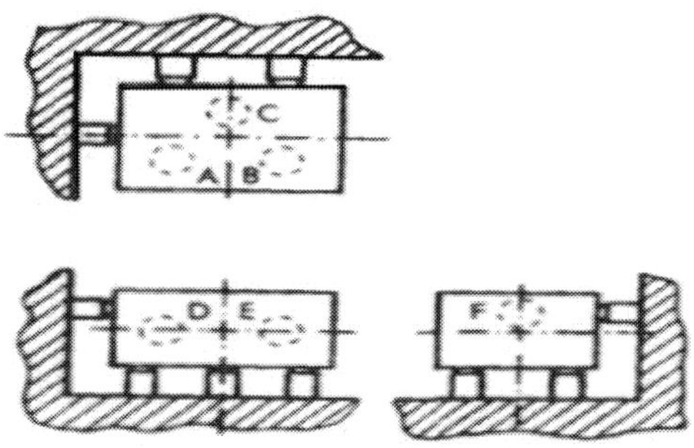

Fig. 2. Six pins arrest nine direction movements.

One of the requirements during a process of the optimizations of position of the pins and clamps on the jig it was that movement of the tool during machining process need to be most continues (Michalewicz, 1995).. By continuity we understand that the tool path need to have as much less curves

and blank movements (movement without contact with material) (Narayanaswami, & Pang, 2003).
As for the utilization of the GA we have been using the Simple GA, prepared in the C++ and rune on the Pentium 4 computer. Basic GA settings are as follow:

- population size = 10000
- max. number of generations = 1 000 000
- crossover = 60%
- mutation = 3%
- bordering cities = 3
- probability of bordering city = 90%

3. SIMULATION OF PRELIMINARY OPTIMIZATION

According to the above proposed method for preliminary optimization we have tested this method with two simple examples. In first example we have made preliminary optimization for the process of face mailing Fig.3. and Fig. 5. and secondly we have made optimization for the jig which will be used in the process of the slot mailing Fig. 4. and Fig. 6.

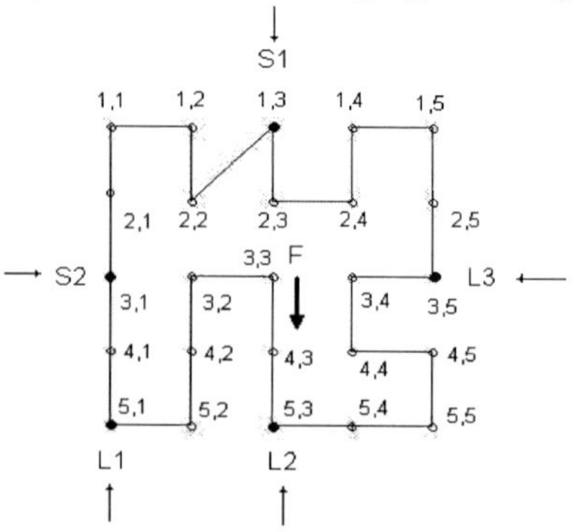

Fig. 3. Obtained position of the pins and clamps for the face milling example

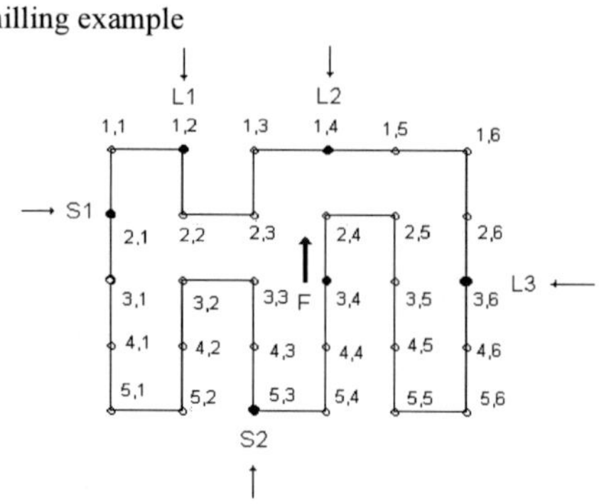

Fig. 4. Obtained position of the pins and clamps for the slot milling example

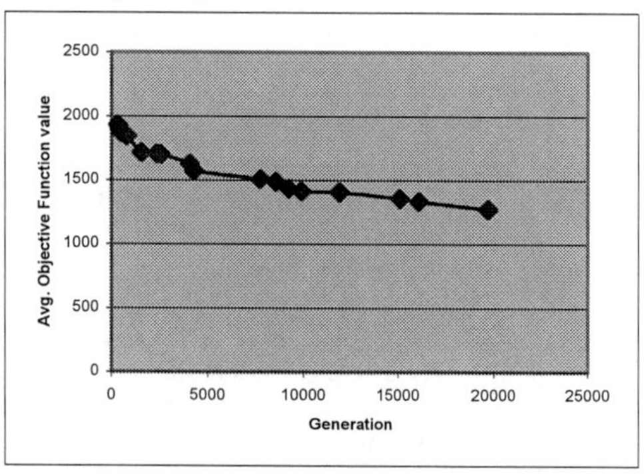

Fig. 5.The convergence of the GA for the first example

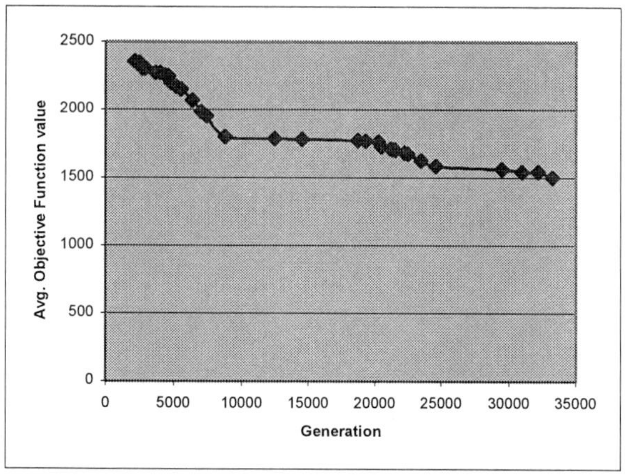

Fig. 6.The convergence of the GA for the second example

4. CONCLUSION

From here presented results we can conclude that with introduction of simple optimization methodology we can obtain good results for the preliminary optimization of design of jig used in the mailing processes. Here we need to say that by the definition of location of the pins and clamps we are able to obtain even distribution of force on the workpiece. By introduction of this method we are able to highly reduce time for the optimization of the preliminary design of the jigs. This is giving ability to the designer to take in the consideration not only one possible layout of the jig, but few in the same moment.
The simulations presented in this paper are only for the 2D problem and for the simple geometrical configuration. As the next task is to introduce this methodology for solution of the complex 3D problems and to introduce knowledge based database.
In this faze this methodology we can only use for the preliminary optimization but the next faze f this research we will introduce this concept through all process of design of the jig.

5. REFERENCES

Society of Manufactoring Engineers (1998). *Fundamentals of tool design*, SME, ISBN 0-87263-650-X, Dearborn, USA

Krishnakumur, K. & Melkote, S.N. (2000). Machining fixture layout optimization using the genetic algorithm *International Journal of Machine Tools & Manufacture*, ISSN Vol. 40 (4), March (2000), 579-598, ISSN 0890-6955

Milfelner, M., Cus, F., Balic, J. (2005). An overview of data acquisition system for cutting force measuring and optimization in milling, *Journal of Materials Processing Technology*, Vol. 162-165, May 2005, 1281-1288, ISSN

Jha, N.K. & Hornik, K. (1995). Integrated computer-aided optimal design and finite element analysis of a plain milling cutter, *Applied Mathematical Modeling*, Vol. 19, No. 6, June 1995, 343–353, ISSN 0307-904X

Michalewicz, Z. (1995). A survey of constraint handling techniques in evolutionary computation methods, in: *Proceedings of the 4th Annual Conference on Evolutionary Programming*, pp. 135–155, San Diego, CA, 1995

Narayanaswami, R. & Pang, J. (2003). Multi-resolution analysis as an approach for tool path planning in NC machining, *Computer-Aided Design*, Vol. 35 (2), February 2003, 167–178, ISSN 0010-4485

Annals of DAAAM for 2006 & Proceedings of the 17th International DAAAM Symposium, ISSN 1726-9679
Editor B. Katalinic, Published by DAAAM International, Vienna, Austria 2006
150th Birthday of Nikola Tesla (1856-1943) Inventor of Mobile Robot Teleoperation

MAX FOR WINDOWS PDM SYSTEM IMPLEMENTATION IN BNT TMIH COMPANY

BEGOVIC, E.; PETKOVIC, D. & VISEKRUNA, V.

Abstract: *Aim of the paper is to present some features of implementation one PDM system in typical BiH postwar factory. MAX for Windows is PDM software package that management of BNT TMiH to bought license for. Paper explores reasons which cause low level of PDM system implementation effects in this company. Some common characteristics of PDM system were the base for this exploration.*
Key words*: Product Data Management, MAX for Windows,*

1. INTRODUCTION

Global village, the term recently very often used, today became unavoidable fact and normal living condition state, in spite of anti-globalization protests around the world. Globalization use very strong and powerful "weapon" - information. Billions of information circulate every day, bringing new ides, new tools, new problems, but at the same time became some kind of *a necessity*, food for soul struggling for information. Definitely, environment has been rapidly changed and is still changing. One of the changes strongly affects company life are - customer demands. Costumers and their demands and wishes are the frame inside which all companies must to work and ensure its survival, continuous development and profit. In that sense company need to build [Petković]:

- New effective business strategy,
- Correctly choose resources used in implementing that strategy,
- Synchronize strategy, employees and flows of information and inventory.

Main tool in synchronization mentioned above are the Information Processing Technologies, or computer technologies. Computer technologies make *dream come true* in area of design product and manufacturing processes. In area of managing companies computer aid is expressed through different software package for MRP, PDM or ERP systems. This paper deals with one PDM software package MAX for Windows, "implemented" in BNT TMiH Company.

2. PDM SYSTEMS

PDM - Product Data Management is organized system for managing product data [Duhovnik] Application of these systems should enable and maximize advantages of concurrent engineering approach to production and reduce time to market period. Automatically distribute data about product to people who need it at the moment when they need it, keeping fully control over it is basic feature of a PDM system [PDM Information Company]

So, manufacturing companies goes from process oriented on product oriented manufacturing. Product life cycle became keyword in manufacturing. PDM systems connect and improve existing CAD/CAM/CAE systems.

Some studies [Gascoine] from period in which BNT TMiH decided to buy MAX for Windows software package, shows that most of engineers (designers) spent only about 20% of their total working time on designing - effective and creative part of work. About 35% of time spends on collecting and retrieving information, drawing parts etc. - non-constructive time. Of course that this data goes in favor of companies that selling PDM software, but no matter of that, it shows also need for some changes in companies.

PDM and Concurrent Engineering seem to be one of possible solution. That was the road BNT TMiH management gone.

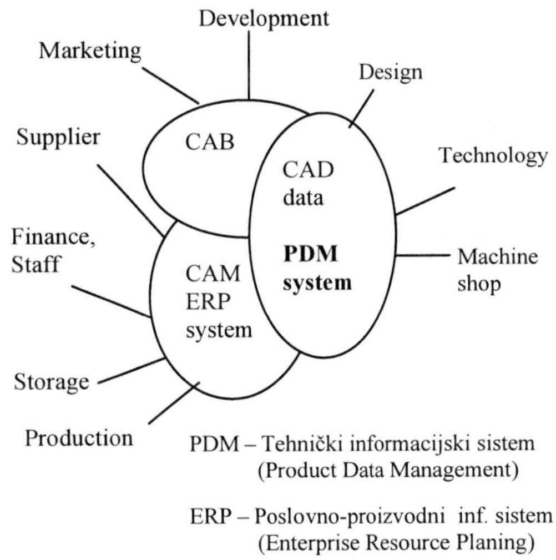

Fig. 1. Position of a PDM in Company (Duhovnik, 2000)

3. MAX FOR WINDOWS IN BNT TMiH

3.1 BNT TMiH company

BNT TMiH Company was a part of a huge military industry of former Yugoslavia. It has been established 1949.[BNT TMiH home page] with orientation on primarily military equipment - cannon howitzers, rocket launchers, guns etc.

Postwar period brought completely different conditions. Main military program has been changed for civil program. Then problems arise due to many political, socio-economical, but also human psychology reasons. Consequently, it was very difficult to adjust existing shopflor to flexible production, with small series of different kind of product. When to all this kind of problems some add state as an owner of factory, with its indolence and inability to do anything helpfully, circumstances in which management have to run factory were pretty confusing.

One of steps management done in order to improve situation and answer effectively on new market demands has been introducing PDM system - MAX for Windows, into factories production routine.

3.2 MAX for Windows

Max for Windows is a product of Micro-MRP Inc [Exact Software]. It is a modular software package that covers all planning and managing functions in one company.
Max for Windows 3.0 , version for which BNT TMiH got the license contain of following modules:

- *Bill of Material*
- *Inventory Control*
- *Master Scheduling*
- *MRP*
- *Purchasing Control*
- *Shop Floor Control*
- *Costing*
- *Multi-Currency*
- *Sales Order Processing*
- *Financial Integration*
- *System Manager*
- *MAX-Access*

So, from list of modules it can be partially foreclosed what are the possibilities that such software offer to company. But, implementation of those modules as well as software in whole has been transferred to human, employees. Moreover, organizational structure, and what is more important production organization has been faced with lacking of clearly defined flows. Some expensive machine laid useless in machine shop, while some were overburden and overloaded. So, machine shop was definitely significant and strong but once projected for mass productive, now seems so inert and long a way from any flexible production.
In such conditions most of MAX for Windows modules implementation has been "frozen" at the start. Some try in software running period occurred, but without any significant results. Technologists (engineers) have been worked as they did earlier, machine operators as well, and other most part of employees also.
So, in technical area of production, PDM system failed, stay without any strong accomplishment. Concurrent engineering, as a powerful manufacturing tool do not exist in this company.

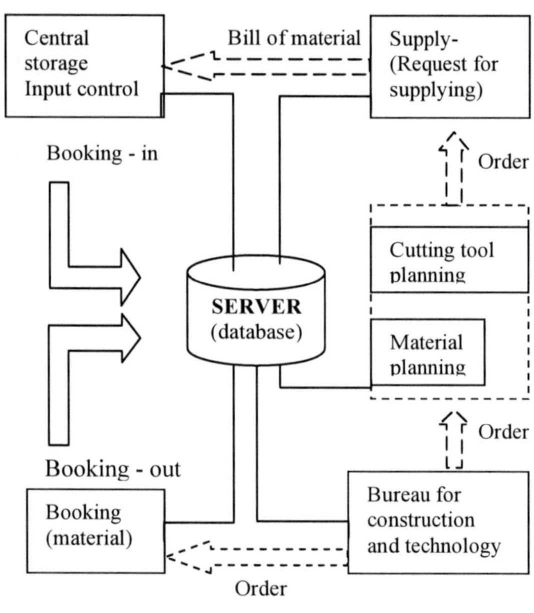

Fig. 2. MAX for Windows in BNT TMiH

Aspects in which MAX for Windows succeeds have been Inventory Control, Financial Integration, Bill of Material. Figure 2. shows current status of PDM implementation in the company.
Obviously, that MAX for Windows in BNT TMiH operate only on level of Inventory, Resource planning, Booking. Hence, implementation of this PDM software package in total became very obscured and restricted. Desired effects in area of production itself failed.

4. CONCLUSION

Globalization forced many companies to implement "new technologies" in order to stay competitive in more and more demanding market. On the other hand, a lot of new technologies have been introduced thanks to this competitiveness and struggling to survive and beat competition. Being although wheel-stone of globalization, IT (Informatical Technologies) are at same time also the edge and main tool for improving competitiveness.
PDM systems are among these technologies with main purpose - to put life into Concureent Engineering, Re-engineering etc., through the linkage of CAD, CAM, CAE and other Computer Aided Technologies.
BNT TMiH management saw that importance of Product Data Management Systems and tries to implement one in their factory. Unfortunately, organizational preconditions for effective implementation of PDM systems were not established at that moment, and this might be a reason for poor effect PDM accomplished in this company. Management overlooked one very important fact - PDM, like any other software solution, is just the tool used as a help, enabling faster exchange of information etc., not "new panacea for manufacturing"
Analysing presented material one can conclude following:

- PDM systems are very powerful tool that enable better performance in managing organizations.
- PDM maximize benefits of CA technologies through the linkage of different software solutions giving real input to Concurrent Engineering, Re-engineering etc.
- Implementation of PDM solutions achieve satisfied results only if it is applied on organization with clearly defined and established organizational structure, production rate or product assortment.
- PDM is an effective tool that gives result only if it is used adequately.
- Success of this implementation depends on those who should implement and run the system; their attitude, wiliness and comprehensiveness.

5. REFERENCES

Petkovic, D.; Introduction to Computer Integrated Designs and Manufacturing Systems, PDS, Zenica, 2002

Duhovnik, J.; Tavcar, J. (2000) Elektronsko poslovanje in tehnicni informacijski sistemi - PDMS Product Data Managemnt Systems, Fakulteta za strojnistvo, Ljubljana, Slovenia

PDM Information Company, Understanding PDM, http://www.pdmic.com/intropdm.html

Gascoine, B; PDM: The Essential Technology for Concurrent Engineering, MCB University Press Limited, Volume 2, November 1, 1995.

BNT TMiH home page, http://www.bih.net.ba/~bnt-tmh

Exact Software,MAX ERP Division, www.micromrp.com

Annals of DAAAM for 2006 & Proceedings of the 17th International DAAAM Symposium, ISSN 1726-9679
Editor B. Katalinic, Published by DAAAM International, Vienna, Austria 2006
150th Birthday of Nikola Tesla (1856-1943) Inventor of Mobile Robot Teleoperation

LASER ALLOYING AND SURFACE HARDENING

BELYI, A.; GORDIENKO, A.; KHEIFETZ, M. & KOUKHTA, S.

Abstract: *The method of laser alloying of control over formation of structures and machine parts surfacing based on principles of synergetics and physico-chemical analysis are considered. Based on the established interconnections between the physical and chemical processes, methods of optimization and control over combined effects have been developed in terms of thermodynamic criteria describing the technological factors of processes through arranging a feedback between the parameters of efficiency and the quality of treatment.*

Key words: *laser alloying, physico-chemical analysis, titanium alloys, technological barrier.*

1. INTRODUCTION

Laser impulse and continual radiation let to focus in small volumes the energy with specific density of power to 10^{15} W/cm^2 (Fig. 1) and can cause acute change of properties in surface layers (Grigor'yants, 1989; Golubev & Lebedev 1988).

Laser surface hardening is using for surfacing processes of steels and alloys and with heating characterizes as uncapacious process. The difference from electrical heating that laser heat hardening is realized from surface into the material therefore it doesn't require to use coolant that simplify the realization of thermo-hardening technology.

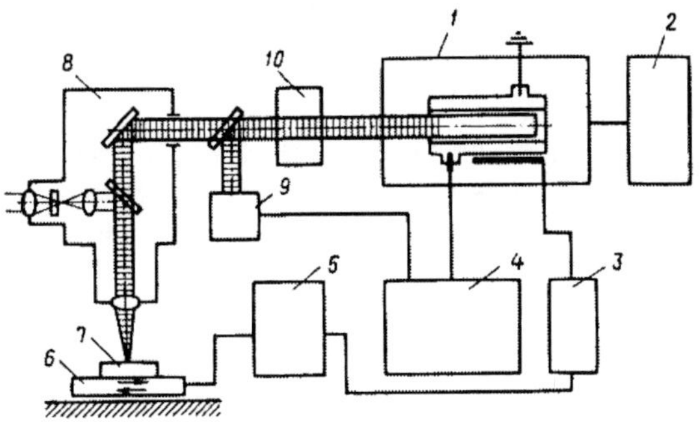

Fig. 1. Scheme of laser technological facility: 1 - radiating source; 2 - cooling system; 3 - unit of lamp incending; 4 - electrical set; 5 - programming unit; 6 - operating place-table; 7 - workpiece; 8 - optical unit; 9 - system of automatic stabilization of exit energy; 10 - unit for energy dosing

Laser thermo-hardening is characterized by the short time of action and provides the absence of treated machine parts deformations. Heat influence during laser hardening is regulated in wide range by changing parameters of laser radiation and conditions of treatment. It allows us to get required structure of surface section and needed properties (Gordiyenko, 1998).

Laser alloying technologies have good perspective in point of view of receiving surface layers with required level of properties. This method is based on puting into the pool with molten alloy elements and allows to saturate surface layers of working parts with many simple and complex chemical compositions. Mutual emplacement of alloy component and material of the basis characterizes the initial conditions of the process. Transfer of alloy substance per zone of alloying is determined by boundary conditions and is realized by convective fluxes arized because of nonhomogeneous heat in a free surface of molten pool. It allows us to form the wide spectrum of structures in solid condition (Grigor'yants, 1989; Gordiyenko, 1998).

2. METHODS OF RESEARCHES

Laser alloying allows considerably to increase such operational characteristics of details as wear resistance, contact strengthing, heat resistance and so on. To the most perspective materials for laser alloying we can concern titanium alloys. Laser thermo-hardening is not so effective for titanium alloys because of high velocity of heating (10^3-10^5 °C/s) and small diffusion coefficient of β-stabilizing elements (Mo, V, Cr, Fe) which in two powers less then diffusion coefficient for carbon into iron.

Complex research of laser alloying processes is performed on an extensive group of industrial α- and (α+β)-titanium alloys (Gordiyenko et al., 1990; Golubev et al., 1991). The powders from different kinds of chemical compositions (combinations of nonmetals, carbides, silicides and boronides of different metals) and some simple elements (B, Si, C, Mo) were used for alloying. High automatizied equipment for transference the ray according programmed contour was used for realization of laser technologies - laser manipulators LTK-1, Unilam, Salar, which can quickly and flexiblely reattune for deciding of concrete tasks (Gordiyenko, 1998).

3. RESULTS AND DISCUSSION

Experimental researches and theoretical calculations show that impulse laser thermo-surfacing (τ~10^{-3} s) doesn't cause hardening of surface layers of titanium alloys VT6 type because of momentary heat-cooling cycle and low diffusion mobility of vanadium. Concentrative redistribution of vanadium between α- and β-phases is able outright to conclude (Fig. 2) during the influence of continuous laser radiation (τ~10^{-2} s) in initial structure with dispersion of high degree presence (the distance between centres of α- and β-phases particles is over 1 μm). That redistribution causes the formation of α'-martensite and alloy hardening from 2.8 to 4.0 HPa. The maximal values of hardness in the zone of thermal influence compile 4-4.5HPa for alloy VT9. It is explained higher degree of α'-martensite alloying then for alloy VT6.

In alloys VT22, VT23 types with more higher concentration of β-stabiliziting elements the full β-phase gomogenization during cooling accomponies by α''-structure formation with hardness 3.2-3.5 HPa.

The experiments show that in laser alloying by boron with argon presence in molten layer in dependence on laser radiation parameters and amount of boron, plated before on the surface, the formation of diboronides of titanium in the form of dendrites and spherical excretions and also cellular or cellular-dendrite structures is possible where boron is in the solid state (Fig. 2).

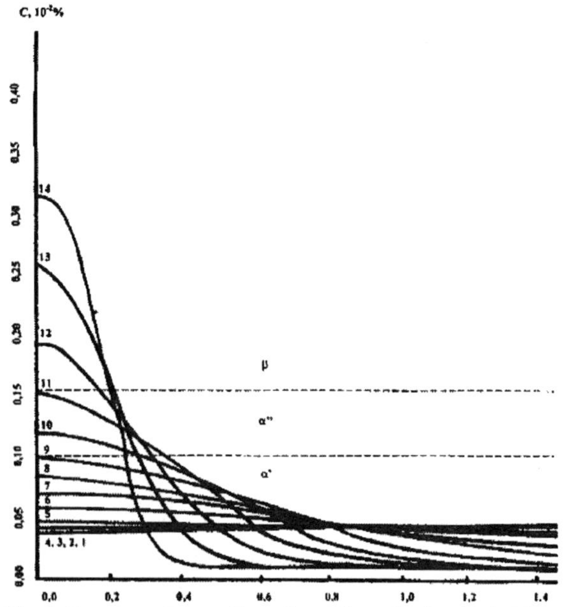

Fig. 2. Concentrative profiles of vanadium distribution in VT6 alloy after laser thermo- surfacing with q=1300 W/cm^2 during τ=0,8 s: 1 - surface; 2 - 100 μm; 3 - 200 μm; ...14-1300 μm

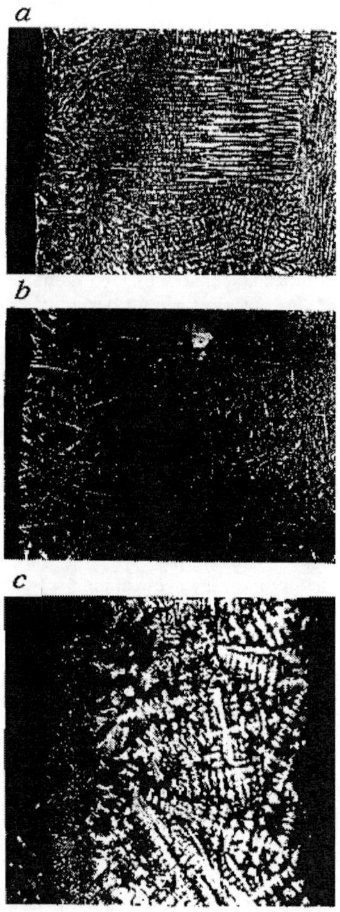

Fig. 3. Microstructure of VT23 alloy after continual laser boroning with q=2·10^4 W/cm^2, γ_0=2 mg/cm^2, v=3 cm/sec (*a*), and nitriding with q=4·10^4 W/cm^2, γ_0=10 mg/cm^2, v=3,5 cm/sec (*b*) (×200) and nitriding with q=4·10^4 W/cm^2, v=l cm/sec (*c*) (×400)

However the achievement of high hardeness (8HPa) in surface layers becomes possible with sputtering in surface before laser meltening the optimal amount of boron γ_0=6-8 mg/cm^2. In this case stable inclusions of titanium diboronides are forming in alloyed zone.

The wear intensity and the total wear intensity of alloy and conterbody are hardly decreasing. The surface roughness decreases top (on 30-50%) with the addition of silicon to boron (Gordiyenko, 1998).

The alloy layer has dendrite structure after laser nitriding of titanium alloys, herewith the composition of dendrites with TiN lattice considerably differs from composition of titanium nitride with lesser concentration of nitrogen.

The subsequent ageing to 700°C doesn't influence on titanium nitride, the disintegration of metastabile α'-, α''- and β-phases is proceeding only. When ageing is higher then 700°C, the structure of zone becomes more homogenous, acute boundaries of dendrites are disappearing of TiN2 and the formation of TiN2 combinations is taking place.

The researches of titanium alloys surface topography during the processes of tribologic tests show that titanium α-alloys wear in initial conditions occurs because of the shear of material with the formation of numerous friction grooves (Golubev et al., 1991). Alloying accompanied with the formation of homogenous dispersive structure in melten layer changes the nature of the wear. The plastic deformation scents appear with grooves of shear. The pining of formed deformation tongues occurs together with shearing. The scents of breakable pining are observed after alloying with incomplete dissolving of particles.

The using of the technology of laser alloying by nitrogen and boron allows to increase the details from VT9 alloy life in 2-3 times in the presence of gasabrasive wear (Gordiyenko, 1998).

4. CONCLUSIONS

1. We can realize the changes in the initial and boundary conditions for processes of the formation of surface layers using both energetic fluxes and substances with varied compaund of components.
2. Technological barriers isolating the nonequilibrium system during the process of treatment are able to move changing the position of system boundaries.
3. The structures and phases of surface layer which formed during the process of treatment are inherited during operation processes and have influence on nature of surface wear.

5. REFERENCES

Golubev V.S., Kindyuk A.P. & Vaganov V.V. (1991), Izv. Akad. Nauk BSSR, Ser. Fiz.-Tekh. Nauk, No.1, 18-22.

Golubev V.S., Lebedev F.V. (1988). *Engineering Foundations of making of Technological Laser*, Moscow.

Gordiyenko A.I. (1998), Izv. Akad. Nauk BSSR, Ser. Fiz.-Tekh. Nauk, No.4, 22-31.

Gordiyenko A.I., Bushik S.V. & Ivashko V.V. (1990), Surface modification technologies, Paris, 18-22.

Grigor'yants A.G. (1989). *Fundamentals of Laser Treatment of Materials*, Moscow.

Annals of DAAAM for 2006 & Proceedings of the 17th International DAAAM Symposium, ISSN 1726-9679
B. Katalinic, Published by DAAAM International, Vienna, Austria 2006
150th Birthday of Nikola Tesla (1856-1943) Inventor of Mobile Robot Teleoperation

FLEXIBLE HYBRID LEAN AUTOMATION

BILBERG, A.

Abstract: *Hybrid LEAN semi automated production cells employed with skilled employees can be designed and operated in a high competitive country with high salaries like e.g. Denmark. The cells are flexible in their design to let employees use their full competence in teams to manufacture a wide variety of products ready to meet changing customer demands. The integration of people, the manufacturing technology and the organization have demonstrated an agile and competitive manufacturing environment which might be inspiring to other industries.*
Key words: *Trends in Manufacturing, Hybrid Automation, Flexible Lean Manufacturing, Organizational Learning, Ergonomics*

1. INTRODUCTION

The future for European manufacturing has pointed in two directions: 1) outsourcing to low salary countries or 2) highly automated production. A few years back, research was focused at CIM – Computer Integrated Manufacturing and the fully automated manufacturing. Since, many companies have moved labor-intensive jobs to Eastern Europe and Asia. These years, Lean Manufacturing is very much on the agenda of western industry. Lean focuses on value adding activities in the companies and removal of any activity that creates no value (waste-Muda). This paper will present a way to combine Lean with automation in what is called "Flexible Hybrid Lean Automation". The example given is based on research within the area and practical implementation in Danish industry, especially the company LINAK A/S producing mechatronic products.

2. THE CONCEPT

Based on theory from advanced manufacturing (Rehg 1994), (Langer & Bilberg 1997) and Lean literature (Bichino 2004), (Rother & Harris 2001), (Monden 1998), different concepts of how to develop a competitive manufacturing environment have been assessed with the objective to get high flexibility according to product types and capacity. This research has resulted in the so-called flexible lean cell which in most cases is formed like a U cell that can be flexibly manned according to the actual demand or product type. The cell combines automated processes and robots with competent employees resulting in very short lead time and high productivity due to a minimum of waste to non-value added activities. The establishment of the cells is not very costly. They are built of modules or components that typically can be reconfigured in new assembly cells. Equipment is not bolted to the floor meaning that they can easily be moved. The power supply for the cells is basically just electrical power and pneumatic air, and the workstation is up running. In this context, the cells are also very flexible and to some extent agile, as they can easily be adjusted to new products and new markets.

3. LEAN AUTOMATION

The flexible Lean cells are often hybrid, i.e. a combination of manual and automated manufacturing processes. One of the challenges in a cell is to find the right level of automation, where automation and employees go hand in hand and with the highest degree of value adding in the manufacturing activities. For typical assembly lines, the level of automation is too high or incorrect because automation of the transport between processes does not add any value to the product. What we do in the flexible cells is to have the machine cycle automated first, and then automate the unloading of the part. To automate the machine cycle is exactly to focus on the value adding process, where the benefits of this kind of automation can be higher quality in the process, avoiding human strain because of non-ergonomic processes, shorter cycle times etc. Generally speaking, the hybrid cells where humans interfere with automated processes might be a very efficient manufacturing concept; because here, you automate processes that might be hard or even impossible to perform by human hand, for example assembly of small mechatronic products and processes that require very critical limits of tolerances, where robots have other capabilities. Human supervision, human interference if an error occurs and other activities might be very difficult to automate; or the automation does not add extra value to the product. Therefore, humans are very suitable for performing these processes.

4. INTEGRATION TO ERP

The flexible Lean cells have been integrated with the ERP system. In the case company Linak, the order management system has been reengineered in such a way that the user interface to the IT system has been made very user-friendly. The operators in the cell get a job list at the PC screen and from here orders can be pulled, kanbans can be released and progress can be reported when a job is started, delayed, finished etc. Inspired by Smalley 2004, communication with suppliers and customers is performed very effectively. The scheduling is performed by the team in the cell but the communication infrastructure is built in the revised ERP system. What is addressed here is actually Lean in the ERP or IT system, which should result in a simple, powerful and user-friendly IT solution with only the functionality that is actually required. In general, a lot of ERP functionalities do not add any value to the user, why it should be considered as waste and probably removed. The Lean IT solution is further documented in (Bilberg 2004).

5. THE ORGANIZATION AND WORK CONDITIONS

At a very early stage and throughout the design phase, a physiotherapist was involved in the project. The "Danish Action Plan against Repetitive Work" focuses very much on better work conditions. In the mature phase, we discussed that the individual employee was actually capable of performing all the processes required to manufacture the entire product by starting at the first process and move from station to station until the product was finished. In such a circle loop, a lot of different working postures are performed. If the work load is limited to a few kilos, the physiotherapist stated that the variability in working postures would be a big advantage to the people in the long run. In the cell, the employees work as a team. They carry out a detailed plan themselves through a person in the team called the waterspider. The waterspider works out the plan and makes it visible to the team on a planning board. Based on the actual day schedule, the actual manning of the cell is decided. Thus, a cell might vary from 3-5 people from day to day, depending on the product mix and the demand of the day. If a team is one worker too many according to the actual demand, workers can be moved to another cell or moved to other Lean activities like continuous improvement activities or 5S activities. The employees of a team have to act as a sports team. They have to move and assist each other. The objective is really not to work very quickly; rather to obtain a continuous, stable and reasonable pace according to the tact time set by the customer. The individual workstations are attempted to be as balanced as possible, but because of the many product variants it is impossible to balance completely. But due to the short walking distance, the close contact between people and the team spirit, it is possible for the team to balance the work load dynamically during operation by helping and assisting each other. The roles of the individuals are well defined and when manufacturing takes place, the individuals work together as a team of players. This involvement and commitment from the employees is also recognized in (Wright & Edwards 2001) based on an example from another industry. It is the team's responsibility to produce according to schedule and therefore the status of the cell is registered each hour on the planning board, which constantly makes the score or status visible to the team and other people of the organization. If the team is lacking behind schedule (red figures on the board), the pace will be improved; and if on the other hand the team is ahead of schedule (green figures), the pace can be reduced.

6. TRAINING THE EMPLOYEES

The human aspect and organizational learning are very important factors when implementing new manufacturing concepts, especially in the Lean cells addressed here with people working together as a team. The human competence, the company culture and the team spirit is very essential and the foundation to get the cells work successfully. Therefore, a Lean Training Academy has been established in order to work with the company culture, competences and team spirit. Here, employees are trained in manufacturing concepts, strategies and the Lean philosophy to obtain the right competence and management attitude. In the challenging task of improving the workstation, we had to engage the mental as well as the physical capacities of the employees. The individuals should understand the concept and the roles to play (get the mindset), they should mentally believe in the concept as a positive and good way to produce and work (the right feeling in their hearts) and finally they should have the right physical skills for the manufacturing processes, and the Lean methods and tools (hands-on experience). The biggest challenge that we have experienced is definitely training the mindset. Here we have applied different production games (Bilberg 2005), e.g. the Buckingham Lean game (Bichino 2004) which makes it possible to illustrate how it is to operate in a lean-based production with focus on flow, team work and elimination of waste. The game has been extended with theory regarding standardized work, balancing of the workstations, continuous improvement activities etc. The idea of the game is to make a turn around in a terrible production company to become a world class company that can deliver on time, satisfy the customer and earn money. During the Lean games, people are challenged and even the most stubborn students should be challenged in such a way that they fail to solve all the problems in the game. Another method applied is to build up full scale models of the workstations to give a more physical understanding of how to operate in the cell. The physical models have been built of simple materials as cardboard boxes and office tables in combination with the physical products. The idea is to build a "quick and dirty" model that the users can see, feel and discuss. This is a pretty amusing process, where lots of ideas for improvements are coming up. The intensive training of the organization including workers, engineers and managers has given a lot of fruitful discussions and has from the very beginning involved the users of the Flexible Hybrid Lean cells, which has proved to be very important for the final result.

7. THE RESULTS FROM DANISH INDUSTRY

The concept has been implemented in the company Linak employing 1,400 people world wide. The results from Linak are among other things an increase in the delivery performance from 85 % to more than 95 %; lead time from 15 days to 6 days, 3 days being the goal; productivity has increased with app. 25 %; the absence due to illness has been reduced from 4 % to app. 3 %. The results have come from the introduction of a productive concept, but also due to better ergonomics and team responsibility.

8. CONCLUSION

The Flexible Hybrid Lean Automation concept has been developed to demonstrate competitive production in Denmark/Europe. The concept has big advantages when it comes to flexibility in volume, product mix and reconfiguration. The ergonomics and operator well-being have been essential in the design phase where repetitive work has been reduced. A responsible and motivated team of operators is able to organize the work to satisfy the customer needs in a dynamic environment. The cells have a combination of automation, e.g. by robots and human activities, which is also the case when interfacing to the ERP system, where Lean has also played a role in designing a simple and efficient user interface to assist the work in the cells. The design of the cells has been decided in co-operation with engineers, university researchers and not least the operators on the shop floor. Training of the organization has been essential to provide the best ideas and to have the users to basically understand the thought and ideas behind the concept.

9. REFERENCES

Bichino J (2004). *The new Lean Toolbox*, PICSIE Books, ISBN 0-9541-2441-3

Rehg J.A.. (1994) *Computer Integrated Manufacturing.* Printice Hall ISBN 0-13-463886-7

Langer G., Bilberg A. (1997). *Architectural Considerations for Holonic Shop Floor Control*, Proceedings of the International Symposium on Manufacturing Systems (ISMS) at the World Manufacturing Congress (WMC97), Auckland, New Zealand.

Rother M, Harris R. (2001) *Creating Continuous Flow – an action guide for managers, engineers production associates*, The Lean Enterprise Institute, June 2001 ISBN 0-9667843-3-2

Monden Y. (1998). Toyota Production System. Engineering and Management Press ISBN 0-89806-180-6

Smalley A. (2004). Creating Level Pull. The Lean Enterprise Institute, April 2004 ISBN 0-9743225-0-4

Bilberg A.(2004): *Implementing Lean in the Supply Chain*, IMS-SIG 9, 2004, Supply Chain Integration Workshop, October 2004, Denmark

Wright M., Edwards P. (2001): *Does Teamworking Work, and if so, Why?* A Case Study in the Aluminium Industry, Economic and Industrial Democracy, 1998 Vol 19: p59-90

Bilberg A. (2005): *LEAN Manufacturing – Part of Business and Manufacturing Strategy*, SMESME 2005 Conference Glasgow 12-15 June 2005

Annals of DAAAM for 2006 & Proceedings of the 17th International DAAAM Symposium, ISSN 1726-9679
B. Katalinic, Published by DAAAM International, Vienna, Austria 2006
150th Birthday of Nikola Tesla (1856-1943) Inventor of Mobile Robot Teleoperation

VALIDATION OF STATOR RESISTANCE IDENTIFICATION ALGORITHM ON A REAL AC INDUCTION MOTOR

BLAHA, P. & VACLAVEK, P.

Abstract: *The aim of this paper is to presents practical validation of novel algorithm for stator resistance identification. The advantage of this algorithm is that it operates online and requires no external excitation. The identification requires steady state operation. The algorithm supposes that the inductances are during operation of vector control algorithm constant when the flux level is kept constant. Both, the stator and rotor resistances are varying with time due to changes in temperature. These changes are slow. The algorithm is not computationally intensive which makes it suitable for real time implementation in reasonable hardware.*

Key words: *AC induction motor, identification, stator resistance, sensor less control*

1. INTRODUCTION

Modern control of AC induction motor requires for its operation the knowledge of parameters of its equivalent circuit. These parameters are used in the estimator for the flux leakage vector position estimation. The rotational speed is estimated based on these parameters as well in sensor-less control scheme. The precision of the estimation is conditioned with the good model, in other words with the precision of motor parameters. The problem is that the motor parameters are not fixed during the operation because they change with the temperature, saturation and with the frequency (Toliyat et al., 2003).

The big amount of identification methods was already published. Probably the most simple and mostly used are the no load test and the locked rotor test (one phase test) (Gastli, 1999). Their disadvantage is the necessity to operate the motor off-line and off-site to lock rotor against movement.

The proposed identification method (Blaha & Vaclavek, 2006) requires steady state operation which partially limits its employment for drives which are designed to run mostly in steady state. It is expected that the load changes slowly or that there are some time intervals when the load is constant.

The article is organized as follows. The second section describes the equivalent circuit of AC induction motor. The next section outlines the algorithm for online stator resistance identification. The fourth section describes practical experiment on a real AC induction motor. The section which follows concludes the results and sketches the future steps in our research.

2. EQUIVALENT CIRCUIT

We expect that the AC induction motor can be substituted with T-model equivalent circuit. This model is composed of stator and rotor resistances (R_s, R_r), stator and rotor leakage inductances (L_{sl}, L_{rl}) and mutual inductance L_m. We also expect that the leakage inductances are equal. The algorithm can be slightly modified also for simplified gamma and inverse gamma model where one redundant leakage inductance is suppressed.

The overall impedance of equivalent circuit is equal to

$$Z_{eq} = R_{eq} + X_{eq} = R_s + jX_{sl} + \frac{jX_m(R_r/s + jX_{rl})}{R_r/s + j(X_m + X_{rl})} \qquad (1)$$

The reactances are given by $X = \omega_{mech} L$. The slip s can be computed using

$$s = \frac{\omega_{el} - z_p \omega_{mech}}{\omega_{el}} \qquad (2)$$

where ω_{el} is electrical speed, ω_{mech} is mechanical speed and z_p is number of pole pairs.

3. STATOR RESISTANCE IDENTIFICATION

When the motor operates with constant rotational speed and the load does not change than the motor is in steady state. The stator voltages and current are then both harmonic and it is possible to compute their RMS values

$$U = \sqrt{\frac{1}{T_{el}} \int_t^{t+T_{el}} u_s^2(t)dt} \qquad (3)$$

$$I = \sqrt{\frac{1}{T_{el}} \int_t^{t+T_{el}} i_s^2(t)dt} \qquad (4)$$

The active power can be computed using equation (5)

$$P = \frac{1}{T_{el}} \int_t^{t+T_{el}} u_s(t) i_s(t)dt \qquad (5)$$

These quantities can be utilized to compute impedance of equivalent circuit. First we compute $\cos\varphi = \frac{P}{UI}$ and $\sin\varphi = \sqrt{1-\cos^2\varphi}$. The real and imaginary parts of equivalent circuit's impedance are

$$R_{eq} = \frac{U}{I}\cos\varphi = R_s + \frac{R_r/s \cdot X_m^2}{(R_r/s)^2 + (X_m + X_{rl})^2} \qquad (6)$$

$$X_{eq} = \frac{U}{I}\sin\varphi = X_{sl} + \frac{X_m(R_r/s)^2 + X_m X_{rl}(X_m + X_{rl})}{(R_r/s)^2 + (X_m + X_{rl})^2} \qquad (7)$$

The inductances L_{sl}, L_{rl} and L_m are known a-priory from some off-line test. They are assumed to be constant during motor operation. This simplification can be used under the condition that the flux magnitude will be kept constant and the saturation of magnetization characteristics will have no effect. By expressing the rotor resistance divided by slip from equation (7) we obtain

$$\frac{R_r}{s} = \sqrt{\frac{(X_{eq} - X_{sl})(X_m + X_{rl})^2 - X_m X_{rl}(X_m + X_{rl})}{X_m - X_{eq} + X_{sl}}} \qquad (8)$$

We can see that this equation depends only on known reactances and therefore it can be evaluated. For sensor-less
When the vector control algorithm works with the speed sensor than it is possible to compute the slip and by multiplying both sides of (8) we obtain rotor resistance estimation.
In the case of sensor-less control (which is of our interest), it is not possible to compute rotor resistance by computing the slip. The estimated speed corresponds to the rotor resistance which is used in estimator so the rotor resistance from the estimator would be obtained instead of real rotor resistance value.
Substituting (8) into (7) and expressing stator resistance R_s gives

$$R_s = R_{eq} - \frac{R_r/s \cdot X_m^2}{(R_r/s)^2 + (X_m + X_{rl})^2} \quad (9)$$

The whole algorithm can be summarized as follows. Zero crossing of voltage is detected. It starts the integration of squared stator voltage, squared stator current and their product. When the next zero crossing is detected then the electrical period T_{el} is computed. It is used to compute RMS voltage, RMS current and the power using (3), (4) and (5). These results are used to compute components of equivalent impedance R_{eq} (6) and X_{eq} (7). Finally, the R_r/s is computed using (8) which helps to compute stator resistance estimate R_s using (9).
The decision on steady state operation is realized as follows. The successive values of electrical period T_{el}, RMS voltage U, RMS current I and power P are compared with their previous values. It they differ for less than 5 percent than the motor is assumed to be running in steady state.

4. PRACTICAL EXPERIMENT

The proposed algorithm was tested in simulations (Blaha & Vaclavek, 2006). Because the results were promising and the estimated resistances coincided with the real values it was added to vector control structure of AC induction motor without speed sensor. The whole algorithm was running on evaluation board powered with Freescale digital controller 56F805. The power of small AC induction motor was 0.25kW. The motor had two pole pairs. The practical experiment was realized as follows. First, the parameters of AC induction motor

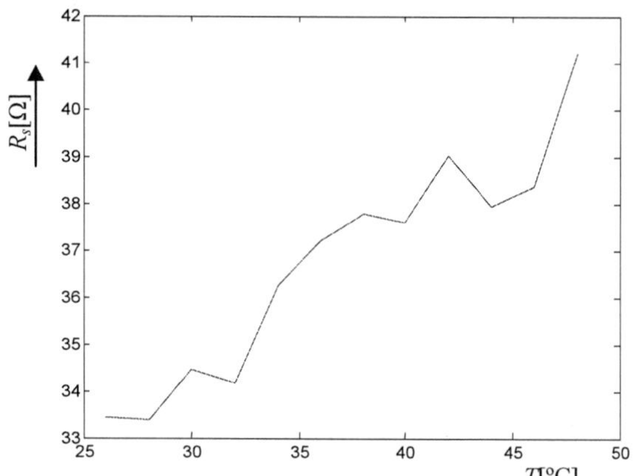

Fig. 1. The dependency of stator resistance on temperature.

equivalent circuit were estimated using no load and one phase test. Vector control algorithm in rotor flux coordinates was employed to drive the motor (Vaclavek & Blaha, 2006). The rotational speed was kept on a constant level of 750rpm with constant nonzero load which caused the increase of motor's temperature. The current and voltage harmonic curves were monitored and stored in files for different temperatures of the motor. The temperature on the top of the motor was measured. The experiment lasted about half an hour. During that time the temperature of the motor changed from 26°C to 48°C.
Measured data files where further treated in Matlab. The algorithm which was described in previous section was used to compute the stator resistance estimate. The obtained data are shown in Fig. 1. The stator resistance measured with the ohmmeter was $R_s = 29.3[\Omega]$. The difference between measured value and estimated one is caused by the fact that the dynamic resistances of the power transistor and diode are also added to stator resistance when the measurement is realized in vector control scheme and the resistance of the leads was also different.

5. CONCLUSION

The proposed paper presented practical validation of novel algorithm for online stator resistance identification of an AC induction motor which does not require any external excitation as it has been already proven in simulations.
Even thought the results are not precise they show good tendency of the estimate which means that the stator resistance increases with the temperature. Probably more precise results can be obtained with more measurements taken during the same temperature and by averaging the resulting stator resistance estimates. It can be concluded that proposed algorithm is working also on a real motor. Our intention is to include this algorithm into digital signal processor as a whole and try to use it for on-line modification of the motor model.

6. ACKNOWLEDGEMENT

The research has been supported by Czech Science Foundation under the project GA 102/06/0949 "Intelligent Control of Induction Machine and Synchronous Machine Electrical Drives" and by Freescale Semiconductor Czech Republic (former division of Motorola).

7. REFERENCES

Blaha, P. & Vaclavek, P. (2006). The algorithm for online stator resistance identification of an AC induction motor. *Proceedings of 10th International Conference on Intelligent Engineering Systems 2006,* Rudas, I. (Ed.), pp. 1-6, June 2006, London Metropolitan University, London.

Gastli, A.(1999). Identification of induction motor equivalent circuit parameters using the single-phase test. *IEEE Transactions on Energy Conversion*, Vol. 14, No. 1, pp. 51–56, ISSN 0885-8969.

Lepka, J. & Stekl, P. (2001). 3-Phase AC Induction Motor Vector Control Using DSP56F80x. *Freescale Semiconductor Application Note*.

Seok, J. K.; Moon, S. I.& Sul, S. K.(1997). Induction machine parameter identification using PWM inverter at standstill. *IEEE Transactions on Energy Conversion*, Vol. 12, No. 2, pp. 127–132, ISSN 0885-8969.

Toliyat, H. A.; Levi, E. & Raina, M.(2003). A review of RFO induction motor parameter estimation techniques. *IEEE Transactions on Energy Conversion*, Vol.18, No. 2, pp. 271–283, ISSN 0885-8969.

Vaclavek, P. & Blaha, P.(2006). Lyapunov function based flux and speed observer for ac induction motor sensorless control and parameters estimation. *IEEE Transactions on Industrial Electronics*, Vol. 53, No. 1, pp. 138–145, ISSN 0278-0046.

Annals of DAAAM for 2006 & Proceedings of the 17th International DAAAM Symposium, ISSN 1726-9679
B. Katalinic, Published by DAAAM International, Vienna, Austria 2006
150th Birthday of Nikola Tesla (1856-1943) Inventor of Mobile Robot Teleoperation

LASER BASED MEASURMENT SYSTEM FOR CALIBRATING MACHINE TOOLS IN 6 DOF

BLEICHER, F.; PUSCHITZ, F. & THEINER, A.

Abstract: *This article is intended to give a brief survey of a measuring device which is based on laser interferometry. The system consists of either three or six plane mirror laser systems and a reflecting reference structure. The laser systems are attached to the moving platform of a machine tool structure, preferably clamped onto the main spindle. The reference structure is fixed in the working area and provides three reflecting plane mirrors which are oriented in a Cartesian manner. Each reflecting plane mirror is either scanned by one or two laser beams. One laser beam delivers information on a translational deviation referring to a starting position. The use of two beams also allows for the determination of the rotational degree of freedom. The measuring system can be used for gathering geometric information of the entire working area of a machine tool, and can especially be used for the calibration of parallel kinematic structures.*

Key words: *Laser interferometry, Parallel kinematic machines, calibration, Error compensation, tooling machine*

1. INTRODUCTION

The accuracy and reliability of machine tools have a decisive influence on the manufacturing accuracy of workpieces. In the past 10 to 15 years, the accuracy of machines has clearly improved (Weck, 2001). Deviations in the accuracy of a machine tool, and consequently in the geometry of the completed workpiece, may be due to a variety of reasons. Causes in the machine tool may include errors in the geometry due to tolerances in the manufacturing and assembly of the machine parts, as well as deformation due to static and dynamic loads or thermal deformation. The newly developed laser measurement system is primarily geared to capturing geometric parameters such as angles and positions.

2. METHODS OF ACCEPTANCE TESTS

To ensure a high degree of machine accuracy, various measurement systems and processes have been created, and standards governing machine tool acceptance tests have been published, such as DIN ISO 230 and VDI/DGQ 3441-3445 for example. Methods of machine acceptance can basically be broken down into indirect and direct processes to identify machine properties.

In the indirect identification of machine accuracy, sample workpieces with defined geometric formal characteristics are manufactured to draw conclusions about machine accuracy on the basis of deviations between the required and actual geometries. The direct determination of machine property allows for the identification of error impacts. Parameters are determined directly on the machine with the help of suited measurement instruments.

The calibration of a parallel kinematic machine tool after assembly is generally significantly more complex compared with serial Cartesian kinematics. It comprises parameter identification, which serves the purpose of accurately determining the geometric parameters, and error compensation. In general, measurement requirements for the identification of the actual machine parameters are usually less complex because it is mostly sufficient to measure only one degree of freedom of a machine tool. For error compensation all degrees of freedom of the machine tool in the working area have to be captured point by point, so that measurement becomes considerably more complex as the number of degrees of freedom rises.

3. MEASUREMENT PROCESSES

Various measurement methods are available to capture the degrees of freedom; in this context, a rough distinction can be drawn between one-dimensional and multi-dimensional measurement systems. Well-known one-dimensional measurement systems include shaft encoders and linear scales working according to inductive, magnetic or photo-electric principles. The laser interferometer has made its mark as the most accurate means of measuring one degree of freedom in machine tool building. However, one-dimensional measurement systems come with the drawback that setting them up and aligning them is a time-consuming process, especially when the entire working area needs to be measured. Multi-dimensional measurement systems are combinations of conventional one-dimensional measurement systems. The double-ball bar is a customary measurement appliance for two-dimensional measurements. Three-dimensional measurement systems and other multi-dimensional systems can largely be broken down into two groups, those of optical and mechanical measurement systems. Optical systems include laser trackers, coordinate measurement appliances like the Metris K-400CMM, photogrammetric and laserscanning devices. These systems however come with the drawback that the degree accuracy is insufficient for the purposes of machine tool building. Examples of mechanical systems include, to name but a few, Omnigage, Unisquare, measurement tripods and the QCM. The QCM or "Quickstep Calibration Machine", including its analysis algorithms, was developed at the Institute of Production Engineering in close cooperation with the Krause & Mauser co. and used for measuring the parallel kinematic machine tool Quickstep©. The concept and algorithms were already described in a publication (Bleicher & Günther, 2004). It is a mechanical system with three passive axes docked onto the main spindle via the tool holding fixture. Comparative measurements using a coordinate measuring machine resulted in a positional accuracy of 12.2μm at a positional variation of 1.7μm and a variation of orientation of 6.3 angular seconds. FEM analyses of the structure and supporting measurements identified room for improvement, but in the end the accuracy could not be increased.

Based on the insights gained in the QCM tests, research for a suited alternative measurement technique, preferably one that uses laser measurement technology and allows for the simultaneous acquisition of values for several degrees of freedom with a high degree of accuracy, started.

4. MULTI-DIMENSIONAL INTERFEROMETER

The basic concept developed for the multi-dimensional laser interferometer envisages a laser source that can be moved with the TCP. A stationary reflector is placed in the working area. There is no mechanical coupling by means of transmission links between the moving and the stationary part of the system. Conventional laser measurement systems with a retroreflector are unsuited for this application because the retroreflector would have to follow the moving laser beam, and would have to be moved mechanically. The new application uses a laser system in which the laser beam is reflected by a front-surface mirror. Due to this laser system it is possible to measure the distance of a point across a reflecting surface as a reference plane. Relevant laser systems using helium-neon lasers are available on the market (e.g. Renishaw RLE10, SIOS SP 2000). To reach the accuracies stated by suppliers, appropriate tolerances in the quality of the reflecting surface must be adhered to. Parameters given for the flatness of the mirror surface are local and global flatness. Local flatness should be around $\lambda/10$ over a distance of 10mm. The wavelength λ of a helium-neon laser is 632.8nm. Global flatness describes distortions over the entire mirror. Deviations of around $\lambda/4$ cause errors of around 158nm (N.N., 2001).
If the reflecting surfaces are mounted appropriately, global deviations can be reduced and deducted from the result by compensation and are of rather secondary importance.
The concept is based on Abbe's comparator principle whereby the distance to be measured should be positioned in alignment with the scale embodiment. A significant benefit of this measuring principle is the method by which the reference body is manufactured as today's manufacturing processes allow for clearly more narrow tolerances to be reached in a plane surface than in spatial geometry.

5. IDENTIFYING A SUITABLE REFLECTOR

The reference body for the analysis of the laser measurement technology was a front surface mirror of high optical quality which had been supplied by the laser manufacturer and could be considered as meeting the requirements of flatness accuracy. First, a hand-polished steel plate was used as a front surface mirror. To keep the guiding movement of the plane mirror laser as regular as possible and within narrow limits in respect of rotation around the axes as well as positioning accuracy, the basic tests were made on a coordinate measuring machine, type Zeiss WMM 850. The plane mirror laser manufacturer stated that the required laser beam intensity is 50%. A software package for the determination of beam intensity is supplied with the laser. In the tests with the polished steel plate, beam intensity was markedly below 50% in many cases so that the measured values cannot be considered as sound. In a further measurement setup, a front surface mirror made of chromium-coated conventional float glass was tested. No special requirements had to be fulfilled in respect of surface flatness. Mirrors of this quality are available from the optical industry as semi-finished goods of almost any dimensions, causing relatively cost-effectiveness. The basic tests showed that the intensity of the reflected laser beam was clearly higher then 50%. In an analysis according to VDI/DGQ 3441 a mean measurement variation $P_{s\text{-mean}}$ of 1.662µm and a measurement accuracy P of 3.427µm were determined along a mirror-parallel path over a measuring distance of 400mm. In the third measurement setup, a front surface mirror made of Borofloat 33©, was tested. The surface was highly polished, sputtered with a layer of aluminium and covered with SiO2 protective coating. The mirror flatness across the entire surface was $\lambda/2$. The laser performance was also constantly at a high level and mean measurement variation $P_{s\text{-mean}}$ of 1.432µm and a measurement accuracy P of 2.153µm were determined along the z-axis. To determine the quality of measurements in beam direction for the plane mirror interferometers, a comparative test with a conventional Michelson-Interferometer was set up. The remaining deviation between the two systems amounts to a maximum of 0.4µm, which is quite normal for a given measurement uncertainty of ±0.7µm/m in a laser interferometer.

6. MEASUREMENT SETUP

To enable simultaneous multi-dimensional measurements the mounting adapter consists of three laser systems and can be directly clamped onto the main spindle or onto the moving platform of the machine tool. The front surface mirrors are mounted immovably in the workspace on the basis of a Cartesian system of coordinates. Two-beam laser systems are well suited for determining angular shifts around the individual axes along with positional measurements. In accordance with the given accuracy of a front surface mirror laser interferometer an angular accuracy of 0.852 angular seconds at a resolution of 0.2 angular seconds can be realized, with the beam distance being 100mm.
First tests of practical application involving a one-axis system were directly made on the Quickstep parallel kinematic machine. The two-beam plane mirror interferometer with a laser beam distance of 12.5mm was directly attached to the tool spindle of the machine by means of an adapter and the spindle was locked in its angular position. The measured results show that the accuracies obtained in the basic tests on the coordinate measuring machine were reproduced in the application test.

7. SUMMARY AND OUTLOOK

Tests carried out in the course of development of the measurement system presented here indicate the potential possibility of covering three to six degrees of freedom simultaneously with a high degree of measuring accuracy. A comparison with systems available on the market shows that there is great potential in respect of accuracy and flexibility. The benefit of greater cost-effectiveness compared with current multi-dimensional measurement systems makes its industrial use seem realistic. Due to the measuring concept used, rotational degrees of freedom can only be covered in a narrow range of angles. In this respect, the measurement system fulfils the requirement of controlling an oriented position via Cartesian movement. This should ensure excellent suitability for many applications in the metrological analysis of machine tool structures, especially of three-axis kinematics

8. REFERENCES

Bleicher, F. & Günther, G. (2004). Automatically Operating Calibration Method for a Three-Axis Parallel Kinematic Machine, *Parallel Kinematic Machines in Research and Practice*, Neugebauer, R. (Ed.), p.165-181, ISBN 3-937524-05-3, Chemnitz, April 2004, Verlag Wissenschaftliche Skripten, Zwickau

N.N. (2001). *RLE10 plane mirror application note*, brochure, Renishaw plc.

Weck, M. (2001). *Werkzeugmaschinen. Bd.5. Messtechnische Untersuchung und Beurteilung - 6.Auflage*, Springer-Verlag, ISBN 3-540-67615-5, Berlin

Annals of DAAAM for 2006 & Proceedings of the 17th International DAAAM Symposium, ISSN 1726-9679
Editor B. Katalinic, Published by DAAAM International, Vienna, Austria 2006
150th Birthday of Nikola Tesla (1856-1943) Inventor of Mobile Robot Teleoperation

ONLINE DETECTION AND ANALYSIS OF CUTTING FORCES DURING THE MILLING PROCESS WITH ROBOTS

BLIEDTNER, J. & MOELLER, K.

Abstract: *The introduction of traditional machining techniques like cutting and milling has proved to be very progressive in connection with robot application. As the stiffness and the exact positioning of robot systems shows some disadvantages against normal production setups an online-process-supervising system has to be adapted for the sake of high production quality and low production waste. Today it is possible to measure complex medical implant geometries and complicated components much quicker and more effectively due to the advantages of robot technology. The process guiding and supervision system for robot application shown here has been integrated in a manufacturing cutting process and can be equipped with many different sensor types. This enables the system to measure a broad range of cutting forces.*
Key words: *Controlling Process, Cutting, Automation, Robots, and Sensor*

1. INTRODUCTION

The manufacturing process with robot technologies has a significant influence on the final parameters as robots are mostly not as stiff as their large industrial tool machine sisters. Because of this deviations occur respective the required cutting line and roughness. These differences have to be regarded in direct connection with high manufacturing forces and low stiffness in the robot system (Hesse 1998).
These are the reasons why cutting techniques with robot technology have proved their reliability just in special fields of material processing where only simple work cycles have to be fulfilled e.g. deburring. The improvement of new cutting technologies also led to techniques like High-Speed-Cutting (HSC). As the cutting parameters were also improved successfully a reduction of work cycle forces was achieved.
These applications are about to recover new application fields for robot technology.
HSC-tool machines show high inertia and are very cost-intensive concerning maintenance and programming.
This makes cheap manufacturing cycles with robot-guided work pieces very interesting in various fields. The important A condition for an improved cutting process is the direct connection of process control and supervision of work forces which can be guided by an adaptive sensor technology. As this guiding system offers various advantages the most important have to be mentioned here: Optimized manufacturing parameters and wear status control of the applied cutting tool / cutting head.
Corresponding to these advantages a low-cost robot application means significantly cheaper production of complex work pieces with high manufacturing tolerances. Examples standing for applications with reduced course accuracy are pattern production, prototype manufacturing, medical implant manufacturing and design model testing in the automotive industry (Tangelder, W.H. 1993).
For each application purpose a proof of documentation is possible with the use of QMS-files in any case. The following paper introduces a cutting regulation setup which improves final manufacturing parameters.

2. EXPERIMENTAL PROCEDURE, RESULTS

The described setup consists of a KUKA robot, a well-known and very reliable model of this company. As clearly visible in Fig. 2 the whole setup is equipped with a cutting head and a cutting force measuring system. Its position accuracy is +- 0.1 mm while this model ran reach any position in the Cartesian coordinate system.
However, for most cutting applications this accuracy is exact enough. The adapted cutting force measuring system is characterized by different sensors of resistive wire strain, piezoelectric and fibre optical (FBG) principle. Within a certain range of tolerance a reliable cutting force data acquisition is possible.

Fig. 1. Robot system with milling unit

Depending on the sensor type (the measuring rate to be achieved is 50 kHz) the range of the cutting force / time regime can be detected comfortably. As soon as the data has been saved the values are transformed in an A/D-transformer. Prepared in that way a PID regulation can easily calculate the signals.

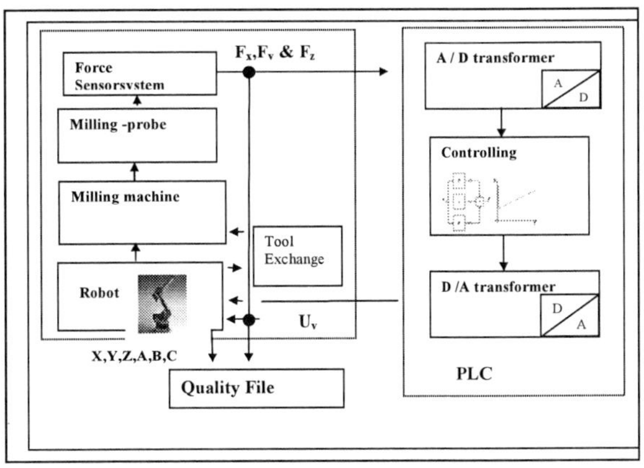

Fig. 2. Diagram of the experimental set up

This reliable PID regulation calculates the cutting parameter for the feed speed and is able to detect early stages of cutting tool wear. The start values are transformed by an A/D-transformer and get transferred to the robot controlling unit.

The data collected by the PID regulative are actively involved in the manufacturing process and lead to an interactive cutting tool replacement operation by the adapted tool exchange system. As all data are present in digital values the signals can easily be used as a QMS- and CAQ-file. The integration of CAQ-systems (Computer Aided Quality Assistance) shows great advantages concerning the immense load capacities for quality documentation which are very important for reasons of product liability. This documentation has to be evaluated and saved for a certain period. Modern CAQ-systems support administrative fields and the methods of modern quality supervision and planning. Also of great importance is quality management, quality improvement and quality guiding as well as the operative field of quality tests.

The force sensors integrated in the system show different technologies and also different handling habits. Three sensor systems were tested during the cutting runs. These three items proved very suitable for this highly sophisticated application purpose:

- Resistive wire strain sensors
- Macro fiber composite piezo sensors (MFC)
- fiber optical Bragg grating sensors (FBG)

The most important properties of these sensors are shown in Table 1.

	Resistive Wire Strain Sensor	Piezo Sensor	Bragg Fiber Grid
measuring rate	0 - 1 kHz	until10 kHz	0 -10 kHz
accuracy	2- 4 % DMS	4.6E+02 pC/N	1pm/µepsilon

Table 1. Parameters of the applied force sensors

Basically, all sensors can be integrated in the complete system. Optical fiber sensors with Fiber Bragg Gratings show interesting advantages in electromagnetic fields.

The following chapter deals with the results of robot cutting research on MFC sensors. For the test runs a standard cutting machine with specified process parameters was used.

Parameter	Bereich
rpm	23000 min^{-1} (0 -27000 min^{-1})
Tool diameter	5 mm
Cutting speed	6,02 m /s
Max. depth of cutting	2 mm
Work piece material	Al

Table 2. Specific process parameters

The crucial point for an optimized data acquisition is the right sensor position on the robot arm. The adaption of sensors for different axes is quite advantageous and enables us to measure axial and cross forces. Also force measurements not depending on the force direction can be saved and calculated for the cartesian coordinate system (Gerke, 2004).

A typical force course is shown in Fig. 3; taken during a non-regulated process. The signal impressively demonstrates the high force gradient during a simple cutting run. The stiffness of the robot clearly reaches its limits. It is obvious that an active force regulation is urgently necessary to achieve a positive result.

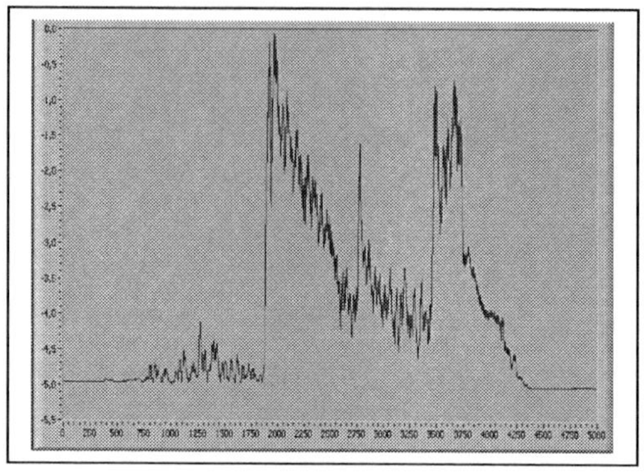

Fig. 3. Cutting force during the milling process

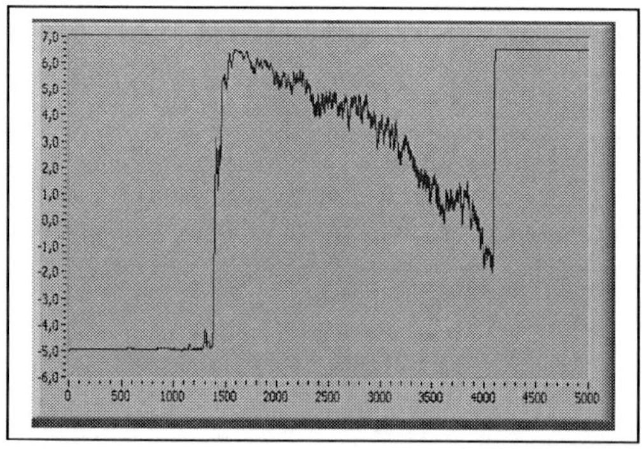

Fig. 4. Cutting force simulated for higher values

Fig. 4 shows up a sudden increase of cutting force simulated during the cutting process for research on the dynamics of the signal processing. The sudden change of force properties, caused by different geometric work piece circumstances and an increase of cutting depth occurs very often during the cutting process and can quickly lead to tool failure. With the customized force supervision system a possible tool failure can be detected in time and therefore regulates the process in the necessary way.

3. CONCLUSIONS

The presented process regulation system enables a close cutting process on complex structures with equal quality features. The application leads to an improvement of product properties, e.g. lower shape deviation and a homogenisation of surface attributes. Judging from this point of view economic and stable production is guaranteed and minimizes production waste. Furthermore, the production process is accompanied by the proof of quality for each product.

4. REFERENCES

Hesse, H: Industrieroboterpraxis- Automatisierte Handhabung in der Fertigung. Braunschweig. Vieweg-Verlag.1998

Tangelder, W.H. Solid Model Representations for Rapid Prototyping with a Milling Robots. Modelling. 2-nd ACM Sohd Montreal, Canada. 1993; ISBN –3-00-003953-8

Schallbroch, Heinrich ; Balzer, Heinrich; Schnittkraft- und Drehmomentmesser für Werkzeugmaschinen; Berlin 1943

Gerke,W. Roboterfräsen von Gießmodellen; VDI Verlag 2004; VDI Bericht; S. 319-327

Annals of DAAAM for 2006 & Proceedings of the 17th International DAAAM Symposium, ISSN 1726-9679
Editor B. Katalinic, Published by DAAAM International, Vienna, Austria 2006
150th Birthday of Nikola Tesla (1856-1943) Inventor of Mobile Robot Teleoperation

PRECISION PROCESSING AND MICROTOPOGRAPHICAL CHARACTERSATION OF TOOLING INSERTS AND THEIR MOULDINGS

BLIEDTNER, J.; BUERGER, W.; ROSENKRANZ, R. & FROEHLICH, M.

Abstract: *Ranges of plastic parts with optical surfaces can increasingly be manufactured in accordance with the users' demands at a high quality level, due to the technical development towards high-precision manufacturing in the field of injection moulding. Optical surfaces of plastic parts have to be of high quality for functional reasons. In the present case, the user of such components demands that the optically effective flat surfaces have a square value of the average surface roughness R_q of ≤ 10 nm.*
Key words: *measurement, plastic component, moulding procedure*

1. INTRODUCTION

In the technological process of injection moulding a process-sure fulfilment of the users' demands is absolutely necessary. Based on concrete users' demands experimental investigations were carried out in the injection moulding process of components with optical surfaces. In these investigations the roughness of the tooling inserts for the injection mould were taken into account according to their respective manufacturing processes (ultra-precision turning, polishing) as well as the roughness of the mouldings from the tooling inserts. Due to the technical roughness demands a high-quality surface testing technology has to be used in order to assure the quality of these components (Duparre´, A.; Notni, G. , 2002).

2. EXPERIMENTAL PROCEDURE, RESULTS

Tooling inserts for injection moulding of plastic optics can be manufactured in accordance with the users' demands at a high quality and precision level, due to the continuous technical development of processing techniques in the field of micro technology. In the technological process of ultra precision turning followed by injection moulding a process-sure fulfilment of the users' demands is absolutely necessary. Based on concrete users' demands experimental investigations were carried out in the ultra precision turning process of tooling inserts with optical surfaces. In these investigations the roughness of the tooling inserts for the injection mould were taken into account according to the technological parameters (rotational speed, feed rate, cutting edge radius, cutting depth) as well as the influence of different nickel-phosphorus layers. Due to the technical roughness demands a high-quality surface testing technology has to be used in order to assure the quality of these components (Fröhlich, M., 2004).
The samples for the roughness measurements and the technological parameters for the experimental investigations were selected in cooperation with the company JENOPTIK Polymer Systems GmbH, Triptis (Jenoptik Group) considering the following aspects: surface hardness, type of layers, processability and quality. 41 nickel-plated (chemically and galvanically) tooling inserts with selected technological parameters (feed speed v_f, rotational speed n, cutting depth a_p, cutting edge radius r_{ϵ}) were processed in the test programme using the ultra precision turning machine Nanoform 200 of the company Precitech. The investigations were carried out with the following range of parts:

- Metallic tooling inserts (different materials and manufacturing procedures/materials: steel, nickel)
- Injection moulded plastic parts (mouldings from the tooling inserts/materials PC Lexan LS 2, COC Topas, PMMA transparent and different PC Makrolon types)

The roughness measurements were carried out by means of the scanning force microscope (SFM) „Dimension 3000" in the tapping mode; a cantilever scanning tip with a tip radius of < 10 nm and an angle of < 25° was used. The measurements were carried out at five different places on the sample surface. The raw surface data were evaluated by means of a roughness analysis of the selected areas, by a cross-section analysis and a 3D-image. The quality of the tooling inserts was evaluated on the one hand by the square value of the average surface roughness R_q and on the other hand statistically by the PSD-function (Power Spectral Density function). The roughness determination of the optical surfaces of ultra precision turned tooling inserts was carried out by means of a scanning force microscope.

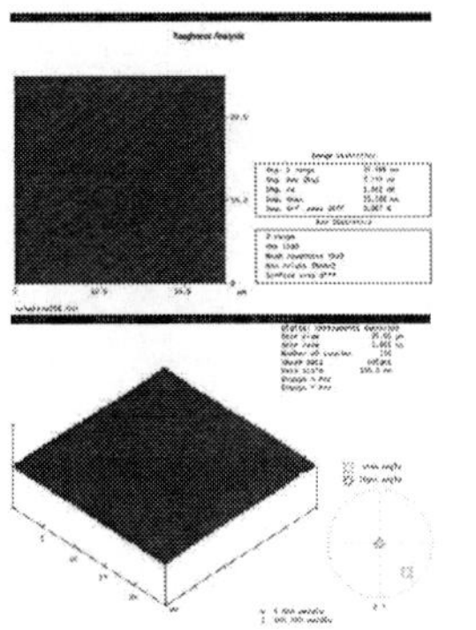

Fig. 1. Roughness analysis with R_q = 2,8 nm and 3D-image of the ultra precision turned tooling insert No. 17

The square value of the average surface roughness R_q (Fig. 1) and the power spectral density function (PSD) (Fig. 2) were determined during the investigations (Rosenkranz, S., 2004).
Basically every precision processing of technical surfaces is characterised by a specific power density distribution (PSD-Power Spectral Density function). According to Beckstette and Roth (see figure 3) three regions can be distinguished in the diagram of the power spectral density distribution as a function of the spatial frequency, which are significant for assessing the quality of the technical surface: roughness (high spatial frequency), waviness (Medium spatial frequency), form error (long spatial frequency).

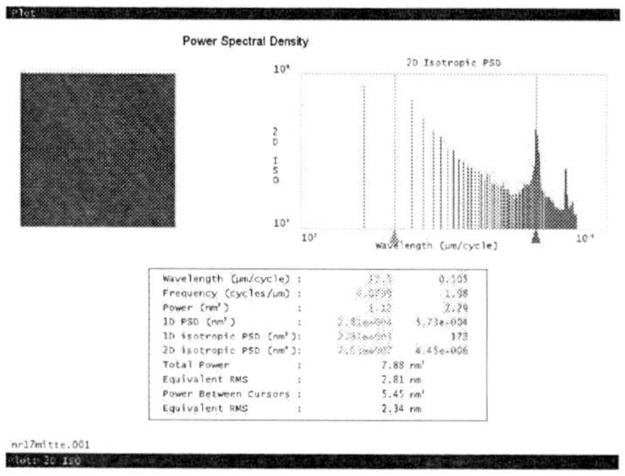

Fig. 2. Two-dimensional PSD-function of the tooling insert No. 17

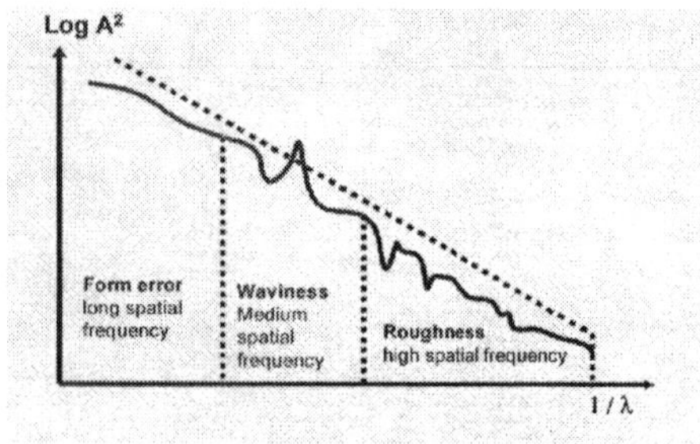

Fig. 3. Regions of the PSD-function of a precision-processed technical surface (Beckstette, K.-F.; Roth, E. 2006)

The PSD-functions for different precision-processed technical surfaces could be determined in own comprehensive examinations. Figure 4 shows results of ultraprecision-turned NiP-plated tooling inserts made of steel with MKD-tools. These examinations led to special results which are too extensive to be dealt with in detail in this article. It is for sure though that ultrprecision turning with increasing feed (all other technical parameters remain constant in this process) results in specific surface manipulations, which can be minimised through subsequent processing. The PSD-function allows a complete description of the surface quality

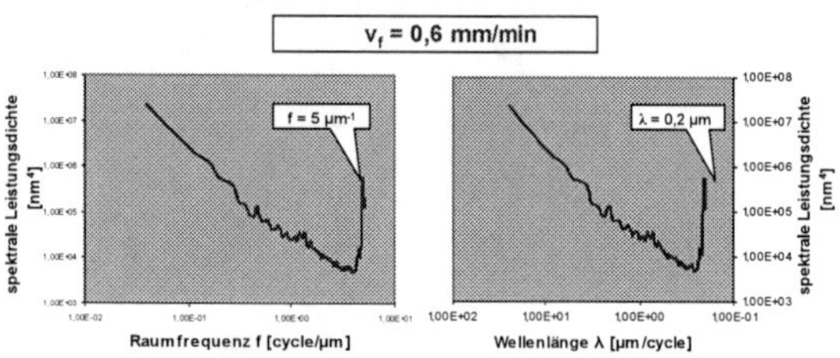

f = spatial frequency λ= wavelength
Feed f = 0.2 µm

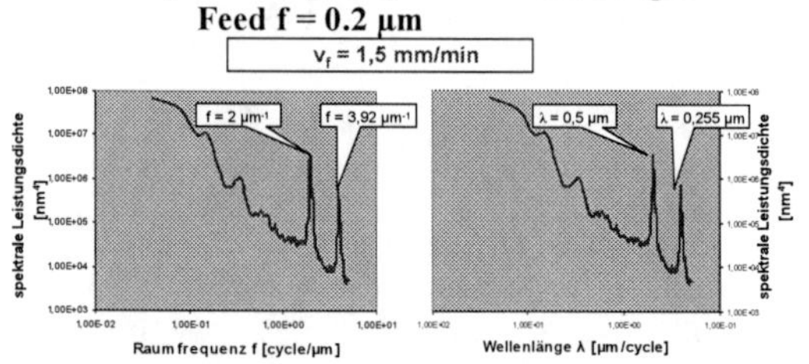

f = spatial frequency λ= wavelength
Feed f = 0,5 µm

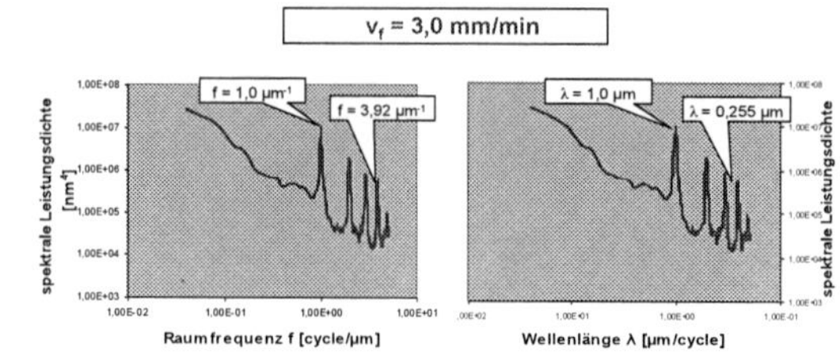

f = spatial frequency λ= wavelength
Feed f = 1,0 µm
Constant Parameters:
n = 3000 min^{-1}, r_ε = 0.5 mm, a_p = 4 µm

Fig. 4. PSD-function of ultraprecision-turned tooling inserts with diamond tools (NiP plated steel)

characteristics and has proved to be particularly suitable for polished surfaces or applications with extreme demands. The two-dimen sional power spectral density function represents, according to Duparee´ and Notni, the relative value of each surface roughness parameter as a function of the spatial frequency.

3. CONCLUSIONS

The extensive experimental investigations have proved valuable for the objectification of production issues in the injection moulding process of components with optical surfaces, for further optimisation of the manufacturing process and for the quality assurance of such components. Selected results of the investigations are talked over e.g. in publications.

4. REFERENCES

Froehlich, M. (2004). Charakterisierung ausgewählter Eigenschaften optischer Funktionsflächen an ultrapräzisionsgedrehten Werkzeugeinsätzen in Abhängigkeit von technologischen Parametern. Diplomarbeit Fachhochschule Jena

Duparre´, A.; Notni, G. (2002). Charakterisierung nanorauher Oberflächen. In: DAKOM 2002: Charakterisierung von optischen und technischen Oberflächen. Darmstadt, 27. Februar 2002

Weckenmann, A.; Ernst, R. (2003). Anforderungen und Randbedingungen für den Einsatz von Meßsystemen in der Mikro- und Nanotechnik. In: VDI-Berichte Nr. 1530 „Sensoren und Meßsysteme 2000“, pp. 297-307, VDI Verlag GmbH, Düsseldorf

Beckstette, K.-F.; Roth, E. (2006): Aspheres & Freeform manufacturing. Ultra precision manufacturing of Aspheres, Freeforms and Microstructures, IOF Jena, May 16th- 17th, 2

Rosenkranz, S. (2004). Charakterisierung ausgewählter Eigenschaften verschiedener spritzgegossener Bauteilsortimente mit optischen Funktionsflächen. Diplomarbeit Fachhochschule Jena

Annals of DAAAM for 2006 & Proceedings of the 17th International DAAAM Symposium, ISSN 1726-9679
B. Katalinic, Published by DAAAM International, Vienna, Austria 2006
150th Birthday of Nikola Tesla (1856-1943) Inventor of Mobile Robot Teleoperation

FORMAL-METHOD-BASED SOFTWARE DEVELOPMENT APPLIED ON EMBEDDED SYSTEMS: BASIC CONCEPTS

BLIZNAK, M. & KOLAR, D.

Abstract: *Programming of embedded systems is a special discipline. In many cases, developers are using only low-level programming technologies and languages so it is obvious that this way of software development can be time-consuming and inefficient. One of the possible ways leading to speed-up of a software development process on embedded systems is application of formal-method-based development processes. The goal of this paper is to point out that some specific formal languages can be effectively used for producing of highly optimized source code that fulfills necessary requirements of embedded systems.*

Key words: *FSM, finite state machine, code generation, ANSI C, optimization*

1. INTRODUCTION

Embedded systems are omnipresent and play significant roles in modern-day life. Programming of embedded systems is a special discipline; developers are often using only low-level programming technologies and languages so it is obvious that this way of software development can be time-consuming and inefficient. One of the possible ways leading to speed-up of a software development process on embedded systems is application of formal-method-based development processes.

One of the best matching formal methods suitable for embedded systems is description of application logic by finite state machines (FSM), which can be used for generation of a source code of developed applications. FSM's general properties ensure not only deterministic behavior of developed software, but they also allow usage of various validating and optimizing algorithms that can significantly improve quality and stability of a software application.

2. FSM VS. APPLICATION LOGIC

A finite state machine (FSM) or finite automaton (FA) is a model of behavior composed of states, transitions, and actions. A state stores information about the past, i.e. it reflects the input changes from the system start to the present moment. A transition indicates a state change and it is described by a condition that has to be satisfied to enable the transition. An action is a description of an activity that is to be performed at a given moment. Generally, there are two groups of FSMs distinguished: Acceptors/Recognizers and Transducers.

Acceptors/Recognizers give a binary output, saying either yes or no to answer whether the input is accepted by the machine or not. **Transducers** generate output based on a given input and/or a state using actions. They are used for control of applications so let us see some details. Two types are distinguished in this case: **Moore machine** and **Mealy machine**.

Moore machine uses only so called *entry actions*, i.e. output depends only on the state. The advantage of the Moore model is a simplification of the behavior. On the other hand, Mealy machine uses only *input actions*, i.e. output depends on input and state. The use of a Mealy FSM leads often to a reduction of the number of states. Mathematical model of transducer FSM is as follows:

A transducer finite state machine is a sextuple $<\Sigma, \Gamma, Q, q_0, \delta, \omega>$, where:

- Σ is the input alphabet.
- Γ is the output alphabet.
- Q is a finite non empty set of states.
- q_0 is the initial state, an element of Q.
- δ is the state transition function: δ: $Q \times \Sigma \rightarrow Q \times \Gamma$.
- ω is the output function.

If the output function is a function of a state and input alphabet (ω: $Q \times \Sigma \rightarrow \Gamma$) then that definition corresponds to the Mealy model. If the output function depends only on a state (ω: $Q \rightarrow \Gamma$) then that definition corresponds to the Moore model .

Figure 1 shows one posible graphical representation of FSM describing two-step control algorithm.

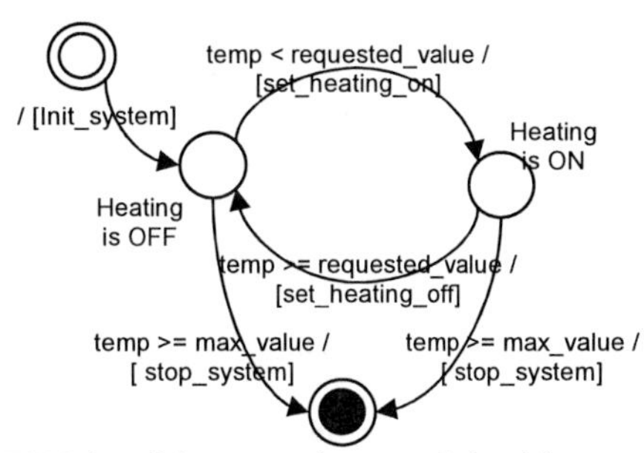

Fig. 1. FSM describing two-point control algorithm

Every program state where a software application can stay in can be interpreted as one separate state of the FSM. Transition paths between these states can be seen as a program flow, which can be driven by set of various conditions (more exactly by conditional statements) or events (for event-driven FSM). Program instructions, commands, methods and functions can be encapsulated into code blocks, which act as executive parts of transition paths called actions.

3. BASIC CODE GENERATION CONCEPTS FOR FSMS

Output source code is a product of one of several possible code generation algorithms. These algorithms differ mainly in an extent of produced source code, used program-flow-control commands and in their transparency. The figure 2 shows differences in the output code produced by two common code generation algorithms, where one of them is based on usage of `switch` command statement for differentiation between considered application states and the second one uses `goto` commands for control of the program flow.

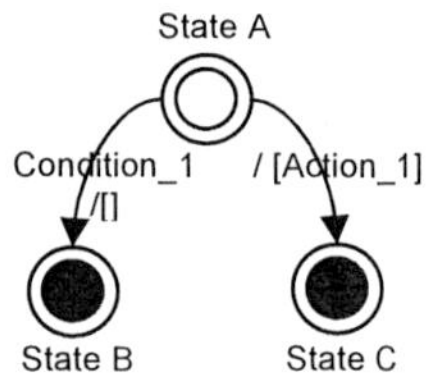


```
/* switch command */
TYPE_STATE
DoSomething(void)
{
  TYPE_STATE state=ID_A;
  for(;;)
  {
    /* Main loop */
    switch( state )
    {
      /* State: A */
      case ID_A:
        if(/*Condition 1*/)
        {
          state=ID_B;
        }
        else
        {
          /* Action 1 */
          state=ID_C;
        }
        break;
      /* State: B */
      case ID_B:
        return ID_B;
      /* State: C */
      case ID_C:
        return ID_C;
    }
  }
}
```

```
/* goto commands */
TYPE_STATE
DoSomething(void)
{
  /* State: A */
  if(!(/*Condition 1*/))
  {
    /* Action 1 */
    goto State_ID_C;
  }
  /* State: B */
  return ID_B;
  /* State: C */
  State_ID_C:
  return ID_C;
}
```

Fig. 2. Output of various code-generation algorithms

As it can be seen from the figure 2, code generation algorithm can significantly impact a structure of output source code. This fact is important because a target platform (embedded system) can have limited system resources. Thus, the source code designed for it should be as efficient as possible.

Therefore, several optimizations should be processed on this code. For example, unnecessary `break` statements or unused code labels should be remved from the code. These optimizations can positively affect not only number of generated program lines but also the program speed.

Another possible basic code optimization deals with the way how the conditions and the actions can be called: their source code (body) can be either inserted directly into source code as it is or it can be encapsulated into functions/methods, which can be called from an appropriate place. The way of invoking of these code fragments can involve both program size and speed. Moreover, especially on embedded systems, excessive usage of function calls can influence program performance and stability due to limited size of a target system's stack.

4. OPTIMIZATION OF SOURCE FSM

This chapter deals with two basic optimization tasks: merging of parallel transition paths with identical actions and linking of states included in direct condition-less automaton branches.

4.1 Merging of Parallel Transition Paths

The basic idea is that all parallel transition paths (i.e. direct paths that have common starting and ending states) including identical actions can be merged into one transition path where all partial conditions can be merged. Sample case is shown in the figure 3.

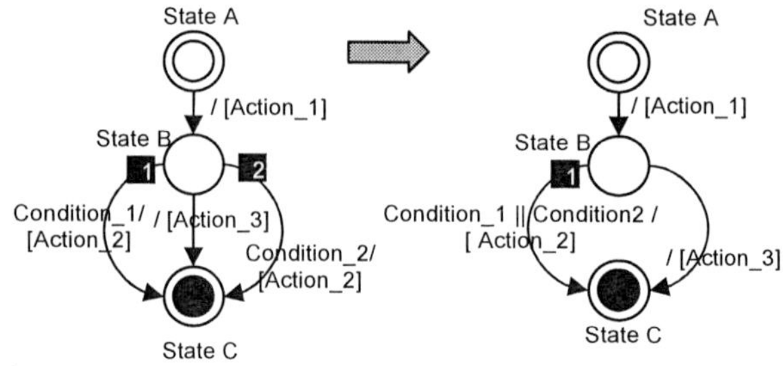

Fig. 3. Optimization by merging of transition paths

4.2 States Linking

Another frequently used optimization step is linking of FSM states that are connected together by direct condition-less transition paths. This is useful in that situation when some application tasks are lined-up one after one for some reason. In this case, all states in the direct branch can be replaced by one state with condition-less transition path where its action consists of all previously used actions in a correct order.

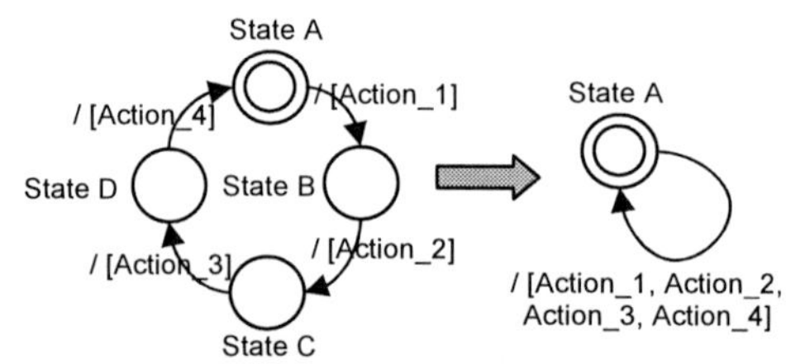

Fig. 4. Optimization by states linking

4. CONCLUSION

Formal-methods-based RAD tools designed for mainframe or desktop systems are commonly known for a quite long time. Nowadays, it is important to bring these techniques to an embedded world as well. As it can be seen from this paper, a quality of a source code produced by code generators based on FSM can be sufficient for nearly all embedded systems' demands.

5. REFERENCES

Carroll, J., Long, D. (1989), *Theory of Finite Automata with an Introduction to Formal Languages*. Prentice Hall.

Barr, M. (1999), *Programming Embedded Systems*, O'Reilly, ISBN: 1-56592-354-5

Qing, Li. (2003), *Real-Time Concepts for Embedded Systems*, CMPBooks, ISBN 1-57820-124-1

Tiziano, V. (1997), *Synthesis of Finite State Machines: Logic Optimization*. Kluwer Acad. Pub., ISBN 0-7923-9892-0

6. ACKNOWLEDGEMENT

This work was supported by the Ministry of Education of the Czech Republic under grant No. MSM 7088352102 and by Czech Academy of Sciences under project No. 1ET400750406.

Annals of DAAAM for 2006 & Proceedings of the 17th International DAAAM Symposium, ISSN 1726-9679
B. Katalinic, Published by DAAAM International, Vienna, Austria 2006
150th Birthday of Nikola Tesla (1856-1943) Inventor of Mobile Robot Teleoperation

ULTRA LOW POWER ECU FOR SMART AIRBAGS

BORBELY, S.; BUMERL-LEXA, R. & KUBINGER, W.

Abstract: *Common automotive airbag Electronic Control Units (ECUs) are connected to the vehicle's power source and communication bus, and continuously evaluate a variety of distributed sensor data as long as the ignition is on. These systems are both sophisticated and complex. In upgrading vehicles without airbags, an autonomous and ultra-low-power battery-driven control mechanism can replace the permanent power and sensor connection, thereby facilitating the upgrade.*
Key words: *embedded systems, ultra low power systems.*

1. INTRODUCTION

Occupant restraint systems undergo continuous research and development. Due to increasingly stringent vehicle safety requirements, the market for this industry is expected to grow. A smart and independent airbag *Electronic Control Units* ECU, based on an embedded ultra-low-power control mechanism, would be a useful method of filling the gap on upgrade systems for vehicles without an onboard occupant restraint system.
Nowadays, airbag ECUs are located outside the airbag unit itself, generally in the passenger compartment. They are connected to the vehicle's communication network and scan sensor data from various built-in and distributed sensors. These up-to-date airbag-based restraint systems (including airbag, ECU, various sensors and cabling) have evolved from countless crash tests over the last 30 years and fulfill relatively strict international crash standards. Recent research has even investigated new methods of passenger detection by image acquisition using stereo cameras to make these systems more failsafe (Faber & Förstner, 2000).
Nevertheless, these systems have already become highly complicated and additionally relatively expensive owing to the widely distributed electronic elements, connected by lengthy, sometimes susceptible wiring (Yu-Chi Ho, 2003; Farmer, 2003). After a frontal crash where airbags are deployed, the car is generally a write-off, even if the crash slightly exceeded 20kmph and the car body itself did not sustain severe damage, because of the costs of re-installing and adjusting the airbag restraint system.
Therefore, airbag-based restraint systems should be reconsidered for a more cost efficient upgrade solution. One way to obtain this is to shrink the whole system (including sensors, ECU and wiring) into an embedded, smart and ultra-low-power control unit. While such a solution would hardly replace the up-to-date systems, it could be an upgrade for vehicles without standard airbags. For that purpose we propose a stand-alone ECU with minimal power consumption for airbag control after a brief discussion of specific crash physics as background.

2. CLARIFYING CRASH PHYSICS

A constant deceleration over the time during the crash is assumed. When calculating the average deceleration, a speed difference and a deformation distance appear as parameters. When the deformation variable is constant, the deceleration is proportional to the square of the speed difference. The level of deceleration is not the only criterion which can be applied. Another significant criterion is the duration of the deceleration process. The longer the period of high deceleration, the more intense the crash. During a crash, the kinetic energy of a vehicle is reduced in a very short time. The car body (and the obstacle) thus have to convert this energy, primarily into deformation and heat. This is done by plastic deformation, whose energy is

$$E_{tot} = \frac{mv^2}{2} \text{ (m=mass, v=velocity). (1)}$$

The transduced energy at the specified point in time t is

$$E(t) = \int_0^t p(t')dt' \quad (2)$$

with the deformation power

$$p(t) = \vec{F}(t) \cdot \vec{v}(t) \text{ (F= force). (3)}$$

The results of the crash are visualized in Fig. 1 and obtained by manual measuring of objects in video frames (Progressive, 2005). Assuming equal time quanta between the frames, it is possible to measure a proportional value of the distance which the car was moving. As a consequence, the speed of the vehicle can be determined and filtered, thereby adjusting the inaccuracy of the manual measurement. It can be seen that the speed reduction is progressive and the crash becomes more intense over time. This is explained by the fact that the deformed car body increases in density after the deformable zone has been exploited. The car and the obstacle hit by the car decreased in length by a total of 1.3 meters.
This non-linear behavior can be difficult to manage because the airbag must be fired approximately after the first 10ms of the crash. Within this early phase, the further progress of the crash cannot be precisely determined. This issue is complicated by the fact that different vehicles have different levels of chassis stiffness.results.

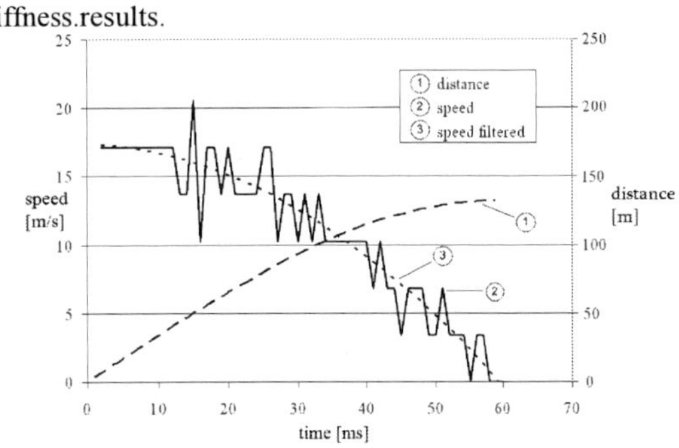

Fig. 1. Smart ECU startup time schedule

3. PROPOSED ACTIVATION AND FUNCTION

The sensor system must trigger in time to initiate the bag deployment, so that a fully inflated airbag is in place to cushion the motion of the passenger. A widely used rule to determine sensor-triggering time for airbags is the so-called *5 inches minus 30 ms rule* (Ching-Yao, 2000). Modern airbag systems need approximately 40 to 50 ms to detect and evaluate an impact and fully inflate the airbag. At this moment the head of the passenger has already moved 12cm (approx. 5 inches) toward the airbag. Following the rule, 30 ms were subtracted and a sensor triggering time of 10 to 20 ms results.

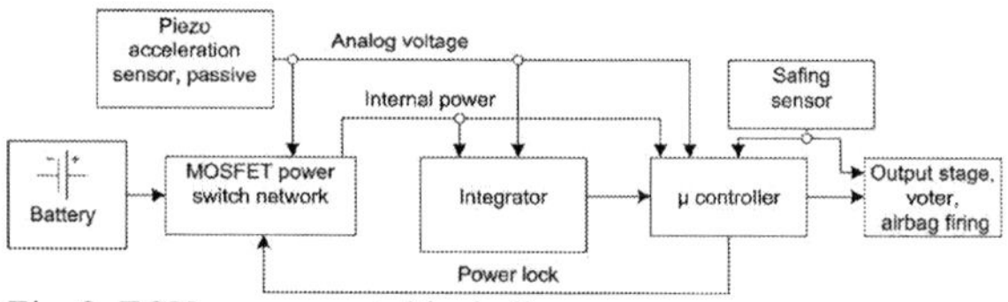

Fig. 2. ECU components block diagram

Fig. 2 shows a block diagram of the proposed system. This system is not powered permanently by the power supply of the vehicle, but has its own battery. If a mechanical force acts on the 3D piezo crystal, a displacement of the electric charge occurs, causing the opposite interfaces of the crystal to be charged differently. Known as the *piezo effect*, this depends on the strength of the mechanical force. The emitted voltage (actually an electrical charge) is proportional to this mechanical force and is used to wake up the system. Fig. 3 shows the startup time-schedule (Ching-Yao, 2000).

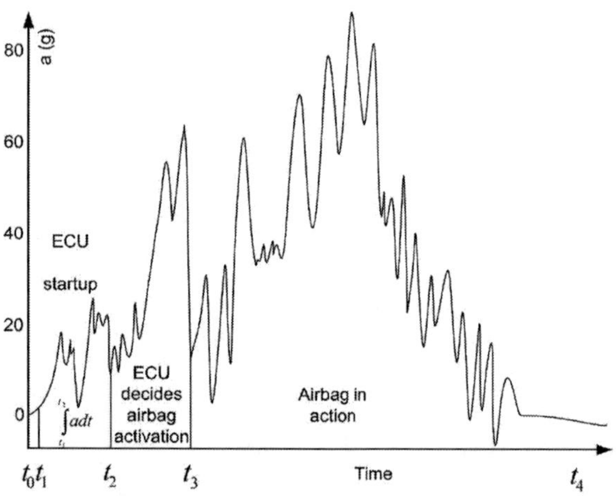

Fig. 3. Smart ECU startup time table.

Utilized on automotive applications, the mechanical force depends on the acceleration of the vehicle. This generated voltage peak drives the *Metal Oxide Semiconductor - Field Effect Transistor* (MOS-FET) power switch network (see Fig. 2) immediately after the impact. The power switch network turns the power for the remaining electronic circuit on at t=t1 and also enables the analog integrator. A microcontroller startup usually requires up to approximately 5 ms (e.g. ATMEL AT87C5). During this time the analog integrator determines an area of the acceleration over the time $\int_{2}^{1} adt$ until t=t2.

This area is a measure of the velocity difference between t1 and t2, which is taken as initial input for the hazard detection algorithm of the microcontroller. As soon as the microcontroller is running (at t=t2), it samples and processes the piezo sensor output voltage directly. Depending on the collision strength, the microcontroller activates its firing output (t2<t≤t3). A second sensor (see Fig. 2) safing sensor realized by an independent mechanical system also provides a firing signal (Ching-Yao, 2000). Only when both systems, i.e. the microcontroller and the safing sensor, enable their firing signals simultaneously does the voter in the output stage fire the airbag. If no crash is detected, the system saves a log of this wakeup event and shuts itself down.

4. SIMULATION RESULTS

The left part of the airbag circuit block diagram that was shown in Fig. 2 was mapped into a Spice model (including piezoelectric sensor, battery, power switch network and analog integrator). As shown in Fig. 4, the piezo voltage V(u_piezo) starts the internal power supply after a few µs. The graph V(vdd) represents the total power consumption of the startup circuit. It starts at a value of approximately zero volts and jumps up to the final value at the power on moment. The graph V(u_integr) displays the instantaneous time integral of the piezo voltage V(u_piezo). This voltage can be captured by the microcontroller as initial value. It is a metric of the violence of the shock which occured before the microcontroller was up and running. This enables the algorithm to estimate the plausibility of a real accident without being in power-on mode the whole time.

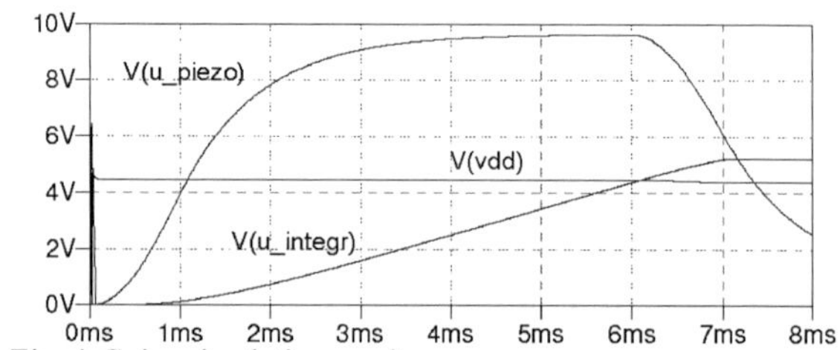

Fig. 4. Spice simulation results.

5. CONCLUSION

A practicable, mixed analog and digital circuit for a smart airbag ECU was presented. A small battery pack is used to power the unit, so that the whole airbag system is not powered permanently and is only started on demand. Power consumption and a lifetime of a minimum of two years were considered. Therefore an appropriate power up mechanism also had to be implemented. While such a solution would hardly replace existing ones, it could be an important upgrade for vehicles without onboard airbags. Car manufacturers could offer steering-wheels with an embedded smart airbag, as an upgrade for their vintage cars.

6. REFERENCES

P.Faber and W.Förstner, 'A System Architecture for an Intelligent Airbag Deployment', Proceedings of the IEEE Intelligent Vehicles Symposium 2000, Dearborn (MI), USA, October 3-5, 2000.

Yu-Chi Ho, Qian-Chuan Zhao and David L. Pepyne, 'The No Free Lunch Theorems: Complexity and Security', IEEE Transactions on Automatic Control, Vol.48, No.5, pp 783-793, May 2003.

Michael E. Farmer and Anil K. Jain, 'Integrated Segmentation and Classification for Automotive Airbag Suppression', Proceedings of the IEEE International Conference on Image Processing (ICIP), pp III-1053-III-1056, September 14-17, 2003.

Progressive Insurance, Mayfield Village, Ohio, USA, Web: http://www.progressive.com/Progressive/history.asp, downloaded Juni 2005.

Ching-Yao Chan, 'Fundamentals of Crash Sensing in Automotive Air Bag Systems', Book, Society of Automotive Engineers, Inc., Published 2000, USA.

Annals of DAAAM for 2006 & Proceedings of the 17th International DAAAM Symposium, ISSN 1726-9679
B. Katalinic, Published by DAAAM International, Vienna, Austria 2006
150th Birthday of Nikola Tesla (1856-1943) Inventor of Mobile Robot Teleoperation

THE ELABORATION OF MULTIPLE PRECESSIONAL GEAR THEORY

BOSTAN, I.; DULGHERU, V. & HEUSCH, C.

Abstract: *The engineering complex study of the triad "gear-technology-transmission" has permitted to elaborate a new type of precessional transmissions with multicouple gear. In this paper, the authors present the mathematic model of the multicouple gear. A computer program for doing this it is also elaborated. It is shown a block-scheme of the algorithm of the program modules including the calculus modalities exposed in the paper.*
Key words*: Precessional transmission, multicouple gear.*

1. INTRODUCTION

The specific character of sphere-spatial (precessional motions of the precessional transmissions pinion makes impossible the utilisation of teeth classical involute profiles. This fact requires the elaboration of new profiles adequate to the sphere-spatial motion of pinion which would ensure high performances to the precessional transmission. Carrying out on the principle of the transfer function continuity and gear based on the principles of the transfer function continuity and gear multiplicity which aims to:

- the elaboration of the gear mathematics model with account of the peculiarities;
- the analytical description of teeth profiles by a system of parametric equations on spherical surface and normal teeth section for inner and plane gear;
- the determination by CAD of geometrical and cinematic parameters influence of the gear upon the teeth profiles shape and the justification of their rational limits of variation;
- the elaboration of the theoretical basis evaluation of teeth gear multiplicity in precessional transmissions;
- the definition of area of gear multiplicity existence by 100% teeth couples.

Constructions peculiarities and high multiplicity of gear create favourable premises for the improvement of precessional transmissions kinematics accuracy. Within these activities we elaborated:

- theoretical basis for the identification of kinematics error generated by various primary error (frontal and radial knocking), on the basis of error independent action principle by fulfilling computer assisted mathematics experiment;
- compensation method for manufacturing and assembling errors;
- method of determination of probable limit error for precessional reducers with account of the stochastic character of manufacturing and assembling errors.

Know-how in the elaboration of the multicouple precessional gear, manufacturing technology and control methods, and a range of precessional transmission diagrams belong to research team from the Technical University of Moldova. During the last 25 years the team patented about 140 inventions.

The elaboration of working machines driving mechanisms is based on the diagram of precessional transmissions, presented in figure 1 (Bostan, Dulgheru & Grigoraş, 1997). The rotating motion of the crank shaft 1 is transformed into sphere-spatial motion of the block pinion 2 with two teethed crowns 3 and 4, which are rolling without sliding on the immovable and driven toothed wheel teeth 5 and 6. Due to the minimum difference between the number of teeth

$$Z_5 = Z_3 - 1, Z_6 = Z_4 - 1, \; Z_3 = Z_4 + 1,2,3...$$

the transmission ratio is

$$i = \pm\frac{Z_3 Z_6}{Z_5 Z_4 - Z_3 Z_6}. \qquad (9)$$

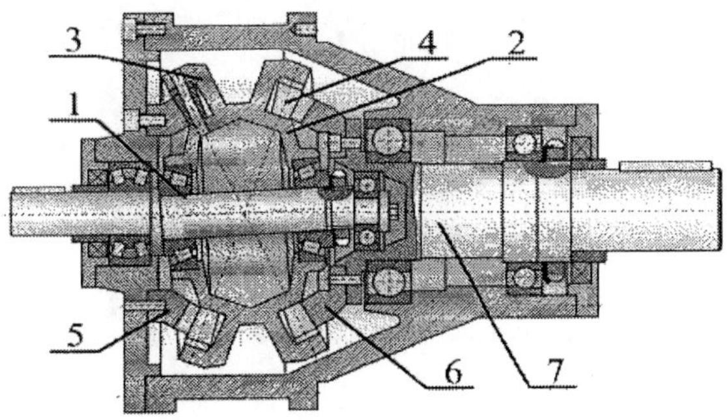

Fig. 1. Planetary precessional

The teeth of crowns 3 and 4 are manufactured in the shape of conical rollers installed on axis having the possibility to rotate round them, and the teeth of central wheel 5 and 6 have non-standard convex-concave profile.

2. ANALYTICAL DESCRIPTION OF THE TEETH PROFILES

In precessional transmission the gear wheel produces sphere-spatial motion round a fixed point. As mentioned the (Bostan, 1992), that the body, which produces spherical motion, has three degrees of freedom. As a rule, in theoretical mechanics, the position of the body, which produces precessional motion, is determined by Euler angles. In this case, the mobile system of co-ordinates $OX_1Y_1Z_1$ is bound rigidly with the gear wheel, as origins of the system of co-ordinates the immobile point O (centre of precession) being chosen (figure 2).

This system of co-ordinates produces jointly with the gear wheel spherical motion related to the immobile system of coordinates $OXYZ$. Analytical relationship between the co-ordinates of gear wheel points, shown in the mobile $OX_1Y_1Z_1$ and immobile $OXYZ$ system of co-ordinates, was obtained analysis of the two pinion positions - initial one, when the system of coordinates axis coincide, and final one, when the axis al removed (displaced) (Bostan et al).

By expressing vectors $\bar{i}_1', \bar{j}_1', \bar{k}_1'$ via basic vectors $\bar{i}_1'', \bar{j}_1'', \bar{k}_1''$ of the immobile system of coordinates $OXYZ$ we obtain

$$\bar{i}_1' = \bar{i}\cos\psi + \bar{j}\sin\psi, \bar{j}_1' = -\sin\psi + \bar{j}\cos\psi, \bar{k}_1' = \bar{k} \qquad (1)$$

The second rotation is produced at angle θ $(0 \le \Theta \le \pi)$ round joint lines, after which the unit vector $\bar{i}_1', \bar{j}_1', \bar{k}_1'$ will move into directions $\bar{i}_1'', \bar{j}_1'', \bar{k}_1''$ (respectively, they coincide with the directions of axis OX'', OY'', OZ''), at the same time vector

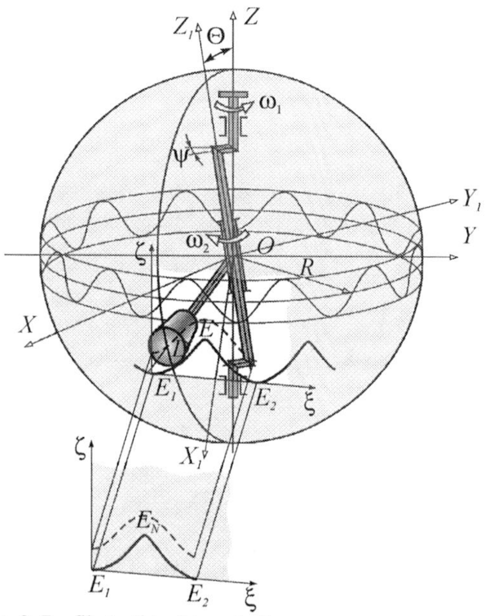

Fig. 2. Profile teeth's determination.

$\bar{k}_1''$, which coincide with vector $\bar{k}_1$, defines the position of axis OZ_1 in final position. By expressing vectors $\bar{i}_1'', \bar{j}_1'', \bar{k}_1''$ via $\bar{i}_1', \bar{j}_1', \bar{k}_1'$ we obtain

$$\bar{i}_1'' = \bar{i}_1'; \; \bar{j}_1'' = -\bar{j}_1' \cos\theta + \bar{k}_1' \sin\theta; \; \bar{k}_1'' = -\bar{j}_1' \sin\theta + \bar{k}_1' \cos\Theta. \quad (2)$$

By operating matrix theory, transition from gear wheel point coordinates is produced (for example, of the roller centre D), given in the mobile system of co-ordinates $OX_1Y_1Z_1$ to the coordinates of the same points in the immobile system $OXYZ$. After some modifications we obtain the coordinates of the modified point D^m:

$$\begin{aligned} X_D^m &= -\sin\delta \sin\left[Y_C^m \sin\theta + Z_C^m (1-\cos\theta)\cos\psi\right]; \\ Y_D^m &= -Y_C^m \cos\delta + Z_C^m \sin\delta\left[\cos^2\psi + \cos\theta \sin^2\psi\right]; \\ Z_D^m &= -Y_C^m \sin\delta\left(\cos^2\psi + \cos\theta\sin^2\psi\right) - Z_C^m \cos\delta. \end{aligned} \quad (3)$$

After some transformations the modified profile (section) of the tooth is described by the equations

$$\begin{aligned} \xi &= X_E'' \cos\frac{\pi}{Z_1} + \left[R_D \cos(\delta+\beta) + Y_E''\right] \sin\frac{\pi}{Z_1}; \\ \zeta &= X_E'' \sin\gamma \sin\frac{\pi}{Z_1} - \left[R_D \cos(\delta+\beta) + \right. \\ &\left. + Y_E'' \cos\frac{\pi}{Z_1} + \left[R_D \sin(\delta+\beta) + Z_E''\right]\right] \cos\gamma. \end{aligned} \quad (4)$$

where

$$\sin\gamma = tg(\delta+\beta) \Big/ \left[\cos^2\frac{\pi}{Z_1} + tg(\delta+\beta)\right]^{1/2};$$

$$\cos\gamma = \cos\frac{\pi}{Z_1} \Big/ \left[\cos^2\frac{\pi}{Z_1} + tg\;(\delta+\beta)\right]^{1/2}.$$

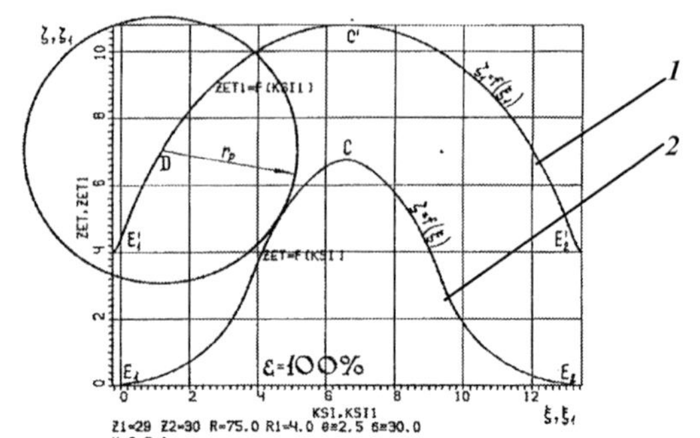

Fig. 3. The profilograms of teeth with convex-concave profile

calculation and generation programme was elaborated. In Figure 3 the profilograms of teeth (curve 2), obtained for the various geometrical parameters of gear are shown. The analyses of profilograms demonstrate the level and direction of influence upon the angle of conical axoid δ, medium radius of wheel R_m, rollers radius R, conical angle of rollers β and correlation between number of teeth Z_1 and Z_2.

On the basis of the parametric equations (4) in the modelling system CAD/CAM/CAE CATIA the mathematic pattern of the gear wheel toothing was obtained (Figure 4).

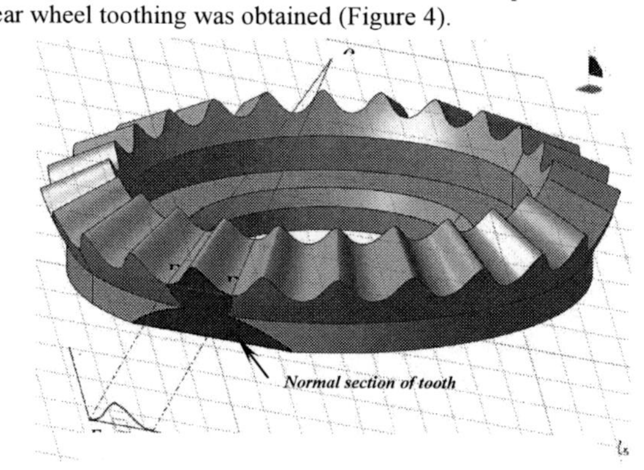

Fig. 4. Mathematic pattern of the gear wheel

In most countries, the mathematic models and algorithms as such, e.g. the model describing the multi-couple gear, as presented herein, cannot be protected by patents. But specific applications of the respective algorithms are patentable, if they meet the basic patentability requirements, that is, if they are novel and inventive. Those who intend to use the present concept(s) will be able to enter into joint development projects with the Technical University of Moldova to develop specific applications. The necessary capital investments for such projects can be protected by filing patent applications together with the present authors.

3. REFERENCES

Bostan, I. (1992). *Precessional transmissions with multicouple gear*, Stiinta, ISBN 5-376-01005-8, Chişinău.

Bostan, I., Dulgheru V.& Grigoraş, S. (1997). *Planetary, precessional and harmonic transmissions*, Tehnica-Bucureşti – "Tehnica",Chişinău, ISBN: 973-31-1069-8.

Bostan, I., Dulgheru V. & Vaculenco M. (2001). *Precessional gear-engagement,* Patent nr. 1886MD.

Bostan, I. & Babaian, I. (1996). *Precessional gear-engagement,* Patent nr.1455094 RU (patent nr. 560 MD).

Bostan, I., Dulgheru V., Vaculenco M. & Bodnariuc I. (2005). *Precessional transmission (variants),* Patent nr. 1886MD.

Annals of DAAAM for 2006 & Proceedings of the 17th International DAAAM Symposium, ISSN 1726-9679
B. Katalinic, Published by DAAAM International, Vienna, Austria 2006
150th Birthday of Nikola Tesla (1856-1943) Inventor of Mobile Robot Teleoperation

DEVELOPMENT OF PROCESSING METHOD FOR NONSTANDARD TEETH PROFILE GENERATION

BOSTAN, I.; DULGHERU, V.; GIURGIUMAN, H. & HEUSCH, C.

Abstract: *The production of non-standard teeth profiles requires a new manufacturing technology. In the complexity of problem "gear-synthesis-profile study- manufacturing " an important role plays the elaboration of efficient methods of teeth manufacturing which ensures a maximum productivity and reduced cost while satisfying the requirements related to the gear with precessional motion.*
Key words: *trajectory, precessional motion, teeth manufacturing*

1. INTRODUCTION

The manufacturing of non-standard teeth profiles needs a new execution technology. In the complexity of the issue *„gearing synthesis-study of the profile-manufacturing"* an important role belongs to the development of efficient teeth manufacturing methods, which might ensure maximum output and low cost under maximum satisfaction of the requirements for precessional motion gearing.

Multicouple teeth gearing and the continuity of the transfer function of precessional transmissions can be performed only by a fundamentally new method for teeth processing that takes into account the peculiarities of the spherical motion of the satellite motion. With this purpose a new technology for teeth processing by precessional tool run method with regard of the revolving semiproduct has been elaborated (Bostan, Bostan *at al*). The method of non-standard profile tooth processing is based on the utilisation of the linkage of a toothing machine-tool and of a special device having a minimum number of mobile parts. The linkage between the tool and the shiftable part of the device is a Hooke joint which generates the non-uniformity of the transfer function in the linkage *„tool-semiproduct"*. This variation will influence the tooth profile.

To elaborate the theoretical basis for tooth profile generation by precessional tool run it was necessary to establish the character of the continuous contact of the tool cutting edge and of the tooth profile of the processed wheel for the duration of a full precessional cycle. In connection with this the mathematic model of tooth generating method by precessional tool run (Bostan *et al*) which gives overall reflection of the teeth interaction in the precessional gearing was elaborated. With regard of this we have described:

- The kinematical linkage of the precessional tool with the semiproduct which provides the consistency of the „tool-semiproduct" linkage transfer function;
- The motion trajectory of the tool centre in the shiftable system of coordinates;
- The motion trajectory of the tool centre in the shiftable system of coordinates linked to the revolving semiproduct;
- The generator contour of the tool in the shiftable system of coordinates and the curve of tool surfaces series which fix the position of the tool contour for a precessional cycle;
- The on plane curve projection in the plane system of coordinates.

2. KINEMATICS OF THE REALIZING MECHANISM FOR THE TEETH GENERATING METHOD

2.1. Qualifications for generating process of the teeth profile

To perform the teeth processing method a mechanism was elaborated which block diagram is shown in Figure 1 (Bostan & Babaian).

In the designed mechanism the joint which involves the tool in precessional motion is fixed from the rotation around the joint axis of the spindle – semiproduct with a mechanism of junction. The rotation of semiproduct 3 and main shaft 1 are coordinated by the indexing gear of the machine tool. Constructional performance of the tool junction mechanism with the housing secures consistency of the transmission function $\omega_1 / \omega_3 = const$ and is defined by point C motion trajectory, which belongs to the movable element.

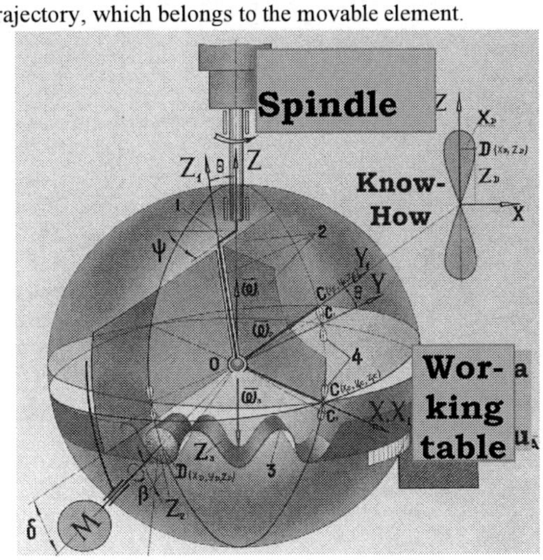

Fig.1 Spatial scheme of the realizing for the teeth processing method by rolling using precessional tool.

In order to establish the dependence of the semiproduct angle of rotation ψ_3 on the main shaft angle of rotation ψ we describe the semiproduct motion composed of the rotational motion of locking in with the main shaft crank ψ_{3e} and the relative rotational motion in relation to the shaft crank ψ_{3r}.

2.2. Analytical description of the motion trajectory of the tool junction mechanism in the fixed system of coordinates

In the compound motion of the semiproduct $\psi_{3e}=\psi$, but ψ_{3r} presents a certain function $f(\varphi)$ of the tool angle of rotation φ, that is

$$\psi_3 = \psi + | f(\varphi) \qquad (1)$$

For real accuracy of the driving mechanism of the machine tool function $f(\varphi)$ will have the form

$$\psi_3 = \psi + \frac{Z_2}{Z_1} arctg\ (\cos\ \theta tg\psi). \quad (2)$$

The momentary ratio of transmission of the device mechanism is obtained by deriving (2) after ψ

$$U_{31} = \frac{d\psi_3}{d\psi} = \frac{\omega_3}{\omega_1} = 1 - \frac{Z_2}{Z_3} \cdot \frac{\cos\theta}{\cos^2\psi + \cos^2\theta \sin^2\psi}. \quad (3)$$

From relation (3) results that the ratio of transmission is not constant and this irregularity will influence the form of teeth profile generated by the tool. Thus the relation of tool with the body introduce a diagram error in the teeth profile

$$\Delta\psi_3 = \psi_3 - U_{31}^{med} = \frac{Z_2}{Z_3}\left[\psi - arctg\left(\cos\ \theta tg\ \psi\right)\right] \quad (4)$$

Figure 2 shows the graph of the diagram error of the tool position ψ_3 at a revolution of the main shaft at point *D* motion in plane *OZY*. In the case of point *C* motion in plane *OZY* this error without distortion is transmitted to the tool and the last generates the teeth profile with the same error.

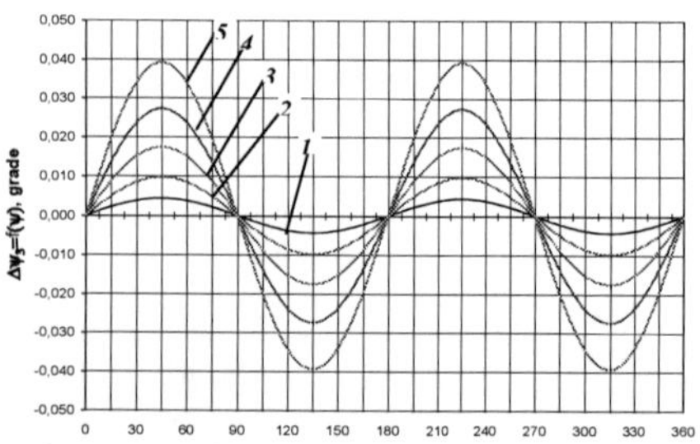

1 - θ=1°; 2 - θ=1,5°; 3 - θ=2°; 4 - θ=2,5°; 5 - θ=3°.

Fig. 2. Dependence of the tool position error on the spindle angle of rotation ψ.

To secure the continuity of the transfer function it is necessary to modify the tooth profile by diagram error value $\Delta\psi_3$ by communicating the tool an additional motion. The rule of additional motion of the tool was set with computer assistance. By settling the junction mechanism position function and the generating wheel 2 equation of motion, and by utilising the matrix instrument for the passage from the movable system of coordinates $OX_1Y_1Z_1$ to the fixed one *OXYZ,* the point *C* coordinates have been defined

$$X_C = R_C(1 - \cos\theta)\cos\psi \sin\psi$$
$$Y_C = R_C\left(\sin^2\psi + \cos\theta\cos^2\psi\right); \quad (5)$$
$$Z_C = R_C \sin\theta \sin\psi .$$

The equations (5) describe the parameter equations of the supporting surface of the mechanism of connection placed in the device housing whose form provides the continuity of the transmission ratio of the „*main shaft-semiproduct*" kinematical linkage.

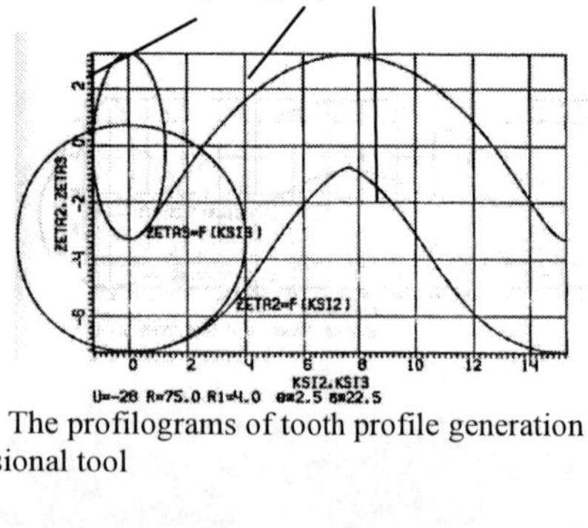

Fig. 3. The profilograms of tooth profile generation by precessional tool

Figures 3 and 4 shows the profilograms of tooth profile generation by precessional tool. On profilogram, curve 1 describes the tool centre motion trajectory in the shiftable system of coordinates *OXYZ* , and curve 2 describes the tool centre motion trajectory in the shiftable system of coordinates $OX_1Y_1Z_1$, curve 3–the precessional tool surface series (tooth profile).

2.3. Determination of the family wrapping of tool surfaces

Tooth profile of the processed wheel represents the family wrapping of the generating contour profiles of the tool. The wrapping is determined by the equations of the working surface of the generating tool and by the relative motion parameters while wrapping.

To make easier the determination of wrapping we pass to the tool centre co-ordinates in the movable system of co-ordinates, bound to the semiproduct

$$\overline{X} = X_D \cos\psi_3 + Y_D \sin\psi_3;$$
$$\overline{Y} = -X_D \sin\psi_3 + Y_D \cos\psi_3; \quad (6)$$
$$\overline{Z} = Z_D.$$

where $\overline{X}_D, \overline{Y}_D, \overline{Z}_D$ are the coordinates of the tool centre in the movable system of co-ordinates; $\psi_3 = \psi / i$ is the rotation angle of the semiproduct; *i* - is the transmission ratio of the cinematic chain "*principal shaft -semiproduct*". Then we determined the equation of wrapping on the sphere (fig.4).

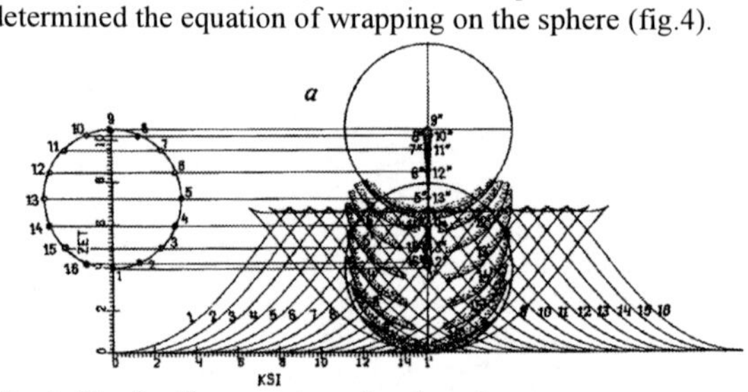

Fig. 4. The family wrapping of tool surfaces.

The analyses of profillograms demonstrate the lever and direction of influence upon the angle of tool insertion δ by the rotation axis of the semiproduct, tool radius *R* and transmission ratio *i* of the cinematic chain "principal shaft - semiproduct " on processed tooth profile. The basic manufacturing technology as such, as presented herein, cannot be properly protected by patents for various reasons, but specific implementations and applications of the respective technology are patentable, if they meet the basic patentability requirements, that is, if they are novel and inventive. Those who intend to use the present concept(s) will be able to enter into joint development projects with the Technical University of Moldova to develop specific applications. The necessary capital investments for such projects can be protected by filing patent applications together with the present authors.

3. REFERENCES

Bostan, I. (1992). *Precessional transmissions with multicouple gear,* Stiinta, ISBN 5-376-01005-8, Chişinău.

Bostan, I., Dulgheru V.& Grigoraş, S. (1997). *Planetary, precessional and harmonic transmissions,* Tehnica-Bucureşti – "Tehnica",Chişinău, ISBN: 973-31-1069-8.

Bostan, I., Dulgheru V. & Vaculenco M. (2001). *Precessional gear-engagement,* Patent nr. 1886MD.

Bostan, I. & Babaian, I. (1996). *Wheel tooth machining method of precessional gear.* Patent nr. 1663857RU.

Bostan, I. & Dulgheru, V. (1996). "*Method of bevel gear pumping and device for its realization*". Patent nr.1758941RU (patent 549 MD)

Annals of DAAAM for 2006 & Proceedings of the 17th International DAAAM Symposium, ISSN 1726-9679
B. Katalinic, Published by DAAAM International, Vienna, Austria 2006
150th Birthday of Nikola Tesla (1856-1943) Inventor of Mobile Robot Teleoperation

THE DEVELOPMENT AND THE IMPLEMENTATION OF THE INTEGRATED MANAGEMENT SYSTEM

BOTEZATU, C.P.; BOTEZATU, C. & CARUTASU, G.

Abstract: *In order to be run, the manufacturing systems need a management system made of subsystems. Nowadays, the Integrated Management Systems are still seen as instruments of the future, a "top managerial technology", although they are increasingly requested. To solve the issue of the Integrated Management System (IMS) design and use, seven institutions interested in this problem signed a Collaboration Agreement, which delegated a team of experts with significant results in this field of activity. This team proposes to create a theory on IMS, to design and produce the components forming the Integrated Management System, to set out a methodology through which such systems can be implemented and used by production and services enterprises, as well as public institutions.*

Key words: *management, manufacturing systems, quality, environment management.*

1. INTRODUCTION

The term of "Integrated System" is used only for the quality and environmental subsystems, but in managerial practice, five of them are used, four more are undergoing the standardization process and other four are mentioned in standards. The management of the manufacturing systems is based on the presence of experts, procedures and endowment. All these elements are part of the enterprise management system. The last years' experience (after 1990) has shown that in fact all organizations need an integrated system. It is should be underlined that the standards show only the way the systems are certified, but not how they are designed.

The system is made of subsystems. In the 1960', the following were considered subsystems: the *decisional subsystem*, the *IT subsystem*, the *financial subsystem* and the *human subsystem*. A new step in the organizational management appeared after 1985, when *quality assurance subsystems* were designed. These subsystems are made of the experts, procedures, equipments, and information used in order to guarantee that products and services quality is achieved. These systems, named nowadays "quality management systems" appeared as a consequence of deep changes in science and technology, initially for the aerospace, nuclear and military industry. After 1990, the quality management systems have been extended to the machine tools industry, electronic, computers, motor vehicles, and after 2000 they were also designed for services, public services included.

After 1990, *environmental management systems* have appeared and these systems allow the enterprise to adopt environmental protective measures. In the meantime, that there are some procedures, which are similar to those of the quality system. Now (2006), the term of IMS is used only with this meaning, defining in ISO standards (from 2000) the audit procedures, corrective and preventive procedures used for quality and environment system. The enterprises in all the European countries (after 1987), USA and Japan (after 1992) are obliged, because of the competition, to build quality and environmental management systems. For this, they aren't vast methodologies, for example the methodologies presented by PBSI, ANCIFAP, ASID, SNAM. The ISO 9000 (for quality) and ISO 14000 (for environment) standards refer only to the way in which the respective systems are certified and not to the way they are build. So far, there aren't published remarkable performances and there aren't significant results in the realization of the Integrated Management System.

In USA there was elaborated a norm for the management system of social responsibility, called SAI 8000 (Social Accountability International dating from 1987) which was analyzed and which has as result the ISO 26000 standard, which will appear at end of the 2006. Through this system it is verified that the free commitments made by the organization referring to the ethic behavior, in environmental area, are true.

This system must be certified and make an audit by a third party, allowing that the behaviors regarding the client, society, employers to be appreciated, assuming the elaboration of a policy in the field and the existence of some procedures of continuous improvement.

The management system of the safety and the employees health has been oriented upon the norm OHS 18000 (Occupational Health and Safety) elaborated by the British Institute of Standards BSI in 1999. ISO started to analyze the possibility of the processing of a standard in the field but some particular interests prevented its elaboration. Despite all this there are enterprises, for example in England and Germany, which elaborated their own procedures, but the system cannot be certified by a third party. On this purpose it must be elaborated a policy for health, an adequate structure and it must be defined the process, the resources and established the procedures.

2. ROMANIAN GROUND RESEARCH

In Romania, there are remarkable performances regarding the designing of the quality management system, numerous enterprises already having a certified system. In European Union, also the enterprises, including SMEs, have such systems. Regarding the other Romanian systems, there are some manufacturing systems which have environmental systems. Recently, food safety systems have been developed. Less than 20% of Romanian enterprises have a certified quality management system, and less than 5% have a certified environmental system. In Europe, many enterprises have environmental systems, food safety systems, but so far the health issue is concerned, only few German enterprises started to take an interest in it. Regarding the social responsibility, except for the USA, theoretical preoccupations of implementing such systems are present only in Italy. The issue, regarding the lack of integration on top management level, is viewed as a source of loss, in general picture of enterprise activity. Supporting the integration of subsystem presented before create the opportunity of an increased overall profit.

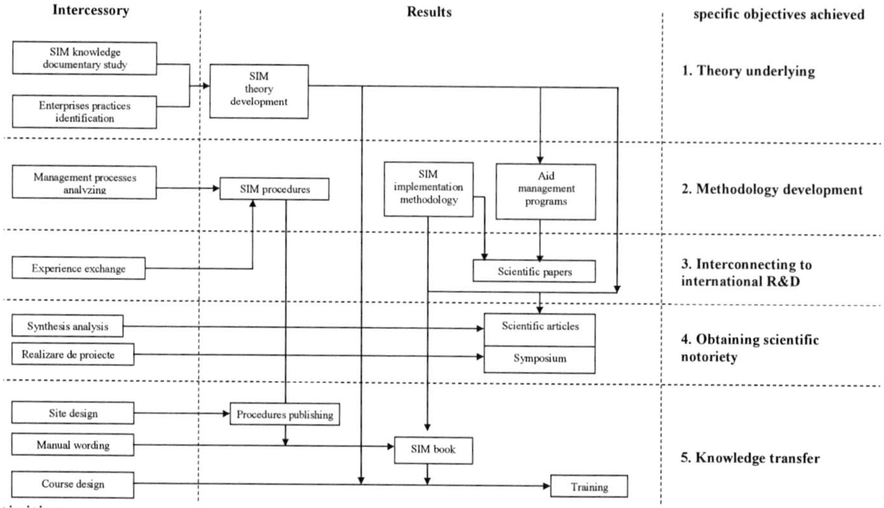

Fig. 1. Project activities

Scientifically speaking, in Romania, specific researches regarding the quality system have been developed only for a number of universities and research institutes.

As for researches on other types of subsystems, the number of published results is rather small; nevertheless, some papers exist. The experts involved in this project established a partnership made from organizations with remarkable performances in managerial subsystems' analysis and design, this also resulting from their list of published papers.

3. DI – IMS PROJECT

An analysis concerning the Romanian enterprises management shows a certain discrepancy between the methods used and the desired performances.

The project aims at generating knowledge that leads to an efficient management, thus allowing the enterprises a smooth integration on the European market. In this presented environment, „Development and the Implementation of the Integrated Management System (DI - IMS)" research project is presented.

The project aims at generating the appropriate knowledge necessary for manufacturing systems (other enterprises and public institutions also) to face the challenges of the European single market. The project activities are presented into figure 1.

The estimated results are: generating an "*integrated systems*" theory, elaborating a number of procedures conforming to the aim of the integrated system and not the subsystems (a global optimization), a methodology for implementation of the integrated systems, a technological transfer site and programs to assist the managerial decision by:

- elaborating a theory on the enterprises management systems, of international scientific value;
- development of operational methods, techniques, procedures that can be assimilated by all kind of organizations;
- transfer of knowledge resulted from the project to the scientific, economic and social environment;
- simulating the virtual enterprise environment, as a result of large collaboration of partners in EU market.

9. CONCLUSION

The present situation on national and international level raises a series of problems, which have to be resolved (the problem is a difference between the existent situation and the desired one). Such problems are, according to the previous analysis:

- the deficiencies in defining the particular organizational management system, while taking into account the enterprise's structure, relations and objectives,
- the need to approach the management system and its subsystems using the General Theory of Systems, measuring the impact of recent scientific and technological changes on the management of organizations,
- drawing the restrictions the management subsystems, defined through different standards, impose upon the integrated management system.

The IMS proposed is including certain social restriction within the enterprises' objectives and is defining a methodology which allows the implementation of efficient management systems in enterprises and institutions, underlining here the importance of all subsystem integration, as a premise for assuring the enterprise efficiency.

The limited awareness of the integrated management systems advantages by organizational decision – making bodies (especially for SMEs).

10. REFERENCES

Abromeit, M. & Sassano L (2006), Integrated Management System For Safety, Health And The Environment: A Continual Improvement Management System, *Available from:* http://www.iapa.on.ca/documents/IMS_Presentation.pdf *Accessed:* 2006-03-20

Botezatu, C.P & Carutasu G. (2006) New actors in business environment, Proceedings of the 2006 International Conference on Commerce , 4pp.583-587 ISBN – 10 973-594-785-4, ISBN – 13 978-973-594-785-, Academy of Economic Studies Bucharest, May 2006, ASE, Bucharest

Breen, K., Pecora, G., Gregerson, D. & McCabe, T (1997), A new direction: Integrated Management Systems, *Available from:* http://www.qualitydigest.com/aug97/html/ims.html *Accessed:* 2006-04-07

Guran, A., Patrascu, G.& Carutasu G (2004), Simulation as a valid tool in project planning, *The 15th INTERNATIONAL DAAAM SYMPOSIUM, "Intelligent Manufacturing & Automation: Focus on Reconstruction and Development"*, 3-6th October 2004, pag. 145-147, ISBN 3-901509-42-9, ISSN 1726-9679.

Ross & Associates Environmental Consulting, Ltd. (2000), Working Guidelines for Integrating Facility Identification Information *Available from:* http://www.epa.gov /oeiinter/pdf/fits2fnl2.pdf *Accessed:* 2005-11-12

Annals of DAAAM for 2006 & Proceedings of the 17th International DAAAM Symposium, ISSN 1726-9679
Editor B. Katalinic, Published by DAAAM International, Vienna, Austria 2006
150th Birthday of Nikola Tesla (1856-1943) Inventor of Mobile Robot Teleoperation

A METHOD FOR AUTOMATED OBJECT CHOICE WITHIN A SET OF AVAILABLE TECHNICAL PRODUCTS USING NEURAL NETS

BOYADJIEV, I.; TOPALOVA, I.; MALAKOV, I. & BACHVAROV, A.

Abstract: *The paper suggests an adaptive method for automated choice of one object from a multitude of technical products on the strength of defined parameters. The method is based on the usage of neural nets. Presented are the outcomes of the method's application for object classifying within a set of pre-existing electrically driven manipulation systems variants.*
Key words: *design, technical products, automated choice, neural nets*

1. INTRODUCTION

Familiar methods for technical product design can be commonly classified as variant referring, adaptive and generative (Spur&Krause, 94). In practice, however, to make a precise distinction between those methods is a tough task since the design of a new item can be accomplished by way of applying already existing solutions. The ever harsh competition between leading manufacturers arousing from the world economies globalization stresses the importance of the first two method groups due to their capability to decrease significantly the design terms. In the case of variation and adaptive methods the design process is reduced to making corrections in pre-existing variants.

One of the main tasks accomplished per such methods is to find the appropriate object among an existing multitude of possible engineering solutions (database) on the strength of parameters defined beforehand. The latter constitute input vectors, which are passed toward the system in order to solve the formulated problem. Normally, it is done manually, by way of using a code classifier. Whenever the amount of familiar solutions grows to huge numbers (frequently this number reaches well beyond 10^6) the process becomes fairly laborious. Often an object with identical number cannot be found. In those cases an item featuring minimum deviation from the basic parametrical values has to be defined. Yet, due to the overlapping of classes, applying traditional methods leads to mistakes based on subjectivity. In addition, modern approaches to WEB based design require an automated setting of parameters of the product and attributing the latter to a certain class.

Subject of the present paper is the creation of a method for automated object choice within a set of available technical products using neural nets.

2. METHOD FOR AUTOMATED CHOICE OF TECHNICAL PRODUCTS

Taking into account that the input vectors describe overlapping classes in the parametrical feature space we choose to apply an adaptive classification method using neural networks.

The usage of neural nets in this case have gained advantage over the traditional methods for recognition and classification, because of their capability to set precisely the boundaries between the classes.

The suggested method comprises following stages:

Stage 1: Defining the technical products classes per providing basic parameters. Coding of the classes per choosing the right code.

Stage 2: Generating a class related database.

Stage 3: Training the neural network with input vectors of the defined classes.

Stage 4: Defining the object's input vector.

Stage 5: Automated choice of classes.

Stage 6: Outcome evaluation.

3. METHOD APPLICATION

The method has been used for selecting an electrically driven manipulation system (Boyadjiev, I.& Malakov, I., 2003).

3.1 Stage 1&2 of the proposed method

The classification system is designed to classify the given customer's parameters for a certain technical product. The classes of objects are determined in a preliminary given set of technical products. Each object is represented by a combination of parameters describing the functioning of the technical product. The parameters are fixed in accordance with the data given in Fig.1 – Shape of the operating area, Load carrying capacity, Speed, Repeatability. Two other parameters – Console/Portal type of the construction and the axis Stroke Length are added to the parameter space. Each parameter combination forms an input vector to the classification system. Taking into account that the input vectors describe overlapping classes in the parametrical feature space an adaptive classification method using neural networks is applied.

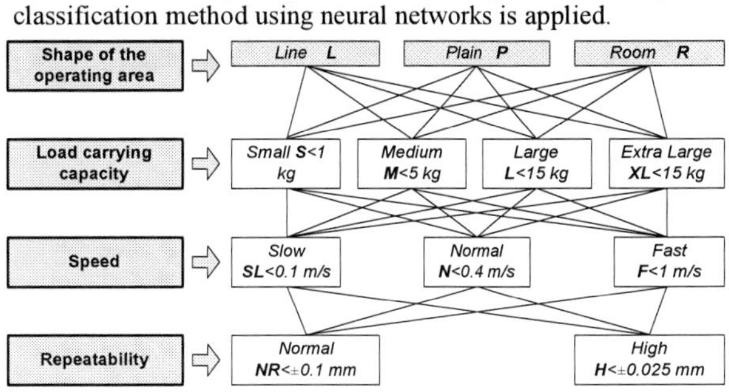

Fig. 1. Definition of the technical parameters for an electrically driven manipulation system.

3.2 Stage 3 of the proposed method

A neural network (NN) of Backpropagation (BPG) type is chosen to be trained with some of the input vectors. We use 35 combinations (classes) of the full network graph on Fig.1. which must be correctly classified.

These 35 input vectors form a subset of NN's input training data (Table 1). Table 1 represents only 10 of the possible input vectors The eight input parameters are attached to the neurons in the input NN layer (Fig.2). We choose a hidden layer with

Shape of the operating area	Load carrying capacity	Speed	Repeatability	Console/Portal	Stroke length-axis S1	Stroke length-axis S2	Stroke length-axis S3	Variant
P	0,98	0,05	-0,08	K	160	320	0	class202A
P	0,9	0,07	-0,09	K	250	1500	0	class202B
P	0,8	0,08	0,08	K	320	1000	0	class202C
P	0,7	0,09	0,09	K	400	400	0	class202D
P	0,6	0,1	0,078	K	160	800	0	class202E
R	4,99	0,38	-0,02	P	160	1000	1000	class335A
R	4,97	0,39	-0,02	P	320	400	400	class335B
R	4,8	0,25	0,024	P	250	500	500	class335C
R	4,6	0,2	0,025	P	400	630	160	class335D
R	4,5	0,37	-0,02	P	160	800	1500	class335E

Table 1. NN's input training data.

doubled number (16) of neurons by comparison with the number in the input layer. That would be a good choice (Kountchev, R.& Topalova, I.,1998) considering the high degree of class overlapping. The output layer contains 35 neurons each representing the corresponding class. The BPG NN training mode is chosen to be "Random Pattern Presentation " and Tangens Hyperbolic is applied as Transfer function for each NN layer.

3.3 Stage 4&5 of the proposed method

50 input vectors representing variations in parameters for the defined classes of objects were tested in query (recognition) mode of NN. These exemplars were not included in the training set aiming to generalize the method.

4. EVALUATION OF THE EXPERIMENTAL RESULTS

Neural network	NN
Number of Layers	3
Neuron count on layer 1	8
Neuron count on layer 2	16
Neuron count on layer 3	35
Transfer Function	Tangens Hyperbolic
Training targets (Classes)	35
Training Error	0,005
Training mode	Pattern random
Learning rate	0,2
Momentum rate	0,8
Epoch count	31646
Training time	00:00:37
% Correctly identified query patterns	96 %

Table 2. NN's parameters and results.

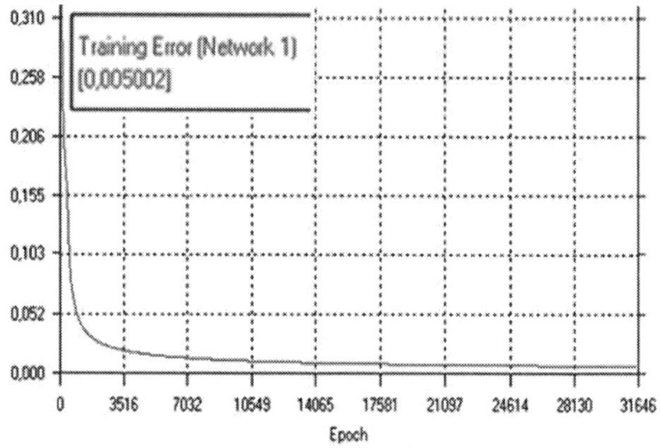

Fig. 3. MSE-Mean Square Error for the trained NN.

The NN is trained with the data in Table 1. with a training error (MSE-Mean Square Error) of 0.005 (Fig. 3.), by learning rate of 0.2, epoch count of 31646 and 37 seconds training time (Table 2.). In query mode 50 exemplars (not included in the training set) of products with different combinations between input parameters were classified with a very good classification rate of 96% (Table 2.). The next step in improving the accuracy of the classification is training the NN with a smaller MSE or developing a more complex structure (Topalova,I., 2005) of the NN classification system applying hierarchical relations between different NNs.

5. CONCLUSIONS

The paper implies an adaptive method for automated object choice from an available multitude of technical products in accordance with pre-defined features. The method refers to the application of neural nets (Skapura,D, 1996). This provides for the classification to be performed under conditions of overlapping classes in the parametrical feature space since neural nets own the capability to set precisely boundaries between the classes. Applying the proposed method will simplify and accelerate the process of solution acquiring, will help eliminating the bias factors and will create preconditions for applying modern approaches to the WEB based design. The method has been tested by way of solving a concrete practical task.

6. REFERENCES

Spur, G., F.,L. Krause (1994), *CAD-Technik*, Carl Hanser Verlag, Wien.

Boyadjiev, I., I. Malakov, A. Batchvarov (2003). Computer-aided selection of an optimal structure of mechatronic drive system. *Annals of DAAAM for 2003 & Proceedings of the 14th International DAAAM Symposium,* Sarajevo, Bosnia and Herzegovina, October,2003, pp 059÷060, ISBN3-901509-34-8 .

Kountchev,R.&Topalova,I.(1998) Invariant recognition system for 2D objects based on three-stage BPNN structure, *Electronics Letters*, 14th May 1998, Vol.34, No.10, pp.974-976, ISSN 0013-5194.

Topalova,I.(2005) Modular Adaptive System Based on a Multi-Stage Neural Structure for recognition of 2D Objects of Discontinuous Production, *International Journal of Advanced Robotic Systems,* Vol.2, Num.1 March, 2005, pp. 045-051, ISSN 1729-8806.

Skapura,D.,(1996) Building Neural Networks, *ACM Press*, N.Y.

Annals of DAAAM for 2006 & Proceedings of the 17th International DAAAM Symposium, ISSN 1726-9679
Editor B. Katalinic, Published by DAAAM International, Vienna, Austria 2006
150th Birthday of Nikola Tesla (1856-1943) Inventor of Mobile Robot Teleoperation

SOFTWARE TOOL FOR WEB-BASED KNOWLEDGE STRUCTURING ON THE EXAMPLE OF MODULAR PRODUCTS CONCEPTUAL DESIGN

BOYADJIEV, I.; TUDJAROV, B. & BACHVAROV, A.

Abstract: *The ability to create, process, support and exchange knowledge assets is the essence of the firms in the new economy. This paper is focused on the firm's knowledge base and especially on the possibility to capture value from their customers whose unique skills, gifts and needs remain still underused. On the example of modular products conceptual design, a systematic approach to elicit customer needs and get customers involved in the design process is combined with the advantages of Extensible Markup Language (XML). A framework of XML-based Computer Added Technologies (CAx) knowledge structuring environment is proposed and a part of created XML description of customers' requests is given.*

Key words: *Knowledge management, Design by Customers (DBC), Extensible Markup Language (XML), Computer Added Technologies (CAx), Modeling*

1. INTRODUCTION

The ability to create, transfer, assemble, integrate, protect and exploit knowledge assets is a main factor for the existence and successful work of the industrial companies. The knowledge base of the firm includes its technological competences, its knowledge of customer needs and supplier capabilities (Teece, 2002).

The approach for the direct customers' participation in the design process by expressing their preferences allows customers to be actively involved in the product definition process rather than passively receive the end product designed by the producer. This approach was applied to support and maintain development of a "virtual family" of modular positioning and handling systems called DriveSets (Boyadjiev, 2005).

Main question here is which kind of technology to choose for the capturing of valuable information from the customers and how to organize knowledge structuring, storing and usage?

In this case, the adoption of new techniques and methods for model representation is a very important issue. In order to resolve the described above problems, we focus on XML document (W3C Recommendation, 2000). XML technology has a lot of possibility for the active document (Igoshi, 1998), such as both: human and machine readable, multi-lingual usage, programming language, multimedia representation and etc. XML allows creation of new languages depending on the needs of designed system, assures extensibility of the models and also we need XML to interoperate with the Web. For our research we adopted to use different XML documents, depending on their functions. We need to incorporate new graphics technologies in a standardized way, so we have to use standardized XML, but for some internal and or new descriptions we have to develop original XML documents. For example Extensible 3D (X3D) language (the successor of Virtual Reality Modeling Language - VRML) is suitable graphical technology (Web3D Consortium, 2006), but for description of customers' requests it is adopted to create original XML document. Real-time 3D communication language X3D enables 3D simulations and web applications to implement features traditionally found only in high-level gaming systems.

A framework of XML-based knowledge structuring environment is proposed and discussed in this paper and an XML description of customers' requests is represented.

2. FRAMEWORK OF XML-BASED KNOWLEDGE STRUCTURING ENVIRONMENT

As outlined above firm's knowledge base can be decomposed on:

- Firms Competences Module (FCM) – technological and other firms' competences module;
- Customer Related Module (CRM) – customer's needs, requests and advices;
- Supplier Related Module (SRM) – supplier capabilities and other information.

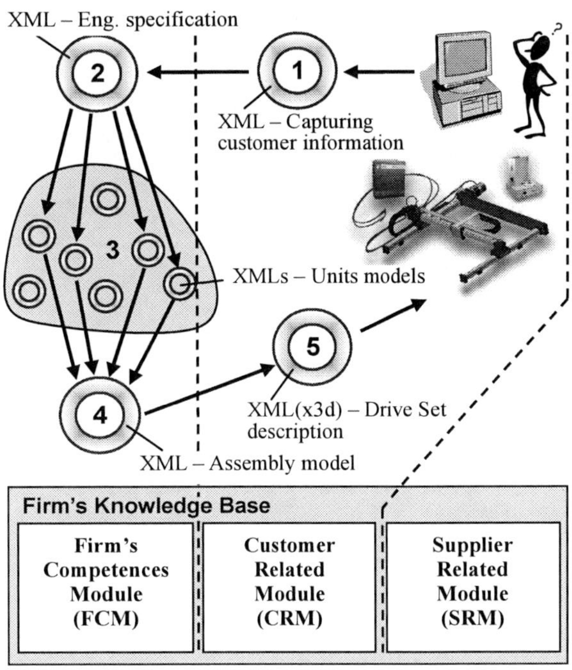

Fig. 1. A framework of XML-based knowledge structuring environment

The proposed framework of XML-based knowledge environment is shown on Fig.1. XML documents used here are marked by ◎ and numbered.

Within **first stage** the Voice of Customers is captured by the web-based assistant and XML1 saves the obtained data. XML1 is processed, analyzed, translated and structured in engineering specification (XML2) - **second stage**. On the base of the obtained results (XML2), predetermined product units (XML3) are chosen, the values of each of their characteristics are identified and assembly model is generated (XML4) - **third stage**. Within third stage a selection of the optimal positioning system for specific requirements and constraints is made. Simulation tests are run to determine whether the intended purposes are satisfied by the selected system configuration. After validation (within **fourth stage**) a graphic model (XML5-X3D) describing the proposed structure is built and proposed system configuration is saved in a database. The technical documentation and HTML-code for web-representation of the new positioning system are generated at the end of the working cycle.

3. EXPERIMENTAL XML DESCRIPTION OF CUSTOMER'S REQUESTS

Customer s' needs identification and structuring for Design by Customer in Product Design stage allows to alter and improve products interactively and so to participate in the product definition process. Customers are presented a wizard, in which a set of product attributes and their possible values are presented for selection and modification (Boyadjiev, 2005). The frame of the parametric range was built by systematic variation and full combination of 6 technical and structure system parameters: load carrying capacity, maximum speed, repeatability, operating area, design type, stroke.

```
<?xml version="1.0" standalone="yes" ?>
<!--   Customers' needs identification file    -->
<CUSTOMER>
<!--   Customers' name, address and other provided information   -->
  <LogInfo>   ...   </LogInfo>
<!--   Short description of the customers' request   -->
  <RequestInfo id="">   ...   </RequestInfo>
<!--   Full description of the customers' request   -->
  <DriveSetInfo>
<!--   Characteristics and possible values and limits   -->
<!--   Operating properties   -->
    <LoadCarryingCapacity>...</LoadCarryingCapacity>
    <Speed>...</Speed>
    <Repeatability>...</Repeatability>
<!--   Structure   properties-->
    <OperatingArea>...</OperatingArea>
    <DesignType>
        <Single>0</Single>
        <Cantilever>0</Cantilever>
        <Gentry>0</Gentry>
    </DesignType>
    <Stroke>...</Stroke>
  </DriveSetInfo>

<!--   Characteristics and chosen and inputted values by the customer   -->
  <CustomerInput>
    <OperatingAreaInput>...</OperatingAreaInput>
    <DesignTypeInput>...</DesignTypeInput>
    <AxesArangementInput>...</AxesArangementInput>
    <LoadPositionInput>...</LoadPositionInput>
    <SingleAxisProperties>
        <OperatingProperties>
            <LoadCapacity>...</LoadCapacity>
            <Speed>...</Speed>
            <Repeatability>...</Repeatability>
            <Stroke>...</Stroke>
            ...
        </OperatingProperties>
  </SingleAxisProperties>
  </CustomerInput>
  </CUSTOMER>
```

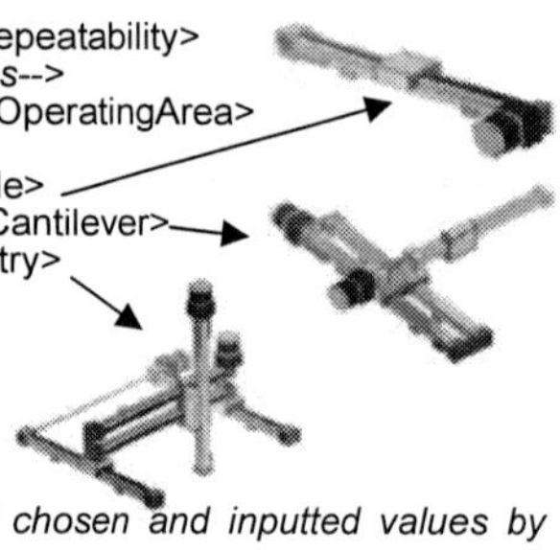

Fig. 2. Customer's needs identification - XML document

On Fig.2 is given a part of developed XML document for the customer's needs identification process (XML1 on Fig.1). Developed XML document is appropriate for the purpose of creation of Web-based knowledge structuring environment, because of its XML syntax, support of structured information and possibilities for unique identification of the single representatives of the DriveSet-family.

4. CONCLUSIONS

Involvement of customers in the process of product design and the development of a web-based knowledge structuring environment represent a perspective solution for increase the efficiency of firms in today's competitive global economy. Here we conclude by pointing following expected results and further research tasks related to proposed Web-based knowledge structuring environment:

- it saves time and costs, reduces the effort in respect to the solving of the design problem and improves the quality of products and their market realization by the involvement of the customers in design process;
- our proposal assures the overcoming of the negative effects of the communicational fragmentation: it is based on XML sets of applications;
- consumers' knowledge and direct participation in the design process facilitates marketing research and brings different customers' attitudes towards buying decision making;
- as a consequence, the traditional cycle of "design – make – sell" is replaced by the new one: "design – sell – make";
- X3D language presentation of results give possibility for attractive, real-time, interactive, high quality graphics, can include audio and video as well as 3D data and allows components to be added to extend its functionality.
- by the use of the idea about rules based protocols for functions recognition (Tudjarov & Igoshi, 2000) it is expected modules for technological knowledge capturing to be developed;
- by attachment of functional information it is expected a X3D based model of the mechanical assembly to be developed for the needs of the further VR immersive representation;

5. REFERENCES

Boyadjiev, I., Bachvarov,A. & Bojkov, D. (2005) *Connecting Customer to the Design - Approach on the Example of the Development of Modular Handling Systems*, 50. Internationales Wissenschaftliches Kolloquium, Technische Univ, Ilmenau, September 19-23, Germany.

Igoshi, M. & Tateno, T. (1998) *Manufacturing System Description by Active Document*, Proceedings of 14th International Conference on Computer Aided Production Engineering, Japan.

Teece, D. (2002), *Managing Intellectual Capital (Organizational, Strategic, and Policy Dimensions)*, Oxford University Press, New York, USA.

Tudjarov B. & Igoshi M. (2000), *A remote value analysis of products by using XML*, 2000 Japan-USA Symposium on Flexible Automation, July 23-26, Michigan, USA.

W3C Recommendation (2000), *Extensible Markup Language (XML) 1.0 (Second Edition)*, W3C Recommendation from 6 October 2000, http://www.w3.org/TR/REC-xml.

Web3D Consortium (2006), *Extensible 3D (X3D)*, ISO-IEC-19774 … 19777, http://www.web3d.org

Annals of DAAAM for 2006 & Proceedings of the 17th International DAAAM Symposium, ISSN 1726-9679
Editor B. Katalinic, Published by DAAAM International, Vienna, Austria 2006
150th Birthday of Nikola Tesla (1856-1943) Inventor of Mobile Robot Teleoperation

OPTIMIZING SMALL BATCH PRODUCTION BY COMBINING VARIOUS RAPID MANUFACTURING PROCEDURE

BRAJLIH, T.; DRSTVENSEK, I.; VALENTAN, B. & BALIC, J.

Abstract: *With increased competition from the global economy, manufacturers face the challenge of delivering new customized products more quickly than before to meet customer demands. During the new product development, the production of several prototypes and test products is often necessary. Various technologies collectively known as Rapid Manufacturing (RM) have been developed, to shorten the design and production cycle, and promise to revolutionize many traditional manufacturing procedures. The following paper presents the possibilities of combining different RM technologies in the production of small batch series.*
Key words: *Rapid manufacturing, Rapid prototyping, Rapid tooling, Silicon rubber moulding.*

1. INTRODUCTION

In production of series of prototypes or test parts, the shortest possible manufacturing time is often the most important aspect to be considered. However, the manufacturing cost can not be completely neglected. With a wide array of various rapid manufacturing technologies available, choosing the optimal one can be somewhat difficult. Often the implementation of a single RM technology in a small batch series production does not yield satisfactory results and the optimal time/cost relation can only be achieved by combining several different technologies (Rapiman, 2006). The following paper presents the advantages of combining PolyJet™ Rapid Prototyping (RP) technology and Silicone Rubber Moulding (SRM) Rapid Tooling (RT) technology in a production of a series of forty antenna housing prototypes.

2. THE ANTENNA HOUSING

The antenna housing consisted of two different parts, the top cover and the backside connector plate (Figure 1). The backside was separated from the rest of the housing, because various different layouts of the connectors were to be tested with the prototype series. The whole series consisted of forty housings and the customer insisted on a shortest manufacturing time possible.

Fig.1. Two components of the housing

3. RAPID MANUFACTURING TECHNOLOGIES

PolyJet™ is a three-dimensional printing technology. Building is done by layers of 16μm. The printing head jets the liquid photopolymer on the work-tray. Solid object is made by polymerization under the influence of the UV light (Figure2).

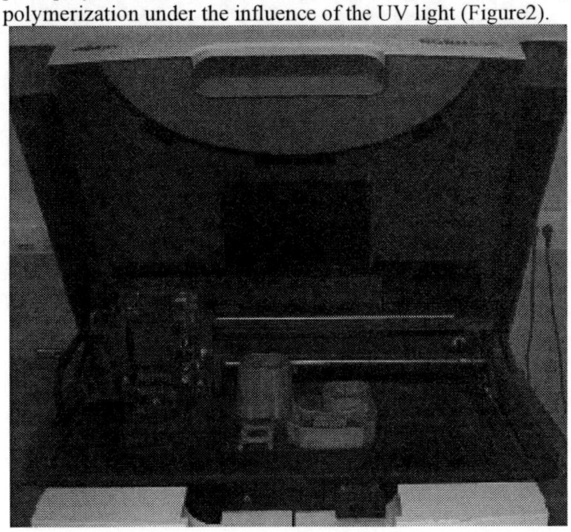

Fig.2. EDEN330™ three-dimensional printer is building objects by PolyJet™ RP technology

Silicone rubber moulding is a Rapid Tooling technology. The pattern (in our case made by PolyJet™) is poured over by a silicone rubber. When the rubber vulcanizes the mould is cut along the parting plane. This mould is then used for vacuum casting of two component resins (MCP, 2006).

4. COMBINING TECHNOLOGIES

Regardless which of both technologies would be chosen for the series manufacturing, at least a single prototype (of both components) had to be build by PolyJet™ RP procedure. If SRM was to be used later, the prototypes already built were to be used as silicone rubber mould patterns. The costs of producing the patters for SRM were calculated in the cost of a mould production. One of the main advantages of using rapid manufacturing technologies is a fact, that the geometrical complexity does not significantly influence the time and cost of manufacturing of a certain part. The most important factors regarding the cost and time are the parts volume and outer dimensions (Drstvensek 2004). As the Figure 2 shows the PolyJet™ rapid prototyping technology is totally inappropriate for manufacturing the cover prototype series in comparison to the SRM procedure. This is largely due to the parts large volume and height (Z axis dimension in the EDEN 330™ RP machine).

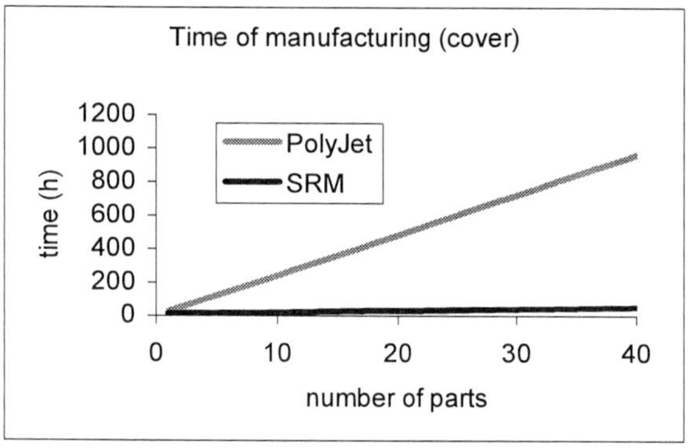

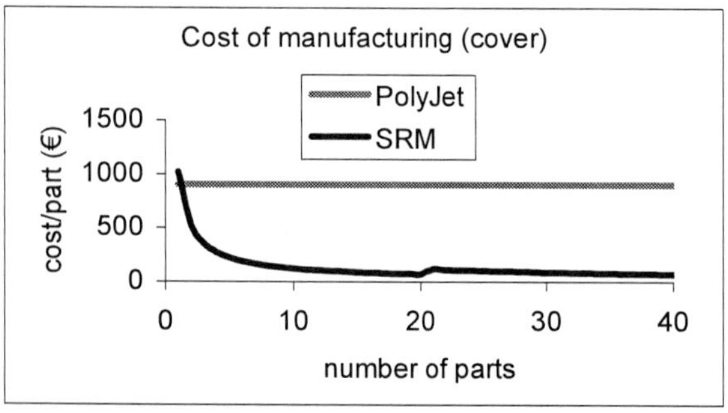

Fig.2. Time and costs of the cover prototype production (SRM and PolyJet™)

However, when considering the manufacturing of the back plate, the PolyJet™ technology yields better result in terms of time and costs as the SRM procedure (Figure 3).

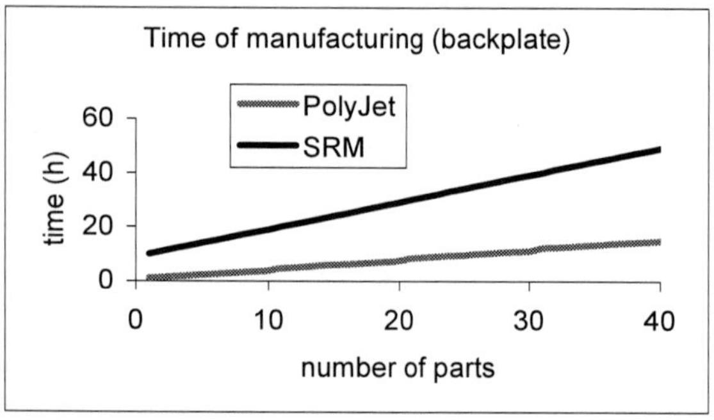

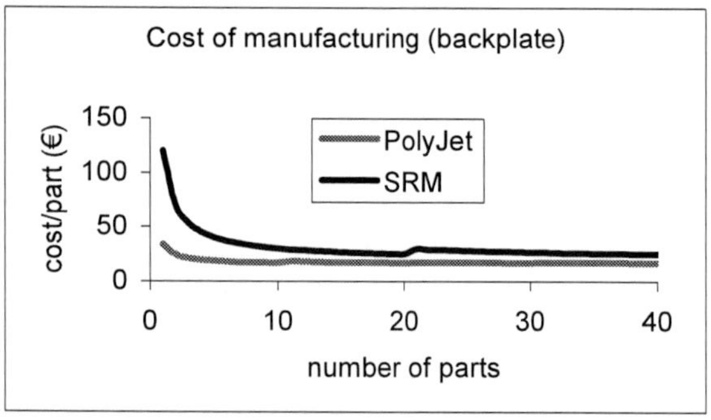

Fig.3. Time and costs of the back plate prototype production (SRM and PolyJet™)

Therefore it was decided; that the whole series of covers would be produced by SRM and the back connector plates by PolyJet™ (Figure 4). This decision had a very interesting side effect. By deciding that the whole series of back plates were to be produced by RP technology, the number of different connector layouts no longer influenced the time and cost of the production. In theory forty back plates, each with a different connector layout could be made in the same time (and at the same costs) as forty identical ones.

This shows an enormous potential of rapid manufacturing in the future, when the individualization of products will play an ever increasing role.

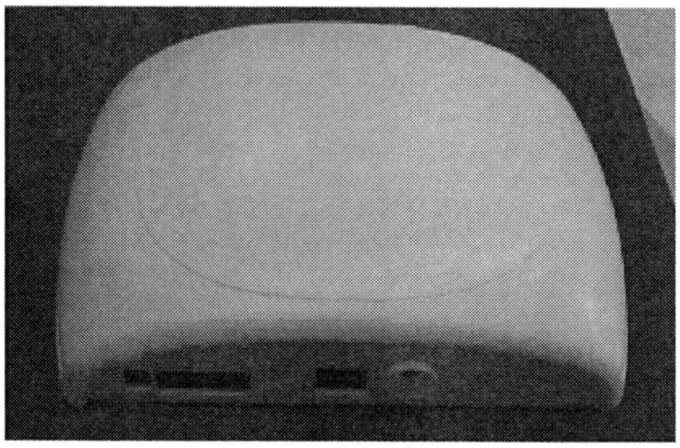

Fig.4. Finished prototype made by combination of SRM (cover) and PolyJet™ (back plate) rapid manufacturing technologies

5. CONCLUSION

This project has shown some of the beneficial effects of combining different rapid manufacturing technologies in production of a small batch series. By using two different technologies the manufacturing time was shortened by 30% comparing to the time of producing everything just with SRM technology. Also, the costs were reduced by 10% (Figure 5).

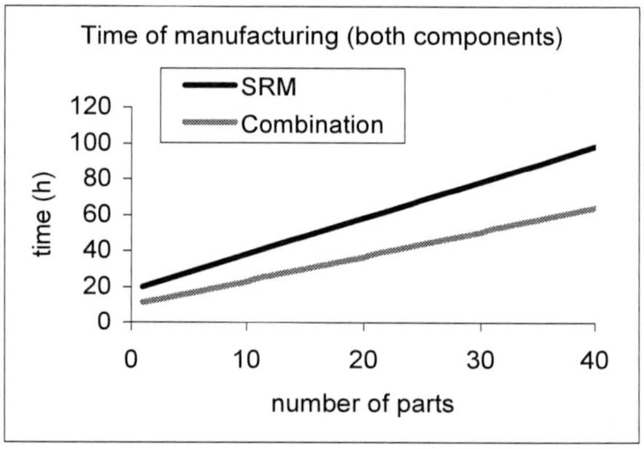

Fig.5. Time of manufacturing of the whole series

The most important aspect was the possibility of manufacturing forty different back plates in the same time/cost frame as forty identical ones. Practically, only four different connector layouts were made, but manufacturing them by PolyJet™ took 60% less time at 50% lesser cost as making them by SRM. This presents the potential of using rapid manufacturing technologies in production of individualized end-products in the future and the way this will revolutionize the traditional manufacturing procedures.

6. REFERENCES

Drstvensek, I. (2004), *Layered Technologies*, Fakulteta za strojništvo, Maribor

MCP Rapid tooling homepage *available at:* http://www.mcp-group.de/, *Accessed*: 2006 5 20

Object geometries homepage *available at:* http://www.2objet.com/home.asp, *Accessed*: 2005 6 22

Rapiman rapid manufacturing homepage, *available at:* http://www.rapiman.mojservis.net/ , *Accessed*: 2006 7 9

Annals of DAAAM for 2006 & Proceedings of the 17th International DAAAM Symposium, ISSN 1726-9679
B. Katalinic, Published by DAAAM International, Vienna, Austria 2006
150th Birthday of Nikola Tesla (1856-1943) Inventor of Mobile Robot Teleoperation

HOLISTIC PLANNING OF MANUFACTURING PROCESS CHAINS BY DESIGNING TECHNOLOGICAL INTERFACES

BRANDES, A. & RABINOVITCH, A.

Abstract: *This paper introduces a methodology for planning manufacturing process chains. The approach is based on modeling and optimisation of technological interfaces between consecutive process steps. The aim is a holistic dimensioning of process chains regarding process spanning dependencies. As reference for validation a manufacturing process chain for gear wheels has been chosen.*
Key words: *Technological Interfaces, Holistic Planning*

1. INTRODUCTION

Usually mechanical components, e.g. gear wheels in the automotive industry, are manufactured in sequential steps. Steel blanks are forged, tempered and hard turned. State-of-the-art in industrial manufacturing is a selective process configuration and optimisation with no regards to holistic appraisal of process chain. Merely the logistic efficiency of workflow is considered processes-spanning. Results are manufacturing planning systems, which configure process chains due to logistic values as adherence to delivery dates, order time and capacity. Room for improvement remains idle, even though it is essential to optimise process chains performance and efficiency continuously in order to assure competitiveness. Thereby a necessity of holistic optimisation of process chains and consideration of process spanning dependencies is recognised in industry and science (Warnecke et al. 1999, Monostori et al. 2001).
In the manufacturing industry the field of production planning stretches across set–up of process chains. In this context, product planning traditionally connects design and manufacturing with technical and logistic information flow (Eversheim, 2002). However currently employed product planning systems are not proper for integrated planning of process chains. Static process models are used, although dynamic process models were required for consideration of process spanning effects, e.g. the impact of dimension variation in raw parts for consecutive machining steps. Defining ranges and limit values is not only sufficient, but additionally, algebraic coherence of input variables, process and output values have to be represented.
The German Collaborative Research Centre 489 (SFB 489), "Process chain for the production of high performance components based on precision forging technology" develops new manufacturing technologies, planning and control methods for holistic optimisation of process chains.
This paper describes a methodology for holistic configuration of process chains by means of a currently developed approach - the so called "designing technological interfaces" (DTI) method.

2. DESIGNING TECHNOLOGICAL INTERFACES

2.1 Procedure of the DTI-method

Designing technological interfaces is a new proceeding for holistic planning, which bases on generic process models, discrete simulation, and multi-criteria optimisation. Generally, the DTI-method bases on a two-stage procedure, the so called qualitative and the quantitative positioning of technological interfaces (Denkena et al. 2003). Qualitative positioning describes the identification of technological interfaces e.g. process elements have to be identified, assessed and chosen for the process chain. The selection of process steps occurs by means of procedures of economic decision making theory considering economical, technological and ecological criteria. Decision maker's preferences are illustrated with fuzzy-sets. Afterwards technological interactions between these processes have to be analysed. If there are conflicting technological needs, an optimisation problem occurs at these interfaces. The quantitative positioning comprises the determination of the most suitable value of technological transfer parameters between single elements of the process chain (Denkena et al. 2005a, b). As a reference for validation of the DTI-method, the process chain for producing automotive gear wheels by means of precision forging technology has been chosen. This process chain begins with separation of blanks, followed by heating, precision forging and hardening. Subsequently the work piece is fine positioned, hard turned and finally continuous grinded.
The quantitative positioning or optimisation stage is divided in four consecutive phases: (1) process analysis, (2) interface modelling, (3) optimisation and (4) implementation (Figure 1).

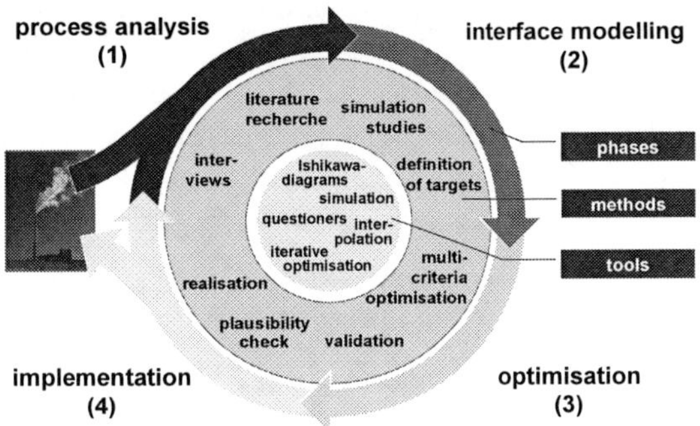

Fig. 1. Phase model of the DTI-method

During the first phase an analysis of the manufacturing processes is accomplished, e. g. information about work piece characteristics, machine tools and machining processes are exploited. Technological and economical dependencies between different manufacturing steps are examined and specified algorithmically.
In the second phase heuristic data (expert knowledge), empirical data (experimentally ascertained cognition), and data models (machine data sheets) are utilised simultaneously. Resulting knowledge is formalised. The implemented optimisation system consists of a process database (MS Excel), a discrete simulation environment (eM-Plant) as well as software for mathematical evaluation of experimental data

(MatLab). The simulation environment reads data of the process models from the database using an OBDC interface. Simulation results are written back to this database. Subsequently, MatLab starts the optimisation and evaluation process. Input data are technological and economical values as well as algorithms which describe the technological interdependencies.
Within mathematical multi-objective optimisation, preference functions of manufacturing targets are prepared and weighted against each other. The outcome is a suitable vector for dependent values between particular processes. In case these values will be chosen, the process chain is designed optimally regarding preference functions and optimisation targets.

2.2 Preliminary implementation
The aim of implementation of the DTI-Method was the evaluation of the developed approach for holistic planning and designing technological interfaces. After implementing the reference process chain in simulation environment (eM-Plant), first simulation studies have been accomplished to test its general functions. Then a debugging process occurs. Manufacturing costs depending on variable lot sizes, collected in a study at different companies, are chosen as validation parameter. After realistic simulation results have been achieved, technological interdependencies are identified. Based on analyses within the first phase, following process spanning correlations are expected:

- Energy consumption and manufacturing costs are addicted to forging temperature (relevant for heating, precision forging and hardening)
- Distortion after heat treatment depends on carbon content (relevant for material costs and treatment effort) and
- Manufacturing costs are addicted to work piece tolerances (relevant for precision forging, hard turning and continuous grinding)

Exemplary, energy costs for heating, hardening and precision forging are depicted in Figure 2.

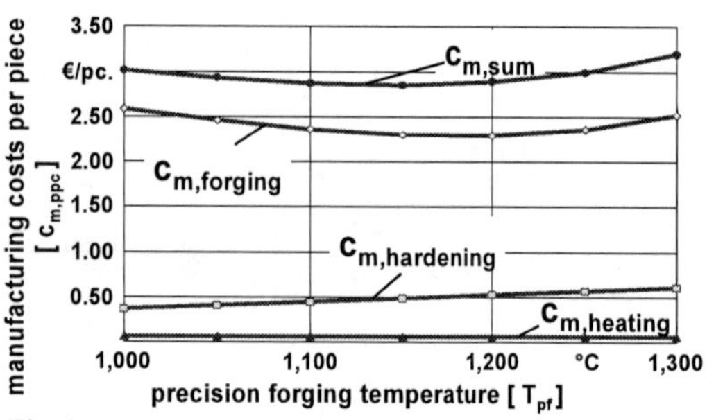

Fig. 2. Energy costs according to forging temperature

Inside an induction oven the work piece is heated up to a temperature of 1,300°C. Energy consumption and thus costs increase with temperature T_{pf} required for forging. The minimum forging temperature is $T_{pf,\,min}$ =1,015°C. Regarding the loss of temperature within this process chain, $T_{pf,\,min}$ can be rated this way: During forging process temperature decreases by 200°C and during transportation by another 15°C. The following hardening process starts with 800°C.
Consideration of manufacturing costs, temperature and energy consumption of the three processes is a typical task for holistic planning. Estimation of forging temperature merely on energy consumption is not sufficient. Manufacturing costs per piece for heating and hardening increase through higher energy consumption with the forging temperature. The higher the forging temperature, the higher the input temperature of the heat treatment process is. With this effect, process chain consumes energy for cooling work pieces down to 800°C. In addition process time increases if work pieces arrive much warmer then 800°C. In contrast manufacturing costs at forging have a minimum at T_{pf} =1,200°C. Mechanical abrasion increases with lower temperature and thermal abrasion with higher temperature. Consequently, forging costs rise because of a decreasing die life time. Aggregate manufacturing costs for these three processes have their minimum at T_{pf} =1,175. Regarding prior and consecutive processes in the whole process chain, the final multi criteria optimisation has delivered an optimal forging temperature of T_{pf} =1,088°C. Considering the process chain not only interdependencies between temperature and manufacturing costs are regarded, but also other correlations e.g. work piece allowance or carbon content are in the focus of studies.

3. RESULTS OF IMPLEMENTATION OF DTI-METHOD

The DTI-Method identifies that technological correlations and dependencies between single manufacturing steps have to be taken into account. Based on a generic approach, single processes were implemented in a simulation environment and analysed. After optimisation manufacturing costs per piece can be compared. In terms of the reference process chain for manufacturing of precision forged gear wheels a potential for rationalisation of about 10% of the manufacturing costs occurs at a lot size of 3,000, as simulation and implementation studies showed.

4. CONCLUSION

These investigations have shown that the optimisation of technological process chains offers substantial potential for rationalisation. The proposed DTI-method allows to set-up manufacturing process chains regarding process spanning dependencies. The described interfaces models enabled holistic design and optimisation of process chains. The perspective is not only to design process chains with consecutive process steps but also analyse dependencies between parallel manufacturing lines. For this purposes approaches from operations research are used to define interfaces between parallel manufacturing steps.

5. REFERENCES

Denkena, B.; Brandes, A. & Apitz, R. (2003). Designing Integrated Process Chains, *Proceedings of XIV Workshop on Supervising and Diagnostics of Machining Systems, "Manufacturing Flexibility Design and Development"*, March 2003, Karpacz, Poland
Denkena, B.; Liedtke, C. & Brandes A. (2005a). Configuration of industrial process chain. At the University of Hannover research engineers investigate new methods for holistic process chain, *wt Werkstatttechnik online,* Vol. 95, No. 11/12, 2005, pp. 866-871, ISSN 1436-4980
Denkena, B.; Apitz, R. & Brandes A. (2005b). Sustainable Process Chain Optimisation based on Simulation of Technological Interactions, Machine Engineering, Vol. 5, No. 1-2, Jedrzeiewski, J. (Ed.), pp. 82-87, Karpacz, Poland
Eversheim, W. (2002). Organisation in der Produktionstechnik-Arbeitsvorbereitung, Springer-Verlag, ISBN 3-540-42016-9, Berlin
Monostori, L. & Viharos, Z. S. J. (2001). Hybrid, AI- and Simulation-Supported Optimisation of Process Chains and Production Plants, *Annals of the CIRP*, Vol. 50, No. 1, 2001, pp. 353-356, ISSN 0007-8506
Warnecke, G.; Eichgrün, K.; Kuge, R. & Zitt, U. (1999). Improvement of Process Reliability within Process Chains by Comprehensive Control Strategies, *Production Engineering, Research and Development*, Vol. 6, No. 1, 1999, pp. 1-6, ISSN 0944-6524

Annals of DAAAM for 2006 & Proceedings of the 17th International DAAAM Symposium, ISSN 1726-9679
Editor B. Katalinic, Published by DAAAM International, Vienna, Austria 2006
150th Birthday of Nikola Tesla (1856-1943) Inventor of Mobile Robot Teleoperation

ADAPTIVE CONTROL MODEL FOR MAINTAINING TOOL WEAR RATE IN THE PREDEFINED CUTTING TIME

BREZAK, D.; MAJETIC, D.; UDILJAK, T.; NOVAKOVIC, B. & KASAC, J.

Abstract: *This paper presents machine tool adaptive control concept for preserving cutting tool durability in the sense of ensuring maximum allowed amount of tool wear rate in the predefined machining time using cutting speed as a control parameter. Additionally, it also needs to maintain the high level of machining process productivity. The proposed model is configured using Radial Basis Function neurocontroller with Modified Dynamical Neural Network filter and is analyzed using analytical tool wear model which parameters are set by the experimentally recorded data.*
Keywords: *Adaptive Machine Control, Neural Networks, Tool Durability Control, Machining Process Productivity*

1. INTRODUCTION

Major research efforts taken in the field of machine process control are related with the machine forces control because of their great importance and influence on entire process. Beside force control, problems concerning chatter suppression, elimination of burr formation, achieving chip discontinuous forms, etc. are also considered. However, a little work has been done in the field of tool wear durability control. Liang et al., 2004, in their recent state-of-the-art work on the machining process monitoring and control are pointing to that problem without referring to any of 126 stated references as a possible solution. There are several advantages of tool wear control such as:

- ensuring scheduled tool change period in a mass production environment along with maximizing process productivity and workpiece surface quality;
- pre-fault tolerance control in the sense of tool breakage avoidance in cases of highly intensive (un)expected wearing when there is a need and possibility to finish a machining operation or providing safely tool exit.

Therefore, it is of great importance to include such criterion in the overall machine control strategy.

2. ADAPTIVE CONTROL MODEL

In this paper an adaptive control model based on Radial Basis Function Neural Network (RBF NN) controller is analyzed and proposed (Fig. 1.).

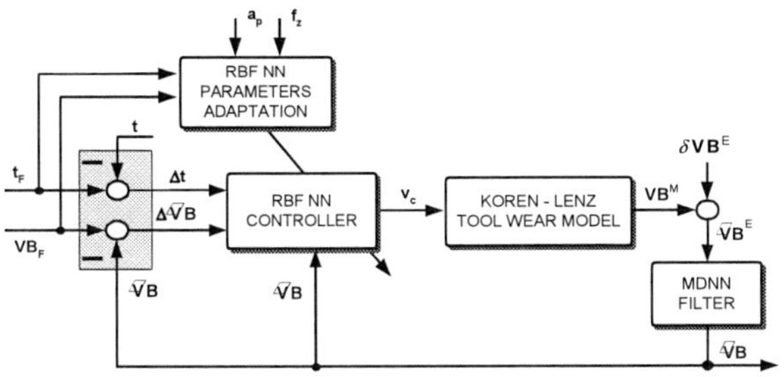

Fig. 1. Tool life adaptive control scheme

The neurocontroller approach is chosen because of neural network capabilities of identification (approximation) of highly nonlinear and partially stochastic processes such as tool wearing. Although, there is prevailing opinion that, due to the previously mentioned characteristics, tool wear process cannot be adequately described using exact mathematical relations, such model was utilized in this work. The model is based on mechanical (abrasive) and thermal (diffusion) wear process mechanisms in turning and is presented in (Koren & Lenz, 1972). Its parameters are taken from (Ulsoy & Danai, 1987) where they have been determined on the basis of experimental analyzes. The only purpose of this model was to replace the dynamic of cutting process together with the tool wear estimator in conducted simulations. In another words, the model was used to generate tool wear curves that would otherwise be determined on the basis of extensive amount of experimentally recorded data that would also be used in configuring tool wear estimator. Hence, the outputs of tool wear model stands for mean values of flank wear parameter for each combination of process parameters (cutting speed, depth of cut and feed). They were additionally weighted with the expected tool wear estimation error in the interval ± 0.03 mm. In order to damp the influence of estimation error a filter based on Modified Dynamic Neural Network (MDNN) architecture has been proposed.

2.1 Estimation error

Koren–Lenz model output in the k–th step depends on the process parameters and output from the previous $k-1$ step

$$VB^M(k) = f\left(v_c(k), a_p(k), f(k), VB^M(k-1)\right), \tag{1}$$

Thus modeled flank wear parameter is additionally weighted with the estimation error

$$\delta VB^E(k) = e(k), \quad e(k) \in \left[-0.03,\ 0.03\right] \tag{2}$$

$$\widetilde{VB}^E(k) = VB^M(k) + \delta VB^E(k), \tag{3}$$

where e is bounded white noise, and $\widetilde{VB}^E$ stands for simulated amount of the flank wear parameter that would otherwise be estimated in shop floor environment on the basis of different types of tool wear features.

2.2 MDNN filter

In order to reduce the influence of estimation noise (error) on neurocontroller's output (cutting speed) tool wear parameter from (3) is filtered using Modified Dynamic Neural Network. Dynamic Neural Network, DNN (Novakovic et al., 1998) is a three layer recurrent neural network with ARMA (Auto Regressive Moving Average) filter integrated in the hidden layer neurons. Every neuron processes past values of its own activity together with its input signals. The input (*net(n)*) and output (*y(n)*) at time instant *n* of the *j*–th hidden layer neuron are defined as

$$net_j(n) = \sum_{i=1}^{I} v_{ji} Z_i, \quad j = 1,2,\ldots, J, \tag{4}$$

$$y_j(n) = e^{-\frac{1}{2}\left(\frac{\bar{y}_j(n)-t_j}{\sigma_j}\right)}, \qquad (5)$$

where $\bar{y}_j$ is ARMA filter output (activation function input)

$$\bar{y}_j(n) = b_{0j}net_j(n) + b_{1j}net_j(n-1) + b_{2j}net_j(n-2) - \\ - a_{1j}\bar{y}_j(n-1) - a_{2j}\bar{y}_j(n-2), \qquad (6)$$

a and b are filter coefficients and i is the number of network input vectors. The original form of DNN network has been modified for the purpose of this problem in the sense of Gauss activation function utilization (5) instead of sigmoid unipolar function. Beside weight coefficients (v, w), widths (σ_j) and positions (t_j) of the Gaussians became also learning parameters. In that way, bias neurons from hidden and output layer can be omitted which speed up the learning process and simplify the network structure. Network k – th output is defined by

$$O_k(n) = \sum_{j=1}^{J} y_j(n) w_{kj}(n) \ , \ k = 1,2,...,K \ . \qquad (7)$$

Instead of using Error Back-Propagation learning method, Resilient Back-Propagation method has been utilized showing faster error convergence and better generalization characteristics. In the learning phase noisy data ($\hat{V}B^E$) were used as network inputs and modeled (approximated) flank wear values as outputs (VB^M). Several tests have been analyzed and the best performance was achieved with 5-5-1 network structure, meaning that the filtered tool wear parameter in the k-th step ($\hat{V}B\,(k)$) is determined on the basis of estimated values from the previous 4 steps and the last k–th step ($\hat{V}B^E(k-4)$, $\hat{V}B^E(k-3)$, $\hat{V}B^E(k-2)$, $\hat{V}B^E(k-1)$, $\hat{V}B^E(k)$).

2.3 RBF NN controller

For the purpose of cutting tool durability control RBF NN controller based on regression algorithm was chosen. The algorithm is based upon well-known RBF NN architecture (Novakovic et al., 1998) with one additional modification. In this work, the Moore–Penrose pseudoinverse of the matrix of hidden layer neurons outputs ($\boldsymbol{H}^+$) for weights calculation ($\boldsymbol{c} = \boldsymbol{H}^+ v_c$, $\boldsymbol{H}^+ = (\boldsymbol{H}^T\boldsymbol{H})^{-1}\boldsymbol{H}^T$) was replaced with recursive least square method (Yu, 2003). That way, the occasional problems with ill-conditioned matrix $\boldsymbol{H}$ that occurred during the conducted simulations have been avoided. From the technological aspect of view it is not convenient to use depth of cut as a variable for tool wear control. Bearing in mind that the process productivity maximization is also a condition that needs to be fulfilled we kept the feed at maximum possible value and used cutting speed as a control variable

$$v_c = \boldsymbol{Hc}\ , \qquad (8)$$

Elements of matrix $\boldsymbol{H}$ are determined from

$$\boldsymbol{H}_{kj} = e^{\frac{-\left(\left\|z_k - \boldsymbol{t}_j\right\|\right)^2}{\boldsymbol{\sigma}_j^2}} \ , \ k=1, ..., N, \ j=1, ..., K \ , \qquad (9)$$

where $\boldsymbol{t}_j$ is a hidden layer neurons position centres vector, $\boldsymbol{\sigma}_j$ is a Gauss (activation) function widths vector, N is the number of patterns and K is the number of hidden layer neurons. The Gauss functions width of the j-th neurone was calculated as geometrical mean value of the Euclidean distances of the centre of j-th neuron and the centres of two of his neighbour neurons $(j-1)$ - d_{1j} and $(j+1)$ - d_{2j}.

$$\sigma_j = \sqrt{d_{1j}\, d_{2j}}\ . \qquad (10)$$

In the learning phase network entrance vector is calculating as $z(k) = \left[\Delta t(k), \Delta VB^M(k), VB^M(k)\right]$. Every time the boundary conditions (VB_F, t_F) change, the RBF NN parameters need to be adapted. In another words, the new learning process for actual a_p and f needs to be conducted. In the recall phase matrix $\boldsymbol{H}$ are calculating for every step separately and entrance vector changes its form into $z(k) = \left[\Delta t(k), \Delta\hat{V}B(k), \hat{V}B(k)\right]$.

3. SIMULATION RESULTS

A series of simulation results were analyzed using different maximum flank wear parameters VB_F and cutting time values t_F. An example of the achieved results, i.e. the comparison between ideal model and controller output (achieved without estimation noise) and the realistic one (influenced by the estimation noise), is presented by Figure 1. From second subplot it can be noticed that there are no filtering in the first several steps since MDNN filter became active from the 5th step of the control process due to its previously explained working characteristics. However, this limitation does not have any relevant influence on the overall control process and the controller successfully maintained tool wear rate inside the set up bounds.

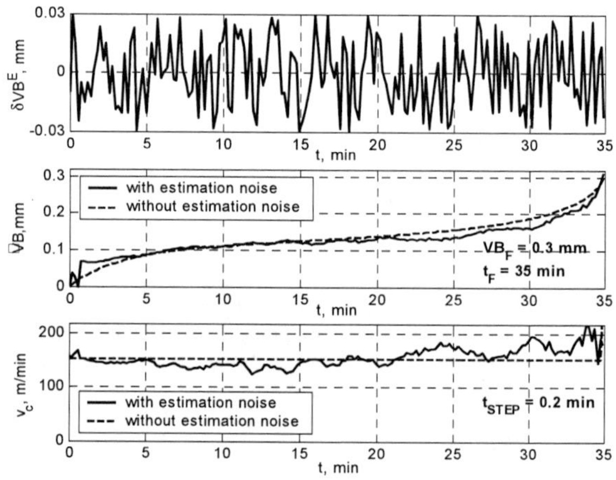

Fig. 2. Simulation results of model and controller outputs (a_p = 2.5 mm, f = 0.35 mm rev^{-1})

4. CONCLUSION

The obtained results from the conducted simulations have shown high accuracy in maintaining desired tool wear rate in the predefined cutting time. In the future research, additional tests will be performed with different types of process disturbances, depth of cut and feed values. Also, a novel model reference adaptive control based model will be analyzed in a predictive control conditions.

5. REFERENCES

Danai, K. & Ulsoy, A.G. (1987) A Dynamic State Model for On–Line Tool Wear Estimation in Turning, *ASME Journal of Engineering for Industry*, Vol. 109, No. 4, November 1987, pp. 396–399, ISSN 0022–0817

Koren, Y & Lenz, E. (1972). Mathematical model for the Flank Wear while Turning Steel with Carbide Tools, *CIRP Proceedings on Manufacturing Systems*, pp. 127–139

Liang, S.; Hecker, R.L. & Landers R.G. (2004) Machining Process Monitoring and Control: The State–of–the–Art, *ASME Journal of Manufacturing Science and Engineering*, Vol. 126, May 2004, pp. 297–310, ISSN 1087–1357

Novakovic, B.; Majetic, D. & Siroki, M. (1998) *Artificial Neural Networks*, Faculty of Mechanical Engineering and Naval Architecture, ISBN 953-6313-17-0, Zagreb, Croatia.

Yu, W. (2003) Recursive Least Square, *Available from*: http://ctrl.cinvestav.mx/~yuw/file/id10.pdf.

Annals of DAAAM for 2006 & Proceedings of the 17th International DAAAM Symposium, ISSN 1726-9679
Editor B. Katalinic, Published by DAAAM International, Vienna, Austria 2006
150th Birthday of Nikola Tesla (1856-1943) Inventor of Mobile Robot Teleoperation

MODEL OF CAPABILITY DEVELOPMENT IN UNCERTAIN CONDITIONS

BUCHMEISTER, B.; PANDZA, K.; KREMLJAK, Z. & POLAJNAR, A.

Abstract: *In the paper our original model of capability development in uncertain conditions is presented. Each uncertainty problem is divided into risk categories and factors. The basic method used in the numerical part is the two-pass Fuzzy Analytic Hierarchy Process. Systematic approach presented by a flow chart and explained in text is composed of three contextual parts: analytical, process and decision-making part. It represents a tool for deciding upon the appropriate activities and instructions.*
Key words: *uncertainty, capability, analytical part, process part, decision-making part, ABC diagram*

1. INTRODUCTION

Manufacturing is and will remain one of the principal means by which wealth is created. However manufacturing has changed radically over the course of the last 20 years and rapid changes are certain to continue. The emergence of new manufacturing technologies, spurred by intense competition, will lead to dramatically new products and processes. New management and labour practices, organizational structures, and decision making methods will also emerge as complements to new products and processes. It is essential that the manufacturing industry be prepared to implement advanced manufacturing methods in time (Trigeorgis, 2002). Managers in production environments face in their decision making a high level of uncertainty due to rapid and big changes defining the environments in which their organisations operate. This means that managers do not possess complete information on future events, do not know all possible alternatives or consequences of all possible decisions. Additionally, flexibility requirements at the enterprise need to be balanced. These issues have not been adequately addressed in published research because the developed models are limited having not covered the entire problem, thereby preventing efficient analysis of realistic problems and decision-making scenarios. Tackling uncertainty involves developing heuristic tools that can offer satisfactory solutions. The problem of decision-making in uncertain conditions is only partially presented in relevant literature (Carpenter & Fredrickson, 2001; Laviolette & Seaman, 1994; Frei & Harker, 1999). In (Buchmeister et al., 2005) our original method of uncertainty estimation is presented. Each uncertainty problem is divided into risk categories and factors. The basic method used in the numerical part is the two-pass Fuzzy Analytic Hierarchy Process (first for the importance and second for the uncertainty of risk factors). The estimates are derived from pairwise comparison, and by using fuzzy numbers we account for the errors of the estimator. In the following stages we use the interval results of this method for calculating the integral uncertainty value, which, in comparison with the boundary value, defines the risk of the process in question.

2. SYSTEMATIC APPROACH TO MANAGING CAPABILITY DEVELOPMENT

Categories and factors of uncertainty are mutually compared according to importance and uncertainty in the diagram. In the diagram we define three areas (low, medium, high) with boundaries around ± 50 % from average share. The position of every category/factor in one of the nine fields of the diagram is defined with an ellipsis, whose boundaries are minimal and maximal shares of category/factor of importance or uncertainty. The diagrams enable selection of factors that need special attention, or vice versa, the factors which can be partially neglected, which is shown in ABC focus diagram (Fig. 1).

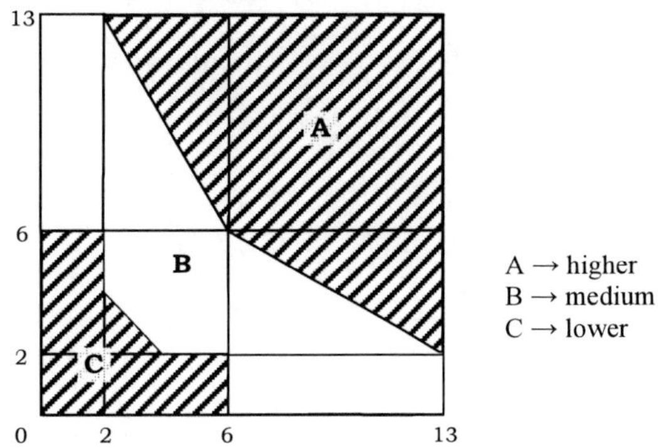

Fig. 1. ABC areas of attention – focus diagram.

The model (Fig. 2) was developed on the case studies of a supplier risk and international military logistics but can be without greater problems transferred in the environment of business and (narrowly) production systems. Systematic approach for directing capability development is developed for easier transfer of the model and its logic between different system environments, and because of the discussion on its applicative value.

2.1 Analytical part

The analytical part resembles the project start-up. Before starting the implementation of the processing and decision part, the basic issues need to be clarified and all info defining the development of discussed capabilities acquired. An expert team should be formed to assess the importance and uncertainty of the factors. The team members should differ with respect to experience, education, formal position and age. The divergence of the team depends on the complexity of the problem.

The analytical part includes a detailed analysis of resources and capabilities. On the one hand we determine which of them are already available and where they could be applied, and on the other, which of them are still lacking. Literature offers a number of systemic approaches to help us with this analysis. Before the expert team analyses the development of capabilities and classifies them into different categories and factors, the types of risks should be determined. Milliken's classification (1987) into the uncertain states, effects and responses helps the team members to come to a uniform understanding of concepts. The analysis of the discussed process into categories and factors of the uncertainty represents a domain which differs with respect to different systemic environments. The analysis represents a decomposition of a complex problem and enables the start of the implementation of the process part.

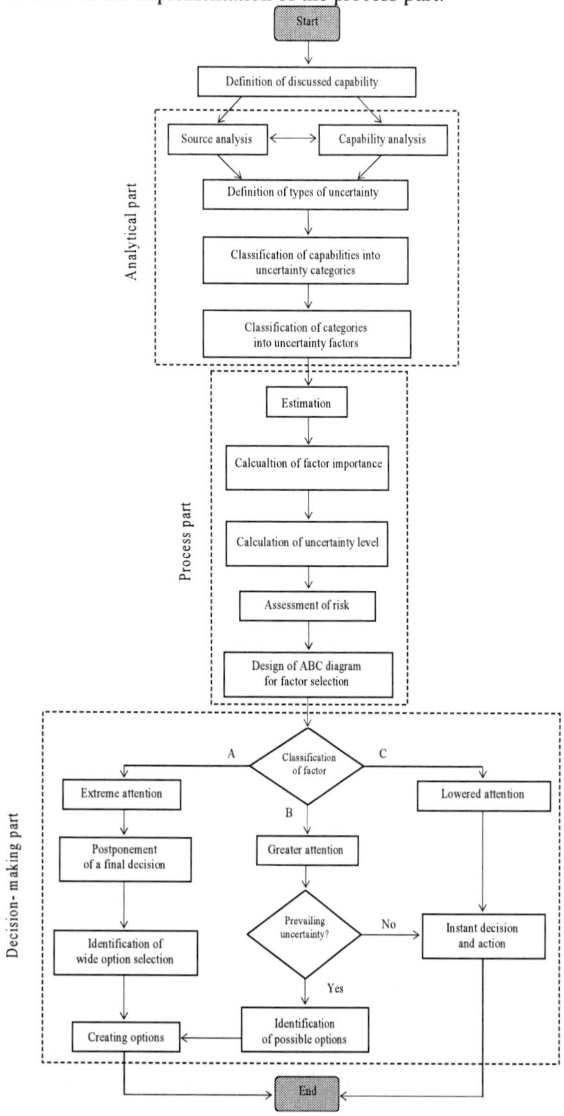

Fig. 2. Flow chart of systematic approach.

2.2 Process part

The process part is intended for evaluation of analysed factors. The basic method used is the Fuzzy Analytic Hierarchy Process. The advantage of this method lies in mutual determining of relationships as the evaluation is much easier with comparison by pairs, and in use of fuzzy numbers when the evaluation itself includes the possibility of the evaluator's error by a class (on either sides) and the calculation takes it into account as results are presented in particular intervals. Relatively large number of evaluations required can be avoided by grouping factors in categories already in the analytical part. The danger of excessively diverging evaluations would require repeating the procedure of comparison and adjustment of selected evaluations.

On the basis of diagrams "Uncertainty/Importance", we create the ABC diagram of attention, which is used for classifying uncertainty factors in the decision-making part.

2.3 Decision-making part

The decision-making part represents the most important part of the approach with regard to the model's applicability criterion. In this part the model's logic was transferred to real-life decisions. The decision-making model begins by classifying factors. *"C"* factors are those, which do not require larger attention with regard to the levels of importance and uncertainty. When such factors require certain decisions or measures, they can be implemented quickly and effectively without worrying how uncertainty may affect the consequences of the decision. *"A"* factors are the opposite pole of factors with regard to importance and uncertainty. These factors require extra attention. The instructions regarding proposing of final decisions should be elaborated in greater detail and should not be interpreted as a proposal for not taking any decisions. The instructions say that in such case no decisions can be taken, which would be final and prevent the flexibility of action. These decisions should be directed towards creating a wide range of options, enabling reaction in case of different development scenarios. Before adopting decisions, which create future options, the wide range of available options shall be identified. *"B"* factors lie between *"A"* and *"C"* factors with regard to their importance and the level of uncertainty. These are by all means factors, which require adequate level of attention. The proportion of attention naturally also depends on the number of factors being defined as *"A"* level factors. If we get a large number of *"A"* level factors, the attention given to *"B"* level factors will be slightly lower than in cases where there are only few *"A"* level factors. In case factors are classified in the *"B"* category due to the fact that they are important but not subject to uncertainty, the same measures as for *"C"* factors can be adopted. In case of increased uncertainty, decisions enabling flexibility of future actions shall also be adopted for *"B"* factors.

3. CONCLUSION

In decision making under uncertainty, risk analysis aims at minimising the failure to achieve a desired result.

The article encompasses the original synthesis of risk management, modelling uncertainty, fuzzy AHP method, and it represents a contribution to the construction of tools for decision-making support in organisational systems.

Based on the "Uncertainty – Importance" relations special ABC focus diagrams are created. These diagrams serve for the classification of uncertainty factors, which provides a decision making part of the systemic approach.

The original contribution in this article is comprised by completion of heuristic approach for effective interpretation of numerical results and their support to decision-making process through the developed model.

4. REFERENCES

Buchmeister, B.; Pandza, K.; Kremljak, Z. & Polajnar, A. (2005). Uncertainty estimation using two-pass fuzzy AHP method. *Proceedings of the 16th DAAAM Symposium*, Katalinic, B. (Ed.), pp. 45-46, Opatija, Croatia

Carpenter, M. A. & Fredrickson, J. W. (2001). Top management teams, global strategic posture, and the moderating role of uncertainty. *Academy of Management Journal*, Vol. 44, No. 1 (533-545)

Frei, F. X. & Harker, P. T. (1999). Measuring aggregate process performance using AHP. *European Journal of Operational Research*, Vol. 116 (436–442)

Laviolette, M. & Seaman, J. W. (1994). The efficacy of fuzzy representations of uncertainty. *IEEE Transactions on Fuzzy Systems*, Vol. 2, No. 1 (4-15)

Milliken, F. J. (1987). Three types of perceived uncertainty about the environment: state effect, and response uncertainty. *Academy of Management Review*, Vol. 12, No. 1, (133-144)

Trigeorgis, L. (2002). *Real options – Managerial flexibility and strategy in resource allocation.* MIT Press, Cambridge

Annals of DAAAM for 2006 & Proceedings of the 17th International DAAAM Symposium, ISSN 1726-9679
B. Katalinic, Published by DAAAM International, Vienna, Austria 2006
150th Birthday of Nikola Tesla (1856-1943) Inventor of Mobile Robot Teleoperation

SADT/STEP/PDM METHODOLOGY USED IN VIRTUAL ENTERPRISE

BUCUR, C.C. & CARUTASU, N.L.

Abstract: *VE is defined as a temporary alliance of enterprises that come together to share skills and resources in order to attend a business opportunity and whose cooperation is supported by computer network STEP it's a standard used in data transfer between these enterprises. SADT is a graphical representation and in VE is used to improve VE management and implicitly to improve data transfer. PDM can be defined as: electronic handling and control of product information throughout the whole product life cycle across system and organization boundaries by means of vaulting, workflow, and product structures. The goal of this paper is to define the task of SADT/STEP/PDM in a VE and also to sketch the SADT for data transfer.*

Key words: *SADT, STEP, PDM, virtual enterprise, data transfer.*

1. INTRODUCTION

The implementation of a supporting infrastructure for Virtual Enterprise (VE) can be based on a number of component technologies and paradigms (Camarintha-Matos & Barata, 2006):

- Interoperability and integration of standards – STEP, EDI, TCP/IP, etc.;
- Integration of Workflow Management Systems;
- Integration of advanced Information Management Systems – PDM;
- Integration of Safety and authentication mechanisms – digital signature, etc.;
- Integration of MAS development environments;
- Integration of legacy systems – PPC/MRP, CAD, CAM, CAE, etc.;
- Infrastructures – Internet/Intranet/Extranet, CORBA, Java, etc.

Virtual Enterprise challenges the way manufacturing systems are planned and managed. Shared virtual environments can allow engineers from different locations to work together and in the same time. These environments give engineers and designers a better understanding of the product, improve quality, reduce time to market and ensure that designs are right from the first time, reducing the need for expensive reworking later in the process. Collaboration can be extended outside a company by sharing virtual product information with suppliers and partners, creating a closer relationship in product development (Popa & Bucur 2004).

2. SADT PRESENTATION

In the 1970's, IDEF0 originated in the U.S. Air Force under the Integrated Computer Aided Manufacturing (ICAM) program from a well-established graphical language, the Structured Analysis and Design Technique (SADT).

The SADT (Structural Analysis and Design Technique) technique really consists of two parts: the structured analysis (SA), followed by the design (DT). SA specifies the requirements using two types of diagrams (actigram and datagram), and then DT explains how to interpret the results. SADT consists of procedures that allow the analyst to decompose software (or system) functions. The diagram includes four factors: inputs, controls, mechanisms, and outputs (Sparling, 2004).

As with many other methods, the diagrams are arranged in a hierarchy to show more detail at lower levels. Each diagram represents a transformation, and at most six diagrams are used to describe a function. If more than six diagrams are needed, the function should be redefined as a set of subfunctions (Sparling, 2004). Inputs are data items that are transformed to outputs. Controls are items such as budget and schedule that constraint the type of degree of the process being described. Mechanisms are external aids to the process, such as tools and techniques used to perform the transformation.

3. STEP PRESENTATION

The ISO 10303 Standard for the Representation and Exchange of Product Data (STEP) provides the means to represent and exchange data on manufactured products in a computer-understandable format.

The objective of STEP is to provide a neutral mechanism capable of describing product data throughout the life cycle of a product that is independent from any particular software application or computer platform. The nature of the description is suitable for neutral file exchange and also as a basis for implementing product databases and for data archiving (Bucur & Ciobanu, 2004).

4. PDM PRESENTATION

Product Data Management (PDM) is a category of computer software that aims to create an automatic link between product data and a database. The information being stored and managed (on one or more file servers) will include engineering data such as CAD/CAM (Computer Aided Design/Computer Aided Manufacturing) models, drawings and their associated documents. The package may also include product visualization data. The central database will also manage metadata such as owner of a file and release status of the components. The package will: control check-in and check-out of the product data to multi-user; carryout engineering change management and release control on all versions/issues of components in a product; build and manipulate the product structure BOM (Bill of Materials) for assemblies; and assist in configurations management of product variants. This enables automatic reports on product costs, etc. Furthermore, PDM (Product Data Management) enables companies producing complex products to spread product data in to the entire PLM (Product Lifecycle Management) launch-process. This significantly enhances the effectiveness of the launch process (Camarintha-Matos, L.M. & Barata, 2006)

5. CASE STUDY

PDM systems are designed to:

- Store, control, managed documents and other information about products;
- Enable engineering teams to share information on products and processes quickly and with consistent accuracy
- Permit the integration of techniques such as CAD/CAM (Computer Aided Design/Computer Aided Manufacturing) into coherent business systems across an entire enterprise.

For CAD/CAM integration and data transfer between different Integrated Design Systems from one enterprise to another we used STEP Standard in our study.

It is the aim of the STEP architecture to standardize the exchange of product model data across three levels: description methods, integrated resources and application protocols (AP). The development and standardization of an AP, is really the development of a standard for the exchange of product data in a given domain. To do this, several intermediate models are set up: the "application activity model" (AAM), the "application reference model" (ARM) and finally the "application interpreted model" (AIM).

SADT is a graphical representation which can transpose in programming language used in AAM (Application Activity Model) defined in norm ISO 10303 (STEP). This standard is utilized in data transfer between different Integrated Design Systems.

The first diagram is presented in figure 1. The purpose of this SADT diagram is to optimize compatibilities between different IDS software (for e.g. CATIA V5R8, UNIGRAPHICS NX, SOLID EDGE and SOLID WORKS). At this case study we work from last year and we don't finish yet but we made some progress from last published paper. This study was presented in a previous paper (Bucur 2005).

The SADT methods has been exploited in order to structure the research oriented towards the realization of the directory scheme, for a better definition of the existing situation and for detaching the functional specifications of the projected system, which has to meet an analysis of the needs.

We choose the actigrammes as tools of shaping in our analysis. We understand to call "actigram" those diagrams of the model, which are graphically represented as boxes designated by verbs, sometimes accompanied by complementation data are represented by arrows and designated by names. This representation allows a description of the series or parallel activities, as well as of feedbacks. The model is hierarchically constructed. At its most general interpretation, the system is perceived as a simple box. I can therefore expand and decompose this box, thus creating a first diagram, which can provide wider information in the field. The boxes of this diagram can be deconstructed in their turn, and can create new diagrams, and so on and so forth, until we reach the proper level of detail.

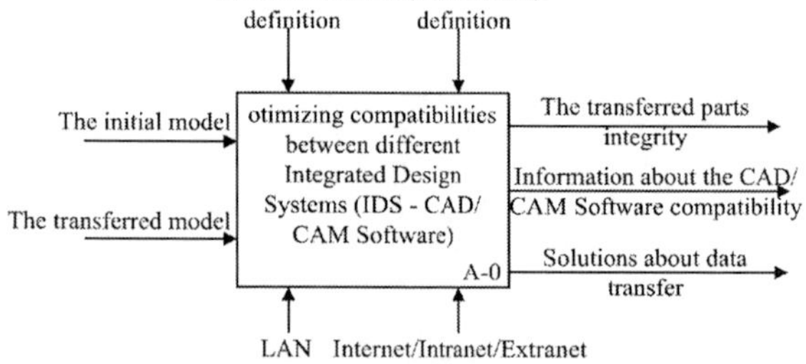

Fig.1. Diagram A-0: The data transfer modeling by using STEP format

We develop this SADT until the "Generating the 2D or 3D parts/assemblies in CAD Module" activity. There are two ways to elaborate the parts/assemblies in IDS: The first one is to design the parts/assemblies directly in actual IDS and the second one is to import the parts/assemblies model realized in a previous IDS. When the parts/assemblies 2D and 3D are generate from another IDS we must take account about transferring standards (STEP, IGES, DWG/DXF) and make the adaptation of the part/assembly in the actual CAD. Whichever are the ways to elaborate the 2D and 3D assembly the final drawings and 3D parts are validating in the present IDS by project director. At the end of this stage the virtual prototype is realized and validated. Technical and functional data, part/assembly geometry are inputs.

6. CONCLUSION

The SADT/STEP/PDM plays a major role in VE.

The advantages of PDM used in Virtual Enterprises are: reduced Time-to-Market; improved design productivity; and improving design and manufacturing accuracy.

All these benefits are achievable only if in the PDM systems are used the STEP Standard for data transfer between enterprise and implicitly is very important to improve this data exchange by using SADT technique.

The techniques for remodeling the manufacturing systems in order to integrate them in the virtual environment can be validated by simulating the manufacturing processes involved in such a project.

The main focus of this paper is to sketch the SADT diagrams of an IDS data transfer model in order to improve CAD/CAM communication. Of course this isn't the final goal of our research. In the future we intend to transpose SADT diagrams in a programming language and to elaborate a CAD/CAM interface. The case study presented in the previous chapter is not finished. The finally results will be published when ready.

7. REFERENCES

Bucur, C.C. (2005). SADT frame of an IDS data transfer model using STEP format. *Proceedings of The 16th International DAAAM Symposium.* Katalinic, B. (Ed.), pp. 47-48, ISBN 3-901509-46-1, *DAAAM 2005,* Croatia, October 2005, DAAAM International Opatija.

Bucur, C.C. & Ciobanu, L.F. (2004). Data Transfer and Simulation for Virtual Enterprise Environment Integration. *Proceedings of The 7th Conference on Management of Innovative Technologies. MIT 2004,* Chircor, M. & Dragoi, G. (Ed.), pp. 19-22, ISBN 973-700-028-5, Romania, October 2004, AIUS, Constanta.

Camarintha-Matos, L.M. & Barata, J. Introduction of STEP/PDM in SMEs, *Available from:* http://www.uninova.pt/~escn/ttt_portugal.html, *Accessed:* 2006-03-18

Popa, C.L. & Bucur, C.C. (2004). Data Transfer between CAD-CAM-CAE Cooperative Systems. *Proceedings of The 7th Conference on Management of Innovative Technologies. MIT 2004,* Chircor, M. & Dragoi, G. (Ed.), pp. 11-14, ISBN 973-700-028-5, Romania, October 2004, AIUS, Constanta.

Sparling, D. (2004). Requirements Specification Methods - Structured Analysis and Design Technique (SADT), *Available from:* http://cctr.umkc.edu/~dsparling/cs457/sadt.html *Accessed:* 2004-11-20

Annals of DAAAM for 2006 & Proceedings of the 17th International DAAAM Symposium, ISSN 1726-9679
Editor B. Katalinic, Published by DAAAM International, Vienna, Austria 2006
150th Birthday of Nikola Tesla (1856-1943) Inventor of Mobile Robot Teleoperation

ECONOMICAL DETERMINATION OF RESIDUAL STRESSES IN CASTINGS

BUDIC, I.; NOVOSELOVIC, D. & BUDIC, H.

Abstract: *After complete cooling off of the castings, residual stresses occur. If stresses exceed tensile strength of the component in production or service, deformation or cracking of the casting will happen. To avoid casting defects, it is necessary to know how big the residual stresses are and if they surpass the allowed values. The paper's presented some of the methods for measurement of residual stresses by DMD 20 and MTS 300 device, and has given comparative analysis of their cost effectiveness.*

Key words*: Residual stresses in castings, determination of residual stresses, DMD 20, MTS 3000.*

1. INTRODUCTION

Residual stresses are those stresses which exist in some machine member, also when there are no external loadings or temperature gradients. Production processes favor by and large creation of residual stresses (for example casting, welding, machining, heat treatment) by capturing residual stresses within the work piece (Heidl & Husnjak, 1987).After complete cooling off of the casting there are residual stresses which equal sum of stresses having different origin. If these stresses exceed tensile strength of the machine member, cracking and distortions of the casting in production or service will take place.

2. CASTING DEFECTS

Some typical casting defects caused by great residual stresses are shown in Fig. 1. (Atlas of Casting Defects, 1965). Fig. 1.a shows gray casting distortion due to unsymmetrical allocation of stresses, as consequence of unequal cooling off of the plate, while Fig. 1.b discloses casting cracks which occurred in warm or cold condition. Hot cracks occur at high temperatures on crystal boundaries. Primary cause for occurrence of hot cracks is first and foremost resistance to changes of volume (resistance to free shrinkage). Cold cracks take place at lower temperatures and due to exposure to loading during operation.

3. DETERMINATION OF RESIDUAL STRESSES

In practice we frequently face the task how to determine size of the residual stresses in the castings. Complexity of constructions, and many influential and relevant parameters quick solution of this theoretical problem, applicable to foundry practice, cannot be expected.

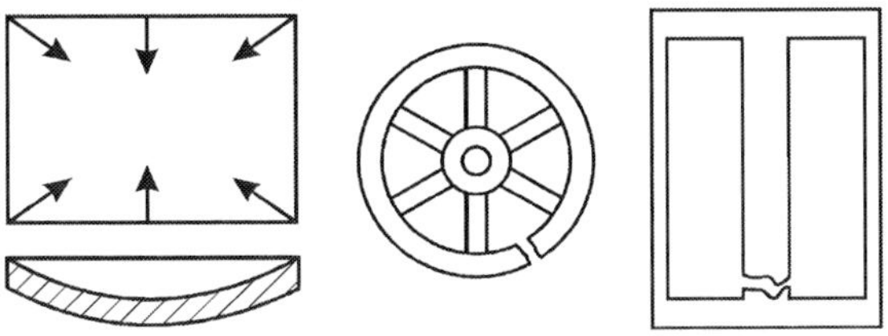

Fig. 1. Casting defects caused by residual stresses

Therefore methods for determination of residual stresses during inspection and manufacturing process of castings are being invented, which are not sufficiently reliable, and what's most important, lack simplicity and cost effectivness. Dependable determination of residual stresses requires measurements on construction itself. It cannot be accomplished on models, because models can only add qualitative impression about allocation of these stresses. That's why measurement devices must be simple so as to be able to apply them on site, frequently on inaccessible parts of the construction. Methods of measurement must eliminate influences arising from change of temperature, humidity and similar influences. Measurements during manufacturing process, when residual stresses come into being, are on the whole impossible due to high temperatures, enormous plastic deformations and similar (Heidl & Husnjak, 1987).

Technically applicable and practically appropriate methods for determination of residual stresses can be classified, according to degree of damage to construction member / casting, into:

- Non-destructive methods of measurement
- Partially destructive methods
- Destructive methods

With non-destructive methods of measurement casting is not damaged. Partially destructive methods can be reduced to non-destructive methods, because areas of damage of the casting are very small. With castings these areas of destruction can be tailored by additions for machining of the section, subsequent machining practically removing damaged areas.

Two most applied methods of partly destructive measurement are cutting of circular groove and hole drilling. Method of cutting of the circular groove is used so as to cut out of a groove around the measuring point, which leads to relaxation of the stresses in the area of the work piece surrounded by the groove. The rest of the core is extracted from equilibrium of forces and moments, easing thus internally available stresses. Thus distortion of component of the work piece is being effected, which is then measured by application of mechanical, electrical resistance-based or photo elastic tension meters. To determine completely state of stress on the surface of the work piece, distortions in at least three directions must be measured. Hole drilling method is very suitable because it can be applied on large pieces, without destruction of the piece itself. Hole diameters range from 1, 5 to 1, 8 mm, with depth up to 4 mm.

Residual stresses around the hole are thus partially relaxed. The accompanying distortions can be measured and be used for determination of residual stresses. Distortions are most frequently measured by tension meters installed at equidistance from the center of the hole.

Destructive methods of measurement are those which totally destroy casting. These methods are also called relaxation methods, because they are associated with cutting and relaxation.

In foundry practice technological probes designed to cause intentional stresses in castings are being used recurrently. There are two such technological probes:

- Grid like
- According to Thomas

Both of these technological probes are designed so as to contain thin and massive profile, which leads to creation of stress during cooling off, due to their mutual resistance to shrinkage. Stress measurement on probes is carried out so as to measure distances of exactly defined positions prior to cutting and after cutting.

4. MEASUREMENT OF RESIDUAL STRESSES ON GRID LIKE PROBE

Technological probes of certain dimensions made of gray casting have been manufactured. After cooling of the casting preparation of measuring points and gluing of measuring tapes in accordance with HBM Instructions (Hottinger Baldwin Masstechnik) (Hoffmann, 1996) has been effected. For determination of residual stresses devices of the company HBM have been used. Measurements on grid like probes have been carried out:

- DMD device for destructive/relaxing method of measurement
- MTS 3000 device for partially destructive hole drilling method of measurement

4.1 Measurement of Residual Stresses with DMD 20 Device

Principle of measurement with DMD 20 is to enable creation of the full Wheatston bridge. Procedure of establishing of the full bridge is enabled in such a way that 3 compensation tapes are glued on 50x50 mm plate made out of the same material as the casting, and 4 active tapes on the casting on which stress is being measured. Prior to measurement itself nullifying of device, calibration and feeding in of the k-factor tape coefficient is carried out. Measurement is conducted so that cables are connected from measuring tapes to device and the real value ε_1 is read out. After that the object is cut through and left for at least two hours prior to second measurements. After that the second value ε_2 is read out. The measured distortion value is:

$$\varepsilon = \varepsilon_2 - \varepsilon_1 \ (\mu m/m) \quad (1)$$

Internal stress is calculated by the equation:

$$\sigma = E \cdot \varepsilon \ (N/mm^2) \quad (2)$$

E – Modulus of Elasticity N/mm^2

Measurement of grid like probe by DMD 20 is shown in Fig. 2.

4.2 Measurement of Residual Stresses by MTS 3000 Device

MTS 3000 device is used for measurement of residual stresses by hole drilling method of measurement. It uses HBM measuring technology. To enable measurement compressed air must be provided for (for example: compressor).

The required pressure is 4 bars, which represents optimal incoming value for air necessary for putting into motion of the turbine within device. Air is introduced through valves and filters into control device for further connection with the hole drilling device. Measuring cables from rosette are jointed onto control device directly or over signal intensifier.

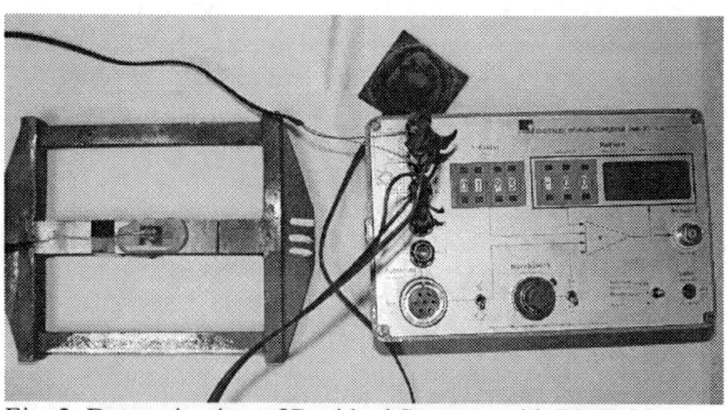

Fig. 2. Determination of Residual Stresses with DMD 20

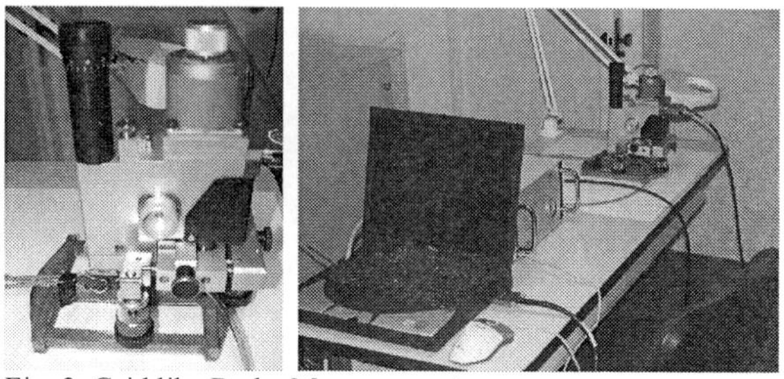

Fig. 3. Grid like Probe Measurement

If connection is effected without signal intensifier, PC or laptop must have a card for signal intensification (i.e. I/O card PCI 6023E or DAQCARD 6024 of the company National Instrument), so that this software implements residual stress measurement. In case that we use the signal intensifier, we don't need cards. The measurement process itself is carried out over computer.

Starting of the Restan program enables adjustments of measurement parameters in the main menu:

- Positioning of the Drill
- Setting of measurement parameters, drill diameter and number of steps.
- Sequencing of steps (lines or curvatures)
- Data on measuring tape/rosette and setting of its values on null (nullifying)
- Data on work piece being tested (Material, Modulus of Elasticity, Poisson's coefficient)
- Selection of the type of intensifier or measuring card
- Record of general measurement data (Date, operator's name, description of measurement ...)

Preparation of measurement - prior to measurement it is necessary to determine drilling speed and the way how to read data. There are two ways how to do it: manual (step by step) or automatic. With manual method, the operator himself stops the device to read out measurement values for certain depth, while with automatic method device itself reads automatically data according to pre-given number of steps. After each step computer reads out distortion values and records them. After measurement data is recorded into software. There, on the basis of obtained results residual stresses are calculated by one of the methods (ASTM E 837-01, INTEGRAL, KOCKELMANN, POWER SERIES).

5. CONCLUSION

If residual stresses are greater then tensile strength they cause casting defects. To avoid defects, it is necessary to know the size of residual stresses, and if they exceed permissible stresses. Casting of probes and testing by destructive methods is suitable for research testing of i.e. influential stress parameters.

Reliable determination of residual stresses demands measurement on construction itself, and cannot be effected on models, because such testing can only add qualitative impression on allocation of residual stresses.

Out of aforementioned it can be concluded that it is more economical to measure stresses on finished pieces, than to manufacture grid like probes.

6 LITERATURE

Heidl, I.; Husnjak, M. (1987):*Tenzometrija*, Tehnička enciklopedija 12, Leksikografskog zavoda "Miroslav Krleža", 688-690

Atlas ljevačkih pogrešaka (1965.), knjiga 1. (sivi lijev), Metalbiro, 1965

Hoffmann K. (1996) *Practical hints for the installation of strain gages*, HBM

Annals of DAAAM for 2006 & Proceedings of the 17th International DAAAM Symposium, ISSN 1726-9679
B. Katalinic, Published by DAAAM International, Vienna, Austria 2006
150th Birthday of Nikola Tesla (1856-1943) Inventor of Mobile Robot Teleoperation

RECONFIGURABLE MANUFACTURING NETWORK: AN EDUCATIONAL TEST-BED

BUTALA, P.; SLUGA, A. & PUTNIK, G.

Abstract: *The paper introduces a concept of a test-bed for studying reconfigurable manufacturing network structures. The test-bed is composed of several experimental autonomous cells (EAC) interconnected in a manufacturing network. The network structure and the structure of EAC are conceptualized. The basic building blocks of each EAC are two desktop CNC work systems. Other elements provide an infrastructure for autonomous control and network operations.*
Key words: *manufacturing network, experimental autonomous unit, desktop CNC work system, ubiquitous manufacturing.*

1. BACKGROUND AND MOTIVATION

Networking of manufacturing enterprises, especially SMEs, is becoming an important issue. Manufacturing networks open new possibilities and potentials in design, development and production of complex high-tech products while they combine good characteristics of large companies with advantages of SMEs. In order to exploit the potential of networking in manufacturing, research into networked working structures and operational principles is essential. A manufacturing network provides a basis for competitiveness, innovativeness, agility, reconfigurability and adaptiveness by enabling interconnected partners to (1) form long-term business coalitions, (2) develop mutual understanding and trust, (3) jointly react to business opportunities, (4) gain synergetic effects by cooperation and collaboration, and (5) share information, knowledge, resources, competencies and risks. Three important concepts of manufacturing networks were proposed recently. The first one is the conceptual framework for collaborative design and operation of networked manufacturing systems (Sluga et al., 2005). It consists of (1) a conceptual model for collaboration among autonomous manufacturing work systems and (2) an ICT platform which supports collaborative operations over the web. The framework improves visibility, understanding and control of the processes involved and on a practical level provides a platform on which geographically distributed operations can be conducted effectively. The second one introduces network building blocks in terms of autonomous work systems (AWS) (Butala & Sluga, 2006). AWS encapsulates functionalities and competencies related to its management and manufacturing operations, as well as its autonomous information system, which supports autonomous decision-making and AWS control, as well as cooperation in the network. The third concept focuses the network architectures and reconfigurability issues in networks on the manufacturing system level (Putnik et al., 2006). These three concepts provide a sound theoretical basis for manufacturing networks design and operations. In the next step of research of the manufacturing network structures and operational principles, an experimental approach should be introduced. Therefore, a corresponding experimental setup, which would enable investigation of different aspects of manufacturing networks, is to be developed.

The paper introduces a concept of an educational experimental test-bed, which will enable in-depth research and education of new concepts, methods and technologies for the next generation of manufacturing systems. The idea is to develop such a test-bed in terms of distributed experimental manufacturing cells. These cells will be installed at several European university laboratories, which are working on the related topics, and interconnected in an experimental manufacturing network.

2. TEST-BED FOR MANUFACTURING NETWORKS

The objective of the test-bed for a manufacturing network is to develop a platform which would enable investigation of various principles of network design and operations, such as changeability, reconfigurability, adaptivity, agility, etc., by experiments. The basic requirement of the test-bed is to provide a manufacturing network structure, composed of several distributed experimental autonomous manufacturing cells and adequate network units, which support and coordinate network operations. Besides, the test-bed should also provide up-to-date ICT solutions, which would enable seamless communication and visibility within the network and, further on, investigation of principles of ubiquitous manufacturing. The concept of the test-bed in depicted in Fig. 1. It consists of several experimental autonomous cells (EAC), interconnected over the internet. An EAC is an experimental version of a so called autonomous work system (AWS), which is defined as a generic building block of a manufacturing network (Butala & Sluga, 2006). According to the AWS definition, EAC encapsulates functionalities and competencies related to its management and manufacturing operations, as well as its autonomous information system, which supports autonomous decision-making and cooperation in the network.

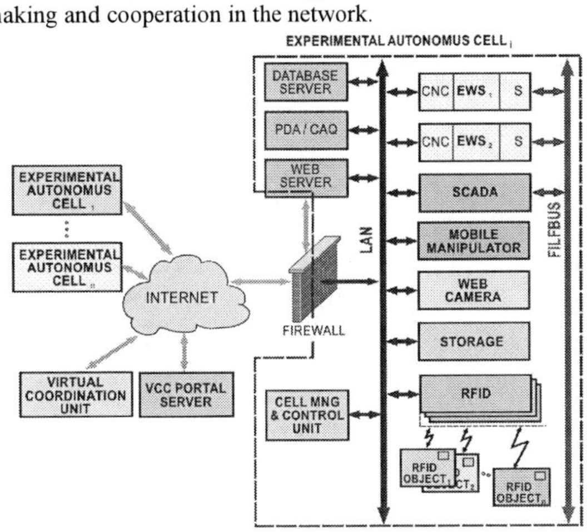

Fig. 1. Architecture of the test-bed for manufacturing networks

The management functionality is implemented in the cell Management and Control Unit (MCU) and is decisive for the autonomy of the EAC. The technological functionality defines the technological competence of EAC and is built upon its core elements, i.e. Elementary Work Systems (EWS), as defined in (Peklenik, 1995). Two identical EWSs, described in Chapter 3, are foreseen for each EAC of the test-bed. Beside the MCU and EWSs, other infrastructural elements (mobile manipulator, storage) and elements of the autonomous information system are included in EAC (see the right part of Fig. 1).

The autonomous information system is briefly described in (Butala & Sluga, 2006). The new element added here is the Radio Frequency IDentification (RFID) system for wireless identification and tracking of physical objects, such as workpieces, human subjects, etc. The role of the EAC information system is (1) to support autonomous decision making within the EAC, (2) to enable seamless communication with other network nodes, and (3) to provide EAC visibility in the network and thus support network coordination.

An important issue of the test-bed is how to assure effective coordination of cooperation and collaboration in the network, where a variety of complex processes occurs, and consequently different experts from different cultural environments are involved.

For these purposes, two special units developed in (Sluga et al., 2005) are integrated in the test-bed, as shown in Fig. 1. Coordination of processes is performed by the Virtual Coordination Unit (VCU). The Virtual Competence Center (VCC) is introduced for communication, teamwork and cooperation support. VCC provides communication among team members, EACs and VCU over the VCC portal. The portal enables access to data and knowledge bases, applications and web services that are intended for common usage. The portal also provides a common electronic work space in order to support communication and asynchronous cooperation.

The implementation of the test-bed will include identical EACs. The reason is to enable research of reconfigurability and adaptivity of the network on one side, and to avoid interoperability problems on other side.

3. DESKTOP CNC MACHINE TOOL LAKOS 150

The key element each EAC are EWSs, while they define the technological functionality and competence of the EAC. The test-bed EACs will include two identical EWSs.

For the experimental set-up, a low-cost but open CNC machine tool was searched. The solution at hand is a desktop CNC engraving machine LAKOS 150. LAKOS 150 is developed for educational purposes in the fields of mechatronics and manufacturing systems. It is a desktop machine capable of performing engraving, milling and drilling operations on small workpieces made of light materials in 2D and 3D. The semi-operational scheme of the machine tool is shown in Fig. 2.

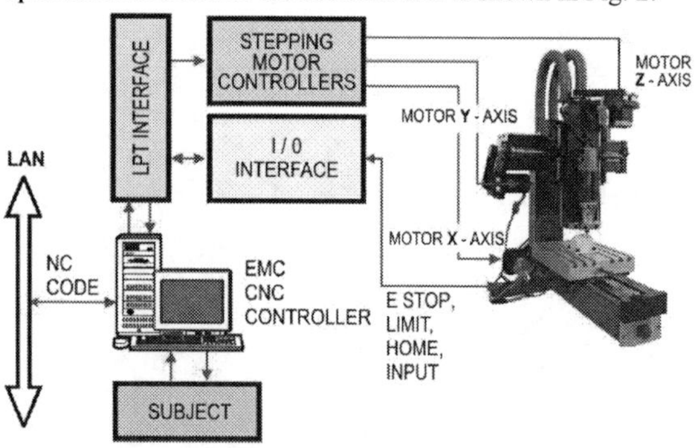

Fig. 2. Desktop CNC engraving machine tool LAKOS 150

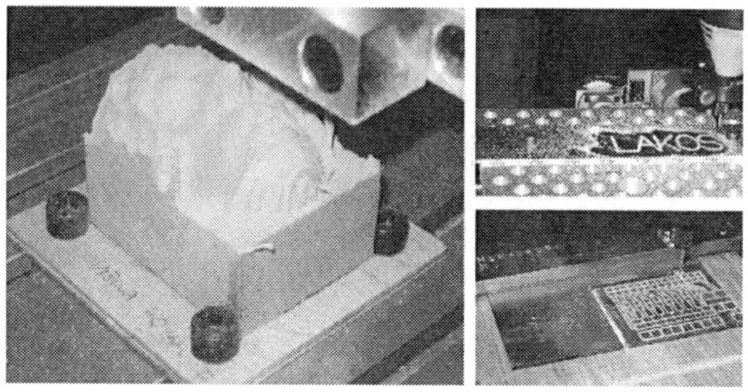

Fig. 3. LAKOS 150 3D and 2D workpiece samples

The mechanical design is a 3 axis Cartesian system with lengths of positioning axes X 200 mm, Y 170 mm, Z 130 mm and speeds up to 3000 mm/min. The positioning accuracy is about 0,05 mm. The main spindle is a universal hand spindle with up to 30.000 rpm. Fig. 3 shows different examples of machining, such as milling of a 3D object, engraving of a sign on a label, and engraving of a pattern on a prototype printed circuit board.

The machine is controlled with a CNC controller based on the open architecture principles and open source. It enables integrability of the machine in the experimental test-bed. The controller is implemented on a PC with the Linux operating system and real time extension. On this basis the solution for using the machine as a web machining service is developed (Butala et al., 2006). This is the key solution for integrating the machine in the experimental network test-bed.

4. CONCLUSION

Research into new manufacturing network structures is getting more and more intensive. But there is the lack of appropriate set-ups for conducting experiments.

The paper proposes the concept and architecture of the platform for studying actual issues of manufacturing network systems, such as reconfigurability and network operations on the network level, and autonomous control and ubiquitousness in on the level of individual experimental autonomous cell. The key building blocks are the open CNC machines LAKOS 150.

5. ACKNOWLEDGEMENTS

The authors acknowledge the support of (1) VRL-KCiP NoE – "Virtual Research Lab for Knowledge Community in Production", Contract No. FP6-507487/2, (2) I*PROMS NoE – "Innovative PROduction Machines and Systems", Contract No. FP6-500273-2, and (3) "Adaptive Distributed Manufacturing Systems / Virtual Enterprises" Inter-Governmental Technology Cooperation between Portugal and Slovenia

6. REFERENCES

Butala, P. & Sluga, A. (2006). Autonomous Work Systems in Manufacturing Networks. *CIRP Ann.* Vol. 55/1: 521-524.

Butala, P.; Vengust, I. & Peklenik, J. (2006). Project oriented education in mechatronics. *Proceedings of the 39th CIRP Int. Seminar on Manufacturing Systems,* Ljubljana, 567-570

Peklenik, J. (1995). Complexity in Manufacturing Systems. *CIRP Journal of Manufacturing Systems* Vol. 24/1:17–25.

Putnik, G.; Sluga, A. & Butala, P. (2006). Reconfigurability of manufacturing systems for agility implementation - part II: two architectures. *Proceedings of the 3rd Int. Conference on Digital Enterprise Technology*, Setubal.

Sluga, A.; Butala, P. & Peklenik, J. (2005). A Conceptual Framework for Collaborative Design and Operations of Manufacturing Systems. *CIRP Ann.* Vol. 54/1:437-440.

Annals of DAAAM for 2006 & Proceedings of the 17th International DAAAM Symposium, ISSN 1726-9679
B. Katalinic, Published by DAAAM International, Vienna, Austria 2006
150th Birthday of Nikola Tesla (1856-1943) Inventor of Mobile Robot Teleoperation

OPTIMALISATION OF TRANSMISSION RATIO FOR THE M-190E GEAR BOX

CAKOLLI, H.; IBRAHIMI, N.; LIKAJ, R. & KYCYKU, A.

Abstract: *The gear box has an important role in the transmission system of the motor power from the engine to the driving wheels. Depending on the type of engine, aerodynamic and outer characteristics and the type of vehicle is determined the velocity rate. In this paper will be contained out the analysis of the gear box with five velocity rates type GL 68/20-B5, used on M190E, in order to reach the optimal transmission values for the first and fifth velocity rate.*

Key words: *Gear box, velocity rate, driving wheels, motor power, vehicle characteristics.*

1. INTRODUCTION

The main outer characteristics of vehicles are: vehicle speed, driving force, dynamical characteristic and ideal driving force (Looman, 1988). The equation of equilibrium for the resistance and driving force is defined as follows:

$$F_{Z,A\max} \geq F_{Z,B} \qquad (1)$$

where:

$F_{Z,B}$ - the overall resistance force of the vehicle,

$F_{Z,A\max}$ - maximal driving force of the vehicle.

The overall resistance force on the driving wheel of vehicle is determined by the following equation:

$$F_{Z,B} = F_R + F_{st} + F_L + F_a \qquad (2)$$

where:

F_R – resistance force of rolling (flip),
F_L – aerodynamic force,
F_{st} – road inclination force,
F_a – acceleration force.

Maximal driving force is determined by the following equation (Cakolli, 2000):

$$F_{Z,A\max} = \frac{T_R}{r_{dyn}} = \frac{T_e . i_A . \eta_p}{r_{dyn}} \qquad (3)$$

where:

T_e - effective engine torque, which is given by equation:

$$T_e = \frac{P_e}{\omega_e} = \frac{P_e}{2.n_e.\pi} \qquad (4)$$

P_e - effective power of the engine, given by equation:

$$P_e = P_{\max}\left[c_1\frac{n_e}{n_{e\max}} + c_2\left(\frac{n_e}{n_{e\max}}\right)^2 - c_3\left(\frac{n_e}{n_{e\max}}\right)^3\right] \qquad (5)$$

i_A - overall power transmission ratio from engine to driving wheels, which is determined by the following equation for the passenger car :

$$i_A = i_n \cdot i_E \qquad (6)$$

i_n - transmission ratio for respective ratio of gearbox,

i_E - transmission ratio for differencial,

η_p - utilisation scale for the power transmission system,

r_{dyn} – dynamic radius of driving wheel.

In this paper will be investigated the two step gearbox with five velocity ratio shown in fig. 1., which is used at passenger car type GL–68/20-B5, M190 E.

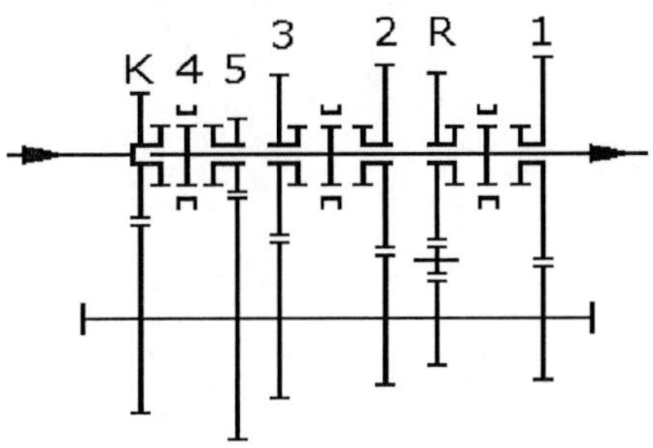

Fig. 1. Two step gearbox with five velocity ratio.

The ideal driving force represents the power proportion between driving wheels and velocity, represented with the following equation:

$$F_{OH} = \frac{P_R}{v} = \frac{P_e.\eta_p}{v} \qquad (7)$$

The model parameters and their respective units are summarised in Table 1.

Symbol	Units	Value
m_F	Kg	1160
P_{emax}	kW	80
n_{emax}	min^{-1}	5500
v_{max}	km/h	185
i_E	-	3,64
i_1	-	3,91
i_2	-	2,17
i_3	-	1,376
i_4	-	1
i_5	-	0,85
r_{dyn}	-	0,308
$c_1 = c_2 = c_3 = 1$	-	1
$c_w \cdot A$	-	0,45

Table 1. Model parameters

The driving force for first, fifth and backward ratio can be calculated with the following equations (Cakolli, 2000):

$$F_{Z,A1} = \frac{T_e \cdot i_{A1}}{r_{dyn}} \cdot \eta_p = \frac{T_e \cdot i_1 \cdot i_E}{r_{dyn}} \cdot \eta_1 \cdot \eta_E = \frac{T_e \cdot i_{12} \cdot i_{34} \cdot i_E}{r_{dyn}} \cdot \eta_{12} \cdot \eta_{34} \cdot \eta_E$$

$$F_{Z,A5} = \frac{T_e \cdot i_{A5}}{r_{dyn}} \cdot \eta_p = \frac{T_e \cdot i_5 \cdot i_E}{r_{dyn}} \cdot \eta_5 \cdot \eta_E = \frac{T_e \cdot i_{12} \cdot i_{910} \cdot i_E}{r_{dyn}} \cdot \eta_{12} \cdot \eta_{910} \cdot \eta_E$$

$$F_{Z,AR} = \frac{T_e \cdot i_{AR}}{r_{dyn}} \cdot \eta_p = \frac{T_p \cdot i_R \cdot i_E}{r_{dyn}} \cdot \eta_R \cdot \eta_E = \frac{T_p \cdot i_{12} \cdot i_{1113} \cdot i_E}{r_{dyn}} \cdot \eta_{12} \cdot \eta_{1112} \cdot \eta_{1213} \cdot \eta_E$$

In fig. 2. are shown the resistance and driving force characteristics, accordingly to velocity ratio.

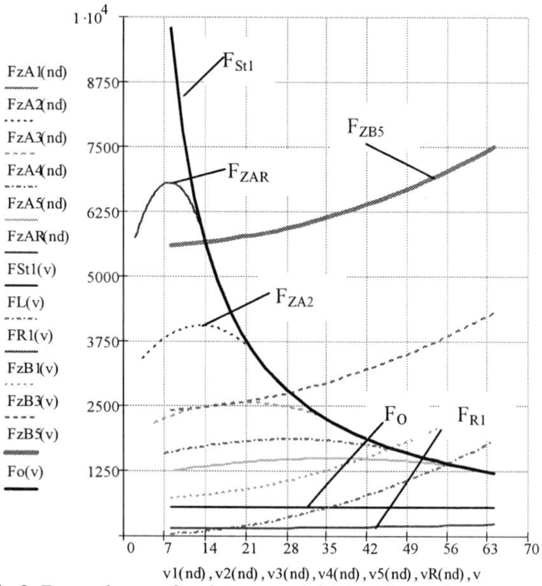

Fig.2. Dependency of resistance and driving forces from velocity ratio.

2. DYNAMICAL CHARACTERISTICS

The dynamical characteristic represents the proportion between the difference of the driving force and air resistance force divided with the overall mass of vehicle, and is given with the following expression:

$$D = \frac{F_{ZA} - R_L}{G} \tag{7}$$

$$D_1 = \frac{F_{Z,A1} - R_{L1}}{G} = \frac{\dfrac{T_e \cdot \eta_{12} \cdot \eta_{34} \cdot i_{12} \cdot i_{34} \cdot i_E \cdot \eta_E}{r_{dyn}} - R_{L1}}{G} \tag{8}$$

The dynamical characteristics for respective velocities of the above selected car type are calculated according to the equation (7), e.g. for the first ratio is given by equation (8), the simulation results are shown in fig. 3. In order to gain the optimal values of transmission ratio for first and fifth velocity ratio, it's used the following interval, thus:

$i_1 = 2{,}5 \ldots 4{,}5$

$i_5 = 0{,}75 \ldots 1{,}0$

while transmission ratio on driving bridge remains constant $i_E = 3.77$.

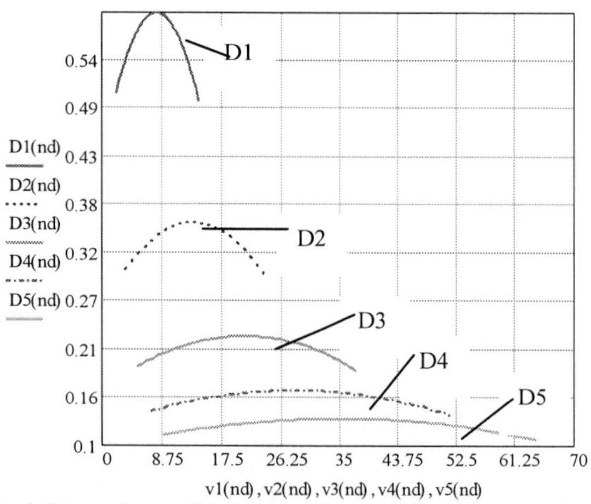

Fig. 3. Dependency of dynamical characteristic from respective velocity of vehicle

Since the dynamical characteristics changes according to transmission ratio for respective velocities, in fig. 4 is shown the dependency of dynamical characteristic from velocity for some of the values of transmission ratio at first and fifth ratio.

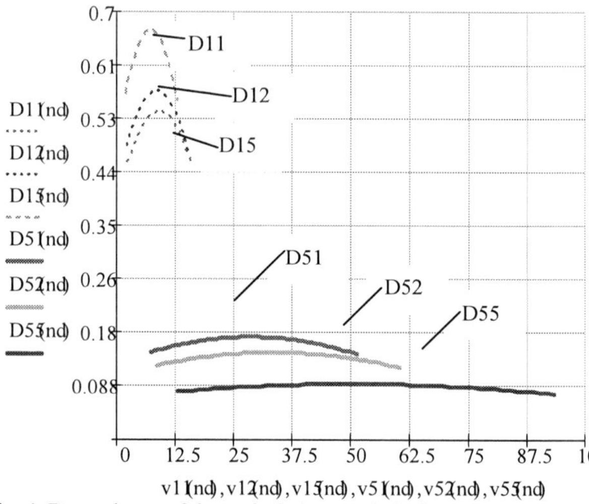

Fig. 4. Dependency of dynamical characteristic from respective velocity for first and fifth ratio.

3. CONCLUSIONS

Based on the results simulation results obtained with it can be concluded that:

- By increasing the transmission ratio at first ratio, the driving force will be increased too, respectively dynamical characteristic, while vehicle speed will be decreased. Therefore, the optimal values of transmission ratio for the first ratio of $i_1 = 3 - 3{,}5$ are the most preferred to use (Ibrahimi & Cakolli, 1997),
- By decreasing the transmission ratio on fifth ratio, the driving force will be reduced accordingly, while vehicle speed will be increased. The minimal value for transmission ratio is limited by a minimal value of dynamical characteristic, which might be negative; this means that the aerodynamic forces are greater than driving force. Therefore for the model used for investigation in this paper the optimal values of transmission ratio for fifth ratio of $i_5 = 0.75 - 0.85$ are the most preferred ones.

4. REFERENCES

Looman, J. (1988). ***Zahnradgetribe grundlagen, konstruktion*** *anwendungen in fahrzeugen,* Springer-Verlag, Berlin Heidelberg-NewYork-London-Paris-Tokyo, Berlin.

Schubert, F.(1983). *Planetengetribe mit verschëenkten planetenradaesen*, PHD, 1983 bericht Nr.66 des Instituts für Maschinenkonstruktion und Getribeban der Universitat Stuttgart, Stuttgart.

Cakolli, H. (2000). *Optimalizimi i karakteristikave të ndërruesit të shpejtësive të kombinuar me transmetues planetar,* PHD, University of Pristina, Pristina.

Ibrahimi, N.; Cakolli, H. (1997). *Shkalla e shfrytëzimit të transmetuesit planetar të tipit 1AI në varësi nga këndi Σ,* Revista Makineria Vol. 2, March 1997 pp. 35-37, Prishtina

Cakolli, H. (1989). *Prilog na istraživanjeto na možnosta na razvojot na planetnite hiperboloidno cilindrični prenosnici,* master thesis, University of Skopje, Skopje.

Annals of DAAAM for 2006 & Proceedings of the 17th International DAAAM Symposium, ISSN 1726-9679
B. Katalinic, Published by DAAAM International, Vienna, Austria 2006
150th Birthday of Nikola Tesla (1856-1943) Inventor of Mobile Robot Teleoperation

ANALYSIS OF OPEN FLAT DIE FORGING PROCESSES IN PLANE STRAIN CONDITIONS BY FEM

CAMACHO, A.M.; MARIN, M.; DOMINGO, R. & GONZALEZ, C.

Abstract: *In open die forging processes several parameters are usually involved. Generally, forces are applied into two flat dies where high contact pressures at the die-workpiece interface are achieved. The friction on this surface is one of the most important parameters to be considered. Additionally, other factors mainly related to the geometrical dimensions of the problem are also relevant. Different cases have been analysed by the Finite Element Method in order to study the influence of geometrical factors on the calculation of the required forces and contact pressure distributions, under plane strain conditions. This work shows that the geometry has a high influence on the required energy to carry out the plastic deformation and also, on contact pressures.*

Key words: *open die forging; plane strain; shape factor; Finite Element Method.*

1. INTRODUCTION

Many variables are involved in compression processes (Camacho *et al.*, 2005a) so different types of them can be distinguished. First, assuming ideal conditions, plane strain or axisymmetrical problems can be studied (Martín *et al.*, 2005). Relating to friction conditions at the die-workpiece interface, cases with and without friction can be also defined (Hartley *et al.*, 1980). But possibly, the geometry of the problem is the most relevant factor in order to define different compression processes (Rowe, 1979). In this sense, the geometry of the dies and the geometry of the workpiece are considered. If dies are assumed to be flat parallel dies, and the workpiece a rectangular billet of rectangular transverse section, different aspect ratios can be defined between the dimensions of both parts. In previous studies (Camacho *et al.*, 2005b; 2006) the ratio between the two dimensions of the billet for plane strain problems is called "shape factor".

2. ANALYZED CASES

Three parameters are considered constant throughout the different cases. The friction at the die-workpiece interface (μ = 0,05), the width of the workpiece (b) and the maximum reduction applied by the forces (5%) defined as:

$$r(\%) = \left(\frac{h_i - h_f}{h_i} \right) \cdot 100 \tag{1}$$

The following parameters are varied in order to solve different cases: the height of the billet, h, and the width of the dies, B. From now on the ratio h/b is used and it is called the "shape factor". Two cases are defined attending to the ratio B/b, where b is the width of the billet: B/b < 1 and B/b > 1. In Figure 1 the two cases considered are represented.

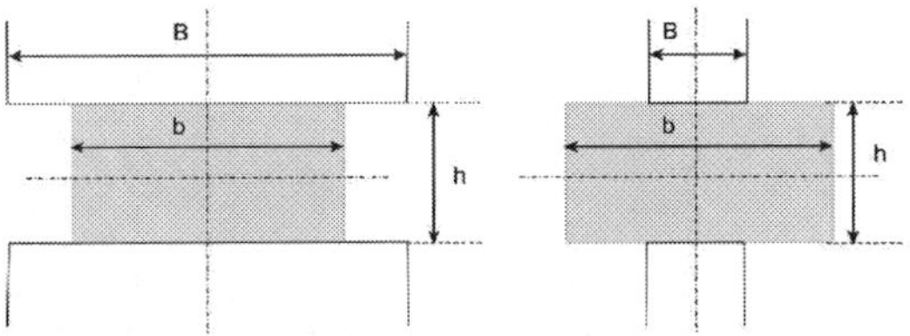

Fig. 1. Typology and geometrical parameters.

In this study three values of the shape factor *h/b* are considered in both types of problem (Table 1), for three values of the height (h_1, $h_2 = 2h_1$ and $h_4 = 4h_1$). Forces and contact pressure distributions are obtained in order to compare both cases.

B/b	**Shape factor *h/b***		
< 1	0,5	1	2
> 1	0,5	1	2

Table 1. Cases of analysis.

3. FINITE ELEMENT MODEL

The Finite Element Method (FEM) is an useful tool for analysing metal forming processes (Rubio *et al.*, 2005; Camacho *et al.*, 2005c; Rowe *et al.*, 1991). A general purpose code of implicit methodology (ABAQUS/Standard) [Hibbitt *et al.*, 2005] is used. The element type of the mesh is CPE4R and consists of a continuous, plane strain, 4-node and reduced integration element.

4. FORCES RESULTS

The first variable analysed is the force required to develop each particular problem. Different values of the reduction are considered (from 1% to 5%). Figure 2 represents the adimensional forces that have been obtained.

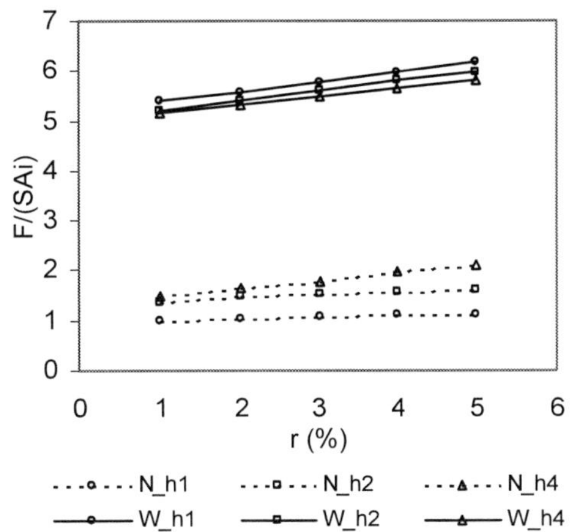

Fig. 2. Forces for B/b < 1 (T) B/b > 1 (W).

Forces values are divided into the initial contact area of the workpiece (A_i) and into the initial yield stress under plane strain conditions (S = 1.155 Y) of the aluminium alloy considered in the simulation. As a general behavior, forces are higher in open die forging processes with B/b >1 (W) than with B/b <1 (N) and the higher the reduction, the higher the required forces. For B/b > 1, the higher the shape factor h/b, the lower the required force. However, for B/b < 1, the higher the shape factor h/b, the higher the required force. On the other hand, results in forces seem to be more sensitive to the shape factor in this last case.

5. CONTACT PRESSURES RESULTS

The next variable to be analysed is the contact pressure at the die-workpiece interface for a reduction of 5%. Figure 3 gathers the obtained results, where the contact length is divided into the initial contact length in order to obtain an adimensional variable. Contrary to forces analysis, contact pressures are higher for B/b < 1 than for B/b > 1. When B/b > 1, the highest contact pressures at the center of the die are obtained for the lowest shape factor, h/b = 0,5. However, at the ends of the die the highest contact pressures are due to the highest shape factor, h/b = 2. Thus, depending on the shape factor, the die should be reinforced in one or another side. When B/b < 1, contact pressure distributions show some peculiar characteristics. First, for h_1, an uniform distribution is observed. On the contrary, for h_2 and h_4 the highest values of contact pressures are at the middle of the half of the die, whereas contact pressures are null form this zone to the end of the die. In both situations, the mean contact pressure is similar for the three cases.

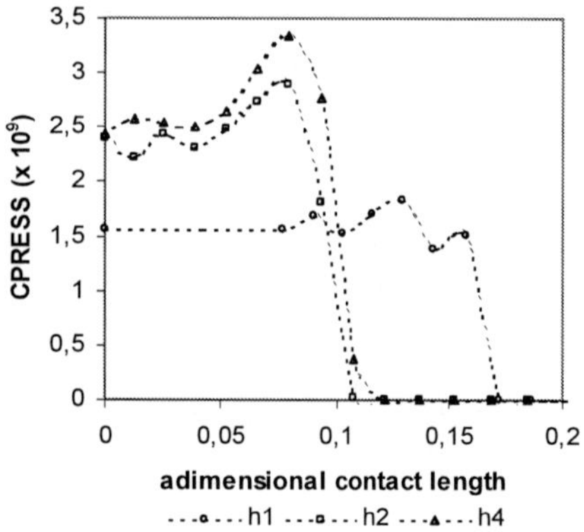

a)

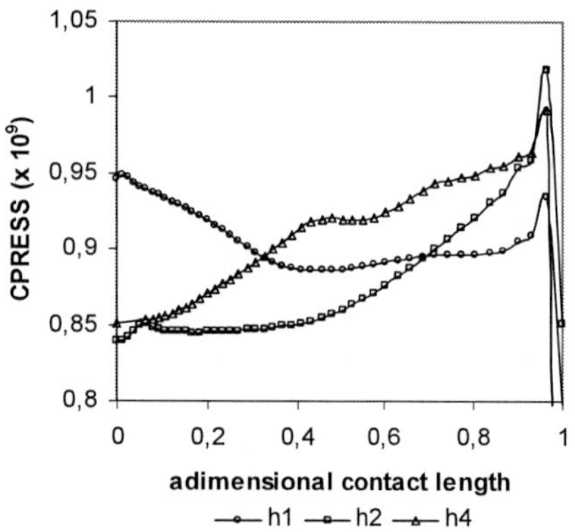

b)

Fig. 3. Contact pressures for a) B/b < 1 and b) B/b > 1.

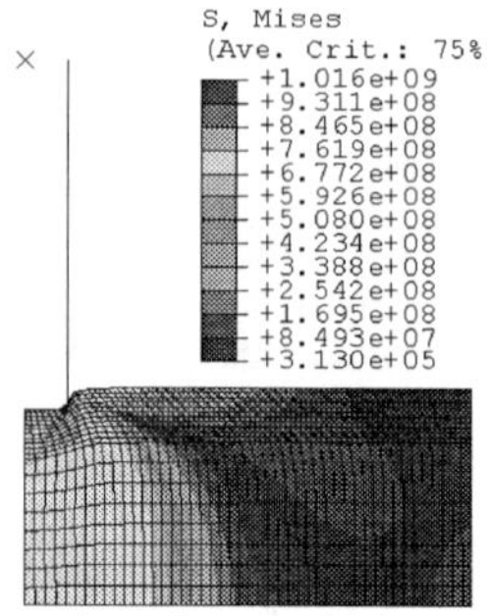

Fig. 4. Stress diagrams for B/b < 1 and h/b = 0,5.

6. CONCLUSION

According to geometrical criteria two main groups of compression processes can be defined: B/b > 1 and B/b < 1, where B and b are the width of the die and the billet, respectively. Different cases are analysed depending on the shape factor h/b . Forces and contact pressure distributions are obtained by the Finite Element Method in order to compare the different cases. Forces are higher when B/b > 1 than when B/b < 1. In the first case, the higher the shape factor, the lower the required force, and in the second case, the higher the shape factor, the higher the required force. Contrary to forces analysis, contact pressures are higher when B/b < 1 than when B/b > 1.

7. REFERENCES

Camacho, A.M.; Marín, M.; Rubio, E.M. & Sebastián, M.A. (2005a). Analysis of forces and contact pressure distributions in forging proceses by the finite element method, *Proceedings of the 16th International DAAAM Symposium*, Katalinic, B., pp 053-054, 3-901509-46-1, Opatija, (October 2005), DAAAM International Vienna, Vienna

Camacho, A.M.; Domingo, R.; Rubio, E.M. & González, C. (2005b). Analysis of the influence of back-pull in drawing process by the finite element method. *Journal of Materials Processing Technologies,* 164-165, (May 2005) (1167-1174), 0924-0136

Camacho, A.M.; Marín, M.; Sevilla, L. & Domingo, R. (2006). Influence of strain hardening on forces and contact pressure distributions in forging processes, *Journal of Achievements in Materials and Manufacturing Engineering,* 15, 1-2, (March-April 2006) (166-173), 1734-8412

Hartley, P.; Sturgess, C.E.N. & Rowe, G.W. (1980). Influence of friction on the prediction of forces, pressure distributions and properties in upset forging, *International Journal of Mechanical Sciences*, 22, (June 1980) (743-753), ISSN 0020-7403

Hibbitt, D.; Karlsson, B. & Sorensen, P. (2005). *ABAQUS v6.5, User's Manuals*, Providence (RI)

Martín, F.; Camacho, A.M.; Marín, M. & Sevilla, L. (2005). Parametrization of analytical and numerical methods in plane strain forging. *Proceedings of the 1st CISIF-MESIC*, pp 1-8, 84-608-0380-5, Calatayud, (September 2005)

Rowe, G.W. (1979). *Elements of metalworking theory*, Edward Arnold, ISBN 0 471 96003 9, London

Rowe, G.W.; Sturgess, C.E.N.; Hartley, P. & Pillinger, I. (1991). *Finite-element plasticity and metalforming analysis*, Cambridge University Press, ISBN 0 521 38362 5, Cambridge

Rubio, E.M.; Camacho, A.M.; Sevilla, L. & Sebastián, M.A. (2005). Calculation of the forward tension in drawing processes. *Journal of Materials Processing Technologies,* 162-163, (May 2005) (551-557), 0924-0136

Annals of DAAAM for 2006 & Proceedings of the 17th International DAAAM Symposium, ISSN 1726-9679
B. Katalinic, Published by DAAAM International, Vienna, Austria 2006
150th Birthday of Nikola Tesla (1856-1943) Inventor of Mobile Robot Teleoperation

DESIGNING INDUSTRIAL ENTERPRISES IN VIRTUAL ENTERPRISES CONCEPT

CARUTASU, G.; BOTEZATU, C.P. & BOTEZATU, C.

Abstract: *Today, the manner in which an enterprise is defined has fundamentally changed: enterprises no longer rely on a hierarchical, vertical structure to develop their products or sell their services, newly accepted business practices, such as out-sourcing and high-profile corporate break-ups. In the present, enterprises are focusing on core expertise processes by applying resources and skills to perform what it is that they do best and leveraging partnerships to complement their own core expertise. Thus, by building higher quality products and better serving consumer needs quickly and at less cost, an enterprise can recruit and retain a stable customer base in an era of global competition, where customer satisfaction is the key to the survival of the enterprise.*
Key words: *e-business, virtual enterprise, management, manufacturing systems.*

1. INTRODUCTION

Globalization has obviating in a great measure the commercial barriers, thus the companies act on market where the products abundance is without precedent. It becomes heavier for a product to impose itself to a more dynamic market, where the concurrent are rapidly adapting their offer. Facing a variety of comparable product, from price and quality perspective, sometimes hardly differentiated, the customer is aiming to appreciate more and more the services affiliated to the product (before sale and after sale) and also the overall relationship quality with the supplier or vendor. The practice of out-sourcing, for example, has led quite naturally to the formation of strategic corporate alliances. In this world, the customer is more sensitive to a personal approach.
The digital era offer new ways to making business. Their success consists in obtaining the needed information, their refinement and the possibility to execute transaction in real time. So far, the companies couldn't entirely capitalize on the business electronic systems advantages. In this moment we assist to a large society transformation, from a paper based to an electronic based fashion.
The virtual enterprises or web-enterprises are also global scale enterprises, acting on planetary level by using ICT. As a matter of fact, enterprises are "based on information", on knowledge, build on advanced information processing technology. In those enterprises, the management structure is simpler, with less bureaucracy and, in conclusion, more flexible. The information has a multiplier role, activating all economical sectors.

2. EXTENDED ENTERPRISE AND INTEREST GROUPS

Starting from a single national company view, two available models can be underlined. First is the extended enterprise (figure 1), concept which brings together the core enterprise, partners, suppliers and customers, in order to improve the economic efficiency and to respond to today's market challenges. This is a centric organizational view, having as objective the activity integration of all organizational categories.

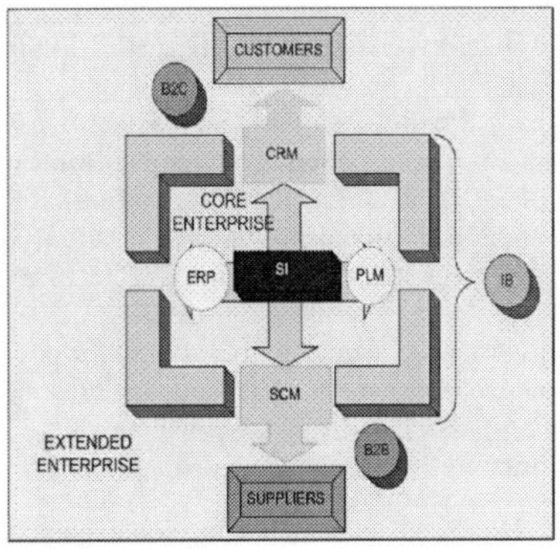

Fig. 1. Extended enterprise

In this model, we can identify the present IT solution. So, for IB relationship, ERP resource centric and PLM product centric are used, considering them as KBS for enterprise. CRM is used to support the B2C, and respectively, SCM for B2B. All those application are related with a SI (System Integrator). The second model presented is the Interest Group (IG), formed as an alliance between independent enterprises (figure 2). The Interest Cluster is defined as area where more IGs, enterprises or other organizations develop their commercial activities (e.g. ITC, construction, healthcare). The IGs' topology can by fixed (grouped by capital nature, such as transnational companies) or dynamic (enterprises can migrate from one IG to another or make their own businesses). The models presented are not opposite, being complementary, seeing the enterprise from two different views, from enterprise centric and alliance centric view. The success of organizational structures presented be is critically affected by e-legal frame and ICT meanings.

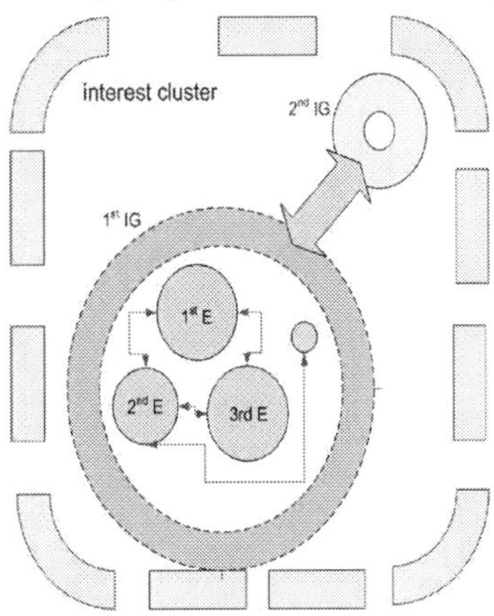

Fig. 2. Interest group

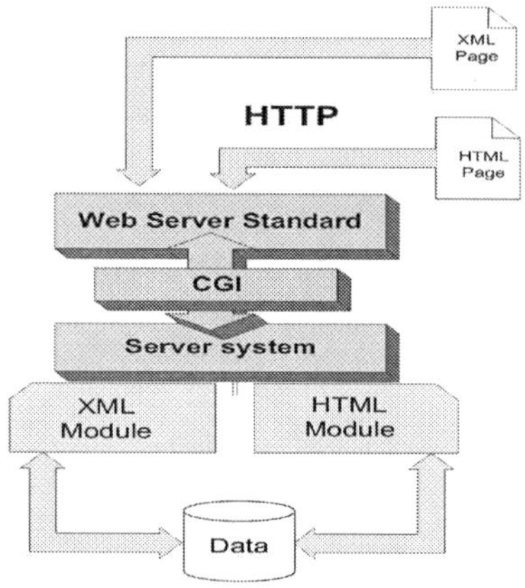

Fig. 3. Client-server architecture

3. SOFTWARE ARCHITECTURE RESEARCH

The client server architecture, which is "a facto" estate for the majority of actual information systems, is offering a large accessibility to DBMS information resources. The Internet expansion, combined with mobility and accessibility needs, has determined the generalization of this architecture in DBMS frame.

The client-server DBMS form the foundation for new solutions, covering financial and economical activities, such as: BI (Business Intelligence), ERP (Enterprise Resource Planning), CRM (Customer Relationship Management) or SCM (Supply Chain Management).

The platform architecture is the physical implementation of application (figure 3), fulfilling the following tasks:

- implementing client-server type information architecture having Internet and WWW technologies as support;
- assuring data and information management by including of Web server, terminal server and application server functions and encoding of numerous Java applet to transfer the client request to the receiver.
- simulating a central server, accessed using Web interface with CGI (Common Gateway Interface) and Java applets;
- being flexible, scalable and to assure components interconnectivity and platform independent;

4. DI – IMS PROJECT

The Research Project CEx nr.83/3.10.2005 "Forming, designing, and developing, in virtual extended organization concept of manufacturing systems" create a community of ideas and employees who use the tools and resources, enabling them to interact on-line with each other, leading to an increased exchange of ideas and business opportunities.

The partners involved can be defined as an on-line platform for services, support, knowledge and experience exchange and cooperation among industry, software suppliers and universities in the area of business development, ICT based applications and e-business services at the national and regional levels.

The other goal of the project is to provide neutral and supplier independent expertise that covers products and services related to ICT applications and e-business that are specific for industry, including training, advice and expertise in this area, facilitating exchange of information, knowledge and experience between companies. This part of the activity is aimed at the creation of centre capable of supplying support services to Romanian enterprises. The specific objectives to be accomplished are:

- consulting services, traditionally offered through the application of expertise in management, ICT and other business advisory services and standard implementation of the best business practices for applications requested by industrial partenrs and other clients
- technical support, updates and new enhancement for daily-use of software products and technical support for network management or Internet connection and use for applications specific to e-business;
- training for staff and managers in using new information and communication technologies, and concepts of competition on the digital economy and global market;
- delivering an environment for e-working and e-learning, combining those two concept in order to achieve the best result for students training.

5. CONCLUSION

The overall business requirements for e-economy environment is directly connected to the market challenges having as result the today's enterprise need in order to satisfy a specific request appeared in market in one time and one place. To do so, specific goals must been defined:

- *changing the enterprise organization fashion*, adopting a matrix organization form, oriented to multi-polar project management, each business being seen as a project, having as result a product or a service;
- *limiting to a core expertise,* the global market impose a maximum efficiency for an enterprise, by out-sourcing all non-profit activities, keeping instead the most competitive activities in comparison with the concurrence;
- *concurrent process implementation*, segmenting and parallel process suite having as result reducing the time-to-market;
- *create multidisciplinary subsystems*, having multidisciplinary subsystems solve various problems that can occur in a project developing;
- *integrating with partners,* develop an IT structure able to maintain the collaborative process with the partners involved in enterprise projects.

6. REFERENCES

Botezatu, C.P (2006) Current demands in economic IR applications design, Proceedings of the 2006 International Conference on Commerce , pp.579-582, ISBN – 13 978-973-594-785-, May 2006, ASE, Bucharest

Carutasu, G. & Aurite, T. (2005) Partners evaluation in virtual enterprise, *Scientific Bulletin of University POLITEHNICA of Bucharest*, U.P.B. Sci. Bull., D Series, Vol. 77, nr. 4/2005, pp. 25-36, ISSN 1454-2358.

Guran, M., Carutasu, G. & Patrascu G. (2005) Virtual workspace in project planning, *Proceedings of International Conference on Integrated Engineering*, pag. 179-180, Editura Politehnica, October Timişoara, ISBN 973-625-259-0

Popplewell, K. & Harding, J. A. (2005) Impact of Moderation through the Virtual Enterprise Life Cycle, *Proceedings of INTEROP-ESA,* , http://interop-esa05.unige.ch/INTEROP/Proceedings/IndustrialPresentations/Popplewell.pdf Inter-operability of Enterprise Software and Applications'05 February, Geneva

Schranz, M., Dustdar, S. & Platzer, C. Building an integrated pan-European News Distribution Network. *6th IFIP Working Conference on Virtual Enterprises*, PRO-VE 2005, 26 - 28 September, Valencia, Spain, Springer-Verlag.

Annals of DAAAM for 2006 & Proceedings of the 17th International DAAAM Symposium, ISSN 1726-9679
B. Katalinic, Published by DAAAM International, Vienna, Austria 2006
150th Birthday of Nikola Tesla (1856-1943) Inventor of Mobile Robot Teleoperation

TOOL WEAR AS ELEMENT OF SURFACE ROUGHNESS MODEL IN THE MODERN MANUFACTURING TECHNOLOGIES

CIELOSZYK, J.

Abstract: *In this study the authors: examine the effect of the tool wear on surface roughness, discuss the tool change during the wear process and its influence on surface speciality in context of a numerical model for calculating surface roughness in the modern manufacturing technologies.*
Key words: *wear, surface, roughness, numerical model.*

1. INTRODUCTION

Several researchers have studied surface finish generation, but not much has been reported in the literature on the tool change during the wear process and its influence on surface speciality in context of a numerical model for calculating surface roughness (Baek et al., 2001). Researchers have studied the influence of tool wear on process outputs with the aim of developing reliable tool condition monitoring system only (Franco et al., 2004).
It must be noted that there are a lot of theoretical models (with the introduction of many additional parameters) that relate surface roughness to cutting conditions (Benardos & Vosniakos, 2003). For all that, there are not possible cooperated real measuring profile with theoretical profile for assuming model. These most models which we have now, are generally not accurate because now give possible to calculation every optional surface parameters too.
The model of surface (which is need) must describe numerically the profiles P (unfiltered actual profile), no only some single standard surface roughness or waviness parameters, or roughness, waviness profiles (Cieloszyk, 2004).
The advanced manufacturing technologies such as Computer Aided Machining (CAM), Integrated Manufacturing Systems (IMSs) required the use of precise, numerical surface roughness model too. It is necessary to develop a model that includes the influence of a lot of factors, but what with the tool nose shape change during the wear process ?
It is generally stated that wear of cutting tools adversely affects the quality of surface. Is it always true and how it may be included in numerical surface roughness models ?

2. EXPERIMENTAL PROCEDURE

2.1 Experimental set-up

The experimental works were performed on a CNC multi-purpose milling machine FYN-50N using typical construction of mills: face mills: 257.1, SECODEX-C R230.19, and solid carbide endmill F4AS0800ADL38 type KC633M.
The body of face mills are fit out in a scrape edges or a scrape edges and one smooth edge (when work with big feed is planed) with cermet CM15 or uncoated carbide S10M. The replace wiper type inserts with the long crowned parallel land are locked like cutting edges. These inserts are normally of precision type and automatically positioned 0,05÷0,07 [mm] below other inserts in the face mill. The investigation of work these tools was conducted on workpiece material, steel 0,55 % C; hardness 25÷35 HRC (face mill) and PSA 1113 (endmill). The cutting speed used during face milling was: for face mill: v_c =111 [m/min] and 176 [m/min]. The feed used were f_r =2, 3, 4 [mm]. These cutters work with depth of cut a_p =0,1÷1 [mm] with smooth edge puts forward 0,07 [mm]. The main depth of cut a_p =0,4 [mm] and width of cut a_e ≈90 [mm]. The cutting parameters for end mill are: cutting speed v_c=35 [m/min], feed v_f =224 [mm/min], depth of cut a_p=1 [mm].
Measurements roughness, waviness and bearing ratio roughness were performed using a Hommelwerke profilometer T8000. All surface finish parameters were defined in the ISO and DIN specifications. Each value surface roughness, waviness and bearing ratio roughness represents the average of five values taken in measurements.

2.2 Results and discussion

The comparison the parameters, profiles of surface machined in milling for the deferent parameters, valuve of tools wear is the fundamental effect of estimate. In the fig. 1 show change the value of surface roughness parameter R_a, R_q during time of work face mill 230.19.

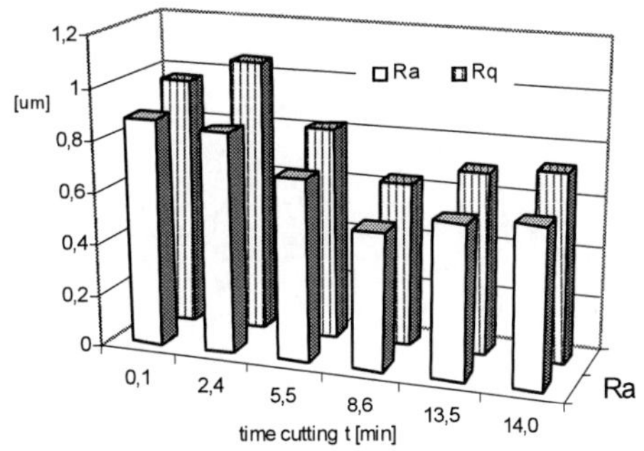

a) v_c=176 [m/min] f_r = 4[mm], a_p =0,4 [mm], (25 HRC)

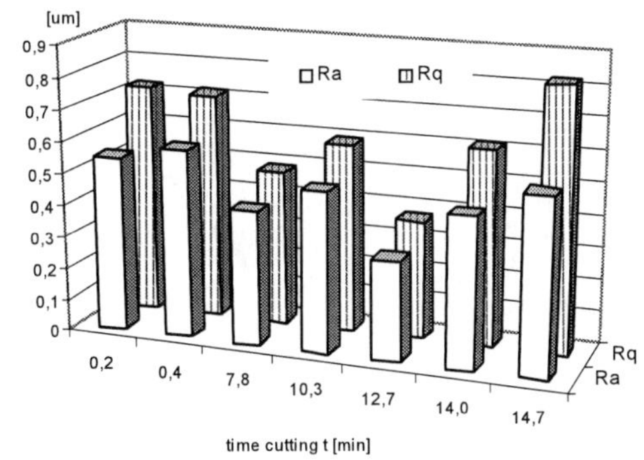

b) v_c=111[m/min] f_r = 4[mm], a_p =0,4 [mm], (35 HRC)

Fig. 1. The average value of surface roughness R_a, R_q for time of machine during face milling; R230.19

Examining this example results, it is seen that the surface change no consistently. During working time of the tool and no the excessive flank wear VB_B, the surface no change reality. It is more important when the effect of wear tool on the generated surface is significant.

In the fig. 2 show change the value of surface roughness parameter R_a, R_q and R_t during time of work face mill 230.19. In the end of working time (fig.2), the excessive flank wear VB_B will lead to poor surface texture and the surface is more heating. For machining more hard material (35 HRC) it is not so significantly because the process must be stop when the edges of tool is sudden failure (fracture).

The next series of milling experiments was made using bad machinability material. Tool live was short and the end of cutting edges frequently was fracture. In the fig. 3 show change the value of surface roughness parameter during time of work solid carbide endmill. In the first pass the roughness surface is bigger. In the next passes (fig.3), the flank wear lead to small surface texture. During working time of the tool and no the excessive flank wear VB_B, the surface no change reality. The last passed of the tool is stopped when the edge is sudden failure like for carbon steel with 35 HRC. This change of surface roughness (for researched cases) during machining and wear procccess of tool is favourable for numerical surface roughness models generaly.

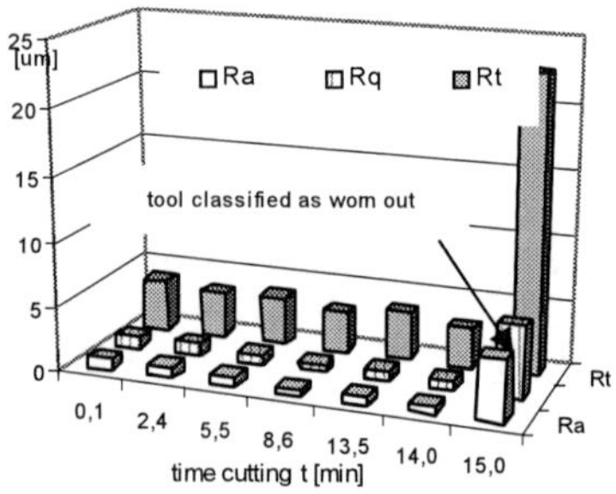

a) v_c=176[m/min], f_r = 4[mm], a_p =0,4 [mm], (25 HRC)

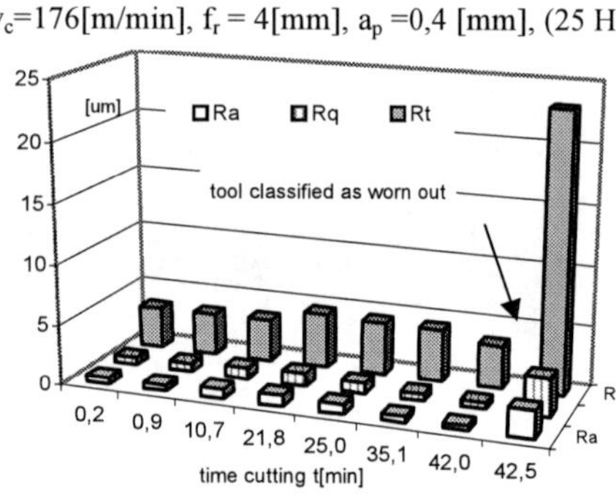

b) v_c=111[m/min], f_r = 4[mm], a_p =0,4 [mm], (30 HRC)

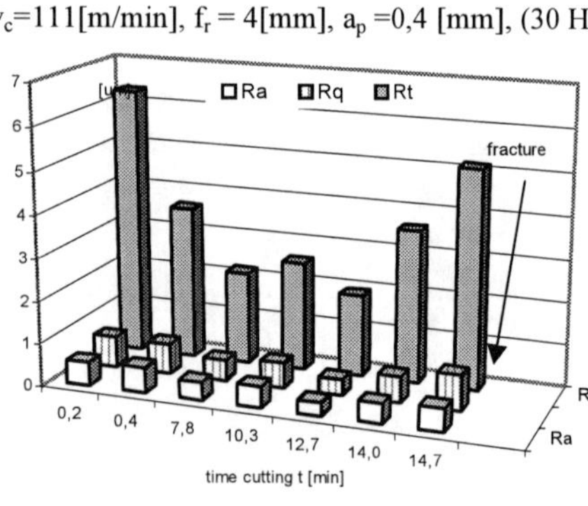

c) v_c=111[m/min], f_r = 4[mm], a_p =0,4 [mm], (35 HRC)

Fig. 2. The average value of surface roughness R_a, R_q, R_t for the end of working time of the face milling; R230.19

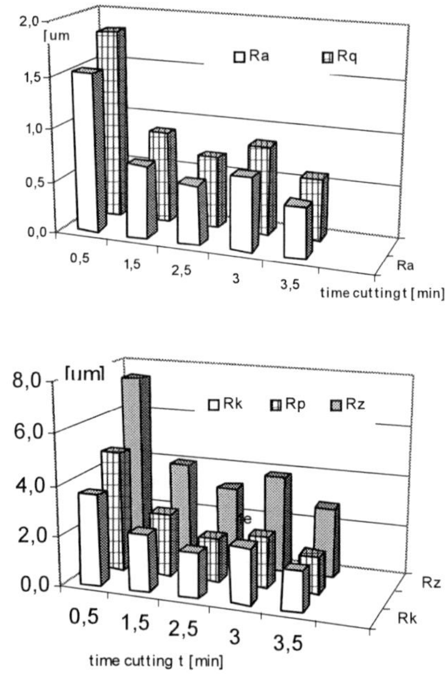

Fig. 3. The average values of surface roughness, for time of milling; solid carbide mill (v_c=35 [m/min], f_r=0,04[mm], a_p =0,4 [mm])

3. CONCLUSION

The presented results as the works of several researches too (Baek et al., 2001, Franco et al., 2004, Gu et al., 2001) show generally, that:

- Wear of cutting tools from start to no excessive, admitted flank wear of tools no influence significantly on the surface and give no worse surface roughness. It is important when the effect of wear tool (from start, stability work, and to finish) on the generated surface is considered. The tool change during cutting (effect of wear) can be frequently no included in numerical model of surface.
- The set up the values of surface roughness for first passes can be consider as a criterion (technological) of wear of the tool.

4. REFERENCES

Baek, D., K; Ko, T., J. & Kim, H. S. (2001). *Optimization of feedrate in a face milling operation using a surface roughness model,* International Journal of Machine Tools & Manufacture 41, pp. 451–462.

Benardos, P.G. & Vosniakos, G.C. (2003). *Predicting surface roughness in machining,* International Journal of Machine Tools & Manufacture 43, pp. 833–844.

Cieloszyk, J. (2004). *The prediction of surface roughness- the methodology and strategy,* Proceedings of 15th International DAAAM Symposium, Katalinic B. (Ed), pp. 073-075, 3÷6th Nov. 2004, Vol.1, Vienna, Austria.

Franco P., Estrems M. & Faura F., (2004). *Influence of radial and axial runouts on surface roughness in face milling with round insert cutting tools,* International Journal of Machine Tools & Manuf. 44, pp. 1555–1565.

Gu J., Barber G.C.; Jiang Q. & Tung S. (2001). *Surface Roughness Model for Worn Inserts of Face Milling, Part I-Factors that Affect Arithmetic Surface Roughness,* pp.47-52, *Part II-An Empirical Model,* pp.142-146, Tribology Transaction, Vol. 44.

Accessed: 2006.09.19

Annals of DAAAM for 2006 & Proceedings of the 17th International DAAAM Symposium, ISSN 1726-9679
B. Katalinic, Published by DAAAM International, Vienna, Austria 2006
150th Birthday of Nikola Tesla (1856-1943) Inventor of Mobile Robot Teleoperation

E-LEARNING IMPLEMENTATION FOR CONTINUOUS HUMAN RESOURCES PREPARATION IN VIRTUAL ENTERPRISES

CIOBANU, L.F. & PARPALA, R.C.

Abstract: *Large global corporations, which are present in different countries or continents, often need to train their staff on certain new processes and procedures. In such cases, face-to-face training would be extremely expensive, or not feasible at all, resulting in employee dissatisfaction and decreased efficiency. The traditional approach of sending instructors to field locations to explain the latest improvement in the invoicing procedures is, in most cases, not feasible anymore. As processes are becoming global, learning processes must also become global. E-Learning can bring all the relevant people together, either synchronously or asynchronously, and facilitate the appropriate training or learning environment in an adequate time frame.*

Key words: *e-learning, virtual classroom, virtual enterprise, continuous education*

1. INTRODUCTION

The terms "distance education" or "distance learning" have been applied interchangeably by many different researchers to a great variety of programs, providers, audiences, and media. Its hallmarks are the separation of teacher and learner in space and/or time, the volitional control of learning by the student rather than the distant instructor, and non-contiguous communication between student and teacher, mediated by print or some form of technology.

The growing popularity of e-Learning (that is, using the World Wide Web and electronic technology for learning) has introduced new terms to education, such as live virtual classroom.

Just as the term *virtual* means a simulation of the real thing, so a *virtual classroom* allows learners to attend a class from anywhere in the world and provides a learning experience that is similar to a real classroom.

When one goes to school or to a university, he has a schedule of classes he must attend. He must arrive on time, and when he enters the classroom, he may find a teacher, fellow learners, a blackboard or whiteboard, an overhead projector, a library, and a television screen with videos.

Likewise, a live virtual classroom is a scheduled, online, teacher-led training session where teachers and learners interact synchronously using computers linked to a network such as the Internet.

A Virtual Classroom offers the following functions:

- Course builder - provides the framework for building outlines, selecting virtual classroom tools, authoring assessments.
- Course scheduling - allows you to schedule courses.
- Notifications - handles course invitations and reminders.
- Course catalogue - lists all available courses, or the courses in which a learner has enrolled

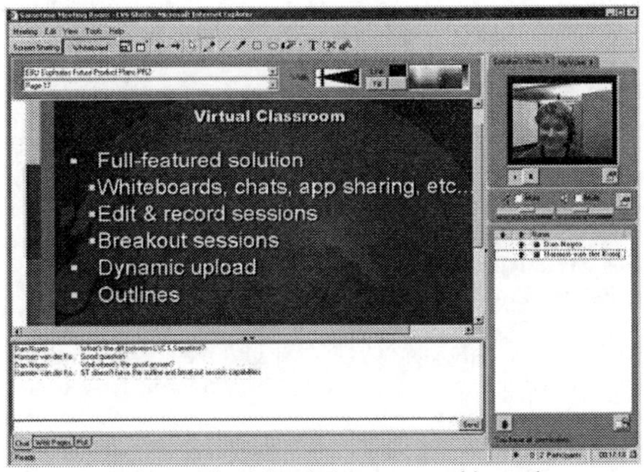

Fig. 1. Overview of live virtual classroom and its tools

- Enrolment manager - administers the enrolment and admission of learners in courses.
- Administrative tools - includes user management, enrolment reports, security settings, and the like.

2. BENEFITS OF VIRTUAL CLASSROOM VERSUS ASYNCHRONOUS WEB-BASED TRAINING

A virtual classroom also carries benefits when compared with an asynchronous course delivered over the Web.

Preparing materials for a virtual classroom session takes much less time than developing an asynchronous Web-based training session. Innovative or complex knowledge that is not yet published can be shared immediately with the rest of the organization. Scheduling a virtual classroom session, creating an outline, and preparing presentation materials take a matter of hours. When training needs to be delivered quickly, the virtual classroom can be the perfect tool for the job.

Since the virtual classroom is a facilitated activity, teachers can respond immediately to learner questions and feedback. In an asynchronous environment, the learner usually has to wait at several hours or more to receive an answer to a question.

Asynchronous environments are notoriously difficult to maintain when learner participation is poor.

When working in a virtual classroom environment, the teacher has a variety of tools and methods to monitor and stimulate participation in real time.

Many types of e-Learning do not offer any communication or collaboration. Asynchronous collaboration can allow learners and teachers to interact, but it is not as direct as in a traditional classroom.

In contrast, a virtual classroom does offer this kind of interaction. Learners can talk to the teacher and to each other, and although this communication is not as intuitive and rich as in a traditional classroom, it still can help learners to learn.

Teachers are key to facilitated learning. They motivate learners, introduce new topics, take care of group processes, and so forth. The virtual classroom allows them to immediately respond to learners' needs.

The virtual classroom uses many metaphors from the traditional classroom with which learners are familiar, including classroom, whiteboard, and hand-raising. Through the use of audio and video, communication is more natural and open.

3. BENEFITS OF VIRTUAL CLASSROOM VERSUS CLASSROOM-BASED TRAINING

A virtual classroom offers benefits when compared with the same course delivered in a traditional physical classroom.

Similar to other e-Learning technologies, a virtual classroom allows learners and teachers to attend a single live training session from any place in the world, as long as they are connected to a corporate network or to the Internet. This solution offers a much broader audience and cheaper solution than a single traditional classroom-based training session; for example, the ability to instantly share the expertise of a research and development department with sales employees across the world.

Traditional learning environments are very resource-intensive and relatively inflexible. They usually require classrooms, furniture, writing materials, and the like. Both under-capacity and over-capacity classes can be very expensive in the traditional environment. For example, near-empty classrooms carry overhead, while overflowing physical classrooms leads to expensive outsourcing. The virtual classroom has a very different resource model. The marginal costs of adding virtual classroom training are relatively insignificant compared to the traditional model.

The speed of virtual classroom session deployment offers the competitive advantage of bringing new training to market a lot faster than before. Training can be organized more quickly than traditional classroom-based training. Classrooms and projectors do not need to be reserved, materials do not need to be distributed, route descriptions do not need to be sent out, and so forth. The sessions are easier to schedule since attendees do not need to travel. If teacher-led training needs to be delivered in other countries, the benefit of the virtual classroom will be even greater. If learners miss a traditional classroom-based training session, they have little opportunity to engage in the learning experience that took place. If a virtual classroom session is recorded, learners or teachers can replay it afterwards. Learners can revisit the session, perhaps making connections and understanding ideas or concepts they had missed at the time of the session. Teachers have the opportunity to review their own or their colleagues' performance. Of course, this should not be considered as a replacement for active participation in a virtual classroom session. For example, once recorded, a virtual classroom becomes an asynchronous activity without the possibility of interaction. Just as the traditional classroom provides learners with a set of communication skills applicable to a specific task, the virtual classroom provides them with new information and technical skills that benefit both individuals and their organizations, such as the usage of Web cams and audio. In general, it will make the participants more familiar with Internet capabilities.

4. CONCLUSION

Virtual classrooms are becoming more and more necessary in our days because of an imperious need for continuous education in the virtual environment.

Beside the benefits already passed in review, there are many others related to courses costs. One area where e-Learning can help a corporation to remain competitive is the reduction of costs associated with learning. With e-Learning one can dramatically improve cost savings.

Some categories of costs that might be reduced are:

- Travel costs

This area of improvement is particularly relevant to geographically dispersed corporations in which travel, living and hotel expenses consume a significant portion of the learning budget. Either instructors must travel to different training locations, or learners must all meet at a particular venue. We must also add the cost of the time invested by both instructor and learner to go to the training venue.

- Instructor costs

In some cases, when the training must be performed simultaneously in different places, or when there is a very tight schedule, companies are sometimes forced to employ more than one instructor (either internal or external), who in turn needs training, in order to provide enough traditional classes.

- Time away from workplace costs

Time away from the workplace is a cost that is often overlooked. While an employee is away taking a course, this person does not produce, and therefore it should be considered an added cost. With e-Learning, training delivery is shortened. Learners can work from their workplace computers. They can choose to run the sessions during off-peak periods when training will not take up production time.

- Venue costs

Whether there is travel involved or not, there are certain costs associated with face-to-face sessions which cannot be avoided, namely: maintenance of the training room or lab; rental of rooms at different locations such as hotels or business centers; provision for refreshments; rental or maintenance cost of required equipment such as computers, overhead projectors, screens, and so on.

- Reuse of materials

When a traditional face-to-face training session has been delivered, it is over. If learners missed it, or had problems in following it because it was about an unfamiliar topic, they will not able to revisit it. With e-Learning, learners can go back to the training environment as many times as required and reread the topics or replay the sound or video tracks. e-Learning also allows one to take modules and reuse them or update them when necessary.

All this costs can be reduced using e-learning and virtual classroom. This gave us the right to say that e-learning is a very important piece in today's economy development and a powerful tool in the hands of virtual enterprises mangers.

5. REFERENCES

Ebbers, M., Balague, D., Ganguly, B., Noyes, D.& Salm, P. (2004), *Using the IBM Lotus Virtual Classroom: A Best Practices Guide to e-Learning,* IBM, ISBN: 0738429414

Ebbers, M. (2004), *Lotus LearningSpace R5.01 Deployment Guide,* IBM, ISBN: 0738426415

Bergland, J., Bravo, A., Chadbourne, G.R., Luz, C., Monson, P., Mueller, H., Savov, T. & Slone. J. (2004), *Lotus Workplace 2.0.1 Products: Deployment Guide,* IBM, ISBN: 0738491934

Niven, P.R., (2005) Balanced Scorecard Diagnostics: Maintaining Maximum Performance, Wiley, John & Sons, Incorporated, ISBN 0471681237

Kaplan, R. et al, (2004) *Strategy Maps: Converting Intangible Assets into Tangible Outcomes*, Harvard Business School Publishing, ISBN 1591391342

Annals of DAAAM for 2006 & Proceedings of the 17th International DAAAM Symposium, ISSN 1726-9679
B. Katalinic, Published by DAAAM International, Vienna, Austria 2006
150th Birthday of Nikola Tesla (1856-1943) Inventor of Mobile Robot Teleoperation

CATIA MODEL MOTION OF ROBOT SCARA IN THE SPACE

COHODAR, M.; KULENOVIC, M.; VOLODER, A. & LIKA, E.

Abstract: *An attention of this paper is focused on problem of simulation of real robot manipulator with CATIA V5 R14. It's pointed out the importance of simulation during design and development of real engineer problems. Simulation allows prediction how mechanism with more degrees of freedom will operate in virtual environment and shows its interaction with environment. Simulation also permits to user easy detection of collision between components – links, joint, etc. during their assembly, as well as analyzing of motion envelopes. Simulation of 'pick and place' tasks of robot SCARA is performed herein and it's shown development procedure of robot mechanism.*

Key words: *simulation, SCARA, 'pick and place', kinematics*

1. INTRODUCTION

Visualization of mechanism motion is challenge task in mechatronics and robotics. This is very efficient way to see out an interaction of robot mechanism with environment, possible collision between components, as well as analyze motion envelopes. CATIA V5 allows completely joining between simulation model and its geometry. This enables see out effects of performed changing of model geometry on resulted motion after re-running of simulation.

There are more software packages used for simulation of different mechanism motion. In (Gnasa, 1999) both CAD-Software Pro/Engineering and Mechanism/Pro (the integrated multibody system analysis module) are applied for modelling and simulation of a hydraulic manipulator. Importance of simulation is subject of paper (Vin, 2001).

2. KINEMATIC MODEL

Kinematics of robot mechanism plays an importance role in simulation of its motion. Kinematics covers investigation of body motion from geometric viewpoint without exploring forces as cause of this motion.

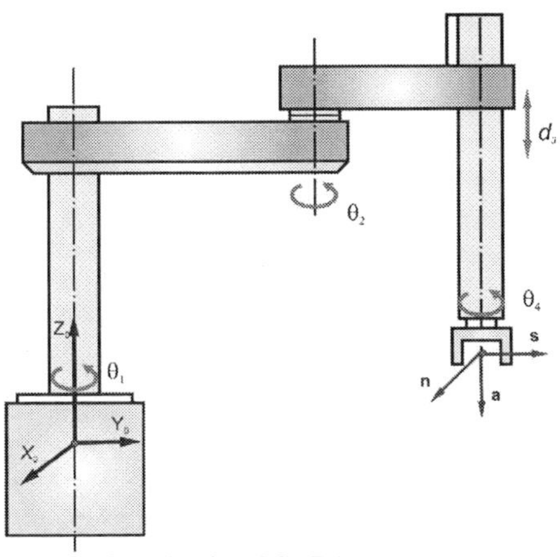

Fig.1. Scheme of mechanism SCARA

Robotic mechanism in more applications can be treated as open kinematics chain of links connected with joints. Many times, rigid body assumption can be accepted because robot operates with short links which relatively large cross-section area. Resulting of robot motion is obtained by composition motions of all links with regards to previous. As robot mechanism achieves interaction with environment via its end-effector that is necessary to know position and orientation of end-effector in the space.

Robot manipulator SCARA, Fig.1, consists of five links jointed with four joints from which first two and fourth joint are revolute, while third one is prismatic. This manipulator structure has high stiffness to vertical loads and compliance to horizontal loads which leads to high mobility in the horizontal plane. Direct and inverse kinematic model of manipulator SCARA are given in (Stadler, 1995).

3. DESIGN OF MANIPULATOR SCARA IN CATIA V5

Design of robot mechanism consists of following steps:

- create separated parts,
- create a mechanism with defining fix parts,
- create joints that means define constrains so that joint degrees of freedom are remained,
- define commands and
- perform simulation.

After that links and the fixed base have been created in module *Part design*, it's necessary to join them in order to create mechanism. It's done in module *DMU Kinematics*, which allows selection of desired joints: revolute, prismatic, cylindrical or screw. Procedure creating of the first revolute joint is shown in Fig. 2. Similar procedures are performed for the other joints. In window *Joint Edition,* it's necessary to fill in constrains consistent with selected joint.

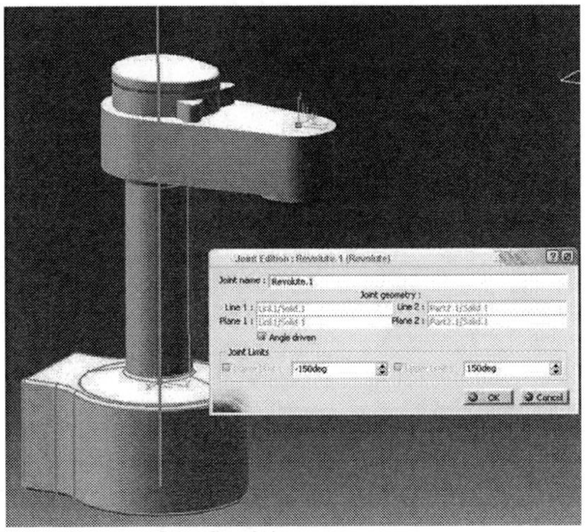

Fig. 2. Creating of the first joint

The specification three updates after each joint has been created (Fig.3.). As we can see from three, next to part title it exists titles of parts in brackets which it joins, for example, *Revolute1* is the first joint of mechanism and it joins *Link1* which is the base part of mechanism with link *Part2.3*.
Final model of robot mechanism SCARA, developed with CATIA V5 R14 is depicted in Fig.4 (Lika, 2006).

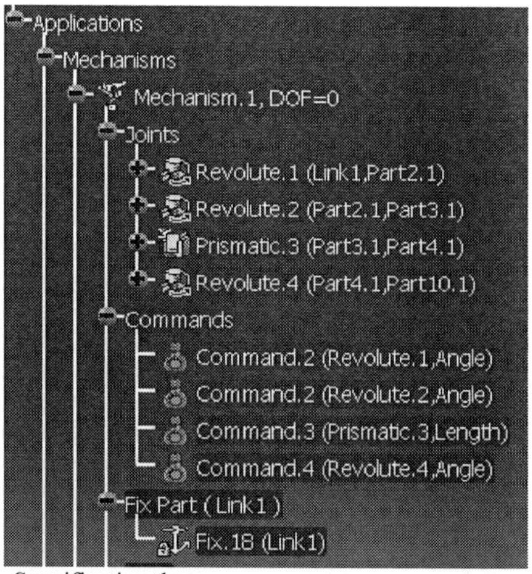

Fig. 3. Specification three

Fig. 4. Final model of robot mechanism SCARA

4. SIMULATION OF 'PICK AND PLACE' TASK

With *Simulation with Commands* it's defined simple trajectory which is frequently used for so-called 'pick and place' tasks and it consists of following steps:

- Move end-effector to some position,
- Down end-effector and pick of workpiece,
- Move end-effector to another position and
- Down end-effector and place of workpiece.

Simulation is performed over 15 s. Coordinates of the last link end-point are tracking and saved during the simulation. This data can be exported in, for example .xls file for processing in MS Excel. Trajectories of the last link end-point in *X-Z* and *X-Y* planes are shown in Figs.5a,b.

5. CONCLUSION

Realistic simulation motion of mechanism allows verify behaviour of the mechanism, as well as checking of validity of its assembly. This paper demonstrates power and the abilities of CATIA V4 R14 - DMU Kinematics module in simulation of a wide variety of engineer problems through simulation motion of robot mechanism. Simulation allows seeing out competed system design in whole, errors detection, assembly collision without performing experiment with real system that leads to less consumption time and less costs. Additionally, it's possible to design of virtual environment according to complex real-world, and further continuance of this investigation would be goes in this direction.

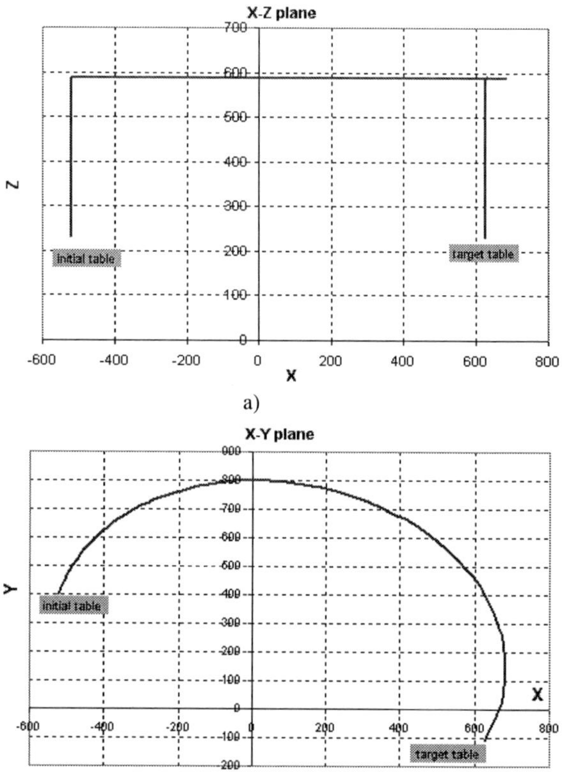

Fig. 5. End-point trajectory in: a) *X-Z* plane, b) *X-Y* plane

6. REFERENCES

De Vin, L. J. & Jagstam, M. (2001). Why we need to offer a modelling and simulation engineering curriculum, *Proceedings of the 2001 Winter Simulation Conference B. A. Peters, J. S. Smith, D. J. Medeiros, and M. W. Rohrer, eds. Available from:* http://www.informs-cs.org/wsc01papers/219.PDF, *Accessed:* 2006-05-20

Gnasa, U.; Thielecke, K.; Modler, K.-H. & Richter, E.-R. (1999). Design and development of hydraulic Manipulator with Mechanism/Pro, *International ADAMS User's Conference Berlin, November 17-18, Available from: http://www.mscsoftware.com/support/library/conf/adams/euro/1999/Wednesday_papers/Parallel%20Papers/Design%20and%20Development%20of%20a%20Hydraulic%20Manipulator.pdf, Accessed:* 2006-05-20

Lika, E. (2006). Simulation of robot mechanisms using CATIA V5, diploma thesis, Faculty of Mechanical Engineering, Sarajevo

Sciavicco, L. & Siciliano, B. (2001). *Modelling and control of robot manipulators (Advance Textbooks in Control & Signal Processing*, Springer-Verlag, ISBN 01852332212

Stadler, W. (1995). *Analytical robotics and mechatronics*, McGraw-Hill Series in Electrical and Computer Engineering, ISBN 0-07-060608-0, San Francisco State University

Annals of DAAAM for 2006 & Proceedings of the 17th International DAAAM Symposium, ISSN 1726-9679
B. Katalinic, Published by DAAAM International, Vienna, Austria 2006
150th Birthday of Nikola Tesla (1856-1943) Inventor of Mobile Robot Teleoperation

DATA GATHERING USING RFID TECHNOLOGY FROM DISASSEMBLY AND RECYCLING SYSTEMS

COSIC, I.; LAZAREVIC, M.; ANISIC, Z. & LALIC, B.

Abstract: *Demands for sustainable development and WEEE directive, placed upon the manufacturers impose the need for optimal material and product flow management at the end of the product life cycle. Till now, these flows didn't have needed intention and these problems were analyzed only in several cases. Disassembly and recycling are the last phase in material and product life cycle. Gathered information is important for optimal management of material flows. Production system, RFID technology, disassembly and recycling systems, designed according to strategies for product end of life management (EoL management), coupled together represent a good way to reach the goal, meaning better management of obsolete products, parts and materials.*
Key words: disassembly, recycling, RFID technology, EoL strategies, information

1. INTRODUCTION

In an attempt to solve the problem of waste and slow down the depletion of non-renewable natural riches, disassembly has become increasingly important. Namely, a conclusion was made that a large number of products which arrive to the waste dump (the last and the usual end phase of a life cycle of most of the products), should now, during the process of disassembly and dismantling to the basic components, be redirected utmost to the systems for renewal, reprocessing and recycling.

WEEE directive (Robbins, 2003), and as its integral part, the product take back strategy support the endeavor to make recycling as efficient as possible. System processes, as disassembly, make it possible that the renewable, non-recycling and dangerous components are selectively separated from those which are possible to recycle. A properly planed process of disassembly, together with recycling and waste disposal, can help a lot to manage the products during their life cycle in the most efficient way.

Information about flows of products, materials, parts/subassemblies and dangerous components/materials out of and in the product processing system, at the end of the life cycle, according to proper strategies, are necessary for something like this to be achieved. By applying RFID technologies to the systems for refining the outdated product, we can improve the management of all the above mentioned flows. Information gathered in this way is reliable, precise and dynamic and they provide optimal management with products that have come to the end of their work lifetime.

2. PROBLEM DESCRIPTION

In the life cycle of the product we can separate partial characteristic phases such as production, sale, use and manufacturing systems for product/material/component (parts/subassemblies) refinement at the end of the life cycle. All possible phases through which products / materials / components (parts / subassemblies) can be processed are shown on Fig. 1., as are the required information paths between certain systems.

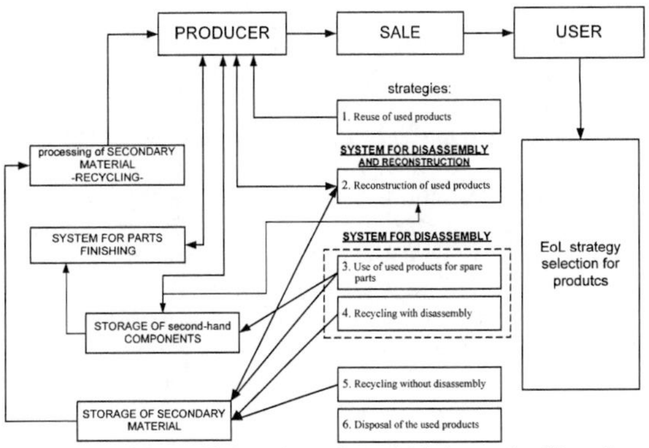

Fig. 1. Required flows of information and material during product life cycle

In this paper, the application of RFID technology will be shown, in certain phases of the product life cycle. A special sugesstion is given for relation to the application of RFID technology in the disassembly systems.

3. THE REALISATION METHOD OF RFID TECHNOLOGY

Studies related to the strategies of the end of the life cycle of a product are very numerous. The most acceptable, and in its character the most comprehensive classification of the end of a life cycle of the product is (Lazarević et al., 2006):

- Reuse of used products;
- Reconstruction of used products;
- Use of used products for spare parts;
- Recycling with disassembly;
- Recycling without disassembly;
- Disposal of the used products.

Reuse of already used products is a strategy which organizes the return of discarded products which are still in function. If such an interest exists, already used products are sold in the market.

Reconstruction of used products is applied for the modernization or to upgrade their performances. The purpose of this strategy is to attain a product which is in quality less or very similar to the quality of the new products. The quality of the reconstructed products depends on the determined depth of disassembly. If a product is disassembled to the level of parts and a control and a replacement of all the parts is conducted, the used products are brought to a high level of quality, required for the new products. Also, it is possible to conduct the modernization of products, by replacing certain modules with contemporary ones (product upgrade), after applying the disassembly.

Using already used products for spare parts is already being applied frequently. In certain companies, out of date products are being collected in an organized manner. The purpose of this strategy is to take a relatively small amount of parts/subassemblys from a used product and use them for the above mentioned strategy, or in another purpose, and the rest will be used for material recycling.
Recycling with disassembly is a strategy used for separating different parts made of different material, before their conversion in the process of disassembly. The purpose of this strategy is to use the materials from the used products and parts, by separating them in the procedure of disassembly into the component parts and with appropriate selection, depending on the determined type of material. These materials can then be used in the production of original or some other products.
Recycling without disassembly is a procedure which is used to compact and compress the product, and then attrition and sorting by type of material.
From the ecological point of view, disposal on the waste is the most inconvenient strategy for managing products at the end of their useful life.
Having the above mentioned strategies in mind it is necessary to design an appropriate production system in accord with them. Systems for refinement planed in a way to consider the EoL strategies in the work result give very complex information regarding the selection of proper components after the completion of disassembly process.
After choosing the appropriate strategy for a given product and a sequential execution of certain operations of disassembly it is necessary to conduct a selection of disassembled components (parts/subassemblies). Basically, we can suggest the next possibilities for the selection of components after the completed disassembly (Lazarevic et al., 2006):

- Dangerous components – materials (H (hazard));
- Material recycling (R);
- Reuse (P);
- Finishing, (D);
- Incineration (I);
- Waste disposal (W).

Components disassembled from a product are identified, measured and sorted in different phases of processes. During the disassembly process it is necessary to place the appropriate RFID tags on the disassembled components or containers for identification of the disassembled components.
A reader placed on the disassembly line is used for identification. Information gathered on the disassembly line imply, for example, groups of products, producers, date of production, disassembled materials, disassembled parts, weight, disassembly instructions, etc.
Disassembled components are stored in the warehouses or proper containers, and RFID infrastructure is well suited for applications where:

- A large number of items needs to be identified individually;
- Additional data about the items needs to be managed.

The advantages of RFID technology in handling of certain parts and materials are shown in (Baudin & Rao 2005):

- Product identification and connection with the producer in any phase of the life cycle of a product;
- The possibility of automatic sorting of products;
- Decentralisation of information and intelligence;
- A list with dangerous materials which are contained in the product;
- Product information on spare parts;
- More efficient use of resources;
- More accurate and faster flow of parts;
- Higher durability than bar codes;
- Part value determination;
- Automation and increased efficiency of information management, etc.

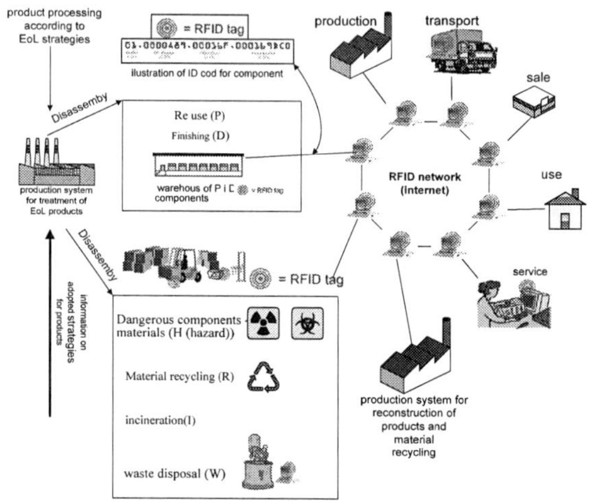

Fig. 2. Data gathering using RFID technology

The application of RFID technology is shown in (Fig. 2) during the life cycle of a product (McFarlane, 2003), with special attention on the production system for the disassembly of products in the end of the life cycle (EOL products) – systems for disassembly.

4. CONCLUSION

There are different reasons why RFID technology has not been in wider use during the life cycle of a product. Some of them are: the widespread usage of barcode, higher expenses due to RFID tags, privacy and security measures, lack of consensus on standards and other reasons.
The application of RFID technologies in the disassembly will be easyly applyed in the next period when certain producers start incorporating RFID tags with unique EPC code in their products (for example, electro- products which are included by the WEEE directive).
In this paper, it is given an overview of possibilities for the use of RFID technologies and information network, for the entire life cycle of the product. Further work will focus on deeper analysis of single application scenarios in disassembly system. The costs and benefits of those applications should be analyzed in detail.

5. REFERENCES

Baudin M., Rao A.(2005). RFID applications in manufacturing, MMTI – Manufacturing Management & Technology Institute. *Available from:* www.wefixfactories.com, *Accessed:* 2005-04-25

M. Lazarević, Ćosić I., Z. Anišić (2005). The Influence of Deciding on Product End of Life Strategy Towards The Strukture of Demanufactoring System, *Proceedings 5th International Conference on Industrial Systems*, Herceg Novi, september 2005, DP Budućnost, Novi Sad

M. Lazarević, Ćosić I., Z. Anišić (2006). RFID Technologies In Product Disassembly Proces, *Proceedings 5th DAAAM ATDC*, Rijeka, june 2006

McFarlane D. (2003). Automating EPC:The Role of EPC in Supply Chain Control, University of Cambridge, *Available from: http://www.ifm.eng.cam.ac.uk, Accessed:* 2006-05

Robbins E. (2003). European Union (Eu)Electrical Andelectronic Products Directives: Directive on Waste Electrical and Electronic Equipment (WEEE), *Available from:* http://www.informinc.org *Accessed:* 2006-07-17

Annals of DAAAM for 2006 & Proceedings of the 17th International DAAAM Symposium, ISSN 1726-9679
B. Katalinic, Published by DAAAM International, Vienna, Austria 2006
150th Birthday of Nikola Tesla (1856-1943) Inventor of Mobile Robot Teleoperation

DISTRIBUTED PRODUCTION – STEP TO THE COMPETIVENESS

COSIC, P.

Abstract: *Research of the most important impact factors for organization distributed production in virtual environment is applied for production specific and original product - bike frame. Process planning and production for selected product, investment costs with eight years period of amortization and the other costs (production materials, technologies, labor, etc) were important in estimation minimum number of bike frame for production costs and business doing without any profit.*
Key words: *distributed production, competiveness, production improvement.*

1. INTRODUCTION

Today dominate often opposed manufacturing criterions: (Bruckner, P. 2004), (Nordstrom, K.A., Ridderstrale, J., 2002):

- inventiveness and innovativeness in a manufacturing,
- fast customize to customer wishes,
- expense optimization,
- flexible manufacturing,
- capacity uniformity manufacturing,
- shorter design cycle,
- faster capital turning,
- uniformity of quality,
- possibility and rate of supply, fast delivery,
- the evaluation life cycle of technology and product, emphasis on core business function,
- define of outsourcing, activities.

Every project in the field of the production improvement for the possible implementation needs companies, present on the global market. These companies need to be: responsible, involved, successful, with own vision and research, and original products manufactured with the new or original technologies.
Today many approaches in improving production are present. Very often, approaches are with marketing impacts on packing use.
Methods of twenty keys, six sigma, lean production, kaizen, reengineering, and virtual manufacturing are only parts of the systematic approach. How to cross over psychological, national, transition borders between researchers and interested companies? This paper tries to give some answers on these very important and complicated questions.

2. ESTIMATION OF TIMES, DELIVERY, COSTS

Now, perhaps more than ever, manufacturing capability – information of the new products, technological processes, systems and services means survival on global market.
Insights by personal contacts, projects, and by work with students and public media, one can gain overall picture of successful companies in Croatia. On one side, regional company with recognizable products that invest in production and which reach respectable export, often is not capable to answer certain inquires (Nordstrom, K.A.; Ridderstrale, J., 2002):

1. cost structure in the product price,
2. how to bond production, development, sales and marketing with existing and future economic environment,
3. which methods, techniques, resources should be used to trace and predict market trends
4. how to, within production strategies, chose production type, perceive behavior: contract manufacturer (CM), original design manufacturer (ODM) original equipment manufacturer (OEM)
5. how to evaluate company competitiveness based on the contradiction demands and constraints.

As a part of manufacturing strategy led by satisfaction before mentioned customer demands, leading unit strategy will be specified who will to be object of research, analysis and appliances:

- product variant as a satisfaction of market demands, various cultures, social degradation,
- product manufacturing variants (Balič, J.; Veža, I.; Čuš, 2006) (impact on lead time, manufacturing expenses, capacity occupation),
- integration and interaction of process planning and manufacturing with projecting and design.

3. ESTIMATION OF COMPANY COMPETITIVENESS

Procedure for the estimation is represented by multilevel, multicriterian and multidisciplinary approach in the process of investigation:

- company information (NACE category, number of employees, company's position within supply-chain, product development, where are products produced, the importance of branding, main customers, etc),
- technological competence (product oriented topics, production oriented topics, business operation oriented topics),
- cost structure in the product price (impacts of variants of production, machine tools, tools, material, workforce, intellectual work quota in the product price, life cycle assessment in the price, etc)
- uniformity of quality (level of automatization, checking in all phases of production, strictly defined procedures of production and control of quality, etc)
- possibility of use some aspects of virtual manufacturing
- business competence (business & economic-related topics, management-related topics).

4. CASE STUDY: PRODUCTION OF BIKE FRAMES

The main idea of this project was to investigate possibilities of production selected product in Croatia with intensively cooperation with companies from virtual environment. Bike frame from Al Zn Mg alloys for mountain bikes is selected as the most important part in definition bike price. The most important criterions in process selection of suppliers of equipment and production materials were:

- experience of the authors project (Lugovic, M; Pihir, M. 2006) as funs of cycling and mountain bikes (one of them is active competitor in triathlon),
- information from web environment,
- tube shape from light material but with sufficient strength of materials and stiffness,
- not too expensive equipment and technologies for small serial production,
- aim market is Croatian and neighboring states; in future e-sale by web application.

Software BikeCAD (Fig. 1) defines all parameters, dimensions, kind of tubes and features. It's very helpfull in design of bike frame. Process planning and production for selected product, investment costs with eight years period of amortization and the other costs (production materials, technologies, labor, etc) were important in estimation minimum number of bike frame for production costs and business doing without any profit (Stutely R. 2002).

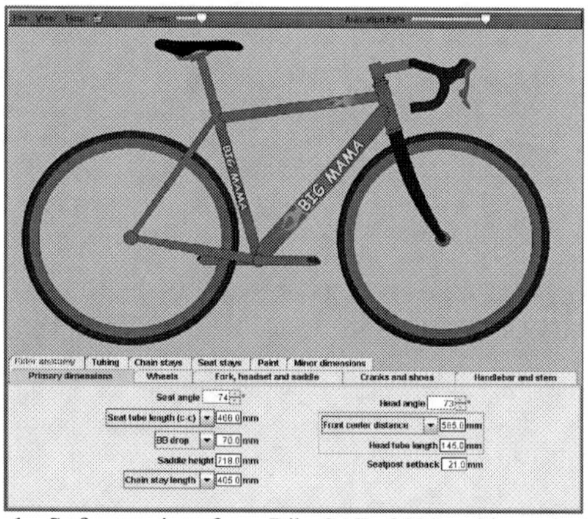

Fig. 1. Software interface BikeCAD 2003 with main bike components

In Table 1 are presented all costs: production materials (tubes), colors, parts (Fig. 2) (bottom bracket, dropout, head tube, seat tube, top tube, chain stay wheel baqse, off set, etc), labour costs, production costs (TIG welding, milling, drilling, cutting, ageing, painting, etc.), investment costs, human resource costs, etc per one bike frame.

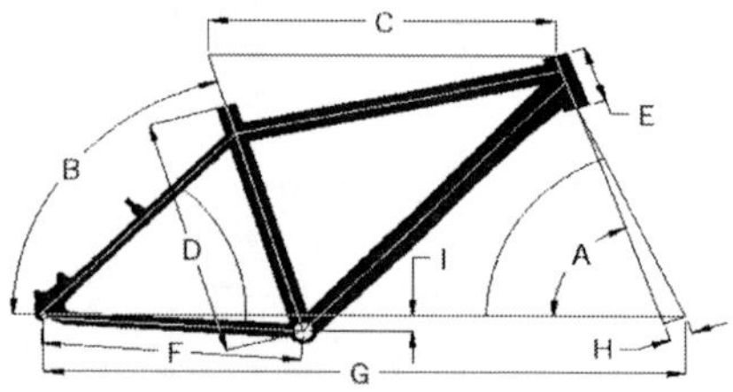

Fig. 2. All important parts of bike frame

Minimal number of bike frames is 14 per day and has importance for minimal profitable production. So, the next phase would be cycle time improvement (reduction) of tools,

	160,00 €	Price/ frame
	Tubes and components + colors	125.47 €
Labor costs	Welding	6.00 €
35,00 €	Machining (Milling, drilling)	15.00 €
Disoxidation		2.00 €
Painting		3.00 €
Ageing in chamber		1.00 €
Additional hand work		8.00 €
	TOTALLY	160.47 €
	Investment costs	76,902.00 €
	Amortization costs (per year)	10,471.50 €
	Business costs (per year)	34,630.00 €
	Costs of human resources (per year)	90,000.00 €
	Minimal number of frames for minimal profitable production	14

Table 1 Distribution production and total costs and minimal number of bike frames estimated times perframe for all types of tools (34), machines

jigs (15)and all operations (92!). So, it would be realised with real manufacturing, economical and business restrictions.

5. CONCLUSION

Project of mountain frame are theme of analyzing from economical, technological, human and labor aspects. For real success of project authors have to improve all aspects of marketing, sale, control and management of production.

Acknowledgement

This project is a part of the scientific project titled Intelligent Process Planning and Reengineering 0120-029 financed by the Ministry of Science and Technology of the Republic of Croatia. We express gratitude for the financial support for the project.

6. REFERENCES

Balic, J.; Veza, I.; Cus, F. (2006); *Napredne proizvodne tehnologije (Advanced production technologies)*, Fakulteta za strojništvo Maribor, Fakultet elektrotehnike, strojarstva i brodogradnje Split, 2006, ISBN 86-435-0776-6, Cosic, P.,: Chapter: *Planiranje procesa u web okruženju* (Planning processes in web environment)

Bruckner, P. (2004). *Bijeda blagostanja (Misére da la prospérité)*, Algoritam, ISBN 953-220-194-7, Zagreb.

Klein, N. (2002). *No Logo,* VBZ, ISBN 953-201-183-8, Zagreb

Lugovic, M; Pihir, M. (2006) Student's project named *"Procedure of mountain bike frame production"*, University of Zagreb, FAMENA, Zagreb,

Nordstrom, K.A., Jonas Ridderstrale (2002), J.; *Funnky Business,* Differo, ISBN 953-96545-0-5Zagreb,

Stutely R. (2002), *The Definitive Business Plan,* Financial Times-Prentice-Hall, ISBN 0 273 65921 9, London.

Annals of DAAAM for 2006 & Proceedings of the 17th International DAAAM Symposium, ISSN 1726-9679
Editor B. Katalinic, Published by DAAAM International, Vienna, Austria 2006
150th Birthday of Nikola Tesla (1856-1943) Inventor of Mobile Robot Teleoperation

INTEGRATION PERSPECTIVES BASED ON THE INTELLECUAL POTENTIAL

COVIC, D.

***Abstract:** The characteristic of the time we are living in is primarily recognized through the phenomenon of globalization - integration processes, liberalization, intensification of new technological ideas and of course limitedness of the resources which surround us and mean life. The recognized trends intervene, with different positive and negative synergic effects. Integration processes create a wish with many national communities to join these processes and they view them as a perspective. The trends of the global additionally in the rough division on the globalizators and the globalized contain the necessity to centralize the knowledge. By building the cells of knowledge on the potential of thousands of educated young people from the whole world of the undeveloped, the centers of economic power concentrate around themselves the centers of intellectual power. The power through the intellectual potential is practically built without risks and costs. To create the preconditions for an acceptable integration process first of all means to build your own national society of knowledge, to network your intellectual potential and knowledge must become the most important investment. By building business systems oriented to intellectual potential and knowledge, by building the networks of knowledge among business systems and the whole social environment, you give yourselves a chance to become wanted, to become a partner. It is obvious that the intellectual power itself is an opportunity for partnership in integration processes.*

***Key words**: integrations, intellectual potential, knowledge*

1. INTRODUCTION

The phenomenon of the global, no matter how limited in time it can be, has created a need for strong integration processes, that are seen by "small" societies as an inevitability without which the future is impossible.

To be a part of European Union or Euro-Atlantic integrations is an imperative to which all is subjected. Many who are in this process do not often see fully what this can mean and how important these changes are. The limitedness of (natural) resources, openness (liberalization) of markets, integration (globalization) and new (highly sophisticated, information and communication) technologies are the crucial characteristics of this moment.

Viewing these characteristics, a need is created for restructuring based on knowledge, intellectual power and potential which will create the opportunity for an internal vision of the strength in the integration process.In order to become a society of negotiators and partnership, we obviously must build, no matter how ambitious this may be, the society of knowledge or the society based on intellectual power, potential and knowledge.

The verification of our processes based on the intellectual resources will follow in the integration process. How successful we are in this intention will be recognized in accordance with role offered to us in the process of admission.

2. INTEGRATION PROCESSES

The basis of integration process lies in the human wish to be a social being. The knowledge of differences encouraged the wish to transfer the message about personal achievements, values and capabilities to the others.Looking at the levels of new knowledge we see the dynamism of this process and the similar ideas about the same on different localities and within different cultures. As the number of these ideas is hard to supervise and monitor in their application, in one moment the need was created to build standards, in fact new behavior which can be acceptable for all of us. Thus we can recognize two parallel pillars on which the integration processes are built or based and they daily complete the network recognizable under the name of globalization.

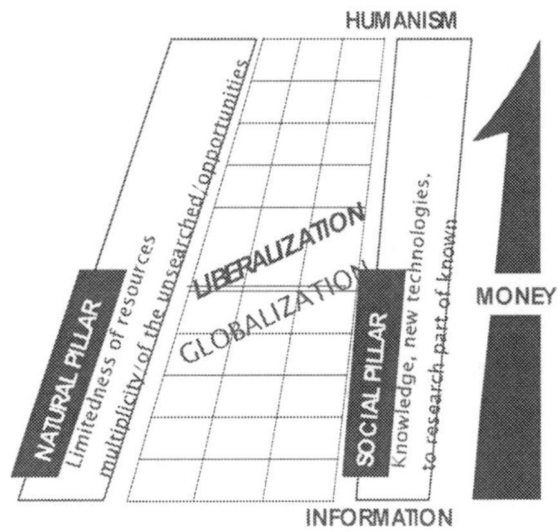

Fig. 1. Basic integration process characheristics

For a whole century it has been attempted to build with these two pillars the networks of the relations of human behavior and natural laws. The numerous small networks which are incorporated in the human life and work and their needs, are daily expanded. They are changing and improving intensively, and are often adjusted to the needs of the others. Out of these small processes, a strong integration process has come out with all its perspectives we are studying today. We will try to show one significant part of these integrations, recognized through the established associations to which Bosnia and Herzegovina itself belongs today or want to belong to.

3. BUSINESS SYSTEMS ORIENTED TO KNOWLEDGE

Business systems oriented to the information and information technologies without focus on the orientation to the activity are threatened by drowning in complexity. This is the answer to the question of many whose business system grew old too soon and falsely followed all the theoretical principles of successful business systems.Information technologies together with communication technologies ensure the accumulation of an enormous amount of information within a business system. The inflation of information makes the process of its selection to the important ones and less important ones for the functioning of a business system more difficult. Our goal can be to direct our organizational structure to the building of a living organism of a community of human networks. In such networks we ensure the continuous acquiring of knowledge and the mutual exchange, without waiting for someone to forward you a new idea, information or knowledge. Human networks are the communities of different capabilities and skills as well as the ideas about certain problems.

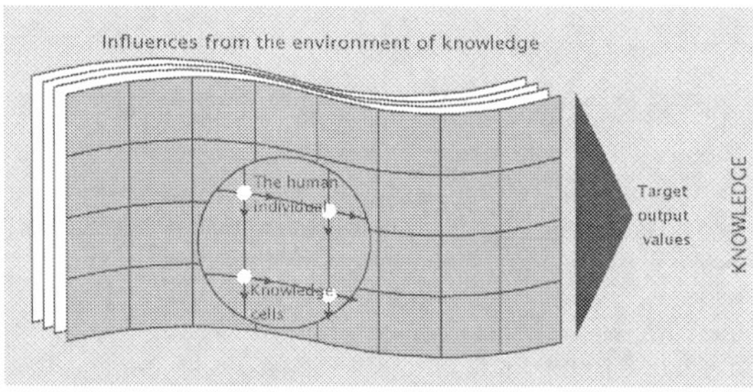

Fig. 2. Knowledge network organization

We cannot ensure the organizational or corporation learning without the knowledge and the intelligence of the basic factor of a business system – the human. The task of the organizational human networks is by all means to raise the level of the organizational knowledge on the synergic incentives of the knowledge of an individual. To ensure that organizational knowledge makes the knowledge of your current but also of your previous human resources means to manage the knowledge as our biggest value, much more important for a business system than the machinery or other material and palpable structures. Organizations learn only through the individuals who learn, but individual learning is not a guarantee to organizational learning. This is the gap created by unused knowledge. The organizations which learn search for the people on all levels who will implement the idea of knowledge in practice, they can observe and work with the forces of change and they do not resist them. Personal conquest is the term for an individual's search for knowledge. The spirit of the learning organization is encouraged by this search and by constant learning. This vision of a learning organization obviously requires the people who will find their personal fulfillment within a business system, for whom the belonging to an organization and work will be one of the beauties of life. The time becomes a partner and an ally to such an organization, and people feel safe in it. If a human is the basis of business system and organization behavior, then we should seek for the organizational models that agree with the human nature.
Since the basis of a modern business system is the orientation to the buyer, then the organization knowledge is not imaginable without a considerable input of the knowledge of the buyer. It is obvious that the organizational networks of people should be moved over the framework of the organization into the environment, by doing which two effects are created:

- Flexibility of the organization to the reactions from the environment
- Organizations gain the new potential of knowledge from the environment which is important in multiple ways and can have considerable effect on the projection of knowledge within the organization.

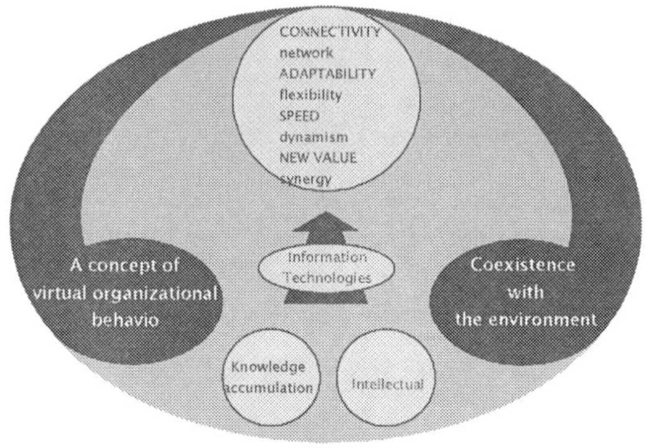

Fig. 3. Features of organically organization structures which basics are knowledge

Business systems attract the buyers with their knowledge – by satisfying the buyer, and the buyers are not aware that they are basically the source of this knowledge.
They gave the basic incentive, information which was sophisticated in the system of knowledge. The capacity of an organization to constantly learn, acquire and efficiently select the important from the unimportant information means the accumulation of knowledge. This process becomes the basic imperative of a business behavior. These approaches provide higher level of the usage of all cells of knowledge and raising the level of the intellectual capital of the organizations of a business system. More and more attention will be paid to this question since the most organized networks do not use more than 10% of the potential of the knowledge of the people in the network.

4. CONCLUSION

The basic planetary characteristic of the space we are living in is more and more recognizable through intensive integration activities. The integration process, directed and carried out, is based on two pillars – natural and social. The unit which is the characteristic of both pillars and their stability is a human, an individual – a man as a social being. The intellectual potential of an individual, current and more importantly accumulated in the cells of knowledge, should be necessarily directed to the transformation from information over money – material value to a human and for the good of a man as a social being.The race for the material on individual and national levels is transformed into a race for the knowledge. The skill of building the networks of knowledge with all their characteristics of business systems towards the society of knowledge will ensure you the respectful place you want in the integration processes – perspectives.

5. REFERENCES

Adizes,I. *Dijagnostika stilova upravljanja*, Prometej, N.Sad, 1994.
Covic, D.*The carasteristic of succesayul business systems – Limitlessuses.* The 12 th International DAAAMSymposium, Vienna 2001.
Covic, D.*Transformation of business systems in Bosnia and Herzegovina for approaching European Union*, Managment, Split, 2006
Hammer, M.*Reinzenjering tvrtke*, biblioteka menadzment u XXI stoljecu, Zagreb, 2004.
Milardovic, A. *Pod globalnim sesirom*, Biblioteka Globalizacija, CPI,Zagreb, 2004.
Sengi, P. *Peta Disciplina*, Asee books, N.Sad, 2003.

Annals of DAAAM for 2006 & Proceedings of the 17th International DAAAM Symposium, ISSN 1726-9679
Editor B. Katalinic, Published by DAAAM International, Vienna, Austria 2006
150th Birthday of Nikola Tesla (1856-1943) Inventor of Mobile Robot Teleoperation

BUSINESS INTEGRATIONS FORERUNNER OF THE SOCIAL ONES FROM THE ASPECT OF BOSNIA AND HERZEGOVINA

COVIC, D.

***Abstract:** The basic characteristic of the time we are living in could be looked at through the prism of the overall integration activities and the resistance to them. The terms related to the new information and communication technologies, globalization and liberalization are the sources of such thinking. The common influence of these recognizable elements of today cannot be stopped by any natural or administrative border, or barrier. If we are aware of this fact, then it means that if we want to make a reasonable concept, we should adjust the space that we "possess" to the trend and to adjust our social and any other structure to the oncoming integrations. The interest of this paper is directed to the strategy of strengthening and accelerating of business integrations, economic connections as a forerunner to the social – legal integrations. Business integrations – connections come by themselves in the expressed interest spheres. However, the wisdom and skill is when we create and direct these processes in the desired direction. It is important to direct the economic groups for which there is no marked interest to partial or total, technological or legal connections. This is all with the aim of strengthening of overall forms of competitiveness and the creation of preconditions for the easier and systematic social connections in the oncoming integration processes.*
***Key words:** business system, integrations, globalization*

1. INTRODUCTION

Our orientation to intensify our admission to European integrations requires the creation of the strategy which can ensure this in as economic way as possible. Taking into account all the standards set in front of the candidates for this integration act, it is very important to protect what today can be called competitive advantage. Regarding the fact that the basis of the standards for the admission is the liberal spirit, it is clear that no national protection or barriers will be put. The way to find the solution is to start and stimulate domestic business systems to a strong business integration process in which these business systems would become the core of the integration activity. This way we protect business systems from the strong competition.
Connecting them to the competition ensures new values for them, and simultaneously the stability which in the network of such activities ensures the macroeconomic stability of the social environment.

2. INTEGRATION PROCESSES AND PERSPECTIVES

Looking with the eyes of an average citizen of Bosnia and Herzegovina or even ex-state, the term of integration processes is necessarily related to the term of NATO or European Union. These are the terms which we expose and promote so much that it is no wonder that we completely neglected what has put us for long within the institutions that have all the global and planetary characteristics.
The multiplicity of integration processes in its most global form have not avoided BiH space, regardless of the fact if these integrations are looked at in the light of social, monetary, fiscal, technological, security, cultural, sport, scientific or some other integrations. But the areas which especially draw our attention are the areas of economic, trade and social associations, which under the influence of the oncoming information and communication technologies develop or adjust so rapidly that by negligent standing by side Bosnia and Herzegovina could be condemned to a long term technological, economic and any other falling behind.
Due to the different standards of living in different parts of the world, the influence of different cultures and traditions, the number of human population and the limits of all resources, the most powerful, the most organized daily offer new integration processes and perspectives.
With all those that have planetary dimension, our attention has been lately and, probably will be in the next ten years, on European Union. "Young" process in terms of time, the project of EU draws the attention and motivates for the membership, but it also brings fear and concern in the closest environment. Twenty-five member countries and two potential members at the end of this year, raise the questions of the possibilities and the reality of the rapid continuation of this process towards other countries and by which dynamics we can achieve that all members of the Council of Europe become EU members, and the multiple integrations that come with the admission process itself.

3. BUSINESS INTEGRATIONS

It is obvious that business integrations are forerunner to social integrations. In which time frame we see it and in how wide geographical, integration space we see it, it is our own decision. All the wisdom of Bosnia and Herzegovina business thinking has to be connected with the incentives to the investment activities, domestic as well as foreign.
The most significant effects expected from these activities are related to the creation of jobs, acquiring new technical and technological ideas.
Business integrations provide multiple opportunities and when it comes to economic system of BiH it is important to emphasize the international dimension. By including and integrating business systems, a global dimension is given the BiH economy. What is important to emphasize at the very beginning is that we have to know how to recognize mutual integration interest of business systems.
The transfers encouraged by the international corporations also mean goods, services, technologies, managerial knowledge and capital.
Which forms of cooperation can be applied by business systems is shown in the figure:

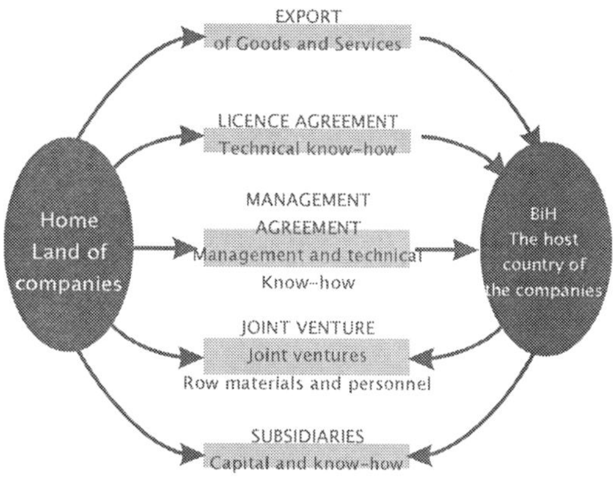

Fig. 1. Forms of business integrations

This rough overview of the potential business integrations is related to a lot of different forms of inter-connections, with multiple different additions and adjustments which do not need to have a clear legal integration dimension.

Namely, all integrations do not need to have for their consequence this legal subjectivity, it is understandable that the result of business connections can be a much "looser" legal connection. Often with the complex economic projects with the international dimension, we have more interested business systems or their parts, with a completely different stress in mutual connections and obligations. As every business enterprise is followed by a financial activity, which is as a service provided by banking institutions, it is clear that they appear as an obligatory logistics of business integrations. The question is how to impose ourselves to these institutions as a wanted and recognizable partner. Since these are big corporations, it is necessary to build a nationally acceptable environment and the business systems as an object of interest.

Tens of ideas that have already been tested in the business practice in the last ten years can be the illustration of what should be our way, how through business integration to become a part of the world which we will join institutionally much later. It is possible, based on a healthy business idea of recognized authors, to create a complex business project which can absorb in itself the capital of those who we want to join, new knowledge and technology, and to use the infrastructure and the potential of the citizens of BiH.

By including the quality capital basis of the businessmen from EU with the necessary instruments of protection, we have made the biggest step towards moving the business idea of EU and everything behind that idea towards the region of Bosnia and Herzegovina.

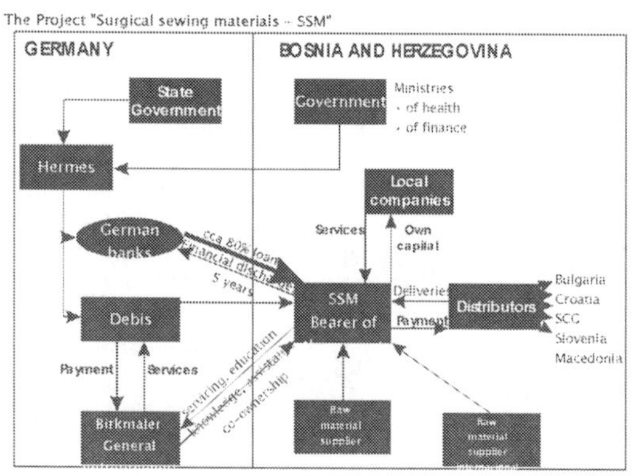

Fig. 2. The concept of business integrations

A concrete example of business integration should be adjusted in its organizational form to the trend of network organization. The core of this network business connection should have in its source a business system with which we connect, as well as the system which we connect in.

With this concept we obtain the system of network of business systems with two cores.

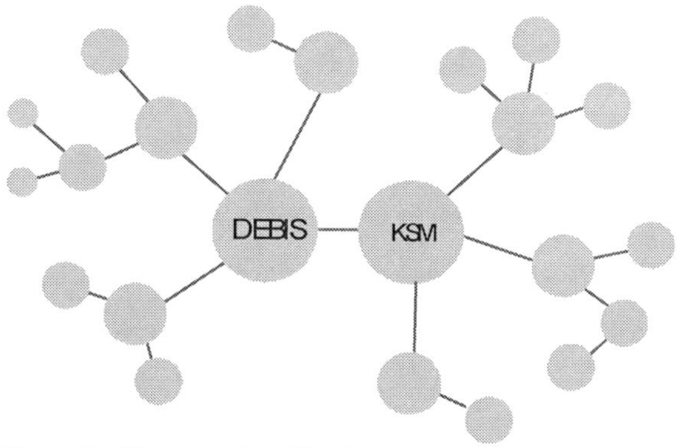

Figure 3: The network of business systems with two cores, based on the concept of business integrations

4. CONCLUSION

Social integrations are today by most, especially the underdeveloped, seen as a perspective. Inclusion in the process of integration by itself means a lot of preconditions, standards to be built. Integration activity often lasts very long, and the unconditional acceptance of the already set standards leads us to a dead end of the neglecting the "national" local reality. Integration processes are followed by the phenomena of openness – liberalization, marked technological novelties and in its broadest sense the insufficiency of resources. To absorb the possibility of going to a "dead end", due to the clumsiness in the inflation of the obligations in front of you, means to involve gradually into the integration processes. The most acceptable way to succeed in this is the way of business integrations. With the business integrations with more or less expressed autonomy of this process you have created the opportunity to learn, to build yourselves, and primarily to be economically sustainable. The standards that you will build with the effects of the economic stability of an integrated business system enable you to transform your business integration into forerunner of social integrations. This is the chance to see the integrations as a perspective finally.

5. REFERENCES

Brekic, I. *Inventurni Managament*, Alineja, Zagreb, 1999.

Covic, D. *Obilježja poslovnog sustava pripremljenog zatržišne izazove unutar Europske zajednice*,International conference „Economic system of European Union and adjustment of BiH", Mostar, 2002

Covic, D. *Transformation of business systems in Bosnia and Herzegovina for approacleing European Union*, Menadzment, Split, 2006.

Hammer, M., *Reinzenjering tvrtke*, Biblioteka Menadzment za XXI stoljece,Zagreb, 2004.

Knezevic, A.*Odrzivi razvoj Bosne i Hercegovine* CETEOR, Sarajevo, 2002.

Milardovic,A. *Pod globalnim sesirom*, Biblioteka Globalizacija, CPI, Zagreb, 2004.

Annals of DAAAM for 2006 & Proceedings of the 17th International DAAAM Symposium, ISSN 1726-9679
B. Katalinic, Published by DAAAM International, Vienna, Austria 2006
150th Birthday of Nikola Tesla (1856-1943) Inventor of Mobile Robot Teleoperation

A THEORETICAL MODEL BASED ON CHIP'S CONTRACTION RATIO FOR CUTTING FORCES EVALUATION

COZMINCA, I. & UNGUREANU, C.

Abstract: *The design of the cutting process and of the technological system's structural elements must be realized based on the real values of the cutting forces. The study of the systems' dynamics can't be accomplished without knowing the average values and the variation laws of the forces that evolve during the cutting process. Now, cutting forces are predicted in different precision conditions, according to the working proceeding. Consequently, this paper presents the theoretical bases of a methodology that allows evaluating the cutting forces level at single-point free cutting, using the chip's contraction ratio values.*

Key words: *orthogonal cutting, cutting forces, chip's contraction ratio.*

1. INTRODUCTION

Analyzing the specialized works we can conclude that the cutting forces evaluation has been made using models that take into consideration the working process type, the theoretical chip's section and the nature of the working material. Some of the models might take account on the wear of the tool's rake face, the approach and the rake angles, the nose angle, the tool's material and, for milling process, also the cutting speed (Albrecht, 1981); (Arshinov & Alekseev, 1976); (Gunay & al., 2004); (Jawahir & von Luttervelt, 1993); (Zorev & Granovski, 1967).

The structure of those models includes a number of constants, exponents and correction coefficients which indicate the influence level of the process parameters. Its values were established long time ago, analytically or by experimental tests, for certain working conditions, but they don't cover all the practical situations and, most of all, ignore the interdependencies between the individual influences exerted by the working parameters [8]. The most utilized general equation for determining the cutting forces is structured as follows:

$$F = C \cdot t^x \cdot s^y \cdot (HB)^n \cdot k_1 \cdot k_2 \cdot \ldots k_n$$

We can also notice that most of the cutting forces models are neglecting some important process parameters, such as the main cutting speed, the throat angle of the cutting edge, the inclination angle of the edges, the cutting medium and the nature of the cutting material. The constants, exponents and coefficients values were established in certain experimental conditions and then were applied in designing for other cutting conditions, materials or tools geometries. So, it is obvious that the errors in cutting forces evaluation will be high, and that should interfere with the actual requisitions on accuracy and quality.

2. THE PRINCIPLES OF THE PROPOSED MODEL FOR CUTTING FORCES EVALUATION

The proposed methodology for cutting forces evaluation is based on the model of the three-dimensional orthogonal cutting, for a single point, referring to the real cutting plane v - v_1, with possibility to extend it to any tool type and working process. The variation of the main parameters v, s, γ, λ is taken into consideration, so one may determine its influence levels considering also the significant interdependencies between the individual influences.

In order to simplify the relations between the cutting forces components F_z, F_x, F_y, we propose to use only the working parameters and specific quantities that include the influences exerted by all the process parameters. This is the reason why we shall use the chip's contraction ratio instead of the friction and shear angles that frequently appear in the structure of actual models, being difficult to determine these with sufficient accuracy. This ratio includes in fact the effects of all the process parameters on the cutting forces developed. Using the experimental values of the chip's contraction ratio, the values of the cutting forces may be appreciated, depending on the working parameters, without their explicit presence into the calculus models (Croitoru, 2000).

In case of the single-point free cutting, in order to develop the models for the cutting force components F_z, F_x, F_y depending on the physical components, F_N, F and F_f, and the tooth geometry, we translate consecutively the physical components of the rake face and then of the side flank in the edge plane, T, and afterwards in the base plane, B. In the plane T we find the F_z component, and in the B plane, the F_x and F_y components (Fig.1).

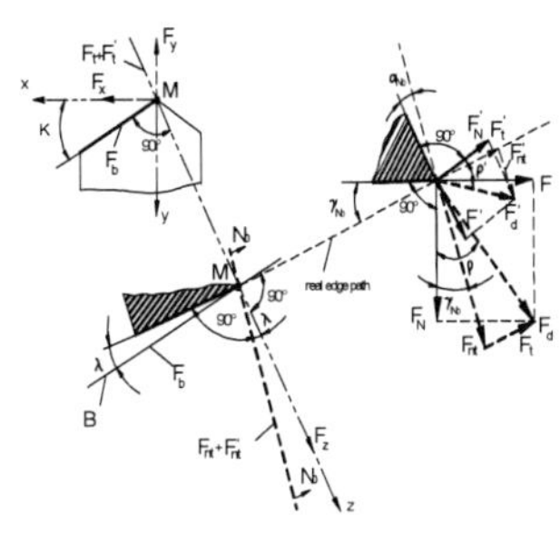

Fig.1. Geometrical model of the forces on the single-point cutting tool

The resultants $\overline{F_d} = \overline{F_N} + \overline{F}$ and $\overline{F_d'} = \overline{F_N'} + \overline{F'}$, corresponding to the rake face and the side flank of the tool, are decomposed in two rectangular convenient directions, i.e. the real edge direction and a direction perpendicular on it. (Fig.1). Finally, the equations (1) result.

$$\overline{F_d} = \overline{F_t} + \overline{F_{nt}}$$

$$\overline{F_d'} = \overline{F_t'} + \overline{F_{nt}'} \quad (1)$$

$$F_t' = F_d' \cdot \cos\left(\rho' - \alpha_{N_0}\right);$$

Depending on the F_N and F_N' components and on the friction angles, ρ and ρ', the resultants F_d and F_d' run into the intermediate equations (2) and (3).

$$F_d = F_N / \cos\rho \quad (2)$$

$$F_d' = F_N' / \cos\rho' \quad (3)$$

Using the relations (2) and (3), the components F_t, F_{nt}, F_t' and F_{nt}' result according to equations (4) – (7) .

$$F_t = \frac{F_N}{\cos\rho} \cdot \sin\left(\rho - \gamma_{N_0}\right) \quad (4)$$

$$F_{nt} = \frac{F_N}{\cos\rho} \cdot \cos\left(\rho - \gamma_{N_0}\right) \quad (5)$$

$$F_t' = \frac{F_N'}{\cos\rho'} \cdot \cos\left(\rho' - \alpha_{N_0}\right) \quad (6)$$

$$F_{nt}' = \frac{F_N'}{\cos\rho'} \cdot \sin\left(\rho' - \alpha_{N_0}\right) \quad (7)$$

By translation into plane T, the components F_t and F_t' were projected on point M, and the components F_{nt} and F_{nt}' constitute the main component, F_z, according to relation (8).

$$F_z = (F_{nt} + F_{nt}')\cos\lambda \quad (8)$$

On the direction of B mark of the base plane (xy), the components F_{nt} and F_{nt}' determine the intermediary component F_b (relation (9)).

$$F_b = (F_{nt} + F_{nt}')\sin\lambda \quad (9)$$

The component F_b, and also F_t and F_t' from the T plane, are translated into the (xy) plane and the design components F_x, F_y result, according to relations (10)-(11).

$$F_x = F_b \cos K + (F_t + F_t')\sin K \quad (10)$$

$$F_y = -F_b \sin K + (F_t + F_t')\cos K \quad (11)$$

Using the relations (4)-(7), the models (8), (10) and (11) become (12)-(14) as follows:

$$F_z = \left[\frac{F_N}{\cos\rho} \cdot \cos\left(\rho - \gamma_{N_0}\right) + \frac{F_N'}{\cos\rho'} \cdot \sin\left(\rho' - \alpha_{N_0}\right)\right] \cdot \cos\lambda \quad (12)$$

$$F_x = \left[\frac{F_N}{\cos\rho}\cos\left(\rho - \gamma_{N_0}\right) + \frac{F_N'}{\cos\rho'}\sin\left(\rho' - \alpha_{N_0}\right)\right]\sin\lambda\cos K + \left[\frac{F_N}{\cos\rho}\sin\left(\rho - \gamma_{N_0}\right) + \frac{F_N'}{\cos\rho'}\cos\left(\rho' - \alpha_{N_0}\right)\right]\sin K \quad (13)$$

$$F_y = -\sin K\left[\frac{F_N}{\cos\rho}\cos\left(\rho - \gamma_{N_0}\right) + \frac{F_N'}{\cos\rho'}\sin\left(\rho - \gamma_{N_0}\right)\right]\sin\lambda + \left[\frac{F_N}{\cos\rho}\sin\left(\rho - \gamma_{N_0}\right) + \frac{F_N'}{\cos\rho'}\cos\left(\rho' - \alpha_{N_0}\right)\right]\cos K \quad (14)$$

With the adopted notations, i.e. relations (15)-(18), the equations (12)-(14) take the (19) form.

$$A = \frac{\cos\left(\rho - \gamma_{N_0}\right)}{\cos\rho} = \cos\gamma_{N_0} + tg\rho \cdot \sin\gamma_{N_0} \quad (15)$$

$$B = \frac{\sin\left(\rho' - \alpha_{N_0}\right)}{\cos\rho'} = tg\rho' \cdot \cos\alpha_{N_0} - \sin\alpha_{N_0} \quad (16)$$

$$C = \frac{\sin\left(\rho - \gamma_{N_0}\right)}{\cos\rho} = tg\rho \cdot \cos\gamma_{N_0} - \sin\gamma_{N_0} \quad (17)$$

$$D = \frac{\cos\left(\rho' - \alpha_{N_0}\right)}{\cos\rho'} = \cos\alpha_{N_0} + tg\rho' \cdot \sin\alpha_{N_0} \quad (18)$$

$$F_z = \left(A \cdot F_N + B \cdot F_N'\right) \cdot \cos\lambda$$
$$F_x = \left(A \cdot F_N + B \cdot F_N'\right) \cdot \sin\lambda \cdot \cos K + \left(C \cdot F_N + D \cdot F_N'\right) \cdot \sin K \quad (19)$$
$$F_y = -\left(A \cdot F_N + B \cdot F_N'\right) \cdot \sin\lambda \cdot \sin K + \left(C \cdot F_N + D \cdot F_N'\right) \cdot \cos K$$

Taking into account the relations (20)-(21) between the real angles and the constructive ones, used for designing, the A, B, C, D constants from equations (19), result from the relations (22)-(25).

$$A = \frac{1 - tg\rho \cdot tg\gamma_N \cdot \cos\lambda}{\left(1 + tg^2\gamma_N \cdot \cos^2\lambda\right)^{0,5}} \quad (22)$$

$$B = \frac{tg\rho' - tg\alpha_N \cdot \cos\lambda}{\left(1 + tg^2\alpha_N \cdot \cos^2\lambda\right)^{0,5}} \quad (23)$$

$$C = \frac{tg\rho - tg\gamma_N \cdot \cos\lambda}{\left(1 + tg^2\gamma_N \cdot \cos^2\lambda\right)^{0,5}} \quad (24)$$

$$D = \frac{1 - tg\rho' \cdot tg\alpha_N \cdot \cos\lambda}{\left(1 + tg^2\alpha_N \cdot \cos^2\lambda\right)^{0,5}} \quad (25)$$

Now one can observe that A, B, C, D values are dependent on the friction angles, ρ and ρ′, and on the constructive angles α_N, γ_N and λ. For most practical situations, the constructive angles take values in a large range, i.e. $\alpha_N = 4°$ … 16° ; $\gamma_N = -40°$ … + 40° and λ = 0° … ± 40° (Arshinov & Alekseev, 1976).

Using the relations (19) for establishing F_z, F_x, F_y values in designing activities involve the necessity to determine with accuracy the values of the intermediary data, i.e. the physical components of the cutting force, F_N and F_N', and the friction angles, ρ and ρ.

3. CONCLUSIONS

In order to apply the models (19) for determining the designing components of cutting force, F_x, F_y, F_z, the values of the chip's contraction ratio, C_d, are necessary, adapted to the specific working conditions.

The C_d values may result from two methods, namely:

- By measuring (weighing) a certain number of chips that result from the cutting process, in the adopted working conditions and then determining C_d with known relations: $C_d = a_1/a$, $C_d = L/L_a$ or $C_d = G_a/\rho \cdot l \cdot t \cdot s$.
- Using analytical models for C_d evaluation, resulting by experimental data processing after cutting the chosen material: $C_d = f(t, s, v, \gamma, \lambda, \ldots)$ or $C_d = f(\sigma_{p02})$.

Applying the proposed models of the cutting process, which are considering all of the physical components of the cutting force and the real cutting plane, may guarantee simplicity, security and maximum precision to the designing components F_z, F_x, F_y evaluation. The experimental tests for C_d values and F_z, F_x, F_y, F_{z1}, F_{x1}, F_{y1} components are strictly necessary only for establishing the variation range of the intermediary data in relations (19).

4. REFERENCES

Albrecht, P. (1981). New development in the theory of metal cutting process, Part II. The theory of chip formation. *Transactions of ASME*, vol.82, no. 4.

Arshinov, V.; Alekseev, G. A.. (1976). *Metal cutting theory and cutting tool design.* Mir. Publishers (Ed.), Moscow.

Croitoru (Cozminca), I. (2000). *Cercetări privind îmbunătățirea metodologiilor de evaluare a forțelor la așchierea metalelor.* Teză de doctorat, Univ. Tehnică „Gh. Asachi", Iași.

Gunay, M.; Aslan, E.; Korkut, H.; Eker, U. (2004). Investigation of the effect of rake angle on main cutting force. *Int. Journal of Machine Tools and Manufacture*, vol.44, no.9, p.953-959.

Jawahir, J. S.; von Luttervelt, C. A. (1993). Recent developments in chip control research and application, *CIRP Annals*, vol. 42, no.2.

Zorev, N.N.; Granovski, G.I. (1967). *Razvitie nauki o rezanii metallov.* Izd. Mașinostroenia (Ed.), Moskva.

Annals of DAAAM for 2006 & Proceedings of the 17th International DAAAM Symposium, ISSN 1726-9679
B. Katalinic, Published by DAAAM International, Vienna, Austria 2006
150th Birthday of Nikola Tesla (1856-1943) Inventor of Mobile Robot Teleoperation

ADDITIVE FABRICATION OF BIOCERAMIC/BIOPOLYMER BONE IMPLANTS

CRUZ, F. & COOLE, T.

Abstract: *This paper presents an overview of the research work carried out in the last few years for the rapid manufacture of hydroxyapatite (HA) based components by means of the SLS (Selective Laser Sintering) technology, to be used for bone tissue implantation, aiming to replace and/or repair bone defects due to traumatized, damaged or lost bone. SLS is an Additive Manufacturing Technology (AMT) that selectively sinters powders of engineering materials by means of a* CO_2 *laser. HA is a bioceramic used since several years for medical applications although being mainly processed by conventional methods (cast, machined, or manually produced). For that purpose CAD/CAM software, Rapid Manufacturing Technologies (RMT) and Medical Imaging Techniques (MRI) were integrated, and this enabled the customized production of anatomical models. This approach provides a direct method of producing HA based prostheses.*

Key words: *Additive Fabrication, Rapid Manufacturing, Direct Bone Implants, Bio- Composites, Selective Laser Sintering.*

1. INTRODUCTION

This paper presents the results of a study, carried out from 2001 to 2004, aiming to investigate the feasibility to directly manufacture bone implants (i.e., without post-operations) from a bioceramic (HA) / biopolymer (PLLA) composite, using a Rapid Manufacturing Technology (Selective Laser Sintering - SLS), rather than a conventional technology as investment casting, slip cast, pressing followed by thermal sintering, or machining. One of the goals of the research work was to prove the capability of AMT to produce medical models, and thus enlarging the field of application of this emergent technology, taking advantage of its main characteristics: to manufacture parts directly from a virtual model; no need of any kind of tools; virtually independent of the geometric complexity of the parts being built; very quick processing; adequate geometric and dimensional accuracy, and a large range of materials used.
The final aim of the study was to manufacture customised bone implants, from medical scanning data captured from patients, having adequate physical and biological properties to allow their direct application in the human body, for bone grafting and bone replacement.
A key requirement in many biomedical applications, including artificial bones, is a property called connected porosity. Green parts produced by the chosen SLS technology inherently have this property. While HA has been used in bone replacement and grafting for more than twenty years, this material was a logical choice for this work. PLLA was chosen as binder because it offers a higher degradation time and a low melting temperature when compared with other acid lactic biodegradable derivates. This biopolymer acts as binder during the sintering operation, and afterwards improves the mechanical properties of the implants.
The study has taken in consideration prior work made at University of Texas, USA, concerning the Selective Laser Sintering of calcium phosphate powders (Lee & Barlow, 1994; Barlow et al., 2000), although that work involved an indirect approach, unlike the direct approach presented in this paper.

2. SELECTIVE LASER SINTERING

The Selective Laser Sintering (SLS) process belongs to the solid-based Additive Manufacturing Systems, which primarily use powder as the basic medium for manufacture. The objects are built layer by layer by sintering / interfusing thin layers of powder. Each layer fuses to the previous one creating a physical object. The main aspects to be considered in the SLS technology are the properties of the powders used in the process and the fabrication parameters. The fabrication parameters depend strongly on the materials used, and have significant influence in the mechanical properties (e.g. tensile strength, surface hardness and density), dimensional accuracy and surface quality (i.e. roughness) of the parts produced. With SLS it is not possible to achieve the best quality in appearance with the best mechanical properties. Maximum density of the parts is only achieved with parameters that result in excess powder sticking to the surface and low speed of construction (which is time-consuming). Using the correct combination of coating and finishing in post-processing, the mechanical properties and surface of SLS parts can be improved. The main SLS process variables are: Part bed temperature, laser power, scans size, scan spacing, slice thickness, and part position and orientation (3D Systems, 2003).

3. EXPERIMENTAL WORK

In this section an overall view of the experimental work is presented.
Based on previous screening trials, a choice of 40% ratio of PLLA by weight has been made. This choice was based on the fact that adding lower than 30% wt. of PLLA to HA will not add significantly for the biodegradation or to enhance the ductility properties, and higher than 50 % wt. the implant could be too plastic and therefore not behave as bone tissue (Cruz, 2004).

3.1 Materials

The materials used in this work (in powder form) were as follows:
- HA: Captal® 120 grade – sintering powder, supplied by Plasma Biotal Limited (UK), with mean particle size (D50) of 111±5µm, melting point of 1250 (-0 +50)°C and 1.30 g/cm^3 of density (~£300 /Kg);
- PLLA: Purasorb® L, supplied by PURAC (The Netherlands), with mean particle size (D50) of 163±5µm (after sieving), melting range of 182.4 – 192.3°C and 0.47 g/cm^3 of density (after sieving) (~£1700 /Kg).

3.2 Equipment

The experiments were performed on a SLS Sinterstation 2000 apparatus installed at CRDM (Centre for Rapid Design and

Manufacture), High Wycombe, UK. This apparatus was modified, namely the feeding envelopes and the part platform, in order to allow operating with non-standard materials, as the standard material is nylon.

3.3 Methods

The aim of the first phase of the experimental work was to identify those factors (process parameters) that have large effects on the response variables (physical properties of the parts produced). In this phase a screening experimental design and analysis, based in the fractional factorial design technique, was applied to determine the most influential factors of interest on fabrication of HA/PLLA models by SLS. A DoE (Design of Experiments) methodology was used (Montgomery, 1997). The results obtained in the tests were focused on three (3) response variables (density, geometric accuracy and surface quality of the parts produced). The results obtained in the screening analysis show that the three factors with most influence on the physical properties of the parts produced by SLS were laser power, scan speed and scan space, i.e., the Applied Energy Density (AED). In fact, these three factors together constitute the Andrew's Equation (Cruz, 2004), corresponding to the energy density applied to the powder bed, which determines the scanning strategy during the SLS operation. After this preliminary stage of the experimental work, the next step was focused on the measurement of the ultimate compressive strength and the elastic modulus in compression of the parts produced, by means of a 2^k full factorial design and analysis, in order to achieve more detailed information about the most influent factors selected in the previous phase. To investigate the degradation behaviour of the HA/PLLA parts produced by SLS, namely its PLLA component, in vitro tests were performed. To study the in vitro degradation mechanism of the HA/PLLA specimens, an aqueous media (physiological fluid) – saline phosphate buffer (SPB, pH=7.4) was used. This media has been taken as a model of biological fluids (Taddei et al., 2002). The study was conducted on their physical properties, namely their mass loss along the time (six months).

4. RESULTS AND DISCUSSION

The main goal of the experimental work was to demonstrate the feasibility of producing HA/PLLA parts by selective laser sintering. The results were successful in proving that feasibility.

In fact, the main results of the trials performed, have shown that:

- It is possible to produce bioceramic parts (HA based) using Additive Fabrication Techniques, namely the Selective Laser Sintering (SLS) process, using a biocompatible/biodegradable polymer as binder;
- For that purpose it is necessary to use a polymeric binder (40% wt.) to promote the agglomeration of the ceramic particles during the laser operation (sintering). The component, in this phase, is the so-called "green body";
- In the case of using a biocompatible binder (e.g. PLLA), the blend HA/PLLA produces a part that can be used directly in the body without removal of the binder (green body). This technique is a direct route to fabricate laser sintered HA based ceramic parts for medical purposes;
- Thus, this route allows the production of parts directly, without the need of post-process operations, therefore reducing the time and cost of production;
- In such a case the mechanical properties of the parts produced constitutes a major limitation, allowing their application only in non-load bearing situations or supported load bearing uses;
- The internal porosity of the HA/PLLA parts produced (mean value >150∝m) is suitable to promote the growth of new tissues in the implant. In fact, this condition is essential to allow extensive blood supply of the new bodies, providing a rich supply of cells, growth factors and other nutrients needed to make bone grow. Moreover, these small holes allow access of bone marrow elements from the host tissues to the bone graft, providing nutrients for bone healing;
- The density of the HA/PLLA sintered parts (mean value = 0.883 g/cm^3) is within the density range values for the cancellous bone (0.14 to 1.10 g/cm^3) as previously referred. The green density in nearly the same of the powder bed. This result is important because, in principle, the higher the density of bone implant, the larger its strength;
- The results obtained by applying the full factorial design have shown that the mechanical properties of the HA/PLLA parts as SLS produced are relatively poor when compared with other biomaterials (σ_c between 2.44 and 4.57 MPa / E_c between 17.19 and 40.80 MPa), but, even so, the Ultimate Compressive Strength and the Elastic Modulus lie in the lower limits of reported values for cancellous bone (1.5 – 38 MPa and 10 – 1570 MPa, respectively). These results confirm the low mechanical strength of the HA/PLLA parts, as SLS produced, thus constituting a limitation for the applications in load-bearing situations. The mechanical properties of the HA/PLLA sintered parts by SLS are expected to rise with the optimisation of the various parameters involved, particularly the energy density applied and the grain dimension of the starting powder.

5. CONCLUSION

The major contribution of this research is the establishment of data and formulation of guidelines for the rapid manufacture of hydroxyapatite based components, by means of the SLS technology, to be used for bone tissue implantation, to replace and/or repair bone defects, due to traumatised, damaged or lost bone. The overall feasibility of producing hydroxyapatite based bone shapes, from SLS, has been established. The research has shown that the Rapid Manufacturing of Biomaterials is possible with a reasonable accuracy, strength and geometry of the models produced. In summary, the research has proved the potential to produce customized components for niche applications, i.e. non-load bearing applications (e.g. facial reconstruction, cosmetic surgery, bone grafting).

6. REFERENCES

Barlow, J. W.; Lee, G. H.; Snyder, T. M.; Vail, N. K.; Swain, C. D.; Fox, W. C. & Aufdlemorte, T. B. (2000). Preparation of Calcium Phosphate Implants, *Available from University of Texas Web Site*: www.utexas.edu. *Accessed:* 2000-06-25.

Cruz, F. (2004). *Direct Manufacture of Hydroxyapatite Based Bone Implants Using Selective Laser Sintering*, Doctoral Dissertation, Brunel University/West London, UK.

Lee, G. & Barlow, J. W. (1994). Selective Laser Sintering of Calcium Phosphate Powders, *Solid Freeform Fabrication Symposium Proceedings – 1994,* September 1994, The University of Texas, Austin, USA, pp. 191-197.

Montgomery, D. C. (1997). *Design and Analysis of Experiments – Fourth Edition*, John Wiley & Sons, Inc., USA.

Taddei, P.; Monti, P. & Simoni, R. (2002). Vibrational and thermal study on the in vitro and in vivo degradation of a poly (lactic acid) – based bioabsorbable periodontal membrane, *Journal of Materials Science: Materials in Medicine*, Kluwer Academic Publishers, Vol. 13, No. 5, pp. 469-475.

3D Systems Inc. *Available from:* www.3dsystems.com. *Accessed:* 2003 -09-10.

Annals of DAAAM for 2006 & Proceedings of the 17th International DAAAM Symposium, ISSN 1726-9679
Editor B. Katalinic, Published by DAAAM International, Vienna, Austria 2006
150th Birthday of Nikola Tesla (1856-1943) Inventor of Mobile Robot Teleoperation

GENETIC ALGORITHM FOR ROBOT PATH ESTIMATION

CURKOVIC, P.; JERBIC, B. & VRANJES, B.

***Abstract**: this paper presents genetic algorithm developed to estimate proper combination of basic incremental moves in order to lead the mobile robot toward the target destination. The robot's initial position and the position of the steady target are given. The robot is considered as simplified model able to move incrementally stepwise front, back, left and right in an environment without obstacles. After approaching the target below the given threshold distance, genetic algorithm stops, returning desired path which is afterwards executed by the robot.*
***Key words**: path planning, genetic algorithm, mobile robot*

1. INTRODUCTION

Genetic algorithms (GA) are a subset of evolutionary algorithms domain. GA is known to be proven optimization tool for different types of problems, especially if the space to be searched is large, and not perfectly smooth and unimodal (Mitchell, 1999). The main procedure is to initially generate random population of possible solutions (candidates) over the search space and to evolve satisfying solution using genetic operators of selection, crossover and mutation pondering the probability of reproducing of the candidates according to their fitness. The candidate's fitness is described by the fitness function, which proper selection has significant impact on the convergence of the algorithm. Proper encoding of solutions and their interpretation is of essential importance for overall performance of the GA, absolute convergence as well as the speed of the convergence. Basic genetic operators can eventually be combined with some operators of higher order such as elitism, which can significantly improve convergence speed of the GA (Koza, 1998). The interaction between parameters of a GA is not linear and their values should be combined simultaneously for finding optimal performance of the specific algorithm. Some common values for crossover and mutation probability are 0.7 and 0.001 respectively, whereby probability of direct copying to the next generation via elitism, size of initial population and some additional parameters are more task specific.

2. PROBLEM FORMULATION

Given the initial position of the robot and the final target position, using GA, the path Γ is to be found from initial to the target position see (2). Robot is modeled as simple mobile device able to move incrementally to front, back, left and right. The corresponding encoding is thus the set $E = \{1,2,3,4\}$, with meaning: 1 - incremental move to front, 2 – incremental move to back, 3 – incremental move to right and 4 – incremental move to left. Since the robot is able to move in discrete manner; generally the positioning error **e** will be generated, depending on the size of the incremental step of the robot and the initial cityblock distance. Therefore necessity for defining the threshold value **t** arises. If:

$$\neg s \in Z \rightarrow e \neq 0 \qquad (1)$$

Where s is the size of a gene, see equation (6).

If **e<t**, the robot has reached the target position and the algorithm stops, returning the desired path. Considering above mentioned constraints we define the next statement:

$$\exists \Gamma \mid \Gamma \subseteq E \wedge |e| \leq t \qquad (2)$$

As the measure of the distance, which has to be minimized and which defines the fitness function, the cityblock measure is selected, because it naturally fits to the discrete nature of the problem.
For cartesian R^2 space, the cityblock distance between two points (x_1, y_1) and (x_2, y_2) is defined as:

$$c = |x_1 - x_2| + |y_1 - y_2| \qquad (3)$$

Initial genome has following structure:

$$\Gamma_{ij} = rand(E) = \begin{bmatrix} 112...334 \\ ... \\ 432...124 \end{bmatrix} \qquad (4)$$

Via decoding, the path is transformed to the total distance traveled by the robot:

$$\begin{aligned} \Delta_1 &= \sum_y \left(\Gamma_{ij} \mid (\Gamma_j = 1 : \Delta y > 0, \Gamma_j = 2 : \Delta y < 0)\right) \\ \Delta_2 &= \sum_x \left(\Gamma_{ij} \mid (\Gamma_j = 3 : \Delta x > 0, \Gamma_j = 4 : \Delta x < 0)\right) \qquad (5) \\ \Delta_{tot.} &= \Delta_1 + \Delta_2 \end{aligned}$$

The size, **s** of the gene depends on the size of incremental step and the initial cityblock distance between the origin and the target position of the robot.

$$s = \frac{|x_{start} - x_{goal}| + |y_{start} - y_{goal}|}{step\_increment} \qquad (6)$$

3. IMPLEMENTATION

The algorithm for simulation was programmed using MATLAB 7.0 platform. At this point, no obstacles in the environment of the robot were considered, but we are currently working on upgrading our algorithm considering this constraint as well as expanding this algorithm for optimization of the paths of two autonomous robots. Algorithm is tested under different GA operators' relations and corresponding results will be presented in Tab. 1. We found that algorithm converges to optimal solution in finite number of generations, providing better results after elitism was introduced. Different crossover operators were considered, but it appeared not to have important influence to the overall speed of the GA. Example of the solution found by the GA is presented by Fig. 1.

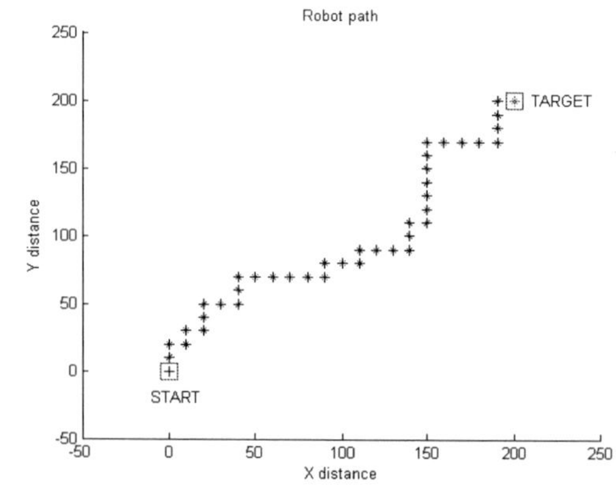

Fig. 1. Robot path found by the GA

The parameters for the solution of the path presented by Fig. 1 are: start position of the robot : (0, 0); target position: (200, 200); incremental step size: 10; population size: 8; crossover rate: 0.7; mutation rate: 0.001; elitism: 10%. It is obvious that with decreasing of the incremental step, the path found by the GA converges to the line connecting the start and the target position.

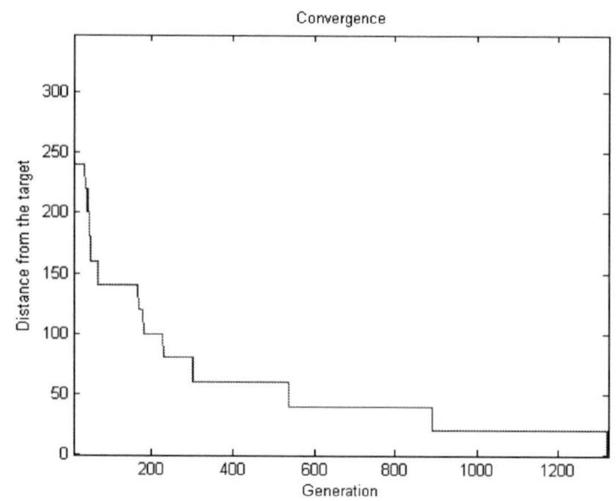

Fig. 2. Convergence vs. generation diagram

Fig. 2 presents the performance of the GA. For the example described above, the GA required 1300 generations to find the exact solution.

The next figure, Fig. 3 presents the same problem, with the step size reduced to 2 mm, which is five times less than the example presented in Fig. 1.

The compensation of better path optimization is done in cost of larger gene size, thus larger search space and more generations needed to find the target.

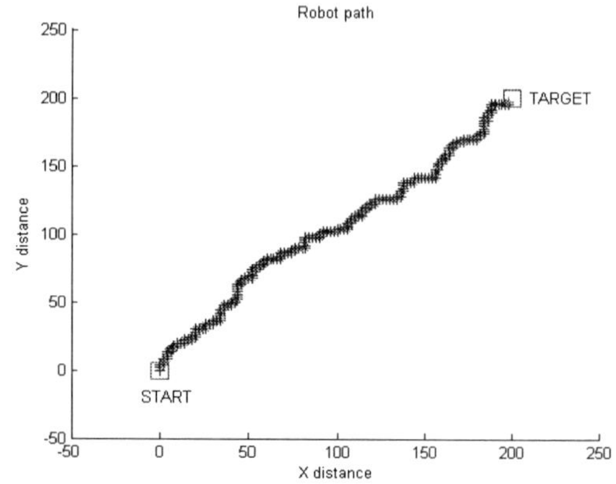

Fig. 3 Optimization with reduced step

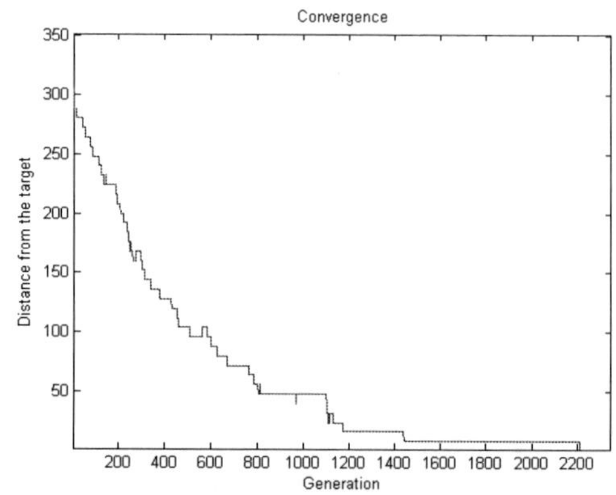

Fig. 4 Convergence vs. generation with reduced step.

Fig. 4 shows that to find the solution with different step size, approx. 2200 generations were needed. Note that the initial distance from the target differs for these two examples, even though the initial position of the robot and the final position are the same. This is due to stochastic nature of the GA, which provided different initial populations for the two solutions.

p_c	p_m	Population	G_{mean}	σ_G
0.7	0.001	8	1547	642
0.7	0.001	16	1054	235
0.7	0.0005	32	1004	145

Tab. 1 Interaction of the GA's parameters.

4. CONCLUSION

Presented results show robustness, reliability and flexibility in mobile robot path estimation using self-organizing GA. The algorithm is also expandable to problems which include environment with obstacles and interactions with other robots, as well as to adapting to real-robots control systems.

5. LITERATURE

Mitchell, M.; *Introduction to Genetic Algorithms (1999),* MIT Press, 0-262-13316-4, London

Koza, J.;*Genetic programming, (1998),* The MIT Press, 0-262-11170-5, Cambridge, Massachusetts.

Farittor, S.; Dubowsky, S.; *A Genetc Algorithm Based Navigation and Planning Methodology for planetary Robotic Exploration*, Proceedings of the 8th International Conference on advanced robotics, 1997.

Annals of DAAAM for 2006 & Proceedings of the 17th International DAAAM Symposium, ISSN 1726-9679
B. Katalinic, Published by DAAAM International, Vienna, Austria 2006
150th Birthday of Nikola Tesla (1856-1943) Inventor of Mobile Robot Teleoperation

CUTTING FORCE GENERATION IN BALL-END MILLING BY EVOLUTIONARY PROGRAMMING

CUS, F.; MILFELNER, M. & GECEVSKA, V.

Abstract: *This paper presents the development and analysis of an efficient, evolutionary, intelligent, genetic modelling method for cutting forces in ball-end milling. An evolutionary method using genetic programming (GP) and evolutionary programming approaches is proposed. In the genetic model, all the influencing parameters influencing on the size of the cutting forces during the milling process are considered. The presented model can be used in flexible manufacturing for reduction of the total machining time, for the increase of accuracy, reliability, productivity and for decreasing the machining costs. The goal of the paper is the development of mathematical equation for the cutting force in ball-end milling process with genetic programming.*
Key words: *intelligence methods, cutting, ball-end milling, evolutionary programming, cutting force.*

1. INTRODUCTION

The paper presents the development of the genetic equation of machining process, which is shown in detail on the process of machining of steels with ball-end milling. The milling process has become a very important and useful procedure for the manufacture of 3D surfaces of different shapes. Due to the widespread use of highly automated machine tools in the industry, manufacturing requires reliable monitoring and optimization models and methods. The cutting forces that are developed during the milling process, can directly or indirectly estimate process parameters such as tool wear, tool life, surface finish, etc. The capability of modelling cutting forces therefore provides an analytical basis for machining process planning, machine tool design, cutter geometry optimization, and on-line monitoring/control. A large amount of work has been carried out on force modelling. Researchers, (Liu et al., 1996), (Yucesan et al., 1996), (Smith & Tlusty, 1991), (Van Luttervelt et al., 1998) have been trying to develop mathematical models that would predict the cutting forces based on the geometry and physical characteristics of the process. However, due to its complexity, the milling process still represents a challenge to the modeling and simulation research effort. By exploring the advantages of the artificial intelligence methods, the genetic equation is developed. The genetic equation is developed for use as a direct modelling method, to predict cutting forces for the ball-end milling operations. The developed genetic equation will be applied into the manufacturing process for the determination of optimal cutting parameters with a few numbers of experiments and maximum cutting power on tool machine.

2. EXPERIMENTAL SETUP

The experiments were made with the system for the cutting force monitoring (Fx, Fy in Fz) in ball-end milling process. The system for monitoring consists of: tool machine (CNC milling machine MORI SEIKI FRONTIER - M), tool material GC 1010), clamping device, workpiece material: (Ck45, Ck45 (XM), 16MnCr5 in 16MnCr5 (XM)), piezoelectric dynamometer (KISTLER 9259A), amplifier (KISTLER 5001), A/D interface board (PC-MIO-16E-4.), computer (PII 350MHz with Windows 2000 software) and programme package LabVIEW.

3. GENETIC EQUATION FOR CUTTING FORCE IN BALL-END MILLING

For the determination of the genetic equation in ball-end milling with the milling cutter type R216.44-10030-040-AL10G and workpiece material Ck45 was selected 45 experimental data. The maximal cutting forces were monitored with different cutting conditions (radial depth R_D = 0,2 - 0,6 mm, axial depth A_D = 0,2 - 0,6 mm, feeding f_z = 0,08 - 0,12 mm/tooth and cutting speed V_c = 125 - 250 m/min). On the base of input and experimental data and with the selection of adequate arithmetic operations the genetic equation for the cutting force in ball-end milling was developed:

$$F_{max} = f(R_D\ A_D\ f_z\ V_c) \tag{1}$$

For the determination of the genetic equation the independent variables must be set - Terminal set: T = {x, y, z, w, R}, where is x - radial depth R_D, y - axial depth A_D, z - feeding f_z, w - cutting speed V_c and R - random floating-point numbers between -10 and 10, arithmetic operations - Functional set: F = {+, -, *, /}, where is "+" - addition, "-" - subtraction, "*" - multiplication and "/" – division and number of arguments: P = {2, 2, 3, 2}.

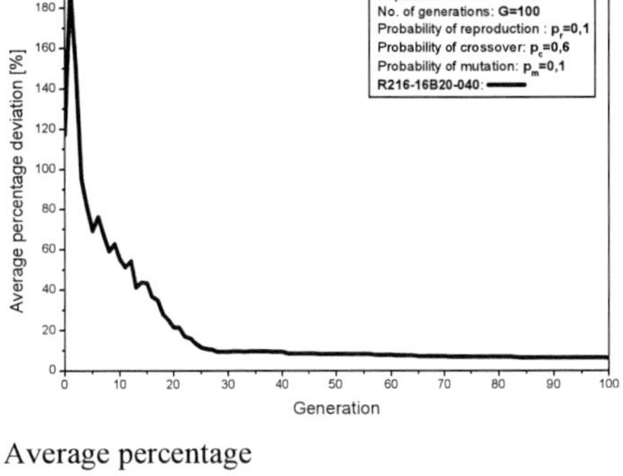

Fig. 1. Average percentage

The evolutionary parameters for the determination of the genetic equation were: population size M=1000 and number of generations G=100 The genetic operations reproduction, crossover and mutation were used. Probability of reproduction was pr = 0,1, crossover pc = 0,6 and mutation pc = 0,1. Figure 1 and figure 3 show the average percentage deviation of the best genetic equation from initial generation to generation 100 where the best genetic equation was reached (equation 2).

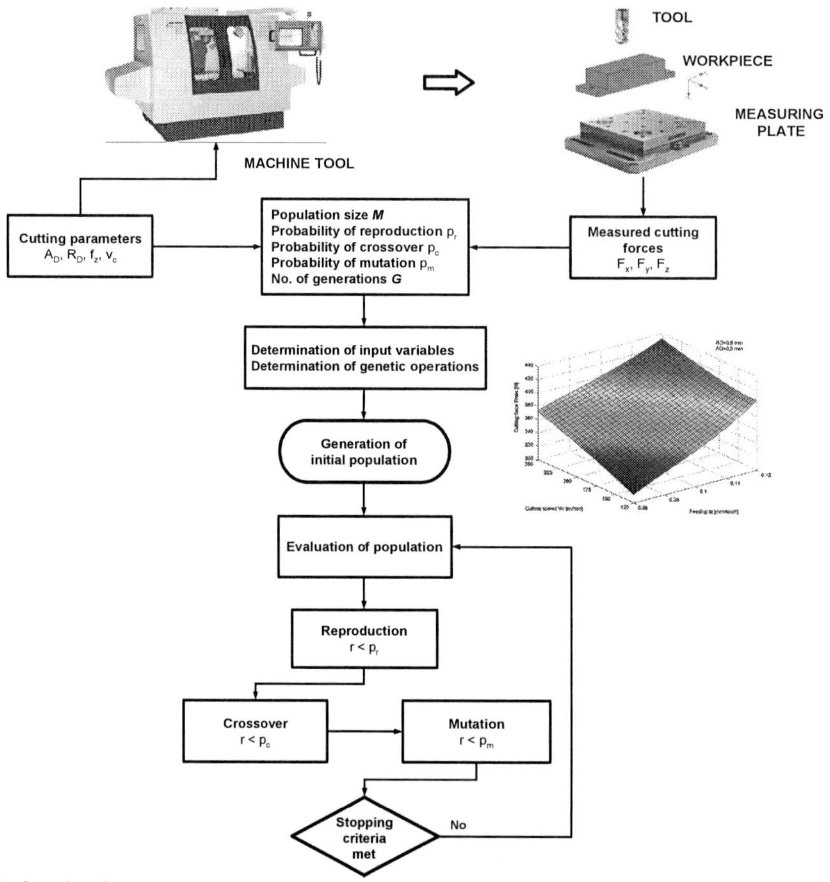

Fig. 2. Genetic model of cutting forces

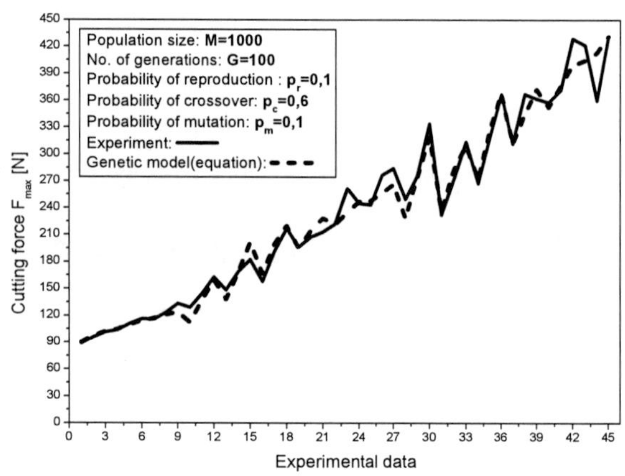

Fig. 3 Experimental and genetic data

$$
\begin{aligned}
F_{max} = {} & 11{,}0486 + R_D - A_D + V_c \cdot R_D \cdot A_D - \\
& 58{,}5012 \cdot (-8{,}2466 + A_D) \cdot A_D + \\
& \frac{58{,}5012 \cdot (-8{,}2524 + A_D) \cdot (0{,}0058 + A_D)}{V_c \cdot (2{,}802 + R_D) \cdot (2{,}7956 - A_D) \cdot f_z^2} + \\
& \frac{V_c \cdot (-0{,}7975 - 0{,}2868 \cdot f_z)}{V_c \cdot (2{,}802 + R_D) \cdot (2{,}7956 - A_D) \cdot f_z^2} + \\
& \frac{V_c \cdot R_D \cdot (2{,}802 + R_D)}{0{,}1012 \cdot V_c - 2 \cdot R_D - A_D + f_z - 0{,}5736 \cdot V_c \cdot f_z}
\end{aligned}
\qquad (2)
$$

4. CONCLUSION

The paper presents the development of the genetic equation in ball-end milling.

The results obtained from the proposed genetic programming approach prove its effectiveness.

The implication of the encouraging results obtained from the present approach is that such approach can be integrated on-line, with an intelligent manufacturing system for automated process planning. Integration of the proposed approach with an intelligent manufacturing system will lead to reduction in production cost and production time, flexibility in machining parameter selection, and improvement of product quality.

This research definitely indicates some directions for future work .

5. REFERENCES

Liu Y., Chen T., Zuo L. & Yang S. (1996). K-L optimization of cutting parameters in machining. *Journal of Huazhong University of Science and Technology*, Vol. 24(5), pp. 50-52.

Mital A. & Mehta M. (1988). Surface roughness prediction models for fine turning. *International Journal of Production Research,*. Vol. 26, pp. 1861-1876.

Smith S. & Tlusty J. (1991). An overview of modeling and simulation of the milling process, *Transactions ASME Journal of Engineering for Industry*, Vol. 113, pp. 169-175.

Van Luttervelt C.A., Childs T.H.C., Jawahir I.S., Klocke F. & Venuvinod P.K. (1998). Present situation and future trends in modelling of machining operations, Progress Report of the CIRP working group on Modelling of machining operations, *Annals of the CIRP*, Vol. 47/2, pp. 587-626.

Yucesan G. & Altintas Y. (1996). Prediction of ball end milling forces, *Transactions ASME Journal of Engineering for Industry*, Vol. 118, pp. 95-103.

Annals of DAAAM for 2006 & Proceedings of the 17th International DAAAM Symposium, ISSN 1726-9679
B. Katalinic, Published by DAAAM International, Vienna, Austria 2006
150th Birthday of Nikola Tesla (1856-1943) Inventor of Mobile Robot Teleoperation

A CONTRIBUTION TO HUMAN-MACHINE COMMUNICATION IN SERBIAN, CROATIAN AND MACEDONIAN LANGUAGE

DELIC, V.; SECUJSKI, M. & TEKIC, Z.

Abstract: *This paper gives a brief review of the development of systems for automatic speech recognition and text-to-speech synthesis in Serbian, Croatian and Macedonian language, at the Faculty of Technical Sciences, University of Novi Sad, Serbia. The systems developed within this project enable two-way communication between humans and machines. These systems are predecessors to many commercial user services such as voice portals and interactive voice response systems. Some original features of these systems, related to certain particularities of south-Slavic languages, will be described as well.*

Key words: *text-to-speech synthesis, speech recognition.*

1. INTRODUCTION

Since speech is the most natural means of communication between humans, people have been trying to develop system that would enable them to extend this interface to communication with machines as well. Using automatic speech recognition (ASR) and text-to-speech synthesis (TTS), humans can talk to devices in their midst, such as household appliances, industry machines, cars, toys or remote computers, even via telephone. One of the important applications of these technologies is also providing independence in computer access to the physically impaired (particularly the visually impaired).

However, speech technologies are language dependent and in some regions they can hardly ever be imported from abroad as most other technologies. They have to be developed for each language separately, especially in case of languages such as Serbian, Croatian and Macedonian, having in mind all their peculiarities. It cannot be expected that some of the world's biggest companies dealing with speech technologies would decide to develop quality ASR and TTS for such a small and closed market in the near future.

A group at the Faculty of Technical Sciences, University of Novi Sad, Serbia, called *AlfaNum* has been dedicated to the development of these technologies for ten years, and their results include phoneme-based automatic speech recognition over the telephone line with accuracy varying between 95% and 99% depending on vocabulary size and sound quality, as well as the first text-to-speech system in Serbian, Croatian and Macedonian taking into account linguistic information and thus greatly enhancing intelligibility and naturalness of synthesized speech. These systems are the basis of a number of applications, such as audio libraries and a speech-enabled web site for the visually impaired, as well as many commercial applications including interactive voice response systems and call centres.

2. ALFANUM SPEECH SYNTHESIZER

During the development of the first high-quality TTS for Serbian language, this team has encountered many problems linked to bridging the gap between plain text and synthesized speech with all its typical features such as intelligibility and naturalness. There is no explicit information in a plain text concerning phone durations, pitch contours nor energy variations. These factors also depend on the meaning of the sentence, emotions and speaker characteristics, which further aggravates the task of attaining high naturalness of synthesized speech (Dutoit, 1997). While Serbian and Croatian are tonal languages, having high-low pitch patterns permanently associated with words, Macedonian is a pitch-accented language with antepenultimate stress on most words, excluding clitics, words of foreign origin as well as some other word groups. Nevertheless, a uniform dictionary-based strategy for lexical stress assignment has been successfully employed in all three cases.

The TTS engine has two main functions: text analysis and synthesis of the speech signal. Text analysis includes text processing such as expanding abbreviations, as well as resolution of morphological and syntactic ambiguities based on a comprehensive accentuation dictionary as well as rule-based syntax analysis. A separate dictionary and syntax analysis techniques were required for each language. Lexical stress assignment, as one of the most important factors influencing the shape of the pitch contour, is based on the algorithm described in (Sečujski, 2005). This approach has proved to produce reasonably correct stress pattern, with word error rate 2,8% for Serbian language. The synthesized speech is highly intelligible and reasonably natural-sounding, much more than any other attempts at speech synthesis in Serbian, Croatian and Macedonian so far. For the development of a multilingual TTS, separate speech databases are generally required for each language, although the Serbian database is at the moment used for Macedonian. This leads to a quite insignificant decrease in speech quality, due to the fundamental similarity between phonetic inventories of these two languages. The Macedonian speech database is expected to be recorded soon. If phonetic inventories are similar enough, as is the case for Serbian, Croatian and Macedonian language, it is appropriate accentuation and appropriate pitch contours that will make synthesized speech sound naturally, almost regardless of the original language of the database.

As to speech signal synthesis, the concatenative approach has been selected as the most promising. The AlfaNum R&D team has recorded a large speech database and labeled it using visual software tools specially designed for that purpose. By keeping score of every phone in the database and its relevant characteristics, use of phones in less than appropriate contexts was avoided, which further contributed to overall synthesized speech quality. This synthesizer is not diphone-based as almost all other speech synthesizers developed for related languages are. The TTS engine can use larger speech segments from the database, according to both phonetic and prosodic requirements, and select them at runtime in order to produce the most intelligible and natural-sounding utterance for a given plain text (Beutnagel et al., 1999). The most significant application of this system so far is *anReader*, a speech synthesizer for the visually impaired that, combined with software known as *screen-readers*, offers them complete independence in computer access. The number of the visually impaired computer users in Serbia has increased significantly since *anReader* was presented for the first time, and its popularity in Croatia and FYR Macedonia is also growing.

3. ALFANUM SYSTEM FOR AUTOMATIC SPEECH RECOGNITION

The goal of ASR is to recognize spoken words in a speech signal, independently of the speaker, the input device, or the environment. A recognized sequence of words W_{ASR} for a given acoustic observation sequence X and all expected word sequences W are usually estimated using Bayes rule:

$$W_{ASR} = \arg_W \max P(W|X) = \arg_W \max P(W) \cdot P(X|W)$$

where $P(W)$ is the *language model* estimated using n-gram statistics and $P(X|W)$ is the *acoustic model* represented by a Hidden Markov Model (HMM), trained using maximum likelihood estimation. HMM encodes the acoustic realisation of speech and its temporal behaviour, while prior probabilities for word sequences $P(W)$ lead to a choice of the word sequence hypothesis with the maximum posterior probability given the models and observed acoustic data. The best word sequence W_{ASR} is computed using a pattern recogniser based on a standard Viterbi decoder. A conventional approach to front-end signal processing of 30 *ms* frames, every 10 *ms*, results in a feature vector X that captures primarily spectral features of the speech signal estimated as cepstrum and energy, along with their first- and second-order time derivatives. A finite vocabulary defines the set of words (sequences of phone units) and phrases that can be recognised by the speech recogniser. The size of the recognition vocabulary plays a key role in determining the accuracy of a system, typically measured in Word Error Rate (WER), including insertion, deletion, and substitution errors.

R&D for Serbian, Croatian and Macedonian ASR has been concentrated on four aspects that define the quality of a speech recognition technique (Gilbert et al., 2005):

- *Accuracy* – WER is less than 5% for small and medium-sized vocabulary continuous ASR; it is achieved by developed acoustic modeling trained with 40 hours of speech databases; good results for large vocabulary continuous ASR in these languages are expected when a more complex language model and more comprehensive post-processing are implemented.
- *Robustness* – channel distortions are compensated by CMS (Cepstral Mean Subtraction), background noise spectrum is subtracted and speaker variations are treated by gender separation and speaker adaptation based on VTN (Vocal Tract Normalization).
- *Efficiency* – long work on software code optimization has resulted in fast decoder and small memory footprint. The ASR engine consumes 2% or more of CPU time on a Pentium IV PC, depending on vocabulary size.
- *Operational performance* – The ASR engine gives a useful confidence scoring and implements barge-in capability, improving operational performance. On the other hand, features such as rejection of out-of-vocabulary speech have not yet been enabled.

Due to the complexity of the problem, a system for isolated word recognition in Serbian language was developed initially. It was later upgraded into a system for connected word recognition in Serbian language. Eventually a system for continuous speech recognition (CASR) was developed, based on recognition of phonemes in particular contexts. Users of this system can define an arbitrary set of words (vocabulary) for each recognition at compiling time (Pekar, 2002). The system takes into account lexical stress (particularly vowel length), giving greater significance to stressed vowels at recognition time. This system can be used for speech recognition in all three aforementioned languages because it is phoneme-based and because of the similarity of phonetic inventories of these three languages. No significant drop in performance for languages other than Serbian has been observed, but actual experiments will be carried out as soon as adequate ASR speech databases in Croatian and Macedonian are available. Even state-of-the-art ASR systems cannot be successful enough if they are based on acoustic features only. In order to achieve natural dialogs in speech applications, AlfaNum ASR has to apply some post-processing such as Spoken Language Understanding (SLU), as well as a lot of experience in both machine learning and design of front-end technology. The goal of SLU is to extract the meaning of recognized speech in order to identify a user's request. Dialog Manager (DM) evaluates the SLU output in context of the call flow specifications, which results in dynamic generation of the next dialog turn. The DM may apply a range of strategies to control dialog flow according to different application tasks. To provide a successful dialog progress, intelligent speech applications have to handle problematic situations caused by system failures or absence of concise or accurate information in a speech utterance. Post-processing makes it viable to adopt natural language dialog applications without having to achieve perfect recognition accuracy and without dictating what a user should say.

4. CONCLUSION

The two speech technologies developed for Serbian, Croatian and Macedonian language have enabled two-way human-machine communication in these languages. This communication can be direct or remote (e.g. via telephone), which introduces the possibility of building speech-enabled intelligent systems. This is a step of a human-to-machine interface in the regions where these languages are spoken from touch-tone prompts toward multimedia and multimodal interface. While both these technologies are still under development, they are already implemented in many commercial applications, and are also in wide use as aids for people with visual disabilities in Serbia, Croatia and FYR Macedonia.

5. REFERENCES

Beutnagel, M.; Mohri, M.; Riley, M. (1999). Rapid Unit Selection from a Large Speech Corpus for Concatenative Speech Synthesis, *Proceedings of EUROSPEECH*, pp. 607-610, ISSN 1018-4074, Budapest, Hungary, September 1999, ESCA, Bonn

Dutoit, T. (1997). *An Introduction to Text-to-Speech Synthesis*, Kluwer Academic Publishers, ISBN 0-7923-4498-7, Dordrecht/Boston/London

Gilbert, M.; Wilpon, J. G.; Stern, B.; Di Fabbrizio, G. (2005). Intelligent Virtual Agents for Contact Center Automation, *IEEE Signal Processing* Magazine, Vol. 22, No. 5, (September 2005) pp. 32-41, ISSN 1053-5888

Pekar, D.; Obradović, R.; Delić, V. (2002) AlfaNumCASR – a system for continuous speech recognition, *Proceedings of 3rd Conference DOGS*, pp. 49-56, Bečej, Serbia, September 2002, Unicersity of Novi Sad, Novi Sad, Serbia

Secujski, M. (2005). Obtaining Prosodic Information from Text in Serbian Language, *Proceedings of IEEE EUROCON*, Belgrade, Serbia, November 2005, *Available from:* http://ieeexplore.ieee.org, *Accessed:* 2006-05-20.

Annals of DAAAM for 2006 & Proceedings of the 17th International DAAAM Symposium, ISSN 1726-9679
Editor B. Katalinic, Published by DAAAM International, Vienna, Austria 2006
150th Birthday of Nikola Tesla (1856-1943) Inventor of Mobile Robot Teleoperation

USING OF TRIBOTECHNICAL DIAGNOSTICS BY WORKING RELIABILITY DETERMINATION OF THE PACKING LINES FOR SPREAD CHEESES

DOBRANSKY, J.; RUZBARSKY, J. & FABIAN, S.

Abstract: *Article deals with working reliability determination of automatic packaging lines for spread cheeses with regard to use of individual sorts of oils, determination of friction regime, machine wear and oils exchange interval estimation. The object of working reliability is the packing line for spread cheeses in dairy. After off-take of oil from lubricating systems of the machine this oil was used for obtaining information about the technical condition of this machine, the friction regime and the wear.*
Key words: lubrication, friction, wearing, tribology

1. INTRODUCTION

Working reliability of packaging lines for spread cheeses is one of the important indicators, which affects their maximum efficiency in using process. The basic problem of working reliability is reliability determination of individual line parts. In generally we can say that the reliability is line property to subserve asked functions according to specified technical conditions. Working reliability was evaluated on packaging lines for spread cheeses from using aspect of individual sort of oils in line, friction regime specification, lines wearing and from determination aspect oils exchange interval by tribotechnical diagnostic methods. Working reliability object was the packaging line for spread cheese in dairy. The specimens of oils which were removed from three packaging lines for packing of spread cheeses KUSTNER YH (Fig.1) were analyzed. The specimens were removed always from the same place of lubricating system – oil box after the end of operation. Ahead of analyze the specimens of oils have to be mix up because the oil particles could be sedimentary.

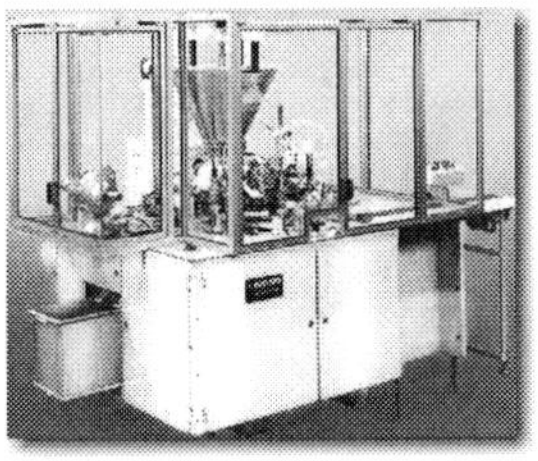

Fig. 1. Packaging line KUSTNER YH

2. TRIBOTECHNICAL DIAGNOSTICS METHODS

2.1 Measurements of total contaminations and H_2O content
Equipment CCT INFRA is using for both measurements. Measuring probe is submerging into the oil and equipment determines quantity of total contamination in oil (%) by photometry. These valuations are showing on the display. A several drops of oil are stash into the equipment saucer. Saucer is warms up on required temperature. Than oil conduct is monitoring. Than is make visual and acoustic analysis of oil.

2.2 Measurements of oil ignition point
Measurements is making on special equipment, which consists of heating and thermometer. Specimen of oil is pouring into the metal container, which is inserted into the equipment. Oil is warmed up and blaze is attached to the oil level. If the blaze is on the oil level, the temperature which is showing on the display is registered. This temperature is oil ignition point.

2.3 Determination of machine wearing regime
For this measurement was used equipment PMA 87 – 2. It uses light tubes which are located at divergent magnetic field of special super magnetic type. Elements situated in oil are segregating and accumulating by the heavy divergent field. Big elements are marked as Dl and they have a dimension bigger than 15 μm and small elements marked as Ds has a dimension approximately 2 μm.
Equation for machine wearing level determination

$$WPC = \frac{Dl + Ds}{1800} \cdot 100 \quad (\%) \tag{1}$$

Dl – big abrasive elements
Ds – small abrasive elements
Equation for machine wearing intensity determination

$$PLP = \frac{Dl.k}{Dl + Ds} \cdot 100 \quad (\%) \tag{2}$$

2.4 Ferrogaphy
For the measurement is using magnetic analyzer (Fig.4). This process enabled separation of abrasive elements which has metal character (especially ferromagnetic metals) from process lubricant and the other alloys which are soluble in organic compounds. This ferrographical process consists of specimen analyze under the scope. The result is displayed as a digital Ferrographical record on the computer by camera.

3. ANALYSIS OF OIL SPECIMENS

As has been mentioned the specimens were removed from three packaging lines KUSTNER YH for packing of spread cheeses, specifically from the lubricating system – oil box. Lubricant is gear oil Shell.
Specimen No.1 - general reparation in 2002
Total contaminations CN – 1,81
H_2O contents – 8 %
Oil ignition point - H_2O contents is more than 5 %, oil ignition point is unknown.
Dl – 310
Ds – 30
Machine wearing level:

$$WPC = \frac{D_l + D_s}{1800} \cdot 100 = \frac{310 + 30}{1800} \cdot 100 = 18{,}888 \ \% \tag{3}$$

Machine wearing intensity:

$$PLP = \frac{D_l \cdot k}{D_l + D_s} \cdot 100 = \frac{97 \cdot 0{,}01}{97 + 12} \cdot 100 = 0{,}892 \ \% \qquad (4)$$

Ferography:

- adhesive elements – 200 - 250 elements in 1 ml of oil,
- abrasive elements – 20 – 30 elements in 1 ml of oil,
- silicium – 50 pieces,
- colour metals.

Evaluation:

- wearing regime – good,
- oil condition – exchange immediately.

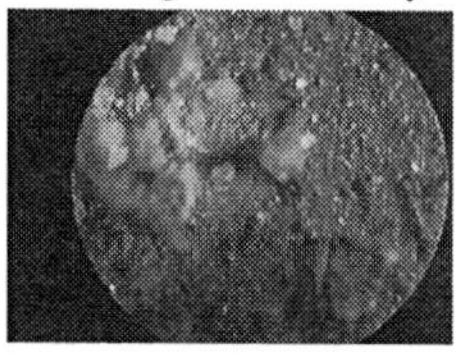

Fig. 2. Ferrographical record of specimen No.1

Specimen No.2 - general reparation in 1997

Total contaminations CN – 1,94

H_2O contents – 8 %

Oil ignition point - H_2O contents is more than 5 %, oil ignition point is unknown.

Dl – 710

Ds – 600

Machine wearing level:

$$WPC = \frac{D_l + D_s}{1800} \cdot 100 = \frac{710 + 600}{1800} \cdot 100 = 72{,}771 \ \% \qquad (5)$$

Machine wearing intensity:

$$PLP = \frac{D_l \cdot k}{D_l + D_s} \cdot 100 = \frac{710 \cdot 0{,}01}{710 + 600} \cdot 100 = 0{,}541 \ \% \qquad (6)$$

Ferography:

- adhesive elements – 700 – 1000 elements in 1 ml of oil,
- abrasive elements – up to 200 elements in 1 ml of oil,
- colour metals.

Evaluation:

- wearing regime – very wrong,
- oil condition – exchange immediately.

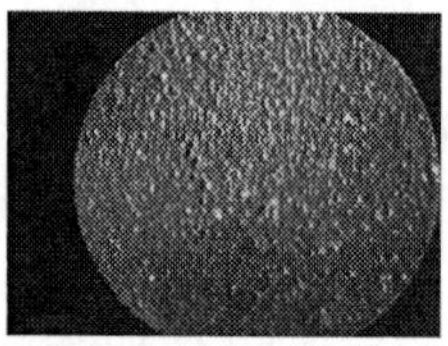

Fig. 3. Ferrographical record of specimen No.2

Specimen No.3 - general reparation in 1996

Total contaminations CN – 1,89

H_2O contents – 10 %

Oil ignition point - H_2O contents is more than 5 %, oil ignition point is unknown.

Dl – 540

Ds – 200

Machine wearing level:

$$WPC = \frac{D_l + D_s}{1800} \cdot 100 = \frac{540 + 200}{1800} \cdot 100 = 41{,}111 \ \% \qquad (7)$$

Machine wearing intensity:

$$PLP = \frac{D_l \cdot k}{D_l + D_s} \cdot 100 = \frac{540 \cdot 0{,}01}{540 + 200} \cdot 100 = 0{,}729 \ \% \qquad (8)$$

Ferography:

- adhesive elements – 400 – 500 elements in 1 ml of oil,
- abrasive elements – up to 100 elements in 1 ml of oil,
- silicium elements,
- colour metals – the most.

Evaluation:

- wearing regime –wrong,
- oil condition – exchange immediately.

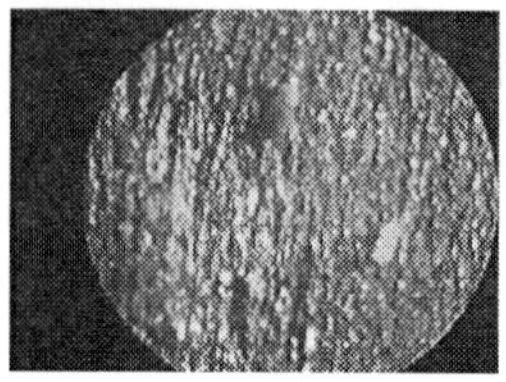

Fig. 4. Ferrographical record of specimen No.3

3. CONCLUSION

At the experiments has been evaluated working reliability of packaging lines for spread cheeses, especially wearing regime and oil condition. It was made individual measurements, which are the important for determination wearing regime and oil condition. The first was detected total contamination (CN). On the specimens was measuring value CN up to 3%, what is the maximum admissible value. We can say that the CN have not the expressive effect on the working reliability. Next was detected the water contents in oil. The all specimens had water contents higher than 5%. Thereat, we are could not detected oil ignition point. Water content is related with washing line, which is the major problem by the machine maintenance. The important will be to improve the machine gasket. Next were detected big and small abrasive elements. Following of these elements was calculated machine wearing level and machine wearing intensity. On the end the specimens was analyzed under the scope. The results were displayed as a digital Ferrographical record on the computer by camera. Following of these measurements we can say that the wearing regime of these machines is wrong and is necessary to remove oil in the all machines.

Acknowledgment

The authors would like to acknowledge the support of Scientific Grant Agency of the Ministry of Education of Slovak Republic, Commission of mechanical engineering, metallurgy and material engineering, for their contribution to project 1/3224/06 – Modelling, simulation and optimization of the influence of technological parameters on products quality parameters at automation packaging manufacturing lines .

4. REFERENCES

Al Hakim, H.; Blazovsky, J. (2002). *Tribolagia a servis strojov,* FVT, ISBN 80-7099-875X, Presov

Bhushan, B. ; Gupta, B.K. (1991). *Handbook of tribology,* McGraw-Hill, Inc., ISBN 0-07-005249-2, New York

Bigos, P.; Pidany, J. (1987). *Prevadzkova spolahlivos strojov,* ALFA, Bratislava

Blaskovic, P.; Balla, J. & Dzimko, M. (1990). *Tribologia,* ALFA, ISBN 80-05-00-633-0, Bratislava

Dobransky, J. (2006). Automatizacia procesu dopravy vstupnej suroviny do linky vtechnológii balenia tavenych syrov, *Proceedings of 7-th International conference: Automation in production planning and manufacturing,* pp. 27-32, ISBN 80-8070-537-2, Žilina, Faculty of mechanical ingineering, Žilina

Fabian, S. (1998). *Zvysovanie spolahlivosti vyrobnych systemov na optimalnu uroven,* Vydavatelstvo Michala Vaska, ISBN 80-7165-174-5, Presov

Havlicek, J. (1989). *Provozni spolehlivost strojů,* Statni zemědelské nakladatelství, ISBN 80-209-0029-2, Praha

Kahancova, E. (1991). *Spolahlivost a diagnostika,* TUKE, ISBN 80-7099-064, Kosice

Kajdas, C.; Harvey, S.S.K. & Wilusz, E. (1990). *Encyclopedia of tribology,* Elsevier, ISBN 0-444-88401-7, Amsterdam

Annals of DAAAM for 2006 & Proceedings of the 17th International DAAAM Symposium, ISSN 1726-9679
B. Katalinic, Published by DAAAM International, Vienna, Austria 2006
150th Birthday of Nikola Tesla (1856-1943) Inventor of Mobile Robot Teleoperation

HIERARCHICAL CONTROL OF THE SELF-EXCITED VIBRATIONS ABSORBER

DOMEK, S.; PARUS, A. & PIETRUSEWICZ, K.

***Abstract:** A Chatter phenomenon is a severe problem in modern machining. Therefore, vibration absorbers are widely applied. In the paper methods to control electromagnetic, self-excited vibration absorber are proposed and compared to each other. The Pole-Zero-Placement (PZP) method is compared with the so-called Model-Following PID 2-DOF Control method (MFC), which uses the nominal model of the controlled plant to evaluate the main control signal, and the corrective signal, which is used to force the process to follow the model. Such a method enables the great robustness of a system. Also the fuzzy-logic PD controller used as a corrective one in MFC system is introduced.*
Key words: *vibration absorber, chatter, self-excited vibration, semi-active eliminator of vibration, model-following control, fuzzy-logic.*

1. INTRODUCTION

Machining process is very often disturbed by high level vibrations. Intensity of vibrations depends on machine tool – cutting process (MT–CP) system e.g. cutting parameters, dynamic properties of cutting tool and the used work-piece. In such systems, under certain conditions, the positive feedback from cutting process can lead to instability and uncontrolled increasing of work-piece vibration. This phenomenon is commonly known as chatter and it is highly undesirable. Poor quality of machined surface is an effect of working with chatter. Moreover, high level of cutting force leads to decreasing of cutting tool life and whole machine tool durability. Chatter is a subject to consider in a many papers. Authors point out that phase shift between inner and outer modulation is the main reason of loosing stability (Olgac & Hosek, 1998).
Figure 1 presents the model of an electromagnetic absorber used to suppress chatter in milling machine. For such system two control methods are proposed.
First is based on the Pole-Zero-Placement approach, the second one involves the robust two-degrees of freedom Model-Following Control system (Skoczowski, 1999). According to this, two new systems are introduced: Hybrid Self-Excited Vibrations Absorber (HSEVA) and HFSEVA based on the fuzzy-logic PD correction, evaluated on the basis of difference between the process and nominal model.

2. VIBRATION ABSORBER

The frequency of the self-excited vibrations is relevant on natural frequency of the mass-damping-spring (MDS) system and cutting parameters. Therefore, chatter frequency may jump in vicinity of one of the MDS natural frequencies and then alter during machining. This is the reason that pure passive absorbers are inefficient in suppressing the self-excited vibrations. The absorber presented in the Fig. 1 exhibits properties of passive and active one. It consists of two electromagnets of appropriate chosen mass m_2 suspended on springs with spring coefficient k_2 and damping coefficient c_2. The natural frequency of absorber is chosen to harmonize with natural frequency of the work-piece determined by mass m_1 and spring coefficient k_1 and damping coefficient c_1. This assures the passive effects of the absorber. Moreover, the electromagnets introduce additional vibration on the work-piece. Properly designed control system enables to shift the absorbers natural frequency.

3. MATHEMATICAL MODEL

The considered here system can be simplified into 2 degree of freedom system composed of two mass moving parallel to the feed direction x presented in Fig. 1.

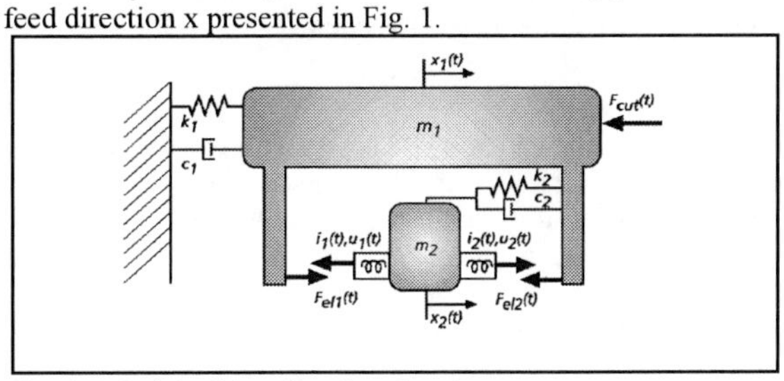

Fig. 1. Model of the vibration absorber.

The system motion is defined analogically to the one proposed in (Chodźko, *et.al., 2004*).

4. VIBROSTABILITY ASSESMENT

Efficiency of proposed control system for electromagnetic, self-excited vibration absorber has been tested in Matlab/Simulink simulation with mechanistic model of the cutting force. The model takes into account geometrical parameters of cutting edge, number of edges and machining parameters, details can be found in (Chodźko *et al.*, 2004). The effectiveness of methods for each gear-box conditioned angular velocity has been evaluated on the basis of the maximal width of cut at which the contact between the cutting tool and the work piece during a given time of machining is lost. Maximal values of the machining parameter width of cut w_0 that enable the machine tool to be operated within the full range of spindle speeds without losing its vibrostability also has been determined.

5. POLE-ZERO-PLACEMENT CONTROL

Vibrostability of machining can be improved by increasing damping factor at f_1 and moving f_2 into higher value. This can be achieved through appropriate pole and zero placement, which cannot be accomplished independently of one another, because of system properties. Transfer function x_1/F_{cut} for system take a general form:

$$G(s) = \frac{N_1(s)}{D(s)} = \frac{(s+\omega_1)(s^2+\omega_2\delta+\omega_2^2)}{\det(s\mathbf{I}-\mathbf{A})} \qquad (1)$$

Increasing of the damping at f_l is achieved trough placement of the conjugate zeros at a frequency closed to f_l by using feedback gain matrix **K**. In view of the fact that the input vector of the model is made up of the supply voltage u and the machine cutting force F_{cut} being in practice beyond reach as stemming from the machining itself, the matrix **K** is determined from equation:

$$\mathbf{N}_i = \mathbf{C}[s\mathbf{I} - \mathbf{A} + \mathbf{BK}]_{ad}\mathbf{b}_i = \mathbf{C}[s\mathbf{I} - \mathbf{A} + \mathbf{BK}_i]_{ad}\mathbf{b}_i \quad (2)$$

where:

$$\mathbf{K} = \left[\mathbf{k}_1 \ldots \mathbf{k}_i \ldots \mathbf{k}_m\right],\ \mathbf{K}_i = \mathbf{K}\big|_{\mathbf{k}_i=[0]} = \left[\mathbf{k}_1 \ldots 0 \ldots \mathbf{k}_m\right] \quad (3)$$

$$\mathbf{B} = \left[\mathbf{b}_1 \ \cdots \ \mathbf{b}_i \ \cdots \ \mathbf{b}_m\right] \quad (4)$$

where m is a number of system inputs.

Hence, there is a possibility to shape the transfer function numerators N_i to a required form, since the i-th row of the gain matrix **K** has no influence on how zeroes of the i-th column in the N_i matrix are placed. Matrix **K** in the form:

$$\mathbf{K} = \begin{bmatrix} 0 & 0 & 0 & 0 & 0 \\ 0 & k_2 & 0 & k_4 & k_5 \end{bmatrix} \quad (5)$$

is used to achieve zero placement for input F_{cut} close to 76Hz and 125Hz. There are two elements in **K** that are not used to zero-placement (Chodźko, *et. al*, 2002). Values of the gains k are experimentally determined and are used to further improvement of controller properties.

6. MODEL-FOLLOWING CONTROL (MFC)

The main idea of the MFC (Skoczowski, 1999) structure is the possibility of acting independently in tracking the reference signal and suppressing effects of disturbances and perturbations. The reference signal *r(s)* is applied to the input of the loop that includes the nominal process model *M(s)* and the controller of the model $R_m(s)$. On the other hand, the nominal model control input is applied to the corrective controller output, and the resulting signal is applied to the process input. The process output is compared with the model output, and the difference is applied to the corrective controller input. Hence, the process and the corrective controller form a second feedback loop. In (Skoczowski, *et.al*, 2003) the robustness was proven in the case of control of the processes with relatively high and varying time-delay.

7. THE HIERARCHICAL VIBRATIONS ABSORBER

The idea of the MFC system was intruduced into the novel structure of the self-excited vibrations absorber (HSEVA).

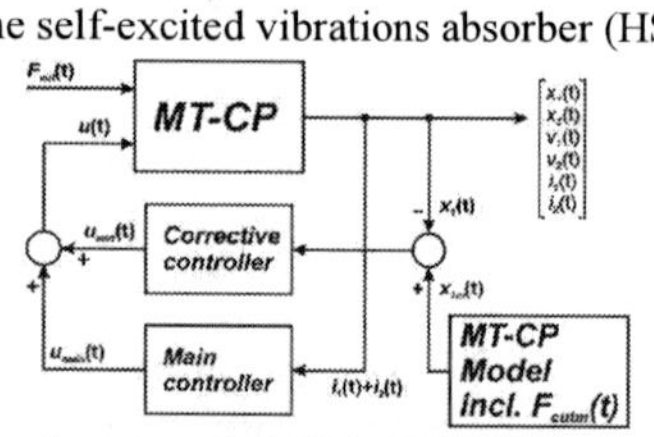

Fig. 2. Proposed system (HSEVA, HFSEVA).

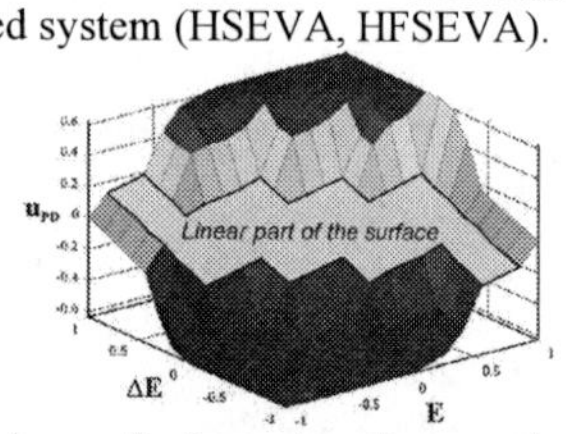

Fig. 3. PD rule base of a fuzzy-based correction.

The modification of the HSEVA system is made with the use of fuzzy-logic PD controller, which control surface (Skoczowski, *et. al*, 2005) with its linear operation is marked in Figure 3.

8. SIMULATION RESULTS

Figure 4 presents one of the simulation tests, did with the use of proposed here approach, compared with the Pole-Zero-Placement approach, known from (Chodźko, *et.al*, 2004).

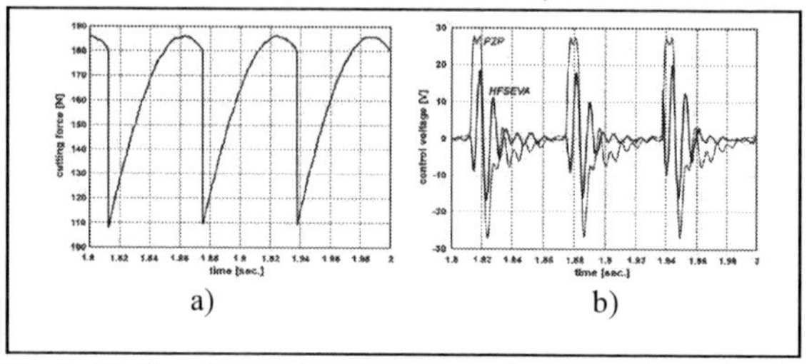

Fig. 4. Cutting force (a) and the control voltage (b) during the test; cutting depth equal to 2.3mm.

The main advantage of the HFSAVE introduced here is the higher quality of the control voltage.

9. CONCLUSION

The proposed in the paper hybrid self-excited vibrations absorber (HSEVA) is a control system that proves its usability for the self-excited vibrations elimination. Fuzzy-logic PD controller introduced here (HFSEVA) into the corrective loop of the proposed system enlarges the robustness, what translates for the greater depth of cutting.

ACKNOWLEDGEMENTS

This work was supported by the Polish Ministry of Science research under Grant 4 T07D 016 29, "The evaluation of active chatter vibration elimination methods in cutting process".

10. REFERENCES

Olgac, N. & Hosek M. (1998). A new perspective and analysis for regenerative machine tool chatter. *International Journal of Machine Tool & Manufacture,* Vol. 38, 783-798

Esma, S. & Marui, E. (2000). Suppression of Chatter Vibration of Boring Tools using Impact Dampers. *International Journal of Machine Tool & Manufacture,* Vol. 4, 1141-1151

Lee, W.: Kim, K. & Sin, H. (2002). Design and Analysis of a Milling cutter with the Improved Dynamic Characteristics. *International Journal of Machine Tool & Manufacture,* Vol. 42, 961-967

Semercigil, S. & Chen, L. (2002). Preliminary computations for chatter control in end milling. *Journal of Sound and Vibration,* Vol. 249, No 3, 622-633

Jang, J. & Tarng, Y. (1999). A study of the active vibration control of a cutting tool. *Journal of Materials Processing Technology,* Vol. 95, 78-82

Chodźko, M.: Domek , S., Marchelek, K., & Parus, A. (2004). Methods of Semi-Active Vibration Absorber Control in Machine Tools. *Proceedings of 10th Methods and Models in Automation and Robotics,* 485-490, August, 2004, Międzyzdroje.

Skoczowski, S. (1999). A robust control system utilizing plant model (in Polish). *PAK,* No. 9, pp. 2-4, ISSN 0032-4110.

Skoczowski S., Domek S., & Pietrusewicz K. (2003) Model following PID control system. *Kybernetes: The International Journal of Systems & Cybernetics,* vol.32, no.5/6, pp. 818-828, ISSN 0368-492X.

Skoczowski S., Domek S., & Pietrusewicz K., Broel-Plater B. (2005). A Method for Improving the Robustness of PID Control. *IEEE Transactions on Industrial Electronics,* vol. 52, No. 6, pp. 1669-1676, ISSN 0278-0046.

Annals of DAAAM for 2006 & Proceedings of the 17th International DAAAM Symposium, ISSN 1726-9679
B. Katalinic, Published by DAAAM International, Vienna, Austria 2006
150th Birthday of Nikola Tesla (1856-1943) Inventor of Mobile Robot Teleoperation

POSSIBILITIES OF RURAL TOURISM DEVELOPMENT IN THE DUBROVNIK AREA

DRAGICEVIC, M.; KRZELJ, Z. & BRAUTOVIC, H.

Abstract: *The objective of this study is to discuss the possibilities of rural tourism in the Dubrovnik area. The area abounds in resources favourable for its development. Due to the lack of projects in the County of Dubrovnik, rural tourism is unable to exploit all of its resources. Its development is in stagnation. Consequently, the Dubrovnik area has only 75 home-style lodgings, out of which only three provide services necessary for this kind of tourism.*

Key words: *rural tourism, Dubrovnik area, development possibilities, conditions for Dubrovnik rural tourism.*

1. INTRODUCTION

In Croatia, not one project for the development of home-style lodgings has been undertaken. The development of rural tourism was broached in the Croatian tourism development strategy as a possible form of tourism. The definition of rural tourism has been the subject of much debate in the literature without arriving at any firm consensus. Several terms like rural tourism, eco tourism, agro tourism, or countryside tourism can be found. The fact is that there are only a few individual examples of a successful development of rural tourism. Istria is an example of a region, which has been well utilized for tourism. The Dubrovnik area has not yet recognized the benefits of rural tourism. Although the area abounds in resources, they have not been exploited. The Dubrovnik area has only three home-style lodgings providing a total service of bread and board and family life. In relation to the total number of possible home-style lodgings in Croatia, the Dubrovnik area is in the second place. According to bed-capacity, it is in the sixth place.

2. RURAL TOURISM IN THE REPUBLIC OF CROATIA

The concept of alternative tourism (AT) is to provide a complement to mass tourism and to ensure that the tourist policy is directed towards demand requirements. (Fennell, 2003)
The existence of various types of tourism enables the tourist season to be prolonged. The high season could be extended throughout the year. It could attract a more diverse and wealthy clientele. (Priporas & Kamenidou, 2003)
An important type of alternative tourism with great development potential is countryside tourism, which includes a “vast pool” of flora, fauna, and cultural profiles of the people. (Koščak, 1995) The development of rural tourism includes all the domains of country life such as architecture, vegetation, culture, national costumes and rural atmosphere. This type of tourism espouses the significance of the rural life-style and represents a retreat to nature.
The Republic of Croatia has done no on-going research of foreign experience regarding development of rural tourism. In 1991, the Croatian Institute for Tourism started a program for researching the possibilities of rural tourism development, which proposed, among other things, foreign experience analysis. It was rejected without valid arguments by the then Minister of Agriculture, who was to have partly financed the cost of the project together with the Ministry of Tourism. (Kušen, 1995)
Although the systematic approach to the rural tourism development was unsupported by the Government, there are some individual successful examples for the development of this type of tourism. The Istrian area is in the forefront of this concept of tourism as it has exploited its attractive, communicative, and receptive factors, as well as its comparative advantages. As a result, Istria today is the only region in the Republic of Croatia who has adequately developed rural tourism.

3. RURAL TOURISM IN THE DUBROVNIK AREA

Almost the entire Dubrovnik area has the possibility to develop this type of selective tourism. Besides the Dubrovnik surroundings and archipelago, the Dubrovnik area includes the river Neretva. The Neretva river mouth has ideal conditions for growing different sorts of fruit and vegetables. The area also has many different species of birds and fish. This is an important condition to the development of a specific type of rural tourism.
To develop the unused tourist resources in the Primorje district, now predominantly inhabited by an older population, would encourage the return of the younger population to their villages as well as create jobs and employment. This would also impede further outflow of the young village population
The land of the Prevlaka district, part of the Konavle region, is becoming private property again. This is a great opportunity to build new and to reconstruct existing buildings for the purpose of rural tourism development. The government has the main role in this because its financial support would enable its success. The Konavle region has many advantages over the Primorje region, as its population has not moved to Dubrovnik. Namely, after the Patriotic War, almost all inhabitants of the Konavle region returned to their homes and reconstructed their houses. This is an important fact for the development of the rural tourism in this area.

3.1. Conditions for the rural tourism development in the Dubrovnik area

The Dubrovnik area abounds with rich natural beauty, cultural and historic heritage. It is a typical Mediterranean climate: the average air temperature is 16,4°C, the average sea temperature from May to September is 17,9°C – 23, 8°C, sunny hours per day average 7,2. The topography of the Dubrovnik area is rocky, and the coast is well indented with numerous coves, beaches, slopes and forested islands. The most beautiful islands in the Dubrovnik archipelago include Korčula, Mljet, and Lastovo. Together with the peninsula Pelješac, these islands have great possibilities for the development of rural tourism. Flora and fauna are abundant and diverse, dominated by cypress, pine and olive trees, vineyards, lemon and orange plantations, as well as the aromatic herbs and flowers like palm trees, cactus, and especially century plants for which this area is famous. The entire Dubrovnik area, including the islands, has great potential for rural tourism development. This area abounds with vineyards and olive groves, important to rural tourism because it is crucial to produce and offer high quality and well-known wines and olive oil. In addition to its natural resources, the entire Dubrovnik area, and especially the city of Dubrovnik, is well known throughout the world for its cultural and historical monuments. Villages are recognizable by their autochthonous architecture and distinctive customs that have survived in some hinterland villages in the Dubrovnik area until now. For rural tourism, village customs are important because the guests can participate and take part in a unique experience. In rural tourism, the most important thing is to give opportunities to the guests to participate actively in the process of the production of a village product such as wine or olive oil, or to help cultivate farmland or feed the cattle. Traffic infrastructure is necessary for tourist mobility. Road traffic at present is inadequate. A highway is planned that would connect Dubrovnik with the rest of Croatia. Most important is to establish air traffic, which is the fastest approach to the area. Although air traffic is quite developed, it is not enough because there are few roads to the villages in the Dubrovnik area. The lack of airports on the islands presents a serious problem. Seldom are guests transferred to the islands by hydrofoils. Receptive factors are lodging facilities to accommodate tourists in the tourist destination. In the Dubrovnik-Neretva County, there are seventy-five home-style lodgings. However, since Croatia has not identified what is needed for tourism, only three out of seventy-five home-style lodgings provide a complete service. Bed capacity in the whole area is seventeen, or five to six beds per location on average. Out of the total number of potential lodgings in the area, there are seventy-two houses for wine tasting. (CCE, 2005) The Dubrovnik area is quite undeveloped with only three lodgings providing the entire service. There is much room for growth.

3.2. The state of rural tourism development in the Dubrovnik area

Past development of rural tourism in the Dubrovnik area has not been satisfactory. There are huge, almost unlimited possibilities for tourist development using the available resources. Only one positive factor influences the development of rural tourism, the natural resource. All other factors are undeveloped. The other problem is that the family-home lodging has not been defined. This is the reason why the Dubrovnik area only has three, out of seventy-five family-home lodgings that can actively participate in the development of rural tourism, and it depicts the existent state of rural tourism in the Dubrovnik area. Another problem is what share of the products is grown domestically and what percentage is imported. Unfortunately, most of the products offered and sold to the tourists visiting home-style lodgings are not of an authentic origin. The products offered in the home-style lodgings are industrially made. The above-mentioned facts explain the reasons for non-existence of the rural tourism development project at the national level. However, neither the Dubrovnik-Neretva County administration has any incentive policy for the development of rural tourism in this area.

3.3. The development possibilities

Analyzing the state and rural tourism development conditions in the Dubrovnik area, the following issues are favourable for the development:

- natural resources
- history, culture and entertainment
- tourist lodging and catering potential; abandoned family homes and villages which could be reconstructed and altered into attractive tourist destinations

There are also many obstacles, such as:

- accommodation capacity underutilized
- lack of apartment/ hotel complex
- unsatisfactory environmental management
- contemporary architecture at the expense of the traditional architecture
- chaos in space planning in rural areas – the disintegration of villages
- disappearance of traditional values – the decoration of the family-style lodgings resembles to those classical catering establishments
- the lack of incentives for shared agricultural and rural tourist activities
- the lack of incentives for traditional building
- insufficient level of education of the local population for providing rural tourism activities
- non-existence of a specific standardization, classification and rural lodging pricing
- rural family-style lodgings are under the purview of higher level departments of inspection
- legally unidentified status of rural tourism
- 10 different Legislations and Regulations regarding the activities of the rural family-style lodgings
- abuse of the term rural tourism

4. CONCLUSION

New, alternative types of tourism are based on travel in which the tourists are active participants. Today tourists want more adventure, cultural and spiritual richness, preserved and protected nature, autochthonous architecture. Only the development of selective types of tourism and selective offerings can meet the challenge of today tourists' needs. Many countries have already prioritized selective types of tourism and have highly ranked rural tourism, considering the great possibilities for development. The Republic of Croatia has not yet established a unified project for rural tourism development. The scientific and professional researches in this area are few. The Dubrovnik area has great development potential, and the richness of resources need to be evaluated for tourism. The present state of rural tourism in the Dubrovnik area is not satisfactory. The support of the Dubrovnik-Neretva County Government is necessary for the development of rural tourism in this area. The strategic importance of rural tourism must become a priority for Croatia, as well as for the Dubrovnik-Neretva County.

5. LITERATURE

Fennell, D., (2003). *Ecotourism*, Routledge, ISBN 0-415-30365-6, London.

Koscak, M., (1995). *Razvoj ruralnog turizmu u Sloveniji. Turizam*, Vol., 18., No., 7-8., (god. XLIII), p. 134-142, ISSN 0494-2639.

Kusen, E., (1995). *Turizam na seljackom gospodarstvu. Turizam*, Vol., 18., No., 7-8., (god. XLIII), p.127-133, ISSN 0494-2639

Priporas, V.C., Kamenidou, I., (2003*). Alternativni turizam kao opcija razvoja turizmu u sjevernoj Grčkoj. Turizam*, Vol., 51., No., 1., (god. LI), p. 49-57, ISSN 0494-2639

Croatian Chamber of Economy Economic Reports, Zagreb, 2005.

Annals of DAAAM for 2006 & Proceedings of the 17th International DAAAM Symposium, ISSN 1726-9679
B. Katalinic, Published by DAAAM International, Vienna, Austria 2006
150th Birthday of Nikola Tesla (1856-1943) Inventor of Mobile Robot Teleoperation

COLLABORATIVE DESIGN SYSTEM FOR THE MECHATRONIC PRODUCTS IN THE CESICED PLATFORM

DRAGOI, G.; ROSU, S.M.; ROSU, L. & COTET, C.E.

Abstract: *This paper presents a solution of a next generation engineering working environments implemented in the experimental platform for training, research and consulting in the new digital economy (CESICED) and describes a core information technology which this systems can be built on. The platform is developed at University POLITEHNICA of Bucharest in the PREMINV center. We emphasize on special skills requirements for collaborative design of the mechatronics products. Within the framework of its activities on collaborative engineering, the PREMINV set up training projects aiming at developing competence's in this field with objectives of education training but also research, and industrial transfers.*
Key words: *collaborative design systems, mechatronics, virtual enterprise, virtual team, outsourcing*

1. INTRODUCTION

Today, integrated product and process development is accomplished by multidisciplinary teams. To support the team approach we have developed the inter-disciplinary communication medium. This paper describes the application of a Collaborative Design System to optimize the Design of a product from a multidisciplinary perspective. Development teams involved in the product development are often geographically and temporally distributed. There is a high level of outsourcing, not only in the domain of manufacturing but also in the actual product development efforts. The cooperation often implies not just a simple made–to-order development but also a real collaboration among the companies involved in the definition of a new product. It is then possible to say that enterprises give rise to a special type of virtual enterprise, in which each company maintains the greatest flexibility and business independence. Each member contributes its specific core know-how to the virtual enterprise. In the virtual enterprising context, it is possible to define the term virtual product development as a temporary cooperative activity of independent and distributed partners concerned with the translation of customer requirements into system functionalities (Dragoi & Cotet, 2004). Also, today, what do you want in an e-teams with multidisciplinary optimization? First, for your enterprise: low risk, low cost, single point of support; reap benefits as soon as possible, ready for unpredictable demand and growth, future capabilities.
Secondly, for your customers: excellent response time, ability to build competitive advantages from a wide variety of applications (use specialized products for unique features, integrate them for seamless customer support etc), ensure optimal customer experience to build loyalty, customer trust (security features build trust: validate who is doing business with whom; secure financial transactions, protect internal assets, halt spread of viruses, protect against hackers), high availability etc.

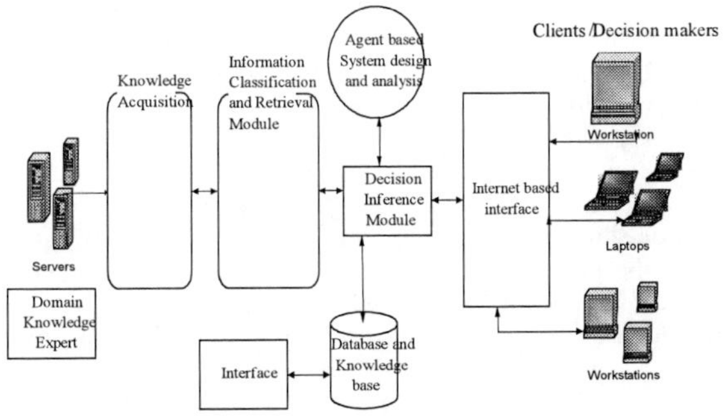

Fig. 1. CESICED: general system architecture for a system geographically dispersed

The CESICED environment (see figure 1) enables the achievement of concurrent engineering goals. It allows their various simulation and modeling tools on the different platforms to inter-operate through the Collaborative Design System of the mechatronic products. One of these goals is to bring engineers from the various disciplines, servicemen, and even customers early in the product development process to access Design of the product concurrently. On the technical level, achieving these goals depends on integrating the modeling and simulation tools used by the different engineering disciplines in the areas of design, analysis, manufacturing etc. The different tools may also reside on heterogeneous platforms geographically dispersed (Dragoi et al., 2004).

2. A MECHATRONIC COLLABORATIVE DESIGN SYSTEM

In order to permit a flexible answer to customer requirements, many products have been modularized in their design to the extend that entire areas of development such as, for example, the design of electrical components can now be outsourced to external suppliers or service providers if required. A majority of consumers and investment goods produced in the mechatronic-engineering domain now consist of a number of modular mechanical, electrical and control software components. For example, in the context of a mechatronics system, the idea is to find the best mechatronics system solution, not the best mechanical solution or the best electronic solution or the best control solution or the best software solution. The solution is the set of values for the variables, which either maximize or minimize, as desired, the value of the objective. In design for competitive advantage the objective would probably be cost, quality or value, which is a combination of the two. Also, both the mechanical characteristics of these modules and the electrical system and software, must be integrated to form a co-operative engineer system. Usually the organizational and systematic division into mechanical and electrical product development is the obstacle to creation of a common virtual product. We propose in the CESICED platform and Internet/Intranet/Extranet based tools (see figure 2) for a collaborative framework for interdisciplinary communication in the conceptual design stage with multidisciplinary optimization. This system is based of the concept of the virtual networks (see figure 2). To support multidisciplinary optimization a base definition must be defined as a common ground among the product development actors (Noel et al., 2003). The base definition contains two major types of information, an entity hierarchy and an entity attributes.

The ability to work with diverse multidiscipline teams members to successfully reach a goal objective. For a multidisciplinary group, the language of the system must do is independent of the languages of the disciplines, but is common to all. Thus, functional language should be a language of choice for multidisciplinary communication. Based on the integration of entirely separated department within the development a mechatronics product it is possible to define an enhanced process for the collaborative development of product properties in partnerships. In order to facilitate effective communication across disciplines, this system argues that an integrated software framework should enable sharing and capture of multi-criteria design proposals, design semantics, critique, and explanation and change notifications.

The product data to be browsed consists first of all of a base definition and engineering views. To support multidisciplinary simulations, a base definition must be defined as a common ground among the product development team members (Szykman et al., 2001). The base definition contains two major types of information (an entity hierarchy and entity attributes): The entity hierarchy describes how the components of the system are grouped together and The entity attributes for a part include, for example, moments of inertia, material properties, etc. The design parameters are considered as attributes of entities in the base definition, and remain with the entities when they are regrouped in engineering views to create assemblies. The feature-based design parameters serve as a common language to support design trade-off across various engineering disciplines where relevant performance of the mechatronics systems is measured. Part of view model is the construction of an entity hierarchy starting with the database definition entity hierarchy according to the needs of the particular engineering discipline at the actors. The actor executes the simulations and analyses in his working environment. The "product model" is composed of a knowledge model (features and production rules) and of data model (components, links and relations).

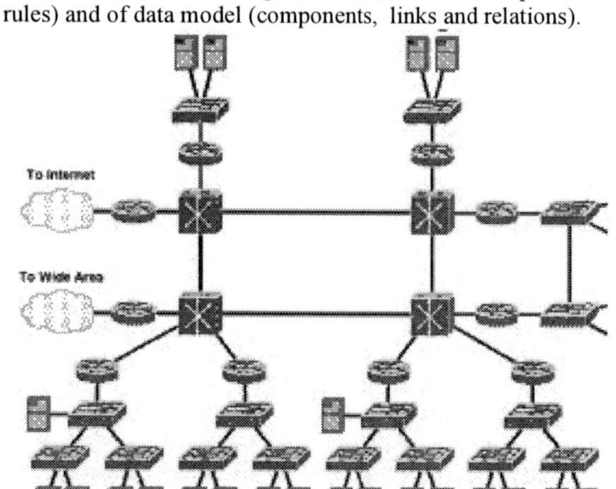

Fig. 2. A hierarchical network design model in CESICED for a system geographically dispersed

The different actors can use a set of predefined features and those can be shared. Some rights are given to specific actors in order to read, to add or modify a particular or universal feature. A main interest in using feature is to be able to declare this feature without to be obliged immediately define the values of their characteristic. Some rights are given to specific actors in order to read, to add or modify the feature. A main interest in using feature is to be able to declare the instance of this feature without to be immediately obliged to define the values of their characteristics. Instead of specifying a value for a characteristic, an actor can use the constraints on this characteristic

3. CONCLUSIONS

This system can be developed in two models alternative:

- In the first model alternative, the different team members will perform varying engineering analyses with local compute resources. These local compute resources include hardware such as workstations and PCs and software such as simulation and analysis tools. Different team members need access to the some of the same computational analysis servers such as an element solver on a supercomputer, besides the process management and server utilities. Their tools should be integrated with the tools of other team members and the shared servers through the common intelligent repository of product and process data and agreed-upon protocols.
- The second model alternative of product development environment considered was a fully distributed development environment (There would not be segregated local computing resources for each team member). All computing resources would be distributed across the network (Through an entry point to the network all simulation and analysis tools may potentially be accessed).

This paper describes the application of a Co-operative Design System to optimize the Design of a product from a multidisciplinary perspective for the mechatronics products. This application demonstrates an approach to solve many of the problems of interoperability, heterogeneity of platforms, and data sharing. An applet provides a convenient user interface capable of being launched from almost any platform with the proper network connections. By being individually interfaced with this system, the various tools and workspaces can inter-operate even from heterogeneous platforms.

4. REFERENCES

Dragoi, G.; Cotet, C.E. (2004). Virtual laboratories for the collaborative design to improve product development, *International Journal of Simulation Modelling-IISJMM*, Vol. 3, number 1, march 2004, pp.29-40, ISSN 1726-4529.

Dragoi, G.; Radulescu, B. & Tichkiewitch, S. (2004). A cooperative system for the design of the mechatronic products with the multidisciplinary optimization in *Methods and Tools for Co-operative and Integrated Design*, pp. 281-290, edited by Serge Tichkiewitch and Daniel Brissaud in Kluwer Academic Publisher, ISBN 1-4020-1889-4,. Netherlands.

Noel, F.; Brissaud, D. & Tichkiewitch, S. (2003). Integrative Design Environment to Improve Collaboration between Various Experts, *CIRP Annals/,* STC Dn, Vol 52, No. 1, 2003, pp. 109-116.

Szykman, S.; Fenves, S..J.; .Keirouz, W & Shooter, S.B. (2001). A foundation for interoperability in next-generation product development systems, in *Computer-Aided Design* Vol 33, pp. 545-559, Elsevier Science.

Annals of DAAAM for 2006 & Proceedings of the 17th International DAAAM Symposium, ISSN 1726-9679
B. Katalinic, Published by DAAAM International, Vienna, Austria 2006
150th Birthday of Nikola Tesla (1856-1943) Inventor of Mobile Robot Teleoperation

A NEW CONCEPT - EDMUS HYBRID MANUFACTURING

DUMITRAS, C.; LUNGU, G. & MIRCEA, D.

Abstract: *The attempt of association of two unconventional machining processes – electrical discharge machining and ultrasonic machining – could be materialized using new machining equipment which is called hybrid machining technology. The equipment is relatively simple to handle and can be used separately or together. These two classical machining technologies do not give birth to a new unconventional technology. Their separate action could be easily emphasized but working together one could observe an improving of the technological performance through much better working conditions given by those two processes.*
Key words: *hybrid machining, electrical discharge machining, ultrasonic machining, Barkhausen phenomenon.*

1. INTRODUCTION

The objective of our research project is to design and build a new equipement for new nonconventional hybrid technology that will combine Die Sinking Electrical Discharge Machining with Ultrasonic Machining (EDMUS). This new type of instrument will open the possibility of research in the field of machining by monitoring surface quality, morphological changes in the outmost layers, the depth of the modifications, and the cost of the machining (Kozak & Rajurkar, 2003)].
By using this new hybrid technology, the Ultrasonic oscillations will allow an increase in the speed of processing through the cavitations effects, and the microcracking of the outer most surface layer of the part that requires machining. We believe that this process will also improve the quality of the processed part by reducing the lateral gap and by creating a higher surface quality (Lungu, a, 2005).
Unconventional technologies were born in fifth decade of the XX century as a natural evolution of the manufacturing technologies development and the turn up of the materials with higher mechanical properties. This was possible as a result of the application of relatively new theories, for that period of time, from the physics domain and inner structures of the matter. One turn up an important number of unconventional technologies manufacturing process, among them electrical discharge machining. As well as the other unconventional technologies, the electrical discharge machining has a spectacular evolution together with the new discovers and achievements in electronic domain.

2. THE ELECTRICAL DISCHARGE-ULTRASONIC HYBRID MACHINING EQUIPMENT (EDMUS)

The basic structure of the hybrid machining equipment, Fig. 1, is obtained by mounting the ultrasonic equipment on the working head (CL) of an electrical discharge machine tool (M). The work-piece, working electrode (EP), is mounted on the machine tool table (ML), while the tool electrode (ES) is fixed on the ultrasonic vibration amplifier connected to the ultrasonic generator (GUS) (Lungu, b , 2005).

An automatic electrical installation (A), command through the (K1) contact the starting of the (GUS) generator in total concordance with the impulse given by the (G) generator of the electrical discharge machine tool; in order to determine the impulses from the (G) generator on use an electrical current transducer (TC). Also, through the (K1) contact one commands the washing installation formed by the command block (BC1D) and electro-valve (EV) which connect the pump (P).

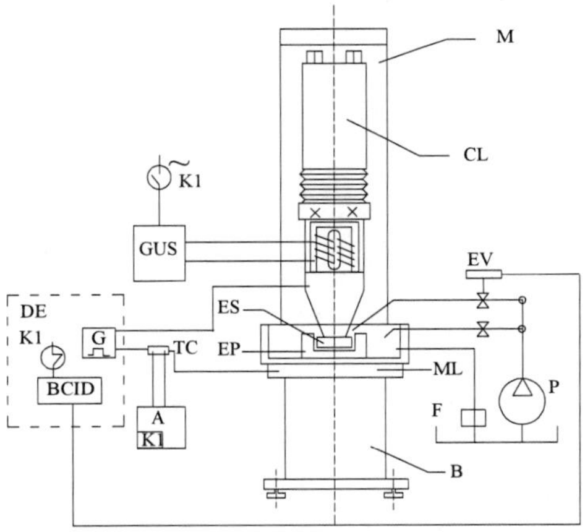

Fig.1. Electrical discharge-ultrasonic machining equipment

One of the major manufacturing parameter is productivity. One can evaluate this parameter based on indirect criteria such as the ratio between effective impulse number and fictitious impulse number or the delay from the launching of the potential impulse and the appearing of the current in working space. In this case, one considers optimum working conditions, for a given situation, when the number of fictitious impulse number is minimum.

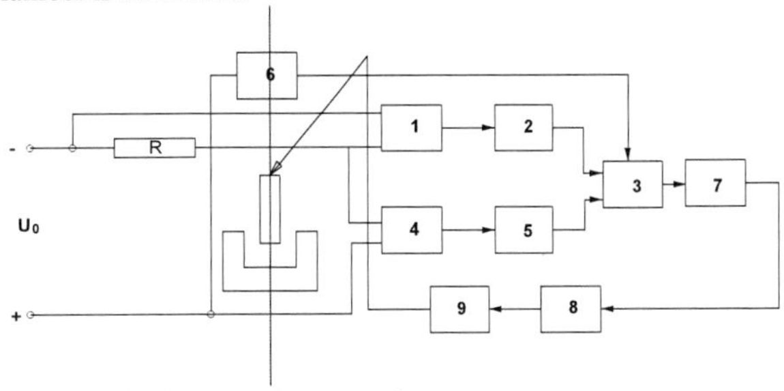

Fig.2 Block diagram of commands

The distance between electrodes in the idle running impulse case must be shorten and in the short circuit case must be increased. The manufacturing optimum regime must be situated between these two extreme situations and the number of the working impulse must have maximum value from the total number of impulses.

In the case of the potential impulse one exist a certain maximum value of the potential which must not be exceed and for the current impulse a certain time which determine the impulse size. The optimum working regime must be situated among above mentioned limits. In Fig.2 on distinguish: 1 and 2 blocks for the current impulse initiation; 3- logical circuit; 4 and 5-blocks of the potential impulse initiation; 6-fictitious impulse detector; 7- workmanship element; 8- automat potentiometer; 9-feed self-regulator. In the case of full potential signal one have only current signal and this means short circuit situation and the logical circuit 3 will send a signal of removing the electrodes. In the case of null current signal one have just potential signal. This means that one have a great distance between electrodes, logical circuit 3 will sent a signal for approaching the electrodes. The optimum working regime is obtained when it exist both potential and current signals. In this period one the fictitious impulse detector 6 is present by approaching the electrodes.

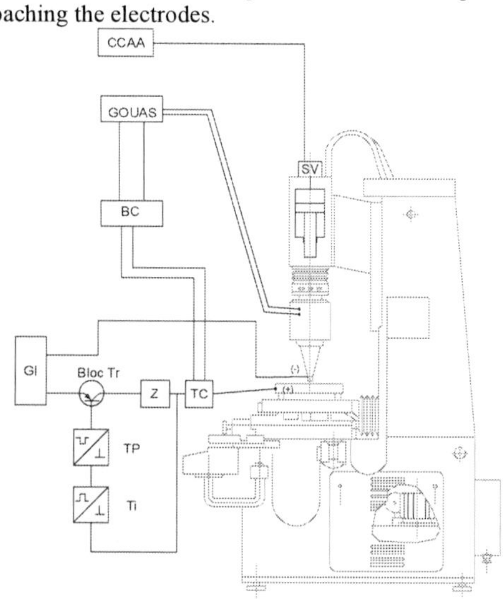

Fig.3 Equipment for hybrid machining

In order to utter the value of the command and optimization scheme of the EDMUS manufacturing process (Fig. 2) is useful to present the relation between the EDMUS working cycle possible impulse types. In Fig. 3 one distinguish the CCAA-command head of the vertical feed; GOUAS-ultrasonic oscillation generator; GI-electrical discharge impulse generator; BC- command block

$$F_i = f_P + f_M + f_G + f_{SC} + f_o \quad (3)$$

where: f_i- impulse frequency; f_P-sampling impulse frequency; f_M-crumbling impulse frequency; f_G-fictitious impulse frequency in working zone (ML); f_{SC}-short circuit impulse frequency; f_o- idle running impulse frequency.

The sampling impulse frequency has a primary sampling impulse component – f_{PP} and a a sampling and crumbling impulse component – f_{PM} with a sampling efficiency η

$$F_P = f_{PP} + \eta \, f_{PM} \quad (4)$$

f_M will be:

$$F_M = f_{SM} + (1-\eta) \, f_{PM} \quad (5)$$

where: f_{SM} is the crumbling mixed impulse at the f_{SM} frequency of secondary impulses of pure crumbling.

The hybrid manufacturing process, named EDMUS (Electro Discharge Machining and Ultrasonic Machining) bring a uniformity of the electrical discharges by modifying the discharge conditions between the electrodes. One can observe after a discharge zone through an simple ED process which include both A1 and A2 zones (the starting of the discharge is not started), a zone where for an interval of 358 microseconds one determine a discharge frequency of 27 discharge per second in which all discharges are with material removal. If one define an filling coefficient as the ratio between the removal discharges f_P and the total number of discharges f_i, one can determine that one have maximum value on this interval and in command scheme one have maximum productivity.

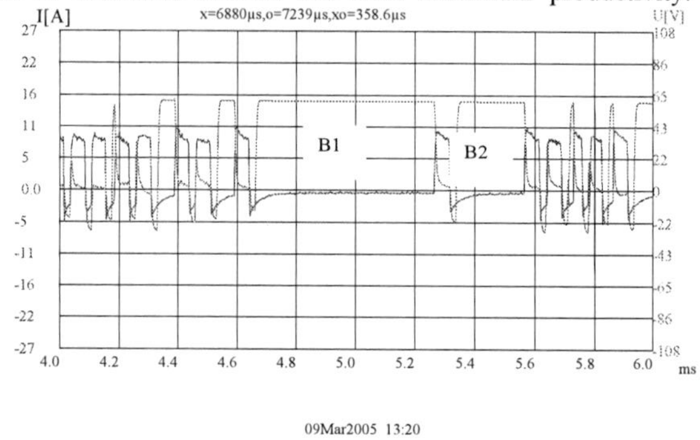

Fig.4 Series of discharges-EDMUS procedure

In figure 4 one considers an EDM type discharging zone where B1 and B2 zones are discharges without starting. For these one do not establish optimum conditions for brake off the space between the electrodes and the machine tool is idle running. The productivity is reduced and the electrical energy consumption is high. As soon as the EDMUS start one determine, after 5...6 microseconds, a calm down of the electrical discharge between the electrodes, maintaining the value of the starting potential in time, and a constant discharge current. Also one has a positive effect of the discharge energy which assures a uniform quality of the manufactured surface (Lungu, c, 2005).

3. CONCLUSION

The attempt of association of two unconventional machining processes – electrical discharge machining and ultra acoustic oscillation machining – could be materialized using new machining equipment which is called hybrid machining technology. The equipment are relatively simple to handle and can be used separately or together. These two classical machining technologies do not give birth to a new unconventional technology. Their separate action could be easily emphasized but working together one could observe an improving of the technological performance through much better working conditions given by those two processes, one for the other. The present paper aim to present such an achievement obtained in the Fine Mechanics Department , Technical University "Gh.Asachi", Iasi, Romania.

4. REFERENCES

Kozak, J., Rajurkar, P.K. (2003), *Hybrid Machining Process Evaluation and Development,* University of Nebraska, Lincoln, USA.

a. Lungu, Gh., (2005) *Non conventional machining technologies,* Ed.Performantica, ISBN 973-7994-84-1 Iasi, Romania.

b. Lungu, Gh. (2005), Some aspects concerning hybrid unconventional technologies, " *Mashinostroienie I Tehnosfera XXI veka",* 12-17 sep, Sevastopol, Ukraine, p.135-138, ISBN 966-7907-19-8.

c. Lungu, Gh. (2005), Some Aspects Concerning the Side Gap in Electrical Discharge Machining, *Intl Conference On Economic Engineering And Manufacturing Systems*, 20-21 oct, Brasov, p.203-209, ISBN 973-635-598-6.

Annals of DAAAM for 2006 & Proceedings of the 17th International DAAAM Symposium, ISSN 1726-9679
B. Katalinic, Published by DAAAM International, Vienna, Austria 2006
150th Birthday of Nikola Tesla (1856-1943) Inventor of Mobile Robot Teleoperation

TECHNOLOGICAL AND METROLOGICAL VERIFICATION OF DE5 MODEL

DVORACKOVA, M.; HRUSKA, F. & SEDLACIK, R.

Abstract: *The system "Integrated automation laboratory – LABI" contents important model DE5 –"Biochemical processes". The model demonstrates and provides some typical processes pro field of biochemistry next to mechanical, electrical and thermodynamic processes. Technical equipment has automation system. During realization there was provided technological and metrological verification of the system. There was verified the measurement and control loops. The paper presents description all activities within it. The results served for process starting too.*

Key words: *measurement, verification, biochemical process, Internet, laboratory.*

1. INTRODUCTION

The system „Integrated Automation Laboratory- LABI" are built at Tomas Bata University in Zlin. The one contents a important model DE5 – biochemical processes (see in fig. 1).

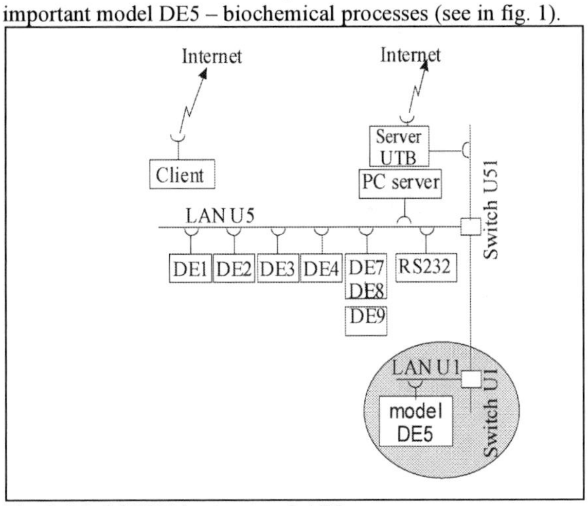

Fig. 1. Model DE5 in structure LABI

The model presents a example process from biochemistry area next to mechanical, electrical and thermo dynamical processes. The automation system was tested technologically and metro logically. There were verified all measurement and control loops. The paper gives a describing of verification results. (Dvorackova, 2005)

The news is a method of testing of the system and measurement loops. In the next period the LABI comes in a regular using, it will use in laboratory teaching and exercising.

2. DESCRIBTION OF MODEL DE5

The model DE5 is used with biochemical processes. Technical equipment is a fermenter of the typ LF20. The loops are designed from a team of specialists. They are (see fig.2):

a)measurement loops : temperature (3a), level (22), pressure (21), pH (8), conductivity (7), disoved oxygen (9), ORP (10), turbidity (11), ratation (19),

b) control loops : rotation of motor (19), heating or cooling (3d or 24), oxidizide (4), dosage by peristaltic pumps (2a, 2b, 2c, 2d), sampling of gas fase (13) and liquid (17).

In the next period will add model for research and development measurement lopps for gas concentration of CO_2, CH_4 and NH_3.

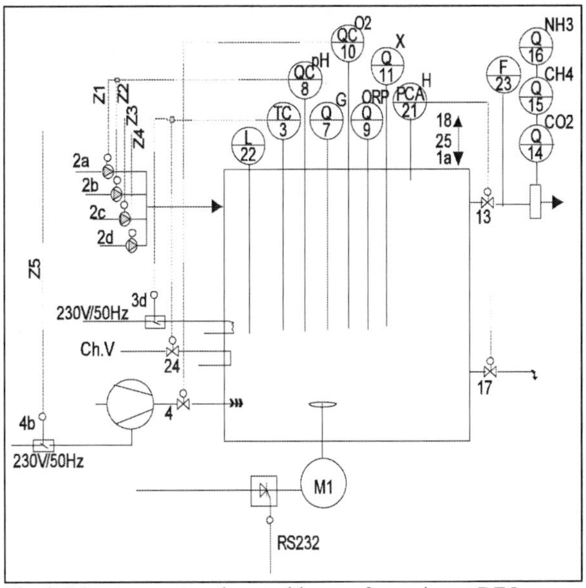

Fig. 2. Measurement and control loops of experiment DE5

3. PROCESSING OF DE5

The automation system of DE5 is used for the functions:

- the automatic measurement and the acquisition of date in set sampling period
- the local archiving
- local and remote access to DE5 in visualization, process and teaching regime.

The model DE5 is processed in three regimes: technological regime, student regime, client regime.

The specific and difference of regimes are:

- In technological regime the employers work, prepare some materials and put it into equipment, define a procedure of technology, set the process parameters, define a process start, control process, stop the process and system.
- The student regime is destined the student, the e-learning in laboratory. The can prepare, start, visualize and evaluate of date.

- The client regime does passive tracing of real process running in technological or student regime. The client can get date from some running process or from before running processes (if it is allowed form the system administrator). Client can be anybody user of LABI system via the Internet.

4. TECHNOLOGICAL VERIFICATION

In the first the fermenter was filled 7 litres of water for technological verification. A change of water parameters was controlled.

The other verification process had filling with seven litres of bio media, 1,6 l active sludge with concentration 7,2 g/l. The solution was aerated through a blower.

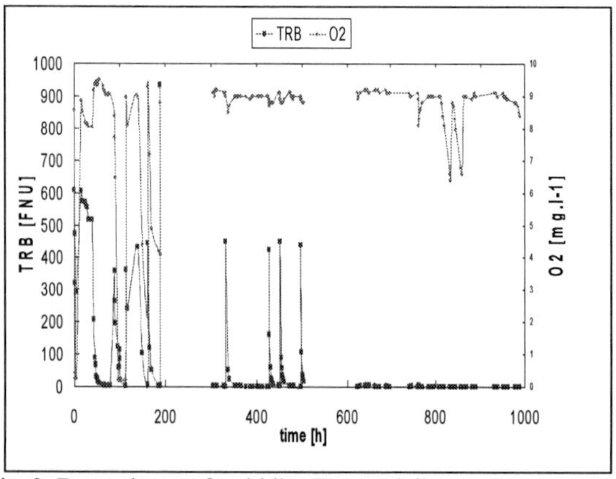

Fig. 3. Dependence of turbidity TRB and dissolved oxygen O2

Graph of dependence of turbidity and dissolved oxygen is in fig. 3. The changes of concentration of dissolved oxygen are in begin of time according to mixing of solution. Identifiably there is graph for ORP.

5. METROLOGICAL VERIFICATION

Verification was done for:

- measurement loops of parameters of water solution (it is described next)
- peristaltic pumps (it is described next)
- measurement of level (it is described next)
- measurement of temperature
- rotation of mixing motor.

	time [h]	Temp. [°C]	Rota. [ot.min^{-1}]	G [µS/cm]	pH [1]	ORP [mV]	O2 [mg.l^{-1}]	TRB [FNU]	press [kPa]	level [cm]	flow [l.h^{-1}]
	42,98	24,9	13	956,0	7,18	127,3	9,16	206,54	11,0	46,0	6,00
	90,25	22,2	12	956,0	7,10	117,2	6,47	195,92	13,0	16,0	2,03
**	114,58	22,0	10	956,4	7,13	149,9	8,97	361,11	13,0	14,0	1,97
**	138,80	21,1	10	956,6	7,10	134,8	9,00	433,90	13,0	7,0	2,00
*	160,75	21,0	16	955,3	7,10	73,9	2,30	4,80	15,0	6,0	2,10
**	160,83	21,0	9	956,7	7,10	118,5	9,30	444,10	12,0	7,0	2,10
*	188,33	21,4	15	955,3	7,10	68,8	4,10	3,30	12,0	7,0	2,10
**	188,41	21,5	10	956,6	7,10	123,5	8,80	934,80	13,0	6,0	2,10
*	260,91	---	---	---	---	---	---	---	---	---	---
**	261,01	---	---	---	---	---	---	---	---	---	---
	762,83	36,0	12	955,3	7,30	63,2	8,10	4,40	12,0	1,0	2,10
	787,58	26,4	11	955,4	7,30	41,2	9,00	0,80	14,0	2,0	2,10
	953,41	24,9	0	1268,1	7,40	368,5	9,00	1,10	1,3	19,0	102,00

Table 1. Example of date from the Internet interface

5.1 Verification of water solution parameters

There were the measurement loops of pH, ORP, turbidity, conductivity and concentration of dissolved oxygen. (Hach Lange, 2004/2005, Klouda, 1996). The digital sensor pHD sc (measuring pH) had two point calibration with puffers pH=4,01±0,01 and pH=7,00±0,01. They were the puffers from firm HAMILTON. It was with temperature correction. The sensor ORP was calibrated according to solution with knew value of ORP (=223,2 mV at 23 °C). The other test was done with sensor SHOTT, 31Rx. The reference electrode was argentochlorid; with 3M KCl solution. The induction conductivity sensor 3700 sc was calibration according to one point verification by KCl solution and for reference value 1377,15 µS/cm. Air calibration was done for measurement loop of dissolved oxygen. The calibration was done automatically according to sample solution. There was selected unit and actual pressure.

5.2 Verification of peristaltic pumps

There was validated peristaltic pumps of DOMET S-31. The hosepipe was used with diameter 2 mm. The was measured: (V) volume, (t) time, (Q) flow. The results in graph are in fig. 4.

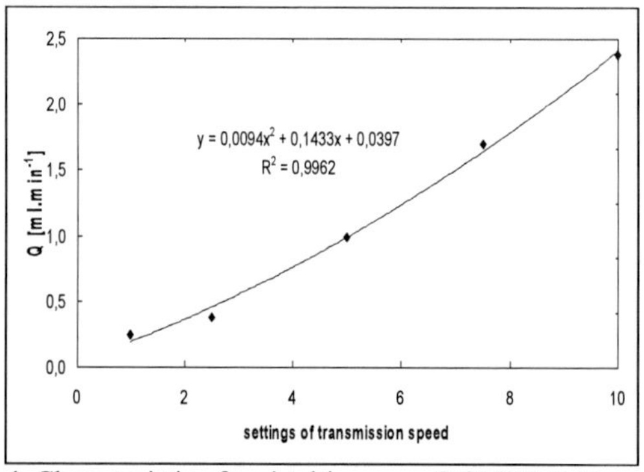

Fig. 4. Characteristic of peristaltic pumps DOMET S-31

5.3 Verification of level

The capacitive level device was calibrated direct at the fermenter according to product guide (Dinel, 2006).

6. CONCLUSION

The bioprocess experiment DE5 is very important part in the project of Laboratory of integrated automation-LABI. It is the next to thermal, electro technical and measurement processes. The automation system was designed according to requirements of technologists. The main criteria for monitoring and control of this processes are concentration of dissolved oxygen, pH value and ORP (Oxidant-Reduction Potential) value, the turbidity and other parameters. This parameters was verified and calibrated for qualitative measurement process.

7. REFERENCE

Dvorackova M.- Hruska F.-Hoffmann J. (2005a). Laboratory biochemical model and WEB communication, *Proceedings of the 16th International DAAAM Symposium,* Opatia, Croatia, pp. 105-106. ISBN 3-901509-46-1.

Dvorackova M.- Hruska F.- Neckar, J. (2005b). Measurement and Control System for Bioprocess Laboratory, *Proceedings of the International Carpathian Control Conference* ICCC' 2006, Rožnov, Czech Republic.

Klouda, P. (1996). *Moderní analytické metody*. Ostrava, Pavel Klouda, 1996. 203 s. ISBN 80-902155-0-5

HACH LANGE (2004), *Katalog.* Düsseldorf, Hach Lange, 2004. 134 s.

Dinel. (2006). *Kapacitní hladinoměry (Capacitive levelmeter) CLM-36.* Zlín: Dinel®, 2006, 6 s.

Annals of DAAAM for 2006 & Proceedings of the 17th International DAAAM Symposium, ISSN 1726-9679
B. Katalinic, Published by DAAAM International, Vienna, Austria 2006
150th Birthday of Nikola Tesla (1856-1943) Inventor of Mobile Robot Teleoperation

THE INVESTIGATION OF CHOSEN PROPERTIES OF CHROMIUM COATINGS AFTER PNEUMATIC BALL PEENING

DZIERWA, A.

Abstract: *The article presents results of the tests concerning chosen properties of chromium coating in relation to specimens made from 40H steel (coating thickness about 25 μm) and finished by pneumatic ball peening. It has been found that pneumatic ball peening of chromium coating can improve level of fatigue strenght and also can assure surface roughness less than Ra = 2 μm.*

Key words: *chromium coating, pneumatic ball peening, roughness, fatigue strenght,*

1. INTRODUCTION

Wear resistance of machine elements can be strongly improved by electrolytic plating. Electrolytic coatings have been used in regeneration technology of the elements being exposed to wear of several tens of micrometers to several millimetres. The most widely used methods of electrolytic plating are iron plating, nickel plating and first of all chromium plating (Przybylski, 1987).

Recently, the range of industrial kinematic pairs coated by various metal platings could be found (in not strengthened state), however their functional properties are not often satisfying. The main reason for such an occurrence is the fact that in the electrolytic plating process internal stresses are created which negatively influence on surface wear resistance (Korzynski, 1994). Besides some methods of electrolytic plating create such defects as pores, microcracks etc. These defects are accountable for decohesion centre of the layers in the process of friction and further intensify corrosion processes (Mlynarczak & Jakubowski, 1998).

Application of burnishing process is an efficient way of removal or considerable decrease of this negative influence on functional properties of the layers. Appropriate selections of the layer material and burnishing conditions can cause increase in surface wear resistance (Korzynski, 1994). At the same time, depending on the requirements the received surfaces may possess additional advantages e.g. increased corrosion resistance.

Among the burnishing methods used for layer processing the following are the most frequently applied: shot peening, micro ball peening and pneumatic ball peening, roller and ball burnishing and slide burnishing (slide planishing, slide diamond burnishing).

The combination of electrolytic processing and burnishing may appear to be more efficient way of surface functional properties improvement than individual application of each of them.

2. PNEUMATIC BALL PEENING

Pneumatic ball peening process is applied mainly to irregular shape machine parts, welded joints or coated elements. This process can be also used to heat treated steel machine parts of low or high hardness alloys (Korzynski, 1994).

Ball peening process is based on making use of stream of small hard bearing balls, which are accelerated by compressed air and circulate in closed loop system, impacting on treated surface and causing its cold working. The process itself is effected by ejector nozzle geometry and: ball diameter, air pressure, distance from working - nozzle to workpiece, ball stream impact angle and peening time.

As the result of the treatment, high strain hardening was achieved together with surface roughness Ra = 0,63 - 4 μm; part shape and dimensional accuracy were not affected. The machined surface have characteristic pricked structure, bearing ball peening created hard surface layer with compressive residual stresses. Such surface layer condition, as generally known, is the best for some properties, particularly fatigue life of machine elements, and it is why bearing ball peening process provides increase in machine parts fatigue strength (and thus much more increase in fatigue life) and tribological wear resistance.

Ball peening process should be done using special machines or special ball peening heads. Developed equipment has many advantages: small overal dimensions, supply from standard compressed-air instalation, repeatability of treatment results, and possibility to treat workpieces of irregular shape and low stiffnes. The process is usefull, especially in small and mean lot production, for strain hardening of parts made from steel working in changing load conditions. Application of bearing ball peening process may cause even elimination of polishing, belt grinding or another finishing technological operations.

Basic advantages of bearing ball peening are as follows (Zarski & Korzynski, 2002):

- As a working medium bearing balls of ø 1-3 mm are used, propelled at limited speed of 3-10 m/sec.
- Low ball and their continouos recirculation cause that only a small ball charge is required and, as air supply source, regular industrial internal compressed air network can be used with a typical oil and moisture trapping equipment.
- After ricochetting, the balls fall down under gravity into the suction zone, thus eliminating ball transportion devices and making bearing ball peening machines much cheaper.

The article describes the results of influence tests of pneumatic ball peening on chromium coating elements properties. The results of fatigue strenght and surface roughness after machining are shown.

3. EXPERIMENTAL

The studies were conducted on steel 40H of hardness 230HB. The base pretreatment included turning and then pneumatic ball peening (Ra = 1.2μm), grinding (Ra = 0.7μm) or polishing (Ra = 0.5μm) depending on the chosen variant. The pneumatic ball peening or polishing was the finishing treatment of the coating (of 25μm thickness). The pneumatic ball peening was done with the following parameters: air pressure – 0,5 MPa;

ball diameter – 2 mm, treatment time – 20 min. The accelerated pendular bending fatigue tests were carried out using an electrodynamic vibrator with resonance frequency of the samples adopting the boundary number of cycles of $2 \cdot 10^6$. The used technique, that consisted in cutting samples out of special shaped cylinders, was described in details in (Lunarski, 1981). Fatigue strength Zgw was calculated using stepped method on the basis of results of experiments on 16-sampled batches. The fatigue experiments results are shown in Tab. 1 and Fig. 1. 3-D surface topography was measured using Rank Taylor Hobson TALYSCAN inductive head, diamond contact tip had 0,002 mm radius, measuring speed was 0,5 mm/s. Profiles were taken from the 4 mm assessment length (Oczos & Liubimow, 2003). The results of the measurements of roughness parameters (in micrometers) are shown in Tab. 2.

The presented treatment variants are as follows: Variant I – the base ground, the coating untreated; Variant II – the base polished, the coating untreated; Variant III – the base ball peened, the coating untreated; Variant IV – the base polished, the coating polished; Variant V – the base ground, the coating polished; Variant VI – the base ball peened, the coating polished; Variant VII – the base polished, the coating ball peened; Variant VIII – the base ball peened, the coating ball peened; Variant IX – the base ground, the coating ball peened.

	Var. I	Var. II	Var. III	Var. IV	Var. V
Zgw [MPa]	341	338	359	354	336
	Var. VI	Var. VII	Var. VIII	Var. IX	
Zgw [MPa]	370	370	407	407	

Table 1. The results of the fatigue strenght experiments according to different treatment variants

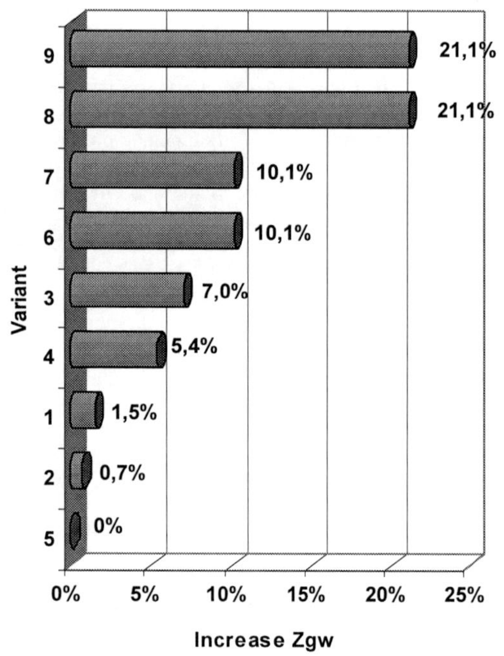

Fig. 1. The increase of the fatigue strength Zgw after different variants of treatment

It is evident from the fatigue examination that one can achieve higher fatigue strenght after two types of machining: the base ball peened, the coating ball peened and the base ground, the coating ball peened. The other variants of chromium coating machining were found to be less beneficial. I also found distinct influence of base machining and on later roughness of chromium coating.

	Ra	Rz	Rt	Rq
Var. I	0,81	4,85	5,89	1,03
Var. II	0,47	2,90	3,68	0,61
Var. III	1,86	7,92	8,88	2,19
Var. IV	0,45	2,65	4,07	0,56
Var. V	0,54	3,68	5,25	0,68
Var. VI	1,41	5,76	8,50	1,68
Var. VII	1,34	6,07	9,26	1,56
Var. VIII	1,78	7,36	8,07	2,09
Var. IX	1,53	6,39	8,15	1,77

Table 2. Selected roughness parameters for particular variants of machining

Coatings from bigger initial roughness height (base surface after pneumatic ball peening) were characterised by greater roughness height than coatings from ground or polished surface. However profiles of coated surface after pneumatic ball peening are distinguished by peaks and depressions of great radius of curvature , made by shot of ball on surface.

4. CONCLUSION

Performed investigations of a chromium plating burnishing process have revealed the possibility of shot peening process realization without fear of cracks, scalings or other defects.
Performed exams proved significant impact of preparatory machining of surface on later roughness of chromium coating. They showed characteristic pneumatic ball peened surface without sharp corners and tops. The results of the realized fatigue experiments revealed the distinctly advantageous effect of pneumatic ball peening on the improvement of the fatigue strength of the elements with chromium coatings. Particularly beneficial findings were obtained with the ground or the ball peened base and with the ball peened coating. The conclusion of the investigations is the necessity of the further studies on this topic.

5. REFERENCES

Przybylski, W. (1987). *Technologia obrobki nagniataniem,* WNT, ISBN 83-204-0742-7, Warszawa

Korzynski, M. (1994). *Nagniatanie powlok chromowych,* Oficyna Wydawnicza Politechniki Rzeszowskiej, ISSN 0209-2689, Rzeszow

Mlynarczak, A & Jakubowski, J. (1998). *Obrobka powierzchniowa i powloki ochronne,* Wydawnictwo Politechniki Poznanskiej, ISBN 83-4143-062-0, Poznan

Zarski, T. & Korzynski, M. (2002). Bearing ball peening influence on surface layer of MgAl18MnZn alloy, *Proceedings of the International Scientific Conference - Mechanics 2002*, pp. 341-345, ISSN 0209-2689, Rzeszow, July 2002, Oficyna Wydawnicza Politechniki Rzeszowskiej, Rzeszow

Lunarski, J. (1981). Porownawcza ocena wlasnosci zmeczeniowych w badaniach technologicznych. *Technika Lotnicza i Astronautyczna,* Vol. 3, Marz 1981 pp. 35-39, ISSN 0040-1145

Oczos K, & Liubimow W. (2003). *Struktura geometryczna powierzchni,* Oficyna Wydawnicza Politechniki Rzeszowskiej, ISBN 83-7199-247-5, Rzeszow

Annals of DAAAM for 2006 & Proceedings of the 17th International DAAAM Symposium, ISSN 1726-9679
Editor B. Katalinic, Published by DAAAM International, Vienna, Austria 2006
150th Birthday of Nikola Tesla (1856-1943) Inventor of Mobile Robot Teleoperation

THE NEGATIVE IMPACTS OF GLOBALISATION ON LOCAL MANUFACTURING

FAHIMNIA, B.; LUONG, L.H.S.; MOHAMMAD ESMAEIL, M.; MOTEVALLIAN, B. & MARIAN, R.

Abstract: *Multinational manufacturers are known as those companies which break the national boundaries and tend to grow globally by investing in other markets directly and/or indirectly. There are a considerable number of papers published in the literature on the reasons why companies choose to go multinational; but no significant research have been conducted on the negative aspects of the multinational manufacturing. This paper presents and analyses four of the major negative aspects of the globalisation on manufacturing.*
Key words: *multinational, exploitation, manufacturing, labour.*

1. INTRODUCTION

Over the past decades many national and individual companies have faced heavy competitions and disappeared from manufacturing arena due to globalisation. On the other hand, multinational giants have grown with alarming rates. This is largely due to the way big companies are organized, operate and choose to grow stateless. Operating globally has obviously great advantages for multinational manufacturing companies, despite the negative effects they may have on local manufacturers and the national markets. This paper analyses and presents some of the negative impacts that multinational manufacturers have on national manufacturing of the country they operate and the local markets they target.

2. IDENTIFICATION OF FACTORS

There are 4 main disadvantages of going multinational particularly in developing countries:

- Labour exploitation
- Law exploitation (Environmental Vandalism)
- No commitment
- Cultural exploitation

3. ANALYSE

3.1 Labour Exploitation

In many cases the negative effect of multinationals in the upward trend of real wage growth can be seen. The outcome of entering a multinational corporation particularly at the production-abroad based stage in developing countries shows a reduction in the real wage growth.

When multinational manufacturers enter their target market in developing countries, they attract the workers normally by paying them considerably higher than their equivalent in the local firms. The rate of unemployment decreases in such a case which sounds to be a positive feature of multinationals' presents, but unfortunately this is not all they want to achieve. After a while, multinationals try to make their business as labour-intensive as possible. Doing so, they can slow down investment on their capital-based business and just simply provide the training for the workers and use what they have learnt for a time period. There are many evidences in the literature showing multinational companies present a negative effect on the real wage growth after a while of their arrival into their target market. Also, hiring children as labours, for instance, is another kind of exploitation, which is common in developing countries.

This can be obviously examined through a case study in Malaysia (Athukorala P., 2000). With a manufacturing share of about 78 %, in 1994 Malaysia was the world's sixth largest exporter of manufactures. The thing is that there is no direct estimate of the share of foreign firms in exports. However, the comparison of data on export and the output share of foreign firms shows that multinationals had provided over three-quarters of total manufactured exports by 1990s.

By the mid 1980s, the rate of unemployment in Malaysia was about 8%. Since then it began to decline falling to just over 2.8% in 1995. Most of the new employment opportunities had come from the rapidly expanding manufacturing sector. Between 1987 and 1994, the direct contribution of manufacturing to total employment increment was more than 60%. With the rapid expansion of export-oriented manufacturing, the structure of manufacturing companies associated with multinationals has become more labour-intensive over the years.

The employment outcome of the export-oriented industrialization is very impressive. In the early 1970s, when multinationals launched to be involved seriously in production, the average wage of the workers in these firms was about 60% higher than their equivalent in local firms. Now, the main point is that the compound annual growth rate of multinational's real wage during 1968 - 92 was 1.52% compared to 3.78% for local firms. By late 1980s, the 60% difference between average wages of the local firms and multinational manufacturers was approximately disappeared. Thus, increased presence of multinationals in export-oriented production has ended with a remarkable slowing down of real wage growth in the foreign-owned segment of Malaysian manufacturing.

3.2 Law Exploitation (Environmental Vandalism)

It has been reported that unfortunately corruptions are widely spread in all levels of governments in developing countries. As many cases proven, ignoring the local law, multinationals can buy their way through. Where there is more benefit for them, they will find out how to penetrate into the governmental levels and do whatever they keen to do by paying the authorities.

This can be substantiated by the case of Union Carbide pesticide factory in their Bhopal plant. The company was established close to the residential area which is against the government policies in all the countries all over the world. This led to the world's worst chemical disaster. On the night of 2nd December 1984, in the worst chemical disaster in history, 40 tonnes of lethal gases released from Union Carbide's pesticide into atmosphere. Safety systems were either malfunctioning, under repair or switched off. The gases overwhelmed the city. Thousands died in their sleep and others gave in the poisons as they ran through the streets. Later on, more than 8,000 died within three days following the disaster. At present, at least 120,000 people, including children born to gas-exposed parents, continue to suffer weakening from exposure-related health effects. At least 30 people continue to die each month from exposure-related illnesses and more than 80,000 are too ill to work (Wenzel C. 2004). More than 95 per cent of the survivors have received US$500 for lifelong injury and loss of livelihood, hardly enough to cover medical expenses for a few years. The Bhopal chapter is now almost closed and all liabilities supposedly have been resolved. The company has grown and is still growing internationally, despite the fact that the Bhopal disaster will never end in social conscious.

3.3 No Commitment

In today's trades, money does not have and recognize boundaries; therefore, multinationals have no nationality. That is because for multinationals money is the main objective and where the return is higher, their money flows, that could be Asia, South America or any other country in the world. The following sections describe the way multinationals enter or leave a country and their targeted market, with no commitment.

3.3.1 Entry Struggles

When a multinational corporation enters a local market, some of the local firms may get lucrative contracts. Although this may be good in the short term for those local firms, in a long run multinationals usually change their suppliers and eventually they replace them with the ones overseas and outsource the contracts. Moreover, when multinationals start their operation in their targeted markets, most of the local and individual firms will close down, since they can not compete as regards price. This is mostly due to 2 main reasons:

- Lack of capital investment and know-how in local and national manufacturing firms do not allow them to compete.
- High volume of output that multinationals produce for international market means that their prices are significantly cheaper than the same product produced by local companies. This is mainly due to the lower fixed cost per item when producing larger quantity as well as cheaper raw materials associated with higher volume.

3.3.2 Leaving Troubles

When the multinationals find another market to supply them for a cheaper price, there is no obligation to continue with the earlier local market. Consequently, they move to another market and leave the earlier country devastated in terms of skilled labour, engineers as well as know-how. Multinationals do not belong to any specific area and they move easily as soon as they find out that there is another market to supply more return on their investment. This brings poverty and increases the rate of unemployment for the country they leave. A true case in Mexico can show the actual outcomes of having no commitment (Miguel A. & Somoza B., 2005). About three decades ago, the multinational industrial plants in the Mexican border with USA started their operations. The most important attractions for the multinational enterprises at the time were the great easy mobility of capital, obtaining high and quick return on investment as well as obtaining more valuable objectives at the international level. The local industrial firms with traditional administration were experiencing severe technological and skilled labour problems. Therefore, none of the local manufacturers could stay in competition when multinationals entered the manufacturing market in Mexico due to their poor traditional technologies and not having enough capital to do so. Very shortly after launching their operation inside the country, multinationals devastated the traditional form of production. Until 1988, multinational manufacturers transferred and implemented some labour exercises that ware being programmed in the headquarters in their original countries. In one side, Mexico needed substantial changes in its manufacturing plants including training qualify workers and professionals in engineering and business areas as well as learning to be adaptive against new technologies coming up. Local manufacturers needed to make a revolution at their work systems in similarity with what multinationals were doing. In the other side, multinationals, which were mostly the ones from US, were never trying to transfer the whole manufacturing technology (know-how) to the targeted market as they did not want to loos the benefits of having Mexico under their direct domination. When in early 90's, the massive market globalization and the operation of the multinational plants in different countries of the world began, the USA in 1993 modified its manufacturing function, disseminating in various geographical areas of the world. The same happened with Japan, Germany and the United Kingdom. It reached its critical level for the year 2000 when most of the multinationals moved to the new markets including US and some parts of East Asia and left Mexico without any care about what they had done there. With the poverty coming from multinationals' presence, it is now easier to be exploited by other interest-hungry companies looking for such environments. It is necessary for Mexican manufacturers to apply new and more sophisticated systems of production and administration in order to compete with other giants coming up.

3.4 Cultural Exploitation

Due to the lower income in developing countries, the purchasing power of majority of the people is week. Once multinationals enter a country, they bring their own values and trends and spread them out in the society. This forces some sections of the society to change their values and priorities by adopting the new trends. As well known, in most developing countries people generally can not afford a modest life style, while multinationals which expand to find market by any means, will push people to change their purchasing habit and consequently pay for non-essential products on credits. In this way the middle and lower middle-classes of the society are exploited, while the wealthy and upper middle class are usually the winners along with multinationals.

4. CONCLUSION

The statistics on the negative side of the multinational manufacturers' presence in their targeted markets, particularly in developing countries, show many evidences of different sorts of exploitation. This paper analysed the negative factors including exploiting work force and local laws, having no commitment as well as devastating the cultural concepts. However, with the serious introduction of globalisation over the last years and positive trends in going multinational, it sounds to be very hard to apply such conceptions while manufacturing corporations find their survival keys in growing stateless.

5. REFERENCES

Athukorala P., (2000). *Multinational Enterprises, Employment and Real Wages in Malaysian Manufacturing*, Australian National University, Australia.

Miguel A. & Somoza B. (2005), *New Expectancy of Administration in Multinational Manufacturing Plants in Mexico*, I.T.R. (Instituto Tecnologico de Reynosa), Mexico.

Rea P. J. & Kerzner H. (1997). *Strategic Planning: A Practical Guide*, John Wiley & Sons, USA.

Singh N. (1996). *Systems Approach to Computer Integrated Design and Manufacturing*, John Wiley & Sons, USA.

Wenzel C. (2004). *The Bhopal Disaster: Briefing*, Greenpeace international, Australia

Annals of DAAAM for 2006 & Proceedings of the 17th International DAAAM Symposium, ISSN 1726-9679
Editor B. Katalinic, Published by DAAAM International, Vienna, Austria 2006
150th Birthday of Nikola Tesla (1856-1943) Inventor of Mobile Robot Teleoperation

A HEURISTIC METHOD TO OPTIMISE MANUFACTURING LEAD TIME

FAHIMNIA, B.; MARIAN, R.; MOTEVALLIAN, B.;
MOHAMMAD ESMAEIL, M. & ABHARY, K.

Abstract: *Manufacturing Lead-Time (MLT) is the total time required to process a given product through the plant. Optimisation of MLT is essential to manufacturing efficiency, because Lead-time reduction decreases operational costs and consequently increases manufacturing effectiveness. This paper analyses and presents the activities, which are required in different stages/phases of manufacturing by identifying the factors of delays and sources of wastes for both value and non-value adding activities and suggestions for improvements are made.*

Key words: *manufacturing, lead-time, value-adding activities, non-value-adding activities, manufacturing process.*

1. INTRODUCTION

Manufacturing lead-time plays an important role in manufacturing efficiency and cost effectiveness of final products, therefore from the business perspective, shorter MLT means quicker turnover, faster growth and higher profits. Thus, from the manufacturing perspective, cycle time is one of the most important variables, which is the time that takes for a unit of production to move from raw material through the process and becomes a finished product. Today due to intense and focussed competitions, world-class manufacturers must shorten cycle times and eliminate sources of wastes. As regards manufacturing planning, short lead-time also reduces the manufacturing plant's dependence on forecasts and allows the plant to operate using a shorter planning and as a result a more accurate master schedule. This paper identifies those factors of delays and sources of wastes and provides suggestions for improvements.

2. IDENTIFICATION OF MANUFACTURING PROCESS

The following steps describe the process that needs to be carried out in order to reduce MLT:

Step1. Recognizing the process scope:
- Find out the natural sequence of the process
- Identify the start and the end points of process
- Identify and number the steps
- Document every step

Step2. Measuring the process speed:
- Follow a single unit throughout the process
- Write down the time required for each step
- Write down the waiting times
- Write down batch size or quantities

Step3. Identify value and non-value-adding activities:
- Recognise the internal and external customers for the process
- Recognise value-adding activities
- Recognise non-value-adding activities

Step4. Collect Lead-time data:
Collection of accurate data would help the accuracy of analyses and accommodate obtaining correct and accurate graphs.

Step5. Graph the obtained information:
- One box for each step which needs to be put in order
- Height of each box represents the time required for step
- Value-adding boxes are set on the left and non-value-adding boxes on the right
- Labelling each box with the activity of that step and time allocated.

Step No.	Time (min)	Value-adding Steps	Non-value-adding Steps	Time (min)
1			Storing	24:00
2	12:00	Processing		
3			Waiting	12:00
4	36:00	Processing		
5			Inspection	12:00
6	12:00	Processing		
7			Waiting	12:00
8			Storing	24:00
Total Time	60			84

Table1. An example of the competed lead-time graph

Step6. Analyse and eliminate non-value-adding activities:
None-value-adding activities include:
- Re-works and inspections,
- Waits and unnecessary moves

Step7. Analyse to reduce or eliminate value-adding activities
Value-adding activities containing factors that contribute to the increase in MLT must be reduced, eliminated or combined.

Step8. Graph the improved process:
By using graphic tools, an accurate graph can be obtained for the improved process, as shown bellow.

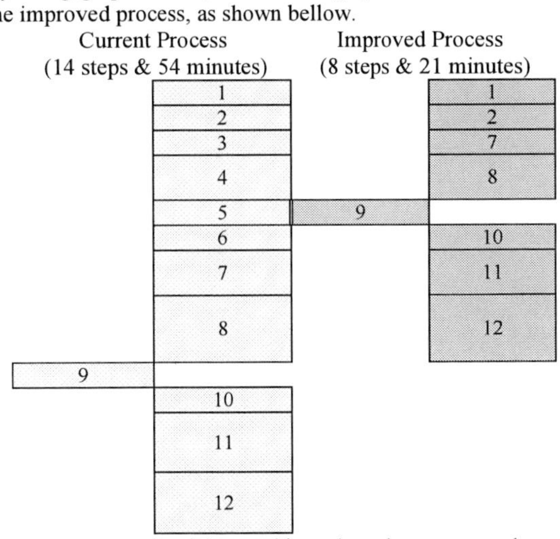

Fig. 1. An example of the graphic tool used to compare the original and improved process

3. ANALYSE & SUGGESTIONS

Due to the limitation of space in this paper, some of the important points of the above steps are analysed and suggestions for improvements are made:

3.1 Non-value-adding activities

3.1.1 Rework & Inspection

❖ **Late changes**

Late changes consume morale, profit as well as time. Most of the wasteful changes occur in product specification reflecting deficiencies in product management skills or processes. Design changes create disruptions and time consuming rework loops at the product development stage. There are, however, a number of measures that can be implemented, to reduce the number of late changes and reduce their negative impacts such as:

- Making strategy and plan explicit at the start stage.
- Improving the quality of market research.
- Having frequent product introductions to reduce changes.
- **Quality Problems**

The cost of production is the sum of costs related to acceptable production, detection of unacceptable production as well as fixing unacceptable production (rework). Low-quality output and/or inappropriate quality of input materials lead to doing more reworks and inspections.

3.1.2 Waits & Unnecessary Moves

❖ **Layout and bottlenecks**

Improving product flow through selecting right layout results in the reduction of waiting times and manufacturing lead time. Equipment and workstations must be arranged in a sequence to support a smooth flow of materials with minimal transport and delay. Process balancing is the best solution to reach this destination. That is because only in a balanced process the product will continually flow and as a result, parts movement is minimized, wait time between operations is reduced, inventory is reduced and productivity increases. There are a number of benefits applying the effective layout including reducing lead times, cycle times and waiting times, Minimizing part movement, Increasing quality performance, maximizing the value added by each worker and finally increasing flexibility. As statistics show, Implementation of a U-shaped design, for instance, can have the following advantages:

- Decrease the material transport time by 80%
- Decrease the WIP by 60%
- Increase the production flow by 50%
- Increase the capacity by 20%

❖ **Long machine setup times**

It is desirable to minimize machine setup time as long setup time means a high amount of waiting time, which is a non-value-adding activity. There are some ways to reduce the setup times some the most important ones are being mentioned here:

- **Adapting cellular manufacturing**

In cellular production machine setup times are minimized because each cell processes only parts that have similar setup and processing requirements. This means the amount of time equipment is available to process parts increases and with fewer machines the company can produce a given level of products.

- **Using multiple machines or automated equipment**

Another approach, if the equipment is available and utilization rates are not a problem, is to use multiple machines that have already been setup for the new task. Alternatively, some of the more advanced and automated equipment can automatically reset itself.

- **Converting internal setup to external setup**

Internal setup requires the equipment to be turned off, whereas external setup time is applied while the machine is still working on the previous part. This needs us to identify those internal setup tasks which can be conducted through external setup work. This might also be done through making some changes in operation. As external setup time is easier to reduce, in the next stage external task times are reduced by applying such techniques as duplicating fixtures and installing roller supports.

3.2 Value-adding activities

There are many factors which contribute to long MLT some of which can be reduced or eliminated. Those include the followings:

- **Errors in different stages of manufacturing:**
- **Complicated planning and control procedures**
- **Poor or complex design**
- **Inefficient utilisation of resources**
- **Fluctuations in the product volumes**
- **Low utilisation of equipment, due to poor job design**
- **High rate of machine idle-time due to inappropriate**

❖ **Errors:**

The major cause of waste in value-adding activities is errors. Going for first-time-right designs should be a part of any manufacturing lead-time reduction plan. The earlier the error, the less the penalty in terms of time and cost can be.

Errors may occur at the product definition stage. These typically result from a poor understanding of the market or an unrealistic belief in the appeal or performance of a new technology. Errors may also occur at the conception stage, particularly if the product involves new technology or design. Conception errors typically result from a lack of skills and experience or from gaps in communications. Furthermore, errors may occur later in the process at the development, engineering, or industrial preparation stages. Such errors reflect insufficient interaction among functions, for example between engineering and manufacturing on product manufacturability. This can be avoided through concurrent engineering. One, however, should pay attention to the following rules to minimise errors:

- **By a formal process to capture and enrich experience**
- **Adopt, learning process by analysing past errors**
- **Implement modern technical documentation systems**
- **Update all technical information regularly**
- **Ensure that people are using latest official update as input to their work**
- **Encourage teamwork and team problem solving**
- **Emphasis on early problem identification & prevention**
- **Implement a formal design review process**
- **Adopt engineering tools and fast prototyping**

4. CONCLUSION

The Lead-time reduction is an important part of the manufacturing process, because it has a direct impact on operational costs and manufacturing efficiency. This paper discussed the activities, which are required in different stages/phases of manufacturing process and analysed the factors of delays and sources of wastes for both value-adding and non-value-adding activities. Conclusion drawn and suggestions for improvements were made.

5. REFERENCES

Cubberly W. & Bakerjian R. (1988). *Tool and Manufacturing Engineer's Hand-book*, Desk Edition, Dearborn.

Marc S. (2004). *Lead Time Reduction.* elsmar.com

Meredith J.R. & Shafer S.M. (2001). *Operations Management for MBAs*, John Wiley & Sons, Inc., USA.

Motevallian, B. (1998). *Reducing Product Lead-time*, School of Advanced Manufacturing & Mechanical Engineering, University of South Australia, A master of engineering thesis, pp1-140, Australia..

Wayne S. (1998). Time Out, John Wiley & Sons. Inc., USA

Annals of DAAAM for 2006 & Proceedings of the 17th International DAAAM Symposium, ISSN 1726-9679
B. Katalinic, Published by DAAAM International, Vienna, Austria 2006
150th Birthday of Nikola Tesla (1856-1943) Inventor of Mobile Robot Teleoperation

ANALYSIS OF ERROR CORRECTION PROPERTY IN DISCRETE SYSTEMS

FILARETOV ,V. & ZHIRABOK, A.

Abstract: *The problem of malfunction analysis for nonlinear and linear discrete dynamic systems is studied on the basis of algebra of functions. A way to analyze an error correction property is suggested. The method of increasing the error correction degree through transforming the initial system is considered. In the linear case, this transformation is based on the Jordan canonical form.*

Key words: *error correction, nonlinear systems, linear systems, discrete systems.*

1. INTRODUCTION

The ability of digital systems to correct malfunctions in its elements can be obtained via error-correcting codes (Peterson & Weldon, 1972), i.e. using certain redundancy. However, in some cases, the system can have error-correcting property due to features of its operation. The problem of interest to us is to investigate this property and to suggest a method to increase the error correction degree.

The problem is initially studied for nonlinear discrete dynamic systems described by difference equation

$$x(t+1) = f(x(t), u(t)) \qquad (1)$$

where $x \in X \subset R^n$ and $u \in U \subset R^s$ are the vectors of state and control respectively, f is nonlinear vector functions. The obtained results are then used for linear dynamic systems

$$x(t+1) = Fx(t) + Gu(t) \qquad (2)$$

where F and G are constant matrices.

Denote the model (1) as $\mathbf{\Sigma}$=(X,U,f). Let $x^1(0)$ and $x^2(0)$ be expected and real (due to an malfunction at t=0) values of the state vector at t=0 and $\varepsilon=x^1(0)-x^2(0)$ is an error. The error ε is said to be self-corrected if $x^1(k)=x^2(k)$ for some k>0 where $x^1(k)$ (or $x^2(k)$) is a state to which the system transfers from the initial state $x^1(0)$ (or $x^2(0)$) under some control $u(\tau)$, $0<\tau<k$.

The problem is to describe a class of all self-correction errors in a given system. To solve the problem, so-called algebra of functions will be used (Shumsky & Zhirabok, 2006).

2. ALGEBRA OF FUNCTIONS

Vector functions are the elements of this algebra, which includes some binary relations, operations and operators.

1. Partial preordering relation $\le$: for any functions α: X $\to$ S и β: X $\to$ W denote $\alpha\le\beta$ if $\gamma(\alpha(x))=\beta(x)$ for some function γ and all $x\in$ X. When $\alpha\le\beta$ and $\beta\le\alpha$, then denote $\alpha\approx\beta$.

2. Operation $\times$: the function $\alpha\times\beta$ is determined as follows:

$$\alpha\times\beta = \max(\gamma \mid \gamma\le\alpha,\ \gamma\le\beta).$$

3. Binary relation Δ: $(\alpha,\beta)\in\Delta$ if the equality $\beta(f(x,u))=$ $=\gamma(\alpha(x),u)$ holds for some function γ and for all $(x,u)\in$ X$\times$U.

4. Operators **M** and **m**: $\mathbf{M}(\beta)$ is a function satisfying the conditions

$$(\mathbf{M}(\beta),\beta) \in \Delta, \qquad (\alpha,\beta) \in \Delta \ \Rightarrow\ \alpha \le \mathbf{M}(\beta);$$

$\mathbf{m}(\alpha)$ is a function satisfying the conditions

$$(\alpha,\mathbf{m}(\alpha)) \in \Delta, \qquad (\alpha,\beta) \in \Delta \ \Rightarrow\ \mathbf{m}(\alpha) \le \beta.$$

3. ERROR-CORRECTION CONDITION

Determine minimum (with respect to the relation $\le$) function φ as follows: $\varphi(x^1(0))=\varphi(x^2(0))$ where $x^1(0)$ and $x^2(0)$ be expected and real (due to an malfunction at t=0) values of state vector at t=0. In our case this signifies that for the state $x(0)$ different from $x^1(0)$ and $x^2(0)$ the inequality $\varphi(x(0))\ne\varphi(x^1(0))$ holds. For example, if ε is an error in the first component of the vector x, then $\varphi(x)=x_2\times x_3\times\ldots\times x_n$.

Theorem. The error ε is self-corrected at the k-th step if and only if $\varphi\le\mathbf{M}^k(e)$. Here $\mathbf{M}^{i+1}=\mathbf{M}(\mathbf{M}^i)$, e is the identity function.

Proof. (Only if). Consider two steps for simplicity. Suppose, the error ε is self-corrected at the second step, i.e.

$$f(x^1,u^1) \ne f(x^2,u^1), \quad f(f(x^1,u^1),u^2) = f(f(x^2,u^1),u^2) \qquad (3)$$

where $x^1=x^1(0)$ and $x^2=x^2(0)$, $u^1=u(0)$, $u^2=u(1)$. Consider two minimum functions φ and ψ: $\varphi(x^1)=\varphi(x^2)$ and $\psi(f(x^1,u^1))=$ $=\psi(f(x^2,u^1))$. Denote $\varphi(x^1)=s$ and $\psi(f(x^1,u^1))=w$ for some u^1. Determine the function γ as follows: $\gamma(s,u^1)=w$, or $\gamma(\varphi(x^1),u^1)=$ $=\psi(f(x^1,u^1))$. This implies the inclusion $(\varphi,\psi)\in\Delta$. By analogy, $(\psi,e)\in\Delta$. Due to the operator **M**, inequalities $\varphi\le\mathbf{M}(\psi)$ and $\psi\le\mathbf{M}(e)$ hold. Therefore $\mathbf{M}(\psi)\le\mathbf{M}^2(e)$ and $\varphi\le\mathbf{M}^2(e)$.

(If). Let $\varphi\le\mathbf{M}^2(e)$, then $\mathbf{m}(\varphi)\le\mathbf{m}(\mathbf{M}^2(e))\le\mathbf{M}(e)$ due to the properties of the operators **m** and **M.** Denote $\psi=\mathbf{m}(\varphi)$, then $\psi\le\mathbf{M}(e)$; since $(\mathbf{M}(e),e)\in\Delta$ the inequality $\psi\le\mathbf{M}(e)$ and property of the operator **M** give $(\psi,e)\in\Delta$. Because $(\varphi,\mathbf{m}(\varphi))\in\Delta$, then $(\varphi,\psi)\in\Delta$ holds. The inclusion $(\varphi,\psi)\in\Delta$ gives the equality $\psi(f(x,u))=\gamma_1(\varphi(x),u)$ for some function γ_1 and all $(x,u)\in$ X$\times$U. If $\varphi(x^1)=\varphi(x^2)$ for some states x^1 and x^2, then $\psi(f(x^1,u))=\psi(f(x^2,u))$. By analogy, the equality $f(x,u)=\gamma_2(\psi(x),u)$ holds. In particular, if $x^{1*}=f(x^1,u^1)$ and $x^{2*}=f(x^2,u^1)$, the equality $\psi(x^{1*})=\psi(f(x^1,u^1))=\psi(f(x^2,u^1))=\psi(x^{2*})$ implies $\gamma_2(\psi(x^{1*}),u^2)=$ $=f(x^{1*},u^2)=\gamma_2(\psi(x^{2*}),u^2)=f(x^{2*},u^2)$, or $f(f(x^1,u^1),u^2)=f(f(x^2,u^1),u^2)$, i.e. equation (3) holds and the error ε is corrected at the second step. Theorem has been proved.

By definition of the function e, the inequality $e\le\mathbf{M}(e)$ holds. Since the operator **M** is monotonic, then $\mathbf{M}(e)\le\mathbf{M}^2(e)$ and $e\le\mathbf{M}(e)\le\mathbf{M}^2(e)\le\ldots$. The equivalence $\mathbf{M}^k(e)\approx\mathbf{M}^{k+1}(e)$ for some k implies $\mathbf{M}^k(e)\approx\mathbf{M}^{k+v}(e)$ for all v=1,2,... This means that if some error is not corrected at the k-th steps, it is not ever correct.

If $\mathbf{M}(e)\approx e$, the system has not self-correction property.

4. INCREASE OF SELF-CORRECTION DEGREE

Self-correction degree can be increased through isomorphic transformation of the initial system.

Recall, the function Φ: $X \to X^*$ is isomorphism $\mathbf{\Sigma} \to \mathbf{\Sigma}^*=(X^*,U^*,Y^*,f^*)$ with $U^*=U$ if $\Phi f=f^*(\Phi\pi_X\times\pi_U)$, where π_X and π_U are projections. In this case, the function Φ^{-1}: $X^* \to X$ has to exist such that $\Phi^{-1}\Phi\approx e$, $\Phi\Phi^{-1}\approx e$, $\Phi^{-1}(f^*(x^*,u))=f(\Phi^{-1}(x^*),u)$ for all $(x^*,u)\in X^*\times U$. Assume that the relationship

$$\mathbf{M}^{k+1}(e) \approx \mathbf{M}^{k}(e) \approx \rho_1\times\rho_2\times\ldots\times\rho_m$$

holds for some k and n-m components of the state vector exist (for definiteness sake - $x_1, x_2,\ldots, x_{n-m}$) such that

$$\Phi = \pi_1\times\pi_2\times\ldots\times\pi_{n-m}\times\rho_1\times\rho_2\times\ldots\times\rho_m, \tag{4}$$

where π_j is projection: $\pi_j(x)=x_j, j=1,2,\ldots,n$.

Consider the case when each function ρ_j contains the only variable different from the ones $x_1, x_2,\ldots, x_{n-m}$. Assume that this variable is x_{n-m+j}; this can be arranged by changing indices at the functions $\rho_1, \rho_2,\ldots, \rho_m$. For the sake of simplicity consider the function ρ_1.

Let $x_{i1}, x_{i2},\ldots, x_{ip}$ be arguments of this function with $x_{ip}=x_{n-m+1}$. Assume that

$$\pi_{i1}\times\pi_{i2}\times\ldots\times\pi_{ip-1}\times\rho_1(\pi_{i1}\times\pi_{i2}\times\ldots\times\pi_{ip})\approx\pi_{i1}\times\pi_{i2}\times\ldots\times\pi_{ip-1}\times\pi_{n-m+1}.$$

Suppose also that the functions $\rho_2,\ldots, \rho_m$ have similar structure and properties. The inverse function Φ^{-1} is in the form

$$\Phi^{-1}= \overset{*}{\pi}_1\times\ldots\times\overset{*}{\pi}_{n-m}\times\mu_1(\overset{*}{\pi}_{i1}\times\ldots\times\overset{*}{\pi}_{ip})\times\mu_2(\ldots)\times\ldots\times\mu_m(\ldots), \tag{5}$$

where the function μ_1 is the inverse one of ρ_1 from the following point of view:

$$\rho_1(\overset{*}{\pi}_{i1}\times\ldots\times\overset{*}{\pi}_{ip-1}\times\mu_1(\overset{*}{\pi}_{i1}\times\ldots\times\overset{*}{\pi}_{ip}))\approx\overset{*}{\pi}_{ip}=\overset{*}{\pi}_{n-m+1}.$$

It is supposed that the functions $\mu_2,\ldots, \mu_m$ have similar properties. In this case, the properties of the projections and operation $\times$ give the following result: $\Phi\Phi^{-1} \approx e$. Thus the function Φ^{-1} from (5) is really the inverse one of Φ defined by (4).

Find properties of the operator $\mathbf{M}$ for isomorphic systems. Let Φ be isomorphism $\mathbf{\Sigma} \to \mathbf{\Sigma}^*$, i.e. $\Phi f=f^*(\Phi\pi_X\times\pi_U)$ and $\beta=\beta^*\Phi$ for some functions β: $X \to W$ and β^*: $X^* \to W$, then $\beta f=\beta^*\Phi f=\beta^* f^*(\Phi\pi_X\times\pi_U)$. By definition of the operator $\mathbf{M}^*$, the inclusion $(\mathbf{M}^*(\beta^*),\beta^*)\in\Delta$ holds. It gives the inequalities $\beta^* f^*\geq\mathbf{M}^*(\beta^*)\pi_X\times\pi_U$ and $\beta^* f^*(\Phi\pi_X\times\pi_U)=\beta f\geq\mathbf{M}^*(\beta^*)\Phi\pi_X\times\pi_U$. Due to the relation Δ, this inequality is equivalent to $(\mathbf{M}^*(\beta^*)\Phi,\beta)\in\Delta$ that implies

$$\mathbf{M}(\beta) = \mathbf{M}(\beta^*\Phi) \geq \mathbf{M}^*(\beta^*)\Phi. \tag{6}$$

Since equation $\beta=\beta^*\Phi$ implies $\beta^*=\beta\Phi^{-1}$, then (6) allows one to write down $\mathbf{M}^*(\beta^*)\geq\mathbf{M}(\beta)\Phi^{-1}$ or $\mathbf{M}^*(\beta^*)\Phi\geq\mathbf{M}(\beta)$ that with (6) gives required relationship $\mathbf{M}(\beta)=\mathbf{M}(\beta^*\Phi)\approx\mathbf{M}^*(\beta^*)\Phi$ or $\mathbf{M}^*(\beta^*)\approx\mathbf{M}(\beta)\Phi^{-1}$. It can be shown that this one is true for the k-th degree of the operator:

$$\mathbf{M}^{*k}(\beta^*) \approx \mathbf{M}^{*k}(\beta\Phi^{-1}) \approx \mathbf{M}^{k}(\beta)\Phi^{-1}. \tag{7}$$

Definitions of the identity functions e: $X \to X$ and e^*: $X^* \to X^*$ and equivalence $\approx$ imply $e\approx\Phi e=e^*\Phi\approx e^*$ if Φ is isomorphism. Then expression (7) with $\beta^*=e^*$, $\beta=e$, and $e^*=e\Phi^{-1}$ can be transformed as follows: $\mathbf{M}^{*k}(e^*)\approx\mathbf{M}^{*k}(e\Phi^{-1})\approx\mathbf{M}^{k}(e)\Phi^{-1}\approx$ $\approx(\rho_1(\ldots)\times\rho_2(\ldots)\times\ldots\times\rho_m(\ldots))\Phi^{-1}\approx\overset{*}{\pi}_{n-m+1}\times\ldots\times\overset{*}{\pi}_n$. Therefore the theorem gives that error in the components $\overset{*}{x}_1, \overset{*}{x}_2,\ldots, \overset{*}{x}_{n-m}$ of the system $\mathbf{\Sigma}^*$ state vector will be corrected within k steps.

5. LINEAR SYSTEMS

In the linear case, when the system is described by model (2), an analysis of self-correction property and increase of its degree can be achieved by more simple way. It can be shown that if $\beta(x)=Bx$ for some matrix B, then $\mathbf{M}(B)=BF$ for system (2). Then $e(x)=Ex$ (E is the unit matrix) and $\mathbf{M}(E)=F$, $\mathbf{M}^k(E)=F^k$. Clearly, if $det(F)\neq0$, then $\mathbf{M}(E)=F\approx E$ and the system has not self-correction property.

Let $det(F)=0$ and $F^k\approx F^{k+1}$ for some k, or

$$rank(F^k) = rank(F^{k+1}).$$

Self-correction condition for the error ε, i.e. inequality $\varphi\leq\mathbf{M}^k(e)=\mathbf{M}^k(E)=F^k$, takes in the linear case the following form: if the matrix F^k contains zero columns with numbers $i_1, i_2,\ldots, i_m$, then errors in components of the state vector with the same numbers will be corrected. To improve the error-correction property, one has to use an isomorphic transformation of the system based on the Jordan canonical for of the matrix F. It is known (Bellman, 1960) that for square $n\times n$ matrix F the nonsingular matrix Φ exists such that

$$\Phi F\Phi^{-1} = \begin{bmatrix} L_{k_1}(\lambda_1) & & 0 \\ & L_{k_2}(\lambda_2) & \\ 0 & & \ldots \\ & & & L_{k_r}(\lambda_r) \end{bmatrix} \tag{8}$$

where $k_1+k_2+\ldots+k_r=n$, $\lambda_1, \lambda_2,\ldots, \lambda_r$ are eigenvalues of the matrix F, $L_k(\lambda)$ is the Jordan block, i.e. $k\times k$ matrix of the form

$$L_k(\lambda) = \begin{bmatrix} \lambda & 1 & \ldots & 0 \\ 0 & \lambda & 1 & 0 \\ \ldots & \ldots & \ldots & \ldots \\ 0 & 0 & \ldots & \lambda \end{bmatrix}. \tag{9}$$

Let $rank(F)=m<n$, then zero is an eigenvalue of the matrix F multiplicity $k\geq n$-m (Bellman, 1960). It follows from (8) and (9) that if all eigenvalues are not equal to zero, then all rows of the matrix $F^* = \Phi F\Phi^{-1}$ are linearly independent. As Φ is a nonsingular matrix and $rank(F)=m<n$, then $rank\,(F^*)=m$-n and the matrix F^* contains n-m Jordan blocks corresponding to the zero eigenvalue of the form

$$L_k(0) = \begin{bmatrix} 0 & 1 & \ldots & 0 \\ 0 & 0 & 1 & 0 \\ \ldots & \ldots & \ldots & \ldots \\ 0 & 0 & \ldots & 0 \end{bmatrix}; \tag{10}$$

if k=1, then $L_1(0)=0$. Therefore, errors in the components of the vector $x^*=\Phi x$ corresponding to the zero columns of the matrix (10) are self-corrected.

6. CONCLUSIONS

A feature of the suggested approach to the error correction analysis is that it allows one to describe set of errors which can be corrected and avoid to analyze each error. The method to increase the error-correction degree through transforming the initial system has been suggested.

7. REFERENCE

Bellman, R. (1960). *Introduction to matrix analysis.* McGraw-Hill Company, Inc. New-York – London.

Peterson, W. & Weldon, E. (1972). *Error-correcting codes.* Mita Press, London.

Zhirabok, A. & Shumsky, A. (2006). Nonlinear diagnostic filter design: algebraic and geometric points of view, *Int. Journal of Applied Mathematics and Computer Science*, V.16, No 1, 115-127, ISSN 1641-876X.

Annals of DAAAM for 2006 & Proceedings of the 17th International DAAAM Symposium, ISSN 1726-9679
B. Katalinic, Published by DAAAM International, Vienna, Austria 2006
150th Birthday of Nikola Tesla (1856-1943) Inventor of Mobile Robot Teleoperation

MEASURING DATA EXTRACTING FROM PBS DETECTOR SIGNAL IN THE NIR MOISTURE METER

FILIC, M. & CORLUKA, V.

Abstract: *The reflectance at two, absorbing and not absorbing wavelengths in the near infrared (NIR) region from several wheat grain samples with different moisture content detected by the precise gravimetric instrument, has been measured. The periodical measuring signal given by the photoconductive PbS detector and amplified with a band pass analogue preamplifier consists of both, beneficial and unfit information. The input radiation was assigned by the mechanical chopper with adequate optical filters and directed to the sample by the mirror system. The frequency analysis of both, the input and amplified output has been conducted by using the Fast Fourier transformation (FFT). The applied approach with all calculation and measurement results are presented in this paper.*

Key words: *photoconductive detector, moisture, NIR, FFT*

1. INTRODUCTION

The optoelectronic measurement system of the moisture meter with one beam which is described and analysed in some authors' previous papers is based on the cooled PbS photoconductive detector, the preamplifier and the mechanical chopper. The applied measuring principle consists of determining the reflected part with respect to the total radiation at two different wavelengths in the near infrared (NIR) region [1]. This was achieved by means of one beam with two pairs of narrow band pass NIR filters, the 1800nm wavelength which does not absorb water, in contrast to the 1940 nm wavelength which absorbs water. Both of them are embedded in the rotational wheel. The mechanical chopper has a double function. First, the NIR radiation at reference wavelengths, which water not absorbs, and narrow band at 1940nm, the so called absorbing for water, is extracted from the totally produced radiation by means of a halogen bulb whereas the lens system directs the incident radiation to the sample. Second, these two domains are mechanically modulated by the exactly chosen frequency [4]. The mentioned moisture measuring system consists of a radiation source, a mechanical chopper with two pairs of optical filters, the photoconductive PbS detector and analogue preamplifier as shown in Fig.1. The chopped radiation from the source is directed by a mirror system to the wheat grain sample. Furthermore, the reflected radiation from the sample is collected and directed by the concave mirror to the photoconductive PbS detector. The continual signal of the detector is AC coupled with the analogue preamplifier and acquired on the output by the PC card. However, the moisture meter is accomplished by one beam [2], the signal modulated by the mechanical chopper is not sinusoidal and it consists of information about radiation values reflected from the sample at both, reference and the absorbing wavelengths. Besides, a large amount of noises as well as some useful information occur in this signal. Henceforth, the measuring data at these two wavelengths have to be extracted from the acquired signal before a subsequent processing is done.

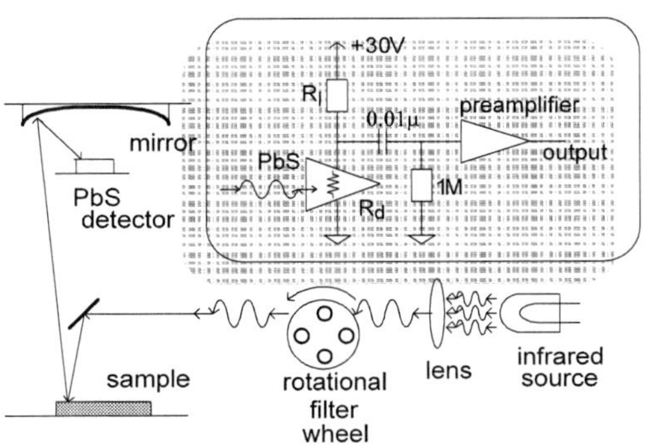

Fig. 1 NIR moisture measuring system

2. NIR SIGNAL MODULATION

The radiation consists of the NIR at 1800nm and 1940nm from the source shown in Fig. 1 which is directed by the achromatic lens to the mechanical chopper. Fig. 2 shows the sight on the wheel surface during its rotation by the angular velocity ω, which cuts the circular shape of the incident beam at the same time. Two pairs of the NIR optical filters are located in the four wheel windows which transmittance characteristics are shown into patterns apart.

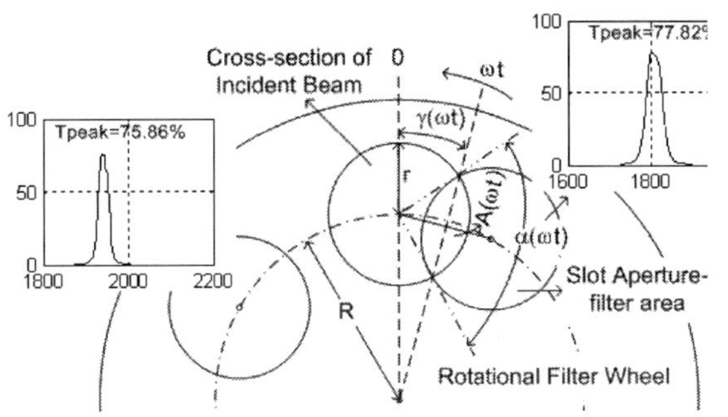

Fig. 2 View on the filter wheel

The energy density of the radiation which penetrates into the sample at 1800nm and 1940nm vary owing to different transmittance coefficients of the remote filters, T_{peak}=77.82% and T_{peak}=75.86% respectively. But, the energy density of the input beam around the corresponding wavelength is particularly constant.

2.1. Input signal evaluation

Some respective units are noted by adequate symbols in Fig.2. Provided that the incident energy of the input beam is constant and the chopper output is proportional to the sector indicated by A(ωt), the following expressions result from Fig.2:

$$A(\omega t) = r^2\alpha(\omega t) \quad (1)$$

$$\alpha(\omega t) = 2a\cos\left(\frac{R\sin\gamma(\omega t)}{r}\right). \quad (2)$$

The rotational wheel changes γ angle twice as slow as ωt from π/8 to 0 corresponding to A changes from 0 to maximum, therefore

$$\gamma = \frac{\pi}{8} - \frac{\omega t}{2}\text{, and} \quad (3)$$

$$A(\omega t) = 2r^2 a\cos\left(\frac{R}{r}\sin\left(\frac{\pi}{8} - \omega t\right)\right) \quad \text{for } \omega t = 0 \div \frac{\pi}{4}. \quad (4)$$

Because the energy density of the radiation is constant and two pairs of the optical filters have different transmittance characteristics when compared with each other, the incident radiation which has the same sample curves as a signal is shown in Fig. 3.

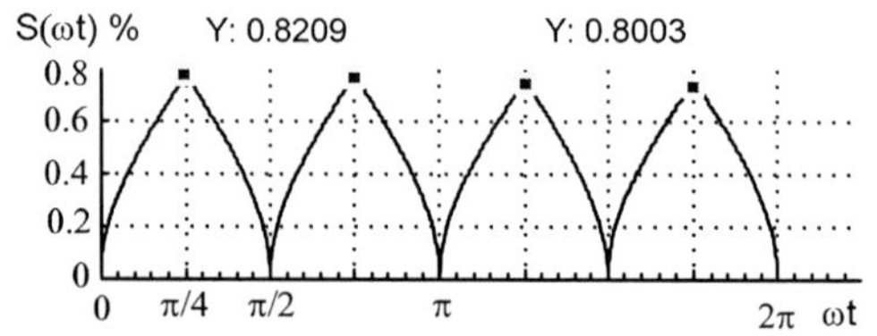

Fig. 3.Curve of the radiation penetrating into the sample

2.2. Fourier series of the reflected radiation

The curves of the reflected radiation from the sample depend on the moisture content and its other features. Linear changing of the reflected energy is assumed by using the reflected coefficient ε_r. The Fourier series of the reflected coefficient,

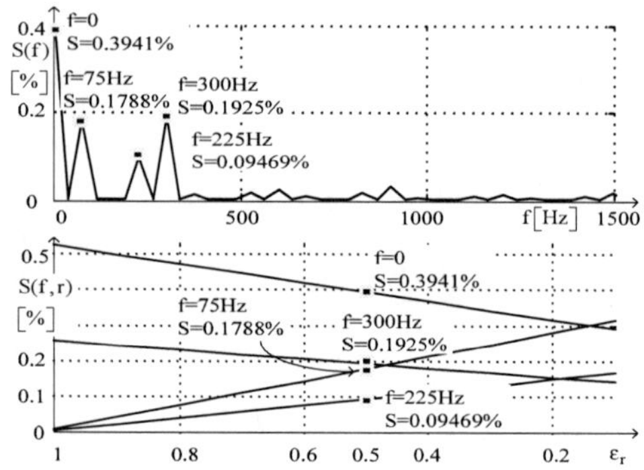

Fig. 4. Fourier series versus different reflection

which were obtained by FFT as well as how some harmonics change with ε_r are shown in Fig. 4. Proportional changing of one part of the reflected signal at 1940 nm [1] only, is being assumed and high band pass of the analog preamplifier circuit. DC (f=0) signal does not pass along the preamplifier which is also shown.

3. EXAMPLES AND EXPERIMENTAL RESULTS

Some of the output signals from the preamplifier are measured by the gravimetric instrument [5] at the constant temperature of the photodetector and ambient 22^0C [3] for the same wheat grain sample with various but known moisture content from 13.36% to 20.96%. One of the time responses and some harmonics compared to the moisture content characteristics which are calculated using the FFT of time curves at the rotary speed of the chopper 4500min^{-1} (75Hz) are shown in Fig. 5. The discrete values which are given by the FFT are marked by dots whereas the appropriate characteristics are indicated by lines.

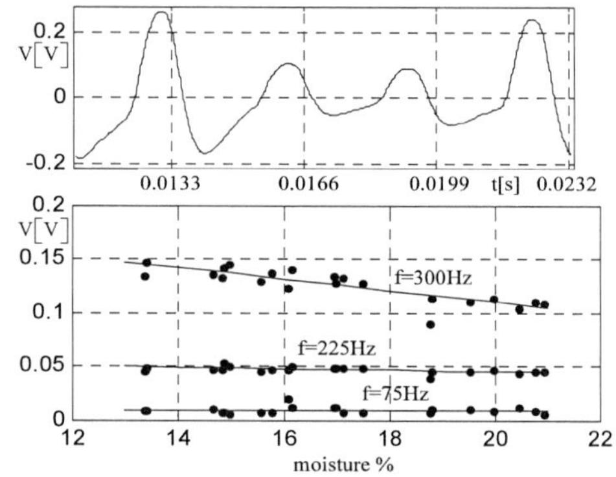

Fig. 5. Time response a) and frequency response b)

4. CONCLUSIONS

The NIR moisture meter measuring system is analysed by the discrete Fourier transformation at different but known moisture contents of the same wheat grain and the constant detector temperature. The detector temperature is measured by the embedded thermistor. The high order frequency components exist in the output signal of the analog preamplifier due to the applied frame of the optical filter windows. The frequency content of the output signal can be changed by the appropriate frame or/and size of the holes chosen in the filter wheel and by the parameter analog preamplifier circuit changing. But, the signals of the reference and absorbing beams are unmistakably measurable in the applied configuration. Linear dependency of several harmonics to the moisture of wheat grain changing was noticed at the process of conducting the analyses, although additional researches have to be carried out in order make a more accurate conclusion. Many problems can be anticipated when the signal of the absorbed beam is small. Noises influence the calculation of the harmonic content, and higher harmonics are distinct if compared to the 3rd or 4th, for example.

5. REFERENCES

Corluka. V.; Filic, M.; Mesic, M. & Valter, Z. (2004). Optoelectronic moisture measurement, *Proceedings of the 3rd DAAAM International Conference ATDC*, Katalinic, B., Veza, I. & Bilic, B. (Ed.), pp. 303-308, ISBN 953-6114-68-2, Split, June 2004, Univ. of Split, FESB, Split

Corluka. V.; Filic, M. & Valter, Z. (2004). Development of one infrared moisture meter, *Proceedings of the 15th DAAAM Int. Symp.*, Katalinic, B. (Ed.), pp. 081-082, ISBN 3-901509-42-9, Wien, November 2004, DAAAM Int. Vienna, Wien

Mesic, M.; Filic, M. & Valter, Z. (2005). Experience with Detectors for Infrared Moisture Measuring, *Proceedings of 6th Int. Conf. on ''Electromagnetic Wave Interaction with Water and Moist Substances''*, Kupfer, K. (Ed.), pp. 535-540, Weimar, June 2005, MFPA, Weimar

Filic, M.; Corluka, V. & Valter, Z. (2005). Frequency Analysis od the measuring system in the NIR Moisture Meter, *Proceedings of the 16th DAAAM Int. Symp.*, Katalinic, B. (Ed.), pp. 117-118, Opatija, November 2005, DAAAM Int. Vienna, Wien

Mesic, M.; Filic, M. & Valter, Z. (2006). Optimal Function-model choice, *Proceedings of 5th Int. Symp. on Humidity and Moisture-ISHM 2006 Brazil,*Julio D.Brionizio, Dr Peter Huang (ur.), pp. 187-190, Rio de Janeiro, May 2006, INMETRO, Rio de Janeiro

Annals of DAAAM for 2006 & Proceedings of the 17th International DAAAM Symposium, ISSN 1726-9679
B. Katalinic, Published by DAAAM International, Vienna, Austria 2006
150th Birthday of Nikola Tesla (1856-1943) Inventor of Mobile Robot Teleoperation

THE ALGORITHM OF THE INTELLIGENT BEHAVIOR OF THE ARTICLE OF CLOTHING

FIRST ROGALE, S.; ROGALE, D.; DRAGCEVIC, Z.; NIKOLIC, G. & BARTOS, M.

Abstract *This paper describes the algorithm of intelligent garment behavior with active thermal protection which changes the value of thermal insulation, depending on the environmental temperature outside the garment and the microclimate inside the garment. Thermal insulation is regulated by activating various combinations of the system of thermo insulating chambers whereby the heat transfer of the human body is regulated by conduction and convection via the thermo insulating chambers into the environment.*

Key words: *algorithm, clothing, intelligent behavior*

1. INTRODUCTION

The research author group of the Department of Clothing Technology of the Faculty of Textile Technology of the University of Zagreb began to engineer, construct and design an intelligent article of clothing with active thermal protection. Thermal protection is realized by activating various combinations of thermo insulating chambers into which compressed air is inflated (Firšt Rogale et al., 2005). Intelligent article of clothing with active thermal protection has two sensors of external and internal temperature, pressure sensors inside the thermo insulating chambers, microcontroller system and actuators, which make use of micro pneumatic elements (Firšt Rogale et al., 2006).

The system control is based on a program support whose fundamental component is an algorithm of intelligent behavior. It enables an appraisal of the external temperature and microclimate temperature inside the garment and to make a decision on necessary optimal thermal protection of garment. The necessary protection is realized in a completely automatic way, independent on the wish of the garment wearer.

2. MEASURING AND CONTROL MICROCONTROLLER PROGRAM WITH THE ALGORITHM OF THE INTELLIGENT BEHAVIOR OF THE ARTICLE OF CLOTHING

After activating the system, a self diagnosis of the state of the technical system of the garment is performed. In the mentioned procedure the proper operation of all temperature and pressure sensors in the chambers, micro compressors, all for inflating and deflating valves and the battery state are examined. After finishing the self-diagnosis procedure of the state the data are displayed, and the external and internal temperature is measured. Afterwards it is possible to select a control mode for the thermo insulating chambers which can be manual or automatic. Provided that the manual control mode has been selected, it is necessary to make a subjective evaluation of thermal comfort. If a feeling of thermal comfort is expressed, the system is directed to the first subprogram (SUB 1) in which the system deflates all the thermo insulating chambers in sequence. If the subjective feeling of a severe cold is expressed, the second subprogram (SUB 2) is activated which activates the sequential inflation of all thermo insulating chambers. By a subjective evaluation of medium cold the third program (SUB 3) is activated which starts the inflation of the shoulder and waist thermo insulating chamber. Subjective feeling of coolness can be evaluated through two states: cooler and slightly cool. In case of a higher coolness the fourth program (SUB 4) is activated and this inflates the shoulder chamber. In case of feeling slightly cool the subprogram 5 (SUB 5) is activated to start the waist chamber. After finishing any of these five subprograms, the external and internal temperature is measured, and the program is reset, i.e. to select a control mode.

In the event that the automatic control mode is selected, it is necessary to enter data on the desirable microclimate temperature. Thereupon the lower microclimate temperature limit and the upper microclimate temperature limit are determined. As the initial condition of the lower and upper limit a temperature difference of ± 2°C is selected. After this adjustment and determination of the initial conditions the internal and external temperature is measured in order to collect parameters for determining the algorithm of intelligent behavior.

The first step in making a decision which can lead to branching in the program flow is the comparison between the internal temperature and the upper limit temperature. In the event that the measurement shows that the internal temperature is higher than the upper limit temperature a conclusion is made that the wearer of the garment is too hot, and it is necessary to activate the first subprogram (SUB 1) and to deflate all the chambers in sequence. In the event that the internal temperature is not higher than the upper limit temperature the algorithm of behavior will continue according to its basic tree. Thereupon a comparison of the value of the internal temperature and the lower limit temperature of the microclimate is made. In the event that the internal temperature is higher than the temperature of the lower microclimate limit the system makes a conclusion that it is within the limit tolerances and hysteresis of ± 2°C of the desirable microclimate temperature, and it makes a decision that the system response is not necessary so that the program is reset to the starting point of the decision-making tree. In the event that the internal temperature is not higher than the temperature of the lower microclimate limit the system makes a conclusion that the wearer of the garment is cold, and the thermal protection of the garment is to be activated.

For the system to make a decision on response intensity of thermal protection, it is necessary to calculate the difference between the desirable microclimate temperature and the measured external temperature.

The response intensity of the technical system of the intelligent article of clothing and the decision on combinations of activated thermo insulating chambers will depend on the above-mentioned difference. If the temperature difference (Δt) between the desirable microclimate temperature and the measured external temperature higher than 15 °C, the system concludes that it is very cold and activates subprogram 2 (SUB 2) in which it inflates all the thermo insulating chambers in sequence. If the temperature difference is lower than 15 °C, the system finds out whether the temperature difference is higher than 10 °C. If the difference is higher than 10 °C, the system concludes that it is medium cold and makes a decision to inflate the shoulder and waist chamber. If the temperature difference is lower than 10 °C, a comparison is made whether the temperature difference (Δt) is higher than 5 °C. If the decision made is positive, the system concludes that it is cooler and activates the shoulder chamber, and if the decision made is negative, it concludes that it is slightly cool and activates the waist chamber by subprogram 5 (SUB 5). After finishing any of five subprograms, an enquiry is made of interrupting the automatic control. If there is no desire to interrupt, the program is reset to the starting point of the branching tree,

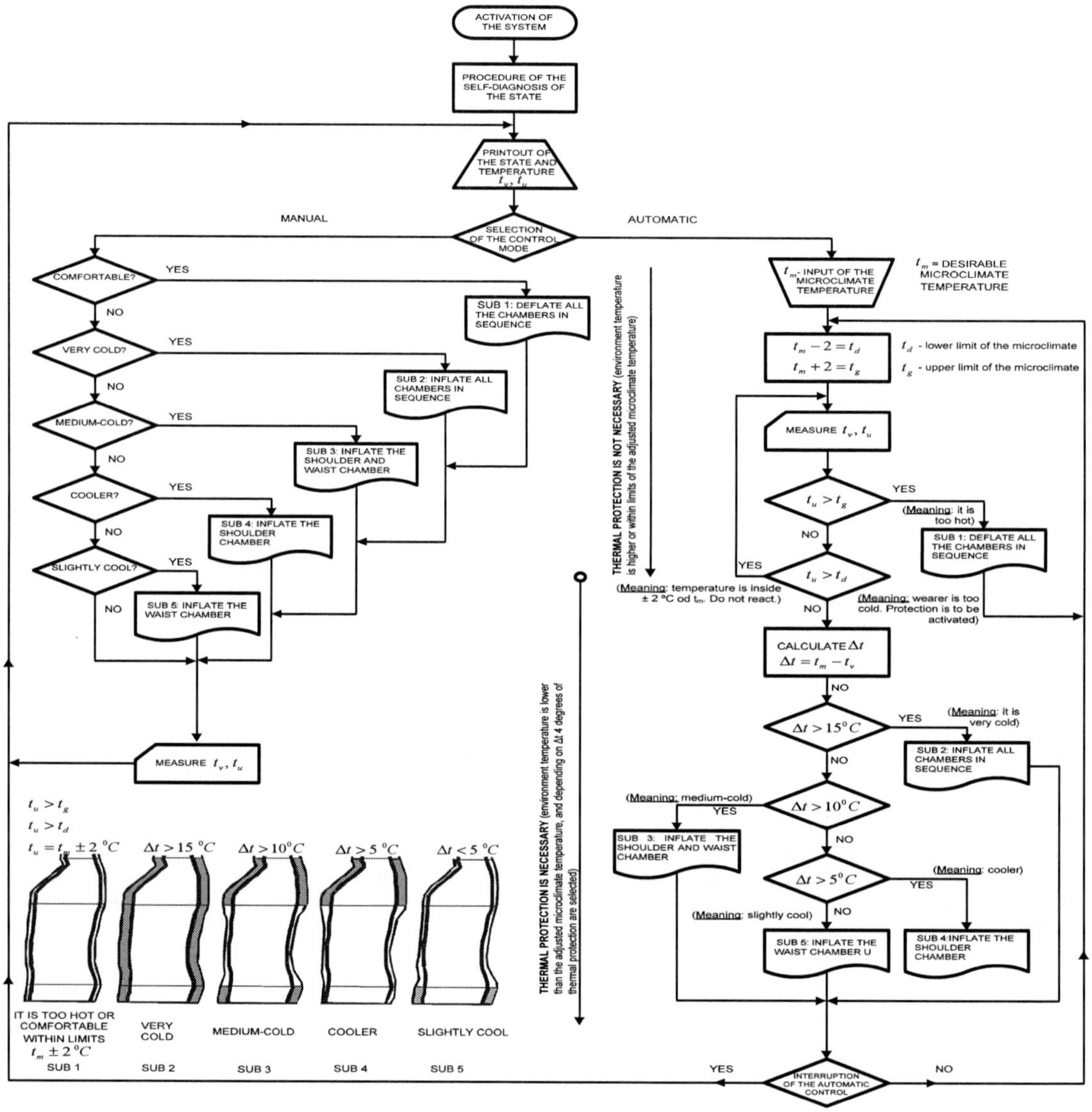

Fig. 1. Operation flow diagram of the intelligent article of clothing with active thermal protection

if there is a desire to interrupt the automatic control, the system is switched to manual control.

3. CONCLUSION

The intelligent article of clothing with active thermal protection, which is patent protected (Rogale et al, 2004; Rogale et al, 2005) is a compound technical system whose operation is controlled by a microcontroller with a program of the algorithm of intelligent behavior of garment.

The described algorithm of intelligent behavior provides a suitable basis for experiments and scientific researches during the introduction of intelligent clothing into human life.

4. REFERENCES

First Rogale, S.; Nikolic, G.; Dragcevic, Z.; Rogale, D. & Bartos, M. (2005). Arhitecture of Clothing with an Active Thermal Protection, Proceedings of 16th DAAAM International Symposium, Katalinic, B. (Ed.), 121-122, ISBN 3-901509-46-1, October 2005, Opatija, Croatia,

Rogale, D. et al. (2004). Intelligent Article of Clothing With an Active Thermal Protection, State Intellectual Property Office of Croatia, P20030727A

Rogale, D. et al. (2005). Intelligent Article of Clothing With an Active Thermal Protection, European Patent Office, Munich, Germany, International application No.PCT/HR2004/000026

Annals of DAAAM for 2006 & Proceedings of the 17th International DAAAM Symposium, ISSN 1726-9679
B. Katalinic, Published by DAAAM International, Vienna, Austria 2006
150th Birthday of Nikola Tesla (1856-1943) Inventor of Mobile Robot Teleoperation

TRENDS IN MANUFACTURING AND INNOVATION IN CROATIA: RESULTS OF THE MANVIS DELPHI RESEARCH

GALETIC, L. & PRESTER, J.

***Abstract:* In early 2004 the MANVIS Delphi research was launched, led by Faunhofer Institute, under the 6th Framework program funded by the European Commission. 22 countries participated, among them Croatia. General results show that Europe is lagging in manufacturing, behind USA and Japan. Innovation is the main issue. One of the suggestions is to enhance funding in R&D. We also investigate Croatian status of manufacturing in comparison to the European average.**

Key words: *Manufacturing, Delphi, MANVIS, Innovation, Croatia*

1. INTRODUCTION

A country can not survive sole on retailing. According to the National Board for Competitiveness in Croatia 45,8% of goods and 10,9% of services are imported, while export is only 24,7% of GDP. Those figures are quite alarming. In this document several strategic propositions are laid out. Two most important suggestions are to increase R&D investment, and to attract students to technical and fundamental science's universities. When Croatia joined the MANVIS survey we thought that the results for Croatia would be very disappointing. But, apart from some problems like relocation, Croatian manufacturing is behind European level, but not as much as we expected. It emerged that whole Europe, in comparison to US and Japan, is struggling with lack of investment in R&D and innovation activities. In this article we will concentrate only on five MANVIS topics, that is, nano-/bio- and advance materials, Microsystems, and information communication technology (ICT). We selected only these topics, because that's were Europe is most behind USA and Japan (Croatia as well). We are interested when those concepts will be in wider use in Europe, that is, when this competitive gap will diminish. In this work we investigate where Croatia stand on these topics, and analyze what are the main barriers for implementation and wider use of these materials and ICT.

2. METHODOLOGY

The MANVIS research uses the Delphi method. The research had more then hundred statements on which experts had to give their opinion. The experts had to predict on a time scale till year 2020. The survey consists of a general part (eg. ecology, management issues, organization...), as well as in depth sectors coverage (eg. machinery, electronics, rubber and plastics, traditional products...). Very similar Delphi survey was launched in USA in 1998 under the name "Visionary Manufacturing Challenges for 2020" and a Korean Delphi research in which 4500 experts were involved (Shint, 1998). The Delphi methodology is only a natural choice for such long range forecasting, and it is only natural that Europe conducts a similar study to check its strong and weak points. Overall 2993 experts, from 22 countries, from industry and academia participated in the study. Croatia was represented by 66 experts. For each statement experts had to give on Likert scales their expertise on the subject, predicted realization of the statement, importance of the question for Europe and potential barriers to implementation.

3. RESULTS

In order to draw conclusions from our data we first checked with t-test do answers from Croatian sample differ from the European sample on the question of the wider use of investigated questions (advanced materials and ICT). The t-test gave $t=0{,}508$ which is lower then the critical $t_{crit}=1{,}721$ (at significance level 5%), which allows us to conclude that there is no significant difference in time of realization of proposed questions between European average and Croatian sample. That on the other hand allows us to draw conclusions for Croatian manufacturing on the basis of data from the overall sample of 22 countries.

The development of new generic technologies and knowledge challenges manufacturing research in two ways. First, it creates a need for manufacturing processes in order to produce the new products and to provide the new services. Secondly, these new technologies and knowledge have to be integrated into the production processes themselves. Basic manufacturing research has to foresee and prepare for the new challenges and applied manufacturing research has to adapt and transform existing technologies and organisational processes. Furthermore, manufacturing research plays a decisive role in combining the long-term horizon in technology trajectories with the short-term need for innovation. This requires a good "timing" of research activities to have solutions and tools ready for industrial adoption. To depict more precisely the problem we show the time scale of possible adoptions of proposed materials and ICT technology.

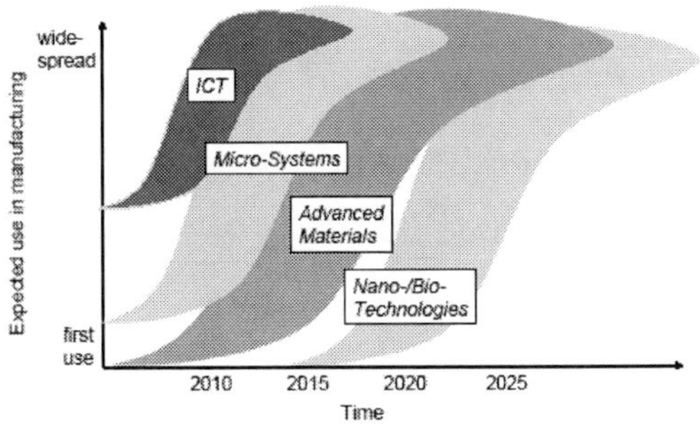

Fig. 1. Expected time of wider use of technology in question

ICT will still play the decisive role in short term perspective; it will play an important role in supporting new forms of cross-company networks (for example SME networks). Hence, targeted applied research for SME's on optimal organisation of networks and its ICT support can be derived as one priority. Microsystems (together with intelligent controllers) are key enablers for plug-and-produce systems aiming at more flexible manufacturing systems as well as for process integration into multi-functional machinery. For this second technology, the necessary link from developing new machinery to creating new business models (although not emphasised by the experts) could be crucial as well as research bringing together equipment suppliers and users. For advanced materials, the problem of making and manipulation of these materials has already been identified as an important research topic by the European Innovation study, also conducted in Croatia. Hence, applied and targeted research on process technology for materials could be a third issue. Finally, basic research on nano- or biotechnology has to be carefully monitored for emerging manufacturing research fields. In addition, cross cutting manufacturing research issues like workplace safety of nano- or bio-based processes etc. may facilitate the basic research activities in this field. The main barriers for implementing the above mentioned innovations, which contribute most to competitiveness of manufacturing, are as follows:

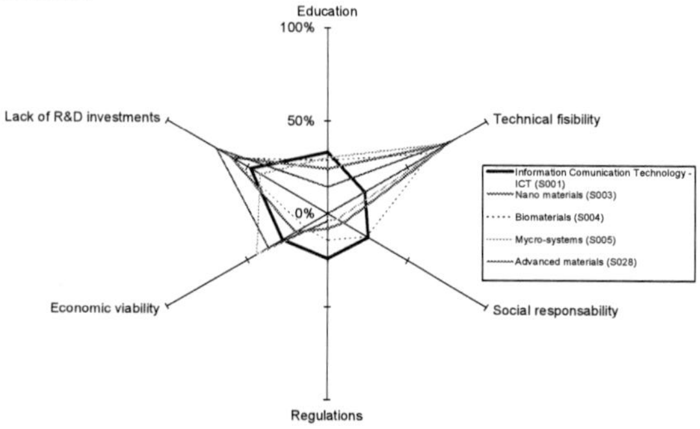

Fig. 2. Croatian barriers to implementing innovative concepts

On the European level the same problem looks in a following manner:

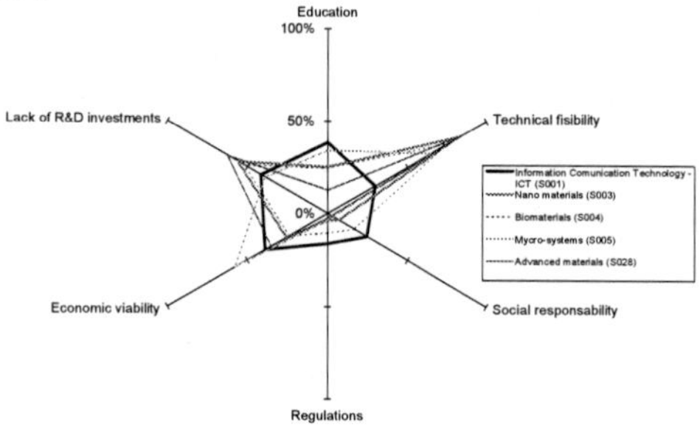

Fig. 3. European barriers to implementing innovative concepts

As we can see the diamonds are almost identical except on the question of economic validity. In Croatian diamond Economic questions are not considered as a barrier (except for micro systems). The main barriers on the Croatian and European level are dominantly lack of R&D investments and technical feasibility, with this technical part being most important. Those slight differences in barriers between European and Croatian experts might be because of Croatian beliefs about its position in comparison to Europe (we have to stress that Croatia has not participated in any such cross-country study before). Croatian experts believe that their existing equipment is unsatisfactory for use of advanced technology, and they believe that with sufficient R&D funding, innovation projects would pay off. In this part European counterparts seem more cautious, and predict that micro-systems might not generate adequate return on investment. Figures 2. and 3. raise an interesting question. Why only on micro-systems there is a difference about economic viability in Croatian (fig. 2) and European average (fig. 3)? The MANVIS research showed that especially in this part (automation, robotics, micro-systems) Japan manufacturing has a significant priority. In field of advanced materials the USA has competitive advantage. We believe that Europe (Croatia included) has sufficient "knowledge capacity" which with higher R&D investments can catch up with US. On the other hand through several past decades Japan highly invested in automation, robotics, micro-systems which is not that easy to catch up now. What we propose here is absolute necessity of higher investment into innovative concepts, especially in the domain of natural sciences and technical universities. But, we also pin point that such investment should not be solely founded from the government but also by industry, following the US model, where industry companies invest highly in R&D.

4. CONCLUSION

Micro electromechanical devices, smart materials, products using nanocoatings – in this timing order – are representing long-term developments in a *new type of products* with a disruptive character for markets. These product challenges offer an opportunity for strengthening competitiveness, which can only be exploited if appropriate manufacturing equipment is available and allowing the use of the technologies in new products. Hence, generic technology development needs complementary manufacturing technology research.

These new technologies and knowledge have to be integrated into the production processes themselves. *Basic manufacturing research* has to foresee and prepare for new challenges, and *applied manufacturing research* has to adapt and transform existing technologies and organisational processes. Furthermore, manufacturing research plays a decisive role in combining the long-term horizon in technology trajectories with the short-term need of firms to innovate successfully. This requires a good "timing" of research activities to have solutions and tools ready for industrial adoption. Good news is that the European Commission, under the 7th research framework intends to invest 3% of GDP for R&D till 2010. That could greatly enhance research activities in Europe, which could in turn result in a better competitive position of Europe.

5. REFERRING

Benson, P., Hill, A. and Hoffmann, T. (1982), "Manufacturing systems of the future – a Delphi study", *Production & Inventory Management*, Vol. 23 No. 3, str. 87-105

Croatian national board for competitiveness: 55 propositions (2005), *Available from:* http://www.vlada.hr/Download/2004/04/13/55_preporuka-uvod-1-7.pdf, *Accessed: 2006-03-18*

European Scoreboard (2005); *Available from:* http://eu-iriscoreboard.jrc.es/docs/scoreboard_2005_key%20findings.pdf *Accessed: 2006-05-12*

National Academy of Sciences, (2004), "Visionary Manufacturing Challenges for 2020", *Available from:* http://books.nap.edu/execsumm_pdf/6314.pdf, *Accessed: 2006-05-12*

Shint T, (1998), Using Delphi for a Long-Range Technology Forecasting, and Assessing Directions of Future R&D Activities, The Korean Exercise, *Technological Forecasting and Social Change*, Vol. 58, 125-154

MANVIS Report No. 6, Policy Summary and Recommendations, *Available from:* http://efzg.globalnet.hr/UserDocsImages/OIM/oim_prester_PolicyPaper_final.pdf, *Accessed: 2006-05-12*

Annals of DAAAM for 2006 & Proceedings of the 17th International DAAAM Symposium, ISSN 1726-9679
Editor B. Katalinic, Published by DAAAM International, Vienna, Austria 2006
150th Birthday of Nikola Tesla (1856-1943) Inventor of Mobile Robot Teleoperation

IMPACT OF TRACTION CURRENT ON THE MONOBLOCK WHEEL OF RAILWAY TRACTION VEHICLES

GAVRILOVIC, B.

Abstract: *This paper provides replies to question of impact of traction current on the abutment of the monobloc k wheel and the railway track by electro locomotives of «Serbian Railways». The paper is analysing the impact of traction current on the abutment of the monoblock wheel and the railway track in transient running regimes of electrotraction vehicles that were designed after the start of trains.*

Key words: traction current, thermal strain, monoblock wheel, railway track.

1. INTRODUCTION

Thermal strains were observed in the monoblock wheel of thyristor locos series ZS 442-002 while the train was trailed along the Beograd – Bar railway track (Fig. 1).

Fig. 1. Thermal strains of the monoblock wheel of thyristor locos series ZS 442-002

The appearance of observed strains on to the monoblock wheel of thyristor locos series ZS 442-002 is better discerned than the apperance of thermal strains issued out of the slipper. Consequently, the folowing questions were put:

- Can the current in transient running regimes of electrotraction vehicles induce the observed strains on the monoblock wheel?
- How shall we be able to protect monoblock wheels from the observed strains?

The replies to these questions follow in the next analysis.

2. ELECTRIC HEATING ON THE ABUTMENT OF THE MONOBLOCK WHEEL AND THE RAILWAY TRACK

Each electric heating originates in the Joule's law. By this law, the generated thermal power in whichever segment of the monoblock wheel is equiavalent to the committed electric power (Vorsic & Bratina, 2000). Therefore, thermal power $P = \frac{dQ}{dt}$ $(\frac{W \cdot s}{s} = W)$ is equivalent to committed eletric power:

$$P = R \cdot I^2 = \frac{U^2}{R} \text{ (W)} \qquad \dots (1)$$

Where:

R - active electric resistance of monoblock wheel;

I - effectual value of current through observed segment of the monoblock wheel;

U - effectual value of tension at the observed segment of the monoblock wheel.

If we register current density J $(\frac{A}{m^2})$ instead of current intensity I (A) and electric field intensity E $(\frac{V}{m})$ instead of electric tension U (V), electric power density is stimulated by warmth at the observed segment of the monoblock wheel which can be defined with the next equation:

$$p_v = \rho \cdot J^2 = \frac{1}{\rho} \cdot E^2 \quad (\frac{W}{m^3}) \qquad \dots (2)$$

Where:

p_v - electric power density $(\frac{W}{m^3})$;

ρ - specific resistance of steel for the monoblock wheel ($\rho = 0{,}09 \cdot 10^{-6} - 0{,}015 \cdot 10^{-6}\ \Omega \cdot m$);

J - current density through the observed segment of the monoblock wheel $(\frac{A}{m^2})$;

E - electric field intensity at the observed segment of the monoblock wheel $(\frac{V}{m})$.

Based on the Joule's law, the thermal energy engendered at a unit of volume of the monoblock wheel is wasted with the progress of temperature and is defined with the next equation:

$$Q = c \cdot \rho_m \cdot (T_l - T_0) = \eta_{th} \cdot p_v \cdot t \qquad \dots (3)$$

Based on the equations (2) and (3), the difference of temperature at the observed segment of the monoblock wheel per unit of time is defined with the next equation:

$$\frac{T_l - T_0}{t} = \frac{p_v}{c \cdot \rho_m} \cdot \eta_{th} \qquad \dots (4)$$

$$\frac{T_l - T_0}{t} = \frac{\rho \cdot J^2}{c \cdot \rho_m} \cdot \eta_{th} \qquad \dots (5)$$

Where:

Q - thermal energy engendered at a unit of volume of the monoblock wheel $(W \cdot s)$, c - specific warmth of the monoblock wheel: $c = 660 \frac{J}{kg \cdot {}^{o}C}$;

ρ_m - specific density of the monoblock wheel: $\rho_m = 7{,}85 \cdot 10^3 \ (\frac{kg}{m^3})$;

T_1 - utter temperature heating of the monoblock wheel ;

T_0 - inchoate temperature heating of the monoblock wheel ;

η_{th} - coefficient of thermal operation: $\eta_{th} = 0{,}9$,

t - heating period (s).

2. HEATING SPEED OF THE MONOBLOCK WHEEL

The equation (5) serves to define heating speed on the abutment of the monoblock wheel and the railway track in stationary and transient running regimes of electrotraction vehicles. Nevertheless, we shall have to consider the possible peak of intensity of traction current, contact geometry of the monoblock wheel and the railway track, profile of the railway strip, scarp and width of the railway track. Besides, we shall have to regard the skin effect of alternative traction current on the abutment of the monoblock wheel and the railway track (Plazinic, 1997). Possible peaks of intensity of traction current in «Serbian Railways» were defined with tehnical characteristics of applied electrotraction vehicles, and profiles of the monoblock wheel and railway strip (type S45, S49 and UIC 60) with clauses of UIC-ORE S10-2, JZS V1.025 and Bulaw 314.

2.1 Impact of skin effect on the abutment of the monoblock wheel and the railway track

Abutment of the monoblock wheel and the railway track is defined with Herz's ellipse. For the profile of the monoblock wheel and the railway strip defined with clauses of UIC-ORE S10-2, JZS V1.025 and Bulaw 314, Herz's ellipse had the size $S \approx 360mm^2$ (Simic, 1997).

Traction current is complex- periodic function of time. Based on the results from literature (Gavrilovic, 2004), traction current of diodic locos series ZS 441 and ZS 461 is inclosed dispared harmonics of Fury series (i.e. harmonics of frequency: 50Hz,150Hz, 250Hz etc.). Traction current of tyristoric locos ZS 444 is inclosed pared and dispared harmonics of Fury series (i.e. harmonics of frequency:50Hz, 100Hz, 150Hz, 200Hz itd.). This traction current through Herz's ellipse tends to gather on the surface of ellipse and consequently the current density tends to become higher on the surface than close to the center of ellipse. If the number of harmonics in traction current thrives, the size of Herz's ellipse dwindles because of skin effect (Plazinic, 1997). Based on the results from literature (Plazinic, 1997), the size of Herz's ellipse of diodic locos series ZS 441 and ZS 461 dwindles within 15% from $S \approx 360mm^2$. The size of Herz's ellipse of tyristoric locos ZS 444 dwindles within 30% from $S \approx 360mm^2$ because these locos have got a bigger number of harmonics in traction current than diodic locos series ZS 441 and ZS 461 .

2.2 Transient traction regimes of electrotraction vehicles

The impact of peak limit of traction current through the observed segment of the monoblock wheel at its heating is of primary significance for our analysis. The value of this current depends on diverse series of the electrotraction vehicle. Besides, the peak limit of traction current was progressing when the electrotraction vehicle moved from the place of departure. Time of the observed abutment of the monoblock wheel and the railway track is the longest in this transient regime of the electrotraction vehicle. This time is dwindled with acceleration of the electrotraction vehicle.

Based on the equation (5), the speed of heating by peak limit of traction current at the observed segment of the monoblock wheel is defined

a) for tyristoric locos ZS 444 with the next equation:

$$\frac{T_1 - T_0}{t} = \frac{\rho \cdot J^2}{c \cdot \rho_m} \cdot \eta_{th} \approx 2462 \frac{{}^{0}C}{s} \qquad ...(6)$$

b) for diodic locos serie ZS 461 with the next equation:

$$\frac{T_1 - T_0}{t} = \frac{\rho \cdot J^2}{c \cdot \rho_m} \cdot \eta_{th} \approx 947 \frac{{}^{0}C}{s} \qquad ...(7)$$

Based on the equations (6) and (7), the difference in temperature at the observed segment of the monoblock wheel per unit of time in tyristoric locos ZS 444 has got an appreciably bigger value than the value in diodic locos serie ZS 461. Consequently, we shall be able to conclude that peak limit of traction current by tyristoric locos ZS 444 can induce the observed strains on the monoblock wheel (Fig 1). Thermal peeps on the monoblock wheel can be unfolded by the fact that the temperature at the observed abutment of the monoblock wheel and the railway track exceeded AC 1 and even AC3 limit. Consequently, the steel on the surface of the monoblock wheel was endangered with austerity transformation. Austerity zone of steel was wading in martensite when the peak limit of traction current did non ran through the observed segment of the monoblock wheel. Martensite was coming in contact with traction current repeatedly after a full circle round wheel when steel traversed in austerity zone. The observed transformation was caused by the reduction of volume of crystal structure of steel, strong thermal strain and micropeeps (Fig.1).

3. CONCLUSION

The difference of temperature at the observed segment of the monoblock wheel per unit of time by peak limit of traction current of tyristoric locos ZS 444 has got an appreciably bigger value then the value in diodic locos series ZS 441 and ZS 461. Consequently, we shall be able to conclude that the peak limit of traction current in tyristoric locos ZS 444 can induce the observed strains and micropeeps on the monoblock wheel

An optimally regulated overstress and overstream shield of electrotraction vehicles are vindicated of uncommited heating of the monoblock wheel. Besides, we shall have to assure a certain course of atmospheric current discharge near the ground.

4. REFERENCES

Vorsic, J.; Bratina, J.(2000): *»Elektrotermija«,* Univerza v Mariboru, Fakulteta za elektrotehniko, računalništvo in informatiko, ISBN 86-435-0414-9, Maribor

Plazinic, M. (1997)*: »Solution of skin effect by method loop«,* University of Nis, Faculty of Electrical Engineering, M.Sc. thesis, Nis

Simic, G. (1997): *»Research of geometry of contact between monoblock wheel and the railway track and its impact at abrasion of system wheel-strip«,* University of Belgrade, Faculty of Mehanical Engineering, Ph.D. thesis, Belgrade

Gavrilovic, B. (2004): *»Impact of traction current on track circuit used in railway signalling – protective devices«,* University of Kragujevac, Faculty of Engineering, Ph.D. thesis, Cacak

Milicevic, Z.; Arandelovic, D.**&** Marjanovic, Pejcic, V. (1999): *»Electro locos JZ 461-000/100«*, Želnid, ISBN 86-7307-085-6, Belgrade.

Milicevic, Z.; Arandelovic, D.**&** Marjanovic, Pejcic, V. (1997): *»Electro locos JZ 441«*, Želnid, ISBN 86-7307-046-5, Belgrade.

Annals of DAAAM for 2006 & Proceedings of the 17th International DAAAM Symposium, ISSN 1726-9679
Editor B. Katalinic, Published by DAAAM International, Vienna, Austria 2006
150th Birthday of Nikola Tesla (1856-1943) Inventor of Mobile Robot Teleoperation

GARMENT APPEARANCE QUALITY AS AESTHETIC FUNCTION OF CLOTHES

GERSAK, J. & ZAVEC PAVLINIC, D.

Abstract: *In the paper the detailed study of the garment appearance quality as garment aesthetic function is presented. On this basis, the intelligent system for prediction of the garment appearance quality was developed. Besides this the garment appearance quality is determined by garment drape, appropriate 3D shape of a garment, quality of the fit, seam appearance quality and garment appearance as a whole. Presented are preliminary research steps of estimation and evaluation of garment quality that were necessary for the development of the prediction system, which is user-friendly and accessible by world wide network.*
Keywords: *garment, fabric, quality, system, knowledge base, prediction*

1. INTRODUCTION

It is known that the processing properties, as well the mechanical and physical properties of fabric, each in their own way, results in the quality of the finished garment. Therefore their mutual dependence should be known as well their influence on the quality grade of the finished garment, i.e. on garment appearance. This is defined by garment drape, appropriate 3D shape of a garment, quality of the fit, seam appearance quality and garment appearance as a whole. The mentioned elements of the appearance quality are of the greatest significance for the evaluation of the aesthetical appearance of a garment.
While there is a connection between the aesthetical appearance and garment quality, the elements for garment appearance quality description should be determined in the first place (Geršak, 2003). Further, the influence of the sort and properties of the material incorporated into the clothing on the appearance quality grade of the finished garment must be investigated. As a result of those studies, the intelligent system for prediction of garment appearance quality, which is the first of its kind in the world, was developed.

2. GARMENT APPEARANCE QUALITY

One of the functions of a garment is to fulfil the desired fashion trend and the aesthetical appearance of the person who wears the clothing. The problem of how to evaluate the aesthetical garment appearance, respectively garment appearance quality, is presented here. Since there exist no standard that determines how to evaluate the appearance of a garment or the methods to be used for this purpose, the elements of garment appearance quality and the criteria for their evaluation must be defined on the basis of the mechanical and physical properties of the fabrics incorporated in the garment. The garment appearance quality is defined by garment drape (fall of the front/back part and sleeves, adjustment of different components and materials), appropriate 3D shape of a garment (quality of a form, its spatial characteristic, fullness, volume adjustment, shoulder and sleeve form), quality of the fit (fitting of the shoulder to the contour of the body, fitting of the front part and collar), seam appearance quality (seam puckering, seam flotation, shear deformed seam) and garment appearance as a whole (Geršak, 2001; Geršak, 2003, Zavec Pavlinić & Geršak, 2004; Zavec Pavlinić, 2005).

2.1 How to predict garment appearance quality?

The previously mentioned elements of garment appearance quality are complex for end-users to understand, while the quality of the clothing, respectively textile products, is non-uniform and non-reliable. How to objective evaluate and predict the non measure able quantities that describe the fabric hand (Kawabata, H.E.S.C., 1980), fabric behaviour (Kawabata et al., 1990; Kawabata et al., 1992; Zavec Pavlinić & Geršak, 2004a) and the appearance quality of high quality garments (Zavec Pavlinić & Geršak, 2004b) is the topic of numerous researches.
The Intelligent system for garment appearance prediction is presented in this contribution. The system is designed using machine learning methods, where the method of k - Nearest Neighbour (k-NN) was used. The algorithm of the k-NN method is implemented in the machine learning system Orange, developed by the Artificial Intelligence Laboratory, Faculty for Compter and Information Science, University of Ljubljana (Demšar & Zupan, 2004). The designing model of the system for prediction of garment appearance is shown on Figure 1.

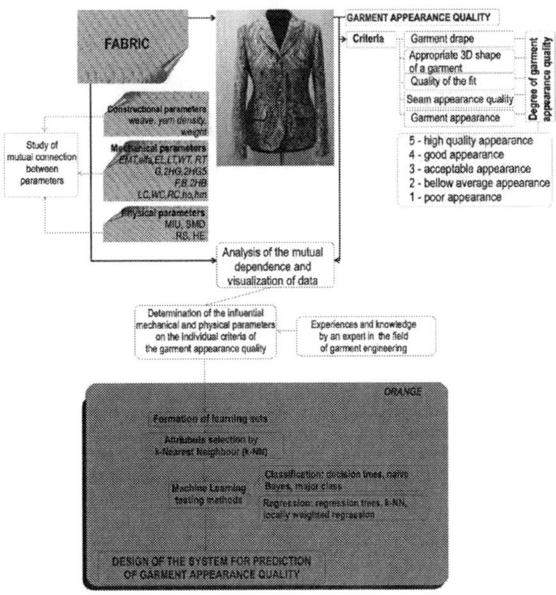

Fig. 1. The designing model of the system for prediction of garment appearance

The designed system works so that the user activates one of the stored data files, using the software installed on his/her computer, Figure 2. The software browses all the data files regarding the selected fabric, opens the appropriate network software and enters the data into the appropriate form, which is showed on the screen.

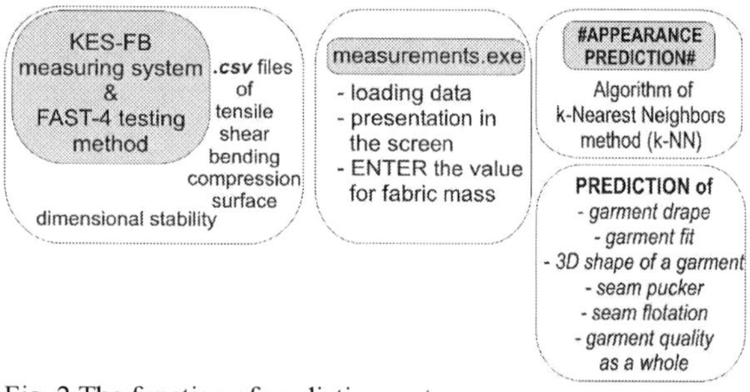

Fig. 2 The function of prediction system

The values of the parameters of the fabric's mechanical and physical properties are checked, and the value for the fabric mass is entered. Clicking the button »APPEARANCE PREDICTION« transfers the data to the processor which, using the Orange software package, with the algorithm of the k-Nearest Neighbour method incorporated, calculates the garment appearance grade prediction.

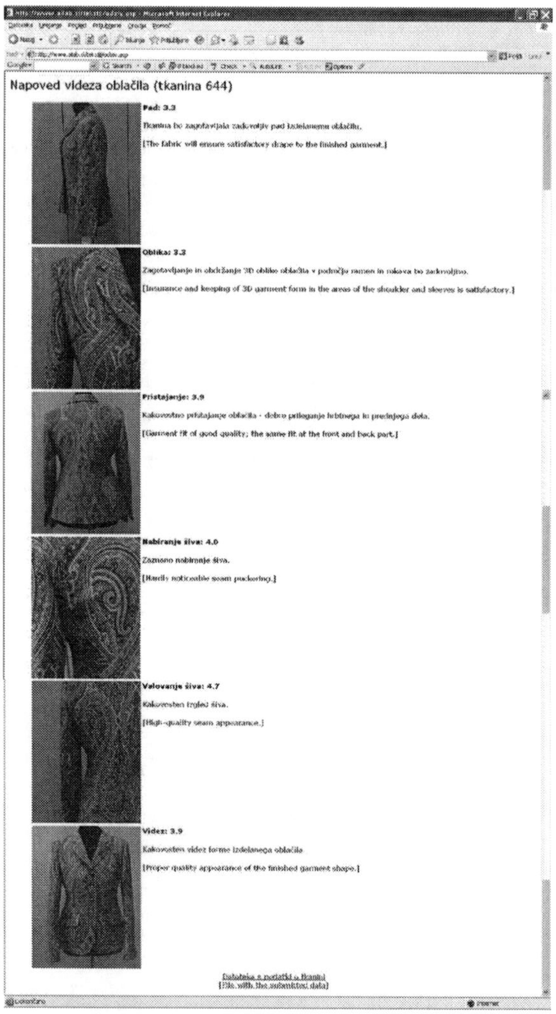

Fig. 3 The example of prediction

The prediction is shown as the quality degree of the five garment appearance quality factors (fall or drape, fit, 3D shape, seam puckering and flotation) and the overall appearance grade. The quality grade is given by estimation from 1 to 5, where grade 1 means poor quality and 5 high quality of the garment. All the criteria were given as descriptive too, and each description contains the appropriate picture view of the finished garment, Figure 3. The description and the picture views refer to the existing knowledge base.

3. CONCLUSION

The developed system enables, on the base of known mechanical and physical properties of the analyzed fabric (measured using KES-FB measuring system and FAST-4 testing method) to predict the quality grade of garment drape, garment fit, 3D shape, seam puckering and seam flotation, as well as the overall garment appearance grade.
The system application can be accessed through the world wide network, which ensures much broader use than any other system so far. The designed system is the first suchlike system in the world; it enables the objective prediction of the garment appearance quality grade and in this way replace, the conventional subjective methods of estimating garment appearance quality.
The system presents an important engineering tool for planning the quality parameters of mechanical and physical fabric properties and for the prediction of the garment appearance quality grade from their known properties. The applicable worth of the system is referred to the prediction of garment appearance quality for a concrete fabric in the real manufacturing processes and for the characterization of material for their specification needs by order as for the planning the finished garment quality grade.

4. LITERATURE

Demsar, J. & Zupan, B. (2004). Programski sistem Orange, http://magix.fri.uni-lj.si/orange/

Gersak, J. (2001). Development of The System for Qualitative Prediction of Garments Appearance, *Proceeding of the 30th Textile Research Symposium at Mt.Fuji*, (Ed.), pp. 135 – 144, ISBN, Fuji, August 2001, Shizuoka, Japan

Gersak, J. (2003). Istraživanje utjecaja mehaničkih svojstava tkanina na izgled odjeće. *Tekstil*, Vol. 52, No. 8, pp. 368-378, ISSN 0492-5882

Kawabata, S., Ito, K. & Masako, N. (1992). Tailoring Process Control, *Journal of the Textile Institute*, Vol. 83, No. 3, pp. 361-373, ISSN 0400-5000

Kawabata, S., Niwa, M. & Yamashita, Y. (1990). A Guide Line for Manufacturing »Ideal Fabrics«, *International Journal of Clothing Science and Technology*, Vol. 11, No. 2/3, pp. 134 - 141, ISSN 0955-6222 2000

Zavec Pavlinic, D. & Gersak, J. (2004). Vrednovanje kakvoće izgleda odjeće, *Tekstil*, Vol. 53, No. 10 (2004), pp. 497 – 542, ISSN 0492-5882

Zavec Pavlinic, D. (2005). *Napovedovanje kakovosti izgleda oblačil*, Doctoral dissertation, University of Maribor, Faculty of Mechanical engineering, Department of textiles, Maribor

Zavec Pavlinic, D., Demsar, J., Gersak, J. & Bratko, I. (2006). Predicting Seam Appearance Quality, *Textile Research Journal*, Vol. 76, No. 3 (2006), pp. 235-242, ISSN 0040-5175

Zavec Pavlinic, D., Gersak, J. (2004). Design of the System for Prediction of Fabric Behaviour in Garment Manufacturing Processes, *International Journal of Clothing Science and Technology*, Vol. 16, No. 1/2, pp. 252-261, ISSN 0955-6222

Annals of DAAAM for 2006 & Proceedings of the 17th International DAAAM Symposium, ISSN 1726-9679
B. Katalinic, Published by DAAAM International, Vienna, Austria 2006
150th Birthday of Nikola Tesla (1856-1943) Inventor of Mobile Robot Teleoperation

DYNAMIC BEHAVIOR OF THE MILLING HEAD ASSEMBLY USED FOR PROCESSING CURVED TEETH IN CYLINDRICAL GEARS

GHIONEA, A.; CONSTANTIN, G.; ANTONESCU, O. & GHIONEA, I.

Abstract: *The paper starts with a kinematic study of the simply planetary mechanism of the multi-cutter device adapted on a milling machine for processing curved teeth cylindrical gears. The means for reducing the generation errors are emphasized. For studying the kinematic and dynamic behavior of the designed device, its modeling and simulation were.*
Key words: *kinematic analysis, hypocycloidal trajectory, milling head, dynamic behavior, simulation.*

1. INTRODUCTION

The processing of the bulged teeth with curved flank lines in cylindrical gears or racks is achieved by milling using multi-cutters tools in mono-block construction or with applied cutters. In most cases, such type of teeth cannot be processed by grinding, hence this implies some careful conditions of milling and abrading. The tooth flak lines are arc of circle, cycloid or hypocycloid (Botez, 1969), symmetrically placed on the gear width. The flank lines are generated kinematically or as trajectory of a point. The involute profile is generated kinematically by rolling (Botez, 1969). The division is discontinuous or continuous. For the generation of the two curves that define the flanks, some conditions regarding the form and position errors are imposed. These determine the accurate achieving of meshing and contact surface of flanks, noise level decrease, increase of durability and gearing precision (Henriot, 1999). For this purpose, certain requirements are imposed: accurate adjustment of tool position in regard with the workpiece, elimination or diminishing of kinematic imprecision of mechanism in the machine structure, increasing of rigidity of clamping and driving systems of the tool and workpiece, improving of dynamic behavior especially during cutting process, adequate choice of the process parameters. The processing method is adapted on machines according to the applied technological method. Most of the cutting tools and methods used have on the basis patents (Ghionea & Oprean, 1995). One specifies data regarding the cutting edge construction, their adjustment in tool assembly, errors arising in processing and their causes, cutting regime parameters, dividing the layer to be processed, forces and moments in cutting, tooth examination and others.

In this the paper the kinematic scheme and dynamic behavior of a cutting tool of changeable milling head type used for processing of flanks of teeth with hypocycloidal lines are presented. The cutting tool is adapted on a machine for cylindrical gear tooth processing of FD Cugir family.

2. KINEMATICS OF THE ORDINARY PLANETARY MECHANISM

For generating kinematicaly a closed hypocycloid, a milling head was designed and achieved having on the basis an annulus ordinary planetary mechanism presented in Fig. 1. It consists of the planetary gear (rolling curve) with the radius R_R – 1, inner teeth gear (base) of radius R_B - 2 that is fixed, and the planetary carrier p. The ratio $R_R / R_B = i_H$ is an integer number.

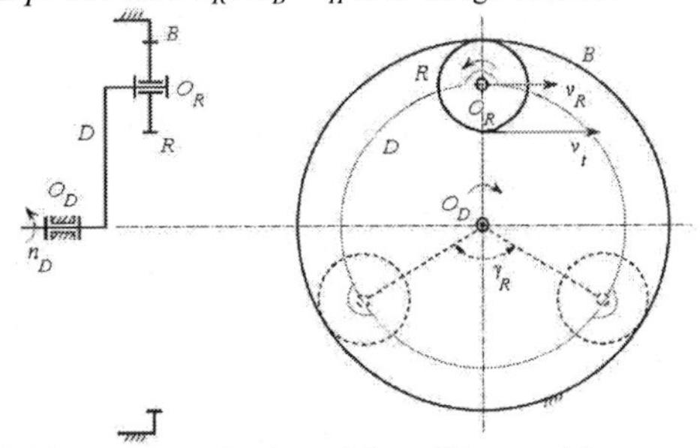

Fig. 1. Planetary mechanism of the milling machine

Analyzing this mechanism by both *analytical* and *graphical methods* (Antonescu 2005) with tangential and angular velocities $v_R = (R_B - R_R)\cdot\omega_D$, $v_t = 2R_R\cdot\omega_R = 2v_1$ where ω_D is disk angular velocity, in rad/s, it results:

$$n_R = \frac{R_B - R_R}{R_R}\cdot n_p, \qquad (2)$$

It was considered the case with gear modulus m = 2 mm. For certain values of n_D = 24; 30; 38 and 47.2 rpm and the ratio i_H / R_R = 3/80; 4/60; 5/48 and 6/40 mm we obtained the relative speed of rollers related to the disc D: n_{R_D} = 72, ..., 283 rpm.

Fig. 2 illustrates the generation of hypocycloidal trajectory of milling head.

On the basis of the vectorial equation of point the M_I position (Fig.2) represented by the milling tool in the orthogonal system $O_DX_DY_DZ_D$ expressed in a parametrical way, we obtain the current point coordinates:

$$\begin{aligned} X_D &= (R_B - R_R)\cdot\cos\varphi_D - R_R\cdot\cos\left(\frac{R_B - R_R}{R_R}\cdot\varphi_D\right), \\ Y_D &= (R_B - R_R)\cdot\sin\varphi_p + R_R\cdot\sin\left(\frac{R_B - R_R}{R_R}\cdot\varphi_D\right), \qquad (3) \\ Z_D &= 0. \end{aligned}$$

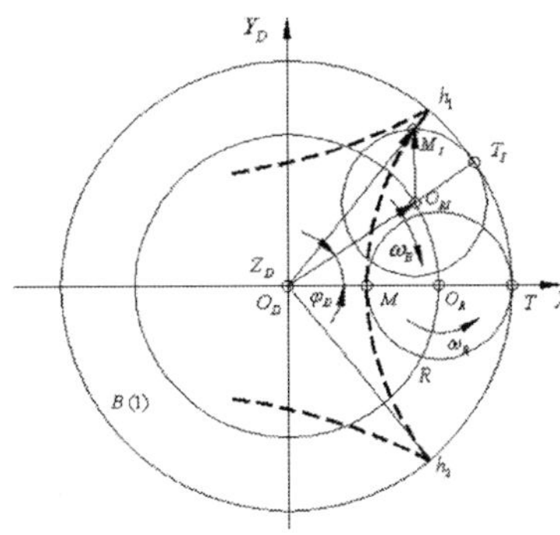

Fig. 2. Hypocycloid generation kinematics

where φ_D is the motion parameter of the rolling circle center. Two possible error cases were analyzed:

- If the rolling circle radius (R_R = 80 mm) has a small error and the real value is 79.9 mm, when R_B = 240 mm, the error on hypocycloid closing trace is 2.0 mm, that is unacceptable concerning the gear flank line shape;
- If the angular dimension between rollers (γ_R = 120 deg) has a small error of 0.1 deg, the angular error between hypocycloid traces will be about 0.111 deg, that is unacceptable concerning the gear flank line shape and pitch too.

These errors were chosen from a value set obtained by KOSIM software. Also, graphical simulations in many possible error cases were achieved.

For diminishing the errors in the design and fabrication of the milling head it was considered the precision level 6. Other constructive measures had in view the backlash reducing and creation of possibilities of compensating for the phase of the hypocycloid (Ghionea & Oprean, 1995).

3. MODELING AND SIMULATION OF THE DEVICE ADAPTED ON MACHINE FD 320 A - CUGIR

The milling head was designed and executed as an experimental model in four variants having i_H = 3, 4, 5, and 6 (Ghionea et al., 2001)). This enables processing gears (precision class 8) with diameters between 50 and 125 mm and modulus m = 1.5, ..., 3.5 mm.

For gearing running in gear testing of the processed cylindrical gearing with polyhypocycloidal teeth, a stand with open mechanical energetic flow was designed and used. The researches emphasized the formation and position of the contact spot and the level of noise. The milling head adapted on the tooth processing machine was modeled as a multi-body system in order to emphasize the kinematic and dynamic behavior of the assembly. All joints concerning the shafts are of revolution type (R). On the main spindle of machine (Fig. 3) a shaft I was mounted, which supports a bevel gear in contact with another one on a perpendicular shaft (driving spindle II, ratio 1:5). This is the shaft that moves the driving disk D carrying the satellite axes (III, IV, and V) on which the satellites rotates - spur gears in contact with the fixed crown. Together with the satellite axes there are the port-tools devices, on which the couples of cutters are mounted and adjusted in position to reduce the error factor.

The bevel gear meshing was modeled as a joint of rolling without sliding type (cone on cone for the bevel gearing or cylinder in cylinder for the cylindrical gearing). The model tree is an open chain. As input at the main spindle an imposed motion was supplied, which corresponds in rad/s to the chosen cutting velocity v_c = 58.3 m/min $\in$ (56.95 m/min, 72.1 m/min). This corresponds to a torque at the main spindle of 70 Nm. The cutting force was considered about 100 N acting about full contact position between tool and workpiece in the range (-3°, +3°).

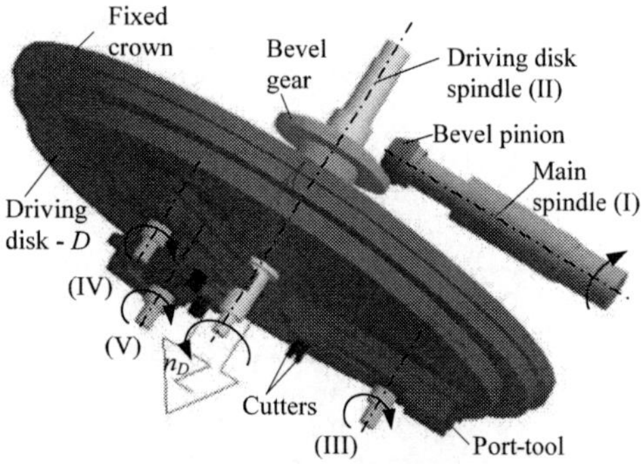

Fig. 3. Milling head with three groups of cutters

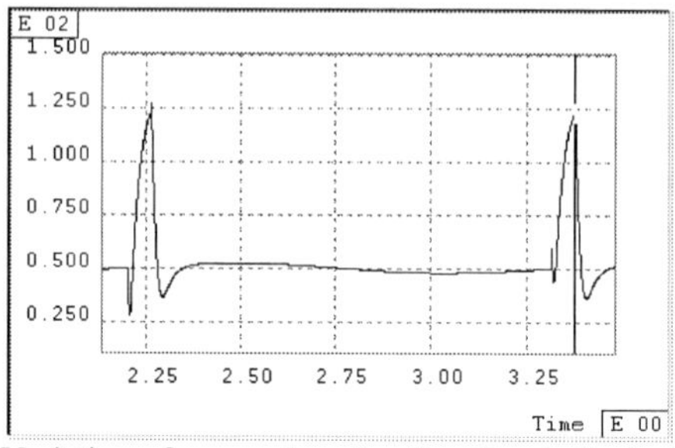

Fig. 4. Variation of moment on the disc D axis (body 1) at the cutting impact

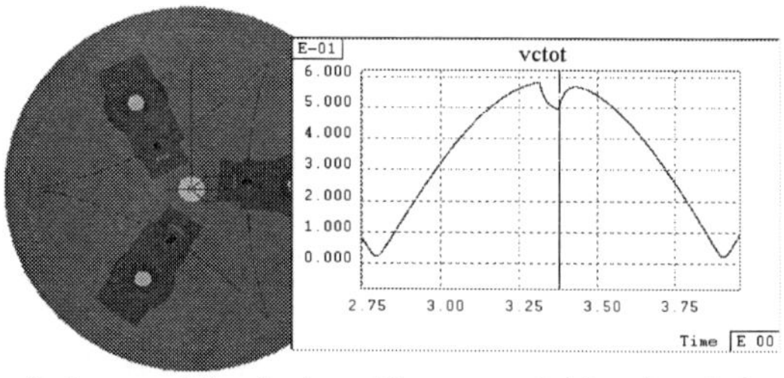

Fig. 5. Image of the device with gypocycloid and variation of cutting speed at the contact tool-workpiece

During simulation, the program provides information concerning positions, velocities, accelerations, point trajectories, the forces and moments applied to the articulations, the energies, as well as other data concerning the system, pre-defined by the software or defined by the user (Ionescu et al. 2003). In Fig. 4 the torque variation on the body 1 (disc D) at the cutting impact is shown. The maximum torque is 126.8 Nm, with an increase of 77 Nm. The variation of cutting speed at the cutting contact (*vctot*) is shown in Fig. 5.

9. CONCLUSIONS

The design of the multi-cutter milling head has on its basis a simple planetary mechanism of high accuracy. For avoiding the generation errors of the hypocycloidal trajectory, the soft KOSIM was used. For diminishing these errors some constructive and adjusting solutions were designed. The 3D model of the milling head was achieved in SolidDynamics programme for obtaining the kinematic and dynamic behaviour of the device under variable loads given by cutting force.

5. REFERENCES

Antonescu, P. (2005). *Mechanism and Machine Science*, Printech Press, ISBN 973-718-293-6, Bucharest, Romania

Botez, E. (1969). *Machine Tools. Fundamentals of Design*, Vol. I, *Cinematics*. Technical Publishing House, Bucharest.

Ghionea, A. (1987). *Contributions to the study of generation process of poyihypocycloidal teeth.* PhD Thesis, Politechnical Institutul of Bucharest.

Ghionea, A. & Oprean, A. (1995). Polyhypocycloidal milling head – *Romanian patent B23F21/12.10.95*, Bucharest.

Ghionea, A.; Constantin, G. & Ghionea, I. (2001). Kinematic structure of the machine tool for machining cylindrical gears with polyhypocycloidal teeth. *Proceedings of the 12th International DAAAM Symposium*, Katalinic, B. (Ed), pp 159-160, ISBN 3-901509-19-4, Vienna, Austria.

Henriot, G. (1999). *Engrenages. Conception. Fabrication. Mise en oeuvre*, 7th edition, Industry Techniques, DUNOD, ISBN 2 10 003903 2, Paris.

Ionescu, Fl., Choynovski, F. & G. Constantin. (2003). Modelling and Simulation of Solid Body Systems. *Annals of ARA 28 Annual Congress*, June 3-7, 2003, Târgu-Jiu, Romania.

Annals of DAAAM for 2006 & Proceedings of the 17th International DAAAM Symposium, ISSN 1726-9679
B. Katalinic, Published by DAAAM International, Vienna, Austria 2006
150th Birthday of Nikola Tesla (1856-1943) Inventor of Mobile Robot Teleoperation

ON VIRTUAL REALITY MODELLING OF AUTOMATED MANUFACTURING SYSTEMS

GOGOUVITIS, X. & VOSNIAKOS, G.-C.

Abstract: *The aim of this paper is to present a project regarding a conceptual framework for modelling automated manufacturing systems in a virtual reality context with embedded intelligence. The focus is on simulating performance of automated manufacturing systems, e.g. FMS, robotic cells, etc. to a level of detail set by the user-analyst. Main areas of research are: smart objects, dynamic simulation, collision detection, and web-based collaboration. The project will make use of semantic modelling, various algorithms, a number of integration tools, and will eventually be tested in three pilot applications.*
Key words: *manufacturing system, virtual reality, smart object*

1. INTRODUCTION

Using virtual reality applications for modelling manufacturing systems has been around for quite some time now, but a number of issues are yet to be examined and resolved (Sung & Ou, 2003). Building of a conceptual framework using both virtual reality tools and embedded intelligence, through smart objects, is an attempt to promote more flexible and accurate simulations of manufacturing systems. At the same time, the desire for better use of the World Wide Web for collaboration among partners is taken into consideration, thus the X3D format is used and extended, in order to accumulate a degree of standardization.

2. VIRTUAL ENVIRONMENTS

Pilot modelling of a robotic cell in VRML was conducted first in order to appreciate the capabilities offered by modelling environments not supporting intelligence of any kind, see Fig.1 Based on the experience acquired, some specifications for a modelling system with embedded intelligence were developed.

2.1 Smart objects

The primary object of research in a dynamic virtual world is the connection of various virtual objects of the world with rules or laws that govern their behaviour, as standalone objects or in connection to each other. Typical examples are kinematics and state transitions triggered by certain events. For instance, if an application aims at interactive modelling of a robotic world, methods that calculate the inverse kinematic transformation are needed in order to calculate the position of every joint of the robot that is compliant with the desirable position of the robot's end. Connection of such information with the objects of the virtual world is what grants them a form of intelligence.
Beyond that, modelling the behaviour of different objects can is associated with complicated state transitions, and/or stochastic changes that cannot be fully represented by normal automata is of interest, too (Pina et al, 2002). In this light, smart objects are introduced. The research field is vast, since different types of smart objects and relations/interactions between them need to be investigated.

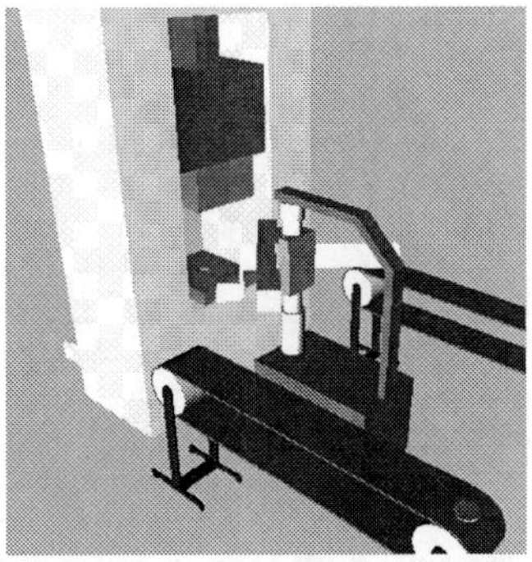

Fig. 1 Pilot modelling of robotic workcell in VRML

2.2 Dynamic simulations

The ability to interactively change the acting scenario in real-time in a virtual environments is of great importance. This, in effect, means that the scenario is not fully pre-decided and executed, but there are features that allow the user to alter scenarios, the simulation of movements is realized through a frame interpolation, where every frame is an instance of the system in a given time point. Therefore, simulation is pseudo-dynamic. In order to achieve a fully dynamic simulation, information regarding the interaction behaviour of objects should be expressed by rules/routines which would be executed in real-time. These options can be realized through the External Authoring Interface or the Script Authoring Interface of the X3D format.

2.3 Collision detection

Collision detection between a moving object and other moving or stationary objects inside a virtual world is necessary in most cases [Zachmann, 1997]. VRML/X3D has internal collision detection algorithms which, for instance,. do not allow the "observer" to pass through a solid object. If that "observer" was connected to the end of a robotic arm, the collision of that arm with any object could easily be detected. Collision detection is based on a bounding box that is connected to every object, but that approach is rather simplistic, especially if the object has a complicated morphology.

2.4 Web-based collaboration

Collaborative development of virtual worlds and scenarios through the World Wide Web, in a way similar to online gaming, is also of great importance. It is generally accepted that a manufacturing system cannot be designed by only one expert, but the cooperation of many is required. These experts need to communicate and the knowledge that each one of them has to contribute into the system is not independent, but is in connection with the information supplied by the others (Shang et al, 2003).
Therefore, procedures that control the information flow and secure parallel or sequential attributes are under examination.

3. PROJECT FRAMEWORK

The desired results of the project are:

- A methodology for the initial design of automated programmable systems.
- A methodology for choosing a formalism for representing knowledge and integrating intelligence into smart objects.
- Software that implements these concepts in a web environment.
- Three pilot applications that present the functionality and mode of using the software developed.

3.1 Semantic modelling

In order to describe the working framework of the smart objects inside the virtual world, a number of semantic models need to be developed. The different kinds of behaviour and their attributes are a main part of this research. These models will, among other things, integrate kinematic and behaviour attributes in a collision. These models will be in full accordance to the X3D format and will differentiate according to whether the model is executed as a stand-alone or in a web environment.

In addition, the formalism of the representation of the objects is of great importance. The use of ontologies (Uschold et al, 1996) as a base system for controlling the objects and their interactions is to be adopted. High level ontologies will be used in order to assign relation attributes to the objects, while low level ontologies will be used in order to provide general and special attributes, such as position, kinematics, interaction to/with the user, collision, deformation etc.

The use of agents will also be considered, in terms of collaborative environments.

3.2 Algorithms

A number of algorithms is needed in order to realize the intelligent behaviour of the objects. Examples of such algorithms include:

- Automatic creation of geometrical model of an object from CAD into X3D, including different LODs and possible parameterization of the modelling.
- Adaptation/development of collision detection algorithms, with special interest in "light" but adequate algorithms for use through a web environment.
- Calculation of direct and inverse kinematic transformation, in order to calculate the positions of the object's parts while moving from one position to another.

Most of these algorithms will be imported from existing libraries, but some should be developed, as necessary. In addition, dynamic algorithms will be developed in order to provide more realism in the representation of certain objects.

3.3 Integration tools

At first, a categorization of the different types of knowledge, that is needed to support the design of automated programmable systems, is to take place. The categorization needs to be done in a convenient way for their use through software in order to enhance the systems' objects with intelligence. Knowledge types will be addressed in full compliance with the semantic modelling of the objects. Next, tools will be developed for the proper representation of these types. The realization of the tools will be done taking into consideration usability in both a web-based environment and in stand-alone applications. The tools that will be examined in order to realize the knowledge are:

- Petri nets, which is a useful and powerful method for modelling discrete event systems.
- Expert systems, which is a proven method for the representation of empirical knowledge.
- Neural networks, which is fast in representing knowledge accumulated from simulations or experiments.
- Fuzzy logic systems, for representing ambiguity.

3.4 Pilot applications

The project aims at building three pilot applications, specifically:

- CNC machine design
- Robotic cell programming
- FMS analysis

The first pilot application refers to an environment for machine design. A press brake will be designed in cooperation with a partner company and it will include a three-axis kinematic chain with simple control. The main focus will be the morphological attributes of the machine and the control of its kinematics and dynamics, as well as the simulation of a CNC program for cutting and bending of a simple product. Also of interest is the use of the model in order to evaluate accuracy of the movements and the effect to the product itself.

The second application refers to an environment for programming robots, focusing on consistency of kinematics and probably dynamics of the model. It will also check the accuracy of the collision detection algorithms, as well as the control of the logical sequence of events. The application will be based on a robotic cell that a partner company is installing.

The third application refers to an environment that will check all the attributes of intelligence of the objects, with respect to the complexity of the manufacturing system that is to be modelled. The manufacturing system will consist of a number of CNC machines, as in another partner company.

4. CONCLUSION

This project aims at examining the usage of virtual environments for a variety of manufacturing subjects. As the demand for faster, lighter, and more accurate simulations of manufacturing systems is increasing, so does the research interest in that field. Intelligence embedding is the key concept of the current project, emphasizing use of 'standard' formats.

5. ACKNOWLEDGMENT

This work is supported by General Secretariat of Research and Technology under project number PENED2003 03EΔ217.

6. REFERENCES

Pina, A; Seron, F.J. & Gutierez, D. (2002). The ALVW system: an interface for smart behavior-based 3D computer animation, *Proceedings of the 2nd International Symposium on Smart Graphics,* 17-20, ISBN 1-58113-555-6, Hathorne, NY, USA, June 2002, ACM Press.

Shang, H.; Zhao, Z. & Thorn, R. (2003). Implementing Manufacturing Message Specifications (MMS) within collaborative virtual environments over the Internet, *International Journal of Computer Integrated Manufacturing*, Vol. 16, No. 2, 112-127, ISSN 0951-192X

Sung, W.-T. & Ou, S.-C. (2003), Using virtual reality technologies for manufacturing applications, *International Journal of Computer Applications in Technology*, Vol. 17, No. 4, 213-219, ISSN 0952-8091

Uschold, M. & Gruninger, R. (1996). Ontologies: principles, methods and applications, *The Knowledge Engineering Review*, Vol. 11, No. 2, 93-136, ISSN 0269-8889

Zachmann, G. (1997). Rapid collision detection by dynamically aligned DOP-trees, *Proceedings of the Virtual Reality Annual International Symposium,* pp. 90-97, ISBN 0-8186-8362-7, Atlanta, USA, March 1998, IEEE.

Annals of DAAAM for 2006 & Proceedings of the 17th International DAAAM Symposium, ISSN 1726-9679
B. Katalinic, Published by DAAAM International, Vienna, Austria 2006
150th Birthday of Nikola Tesla (1856-1943) Inventor of Mobile Robot Teleoperation

TRAVERSE DIRECTION INFLUENCE EVALUATION ON AVERAGE ROUGHNESS AT AWJ CUTTING

GOMBAR, M.; HLOCH, S. & RADVANSKA, A.

Abstract: *Article deals with the experimental research of solid phase distribution (abrasives) influence in waterjet to quality of surface roughness measured in the depth of 9 mm in 10 mm thick cast aluminium. As an experimental method full factorial design was applied to study the effects of submitted selected factors such as abrasive mass flow rate [g.min^{-1}], traverse speed [mm.min^{-1}], traverse direction [°], and pressure [MPa] on the surface profile parameter the average roughness pressure [μm] as a dependent variable. It has been found that new factor submitted to the experimental study influences significantly the surface profile parameter – average roughness in soft material cutting with connection to the solid phase distribution in waterjet.*

Key words: *waterjet, cast aluminium, traverse direction*

1. INTRODUCTION

With the development of technology, the scientists and the technologists in the field of manufacturing are facing more and more challenging problems. The demand for the highest accuracy and surface finish, the challenge to produce critical surfaces and complex shapes have necessitated the use of non-traditional machining techniques such as abrasive waterjet machining (AWJ) (Valicek et al., 2005). AWJ technology has greatly altered the tooling and manufacturing industry, resulting in the dramatic improvement in accuracy, quality, and productivity. Presently, this AWJ cutting process is being used for many applications. However, such techniques are not favourably nourished due to the difficulties and complexities involved in setting their process factors which enters to the cutting process date (Hires, 2004).

2. RELATED AND PREVIOUS WORKS

The updated model of AWJ cutting presented in (Hloch 2006) contains new factors - traverse direction and abrasive feeding direction. The influence of these factors has not been exactly explained. It is assumed that these factors cause the asymmetry roughness values due to the feeding direction of solid phase, distribution of abrasive particles in the waterjet, and traverse direction. The quality of surface roughness depends on AWJ tool and velocity conditions and structure that were observed (Jurisevic et al., 2005) by high-speed photography to measure the velocity conditions of the AWJ. It has been found by means of Laser Doppler Anemometry (LDA) (Chen & Siores 2001) that the kinetic energy distribution of the abrasive particles in the AWJ has the double slope distribution shape. This method has been used by (Jurisevic et al., 2005) to measure velocities of abrasive particles and water droplets. Separation methods were used by (Jurisevic et al., 2005) to define distribution in the AWJ radial direction. In previous works it has been found that distribution of abrasive mass flow rate may cause the asymmetry of roughness values. Traverse direction has been added to the experiment to explain the significance with connection of the selected factors that influence the quality of the hydrodynamic action of the abrasive waterjet (Hloch et al., 2005).

3. EXPERIMENTAL METHOD AND SET UP

For investigation of the influence of the traverse rate and the solid phase distribution in waterjet the samples created for this purpose have been cut in two directions +180° and −180°. Traverse direction has been submitted to the experiment to explain the significance with interaction of the next selected factors. In order to investigate the influence of above mentioned abrasive waterjet process factors on surface profile parameter average roughness Ra, the two full factorial designs for four independent variables has been designed. Full factorial analyses were used to obtain the combinations of values that can optimize the response, which allows to design a minimal number of experimental runs (Montgomery, 2001), (Toutenburg, 2002). Among the many process variables that influence the cutting results, four factors have been selected and considered as factors in the experimental phase. Four factors – independent variables submitted for the analysis for the two factorial designs of each constituent at levels [–1; +1] are listed in the Table 1.

Factors		Factor level	
Var	Terminology and dimension	-1	+1
x_1	Abrasive mass flow rate [g.min^{-1}]	200	400
x_2	Pressure [MPa]	300	350
x_3	Traverse rate [mm.min^{-1}]	100	200
x_4	Traverse direction [°]	-180	180

Table 1 Coded independent variables at defined levels

The experiments were carried out based on the analysis using Statistica 7.0 and Matlab to estimate the responses of the dependent variables. A two dimensional abrasive waterjet machine Wating was used with following specification: work table x-axis 2000 mm, y-axis 3000 mm, z-axis discrete motion, with maximum traverse rate of 250 mm.s^{-1}. The high-pressure intensifier pump Ingersoll-Rand Streamline model with maximum pressure of 380 MPa was used. An Autoline cutting head from Ingersoll-Rand has been used with two focusing tubes of 1 mm and 1.2 mm diameters. The target material was cast aluminium, the properties of each sample were: length 35 mm, width 8 mm, and height 10 mm. A digital surftest Mitutoyo 301 has been used to calculate the average roughness with 0.01 mm measurement precision.

4. RESULTS AND DISCUSSION

The quantitative description of the conditions effects on average roughness was performed. The results were analyzed using the analysis of variance as appropriate to the experimental design used. The normality of experimentally measured data has been tested according to Shapiro-Wilkson test. The significance of factors is interpreted in the Pareto Charts (Fig. 1, 2), which graphically displays the magnitudes of the effects from the obtained results. The Pareto chart shows the significance of the traverse feed in affecting the average roughness.

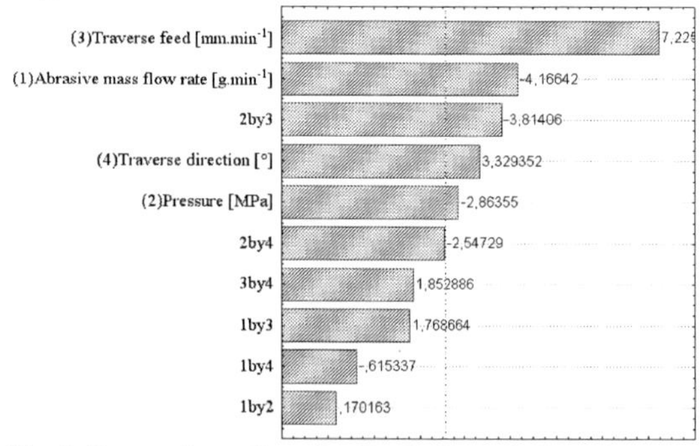

Fig. 1. Pareto chart of standardized effects (ϕ f_t = 1.2mm)

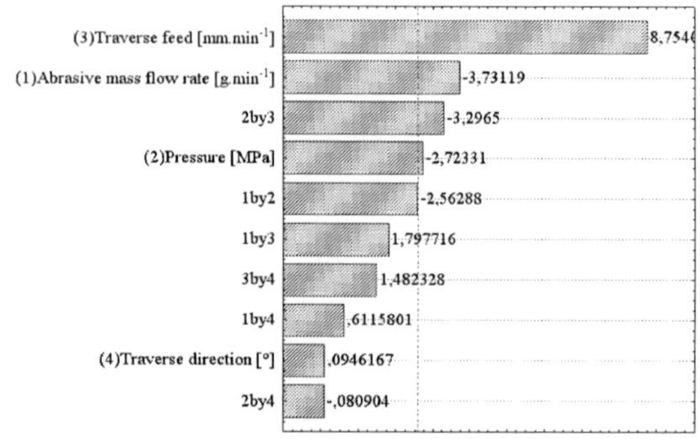

Fig. 2. Pareto chart of standardized effects (ϕ f_t = 1mm)

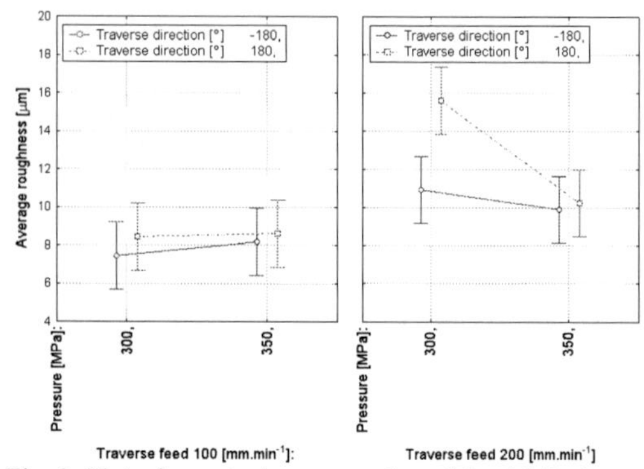

Fig. 3. Plot of marginal means and confidential limits

With the raising traverse feed and depth Ra is affected mainly by pressure abrasive mass flow rate. This phenomenon is connected with the deformation of the abrasive waterjet as a tool. Traverse direction was significant only in the case (Fig. 1), when the focusing tube of 1.2 mm diameter was used. When the focusing tube with diameter 1mm was used the significance of traverse direction conspicuously decreased. It is assumed that numeric values of Ra are affected by traverse direction that is significant at the high diameter of focusing tube. The influence of the traverse direction is connected with the abrasive particle concentration distribution in the jet; it was not uniform and a peak appeared around 1/3–2/3 of the radius of the jet (Chen & Siores 2001). The J/T abbreviation and pressure belongs to the hydrodynamic factors that create the AWJ tool. Generated tool through these factors enters to the cutting technology process of material at the large number of localities, by means of direct factors. Through cutting factors, created tool hits the workpiece at upper erosion base, where erosion process begins. These factors create surface area as the area of abrasive waterjet trajectory. It is the specific way of material machining because particles with more edges are used; those cutting edges are made by abrasive particles that are random oriented in the waterjet liquid phase. This random position and different shape of abrasive particles causes irregular removal mechanism of material and may cause the asymmetry of the roughness values due to the feeding direction of solid phase that is connected also with wavy particle distribution and consequent distribution of abrasive particles in the waterjet. Therefore, the surface quality improves with increasing pressure and smaller diameter because in such manner the abrasive water jet disposes higher energy concentrated to smaller area of the workpiece. Machined surface is created by a trajectory of working abrasive waterjet tool movement. There is a difficulty and specificity of such way of machining, as particles with a lot of clines ordered irregularly at the stream are applied in a process.

5. CONCLUSION

The problem analysed in these pages is the study of the influence of solid phase distribution in waterjet to quality of surface roughness. Full factorial design has been used as a statistical method to study the effects of selected factors. Experiments also proved an asymmetry of the roughness values Ra due to the feeding direction of solid phase - abrasive at the entry of the mixing chamber at the rate of speed 100 mm^{-1} at the depth of 1 mm. Observations proved the relationship between the influence of the orifice diameter and traverse rate related to solid phase distribution on average roughness. Exact conclusions will be done at another experiment where all AWJ factors will be evaluated according to Taguchi design.

6. REFERENCES

Hires, O. The study of cutting surfaces quality. In: *Funkcne povrchy 2004*: Trencin, 2004. Trencianska univerzita, 2004. ISBN 80-8075-021-1.

Montgomery, D., C. *Design and analysis of experiments. 5th. edition*, Hamilton Printing Company, 2001. ISBN 0-471-31649-0.

Toutenburg H. *Statistical analysis of designed experiments*. Springer-Verlag New York, Inc. 2002, ISBN 0-387-98789-4.

Chen, F. L.; Siores, E. The effect of cutting jet variation on striation formation in abrasive water jet cutting, *International Journal of Machine Tools & Manufacture* 41 (2001) 1479–1486.

Jurisevic, B., Coray, P. S., Heiniger, K. C., Junkar, M.: Tool Formation Process in Abrasive Water Jet Machining. In: *Management of Innovative Technologies MIT'2003* – post-conference edition. Ljubljana: University of Ljubljana. 2003. s. 73–85.

Valicek, J., Madr, V., Drzdik, M., Luptak, M., Slivecka, L. Method of the determination of superposition relation of changes in surface geometry to mechanical parameters of material due to corrosion. In. *Acta Mechanica Slovaca*. 3-A/2005. ISSN 1335-2393. p.171-176.

Hloch, S., et al. Factor analysis of abrasive waterjet process factors influencing the cast aluminum surface roughness. In. ICOMAST 2006, Melaka, Malaysia., ISBN 983-42051-1-2

The authors would like to acknowledge the support of Scientific Grant Agency of the Ministry of Education of Slovak Republic, Commission of mechanical engineering, metallurgy and material engineering, for their contribution to project 1/1095/04 - Optimization of abrasive waterjet cutting technology process, by mathematical and experimental planning.

Annals of DAAAM for 2006 & Proceedings of the 17th International DAAAM Symposium, ISSN 1726-9679
B. Katalinic, Published by DAAAM International, Vienna, Austria 2006
150th Birthday of Nikola Tesla (1856-1943) Inventor of Mobile Robot Teleoperation

ABRASIVE WATERJET CUTTING PROCESS RISK ANALYSIS

GOMBAR, M.; RADVANSKA, A. & HLOCH, S.

Abstract: *The paper deals with the evaluation of safety and noise level of technological system with abrasive waterjet cutting technology. The safety has been evaluated by means of Failure Modes and Effects Assessment Method. Noise as a negative phenomenon of abrasive waterjet cutting has been evaluated by design of the experiment. Further, the manufacturing system with abrasive waterjet machining and the cutting process was evaluated. Results show the abrasive waterjet factors significance and their effect to the noise environment. It has been found that the most significant is the focusing tube diameter.*
Key words: *abrasive waterjet, safety, risk analysis, noise.*

1. INTRODUCTION

Abrasive waterjet cutting machining (AWJM) represents cold, precise, computer controlled shape cutting without any strain. AWJM is for to-date high requirements on quality and productivity applied in full-automated workplaces with automatic CNC control. Flexible and smart automated technique application does not exclude the human being from the manufacturing process; just moves his working activities from strenuous jobs and jobs in malign environment to the areas of control, maintenance and operation management.

2. AWJ PRODUCTION SYSTEM

The following Figure 1 shows the manufacturing systems and subsystems of abrasive waterjet machining technology.

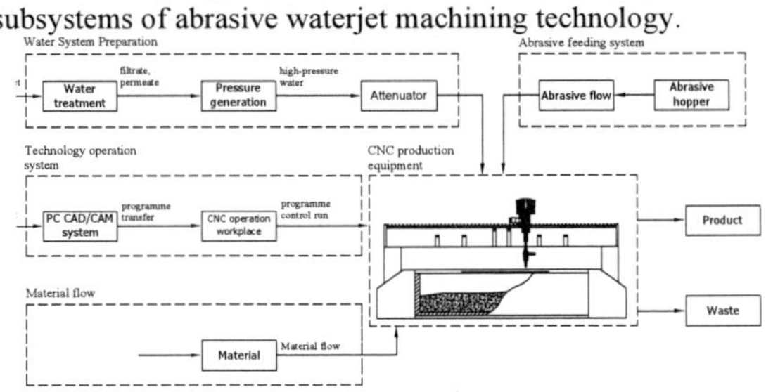

Fig. 1 Division of assessed subsystems of AJW production system

Manufacturing system of abrasive waterjet cutting consists of subsystems, through the cutting tool is created. Waterjet with purity of redistilled water flows from treatment system and enters the system where pressure is generated and flows through the attenuator to the cutting head. In cutting process high pressurized water is transformed by orifice to the high-speed water.

3. AWJ TECHNOLOGY PROCESS RISK EVALUATION

As a risk assessment method the Failure Mode and Effects Assessment (FMEA) method was used to evaluate the risk size. The method is semi-quantitative and evaluates the accident probability (P) and effects (E) seriousness and can be expressed as a combination of both factors as follows:

$$R = PxE \tag{1}$$

The total risk (R) is expressed in risk matrix listed in (Sinay, 1997).

Low pressure circuit		
Danger source	Danger	Effect
Electrical system, Pressure accessories, Hydraulic fluid.	Electric isolation damage, Fluid outflow, Fluid outflow in form of aerosol, Part break.	- falling, - killing, - bruising, - poisoning, - lungs damage.
High pressure circuit		
Danger source	Danger	Effect
High pressure accessories, High pressure fluid (permeate)	Damage of tubes, Outflow of permeate, Hit of operator by high pressure tubes breakage	- killing, - falling, - cutting, - amputation, - bruising.
Filtration system		
Danger source	Danger	Effect
Deficient of water treatment, Impurities sedimentation,	Mechanical danger, Damage of device	- killing, - cutting, - bruising.
Abrasive feeding system		
Danger source	Danger	Effect
Abrasive hopper, Abrasive flow, Recycling system	Impurities penetration, damage of abrasive cutting head.	- Production losses
Material machining		
Danger source	Danger	Effect
Abrasive waterjet Workpiece	Abrasive particle rebounding Abrasive waterjet rebounding Noise Workpiece fall, sharp edge, Physical and chemical composition of workpiece	- amputation, - hearing damage, - cutting, - lungs damage, - poisoning.
Operation system		
Danger source	Danger	Effect
Operation system, Manipulation system, Motion system, Abrasive waterjet.	Collision with manipulation and motion system, Affection of abrasive waterjet to operator, Dispersion of abrasive waterjet.	- amputation, - hearing damage, - cutting, - lungs damage, - poisoning.

Table 1. Overview of possible dangers and their effects

Abrasive waterjet machining contributes to the manufacturing process by number of incipient dangers, which mostly contribute to the risk origin of the AWJ technology system operation. Effects of possible personal accidents are in general considered to be most serious; therefore the assessment of the possible risk is necessary. For general assessment of the defect cause it is necessary to name the systems that can significantly influence the work safety at technology of abrasive waterjet by their operational properties and performance function. Each estimated system is assessed independently (Tab. 1). At the selection of assessed systems not only common working process is taken into account, but also some exceptional activities. Very important part of the method is estimation of assessed systems parameters. The two elements, high speed waterjet and workpiece take part at the cutting process. In manufacturing automation of the AWJ cutting technology a number of factors influences the creation of defects and risks. In the Table 1 the risks for each subsystem of AWJ technology are assessed. Basing on the properties of evaluated systems risk size has been estimated according to FMEA method (Tab. 2).

Danger	Accident probability	Effect seriousness	Risk size	Verbal expression of risk size
Electric current	D	I	8	Undesirable
Falling of persons	A	III	7	Undesirable
Water pollution	E	II	15	Acceptable with inspection
Oil dispersion inhalation	D	II	10	Undesirable
Hit by AWJ	C	II	6	Undesirable
Noise	A	III	7	Undesirable
Sharp edges	A	III	7	Undesirable

Table 2. Qualitative risk expression of the assessed system

5. NOISE EVALUATION

The AWJ velocity reaches up to 1000 $m.s^{-1}$ and disposes the very high kinetic energy thus it is the source of the majority of risks. At AWJ focusing tube outlet, where the elasticity interface is, jet interferes with the outer environment molecules.

Fig. 2. Abrasive waterjet rebounding at the cutting of 20 mm thick stainless steel AISI 304 (Wating, Ltd.)

This induces the elements oscillation and results from the energy change that forms the acoustic field. This is manifested by noise, predominantly of high frequency, which has a negative impact on central nervous system of the CNC motion operator. To reduce this negative phenomenon and enhance operators' safety it is necessary to recognize the most potential noise sources in water jet machining system. Results show the abrasive waterjet machining factors significance and their effect to noise environment. Abrasive waterjet manufacturing system consists of subsystems, by which the initial tool is created. Technological cutting process by hydroabrazive erosion is performed by means of cutting tool – abrasive waterjet – properties of which are not reduced due to the operation unlike it is at the conventional cutting knife. Noise is initially produced by vibration of solid objects, by turbulent motion of fluids, by explosive expansion of gases, or by other means. The pressures, amplitudes, and velocities of the components of the sound wave within the range of hearing are quite small. With a proper catcher, the abrasive waterjet is as loud as 90 dB (Hloch, et al, 2005). In the open air the noise level can reach 110 dB or more (Slattery 1991). Another disadvantage is mixing nozzle wear - the mixing nozzle needs to be replaced every 2-6 hours depending on the abrasive type being used and abrasive flow rate (Hitchox, 1986). Machined material participates on danger origin by many ways. Here belongs target material weight, its chemical composition and physical properties, material handling. Danger may be caused at material handling before or after manufacturing operations. The next dangers arise during the cutting process itself, at improper material placing or work clamping. Attention has to be given to the chemical composition of the material. The impact of dispersed particles to organism is different in case of allergens. The effects of the particles to the skin is different too, if the material is brittle or ductile. The possible dangers are drop of the workpiece, cuts at material handling, dispersed particles inhalation.

6. CONCLUSION

Abrasive waterjet technology risk analysis proved the existence of numerous dangers that expose the operators at workplace. Except of acceptable risk the FMEA has shown the existence of undesirable and unacceptable risks. According to risk assessment, the most severe are risks caused by personal protective equipment omission, workplace noise, waterjet and rebounded abrasive particles hitting. Undesirable and unacceptable residual risks have to be delimitated or eliminated in given working system to acceptable level and it is necessary to create acceptable working conditions for stuff operators and hence raise the effectiveness of AWJ technological devices, their reliability, working safety. At the experiment from the occupational safety and health point of view, the limits of acoustic sound pressure level were exceeded. For audio frequency noise the overrun was 7 dB, for high frequency noise up to 26 dB. The noise elimination at abrasive waterjet machining can be achieved by reduction of the sources of acoustic sound pressure level.

7. REFERENCES

Hitchox, Alan L., "Vote of Confidence for Abrasive Waterjet Cutting", *Metal Progress*, Vol. 130, July 1986, pp. 33-34.

Hloch, S. Fabian, S. Rimár, M.: Design of Experiments Applied on Abrasive Waterjet Factors Sensitivity Identification. In: *Nonconventional Technologies Review.* 1454-3087 2005. 119 - 126.

Hloch. S., Fabian, S., Radvanská. A., Ragan. E. Acoustic environment evaluation of manufacturing system with abrasive waterjet technology In. *Advances in Materials, Product Design & Manufacturing Systems 2005,* Bannari Amman Institute of Technology, Sathyamangalm, Tamil Nadu, India,

Radvanská, A., Hloch, S. Riziká technológie delenia vysokorýchlostným hydroabrazívnym prúdom. In: *Bezpečná práca.* roč. 36, č. 2 (2005), s. 7-12.

Slattery, Thomas J., "Abrasive Waterjet Carves Out Metalworking Niche", *Machine and Tool Blue Book*, July 1991, pp. 31-34.,

Sinay, J.: *Riziká technických zariadení*, OTA, Košice 1997, ISBN 80-967783-0-7,

The authors would like to acknowledge the support of Scientific Grant Agency of the Ministry of Education of Slovak Republic, Commission of mechanical engineering, metallurgy and material engineering, for their contribution to project 1/1095/04 - Optimization of abrasive waterjet cutting technology process, by mathematical and experimental planning.

Annals of DAAAM for 2006 & Proceedings of the 17th International DAAAM Symposium, ISSN 1726-9679
B. Katalinic, Published by DAAAM International, Vienna, Austria 2006
150th Birthday of Nikola Tesla (1856-1943) Inventor of Mobile Robot Teleoperation

THE IMPORTANCE OF OPTICAL DETECTION OF SURFACE QUALITY CREATED BY ABRASIVE WATERJET

GOMBAR, M.; VALICEK, J.; HLOCH, S. & RADVANSKA, A.

Abstract: *The paper deals with a newly developed optical no-contact measurement method and its importance in comparison with classical measurement by contact profilometer used for quality measurement of the surfaces created by abrasive waterjet technology. The main emphasis is put on the analysis of results for defining the process of creation of a new surface generated by the stream of abrasive water jet, including its geometric parameters and mechanisms of cutting tool-material interaction. New possibilities for surface quality evaluation and optimization, the choice of technological parameters of the cutting process are developing.*

Key words: *abrasive waterjet, optical measurement, surface quality*

1. INTRODUCTION

From the manufacturing and technical point of view, we are interested in the technological property of work pieces. This property influences the durability and reliability of components, (waste of energy), wear resistance and the economics of machines and equipment (Hloch 2005). The scale of the basic metrology operation is defined by the standards determining the required quality class. These operations lead to correct results for surfaces created by traditional technologies. The physical basis of new surfaces formation by non-conventional technologies is different; therefore the geometrical characteristics of the created surface will differ (Valicek, et al, 2001). Good agreement has been confirmed by comparison of the results obtained using the newly developed measurement method with the commercial method. The principle of the measurement shadow method (SM), design and optical scheme adjusted for measurement of cutting surfaces by abrasive waterjet have been published (Bumbalek, et al, 1989), (Brezina 1991). Knowledge of the surfaces topography and their classification is very important for machining techniques. Measurement of surfaces generated by water jets and abrasive water jets is quite difficult. It is caused by specific surface structure. In contrast to classical methods, the surface created by water jets and abrasive water jets is rather diffuse. As main surface geometry parameters have been proposed average roughness Ra [μm], stream deflection Y_{ret} [mm] and deviation angle D [°].The character of the distribution of topographical elements divided surfaces generated by AWJ into cutting wear area h_c [mm] and deformation wear area h_d [mm].

2. MEASUREMENT METHOD

The specific characteristics of the surface profile of samples cutted by WJ and AWJ (Vasek, et al, 2002). Further parameters for the quantification of surface quality are defined in the engineering literature (Bumbalek, et al, 1989) and standards EN ISO 4287 and EN ISO 4288. Nevertheless, many of standards are derived especially for the measurement of surface quality of workpieces by a contact profilometer. Most of them can be evaluated by optical measurement of surface quality. In the case of measurement by contact profilometers we use direct measurement and the indirect measurement is used for optical methods. The surface quality of each side of samples has been measured by an optical method using a CCD camera on 22 geometric lines with a vertical step of 0.364 mm (Fig. 1) (Valicek, et al, 2001). The optical signal of light and shadows surface distribution have been analysed by means of the Fourier transformation analysis, spectral decomposition and frequency band filtration with the aim of obtaining RMS (Root Mean Square) light reflection intensity from the surface and transforming equations between RMS and surface roughness parameters, particularly the surface roughness profile parameter, the average roughness Ra [μm]. The basic set-up of the shadow visualisation method of shadow visualisation by CCD camera (version A) is shown in (Fig. 1).

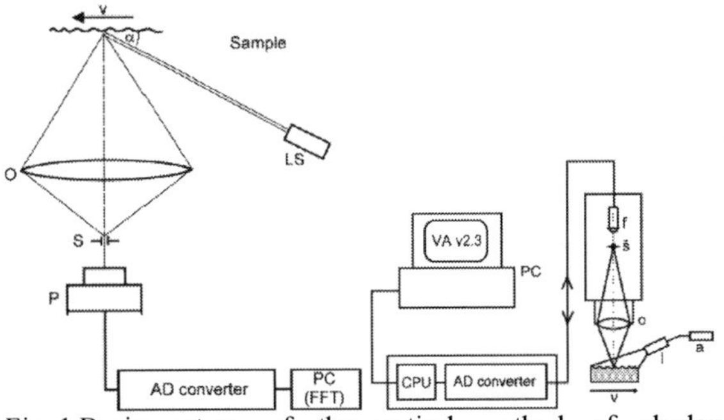

Fig. 1 Basic set-up of the optical method of shadow visualization by CCD camera (version A)

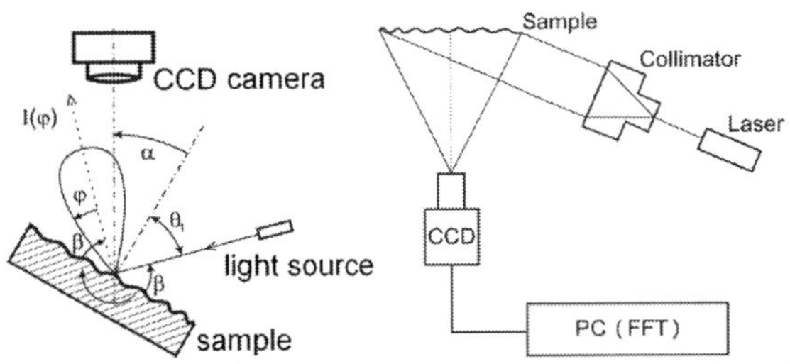

Fig. 2 Optical manner of optical signals calibration. (version B)

The observed surface is illuminated at a small oblique angle by a beam of collimated light. The shadow visualization effect arises from the change in surface reflectivity with the illumination angle. Using this kind of illumination, hills, dimples and other defects are easily observable and the typical waviness created by abrasive water jets. RMS parameter fluctuation in dependence on topographical properties and the depth is shown on Figure 3. For comparison and verification of the fluctuation character on topography the contact profilometer HOMMEL TESTER T8000 and optical profilometer MicroProf (FRT) have been used. The following figure 3 shows the results of Ra measurement obtained by contact profilometer. Figure 4 shows the results of surface roughness profile parameter - average roughness Ra obtained by no contact optical method. According these results shown on Figure 3 the initial zone is not identified by contact profilometer, because measurement principle by means of contact point do not allow measurement in close distance of the workpiece edge.

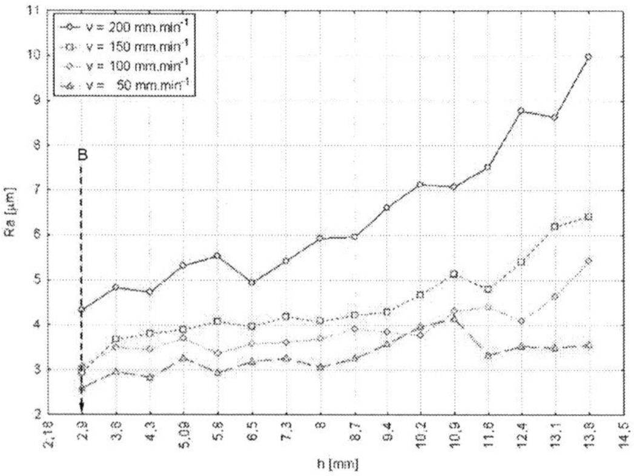

Fig. 3. Results of Ra measurement obtained by contact profilometer

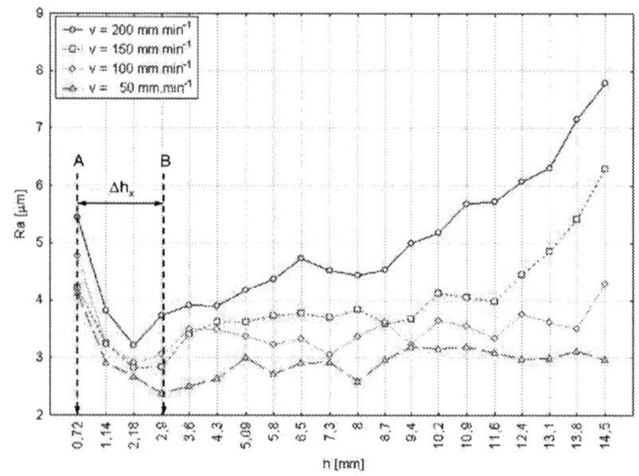

Fig. 4. Results of Ra measurement obtained by optical method

3. RESULTS AND DISCUSSION

Based on the favorable results measured by a new optical method shadow method in comparison with the data obtained by a commercial optical set and by a contact profilometer, the Physical Institute Laboratory possesses an optical measurement device that is suitable for basic measurement of topographical geometrical properties of a surface created by abrasive waterjet. The main optical quantity RMS has been estimated and defined and, besides the cutting depth h, the main interpreted parameters Ra, Y_{ret} and D. These are new structured geometrical properties of the surface that will be systematically stored in the data-base after each measurement. In terms of the development of the WJ/AWJ technology, the above described procedure lays only foundations for solution of much difficult problems relating to interaction between a flexible cutting tool and the target material, and for knowledge of the theoretical relations to the technological factors of the process. Systematic drafting data-bank with important input and output data about provided cut will have an irreplaceable role. From statistical, physical and analytical regularity evaluation of the relationship between input and output data, it is possible to proceed to the mathematical generalization of these regularities and derive an equation for prediction and project calculation of the concrete cuts. By such calculation it could be theoretical based technology factors selection that will be optimal for the given machined material on required quality, the performance parameters and the total machining economy. We are currently solving the problems of quantification of the cutting process prediction, stress and strain state in the machined material and their influence on development of geometrical surface parameters. In our opinion, we have achieved some remarkable results in the field of calculation of static and tensometric properties of a workpiece, in describing the development of the stress and strain state in the elastic and plastic zone of the material being stressed, as well as in describing the hydrodynamics of the disintegration process using the AWJ tool.

4. CONCLUSION

For basic geometrical properties topography measurement of new created surfaces by means of abrasive waterjet cutting technology the main optical quantity RMS, and main interpreted geometrical parameters Ra, Y_{ret} and D have been estimated and defined. The properties, mainly if dealing with average roughness Ra, are considered to be very important from the theoretical and practical known reasons and hence they have to be controlled. Surface quality output control has to be provided by adequate equipment. Optical methods, in contrary to the classical surface measurement, represent the substantial objectives to the output control in AWJ technology.

5. REFERENCES

Bumbálek, B., Obvody, V., Ošťádal, B.: Surface roughness. Praha, SNTL 1989. (in Czech)

Brezina, I.: Surface roughness. Fine mechanics and optics, No. 7, 1991. (in Czech)

Pernikář. J., Tykal, M., Vačkář, J.: Metrology and quality. Brno, CERM 2001. (in Czech)

Vašek, J., Martinec, P., Foldyna, J., Sitek, L., Ščučka,J.: Abrasives for AWJ cutting. Academy of Sciences, Ostrava, 2002.

Sigmund, M., Brychta, J., Čep, R. Quality control at high speed machining. Technological Engineering, 2005, roč. II., č. 1, pp. 20 – 21.

Valíček, J., Držík, M., Ohlídal, M., Mádr, V., Hlaváč, L.M.: Optical method for surface analyses and their utilization for abrasive liquid jet automation. In Proceedings of the 2001 WJTA American Waterjet Conference, M. Hashish (ed.), WJTA, Minneapolis, Minnesota, 2001, pp. 1 – 11.

Valíček, J., Louis, H., Schenk, A., Držík, M., Hlaváč, L. M., Chlpik, J.: Utilization of the optical methods for analyses of cutting edges. BHR Group, 2004, p. 487 – 501

Hloch, S. Evaluation of abrasive waterjet factors influencing the surface quality. In: Transactions of the Universities of Košice. No. 2 (2005), pp. 12-21, ISSN 1335-2334.

Hashish, M.: Pressure Effect in Abrasive – Waterjet (AWJ) Machining. Trans. of the ASME, Journal of Eng. Mat&Tech., Vol. 111, No. 7, 1989.

Guo, N. S.: Schneidprozess und Schnittqualität beim Wasserabrasiusstrahl – schneiden. VDI Verlag, 1994.

Kadnár, M., Rusnák, J.: Teoretický výpočet hrúbky mazacieho filmu pri elastohydrodynamickom mazaní. In: 11. Medzinárodného strojárskeho veltrhu v Nitre, Nitra, 2004, s. 55-58.

ACKNOWLEDGEMENTS

The authors would like to acknowledge the support of IGS-HGF VŠB TUO-2005-516/6 grant and VEGA 1/1075/04 and to 1/1095/04 - Optimization of abrasive waterjet cutting technology process, by mathematical and experimental planning

Annals of DAAAM for 2006 & Proceedings of the 17th International DAAAM Symposium, ISSN 1726-9679
Editor B. Katalinic, Published by DAAAM International, Vienna, Austria 2006
150th Birthday of Nikola Tesla (1856-1943) Inventor of Mobile Robot Teleoperation

INDUCTION HEATING OF DEFORMED SURFACE OF TITANIUM ALLOYS

GORDIENKO, A.; BELYI, A.; KHEIFETZ, M. & KOUKHTA, S.

Abstract: *The rapid induction heating method of control over formation of structures and machine parts surfacing based on principles of synergetics and physico-chemical analysis are considered. Based on the established interconnections between the physical and chemical processes, methods control over combined effects have been developed.*
Key words: *rapid induction heating, physico-chemical analysis, titanium alloys, technological barrier.*

1. INTRODUCTION

Induction heating of a deformed surface is the technology of fine-grain structure formation in metal characterized by improved physico-chemical properties. Great loads under strain, non-uniform distribution of temperature and strain in the surface layer restrict the application of intensive deformation in β–region with subsequent 50%-80% deformation in α+β region, when machining two-phase titanium alloys. Therefore, to make grains much smaller in β–structure of titanium alloys, moderate strain is applied – up to 30% – in α+β region, combined with a short-term heating in β–region. This short-term heating aimed at complete weakening of two-phase titanium alloys changes initial conditions of the process (Bodyako et al., 2000).

2. METHODS OF RESEARCHES

First of all, depth of electric current penetration was calculated on the basis of the data on induction heating modes (Bodyako et al., 2000). Then calculation results were compared to temperature distribution in surface layers and microstructures resulted from the electric current heating effect, and to phases. Radiographic and structural analysis data, as well as electronic microspectroscopy data were processed on the basis of methods described in (Gordiyenko & Shipko 1983, Pushinskii & Plokhov 1986). To investigate the processes of polygonization and recrystallization, cold and hot deformation of samples with constant cross section was performed on the milling machine. Milling of wedges made it possible to obtain variable distribution of strain along the sample length ranging from 10% to 50-60%. To study the processes of structure formation and provide grounds for selecting the strain and heating treatment modes making β–grains as small as possible, analysis was done to the following: the effect of initial structure parameters in two-phase titanium alloys, degree of deformation and speed of heating on recrystallization temperature range under conditions of high-speed heating in furnaces.

3. RESULTS AND DISCUSSION

Initial recrystallization was registered to start at 880°-920°C in VT9 and VT18 alloys with a coarse grain structure (ε_d =20%) when the latter were heated in the furnaces. The first recrystallized grains 1-2 μm in size appear within the range of α–phase, skirting the boundaries of β–grains. As the temperature rises to 950°-970°C, the degree of recrystallization increases, however, the process does not complete even in 60%-deformed alloys. In radiograms obtained using the technique of reflection from the sample surface, in addition to point reflections from recrystallized grains, one can notice dim spots testifying to the presence of deformed sections in the structure. Recrystallization process is complicated by the presence of closely located laminated formations of α- and β–phases, since in this case formation of major part embryos of «critical» size is connected with redistribution of dislocations within the range exceeding the distance between the plates; phase interfaces become, in fact, insuperable barriers for thermally activated dislocations and high-angular boundaries. Therefore, the development of primary recrystallization is closely connected with phase coagulation and coalescence processes, which provide the increase of particle size or interphase space to values commensurable to/with the size of the «critical» embryo. Depending on the temperature of annealing, primary crystallization can take place both in α–phase and in β–phase. In the course of heating of a deformed (ε_d = 60%) VT9 alloy to 850°-900°C, initial recrystallization usually proceeds within the range of α–phase. β–phase plates, whose content amounts to 20-25%, prevent the process from completing in full volume. With further curing, development of recrystallization in α–phase is closely connected with the division and coagulation processes of deformed β–plates.
If heated to temperatures approaching the boundary of α→β transformation (970-980°C), when the β–phase amount rises sharply, new grains appear within the β–range. The remaining undissolved α–plates serve as barriers impeding rapid recrystallization development and β–grains growth. Percentage of phases is approximately the same within the 930-960°C-temperature range and initial recrystallization proceeds both in α– and in β–phases. In VT18 alloy β–phase recrystallization is 20°C higher than in VT9 alloy. After the alloys have been heated in the furnace beyond the boundary of α→β transformation, and after the residual α–phase has been dissolved, intensive development of initial and collective recrystallization takes place in all the alloys. Within one hour the size of β–grains approaches 500-600 μm, i.e., it becomes commensurable to the initial size of β–grains before deformation. In VT9 and VT18 alloys, which originally have fine grain structure in the initial state, initial crystallization, like in the previous case described above, does not complete in full volume after they have been subjected to strain (ε_d = 10-30%) and heated in the furnace to 900-970°C. In addition to separate recrystallized 1-3 μm grains, one can observe initial deformed β–grains with a laminated structure. Compression degree 40-60% increased, recrystallization kinetics becomes more intensive. The process of replacing old grains by the new ones completes after heating to 960-980°C. Cooling in the furnace results in the formation of homogeneous fine grain structure, the dimension of α–grains being 2-4 μm; coagulated β–phase

grains are located along their boundaries. It has been found that deformation inside the grains of β–converted laminated structures proceeds by means of irregular sliding, which spreads within the range of a separate colony of α–plates. As the average size of α–plates colony increases, the sliding paths become more irregular, their number decreases and the intensity of local cold hardening is several times higher than in alloys with small colonies. Therefore, α–plate and α–colonies size reduction during thermal treatment decreases the degree of structural anisotropy in β–crystallites, which contributes to increasing homogeneity of grain microdeformation and reduces the tendency of the alloys to intercrystillite destruction. Besides, alloys with reduced grain size structure do not produce obvious α–phase extractions at grain boundaries. Such a structural state impedes the development of deformation between the grains, making processes of multiple sliding inside the grains more intensive. Availability of a highly developed interphase surface, in addition to multiple sliding inside β–plates, considerably accelerates the processes of plate division, when subjected to heating in two-phase region, as well as their coagulation and spheroidizing, thus contributing to rapid development of initial recrystallization. The above is also proved by high-speed annealing tests. High speed heating of deformed alloys introduces considerable changes in the temperature kinetics of structure formation. First, when tested alloys are heated at 10-300°C/s to the temperature of α→β conversion, diffusion processes of phase coagulation and spheroidizing do not develop sufficiently, which ensures/provides the retaining of non-crystallized structure in the two-phase region. Secondly, a high temperature β–phase inherits deformation defects of the low- temperature α–phase during the α→β transformation process. All the above results in the fact that the start and completion of initial recrystallization of alloys take place in β–phase, after the phase α→β conversion has been completed. Rapid dissolution of laminated α–phase contributes to simultaneous emergence of new grains, providing efficient size reduction of original coarse structure at the stage of initial recrystallization completion. Phase cold hardening does not play an important part in the above process of β–grains size reduction. This is proved by the fact that the original coarse structure does not change in undeformed alloys when subjected to rapid heating in β–region, only the alloy intragranular structure (α–colony dimension, thickness of α–plates) undergoes changes resulting from cooling modes. Temperature level of the start and completion of initial alloy crystallization depends on heating speed. When a 40% deformed VT18 alloy is heated at a speed of 10°C/s, initial recrystallization takes place in β–region within a narrow temperature range. Recrystallization start is registered with heating to 1000°C and its completion – at 10°C and higher. As heating velocity is increased to 300°C, the temperature range, in which recrystallization takes place, is extended to 50°C, and the boundaries of recrystallization start and completion move upwards to 1050°C and 1100°C, respectively.

In VT9 alloy dependence of the temperature of the initial recrystallization start and completion on the heating modes is less evident than in BT 18 alloy. Thus, if heating temperature is raised from 10°C/s up to 300°C/s, recrystallization start is displaced from 1000°C to 1020°C and completion of the process is registered at 1010°C and 1050°C, respectively. The results of the experiment show that linear dependence is evident between the temperatures of recrystallization start, completion and the logarithm of the heating speed.

Initial alloy intragranular structure is a vital parameter affecting the temperature boundaries of recrystallization and the kind of structure obtained at high speed heating. The above feature is particularly evident at high speed recrystallization annealing of alloys with a small amount of strain. For example, when VT9 alloy with initial coarse structure is heated beyond the boundary of α→β transformation, the first embryos appear in maximally distorted areas, which are often the boundaries of the old grains or colonies, and, also, irregular sliding lines. Due to inhomogeneous distribution of defects during the process of deformation of alloys with a coarse intragranular structure, local difference in speed of formation and growth of embryos is great. All the above results in forming diversified granular structures at initial recrystallization stage. As the compression degree decreases, the number of newborn grains decreases greatly, and after a 5-7% deformation all the tested alloys, in fact, do not recrystallize. When heated to 1200°C at 25°C/s, initial coarse β–grains of these alloys are preserved, however, the old boundaries of β–grains assume a winding shape. The emergence of winding boundary lines results from the inhomogeneous migration, stipulated by the inhomogeneous distribution of defects in sections lying in the vicinity of the boundaries. If deformation degree is raised from 20 to 60%, general laws of recrystalization at high speed heating remain valid: the amount of embryos grows and after initial recrystallization completion smaller grains are formed; temperature of recrystallization start and completion drops. However, it should be noted that it is impossible to obtain sufficiently homogeneous transformed 10-20 μm β–grains in alloys with coarse α–colonies and α–plates, even at high speed heating after the 35-60% deformation, because the centers of the new grains do not emerge homogeneously. Grain diversity is only eliminated at the stage of collective growth and becomes complete when average size of β–grains reaches 30-40 μm.

Preliminary intragranular structure size reduction in alloys ensures at subsequent strain (ε_d = 10-15%) a homogeneous distribution of defects across the grain, and, consequently, homogeneous emergence of new grain embryos, and, thus, effective β–grain size reduction in the course of high speed recrystallization annealing. In VT9 and VT18 alloys with initial coarse intragranular structure with ε_d= 10-20%, the recrystallization process takes place within a more narrow temperature range, compared to the initial intragranular structure, which has undergone size reduction, and β–grains formed by the completion of primary recrystallization are homogeneous and smaller.

4. CONCLUSIONS

1. It is possible to considerably reduce the effect of the diverse grain size resulting from high-speed annealing of two-phase titanium alloys stipulated by inhomogeneous hot plastic deformation of grains, if the intragranular structure is changed after preheating.
2. The operation can perform the role of a barrier in the technological sequence, with some of the factors being unable to overcome it; in order to ensure complete weakening of two-phase titanium alloys which have undergone plastic deformation in α+β–region, it is necessary to apply a short-term heating in β–region.
3. One can observe the following heredity at high-speed treatment: defects at low temperature α–phase introduced by plastic deformation transfer to a high temperature β–phase in the course of α→β phase transformation and, as a result, emergence and growth of new grains takes place in β–region.

5. REFERENCES

Bodyako N.M., Gordiyenko A.I. &. Ivashko V.V, (1982). Izv. Akad. Nauk BSSR, Ser. Fiz.-Tekh. Nauk, No.9, 27-32.

Gordiyenko A.I. & Shipko A.A. (1983). *Structural and Phase Transformations in Titanium Alloys at High Speed Heating*, Minsk.

Pushinskii L.I. & Plokhov A.P. (1986). *Investigation of the Structure and Physicomechanical Properties of Coatings*, Novosibirsk.

Annals of DAAAM for 2006 & Proceedings of the 17th International DAAAM Symposium, ISSN 1726-9679
B. Katalinic, Published by DAAAM International, Vienna, Austria 2006
150th Birthday of Nikola Tesla (1856-1943) Inventor of Mobile Robot Teleoperation

PROMOTIONAL ACTIVITIES IN TOURISM

GRGONA, J.

Abstract: *Promotional activities in tourism are part of the joint framework of marketing efforts aimed at satisfying tourists' needs. This paper highlights different forms of promotional activities by which agents in tourism system participate in communication processes on the tourism market. Sports and activity play a significant role in promotional activities. The paper stresses the importance of promotional activities coordinated by all those involved in tourism sector and on all levels so that, based on mutual interest and marketing concept, the tourism product is successfully presented and business targets are achieved. Promotion therefore bears particular significance in the activities of agents in tourism system and is also very important for the development of tourism.*
Key words: *promotion, development of tourism, promotional activities in tourism, sports and activity*

1. INTRODUCTION

From the theoretical and practical point of view marketing basics in tourism do not differ from those in any other business. In other words, all activities are eventually concentrated on the final user, i.e., the tourist. Tourism marketing has its own specific features that should be respected by entities active in tourism. However, they will act in accordance with general principles of marketing and adjust them to the specifics of tourism marketing. Marketing as a business concept takes into consideration particular characteristics of a business activity and adjusts to them accordingly. It is necessary to respect the characteristics of a business activity as they differ in areas such as: service features, types of distribution and sales channels, business strategy, pricing, markets and market segments, use of natural resources and their availability, business environment, various external factors, etc. Let us not forget that tourism product, as an amalgam of different products and services, goods traded on the tourism market. The majority of tourism services follow general principles of service industry marketing. But in some aspects the implementation of these principles is different from other services. The differences that exist arise from the specific nature of tourism and tourism market and also because in tourism we sell things that do not bear features of goods and services but nevertheless have significant influence on the sales of goods and services that satisfy the needs of tourist. We primarily have in mind various benefits that a tourist area offers. Frequently, this is the main reason why tourists come and spend time and money in that area. But in order to achieve this and a certain level of income it is necessary to research relevant aspects of tourism market beforehand. It is particularly important to research different motivation and behaviour of tourists, as they are the ones that indicate their needs & demand. By identifying and getting to know different aspects of motivation we are able to tailor the offer so that it adequately responds and ultimately satisfies these motivations. Tourism offer and demand are formed on the market under the influence of many factors. Tourism offer is static and inelastic while tourism demand is characterized by mobility and elasticity. Dispersion and a physical separation of both ends of tourism market is a common and very important feature for the functioning of tourism market.

2. THE BASICS OF THE MARKETING CONCEPT IN TOURISM

Marketing concept is not only essential for tourism development of a country, specific tourist destination, region and tourist resort but also for companies active in the tourism sector. Due to many different activities that directly or indirectly partake in creation of a tourism product of a region or a country this represents a complex system within national economy. The complexity of tourism product and equally important economic effects from tourism necessitate the creation of a specific tourism policy. This is particularly true of the receiving tourism countries. Physical separation of tourism product can be viewed in two ways, from the geographical point of view and from the immobility of providers of tourism product aspect. As a rule, the tourist generating areas are geographically separated from the receiving areas regardless of the real distance between the two. A clear feature of the physical distance between tourism offer and tourism demand is that tourism product can be consumed by traveling and arrival of tourists and from the generating to receiving areas. This feature makes the application of marketing in tourism somewhat specific. Senečić emphasizes that marketing in tourism is the implementation of the marketing concept in tourism companies and other providers of tourism products but not limited to them only as the implementation is coordinated on different levels where interests of providers of tourism products and aims of tourism policy of a country, region or local community are agreed. However, tourism companies must also create their own marketing policy and management system. They should and must build their own business concept based on marketing and take into consideration common interests that link the entire system of tourism offer and enable its functioning (Senečić, 1998). In the context of what has already been mentioned, the link between agents in tourism system and implementation of tourism policy on the level of a country and particular regions is clearly evident. Specific relations on the tourism market, complexity of tourism product, characteristics and behaviour of tourist on the market are fundamentally important for the implementation of marketing in tourism. According to the findings of a recent research (TOMAS 2004) sports and sports-related activities are among the key reasons why people travel and stay in a destination. Bartoluci clearly indicates the links between sport and tourism and stresses that sport in contemporary tourism does not only imply stay in a certain destination but is frequently the main reason for travel to a destination. The relationship between sport and tourism leads to the development of a new type of tourism: sports and activity tourism. The same author emphasizes that interdependence between sport and tourism rests on the factors that lead to the development of tourism and sport and of functions that generate tourism and sport (Bartoluci, 1998). It is evident that development of sports and possibilities for sporting activities leads to the development of tourism. The complementary nature of these two phenomena is evident in promotional activities. The promotion of different sporting events in a destination is also the promotion of the destination. Promotional activities aim to present overall tourist potentials of a destination. Sports and activities should be considered an essential element of tourism offer. They attract tourists and contribute to the development of tourism.

3. THE IMPORTANCE OF PROMOTIONAL ACTIVITIES IN TOURISM

As one of the controllable marketing variables the role of promotion in tourism and its development is particularly stressed in this paper. Without ignoring other aspects of marketing mix, promotion on tourism market is the key link between tourism offer and demand. Let us define tourism promotion as a group of planned and continuous activities that form the communication process on the tourism market between suppliers and customers with the aim to create positive attitude towards a tourism product. The opinion to which potential tourist arrives in the course of communication process, from opinion, via emotion to behaviour, ultimately results in the selection of a destination. We can hence talk about a model of opinion influencing the behaviour that works in the following pattern. Firstly, a potential tourist, via communication, within his or her environment acquires knowledge and perception of a destination. Then he looks for additional information (again via communication) and firms up his/her opinion about a destination and its abilities to satisfy his/her needs and expectations. These opinions are integrated and valued thus creating an emotion towards the destination. Based on the opinion and emotion a tourist ultimately expresses the intention, or lack of it, to travel. It is clear that this model of opinion formation is based on the communication process (Grgona, 2003). Tourism promotion implies activities aimed at attracting a tourist/customer to a destination and at persuading tourists to buy products and services when travelling and when in a destination (Senečić, 1998). Therefore, promotion plays a significant role in the marketing policy and activities of agents in the tourism system. Let us mention the main tourism promotional activities that are almost the same as in other business activities although they are slightly different in their application. In the first place, advertising is the most widely spread form of promotion, followed by direct marketing, sales promotion, public relations plus other forms of promotional activities characteristic for tourism such as trade fairs, exhibitions and events where different elements of tourist offer are presented. According to *Rječnik turizma [Dictionary of Tourism]* the importance of advertising in tourism as a form of promotion stems from the fact that advertising in tourism carries several critical communicational functions: information, persuasion, reminding, increasing the value of product (service), support for other marketing efforts (Vukonić & Čavlek, 2001). When talking about advertising of tourist countries we must single out a feature different to advertising in other sectors. Namely, advertising acts as general tourism advertising, on the country, region or destination level with the aim of presenting these areas and attracting tourists. Direct marketing in tourism is a form of promotion where direct communication with tourists is established through personal contact thus getting to know better the tourist's preferences, wishes and demands. That is the main advantage of this form of promotion. With direct communication it is easier to adjust to different tourists' needs and at the same time to receive an adequate level of feedback. The other advantage is the possibility to approach target groups more directly and with greater intensity. A systematic activity aimed at forming positive public image creates a level of trust, understanding and empathy for tourists, suppliers, own employees, banks, sub-contractors, businesses and experts. The aim is to create a positive image in the public and to integrate with it, so as to build a long-term trust and empathy. In their efforts public relations agents have commitments towards the public. One of them is the social obligation requiring the explanation of activities to all relevant people and social groups so as to avoid any potential conflict situations.

4. CONCLUSION

Tourism and its development should be studied in the context of general development of a society, region or country. Explanations of tourism phenomenon can be found in the works of many theoreticians. All of them and without exception stress that development of tourism in an area is primarily triggered by economic reasons that presume the overall economic development in that area. We can only guess whether these arguments are justified and what possible consequence tourism development might have on the economy of a place or a country. However, there should not be any dilemma as to the role marketing and marketing communication via promotional activity play in the development of tourism. When talking about the role of promotion in tourism development we again must stress the importance of the systematic and consistent communication on the tourist market aimed at attracting tourists and creating positive attitude towards tourism product, company and destination. Promotion and promotional activities in tourism as a part of marketing mix are a relevant source of information that aids the existing and potential tourists in decision-making process and selection of a holiday destination. Through this communication process tourists firm up their attitudes, opinions and convictions about destinations they know or have already visited. They also discover new destinations about which they know little or nothing at all. If we measure the effects of tourism on the economy of a region or a country through economic indicators we realistically assess the possibilities for the development of other sectors. In other words, development of tourism can instigate or improve development of other sectors. By creating the conditions for tourism consumption we indirectly stimulate the development of tourism in an area. Promotion plays a crucial role here. Tourists need to be acquainted with a tourist destination and its product and they must be provided with timely information. They need to get a realistic and a clear image of all positive attributes that will represent the factors of attraction for a destination. As this is all about communication on the tourism market it is important to co-ordinate the activities on different levels, from agents of tourism system to relevant state institutions tasked with the implementation of tourism policy.

5. LITERATURE

Bartoluci, M., & Cavlek, N. (Ed.) (1998). Tourism and sport. Zagreb: Fakultet za fizičku kulturu Sveučilišta u Zagrebu, Ekonomski fakultet Zagreb, Zagrebački velesajam. ISBN 953-6378-18-3.

Grgona, J. (2003). Image of a Tourist Destination: A Contribution to a Scientific Debate. *Acta Turistica*, UDC/UDK 338.48 No.1., Vol 15/2003., Ekonomski fakultet Zagreb.

Senecic, J. (1998). Promotion in tourism. Mikrorad. ISBN 953-6286-28-9, Zagreb.

TOMAS 2004, Attitudes and Expenditures of Tourists in Croatia. Institut za turizam. UDK 338.484(497.5). ISBN 953-6145-11-1., Zagreb

Vukonic, B., & Cavlek, N. (Ed.) (2001) Dictionary of Tourism. Masmedia UDK 338.40(03). ISBN 953-157-111-2. Zagreb.

Annals of DAAAM for 2006 & Proceedings of the 17th International DAAAM Symposium, ISSN 1726-9679
Editor B. Katalinic, Published by DAAAM International, Vienna, Austria 2006
150th Birthday of Nikola Tesla (1856-1943) Inventor of Mobile Robot Teleoperation

COMPARISON OF RESULTS OF PREDICTION OF CUTTING FORCES WITH REAL QUANTITIES – PART ONE: METHOD OF RESEARCH AND RESULTS OF ANALYSIS

GRZESIAK, D.D. & KROLIKOWSKI, M.A.

Abstract: *Referring to prior research work (Grzesiak & Królikowski 2005) improved model for prediction of the tool deformation during profile milling is presented in paper. Model applies for slim rotary milling tools. According to conclusions driven from prior experimental work, the FEM methodology was applied. The FEM results were compared with classic analytical strength resistance calculation results and experimental investigations. Presented analysis is a first part of the investigations of cutting forces influence on cutting tool deformations. Conclusions will be applied to experimental work, with application of cutting forces measurement equipment.*
Keywords: *milling, cutting forces, cutting tool deformation.*

1. INTRODUCTION

Profile milling with application of slim mill is a common, widely used method of machining. Many contouring operations can be carried out only with application of this machining method without replacement methods available. In order to fully benefit from profile milling, there is a need to predict the forces acting on the cutting tool, and the same deformations influencing on machining accuracy. The cutting tool deformation risk should be than applied to tool path during CAM processing. Referring to prior research on this subject (Grzesiak & Królikowski 2005), improved model for prediction the cutting tool deformations will be presented. New model includes conclusions driven from prior investigations. Results will be compared with recently obtained deformation measurements, basing on machining errors at the workpiece. This comparison will justify new model presented for further investigations. Cutting tool deformations would be then applied as correction of the tool path in critical moments, where machining errors are foreseen to occur.

2. ANALYTICAL WORK

2.1 Correlation with prior research

In order to have a base for comparative analysis, the same sample model was used. (fig. 1). Model consists of different tangent second-level polynomial surfaces: two elliptical, two parabolic and sinusoid. Curvilinear contour of the model was divided regularly by points located at the surface. In order to have comparative results, the same cutting tools and the same machining parameters were applied, like in prior experimental work (Grzesiak & Królikowski 2005).

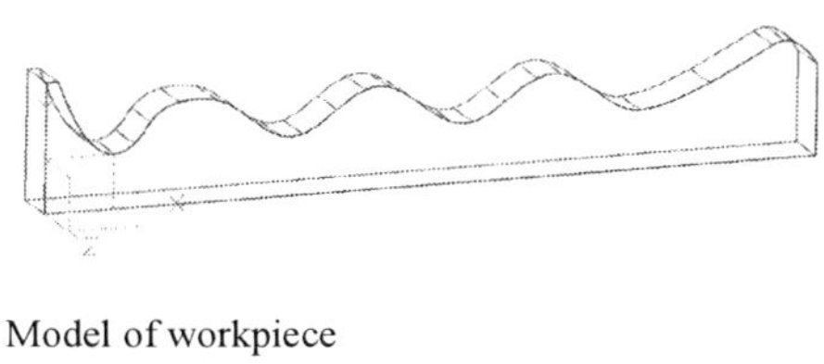

Fig. 1 Model of workpiece

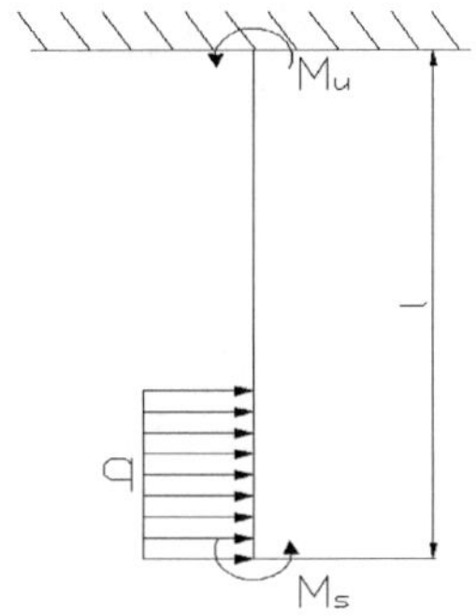

Fig. 2 Classic strength resistance calculation model

The equation for estimating the cutting force values was also the same (1).

$$F_{c\Sigma \acute{s}r} = k_{c1.1} \cdot a_p \cdot \frac{\psi}{\delta} \cdot f_z^{(z+1)} \cdot \left(\frac{a_e}{D}\right)^{\frac{z+1}{2}} \qquad (1)$$

Following simplifying assumptions were made:

- Vibrations of system Machine toll – fixture – workpiece-cutting tool were not taken into account.
- A play between moveable elements in machine tool was omitted.
- Values of coefficients „k_{c1}" and „z", used in equation were taken from publications.
- Cutting force variations, caused by cutting tool geometry were omitted as well.

According to conclusions driven from prior research and analysis, Finite Element Method was applied for improved results of prediction of cutting tool deformation. Cutting tool model (fig. 3) was meshed optimally, in order to obtain reliable results and shorten the calculation time of passes (fig. 4).

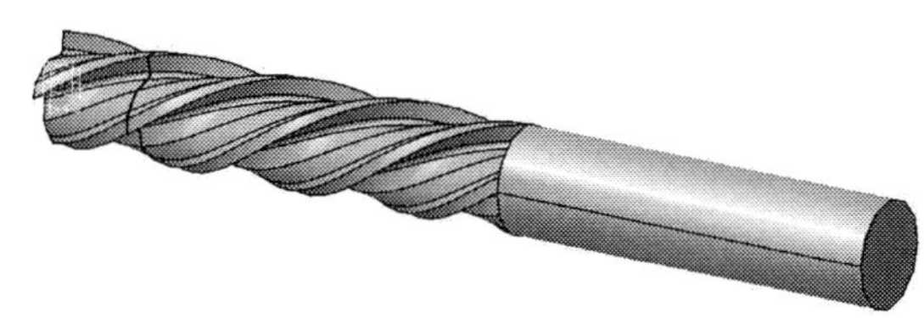

Fig. 3 CAD model of mill

2.2 Calculation with application of classic equations

In first stage of calculations, classic strength resistance model was applied (fig. 2). Tool is represented as a round bar with complex cross-cut contour, fixed one side on the shank and loaded by torque moment and continuous load on the length of cut (20mm).

2.3 Calculations with application of FEM

Second stage of calculations was carried out using FEM analysis. As foreseen FEM method, based on model made in CATIA CAD/CAM system (fig. 3) should provide more accuracy and reliability in prediction of cutting tool deformation. FEM analysis was carried out using COSMOS system. Load values were applied from equation (1). Model of fixing the cutting tool at the shank was realised by taking away DOF in all directions, by friction forces, on the cylinder part of the shank.

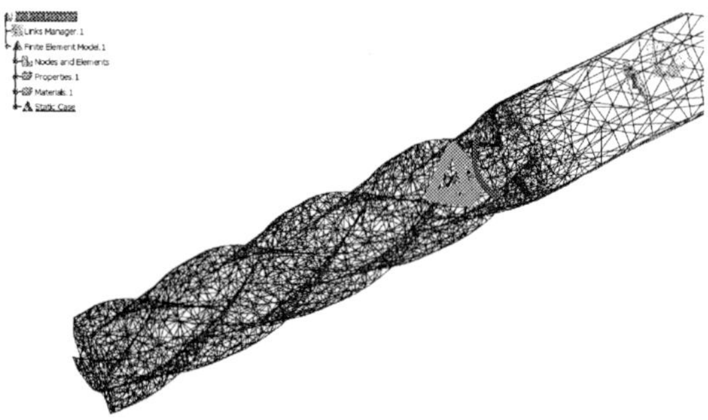

Fig. 4 Meshed model of mill

Results of calculations with application of FEM method were compared with results obtained by classical method. It seems that good convergence of results was obtained. The differences of deformations obtained by different methods were less than 0,01 mm. The deformation values are show at the fig. 5. At lower part of this chart, the sample contour is presented. Difference values between different calculation methods could be therefore analysed in respect to profile curvature. It can be observed, that in places where the clasp angle increases, the difference increases as well, respectively. Difference values increase in accordance to increment of value of cutting forces acting at tool. It can be observed that difference values "follow" the curvature change of the profile. It points on the need to include however the cutting force variations what was omitted in assumptions, in order to simplify the task. Comparison of results obtained by FEM and results obtained in first research work (Grzesiak & Królikowski 2005) shows, that assuming simplifications presented in former analysis and in chapter 2, there is no possibility to approach to real tool deformations. As it is presented at the fig. 6 the differences increase up to 0,6 mm. It should be however taken into account that this maximal values are observed at the beginning of cutting process, where the cutting process is not stable and depends on many factors – in this case most probably by strong vibrations. The difference values in the rest of profile are oscillating around 0,1 mm.

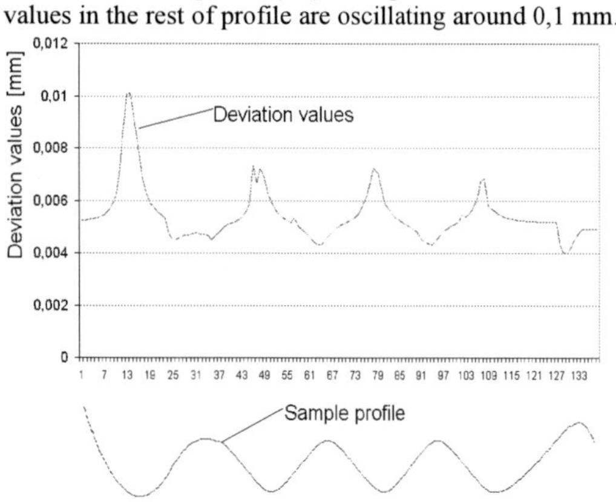

Fig. 5 Deviations from deformations calculated with application of classical methods and FEM analysis

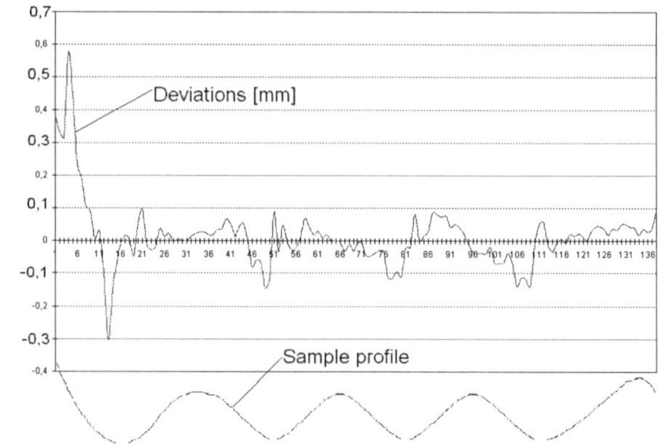

Fig. 6 Differences between real machining errors and deformations calculated with application of FEM

3. CONCLUSIONS

Next approach to examination of deformation of cutting tool during profile milling, in comparison to first, basic approach presented (Grzesiak & Królikowski 2005) allowed to draw some new conclusions. First of all it was shown, that application of more accurate calculation method – FEM – did not provide expected improvements in prediction of cutting tool deflection and its influence on machining accuracy at profile milling (for comparative analysis prior experimental results gathered by measurement of real workpieces after machining were used). Secondly, the research work on real cutting force measurement must be carried out in order to obtain results for improvement of the methodology of cutting forces estimation and analysis in this case of machining. In the case of XY millin, previously assumed classical method of strength resistance analysis was good enough, without the need of complex FEM analysis, what is proved in presented research work. It can be observed that deviations in calculations by comparison of two methods are lower than deviations from calculated and real deformation observer by workpiece measurements.

Concluding, following experimental research is foreseen respectively: 1) measurement of cutting forces applying the same workpiece and machining conditions, 2) analytical research driving to correlate real cutting forces, deformations and calculations, 3) in case of unsuccessful results an approach to more complex calculations and finally an approach to estimate experimental coefficient values.

Further investigations are foreseen to 3D surface sculpturing in milling operations.

4. REFERENCES

Dyląg, Jakubowicz, Orłoś: *Podstawy wytrzymałości materiałów*. WNT Warszawa

Górski E. (1999). *Poradnik frezera*, WNT, Warszawa

Grzesiak D. D., Królikowski M. A. (2005) *Analysis of influence of variation of cutting forces direction on accuracy of curvature representation by profile milling operation*

Grzesik W. (1998). *Podstawy skrawania materiałów metalowych,* WNT, Warszawa

Praca zbiorowa (1990). *Poradnik Inżyniera – Obróbka Skrawaniem t. II*, Praca zbiorowa, WNT, Warszawa

Olszak W.(2005). *Obróbka Skrawaniem, Rozdział 11 – Frezowanie*, Wydawnictwo Politechniki Szczecińskiej, Szczecin

Olszak W. (2001). *Obróbka Skrawaniem i Narzędzia, Rozdział 5 – Proces Skrawania*, Wydawnictwo Politechniki Szczecińskiej, Szczecin

Annals of DAAAM for 2006 & Proceedings of the 17th International DAAAM Symposium, ISSN 1726-9679
Editor B. Katalinic, Published by DAAAM International, Vienna, Austria 2006
150th Birthday of Nikola Tesla (1856-1943) Inventor of Mobile Robot Teleoperation

MECHATRONICS: THE COMING REVOLUTION

GUPTA, P.; TIRTH, V. & SRIVASTAVA, R.K.

Abstract: *Mechatronics related issues are often overlooked in engineering systems. Most of these overlooked issues are vital in design of integrated systems. After giving a brief account of mechatronics, the paper explains the interaction between mechatronical integration and complexity involved, along with logical integration of different activities in a manufacturing system. It also gives an idea for the formation of an efficient and effective integrated design team of mechatronic engineers. The paper looks at how organizations can benefit from learning to apply the ideas and principles that mechatronics embodies.*
Keywords: *Mechatronics, automation, communication, Manufacturing,, Integration.*

1. INTRODUCTION

Mechatronics a new technology coined by Japanese, is the combination of mechanical, electronics and information technology in design of products and systems to give better value. Engineering design and product development are complex processes involving an interaction between many skills and disciplines. The success of industries and selling goods in a competitive world market increasingly depends upon an ability to integrate electronics and computing technology into a wide range of primarily mechanical products and processes. This results in the systems, which are cheaper, simpler, and more reliable with greater flexibility of operation.

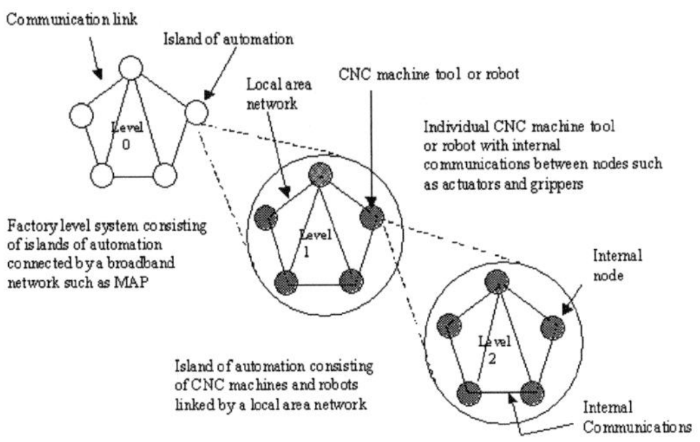

Fig. 1. A Mechatronics Hierarchy within Manufacturing

2. CONCEPTUAL DEFINITIONS

1. Mechatronics has been described as the design and manufacture of products and systems processing both a mechanical functionality and an embedded algorithmic control. Mechatronics is the synergistic combination of precision mechanical engineering, electronics control and system thinking in the design of products and manufacturing processes.
2. Mechatronics is a methodology for designing products that exhibits fast precise performance. These characteristics can be achieved by considering not only the mechanical design but also the use of servo control, sensors and electronics.
3. Mechatronics can be described as an integration of mechanical and electronics engineering systems, which help to multi-disciplinary approach to product and manufacturing system design.

3. MECHATRONIC SYSTEM AND ITS EVOLUTION

3.1 Interdisciplinary nature

Mechatronics is a combination of several areas of specialized knowledge. Generally it includes (i) Mechanical (ii) Electronics (iii)

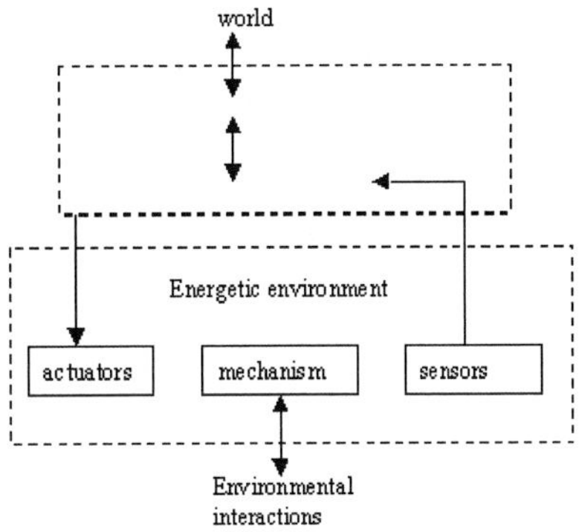

Fig. 2. General structure of a mechatronic system

Computer (iv) Information technology (v) Opto-Electronics (vi) Control Engineering (vii) Bio-Medical technology (viii) Robotics and automation (ix) Manufacturing Science (x) Micro-devices (xi) Vibration and noise control (xii) System and Cybernetics (xiii) Intelligence and expert system (xiv) Actuators and sensors (xv) Motion control and trajectory planning (xvi) Modeling and design etc.

3.2 Merits / Benefits of Mechatronic systems

- Increased quality and reliability in performing the work
- Reduced cost with improved specifications and better design
- Reduced assembly cost, warranty cost etc.
- Improved customer specifications
- Increased sales and thus more profits
- More functional products at reduced cost
- Designing of complex system with high level of integration
- Deployment of multi sensor environment, multi program structure for user selection, real time

structure involving distributed and developed intelligence etc.

3.3 Comparative study (See Table no. 1)

HUMAN BODY PARTS	ELEMENTS OF MECHATRONICS
1. Muscles : Provide force and energy to do the work	Mechanical Engineering: Provides analysis of mechanism, motion planning, energy transformation etc.
2. Brain : To think and take decisions	Microprocessor/Computer : • To take decision • Speedy processing of data • Repetitive processing
3. Nervous system: Linking and coordination of the various parts of the body including brain.	Information Technology : • Communication • Data Processing • Linking various module of the system
4. Integration of all parts nervous system, and brain is necessary for proper functioning of human body	Integration of Mechanical engineering, microprocessor / computer, electronics, control and information technology is necessary for synergistic effect in Mechatronics

Table 1. Human body parts and Mechatronic elements

4. THE HOW OF MECHATRONICS

Achieving a successful mechatronic design environment depends on three factors:

- Communication
- Collaboration
- Integration

(i) Communication Mechanisms: Generally these will comprise computer hardware and software and associated communication equipment and will function to achieve the interchange of digital data between distributed and heterogeneous computer systems.

(ii) Collaboration and Integration: The main aim of any company is to maintain, develop and improve its competitive position in the market, hence to be effective; the adoption of a mechatronic approach to product development requires the collaboration of all members of an integrated product-development team towards that goal.

5. INDUSTRIAL APPLICATIONS

A typical mechatronic system picks up signals from the environment, process them to generate output signals, transforming them into forces, motions and / or actions.

Few specific examples, which have used mechatronics technology, are: CNC machine tools, Washing Machines, Computer Disc Drives, Electronic Watches, Robotics and AGVs, Servo Technology, Temperature control in air conditioners and refrigerators, Automobiles, printing machines, robotics, automatic systems and computational devices, Power provides a viable candidate for shifting to mechatronic technology.

6. COMPARISON WITH CONVENTIONAL SYSTEMS

Conventional System	Mechatronic System
Bulky	Compact
Complex mechanics	Simplified mechanism
Constant speed drives	Variable speed drives
Mechanical synchronization	Electronics synchronization
Rigid heavy structures	Lighter structure
Non adjustable movements	Programmable movements
Accuracy determined by mechanism	Accuracy achieved by tolerance of feedback
Manual Control	Automatic and programmable controls

Table 2. Mechatronic Systems Vs Conventional

7. CONCLUSION

Many current products and system depend for their success on a mechatronic approach to engineering design and product development. Manufacturing projects has undergone revolutionary changes in the past decade in its objectives, philosophy and applications. The widespread use of digital computers technology and microelectronics is one of the most important aspects for futuristic projects. It is evident that application of mechatronic aspects to manufacturing projects has remained a major task for the automation engineers and planners. To be truly mechatronic, however, it is not enough simply to consider the technical and technological aspects of the design; it is also necessary to adopt an integrated approach to product development.

8. REFERENCES

P. M. Taylor and M. B. Gunner, "Mechatronics in automated garment manufacture," in Mechatronic Design in Textile Engineering, M. Acar, Ed. Kluwer, 1995, pp. 271-289.

Parul Gupta, Vineet Tirth, Mayank Gangwar, "Global Trends and Best Practices in Mechatronics" 'National conference On Recent trends in mechatronics, nanotechnology, and robotics', Feb 2006 in NIT Rourekela.

Tähemaa T. Assurance of Synergy and Competitive Dependability at Non-Safety-Critical Mechatronic Systems Design, Doctoral degree thesis, Tallinn 2002.

S.S.Rao. Reliability-Based Design, School of Mechanical Engineering, Purdue University,1992.

www.cordis.lu/trendchart.S.

Annals of DAAAM for 2006 & Proceedings of the 17th International DAAAM Symposium, ISSN 1726-9679
B. Katalinic, Published by DAAAM International, Vienna, Austria 2006
150th Birthday of Nikola Tesla (1856-1943) Inventor of Mobile Robot Teleoperation

MEASUREMENT OF A SPUR GEAR ON A GHIBLI TRAX MACHINE USING QUINDOS

GYENGE, C.; BOB, M. & BOB, D.

Abstract: *This papers purpose is to solve the problem of measuring a spur, gear defined according to AGMA or DIN standards, gears , without using special gear-measuring devices or rotary table. We will present the case of two particular gears, measured on a GHIBLI TRAX universal measuring center, commanded by QUINDOS, using special gear-measurement soft modules.*
Key words: *spur gears, measuring, CMM.*

1. INTRODUCTION

Measuring gears, especially measurements that involve profile deviations, can be very difficult, usually requesting special gear-measuring devices, not always available. The problem is even more difficult when profile modifications are involved, or the gear definition is made according to AGMA standard, considering that the traditional gear-measuring devices are usually working according to DIN standard (Dudley, Darle W. (1994)).
Using the opportunities offered by the GHIBLI TRAX measuring machine (fig.1), we can very simply solve the problems, simply using the QUINDOS command MEGEAR.
The ultra-precise mode of operation and the QUINDOS application-oriented modules make this machine a perfect choice for this particular case (Brown & Sharpe DEA Spa 1999).

Fig.1. GHIBLI-TRAX Measuring Machine

The logic of the QUINDOS measurement program and the results will be described below, considering two particular cases:

- standard gear according to DIN 3960, without file modification (m = 2,5; z = 34),
- pinion shaft, with both profile and trace modification, defined according to AGMA, (fig.2).

In both cases, the logic of the measurement process is described by the same logic scheme (fig.3):

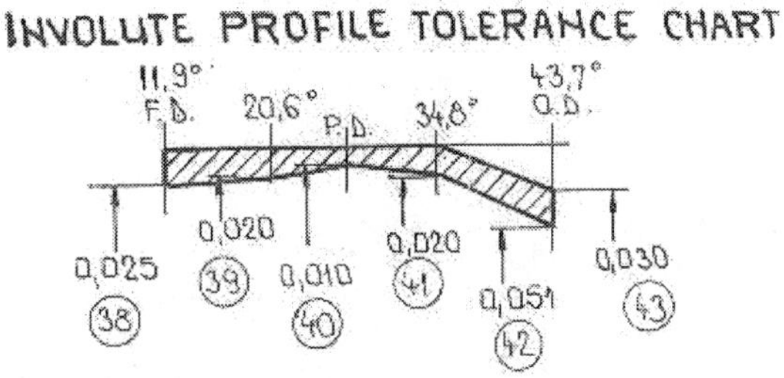

Fig.2. The K chart of planet pinion

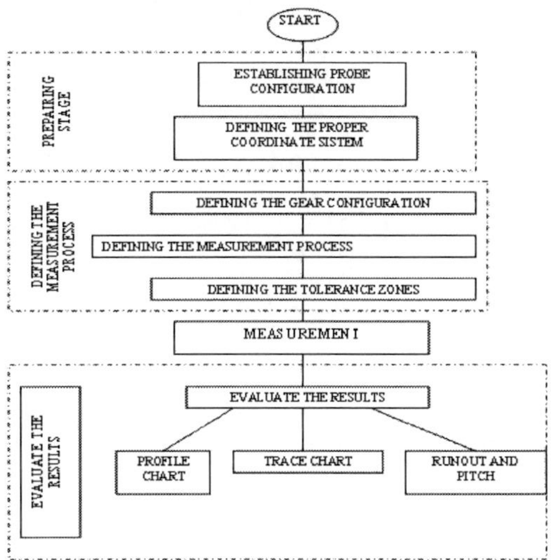

Fig.3. Logic of measuring process

2. MEASURMENT

For the pinion shaft, the profile modification will be calculated considering the diametral pitch, the pressure angle and the number of teeth.
The parts should be cleaned and adjusted before the measurement, and should be kept at least 24 hours in the measuring environmental conditions.
These probe configurations allows all the necessary measurements, both profile and trace measurement and the run out and pitch measurements. The great mobility of the GHIBLI TRAX machines offers the possibility of probe movement around the part, and thus, through a proper placement of the part in the measuring volume combined with a proper alignment, gives access to all the necessary measuring points, without the need of a rotary table. The probes must be calibrated before measurement.
For this kind of measuring machine, establishing the proper coordinate system, linked to the part and its characteristics, is

an essential mater, which can influence decisively the measurement results (Brown & Sharpe (1996).
The direction of the internal bore will give the Z axis direction, while in the case of the pinion, the shaft direction is the Z direction (fig. 4).

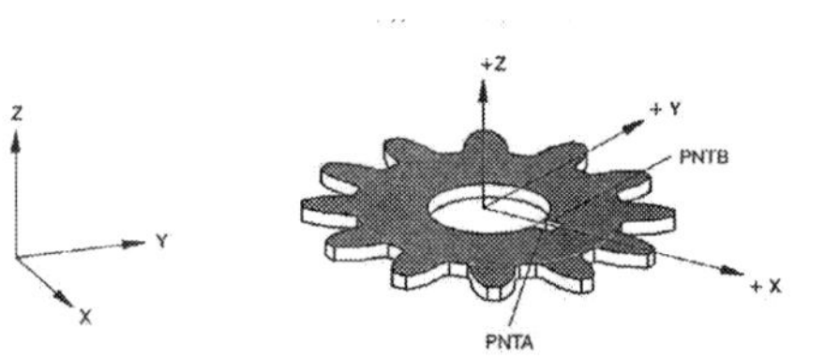

Fig. 4. Alignment scheme for the gear

For the parts we considered, a planar alignment is not necessary, because the symmetry of the splines. During the gear measurement procedure, a special planar alignment is made by QUINDOS, probing a reference tooth or gap, but in order to avoid the danger of running into a collision, we shall place a gap approximately on the X-direction.
For the AGMA defined profile, the addendum modification coefficient will be at first calculated considering the number of teeth, the pressure angle and the diametral pitch.
In case of the gear, defined according to DIN, the quality number according to DIN can be introduced, and all the corresponding tolerances will be taken into account. In case of the pinion shaft, the problem is a little more complicated. The involute is defined trough a tolerance zone included in the K-chart. The MEGEAR command, offer the opportunity to define such a K chart, introducing the values from the drawing, converted in roll length. (see fig. 2) Also, for trace evaluation, the crowning must be considered. As special measurements for pitch and run out, for the gear, the dimension over five teeth must be evaluated. In case of the pinion shaft, the dimension over balls of Φ14.630 must be evaluated. For the run out and pitch measurement a point on the pitch diameter will be probed for each flank. The clearance points and the probing points will be calculated by the program. The profile and trace diagram for the gear is presented in fig. 6

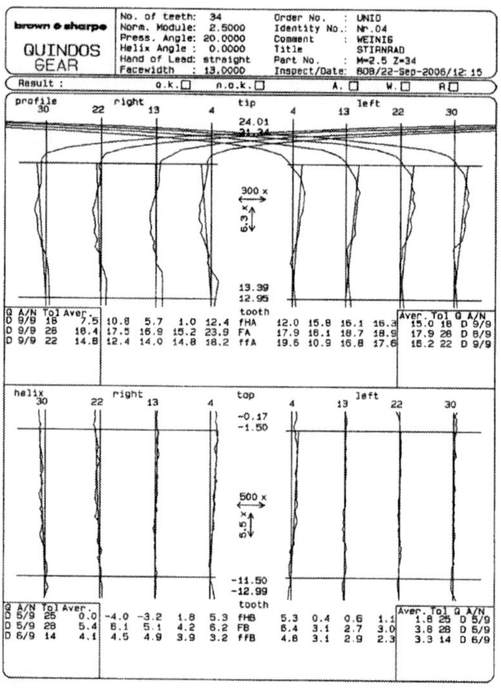

Fig.6. Profile and trace diagram for gear

The profile and trace diagram for the pinion shaft is presented in fig. 7

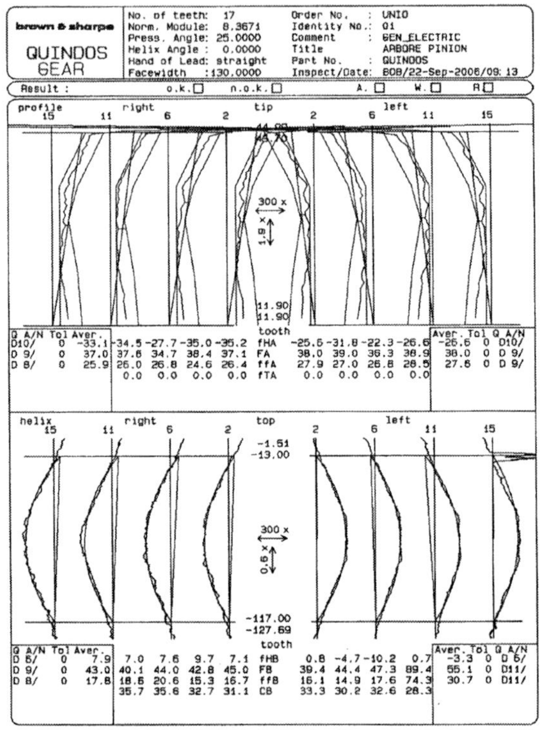

Fig.7. The profile diagram

For the pinion shaft, the pitch diagram also was realized.
We notice that the gear is also DIN9, and in the bottom of the diagram we have the transverse span over 5 teeth, calculated from the points collected, with the average value inside the tolerance zone.

3. CONCLUSIONS

- Using the GHIBLI TRAX measuring machine, the measurement of the gears was possible without rotary table or special devices
- All the measurements were performed inside a single command, including special measurements and the profile tolerance zone of the gear
- The precision for the pitch measurement and for the special measurements like dimension over balls and the span over teeth is better due to the way in which they are obtained (calculated, without probing points with special devices)
- The measuring time was considerably reduced.

4. REFERENCES

Brown & Sharpe (1996). ***Quindos .Users Manual and Tutorial. 3rd edition***, Wetzlar, Germany

Brown & Sharpe (1992). ***Quindos .Measurement of Cylindrical Gears. Reference Manual,*** Wetzlar, Germany

Brown & Sharpe DEA Spa (1999). ***Ghibli Trax Measuring Machine. User's Manual,*** Grugliasco, Italy

Dudley, Darle W.(1994). ***Handbook of Practical Gear Design,*** Technomic Publishing Company, ISBN 1-56676-218-9, Lancaster, Pennsylvania, USA

DIN 3960, (1976). Concepts and parameters for Cylindrical gears ,Germany

Annals of DAAAM for 2006 & Proceedings of the 17th International DAAAM Symposium, ISSN 1726-9679
B. Katalinic, Published by DAAAM International, Vienna, Austria 2006
150th Birthday of Nikola Tesla (1856-1943) Inventor of Mobile Robot Teleoperation

SOME CHARACTERISTICALLY ASPECTS OF RECYCLING THE HUGE AGGREGATE FROM PAPER INDUSTRY

GYENGE, C.; ROS, O.; FULOP, I.; ITU, C.; TODEA, M. & COSTEA, A.

Abstract: One of the main conditions regarding Romania's integration in the European Community needs some qualitative changes in most of the areas of the industry. Therefore, in paper and cellulose industry is a necessary condition to modernize the technology and the equipments. In this program one specific technical problem is related to the disassembly and the recycling of the drying cylinder from paper drafter unit. On this paper we'll present the specific problems which should be solved, some adopted solutions and in the end the concrete achievement of the disassembly.
Key words: *paper machine, drying cylinder, milling device*

1. DESCRIPTION OF THE PAPER MANUFACTURING AGGREGATE

The utilized aggregate is composed by the following main parts (fig.1.):

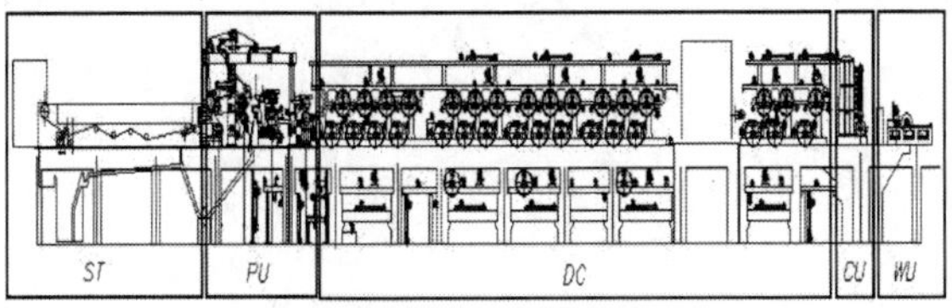

Fig 1. The basic scheme of the paper aggregate

- ST - the table of the sieve
- PU - press unit
- DC - drying cylinders
- CU – calender unit
- WU – wrapping unit

The paper sheet is passing through the drying cylinders heated with steam. In the calender unit **CU** the paper is pressed in continuation in order to obtain gloss and then the paper finalized in this way is rolled up in wrapping unit **WU**

2. THE NECESSITY OF DISASSEMBLY THE DRYING CYLINDER TYPE "YANKEE"

Paper machine from the firm SC. Somes Dej was designed to function at a nominal speed between 380 m/min and 450 m/min. To increase the working speed of the machine it is was necessary to replace some components that can limit the possibility of increasing the speed. To change this drying side of the machine implies the change of the drying cylinder with 5 smaller drying cylinders.

As a result, it is necessary the disassembling of the actual drying cylinder and the preparation of the machine for the assembly of the new five cylinders.

Taking into account the weight of the drying cylinder (fig.2) of appreciatively 70 tones, there wasn't possible the elevation of the drying cylinder through classical methods, for example like usual cranes, hoists, bridge-cranes, because the maximum elevation capacity of the bridge-crane in the assembly room was of 32 tones. Disassembly of the drying cylinder implies the necessity of finding other solutions.

Fig.2. Side view of the drying cylinder

2.1. Analyzed solutions for the disassembly of the cylinder

A. The total disassembly of the cylinder without any dismantling

First alternative of removal of the drying cylinder took it into account was the disassembly if the cylinder without dismantling.

Adopted solution was abandoned due to some of its disadvantages as following:

-Expensive costs of the work (180.000 €). Working time of the work exceeds 3 weeks.

-Consolidation of the platform floor assumes the deviation of some technological pipes routes and some bridges for supporting electric cables, work that may take 1 week.

-Manufacturing a special base, a metallic structure on which it will be laid the drying cylinder with adapted bearings.

-After the setting of the drying cylinder on the metallic base and its translation to the charging platform, the loading on a special transport means it could be done only with a handling crane of 350 tones; equipment renting costs are also very high.

B. Disassembly of the drying cylinder by plasma-jet cutting

This method was unfavorable of all the analyzed solutions due to the very imminent danger of producing fire burnings.

The method consists in the cutting of the drying cylinder casing in many pieces so that the necessary load for lifting the bridge crane not to exceed 32 tones.

C. Disassembly by drilling.

The method consist in applying on the cylinder casing of some holes of 10 mm in diameter (d=10 mm) with the pitch of 15

mm in 3 directions obtaining in the end 3 spherical cap from the drying cylinder casing with a weight that allows an easy handling by the help of the bridge crane.
The solution consists in manufacturing a device on which should be located the drilling machine, using in this way, a total of 1500 number of holes of d=10 mm.
The problem arise here in applying this solution is the possibility of breaking the contact bridges resulted as a consequence of the drilling process and taking into account the fact that the thickness of the casing is of 40 mm.

D. Combined disassembly by drilling and milling.
We have analyzed 2 possibilities of milling:
a). by rounding cutter with alternative teeth with the diameter of 350 mm from 2 passes;
b). by end-milling cutter with nicked teeth with diameter of 18 mm from rapid steel covered by TiNC.
Analyzing different alternatives of milling devices in order to be mounted on the cylinder and in order to assure the possibility of the mill feed we jumped the conclusion that only the last solution, the one made by end-milling cutter it could be applied.
Milling device was fixed by 4 screws M 10 in the cylinder holes made before, and it was moved step by step along the canal in the milling cut on the length of 250-300 mm. The required end-milling cutters with nicked teeth covered by TiNC is of 5 pieces, purchased at the price of 130 € / piece.
At the first tests it could be remarked the appearance of some big oscillatory movements in cross direction and it was necessary the improvement of the device by using 2 guide bars. (fig.3)

Fig. 3. Improved device by 2 guide bars

3. RESULTS OBTAINED

Using the combined process of drilling and milling with the presented device (fig.3) and the following technological parameters (which was estblished in cooperation with tools firm producer EMUGE FRANKEN TOOLS):

a) Drilling ϕ 10 mm (to assure the mill entrance):
$v = 15\ m/min,\ f = 0,3\ mm/rot$

b) Milling with end-mill cutter ϕ 10 mm;
$v = 20\ m/min,\ v_f = 34\ mm/min.$

It was not possible to use some faster cutting conditions because the cylinder has over 3000 mm in diameter and 3.5 m in length, on a height of 7 m, and the device used can not assure a great stability.
Using these cutting conditions we achieved the total manufacturing time:

Drilling: $t_m = 3$ h
Milling: $t_m = 33{,}5$ h

Total costs of the dissasembling and transport of the drying Cylinder: C=12.328.8 Euro
The cylinder was disassembled in 3 cylindrical sectors (fig.4) and in 2 flanges and the axle (fig.5).

Fig. 4. Cylindrical sectors after disassembly

Fig. 5. Front view of the cylinder after disassembly

4. CONCLUSION

After a long study and some practical tests it could be accomplished the disassembly of the cylinder, on the impose requirements and also creating the conditions of the recycling and modernizing demands.

5. REFERENCES

Boothroyd, G, Dewhurst, P., knight, W., (1994), *Product Design for Manufacture and Assembly*, Marcel Dekker inc

Gyenge Cs., Fulop I., Hosu A. & Costea A.(2005): *Some Characteristics Aspects regarding the Maintenance of Equipments for paper Industry.* Proceedings of microCAD 2005 International Scientific Coference 10-11 March,2005,Section M, Miskolc. ISBN 963 661 646 9 0

Gyenge Cs., Fulop I., Hosu A. & Costea A. (2005): *Main aspects regarding the maintenance of equipments for paper and pulp industry.* Proceedings of 9th International TMTM 2005 Conference, Antalya, Turkey. 26-30 sept.2005

Grigoras, St. (1991) *Service life of the reconditioned line contact versus classic one*, Bulletin Inst. Politehnic, Iasi,

Smith, D. (2000) *Reliablity, maintainability and risk*, Practical methods for Engineering- fifth edition

Annals of DAAAM for 2006 & Proceedings of the 17th International DAAAM Symposium, ISSN 1726-9679
B. Katalinic, Published by DAAAM International, Vienna, Austria 2006
150th Birthday of Nikola Tesla (1856-1943) Inventor of Mobile Robot Teleoperation

PROCESS MODELING AND DEVELOPMENT IN SMALL AND MEDIUM SIZED COMPANIES IN FINLAND

HAAPALAINEN, P. & MAEKIRANTA, A.

Abstract: *Process modeling is a useful tool that can be utilized in process development. However, it is not commonly used in smaller companies. In this paper we introduce first the concept of process modeling and the connection between modeling and developing processes and then some empirical findings about the situation of process modeling in the small and medium sized production companies in Finland.*
Key words: *process modeling, process development.*

1. INTRODUCTION

Our research group was involved in process modeling in a public organization case and after that we wanted to find out what is the situation of process modeling in process development in Finnish small and medium sized enterprises. Process modeling is one of the tools that can be used in process development. It seems to be typical that in many larger organizations processes are developed and specially preparing quality handbooks has forced companies to also model their processes. However, we suspected that situation in smaller companies might be different.
Thus our research questions were:

1. Have the companies modeled their processes and has the modeling been used in process development?
2. Is process modeling considered to be useful?

A survey was made to gather research material. In chapter 3 there is a full presentation of the empirical part of out research. In chapter 2 there is some theoretical background for this research.

2. THEORETICAL BACKGROUND

2.1 Process modeling

Process is a set of causes and conditions that repeatedly come together to transform inputs into outcomes. (Moen et al., 1991). Bergman and Kelfsjö (1994) see that a process is a transformation of inputs into outputs with help of careful planning. Scherkenbach (1991) says that general resources: people, method, material, equipment and environment are both inputs and outputs. All resources transform, or change their state, over time. That is why one should not simplistically focus, for example, on material as the process output.
There are always one or more customers for the final results of the process. The customer satisfaction of high level by using as small an amount of resources as possible is the main goal of each process. The suppliers of the process have to be identified, as well as, the owner of the process, who is responsible for the improvement work of that process. Horizontal processes, traditionally, do not have any owners. An owner should be appointed, however (Bergman & Klefsjö 1994).
According to Bergman and Klefsjö (1994) the process is a link between history and future. The process data and the measurements of the results indicate the customer satisfaction.

There are several ways to model a process, for example, flowchart, action chart and cause defect analysis. These are all useful ways to show in paper how a process works and what happens in it. In Fig 1. is one example of modeling process.
The process inputs and outputs as well as the most important actors (process owner, the customer of the process, suppliers etc.) related to process are usually presented in process models. Also the relations between the process and sub processes are often depicted. Modeling has an important role in process development because it reveals where possible problems could lie.

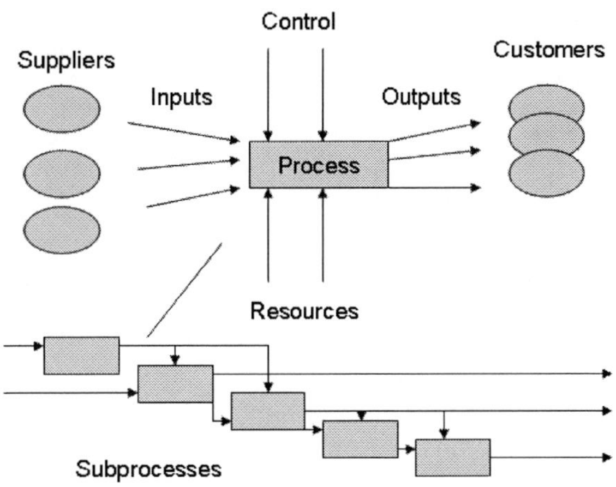

Fig. 1. *Modeling the process. (verify Marsh, IDEF standard)*

2.2. Process development

Process improvement can be seen as a continuous effort to learn about the cause system in process and to use this knowledge to change the process to reduce variation and complexity and to improve customer satisfaction. (Moen et al., 1991).
Before you can improve the process you have to know the current state of the process. That is why you need to model processes. After modeling you need to analyse the process to choose the way to develope it.
The most important success factors for process development according to Lager & Hörte (2005) are:

- Changes in the production process resulting from successful process development work are fully accepted by the production organisation.
- The development organisation includes individuals with suitable personal qualifications for process development work.
- Development results are measurable and trustworthy.
- There are good incentives and driving forces for process development.
- Well formulated and measurableproject objectives.

There are several ways to improve process. Sometimes small changes can be enough. In other case it can be essential to renew the whole process. As an extreme case can be seen closing down of the process and outsourcing or in the other hand the expanding of process by integrating supplier and customer processes to it.

3. EMPIRICAL FINDINGS

3.1 Research method

In order to find out the current state of process modeling and process development in Finnish companies, a survey was made. The target group was small and medium sized production companies. The survey was sent to about 200 companies around Finland in May 2006. The respondents were CEOs, quality managers or production managers of the companies, however, in small companies the positions and tasks are not necessarily so divided. The respondents received an e-mail that contained a link to web page containing the questionnaire.

The questions were kept short and simple to get as many answers as possible, however, it is likely that the used method caused that some people didn't want to answer. The following questions were included in the survey:

- Have you modeled your processes and if you have, which methods have you used?
- Have you developed you processes and if you have, have you utilized process modeling in process development?
- Do you feel that process modeling is / could be useful in process development?
- Do you plan to model / develop your processes in the future?

The questions were formulated so that information about both process modeling and process development and their relationship would be gained.

3.2 Results

The answering rate to the survey was about 10 %, which can be considered to be quite good for this kind of information gathering, Most of the answering companies were located less than 100 km away from Vaasa, where the university submitting the survey is situated. The answering companies represent the following producing industries: metal, plastic, electronics, textile and furniture. The sizes of the companies are in the table 1.

Employees	< 10	11-30	31-50	51-100	> 100	
Number of answers	2	4	4	5	1	
Turnover (million €)	< 1	1-2	2-5	5-10	10-20	> 20
Number of answers	1	3	4	5	1	1

Table 1. The size of the answering companies.

50 % of the answering companies said that they have both modeled and developed their processes. They had utilized process modeling in process development and all of them felt that modeling had been useful and they intended to use it also in the future. The modeling techniques used included the following: PDCA circle, flowchart (the most common), wall posting technique, and simulation. The reasons for using these methods were that it was noticed that they worked well or a consultant had recommended them.

Processes had been developed but not modeled in 25% of the companies that answered the survey. However, three out of four of these companies felt that modeling might be useful in process development. The common reason that these companies had not modeled their processes and were not certain that they would like to do it in the future was that they didn't know enough of modeling processes.

25% of the answering companies said that they had neither developed their processes nor modeled them. Half of these companies felt that both modeling and developing processes should be done in the future. Reasons for not developing processes were that they were not having enough time or not having enough knowledge of how to do modeling. These reasons reflect well what we expected to be found.

4. CONCLUSION

Today's fierce competition forces companies to improve their processes to make them more effective and more efficient. Production should be faster and cheaper and also the other processes can usually be improved. If the current state of the process is not known, it will be difficult to try to make it better. That's where process modeling comes into the picture.

The process input and outputs as well as the clients and the suppliers of the process are often depicted in the process model. Several different methods can be utilized when modeling processes but as out survey suggests these methods are not very well known in smaller companies and process modeling is considered to be quire complex and time consuming task.

Both process development and process modeling are considered to be important but in spite of this only half of the respondent companies had done them both. It also seems that it is known in companies that process modeling is a useful tool in process development but the fear of modeling methods and lack of time cause that it is not utilized more.

In many answers it was shown that consultants were used in the process modeling process. Hiring an outsider consultant may be too expensive for some of the smaller companies. For the future research, there is a need for spread the knowledge about simple process modeling methods that could be used in the companies by the personal themselves. Also the role and the benefits of process modeling for process development should be clarified.

5. REFERENCES

Bergman, B. & Klefsjö, B. (1994) *Quality from Customer Needs to Customer Satisfaction,* McGraw-Hill.

IDEF. A Structured approach to enterprise modeling and analysis. Via the internet at http://www.idef.com Knowledge based systems, inc. (KSI). 1998.

Lager, T. & Hörte, S-Å. (2005*) Success factors for the development of process technology in process industry.* Part 1: a classification system for success factors and a rating of success factors on a tactical level, International Journal Process Management and Benchmarking, Vol. 1, No. 1, 2005, Inderscience enterprises.

Marsh, J. (1993) *Quality Toolkit, An A-Z of Tools and Techniques*, IFS ltd.

Moen, R. ; Nolan, T. & Provost, L. (1991) *Improving Quality Through Planned Experimentation,* McGraw-Hill.

Scherkenbach, W. (1991) *Deming`s Road to Continual Improvement,* SPC Press, Inc.

6. ACKNOWLEDGEMENT

We would like to thank Jukka Rohila and Miiled for providing us the Internet survey environment that was used when gathering the empirical data for this research.

Annals of DAAAM for 2006 & Proceedings of the 17th International DAAAM Symposium, ISSN 1726-9679
Editor B. Katalinic, Published by DAAAM International, Vienna, Austria 2006
150th Birthday of Nikola Tesla (1856-1943) Inventor of Mobile Robot Teleoperation

MATHEMATICAL MODELING OF FEATURE CONVERSION FOR INTEGRATION OF CAD AND ASSEMBLY PLANNING

HAMIDULLAH, H.; BOHEZ, E. & IRFAN, M.A.

Abstract: *Feature conversion has been an active research area in feature-based modeling for many years. It is considered to be a critical component for concurrent integration of CAD with various down stream applications. Features are application dependent. Each application has its own feature model that contains features relevant for that application. That's why data in design domain can not be directly used in the down stream applications, like assembly planning, etc. To create feature models for down stream applications, feature conversion plays an important role. This paper addresses the problem of how to convert the design features into assembly features, with the help of a mathematical model for feature conversion. The mathematical model is represented by a mapping function which is a function of two other mapping functions. The first mapping function extracts the form features on the mating parts from the design model. The second mapping function groups these extracted form features into assembly features, according to the assembly relationship between these features, and matches them with the pre-defined assembly features.*
Keywords: *Design features, assembly features, mathematical modeling, feature mapping, feature-based design.*

1. INTRODUCTION

Features are used in different areas in mechanical engineering and are considered as a main factor in the integration of CAD with downstream applications. Because various design, engineering, manufacturing, and assembly data can be associated with a feature. Feature-based design means building a product model with a predefined set of design features. Feature-based design provides an explicit product data in a neutral format which can be used directly with down stream applications (Amaitik & Kilic 2005). Feature recognition is the interpretation of a geometric model of an object to identify the relevant features and associated information. Feature conversion, also called feature mapping is the process of deriving a new feature model for some view from an existing feature model for another view.
This paper presents mathematical model for feature conversion from design model to assembly model after incorporating necessary modifications in the model developed by (Gao J. et al. 2004). The model of Gao J. et al. used to convert design features into machining features is modified for conversion of design features into assembly features. Also, their model based on extraction of low-level geometric entities (faces) is modified for extraction of meaningful entities (features). Mathematical modeling for feature conversion from design feature model to assembly feature model can be the future research target.

2. FEATURE REPRESENTATION

2.1 Representation of features in design domain

Design features are defined as a combination of abstract form features and design intents. Design feature can be represented mathematically as:

$$DF^i = F_d^i \cup I_d^i \tag{1}$$

Where, DF^i shows an ith design feature, F_d^i represents an abstract form feature in design domain, and I_d^i shows design intents. I_d^i includes material features F_m, precision features F_p^i, and functional features F_f^i. With this definition of design intents, the design feature DF^i can be represented as:

$$DF^i = F_d^i \cup \left(F_p^i \cup F_f^i \cup F_m\right) \tag{2}$$

2.2 Representation of features in assembly domain

Assembly feature is defined as, an association between two form features present on different parts (Shah & Mantyla 1995). Assembly features are used as an abstraction to specify several elementary relations. By combining such elementary relations, higher-level relations called assembly features can be generated. These assembly features can then be extended with additional assembly-specific information, such as degrees of freedom and fit information. Using this definition, assembly feature can be mathematically represented as:

$$AF^i = F_u^i \cup I_a^i \tag{3}$$

Where, AF^i shows an ith assembly feature, F_u^i represents union of form features in assembly domain, and I_a^i is the assembly intents. Assembly intents include assembly operations A_o^i (e.g., mating, fastening, fusion), assembly type A_t^i (e.g., align, against, fit, contact, insert), and other assembly attributes A_a^i (e.g., feature's position and orientation or degree of freedom). Therefore, assembly feature is mathematically represented as:

$$AF^i = F_u^i \cup \left(A_o^i \cup A_t^i \cup A_a^i\right) \tag{4}$$

The geometry of an assembly feature F_a^i can be represented as, an association between two different form features on different parts or components, and their assembly relationship.

$$F_a^i = \left(F_1^i \cup F_2^i\right) \cup R^i = \left(\bigcup_{u=1}^{2} F_u^i\right) \cup \left(R^i\right) \tag{5}$$

Where, F_u^i represents union of feature F_1^i and feature F_2^i. R^i shows assembly relationship between the mating features F_1^i and F_2^i at the ith mating pair. Examples of assembly features are: rib-slot, pin-hole, and dove-tail etc.

3. MATHEMATICAL REPRESENTATION OF FEATURE CONVERSION

Suppose U is a set domain of form features in the design domain, and V is a set domain of form features in the assembly domain, then $P(U)$ and $P(V)$ are the index set of the domain U and of the domain V. Apparently, a feature set SF_d in the design domain and a feature set SF_a in the assembly domain should be a point set of U and V, i.e., $SF_d \in P(U)$, and $SF_a \in P(V)$.
If there is a rule, $g_{d\text{-}a}$ by which for every element (feature) $x \in SF_d$, an element (feature) $y \in SF_a$ can be created correspondingly, then this rule $g_{d\text{-}a}$ is called form feature

mapping function from SF_d in the design domain to SF_a in the assembly domain, and is represented as:

$$g_{d-a} : SF_d \rightarrow SF_a \quad (6)$$

Where, SF_d is known as "definition" domain of g_{d-a} and is represented by *dom* $g_{d-a} = SF_d$; while SF_a is called "value" domain and is represented by $val\ g_{d-a} \subseteq SF_a$. If an element y is an image of an element x under the action of g_{d-a}, then,

$$y = g_{d-a}(x) \quad (7)$$

$$g_{d-a} : x \rightarrow y \quad (8)$$

In other words, the mathematical representation of feature mapping from the design domain to the assembly domain is represented as:

$$g_{d-a} : SF_d \rightarrow SF_a \quad (9)$$

$$x \mapsto y = g_{d-a}(x) \quad (10)$$

3.1 The mapping function g_{d-a} of form features

The process of feature conversion from design domain to assembly domain is divided into two stages. The first stage is carried out through a process mapping function g_1 from an abstract form feature set F_d to a transition feature set F_g. The second stage is carried out through a process mapping function g_2 from the transition feature set F_g to assembly feature set F_a. Consequently, the form feature mapping function g_{d-a} is a composition of the function g_1 and g_2 represented as:

$$g_{d-a} = g_2 \circ g_1 : F_d \rightarrow F_a \quad (11)$$

Where, $\circ$ represents composition mapping by g_1 and g_2.

3.2 Process mapping function g_1

Suppose F_d be the abstract form feature set, and F_g be the feature set corresponding to F_d in a mating pair. F_d is basically a form feature F, for example, rib, slot, or hole and will therefore be represented as F when assigned to F_g. If *dom* $g_1 = F_d$, $val\ g_1 \subseteq F_g$, then, g_1 is a process mapping function from F_d to F_g, represented as:

$$g_1 : F_d \rightarrow F_g \quad (12)$$

$$x_1 \mapsto y_1 = g_1(x) \quad (13)$$

Where, $x_1 \in F_d$ *and* $y_1 \in F_g$

$$g_1(x_1) = g_1(F_{d1}^{12} \cup F_{d2}^{12} \cup F_{d1}^{13} \cup F_{d3}^{13} \cup ... \cup F_{dp}^{pq} \cup F_{dq}^{pq})$$

$$= ((F_1^{12} \cup F_2^{12}) \cup (F_1^{13} \cup F_3^{13}) \cup .... \cup (F_p^{pq} \cup F_q^{pq}))$$

$$= F_g^{12} \cup F_g^{13} \cup .... \cup F_g^{pq} \quad (14)$$

Where, F_g^{pq} is the association of the form features F_p and F_q at the pq^{th} mating pair.

3.3 The process mapping function g_2

Let F_g be feature set obtained from the process mapping function g_1, showing association of form features at mating pairs in a given assembly. Let F_a be a set of assembly features. If dom $g_2 = F_g$, $val\ g_2 \subseteq F_a$, the function g_2 is called the process mapping function from F_g to F_a and is written as:

$$g_2 : F_g \rightarrow F_a \quad (15)$$

$$y_1 \rightarrow y_2 = g_2(y_1) \quad (16)$$

On the basis of the abstract form features derived by the function g_1, the process mapping function g_2 is applied to search and deduce the features attributes and their assembly relationships, and to deduce the description of assembly feature similar to that of the pre-defined assembly features.

Using this explanation, g_2 can be represented as:

$$g_2 = g_{22} \circ g_{21} : F_g \rightarrow F_a \quad (17)$$

$$y_1 \rightarrow y_2 = g_2(y_1) = g_{22}(g_{21}(y_1)) \quad (18)$$

Where, g_{21} is searching function used to deduce the feature attributes and assembly relationships, and g_{22} is constraint function used to re-associate the features and inspect their relationships.

Let REL be a set of assembly relationships, represented as:

$$REL = \{R_{align}, R_{against}, R_{fit}, R_{contact}, R_{insert}\} \quad (19)$$

$P(REL)$ is an index set of REL, which includes all subsets except the empty set, namely, $P(REL) \neq \phi$. Hence, the result of the searching function g_{21} is "a union of mating form features and their corresponding assembly relationships".

$$g_{21}(y_1) = \bigcup_{mn=12}^{pq} \left(\bigcup_{i,j=1}^{p,q} F_{i,j}^{mn} \cup P(REL) \right) \quad (20)$$

Putting equation (20) into equation (18), function g_2 becomes:

$$g_2(y_1) = g_{22}(g_{21}(y_1)) \quad (21)$$

$$= g_{22}\left(\bigcup_{mn=12}^{pq} \left(\bigcup_{i,j=1}^{p,q} F_{i,j}^{mn} \cup P(REL) \right) \right)$$

$$= g_{22}\left(\left(\bigcup_{i,j=1}^{1,2} F_{i,j}^{12} \cup P(REL) \right) \cup \left(\bigcup_{i,j=1}^{1,3} F_{i,j}^{13} \cup P(REL) \right) \cup .....\cup \left(\bigcup_{i,j=1}^{p,q} F_{i,j}^{pq} \cup P(REL) \right) \right)$$

$$= (F_1^{12} \cup F_2^{12} \cup R^{12}) \cup (F_1^{13} \cup F_3^{13} \cup R^{13}) \cup .... \cup (F_p^{pq} \cup F_q^{pq} \cup R^{pq})$$

$$= F_a^{12} \cup F_a^{13} \cup ..... \cup F_a^{pq}$$

$$= \bigcup_{mn=12}^{pq} F_a^{mn} \quad (22)$$

Where, $i = 1, 2, 3...p$, $j = 1, 2, 3...q$, $m = 1, 2, 3...p$, $n = 1, 2, 3...q$, $i \neq j$, $m \neq n$. R^{pq} shows assembly relationship at the pq^{th} mating pair. Above equation (22) shows how the feature mapping process from F_g to F_a is carried out through the process mapping function g_2.

4. CONCLUSIONS

Using a mathematical description of the feature mapping process, assembly features are deduced and formed by Set Operation, and the procedure of converting design features to assembly features is represented by two process mapping functions g_1 and g_2. Function g_1 extracts form features from assembly model in design domain and assigns them to a transition feature set F_g which represents association of two form features at a mating pair in a given assembly. Function g_2 defines assembly relationship for each pair of mating features by a function g_{21}. It combines the mating features and checks the assembly relationships corresponding to the pre-defined assembly features by another function g_{22}. Thus, the research presented in this paper plays an important role in the integration of CAD and assembly planning through feature conversion.

5. REFERENCES

Amaitik, S.M. and Kilic S.E. (2005), STEP-based feature modeler for computer-aided process planning, *International Journal of Production Research*, 43, 15, 3087-3101.

Gao, J., Zheng, D. and Gindy, N. (2004), Mathematical representation of feature conversion for CAD/CAM conversion system integration, *Robotics and Computer Integrated Manufacturing*, 20, 457-467.

Shah JJ and Mantyla M. (1995), *Parametric and feature-based CAD/CAM: concepts, techniques, and applications*, John Wiley & Sons Inc, 0-471-00214-3, New York.

Annals of DAAAM for 2006 & Proceedings of the 17th International DAAAM Symposium, ISSN 1726-9679
B. Katalinic, Published by DAAAM International, Vienna, Austria 2006
150th Birthday of Nikola Tesla (1856-1943) Inventor of Mobile Robot Teleoperation

VISUALIZATION OF INFORMATION ON TRAIN POSITION MONITORING IN RAILWAY CONTROLLING

HARAMINA, H.; TOS, Z. & JURUM-KIPKE, J.

Abstract: *Participation of humans in the tasks regarding surveillance and control of the railway traffic process represents an important factor for safe and technologically optimal flow of this process. New solutions in this field are provided by automation of these processes. For a human to perform the tasks regarding surveillance and control of traffic as reliably and efficiently as possible, the relevant information about the traffic condition need to be visualized in an appropriate manner. This paper tends to analyze the visualisation of information at the interface between the human and the technical means, realizing the activities in the process of surveillance and control of railway traffic within railway station area of minor and mid-size stations.*
Key words: *railway traffic control, train routing system, reliability*

1. INTRODUCTION

The development of railway traffic meant also the changing in the way of controlling train traffic in the railway traffic process. All the methods of control, regardless of the level of technical development have a characteristic in common, that in the existing traffic flow conditions they tend to achieve the highest possible safety level at the same time optimizing the technological operation in the railway traffic process. Railway scheduling of trains as an operation plan at the railways depends among other things also on various types of systems of signal-safety instruments which also affects the method of traffic control. This means that the technically more advanced solutions regarding railway traffic control system allow greater efficiency of railway traffic within its intended function. There are systems with different control levels that have been classified according to hierarchy. In such systems the decisions are made primarily in the highest level of control whereas lower levels with authorisation from higher levels may control traffic in certain segments of the overall line section. The development of new solutions is stimulated by the tendency to reduce the human activities requirements in the control procedures in order to avoid the possibility of human error, i.e. in order to achieve greater reliability in the operation of the system which includes the participation of humans. Today numerous informatics solutions are being developed, in the form of expert systems which help humans in processing the collected information and in making valid decisions related to the process of railway traffic control. Thus, the volume of human activities in the control process would be reduced and the content of the encountered problem changed, providing also the opportunity to increase the reliability of work as well as of the entire system of control in which humans participate as an inevitable component.

2. PROBLEM OF SAFETY AND OPTIMIZATION IN RAILWAY TRAFFIC

Optimization of the traffic process along with the safety of railway traffic represents an important factor in finding new solutions in the control process of the train traffic through the railway network. With the use of new technical solutions, new control methods have also appeared. The railway traffic safety does not depend, namely, only on the investment into new technical means but it is rather much more important to set the rules in the control process and to strictly comply with them. In this way the probability of undesired occurrences in traffic is reduced. Optimal flow of railway traffic mainly means the realization of the planned schedule for a certain railway network area. It is often the case that trains cannot operate as planned by the schedule. Such cases occur due to frequent disturbances in the network which expand depending on the schedule of the movement of railway vehicles and their hierarchy according to which higher-priority trains have the right of way to operate compared to the lower-priority trains. The propagation of disturbances through the railway network results in the delays of certain trains regarding the schedule. This delay is more pronounced in lower-priority trains, which are mainly cargo trains, which are sometimes said to be "pulling through" the railway network. The dispatcher controlling the traffic on a certain line section, by means of the central control system, attempts in this way to balance the actual situation of train operation with the one planned by the schedule. Since this is not always entirely possible, the dispatcher needs to try to optimize the train operation in real time which depends on their ingenuity, as well as experience in the job. The dispatcher controls traffic on a relatively long line section and deals mainly with the optimization of traffic flows by guiding trains through the line section for which they are in charge. The jobs related to the technological process of station operation are mainly entrusted to the lower level of management (regarding the station level) so that activities with trains determined by the technological process of station operation would be performed from that level. However, in case the line section with central management accommodates only minor or mid-size stations in which the technological process is not so complex nor comprehensive as at major railway stations, such as arranged station and shunting yards, then the dispatcher could be allowed to control the technological process at stations on their respective line section. In order to realize this, it is necessary to provide the dispatcher with the required collected information from the traffic process in such a way as not to exceed the volume of information that could result in concentration decline, i.e. reduction in the reliability of the railway traffic management system.

3. ADVANCED TECHNOLOGIES AS THE BASIS FOR THE DEVELOPMENT OF RAILWAY TRAFFIC CONTROL SYSTEM

The development of computer technologies as well as the development of information sciences has had great impact also on the development of the visual interface for the dispatcher who centrally controls the railway traffic. The possibilities offered by advanced computer technology allow skilful way of adopting information necessary to control the traffic process at the railways. This does not only refer to the optimization of the traffic flow of trains and setting of routes on the open tracks or at stations but also to the performance of technological operations at stations done according to precisely defined instructions for which good knowledge is required about the technological work process at a certain railway station. Harmonization of train operation according to a given schedule which is contained in the timetable, imposes the dispatcher's requirement to focus on the task of optimizing the traffic process, particularly in case of major deviations of trains from the given routes in the schedule, where the dispatcher's expert skills may become most prominent. In such complicated situations, in order to alleviate the burden of some of his tasks, the dispatcher transfers a part of his responsibilities related to the technological operation process at a certain railway station to the lower level of station management field. In such situations often problems occur related to the communication between various levels of management. Because of more reliable communication between different managerial levels, the dispatcher's interface should be provided with graphical presentation of relevant information that are simultaneously transferred in another way as well. For instance, everything transferred by special telephone line could be also presented in writing on an interface, so that the dispatcher can be sure that he has understood correctly the message given from some lower management level. Besides, the dispatcher's confirmation would unambiguously confirm the receipt and understanding of the contents of the spoken and textually written message.

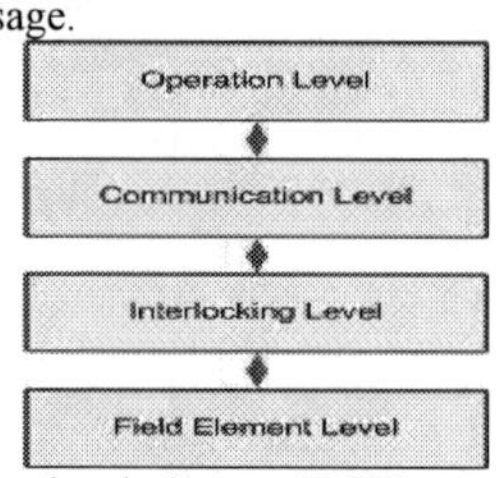

Fig. 1. System levels (source IRSE)

In order to provide the dispatcher with the relevant conditions in traffic in a satisfactory way, it is necessary to equip his interface with adequate hardware equipment (Woodson 1992). Such equipment includes different tools and reliable monitors for graphical presentation of the conditions of stable and mobile elements of traffic process that should be designed so as to best match the presentation of the used software. Using advanced computer technologies it is possible to develop flexible software that in semi-automatic railway traffic control could serve as the basis for the upgrade of various auxiliary applications that can additionally improve the work of the central dispatcher in different circumstances of traffic process control. Flexible software means that it can be adapted to the conditions that characterize the situation regarding the technical condition of the tracks, primarily the instrument type of security, surveillance and traffic management. The flexibility of such software is also necessary for the very operation of the traffic process which would, regarding its characteristics, emphasise the need to including individual software functions as well as their necessary modification. It is namely the detailed research of each line section that would be used to select the priorities with the emphasis on safety and optimization of the traffic process on that line section, and modifications would then be performed on the standard type of software by subtracting or adding certain functions it offers. For example, it would not be always necessary for the system to visualise to the dispatcher all the data related to trains that intersect at one of the traffic places of work within his control area. The dispatcher would for instance present the data on the exact time of stopping of the first train on the tracks at the crossing station only in case the overlap safety way of that train needs to be set so that it threatens the safe entry or passing through of the next train from the open line into or through the railway station. Thus, if the overlap safety way at traffic places of work on that line section for the first train could be set on any of the tracks where there would be no risk at all for the traffic safety, e.g. if there is within the traffic place of work a turn-out track which can be used as the overlap safety way, then the dispatcher should not need this function embedded in the software for such traffic places of work, so that he could safely and optimally carry out the scheduling operation of the trains on that line section. Apart from the mentioned, the software should adequately present relevant data from the traffic process in such a way as to be easiest for the dispatcher to adopt without any technical or other limitations. E.g. in some solutions there is the problem that the visualised display of the line section is too reduced or divided into lines so as to fit on the monitor as a whole, but not in one direction. Such cases are stipulated by the limitations of the technical equipment and they are not completely adapted to the human who participates in controlling the railway traffic. Apart from the mentioned, the dispatcher needs to have adequate ergonomic conditions of his working environment required for the adequately smooth and reliable work.

4. CONCLUSION

It is necessary to allow an optimally designed visual presentation of information on the condition in the railway traffic which would at the same time provide reduced risk of dispatcher's error, thus increasing his reliability in work, as well as the safety of railway traffic operation. High-quality presentation of the necessary data would increase the reliability of operation of the entire management system in the railway traffic by reducing the dispatcher's efforts who controls centrally the traffic in making valid decisions for the sake of safety and optimization of the traffic process. Using advanced computer technologies it is possible to design flexible software that could, in semi-automatic railway traffic management, be used as the basis for the upgrade of different auxiliary applications that may additionally improve the work of the central dispatcher in different circumstances where it is necessary to realize the railway traffic process management in a safe way. The result of this work is to emphasise the flexibility in the railway traffic management system in the segment of visualising the display on the interface by means of which the dispatcher of the central control communicates with the lower levels of the railway traffic control system.

5. REFERENCES

Tos Z.;(2005). *Signalizacija i automatizacija željezničkog prometa,* Fakultet prometnih znanosti, Zagreb

Woodson, W.E. Tillman, B, Tillman, P. (1992). *Human Factors Design Handbook.* (2nd ed.) New York: McGraw.

European Railway Signalling, Institution of Railway Signal Engineers, London, 1995.

Törnquist J. and Persson J.A. (2006): *N-tracked railway traffic re-scheduling during disturbances,* Blekinge Institute of Technology, Sweden.

Higgins A., Kozan E. and Ferreira L. (1996): *Optimal scheduling of trains on a single line track,* Transportation Research Part B: Methodological, Volume 30, Issue 2, April 1996, Pages 147-161.

Annals of DAAAM for 2006 & Proceedings of the 17th International DAAAM Symposium, ISSN 1726-9679
Editor B. Katalinic, Published by DAAAM International, Vienna, Austria 2006
150th Birthday of Nikola Tesla (1856-1943) Inventor of Mobile Robot Teleoperation

OPTIMIZATION OF DISTRIBUTING OF THE FUNCTIONS BETWEEN A MAN AND MACHINE IN COMPUTER-AIDED DESIGNS

HARBARCHUK, V.

Abstract. *Approach to formalization and automation is offered planning of the difficult man-machine systems on the basis of mathematical decision of tasks of optimization of distributing of functions between a man (by an auxiliary personnel) and hardware's (machine) with simultaneous optimization of choice of concrete personnel and machine of the designed systems.*
Keywords : *system. optimization, automation, planning, man- machine*

1. INTRODUCTION

The modern process of construction of different practical systems requires creation of the integrated intellectual automated systems consisting of two main the subsystems:
1) subsystem of computer-aided design (CAD),
2) subsystem of automation of preparation of production (CAM).
Therefore, speaking about the computer-aided design we imply that this process includes all complex of tasks and problems – from an idea and goal setting – to production of a product according the put purpose. Such approach and researches on creation of bases of methodology and theory of computer-aided design by an author develops from the 80-th years of the 20-century (Harbarchuk, 1982). And this not simple connection of two parts CAD & CAM.
We consider the systems of computer-aided design how the intellectual man-machine systems in which intellectual functions are executed by a man only. But in what then essence of the considered problem of distributing of functions?
And that nonintellectual functions in the system can be carried out by a man and hardware's (a machine). And these "performers " are considered how competitors are on the great number of criteria of efficiency.
The scientific novelty of the article is that the mathematical models of optimization are offered, it is enough simple and it is adequate describing essence of practical tasks of planning and allow to promote the level of automation of planning processes of different man-machine systems. At the same time there are some new mathematical and algorithmic problems touching multi criterion optimization of decisions.

2. RAISING OF TASK

In accordance with developed by us methodology and theory of computer-aided design, after the decision of tasks of the general system planning, tasks of choice of optimum structure of the system, it is necessary to decide the great number of different other tasks which it is possible to name the tasks of "optimization of choice ".

3. MODEL OF TASK

For the mathematical design and optimization of this task by us and by other authors the great number is developed effective approaches, methods and decisions (Harbarchuk, 1982, 1985, 2003, Mikhalevich &Volkovich, 1982, Sergienko & Shylo, 2003).
One of most simple, but effective (in sense of providing of adequacy and exactness of decision) there is a vehicle on the basis of models of the mathematical programming with [1-0] variables. In addition by us the methods of the vector game theory, (Harbarchuk, 1985) allowing to decide the given tasks as a multi criterion and taking into account the competition decisions at the market of the proper systems, are developed.

3.1. Raising of the task 1.

From some great number, "candidate" $a_i \in M_1$ into every place ***of performer*** of concrete functions in the designed system *S* is required ***to choose*** only on one of these applicants optimum on the great number of criteria, for all system *S*. In all subset applicants there are all type of machine and competition for them specialists (people).

3.2. Raising of the task 2.

It is present great number of destinations $k = \overline{1,r}$. – functions in the system. Into every place there is subset of applicants ., - performers of these functions.
It is necessary from subsets M_1^k to choose on one applicant on one of great number of destinations. So that into every place a sole applicant was appointed and an optimum on the great number of criteria for all system S was achieved. Hardware is applicant only.

3.3. Raising of the task 3.

In the subsets of applicants on all, or some places are only specialists of a different level of qualification. In each of these tasks criteria can make such variants:

1) all criteria are minimized,
2) all criteria are maximized,
3) part of criteria is minimized, and part – is maximized.

In addition, different limitations are set for the criteria. In practical multi criterion tasks the criteria have incomparable physical to the dimension (mass, power, speed and etc.). Therefore it is necessary originally to decide a task about taking of all criteria to the comparable form.

Simpler than all it is done by normalization of size parameters in dimensionless.

4. MATHEMATICAL MODELS

We will enter a variable $x_i \in \{0,1\}$ such, that

$$xi = \begin{cases} 1\text{- } ai \text{ is chosen into a place in } Ao, \\ 0 \text{ – in the opposite case.} \end{cases}$$

Then for the cases with one-sided criteria we have a model

$$\sum_i \omega_j(a_j) \cdot x_i \to \min,(\max),$$

$$\sum_i x_i = 1,$$

$x_i \in \{0,1\} \quad \forall i \in \overline{1,m}$.

where *wj*- criterion dimensionless estimations - *aj* applicants. For the case with the mixed criteria a model has a kind

$$\sum_{i=1}^{m} \omega_{j_1}(a_i)x_i \to \min,$$

$$\sum_{i=1}^{m} \omega_{j_2}(a_i)x_i \to \max.$$

$$\sum_{1=}^{m} x_i = 1,$$

$$\forall i \in \overline{1,m} : \overline{1,n} \ni (j_1, j_2).$$

If limitations determined by the type of the designed system are set by some recommendation of applicants, for example, limitation on a cost, sensitiveness, weight and etc., the model of the first task will have a kind

$$\sum_{i=1}^{m} \omega_{j_1}(a_i)x_i \to \min,$$

$$\sum_{i=1}^{m} \omega_{j_2}(a_i)x_i \to \max.$$

$$\sum_{i=1}^{m} \omega_{j_3}(a_j)x_i \le \omega_j^*,$$

$$\sum_{i} x_i = 1,$$

$$x_i \in \{0,1\} \quad \forall i \in \overline{1,m} : \overline{1,n} \ni (j_1, j_0, j_3).$$

In these models the variables x_i are definite before, and - great number of applicants, - *wj** values of limitations, *wi*- criteria from the great number W.

The models of the second task can be built by a similar appearance, taking into account the basic data indicated in raising of her and disposing by data about the values of descriptions of applicants which limitations are set on.

Here also there will be three types of models. For one-sided criteria a model has a kind

$$\sum_{k}\sum_{i} \omega_j(a_i^k)x_{ki} \to \min,(\max).$$

$$\sum_{i} x_{ki} = 1.$$

$$\sum_{k} a_i^k = r, x_{ki} \in \{1,0\} \quad \forall k \in \overline{1,r}. \ \forall i \in \overline{1,m}.$$

For the mixed criteria:

$$\sum_{k}\sum_{i} \omega_{j_1}(a_i^k)x_{ki} \to \min,$$

$$\sum_{k}\sum_{i} \omega_{j_2}(a_i^k)x_{ki} \to \max,$$

$$\sum_{i} x_{ki} = 1,$$

$$\sum_{i}\sum_{k} a_i^k = r,$$

$$x_{ki} \in \{1,0\} \ \forall k \in \overline{1,r}, \ \forall i \in \overline{1,m}, \quad (j_1, j_2) \in \overline{1,n}.$$

For a task with limitations by some recommendation of applicants (criteria mixed):

$$\sum_{k}\sum_{i} \omega_{j_1}(a_i^k)x_{ki} \to \min,$$

$$\sum_{k}\sum_{i} \omega_{j_2}(a_i^k)x_{ki} \to \max,$$

$$\sum_{k}\sum_{i} \omega_{j_3}(a_i^k)x_{ki} \le \omega_{j_3}^*,$$

$$\sum_{i} x_{ki} = 1,$$

$$\sum_{i}\sum_{k} a_i^k = r,$$

$$x_{ki} \in \{1,0\} \ \forall i \in \overline{1,m}, \ \forall k \in \overline{1,r}, \ (j_1, j_2, j_3) \in \overline{1,n}.$$

4. DECISION OF TASKS

After the task of the concrete given values of the descriptions used in criteria and limitations, the decision of these tasks is produced by the methods expounded in (Harbarchuk,1985, 2003, Mikhalevich & Volkovich, 1982, Sergienko & Shylo, 2003).

Some new algorithms of decision of the adopted tasks with ***unclear*** limitations are developed by us. Such situations arise up at the beginning of tasks raising, when it is very difficult to set concrete limitations for a designer.

Then we set some really possible lower or overhead (depending on the type of criterion of optimization) ***scope values***. Then, by the frequent decision of tasks and ***narrowing (expansions)*** of limitations we come to the optimum decisions.

Thereon ***the stage of planning*** (preliminary, technical or works) on ***the life cycle*** of the system ***is*** closed and ***begun preparation of technology of making*** of the projected system. Here also there are the tasks of distributing of functions between a man and machine, for example, at implementation of functions of control of quality, control after the state of technological processes and etc. For the decision of such tasks the offered method of design and optimization of decisions can be used also.

5. CONCLUSION

An approach developed by us enables effective application of mathematical methods of project decisions optimization and optimum choice for the man of such functions at implementation of which he is a competitor for a machine. After the simplest, model, well formalization procedures of raising of purposes (tasks) will be passed to the machine, so the first computer-aided designs with artificial intelligence will appear.

6. LITERATURE

Garbarchuk V. (1982) Mathematical construction of the big ships systems, Shipbuilding, Leningrad

Garbarchuk V. (1985) *Design of conflict project situations and acceptance of decisions by the methods of vector games.*- In a book : Design of the big processes and systems, p.p.. 221-230, Naukova dumka, Kiev

Mikhalevich V, Volkovich V. (1982) *Calculable methods of research and planning of the difficult systems.* Science, Moscow

Garbarchuk V., Zinovich Z., Svits A. (2003) *Cybernetic approach to planning of the systems of defence of information,* Ukrainian academy of informatics, ISBN 966-8064-84-4, Kiev-Lublin

Sergienko I, Shylo V. (2003) *Tasks of discrete optimization. Problems, methods of decision, research.* Naukova dumka, ISBN 966-00-0114-2, Kiev.

Annals of DAAAM for 2006 & Proceedings of the 17th International DAAAM Symposium, ISSN 1726-9679
B. Katalinic, Published by DAAAM International, Vienna, Austria 2006
150th Birthday of Nikola Tesla (1856-1943) Inventor of Mobile Robot Teleoperation

BUILDING AN INTELLIGENT ROBOT ARM TO DISTINGUISH COLOR OBJECT

HASSAN, K.; BASHER, T.; RAHMAN, M.; RAHMAN, I.; AKHTER, S. & RAHMAN, A.

Abstract: *This paper describes the implementation of a robot arming system that is designed for distinguishing color objects (ping pong ball). It is planned to implement the whole system with 7 different phases. These are 1) User input, 2) Move left/right, 3) Scan image, 4) Process image, 5) Move up/down, 6) Grip/release the object, and 7) Go to initial phase. Three stepper motors are used for stirring this robot arm. User will provide instructions to the robot to select one specific color ball through computer interface, and then these instructions will be passed to the robot arm. In this research, we are going to describe the Robot Arm building technologies with the capability to distinguish color objects.*

Key words: *Robot, artificial intelligence, parallel port, web cam, color object, stepper motor, circuit.*

1. INTRODUCTION

While scientific research usually takes credits as the inspiration for science fiction, in the case of AI and robotics (Russell & Norvig, 2003), it's possible that fiction led the way for science. At present, industrial robot arms have been widely adopted in many areas. Unfortunately, operation of industrial robot arms is not easy to master for workers due to complex architectures as well as their various control patterns. Therefore, if there is some intelligence in robot that will help the workers to operate the robot easily, it will remarkably reduce the difficulty of the usage of industrial robot arm.

Today, Building a humanlike robot arm is a formidable engineering task (Rothbart, 1986) that requires a combination of mechanical engineering, electrical engineering, computer architecture, real-time control, and software engineering. Research issues from each of these fields as well as issues particular to integrated systems and robotics must be addressed to build a robot: What types of sensors should be used and how should the data be interpreted? How can the motors be controlled to achieve a task and remain responsive to the environment? The aim of this research is to develop a robot arm that can do things by itself; it will take the order and execute it correctly.

The system is designed as a preliminary step towards more difficult task of the coordinated targeting & tracking with all the parts of robots synchronized at the same time. The primary goal of this initial research is to create a system, a robot arm that would locate the objects and then pick & put to its desired location. Sub goals included working in possible location & as well as optimal placement of the object and other components of future designs.

The objective of this research is to focuses on building autonomous robot arm that is not under human control or supervision. Unlike industrial robots that operated in a fixed environment on a small range. Because we require the system to operate without human control, we must address issues such as controlling robot arm & taking images from environment. However, it is believed that autonomous systems provide robustness & flexibility that task specific systems can never achieve.

2. METHODOLOGY

2.1 Focus

The ability to control a robot arm relies upon many perceptual, cognitive & motor capabilities. Many of these requirements are precursor skills which are necessary before attempting any task of this complexity. For example, this robot arm will require a system for basic motor behaviors, perceptual ability to detect color, objects, postural control, and manipulative abilities such as reaching for target & acquire new objects in the memory.

Even if we were to construct a system which had all the requisite skills, controlling also presents its own unique set of research questions. Those are describing below

- How does the robot arm know, which is the right object?
- How does robot arm moves toward its object?
- How it grab the object?
- How can it put the object into desired destination (ball-been)?

2.2 Circuit Design

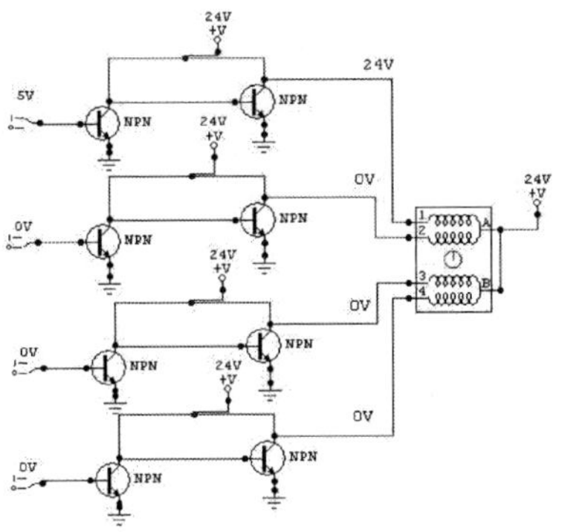

Fig. 1. stepper motor circuit

The following circuit diagram can be used to control a stepper motor (Fig.1). Power is supplied by a DC transformer or DC power supply. The voltage can be anything from 5 to 30 volts, depending upon the rating of the stepper motor. Following are the basic movements for the robot arm movement a) move left, b) move right, c) go up, d) go down, e) grab object, f) release object.

Sending bits for action

If left move then send bit '00000001; 00000010; 00000011; 00000100'
If right move then send bit '00000100; 00000011; 00000010; 00000001'
If going up then send bit '00001000; 00000111; 00000110; 00000101'
If going down then send bit '00000101; 00000110; 00000111; 00001000'
If grabbing object then send bit '00001001; 00001010; 00001011; 00001100'
If releasing object then send bit '00001100; 00001011; 00001010; 00001001'

Received bits action
If ('00000001; 00000010; 00000011; 00000100') Move one step left;
Else If ('00000100; 00000011; 00000010; 00000001') Move one step right;
If ('00000101; 00000110; 00000111; 00001000') Move one step down
Else If ('00001000; 00000111; 00000110; 00000101') Move one step up;
If ('00001001; 00001010; 00001011; 00001100') Grab object;
If ('00001100; 00001011; 00001010; 00001001') Move one release;

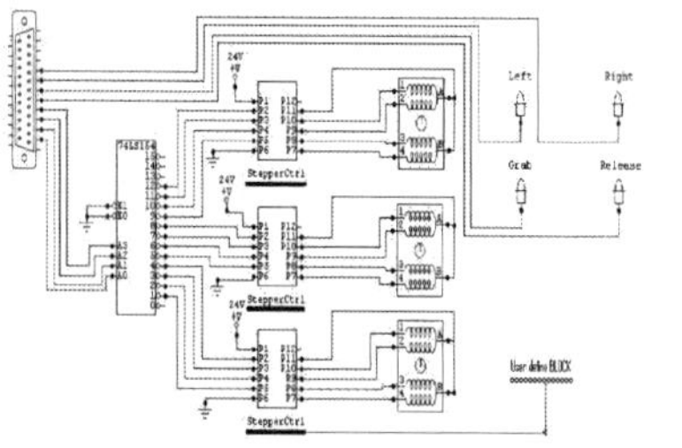

Fig. 2. Main Circuit

The step angle of our mechanical part is designed as the design step angle of the stepper motor. The circuit to control is not complex and is cheaper than many dedicated driver / controller devices and the parts are easy to find. In our case, we send bit sequence from the computer parallel port. That's why we don't need the shift resistor, timer circuit. To control our robot arm, we just arrange some switching circuit with supporting IC's. The direction and speed is controlled by the computer interface. Though we need to control 3 stepper motor (3x4=12 output line) one Decoder (4-line-to16 line) IC is used to give appropriate output line. Decoder is also used to reduce the input line. Inverter is used to invert signal from the decoder IC and is send to the switching circuit. The following diagram is for the main circuit of our robot arm. LED's are used to indicate how the robot arm acts.

2.3 Image Processing

For image processing (Gonzalez & Woods, 2001) part we use standard web-cam that's available in the market. By using the MATLAB7 tools (http://www.utexas.edu/math/ Matlab/Manual /) we capture the image of the object. Then our image processing part starts to analyze the image (Sonka et al., 2001). At first it resizes the image into (320 * 240) resolution. Then it will take the colors (RGB) from 256 points. At that time it will just add the new value of colors (RGB) with the old values. After we get all 256 points color values then we find the average of these three colors (RGB). Now which ever color has the highest value will be consider as the color of the object. This process can handle if there are noise in the image (spot on the objects, light reflect from background).

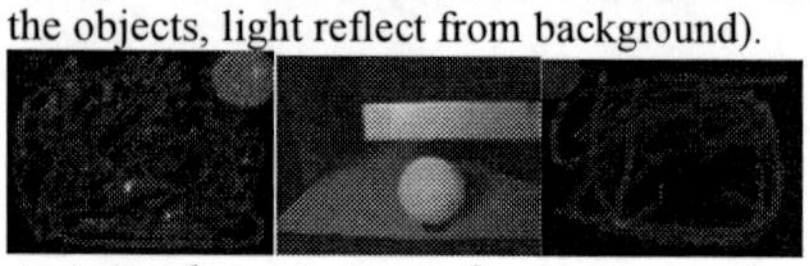

test case-1 test case-2 test case-3
Fig.3. Image Processing

2.4 Structure Design:

For the robot arm, one of the first tasks is to locate the object; so it's important it should do its part easily. As our robot arm is designed to especially for grab an object at a time, its structure also is quite simple. In first part, its have right-left movement capability, then its moving capability for up-down, and pick up that object through grab and release command. As its primary target is to acquire picture of that object, there is a web-cam with the grabber.

To control mechanical part it needs circuit to run accordingly & a computer is use for the do this job and sends the command consequently. It can observe the whole system with fractionally. As object is to work autonomously the robot arm we use a computer design interface with the support of MATLAB7. After running the program by itself any user can say which ball to choose and then it can decide autonomously from which side it is going to search the object by itself. As an example, clicking the option pick red ball (user input for searching desire colored ball), as the whole robot arm is in its initializing stage. It can either move to left or right. Then the web-cam will be activated to take the relevant ball image. And then in PC the image will be analyzed with MATLAB7 toolbox. Processing the images (800x600 in RGB format, and the resize image is 320x320) using MATLAB7 is quite compressively. When the desire colored ball is found, the robot arm starts moving down and grabs the ball then again moves up with the ball, then put the ball in a specific place and release it there.

Fig.4. Robot Arm

3. CONCLUSION AND RECOMMENDATION

In the last couple of years, robotics has become the focus of many research groups, conferences, and special issues. While all projects must address many of the same fundamental problems of motor control, sensitivity, and general architecture, and our group has also focused on various additional aspects. We are committed to building a robot arm that can grab an object and acquire its image in right time & move towards its goal. Finally, we believe that constructing systems that can lead to simpler techniques for machine learning and human-computer interfaces. While it may be difficult for us to outpace the imaginations of science fiction writers, though our work does indicate one possible future, movement of robot & its taking decision of by its own.

In future, we have a plan to make this project with wireless system. To fulfill these criteria two proposed options are given below:

Option 1: Two Bluetooth (Mahgoub & Ilyas, 2004) sending and receiving devices can be used to merge wireless system with our Robot Arm system. These criteria will support remote controlling system. Mobile phone can also be supported to control the Robot Arm form Bluetooth distance range (one Bluetooth mobile will act as a sender and another will act as a receiver). We just need to develop mobile software for both sender and receiver mobile for controlling activities. Microcontroller will provide us the embedding power to develop the codes inside Robot Arm.

Option 2: Another option is to develop a new wireless system to replace the parallel port cable connecting the Arm and the PC. This system will provide the PC to operate multiple robot arms at the same time.

4. REFERENCES

Gonzalez C. R. & Woods E. R. (2001). *Digital Image Processing,* Addison Wesley Longman,81-7808-087-7, Singapore.

Mahgoub, I. & Ilyas, M. (2004). *Mobile Computing Handbook,* CRC Press, 0849319714, USA.

Rothbart H.A. (1986). *Mechanical Design and System Handbook (2nd Edition),*McGRAW-HILL Book Co, 0-07-100272-3, Printed and Bounded in Singapur.

Russell, S. & Norvig, P. (2003). *Artificial Intelligence: A Modern Approach (2nd edition),* Prentice Hall, 0137903952, USA.

Sonka, M.; Hlavac, V. & Boyle, R. (2001), *Image Processing, Analysis and Machine Vision (Second Edition),*Brooks/Cole Publishing Company, 981-240-061-3, USA.

Annals of DAAAM for 2006 & Proceedings of the 17th International DAAAM Symposium, ISSN 1726-9679
B. Katalinic, Published by DAAAM International, Vienna, Austria 2006
150th Birthday of Nikola Tesla (1856-1943) Inventor of Mobile Robot Teleoperation

REAL-TIME FAULT DIAGNOSIS USING PROBABILITY DISTRIBUTION OF FAULTS

HOFREITER, M. & GARAJAYEWA, G.

Abstract: *The paper focuses on fault diagnosis (FDI) in stochastic nonlinear systems using Bayesian statistics. The main concept of this approach relies on a probabilistic model to estimate the occurrences of faults by probabilities. The methodology consists in probabilistic mapping of the measured data into a fault variable, which acts as an indicator of considered modes of the systems. One of the main advantages of the proposed FDI approach is the ability to perform real-time supervised training in which the parameters of the fault probability table are updated in real time. The suggested methodology is very simple for numerical calculations and enables one to include heuristic knowledge about the faults. The practical aspects of the proposed FDI algorithm were successfully tested in real time using a laboratory heating system.*

Key words: *Fault, diagnosis, Bayesian statistics, nonlinear system, probabilistic model.*

1. INTRODUCTION

The paper focuses on fault diagnosis (including fault detection and isolation, (FDI)) for non-linear stochastic dynamic systems. Fault diagnosis is an important subject in modern control theory and practice which allows an increase in the reliability of safety-critical systems and improves plant efficiency, maintainability, availability and reliability.

A traditional approach to fault diagnosis is based on "hardware redundancy". But this concept requires additional costs for adding more hardware. Therefore, the concept of "analytical (functional) redundancy" using model–based FDI is more and more used, as no additional hardware components are needed.

Model-based fault diagnosis methods are mostly based on linear system models (Chen & Patton, 1999). But this approach fails for systems with high nonlinearity and a wide dynamic operating range. To overcome this problem, it is desirable to find a "universal" approximate model to represent any non-linear system approximately. The neural network is such a powerful tool for handling non-linear problems. But the neural networks are mostly used for the FDI of steady-state processes and deterministic systems only. In this paper, a different model for fault diagnosis of non-linear dynamic systems is suggested, which is even able to respect system uncertainty. The probabilistic model determined by the fault probability table is the essence of this methodology, which is also used in Markov chain applications (Hofreiter, 2000).

2. PROBABILISTIC DESCRIPTION OF THE FDI

Let us assume that we have a stochastic non-linear dynamic system, where we want to detect a finite number N_f of known faults. We set up a random variable f_t to serve as a pointer of faults, $f_t \in S_f$, $S_f \equiv \{0,1,\ldots,N_f\}$, which detects the faults at the discrete time $t \in t^*$, $t^* \equiv \{1,2,\ldots,\bar{t}\}$, $\bar{t}$ is a natural number. If $f_t = 0$ the stochastic system at the time t is without faults. The sequence of discretized data observed at the discrete time t is denoted as D_t and the sequence of all data from the beginning of observation up to the discrete time instant t is denoted by $D^{(t)} \equiv \{D_1, D_2, \ldots, D_t\}$.

In a real case we may rely only on analytical redundancy, which is contained in the available measurements of the system and in the prior information about it. As the considered system is uncertain and the measured data are affected by noise the concept of probability is used for the FDI.

We will assume that the information about the particular faults is contained in the observed data vector x_t, $x_t \in S_x$, where S_x is the set of possible observed data vectors with μ_x elements that may contain the finite number of inputs and outputs. From this point of view, our aim is to determine $p(f_t \mid x_t)$ at discrete time $t \in t^*$, where $p(.\mid.)$ denotes a conditional probability mass (density) function. Since the probability mass function $p(f_t \mid x_t)$ is not available, we have to use the parametrized time-invariant model. This model defines the set of probability mass functions $p(f_t \mid x_t, \Theta)$, for $t \in t^*$, where the parameter Θ is the fault probability table to define the relationships of observed data into faults.

The matrix Θ may be estimated using Bayesian statistics where the probability is interpreted as a subjective measure of one unit of the statistician's belief distributed over the set of values which random variable could possibly take. With this approach using the natural conditions of control (Peterka, 1981), we can determine the posterior probability density function $p(\Theta \mid D^{(t)}, f^{(t)})$ conditioned by the observed data and information about faults.

Using the probability density function $p(\Theta \mid D^{(t)}, f^{(t)})$ and under the previous mentioned assumptions we may derive:

$$p(f_{t+1} = \varphi \mid x_{t+1} = \xi, D^{(t)}, f^{(t)}) = \frac{n_{\xi,\varphi}(t)}{\sum_{\varphi_p=0}^{N_f} n_{\xi,\varphi_p}(t)}, \qquad (1)$$

where $n_{\xi,\varphi}(t)$ is the number of events $f_\tau = \varphi, x_\tau = \xi$ for $\tau \in \{1,2,\ldots,t\}$, see (Hofreiter & Garajaÿewa, 2006).

According to formula (1) the data $D^{(t)}, f^{(t)}$ must be available for the probability estimation of the particular possible values of the random variable f_{t+1}. But the true values of the quantities $f^{(t)}$ are known only for the discrete time $\tau = 1,2,\ldots,t_r$, where t_r is the time of the last observed training sample $(t_r < t)$, which we can use for the parameter Θ estimation. Therefore instead of formula (1) for the estimation of f_t at discrete time $t > t_r$ formula (2) may be used for $t = t_r + 1, t_r + 2, \ldots, \bar{T}$:

$$p(f_{t+1} = \varphi \mid x_{t+1} = \xi, D^{(t_r)}, f^{(t_r)}) = \frac{n_{\xi,\varphi}(t_r)}{\sum_{\varphi_p=0}^{N_f} n_{\xi,\varphi_p}(t_r)} . \quad (2)$$

Formula (2) is very simple from a numerical point of view, since it only contains trivial operations of counting and adding. This fact enables one to sequentially update information regarding the matrix Θ in real time.

Further improvement of the fault diagnosis can be done for faults with time duration lasting longer than n sampling periods, when $f_{t+1} = f_{t+2} = \ldots = f_{t+n}$. At least n probability distributions $p(f_\tau \mid x_\tau, D^{(t_r)}, f^{(t_r)})$, $\tau = t+1, t+2, \ldots, t+n$ can be used for the fault diagnosis in this case, e.g. by the application of the arithmetic or geometric means of these probability functions.

From numerical point of view, this approach to fault diagnosis is very simple but the dimension of the sufficient statistic $n(t) = \left[n_{x,f}(t) \right]$, $x \in S_x$, $f \in S_f$ is extremely large even for small cardinality of data-value sets of f_t and x_t, $t \in t^*$. To overcome the above-mentioned drawback the approximation of the Markov chain-based prediction algorithm (Hofreiter, 1997; Garajayewa, 2005) was adapted for the estimation $p(f_{t+1} \mid x_{t+1} = \xi, D^{(t_r)}, f^{(t_r)})$, where $\xi \notin S_x^{(t_r)}$, $S_x^{(t_r)}$ is the set of the observed data vectors x_τ, $\tau = 1, 2, \ldots, t_r$.

3. APPLICATION

This suggested way of fault diagnosis was tested on a laboratory heating system (Fig.1) designed and built by the Centre for Applied Cybernetics at Czech Technical University in Prague.

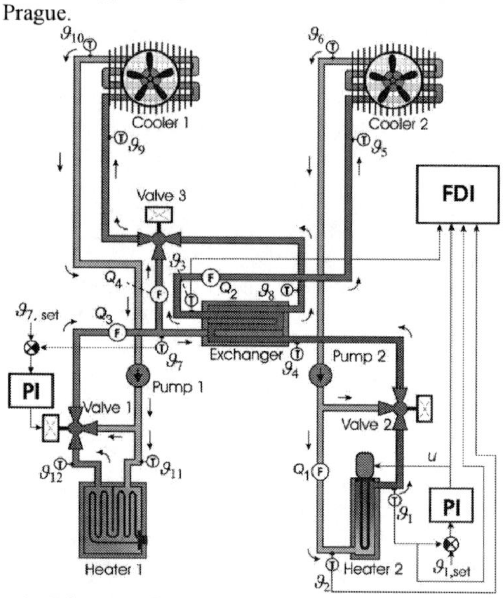

Fig. 1. Diagram of the laboratory heating system

The heating system consists of two independent heating circuits with the circulation of the heat medium (water) performed by two pumps. The water is heated in 2 heaters and cooled down in 2 coolers. From the temperatures measured in the right circuit, only the temperatures $y_1 = \vartheta_1$, $y_2 = \vartheta_2$ and $y_3 = \vartheta_3$ were chosen to monitor the behavior of the system. Besides temperatures, the control signal u assigning the performance of Heater 2 is used as an input for fault detection.

Several possible modes of the system behavior were taken into consideration (e.g. the changes of the heads of the pumps, the performances of the coolers and the heaters, the setting of the opening of the valves etc.). Fig.2 displays the obtained results of the introduced fault diagnosis for one experiment. One can see the good matching between the real fault evolution (Line 1) and the fault estimation (Line 2).

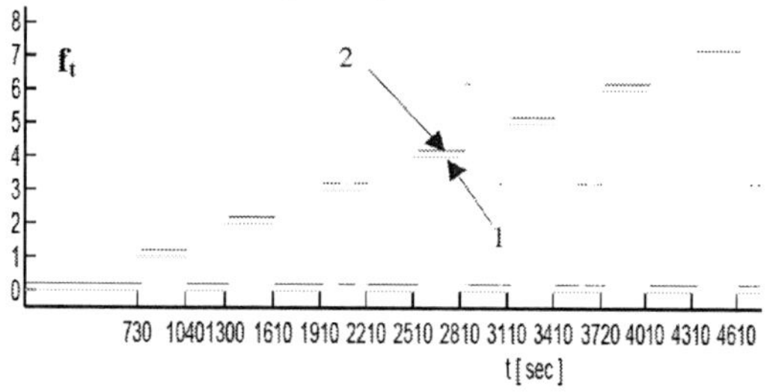

Fig. 2. Real fault evolution and fault estimation

4. CONCLUSION

The aim of this paper is to inform the reader about the proposed methodology for real-time fault diagnosis of stochastic nonlinear dynamic systems, which is based on Bayesian statistics and Markov chains and consists in probabilistic mapping of the measured data into a fault variable that acts as an indicator of the considered modes of the system. The methodology is very simple for numerical calculations and enables one to include heuristic knowledge about the faults. The suggested method and the presented approximation, which helps to overcome the dimensionality problems, were successfully tested on a laboratory heating system. The experimental results confirmed the promising properties of the introduced theory and related algorithms and the ability to perform real-time supervised training by which the parameters of the fault probability table are updated in real time. The FDI methodology is also applicable for time-variant stochastic systems with slowly varying parameters if the probabilistic model is gradually rebuilt by newly measured data.

Acknowledgement: This research is supported by the Ministry of Education of Czech Republic under Project 2B06023.

5. REFERENCES

Chen, J. & Patton, R.J. (1999). *Robust Model-based Diagnosis for Dynamic Systems,* Kluwer, ISBN: 0-7923-8259-5, Boston, USA

Garajaÿewa, G. (2005). *Bayesian Approach to Real-Time Fault Detection and Isolation with Supervised Training, CTU in Prague,* Ph.D. Thesis, Fac. of Mechanical Eng., Prague, Czech Republic

Hofreiter, M. (1997). Approximate identification of stochastic systems, *International Journal Automation Austria,* Vol.5, No.2, pp. 51-63, Vienna, Austria

Hofreiter, M. (2000). On-line fault detection and isolation by Markov chains, Proceedings *of the 11th International DAAAM Symposium,* Katalinic, B. (Ed.), pp. 184-184, ISBN 3-901509-13-5, 2000, DAAAM International, Vienna, Austria

Hofreiter, M. & Garajaÿewa, G. (2006). Real-time fault detection and isolation with supervised training, *Proc. of 6th IFAC Symposium SAFEPROCESS'2006,* Korbicz, J. (Ed.) Beijing, P.R.China, *(accepted for presentation, Paper No.1187)*

Peterka, V. (1981). Bayesian approach to system identification, In: *Trends and Progress in System Identification,* Eykhoff, P. (Ed.), pp. 239-304, Pergamon Press, Oxford, GB

Annals of DAAAM for 2006 & Proceedings of the 17th International DAAAM Symposium, ISSN 1726-9679
B. Katalinic, Published by DAAAM International, Vienna, Austria 2006
150th Birthday of Nikola Tesla (1856-1943) Inventor of Mobile Robot Teleoperation

SOLAR HEATING CONTROL SYSTEM

HRUSKA, F.

Abstract: *Control system of solar heat source used for industrial utilizing has to have other and specific solving. This is given requirements for industrial application. There is control according to optimum criteria as a basic requirement. Maximum of solar energy, accumulation of surplus of energy, minimum of utilizing of secondary source. Other specific is utilizing of extreme control of position of collector and its rotation according to sun.*
Key words: *solar system, heating system, control.*

1. INTRODUCTION

Utilization of solar energy for industry technology requires to solve, design and use specific control system of solar supply. This is given requirements for industrial application. There is used the control according to optimum criteria as a basic requirement (Hruska 2005). Maximum of solar energy, accumulation of surplus of energy, minimum of utilizing of secondary source are requirements too. Other specifically is utilizing of extreme control of position of collector and its rotation according to sun. The house application has only simple solving of temperature control.
The sun has surface temperature around 5800 K. The solar energy is radiated as electromagnetic energy with wave length from 100.10^{-9} do 15.10^{-6} m and with energy intensity 75 MW/m2 of sun surface. The solar constant is 1360 +/-20 W/m^{-2} in the boundary of atmosphere. This intensity is decreased by getting trough the Earth atmosphere according to water stream, CO2, methane, ozone absorption and by scattering.
The solar energy presents a big energy source on the Earth. Therefore is a question, how to take it in maximum value and to use it for utilizing of people. The important is energy efficiency of solar heating system.
In the industry there is much areas, where it is possibility to use heating liquids in range about 40 to 95 °C. In some application there is to use an air heating system. This specific solving about control system is described in this paper.

2. UTILIZING OF SOLAR ENERGY IN INDUSTRY

In industry there is possibility to use solar heat energy in hot water or hot air systems. It is e.g. for hot water:

- direct heating with exchanger
- system with accumulation
- large capacitive industrial system
- outdoor industrial bath.

The hot air system products the hot air up 100°C. The run of air is with pumps and the intern air is heated according to temperature intern hot wand. The hot air solving is used for drying, for sub heating of air condition, heating for oven.
The solar system for family houses is solved for fixing position of collector again sun in the 100% using. Only in any realization is for summer a fixing position of 5 to 30 °elevation and for winter of 60 – 70°. This projects count with a smaller power. The using of solar energy is only during summer, in winter sun shines a little.
The using for industry has more intensive motivation. If the sun shins, the solar energy has to use for 100%. Therefore the control system is simple for house applications. It has only temperature control without optimal algorithm.

3. I-SOLAR-CONTROL SYSTEM

The scheme a new control system for industrial solar energy plant is in fig. 1.

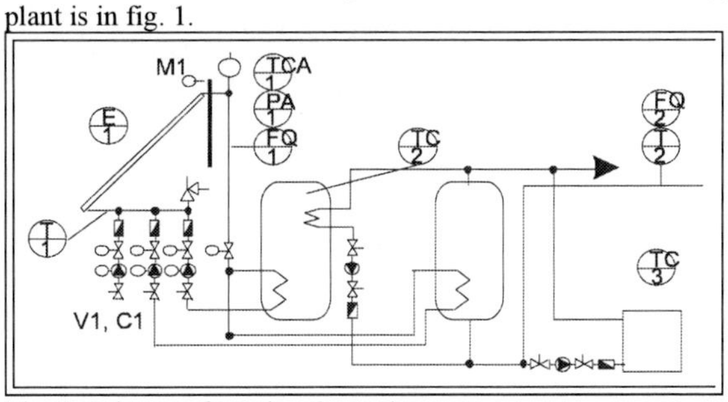

Fig. 1. Scheme of I-Solar_Control system

The solar collector (1) heats amount of accumulation tanks (2). The temperature loop (TCA1) controls temperature in output of collector via change of flow through collector. The action units are valve (V1) and pump (C1). In the situation, when is surplus of energy there is heated amount in the accumulation tanks. This is controlled via loop (TC2).
The collector has other control loop for control of position of collector. Motor (M1) serves for the control loop. Measured value of the loop is solar radiation (E1). The algorithm is used according to extreme control, for maximum solar energy.
The collector has measured parameters: temperature of input water (T1), pressure in collector with alarm function (PA1), flow of hot water from collector (EQ1).
The secondary (bivalent) hot water source (3) can be electrical, water-water or stream-water exchanger or gas unit. The temperature control does the loop of TC3 for output of this source.

4. CENTRAL UNIT

The I-Solar-Control system has central unit of microcontroller element. The decision about the using of microcontroller was done according to new trends automation technique. The element ADuC was selected from Analog Devices. The microcontrollers ADuC combine some precision analogue function and industrial standard microcontroller of type 8051. The have a high resolution of ADC (up 24 bits) and DAC part. The precondition of ADuC is next: multi channel subsystem, weight measurement, intern temperature sensor etc.
The microcontroller ADuC 845 was selected from family of ADuC. (Palka et al, 2005). It has:

- up 10 channels multiplex for analogue inputs
- programmable gain amplifier
- full integrated dual Sigma-Delta-ADC up 24 bit resolution
- on-chip integrated 8 bit microcontroller compatible 80x51
- 32 kHz extern crystal
- 2 channels DAC
- 24 digital inputs and outputs
- 2 date communication channels RS232 and SPI
- U_{ref} and I_{ref} for extern using
- intern temperature sensor
- RAM and Flash memory intern and extern
- watchdog circuit
- two PWM channels
- three channels of 16 bits timer and counter
- 11 channels of interrupts
- dual date pointer, extern 11 bits stack pointer.

Scheme of the ADuC is in fig. 2.

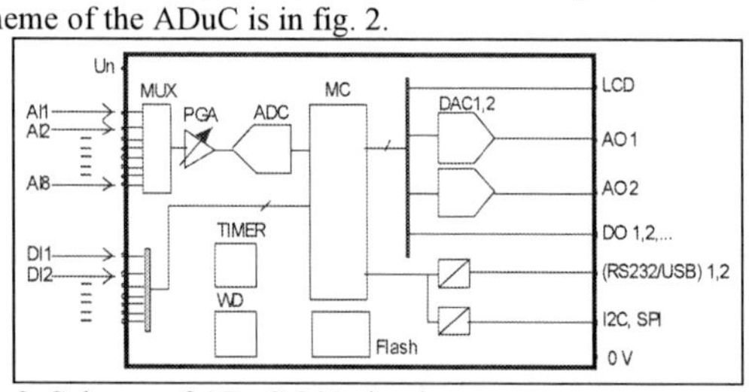

Fig. 2. Scheme of ADuC 845 circuit

The ADuC is a base of central unit of I-Solar_Control. It is connected with inputs of sensors and actuator units. The scheme is in fig. 3.

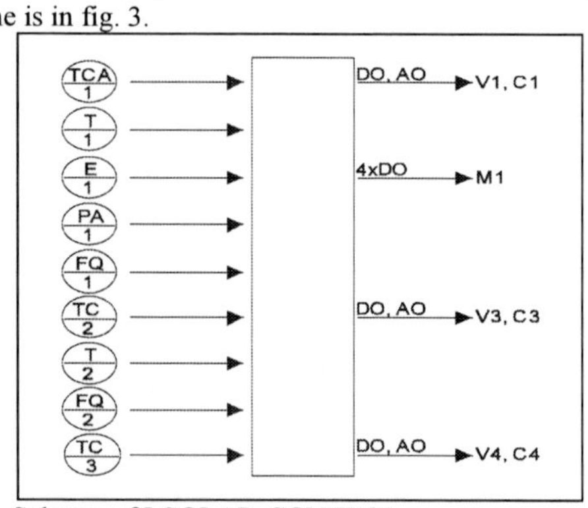

Fig. 3. Scheme of I-SOLAR-CONTROL

The signals of input channels are evaluated and achieved. Some of them go into algorithm of control function. Output of this SW evaluation are some signals for actuator unit.

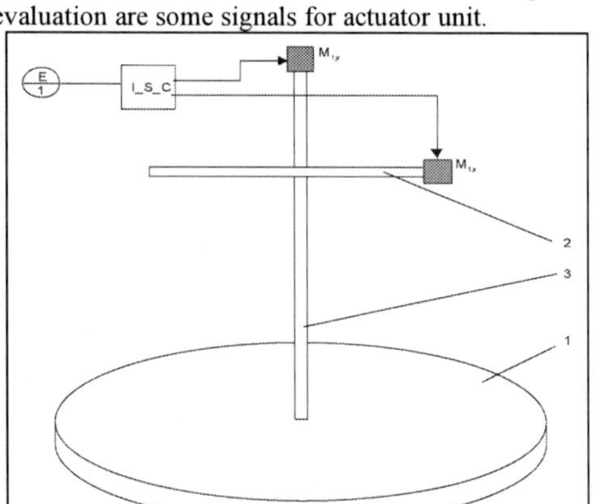

Fig. 4. Scheme of position control of collector.

The control algorithms are for temperature control (TCA1, TC2, TC3) and position control according to maximum radiation (E1).

The position control loop is new and specific. The rotation system (fig. 4) has a base (1) that is storm-proof and wind-proof. The setting of position is in two axels: x and y. The construction allows rotation in range: x=from 0 to 90°, y=from -90° to +90°.

The measured value is parameter (E1). The I_Solar_Control system controls the position by step motors $M_{1,x}$ and $M_{1,y}$. The scheme of position algorithm is in fig. 5. The algorithm needs some sensors of position and a digital system clock.

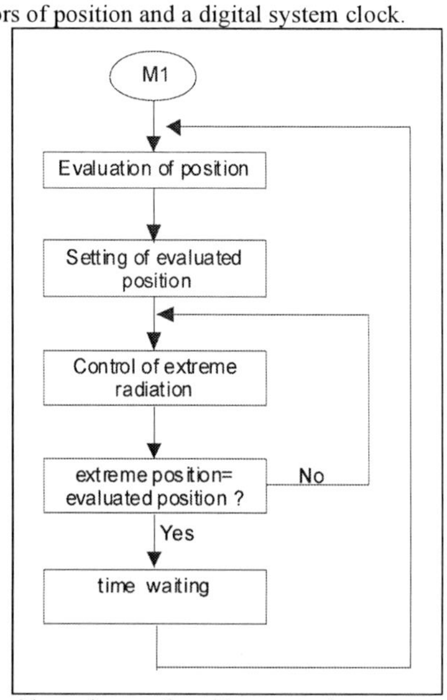

Fig. 5. Algorithm scheme of position control

5. CONCLUSION

The I-SOLAR-CONTROL system presents new way to control of solar industrial system. It use microcontroller with analogue digital interface from family ADuC. The position control of collector is a new solving.

The theory helps to solving of solar supply for industry (Hruska 2005). The help is used for project of solar supply equipments and for analysis and synthesis of control system..

6. REFERENCE

Hruska F. (2005). *Control system specifics for solar supply of heat energy*. In: Proceedings of the 9th International Research/Expert Conference "Trends in the Development of Machinary and Associated Technology" TMT 2005, Antalya, Turkey, pp. 1557-1560. ISBN 9958-617-28-5.

Palka, J. – Hruska, F. (2005). E*mbedded systém pro vyhodnocování tepelné pohody*. (Embedded system for thermal comfort evaluation). In: VII Ogólnopolska Konferencja Naukowo-Techniczna „Komputer W Ochronie Œrodowiska" Gniezno, 14-17.9.2005

Palka, J. – Hruska, F. (2005). *Vyhodnocování tepelné pohody pomocí embedded systému ADuC845*. (The thermal comfort evaluation with ADuC 485 system.) In: Projekt Talent, Brno, 6-7.10.2005.

ACKNOWLEDGEMENT

This work was supported under grant No: MSM 281100001 "Automation of technologies and product processes" and MSM 7088352102 : "Modeling and control of processes of natural and synthetic polymers".

Annals of DAAAM for 2006 & Proceedings of the 17th International DAAAM Symposium, ISSN 1726-9679
B. Katalinic, Published by DAAAM International, Vienna, Austria 2006
150th Birthday of Nikola Tesla (1856-1943) Inventor of Mobile Robot Teleoperation

INCREASING AND DECREASING OF DIMENSION FACTOR BY ANGLES FOR CENTRAL GEAR OF CYLINDRICAL HYPERBOLICAL PLANETARY TRANSMITTER

IBRAHIMI, N.; CAKOLLI, H.; LIKAJ, R. & KYCYKU, A.

Abstract: *In this paper is given the basic geometric concept of cylindrical hyperbolical planetary transmitter type 1AI, which are the gears with different line side angles of tooth threads. Particularly the increasing and decreasing dimension factor for the central and planetary gears has been investigated, when the inclination angle and direction of side lines of the gear change. This is because those angles are dependent on each other and the cross angle (Σ) for rotational axle of central and planetary gears.*
Key words: *Dimension factor, angle (Σ), central gear, planetary transmitter.*

1. INTRODUCTION

The planetary transmitters occupies a very important place among the toothed transmitters and have a broad technical application, especially for transmitting a great powers since here is used the principle of power branching. A simple cylindrical planetary transmitter type 1AI is shown in fig. 1 (Cakolli, 2000 and Schubert, 1983).
To the cylindrical planetary transmitters the geometric axes of all gearings are parallel, while the inclination angles of teeth are the same. The power which is carried from central gear (1), it is branched in three parts (planetary gears 2) and again turns back to the planetary gears holder (s). During this process, the gear couple of cylindrical planetary transmitter will be loaded only with 1/3 of total load that comes from stationary geared transmitter which is under the same amount of power load, (Ibrahimi, 1987).

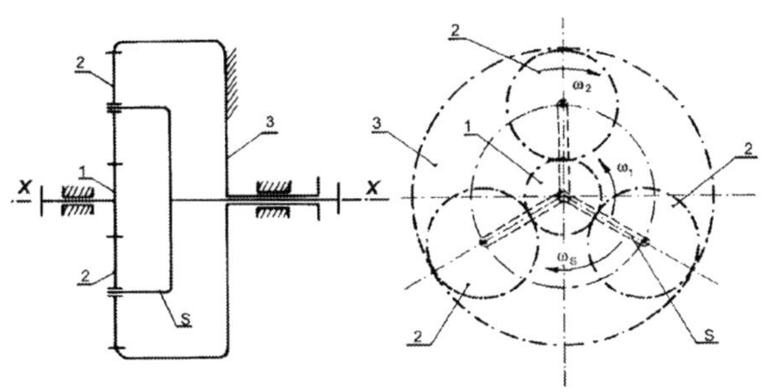

Fig. 1. Cylindrical planetary transmitter 1AI

2. MAIN ISSUES OF CYLINDRIC HYPERBOLICAL PLANETARY TRANSMITTERS HPT

As a distinction from cylindrical planetary transmitters (CPT), the geometric axes of planetary gearings intersect the axes of central gearings, while the inclination angles are different to each other for all gears. Transmission ratio of HPT (Cakolli, 2000), is given as:

$$i_{1s} = 1 - i_{13} = 1 + \frac{d_3 \cos\beta_3}{d_1 \cos\beta_1} \qquad (1)$$

Intersection angle for rotation axes of planetary and cylindrical gearings, of HPT type 1AI is given with the following formula:

$$\Sigma = \beta_1 + \beta_2 = -(\beta_2 + \beta_3) = \frac{\beta_1 - \beta_3}{2} \qquad (2)$$

The diameter of separator circle for the central gear with external tooth's of HPT is determined with following expression:

$$d_{1TPH} = \frac{d_3 \cos\beta_3}{(1 - i_{1s})\cos\beta_1} \qquad (3)$$

The increasing and decreasing factor for central gear with external tooth's of HPT F_R, (by Cakolli, 2000), is determined by the following expression and its dependent on angles β_1 and β_3:

$$F_R = \frac{d_{1TPH}}{d_{1TPC}} = \frac{\cos\beta_3}{\cos\beta_1} \qquad (4)$$

Similarly, the increasing and decreasing factors can be determined for the planetary gears F_p, as follows:

$$F_p = \frac{d_{2TPH}}{d_{2TPC}} = \frac{i_{1s} - (1 + F_R)}{i_{1s} - 2} \qquad (5)$$

In fig. 2. is shown an cylindrical HPT, with the corresponding values of parameters, based on the calculations are derived the dimensions of central and planetary gearings for certain axle distance, as shown in table form.

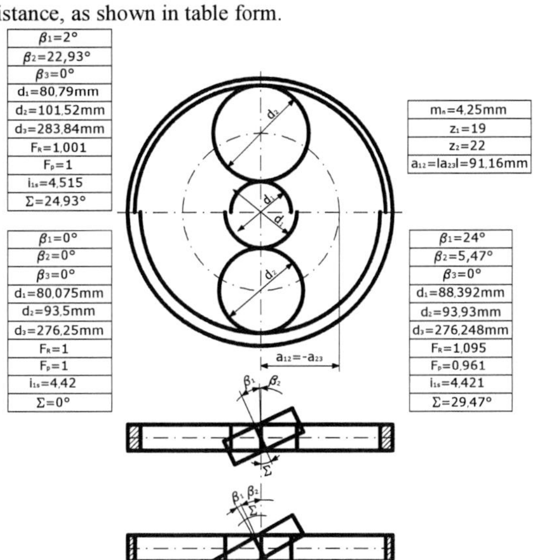

Fig. 2. Gear dimensions, given for different angles of side lines and equal axle distance.

Based on equations (4) and (5) are calculated the increasing respectively decreasing factors for the dimensions of central and planetary gears, based on inclination angles and transmission ratio and are graphically represented in fig. 2. and fig. 3.

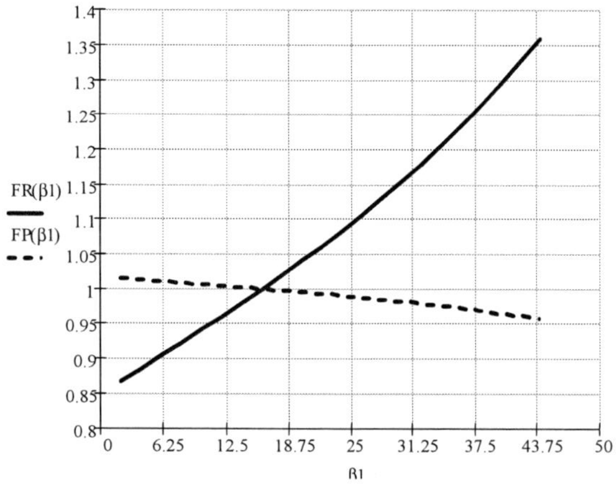

Fig. 3. Dependence of factors F_R and F_p from angles
β_1=2.80... 42°, β_2 = -16°, $\beta_3 = \Sigma - \beta_1$ for i_{1s} = 10,428

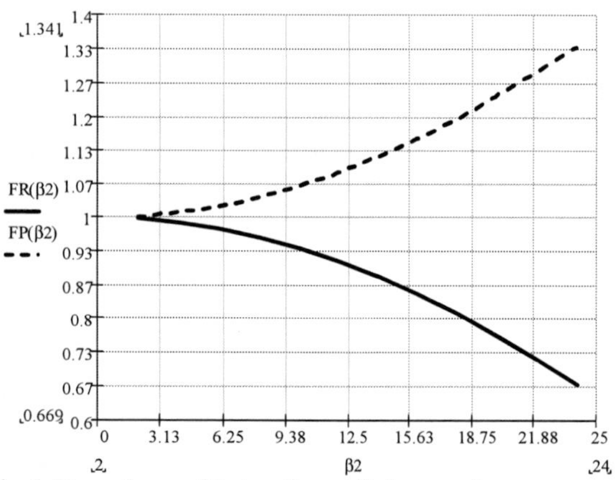

Fig. 4. Dependence of factors F_R and F_p from angles
β_2=2.6... 26°, β_1 = 0, β_3 = -2 β_1 for i_{1s} = 2,9789

By increasing the inclination angle of side lines for central gear 1, the dimension factor of central gear with external tooth's decreases F_R = 0,85 ...0,35, while the dimension factor of planetary gear increases F_p = 1,02...1.09, shown in fig. 5.

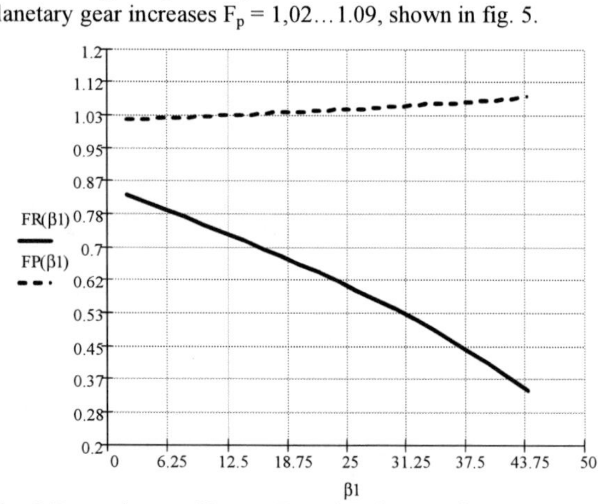

Fig. 5. Dependence of factors F_R and F_p from angles
β_1=2... 45°, β_2 = 16°, $\beta_3 = \Sigma - \beta_1$ for i_{1s} = 10,428

Similarly, by increasing the inclination angle of side lines (left or right) for planetary gear 2, the dimension factor of central gear with external tooth's increases F_R = 0,67 ...1.0, for the angle β_2=-25,-24...0, while the dimension factor of planetary gear decreases F_p = 1.35...1.0. But, at interval β_2=0 to $\beta 2$=10 the dimension factor of central gear with external tooth's decreases F_R = 1.0,...,0.93, while dimension factor for planetary gear increases F_p = 1.0...1.07, as is shown in fig. 6.

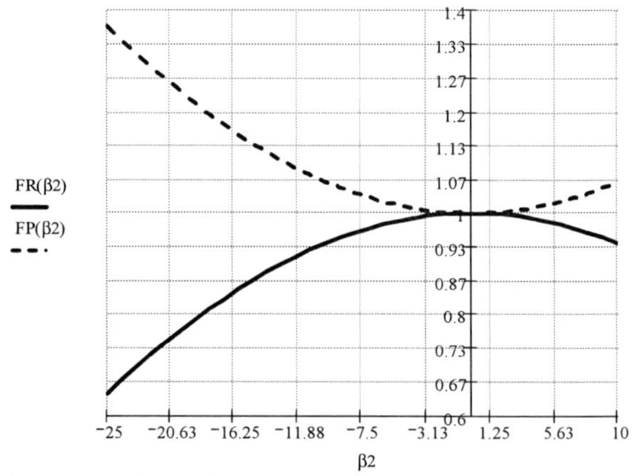

Fig. 6. Dependence of factors F_R and F_p from angles
β_2=-25,-24... 10°, β_1 = 0, β_3 = -2 β_1 for i_{1s} = 2,97

3. CONCLUSIONS

Based on the obtained and graphically presented results for the Cylindrical Hyperbolical Planetary Transmitter HPT it can be concluded that:

- When the central gear has external straight (spur) tooth's the respective factor increases very slowly and its value is always below 1, while the dimension factor for planetary gearings decreases,
- By increasing the inclination angle of side lines for the central gear, in the same time increases the dimension factor for the central gear with external tooth's, but dimension factor for planetary gearings and central gear with internal tooth's decreases,
- For practical application, the appropriate value for inclination side lines for central gear with external tooth's is in interval of $\beta_1=(10^0-30^0)$, while intersection angle $\Sigma=(5^0-15^0)$.

The model proposed by Ibrahimi and Cakolli is practically realised and tested in Strojna, Maribor, Republic of Slovenia.

4. REFERENCES

Cakolli, H. (2000). *Optimalizimi i karakteristikave të ndërruesit të shpejtësive të kombinuar me transmetues planetar,* PHD, University of Pristina, Pristina.

Ibrahimi, N. (1987). *Prilog istraživanju mogućnosti razvoja planetarnih prenosnika,* PHD, University of Pristina, Pristina.

Cakolli, H. (1989). *Prilog na istraživanjeto na možnosta na razvojot na planetnite hiperboloidno cilindrični prenosnici,* master thesis, University of Skopje, Skopje.

Schubert, F. (1983). *Planetengetribe mit verschëenkten planetenradaesen,* PHD, 1983 bericht Nr.66 des Instituts für Maschinenkonstruktion und Getribeban der Universitat Stuttgart, Stuttgart.

Lechner, G. (1994). *Fahrezug getriebe,* Springer Verlag, Berlin.

Annals of DAAAM for 2006 & Proceedings of the 17th International DAAAM Symposium, ISSN 1726-9679
Editor B. Katalinic, Published by DAAAM International, Vienna, Austria 2006
150th Birthday of Nikola Tesla (1856-1943) Inventor of Mobile Robot Teleoperation

THE INFLUENCE OF COEFFICIENT OF ELASTIC SLIPPING IN TRANSMITTAL CAPABILITY OF TRAPEZIUM BELTS

IBRAHIMI, N.; KYCYKU, A. & CAKOLLI, H.

***Abstract**: The coefficient of elastic slipping is an important factor for the dynamic analysis of belt transmitters, because it clearly determines the level of transmittal capability for a given force of preliminary tightening. By laboratory experiments is researched the influence of the transmitted power and the preliminary tightening in the coefficient of elastic slipping on purpose to determine the optimal limit of this coefficient.*
***Key Words**: Experiment, belt transmitter, trapezium belt, elastic slipping,*

1. INTRODUCTION

The transmitters with belts take part in the group of mechanical indirect transmitters, which have a great application for power transmission in all kinds of machines and devices. During power transmission a specific problem appears to be the non-constant value of transmission ratio because of the elastic slip of the belt. The elastic slipping appears on a part of the angle where belt embraces the pulley. The elastic slipping is a result of the action of forces with varying intensities on the belt embranchments.

2. DETERMINING ELASTIC SLIPPING COEFFICIENT

As the elastic slipping angle cannot be determined experimentally, then the phenomenon of elastic slipping is analyzed through *the coefficient of elastic slipping K_{rre}*. This coefficient presents the ratio between the difference of the belt velocities in pulling embranchment and the free embranchment toward the speed in the pulling embranchment.

$$K_{rre} = (v_1 - v_2)/v_1 \quad (1)$$

The mathematical model of the influence of elastic slipping coefficient in the transmission ratio is given through the expression (2).

$$i = \frac{n_1}{n_2} = \frac{d_{p2}}{d_{p1}(1 - K_{rre})} \quad (2)$$

The expression (2) shows that for a certain number of driving pulley rotations n_1 depending on the value of elastic slipping coefficient K_{rre}, can be gained different numbers of driven pulley rotations n_2. This makes that during the work to have an inflection of transmission ratio,
depending on dynamic conditions created in the transmitter.

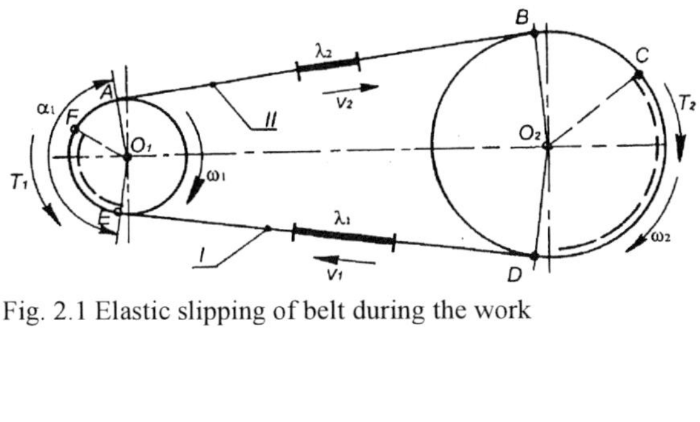

Fig. 2.1 Elastic slipping of belt during the work

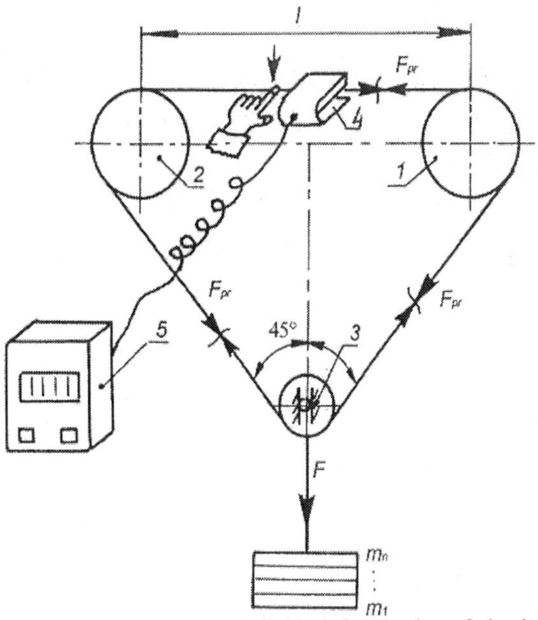

Fig. 2.2 Measurement of individual frequencies of the belt transmitter

An element of belt during the work of transmitter, in the pulling embranchment has the length λ1 and the speed v1, whereas in the free embranchment will have the length λ2 and the speed $\mathbf{v_2}$ (Fig. 2.1). The elastic slipping ground is determined by the elastic slipping angle. With the increase of changing forces in belt embranchments the angle of elastic slipping will b increased as well.

3. DETERMINING THE ELASTIC SLIPPING COEFFICIENT K_{rre} EXPERIMENTALLY

Experiments have been conducted in the laboratory for dynamic research to better analyze the influence of this coefficient on the work of transmitter and the dynamic conditions of inflection of this coefficient,. From the research done is ascertained that this coefficient depends on the power being transmitted P and by the force of preliminary tightening of belt F_{pr}. The frequence of individual oscillations of the agitated belt is measured by the digital tensiometer. The force of preliminary tightening based on the frequency individual oscillations of the belt is calculated by the expression:

$$F_{pr} = 4m_{rr}(lf_L)^2 \quad (3)$$

Experiments are done on transmitters with three pulleys. Also the research is done with elastic slipping coefficients for the belts produced with wrapping and cut. The research device has had the possibility of transmitting power ranging from 5 kW till 20 kW, and the force of preliminary tightening.

The increase of transmitting power is done by burdening the driven pulley, whereas the increase of preliminary tightening force is done by burdening the tightening pulley.
The results gained by experiments are presented in tables 3.1 and 3.2, respectively in diagrams in Fig. 3.1 and 3.2.

Transmitting ,power P(kW)	F (daN)	68	56	48	40	32
	f_L (Hz)	123	112	102	96	86
	F_{pr} (daN)	53.28	44.18	36.64	32.46	26.06
5		0.25	0.25	0.32	0.32	0.34
6		0.35	0.38	0.50	0.55	0.65
7		0.47	0.50	0.70	0.75	0.80
8		0.61	0.64	0.90	1.0	1.2
9		0.77	0.82	1.2	1.5	6.0
10		1.0	1.1	1.6	4.7	
11		1.2	1.4	2.2	10	

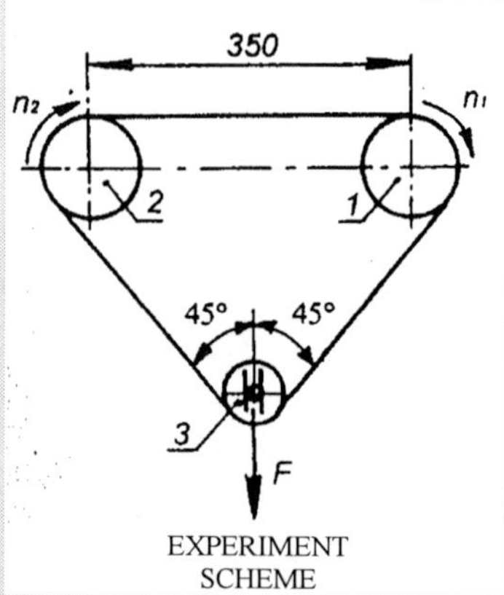

AV10x1200 L_a (production of ''Ballkan'' factory) $d_{p1}=d_{p2}=117.3$ mm, $d_{p3}=57$mm $n=4900 min^{-1}$,$t=20^0C$, $\varphi=60\%$ $P_{EM}=20$ kW, $n_M=3000$ min^{-1}

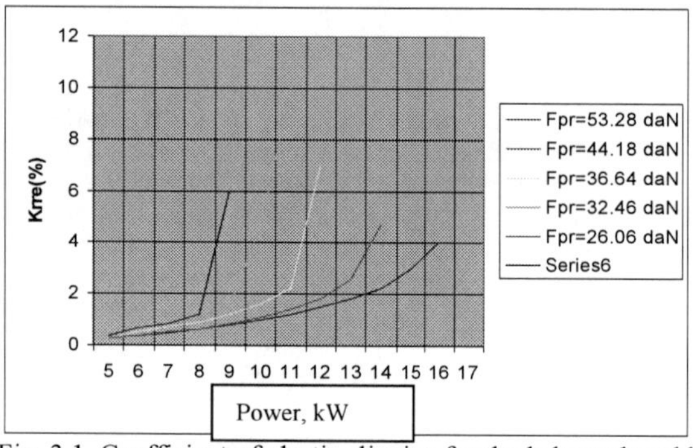

Fig. 3.1. Coefficient of elastic slipping for the belt produced by wrapping.

From table 3.1. and from diagram depicted in Fig 3.1. for the belts produced by wrapping, it can be seen that the coefficient of elastic slipping increases with the increase of transmitting power P. Also by decreasing the preliminary tightening force F_{pr} this coefficient increases. After the detailed analysis of results gained and presented in the table 3.1. and in the diagram in Fig. 3.1., it can be seen that value $K_{rre}=2\%$ represents the optimal limit of stable work of the transmitter with three pulleys.
From the table 3.2. and from the diagram depicted in Fig. 3.2. for belts produced with cutting, it can be seen that the same definition is valid on the influence of power being transmitted and preliminary tightening force in coefficient of elastic slipping, but it has lower values.

Table 3.2. Coefficient of elastic slipping K_{rre}[%] for the belt produced with cutting.

Transmitting power P(kW)	F (daN)	68	56	48	40	32
5		0.20	0.20	0.20	0.22	0.22
6		0.35	0.35	0.35	0.40	0.40
7		0.50	0.45	0.45	0.55	0.55
8		0.60	0.60	0.55	0.8	0.85
9		0.75	0.70	0.70	1.10	6.0
10		0.90	0.90	1.0	4.0	
11		1.1	1.2	1.4		

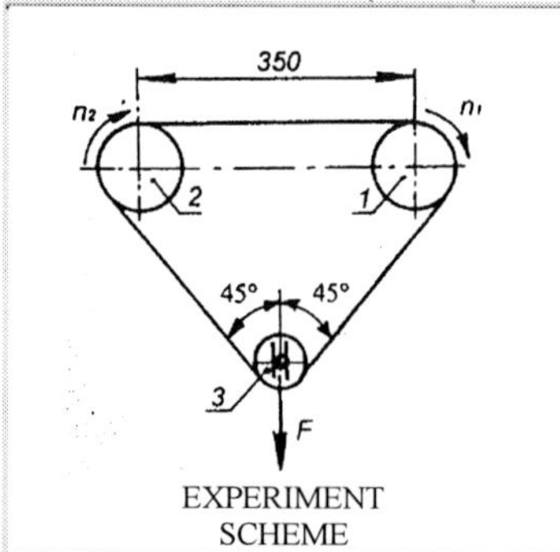

Belt AV10x1200L_a (produced with cutting) All other conditions are like in the first case

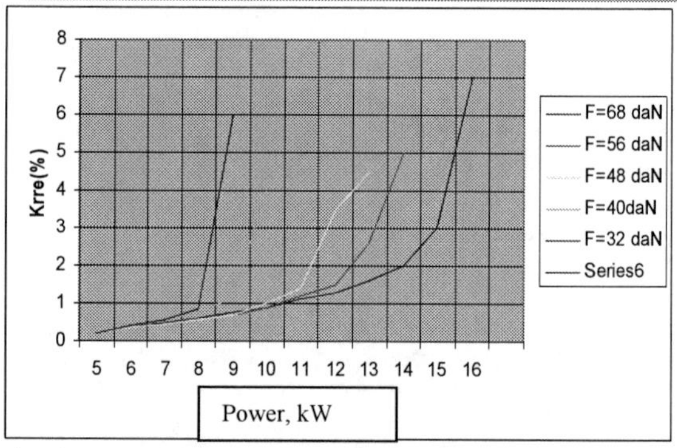

Fig. 3.2. Coefficient of elastic slipping for the belt produced with cutting.

4. CONCLUSION

By analyzing the results gained experimentally, which are presented in tables and graphically, it can be concluded that:

- the coefficient of elastic slipping increases with the increase of power that the transmitters transmits
- this coefficient increases by decreasing the force of preliminary tightening
- belts produced by cutting have a lower coefficient of elastic slipping than belts produced by wrapping

5. REFERENCES

Kyçyku A. *"A contribution to optimalization of transmitting capabilities of trapezoid belts"* work of magistracy, Pristine 2003
STANDARDS for trapezoid belts research
Technical documentation of "Ballkan" factory, Suhareka.
G.Gerbert: "Load Distribution in Timing Belts", Technical University, Sweden.

Annals of DAAAM for 2006 & Proceedings of the 17th International DAAAM Symposium, ISSN 1726-9679
B. Katalinic, Published by DAAAM International, Vienna, Austria 2006
150th Birthday of Nikola Tesla (1856-1943) Inventor of Mobile Robot Teleoperation

BIFURCATION ANALYSIS OF ELASTIC SYSTEMS BASED ON FREQUENCY SPECTRUM OF LARGE VIBRATIONS

ISIC, S.; DOLECEK, V. & KARABEGOVIC, I.

Abstract: This paper presents the analysis of bifurcation of elastic systems, where character of a bifurcation is identified from the analysis of large vibration of these systems around their primary equilibrium state. It is identified that content and change of frequency spectrum for large amplitudes show on post-critical behaviour of the system. Analysis is demonstrated on the double pendulum supported by angular springs. Frequencies of vibration and their change caused by large amplitudes are obtained analytically using perturbation method, and using FFT transformation of the results obtained by numerical simulation of the vibration.

1. INTRODUCTION

Bifurcation analysis of elastic systems is commonly based on detection of the stationary or critical points using linear analysis, and then the characteristics of these points, i.e. postcritical behaviour, are determined using nonlinear analysis. Nonlinear analysis may be either statical or dynamical. In use is often statical approach to nonlinear analysis, which is based on determination of equation of post-critical path in shape of power series (Thompson, 1973). A sign of coefficients of odd and even powers define character of bifurcation, which may be, in case of conservative systems, stable or unstable. In case of dynamic approach, large number of higher order terms is necessary to determine if a system has stable or unstable bifurcation, since, for conservative systems, eigenfrequency becomes imaginary and causes exponential departure from initial equilibrium position (Thomsen, 1994).

It is already known that analysis of eigenfrequencies of the small vibration around primary equilibrium position may be used to define stability of the system (Ziegler, 1968). Analysis of frequencies of vibrations with large amplitudes around primary equilibrium position, considering stiffness nonlinearity, shows change of the frequency, depending on the stiffness characteristics (i.e. is system hardening or softening) (Cunningham, 1958; Thomsen, 1994). In the literature less attention is given to the change of frequency caused by large displacement and load, and possibility of determination of post-critical behaviour on the basis of this analysis.

This paper presents the analysis of bifurcation of elastic systems, where character of the bifurcation is identified from the analysis of large vibration around their primary equilibrium state. It is identified that content and change of frequency spectrum caused by large amplitudes and load shows on post-critical behaviour of the system (i.e. on the character of the bifurcation). Analysis is demonstrated on the double pendulum supported by angular springs with linear stiffness, for which stable symmetric bifurcation characteristic is already known (Isic, Dolecek, 2006). This system is still simple for mathematical description, but despite its simplicity exploits all dynamical characteristics of more complex beam structures. Frequencies of vibration and their change caused by large amplitudes are obtained using perturbation method, in case of analytic approach, and by FFT transformation of the results obtained using numerical simulation of the vibration.

2. CALCULATION OF CRITICAL POINTS

2.1 Nonlinear equation of motion

Equation of motion for double pendulum shown on Fig. 1 is

$$2ml^2\ddot{\theta}_1 + ml^2\ddot{\theta}_2\cos(\theta_1-\theta_2) + ml^2\dot{\theta}_2^2\sin(\theta_1-\theta_2) + k\theta_1 - k(\theta_2-\theta_1) = Pl\sin\theta_1 \qquad (1)$$

$$ml^2\ddot{\theta}_2 + ml^2\ddot{\theta}_1\cos(\theta_1-\theta_2) + ml^2\dot{\theta}_1^2\sin(\theta_1-\theta_2) + k(\theta_2-\theta_1) = Pl\sin\theta_2$$

where k is stiffness of springs, m is attached mass, l is length of massless rods, and P is applied axial load.

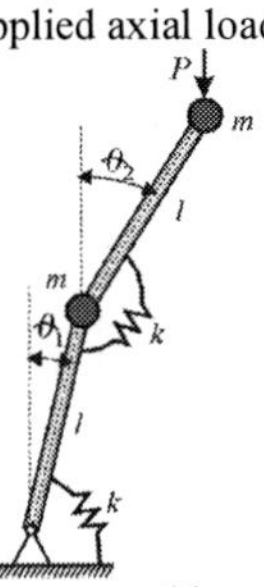

Fig. 1. Double pendulum supported by angular springs.

Limiting to cubic powers of displacements θ_1 and θ_2, approximation of the exact equation (1) may be written in the matrix form

$$[\mathbf{M}]\{\ddot{\boldsymbol{\theta}}\} + [\mathbf{K}]\{\boldsymbol{\theta}\} + [\mathbf{M}_1]\{\ddot{\boldsymbol{\theta}}\} + [\mathbf{K}_1]\{\boldsymbol{\theta}\} + [\mathbf{C}_1]\{\dot{\boldsymbol{\theta}}\} = 0 \qquad (2)$$

in which mass matrix $[\mathbf{M}]$, stiffness matrix $[\mathbf{K}]$, displacement vector $\{\boldsymbol{\theta}\}$ are given by

$$[\mathbf{M}] = ml^2\begin{bmatrix} 2 & 1 \\ 1 & 1 \end{bmatrix}, [\mathbf{K}] = \begin{bmatrix} 2k-Pl & -k \\ -k & k-Pl \end{bmatrix}, \{\boldsymbol{\theta}\} = \begin{Bmatrix} \theta_1 \\ \theta_2 \end{Bmatrix},$$

and $[\mathbf{K}_1]$, $[\mathbf{M}_1]$ and $[\mathbf{C}_1]$ are matrices of nonlinear effects

$$[\mathbf{K}_1] = \frac{Pl}{6}\begin{bmatrix} \theta_1^2 & 0 \\ 0 & \theta_2^2 \end{bmatrix}, \quad [\mathbf{M}_1] = \frac{1}{2}ml^2(\theta_1-\theta_2)^2\begin{bmatrix} 0 & 1 \\ 1 & 0 \end{bmatrix},$$

$$[\mathbf{C}_1] = ml^2(\theta_1-\theta_2)\begin{bmatrix} 0 & \dot{\theta}_2 - (\theta_1-\theta_2)^2\dot{\theta}_2/6 \\ \dot{\theta}_1 - (\theta_1-\theta_2)^2\dot{\theta}_1/6 & 0 \end{bmatrix}.$$

2.2 Eigenvalues of linearised equation

Considering small vibration around vertical position for load P less then critical P_{cr}, we get eigenfrequencies of the system and first vibration eigenmodes in the form

$$\omega_{1,2}^2 = (6k - 3Pl \pm \sqrt{5P^2l^2 - 24kPl + 32k^2})/(ml^2) \qquad (3)$$

$$\theta_2 = \frac{2(4k - 2Pl - \sqrt{5P^2l^2 - 24kPl + 32k^2})}{3Pl - 8k + \sqrt{5P^2l^2 - 24kPl + 32k^2}}\theta_1 \qquad (4)$$

Increasing of the axial load P causes decreasing of eigenfrequencies. For $P = P_{cr}$, the lowest eigenfrequency becomes zero, what indicates possibility of bifurcation and first buckling and vibration eigenmodes become coincident, i.e.

$$P = P_{cr} = \frac{3-\sqrt{5}}{2}\frac{k}{l}, \quad \theta_2 = \frac{1+\sqrt{5}}{2}\theta_1 \qquad (5)$$

3. FREQUENCY CHANGE FOR LARGE AMPLITUDES

3.1 Numerical calculation of frequency change

Equation of motion (1) is solved for unknown accelerations

$$\ddot{\theta}_1 = g_1(P,\theta_1,\theta_2,\dot{\theta}_1,\dot{\theta}_2), \; \ddot{\theta}_2 = g_2(P,\theta_1,\theta_2,\dot{\theta}_1,\dot{\theta}_2), \qquad (6)$$

and transformed to system of first order differential equations

$$\{\dot{\mathbf{x}}\} = \{\mathbf{g}(\mathbf{x})\}, \qquad (7)$$

in which $\{\mathbf{x}\} = \{\theta_1 \;\; \theta_2 \;\; \dot{\theta}_1 \;\; \dot{\theta}_2\}$ and $\{\mathbf{g}\} = \{\dot{\theta}_1 \;\; \dot{\theta}_2 \;\; g_1 \;\; g_2\}$.

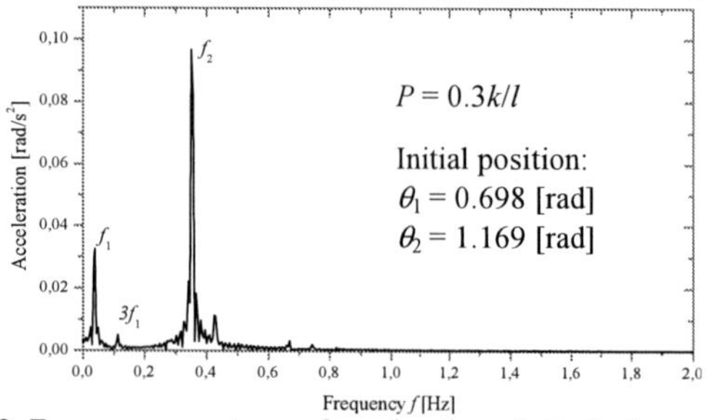

Fig. 2. Frequency spectrum of acceleration of attached mass m_1 ($f = \omega/2\pi$).

The equation (7) is solved numerically, for specified initial conditions and value of axial load, using forth order Runge-Kutta method. Simulation is performed for axial load near critical, $P = 0.3\ k/l$, where m, k and l have unit values. Initial position of pendulum $\{\theta\}_0$ is chosen to be near to first buckling eigenmode, i.e. $\theta_2/\theta_1 \approx 1.618$. Transformation of solutions in time domain to frequency domain using FFT transform is performed, and then change of frequencies is analyzed, Fig. 2. Increasing of the base frequency under increasing amplitude of oscillations is observed. It is concluded that nonlinear terms of geometric stiffness cause increasing of eigenfrequencies ω_1, where $\Delta\omega_1(\{\theta\}_0) = \Delta\omega_1(-\{\theta\}_0)$, which points out to stable symmetric bifurcation of equilibrium position. Symmetric bifurcation is also indicated by presence of the frequency $3\omega_1$.

3.2 Analytic calculation of frequency change

Supposing vibration in shape of the first buckling eigenmode (5) and premultiplying by eigenvector (Isic, Dolecek, 2006), equation (2) may be transformed in the following differential equation

$$\ddot{\theta}_1 + \omega_0^2\theta_1 + h\theta_1^3 = 0 \qquad (8)$$

in which

$$\omega_0^2 = \frac{(5-\sqrt{5})k-(5+\sqrt{5})Pl}{3(3+\sqrt{5})ml^2}, \quad h = \frac{5+\sqrt{5}}{18(3+\sqrt{5})ml}P.$$

A solution of the equation (8), for initial condition $\theta_1(0) = A$, $\dot{\theta}_1(0) = 0$, is obtained using perturbation method, considering solution in the form

$$\theta_1 = \theta_{10} + h\theta_{11} + h^2\theta_{13} + \dots, \qquad (9)$$

where change of base angular frequency is considered as

$$\omega^2 = \omega_0^2 + hb_1(A) + h^2b_2(A) + \cdots. \qquad (10)$$

The obtained solution has the form

$$\theta_1 = (A - \frac{A^3}{32\omega^2})\cos\omega t + \frac{\omega t}{2}(b_1A - \frac{3A^3}{4})\sin\omega t + \frac{A^3}{32\omega^2}\cos 3\omega t, \qquad (11)$$

while first order approximation of the change of base angular frequency for large amplitudes is given by

$$\omega_1^2 = \omega_0^2 + \frac{3h}{4}A^2 = \omega_0^2 + \frac{3(5+\sqrt{5})P}{72(3+\sqrt{5})ml}A^2 \qquad (12)$$

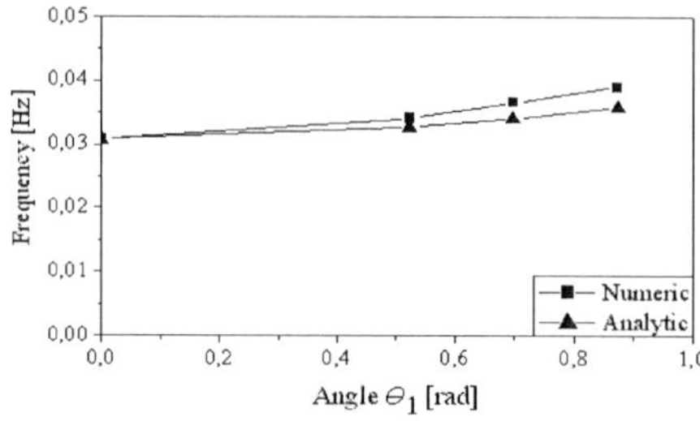

Fig. 3. Numerical and analytical results of frequency change for large amplitudes ($f_1 = \omega_1/2\pi$).

Equation (12) shows the increasing frequency for large amplitudes. As in the numeric solution, only odd multiple of base frequency exists in the analytic solution. This result, as numeric results also, leads to conclusion that system has stable symmetric bifurcation for load greater then critical.

On the Fig. 3 is presented comparison of results obtained numerically and results given by (12).

4. CONCLUSION

The analysis of bifurcation type on the basis of frequency change of vibrations with large amplitudes is presented in this paper. Analysis is demonstrated on the double pendulum. Frequency change for large amplitudes is calculated numerically and analytically. Numerical solution is determined by simulation of vibration and FFT transformation of results. Analytical solution is determined by perturbation method, where instead of original equation of motion, which is system of differential equations, one differential equation derived from the original equation is analyzed. Analysis of frequency change shows the increasing of frequency if large amplitudes is present, which points to the same effect of nonlinear terms of geometric stiffness as stiffness nonlinearity, and also shows stable symmetric bifurcation of the considered system for load greater then critical. An aacceptable agreement between numerical and analytical results shows possibility of avoiding of numerical analysis for variety of initial condition and load intensity in order to determine post-critical behaviour of the system, and replace it by an analysis of single differential equation using some of perturbation methods.

5. REFERENCES

H. Ziegler, "Principles of Structural Stability", Blasidell Publishing Company, Watham, Massachusetts, Toronto, London, 1968.

J.M.T. Thompson, G.W. Hunt, "A General Theory of Elastic Stability", John Wiley & Sons, 1973.

J.J. Thomsen, "Vibration and Stability – Order&Chaos, Technical University of Denmark, Lyngby, 1994.

W.J. Cunningham, "Introduction to Nonlinear Analysis", McGraw-Hill, New York, 1958.

S. Isić , V. Doleček, "A Dynamical Approach To The Nonlinear Analysis Of The State Stability Of Elastic Systems", 10th International Research/Expert Conference "Trends In The Development Of Machinery And Associated Technology" TMT 2006, Barcelona-Lloret de Mar, Spain, 11-15 September, 2006.

S. Isić, V. Doleček, I. Karabegović, "An Identification Of Bifurcation Type Using Postcritical Motion Analysis", 5th International Congress of Croatian Society of Mechanics, Trogir/Split, Croatia, September, 21-23, 2006.

Annals of DAAAM for 2006 & Proceedings of the 17th International DAAAM Symposium, ISSN 1726-9679
Editor B. Katalinic, Published by DAAAM International, Vienna, Austria 2006
150th Birthday of Nikola Tesla (1856-1943) Inventor of Mobile Robot Teleoperation

SURFACE PROFILE MEASUREMENT SYSTEM FOR TIMBER

ISLAM, N.; PARKIN, R.; MUELLER, P.; JACKSON, M. & FAN, K.

Abstract: *One of the most common and primitive methods of surface quality or defect detection in process industries is the visual approach, which to this day remains a valid procedure for surface inspection. This is a very subjective way of looking into product finishes and in most cases fails to ensure a reliable and consistent minimum standard of end product finish. This has inspired the invention of mechanical stylus based profilometers. However, this is a slow and destructive process of obtaining profile information despite very well-established standards and cost-effectiveness. As a result, non-contact techniques are emerging as the leading candidate for profile measurement of some materials. In this paper, a surface profile measurement system for timber surface inspection based on the pickup head of a CD/DVD player has been proposed. Design ideas and preliminary test results have been presented.*
Keywords: *Woodworking, optical method, DVD pickup head, surface profile measurement.*

1. INTRODUCTION

The most commonly employed technique until now to evaluate machined surfaces is human inspection (Sandak & Tanaka 2003). Not only is it a highly subjective way of looking into finish quality, it is also prone to errors due to human fatigue and judgment. Hence, instrumentations to accurately analyze the finish quality had become a necessity. Thus, mechanical stylus based instruments have been widely employed in various process and manufacturing industries in the past few decades.
Despite well-established standards and cost-effectiveness, this technique is becoming unsuitable for some applications. Most notably, mechanical stylus based instruments can only be used in high-speed woodworking with limited success (Parkin & Jackson 1996). The method is destructive to timber surface due to the loading of stylus tip as well as slow and yielding parameters of little importance to indicate the surface form defects due to timber machining. Therefore, non-contact methods are proving to be the more suitable instrument in the wood machining scenario (Cutri *et al.* 1991).

2. BACKGROUND

Various non-contact methods for surface profile measurement have been developed and tested. However, most of the methods have been used to obtain surface texture information of metal or similar products (Parkin & Jackson 1996).
Very little or almost no research has been carried out to successfully determine the finish quality of timber. Among these few researchers, Yang *et al.* (2005), Jackson *et al.* (2002), Sandak *et al.* (2003)) have successfully explored non-contact methods for surface profile measurement of timber.
As stated before, various non-contact measurement methods are being developed and investigated around the globe. Ultrasonic and optical probes have become the most suitable candidates for many present day non-contact measurements. In this section of the paper, some of these non-contact techniques will be discussed in certain detail.
Ultrasonic technique mainly includes the pulse echo and standing wave resonance methods. In principle, ultrasonics may still be regarded as a contact method as good acoustic coupling is essential for the electromagnetic waves to travel from one medium to the other (Blessing *et al.* 1993). Thus, the surface being measured must be immersed in an oil bath along with the transducer. Water jets are also used as a coupling medium.
Optical methods include optical profilometers (mostly laser based), microscopes, image analyzers, imaging spectrographs, interferometers, fiber-optic transducers, white-light speckles, laser scattering, optical light sectioning systems, etc. Some of the techniques e.g., microscopes, suffer from the fact that measurement speed is rather slow and thus they are not suitable for in-process measurements (Lee *et al.* 2002). Also, the cost is prohibitive notwithstanding the great accuracy and ultra-fine resolution of these probes. Image analyzers, spectrographs etc. are highly costly approach and only suitable for certain measurement tasks and environments.
Another common precision method used for surface measurement is the fiber-optic transducer (Persson 1999). Very fine measurements can be carried out using this method with a high resolution. The drawbacks seem to be use of complex hardware as well as complicated signal processing algorithms. The fiber is very delicate in nature and thus must be shielded very carefully in the measurement setup.
It has been observed that, most of the techniques currently being used or investigated require complex hardware, high capacity image/signal processing, and complicated arrangement leading to high cost and low flexibility. Therefore, scope remains to develop low-cost, easy-to-use and flexible system for non-contact surface profile measurement.

3. THE PROPOSED METHOD

An alternative to the aforementioned non-contact methods has been investigated within the Mechatronics Research Group at Loughborough University and reported in this paper. The sensor used for this measurement purpose is the optical pick-up head of a Compact Disc (CD) / Digital Versatile Disc (DVD) player.
The research work presented in this paper mainly focuses into the possibility of measuring materials having lesser reflectivity and integrate such measurement technique into a woodworking machine currently under development within the research group.
All the previous research have used micro-precision stage to position the optical head over the specimen, thus obtaining accurate positioning information. This approach is effective for laboratory based nano or micro metrology but would be prohibitively costly for viable industrial application.
To overcome this drawback, a low-cost optical mouse sensor based linear encoder has been developed. This would eliminate the need for an expensive stage and the scanning motion can be provided through any motor driven linear stage.

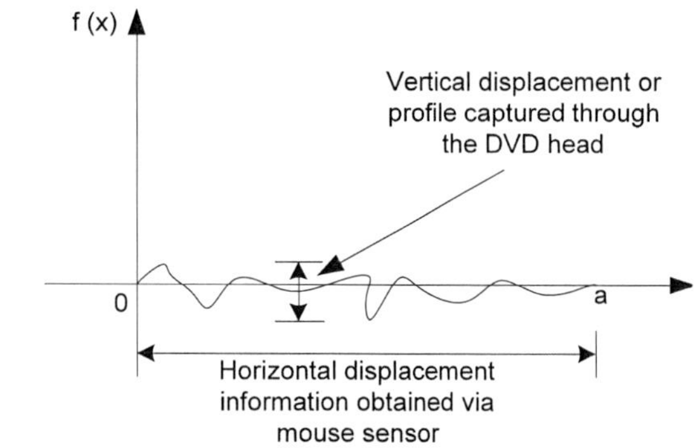

Fig. 1 Surface profile measurement idea schematics

The measurement idea has been depicted in figure 1. The optical pickup is being used to measure the profile information, while the optical mouse based encoder provides the positional information of the sensor over the work-piece under test

4. RESULTS AND DISCUSSION

Result obtained from a previous research has been shown in figure 2 (Fan *et al.* 2001). The profile of a CD has been measured with the system. The track distance was measured to be in the range of 1.4 to 1.7 µm, which compares well with the nominal distance of 1.5 µm. Also, the average height was found to be 0.15 µm, which is slightly less than the manufacturer's value of 0.20 µm (Fan *et al.* 2000, 2001).

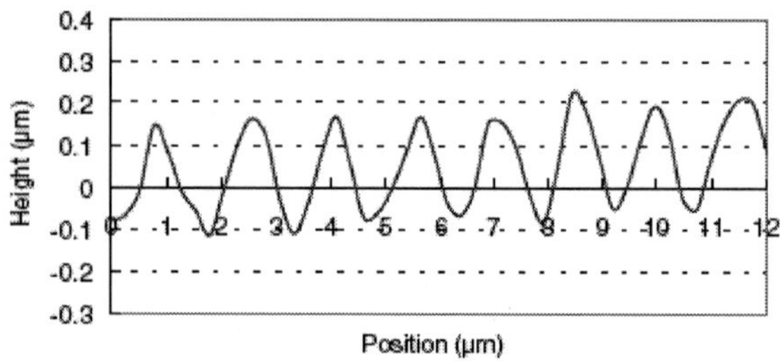

Fig. 2 Measured profile of a CD (Fan et al. 2001)

It is acknowledged that, the measurement results will be significantly different in the case of obtaining surface profile of wood or similar material as opposed to a CD. Construction of the test rig for such measurement purpose is currently underway and test results on timber will be presented in future publications.

Some testing has already been conducted to establish the suitability of optical mouse sensor based linear encoder in the case of timber. Repeatability error of this linear encoder has been reported in figure 3.

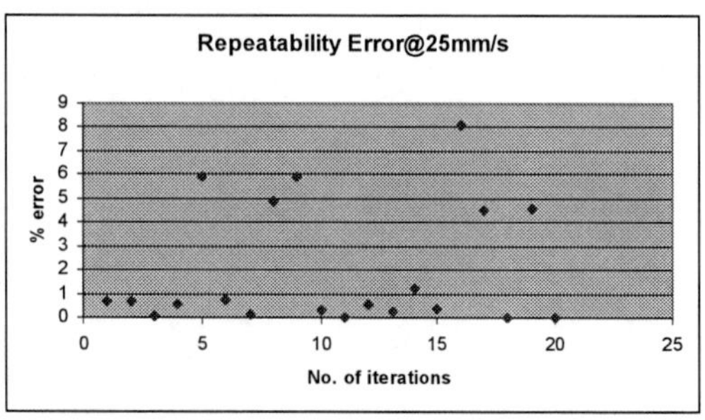

Fig. 3 Repeatability error of the linear encoder at 25 mm/s

An industrial wood machining environment was simulated in the lab, by mounting a light colored timber on to a linear stage and passing it through the sensor assembly at 25 mm/s.

From figure 3 it is evident that, in almost 80% of the cases, the measurement error was lower than 1%. However, many more iterations will have to be carried out to determine the actual average repeatability error of the encoder.

5. CONCLUSION

A research carried out on the possibility of using DVD pickup head as a surface profile measurement sensor has been presented in this paper. Also, a novel linear encoder to obtain measurement position on the specimen has been discussed. Great potential lies with such a system combining the two sensing system in terms of cost-effectiveness, smaller footprint and seamless integration into existing process machineries. Some preliminary results have been reported in the paper, showing a good potential of the system being able to meet the measurement requirements in wood machining industry.

6. ACKNOWLEDGEMENT

The lead author would like to gratefully acknowledge the research funding provided by Loughborough University through its Innovative Manufacturing and Construction Research Center.

7. REFERENCES

Blessing, G. V., Slotwinski, J. A., Eitzen, D. G. and Ryan, H. M. (1993). Ultrasonic Measurements of Surface Roughness. *Applied Optics* 32(19), pp: 3433-3437. ISSN: 0003-6935

Cutri, F., Maycock, K. and Parkin, R. (1991). Surface Measurement of Planed and Moulded Timber Products. *Sensors and Actuators, A: Physical* 26(1-3), pp: 433-437. ISSN: 0924-4247

Fan, K. C., Chu, C. L. and Mou, J. I. (2001). Development of a Low-Cost Autofocusing Probe for Profile Measurement. *Measurement Science and Technology* 12(12), pp: 2137-2146. ISSN: 1361-6501

Fan, K. C., Lin, C. Y. and Shyu, L. H. (2000). Development of a Low-Cost Focusing Probe for Profile Measurement. *Measurement Science and Technology* 11(1), pp: 1-7. ISSN: 1361-6501

Jackson, M. R., Parkin, R. M. and Brown, N. (2002). Waves on Wood. *Proceedings of the Institution of Mechanical Engineers, Part B: Journal of Engineering Manufacture* 216(4), pp: 475-497. ISSN: 0954-4054

Lee, S., Miyoshi, T., Takaya, Y. and Takahashi, S. (2002). Non-Contact 3d Edge Profile Measurement for Die and Mould Model Surface. *Journal of Materials Processing Technology* 127(2), pp: 286-291. ISSN: 0924-0136

Parkin, R. M. and Jackson, M. R. (1996). A Mechatronic Approach for Analysing Timber Surfaces. *Mathematics and Computers in Simulation* 41(5-6), pp: 445-450. ISSN: 0378-4754

Persson, U. (1999). Fibre-Optic Surface-Roughness Sensor. *Journal of Materials Processing Technology* 95(1-3), pp: 107-111. ISSN: 0924-0136

Sandak, J. and Tanaka, C. (2003). Evaluation of Surface Smoothness by Laser Displacement Sensor 1: Effect of Wood Species. *Journal of Wood Science* 49(4), pp: 305-311. ISSN: 1611-4663

Yang, D., Jackson, M. and Parkin, R. (2005). Measuring Cutter Marks on Wood Surfaces with Machine Vision Techniques. *17th International Wood Machining Seminar*, 26-28 September 2005 Rosenheim, Germany. pp: ISBN: 4-9901767-0-7

Annals of DAAAM for 2006 & Proceedings of the 17th International DAAAM Symposium, ISSN 1726-9679
Editor B. Katalinic, Published by DAAAM International, Vienna, Austria 2006
150th Birthday of Nikola Tesla (1856-1943) Inventor of Mobile Robot Teleoperation

Z8 ENCORE MICROCONTROLLER BASED DATA LOGGING SYSTEM

ISLAM, N.; WATSON, R. & MOORE, P.

Abstract: *Data logging systems find numerous important applications in fields ranging from sport science to automotive engineering. The need for accurate, robust and low-cost data logging solution has been there for a while now as machine, process and product inspections are becoming critical issues. In this paper, the design and construction of a novel Z8 Encore based embedded data logging system has been discussed in detail. A GPS device has been used to provide the timing information of the logged data. This particular data logger provides the user with the desired accuracy, flexibility and cost-effectiveness rarely seen in commercially available such devices.*
Keywords: *data logger, Z8 encore microcontroller, embedded system, reliability, cost-effectiveness.*

1. INTRODUCTION

A data logger is an electronic instrument (or specialized computing device in some cases) that records digital, analogue, frequency or smart protocol based measurements over time (Wikipedia 2006). There is a greater need for monitoring of equipments and systems in industrial environments due to their increased complexity and safety related issues. Thus, the monitoring systems such as the data loggers have become more complex as well as costly. This adds to the overall cost of operation of any plant or automation line. So the urgent need for cost effective data logging solution has increased over the last few decades.

2. APPLICATIONS

The main areas in which data loggers are widely used are in automotive engineering (Illic *et al.* 2004), biomedical instrumentation (Cifrek *et al.* 2004), mass spectrometry (Payne and Gannon 1993), power quality measurement (Yeary *et al.* 2001), machine health monitoring (Babb 2005) and so on. In most cases, the data logging system provide the much-needed real time data for the analysis of a particular fault. It can also provide the required data to optimize a control system in a given process industry.

3. RESEARCH CONTEXT

Non-invasive measurement of partial discharge (PD) from the overhead electrical transmission line is a field where the data logging system can play a typical and very important role. The detection and measurement of partial discharge has become very important non-invasive method of checking the quality and technical integrity of the high-voltage (HV) electrical systems (Moore *et al.* 2003). Moore *et al.* (2004) has proposed and demonstrated mobile remote monitoring of substation equipments by using a properly equipped vehicle. The common practice is to use helicopters to measure PD from overhead transmission cables. But the use of mobile van to accomplish the same task gives the flexibility and cost effectiveness desired in the power generation industry.

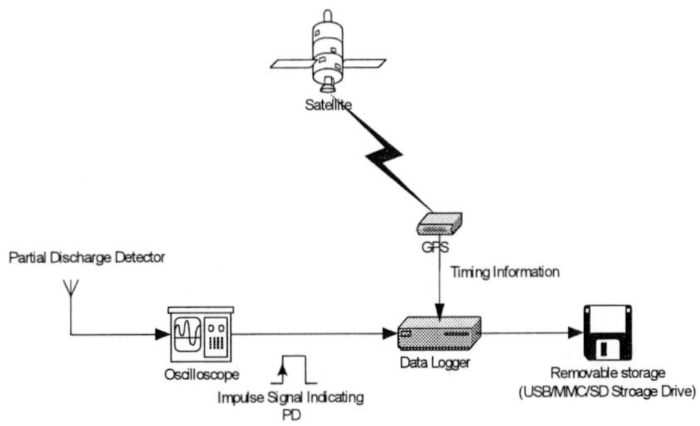

Fig. 1 Integration of the data logging system into PD

The data logging system presented in this paper can be successfully integrated in the aforementioned mobile PD measurement practice, thus enhancing the experimental setup. The lower power consumption of the Z8 encore development board ensures that the data logging system can be easily operated in the vehicle. Since GPS based logging system is being used, the timing of the event of partial discharge can be kept in a very accurate manner. Also the data stored in a structured fashion can be used easily at a later stage to do the required off-the-site analysis of the measurement carried out. Figure 1 gives a good schematic idea of the proposed setup.

4. IMPLEMENTATION

Development of the Z8 Encore microcontroller based data logging system can be divided into two parts i.e., hardware and software. Since the development board comes with well-defined ports and communication interfaces, hardware implementation did not turn out to be a time consuming task. Most of the development time went into developing required routines to drive the board.
Hardware implementation of the data logger involved the interfacing of the analog impulse signal obtained from PD measurement sensor along with the GPS serial link. The schematic hardware connection diagram is given in figure 2. The analog sensor signal can be interfaced to any of the analog ports via the expansion slots onboard the development board.
It can be noted that, there is also a provision for a 1 PPS signal to the development board. This signal is in fact obtained from the GPS receiver, which acts as a reset to the internal counter as discussed in the later part of the paper.

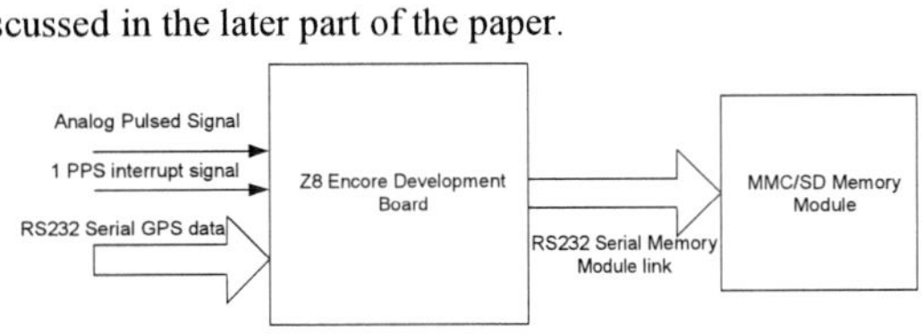

Fig. 2 Hardware interface

As in the proposed setup of figure 2, a serial memory module link was interfaced to the board in order to store the logged data. The need for this device is two-fold. As the memory onboard the microcontroller is very small, a realistic amount of data cannot be stored, resulting in loss of recorded data. Also, this memory card can be detached from the measurement setup and the stored data can be analyzed at a convenient time and place with the help of a PC. uMMC Serial Data Module from Rogue Robotics Corporation Inc. was used for this purpose.
The GPS timing data has a resolution of 1 s. Thus, in order to increase the resolution of the measured data, an internal counter was also setup to effectively give a resolution of 10 µs. A counter reset was devised for each second. This reset process can be carried out either by the 1 PPS signal obtained from the GPS and interfaced via the analog input ports or the onboard timer of the microcontroller.

5. RESULTS AND DISCUSSIONS

The error analysis of the system was carried out in two phases. At first, the error in the recorded data obtained from the GPS was evaluated. Nearly 400 trials revealed that there were no errors in logging the GPS timing information (figure 3). Thus, it was ascertained that the timing information was correctly logged by the system. Figure 4 shows the counter data errors for a fixed interval interrupt. The jittery curve for almost 1800 trials signifies the fact that the counter readings were erroneous in many cases. This is because of the fact that the counter is being reset in this case with the help of the internal timer of the board. Due to the drift in the onboard oscillator, the internal timer has been inaccurate and the counter was not reset accurately on the 1-second mark. This problem can be overcome by using the 1 PPS signal from the GPS instead of the internal timer to reset the counter. Also, ultra-precision external oscillators can be used to drive the timer instead of the onboard inaccurate oscillator.

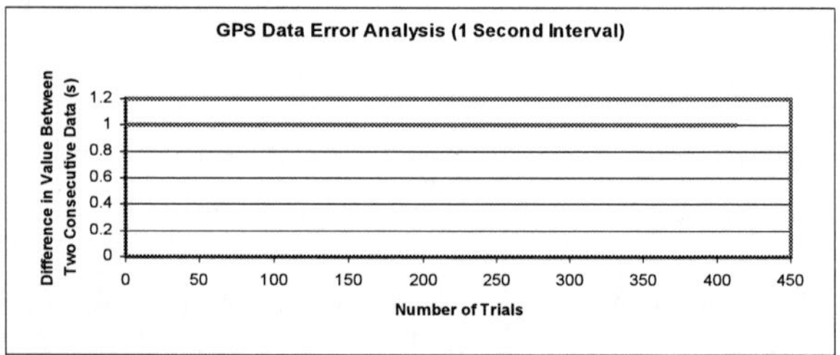

Fig. 3. GPS data error analysis

It is worth mentioning here that, this is the exact reason why a GPS device has been used for timekeeping purpose. As the time obtained from the GPS is synchronized with the atomic clock onboard the satellites, it is extremely accurate. Thus, the error in the timestamp was successfully avoided.

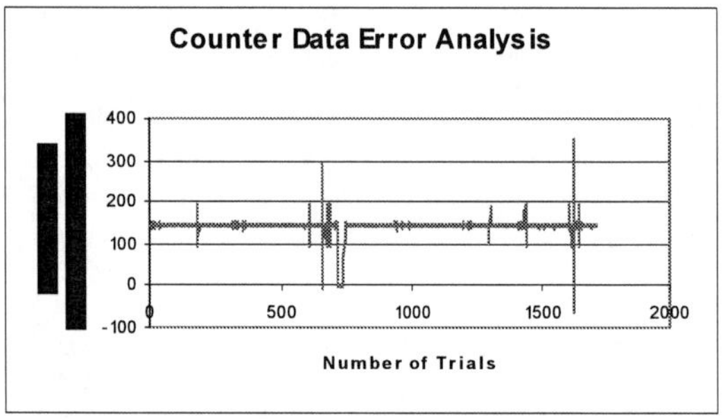

Fig. 4 Error in counter data reading

6. CONCLUSION

A data logging system using Z8 Encore microcontroller and a GPS receiver has been presented in this paper. A detailed construction of the system has been discussed and performance analysis has also been carried out.

Trial data collected over a period of two weeks have indicated that the system was accurate to an acceptable level provided the counter reset issue has been addressed as discussed in the previous section.

Despite some robustness related drawbacks, overall performance of the system was satisfactory and the goal of a cost-effective, flexible data logger was achieved. Once this issue is addressed, this system can be safely used in real-world industrial applications.

7. REFERENCES

Babb M.; 2005. Machine Health Management: A New Role for Process Automation Systems. *IEE Computing and Control Engineering Magazine*. June/July 2005, pp: 29 – 33.

Cifrek, M.; Mrvos, S.; Zufic, P.; 2004. Portable data logging system for long-term continuous monitoring of biomedical signals. *Proceedings of the 12th IEEE Mediterranean Electrotechnical Conference (MELECON 2004)*, 12-15 May 2004, Volume 1, pp: 399 – 402

Illic, S.; Katupitiya, J.; Tordon, M.; 2004. In-vehicle data logging system for fatigue analysis of drive shaft. Proceedings of the 2004 International Workshop on Robot Sensing (ROSE 2004), 24-25 May 2004, pp: 30 – 34

Moore, P.J.; Portuguese, I.; Glover, I.A.; 2003. A nonintrusive partial discharge measurement system based on RF technology. *Proceedings of IEEE Power Engineering Society General Meeting,* 13-17 July 2003. Volume 2, pp: 628-633

Moore, P.J.; Portuguese I.E.; Glover, I.A.; 2004. Remote diagnosis of overhead line insulation defects. *Proceedings of IEEE Power Engineering Society General Meeting*, 6-10 June 2004. Volume 2, pp: 1831 – 1835

Payne, J.; Gannon, J.; 1993. Leak checker data acquisition system. *Proceedings of the Particle Accelerator Conference, 1993.* 17-20 May 1993, Volume 5, pp: 3870-3872

Wikipedia; 2006. Data Logger. http://en.wikipedia.org/wiki/Data_logger [Accessed March 31, 2006]

Annals of DAAAM for 2006 & Proceedings of the 17th International DAAAM Symposium, ISSN 1726-9679
Editor B. Katalinic, Published by DAAAM International, Vienna, Austria 2006
150th Birthday of Nikola Tesla (1856-1943) Inventor of Mobile Robot Teleoperation

ROLE OF FOURTH PARTY LOGISTICS IN MINIMIZING THE "BULLWHIP EFFECT"

IVAKOVIC, C.; SAFRAN, M. & BABIC, D.

Abstract: *The system which connects all the elements between the producer and the consumer is called the supply chain. This definition of the supply chain shows its complexity, dynamics and uncertainty. Successful operation of supply chains requires the best possible coordination of its elements, which obviously shows that the information flow along the supply chains is of utmost importance. The work analyses the processes within the supply chains and the occurrence of the bullwhip effect. The main causes of the bullwhip effect, that we encounter daily in the supply chains, have been clearly defined. The paper also presents the main negative consequences of such a disturbance within the supply chains, as well as the possible solutions to avoid the bullwhip effect, in the development of 4PL operators.*
Key words: *logistic, distribution, supply chains, Fourth Party Logistics*

1. INTRODUCTION

Supply chains include all the participants and processes in satisfying the demand of the end users: transport, inventories, wholesalers, retailers, distributors and producers. A great number of participants results in the diversity of interconnections and processes, dynamic characteristics, uncertainty, and stochastic flow of goods and information, and many control points prove that the supply chains have to be considered as complex systems in which coordination plays the key role of successfulness.
The need for continuous improvement of company's business operation (since only permanent improvement means no regression) sets increasingly complex requirements in the resource management system – logistic management, which together with the outsourcing trend and the sudden increase in the application of information technologies opens up a new space for the development of higher level services i.e. some kind of supermanagement which includes logistics, know-how and information system of a company. Such supermanager represents at the same time also a unique point for the contact between the respective company and all their subcontractors (suppliers, 3PL, IT, intellectual and other services). **3PL *(Third Party Logistics)* - Provider** is an external logistic service provider, i.e. a company specialized in providing complete warehouse, transhipment-manipulation and transport services, which takes over the organization and implementation of a certain part of the user's supply. chain. **IT *Information Technology)* – Provider** is an external provider of information services i.e. a company specialized in providing services of computerization and maintenance of information technology systems.

2. "BULLWHIP EFFECT" IN SUPPLY CHAINS

"Bullwhip effect" was noticed for the first time at the "Procter&Gamble" (P&G) company. While testing order samples for one of the most sold products, their logistic experts noticed certain fluctuations in the flows of goods, but these variances were not specially excessive. Nevertheless, the experts were surprised by the increased level of variability in the distribution of orders. When they checked, namely, the P&G data against their raw material suppliers, they discovered even greater changes of variables. At first glance the obtained results showed no sense. While the consumers used the product at a constant pace, the order variables (demand) in the supply chain were increasing continuously as we move along the supply chain. The P&G experts called this phenomenon the "bullwhip effect", and it was proven that this phenomenon occurs in almost all the organizations whose products or services include a multi-level supply.
When the bullwhip effect occurs in the supply chains, and when the demand information along the supply chain is distorted, then the following are the undesired consequences:

- increase in inventories,
- poor forecast of future conditions,
- reduction or increase of delivery possibilities,
- loss of revenue,
- poor planning of the future capacities,
- passive transport and logistics,
- disturbances in the production schedule,
- low level of service,
- uncertain and expensive production,
- high costs of corrections (e.g. shipments, overtime, etc.).

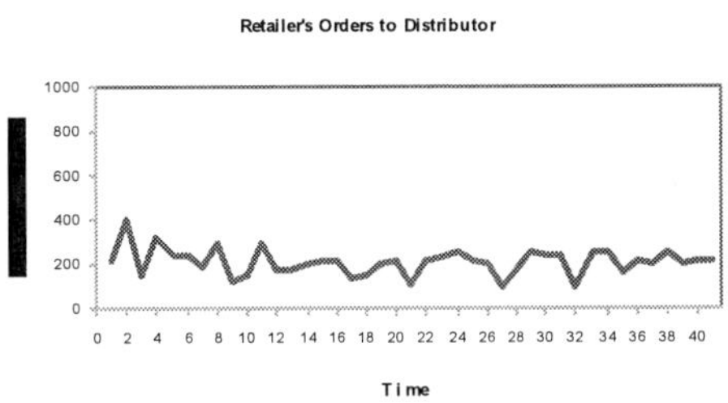

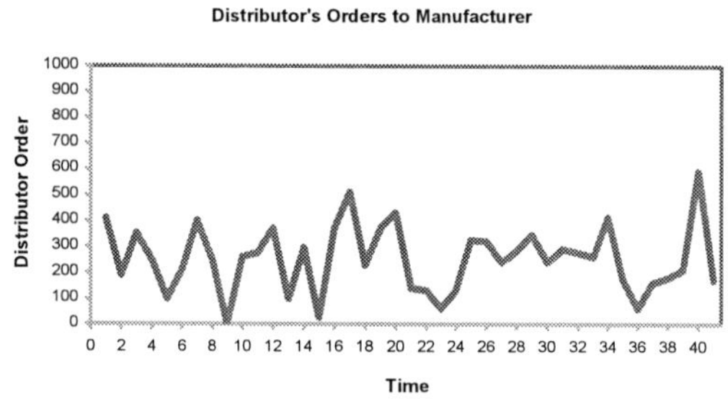

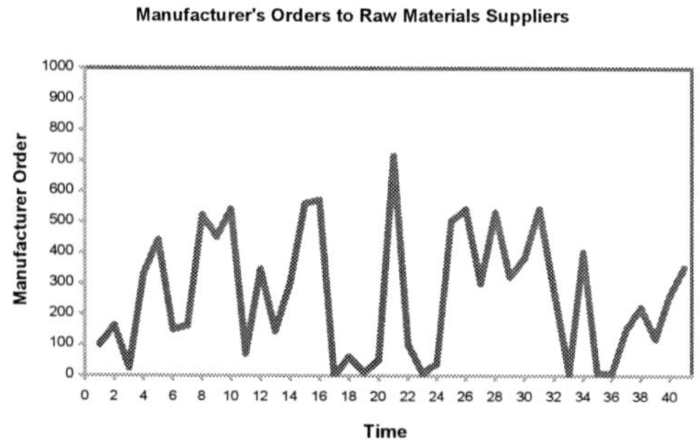

Fig. 1. Bullwhip Effect

3. ROLE OF FOURTH PARTY LOGISTICS (4PL) IN ORGANIZING SUPPLY CHAINS

4PL *(Fourth Party Logistics)* is an integrator that functionally connects resources, capabilities, and the technology of one's own organization and other organizations in order to design, construct and implement complete solutions for supply chains.
The 4PL concept is based on intellectual logistic capital and IT systems, and not on owning the infrastructure (terminals, warehouses, etc.) and other material means (transport means, handling machinery, etc.), for the purpose of which, in order to provide services to their clients, they use various 3PL suppliers. The role of 4PL represents therefore a unique point of integration of the information flows and executive responsibility.
The competencies of 4PL are basically found in the following three areas:

1. Surveillance and management of operation of several 3PL suppliers, both on the strategic and operative levels;
2. Surveillance and management of know how regarding the accessibility and practical implementation of new achievements;
3. Surveillance and management of IT sector, with the emphasis on the integrated IT systems with operative sector.

The increasing connectivity at all levels of the world economy makes the sophisticated technologies universally accessible, thus allowing 4PL suppliers not only to efficiently manage the complex supply chains, but also to modify and improve them. Here, the outsourcing in the supply chain passes from the cost centre into a potential profit centre.

4. MINIMIZING THE “BULLWHIP EFFECT”

To minimize the bullwhip effect it is necessary to define and understand the "forces" that give incentive to the consumer demand planning, since they trigger the supplementation of the orders at different points in the supply chain. The most efficient way to mitigate the oscillations caused by the bullwhip effect is for the distributors and suppliers to understand what makes the demand and supply work, and then make improvements in the quality of the exchanged information and reduce the duration of activities along the entire supply chain.
It may be said that for the reduction of the bullwhip effect, certain or all of the following activities need to be carried out:

- minimize the cycle from receiving the planned and actual information about consumption,
- understand the demand for a product at every level of the supply chain at the same time connecting the logistics service providers,
- increase frequency and the quality of information among the subjects in the supply chains,
- minimize or eliminate "waiting" of information that causes delay in the flow of goods,
- eliminate varying of the inventories that cause demand lumps in the supply chains
- offer products at relatively acceptable prices in order to minimize the purchase of the products at promotional discounts,
- recognize and eliminate the causes for the reduction and cancellation of consumers' demand,
- allow the inventories to be managed by the retailer so that the inventories are planned in cooperation with the consumers who project the final demand.

5. CONCLUSION

One of the biggest negative consequences that may occur in the supply chain is the bullwhip effect. This is an undesired occurrence of variability in the quantities and flows of goods in the supply chains, as a consequence of small variations in the end-users' demand. This phenomenon depends mainly on the fact that decisions are made that are based on the limited amounts of information. It is precisely the distorted information from one end of the supply chain to the other end of the supply chain that can result in very damaging consequences.
The enterprises can successfully counter the bullwhip effect if they understand fully its main causes. Big companies implement new strategies that set new challenges: integration of a new information system, defining of new organizational connections, implementing of new incentives and control systems.
The 4PL services market should be considered as a new market segment, which expands the market of logistic services. The demand for 4PL services directly corresponds to the development of technology (e.g. wireless communication, RFID, on-line tracking and tracing systems) and with globalization processes in the world economy (elimination of administrative barriers, harmonization of commerce, tax and customs systems, elimination of custom duties, etc.).

6. REFERENCES

Koether, R. (1993). *Technische Logistik*, Carl Hanser Verlag, ISBN 3-446-17139-8, Munchen Wien

Martin, H. (1995). *Transport – und Lagerlogistik*, Vieweg Verlag, ISBN 3-528-0494, Braunschweig/Wiesbaden

Robeson, J. F.; Copacino, W. C. (1994). *THE LOGISTICS HANDBOOK*, Andersen Consulting, ISBN 0-02-926595-9, New York

Zelenika, R.; Jakomin, L. (1995). *Suvremeni transportni sustavi*, Ekonomski fakultet, ISBN 953-6148-05-6, Rijeka

Ghiani, G.; Laporte, G. & Musmano, R. (2004). *Introduction to Logistics Systems Planning and Control*, John Wiley & Sons Ltd, ISBN 0-470-84916-9, West Sussex

Annals of DAAAM for 2006 & Proceedings of the 17th International DAAAM Symposium, ISSN 1726-9679
B. Katalinic, Published by DAAAM International, Vienna, Austria 2006
150th Birthday of Nikola Tesla (1856-1943) Inventor of Mobile Robot Teleoperation

INTELLIGENT MANUFACTURING IN PRODUCTION NETWORKS BY THE VALUE ADDING-PROCESS RELATED EVALUATION OF PERFORMANCES

JAEHN, H.

Abstract: *The production of goods in networked organisation structures gains significance. Thereby economically and legally independent enterprises form order-specific networks in order to compete with the large companies on the global market. This paper focuses the problem of performance measurement and evaluation in production networks. In that context a new approach for value adding-process related evaluation of enterprise-specific performances is introduced.*
Key words: *Evaluation, Production Network, SME, Network Controlling, Performance Measurement*

1. MOTIVATION

In order to bear the challenge of global competition, the realisation of value adding processes within networked production structures gains significance. To optimise the supply chain in production networks hereby is one major challenge in order to assure success. This contribution focuses the procedure of performance evaluation in production networks as one part of the operative network controlling. Thereby the specific model of an order-specific (dynamic) production network based on small and medium-sized enterprises (SME) serves as the theoretical background.
The evaluation of performances of the involved enterprises is one important task in that context. It is realised during the final phase of an ideal-typical life-cycle of a production network. In order to quantify the evaluation data, suitable performance attributes need to be identified. The evaluation process is realised by an adapted value benefit analysis. As the final result a measure for the valuation of the performance of participating enterprises is ascertained. In case of an unsatisfying performance the individual profit shares can be shortened. Thereby a specific approach for the distribution of profit in networks serves as the basis (Jähn, 2005).

2. PRESENT SITUATION

2.1 Problem formulation and classification

The terms performance measurement or performance evaluation are very popular in literature. However, it has to be exactly differentiated which relation this performance measurement is in. Therefore, three perspectives are practicable:

- the network perspective,
- the static enterprise perspective and
- the dynamic enterprise perspective.

For evaluating the performance of the entire network (variant 1), only the performance parameters for the manufacture of the final product are evaluated. Therefore, the date of delivery of the final product to the customer, the final quality or the sales price is for example significant. Thus, it is an evaluation of the performance of the entire network without taking a closer look on the partial performances of the single value adding units.
This is the content of the enterprise-perspective. Thereby, a static (variant 2) and a dynamic (variant 3) consideration can be differed. The static consideration is geared to long terms. The (evaluation) data of several value adding processes are collected and aggregated. This procedure results in the possibility to observe the performance of enterprises for a longer period of time and to critically evaluate it. The dynamic perspective is relevant with regard to the ascertainment of sanctions as for example shortage of profit shares. Thereby, simply the data of the performance parameters of all the enterprises participating in a certain value adding process are analysed. This paper focuses on the third perspective.

2.2 Literature Study

Numerous publications focussing on the measurement of performances have been published. Most of them however focus on the static enterprise-related perspective or respectively the network perspective. *Erdmann* (Erdmann, 2003) gives a comprehensive approach for „performance measurement“ in supply chains. Thereby, performance measurement is interpreted as an operative or respectively strategic controlling instrument for the support and improvement of management tasks. The order-related performance evaluation thereby is hardly considered. On the basic principles of performance measurement focuses *Klingebiel* (Klingebiel, 1999) However he concentrates on the transparency of the measured data.
The SCOR-model (SCOR, 2006) of the Supply-Chain Council (SCC) is another relevant model for the controlling of supply chains. Within that approach, performance attributes which can be transferred on single value adding units as well as on the entire supply chain are introduced. The Balanced Scorecard (Kaplan & Norton, 1996) is the most famous approach for the measurement and comparison of performances. However, that approach is not network-specific.So far, a performance evaluation has almost exclusively been applied in the field of information technology for the evaluation of the performance of computers. As oppose, an approach for networks is new. The approach for the performance evaluation of enterprises in SME-based production networks is introduced in the following. Thereby this contribution deals with matters beyond a simple measurement: it focuses the evaluation process.

3. MEASUREMENT AND EVALUATION

3.1 Selection of relevant performance attributes

In the first step of the procedure of performance evaluation the performance attributes that are to be considered have to be determined. Thereby quantitative as well as qualitative performance attributes are relevant. A finite number of performance attributes is to be included in the analysis. A wide range of performance attributes, which can be included in a performance measurement approach, are suggested within the scope of the SCOR-model. Those for example include: delivery performance, fill rate, perfect order fulfilment, order fulfilment lead time, supply-chain response time, production flexibility, supply chain management cost, cost of goods sold, value adding productivity, warranty cost or returns processing cost, cash-to-cash cycle time, inventory days of supply and asset turns. Several questions arise concerning the further procedure. Thus it has to be described:

- How can the performance attributes be determined / measured?
- Which performance attributes are SME-related?
- Which performance attributes are relevant for the calculation of sanctions (profit shortage)?
- Which weightings should the single performance attributes be given?

The price, the date of delivery and the specification of the quality for example are the most prominent performance attributes. However, there are also parameters that can hardly be expressed as figures besides the classic quantity-orientated performance attributes. Those for example include the quality of the cooperation and the climate of confidence within the network. Such so-called soft-facts however are very important for the long-term success of a network. For the evaluation of these parameters a specific approach is required.

3.2 Determination of the weightings

Subsequently for every attribute a weighting has to be determined according to the estimated significance within the performance evaluation process. This is a singular occurring task. In decision theory literature several approaches for the fixing of weightings are introduced: the trade-off method, swing weighting and direct-ratio-method. Thereby the application of an approach depends on the specific structure of a problem. In the present case the determination of the weightings is based on the controlling target system. However that process is highly subjective. For limiting the danger of an abuse, the determination ideally is carried through by several decision makers. The weightings should not be changed too often in order to keep them continuously comparable within several value adding processes. In the ideal case, they are selected in the beginning of the evaluation activities. Therefore, rankings or empirical investigations might serve as aiding means. For achieving a standardisation of the evaluation, the sum of all weightings must be 1. This guarantees the comparability of the parameters concerning their significance.

3.3 Determination of the target performance

In order to allow a comparison of planned and realised performances for all attributes, the highest achievable number of evaluation credits for every attribute has to be calculated and compared to the actual achieved evaluation credits of each parameter and each enterprise. This task is fulfilled after every value adding process in the order-specific network. The highest achievable number of evaluation credits for one parameter results from the product of the maximal achievable number of credits and the corresponding parameter weighting. It is identical for all the attributes. Its value results from the degree of differentiation of the consideration, which means the number of gradations. Usually, a value of 10 is chosen. This means, that a later evaluation of a performance between 0 and 10 credits is made available, whereby 10 credits are awarded in case of complete satisfaction or accordance of the target and actual performance while 0 credits refer to an absolutely unsatisfactory or respectively not performed service. The target value finally can be ascertained by summing up the target values for all involved attributes. The resulting value is the comparison parameter for all enterprises in a static network (pool of resources) and it is valid for a longer period of time. In a subsequent step, the comparison parameters for the evaluation of the performed services of the single enterprises for a certain value adding process must be ascertained. That process is called performance measurement.

3.4. Determination of the actual performance

After the credit evaluation process resulting from the analysis of the coherences between credit evaluation and the considered performance attributes, the calculation of the actual performance value follows. It results from the multiplication of the achieved credit evaluation and the corresponding weighting for an attribute. Subsequently, the aggregated evaluation of an enterprise can be ascertained by summing up the actual values for all performance attributes. This ascertained parameter then serves as basic parameter in the further analysis.

3.5. Network conformity

In the next step the achieved actual value and the target value of every enterprise are compared in order to ascertain the measure network conformity. That measure represents the level of fulfilment of the performances because the aggregated actual value and the aggregated target value are proportionated for the calculation. Therefore, this variable can also be interpreted as the parameter of the quality of fulfilment of performances.

It must be the aim of every enterprise participating in a network to realise a perfect performance according to the given targets in order to get a satisfying evaluation.

3.6. Possible consequences

The final step of performance evaluation focuses on the quantification of possible sanctions in case of an unsatisfactory performance. One example could be the shortage of profit shares. Sanction mechanisms are a valuable instrument of the network controlling. In this case sanctions can result in a shortage of the profit share. However sanctions must not lead to the fact that a future cooperation is made more difficult or even does not take place anymore. For quantifying sanctions, it is necessary to observe the behaviour of the enterprise with regard to the fulfilment of contracts.

4. CONCLUSION

After finishing a production process in production networks, the question arises, how the performances of the single enterprises involved in a value adding process can be measured and evaluated. The introduced model offers a new approach for dealing with that topic using an adapted value-benefit analysis. The future work will concentrate on the formulation of interdependencies for further attributes as well as on the information-technical implementation. Thereby this modelling aims at making a valuable contribution to the operative network controlling.

5. REFERENCES

Erdmann, M. (2003). *Supply Chain Performance Measurement*, Josef Eul, ISBN 3-89936-082-6, Köln und Lohmar.

Jähn, H. (2005). Profit Ascertainment and Distribution in Production Networks, *Proceedings of the 16th International DAAAM Symposium 2005*, Katalinic, B. (Ed.), pp. 169-170, ISBN 3-901509-46-1, Opatija, Croatia, October 2005, DAAAM International, Vienna.

Kaplan, R.S. & Norton D.P. (1996). *The Balanced Scorecard: Translating Strategy into Action*, Harvard Business School Press, ISBN 0-87584-651-3, Boston.

Klingebiel, N. (1999). *Performance Measurement*, Gabler, ISBN 3-85496-007-7, Wiesbaden.

SCOR (2005), Name of the Paper, *Available from:* http://www.supply-chain.org/slides/SCOR5.0OverviewBooklet.pdf *Accessed:* 2005-11-24.

Annals of DAAAM for 2006 & Proceedings of the 17th International DAAAM Symposium, ISSN 1726-9679
B. Katalinic, Published by DAAAM International, Vienna, Austria 2006
150th Birthday of Nikola Tesla (1856-1943) Inventor of Mobile Robot Teleoperation

CALCULATION OF CUTTING FORCES BY GRINDING

JANAC, A.; LIPA, Z. & STEFANEK, M.

Abstract: *Within area of machining there were devoted relative enough space to cutting forces research. But not all problems were solved. By theoretical determination of cutting forces by grinding the problems are: uncertain number, shape and distribution of cutting wedges. It is fairly difficult to take all this into the account. Our contribution sets tangential component of cutting force by grinding falling on one abrasive grain, also regarding shapes and distribution of grinding grains. Our determination of cutting forces by grinding is more detailed than relations utilized till now.*

Key words: *cutting force, specific cutting resistance, grinding, abrasive grain.*

1. INTRODUCTION

In machining theory has very important position the theory of cutting forces. We can say, that cutting force is force, by that the cutting tool influences the workpiece in process of cutting and chip removal. According to action and reaction law the workpiece backwards influences the cutting tool, by its cutting resistance, in size tantamount to cutting force and being in opposite direction. Mutual influence of tool and workpiece invokes change of mechanical motion of workpiece part (chip), called mechanical influence. If result of this mechanical influence is removal of chip from workpiece and elastic and plastic deformations in both chip and workpiece, then such force we name the cutting force. By influence of cutting tool on workpiece is creating force field. This field has potential U. From this potential is possible to deduce some partial forces as potential gradient of force field and partial forces remains non potential. Strain energy uprised by influence of tool on workpiece is potential energy. To non potential forces belong frictional forces.

By grinding the radial component of cutting force predominates the tangential component, because deformation of material consumpts more energy than cutting. This is the consequence of negative geometric conditions of cutting wedges (abrasive grains). For grinding of hardened steels is usual to introduce this two components in form – for surface grinding (however, there are also possibilities for other expressions), e.g. in [Maslov, J.N.: (1976)]:

radial:

$$F_p = 50 \cdot v_{fw}^{0,7} \cdot f_r^{0,7} \cdot a_p^{0,6}; \qquad [\mathrm{N}] \qquad (1)$$

tangential

$$F_c = 22 \cdot v_{fw}^{0,7} \cdot f_r^{0,7} \cdot a_p^{0,6}; \qquad [\mathrm{N}] \qquad (2)$$

where v_{fw} ... workpiece velocity of feed [mm.min^{-1}]
f_r ... transversal feed [mm]
a_p ... depth of cut [mm]

Tangential component of cutting force can be written in form:

$$F_c = k_c \cdot S_r = k_c \cdot S_{r1} \cdot z = k_c \cdot S_{r1} \cdot N_c \cdot S_S = k_c \cdot S_{r1} \cdot N_k \cdot V_\Delta = k_c \cdot S_{r1} \cdot N_c \cdot l_k \cdot b_S \qquad (3)$$

where k_c ... specific cutting resistance [MPa]
S_r ... overall cross-section of grinded layer [mm^2]
S_{r1} ... chip cross-section falling on one abrasive grain
z ... number of concurrently working abrasive grains
N_c ... number of active grains on 1 mm^2 of grinding wheel surface [mm $^{-2}$]
S_S ... actual active surface of grinding wheel [mm^2]
N_k ... grain density in grinding wheel [mm^{-3}]
V_Δ ... actual volume of grinding wheel active layer [mm^3]
l_k ... length of contact arc between tool and workpiece [mm]
b_S ... active width of grinding wheel [mm]

Sometimes is also introduced (specific tangential component of cutting force)

$$F'_c = C_F \cdot a_{p\ ekv}^n \qquad (4)$$

$$F'_c = \frac{F_c}{b_S} \qquad (5)$$

where $a_{p\ ekv}$ ... is equivalent chip thickness [mm or μm],
C_F, n ... are constants.

2. CUTTING FORCE FALLING TO ONE ABRASIVE GRAIN

We will go out from relations:

$$F_{c1} = k_c \cdot S_{r1} = \frac{C_{kca}}{a_{z\ str}^{\frac{1}{3}}} \cdot S_{r1} = C_{kca} \cdot k \cdot a_{z\ str}^{\frac{2}{3}} \cdot b_{z\ str} \quad [\mathrm{N}] \qquad (6)$$

$$F_{c1} = k_c \cdot S_{r1} = \frac{C_{kcs} \cdot k^{0,47}}{S_{r1}^{0,47}} \cdot S_{r1} = C_{kcs} \cdot S_r^{0,53} \cdot k^{0,47} [\mathrm{N}] \qquad (7)$$

In relation (6) we can consider

$$S_{r1} = \frac{1}{2} a_{z\ str} \cdot b_{z\ str}$$

if the abrasive grain profile will be triangular, or

$$S_{r1} = a_{z\ str} \cdot b_{z\ str}; \ (k = \frac{1}{2} \text{ or } 1) \text{ by rounded profiles.}$$

In our contributions [4] and [5] we introduced attributes

$$a_{z\ str}, b_{z\ str}, S_{r1}.$$

In introduced formulas (6), (7) we have

F_{c1} ... tangential component of cutting force falling on one abrasive grain [N]

C_{kca} … constant

C_{kcs} … constant

$a_{z\ str}$ … average chip thickness falling on one abrasive grain [mm]

$b_{z\ str}$ … average chip width falling on one abrasive grain [mm]

k … constant

As a form of abrasive grains we will consider spherical (circular cross-section with radius ρ [mm]), triangular with vertex angle 90° and 120°, trapezoidal with side angle 15°, above introduced triangular profiles with vertex rounded by radius ρ.

In contributions [Janac, A., Lipa, Z.& Stefanek, M. (2005)] and [Janac, A., Lipa, Z.& Stefanek, M. (2005)] we have introduced average thicknesses, average widths and cross-sections; also for the case of uniform distribution of all these abrasive grain forms and we wil utilize them for relations according to (6) and (7). Now we introduce some simplifications of relations in Janac, A., Lipa, Z.& Stefanek, M. (2005)] a [Janac, A., Lipa, Z.& Stefanek, M. (2005)]. For equivalent chip thickness by surface grinding is valid the Peters relation:

$$a_{p\ ekv} = \frac{v_{fw} \cdot a_p}{60 \cdot v_c} \quad ; \quad [\text{mm}] \qquad (8)$$

For average chip thickness falling on one abrasive grain is valid the Maslow relation:

$$a_{z\ str} = \frac{v_{fw} \cdot a_p \cdot l_z}{60 \cdot v_c \cdot l_k} = a_{p\ ekv} \cdot \frac{l_z}{l_k} = \frac{a_{p\ ekv}}{z_L} ; \qquad (9)$$

where … l_z … is distance of two consecutive abrasive grains on grinding wheel arc,

z_L … number of concurrently working grains on contact arc.

In formulas introduced in our contributions [Janac, A., Lipa, Z.& Stefanek, M. (2005)] and [Janac, A., Lipa, Z.& Stefanek, M. (2005)] occurs the specifyable factor:

$$\frac{v_{fw} \cdot a_p}{v_c \cdot l_k \cdot N_k},$$

where N_k … is density of abrasive grains in grinding wheel $[\text{mm}^{-3}]$

$$\frac{v_{fw} \cdot a_p}{v_c \cdot l_k \cdot N_k} = \frac{v_{fw} \cdot a_p}{60 \cdot v_c} \cdot \frac{60}{l_k \cdot N_k} =$$
$$= a_{p\ ekv} \cdot \frac{60}{l_k \cdot N_k} = V_{red} \qquad (10)$$

unit of this factor is mm^3, let us to denote it as reduced volume of grinding V_{red} $[\text{mm}^3]$ and denote also the expression occurred in formulas [Janac, A., Lipa, Z.& Stefanek, M. (2005)] and [Janac, A., Lipa, Z.& Stefanek, M. (2005)]

$$\left(\frac{v_{fw} \cdot a_p}{v_c \cdot l_k \cdot N_k}\right)^{0,33} = \sqrt[3]{\frac{v_{fw} \cdot a_p}{v_c \cdot l_k \cdot N_k}} = a_{red} \qquad (11)$$

as reduced thickness of grinding a_{red} [mm], an also further occurred expression

$$\left(\frac{v_{fw} \cdot a_p}{v_c \cdot l_k \cdot N_k}\right)^{0,67} = \left(\frac{v_{fw} \cdot a_p}{v_c \cdot l_k \cdot N_k}\right)^{\frac{2}{3}} = a_{red}^2 = S_{red} \qquad (12)$$

as reduced cross-section of grinding S_{red} $[\text{mm}^2]$.

From formula (6) by using relations from [Janac, A., Lipa, Z.& Stefanek, M. (2005)] and [Janac, A., Lipa, Z.& Stefanek, M. (2005)], with above introduced denoting we get (for k=1):

$$F_{c1} = C_{kca}\left(0{,}013\, V_{red}^{0,4} \cdot \rho^{-0,2} + 0{,}069\, a_{red} - 0{,}003\, \rho\right) \cdot \left(0{,}042 \cdot V_{red}^{0,2} \cdot \rho^{-0,4} + 0{,}177 \cdot a_{red} + 0{,}215\, \rho\right) \quad [\text{N}] \qquad (13)$$

Simpler expression we get from formula (7) for k=1 ; :

$$F_{c1} = C_{kcs} \cdot S_r^{0,53} =$$
$$= C_{kcs}\begin{pmatrix} 0{,}002 \cdot S_{red} + 0{,}003 \cdot V_{red}^{0,6} \cdot \rho^{0,2} \\ + 0{,}008 \cdot a_{red} \cdot \rho - 0{,}001 \cdot \rho^2 \end{pmatrix}^{0,53} \cdot k^{0,47} =$$
$$= C_{kcs}\left(0{,}003\, V_{red}^{0,6} \cdot \rho^{0,2} + 0{,}002\, S_{red} + 0{,}008\, a_{red} \cdot \rho - 0{,}001\, \rho^2\right)^{0,53} \quad [\text{N}] \qquad (14)$$

This relation is conformable with experimentally founded facts, is addition to theory of cutting forces by grinding.

3. CONCLUSION

Although there were devoted relevant attention to theory of cutting forces by grinding [Peters, S., Snoyes, R. & Decneut, A. (1976)] , after all some problems stayed not solved. By theoretical determination of cutting forces by grinding was unsuficiently considered number, form and distribution of abrasive grains in grinding wheel. Although it is challenging task, it is necessary to solve it. Our contribution to determination of cutting forces by grinding respects all this facts and gives more detailed relations than actual.

Acknowledgement

The contribution was elaborated within grant project 1/1079/04 Theory of finishing processes in machining.

4. REFERENCES

Maslov, J.N.: (1976), Teorie broušení kovů (Theory of metal grinding, SNTL, Praha

Buda, J.& Békés, J. (1977), Teoretické základy obrábania kovov, (Theoretical fundamentals of metals machining), ALFA, Bratislava

Přikryl, Z. et al. (1957), Obrábění (Machining), SNTL, Praha

Janac, A., Lipa, Z.& Stefanek, M. (2005), Average chip thickness by grinding, Proceedings of ICPM 2005, Afjehi-Sadat A. et al. (Ed.), pp.165-168, ISBN , Vienna, Oct. 2005,

Janac, A., Lipa, Z.& Stefanek, M. (2005), Average cross-section of chip by grinding, Proceedings of 16th Int.Symp.DAAAM 2005, Katalinic, B. (Ed.), pp.171-172, ISSN 1726-9679, Opatija, Croatia, Oct. 2005

Peters, S., Snoyes, R. & Decneut, A. (1976), The proper selection of grinding conditions, Proceedings of 16th Int. Mach. Tool Conf. London, McMilan Press

Annals of DAAAM for 2006 & Proceedings of the 17th International DAAAM Symposium, ISSN 1726-9679
B. Katalinic, Published by DAAAM International, Vienna, Austria 2006
150th Birthday of Nikola Tesla (1856-1943) Inventor of Mobile Robot Teleoperation

COMPARISON OF EFFICIENCY OF FLUID MACHINE AND DRY MANUAL FEEDING

JANZEKOVIC, M.; BRUS, M. & MURSEC, B.

Abstract: *The objective of our research was to compare different methods of feeding of pigs. During the research the fluid machine feeding and the dry manual feeding were compared. The comparison was made on two groups of pigs divided into three groups depending on the input mass. The results show that the fluid feeding proved to be the best on the 6th group, whereas dry feeding proved to be the best on lighter groups of pigs.*
Key words: *Fluid machine feeding, manual feeding, pig, gain*

1. INTRODUCTION

The purpose of correct feeding of pigs is full utilization of their genetic potential for the gain and meat production. In today's fattened pig production the bases of economic production are the daily gain, feed efficiency, meat production percentage of about 59 % and losses. In practice, too little emphasis is put on the feed efficiency. The change for only 0,1 kg in exploitation of the feed changes the feed costs for about 1,5 € or approximately 8,5 kg of feed (additional fattening from 85 kg). In case of 1000 fattened pigs that quantity amounts to 8500 kg of the feed. Therefore, precise consideration of efficiency of feed concentrates pays (Garant, 2006). Also Stock (2006) states a saving of up to 12 € per fattened pig when using fluid feeding and considering the gain curve for reaching the optimum extent of additional fattening. The trial period was since putting into pigsty (33 kg mass) till slaughter (110 kg) as researched in detail by Andersson et al. (1997) from 26 to 105 kg on the average. There were 8 to 10 animals in the box with 0,6 m^2 of lying area per animal. The use of automation in everyday life implies also its introduction into agriculture as an aid facilitating the everyday routine work requiring much time and effort. Ernst et al. (2005) and Janzekovic et al. (2005) state that modern intensive raising systems have the tendency towards introduction of newer technical aids ensuring interaction with animals and, consequently, their better feeling. Thus we were finding out whether introduction of the automatic fluid feeding (f. m.) influences the gain itself of pigs in comparison with manual dry feeding (d. m.) of complete feed concentrates.
At least two-stage fattening is recommended (smaller loadings because of exceeding the quantities of raw proteins). The advantages are as follows:

- it prevents surplus supply of nutritive and mineral substances to pigs (raw proteins) during fattening,
- the costs of feed concentrates are reduced,
- secretion of N and P with liquid manure is reduced,
- the climate in the pigsty is improved (less ammonium),
- the fattened pigs are less stressed and the gain and health of animals are improved.

2. FLUID FEEDING

High economic shortfalls occur in fattening of pigs (due to losses and unfavourable daily gain) because of defective feed hygiene in fluid feeding. The central component in fluid feeding is the mixing tank. Its size, shape and construction influence the precision and quality of the mixture. In the bottom conical part smaller quantities of the feed can be prepared (Fig. 1).

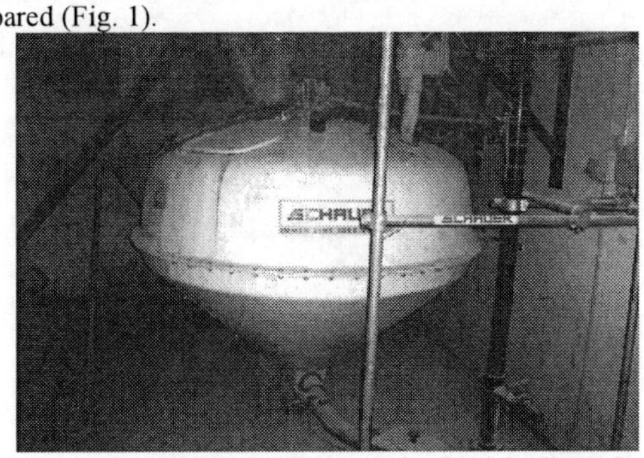

Fig. 1. Part of automatic feeding unit for fluid feeding of pigs.

In tanks the proper hygiene is reached by incorporation of components having automatic shut-off flap and by using a strong cleaner for collectors. As individual units the collectors are portable. Some systems use high-pressure water cleaners to maintain proper cleanness. Farzan et al. (2006) and Stock (2006) state that cleaning must be performed on one individual unit of the control system after another. From the point of view of health condition of animals significantly more sero positive samples with *Salmonella* are established in farms, where dry feed concentrates are used than in farms with fluid feeding. *Salmonella* was isolated in 6 % of samples in case of dry feeding and in 0,8 % in case of fluid feeding (Farzan et al., 2006).

2.1 Distribution of feed

Fluid feeding considers all rational assumptions of the uniform group of animals, feeding space for each animal and simultaneous dosing of feed. In case of fluid feeding system electronic sensors, ensuring precise dosing of the feed in each feeding place, are used (Rasmussen et al. 2006). The motor-driven movable valve controls the flow and precisely ensures reaching the desired rate of the feed jet, which is particularly important when dosing small quantities of feed. The control valve in combination with the rotary pump is a cheap alternative controlling the number of revolutions. The valve, specially designed, for adjusting the feed rate is without cavities in which the feed remains would accumulate; it has a robust membrane for long-standing trouble-free operation (Schauer, 2004). To reach optimum gains, a sufficient number of feeding places is required. In case they are scarce, aggressive behaviour of pigs appears. When studying the influence of different technological solutions of feeding and type of feed (dry, fluid) on the competitivity of animals during feeding, Andersen et al. (1999) established that the frequency of crowding, bites and other disturbances is reduced in the group, fed with fluid feed in case of simple way of feeding without isolating bulkheads or with short bulkheads in the area of the front part of the trunk.

3. INTENSITY OF GROWTH

Fig. 2. shows the intensity of growth of the individual group of animals tested in the research. It can be seen that the growth of the group of fluid-fed medium-weight pigs (f.m.m.) is the most intensive; the light-weight pigs fed manually with dry feed (d.m.l.) grew with the same intensity; a less intensive gain was characteristic of the group (f.m.h.), but with higher initial weight. For light-weight (d.m.m.) and heavy-weight (d.m.h.) pigs fed manually with dry feed and for the light-weight pigs fed with fluid feed (f.m.l.) high intensity of growth is not noticeable in this period of fattening by the applied way of feeding.

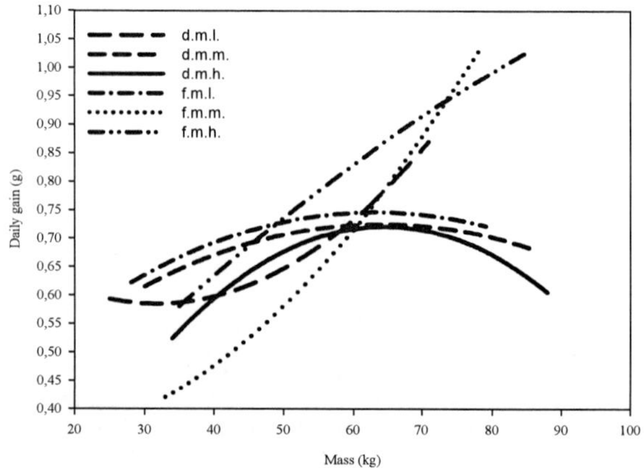

Fig. 2. Intensity of growth of pigs of differently fed groups.

4. REACHED MASS AND DAILY GAINS

With different systems of feeding the animals reached different masses (Table 1) and different daily gains (Table 2) during successive weighings.

Group	1. mass (kg)	2. mass (kg)	3. mass (kg)
d.m.l.	28,43±1,51[a]	65,43±4,68[ac]	107,71±9,40[a]
d.m.m.	32,00±1,73[b]	71,71±7,76[ab]	109,43±10,97[a]
d.m.h.	36,83±3,12[c]	72,17±11,27[ab]	108,33±16,44[a]
f.m.l.	30,86±1,68[db]	71,29±5,70[ab]	110,29±8,57[ab]
f.m.m.	33,86±0,69[eb]	61,14±10,40[abc]	100,29±23,13[a]
f.m.h.	37,50±2,07[fc]	76,17±9,45[b]	126,67±15,13[b]
p	0,000	0,035	0,079

Table 1. Resulting average mass (kg) and standard deviation (sd) during individual weighing of the animal depending on method of feeding.

Group	1. daily gain	2. daily gain	3. daily gain (b-e)
1	0,59±0,07[a]	0,80±0,12[ab]	0,68±0,07
2	0,63±0,10[a]	0,71±0,15[a]	0,67±0,09
3	0,56±0,16[ab]	0,68±0,25[a]	0,62±0,14[a]
4	0,64±0,08[a]	0,74±0,12[a]	0,68±0,07
5	0,43±0,16[b]	0,74±0,28[a]	0,57±0,20[a]
6	0,61±0,13[a]	0,95±0,13[b]	0,77±0,12[b]
p	0,032	0,170	0,117

Table 2. Average daily gain and standard deviation of differently fed pigs.

During the first weighing the statistical analysis showed significant differences between groups as indicated in Table 1. The individual groups mutually differ significantly ($p \leq 0,05$); this was marked by different designation (a, b and c) in the power of the individual average value; the same applies to Table 2. During the second weighing the f.m.m. had the significantly lowest gain out of all groups in test with the exception of the group d.m.h. The other groups did not differ in the reached gain during that weighing. During the second finding out of the gain the group f.m.h. reached the significantly ($p \leq 0,05$) highest gain in comparison with other groups except the group d.m.l., whose gain did not differ significantly. Group f.m.h. reached the highest average gain which significantly ($p \leq 0,05$) differed from group f.m.m. and from group d.m.h. The applied method of feeding with other controlled conditions of raising has a significant ($p = 0,032$) influence on the gain. That influence is slightly reduced in the second gain and remains insignificant in the average gain.

5. CONCLUSION

- Fluid feeding is most fit for medium-weight (f.m.m) and heavy-weight (f.m.h.) fattened pigs.
- Dry feeding is most fit for light-weight (d.m.l.) pigs.
- Two-stage feeding of pigs being fattened is recommendable.
- Fluid feeding is a very flexible method, particularly, when a great number of different feed concentrates are used.
- Animals are fed with optimized ration meeting the food needs of animals of individual masses and ages.
- The system allows savings of time, energy, better gains and better utilization of feed and has favourable influence on the economy.
- This method of feeding allows to save up to 12 € per fattened animal.

6. REFERENCES

Andersen, I.L. (1999). The influence of different feeding arrangements and food type on competion at feeding in pregnant sows. *Applied Animal Behaviour Science,* No. 65, pp. 91-104

Andersson, K. (1997). The effects on feeding system, lysine level and gilt contact on performance, skatole levels and economy of entire male pigs, *Livestock Production Science,* No. 51, pp. 131-140

Ernst, K. (2005). A complex automatic feeding system for pigs aimed to induce successful behavioural coping by cognitive adaptation, *Applied Animal Behaviour Science,* No. 91, pp. 205-218

Farzan, A. (2006). Prevalence of Salmonella spp. on Canadian pig farms using liquid or dry-feeding, *Preventive Veterinary Medicine,* No. 73, pp. 241-254

Garant. Garant high-quality feed concentrates for fattened pigs, Available from: http://www.raiffeisen-trgovina.si/pic/slike/Brosura%20Svinje%20za%20Web.pdf *Accessed:* 2006-06-14

Janzekovic, M.; Janzekovic, I.; Mursec, B. (2005). Following up heart rate in cattle by noninvasive method. *DAAAM International scientific book 2005,* Vienna pp. 291-298 (chapter 22)

Rasmussen, D.K. (2006). Effects of animal/feeding-place ratio on the behaviour and performance of fattening pigs fed via sensor-controlled liquid feeding, *Applied Animal Behaviour Science,* No. 98, pp. 45-53

Schauer. (2004). *Fluid feeding, innovative ideas for "current" topics.* Prospectus

Stock, M. (2006). Mit der Hygiene steht und fällt der Erfolg. *DGS,* 22, S. 41-44

Annals of DAAAM for 2006 & Proceedings of the 17th International DAAAM Symposium, ISSN 1726-9679
Editor B. Katalinic, Published by DAAAM International, Vienna, Austria 2006
150th Birthday of Nikola Tesla (1856-1943) Inventor of Mobile Robot Teleoperation

WORKPIECE CLAMPING AND ITS DIAGNOSTICS

JAVOROVA, A.; KATALINIC, B. & ZVOLENSKY, R.

Abstract: The one of ways to supplementary time decreasing is suitable fixture using and supplementary time operations automation. Fixture using increasing the production quality, the productivity and decreasing the production costs. Development of new generation production machines and systems demand that they are equipped by adequate fixture devices. Efficiency of fixture device in automatized production systems underlie using efficiency of entire production systems. In time of work cycle these fixture devices works automatically, without operator intervention and cooperate with other devices of production system.

Key words: fixture device, clamping, positioning, automation, robot

1. INTRODUCTION

The one of ways to supplementary time decreasing is suitable fixture using and supplementary time operations automation.
The fixture provide clamping of workpiece on machine desk, so as the workpiece have the right position toward tool. This position must be retain in machining time too.
Fixture using increasing the production quality, the productivity and decreasing the production costs. In some cases are fixture using necessary. The fixture design are dependent on the batches of production. In small batches we use the fixtures designed from modular system and in large batches we can use the dedicated fixtures.
The clamping fixture providing these basic functions:

- Workpiece positioning on the desk of machine
- To prevent of workpiece deformation when cutting and clamping forces are acting
- In some cases tool support

These Basic functions are provided by positioning, clamping and supporting elements (active elements) of fixture. These elements can be placed on standalone unit or on some units or can bee mounted to machine or some its part.
Each technological process aim is produce required quality product. Required quality workpiece realization in automated technological system (ATS) is different as in conventional technological system. This different are caused human acting elimination to production. Automated technological system must ensure required product quality by theirs properties and parameters without human action. Consequently all subsystems automated technological systems are participated to provide for required quality with different but function dependent deals. For automated technological systems production quality requirements assurance are needed correct decomposition functions between individual subsystems ATS, their time and position synchronisation. In this process are needed respect dynamic shows and compatibility bilateral inputs and outputs of technological process in ATS.

2. PRODUCTIVITY AND QUALITY TECHNOLOGICAL PROCESS

ATS productivity p_c is standard expressed by operating cycle time t_c and number of quality suitable products. This functionality is affectioned by out of cycle idletime, which are expressed operating ratio η_v.

$$p_c = \frac{1}{t_c} \cdot \eta_v \tag{1}$$

Basically this relation we can formulate a fie principles to compose conception of operations ATS:

- principle optimal accuracy
- principle maximum efficiency
- principle of prevention.

These principles relate elimination possible systematic errors and optimalization influence of accidental errors for concrete ATS.

Toleranced dimensions w_v of product realized by ATS must meet condition

$$w_v = \delta_d \tag{2}$$

δ_d is specified permissible deviation.

Both inscribed relations are realized in system: machine-tool-fixture-workpiece. This realization we can decompose to tasks:

- constitute systems
- positioning workpiece in production device
- clamping workpieces
- static set up system machine-tool-fixture-workpiece
- dynamic tooling

Each of these tasks is potencional systematic and accidental errors source . These errors are result in dependency of variety factors, which are accumulated in parameter w_v. This is a count a few components :

$$w_v = w_p + w_s + w_d \tag{3}$$

w_p - errors of constitute/positioning workpieces

w_s – errors of static set up system machine-tool-fixture-workpiece

w_d – errors of dynamic set up system machine-tool-fixture-workpiece

This relation obtain in automated technological system in which industrial robot manipulate with workpieces and all process is automated too.

3. CLAMPING WORKPIECE PROBLEM IN AUTOMATED MANIPULATION

Parameter w_p diagnostic is based on technical system machine-tool-workpiece-fixture analyze, in which is a new item

– industrial robot and than this system we can define as system machine-tool-workpiece-fixture-robot. Analyze realization is on base space analyze of geometry – kinematics relations three function items : fixture – workpiece – gripper. In this case is w_p dedicated as a :

$$w_p = f(w_{pf}, w_{pw}, w_{pg}) \tag{4}$$

w_{pf} – errors fixed on fixtures

w_{pw} – errors fixed on workpiece

w_{pg} – errors fixed on gripper

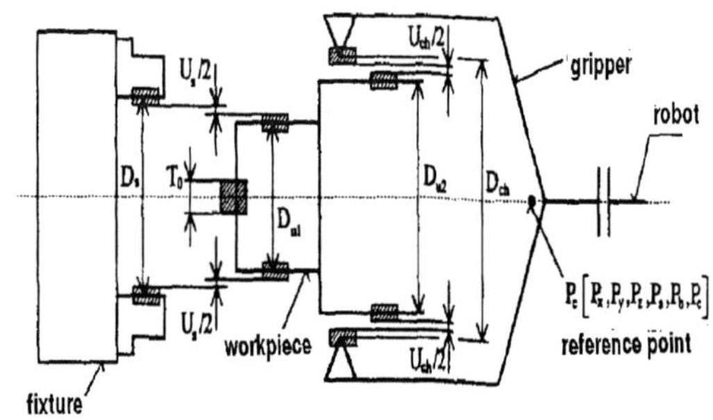

Fig. 2. Components of system: fixture- workpiece-gripper

Reason of variance analyze and ratio participation of individual parameter to their we must regard especially participation of function industrial robot on manipulate . We can simplify describe by these tasks:

- grasp workpiece from buffer
- hand in workpiece to production device
- clamping workpiece

Functional determining factor systems machine-tool-workpiece-fixture-robot which eliminate w_p is clamping latitude u_s , which describe following function :

$$u_s = f(w_{uf}, w_{uw}, w_{ug}) = F(s) \tag{5}$$

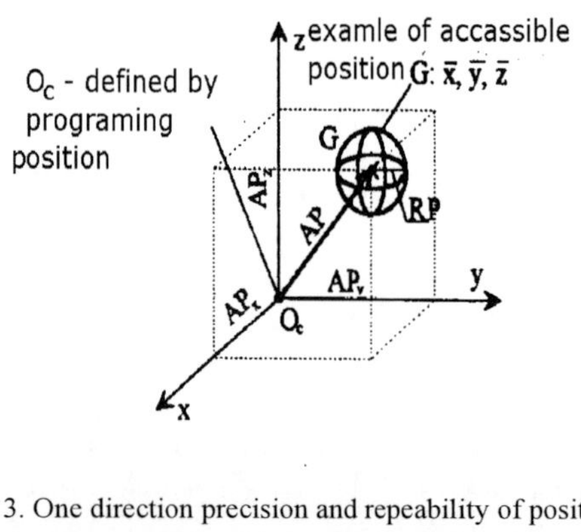

Fig. 3. One direction precision and repeability of positioning

components w_{uf}, w_{uw}, w_{ug} represents actual state fixed on:

- precision opening fixture of production devices
- accuracy to size of workpiece
- robot repeatability precision opening gripper

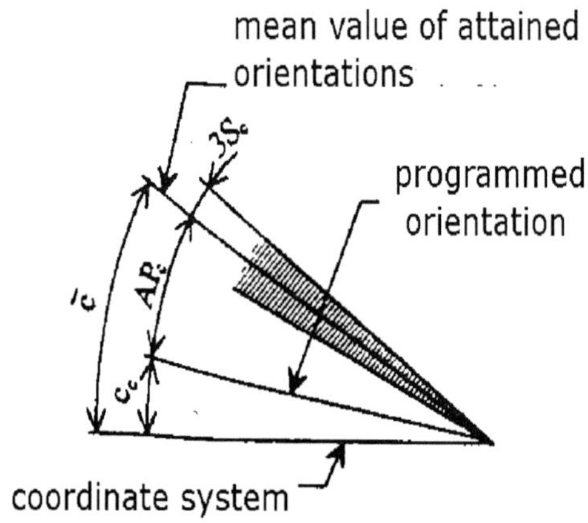

Fig. 1. One direction precision and repeability of orientation

4. CONCLUSION

Algorithmization of present access allow quarantee certain and safer workpieces clamping on design stage of automated technological system. At design gripper stage allow specify its assignment, which advance quality of input parameters for technical solve

5. REFERENCES

Buda, J., Kováč, M.: Metodika projektovania výrobných procesov v strojárstve. *Alfa,* 1990, Bratislava

Alpek, F., Velíšek, K.: Intelligent fixturing in production automation. *Procedings of DAAAM symposium*, Katalinic, B. (Ed.), Austria, 2002, Viena

Zuperl, U., Cus, F.: Analytical model for analysis of fixtures for clamping thin-wall workpieces., *Procedings of DAAAM symposium,* Katalinic, B. (Ed.), Austria, 2001, Viena,

Košťál, P. Matúšová,M.: Teoretické otázky ustavenia a upnutia, *Procedings of Vedecké práce MtF STU*, Zväzok 10., pp.53-56,Bratislava, 2001,

Smrček, J.,Fotopulos, J., Benča, A.: Testing of industrial robots, Selection of performance criteria for different applications, *Proceding of 6th International Symposium on measurement and control in robotics ISMCR 96,* pp. 576-579, Brussels, 1996

Matúšová,M. (2003). Postup návrhu upínacieho prípravku, Procedings of 9. Vedecká konferencia s medzinárodnou účasťou Akademická Dubnica 2003, pp. 219, ISBN: 80-2227-1915-3, Dubnica nad Váhom, 18-19.6.2003.

Annals of DAAAM for 2006 & Proceedings of the 17th International DAAAM Symposium, ISSN 1726-9679
Editor B. Katalinic, Published by DAAAM International, Vienna, Austria 2006
150th Birthday of Nikola Tesla (1856-1943) Inventor of Mobile Robot Teleoperation

RANKING OF THE VISUAL INFORMATION ON GRAPHIC DESIGN OF PACKAGING WHICH MOSTLY MOTIVATE THE PURCHASING OF THE PACKET PRODUCT

JURECIC, D.; BABIC, D. & VANCINA-KROPAR, V.

Abstract: *This paper describes the visual information displayed on packaging and their importance in the process of getting to users. From the economic and designing point of view, the product, as well as the medium by which the information is transferred, must be valorised. The use of colours, text information and other graphic solutions are explained in this paper. Whether customers will buy a certain packaged product or not depends mostly on their evaluation of the visual information displayed on the packaging. It applies primarily to the selection of colours that represent the packed product. Secondly, it also applies to the visual information (example: pictures, graphics, symbols, codes etc).*

Key words: *visual information, product packaging, design, evaluation*

1. INTRODUCTION

Text, image and colour are basic visual elements containing information that are used in manipulating the buyer. The topic of our research is the medium which transfers that information, i.e. packaging. Packaging needs to meet two basic design criteria: it has to satisfy technical and physical demands (i.e. protect the product during transport and distribution) and printing demands when decorating the surface, informing consumers and motivating buyers to make the purchase. Before we start creating a possible packaging design, it is necessary to take into consideration the product itself. A typical supermarket or self-service can offer more that ten different varieties of a single product (such as sweets or cosmetic products) produced by different manufacturers. The challenge which a designer needs to meet is how to make customers notice the aspect of his product (which might be the eleventh product on the shelf) and, when customers notice it, how to convince them to buy this very product instead of any of the other ten. The only remaining medium which can influence customers is the packaging itself. It is thought that, depending on the product type and the source of information on the packaging, 68 - 80% of decisions on purchasing depend on the decision made by the customer in the shop while looking at the product, i.e. at the graphically equipped packaging on the shelf. Typical consumers most frequently walk through the store and buy what they like. A particular product attracts their attention for only a few seconds. During such a short period of time the medium (packaging) must send the message which motivates potential customers to purchase the product. In those few seconds, a complicated graphic design cannot send the message "BUY ME" and neither can contradictory and ambiguous messages. A potential buyer does not notice such products, and that what is not noticed remains unsold.

2. EXPERIMENT

The examiners ranked the image information by giving it marks from 1 to 5. The method is based on the procedure by which each of the experts (m) participating in the examining, i.e. evaluation of the objects (n), gives each object a number as the rank of that parameter. The object of the highest ranking gets the number 1; the next one gets the number 2, etc. The criteria of the technological process optimality appear in such evaluations as the object. The degree of the experts' agreement in the observed example is expressed by M. G. Kendall's variant. The conformity of the experts' ideas, according to Kendall, is determined by the expression (1):

$$W = \frac{12 \times s}{n^2 \times \left(n^3 - n - b\right)} \qquad (1)$$

where *m* is the number of experts and *n* the number of objects (questions) presented in the table. As the answer to a given question, each of the mentioned *m* experts (examiners) in the conducted poll gives a number *uij* as a ranking parameter in relation to other questions. It is possible, of course, that the same expert gives the same rank to more than one object. According to the agreement, the best-ranked object gets the parametric value 1, that is rank 1, followed by the next rank 2 etc. In the observed model, $k<n$ and the objects are given a so-called standardized rank. The standardized rank is defined as the analytical middle of the total number of places occupied by the objects of the same rank k. In the standardized table, the column T_i is calculated. The element of that matrix is calculated according to the expression (2):

$$T_i = \sum_{t_j} \left(t_j^3 - t_j\right) \qquad (2)$$

where t_j is the number of times the *j-ti* rank has appeared in the *i*-line of the matrix. The expression $0 \leq W \leq 1$ is valid for Kendall`s coefficient *W*. As this coefficient is probability, it means that if *W* is nearer to one, the experts agree.

3. RESULTS AND DISCUSSION

Diagram 1 shows the results of the evaluation of elements containing visual information on graphically designed packaging, which is the main motivating factor in the purchase of some packed products. The examiners almost agreed that the colour of the packaging is the most important motivating factor in the purchase of the product and it was marked *excellent* by 60% of the examiners. The mark *very good* was given by 30% of the examiners and less than 10% thought that the colour was not an important motivating factor and marked it *good* and *sufficient*.

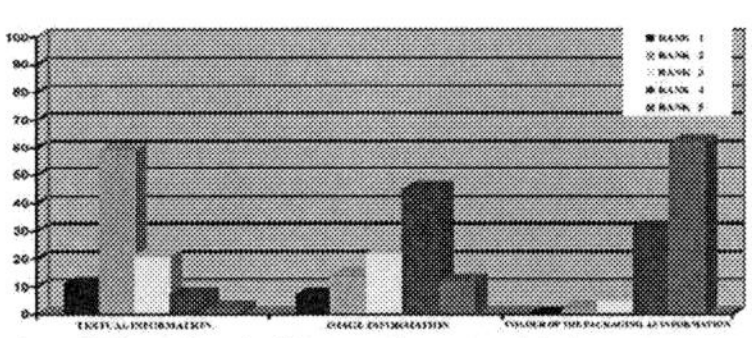

Diagram 1. Visual information which most frequently motivates the purchase of packed products

Experti		Objecsi		
		Standardized ranks according to the formula		
Ordinal number	Randomly taken samples from the questionnaires	Textual information	Image information	Colour of the packaging as information
1.	7	5	4	3
2.	12	4	5	1
3.	18	5	4	4
4.	21	3	4	5
5.	38	2	3	5
6.	45	4	3	5
7.	54	3	5	4
8.	72	1	3	5
9.	88	5	4	2
10.	99	2	5	5

Table 1. Evaluation of the visual information which most frequently motivates the purchase of packed products

Experti		Objecsi				
		Normirani rangovi izvedeni prema formuli				
Ordinal number	Randomly taken samples from the questionnaires	Textual information	Image information	Colour of the packaging as information	Σ	T
1.	7	1	2	3	6	0
2.	12	1	2	3	6	6
3.	18	1	2,5	2,5	6	0
4.	21	1	2	3	6	0
5.	38	1	2	3	6	0
6.	45	1	2	3	6	0
7.	54	1	2	3	6	0
8.	72	1	2	3	6	0
9.	88	1	2	2	6	0
10.	99	1	2,5	2,5	6	6
	Σ	10	21	29		

Table 2. Standardized ranks of the evaluation of the visual information which most frequently motivates the purchase of packed products

The mark *insufficient* was not given at all, which proves that there was nobody who thought that the colour was completely unimportant when deciding what to buy. Such results show that colours attract attention, influence human emotions, convey information about the product origin and quality, stir up associations and create satisfaction when purchasing a product. According to the poll, the image information displayed on a graphically designed packaging is the second most important motivating factor for making a purchase. More that 40% of the examiners marked that kind of visual information *very good* and a little more that 10% of them thought that the visual information deserved the best mark. The mark *good* was given by 20% of the examiners. Less than 20% of them thought the image information was unimportant when deciding whether to buy a product and marked it *sufficient* and *insufficient.* The third element containing visual information was the text, which is thought to be the least important motivating factor. Only 10% of the examiners marked it *excellent* and *very good,* 20% marked it *good* and 60% *sufficient.* This motivating factor was considered quite unimportant by 10% of the examiners. Low marks given to the textual information are surprising, considering the results of the poll on the importance of textual information displayed on graphically designed packaging, which proved the price to be the most important textual information. In this sense, the price should be a motivating factor influencing the decision to purchase but the examiners did not think of that. One should agree with those psychologists who claim that there are two people in each person: one is the buyer himself as he really is and the other is the person he would like to be. The calculated correlation coefficient of the ranks R = 0.96 shows that the productivity of the examiners' ideas is high. Tables 1 and 2 show the evaluation marks given by the examiners, on the bases of which the standardized ranks were calculated (Calculation 1).

Calculation 1. Evaluation of the correlation ranks of the visual information which most frequently motivates the purchase of packed products

$$T_1 = \sum_{t_j} (t_j^3 - t_j)$$

$$T_1 = (2^3 - 2) + (2^3 - 2) = 12$$

$$T_2 = (2^3 - 2) + (2^3 - 2) = 12$$

$$T_3 = (1^3 - 1) + (1^3 - 1) + (1^3 - 1) + (1^3 - 1) = 0$$

$$T_4 = (1^3 - 1) + (3^3 - 3) = 24$$

$$T_5 = (2^3 - 2) + (2^3 - 2) = 12$$

$$T_6 = (3^3 - 3) + (1^3 - 1) = 24$$

$$T_7 = (1^3 - 1) + (2^3 - 2) + (1^3 - 1) = 6$$

$$T_8 = (1^3 - 1) + (1^3 - 1) + (2^3 - 2) = 6$$

$$T_9 = (1^3 - 1) + (2^3 - 2) + (1^3 - 1) = 6$$

$$T_{10} = (1^3 - 1) + (1^3 - 1) + (1^3 - 1) + (1^3 - 1) = 0$$

$$s = \sum_{j=1}^{n} u_{ij} (-\frac{1}{2} \cdot m \cdot (n+1))^2$$

$$s = \frac{1}{2} m \cdot (n+1)$$

$$s = \frac{1}{2} \cdot 10 \cdot (n+1)$$

$$s = 0{,}96$$

$$s = (10 - 20)^2 + (21 - 20)^2 + (21 - 29)^2$$

$$s = 182$$

$$b = 10 \cdot m \cdot \sum_{j=1}^{n} u_{ij}$$

$$m = 10$$

$$b = 10 \cdot 12$$

$$b = 120$$

$$R = \frac{12 \cdot 182}{10^2 \cdot (3^3 - 3) - 120}$$

$$R = 0{,}96$$

4. CONCLUSION

A high coefficient obtained from the answers of some examiners proves a high qualitative ranking in this case. The hypothesis was proven by 96%. The main motivating factor influencing the decision to purchase a packed product is colour, then image information and, at the end, textual information. As the work proves that colours are the most important motivating factor, they must undergo some changes in order to minimize their positive and negative qualities. The designers of the product must make the message more appealing to potential buyers and attract their attention. They must make the message legible and understandable, clear and distinct and shape it in such a way that it has an impact on potential buyers.

5. REFERENCES

Bessen, A. H.: *Design and Production of Corugated Packaging and Displays,* Jelmar Publishing Co., Planview, 1994

Favre, J. P.: *Color Sells Your Package,* ABC Edition, Zurich, 1969

Jackson, C.: *Colorme Beautiful Ballantine Books,* Jelmar Publishing Co., New York, 1991

Jurecic, D.: *Evaluacija elemenata vizualne informacije na grafičkoj opremi ambalaze, FOI, Varazdin, 2004*

McCaughey, D. G.: *Graphic design for corrugated packaging,* Jelmar Publishing Co., Inc., New York, 1995

Annals of DAAAM for 2006 & Proceedings of the 17th International DAAAM Symposium, ISSN 1726-9679
B. Katalinic, Published by DAAAM International, Vienna, Austria 2006
150th Birthday of Nikola Tesla (1856-1943) Inventor of Mobile Robot Teleoperation

THE EXPERIMENTAL DEPENDENCES OF VARIABLES IN GRINDING PROCESS

JURKO, J.

Abstract: *The use of SG grinding wheels with various content of abrasive grain in grinding process of roller-bearing provides new knowledge compared with the use of grinding wheels Al_2O_3. This paper is result of the realized experimental measuring. The grinding diagrams represent complex interaction of grinding wheel and workpiece under various conditions of grinding.*
Key words: *grinding wheel, grinding, grinding parameters*

1. INTRODUCTION

The surface roughness measurement, the measuring of cutting force components and also the measuring of specific grind size allow to create the grinding diagrams.
The optimal cutting conditions can be specified relatively simple way in grinding process by tools with defined geometry of cutting edge. The situation in grinding is more difficult by reason of influence of many variability parameters. If, in generally, we designate the examined variable by ***A*** letter (e.g. tool life, cutting force, specific grind, surface roughness), then the mentioned variable is function of many variables as are machined material, the kind of abrasive, bond, grinding wheel structure, grinding wheel hardness, accuracy and rigidity of grinding machine, quality of truing, grinding width, feed, grinding depth, cutting speed etc.

2. GRINDING PROCESS

The experimental determination of functional dependence of variable ***A*** on individual mentioned variables is very complex and difficult to realize with respect to extensity of work. From mentioned reasons, the single parameter was proposed. This parameter is so-called grinding depth equivalent h_{eq} which characterize the grinding process very suitable and at same time it allows to determine the dependence of grinding conditions on grinding process side effects, on quality and on grinding economy. Plotting of the above-mentioned dependences creates the grinding diagrams. The diagrams describe grinding wheel and workpiece interaction in varying grinding conditions. The diagrams can be used for material machinability analysis, cutting property of grinding wheel analysis, cutting fluid selection etc. In order to determine the grinding depth equivalent, the continuity of material volume entering into cutting zone by speed v_f and leaving it by speed v_c is the basis. On base of kinematical conditions, the grinding depth equivalent can be determined for surface grinding (Fig. 1) as follows:

$$h_{eq}.v_c.b = a_p.v_f.b \qquad (1)$$

$$h_{eq} = a_p.\frac{v_f}{v_c} \quad [\mu m] \qquad (2)$$

The grinding diagrams are elaborated for one grinding technique of one kind of material by one kind of grinding material and by one type of grinding wheel. The values of grinding depth equivalent are at horizontal axis; the components of cutting force, the surface roughness and the specific grind are at vertical axis. The upper part presents basic information on grinding operation, machined material, used grinding wheel, cutting conditions of grinding, of wheel truing and others.

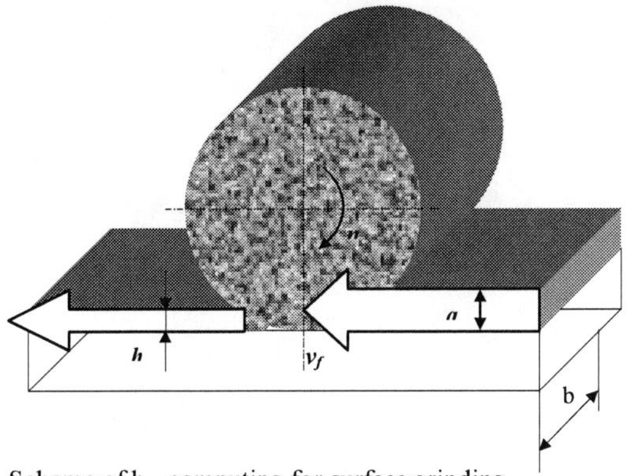

Fig. 1 Scheme of h_{eq} computing for surface grinding

The experimental studies of different materials grinded by various grinding wheels confirmed functional dependence of grinding depth equivalent and examined input grinding parameters. These dependences have a character that can be describe in following form:

$$X = X_1.h_{eq}^{x} \qquad (3)$$

If we take a logarithm, the linear dependence is given:

$$\log X = \log X_1 + x.\log.h_{eq} \qquad (4)$$

G, Ra, F are parameters established for X value,
X_1 is y-intercept if h_{eq}=1,
X is tangent of inclination of line.

The examined parameters in grinding diagram are formulated in following form:

Tangent cutting force component	$F_c = F_1.\ h_{eq}^{f}$	[N]*	(5)
Radial cutting force component	$F_p = F_2.\ h_{eq}^{g}$	[N]*	(6)
Surface roughness without sparking	$Ra = R_1.\ h_{eq}^{r}$	[µm]	(7)
Surface roughness with sparking	$Ra^d = R_{1,d}.\ h_{eq}^{rd}$	[µm]	(8)
Specific grind	$G = G_1.\ h_{eq}^{q}$	[$mm^3.mm^{-3}$]	(9)

**Note*: cutting force component per 1 mm grinding width

3. EXPERIMENTS

As was mentioned above, the grinding diagrams are used for examination of cutting property of grinding wheels in relation to cutting conditions (expressed by h_{eq}). While the left side of diagram ($h_{eq}<0,1\mu m$) represents surface finishing grinding operations, right side ($h_{eq}>0,1\mu m$) represents stock opera-tions. Resulting from computed equations for conventional Al_2O_3 grinding wheel, the rising of surface roughness Ra and specific grind G and h_{eq} is low marked compared with components of cutting force F_c and F_p (equations 10–16). The higher abrasive grain SG content in grinding wheel, the higher rising of surface roughness and specific grind, the less intensive components of cutting force compared with conventional grinding wheel Al_2O_3 (equations 17–37). The mathematical interpretation of grinding diagrams according to relations 1–9 is following for various wheels: (subscript "o" with components of cutting force and surface roughness refers given parameter to V_o=1000 mm^3).

Grinding wheel Al_2O_3: (Fig.2) * per 1 mm grinding width

$F_c = 32.1\ h_{eq}^{1.05}$	[N]*	(10)
$F_{co} = 26{,}6\ h_{eq}^{0.78}$	[N]*	(11)
$F_p = 46.1\ h_{eq}^{1.06}$	[N]*	(12)
$F_{po} = 41.2\ h_{eq}^{0.71}$	[N]*	(13)
$Ra = 1.7\ h_{eq}^{0.15}$	[μm]	(14)
$Ra_o = 1.6 h_{eq}^{0.1}$	[μm]	(15)
$G = 84.4\ h_{eq}^{0.1}$	[$mm^3.mm^{-3}$]	(16)

Grinding wheel 3SG: * per 1 mm grinding width

$F_c = 24.1 h_{eq}^{0.94}$	[N]*	(17)
$F_{co} = 29.8\ h_{eq}^{0.83}$	[N]*	(18)
$F_p = 30.3\ h_{eq}^{0.88}$	[N]*	(19)
$F_{po} = 55\ h_{eq}^{0.81}$	[N]*	(20)
$Ra = 2.4\ h_{eq}^{0.24}$	[μm]	(21)
$Ra_o = 2.3\ h_{eq}^{0.35}$	[μm]	(22)
$G = 1780\ h_{eq}^{1.32}$	[$mm^3.mm^{-3}$]	(23)

Grinding wheel 5SG: * per 1 mm grinding width

$F_c = 24.4\ h_{eq}^{0.93}$	[N]*	(24)
$F_{co} = 16.3\ h_{eq}^{0.56}$	[N]*	(25)
$F_p = 31.2\ h_{eq}^{0.85}$	[N]*	(26)
$F_{po} = 30.1\ h_{eq}^{0.57}$	[N]*	(27)
$Ra = 2.5\ h_{eq}^{0.24}$	[μm]	(28)
$Ra_o = 2.7\ h_{eq}^{0.41}$	[μm]	(29)
$G = 3756\ h_{eq}^{1.57}$	[$mm^3.mm^{-3}$]	(30)

Grinding wheel 100SG: * per 1 mm grinding width

$F_c = 19.5\ h_{eq}^{0.82}$	[N]*	(31)
$F_{co} = 17.1\ h_{eq}^{0.60}$	[N]*	(32)
$F_p = 20\ h_{eq}^{0.65}$	[N]*	(33)
$F_{po} = 12.5\ h_{eq}^{0.18}$	[N]*	(34)
$Ra = 3.3\ h_{eq}^{0.30}$	[μm]	(35)
$Ra_o = 3.8\ h_{eq}^{0.52}$	[μm]	(36)
$G = 4738\ h_{eq}^{1.63}$	[$mm^3.mm^{-3}$]	(37)

Emulsion H – 2% concentration

4. CONCLUSION

On the present, the one of the most important tendency in the field of grinding wheels SG represents the development of new kinds of bonds that are specially developed for abrasive grain SG. It indicates that for achieving the specific properties of abrasive grain SG compared with conventional abrasive grain Al_2O_3 is necessary to modify also the bond, which can markedly increase cutting properties of grinding wheels SG and that is way how to widen the field of usage as the replacement of conventional abrasive Fig. 3.

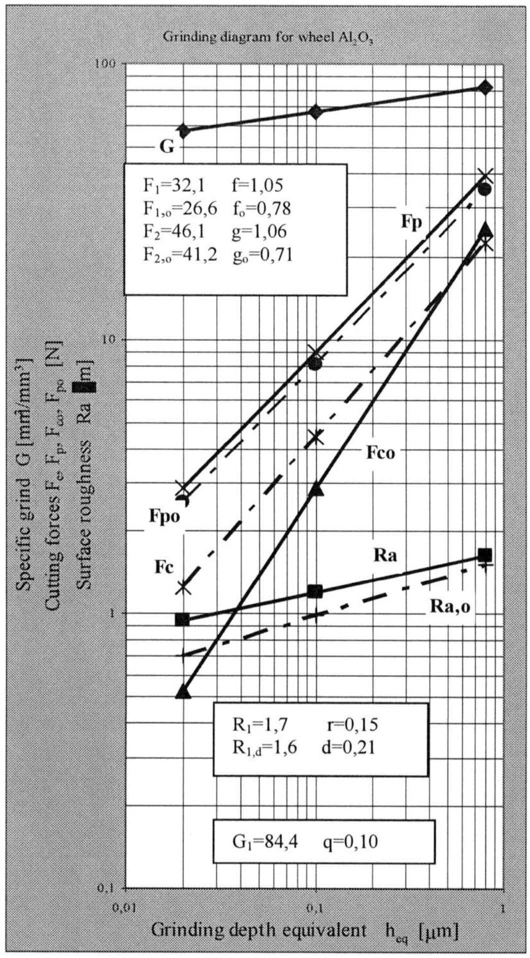

Fig.2 Grinding diagram

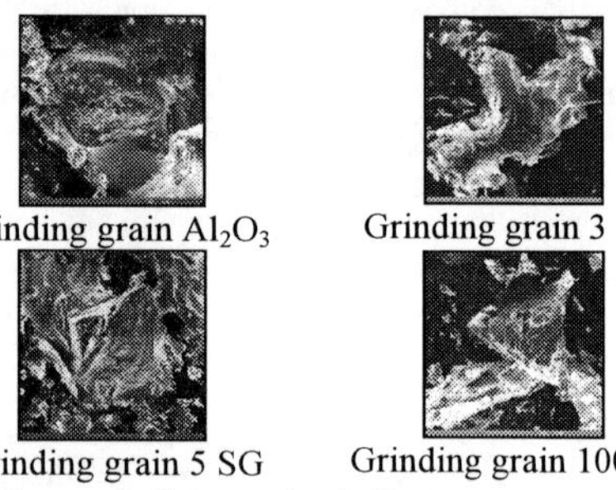

Fig.3 Grinding grain form past grinding

The grinding diagrams allow optimising the cutting process from technical and economy point of view. They provide simple and quick tool to find information on size of specific grind, components of cutting force as well as surface roughness depending on cutting conditions.

5. REFERENCES

Jurko, J. (2005). *Opotrebovanie rezných nástrojov*. Prešov, D.A.H. Prešov, ISBN 80-8073-255-8, Prešov

Jurko, J. (2000). *Technológia zmeny rozmerov*. FVT Prešov, ISBN 80-7099-617-X, Prešov

Annals of DAAAM for 2006 & Proceedings of the 17th International DAAAM Symposium, ISSN 1726-9679
B. Katalinic, Published by DAAAM International, Vienna, Austria 2006
150th Birthday of Nikola Tesla (1856-1943) Inventor of Mobile Robot Teleoperation

NEED OF APPLYING ERGONOMIC METHODOLOGIES FOR THE TRAIN DRIVERS CAB DESIGN

JURUM-KIPKE, J.; TOS, Z. & VIDAKOVIC, I.

Abstract: *The objective of this paper is to determine the optimal location for the Man-Machine Interface (MMI) equipment, based on the analysis of position of the engine driver in the driver's cab. The basic data to determine the scope of driver's activities within the cab were linear anthropo-measures for the persons of our population.*
The redesign of the driver's cab has to allow optimal position of the CDU regarding the driver's field of sight, access and visibility of all the instruments within the cab. Since it may be expected that the population of drivers will include participation of women as well, this has been taken into consideration through the analysis of a number of situations which correspond to everyday work.
Key words *MMI, virtual reality, cab fitment, ergonomics methodology, anthropometry*

1. INTRODUCTION

The purpose of this work is to include the human factors in designing the modifications of the existing engine driver's cabs, as well as in the construction of new or upgraded engines. The analysis of human factors presented in this document represents the beginning of the design process. Full understanding of the human factor problems certainly affects the engine design, since the work of the engine driver has not been completely studied yet. In preparing this document many problems have been observed, in introducing the new information technologies that is yet to be discussed quite a lot. The location of the Man-Machine Interface (MMI) equipment, which consists of the control display unit (CDU), manual commands, manual telephone, and loudspeakers, is defined by a set of human factors. The role of human factors in the management process is very important in solving these questions. However, the currently available literature does not analyze often the skills of humans for these new technologies (Rogers, Myers, 1993).

2. METHODS AND EXAMINEES

Basic data to determine the dynamic anthropo-characteristics in the people of Croatian population are based on linear anthropo-measures (Rudman). These data can be found for the adults of Croatian population in Tables 1 and 2. Further step in determining dynamic characteristics of the Croatian population was to define the biomechanical model of examinees in the assumed posture in the vehicle. In accordance with the modeling method, regular geometric solids have been selected for the segmental body parts, such as e.g. cylinders, parallelepipeds and ellipsoids, for which it is possible to simply calculate the dynamic inertia moments with appropriate assumptions on the density, etc. Eventually they can be reduced to an assumed central coordinate system as presented in Figure 1. In this way it is possible to determine the dynamic inertia moments for the static subjects for the indicated axes x,y,z.

	Percentiles				
Variable in mm	5%	25%	50%	70%	95%
Body height	1592	1620	1648	1724	1801
Seating height	820	848	882	908	941
Length of upper arm	288	303	316	326	341
Length of forearm	241	260	270	278	298
Length of arm	708	743	759	784	821
Length of upper leg	474	502	523	540	570
Length of lower leg	333	362	376	390	414
Length of leg	881	935	965	996	1045
Bichromial range	358	377	393	404	432
Pelvis width	269	283	293	302	318
Body mass in kg	56,5	64,9	71,5	80,2	95

Table 1: Anthropo-measures for male persons (157 subjects), Rudan.

	Percentiles				
Variable in mm	5%	25%	50%	70%	95%
Body height	1468	1531	1572	1617	1684
Seating height	775	803	828	854	891
Length of upper arm	263	280	290	298	315
Length of forearm	222	231	242	252	267
Length of arm	640	676	692	712	745
Length of upper leg	450	478	491	508	534
Length of lower leg	311	334	348	362	388
Length of leg	830	870	896	925	973
Bichromial range	333	353	365	379	393
Pelvis width	264	283	298	306	324
Body mass in kg	52	57,1	62,2	70,2	83,5

Table 2. Anthropo-measures for female persons (155 subjects), Rudan

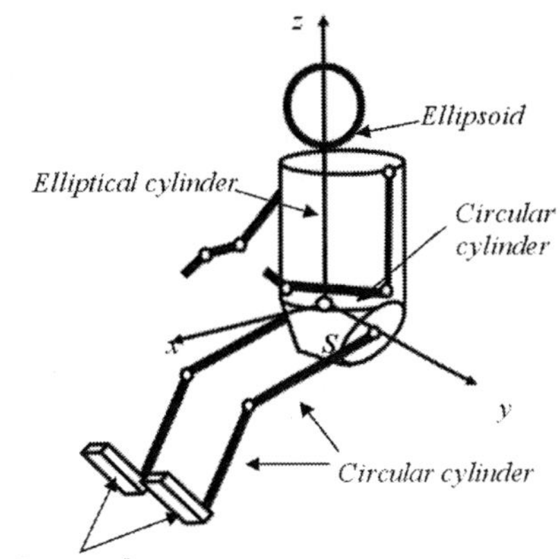

Fig. 1. Biomechanical model of the engine driver with indicated axes of coordinate systems

3. DRIVER'S CAB DESIGN

The engine driver's cab represents the main operating environment of the driver. Direct operating requirements he has to perform are done in the driver's cab. However, there are environmental requirements for long-term accommodation of the engine control crew. Since the train crew changes periodically, it is necessary to plan enough space for work

depending on the anthropometric measures of humans. The height of the cab ceiling should be sufficient for the accommodation of 95% of the engine drivers of the male population plus 5% of engine drivers from the group of female population, as a relation that can be predicted in the overall population (Muftic 1992). The implementation of a wider range of anthropometric sizes for the engine drivers is stimulating, but it can exceed the point of economic interest of the users and the producers. Maximal acceptance of the users' requirements means that the user has to participate during the design in the earliest possible stage of the beginning of the development process, when system requirements are set.
The previous claim is a good illustration of another important problem typical of the process of engine design. By adding new technologies and by changing the designs of the existing ones, no new engine could be produced to be better than the previous version which lacked new technologies and new forms. The design of the complex Man-Machine system and its interface (MMI), as well as the engine driver's cab often includes the development, changes and compromise solutions, but only after the customer's requirements are known, which are usually stipulated by requiring that the price remains the same or by some other limitations.

4. COMMANDS AND ELECTROMECHANICAL DISPLAYS

The position of commands to control the movement has to be located directly in front of the engine driver with the braking modulus opposite to the accelerator. The position of the radio should allow manipulation by left hand with the possibility that allows the engine driver to act using his right hand after a certain time. The commands for the horn, light and radio microphone should be located within the reach zone presented in Figure 3 or, even better, within the comfort zone if possible. Commands should be located so that the position of the body does not have to be changed when activating them. All the commands need to be set so that, with normal handling, the operators do not have to shift significantly their field of vision, resulting in overseeing an important event outside the engine or on the main internal indicator display (Woodson, 1992). The layout of commands should be according to the level of expected usage, all in accordance with the normal movements of the extremities. This means that where a hand waves the control movement should be forwards and backwards, and certainly not sideways. The commands that have similar functions or purpose need to be grouped together.

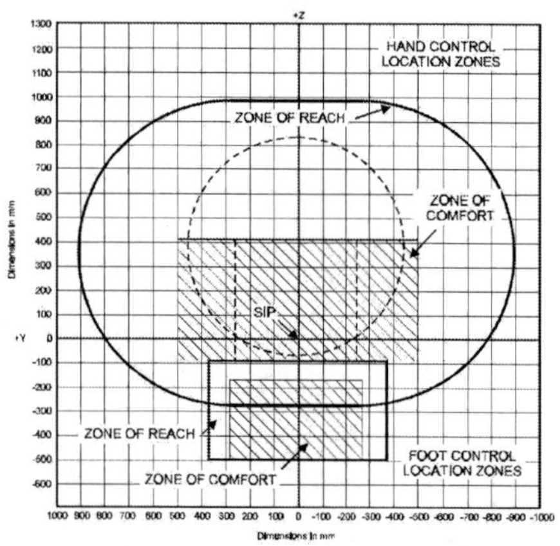

Fig. 3. Front view - Zones of Comfort and Reach (SAE Standard J898© 1987 Society of Automotive Engineers, Inc.)

If the displays are usually read in a sequence, then they also have to be organized either horizontally or vertically. When the circumstances include the layout of the displays which have to be controlled regularly, those conditions need to be determined which are almost "normal", with the arrangement of the display in such a way that all the indicator-hands are arranged equally when the instruments indicate normal operation. In case of harmonizing the needs of the customers and production possibilities the work in design development and production become easier for the producer. It also needs to be taken into account to reduce the probability of engine driver's error while controlling different engines when they move from one driver's cab into another. The attempt should be made always to realize all the requirements related to the safety, health, productivity and satisfaction of the engine driver. Electronic displays with software type of control and the nature of information display offer potential for a greater diversity. The ergonomic conditions, including also anthropometric elements for the users' population, are often outside the producer's range. The example of a layout of commands and displays in the modern engine driver's cab of a passenger diesel-motor set without anthropo-measures is presented in Figure 4.

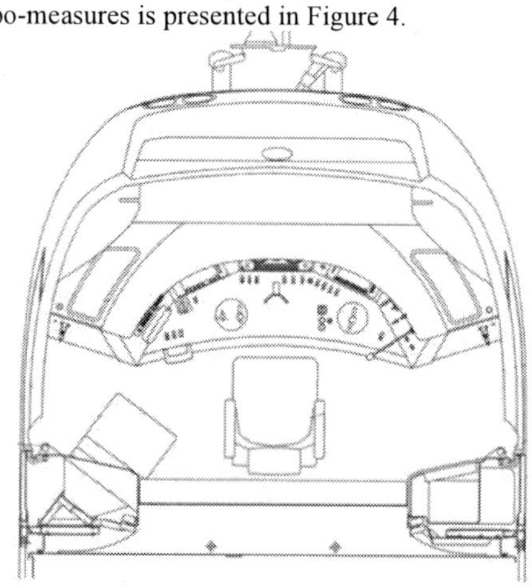

Fig. 4. Arrangement of commands in the engine control cab HZ 7123

5. CONCLUSION

During design, modifications of the existing engine driver's cabs and in construction of new engines it is necessary to analyze human factors. Therefore, it is necessary to use linear and dynamic anthropometric measures. The type of control will affect the reach requirements due to the part of the body that operates the control. The work uses anthropo-measures for male and female persons of the Croatian population. Regarding these measures, optimal positions of commands and electromechanical displays are recommended.

6. REFERENCES

Muftic, O.; Jurcevic T.(1992): *Analiza promjene tlaka trbušne preše u različitim radnim položajima* (in Croatian), Sigurnost, Zagreb

Multer J.; Rudich R.; Yearwood K.(1998): *Human Factors Guidelines for Locomotive Cabs*, U.S. Department of Transportation

Rogers, S.P. and Myers, L.D. (1993). Development of an intelligent system to aid in avionics display design. *AIAA/IEEE Digital Avionics System Conference*. New York: American Institute of Aeronautics and Astronautics.

Rudan, P.: *Dimenzije tijela i tjelesni položaj pri radu* (in Croatian), Medicina rada,pp 87 – 92, Sarajevo, 1979.

Woodson, W.E. Tillman, B, Tillman, P. (1992). *Human Factors Design Handbook*. (2nd ed.) New York: McGraw.

Annals of DAAAM for 2006 & Proceedings of the 17th International DAAAM Symposium, ISSN 1726-9679
B. Katalinic, Published by DAAAM International, Vienna, Austria 2006
150th Birthday of Nikola Tesla (1856-1943) Inventor of Mobile Robot Teleoperation

AUTOMATED REAL-TIME ANALYSIS AND PROCESS CONTROL USING PROCESS DATA MANAGEMENT AND MULTIVARIATE CALIBRATION MODELS

KANDELBAUER, A.; KESSLER, R.W.; WUZELLA, G. & ZANDER, J.

Abstract: *The paper presents calibration models for an automated real-time online analysis of process data by multivariate statistics in the manufacturing of hard boards. It is shown, that chemical and morphological information from intermediates and final products can simultaneously be observed by a continuous and automated measurement of more than 500 different process variables. These data can also be used to monitor and control in real time the quality of mass production of wood based panels.*
Key words: *process data management, real time analysis, multivariate calibration, wood based panel industry.*

1. INTRODUCTION

The wood based panel board industry is a typical example for an industrial branch with high technological input but low added value products as an output. Its profitability relies very strongly on "High Tech" production processes with expensive manufacturing equipment. The only means being profitable for such a plant is to manage efficiently a "high throughput" mass production. It is therefore of substantial interest to implement strategies which allow to manufacture wood panels of guaranteed and specified quality just in time. This is only possible by means of a fast control of optimum manufacturing conditions and an adaptation of the process to variable raw material qualities. One possible strategy is to implement online measurement systems which provide direct information about material properties even on a molecular level. Especially optical spectroscopic methods have been shown to be very suitable for the characterization of single (Schimleck et al., 2001) or multiple (Via et al. 2003) properties of wood surfaces. In the case of the wet hard board manufacturing, it has been shown earlier that optical spectroscopy in the UV/Vis as well as in the NIR spectral range is a very powerful tool for the real-time classification of raw materials, product intermediates and final product qualities at various stages of production (Dolezel-Horwath et al., 2005). However, the use of online measurement systems is accompanied by a number of principal drawbacks: (1) ideally, a series of expensive spectrometers need to be installed, (2) maintenance requires trained staff, (3) environmental conditions for on-line equipment are demanding and optimum working conditions are sometimes difficult to meet. Another possible strategy is to exploit the information which is documented in process data management systems. The advantage here is, no additional equipment is necessary and trained staff is in most of the situations also available.
In the present publication we demonstrate how morphological and chemical information can also be extracted indirectly using process data only. We show furthermore that process data gathered in real-time can also be used to monitor material properties. After calibration by means of designed experiment, these data can also be used to create a causal model of the production process.

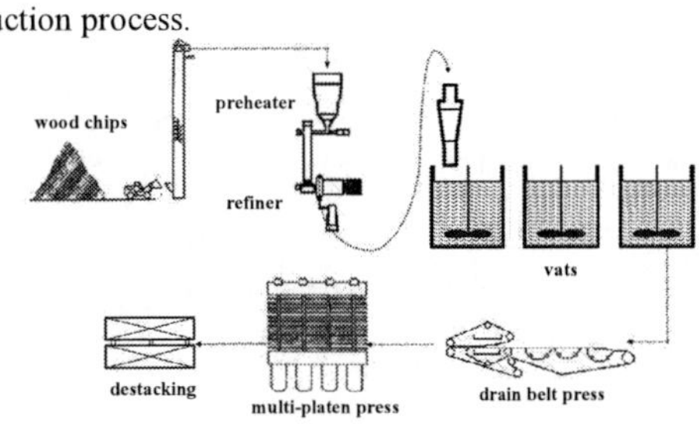

Fig. 1. Schematic representation of the wet process for the manufacturing of hard boards.

2. METHODOLOGY

Raw process data were collected on-line using the Promacs system of ATR, Viersen. A design of experiments was performed on the fibre board plant where the following parameters were varied: wood species mixture (spruce/beech, spruce with/without bark), refiner plate gap (RPG), severity factor chemistry (SFC), aeration time t_a and pressure p_a).
Technological parameters of boards such as swelling, density profile, water uptake, flexural strength, modulus of elasticity and internal bond strength were determined for each experiment. Table 1 summarizes some parameter variations of the experimental design. On-line UV/Vis spectroscopic measurements were performed (1) on the raw material, (2) in the blowline after digestion in the refiner, (3) after sheet formation and (4) on the final board after hot pressing (Dolezel-Horwath et al., 2005). Figure 1 schematically depicts the wet process for hard board manufacturing. All data were subjected to multivariate statistical analysis (Principal Component Analysis, PCA and Partial Least Square, PLS regression) using the software package Unscrambler 9.0 and Statistica. Calibration models were established and validated for each, process and on-line spectroscopic data sets.

#	Wood type	SFC	RPG	t_a	p_a
1	spruce / beech	2.5	large	120	65
2	spruce / beech	2.9	large	180	45
3	spruce / beech	2.9	large	240	55
4	spruce / beech	2.9	large	120	55
5	spruce / beech	2.9	large	180	65
6	spruce without bark	2.1	small	180	45
7	spruce without bark	2.1	small	120	55
8	spruce without bark	2.1	small	240	55
9	spruce without bark	2.1	small	180	65
10	spruce without bark	2.1	small	240	45

Table 1. Some parameter variations in the designed industrial experiments (for detailed explanation see text)

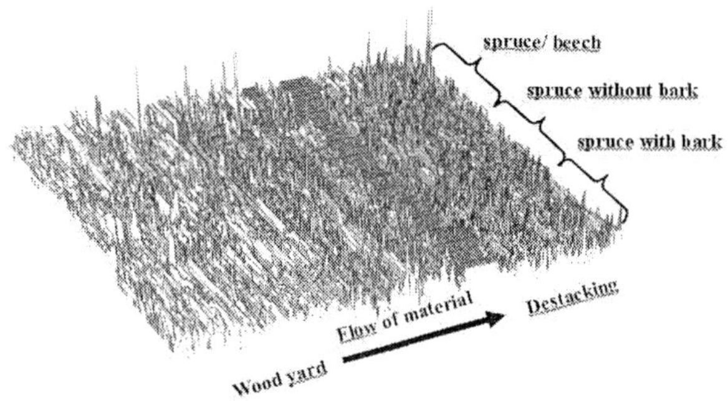

Fig. 2. 3D-Plot of the industrial plant process data during the designed experiments at all stages of production from the raw material to the final product

3. RESULTS AND DISCUSSION

The time course of more than 500 different process data throughout the industrial experimental design is shown in Figure 2. Application of PCA and PLSR models reduces the complexity of the data set. The correlation with material parameters creates information about the process at various stages of production.

The multivariate analysis of process data has been studied in detail earlier (Bernardy et al., 2004) for discrete batch processes (see for example Westerhuis et al., 1999) and for continuous production processes such as the manufacturing of medium density fibre (MDF) board (Hasener, 2004).

In Figure 3, an example of a PCA analysis of a subset of process data is presented. It is evident that the process data correlate well with the morphological properties of the refined fibres in the first principal component: the clustering of data points reveals the classification of refined fibers as either course or fine. This reflects the variable "plate gap" in the refiner which was varied in the experimental design on 2 levels ("small"/ "large").

Even more interesting is the information illustrated by the second principal component in Figure 3 which is interpreted as the severity factor chemistry (SFC). SFC is the combination of temperature, pressure and incubation time during refining which governs the severity of the thermo-mechanical refining conditions. The severity of the refining conditions from low (SFC 2.1) to high (SFC 2.9) is obviously reflected by the process data and manifests the temperature – time variation of the process. Different SFC causes selective degradation of lignin thus defining the chemical properties of the inherent glue.

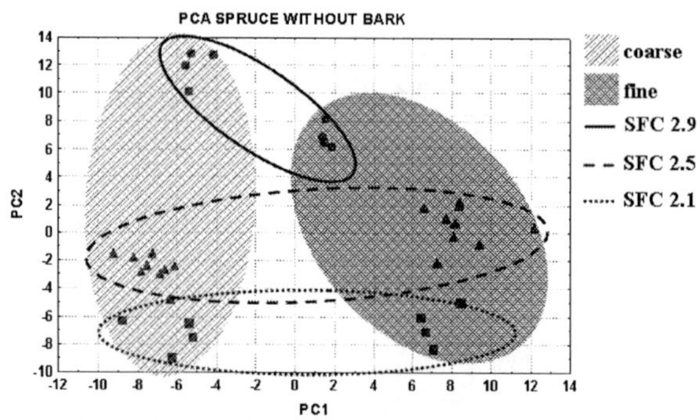

Fig. 3. Principal Component Analysis (PCA) of process data reflecting the correlation with morphological properties of the intermediate material

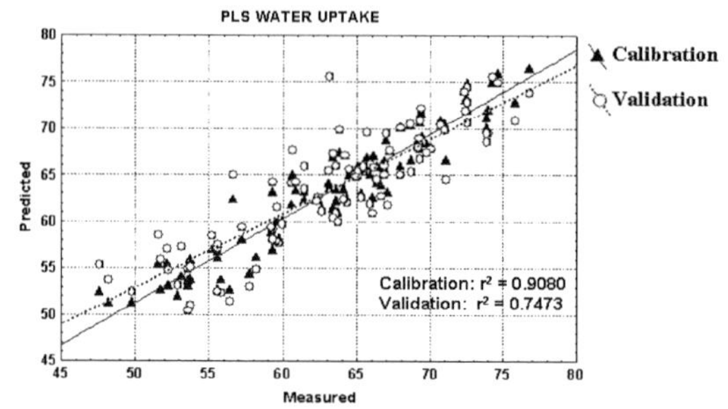

Fig. 4. PLS regression model of correlation between process data and water uptake of manufactured hard boards

While Figure 3 illustrates the correlation of the process data with intermediate material properties in the blowline, Figure 4 shows an example of the quantitative correlation between the process data and the technological performance (water uptake) of the final hard boards.

4. CONCLUSIONS

Process data can be used for the real-time monitoring of the continuous production of hard boards in a similar way as online spectroscopy. They can reflect material properties of intermediates as well as of end products quality during the whole production process. Using experimental design, causal models of the manufacturing process can be deduced. The mathematical models can be implemented to allow an automated process control system for the manufacturing of hard boards in the wet process.

5. REFERENCES

Bernardy, G.; Lingen, A. & Sandweg, T. (2004). Process Data-Based, Online Quality Control in Everyday Production Situations, *Proceedings of 38th International Wood Composites Symposium*, Tichy, R. J. & Yadama, V. (Eds.), pp. 155-163, ISBN 2004010184, Washington State Univ., Pullman, Washington, April 6-8, 2004, Washington State University, Pullman

Hasener, J. (2004). Statistische Methoden der industriellen Prozessmodellierung zur Echtzeitqualitätskontrolle am Beispiel einer kontinuierlichen Produktion von Faserplatten, *Available from:* http://www.sub.uni-hamburg.de/opus/volltexte/2005/2346/pdf/Dissertation_Hasener.pdf *Accessed:* 2006-09-28

Dolezel-Horwath, E.; Hutter, T.; Kessler, R. W.; Wimmer, R. (2005). Feedback and feedforward control of wet-processed hardboard production using spectroscopy and chemometric modelling. *Analytica Chimica Acta,* Vol. 544, pp. 47-59, ISSN 0003-2670

Schimleck, L.; Evans, R.; Ilic, J. (2001). Application of near-infrared spectroscopy to a diverse range of species demonstrating wide density and stiffness variation. *International Association of Wood Anatomists Journal.* Vol. 22, Issue 4, pp 415-429, ISSN 0928-1541

Via, B.K.; Shupe, T.F.; Groom, L.H.; Stine, M.; So, C.-L. (2003). Multivariate Modelling of density, strength, and stiffness from near-infrared spectra for mature, juvenile and pith wood of *Pinus palustris. Journal of Near Infrared Spectroscopy.* Vol. 11, pp. 365-378, ISSN 0967-0335

Westerhuis, J.A.; Kourti, T.; MacGregor, J.F. (1999). Comparing alternative approaches for multivariate statistical analysis of batch process data. *Journal of Chemometrics.* Vol. 13, Issue 3-4, pp397-413, ISSN 0886-9383.

Annals of DAAAM for 2006 & Proceedings of the 17th International DAAAM Symposium, ISSN 1726-9679
B. Katalinic, Published by DAAAM International, Vienna, Austria 2006
150th Birthday of Nikola Tesla (1856-1943) Inventor of Mobile Robot Teleoperation

KNOWLEDGE MANAGEMENT FOR NETWORK OF ENTERPRISES

KARAULOVA, T; KRAMARENKO, S. & SHEVCHENKO, E.

¶

Abstract: *Rapid advances in technologies such as networks, database analysis techniques and high speed processing helped the growing importance of organizational decision support systems (DSSs). Today's organizations invest in various knowledge management (KM) systems and tools to enable seamless integration of the constantly increasing volume and sources of information.*
In the paper are offered an approach to integrating various sources of information and knowledge for companies to enhance their decision-making process.
Key words: *Knowledge management, data warehousing, data mining.*

1. INTRODUCTION

Small and medium enterprises (SME) with limited resources are slow in grasping the opportunities of business change in the digital era. The exploitation of information technology (IT) and exploration in the use of information systems (IS) is the key for change in transforming the existing business models to those embracing digital enterprises (Lun et al., 2001). The transfer of knowledge between industrial firms is important since industry is relying more heavily on these types of relationships to broaden and enhance their existing knowledge base (Lee et al., 2000). The paper highlights different perspectives on collaboration: management of the SME, the SME as a part of a network of regional – international SMEs and collaboration in the daily work (maintenance, engineering).
These aspects can be seen as stages in a life cycle development of SME network evolution. At the same time collaboration is seen as to evolve naturally from the needs of the market and the unique and complementary skills and capabilities of partners. The first place in this process belongs to information and knowledge collaboration.
An organization's most valuable business asset flows through its enterprise every day. It's the data from customer orders, purchase receipts, operating statistics and more. Analyzing that data can provide business insight an organization can't get any other way—insight executives and managers can use to guide the tactical and strategic decisions that drive company's performance (Far, 2005).
The requirements of an enterprise wide Decision Support System (DSS) can only be met by a system optimized for data access, quality and integrity. Knowledge is a key organisational resource. Acquiring knowledge is a concern of organisations as most are significantly knowledge–dependent for their success (Edwards et al., 2005).

2. KNOWLEDGE MANAGEMENT

There is a difference between data, information and knowledge. Data is defined as raw facts. Information is organised data that results from the processing of data for a purpose but without any interpretation (for example, its summarisation and reporting). The progression shown in Figure 1 illustrates the value chain of information.

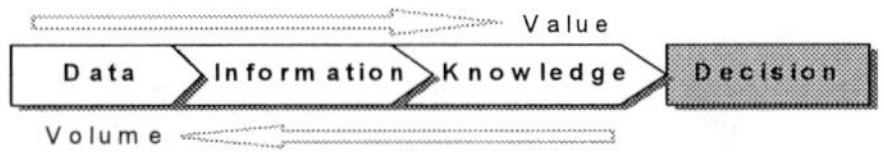

Fig.1. Value chain of information

Knowledge management has been described as a process of creating, capturing and using knowledge to enhance organisational performance at the organisational, rather than the individual, level of analysis.
Enterprise knowledge management (KM) entails formally managing knowledge resources in order to facilitate access and reuse of knowledge, typically by using advanced information technology. The overriding purpose of enterprise KM is to make knowledge accessible and reusable to the enterprise. Knowledge resources vary for particular industries and applications, but they generally include manuals, letters, summaries of responses to clients, news, customer information, competitor intelligence, and knowledge derived from work processes (O'Leary, 1998). A wide range of technologies are being used to implement KM systems: e-mail; databases and data warehouses; group support systems; browsers and search engines; Intranet and Internet; expert and knowledge-based systems and intelligent agents. Knowledge management processes are a combination of acquiring, sharing, retaining and utilizing knowledge (Edwards et al., 2005):

- knowledge acquisition relates to obtaining knowledge that an organisation does not currently have;
- knowledge sharing refers to the availability of knowledge that is already held by the organisation;
- knowledge retention is concerned with not losing knowledge as people leave or move around the organisation;
- knowledge utilisation is concerned with the value that can be added through the effective use of organizational knowledge.

3. DATA WAREHOUSING

Since the introduction of the term "data warehousing", companies have explored the ways they can capture, store and manipulate data for analysis and decision support. Data warehouses differ from traditional transaction databases in that they are designed to support decision making rather than simply efficient capturing of transaction data.
For the largest companies today Enterprise Resource Planning (ERP) serves within the core set of day-to-day transaction processing applications. Even when data is efficiently captured and stored in ERP systems, it may remain relatively useless for reporting and decision making purposes ERP systems do central work of running, tracking and reporting on business data processing.

Providing information requires all the operational data be available for analysis, including the vast amounts held in ERP, CRM (Customer Resource Management) and SCM (Supply Chain Management) applications. Each business is unique; each solution must be tailored to the company's own requirements and resources.

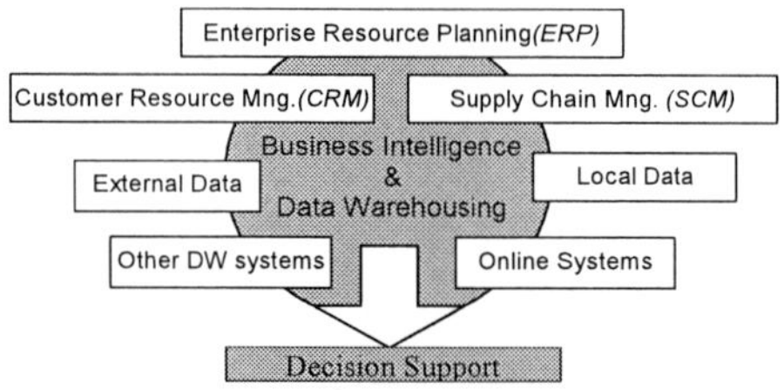

Fig. 2. Information collaboration for decision support

Data Warehousing (DW) is a concept. It is a set of hardware and software components that can be used to better analyse the massive amount of data that companies are accumulating to make better business decisions (Fig. 2).

Data Warehousing is not just data in the data warehouse, but also the architecture and tools to collect, query, analyse and present information.

4. KNOWLEDGE DISCOVERY FOR DECISIONS MAKING

Organizations are gathering and storing more and more data. Every year the amount of data in the world is approximately doubling. This data is of little benefit unless it can be turned into useful information and knowledge (Muehlen, 2002).
Recognizing that needless data is found within most enterprises, organizations have implemented some form of data warehousing to bring this data together and allow efficient enterprise reporting and analysis. Key information needs to reach multiple decision makers within the organization. Inconsistencies and inefficiencies result when information is not distributed and shared across the organization.

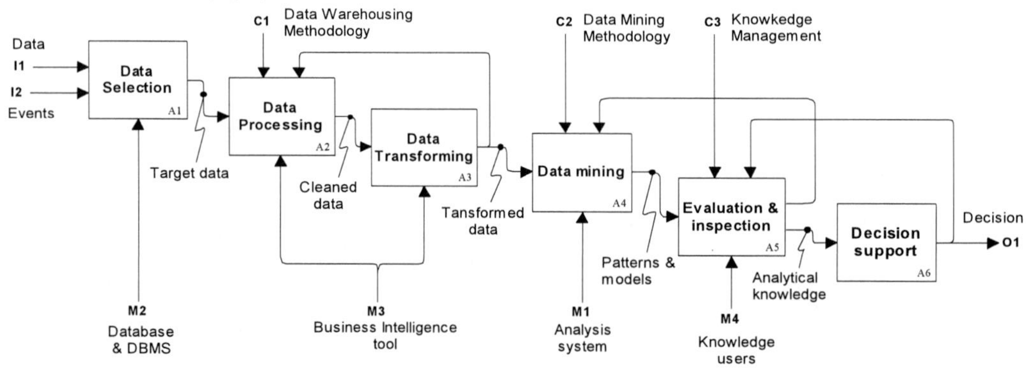

Fig. 3. Knowledge discovery for decisions making

The steps taken to discover the knowledge for decision making in the IDEF0 language are given in figure 3.
Data Selection: The goal of this phase is the extraction from a larger data store of only the data that is relevant to the data mining analysis. This data extraction helps to streamline and speed up the process.
Data Processing: This phase of knowledge discovery is concerned with the data cleansing and preparation tasks that are necessary to ensure correct results. Eliminating missing values in the data, ensuring that coded values have a uniform meaning and no spurious data values exist are typical actions that occur during this phase.
Data Transformation: This phase of the lifecycle is aimed at converting the data into a two-dimensional table and eliminating unwanted or highly correlated fields to attain the valid results. This process includes replacing codes with actual values.
Data Mining: The goal of the Data Mining (DM) phase is to analyse the data by an appropriate set of algorithms in order to discover meaningful patterns and rules and produce predictive models. This is the core element of the knowledge discovery cycle.

5. CONCLUSION

The proposed solution brings together benefits of the available technologies of data querying, warehousing, analysis and visualization, as well as information retrieval and sharing knowledge with its users. Those who have access will be able to make a fact-based and timely decision.

Acknowledgements Hereby we would like to thank the Estonian Science Foundation, for the grant G6795 enabling us to carry out this work.

6. REFERENCES

O'Leary, D. (1998) Enterprise knowledge management. *Computer* Volume 31, Issue 3, March 1998

Lee, G., Bennett, D., Oakes, I. (2000) Technological and organizational change in small- to medium-sized manufacturing companies: A learning organization perspective, *International Journal of Operations & Production Management*, Vol. 20, No. 5

Lun, E., Yuen, K., Chan, M., Chung,W. (2001) Enterprise Transformation for Management Succession *Proceeding of The 6th International Manufacturing Research Symposium*

Muehlen, M. (2002) *Workflow-based Process Controlling*, Logos Verlag Berlin ISBN 3-8325-0388-9, ISSN 1611-3101

Far, B., (2005) Intelligent project lifecycle knowledge management and decision support, *http://www2.enel.ucalgary.ca/People/far/*

Edwards, J., Collier, P., & Shaw, D., (2005) Knowledge Management and Its Impact on the Management Accountant, Research Report, ISBN 1-85974-569-9, The Chartered Institute of Management Accountants.

Annals of DAAAM for 2006 & Proceedings of the 17th International DAAAM Symposium, ISSN 1726-9679
B. Katalinic, Published by DAAAM International, Vienna, Austria 2006
150th Birthday of Nikola Tesla (1856-1943) Inventor of Mobile Robot Teleoperation

FEM ANALYSIS IN REDESIGN OF CONTINUOUS PROFILING UNIT

KERAN, Z.; SKUNCA, M.; MATH, M. & MEDVED, T.

Abstract: *The intention of this paper is to analyze the technology of continuous profiling of garage doors. According calculations process includes ten phases of continuous roll metal forming. Calculation of distance between two phases was regularly made according Shulze and was improved using FEM based software. The results that were obtained have been brought into comparison and presented in the paper.*

Key words: *Continuous profiling, FEM, Shulz, UBECO Profile.*

1. INTRODUCTION

Rollforming is a continuous bending operation in which sheet or strip metal is gradually formed in tandem sets of rollers until the desired cross-sectional configuration is obtained. One of the important questions in process design is determination and calculation of distance between two pair of rollers or between two phases in a way that those rolling processes do not influence on each other. Regularly, this calculation is made by Shulze (Shulze, 1959), or some of other authors who determine equations for calculation of edge stresses. But there are certain inconveniences that designer has to deal with when using standard way of calculation – each author gives some restrictions in use. Profil is one of the rollform design softwares for manufacturer of cold roll-formed profiles from sheet metal and for designers of rollformers and tube forming machines. It enables quicker working and cost reductions in planning, design, calculation and drawing of the profile, the flower pattern (bending steps) and the rolling tool. Profile Stress Analysis (PSA) and Virtual Rollforming Machine (VRM) are steps of process design. The sheet metal strip is divided into small shell elements and the geometrical deformation while passing through the roll forming machine is calculated. From it the strain and stress in the whole profile is obtained and displayed by colors. The FEA simulation of the roll forming process enables the designer to validate and optimize his rollform design at an early stage before the rolls are manufactured to ensure that the final product meets the particular needs. To proceed this very precise calculation of the stress and the longitudinal deformation within the whole profile, an interface has been provided to the loading FEA system ABAQUS/Explicit.

2. SHAPE GRADUATION

Procedure of shape for each phase always starts from final profile shape until it becomes consisted of only straight elements. As a help in approximate shape defining completed examples solved in practice can be used. Number of phases is determined using AUTOCAD to 10 phases.

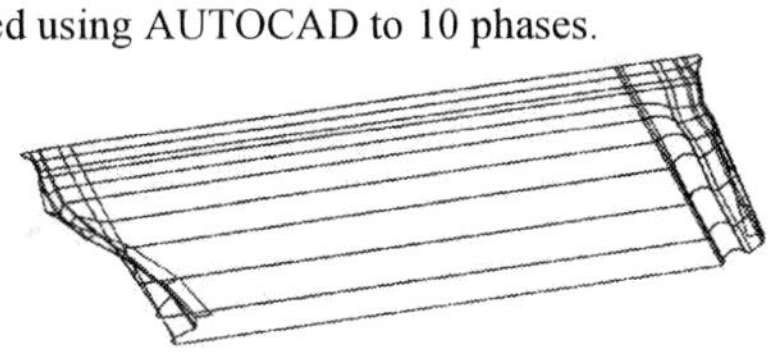

Fig. 1. Shape graduation.

3. DISTANCE BETWEEN PHASES

Once the shapes for 10 phases have been defined, the distances between the phases have to be set up. It was done using the empirical expression after Schulze and using UBECO profile – FEM based software. Restriction of Shulze's expression to the bending around sharp corners disables the calculation of the distances when bending around round corners occurs. Therefore Profile was used to determine all the distances appearing in continuous profiling. Results of both methods used are given in comparison in Table 1.

	Distances between phases					
	Left side			Right side		
Phase numbers:	Sulze's expression [mm]	UBECO Profile [mm]	Difference regarding Schulze	Shulze's expression [mm]	UBECO Profile [mm]	Difference regarding Schulze
0-1	150	130	-13%	150	130	-13%
1-2	150	130	-13%	150	130	-13%
2-3	100	80	-20%	100	80	-20%
3-4	100	160	60%	-	160	-
4-5	110	190	72%	-	150	-
5-6	-	300	-	-	100	-
6-7	-	350	-	-	350	-
7-8	-	460	-	-	100	-
8-9	-	450	-	90	150	67%
9-10	-	360	-	60	130	117%

Table 1. Distance between phases obtained by UBECO Profile and empirical methods

4. TOOL DESIGN

In order to correctly define the geometry of the rolls, springback effect has been calculated. It was done only for the phases 5 and 10 of the left and for the 8 and 10 of the right side where partial and final bending onto the final shape geometry appears. The calculation was performed using empirical approach after and using UBECO Profile calculation. Comparison of the results is given in Table 2.

		Change of the bending angle Δα			Change of the bending radius Δr		
Ph. Nr.	Side/zone	UBECO Profile, [°]	Empirical expression,[°]	Difference	UBECO Profile, [°]	Empirical expression,[°]	Difference
5	Left	0.83	1.41	69%	0.015	0.025	67%
10	Left	2.89	3.25	13%	0.045	0.050	11%
8	Right – 1	1.08	1.84	70%	0.015	0.025	67%
	Right – 2	5.23	7.50	42%	0.261	0.364	39%
	Right – 3	1.66	1.84	11%	0.051	0.056	10%
10	Right	0.83	1.07	29%	0.034	0.043	26%

Table 2. Change of bending angle and bending radius obtained by UBECO Profile and empirical methods

Right side of the 8th phase was divided into 3 zones since three bending angles with adjacent radii appear. Each one of them requires a separate calculation.

5. CALCULATION OF FORCE

To obtain force within the greatest interval of confidence, three empirical expressions were used. Authors are Mäkelt (Mäkelt, 1954), Griffin (Griffin, 1956), and Kretzchmar (Schulze 1959). All of these methods are based upon the basic geometrical relations who make them easily applicable in praxis. The information needed on rolled material is also basic: tensile strength, bending stress and yield stress. This makes the mentioned methods very swift on giving arbitrary information, but not as swift as Profile with its own implemented codes. The results obtained in all four ways are on Fig. 2 and Fig. 3.

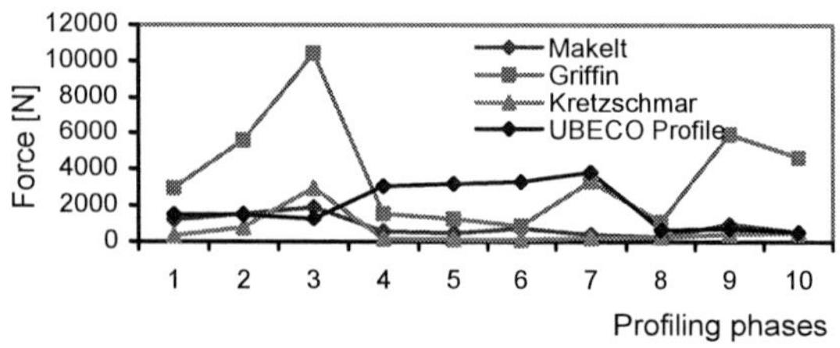

Fig. 2. Profiling force over the phases – Right side.

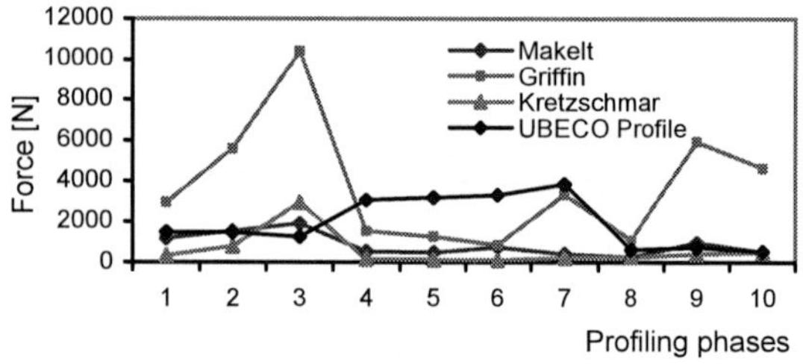

Fig. 3. Profiling force over the phases – Left side.

6. FEM SIMULATION

The simulation for the right side was performed. It consisted of approximately 4000 shell elements whose dimensions are 1x3 mm. Although no forces, momentum and springback effects were analyzed, simulation undisclosured some very interesting phenomenon. In Fig. 4 deformation at the spot designated with ¨A¨ happened because all the right sides of the forming rolls were lying in the same plane. Complanarity of the right sides of the roll induces an extra hardening of the material along the line inducing the final error at the spot ¨A¨.

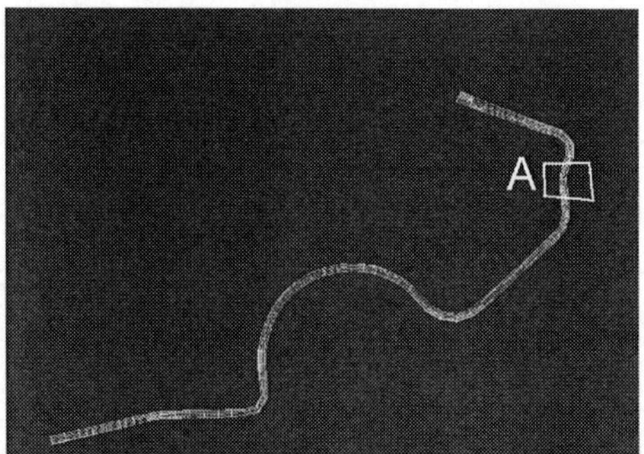

Fig. 4. Occurrence of an error specific to profiling technology captured by FE method.

Forming error at spot ¨A¨ shown in Fig. 4. was corrected by changing the roll geometry in order to break the complanarity by distributing their right sides around an arbitrary longitudinal plane along the right side of the profile. The result is correction of the error as it can be seen in Fig. 5. This phenomenon could have been only captured by FE method while all the other methods can not give this kind of data.

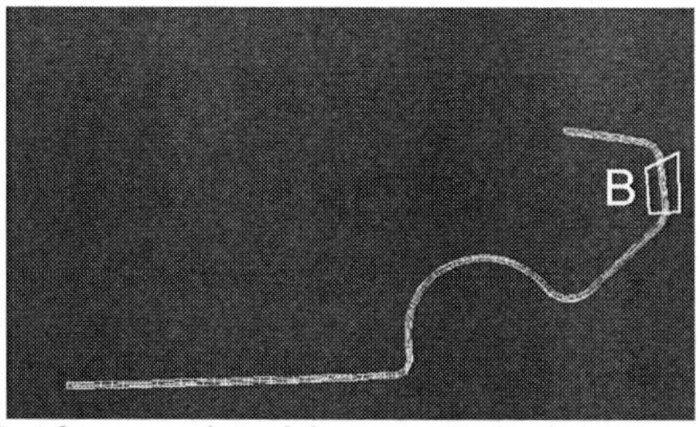

Fig. 5. After correction of the error FE simulation has shown the absence of wrinkling at the spot ¨B¨.

7. CONCLUSION

Calculation of continuous profiling using standard numerical methods suggested of different authors gives results that much differ from each other. The only certain way of determining all the parameters is creation of prototypes and gradual declining of possible errors. Such way is unimaginable and unacceptable because of too long time that has to pass from idea to final product. Because of that reasons FEM based software offers a great help in solving design problems.

Described problem was successfully solved using FEM based software. It has been also successfully implemented in production process. Time of designing of improved process is spared. It can be noticed that combining the possibility of capturing phenomena with the future possibility of obtaining the force and momentum from FE simulation, gives a very good tool for everyday engineering praxis.

But, one critical statement toward specific software has been formulated during designing this process. For simulation of process on a model consisted of approximately 4000 shell elements, about 48 hours was needed. This means that specific program is not optimised on calculation time.

7. REFERENCES

Anderson J. M., Berc L. M., Dean J., Continuous Profilling: Where have all the cycles gone, *ACM Transactions on Computer Systems*, Vol. 15, (November 1997), 357-390.

Griffin E. (1956). Cold roll forming and manipulation of light-gauge sections, *Journal of the institute of metals*, Vol. 84, (1956), 181-197.

Lange K. (1990). *Umformtechnik Handbuch für Industrie und Wissenschaft, Band3: Blechbearbeitung*, Springer-Verlag, 3-540-50039-1, Berlin.

Mäkelt H. (1954). Das Kaltwalzen von Profilen aus Bleich, *Industrie – Anzeiger*, Vol. 76, (1954), 1034-1038.

Barisic B., Cukor G., Math M., Estimate of Consumed Energy at Backward Extrusion Process by Means of Modelling Approach, *Journal of Materials processing Technology*, Vol. 153-154, (2004), 907-912.

Ranatunga V., Gunasekera J.S., Vaze S. P., De Souza U., Three-Dimensional UBET Simulation Tool for Seamless Ring Rolling of Complex Profiles, *Journal of Manufacturing Processes*, Vol. 6, (2004), 179-186.

Shulze G. (1959). Kenngrößen für die Enwicklung und den Einsatz von Profiliermaschinen, *Machinenbautechnik*, Vol. 8, (1959), 181-191.

Annals of DAAAM for 2006 & Proceedings of the 17th International DAAAM Symposium, ISSN 1726-9679
Editor B. Katalinic, Published by DAAAM International, Vienna, Austria 2006
150th Birthday of Nikola Tesla (1856-1943) Inventor of Mobile Robot Teleoperation

THE STRATEGY OF CONCENTRATION IN CROATIAN RETAIL TRADE

KNEGO, N. & KONJHODZIC, A.

Abstract: *The topic of our interest is the process of concentration in European retailing and factors that either aid or inhibit this process. Factors that either stimulate or de-stimulate the processes of business concentration can be divided into a few groups. We take a look at this by using the case of Croatian retailing.*
Key words: *strategy, concentration, retail trade, Croatia.*

1. INTRODUCTION

We will place an emphasis on the concentration in the Croatian retail trade. Our objective is to point out in which measure this is and to determine whether it is specific to Croatia, or if it involves something else. If something else, it could be s certain legitimacy that could be anticipated by knowing and monitoring the issue in question. Concentration is a result of the internationalization process. Numerous factors stimulate or slow down the internationalization process in Croatian retailing.

2. CONCENTRATION IN CROATIAN RETAIL TRADE

2.1. An explanation of the phenomena

Concentration is one of the phenomena that characterized the economic progress of a number of countries, starting with the nineteenth century. Its affect on modern economic progress has intensified particularly over the last decades, influenced by scientific-technological progress.
The concentration processes in production affected the appearance of like processes in social production. Concentration, as a phenomena and a process, manifests itself through the creation of large companies, up to those gigantic.
In clarifying the role of trade, one starts from the observation that production is increasingly concentrated, while consumption is dispersed spatially. The concentration of production means that products are manufactured in a smaller number of places. The spatial dispersion of consumption results in its spatial, time and personal withdrawal from production.
The practical approach towards evaluating the level of concentration achieved in trade, especially in retailing, are generally based on data that concern the relative share of a small number (few) of leading trading companies in the total turnover realized. An identical approach is used in investigating the concentration of production (Knego, 1998).
Depending on the number of companies compared to the total (4, 5, 8 or some other), we can speak of the application of methods K4, K5, K8 or Kn. K is the sign for concentration, and the index n determines the number of leading companies whose share in the overall phenomena is being monitored or investigated (Rugman, 2000).
Generally, the claim that the larger the share of fewer leading companies in realized turnover, the larger the level of concentration, and vice versa. It is also possible to identify the level of concentration achieved on the basis of similar variations of indicators, such as: stores and their number, the number of employees, available resources, etc.

2.2. The segmentation of Croatian retailing

An analysis of the data on the number of business entities according to their main activity in the distributive trade of the Republic of Croatia shows the pronounced segmentation of Croatian trade (applicable to both wholesale and retail trades) and the hypertrophies of the Croatian wholesale trade.
There were slightly less than two retailing business entities for each wholesale business entity. There is pronounced hypertrophies of the Croatian wholesale trade as compared to retailing. Our research shows that wholesale entities cannot base their market survival on being the intermediaries between the manufacturer and the retailer. This claim arises from their exaggerated number as compared to retail business entities. It is enough to mention that each business entity in Croatian retailing had less than two stores on average (1,3), and that it employed about four workers on average. Simultaneously, the business entities in the Croatian wholesale trade had less than one store (0,6), with an average of five employees. It is realistic to suppose that the impossibility of any long-term survival in the relationship of many Croatian wholesale business entities will mean their disappearance from the market scene.

3. CONCENTRATION IN RETAIL TRADE: CHARACTERISTIC OF CROATIA OR SOMETHING ELSE

Concentration in trade is not just characteristic of Croatia. Simply said, it is a rule that is more pronounced the more developed a country is. People who are experts in trade and who manage companies should have anticipated and prepared for the challenges presented by it.
The events on the Croatian market are only a time-delayed picture of events that have already occurred on the markets of developed countries. The processes mentioned have to be monitored and anticipated, their influence evaluated and preparations should be made for them. It is obvious that this has not been done, keeping the Croatian case in mind. Many are caught by surprise, and they try to frantically think up fast protection mechanisms and to animate the public regarding how many have been «swept» away or will be from the market scene.
The most direct relationship of numerous small independent traders is mentioned. It is true that many have disappeared and that many will disappear yet. It is true that developed countries, especially those European, did not anticipate this event. This is a process that they have already gone through and through which they are still going. If it had been understood today there would have been an association of independent traders in Croatia. Unfortunately, we are witness to the rare operational business ties that are connected to the more rational way of doing business over recent years.
Which are the processes that are present in developed European retailing and which represent elements of a scenario that is already penetrating the Croatian market softly, and which will come on strong in due time. We note the following processes:

a) Less growth, more segmentation; b) Fewer national operators, more international operators; c) Fewer shops, more sales area; d) Less stock, more customer service; e) Fewer independent retailers, more affiliations and f) Lower turnover per square meter, greater margin differentials (Alexander, 1995). The processes mentioned under b, c and e directly affect the greater significance of the concentration process. The process under f is a certain consequence of the processes previously mentioned. Everything that has been mentioned can be documented, thereby confirming that the events on the Croatian market are not specific to us, rather that it is something that competitive markets come across either sooner or later, with either lesser or greater intensity. Accordingly, the Croatian market is no exception. In comparison to some other markets, it can be differentiated only by the extent of the phenomena mentioned and by their intensity. The above is dependent primarily on the level of economic development in Croatia.

An inadequate knowledge of this issue would give the feeling of a catastrophic event, when faced with the data noted. A question could be raised as to the number of workplaces that were terminated with the disappearance of the stores.

Those with a better knowledge of this issue would approach the process that is unraveling in European retailing more carefully, and we mentioned it under c) less stores, more sales space (Segetlija & Lamza-Maronic, 1999).

4. WHAT DOES THE PROCESS MEAN: LESS STORES, MORE SALES SIZE?

The process mentioned deals with the expansion of stores. This means that the average store size is on the rise. This should be supported, as stores with large surface areas are the key to introducing modern retail technology. The growth in sales surface size influences the higher value of one of the qualitative indicators for the progress of retailing structures, such as: sales surface size per capita. Modern sales technology and the growth of sales size per capita help increase: capacities of stores, the speed of shopping and the better quality of retail service (Walters & Hanrahan, 2000). So, the results of less stores did not mean a lower quality of service, and a lowered capacities of the store, rather, it was contrary. A research of the indicators on the development of retailing structures, such as the number of inhabitants per store in Croatian retailing, in developed European countries, and some developed non-European countries shows that we are very similar to them. We will take the case of such countries as: Austria, Sweden or the USA. A layman could say, "Look, we are not so bad. We are almost the same as they are. What do you want?" What is it about or what could be the misunderstanding? The misunderstanding could be not understands the cause of such a state. This situation, the mentioned indicator of the development of the retail structure is, using the case of Croatia, the result of a poorly developed retail network and its segmentation. For the three countries mentioned, this is the result of a highly developed retail network, in which powerful retail buildings dominate, up to gigantic proportions. The above is no coincidence. This is a rule. This rule in the number of trading companies and the number of stores is closely tied to the level of economic development. It could be presented graphically with a coordinate system, so that y represents the number of companies or stores, and x represents the level of economic development (GDP; GDP per capita). The graphical illustration of this relationship is in the shape of a semi-circle whose base faces upwards. See Figure 1. The curve shaped like a semi-circle, above-mentioned graphic illustration, was divided into two. The first part that is found left of the angled direction represents the situation of countries in transition, as is Croatia. There is stillroom on the Croatian retail market for new companies and new stores.

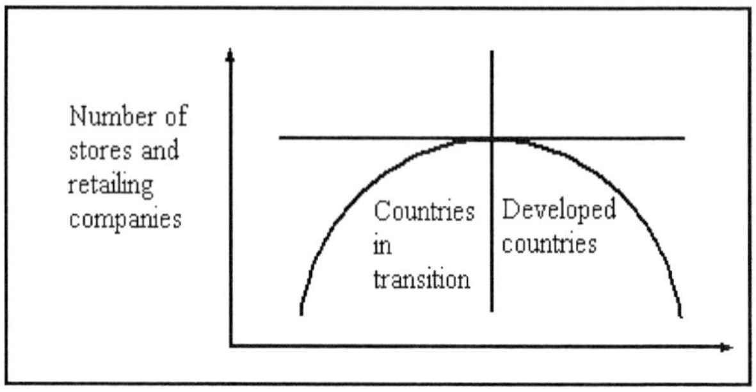

Fig. 1. The affect of economic development on the member of stores and retailing companies

A similar situation can be seen with other transition countries. There is no need to despair and to say that there is no more room for new players. The key question is whether the new ones can do their job differently, better and cheaper than existing ones. Those that can still have open possibilities on the Croatian retail market. It is still significantly distant from the base point of the mentioned semi-circle, in which significant changes will occur that will benefit the better quality of retail services. This will be the moment when the Croatian retail market will experience what has already been a process in developed European retail markets. This means the absolute decrease in the number of stores, followed by the growth in store sale size, and a growth in the average store size. Passing the base of the curve mentioned, we enter into a field of new quality. The area right of the angle is the area of developed countries.

5. CONCLUSION

We can conclude that trade is an exceptionally important economic activity, measured by its contribution to BDP and employment, through the use of a rich analytical base, and by the application of comparative methods on a few transition countries. This is not something particular to Croatia.

The processes that are present in Croatian trade today are not coincidental or something that is happening in Croatia and nowhere else. We show that by listening carefully, monitoring and knowing the expert issues, we can be ready for the approaching processes. Nothing happens by chance. This is not peculiar to us.

6. REFERENCES

Knego, N. (1998). Moguce ulazne strategije u procesu internacionalizacije maloprodaje, Potential Entry Strategies in the Internationalization of Retail Processes. *Ekonomska misao i praksa*, Vol. 7 , No. 2., 1998, pp. 105-120, ISSN 1330-1039

Rugman, A. (2000). *The End of Globalization-A New and Radical Analysis of Globalization and What It Means for Business*, Random House, ISBN 0 7126 8495 6, London

Alexander, N. (1995). Internationalization: Interpreting the Motives, In: *International Retailing-Trends and Strategies*, Pitman Publishing, Ed. McGoldrick, P.J. & Davies, G., pp. 80-95, ISBN 0 273 61183 6, London

Segetlija, Z. & Lamza-Maronic, M. (1999). Internacionalizacija maloprodaje kao vazan cimbenik razvoja turisticke ponude, Internationalization of Retail Trade as an Important Factor in Tourism Supply Development. *Ekonomska misao i praksa*, Vol. 8, No. 2., 1999, pp. 177-195, ISSN 1330-1039

Walters, D. & Hanrahan, J. (2000). *Retail Strategy-Planning and Control*, MacMillan Business, ISBN 0-333-75235-X, London.

Annals of DAAAM for 2006 & Proceedings of the 17th International DAAAM Symposium, ISSN 1726-9679
B. Katalinic, Published by DAAAM International, Vienna, Austria 2006
150th Birthday of Nikola Tesla (1856-1943) Inventor of Mobile Robot Teleoperation

THE MEASURMENT OF BIOGAS MICRO FLOW DURING AMARANTH BIOMASS FERMENTATION

KODRIKOVA, K. & KOLOMAZNIK, K.

Abstract: *The aim of the article is oriented at a measurement of biogas micro flow which springs up while amaranth biomass decomposition. The fermentation runs strictly under anaerobic conditions, the possibility of biogas production is primarily tested as four-level mathematical model. The mechanism of four-level decomposition is described by a vector equation, which presents a first order mechanism with regard to all reaction components. For analytical solution is Laplace transformation used. The designed mathematical model is verified by an experimental measurement. Experimental device is supplemented by visualization program which is designed in programming environment Control Web. Archived data are compared with actual measured values.*

Key words: *amaranth, anaerobic decay, amaranth mass exploitation, modeling*

1. INTRODUCTION

Amaranth [Amaranthus hypochondriacus, A. cruentus (grain), A. tricolor (vegetable)] is a plant with an up-right growth habit, cultivated for both its seeds, which are used as grain, and its leaves, which are used as a vegetable. Both the leaves and seeds contain protein of an unusually high quality. The grains are milled for flour or popped like popcorn. The leaves of both the grain and vegetable types may be eaten raw or cooked [AMR Amaranth, 2006].

Amaranth produces many thousands of tiny seeds having very valuable contents. The crop has high nutritional values, contains remarkable amounts of protein with high portion of essential amino-acids and other considerable substances, like squalene and flavonoids (rutin). As a result, amaranth starts to be exploited especially in the food industry to produce new products, conducive against civilization diseases. Amaranth has a wide variety of applications in the food industry, it can be used in a number of food products including breakfast cereals, confectionery products, salad condiments, baked products, etc. This utilization is dominant [Kodrikova, 2004].

On the other hand, amaranth starts to be used also in medicine, pharmaceutics and cosmetics. Research is in progress, for example in the area of amaranth biomass treatment where gaps in present amaranth exploitation are. Fermentation of amaranth mass runs in anaerobic decay, however this way of amaranth exploitation is not often used. Therefore we are interested in experiments which prove the possibility of amaranth biogas production.

It is known, that the description of fermentation is possible by mathematical equations of the decay process. We assumed the mechanism of four-level decomposition, which is depicted by a vector equation. The formula presents a first-order mechanism with regard to all reaction components. For analytical solution of equation system we used Laplace transformation [Zábranská, 1997]. The experimental measurement runs in experiments for gas production monitoring we compiled in laboratory conditions. As a source of anaerobic bacteria we used anaerobic sludge from the sewerage plant. The purpose of our present experiments is to prove that amaranth biomass is suitable for anaerobic decomposition and that the produced gas can be successfully employed in practice.

2. MATHEMATICAL MODEL OF ANAEROBIC DECAY

The purpose of the article is to check out the anaerobic decay efficiency of amaranth in laboratory conditions. Generally, anaerobic decay proceeds in four stadiums, therefore mathematical model of four-level decomposition is used.

Mathematical simulation of anaerobic decay of amaranth biomass will progress in these presumptive steps:

•degradation
•hydrolysis
•acetolysis
•methane genesis.

This could be expressed in the following scheme:

$$A \xrightarrow{k_1} B \xrightarrow{k_2} C \xrightarrow{k_3} D \xrightarrow{k_4} E$$

where k1, k2, k3, k4 are speed constants each subsequent reaction. The constants dependent on the sludge activity. In the first approach the mechanisms of the first degree was contemplated. In this presumption the following system of differential equations will be applied:

$$\frac{dc_A}{d\tau} = -k_1 c_A \tag{1}$$

$$\frac{dc_B}{d\tau} = k_1 c_A - k_2 c_B \tag{2}$$

$$\frac{dc_c}{d\tau} = k_2 c_B - k_3 c_c \tag{3}$$

$$\frac{dc_D}{d\tau} = k_3 c_C - k_4 c_D \tag{4}$$

$$\frac{dc_E}{d\tau} = k_4 c_E \tag{5}$$

$$\dot{C} = A \cdot C \tag{6}$$

where $C_{A,B,C,D,E}$ are concentrations of particular compounds. Equation (6) is a vector differential equation with consecutive initial conditions:

$$\begin{bmatrix} c_A(0) \\ c_B(0) \\ c_C(0) \\ c_D(0) \\ c_E(0) \end{bmatrix} = \begin{bmatrix} c_{A0} \\ 0 \\ 0 \\ 0 \\ 0 \end{bmatrix} = C(0) = C_0 \tag{7}$$

Consequently Laplace transformation is solved, which leads to these results:

$$c_A = c_{A0} e^{-k\tau} \tag{8}$$

$$c_B = \frac{c_{A0} k_1}{k_1 - k_2} \left(e^{k_L \tau} - e^{-k_1 \tau} \right) \tag{9}$$

$$c_C = c_{A0} k_1 k_2 \left[\frac{e^{-k_1 \tau}}{(k_3 - k_1)(k_2 - k_1)} + \frac{e^{-k_2 \tau}}{(k_3 - k_2)(k_1 - k_2)} \right] \tag{10}$$

$$c_D = k_3 k_2 k_1 c_{A0} \left[\frac{e^{-k_4 \tau}}{(k_2 - k_4)(k_1 - k_4)(k_3 - k_4)} + \frac{e^{-k_2 \tau}}{(k_4 - k_2)(k_3 - k_2)(k_1 - k_2)} \right] + \left[\frac{e^{-k_3 \tau}}{(k_4 - k_3)(k_2 - k_3)(k_1 - k_3)} + \frac{e^{-k_1 \tau}}{(k_4 - k_1)(k_3 - k_1)(k_2 - k_1)} \right] \tag{11}$$

$$c_E = 1 + k_4 k_3 k_2 k_1 c_{A0} \left[\frac{e^{-k_4 \tau}}{-k_4 (k_3 - k_4)(k_2 - k_4)(k_1 - k_4)} + \frac{e^{-k_3 \tau}}{-k_3 (k_4 - k_3)(k_2 - k_3)(k_1 - k_3)} \right] + \left[\frac{e^{-k_2 \tau}}{-k_2 (k_4 - k_2)(k_3 - k_2)(k_1 - k_2)} + \frac{e^{-k_1 \tau}}{-k_1 (k_4 - k_1)(k_3 - k_1)(k_2 - k_1)} \right] \tag{12}$$

3. EXPERIMENTAL SET-UP

The mathematical model of anaerobic decay was verified by the help of experimental equipment which is shown in Figure 1. The reactor with a magnetic stirrer was filled with the mixture of anaerobic sludge and biomedium, then the substrate was added and saturated with water so the inlet of inert gas was dipped. The reactor was hermetically closed and aerated with nitrogen for 15 minutes, which provided oxygen expulsion from water liquor and performance of anaerobic conditions. Produced gas was collected and drained through the bubble flow meter to ambient air. The sensor was composed of a glass tube, which was encased in a chuck from organic glass. The inlet of gas measuring through a nozzle needle was fixed on the chuck and leads below the liquid level. The bubbles were formed by the low flow rates in the issue of the jet. The needle was placed to the gas inlet hence regular separation of commensurate bubbles. These were photoelectric detected by virtue of a photodiode and phototransistor and added to figures on the counter. The bubble flowmeter was volumetric gauged. The scaler was connected to a computer, where the process of the measurement was registered via program Control Web. The bioreactor was maintained with constant temperature 23°C and the content was stirred with 50 rpm.

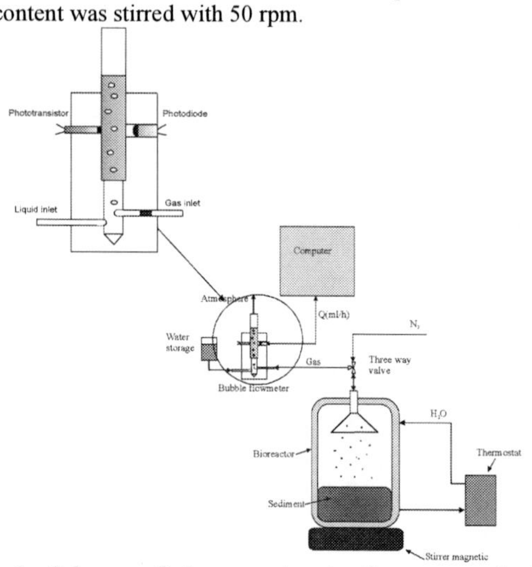

Fig. 1. Scheme of the experiments for gas productions monitoring

In the mathematical model in Figure 2, which was designed in computer program Mathematica 4, the time dependence of concentration by the four-level model can be seen. Methanogenesis (the last step of anaerobic decay) represents the curve c_E. It was proved as the slowliest part of the decomposition. Therefore the speed of methanogenesis is the most important part of the reaction.

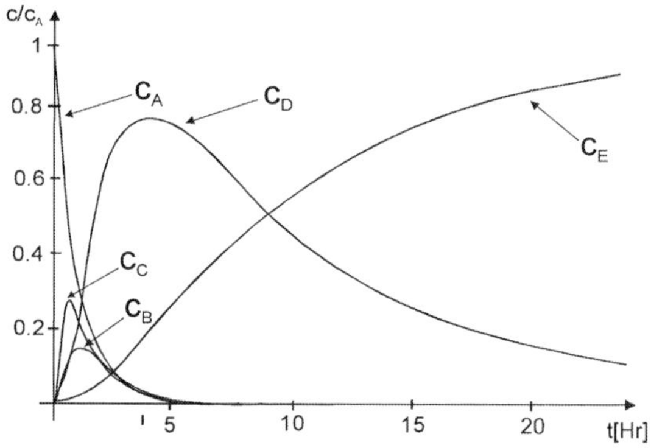

Fig. 2. Time dependence initial substance, intermedi-ate product and product concentration by the four-level model, where constants were chosen as k1= 1, k2= 2, k3 = 3, k4 = 0,5

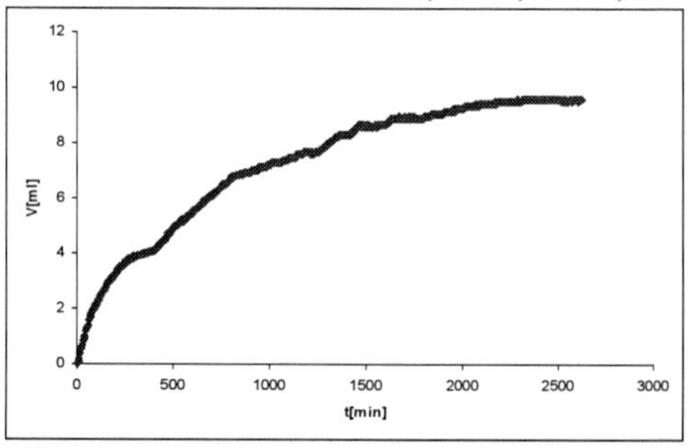

Fig. 3. The time depandance of amaranth mass anaerobic decay

As can be seen in the graph above, the decomposition of amaranth biomass grows with time. At the beginning the speed of methanogenesis is higher, than it slows down according to activity of the anaerobic methanogene bacteria. After the sludge is exhausted, the biogas production stops. This process takes several days.

4. CONCLUSION

In this paper, the efficiency of anaerobic decay of sodium acetate and amaranth mass and its time behavior were measured using the photodiode and phototransistor. A constant momentary reaction speed during all the reactions, including amaranth mass, was proved with the same value. This fact shows that the speed of methanogenesis is probably the most important part of the reaction. Finally, the experiments verified that amaranth biomass is suitable for anaerobic decomposition and therefore to biogas production.

5. REFERENCES

AMR AMARANTH a.s. [online], 2006-[cit. January 10 2006], available on WWW: http://www.amaranth.cz

KODRIKOVÁ K. 2004. Evaluation of amaranth components and research of the polysaccharide component fermentation to fractional fission products, Master thesis, UTB Zlín, 2004.

ZÁBRANSKÁ J. 1997. Laboratory methods in water technology. VŠCHT Prague, 1997

Acknowledgment

The work has been supported by the Ministry of Education of the Czech Republic under grant MSM 7088352102. This support is very gratefully acknowledged.

Annals of DAAAM for 2006 & Proceedings of the 17th International DAAAM Symposium, ISSN 1726-9679
B. Katalinic, Published by DAAAM International, Vienna, Austria 2006
150th Birthday of Nikola Tesla (1856-1943) Inventor of Mobile Robot Teleoperation

FUZZY MODELS FOR CAPACITATIVE VEHICLE ROUTING PROBLEMS IN UNCERTAINTY

KONDRATENKO, G.; KONDRATENKO, Y. & ROMANOV, D.

Abstract: *This paper deals with solving capacitate vehicle routing problem (CVRP) in uncertainty. The problem concern the bunkering tankers operations in the various ports and open sea for fuzzy demands of served ships and for restricted capacity of bunkering tankers. According to optimisation of routes planning each tanker should serve all ships on the its route with minimised the total length of route and maximised the total loading cargo (fuel) for served ships in real marine conditions as in fuzzy environment. It is suggested to use relatively input data for fuzzy logic algorithm with was designed for solving CVRP with fuzzy demands. Modelling results confirm the efficiency of suggested algorithms for CVRP decision-making based on the special fuzzy expert system.*
Key words: *fuzzy demands, bunkering CVRP, expert system*

1. INTRODUCTION TO BUNKERING CVRP

One of the most important vehicle routing problems in marine industry is routing problem for bunkering tankers (in further, tankers). Tankers should provide bunkering (transportation and unloaded) operations for various served (ordered) ships which can be located in different marine ports and in different open sea points. Marine practice shows that very often the information about fuel demands of served ships and ports is uncertain. As usual ship's customer sends to bunkering company the order about quantity of fuel supplying as approximate value. It is possible to represent such kind of orders as fuzzy demands, for example, as fuzzy numbers with triangular membership function. Taking in to account the restricted fuel capacity of each tanker and fuzzy demands of ordered ships the well-known vehicle routing problem will transfer to capacitate vehicle routing problem in uncertainty. The efficiency of preliminary bunkering operations planning as efficiency solving CVRP in uncertainty is evaluated by: a) possibility to serve all ordered ships with maximum possible quantity of unloaded fuel; b) minimum consumption of own tankers' fuel on the all routes which depends on the total length of tankers' routes.

The vehicle routing problem (VRP) with fuzzy demands at nodes is considered in (Teodorovic & Pavkovic, 1996) where decision-making is based on the heuristic "sweeping algorithm", on the set of fuzzy rules formed using fuzzy logic and on the use of preference value for comparing route planning results. Such approach is efficient if VRP don't consists any requirements, especially, to provide the maximum unloaded cargo at the all routes as in the case of bunkering CVRP. The solving CVRP for bunkering operations with modelling results was successfully considered in (Malakhov,V. & Kondratenko, G., 2002) and (Kondratenko, Y. & Werners, B., 2001). But in this published works the corresponding decision-making expert systems based on the fuzzy logic did not take in to account (on the last steps of each route planning) some relationships and dependencies between very important for decision-making parameters. First of all, it concerns such kind of parameters as: a) order of next node; b) remain fuel quantity at the tanker; c) tanker capacity and lengths of alternative routes and others. The new approach suggested by authors deals with the application of relative input signals for above-mentioned parameters as input data for developed decision-making expert system. Modelling results obtained by authors confirm the efficiency of suggested structure of expert system, high information level of relative input data and influence of preference value to current decision-making process. It is expediency from the authors point of view to consider in future a combination for general and relative signals as input data with corresponding correction of fuzzy rules for developed decision-making expert system.

2. STRUCTURE OF FUZZY EXPERT SYSTEM FOR SOLVING CVRP IN UNCERTAINTY

2.1 CVRP with fuzzy demands

Let's consider depot P_0 as port where bunkering company is located. The decision-maker of such bunkering company should solve tanker's CVRP before beginning of any bunkering operations. At the deport P_0 (Fig. 1) bunkering company have m tankers for bunkering operations at the served

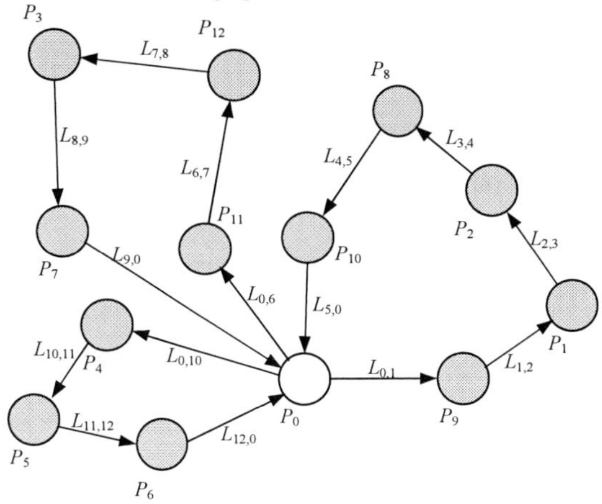

Fig. 1. CVRP with fuzzy demands of ships

marine region $(P_1 ... P_{12})$ and warehouse with satisfied quantity of fuel for any real complete set of demands of n ships which need fuel supplying for their further performance and mission continuation. Each i-th tanker (i = 1,.., m) is special and has the varies fuel capacity Q_i. The ships, which must be served by tankers, have various capacities S_j and various demands q_j $(j = 1, ..., n)$, respectively. Taking into account that above-mentioned customer's orders for all bunkered ships are uncertain each ship's demand can be presented as fuzzy number of triangular form (Malakhov, V. & Kondratenko, G., 2002)

$$\tilde{q}_j = \left(\underline{q}_j, \hat{q}_j, \overline{q}_j\right), \qquad (1)$$

where $\hat{q}_j$ is a value of demand with value of membership function equal 1; $\underline{q}_j$ and $\overline{q}_j$ are the lowest and highest possible values of demand, respectively.

2.2 Fuzzy algorithm for decision-making expert system

The suggested decision-making fuzzy expert system based on the various types of classical and modified algorithms for VRP (Laporte, 1992) may determine a current value of parameter λ. The criteria for positive decision about continue bunkering operations (about including next s-th port to current planning route) is (Kondratenko and Werners, 2001; Teodorovich and Pavkovich, 1996)

$$\lambda \geq \lambda^* \qquad (2)$$

where λ^* is desired value of satisfaction level (preference value); λ is a current value of satisfaction level which is calculated at the each served port as possibility level for next port serving taking in to account its fuzzy demand.

The developed fuzzy logic decision-making algorithm can be presented as follow

$$\lambda_j = FIE\left[FRB\left(\underset{\sim}{x}_1, \underset{\sim}{x}_2, x_3\right)\right], \qquad (3)$$

where λ_j is satisfaction level for each alternative decision-making; $\underset{\sim}{x}_1 = \frac{\tilde{q}_s}{\Delta\tilde{Q}_i}$ is relative unit between fuzzy demand of next s-th port $\tilde{q}_s$ (s=1,2,...,12) and fuzzy value of remain tanker cargo $\Delta\tilde{Q}_i = \left(Q_i - \sum_{j=1}^{k} \underline{q}_j, Q_i - \sum_{j=1}^{k} \hat{q}_j, Q_i - \sum_{j=1}^{k} \overline{q}_j\right)$ with k – number of served ships on the i-th route before decision-making process; $x_2 = \frac{\Delta\tilde{Q}_i}{Q_i}$ – relative unit between fuzzy numbers $\Delta\tilde{Q}_i = (\Delta\underline{Q}_i, \Delta\hat{Q}_i, \Delta\overline{Q}_i)$ and tanker capacity Q_i; $x_3 = \frac{L_1}{L_2}$ is relative unit between length of two alternative routes: route with decision "return to the depot" and route with decision "continued movement"; *FIE* – fuzzy inference engine; *FRB* – fuzzy rules base.

The structure of fuzzy rules is, for example, like this:

IF (input signal $\underset{\sim}{x}_1$ is *L*) **AND** (input signal $\underset{\sim}{x}_2$ is *M*) **AND** (input signal x_3 is *H*) **THEN** (output signal λ_j is *H*),

where *L*, *M* and *H* are the names of corresponding linguistic terms (Low, Middle, High).

2.3 Modelling results for CVRP solving

The modelling results for decision-making by developed expert system based on the fuzzy-logic structure as dependence between satisfaction level λ_j and third input variable x_3 for fixed fuzzy triangular number $\Delta\tilde{Q}_i$= (150, 300, 380) and various fuzzy triangular numbers $\tilde{q}_j = \left(\underline{q}_j, \hat{q}_j, \overline{q}_j\right)$ are presented in Fig. 2. Different cases are considered in Fig.2, in particular: *Line* 1 - for $\tilde{q}_s$ = (100, 200, 300); *Line* 2 - for $\tilde{q}_s$ = (200, 300, 400); *Line* 3 - for $\tilde{q}_s$ = (300, 400, 500); *Line* 4 - for $\tilde{q}_s$ = (400, 500, 600); *Line* 5 - for $\tilde{q}_s$ = (500, 600, 700).

The characteristic surface for determination of satisfaction level λ_j (by fuzzy expert system) for CVRP with fuzzy demands (for fixed value $x_2 = 0{,}8$) is presented in Fig.3.

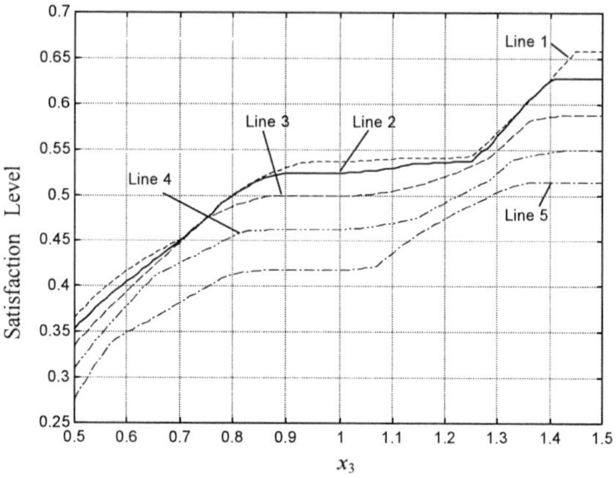

Fig. 2. Current value of satisfaction level λ_j as dependence from parameters $\underset{\sim}{x}_1$ and x_3 : modelling results

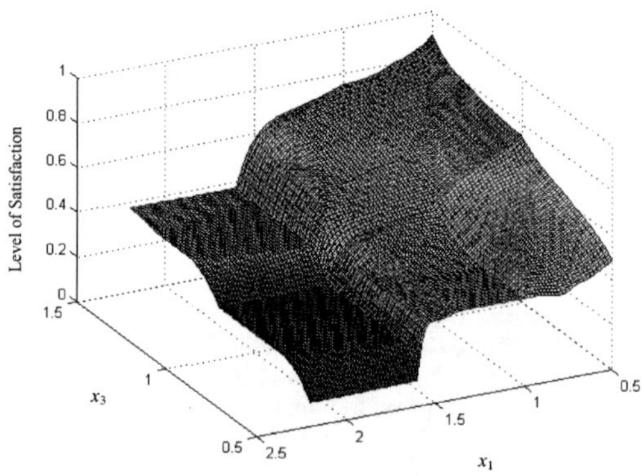

Fig. 3. Characteristic surface of fuzzy expert system

The main criteria for evaluation of bunkering process (BP) efficiency based on CVRP solving with capacity constraints and fuzzy demands of ships can be presented as

$$\max E_{BP} = \frac{1}{r}\sum_{i=1}^{r}\left(1 - \frac{\Delta Q_i^{remain}}{Q_i}\right), \qquad (4)$$

which can be transferred to multicriteria approach as follows

$$\Delta Q_i^{remain} \to \min, \quad (i = 1, 2, \ldots, r); \quad r \to \min,$$

where ΔQ_i^{remain} – remain fuel capacity before tanker returning to depot P_0 ; r – number of tanker's routes.

3. REFERENCES

Kondratenko Y. & Werners B. (2001). Solving routing problem in marine environment using fuzzy logic. *Proc. of Intern. Conf. on Modelling and Simulation* MS'2001, pp.109-110, 23-26 May, Lviv, Ukraine.

Malakhov, V. & Kondratenko, G. (2002) Simulation models and algorithms for planning routes and trajectories based on fuzzy logic. *Trans. of Odessa Polytechnic Univ.*, Vol.2 (18), pp.72-80, (in Ukrainian).

Laporte G. (1992). The Vehicle Routing Problem: An overview of exact and approximate algorithms. *European Journal of Operational Research*, Vol. 59, pp.345-348.

Teodorovic, D. & Pavkovich, G. (1996). The fuzzy set theory approach to the vehicle routing problem when demand at nodes is uncertain. *Fuzzy Sets and Systems*, Vol.82, pp.307 – 317.

Annals of DAAAM for 2006 & Proceedings of the 17th International DAAAM Symposium, ISSN 1726-9679
Editor B. Katalinic, Published by DAAAM International, Vienna, Austria 2006
150th Birthday of Nikola Tesla (1856-1943) Inventor of Mobile Robot Teleoperation

FLEXIBLE ASSEMBLY CELL'S SHELF STORAGE

KOSTAL, P. & VELISEK, K.

Abstract: *In this paper we want introduce shelf storage design for our manufacturing cell. This shelf storage contains from shelf and manipulator. This shelf storage is realized on base of our design. Now a day we are working on control system of this device. This control system will be at base of industrial PC equipped by several I/O cards with digital and analog ports.*

Key Words: *shelf storage, flexible manufacturing, assembly, manipulator, industrial robot*

1. INTRODUCTION

In industry are used several types of manipulators very often. They can change a lot of manpower with high efficiency on industrial production.
Today electronic systems are less expensive and have a lot of possibilities in controlling. One of these extended possibilities of electronic control systems opposite to mechanical controlling are:

- Modularity,
- Flexibility,
- Lower buying cost,
- Lower cost to program change,
- Easy reprogramming,
- Higher precision and accuracy.

he basic division of manipulators we can doing on base of its functionality:

- **One-purpose manipulators** – these manipulators are part of machine. The functions of these manipulators are usually very simply. By these manipulators we can realize only material feeding for machine.
- **Universal manipulators** – these manipulators are designed as standalone device. The power supply and control system are isolated from machine, but can communicate with machine control system.

The other important data about a manipulator are number, type and location of his working axis. On base of kinematic structure we can divide manipulators to four basic groups:

- One rotational and two translation motions – cylindrical workspace,
- Two rotational and one-translation motions – spherical workplace,
- Tree rotational motions – angular workspace,
- Tree translation motions – Cartesian workspace

2. SHELFSTORAGE'S MANIPULATOR

Shelf storage manipulator has Cartesian workspace and tree DOF. The manipulated workpiece together with system palette will has less than 3 kg. The manipulator will take this pallets (together with workpiece) to rotational feeding device (interface device between assembly cell and shelf storage) and giving its to shelf storage cells. Movement along X and Y-axis are realized by pneumatic cylinders. Movement along Z axis is realized by electrical stepper motor with possibility of usage preprogrammed positions addressed by 4 bits.
In the shelf storage design time we must accept give conditions, which determine its structure. Space for shelf storage and its kinematic structure wee can see at Fig. 1.
The shelf contain 3 rows and 4 columns, we have 12 cells. The assembled shelf storage and manipulator we can see at Fig. 4.

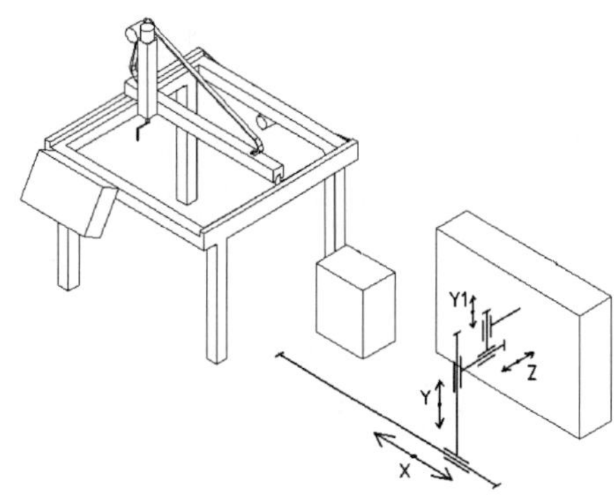

Fig. 1. Kinematic structure of shelf storage

The shelf storage is assembled from extruded aluminum profiles. This type of design has some advantages:

- elimination of mechanical and thermal stresses after welding,
- high strength of construction,
- modular construction, easy and quick assembly,
- don't need surface finishing (painting),
- smaller weight as steel construction,
- easy application several types of sensors without drilling to construction.

Fig. 1. Assembled shelf storage with manipulator

2.1 Sensors in shelf storage

The shelf will be equipped by sensors for palette detection. These sensors will detect not only "the palette is present or not", but detect which palette is present. This detections are realized by four switches at each cells of shelf. Each pallets has a binary combination of four pins. The system know 15 pallets, the combination of four zeros means "palette is not present". This scheme of sensors will be assembled in manipulator's end effector and in rotational feeding device too. The pallets addressing space can be expanded to 5 bits. After expansion of address range the system will know 31 pallets.
The control system of flexible manufacturing cell will controlling the shelf storage and its manipulator too. Control system has actual informations about each pallets contents every time.
The all pallets has justified pins. These pins allows the exact position of pallets at shelf, at manipulator's end effector and at feeding device too.
Motion along X axis is realized by pneumatic driver. This motion must be able stopping at 5 position (5 columns of shelf) with high accuracy. This task are solved by pneumatic stopping cylinder and 5 "**V**" shape undercuts. All these undercuts are between two magnetic sensors. These sensors says to system that position of manipulators column is near to needed position and motion sped must be down creased.
Motion along a vertical axis (Y axis at Fig. 1) is realized by electric stepper drive which controller can be programmed by 4 bits address to achieve up to 16 exactly defined positions.
At all needed positions along Y axis will place the optical sensors as feedback to position achieving.
Motion along a second horizontal axis (Z axis at Fig. 1) will be realize by simply pneumatic cylinder. This motion has only two end points. At these endpoints will be assembled the magnetic sensors as feedback .

2.2 End effector of manipulator

The manipulator has passive end effector – gripper. This gripper has two holes for justified pins of pallets. In loading and unloading time of pallets, the gripper realize some micro movement along a vertical axis (Y1 at Fig. 1). This movement unlock the pallets from its position (giving out justified pins from its holes). This moving will realize by pneumatic driver.
The gripper is equipped by switches for pallets identification too.
The whole pallets with its contents will be fold by manipulator from shelf to feeding device and backward. The feeding device give the pallets from shelf storage manipulator workspace to workspace of manufacturing cell.
Model of end effector is at Fig. 3.

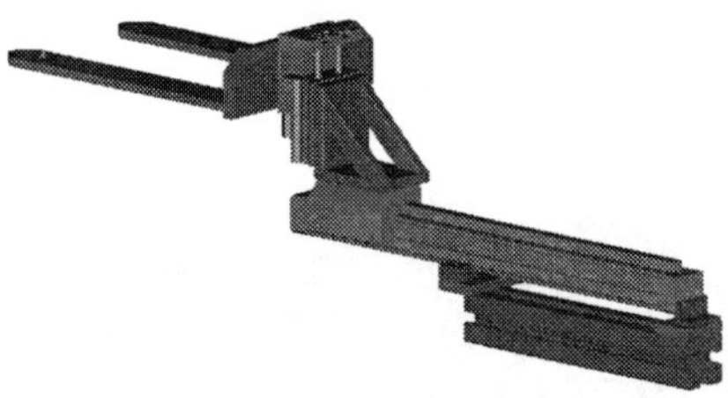

Fig. 3. Model of storage manipulator's end effector

3. COMPLETE MANUFACTURING SYSTEM

The manufacturing cell is based on industrial robot by Cartesian coordinate system. Frame of robot are made from extruded aluminum profiles. Electric stepping motors realize drives of axes. Repeatability of positioning is better than 0,0125 mm and loading may be up to 100N. Workspace is c 1000x1000x350 mm.
Preliminary CAD model of manufacturing system is at Fig. 4, Photo of current stay of building is at Fig. 5
The control system of this robot is realized by PLC from Mitsubishi. In this time we can control 6 axes. Number of controlled axis are sufficient for industrial robot, but we want

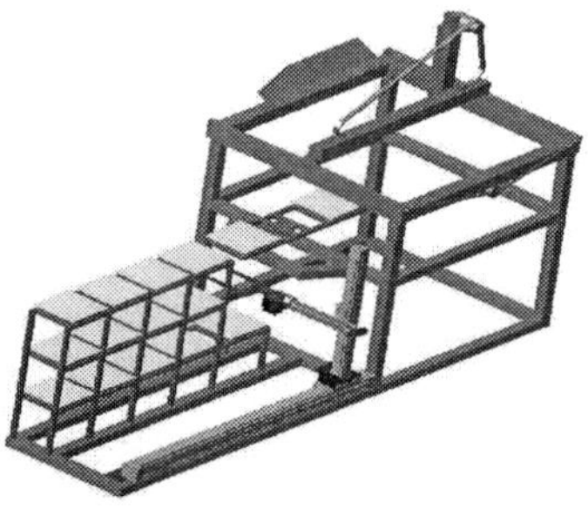

Fig. 4. Model of flexible manufacturing system and shelf storage with manipulator

use this control system to control of all devices (shelf storage, storage manipulator, feeding device, fixture, ...). That means we must change this control system.
The new control system will be at base of industrial PC equipped by industrial I/O cards with lot of input and output channels.

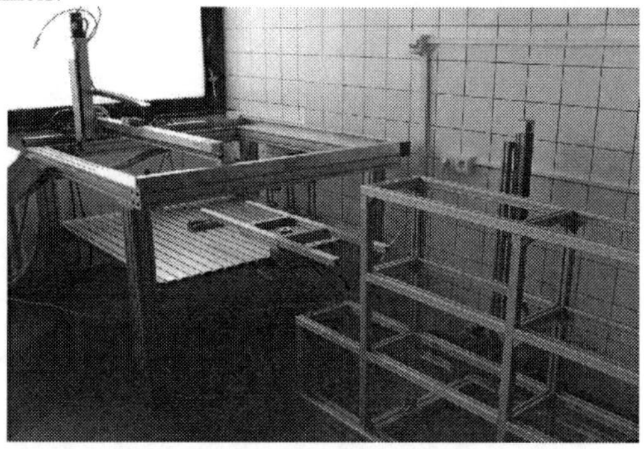

Fig. 2. View to manufacturing cell with shelf storage and feeding device

4. CONCLUSION

Now days are shelf storage and storage manipulator assembled. We working on control system which will be able to control the whole manufacturing system.
In future we want integrate the control system of shelf storage and control system of assembly cell to one control system based on industrial PC.

5. REFERENCES

Javorova,A., Katalinic, B. (2004). Flexible assembly workstation, *Proceedings of 15th International DAAAM Symposium*, Katalinic, B. (Ed.),pp. 191-192, ISSN 1726-9679, 3-6th November 2004, Vienna, Austria

Košťál, P. Matúšová,M. (2001). Teoretické otázky ustavenia a upnutia, *Proceedings of Vedecké práce MtF STU*, Zväzok 10., pp.53-56, Bratislava, Slovakia

Matúšová,M.(2002). Činitele pre výpočet upínacích síl, *Proceedings of 8. Vedecká konferencia s medzinárodnou účasťou Akademická Dubnica 2002*, pp 229, November 2002, Dubnica nad Váhom , Slovakia

Danišová, N. Zvolenský, R. (2006). Automated flexible manufacturing systems, *Proceedings 4th International symposium KOD 2006*, pp.253-256, ISBN 86-85211-92-1 Palič, Serbia & Monte Negro.

Annals of DAAAM for 2006 & Proceedings of the 17th International DAAAM Symposium, ISSN 1726-9679
Editor B. Katalinic, Published by DAAAM International, Vienna, Austria 2006
150th Birthday of Nikola Tesla (1856-1943) Inventor of Mobile Robot Teleoperation

CASE STUDY OF PROJECT MANAGEMENT PROCESS IMPROVEMENT

KRAMARENKO, S. & KARAULOVA, T.

Abstract: *This paper discusses project management process and its improvement through process modelling, simulation and analysis of processes. A process model contains representations of individual facts, objects, and the relationships that occur within an enterprise. We will focus our efforts on the management of an investment project on one of the biggest enterprises in Estonia BLRT Grupp AS.*
Key words: *process modelling; project management; business process; investment project.*

1. INTRODUCTION

In today's business environment managers are trying to find new approaches to improve their organisation's performance. Therefore they need to manage their sources of competitiveness effectively.
The objective of this research was to improve management of investment projects and to define the main criterions of its estimation. The problem was studied on the *BLRT Grupp AS*, where a great deal of works haves the status of projects. The term "improvement" implies: minimizing project lead-time (for separate works) and decrease in the expenditures on the process (cost of working time of each participant) with the same (or even higher) level of productivity.
BLRT Grupp AS development centre *BLRT Kinnisvara (*KV) is engaged in investment project management. Investment Projects (IP) is deployed in order to solve any technical problems, which need centralized financing in assets. The investment project management means planning, implementation and analysis of a project's effectiveness. The principle scheme of project relationships within an enterprise is shown in Fig. 1.

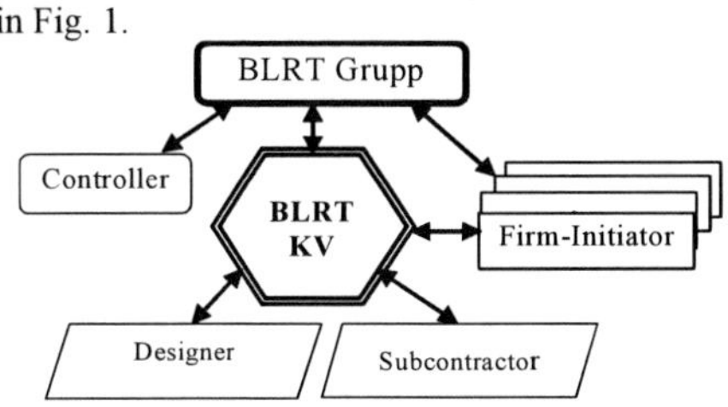

Fig.1. Project relationships at the BLRT Group

2. OBSERVATION OF PROBLEMS

BLRT Grupp AS is one of the biggest enterprises not only in Estonia, but also among countries of the Baltic Sea basin. It combines over 20 daughter enterprises. Every branch establishment conducts its own business activity: ship repair and shipbuilding, machine building, fabrication of steel structures, stevedoring, rolling stock repairing and etc. And each of these companies has its own outcomes. One company, for example machine building, with hundreds of workers could achieve the balance between loss and profit only because of lack of metal. Another one, a relatively small metal trading company was not informed about the need of materials by that company. Hence, the prosperity of a group depends not only upon a successful activity of each concern's firm, but also upon the level of co-operation between all of them. The level of an enterprise's collective work is characterised by the degree of its infrastructure development, especially informational. It means that today's development strategy for any enterprise must include the development of whether an ideal or at least ensuring competitiveness information infrastructure. That depends neither on the extent of the enterprise nor on its position at the world market. Managers of the BLRT Grupp AS take this fact into account. That's why the investments directed to the development of information systems increase from year to year. These features (obstacles) cause the following problems in project management:

- Projects are usually out of the scheduled execution time (a lot of routine work);
- Projects exceed their budgets;
- A project (portfolio of projects) performance decreases.

Sometimes these problems cause even collapse of projects. Standardization and integration was the key to the successful future operation, especially with respect to project work breakdown structure and internal workflows (Pitagorsky, 2003). The project management objective, contribute to the optimisation of the business case of the investment, initialised by the project, is of great concern in product development and capital investment projects and of less concern in contracting projects (Morris & Pinto, 2004).

3. MODELLING OF THE PROJECT PROCESSES

Process modelling involves modelling of processes in a project and can often include the data and material that flows between them. Visual models are commonly produced through mapping conceptual data and process models.
Models provide "blueprint" of functions and their interfaces that must be captured and understood in order to make systems engineering decisions that are logical, integratable, and achievable, to provide an approach to (Morris & Pinto, 2004):

- performing system analysis and design at all levels, for the systems composed of people, machines, materials, computers, and information of all varieties – the entire enterprise, a system or a subject area;
- producing the reference documentation concurrent with development to serve as a basis for integrating new systems or improving the existing ones;
- communicating among analysts, designers, users, and managers;
- allowing team consensus to be achieved by shared understanding;
- managing large and complex projects using qualitative and quantitative measures of progress;
- providing reference architecture for enterprise analysis, information engineering and resource management.

For process modelling we used the IDEF standard. The modelling language makes itself the purpose of particular activity explicit.

The IDEF0 model, introduced in Fig. 2 was used to improve the investment project management through a detailed description of each activity included in the process of the project management. On the base of the IDEF0 model it is possible to actualise the simulation of the process.

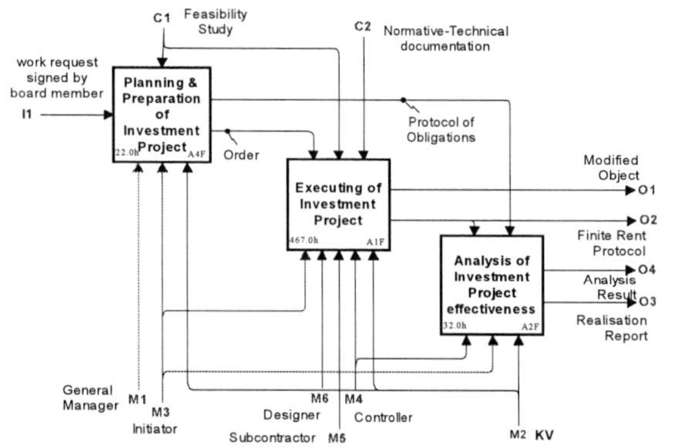

Fig.2. The main function of the investment project management

The ability of the simulation model to analyse real systems yields a far more useful tool than simple analytical techniques. This step is also important during system design to assist decision-making by providing performance measures, presented in simple tables and charts, for each alternative solution (Perera, 2000). The amount of data that must be stored and manipulated for most real-world systems dictates discrete-event simulations (Law, 2000).

The analysis of the model AS-IS shows that the most expenses of time and cost have information collecting, analysing and releasing. Information treatment is an integral part of whatever processes, especially project management, and it is vitally important to help a project manager to get all the needed information. The present business technology facilitates information management by automating such procedures like planning, scheduling and different paperwork. There are a lot of assistance tools for this purpose. And one of the most widespread applications is Microsoft Project.

In comparison with the existing (AS-IS process) TO-BE process have two new main activities: *Preparing of Investment Project* and *Validation of the Project Request* before process *Planning & Preparation of Investment Project* (see Fig. 3).

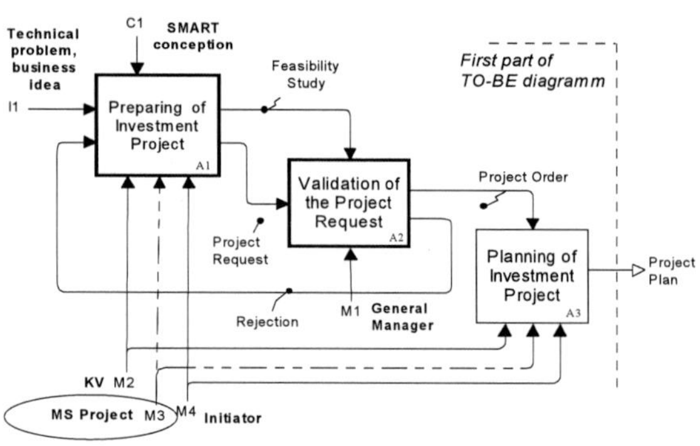

Fig. 3 A fragment of a TO-BE diagram

4. RESULTS OF IMPROVEMENT

Comparing the results of AS-IS and TO-BE simulations it is seen, that the total number of the implemented projects within a year has increased. From the results of the simulation it can be noticed: 6 projects have already achieved a final phase – works acceptance; General Manager is considering the 9th project.

The duration and cost levels of both models were:

- The average cost of the TO-BE model equals €71 compared with the €110 of the AS-IS model.
- The average duration of the TO-BE model equals 5 hours, whereas for the AS-IS model it was 7 hours.

The summary of both models is introduced in Table 1. The difference and percentage of the improvement demonstrate the expediency of the improvements performed in this work. All the data in Table 1 is calculated for one investment project.

Name of parameter	*AS-IS*	*TO-BE*	*Difference, +/-*
Total project time, hours	519	484	- 35
Cycle time, hours	260	256	- 4
Cost of the project, euro	6390	5552	- 838
Work-time of each member, %			*Improved, %*
Project manager	48	47	2.1
General Manager	4.67	2.15	54.0
Initiator	28.67	28.85	0.5
Controller	8.13	4.92	39.5

Table 1. Summary table of the basic process parameters

Collaboration is the business model making process-orientation effective. This optimises joint planning, managing, monitoring and controlling of all business processes. The key principle is: "You can only manage what you can measure!" Business strategy determines which business processes are to be executed and managed by the enterprise (Wolfgang, 2003).

5. CONCLUSION

A real model of the investment project management of a living and well developing company was represented in the paper. All the activities of the BLRT Grupp AS and participants of the project, as well as their functions, were defined, which enabled to create a functional model (AS-IS) of the process (investment project management) by means of *Meta Software*. The static AS-IS model was reorganized into the simulation model, which provided the structural analysis of the process. The modelling and simulation techniques, based on the IDEF standard, are proper tools for the early estimation of the system performance and may be used to support solution of practical problems in industry. In the paper we introduced the way to improve the management process in the BLRT Grupp AS by integrating into it the MS Project Management System. The process analysis is performed by the same criterions: resource utilization, the work-time allocation, cost, duration and effectiveness of the process. The estimation of improvements (results) could make a beneficial effect on the *BLRT Grupp AS*.

Acknowledgements Hereby we would like to thank the Estonian Science Foundation for the grant 6795 enabling us to carry out this work.

6. REFERENCES

Morris, P. & Pinto, J. (2004) *The Wiley Guide to Managing Projects*, ISBN 0-471-23302-1, John Wiley & Sons, Inc.

Perera, T. & Liyanage, K. (2000) – Methodology for Rapid Identification and Collection of Input Data in the Simulation of Manufacturing Systems, *Simulation Practice and Theory, Volume 7, Issue 7*, pp. 645-656

Law, A. M. & Kelton, W. D. (2000) – Simulation Modeling and Analysis. *ISBN 0-07-116537-1 McGraw-Hill Higher Education, Inc. New York, 2000*

Wolfgang, M. *Business Performance Management*, IT Research Note, 2003

Pitagorsky, G (2003) – Managing projects the right way: Key principles for successful projects.

Annals of DAAAM for 2006 & Proceedings of the 17th International DAAAM Symposium, ISSN 1726-9679
B. Katalinic, Published by DAAAM International, Vienna, Austria 2006
150th Birthday of Nikola Tesla (1856-1943) Inventor of Mobile Robot Teleoperation

MODEL FOR SURFACE ROUGHNESS IN HIGH SPEED FACE MILLING

KULENOVIC, M.; CEKIC, A. & HALILAGIC, D.

Abstract: *This paper presents the mathematical model for predict roughness of machined surface during high speed face milling. In order to find the most suitable mathematical model to describe the depending factors of the surface roughness the following cutting parameters have been taken into consideration: the cutting speed V_c and the feed per tooth f_z. The experiments are performed during machining the carbon steel Č.1737 on the machining system VBS. By combining experimental results and the adequate mathematical model the dependence of surface roughness of the cutting parameters in analytical form is done. Experimental results are compared with calculating results.*
Key words: *High-speed face milling, surface roughness, cutting speed, feed per tooth, mathematical model.*

1. INTRODUCTION

High speed machining is an advanced manufacturing technology with a great future potential (Schulz, 1996). Since, there is no unique definition of high speed machining because it is defined differently from the different authors, high speed machining can be defined as machining with the feed rate and cutting speed having a high material removal ratio without degrading the quality and precision of the machined surface (Kim et al., 2002).
The significance for high speed machining and especially high speed milling in production has increased since new machines and tools enable the possibility to reduce process time on one hand and to improve work piece accuracy and work piece surface on the other hand (Dudzinski et al., 2002).
The main purpose of this paper is that the suited planning of experiment and applying of regression analysis to determinate the dependence of the surface roughness from cutting speed and feed per tooth during high speed face milling of the carbon steel Č.1737.

2. EXPERIMENT CONDITION

The milling investigations were conducted on a machining system type VBS constructed and built by Laboratory for Cutting Metal and Design Tool of Mechanical Faculty in Sarajevo (Fig. 1). Motor spindle HS type HSM 090-40000 allows maximum revolutions up to 40000 rpm. The worktable is driven about 2 axes by two "SIEMENS" linear motors and monitored by SINUMERIK 840D. For implementation the experimental investigations as the cutting tool using a cylindrical milling tool R390-020A20-11M with exchangeable inserts of hard metal – manufacturer SANDVIK Coromant (www.sandvik.com). The cutting tool is balanced for speed up to 34600 rpm. The cutting tool is clamped to the motor spindle HSM by a tool interface HSK 40E, and it is clamped to the tool interface HSK by hydraulic pump providing a pressure of 700 bar.
The machining process was carried out such as always just one tooth of cutting tool was in engagement with work piece.

Fig. 1. Machining system VBS

Width of cut and depth of cut were constant, 8 mm and 0,1mm respectively. Cutting tests were carried out under dry machining conditions. A new cutting edge was used for each machining experiment. The experiments were carried out in down-cut milling of carbon steel Č.1737 (internal mark G70). Mechanical properties and chemical composite of the examined steel (Tab.1). For measuring roughness of machined surface the modern device Perthometer Concept is used.

Mechanical properties							
Tensile strength (N/mm^2)			Yield strength (N/mm^2)		Hardness HRc		
734,71			506,13		30		
Chemical composite, %							
C	Si	Mn	Cr	S	P	Cu	Ni
0,68	0,25	0,78	0,10	0,02	0,02	0,08	0,03

Table 1. Mechanical properties and chemical composite of the carbon steel Č.1737 (G70)

3. MATHEMATICAL MODELING OF PROCESS

Mathematical model for calculating of surface roughness is approximated with second order polynomial expression:

$$Y_i = R_a = b_0 + b_1 X_1 + b_2 X_2 + b_{12} X_1 X_2 + b_{11} X_1^2 + b_{22} X_2^2, \quad (1)$$

where are:

R_a, μm - arithmetical mean deviation,
$b_0, b_1, b_2, b_{12}, b_{11}, b_{22}$ – model coefficients which would be determined after experiment was executed,
Xi ($X_1 = V_c$; $X_2 = f_z$) - variables.

Namely, bellow parameters, which significantly influence to surface roughness during high speed milling, are:

- cutting speed V_c, m/min and
- feed per tooth f_z, mm

Whereas experiment is executed with two factors, total number of experiments which are needful (based on central composition plan) is (Jurković, 1999):

$$N_e = n_k + n_\alpha + n_0 = 2^k + 2k + n_0 = 2^2 + 2 \cdot 2 + 5 = 13 \quad (2)$$

where are:

N_e – total number of experiments,
k – number of basic factors, k = 2
n_k – number of first order plan points, $n_k = 2^k$
n_α – number of plan points on central axis, $n_\alpha = 2k = 4$
n_0 – number of plan middle points, $n_0 = 5$

Previous analysis of the significance estimation of coefficients from the model (1) shows that coefficients b_2, b_{12} are insignificant. In this reason new model for surface roughness is defined by:

$$Y_i = R_a = b_0 + b_1 X_1 + b_{11} X_1^2 + b_{22} X_2^2 \quad (3)$$

Mathematical model (3), based on calculated coefficients, can be written as:

$$R_a = 0{,}6872 - 0{,}00108 \cdot X_1 + 4{,}907 \cdot 10^{-7} \cdot X_1^2 + 20{,}223 \cdot X_2^2 \quad (4)$$

Results of investigation of roughness parameter R_a for some machining parameters (cutting speed and feed per tooth) are given in table 2.

N_e	Variables		Results of research	
	V_c (m/min)	f_z (mm)	R_a^E (µm)	R_a^M (µm)
1	750	0,05	0,199	0,199
2	1250	0,05	0,188	0,148
3	750	0,15	0,649	0,604
4	1250	0,15	0,570	0,552
5	1000	0,10	0,290	0,294
6	1000	0,10	0,250	0,294
7	1000	0,10	0,30	0,294
8	1000	0,10	0,280	0,294
9	1000	0,10	0,270	0,294
10	650	0,10	0,370	0,391
11	1350	0,10	0,287	0,318
12	1000	0,03	0,150	0,110
13	1000	0,17	0,670	0,676

Table 2. Testing results of roughness parameter R_a (R_a^E - experimental value; R_a^M - model value)

Coefficient of determination R^2 = 0,986 shows very good interactive dependence between variables X_i and roughness parameter R_a.

Graphical interpretation of computed values of roughness parameter R_a for determined mathematical model (4) is done in figure 2. Graphical presentation is done mathematical dependence of roughness parameter R_a from cutting speed V_c and feed per tooth f_z.

On this way, mathematical model (4) with good accuracy and reliability describes of surface roughness in function of cutting speed and feed per tooth during high speed face milling of carbon steel Č.1737 in experimental domain, figure 3.

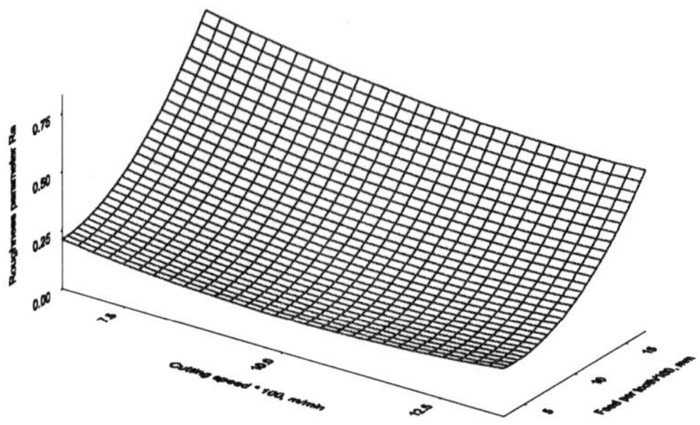

Fig.2. Graphical representation of mathematical model (4) for the surface roughness in high speed face milling of steel Č.1737

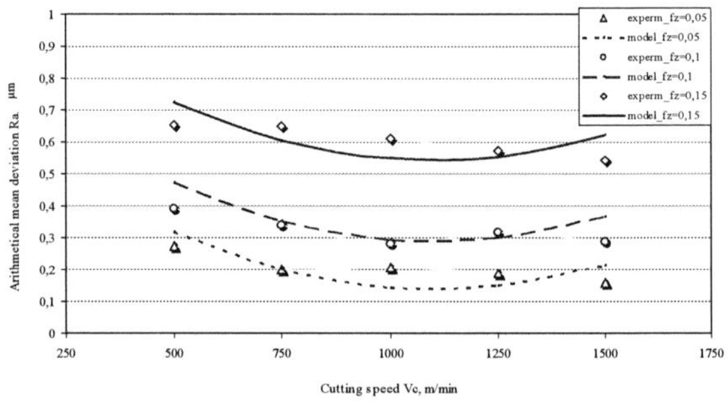

Fig.3. Comparison of the experimental results by the calculation results of the mathematical model (4)

4. CONCLUSION

Presented mathematical model to adequate describe dependence of roughness parameter R_a from cutting speed and feed per tooth. According to the given results in Tab. 2 and diagrams in fig. 2 and 3, it is possible to derive the following conclusions:

- Surface roughness decreases with increasing the cutting speed.
- Feed per tooth has an important impact on the surface roughness. Surface roughness increases when feed per tooth increases.
- Mathematical model (4) good describes of surface roughness in high speed face milling of carbon steel Č.1737 in experimental domain.

Authors of this paper are planning development of mathematical model for roughness surface as function of machining parameters, materials of workpiece and tool and etc.

5. REFERENCES

Catalog Sandvik Coromant, *Available from*: http://www.sandvik.com, 2002

Dudzinski, D.; Molinari, A. & Schulz, H. (2002). *Metal Cutting and High Speed Machining*, Kluwer Academic/ Plenum Publishers, ISBN 0-306-46725-9, New York

Jurkovic, M. (1999). *Matematicko modeliranje inzenjerskih procesa i sistema*, Mašinski fakultet Bihac, ISBN 9958-624-04-4, Bihac

Kim Kyoon, K.; Kang Chang, M.; Kim Suk, J.; Jung Ho, Y. & Kim K. (2002). A study on the precision machine ability of ball end milling by cutting speed optimization. *Journal of Materials Processing Technology*, pp (357-362)

Schulz, H. (1996). *Hochgeschwindigkeits-bearbeitung*, Carl Hanser Verlag Munchen,ISBN 3-446-18796-0, Wein.

Annals of DAAAM for 2006 & Proceedings of the 17th International DAAAM Symposium, ISSN 1726-9679
B. Katalinic, Published by DAAAM International, Vienna, Austria 2006
150th Birthday of Nikola Tesla (1856-1943) Inventor of Mobile Robot Teleoperation

DESIGNING EXPERIMENTAL FRAMEWORK FOR MULTI-AGENT AUTONOMOUS MECHATRONIC PROCESSES

KUNICA, Z. & ZORC, D.

Abstract: *Inventions and technological progress allow technical systems to deal with more and more complex industrial tasks with increased autonomy, enabling profit to companies and liberation of humans of tedious work. At the same time, it means that the technical systems become more and more complicated, both on hardware and control levels, requiring enlarged planning effort. The paper discusses a concept of a mechatronic environment and experiments suitable for research and education in the field of automatic industrial processes, such as assembly, disassembly and packaging. The concept covers planning and execution issues, and it is based on a multi-agent approach, i.e. involved mechatronic pieces of equipment and processes are thought as autonomous agents.*
Key words: *Mechatronics, assembly, disassembly, packaging, multi-agent approach.*

1. INTRODUCTION

For more than last 25 years, proclaimed product diversification and customisation to meet customer needs (meaning decreased production volumes), very often has not been followed with automation achievements. Yet still, mass consumption allows hard or just slightly softened automation that can deal with one product in large production volumes only. Moreover, in fact, nowadays automation desperately tries to deal with rapidly changing production environment, to be applicable to wider range of products, in smaller production volumes, giving appropriate satisfaction to automation labourers (equipment manufacturers and system integrators) and decent profit to stockholders. The cost of robotic insertion dropped down 50 % in last six years (Bailey & Valadez, 2005). The globalised and unbalanced world market and corresponding phenomenon of outsourcing, driven with a wish for short-time profits, lead to neglecting of real technical and technological issues (Onori et al., 2003). But, the history wheel cannot stop, and it is only a matter of time, when all ignored will become relevant more than ever. At the same time, inventions and technological progress allow technical systems to deal with more and more complex industrial tasks with increased autonomy, enabling potential profit to companies and liberation of humans of tedious work. It means also that the technical systems become more and more complicated, both on hardware and control levels, requiring enlarged planning effort. The planning of technical systems includes several aspects, such as their design, process planning and programming. This paper depicts design of experiments of mechanical assembly and disassembly of a simple product, including packaging, and covering planning and execution using CAD/CAE/CAM software (commercial and self-developed), vision, force/torque sensor, and robotic and assembly cells (Kunica et al., 2004). The experiments are aimed at the research of possibilities of knowledge acquiring and structuring to allow a creation of a multi-agent tool, in order to get more efficient and reliable planning and operation of industrial systems such as those for assembly, disassembly and packaging.

2. MULTI-AGENT APPROACH IN A TECHNICAL SYSTEM

A technical system is an artificial system manufactured by humans. It has a purpose intended by human designers, to act in an environment in a predicted time, under human supervision and support. A complexity of nowadays and emerging technical systems implies increased requirement for symbolic reasoning. In that sense, the multi-agent approach seems to be useful. Besides, it is necessary to anticipate that future systems would be required to have much more autonomy in their acting, so the humans would be only first “ancestors” of, let us say, several generations of a system. The autonomy or large-scale automation, implies comparable mechanical intelligence that would eventually evolve to artificial life. Human survival, playful intellectuality, conscience of its own limitations, loneliness and imperfection have always been mirrored to creation of “deux ex machina”. Many authors stress the importance and similarity of behaviour of multi-agent systems as human-like groups (Wooldridge, 2002), somewhat ignoring complexity of aspects such as motivation and success validation criteria, which are always a source of misunderstandings in human groups. Even trivial communication that can be treated as a “technical” problem, often generates very serious failures in human behaviour. Emotions and irrational assumptions not to mention... Maybe mechatronic applications with included advanced sensorics does not imply so “intelligent” artificial systems, but rather “emotional”? Even trivial technical system may show very complex behaviour and impress uninformed observer (Breitenberg, 1998). The question is: what system of values to use? Can we expect from artificial systems to resolve our own human confusions and illusions? Why would the multi-agent systems be better than humans would? Nothing other can be concluded but the future systems will constantly have some inherited system's errors just as we humans – their creators – have (Capek, 1936). In the worst case, emerging technical – multi-agent – systems, occur as means of human fierce competition, conflict, alienation and escape from life (Fromm, 1968). Unfortunately, too often the use of advanced technology is presented as all-problem-solving facility, unreasonable individual consumption, self-forgetness and decorative symbol of status (as bones-in-nose and shells-in-ears in so called “primitive” tribes), meaning that scientists and engineers, academic and research communities failed to emphasize their role – that survival and progress of individuals and societies are, beside the others, universal moral categories, which include improving of production relationship and production forces, aimed at: benefit (profit) for all, maturity and creativity, and right of free research of own life phenomenon (freedom). Too often, researches and academic institutions are an isolated class of/for their own kind, without sufficiently expressed creative criticism that would strengthen their work to improve results regarding influence and benefits to particular and global society. Instead, sometimes they appear as nothing more than another tempocentric consumer group, which is satisfied with lean transfer of terms, ideas and technology generated from time to time by a few globally accepted paramount academic institutions; shaded with fake-work controversies.

Further, ubiquity of mechatronic widgets and gadgets would obviously generate many "overpopulation"/"life-space" physical and wireless issues with consequences to their functioning and human safety. Perhaps the main feature of life is perpetuation and it is not hard to believe that we should make love-driven systems (Breitenberg, 1998), knowing that each "mechanisation" of love may result with great damages.

Certainly, it is a tough task to develop such systems that would have our best capabilities adopting at least "honest" system of values, while acting, communicate, sense and process information. Fortunately, industrial domain (aside existing greatly virtualised, mostly non-materialised computer service/networks domains) is pretty constrained and therefore grateful environment regarding expectations, so the first applications of multi-agent autonomous systems can be developed there. Limited tasks allow step-by-step development, and man-machine interaction can be very inspiring experience in terms of shared control and levels of mechanisation-automation, such in case of Cobots (Peshkin et al., 2001).

3. THESIS AND EXPERIMENTS

Humans are sufficiently efficient, because they still exist. Industrial production, products and technical systems are materialisation of humanity. In that sense, it could be assumed further that structure and behaviour of technical systems may have features that are inherent to humans. If the majority of people behave as robots, then really it would not be hard to produce robots similar to humans (Fromm, 1968). According to commonly adopted approach to human mental structure and behaviour (Berne, 1964), a multi-agent technical system may possess three levels: ego, super-ego, and id. All three levels include planning (data processing) and operating/acting. It is believed that such structure allows us to deal more efficiently in our environment. Further, the next analogy may be established: ego – operating and low-planning level; super-ego – higher planning and upgrade/maintenance level; and id – fully automatic and entropy level. All levels imply sensing (inputs) and communication (inputs and outputs). The main issue is the development of criteria of desirable and undesirable behaviour of the multi-agent system, because they depend on current situation which may be difficult to anticipate. Instead, common technical criteria of effectivnes (productivity, costs etc.), should be considered as dialectical criteria of construction and deconstruction and then applicable to the multi-agent system. Seductive and popular "darwinistic" rule "survival of the fittest" simply does not apply because it is not true, and solely applied may lead to unproductive and therefore unwanted unification. On the other side, does it mean that solutions that are harmful at one particular moment may be tolerated? The experiments aimed at identification of features and relationships within mentioned knowledge, as well as development of the appropriate multi-agent system, would contain answers to questions: How are resources represented (virtually and physically, CAD-data, status of signals, etc.)? How to assemble (the parts into) the product? How to disassemble and how to pack the product? How to pack dismounted parts? Which are the prospective resources in the particular system's state? What positions and orientations have parts and assemblies? What are the initial states of resources? What are the assembly/disassembly/packaging actions? What are validation criteria (of success, efficiency etc.)? The vision issues cover: overlapping parts, choice of acting based on several possible natural resting aspects (orientations) of parts, and collision-free grasping. CAE planning procedure and data are primary related to off-line arrangement of system, resulting with assembly plan and fixed parts positions. In contrary, acting using vision is related to on-line situation, with randomly positioned parts, including recognition of parts' positions and orientations. Therefore, it is interesting to consider planning at execution level (on-line planning), which may be in collision with that obtained off-line. The next issue is that different pieces of equipment allow execution of the same tasks, so there is a potential conflict based on equipment jurisdiction. Experimental apparatus contains the following: the simple product; CAD/CAE/CAM software – commercial (CapePack; Catia, Delmia) and self-developed (Kunica & Vranjes, 1996; Kunica et al., 2003); vision system (HexSight); force/torque sensor (ATI-IA); robotic cell (Adept); assembly cell (Festo); mobile robots (Pioneers).

4. CONCLUSION

The mentioned experiments would be aimed at the research of possibilities of knowledge acquiring and its structuring to allow a creation of a multi-agent tool, to achieve more efficient and reliable planning and operation of industrial systems. It is expected that such approach should produce appropriate benefits in terms of design and exploitation of future industrial systems and processes. However, the effort to develop such systems cannot be treated solely as a technical issue, or technical systems must be treated differently, since many risks and threats may arise by neglecting past experience and some today's obviously concerning trends. Although the engineering class has its own limitations to deal with the above content, engineering science and praxis should always be human-centric avoiding indifference. Less dramatically, the psychology paradigm of human mental structure and behaviour may be used for design and operation of emerging technical systems regarding extended information content required and smoother work choreography.

5. REFERENCES

Bailey, K. & Valadez, A. (2005). Adept Update, Technical presentation, Adept, September 2005

Berne, E. (1964). *Games People Play: The basic handbook of transactional analysis,* Ballantine Books, Reissue edition (August 1996), ISBN 0345410033

Braitenberg, V. (1998). *Vehicles, Experiments in Synthetic Psychology*, 6th printing, The MIT Press, Cambridge, MA.

Capek, K. (1936). *War With the Newts*, Catbird Press (February 1990), ISBN 0945774109

Fromm, E. (1968). *The revolution of hope, toward a humanized technology*, Harper & Row, New York.

Kunica, Z. & Vranjes, B. (1999). Towards automatic generation of plans for automatic assembly, *Int'l J. of Prod. Research*, Vol. 37, No. 8, pp. 1817-1836, ISSN 0020-7543

Kunica, Z.; Vranjes, B. & Hrman, M. (2003). Some New Algorithms for CA Assembly Planning, *Journal of Mech. Engineering*, Vol. 49, No. 1, pp. 41-51, ISSN 0039-2480

Kunica, Z.; Vranjes, B.; Zorc, D. & Jerbic, B. (2004). *A Framework for Laboratory Modelling of Intelligent Production Processes*, Proc. of 15th Int'l DAAAM Symp., Katalinic, B. (Ed.), pp. 243-244, ISBN 3-901509-42-9, DAAAM Int'l Vienna, Vienna, November 2004.

Onori, M.; Camarinha-Matos, L. M. & Barata, J. (2003). European assembly: opportunities or threats?, *Proceedings of ISATP 2003*, pp. xiii-viii, Besancon, July 2003.

Peshkin, M. A.; Colgate, J. E.; Wannasuphoprasit, W.; Moore, C. & Gillespie, B. (2001). Cobot architecture. *IEEE Trans. on Robotics and Automation*, Vol. 17, No. 4, pp. 377-390.

Wooldridge, M. (2002). *An Introduction to MultiAgent Systems*, John Wiley & Sons, ISBN 047149691X

Annals of DAAAM for 2006 & Proceedings of the 17th International DAAAM Symposium, ISSN 1726-9679
B. Katalinic, Published by DAAAM International, Vienna, Austria 2006
150th Birthday of Nikola Tesla (1856-1943) Inventor of Mobile Robot Teleoperation

DYNAMIC CLASSIFICATION IN CAPP

KURIC, I. & KUBA, J.

Abstract: *The paper deals with aspect of classification systems in Computer Aided Process Planning (CAPP) systems based on Group Technology (GT). The paper presents possibilities of dynamic classification systems. The dynamic classification will be introduced as new method utilised especially for non-cutting technologies such as casting and forging.*
Key words: *classification, GT, CAPP*

1. INTRODUCTION

Group Technology /GT/ has a great signification in engineering industry. There is the greatest utilisation of GT in planning activities, especially in process planning. GT emerges as one of the prime forces that will integrate the engineering and manufacturing processes.

The GT methods are especially utilise in process planning for machining processes. The classification system is very important part of the CAPP /Computer Aided Process Planning/ system based on GT. The classification system for machining has static character. It is not needful to change classified parts into individual groups. However there is a proposal to utilise GT methods also for non-cutting technologies such as casting and forming. The static classification system is not suitable for process planning of non-cutting operations. Therefore there is a concept design of dynamic classification system oriented especially for non-cutting technologies.

2. STATIC CLASSIFICATION METHODS

A company may make hundreds and thousands of different parts. Because the parts are made in a concrete manufacturing environment, many parts are similar in some way /Fig.1./. Therefore many process plans must be also similar. If similar parts are situated in one group afterward their process plans likely are similar. It is possible to create some groups of parts with similar characteristics. If similar parts have similar processes afterwards utilising this approach get very good economic benefit. The above describing principle of grouping is a basis for GT philosophy.

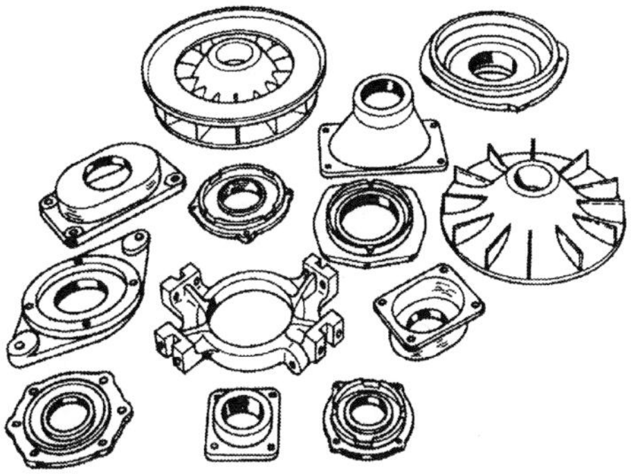

Fig. 1. Similar parts situated in one group

Process plans utilising is a main idea of utilising of grouped parts. The important task is to determine a similarity of characteristics among the parts. A wide variety of statistical, numerical and optimisation techniques solve the problem of grouping parts into families and machines into cells.

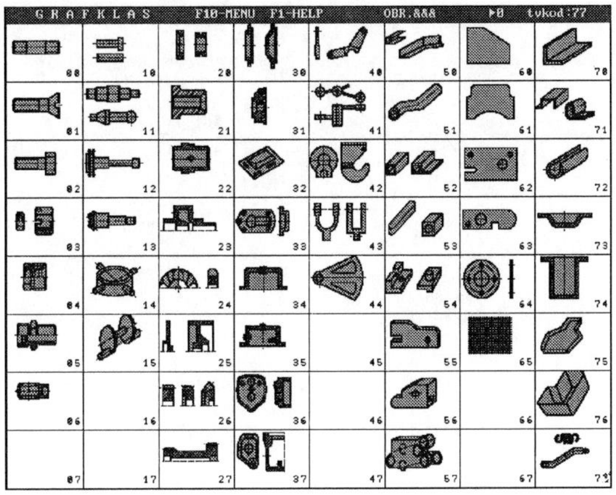

Fig. 2. Static classification system of Sysklass system

GT principle is one of principle used in CAPP methodology (Chang, 1990). The standard plans are created for a family of similar items. In the GT CAPP systems human retrieves the plan for similar components using coding and classifications of parts. The planner edits the retrieved plan to create a variant to suit the specific requirements of the component being planned.
Many of classification systems especially based on graphical classification (Fig. 2.) are utilised in GT CAPP systems in SK, CZ and PL. All systems are based on static classification method and they are utilized especially for planning of cutting operations.

3. CAPP SYSTEMS AND DYNAMIC CLASSIFICATION

The classification system is very important part of the CAPP system based on GT. Majority of CAPP systems based on GT is intended for manufacturing process planning. It is sufficient for the manufacturing process to create classification system which will be only fill up. There is no need for changing the number of groups, change localisation of individual engineering parts in individual groups. Therefore it is possible to consider these classification systems as static system.
However there is a big demand to utilise the group technology also for other technologies and not only for machining process planning. As the characteristic of non-cutting technologies (such as forging and casting) are different as cutting technologies, there is need to take other view on utilisation of GT in this area (Varga, 2000).

As the static classification system is not suitable for process planning of non-cutting operations, therefore there is a concept design of dynamic classification system oriented especially for non-cutting technologies (Sugar, 2000).
The dynamic classification is based on *flexible classification system* (Kuric, 2006). The engineering parts are *dynamic grouped* to the individual groups according to classification aims. For example the engineering parts will be dynamic grouped to the family groups according the total costs or operational total times, number of produced parts, series, etc..
There is a mathematical method - *cluster analysis* - which seems to be a very good candidate for support of dynamic classification system creation (Marcincin, 2006). *Clustering techniques* have been applied to a wide variety of research problems. The term cluster analysis actually encompasses a number of different classification algorithms.

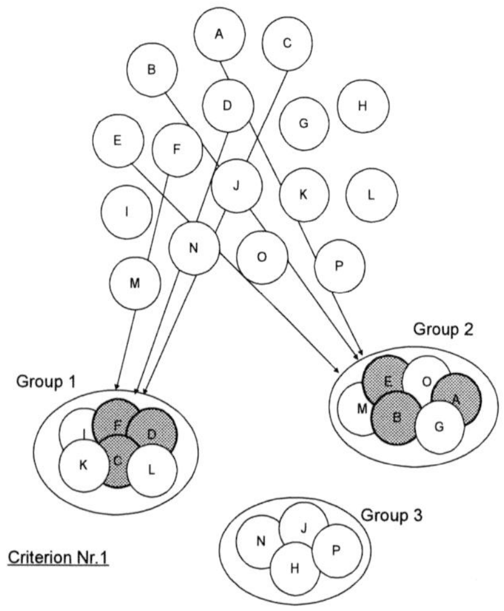

Fig. 3. Dynamic classification according 1st criterion

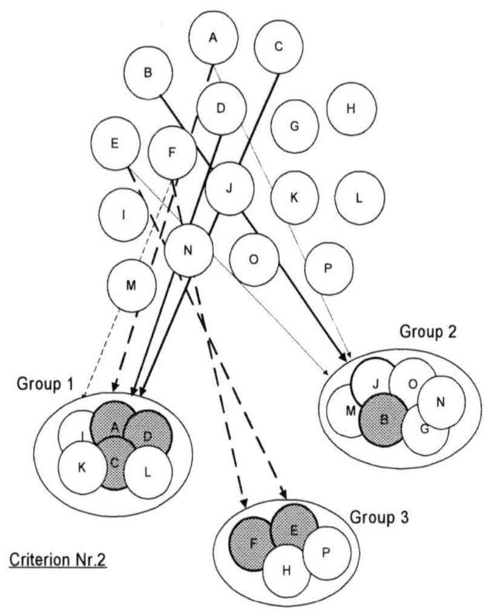

Fig. 4. Dynamic classification according different criterion

The principle of dynamic classification is evident on Fig.3. and Fig.4. The parts are flexible and dynamic grouped according selected criterions. It is still appropriate to utilise the visual classification as it is very simple and effective method however with flexible possibility the grouping the parts according actual demands. There is a development of dynamic classification systems and GT CAPP SW on the University of Zilina (Fig. 5.).
Disadvantages of current CAPP systems based on GT consist in their static classification systems which are not suitable for flexible change of GT representatives. There is no support to serve it in this systems. New approach consists in applied methods which making possible dynamically to group the engineering parts in the individual groups according selected criterions (e.g. cost, precision, equipments, level of automation, etc.).

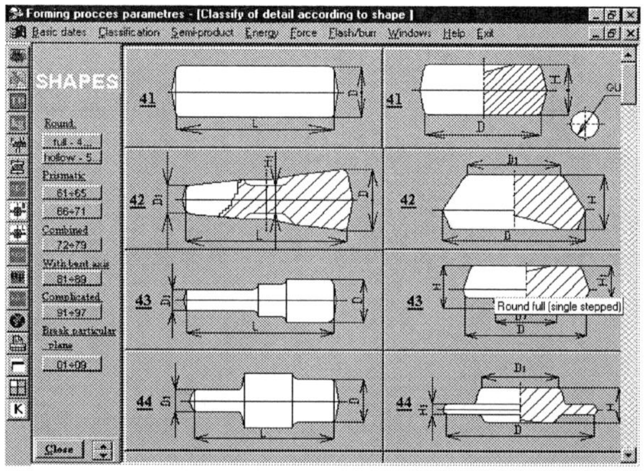

Fig. 5. Developed dynamic visual classification

4. CONCLUSION

The above mentioned topics are elaborated as PhD thesis. The building of GT CAPP system is time demanding and very labour task. The CAPP tasks require the theoretical elaborating, working out the serious methodology of process planning and used advanced programming technique. It seems that dynamic classification method is very effective and flexible method of part grouping for casting and forging process planning.

Acknowledgement: The article was made under support of Grant Agency VEGA – project Nr.1/0046/03 and the project of MS - Applied Research Nr. 4/0002/05.

5. REFERENCES

Chang, T.C. (1990). *Expert process planning for manufacturing*. Addison-Wesley Publ., Massachusetts, USA

Varga G.; Dudas, I. (2000). Intelligent Manufacturing System for Productions of Sophisticated Surfaces, *Proceedings of the Int.Computational MicroCAD 2000*, pp.99-104, Miskolc, 23-24.2.2000

Marcincin, J.N. (2006). Application of Group Technology theory for creating of NC programs. *Proceedings of 11th Int.DAAAM Workshop "CA Systems and Technologies"*, pp. 119-123, ISBN 3-901509-56-9, Cracow, 2.-4.10.2006, Cracow

Kuric, I. (2006). Static and dynamic classification systems. *Proceedings of 11th Int.DAAAM Workshop "CA Systems and Technologies"*, pp. 113-118, ISBN 3-901509-56-9, Cracow, 2.-4.10.2006, Cracow

Kuric, I.; Matuszek, J. & Debnár, R. (1999) *Computer Aided Process Planning in Machinery Industry*, Politechnika Lodzka, ISBN 83-87087-00-9, Bielsko Biala

Sugar, P. (2000) *Similarity of objects and processes of machine production*, Publishing center of TU Zvolen, Zvolen

Annals of DAAAM for 2006 & Proceedings of the 17th International DAAAM Symposium, ISSN 1726-9679
Editor B. Katalinic, Published by DAAAM International, Vienna, Austria 2006
150th Birthday of Nikola Tesla (1856-1943) Inventor of Mobile Robot Teleoperation

THE IMPLEMENTATION OF HACCP SYSTEM IN CROATIAN FOOD INDUSTRY

LAZIBAT, T.; VOUK, R. & BAKOVIC, T.

Abstract: *The quality of food products is a theme that deserves to be closely elaborated. Since HACCP system is considered to be one of the most efficient ways of building food quality management system; its history, basics and application are presented in this paper. The most recent developments in food quality management systems, regarding the appearance of ISO 22000:2005 international standard, are also described together with results from a survey conducted among Croatian food industry enterprises.*
Key words: *HACCP quality management system, ISO 22000, food safety, HACCP principles.*

1. INTRODUCTION

HACCP is an acronym for the Hazard Analysis and Critical Control Points system, which is a preventive food safety management system. It is recognized internationally as the most effective way to produce safe food. The application of HACCP is possible throughout the food supply chain from primary productions to the consumers. Because it is a step by step approach it is less likely that hazards will be missed. HACCP therefore, offers increased confidence to the food business and its consumers.

2. THE DEVELOPMENT OF HACCP

Initially developed as a means of increased confidence over the microbiological safety of food used in space program, the potential of the technique for ensuring control of food safety hazards in a broad range of foods in manufacturing, distribution, food service and retail situations was soon realized. Although the HACCP concept continued to develop, it did so in a somewhat piecemeal fashion and it was not until the mid to late 1980s that HACCP development begun to accelerate in a coordinated way. The HACCP concept was endorsed by WHO/FAO on an effective way of controlling food borne disease in 1983 when the Joint WHO/FAO Expert Committee on food safety advised that HACCP should replace the traditional end product testing approach to food safety assurance (Mayers & Martimore 2001).

The most important document in HACCP history is the CAC (Codex Alimentarius HACCP) document from 1997, since it is the first truly authoritative, internationally agreed document that addresses not only the agreed HACCP principles but also guidelines for its application. This document clearly identifies that the HACCP System consists of seven principles listed in Table 1. Most people today regard this Codex document the international benchmark for HACCP.

The primary area where HACCP will help is in the processing of safe food. It helps people to make informed judgments on faulty matters and removes bias, ensuring that the right personnel with the right training and experience are making decisions connected to food safety.

As HACCP is a universal system, it can be passed on to your suppliers to assist in assuring row material safety and will also help to demonstrate effective food safety management through documented evidence which can be used in the event of litigation (Martimore & Wallace 1998).

Principle 1.	Conduct a hazard analysis
Principle 2.	Determine the critical control points (CCP)
Principle 3.	Establish critical limit
Principle 4.	Establish system to monitor control of HACCP
Principle 5.	Establish corrective action to be taken when monitoring indicates that a particular CCP is not under control.
Principle 6.	Establish procedures for verification to confirm that the HACCP system is working effectively.
Principle 7.	Establish documentation concerning all procedures and records appropriate to these principles and their application.

Table1. Seven principles of the HACCP System

The guidelines for the application of HACCP system are also clearly outlined in 12 tasks, which are as follows:

1. Assemble the HACCP team,
2. Describe Product,
3. Identify intended use,
4. Construct flow diagram,
5. On-site verification of flow diagram,
6. List all potential hazards, conduct a hazard analysis, determine control measures,
7. Determine CCP-s,
8. Establish critical limits for each CCP,
9. Establish a monitoring system for each CCP,
10. Establish corrective action for deviations that may occur,
11. Establish verification procedures,
12. Establish record keeping and documentation.

HACCP users will almost certainly find that there are additional benefits in the area of product quality. This is primarily due to the increased awareness of hazard in general and the participation of people from all areas of the operation. Many of these mechanisms that are controlling safety are also controlling product and service quality.

New European Union hygiene regulations which will apply to member states from 2006, and very soon the Republic of Croatia, will require that all sectors of food chain, except primary producers, operate food safety procedures based on HACCP principles. The Food and Drug Administration (FDA) has estimated that the current level of compliance of food business is 30 per cent, of these being mostly manufacturers and larger retailers and caterers. (Worsfold 2006).

3. THE USE OF HACCP IN CROATIAN FOOD INDUSTRY

For the needs of this paper it was most important to investigate how did the introduction of HACCP system influence Croatian food industry export? The survey was conducted among 87 Croatian companies that had HACCP system implemented in their business. After sending the questionnaire twice in February 2006, 25 answers were collected. The summary of results is presented in extension.

Export existence	No. of enterprises	Percentage
Yes	20	80
No	5	20
Total	25	100

Table2. Do you export your products?

As we can se according to the data from Table 2. 80% of enterprises from Croatian food industry had exported their products. Another question that arises from this fact is to investigate weather foreign partners made these companies to implement HACCP system. According to the data from Table 3. HACCP was indeed a main precondition for exporting food products for 80% of the analyzed enterprises. It is obvious that implementing HACCP is the unexceptional term for participation in international food products trade.

Export precondition	No. of enterprises	Percentage
Yes	20	80
No	5	20
Total	25	100

Table3. HACCP as a basic export precondition

Table 4. Shows us that 64% of food exporters accomplished better export results after implementing HACCP food safety management system.

Export impact	No. of enterprises	Percentage
Raises export	16	64
No influence	3	12
No reply	6	24
Total	25	100

Table4. HACCP export impact

According to Table 5. 60 % of analyzed companies possessed ISO 9001:2000 certificate. That number can not be considered satisfying because ISO 9001 certificate is also one of main preconditions for participation in international trade. One of the reasons why so little companies possess this certificate is that in food business HACCP is considered much more important.

ISO 9001:2000	No. of enterprises	Percentage
Yes	15	60
No	10	40
Total	25	100

Table5. Existence of ISO 9001:2000 QMS?

4. THE FUTURE OF HACCP-ISO 22000: 2005

If we consider the fact that ISO 9001:2000 and HACCP had many common points, it is understandable why ISO (International Organization for Standardization) created a new standard that united these to into a new standard called ISO 22000:2005.

ISO 22000:2005 published on 1 September 2005, is a new International Standard designed to ensure safe food supply chains worldwide and the first of a family on food safety management systems. ISO 22000 is therefore designed to allow all types of organizations within the food safety management system. These range from feed producers, primary producers, food manufacturers, caterers, transport and storage operators and subcontractors to retail and food service outlets. In addition food safety management systems that confirm to ISO 22000 can be certified- which answers the growing demand in the food sector for the certification of suppliers-although the standard can be implemented without certification of conformity, solely for benefits it provides.

Developed with the participation of food sector experts, ISO 22000 incorporates the principles of HACCP, and covers the requirements of key standards developed by various global food retailer syndicates, in a single document. Another benefit of ISO 22000 is that it extends the successful management system approach of the ISO 9001:2000 quality management system standard which is widely implemented in all sectors but does not itself specifically address food safety. The development of ISO 22000 was based on the assumption that the most effective food safety systems are designed, operated and continually improved within the framework of a structured management system, and incorporated into the overall management activities of the organization. While ISO 22000 can be implemented on its own, it is designed to be fully compatible with ISO 9001:2000 and companies already certified to ISO 9001 will find it easy to extend this certification to ISO 22000.

5. CONCLUSION

Certified quality management system represents a comparative advantage and serves as a base for building trust between buyer and seller. (Lazibat & Kolaković 2004). Food industry makes no exception to this rule. A certified ISO 22000 or HACCP company proves not only company products are of the highest quality, but also that there is no risk for human health from consuming these products. HACCP was for a long time considered as the most obvious and truthful instrument of insuring food safety, but soon it became obvious it is necessary to link it with ISO 9001:2000. The areas of work for these two standards lapped in many points so ISO developed a new standard called ISO 22000 to solve this problem. Results from a survey conducted for the needs of this paper pointed out that Croatian food industry companies had to implement HACCP because it was a basic precondition for exporting (80%).

6. REFERENCES

Mayers, T. & Martimore, S. (2001). *Making the most of HACCP*, CRC Press, ISBN: 0-8493-1218-3, New York.

Martimore, S. & Wallace, C. (1998). *HACCP A practical approach (2E)*, Aspen Publishers, ISBN: 0-8342-1932-8, New York.

Lazibat T., & Kolalović M., (2004). Međunarodno poslovanje u uvjetima globalizacije (International business in terms of globalization) , Sinergija, ISBN 953-6895-19-6, Zagreb

Frost, R.; (2005). ISO 22 000 is first in family of food safety management system standards, *Available from:*http://www.iso.org, *Accessed:* 2006.08.25.

Warsfold. D, (2006). HACCP workshops- practical guidance for small fast food businesses, Nutrition & Food Science, Vol. 36, No.1, p. 32-42, ISSN:0034-6659-91.

Annals of DAAAM for 2006 & Proceedings of the 17th International DAAAM Symposium, ISSN 1726-9679
B. Katalinic, Published by DAAAM International, Vienna, Austria 2006
150th Birthday of Nikola Tesla (1856-1943) Inventor of Mobile Robot Teleoperation

A MODULAR CONTROL ARCHITECTURE FOR WATER SYSTEM MANAGEMENT

LENDAK, I.; ERDELJAN, A. & VUKMIROVIC, S.

Abstract: *The aim of this paper is to present a modular architecture able to communicate with various commercial and open source solutions. Its primary use would be planning, simulation and control in water management systems or theoritecally in any similar system. The solution allows an effortless integration of existing and future tools into a complete WMS as well as their easy replacement.*

Key words: *Water System Management, system planning and modelling, EPANET, high level software design.*

1. INTRODUCTION

There is wide variety of software solutions in use in water management systems (WMS). There are acclaimed and highly esteemed simulation software solutions for WMS, which are published under some forms of open source licenses (EPANET, Modflow). In the course of our research we needed to use these solutions and integrate them into a WMS prototype. We faced various problems during this process, and therefore decided to develop a modular software architecture which could accommodate these existing solutions, as well as any newly emerging similar products.

In the process of our work, we proposed a unified object model for water distribution and control systems, developed a prototype application hosting the model and communication modules for communicating model data with EPANET, Matlab, a relational database management system (RDBMS) and a number of other data sources. The aim of this paper is only to discuss the modular architecture which enables this solution to communicate with the above mentioned data sources/sinks.

2. DISCUSSION

There are fewer standards and recommendations available for water management software systems (WMS) as there are for energy/distribution management systems (EMS/DMS) (e.g.: CIM, Multispeak). This fact is a major hindrance for projects aiming to deliver a complete solution for WMSs.

Another important fact is that there are freely available tools of high quality, like EPANET for distribution network planning and modeling or Modflow for modelling water sources with their major drawback being that they do not contain standardized interfaces which would allow their integration into a complete WMS solution. Our goal is to propose a software architecture which would rest on an object oriented model of water systems, and communicate with other software solutions in a standardized way, through a unified software bus.

2.1 The OO model

The multitude of standards for EE systems was a great basis based on which an object model of water management systems was developed (Vukmirović, 2006). Our work on the model was based on the Common Information Model for EE systems and on the matrix based model contained within EPANET (EPA, 2001). The model's name (WCIM), the package hierarchy and the base classes were borrowed from CIM and then the WMS specific classes were identified based on EPANET data. The discussion of this model is outside the scope of this paper and is mentioned here as it will be used as the central part of the proposed architecture, containing the data accessed by other modules.

2.2 Data exchange

The first few communication tasks, identified during our work, in which the WCIM model is accessed by an external software system are the following:

- Populating WCIM from a database
- Populating WCIM from an EPANET input file
- Setting simulation results calculated in EPANET
- Setting simulation results calculated in Modflow
- Setting control schedules calculated in Matlab

In order to accommodate these communication tasks, we propose the architecture shown in figure 1.

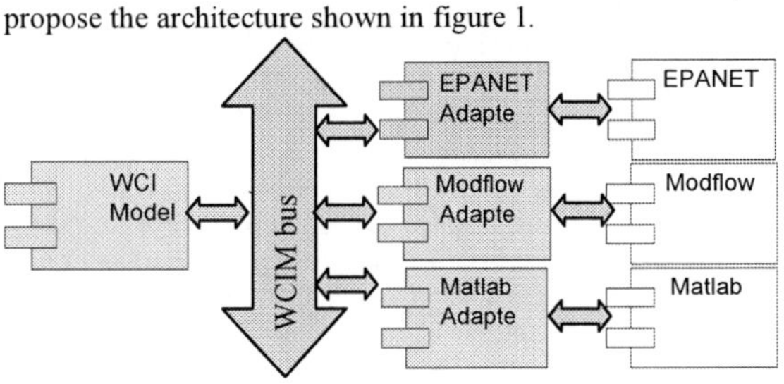

Fig 1. WCIM adapters

The basis of the architecture is the WCIM model on the left. Data contained within this model is accessible through the WCIM bus. The interfaces used for communication on this bus are outside the scope of this document. On the right we have external software solutions which need to exchange data with our WCIM model. As these do not have standard interfaces, they can not be directly plugged into the WCIM bus. There had to be certain components which will marshal data between the WCIM bus and these external applications. These components were named adapters. One such component has to be written for every external application. These adapters connect to the WCIM bus in a unified way, while their other end facing the external applications can contain various communication technologies (eg. direct DLL calls for EPANET, COM client-server for Matlab). The shaded components in Figure 1. are part of our solution, while the non shaded ones are external applications. It is important to mention at this point that the number of modules is not limited to the few shown in this figure. Should there be a need, the proposed solution allows the effortless addition of other adapters.

A not so obvious, but nevertheless very important fact is, that the adapters allow the complete solution to be proprietary, even

though it uses open source components. This is possible even if those have stricter licenses, as the source code of the adapters could be separately made open, while keeping the complete solution's code closed.
In order for the proposed solution to be scalable and modular, the technology behind the WCIM data bus had to be chosen well. This choice had an impact on the ease of development, testing and above all on system performance.

	Strengths	Weaknesses
.NET Remoting	- Quick results	- Windows only - Not used in the industry - Relatively immature
CORBA + XML	- Multi platform - Performance - Processor util.	- No widespread standard
(D)COM/ OPC	- Standard (OPC) - Performance - SCADA support	- Windows only - Delayed results

Table 1. Communication protocol choice

We have conducted a simplified SWOT analysis of three widespread communications methods in order to identify which could be best used for the task at hand (Table 1.). All of the three candidates had their advantages and drawbacks, but we deemed CORBA's ability to work on various operating systems important and therefore suggest its use in combination with XML messaging. The format of XML messages traveling on the WCIM bus is outside the scope of this document. CORBA's multiplatform capability is even more important, as the external components used during our work are either open source and platform independent (EPANET, Modflow), or they have versions for Unix based operating systems as well (Matlab).

2.3 Additional communication scenario

Let us suppose the following scenario: Matlab is chosen for optimal pump speed schedule calculation, applying a genetic algorithm. For the successful generation of these schedules, Matlab needs simulation results from EPANET. In this case, two external components need synchronized access to WMS data. This is not an easy task to solve, as if we directly connect, the two external components we contaminate our architecture's modularity. On the other hand, which component should oversee the communication between these two adapters?
The solution we propose is the introduction of higher level functions, which in a way similar to the above discussed adapters, plug into the system via the WCIM data bus. They are not necessarily connected to other such functions or adapters, as they may only need model data for their calculations. If a function needs data from multiple adapters (eg. simulation results) it has to be able to call those adapters. For these calls we introduced the control bus, which enables the higher level functions to control the adapters.

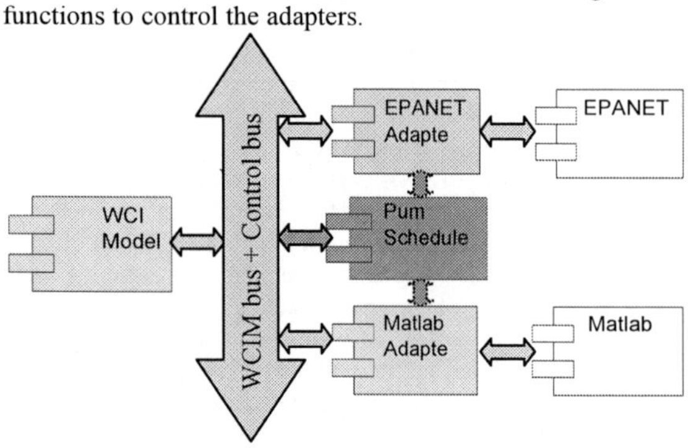

Fig 2. WMS function Pump Curves

On first glance figure 2. showing the Pump Schedule function seems very similar to the previous figure. The most important component here being "Pump Schedule", communicating with two adapters and delivering the optimal pump speed schedules to the system. The arrows between this functional block and the other two adapters are dotted as this communication is realized through the control bus. In this case Pump Schedule (PS) sends its synchronization messages through this bus, enforcing a predefined sequence of calls between the EPANET and Matlab adapters. These adapters do not communicate with each other, they just access model data through the WCIM data bus on the PS function's demand.
The simplified sequence of events carried out by the PS function is shown in the sequence diagram in figure 3.

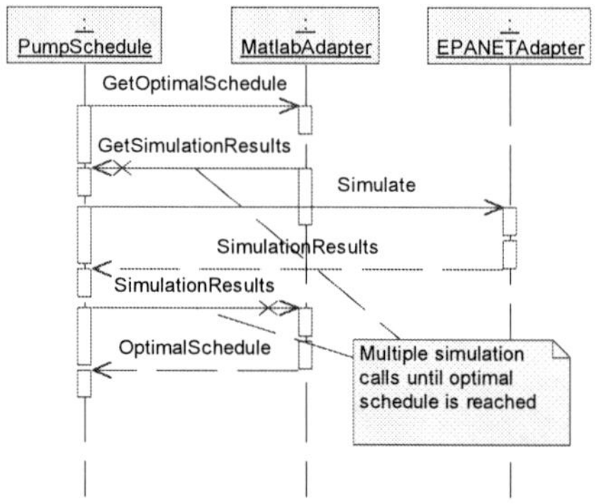

Fig 3. Pump schedule calculation sequence

The calculation of optimal pump schedules is by no means the only function identified during our work on IT systems for WMS. Another example could be fault detection, which could be achieved by an additional module, which would use EPANET to simulate pressures and flows, compare those results with the measured values from the field and decide whether there are faults (eg. spillage) in the distribution network. Further function examples: water source planning (requiring Modflow simulation results), topology analysis, load calibration, pressure losses, etc. All of these could be separately written modules, plugged into the system just as the "Pump Curves" function is in Figure 2.

3. CONCLUSION

The main advantages of the proposed solution are its modularity, the effortless replacement of modules and its ability to use open source solutions with various licenses without it being under an open source license. Another advantage is that this system could be sold to clients with only those functions which are needed by clients. The authors would like to express their thanks to the Ministry of Science of Serbia, which allowed them to conduct their research within its National Project "Information system of water systems in large settlements".

4. REFERENCES

Booch G., Jacobson I. & Rumbaugh J. (1999), *The Unified Modeling Language User Guide*, Addison-Wesley
EPA (2001), *Environmental Protection Agency NET software, http://www.epa.gov/nrmrl/wswrd/epanet.html*
EPRI (2001), *Common Information Model (CIM): CIM 10 Version* (2001), Palo Alto, CA, 1001976
IEC 61968-1 (2001), *Power system control and associated Communications*
Vukmirović S, Erdeljan A. & Lendak I. (2006), *Architecture of SCADA software for water distribution systems*, 50th ETRAN Conference, Belgrade, Serbia

Annals of DAAAM for 2006 & Proceedings of the 17th International DAAAM Symposium, ISSN 1726-9679
Editor B. Katalinic, Published by DAAAM International, Vienna, Austria 2006
150th Birthday of Nikola Tesla (1856-1943) Inventor of Mobile Robot Teleoperation

A MORE SIMPLE G-CODE PROGRAMMING WITHIN AN OPEN SOURCE CNC SYSTEM

LICARI, R.; VANOLI, R. & PIACENTINI, M.

Abstract: *In this paper we introduce the utilization of a new CNC programming language, named O-Word, a way to implement several branching routines into the EMC interpreter. It opens new interesting possibilities to programmers who can build subroutines, loops, and conditionals while taking advantage of the interpreter's use of variables and computation. Therefore, it allows us to make pieces describing theirs features instead of tool's movements. We will demonstrate a new complex routine in order to show both the language syntax and its potential for advanced applications*
Key words: *CNC, G-Code, O-Word, Open Source .*

1. INTRODUCTION

Thanks to low cost and Open Source CNC system availability, it is currently possible to use CNC technology even in very simple and small production environment. Yet, the generation of part program often requires long time or expensive CAM system.
Many companies use cutter location (CL) data output from CAD/CAM systems for machine tool control. The CL data must be post-processed, or translated, into machine language for a specific controller in order to be executed. If the job has to be transferred to a different machine, the code has to be processed again, translating it into the new machine language.
Translating the source programming language into the various dialects can become a time consuming task. Developing and maintaining a postprocessor for each controller dialect can become extremely expensive. Programming all jobs with one language would allow code reuse and immediate portability.
This problem could be solved using the most interesting characteristic of open source CNC systems: they can be programmed.

2. THE O-WORD LANGUAGE

This paper demonstrates how to simplify the NC code generation using O-Word, a C-like programming language. It is named O-Word because each subroutine and function is introduced by an "O" followed by an identifying number (a label).
The O-Word language, created by Kenneth Lerman, allows us to write tree-shaped routines in EMC.
EMC is a Linux CNC-executable control program, freely available on the Internet, able to control servo-motors, stepper motors, robots and other automated devices. It allows us to manage a numeric control machine from the parallel port even of a low-performing PC. EMC runs on Linux using real time extensions.
By O-word, a programmer who uses G-Code can build with less effort subroutines, loops, and conditional structures (containing either "while" or "if") which can be used by EMC.
The programming of subroutines in this language is still in embryo, but we believe its development is desirable, for the following reasons:

- it grants us an opportunity to write ISO programs based on the description of manufacturing rather than movement;
- it determines a drastic reduction of ISO command lines;
- it allows us to make relatively complex works without using any CAM program.

2.1 O-Word applications

ISO programming in RS276NGC language, containing all movement information and all working parameters, allows us to describe a trajectory only by either linear (G00, G01) or circular interpolations (G02, G03). With O-Word it is possible not only to "build" such more complex interpolations as ellipses and spheres, but also to more easily read and modify the program, which will be composed of just few dozens of lines, rather than the hundreds – or even thousands – of lines which would be generated by a CAM software. Notice, moreover, that this approach is very useful in case of EMC, which executes – and does not generate – ISO programs. This is why EMC needs a CAM software. Since most of these software run on a Windows platform, it is necessary to buy a Windows licence, which increases expenses. There are also some Linux CAM software, but they are not so much common and are usually quite expensive. In both cases, an increase of expenditure would be taken into account.
As an example of how does O-Word simplify NGC, we have made an easy comparison of NC programmes for the realisation of 5 elliptic grooves, with the following results:

- Length of the NC file created by EdgeCam (a commercial CAD/CAM software): 600 lines;
- Length of the NC file created by O-Word: 21 lines.

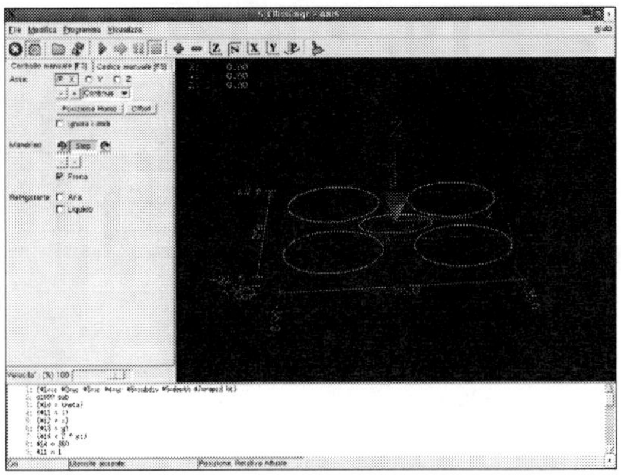

Fig. 1. Five elliptic grooves in EMC preview

It is evident that it is easier to read and modify a short file then a long one, even acting directly on the machine.

2.2 O-Word development

We developed a subroutine for both face milling and rectangular/square bosses. The algorithm is composed of two

symmetrical "while" cycles and one nested "if" function.
In order to create this program we will define:

- the parallelepiped dimensions (x_box_length; y_box_length
- the boss dimensions (Xr ;Yr)
- the depth
- the tool radius
- the overlapping percentage of passages (% diameter_step)

The steps performed by the following program will be parallel to X and as long as the piece length, at the distance of one step from each other. When Y becomes equal to Yr of the boss, the program will perform a series of parallel passages which length will be equal to the half of the piece's length minus Xr. The distance between the passages will be equal to the step.
This subroutine has resulted to be useful and flexible: we succeeded in using it for a great number of different operations, by merely substituting either the kind of call or its values. In compliance with each kind of call, we have been able to perform face millings and bosses by one or more subsequent passages, as it will be explained by the following applications.
In order to make a face milling, it will be sufficient to attribute a null value to Xr and Yr (respectively the third and the fourth) parameters, as evidenced in bold in the following line:

o1000 call [80] [50] **[0] [0]** [200] [-2.5] [2] [4] [0.5]

By the following call, for example, a rectangular boss sized (40 x 24) mm^2 , equal to (2Xr x 2Yr), 2.5mm high (depth of cut) will be realised by only one passage.
o1000 call [75] [55] **[20] [12]** [200] **[-2.5]** [2] [4] [0.5]

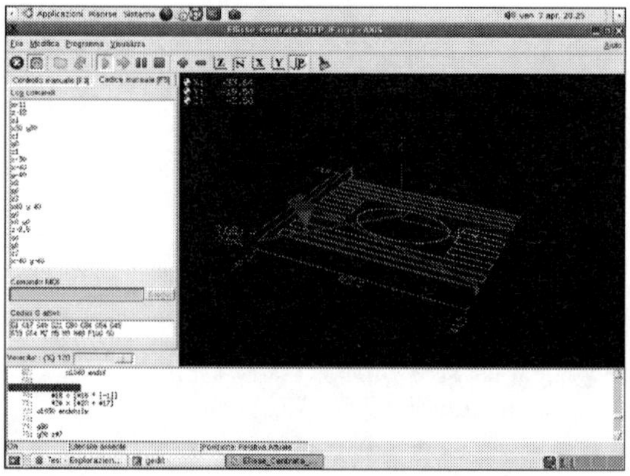

Fig. 2. O-Word example routine

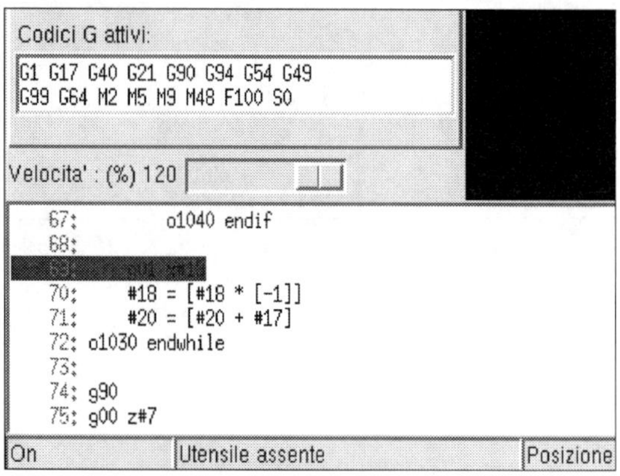

Fig. 3. O-Word routine example detail

By calling the subroutine again and again, and by changing each time the depth of cut, we will obtain a boss with the desired height.

The following example shows a 11.2 mm high square boss, obtained by four passages at various depths, equal to 3, 6, 9, 11.2 mm.

o1000 call [80] [50] [13.7] [13.7] [200] **[-3]** [2] [4] [0.5]
o1000 call [80] [50] [13.7] [13.7] [200] **[-6]** [2] [4] [0.5]
o1000 call [80] [50] [13.7] [13.7] [200] **[-9]** [2] [4] [0.5]
o1000 call [80] [50] [13.7] [13.7] [200] **[-11.2]** [2] [4] [0.5]

O-Word programming is modular, so that it is possible to join different programmes to realize more complex works.

3. CONCLUSIONS

The most interesting feature of this approach is that, by using this technology, we can describe the piece's shape rather than the tool's movements, as the STEP NC language technology does.
In fact STEP-NC replace the traditional G and M codes used to program CNC machines with a data structure containing information on how to build a part.
In our opinion, it is also possible to describe such complex curves as spline and B-spline (NURBS), even if we are aware of the difficulties in writing an efficient generation algorithm by using the O-Word language. The benefit of using NURBS is that the amount of space occupied by part programs and the time taken to execute them can be significantly reduced. Large sections of part programs often consist of many small linear moves. With our approach each of these moves can be on-line computed rather than previously defined.

4. ACKNOWLEDGEMENTS

This work was performed with the financial support of M.I.U.R. of Italy.

5. REFERENCES

Borsellino, C.; Licari, R.; Lo Valvo, E. & Piacentini M. (2004). *An XML Interface to STEP-NC Systems for Remote Collaborative Concurrent Engineering*, ISOMA.

EMC official home page, *Available from:* http://www.linuxcnc.org

NIST RS274NGC programming language, *Available from:* http://www.isd.mel.nist.gov/personnel/kramer/pubs/RS274NGC_3.web/RS274NGC_3TOC.html

Olszta, M. W. & Klemm, K. *Interactive Qualifying Project Report: The Transfer of Technology to Small and Medium Sized Manufacturers*, Worcester Polytechnic Institute, *Available from:* http://www.linuxcnc.org/Dropbox/iqp.PDF

O-Word programming language home page *Available from:* http://wiki.linuxcnc.org/cgi-bin/emcinfo.pl/emcinfo.pl?Oword

Proctor, F.; Damazo, B.; Yang, C. & Frechette, S. *Open architectures for machine control*, National Institute of Standards and Technology Gaithersburg, Maryland 20899, *Available from:* http://www.isd.mel.nist.gov/ documents/ proctor/sim/sim.html

Proctor, F.; Shackleford, W. & Yang, C. *Simulation and Implementation of an Open Architecture Controller*, National Institute of Standards and Technology.Gaithersburg, MD 20899, *Available from:* http://www.isd.mel.nist.gov/documents/proctor/315openarchit.pdf

Yang M.-Y., Hong W.-P., *A PC–NC milling machine with new simultaneous 3-axis control algorithm*, Int. Journal of Machine Tools & Manufacture 41 (2001) 555–566

Annals of DAAAM for 2006 & Proceedings of the 17th International DAAAM Symposium, ISSN 1726-9679
B. Katalinic, Published by DAAAM International, Vienna, Austria 2006
150th Birthday of Nikola Tesla (1856-1943) Inventor of Mobile Robot Teleoperation

MODEL VALIDATION PROCEDURE AND DAMAGE DETECTION ALGORITHM BASED ON DYNAMIC PROPERTIES OF STRUCTURES

LOZINA, Z.; VUCINA, D. & SEDLAR, D.

Abstract: *Finite element methods are extensively used to model the structures in order to foresee their behaviour under loading condition as well as to extract information on the respective dynamic properties such as frequencies and mode shapes. The simulation of behaviour as well as dynamic properties heavily depends on model quality. This fact is used to in two directions: model validation and damage detection.*

Key words: *dynamics properties, modal analysis, changed elements, finite element.*

1. NUMERICAL BACKGROUND

Modal properties of structure can be measured experimentally or simulated using finite element method. When modal properties for structures are known in their original state and changes are detected, changed regions can be located.
Motion equation of undamped dynamic problem is:

$$\mathbf{M}\ddot{\mathbf{x}}+\mathbf{K}\mathbf{x}=\mathbf{f}(\mathbf{t}) \tag{1}$$

where $\mathbf{M}$ is the mass matrix, $\mathbf{K}$ is the stiffness matrix, and $\ddot{\mathbf{x}}$ and $\mathbf{x}$ are vectors representing accelerations and displacements, $\mathbf{f(t)}$ is excitation force vector. Free vibration are characterized with force $\mathbf{f(t)}$ equals zero. To compute the natural frequencies and mode shapes eigenvalue problem have to be solved

$$(\mathbf{K}-\lambda_i\mathbf{M})\varphi_i=0 \tag{2}$$

where λ_i is the *i*th mode eigenvalue, and φ_i is *i*th mode shape. The λ_i and φ_i can be considered *i*th modal property.
To locate changed regions, difference in modal properties between two states has to be determined. Assuming eigenvalue problem for changed state can be written as

$$(\mathbf{K}_c-\lambda_{ci}\mathbf{M}_c)\varphi_{ci}=0 \tag{3}$$

where subscript c indicates changed state. Stiffness and mass matrices in changed state $\mathbf{K}_c$ and $\mathbf{M}_c$ are defined as

$$\begin{aligned}\mathbf{K}_\mathbf{c}&=\mathbf{K}+\Delta\mathbf{K}\\ \mathbf{M}_\mathbf{c}&=\mathbf{M}+\Delta\mathbf{M}\end{aligned} \tag{4}$$

where $\Delta\mathbf{K}$ and $\Delta\mathbf{M}$ is change of original stiffness and mass matrices $\mathbf{K}$ and $\mathbf{M}$. Substituting the last equation into eigenvalue problem and rearranging residual force vector is

$$\mathbf{R}_i=\Delta\mathbf{K}\varphi_{ci}-\lambda_{ci}\Delta\mathbf{M}\varphi_{ci}=(-\mathbf{K}+\lambda_{ci}\mathbf{M})\varphi_{ci} \tag{5}$$

Residual force vector $\mathbf{R}_i$ can be simple determined if is known natural frequencies and mode shapes. Subscript *i* denote *i*th degree of freedom. From the last equation it is observed that residual force for *i*th degree of freedom is zero if changed in stiffness and mass matrices are zero. Furthermore, only members of residual force which belong changed elements are different from zero.
Theoretically, only one natural frequency and mode shape are needed to locate changed element. In real world, measured data usually include certain amount of errors which may lead to inaccuracies in residual force vector. Therefore, it is needed several natural frequencies and mode shapes. If it is obtained *p* measured frequencies and mode shapes, pseudo residual force vector may be expressed as

$$\mathbf{R}=\{r_1,r_2,\ldots,r_n\}^{\mathrm{T}} \tag{6}$$

where r_j is

$$r_j=\left[\prod_{i=1}^{p}\left(|r_j|_i\right)\right]^{\frac{1}{p}} \tag{7}$$

in which $|r_j|_i$ is the absolute value of the *j*th entry of $\mathbf{R}_i$. In obtained pseudo residual force vector the real signals are amplified, and pseudo residual force vector is more accurate.
To determine the amount of changes, natural frequencies and mode shapes in changed state are defined as

$$\begin{aligned}\lambda_{ci}&=\lambda_i+\Delta\lambda\\ \varphi_{ci}&=\varphi_i+\Delta\varphi\end{aligned} \tag{8}$$

Now, the eigenvalue equation can be updated as

$$(\mathbf{K}-\lambda_i\mathbf{M})\varphi_i+(\mathbf{K}-\lambda_i\mathbf{M})\Delta\varphi_i+(\Delta\mathbf{K}-\lambda_i\Delta\mathbf{M})\varphi_i-\Delta\lambda_i\mathbf{M}\varphi_i+$$
$$+(\Delta\mathbf{K}-\lambda_i\Delta\mathbf{M})\Delta\varphi_i-\Delta\lambda_i(\mathbf{M}\Delta\varphi_i+\Delta\mathbf{M}\varphi_i+\Delta\mathbf{M}\Delta\varphi_i)=0$$
(9)

The first term is equal to zero, moreover, neglecting higher order terms last two terms are equal to zero. Pre-multiplying all remaining terms by φ_i^{T}, second term becomes equal to zero. Furthermore, if φ_i is normalized so that $\varphi_i^{\mathrm{T}}\mathbf{M}\varphi_i=1$, last equation can be reduced to

$$\varphi_i^{\mathrm{T}}\Delta\mathbf{K}\varphi_i-\lambda_i\varphi_i^{\mathrm{T}}\Delta\mathbf{M}\varphi_i=\Delta\lambda_i \tag{10}$$

Values of $\Delta\mathbf{K}$ and $\Delta\mathbf{M}$ matrices are nonzero only at location where degrees of freedom are associated with changed elements. Reducing $\Delta\mathbf{K}$ and $\Delta\mathbf{M}$ to $\Delta\mathbf{K}'$ and $\Delta\mathbf{M}'$ matrices that contains only nonzero rows and columns associated with changed elements, and reducing φ_i to φ_i' that contains values only at corresponding changed elements, the last equation can be expressed as

$$\varphi_i'^{\mathrm{T}}\Delta\mathbf{K}'\varphi_i'-\lambda_i\varphi_i'^{\mathrm{T}}\Delta\mathbf{M}'\varphi_i'=\Delta\lambda_i \tag{11}$$

This can be used to solve for changes in original stiffness and mass matrices. To reduce the number of unknowns that need to be solved from the last equation, matrices $\Delta\mathbf{K}'$ and $\Delta\mathbf{M}'$ are expressed as function of original element stiffness and mass matrices by following equations

$$\begin{aligned}\Delta\mathbf{K}'&=\sum_{j=1}^{n_c}\mathbf{K}_j^e\delta k_j\\ \Delta\mathbf{M}'&=\sum_{j=1}^{n_c}\mathbf{M}_j^e\delta m_j\end{aligned} \tag{12}$$

where $\mathbf{K}_j^e$ and $\mathbf{M}_j^e$ are stiffness and mass matrices of the *j*th changed element, n_c is the number of changed element and δk_j and δm_j are stiffness and mass proportional modification factors for changed element *j*, on this one way number of unknowns is reduced to two per changed element. Now we have:

$$\mathbf{A}\delta\mathbf{k} + \mathbf{B}\delta\mathbf{m} = \Delta\boldsymbol{\lambda} \tag{13}$$

where **A** and **B** are *p* x n_c matrices, δk_j and δm_j column vector of size n_c, and $\Delta\boldsymbol{\lambda}$ iz column vector of size *p*, given by

$$\mathbf{A}_{ij} = \boldsymbol{\varphi}_i'^{\mathrm{T}} \mathbf{K}_j^e \boldsymbol{\varphi}_i' \qquad \mathbf{B}_{ij} = -\lambda_i \boldsymbol{\varphi}_i'^{\mathrm{T}} \mathbf{M}_j^e \boldsymbol{\varphi}_i' \tag{14}$$

$$\delta\mathbf{k} = \left[\delta k_1, \delta k_2, ..., \delta k_{n_c},\right]^{\mathrm{T}} \qquad \delta\mathbf{m} = \left[\delta m_1, \delta m_2, ..., \delta m_{n_c},\right]^{\mathrm{T}} \tag{15}$$

$$\Delta\boldsymbol{\lambda} = \left[\Delta\lambda_1, \Delta\lambda_2, ..., \Delta\lambda_p\right]^{\mathrm{T}} \tag{16}$$

Values $\boldsymbol{\varphi}_i', \mathbf{K}_j^e, \mathbf{M}_j^e$ and $\Delta\lambda_i$ can be obtained by measuring or modelling using finite element method structures in their original state. For unique solution only $p = 2n_c$ eigenvalues needed to be determined. As far as measured data usually include certain errors it is necessary to have more natural frequencies than $2n_c$.

Further simplification of the equation can be:

$$\mathbf{L} \cdot \mathbf{D} = \Delta\boldsymbol{\lambda} \tag{17}$$

where $\mathbf{L} = [\mathbf{A}, \mathbf{B}]$ and $\mathbf{D} = [\delta\mathbf{k}, \delta\mathbf{m}]$.

To solve equation this equation an optimization procedure is needed. The least square method is used to obtain acceptible solution. If the error iz denoted as

$$\mathbf{E} = \mathbf{L} \cdot \mathbf{D} - \Delta\boldsymbol{\lambda} \tag{18}$$

the condition from least square method that is needed to be satisfied is

$$\frac{\partial \mathbf{E}^T \mathbf{E}}{\partial d_k} = 0 \tag{19}$$

where d_k is the *k*th entry of vector **D**. Combining the last two equations one can obtaine

$$\mathbf{D} = \left(\mathbf{L}^T \mathbf{L}\right)^{-1} \mathbf{L}^T \Delta\boldsymbol{\lambda} \tag{20}$$

The last equation can be used to solve for all the unknown stiffness and mass proportional modification factors which indicate amount of changes.

2. NUMERICAL EXPERIMENTS

The numerical example is performed on steel frame of 8 element.

Results are obtained using finite element method, each node has three degrees of freedom – two translational and one rotational, stiffness and mass element matrices are given by Chandrupatla and Belegundu. For computations MATLAB written code is developed and used to present results.

A five-story steel frame shown in Figure 1 is analysed. It has 25 elements and 18 nodes, the first node degrees of freedom are labeled as 1, 2, 3, for second 4, 5, 6 and etc. All 25 elements are S12x35 type I beam, elastic modulus used for steel was 210 GPa. Three different types of change are investigated on element 8 and the same change on element 12:

- 15% stiffness reduction
- 15% mass reduction
- 15% stiffness and 15% mass reduction

Obtained pseudo force vector for degrees of freedom which belong change element are in range 10, for other degrees of freedom pseudo force vector is less than 10^{-8}.

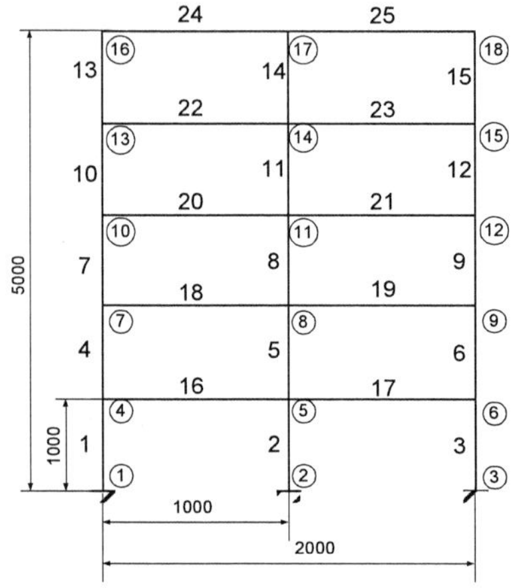

Fig. 1. Model of the structure

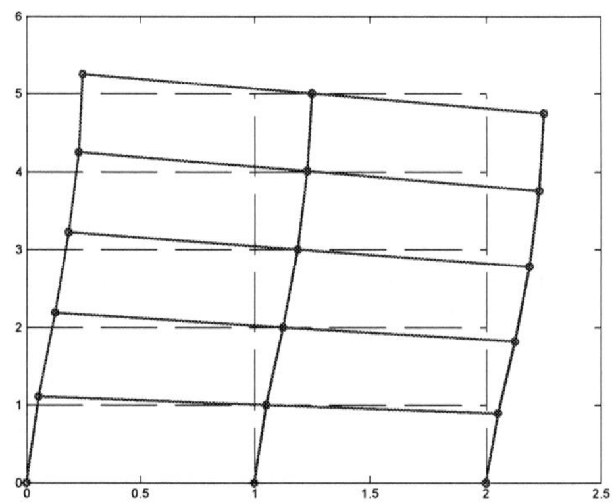

Fig. 2. The first eigen mode

So, we find the method presented very reliable and efficient.

3. CONCLUSION

The measured modal properties of structure can efficiently be used in validation of finite element model. When modal properties for structures are known in their original state and changes are detected, changed regions can be located and change evaluated.

The procedure is reliable and accurate. The error range is under 1.0% changed property.

4. REFERENCES

Chandrupatla, T.R. and Belegundu, A.D., *Introduction to Finite Elements in Engineering*, Prentice-Hall Inc., New Jersey, 1991

Farrar, C.R. and Doebling, S.W., An Overview of Modal-Based Damage Identification Methods, *Euromech* 365 *International Workshop*: DAMAS 97, Sheffield UK, 1997

Ge, M. and Lui, E.M., "Structural damage identification using system dynamic properties", *Computers and Structure*, 83, 2005, 2185-2196.

Hatch, M.R., *Vibration Simulation using MATLAB and ANSYS*, Chapman & Hall, 2001.

Meirovitch, L., *Elements of Vibration Analysis*, McGraw-Hill, 1986.

Annals of DAAAM for 2006 & Proceedings of the 17th International DAAAM Symposium, ISSN 1726-9679
Editor B. Katalinic, Published by DAAAM International, Vienna, Austria 2006
150th Birthday of Nikola Tesla (1856-1943) Inventor of Mobile Robot Teleoperation

SIMPLORER/SCHEMATIC BASED MODELLING AND SIMULATION OF A FOUR-QUADRANT CONVERTER IN TRACTION APPLICATIONS

MACAN, M.

Abstract: *This paper describes the development of a simulation model of a four-quadrant converter in AC traction drives. The functions of each of the major components of the system have been analyzed. The parameters of the model have been chosen in a way that it will be possible to build the prototype using IGBT modules with blocking voltage of 3.3kV. The model is constructed using SIMPLORER/Schematic graphical simulation platform.*

Key words: *Traction Application, Four-Quadrant Converter, Modelling and System simulation.*

1. INTRODUCTION

Modern AC traction vehicles use pulse-width modulation (PWM) controlled converters on the line-side refereed to as four-quadrant converter (4QC). The 4QC basically operates as a boost chopper (often called a PWM rectifier) with bipolar voltage (AC) at the input, but unipolar voltage (DC) at the output, maintaining sinusoidal line current. The example of traction configuration of a 4QC is shown in Fig. 1. The units are supplied from a single-phase traction transformer with multiple secondaries of high leakage reactance, allowing the switching instants to be phase-shifted. The output DC-link voltage is fed into a three-phase voltage source inverter (VSI) that drives the traction. In normal motoring mode of the drive, power flows to the motor, and the line-side converter operates as a single-phase PWM rectifier, whereas the load-side converter operates as an inverter. In regenerative braking mode, their roles are reversed, that is, the load-side converter operates as rectifier, whereas the line-side converter operates as an inverter. In many applications, such as subway and locomotive tractions, several identical induction motors are required to operate in parallel with a single VSI, and each machine drives an axle of the locomotive.

2. MODELLING OF TRACTION SYSTEM

2.1 25kV traction system

The overhead contact lines distribute the energy to the trains running on the route and they are placed above the upper limit of the vehicle gauge and supply vehicles with electric energy through roof-mounted current collection equipment referred to as pantographs. Return current circuit and return conductors feed back the traction current to the substations. Electric locomotives move continuously along the contact line, rendering the topography of the supply system a time-variable. Under these conditions, the resonant impedances vary as the traction vehicle moves along the track.

Although the latest generation of the locomotives use AC traction motors powered through sophisticated PWM AC drive control system, many of the locomotives, today still in service, are based on the DC traction motors. These older types of locomotive use thyristor-based rectifier converters for speed control, which draw current rich in harmonics. The flow of this harmonic current through the feeder impedance can result in a distorted pantograph voltage waveform, which in return, can constrain the amount of the power delivered to the locomotives.

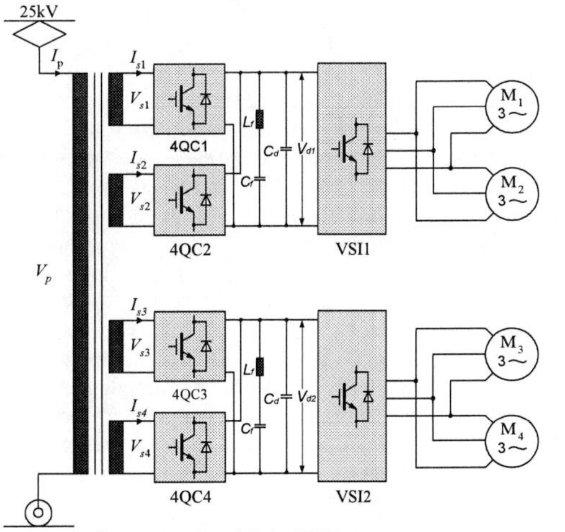

Fig. 1. Block diagram of multiple PWM converters.

The traction system considered in this simulation consist of a 30km single-phase contact feeder section with a longitudinal impedance of (0.169+j0.432)Ω/km at 50Hz and a shunt capacitance of 0.011μF/km, fed from a substation step-down transformer at 25kV, as shown in Fig. 2 (Tan, et al., 2005). The 30km contact feeder is modelled as three 10km pi-sections, and the substations transformer is represented by its equivalent series inductance and resistance. Up to four locomotive loads, each nominally rated as 2MW at 25kV, are considered. Each of the three locomotives is modelled as a pair of serially-connected half controlled thyristor bridges and fourth locomotive uses PWM controlled converter on the line-side, i.e. 4QC. The used system parameters are as follows: R_s=1000Ω, L_s=27.1mH, R_L=1.69Ω, L_L=13.8mH, C_L=0.11μF.

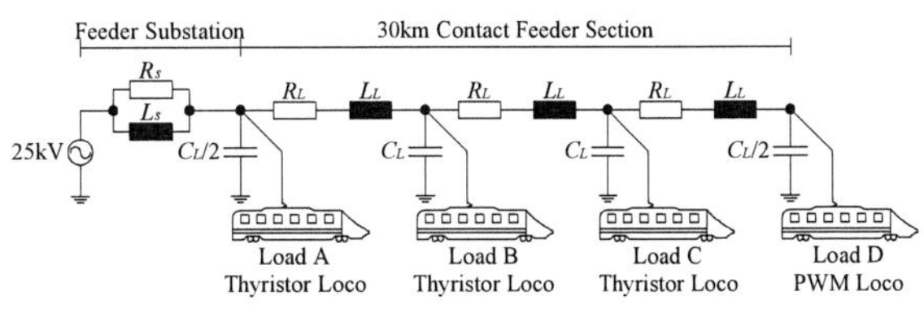

Fig. 2. Simulation model of the system representing a 30km section of a traction feeder.

2.2 Transformer model

The traction transformer receives single-phase overhead contact line AC power system on its primary side and size it down to required voltages. The secondary windings are used to supply converters for the regulated DC-link used to supply both the propulsion system and the auxiliary power system.

The most general approach to introduce all terminals of the transformer is the polygon representation, which comprises of impedances from each terminal to every other terminal (Seitlinger, et al., 1995). For m winding there are $m(m-1)/2$ impedances.

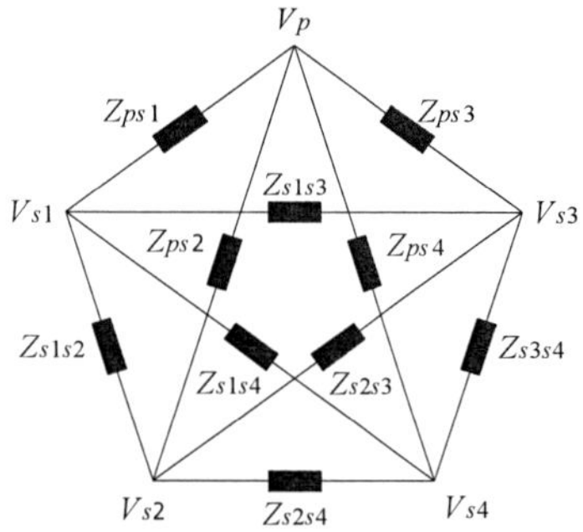

Fig. 3. Low frequency model of transformer.

The polygon impedances Z_{ij} are identified by short-circuit measurements from each winding to every other winding, e.g. *i-j* means, *i* terminals supplied and voltage/current calculated or measured, while *j* terminals are short-circuited, with $i, j=1\ldots m$, $i\neq j$ (Runge, 1997).

The rail vehicle transformer on the Fig. 1 consists of four secondary windings and one primary winding. Each secondary winding is connected to 4QC. The transformer model of the rail vehicle is implemented as a net equivalent circuit as shown in Fig. 3. Each link between two windings comprises of one resistive and one inductive part. The magnetizing current and the iron losses are nonlinear parameters, but they can be neglected in short circuit conditions and are not part of the equivalent circuit. Also no capacitive couplings are considered and saturation and skin effect can be neglected.

2.3 The four-quadrant converter

The four-quadrant converter units are supplied from a single-phase traction transformer with multiple secondaries of high leakage reactance. Naturally sampled PWM is used with PWM switching instants being determined by the intersection of a reference waveform with a symmetrical triangular carrier waveform. The control signals are phase-shifted with respect to each other to produce ripple cancellation, so the harmonic spectrum of the primary current I_p is characterized by dominant harmonic of the frequency $f_r=n_i f_s$ and by multiple line frequency side bands. For configuration (Fig 1.) $n_i=4$ and carrier frequency f_s of 1000Hz, the dominant harmonic of frequency is 4kHz. Each current control module regulates the current drawn by one converter from one secondary winding and the control principle is shown in Fig. 4 (Weem, et al., 2003).

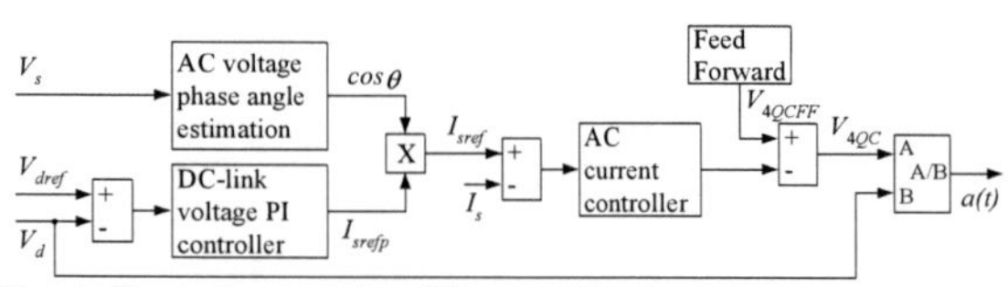

Fig. 4. Control principle of four-quadrant converter.

The command DC-link voltage V_d is compared to the actual voltage V_{dref} and the error signal trough the PI-controller is multiplied with the phase angle signal $cos\Theta$ to generate the line current command I_{sref}. The rectifier line current I_s is compared to the command line current I_{sref} and controlled by the AC current controller. An additional feed forward controller compensates the influence of the line voltage and the voltage drop across the leakage inductance. Its output is the required converter input voltage V_{4QC} which, divided by the actual DC-link voltage V_d, results in the duty cycle *a(t)*.

2.4 The DC-link

The DC-link consists of a link capacitor and of a serial resonant filter. The DC-link capacitor must be large enough to limit the ripple at the input of the inverter to an acceptable level during steady state operation, and to limit the link voltage variation as the load current changes. The filter is tuned to the second harmonic of the line frequency to eliminate strong harmonic produced by the converter (Shen, et al., 1995).

2.5 Traction motor and converter

The propulsion system traction motors and converter are represented by resistance. The value of resistance is set to 4x4.5Ω to adjust the DC-link load, approximately 4x500kW. To test the transient response, simulations were performed under a gradual load change conditions of 4x5kW/ms.

3. SIMULATION RESULTS

Simulation results are presented to illustrate the efficiency of the simulation system (Fig. 5).

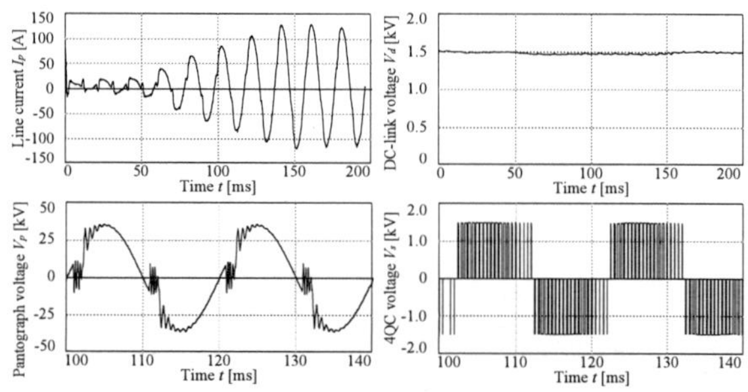

Fig. 5. Simulation results of the four-quadrant converter model (Line current I_p, DC-link voltage V_d, Pantograph voltage V_p and 4QC voltage V_s).

4. CONCLUSION

This paper describes modelling and simulation of the four-quadrant converter in traction applications using SIMPLORER/Schematic tool. The model is useful for studying the variations of different system parameters under varying conditions as well as for performing measurements of the system component peak stress levels under regular and irregular operation. It is important to point out that this paper should actually be followed by construction and testing of a real prototype of 4QC for traction application and hopefully obtain results which will support this simulation model.

5. REFERENCES

Tan, P.-C.; Loh, P.C. & Holmes, D.G. (2005). Optimal impedance termination of 25-kV electrified railway systems for improved power quality, *IEEE Transactions on Power Delivery*, Vol. 20, No. 2, (April, 2005), pp. 1703-1710.

Seitlinger, W.; Bachner, T. & Schrödl, M. (1995). Traction transformer for locomotive 1014-optimizing various requirements. *ELIN*-Zeitschrift, Heft ½, (February, 1995), pp. 45-52.

Runge, W. (1997). Control of line harmonics due to four-quadrant-converter in AC tractive stock by means of filter and transformer, *EPE 97*, pp. 3.459-3.464, Trondheim, Norway, September, 1997.

Weem, J.; Song, H.-S. & Mutschler, P. (2003). New generation IGBT four-quadrant-converter for multi-system rail vehicles using a novel control strategy, *EPE 2003*, ISBN: 90-75815-07-7, pp. P1-P10, Toulouse, France, September, 2003.

Shen, J. & Mansell, A.D. (1995). The simplified analysis and design of a PWM converter system for a three-phase traction drive, *Supplied by the British Library*, C478/6/043

Annals of DAAAM for 2006 & Proceedings of the 17th International DAAAM Symposium, ISSN 1726-9679
B. Katalinic, Published by DAAAM International, Vienna, Austria 2006
150th Birthday of Nikola Tesla (1856-1943) Inventor of Mobile Robot Teleoperation

MODELLING AND CONTROL OF CHEMICAL BATCH REAKTOR

MACKU, L. & ADAMEK, M.

Abstract: *The paper presents a chemical batch reactor control mechanism design. A program for a computer simulation using a mathematical model of the real reactor was created. An experimental based data were filled to the program to trace the real process state variables dependencies. Different process control approaches were simulated and the results were discussed. The components dosage influence and the risks related to the process were also discussed.*
Key words: *chromium waste, process modelling, batch reactor, temperature control*

1. INTRODUCTION

The alternative waste free method of dealing with the chromium waste is an enzymatic hydrolysis (Kolomazník at al. 1996, Cabeza at al. 1998). This technique separates the chrome from the protein in the form of a chromium filter cake (chromium sludge). The chromium sludge can be processed in a chemical reactor by an exothermic chemical reaction with chrome sulphate acid. The process output chemical is reusable as a tanning salt again. During the exothermic reaction a considerable quantity of heat is developing so the reaction control is necessary. The reactor temperature has to stay under a critical temperature; otherwise the reactor could be destroyed. On the other side we need to use the maximum reactor capacity to process the maximum waste amount in the shortest possible time (request on higher temperature during processing).

2. THE SYSTEM MODEL

The chemical reactor mathematical model was derived to simulate the real process (Macků & Adámek, 2005). We can see a chemical batch reactor scheme in Fig.1. The initial reactor filling is the chemical solution without the chromium sludge (filter cake) and its weight is symbolized by m_P*[kg]*. The chromium sludge dosage $\dot{m}_{FK}$ *[kg.s⁻¹]* is used to the developing heat control. The process balance equations are shown in equation (1). The first equation expresses the total mass balance of the chemical solution in the reactor. The symbol $\dot{m}_{FK}$ *[kg.s⁻¹]* expresses the mass flow of entering chromium sludge, $m'[t]$ the accumulation of the in reactor content.

$$\begin{aligned}
&\dot{m}_{FK} = m'[t] \\
&\dot{m}_{FK} = m[t]\ a'_{FK}[t] + k\, m[t]\, a_{FK}[t] \\
&\dot{m}_{FK} c_{FK} T_{FK} + \Delta H_r k\, m[t] a_{FK}[t] = \\
&\qquad = K\, S\,(T[t] - T_V[t]) + m[t]\, c_R\, T'[t] \\
&\dot{m}_V c_V T_{VP} + K\, S\,(T[t] - T_V[t]) = \dot{m}_V c_V T_V[t] + m_{VR} c_V\, T'_V[t]
\end{aligned} \tag{1}$$

The second equation expresses the chromium sludge mass balance. The input is $\dot{m}_{FK}$ *[kg.s⁻¹]* again, the accumulation is $m[t]\ a'_{FK}[t]$ *[kg.s⁻¹]* and the express $k\, m[t]\, a_{FK}[t]$ *[kg.s⁻¹]* means the chromium sludge extinction by the chemical reaction.

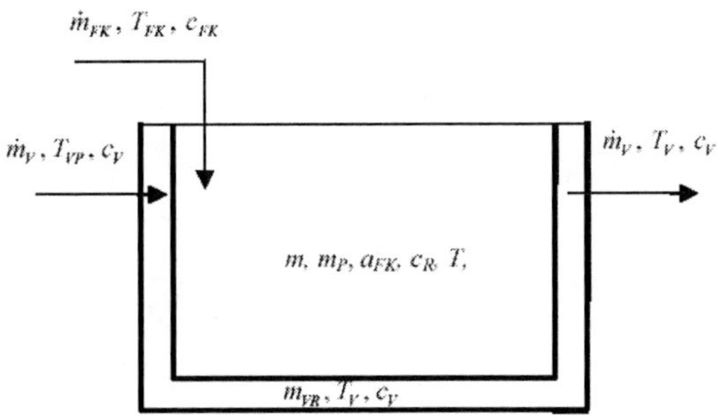

Fig.1. Scheme of chemical reactor

Symbol k *[s⁻¹]* means reaction rate constant expressed by an Arrhenius equation (2).

$$k = A\, e^{-\frac{E}{RT[t]}} \tag{2}$$

$a_{FK}[t]$ means mass concentration of chromium sludge in the chemical reactor, $m[t]$*[kg]* weight of reaction components in the reactor.
The third equation describes enthalpy balance. The reactor input heat is expressed by $\dot{m}_{FK} c_{FK} T_{FK}$, the heat rising from the chemical reaction is $\Delta H_r k\, m[t] a_{FK}[t]$ and the reactor wall heat trans-mission is $K\, S\,(T[t] - T_V[t])$. The individual symbols mean: c_{FK}*[J.kg.K⁻¹]* – chromium sludge specific thermal capacity, T_{FK}*[K]* – chromium sludge temperature, ΔH_r*[J.kg⁻¹]* – heat of reaction, K*[kg.s⁻³.K⁻¹]* - overall heat transfer coefficient, S*[m²]* – heat transfer surface, $T[t]$*[K]* - temperature of reaction components in the reactor, $T_V[t]$*[K]* – temperature of a coolant in the reactor double wall.
The last equation describes a coolant balance. The coolant input heat is expressed by $\dot{m}_V c_V T_{VP}$, the heat entering to the coolant from the reactor by the reactor wall is $K\, S\,(T[t] - T_V[t])$, the heat going out with the coolant is $\dot{m}_V c_V T_V[t]$ and the heat accumulated in the double wall coolant expresses $m_{VR} c_V\, T'_V[t]$. The symbols mean: $\dot{m}_V$ *[kg.s⁻¹]* – coolant mass flow, c_V*[J.kg.K⁻¹]* - coolant specific thermal capacity, T_{VP}*[K]* – input coolant temperature, m_{VR}*[kg]* – weight of the coolant mass in the reactor double wall. The symbol $[t]$ after some symbols indicates that the symbols are function of time.

3. SIMULATIONS

To the theoretical reaction heat authentication and to the reaction kinetic determination calorimetric experiments were done. The experimental heat of reaction varied between *1624,44–1961,29kJ.mol⁻¹*, the average value was *1747,43kJ.mol⁻¹*. The velocity constant on the temperature dependence expressed by Arrhenius equation was found as:

$$k = Ae^{-\frac{E}{RT}} = 219{,}588e^{-\frac{3604{,}4634}{T}} \qquad (3)$$

Simulations with these constants filled in the system model (1) were done in the "Mathematica" software: $A = 219{,}588\ s^{-1}$, $E = 29967{,}5087\ J.mol^{-1}$, $R = 8{,}314\ J.mol^{-1}.K^{-1}$, $c_{FK} = 4400\ J.kg.K^{-1}$, $c_V = 4118\ J.kg.K^{-1}$, $c_R = 4500\ J.kg.K^{-1}$, $m_{VR} = 220kg$, $\Delta Hr = 1392350\ J.kg^{-1}$, $K = 200\ kg.s^{-3}.K^{-1}$, $S = 7{,}36\ m^2$, $\dot{m}_V = 1\ kg.s^{-1}$, $T_{FK} = 293{,}15K$, $T_{VP} = 288{,}15K$. The $\dot{m}_{FK}$ dosage varied from *0* to $3kg.s^{-1}$. Initial conditions: the initial weight corresponded to the weight of initial reactor filling and was $m[0] = m_P = 1810\ kg$, the initial chromium sludge mass concentration in the chemical reactor was $a_{FK}[0] = a_{FK0} = 0$, initial reactor filling temperature $T[0] = T_0 = 323{,}15K$ and initial coolant temperature $T_V[0] = T_{V0} = 293{,}15K$. The desired in-reactor temperature value was set as *370K* (the safe in-reactor temperature was determined as *373,15K*). Because of the velocity constant influence, the developing heat doesn't follow the actuating signal (the chromium sludge dosing) but it is delayed. The in-reactor chromium sludge concentration accumulates quickly but the in-reactor temperature rises very slowly until the chemical reaction starts up. This reaction feature can caused uncontrollable temperature developing and reactor damage as a result. This phenomenon has to be considered with care during the process control design.

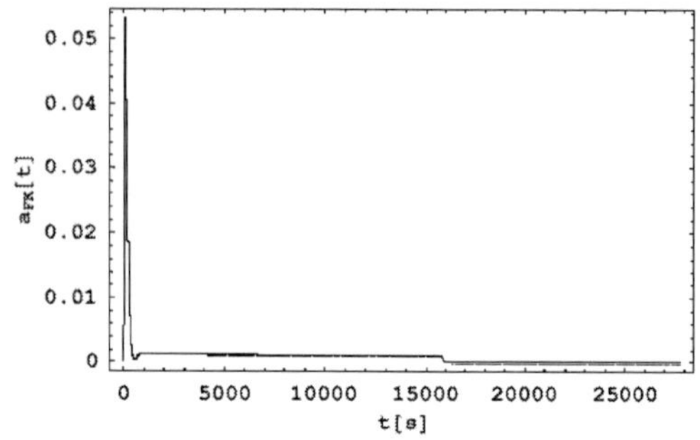

Fig. 2. The in-reactor temperature development – predictive control

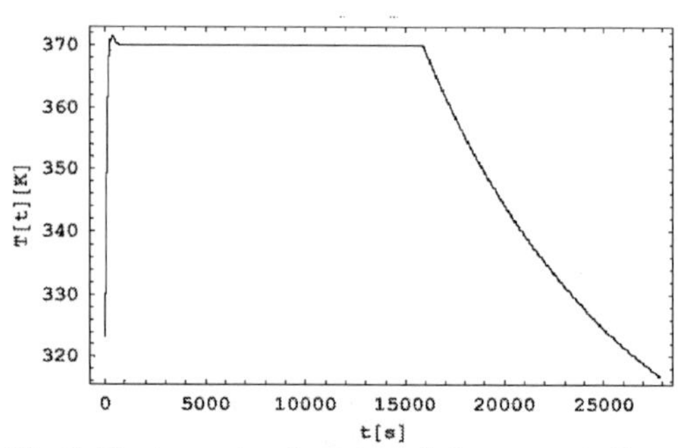

Fig. 3. The in-reactor chromium sludge concentration development – predictive control

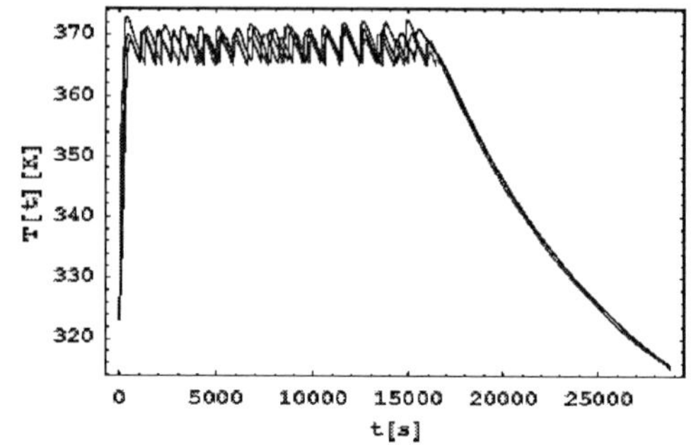

Fig. 4. The in-reactor temperature development – two step control with the penalty

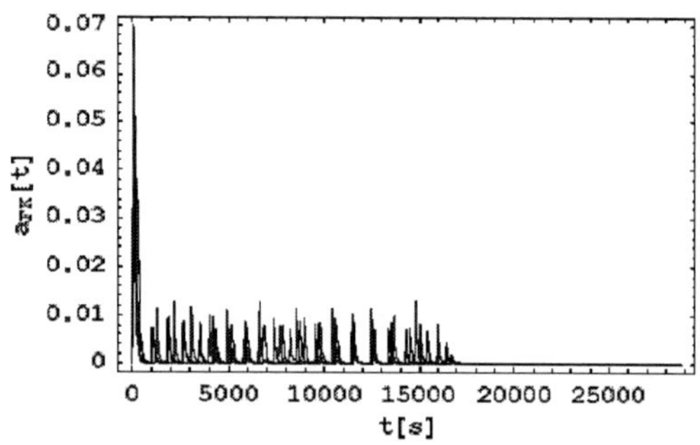

Fig. 5. The in-reactor chromium sludge concentration development – two step control with the penalty

Results of the predictive control you can see in (Fig. 2.) and (Fig. 3.). The two-step control was not suitable because the temperature got over the safe value, but the two-step control with the penalty showed very good results (Fig. 4., Fig. 5.).

4. CONCLUSION

On the basis of the experiments and simulations was found out that industrial application of the chemical reactor using two-step control with the penalty is suitable enough and the costs of such an application should be less expensive than the predictive control application costs (Macků 2003). Some more details could be found out in a next research.

5. ACKNOWLEDGMENTS

The work has been supported by the Ministry of Education of the Czech Republic under grant MSM 7088352102.

6. REFERENCES

Kolomazník, K.; Mládek, M.; Langmaier, F.; Taylor, M.; Diefendorf, E. J.; Marmer, W. N.; Brown, E. M. & Tribula, L. (1996). *CR Patent 280 655*, Februar 2, 1996.

Cabeza, L.F., Taylor, M.M., Dimaio, G.L., Brown, E.M., Mermer, W.N., Carrio, R., Celma, P.J. & Cot, J. (1998). Processing of leather waste: pilot scale studies on chrome shavings. Isolation of potentially valuable protein products and chromium, *Waste Management,* Vol 18, Iss 3, pp. 211-218

Macků, L.; Adámek, M. (2005): Model of chromium salts regeneration process, Proceedings of the 16th International DAAAM Symposium, Vienna, Austria, 2005.

Macků, L. (2003): *Control design for the preparation of regenerate for tanning*. PhD. Thesis, UTB in Zlin, 2003

Annals of DAAAM for 2006 & Proceedings of the 17th International DAAAM Symposium, ISSN 1726-9679
B. Katalinic, Published by DAAAM International, Vienna, Austria 2006
150th Birthday of Nikola Tesla (1856-1943) Inventor of Mobile Robot Teleoperation

DEVELOPMENT OF A BIOMECHANICAL MODEL FOR A ROBOTIC MANIPULATOR

MADHAB, G.B. & KUMAR, R.P.

Abstract: *When compared with the human arm, the low performances of robotic manipulators motivated the development of new mechanical structures. Joint-driven manipulators, with standard actuators, are not well adapted to the transients imposed by the robotic applications. The kinematic analysis of these systems highlights its main properties and constitutes a step towards the design of new mechanical biological-like robotic structures adopting linear actuators replacing the standard rotational joint-driving motor systems. PID tracking control of the arm was studied by implementing the biomechanical model of the forearm and hand. This model emphasizes the role of the visco-elastic properties of the musculo-skeletal system of the actual limb in controlling its net configuration and movement.*

Key words: *Antagonist actuators, biomechanics, biorobotics, PID control.*

1. INTRODUCTION

In recent years, the interface between biology and robotics is carried out by biorobotics researchers. Each component of a biorobotic system must incorporate the knowledge of diverse areas as neuromuscular physiology, biomechanics, and neuroscience to name a few, into the design of sensors, actuators, circuits, processors, and control algorithms (Garcia-Cordova et. al., 2001). It is found that some dexterous motions of human articulations are due to the capability of regulating the stiffness in accordance with a task that he/she is going to do. The skeleto-muscular system of human articulations is able to regulate its stiffness mechanically rather than by efferent command from the CNS using exteroceptive force feedback (Ito et. al., 1994). The key mechanism for regulating the stiffness is the antagonistic structure of the skeleto-muscular system; one agonist and its antagonist muscles counteractively drive one articulation as shown in Fig.1. Simultaneous stretching of both muscles provides high stiffness of the articulation and both relaxing gives us its low stiffness. It is notified that the nonlinear elasticity of the muscles is prerequisite to the agonist-antagonistic structure for regulating the stiffness (Hogan, 1984).

The visco-elastic properties of the arm play an important part in controlling its configuration and movement (Akazawa and Kusumoto, 1992). Since muscles are configured as an agonist-antagonist pair such as the flexor and extensor muscles, a limb's posture is maintained when the forces exerted by the agonist and antagonist muscle groups are equal and opposite (Koganezawa,2005).

By adjusting the visco-elasticity skillfully, we handle soft and easily breakable objects or carry out smoothly biped walking and arm movements restrictive to the environments. This paper investigates the arm biomechanics and evaluates its influence upon the driving actuators. In this perspective the article is organized as follows. Section two formulates a biomechanical model of the human arm with muscles as actuators. Section three discusses the results of simulation and PID control of system. Finally, the main conclusions are outlined in section four.

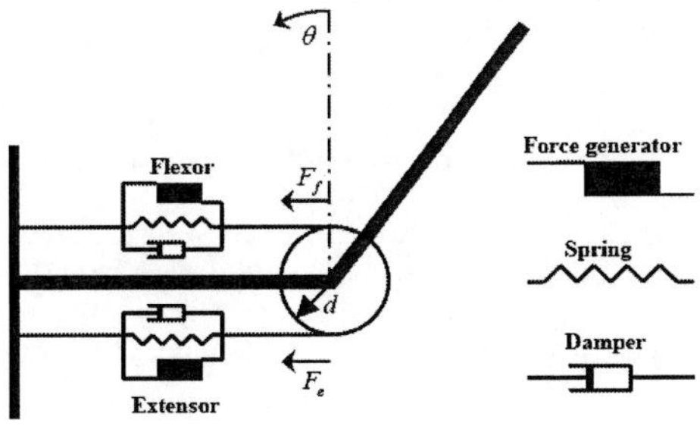

Fig. 1. Schematic diagram of muscle-arm system.

2. BIOMECHANICAL MODEL

The system consists of a planar one-link arm hinged to an inertial reference frame and acted upon by two muscle-like actuators: an extensor that rotates the arm clockwise and a flexor that rotates the arm counterclockwise as shown in Fig.1. The two states of the system are the angle θ, measured from the vertical and the angular velocity $\dot{\theta}$.

The muscle force can be assumed to be proportional to the level of activation ($0 \le \alpha \le 1$, normalized). The muscle force is given by

$$F = \alpha\, g(L,V), \tag{1}$$

where $g(L,V)$ is a nonlinear function under the maximum level of activation. Linearizing $g(L,V)$ around the rest length $L=l_0$ and the velocity of contraction, $V=0$, and neglecting the second and higher terms, the following linear relation is obtained.

$$g(L,V) = f_{\max} - k_1 x - b_1 \dot{x}, \tag{2}$$

where $f_{\max}$ is the maximum tension at isometric contraction ($V=0$), x is the relative length of the muscle ($x=0$ at rest and $x>0$ is shortening), $\dot{x}$ is the velocity of shortening, and k_1 and b_1 are positive constants. Substituting Eq.(2) into Eq.(1) yields

$$F = \alpha(f_{\max} - k_1 x - b_1 \dot{x}),$$

$$\text{or,}\quad F = u - k' u\, x - b' u\, \dot{x}, \tag{3}$$

where $u = \alpha f_{\max}$, $k' = k_1 / f_{\max}$ and $b' = b_1 / f_{\max}$.

Eq.(3) represents the visco-elastic model of a muscle, where the viscous and elastic coefficients are not constant but proportional to the contractile force u.

The musculo-skeletal system of the limb is given an equivalent representation of a parallel force generator-spring-damper combination. Based on Fig.1, the muscle torques T_f and T_e about the joint are given by

$$T_f = F_f d = (u_f - k' u_f x - b' u_f \dot{x}) d \text{ and}$$
$$T_e = F_e d = (u_e + k' u_e x + b' u_e \dot{x}) d \quad (4)$$

where, d is the radius of the cylindrical joint. Using $x = \theta d$ and defining $k = k'd$, $b = b'd$, we get

$$T_f = (u_f - k u_f \theta - b u_f \dot{\theta}) d \text{ and}$$
$$T_e = (u_e + k u_e \theta + b u_e \dot{\theta}) d \quad (5)$$

Here, θ is defined as zero in the upright position and is positive toward flexion. u_f and u_e are the contractile forces of the flexor and extensor.

From the above expressions, the dynamic model equation of the forearm and hand can be expressed as,

$$I\ddot{\theta} = T_f - T_e$$
$$= (u_f - u_e) d - (u_f + u_e) k\theta - (u_f + u_e) b\dot{\theta} \quad (6)$$

where I is the moment of inertia of the forearm. The above equation shows that the driving torque about the joint and the visco-elasticity of the joint can be controlled independently through the sum and difference of the controlled forces. Since the difference $u_f - u_e$ controls the input while the sum $u_f + u_e$ controls the system parameters, Eq.(6) represents a bilinear system. The constraints on the system is represented as,

$$0 \le u_f \le u_{fmax} \text{ and } 0 \le u_e \le u_{emax} \quad (7)$$

3. SIMULATION

Numerical simulations are done for a smooth angular position trajectory using PID control. The desired and actual trajectories are shown in Fig.2. In this simulation the sum of contractile forces ($u_f + u_e$) is assumed to be constant and is equal to $u_{fmax} = u_{emax}$. Control effort will be the difference ($u_f - u_e$). Fig. 3 shows the variation of contractile forces with respect to time. The control input shown in Fig. 4 can be positive or negative but the contractile forces cannot be negative because of the constraint that a muscle cannot apply a pushing force, but always pull.

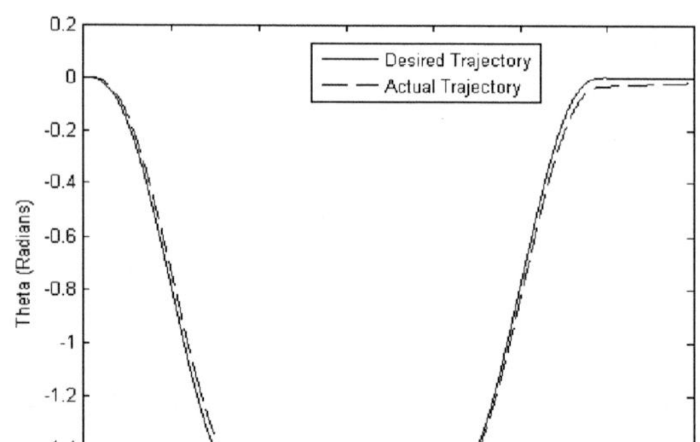

Fig. 2. Desired and actual trajectories.

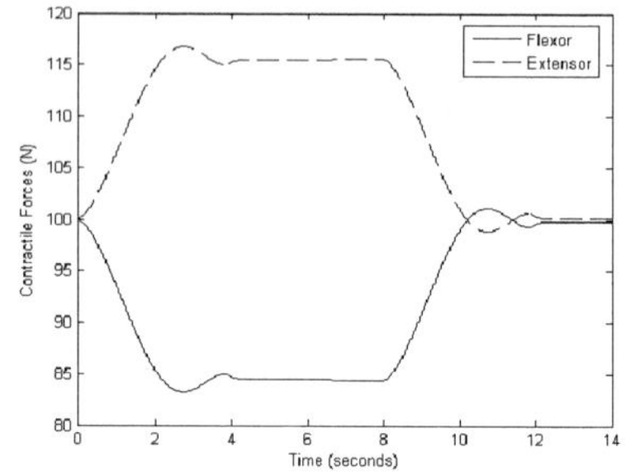

Fig. 3. Contractile forces produced by the flexor and the extensor force generators.

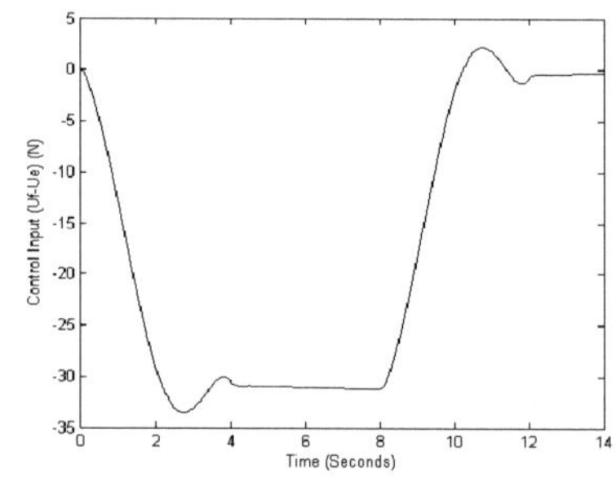

Fig. 4. Control input to the system.

4. CONCLUSIONS

A PID trajectory control scheme for forearm movement has been designed and simulated using MATLAB/SIMULINK. The advantages of antagonist muscle pair actuators in regulating the arm movement are shown in the results in terms of differential control input and individual activation of muscle actuators. With the development of new muscle-like actuators (e.g. EAP, fluidic, SMA etc.), the human and animal bio-mechanisms can be implemented in real robotic devices (e.g. manipulators, walking robots, swimming robots etc.). Thus a detailed biorobotics study of these mechanisms will enhance the overall performance and applicability of these devices.

5. REFERENCES

Akazawa, K. & Kusumoto, H. (1992). Cybernetic actuators: muscle mechanics and hardware model, *Proceedings of IEEE Int. Conf. on Biomedical Engineering Days*, pp. 48-57, ISBN: 0-7803-0743-7, Istanbul, Turkey, August 1992.

Garcia-Cordova F.; Guerrero-Gonzalez A.; Pedreno-Molina J. L. & Moran J. C. (2001). Emulation of the animal muscular actuation system in an experimental platform. *Proceedings of IEEE Int. Conf. on Systems, Man, and Cybernetics*, pp. 64-69, ISBN: 0-7803-7087-2, Tucson, AZ, USA, October 2001.

Hogan; N. (1984). Adaptive control of mechanical impedance by coactivation of antagonist muscles. *IEEE Trans. on Automatic Control*, Vol. AC-29, No. 8, pp. 681-690, ISSN: 0018-9286.

Ito K.; Pecson M.; Luo Z. W.; Yamakita M.; Kato A.; Aoyama T. & Ito M. (1994). Compliance control of an EMG-controlled prosthetic forearm using ultrasonic motors. *Proceedings of Int. Conf. on Intelligent Robots and Systems*, pp. 1816-1823, ISBN: 0-7803-1933-8, Munich, Germany, September 1994.

Koganezawa, K. (2005). Mechanical stiffness control for antagonistically driven joints. *Proceedings of Int. Conf. on Intelligent Robots and Systems*, pp. 1544 – 1551, ISBN 0-7803-9178-0, August 2005.

Annals of DAAAM for 2006 & Proceedings of the 17th International DAAAM Symposium, ISSN 1726-9679
Editor B. Katalinic, Published by DAAAM International, Vienna, Austria 2006
150th Birthday of Nikola Tesla (1856-1943) Inventor of Mobile Robot Teleoperation

INFLUENCE OF THE CHEMICAL PROCESSING ON THE CTP PRINTING FORMS

MAHOVIC POLJACEK, S.; AGIC, D. & GOJO, M.

Abstract: *The aim of this research was to make a comparison between the ageing process of the developing solution and the adsorption of the fountain solution on the CtP printing forms. Different adsorption is mostly caused by variations in chemical processing conditions. According to that, physical and chemical characteristics of the fountain and developing solutions were measured within different processing conditions. The investigation has showed that during the ageing of the developer solution certain physical – chemical changes have appeared. These changes can considerably influence the application of the fountain solution on the printing form and can result with incorrect water-ink balance during the printing process.*
Key words: *offset printing, CtP plates, chemical processing, fountain solution*

1. INTRODUCTION

During the production of the printing forms aluminium oxide base with photoactive layer must be exposed with defined laser and than processed in the certain developing solution. By the processing of the direct computer controlled imaging of the printing forms (CtP), there is the trend of creating the "processless" printing forms, i.e. creating of such forms which do not require the process of chemical developing. The reason for that is a great number of parameters that can easily vary in the developing process and that can influence the obtaining of printing forms of lower quality (Mahovic et al., 2005). Some of these parameters are: concentration of the developer components, physical and chemical characteristics of the developing solution, after treatment of printing form like gumming, developing speed, developing temperature etc. (Lovrecek et al., 1999).
In this paper the ageing of the developing solution during the processing of the printing forms was observed. It was presumed that chemical interaction between the developer and the surface of the printing forms can influence on the changes in the chemical characteristics of the developing solution (Shriver & Atkins, 1999). This kind of instability of the process could cause changes in the adsorption of the fountain solution during the printing process and could cause obtaining the imprints with lower quality (Gojo et al., 2004).

2. BACKGROUND

In order to achieve the offset printing process, the printing and nonprinting areas of the printing form must differ in their physical–chemical properties. The nonprinting areas are hydrophilic and they attract water. On the contrary, the printing areas are oleophilic i.e. hydrophobic because they absorb printing ink which is produced on the basis of oil and resins. Hydrophobic ability is not much expressed and that is the reason why the fountain solution is applied first on the printing form during the printing process. Stability of the interaction between the fountain solution and the printing ink during the printing process is very important because any kind of their emulsification could cause scumming on the imprints.

3. EXPERIMENTAL

Investigation included the imaging and developing of CtP aluminium forms with silver halide coating. Silver halide-coated printing forms have a considerably high sensitivity to UV light. Imaging and chemical processing of the silver halide coating on the printing form surface resulted in forming the ink-repelling, i.e. hydrophilic and ink-accepting areas.
In this research the first sample of the developing solution has been taken after development of the first plate in the newly prepared developing solution. The other samples of the solution were taken in an interval after a development of 250 printing forms. After 250th sample processed printing forms were unusable for printing process. pH values of the samples were measured by pH meter 330/SET from the company WTW GmbH with standard electrode with previous calibration. Electric conductivity of the solution samples were measured by the conductometer LF 330/SET by WTW GmbH. Samples of the fountain solution were prepared by mixing the concentrate with the demineralized water in the concentration of 10 vol%. Measurements of the contact angle were performed by goniometer NPL C.A. model N A-100 of the firm Rame-Hart. The solution samples were measured by means of the immersing method (Dragcevic et al., 2002). Surface tension of the prepared samples was measured by the stalagmometer method. This comparative method measures indirectly the mass of the drop sample in comparison with the sample of the known surface tension. The distilled water with the surface tension 72.75×10^{-2} mNm^{-1} at 21°C was used as the reference solution.

4. RESULTS AND DISCUSSION

Measuring results of mentioned physical-chemical parameters showed notable changes of the developing solution. Results are performed in Table 1. The decrease of electric conductivity and pH values of developing solution could have a negative influence on the adsorption of the fountain solution.

No. / Samples	Conductivity (χ / $mScm^{-1}$)	pH value
0	87.13	12.550
1	88.16	12.550
10	87.80	12.520
20	88.06	12.520
30	87.96	12.526
40	87.23	12.486
50	87.20	12.466
60	86.66	12.446
70	86.83	12.436
80	86.80	12.420
90	86.66	12.413
100	86.70	12.400
150	86.26	12.326
200	86.23	12.276
250	86.20	12.246

Table 1. Measured results of developing solution.

Contact angle of the fountain solution were measured on the printing forms separated from the processing within the interval of 50th form. It can be seen in Fig. 1 that there were certain differences in the contact angle values. These results were expected. It was presumed that any change of electric conductivity and pH values of the developing solution could cause the differences in adsorption of fountain solution. Fig. 1 shows these changes. The differences were most significant between values measured on the 1st and values measured the 250th printing form sample and were obviously caused by ageing of developing solution.

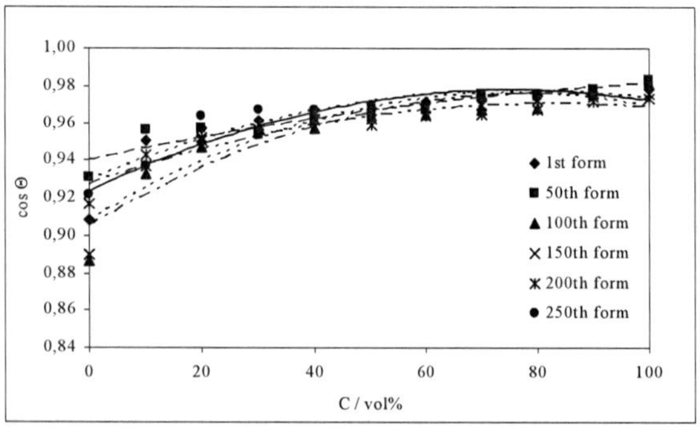

Fig. 1. Measured results of contact angle.

Further analysis included comparison of the physical and chemical characteristics of fountain solution on the 1st and 50th, on the 100th and 150th, and on the 200th and 250th printing form sample (Fig. 2, 3, 4).

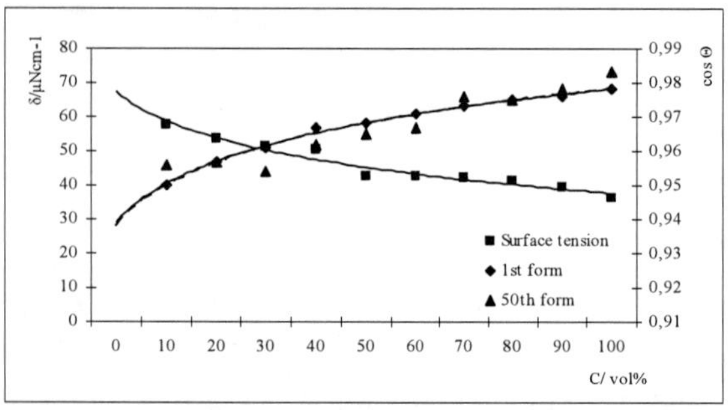

Fig. 2. Correlation between surface tension and contact angle.

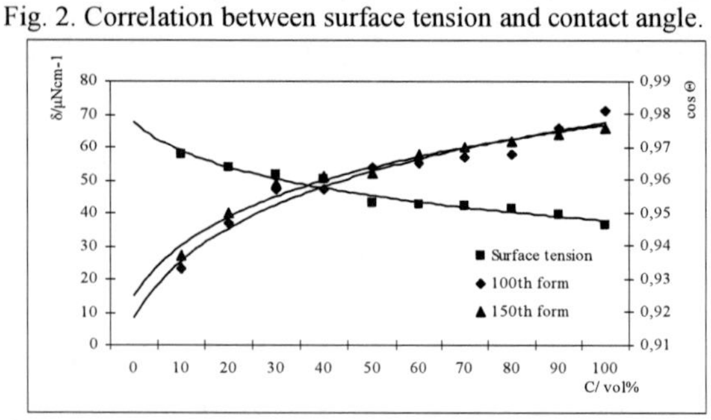

Fig. 3. Correlation between surface tension and contact angle.

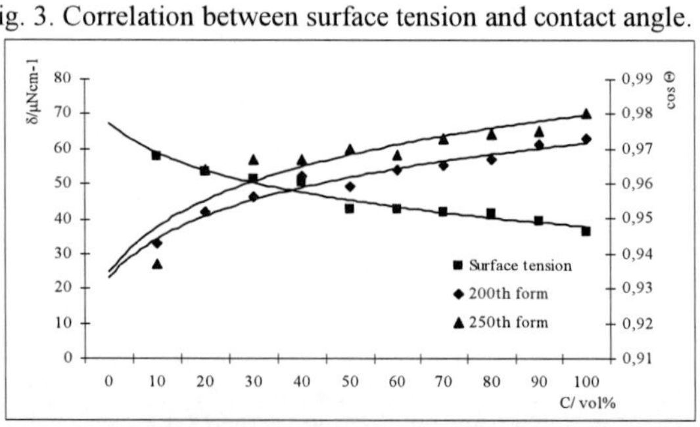

Fig. 4. Correlation between surface tension and contact angle.

The most significant differences were measured between the 1st and 250th sample (Fig. 5). The linearity between the surface tension and contact angle of fountain solution can be seen. According to the previous results, printing forms which were developed in the newly prepared processing solution have shown lower values of contact angle (higher values of cos Θ). Further results have shown that printing forms which were developed between the 200th and 250th samples have shown certain increase of contact angle values (decrease of cos Θ).

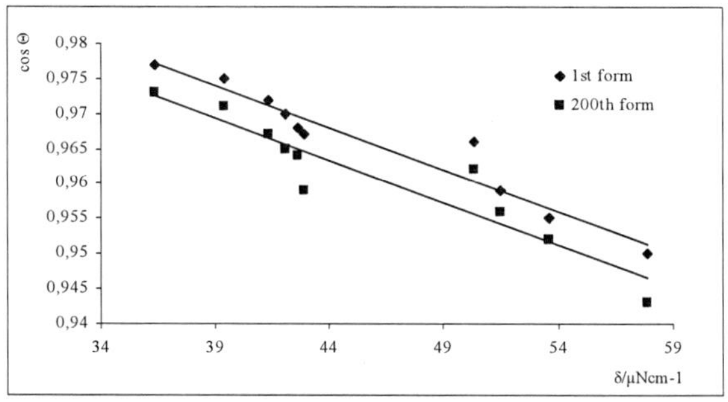

Fig. 5. Correlation between surface tension and contact angle.

5. CONCLUSION

Performed investigations were based on the fact that ageing of developing solution can cause differences in fountain solution adsorption on the nonprinting areas of the printing forms. Results have shown significant differences in electric conductivity of developing solution which have probably happened because of dissolving of the organic nonconductive particles from silver halide coating in the processing solution. pH value of developer has been changed according to the number of processed forms and with further development the quality of printing forms decreased.

Ageing of developing solution has caused a different adsorption of the fountain solution on the printing form surfaces. Increase of the contact angle can cause weaker wetting of the nonprinting areas (aluminium oxide surfaces) and additional problems in the offset printing process.

These problems are most often expressed by disturbance of the optimal balance of fountain solution and printing ink during the printing process and by appearance of scumming on imprints.

6. REFERENCES

Dragcevic, K.; Gojo, M. & Agic, D. (2002). Investigations of Physicochemical Properties of Fountain Solution in the Function of Printing Quality Prediction, *Proceedings of 13th International DAAAM Symposium*, Katalinic, B., 141-142, ISBN 3-901509-29-1, Austria, October 2002, Daaam Int., Vienna

Gojo, M., Mahovic, S., Agic, D., Mandic, L. (2004). The Influence of Paper on Physical-Chemical Characteristics of Fountain Solution, Chapter 22, In: *DAAAM International Scientific Book*, B. Katalinić (Ed.), pp. 219-230, Published by DAAAM Int., ISBN: 3-901509-38-0, Vienna, Austria

Lovrecek M., Gojo M., Dragcevic K. (1999). Advances in Printing Science and Technology, Vol. 25, Pira International, Surrey UK

Mahovic, S.; Gojo, M. & Mahovic, S (2005). Analysis of the Surface Properties of Thermal Printing Forms, *Proceedings of 16th International DAAAM Symposium*, Katalinic, B., 235-236, ISBN 3-901509-46-1, Austria, October 2005, Daaam Int., Vienna

Shriver, D. F. & Atkins, P. W. (1999). *Inorganic Chemistry*, 3rd Edition, W. H. Freeman and Company, NY

Traber, K. (1996). *Fogra No 32.*, 107, Munchen

Annals of DAAAM for 2006 & Proceedings of the 17th International DAAAM Symposium, ISSN 1726-9679
Editor B. Katalinic, Published by DAAAM International, Vienna, Austria 2006
150th Birthday of Nikola Tesla (1856-1943) Inventor of Mobile Robot Teleoperation

A CONTRIBUTION TO INSTITUTIONAL DEVELOPMENT OF QUALITY SYSTEM AT UNIVERSITY OF MOSTAR

MAJSTOROVIC V.; PETKOVIC, D. & VISEKRUNA, V.

Abstract: *University in Mostar is at its turning point in a process of fulfillment conditions stated in Bologna process. Proposal of Law on Higher Education in BiH which came from this University tries to fulfill modern achievements of European Universities. Very special part of that proposal is development of appropriate quality system, which will make structures and develop the quality management in the future development in clear and recognizable way. Even developed Europe is not very experienced in this field, so the solutions given in this papers present universal principles in modern quality system development.*

Key words: *quality, system, university, strategy, education*

1. INTRODUCTION

Development of modern quality management system means creation and implementation of internal system of organization, that is, quality management system (QMS). That means arrangement of business system which treats quality through all aspects of business operations and through which an organization is directed toward demands of its users. This also comes from the very definition of quality as a group of all properties and characteristics of products and services which meet needs of users in their best way. In other words, traditional opinion that quality is defined "from upwards" and that it expresses opinion of experts is a matter of past in all aspects of work and business operations. Quality has to be in function of users, that is, its function is to meet their satisfaction. Higher education institutions are not excepted from this trends, as Bologna model of higher education organizations dedicate its very important part to quality insurance. A number of reports, analysis and discussions given as a part of actual reform show how much radical changes of current situation concerning organization and higher education quality development are necessary. In accordance to that, solutions, given in this work, present universal principles in modern quality system development.

2. QUALITY IN HIGH EDUCATION IN BiH

It is known that, so far, higher education institutions in this area have not taken enough care of users of their services. Trying to create wider offer of higher education services, higher education institutions were led only by their own understanding of quality, and they often interpreted their users' needs independently, without thinking about their real wishes. In that way, different business systems of higher education institutions, were created (universities, faculties, colleges) and they reacted very slowly to actual society courses as well as other background impulses. Because of that ignoring attitude and neglecting of users demands, a disadvantageous ambient for higher education services development was created. That is why it can be concluded, with certainty, that it was impossible to get positive changes in higher education work, or if some appears, from time to time, they were completely wrong. It is obvious today that this approach to design and providing of higher education services was completely opposite to modern theory of quality management. Because of that, led by experiences of production organization, and faced with incoming changes and practical problems in their function, higher education institutions all over the world tried to find the answers on trends and incoming problems in focusing on quality. That meant initiation of programs for quality system creating in accordance to some of existing models.

BiH does not have quality insurance system which is based on law because of the reason that existing system does not mention existence of such a system. On the other side, existence of such a system is foreseen and described in the Law on Higher Education (version made in 2005), which is prepared and submitted to the authorities, but at the moment there are no signs when the law will be accepted. University and all university members should answer the tasks which are at the same time strategic aims of the organization through corresponding policy of quality of each organization. Quality policy consists of the following vision, mission, policy, aims and strategy. The first and minimal precondition for quality culture establishment at the university is, for sure, building of appropriate and efficient quality management system. In that way, strategic steps in quality culture establishment correspond to strategic steps in development of quality insurance system at universities in EU. Creation of procedures and directions is based on existing reports, recommendations and standards.

3. PROJECT FRAME

It is natural that in national higher education systems of all countries in the world there are clear legislative regulations which provide creation of certain elements of quality systems and criteria necessary for accreditation of higher education programs and institutions. But even partial analysis of it in BiH shows that there are no key principles of quality management based on systematic and procedural approach, programs of improvement and users' satisfaction. European quality management standards in higher education are divided in three different parts which cover internal quality insurance of higher education institutions, external quality insurance of higher education as well as quality insurance of agencies for external quality insurance.

University in Mostar is in its first phase of institutional solution of creation of proper model for quality management. Within the project led on level of all BiH universities (Tempus project Quality System Improvement at BiH Universities) a manager for University quality has been appointed, and the Office for University quality will be opened in the near future. It is, of course, just a small step in appropriate quality system establishment, which is tried to be developed at Faculty of Computing and Mechanical Engineering in Mostar, a University member which tries to be real leader in this project. In that sense, the project task which contains the following parts has been developed:

1. Introduction
2. Reasons for quality system implementation at University of Mostar
3. Planned ways and strategic steps within goals fulfillment
4. Structure and organization of institutional quality management system
5. Development of quality culture
6. Development of manpower
7. Financing of institutional quality management system
8. Co-operation with other institutions
9. Program of work and schedule
10. Control of system success
11. Conclusion

4. COMPARATIVE REQUIREMENTS OF BOLOGNA PROCESS AND ISO 9001:2000 MODEL

Through a partial review of development of idea about establishment of unique European area of higher education, some key terms of reference important for choice of model and quality management system establishment can be made. They are contained in the following:

- quality insurance and management is one of key high-positioned goals of Bologna process and EHEA (European Higher Education Area) implementation,
- with the aim of quality insurance and management each institution of higher education should define, establish and further develop their own, internal mechanisms and programs.,
- the established internal quality management system has to be directly connected to the external evaluation and judgments made by independent bodies and agencies.

World experiences show that choice and implementation of quality management model at Universities in accordance to ISO:2000 is in complete harmony with Bologna process provisions in segment of quality insurance and management, and that is why it is our proposed term of reference at this moment of quality management system development at University in Mostar. At the same time, even partial and very brief comparative analysis of some elements which are the subject of both Bologna process and ISO 9001:2000 standards requirements shows many similarities even their complete overlapping.

Above stated comparative review of similarities of some elements related to both, Bologna process and the International standard ISO 9001:2000, shows that they have the same final goal. Simply talking, that goal is to achieve high quality process of work at higher education institutions through organized, systematic and standardized approach. Because of that, and also because of numbers of other reasons, proposed model, as known and acknowledged methodology is, for sure, a good choice, and reliable and efficient method which is able to meet users' requirements and modern trends in higher education.

BOLOGNA PROCESS	ISO 9001:2000 STANDARDS
- to establish EHEA and clearly define position and role of each higher education institutions within the same	- to define vision and mission of higher education institutions as well as quality policy
- to provide mobility of students, teachers and administrative staff through compatible curriculums and clear definition of inputs and outputs and mutual connections within both the institution and EHEA	- to define key and other own processes, their inputs, outputs and mutual connections within higher education institutions as well as toward the background
- to harmonize standards and outputs of higher education institutions with the aim of their mutual recognition	- to define goals of quality, plan their fulfillment and continually review the same
- to provide corresponding resources for realization of Bologna process principles	- to provide necessary resources for realization of defined goals and processes
to define and implement obligations, responsibilities and rules of behavior for all participants in higher education through corresponding curriculums and credit (ECTS) system of evaluation, as well as appropriate models for observation of development of students, teachers and administrative staff	-to define and establish authorizations, responsibilities and rules of behavior for all participants within each identified process through appropriate documentation
-to provide equal standards, easy understandable and comparable diplomas for providing passable study and equal quality at higher education systems	-to establish system for collecting feedbacks from the process and service users, as well as appropriate methods to measure their satisfaction with the aim to improve overall higher education activities continually
-to provide and promote study for all life	-to plan and implement continual qualifying and additional education with proof about their conducting
-to accredit curriculums and higher education institutions	-to certificate quality management system.

Figure 1. Comparative Review of Similarity of more Important Elements of Bologna Process And ISO 9001:2000 For Higher Education Institutions

5. CONCLUSION

Quality insurance and management system means establishment of standard and transparent work procedures created to contribute fulfillment of certain goals and missions of any organization, as well as higher education institution. That is why establishing of quality management system can be considered as basement and the first step in promoting, improvement and further development of quality infrastructure within certain higher education institution. Universality of a subject model and practical experiences of its application are key factors for its implementation at integrated university also, which is promoted by Bologna process. At the same time, such an approach, beside the fact that it meets nominated principles of Bologna process, provides simpler application and easier implementation of quality management principles, standardization and harmonization considering the level of realized quality insurance and management efforts at entire university, that is, all its members, and also more and more necessary rationalization of organization and decreasing of total resources necessary for any development activities within higher education

5. REFERENCES

Petkovic,D.,Jasarevic,S.&Plancic,I.: (2005)Razvoj sustava upravljanja kvalitetom na Univerzitetu u Zenici kroz prizmu Bolonjskog procesa; *Kongres Kvalitet 2005, Fojnica, 2005.* ISSN 1512-9268 (Ed.Brdarevic S

Prirucnik za uspostavu sustava upravljanja kvalitetom u visokom obrazovanju, (2001), *Tempus projekat UM_JEP 16015-2001*, Sveucilište u Zagrebu, Zagreb, srpanj 2004.

Injac, N.: (1999), *Sustavi kvalitete2000, velika revizija normi ISO 9000*; Oskar, Zagreb, 1999.

Maksimovic, R, & Cosic,I.(2005): Iskustva u primjeni standarda ISO 9000 u obrazovnim institucijama, *XI Skup Trendovi razvoja "Sta donosi novi zakon o visokom obrazovanju"*, Kopaonik, 07-10. 03. 2005.

Tunjic, Dj., Kljajin, M. & Maglic, L.(2005): Uvođenje sustava upravljanja kvalitetom u visokoškolske ustanove, *Tehnički vjesnik 12*, str. 39-47, 2005.

Annals of DAAAM for 2006 & Proceedings of the 17th International DAAAM Symposium, ISSN 1726-9679
B. Katalinic, Published by DAAAM International, Vienna, Austria 2006
150th Birthday of Nikola Tesla (1856-1943) Inventor of Mobile Robot Teleoperation

AN APPLICATION OF INDUSTRIAL ROBOTS AND AN INTELLIGENT CONVEYER SYSTEM BASED ON CHAOS TECHNOLOGY IN CATERING INDUSTRY

MALISA, V. & ZUPERL, U.

Abstract: *The paper presents the application of three industrial robots and an intelligent logistic transport system for delivering beverage during the realization of fully automated ROBOTIC bar. By using hand terminals and vision system, the customer can communicate with robots and select/order drinks. The system includes a clever innovative transport system Montrac which ensures that the right drink is served to the right customer, even without controls, thanks to new Chaos technology. The ROBOTIC bar demonstrates how efficient and safe the applications of industrial robots can be in catering industry. It presents an example how to employ industrial robots in completely new sphere of activity.*

Key words: *mechatronics, robotics attendance, industrial robots, transport system.*

1. INTRODUCTION

In this contribution a fully automated bar is presented. The bar consists of two robot cells, an intelligent transport system for delivering beverage to the customers, four order terminals and main control unit (Malisa & Katalinic, 2004). The schematic structure of the whole system is shown on Fig. 1. The robots in the first cell are used for preparing and serving fresh beer on the visitors demand while the robot in cocktail cell is used for mixing cocktails. A clever transport system based on Chaos Technology ensures that the right beverage is served to the right costumer. By using four order touch screen terminals, the visitor can order beer, cocktail or combination of both. The order is sent to the main computer (soft PLC) via Profibus. The main computer ensures the optimal coordination and communication between all units. The transport system is connected to main PLC via serial communication.

2. COCKTAIL CELL

The cocktail cell consists of an Epson E2S551S scara-robot, controller RC 420, gripper GPL-40-1, cup magazine with cup dispenser, dispenser system (DRINKOMAT liquor dispensers) and ice dispenser (gravity-pick system). The cocktails are prepared by an Epson robot. The cups are taken from one of the cup dispensers which are automatically switched when one is empty. As soon as the cup is in the griper, it is controlled by an infrared sensor. The next step is to get the ice from the gravity-pick ice dispenser, which work like a lotto machine. Then the electronic dispensers fill the cup with the ordered cocktail, and waits for the shuttle to pass in order to place the cocktail into the plexiglass fitting. The innovation of the cocktail cell is that the drink dispensers and the pneumatic cylinders from the ice dispenser and the cup separator are controlled by the main computer. The valves, sensors and the dispensers are wired to a Profibus I/O block. The capacity of cell is 60 cocktails/hour.

3. BEER CEEL

The beer cell consists of two industrial robots Motoman HP-20 6 controlled by NX100 controller with direct connection to the pneumatic valves and sensors. Integrated NX100 robot controller coordinates the synchronized movement of both robots while automatically preventing collisions. The Multiple Robot Control capability of the NX100 allows the ability to operate up to four robots (36 axes of motion) from a single controller – offering a patented method of Multiple-Cooperation Robot Control, as well as a single point of control from one programming pendant to maximize productivity. Multi Robot Synchro Function, Robot Link, Robo Team are software control functions for this concept of linking robots to work in synchrony. Such multi-robot systems bring additional benefits to a range of sectors – cooperating robots working together not only complete a task faster, but they can also be used singly for other functions. The advantages of multi-robot cooperation technology have also attracted interest from other sectors. The food industry and lifting heavy loads using two smaller robots instead of one large robot, for example, are realising the benefits to be gained from having two or more robots working together. Such jobs require a very well synchronised process. Communication with the main control unit is achieved by Profibus. Based on the customer order, the first robot picks the bottle from one of beer crates and passes it to the other robot, so the first robot is able to hold the bottle. Then the second robot opens the bottle with the opener mounted on the gripper. An inductive sensor checks if the bottle has been opened properly. If the bottle was not opened in the first attempt, the robot tries to open it again. During this operation the second robot takes clean cup from the cup dispensers. Next, the beer is filled into the cup, and waits for the shuttle. The cup and the bottle are then placed into the Plexiglas fitting on the shuttle. Finally the shuttle is released, so it can deliver the beer to the terminal where the order was made. The capacity of this cell is 40 beers/hour.

4. INTELLIGENT TRANSPORT SYSTEM BASED ON CHAOS TECHNOLOGY

The beverage is served to the right customer by an intelligent conveyer system Montrac. This system is linked to the main control unit by a serial interface. It works without any controls thanks to the new Chaos Technology. It has the following advantages over traditional classical conveyor systems (Montech, 2001):

- Simplicity; It is comprised of very few components, making design and assembly easy. The components are quick to assemble. The exclusive use of electronic components eliminates the need for pneumatic connections and complicated wiring. Chaos Technology greatly reduces the cabling and programming. The greatest benefit of this simplicity is not only during assembly but also when changing or expanding the whole system.

- Reliability; A shuttle can always be removed and additional shuttles can be added. The track itself is a passive element and not subject to failure. The shuttle was designed to be maintenance free.

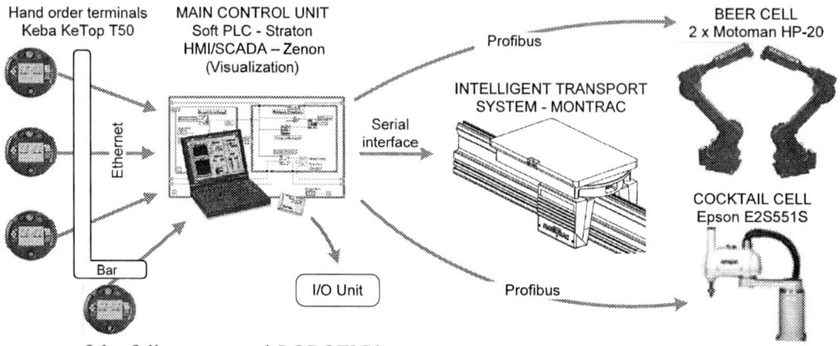

Fig. 1. Schematic structure of the fully automated *ROBOTIC* bar.

- Flexibility; The conveyer system is especially well suited for chaotic manufacturing, planned expansion, phased integration, short product life, and many other applications that are not generally considered suitable for transport systems. The transport system is flexible and it can be expanded, modified or adapted to different system configurations almost effortlessly. It has modular construction with very few interchangeable mechanical parts, so it can easily be reconfigured.
- No Maintenance; Mechanical components constitute a tiny portion of the overall system and are very robust. The track itself is a passive element and is not subject to wear. The shuttle is self-propelled by a brushless DC motor. All bearings are sealed and lubricated for life. Modular design means that the unlikely minor repair can be made easily.
- Efficiency; System consumes 95% less electricity and no compressed air when compared to traditional systems. The simplicity of conveyer system means considerably less time spent designing, installing, and programming. Travel time becomes production time.

Montrac virtually excludes the possibility of a system failure (Sallez et al., 2004). Montrac is an efficient solution for complex and expandable production systems, but it is also competitive in smaller/simpler layouts.

4.1 Chaos Technology

When making a variety of products (drinks in our case), some process may be common to all products while other processes are dedicated to a specific product. It may also be that different products require processes in different order. This makes writing a centralized logistic program nearly impossible. With Chaos Technology, products can easily navigate through chaotic production processes without a centralized control system. This is the main reason why the Montrac transport system is used for this project. Chaos Technology is based on the simple principal that, upon completing a process, that process station (Cocktail and Beer cell) communicates to the shuttle what the next process station will be. At each place that a routing decision needs to be made, the Trac asks the shuttle where it is headed and directs it accordingly to pneumatically controlled turnouts. In addition to providing chaotic routing capability, Chaos Technology reduces the need for I/O hardware, programming, and cabling; and localized controls make the system easy to reconfigure.

5. MAIN CONTROL UNIT

The task of the main control unit is to manage the orders from customers, to control the valves of the cocktail cell and to visualize the error codes from both cells and the conveyor system (Malisa, 2005). The main control unit is a Windows based PC with installed Zenon and Straton software. The Straton works as a soft PLC, while the Zenon 6.2 SP3 is used for process visualization. Zenon is a HMI/SCADA (Human-Machine-Interface/ Supervisory Control and Data Acqusition) software.

6. ORDER TERMINALS

The drink order can be performed on four touch-screen terminals KeTop T50. They are used as order terminals. The order is then sent to main control unit via Profibus.

7. CONCLUSION

In this project the University of Applied Sciences Technikum of Vienna worked together with Epson, Motoman, Keba, Montech and Copa-Data to create the ROBOTIC bar. The bar was successfully presented and tested by the guests at the Automatica Fair in Munich (2006), where it has shown that the industrial robots can be successfully, efficiently and safely used in bar attendance (Malisa, 2006). The robots were preparing and serving the visitors with 18 different kinds of drinks.

Extensive testing confirms that the bar has a capacity of 40 drinks/hour. The main effort during realization of the project was how to achieve efficient and reliable communication between two robot cells, transport system and order/hand terminals. The innovation of this project is also the design and optimization of beverage transport logistic.

8. REFERENCES

Malisa (2006). Mechatronik- mehr als ein Modewort, *Available from*: http://www.handling.de *Accessed: 2006-06-30*

Malisa (2005). Tag der Bildverarbeitung, Panasonic Electric Works Austria GmbH, pp. 1-6, September 2005.

Malisa, V. & Katalinic, B. (2004). Next Generation of Production Systems: Original Concept of Selforganizing Production Systems, *Eight International Conference on Manufacturing & Management Proceedings*, pp. 1-14, 0-9578296-1-2, Queensland, Australia; October, 2004

Montech (2001). Powered pallets on track replace conveyors, *Available from*: http://www.manufacturingtalk.com. *Accessed: 2006-05-2,* Edited by the Manufacturingtalk Editorial Team

Sallez, Y.; Trentesaux, D.; Berger, T. &. Tahon, C. (2004). Product-based and resource-based heterarchical approaches for dynamic FMS scheduling. *Computers & Industrial Engineering*, Vol. 46, No. 4, July 2004, pp. 611-623.

Annals of DAAAM for 2006 & Proceedings of the 17th International DAAAM Symposium, ISSN 1726-9679
B. Katalinic, Published by DAAAM International, Vienna, Austria 2006
150th Birthday of Nikola Tesla (1856-1943) Inventor of Mobile Robot Teleoperation

FMS CONTROL WITH ARENA

MALOPOLSKI, W. & SLOTA, A.

Abstract: *This paper describes the use of Arena simulation software for a manufacturing system control in real time mode. To integrate Arena with real system Message Dispatcher application is built – it translates and transfers messages between Arena and local controllers of the system. Real time Arena model of educational EMCO system was built and the proposed control solution was tested for this system.*
Key words: *Arena, simulation, control, real time.*

1. INTRODUCTION

Today's manufacturers require manufacturing systems which may be quickly adapted to market demands (Mazziotti & Horne, 1997). Therefore manufacturing systems' structure and control systems change very often. Before starting a procedure of the system rebuilding the changes should be proved correct. Due to complex structure of real manufacturing systems the way the FMS works is often analysed by simulation. Using simulation techniques the impact of designed changes on the system behaviour may be quickly assessed and a suitable solution may be found. Simulation is an execution of a virtual process which corresponds to processes execution in the real system. Data crucial for analysis of FMS operation are included in the FMS model which is processed during simulation experiment. For FMS control some kind of a mathematical model of the system is also required. Such model, processed in real time, is used to calculate control decisions. Thus it comes natural to make use of the model proved to be correct in simulation for control of a real system. The second advantage would be using simulation software, which processes the model during simulation, for processing the same model during control.
Arena simulation software is a very flexible tool for modelling and simulation (Eldabi & Paul, 1997). It provides an intuitive environment for model creation and implementing the model with real-world data like processing times, resource requirements, etc. Such models are simulated and then reports on specified criteria are generated. RT Arena extension enables running a simulation experiment in execution mode. In this case Arena can coordinate the model with real-world processes by exchanging messages with external applications. This feature may be used for monitoring and control of real systems (Kelton et al., 2002). The main goal of this work is to integrate Arena with local controllers of EMCO system in the laboratory at Cracow University of Technology so that it can be used for control of this system.

2. PROPOSED SOLUTION

The diagram of the Arena based FMS control system is presented in the figure 1. Message Dispatcher is a piece of software which integrates Arena with local controllers of FMS. It receives commands of activity execution from Arena and dispatches them to local controllers which are responsible for the activity execution. It also receives messages confirming activity execution form local controllers and sends them back to Arena.

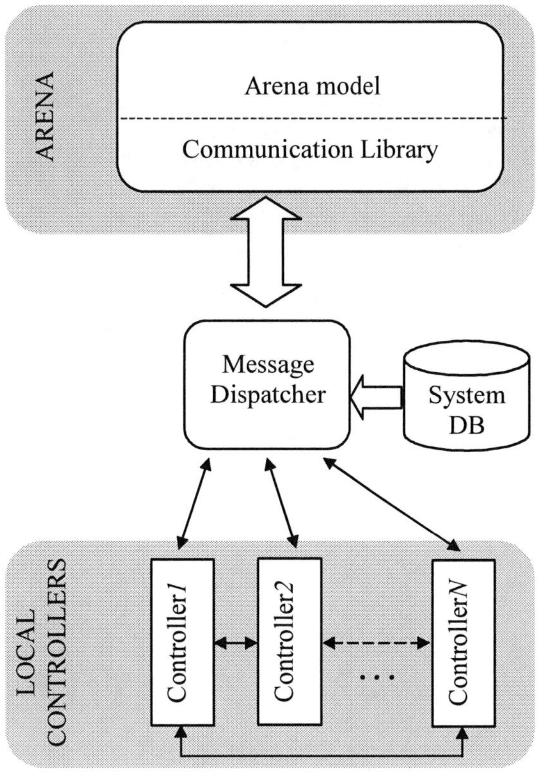

Fig. 1. Arena based FMS control system.

Communication between Message Dispatcher and Arena uses socket mechanism. The first step of running Arena model in execution mode is establishing a connection with Message Dispatcher. Arena goes into a listen state and waits for connection from Message Dispatcher. After a correct connection request has occurred it is accepted and socket communication is established. Any time Arena model needs to send an activity execution command to Message Dispatcher writeIPCQueue function, which sends messages through the socket, is called. When Message Dispatcher sends a message to Arena readIPCQueue function, which reads data form the socket, is called. These functions are located in Communication Library (dynamic linked library) provided with Arena (Fig. 1).
Since local controllers may use different communication protocols and data format the part of Message Dispatcher responsible for communication with local contollers is dedicated to a given system.
Message Dispatcher extracts data from a message received from Arena and transforms it into the format required by local controllers. For this purpose System DB is used. System DB

defines also to which local controllers a given activity execution command should be sent. The same controllers are expected to send back to Message Dispatcher confirmation of activity execution.

3. EXAMPLE OF ARENA BASED CONTROL SYSTEM

3.1 EMCO system description

The proposed solution of using Arena for FMS control will be presented for a simple educational production cell EMCO. The layout of the cell is shown in figure 2. The cell consists of two identical machine tools: LATHE1 and LATHE2, TABLE and ROBOT.

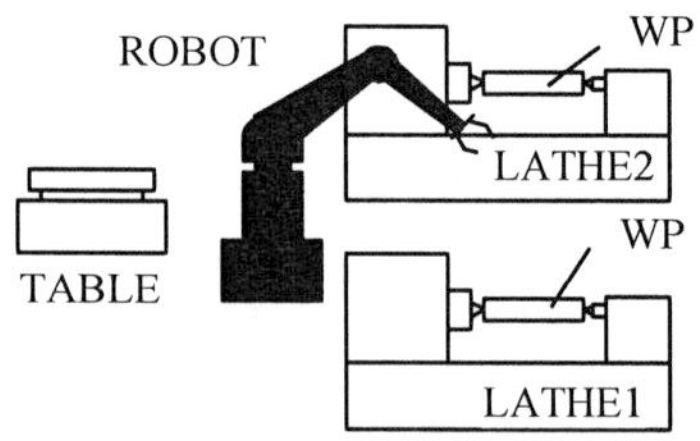

Fig. 2. Educational production cell EMCO

3.2 Arena model of EMCO system

Arena model of EMCO system is presented in figure 3. In the model three stations are defined: TABLE, LATHE1 and LATHE2. Routing is added to each generated workpiece (LATHE1 and then LATHE2) and workpieces are moved to TABLE station. Then single workpieces are processed in the system according to their routes.

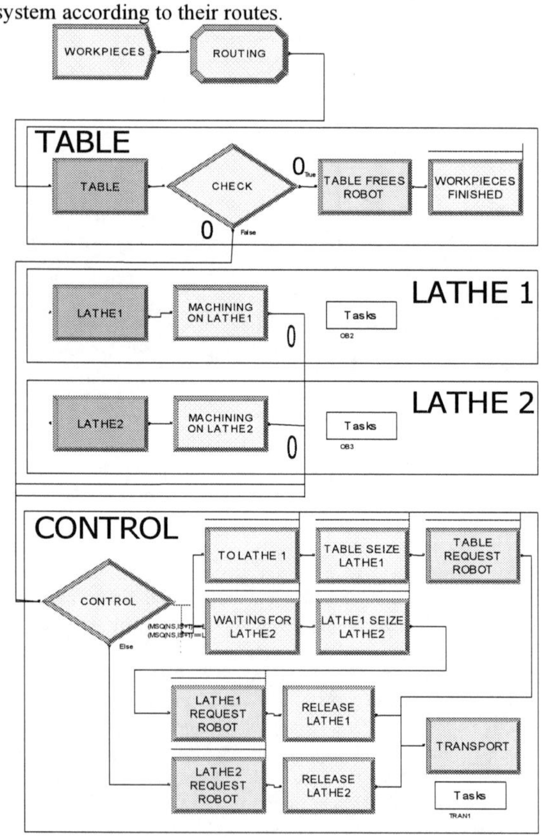

Fig. 3. Real time Arena model of EMCO system

For communication purposes three tasks elements are defined in the model. They are assigned to process modules: MACHINING ON LATAHE1 and MACHINING ON LATAHE2 and to transport module TRANSPORT. Tasks elements define data and format of messages which Arena sends to Message Dispatcher. Initial and final positions of a workpiece for different transport activities are calculated in CONTROL part of the model on the basis of the workpiece's current state and its route.

3.3 Control system description

As local controllers of LATHE1, LATHE2, TABLE and ROBOT PC based controllers, prepared for Petri Net based control system are used (Slota, 2003). Local controllers are installed and run on three computers:

- computer1: controller of LATHE1,
- computer2: controller of LATHE2,
- computer3: controllers of ROBOT and TABLE,

To integrate Arena with local controllers of EMCO system computer program Message Dispatcher, characterized in chapter 2, has been worked out. Communication between Message Dispatcher and Arena is implemented as described in chapter 2. For exchanging messages between Message Dispatcher and local controllers DCOM technology is used (Zajac, 1998).

4. SUMMARY

Building FMS control software is a laborious, time consuming and error prone process. Simulation is a basic technique for analysis and verification of FMS control systems. The possibility of using available simulation software and verified simulation models for FMS control eliminates all the work necessary to build special control systems and adapt control models. What is necessary is to integrate simulation software with existing local controllers of FMS. In the paper FMS control system based on Arena simulation software is presented. Arena provides communication mechanism with external processes by means of a dynamic linked library (dll). In the proposed solution, instead of customizing Arena provided dll, an external application – Message Dispatcher – was built. This application is designed to integrate Arena with EMCO system but it may be modified so that it can be used for other systems.

5. REFERENCES

Eldabi, T. & Paul, R. J. (1997). Flexible modelling of manufacturing systems with variable levels of detail, *Proceedings of the 1997 Winter Simulation Conference,* S. Andradottir, K. J. Healy, D. H. Wither, and B. L. Nelson (Ed.)

Kelton, W. D. & Sadowski, R. P. & Sadowski, D. A. (2002). *Simulation with Arena*, McGrow-Hill Companies, ISBN 0-07-112239-7, New York

Mazziotti, B. W. & Horne, R. E., Jr. (1997). Creating a flexible, simulation - based finite scheduling tool, *Proceedings of the 1997 Winter Simulation Conference,* S. Andradottir, K. J. Healy, D. H. Wither, and B. L. Nelson (Ed.)

Slota, A. (2003). Application of Object Observable Petri Nets in simulation and operational control of discrete manufacturing systems, *Proceedings of the 14th International DAAAM Symposium*, Katalinic, B. (Ed.), pp. 431-432, ISBN 3-901509-34-8, Sarajevo, October 2003, DAAAM International, Viena

Zajac, J. (1998). Interobject communication in distributed manufacturing using COM technologies, *Proceedings of the 9th International DAAAM Symposium*, Katalinic, B. (Ed.), pp. 513-514, ISBN 3-901509-08-9, Cluj-Napoca, October 1998, DAAAM International, Vienna

Annals of DAAAM for 2006 & Proceedings of the 17th International DAAAM Symposium, ISSN 1726-9679
Editor B. Katalinic, Published by DAAAM International, Vienna, Austria 2006
150th Birthday of Nikola Tesla (1856-1943) Inventor of Mobile Robot Teleoperation

PROPERTIES OF FIBER REINFORCED POLYESTER LAMINATES

MARUSIC, V.; KRUMES, D. & DZEPINA, I.

Abstract: *Mechanical properties and damages of fiber reinforced polyester laminates have been examined, depending upon number of reinforcing layers, their layout and thickness. Different layers of glass fibers ("mat"), and "roving", as well as matrix of polyester rosin with additives and calcite have been used as reinforcing links. It has been concluded that mechanical properties of polyester laminates depend not only upon composition but also on homogeneity of structure, i.e. presence of larger number of cavities and holes, as indicators of non-homogeneity.*

Key words: *polyester laminates, mechanical properties, composition, structure*

1. INTRODUCTION

Research of properties and shapes of damages of polyester laminates is considered to be very complex and challenging field of study (Rossmanith,1995), (Kunej,2001). Adhesion between fibers and matrices is the most important precondition for transfer of strains in composites. Change in number, layout and thickness of reinforcements result in optimal improvement of properties (Semenski,1998), (Španiček,1998).

2. MATERIAL, METHODS, TEST REZULTS

Seven plates of polyester laminates have been fabricated for the purpose of testing. Cross section of plates, reinforcement layout within the matrix is shown in Fig. 1.a (schematic). Review of plate composition has been shown in Table 1.
Plates with size 500x700 mm have been fabricated by soaking of the reinforcement in polyester resin, and differed in thickness, type and content of reinforcement and additives. There are two sides of plate: "A"- coarse side with polyester resin; "B"- smooth and shiny side with gel coat additive so as to obtain esthetic effect.
Thicknes of the plate, test results of tensile strength (DIN53455) and module of elasticity (DIN53457) are shown in Table 2

No. of the plate	Thicness, mm	Tensile strength, N/mm^2	Tensile module of elasticity, N/mm^2
1	3,8	126,6	8803
2	4,1	117,4	12073
3	4,2	105,9	10916
4	3,9	without fracs	12791
5	4,2	106,1	8734
6	4,0	130,7	10101
7	4,8	100,3	6083

Table 2. Arithmetic mean of tensile strength and module of elasticity (3 specimen).

Test results were used for calculation of arithmetic mean: mass content (DIN52330) of inorganic contents (glass fibers and fibers) and bending strength (DIN53452) are shown in Table 3.; interlayer strength (BS2782) and tensile strength by Charpy ,4J, 4x6mm sample with groove (DIN53453) are shown in Table 4. Tests were performed separately for side "A" and for side "B" of laminates.

No. of the plate	Content of inorg. matt. %	Bending strength, N/mm^2	
		Side "A"	Side"B"
1	49,49	239,8	242,4
2	44,76	251,1	239,5
3	42,06	202,7	233,5
4	47,85	247,1	291,0
5	51,49	157,3	212,0
6	44,79	209,0	175,2
7	44,65	156,6	175,3

Table 3. Arithmetic mean of mass content of inorganic substance in plates and bending strength (5A+5B specimen).

Compositon						
1	2	3	4	5	6	7
Side "A" resin	Side "A" resin	Side "A" resin	Side "A" resin	Side «A» resin with addition of calcite	Side «A» resin with addition of calcite	Side «A» resin self extinguishing
3 layers of mat (450 g/m^2)	2 layers of mat (450 g/m^2)	2 layers of mat (300 g/m^2)	1 layer of mat (450 g/m^2)	3 layers of mat (450 g/m^2)	2 layers of mat (300 g/m^2)	3 layers of mat (450 g/m^2)
2 layers of mat (450 g/m^2)	1 layer of textile (300 g/m^2)	"Combimate"	2 layers of textile (300 g/m^2)	2 layers of mat (450 g/m^2)	2 layers of textile (300 g/m^2)	2 layers of mat (450 g/m^2)
Side "B" gel coat	2 layers of mat (450 g/m^2)	1 layer of mat (450 g/m^2)	2 layers of mat (450 g/m^2)	Side "B" gel coat	1 layer of mat (450 g/m^2)	Side "B" gel coat
	Side "B" gel coat	Side "B" gel coat	Side "B" gel coat		Side "B" gel coat	

Table 1. Review of composition of test plates.

No. of the plate	Interlayer strength N/mm^2		Tensile strength, kJ/m^2	
	Side"A"	Side"B"	Side"A"	Side"B"
1	21,9	23,7	62	77
2	23,7	24,5	76	92
3	16,8	19,8	87	88
4	18,7	21,8	77	115
5	24,9	25,1	71	75
6	19,3	22,8	77	89
7	15,0	18,2	62	73

Table 4. Interlayer strength and Tensile strength (calculated from impact strength)(5A+5B specimen).

Recording of plate cross sections, Fig. 1. (b, c, d, e, f, g, h) has been conducted on light microscope under polarized light (with magnification 100X), for the purpose of linking of mechanical property test results with structure of laminates. These recordings should contribute to detection of defects, i.e. irregularities, showing eventual possibilities for further damage of laminates.

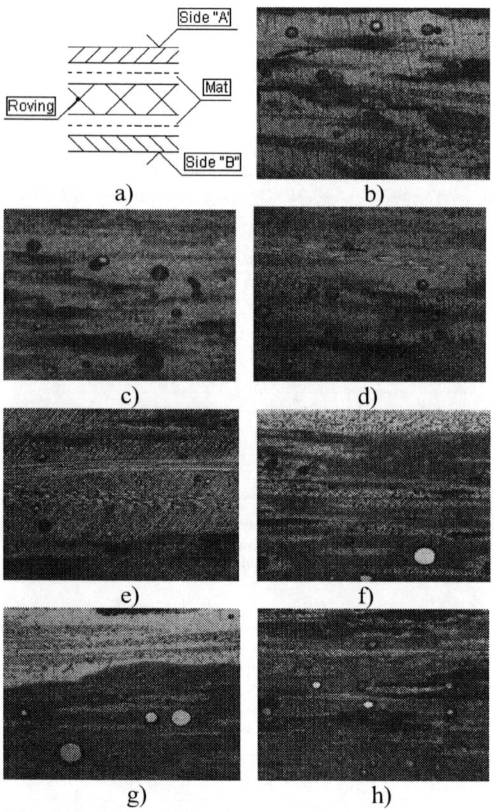

Fig. 1. Structure of plate cross sections:
a) Schematic layout of reinforcement in the matrix;
b) Plate 1; c) Plate 2; d) Plate 3;
e) Plate 4; f) Plate 5;
g) Plate 6; h) Plate 7.

3. RESULT EVALUTATION AND CONCLUSION

Focus of research has been laid upon interdependency between tensile, bending and interlayer strength and tensile strength in mass volume fraction of reinforcement, on example of polyester laminates with different composition. According to tested mechanical property results (Side "A"), laminate plates (designated by serial numbers) have been classified (Roman number designations) from the best quality plates (I; II…) to the worst quality plates (VI, VII), Table 5 (*from greater reinforcement content (%) to the lesser).

Property	Classes of laminate plates(1-7) from the best to the worst						
	I	II	III	IV	V	VI	VII
Tensile strength	6	1	2	-	5	3	7
Tensile module of elasticity	4	2	3	6	1	5	7
Tensile strength by Charpy	4	2	6	3	1	5	7
Interlayer streng.	5	2	1	6	4	3	7
Content of inorganic matter*	5	1	4	6	2	7	3

Table 5. Classification of laminate plates of polyester composites according their tested properties.

According to comparison of test results, structure and composition of plate cross sections of tested polyester laminates the following conclusions may be reached:
- Average high classification of plate 4 is most probably result of homogenous structure (Fig. 1.e), and the biggest number of textile layers (two layers) in the middle of the plate.
- Difference in plate classification 1, 2 and 5 (having also two layers of textile in the central part) is due to greater porosity of material (Fig. 1.b, 1.c and 1.f).
- Average classification of plate 7 is due not only to lesser content of reinforcement (no textile) but also to evident lack of homogeneity of structure (Fig. 1.h).

Out of comparison of test results it can be concluded that due to strains different aspects of damages of polyester laminates may occur. Thereby different factors exert influence on these properties, not only sort and fraction of reinforcement, but also condition on boundary between reinforcement and matrix. In production by soaking (applied in production of these test plates) it is of the utmost importance to force out the air (i.e. by roller..), and so to decrease number of present holes in the adjacent surfaces.

5. REFERENCES

Kunej W. (2001). Analiza stanja poliesterskih smola i kompozita na njihovoj osnovi, Polimeri, 22(1), 14-28,0351-1871

Rossmanith H.P.; Knasmilner R.E. & Semenski D.(1995). Crack - tip Caustics in Mechanically Anisotropic Materials, Experimental Mechanics, 35, 31-35

Semenski D., Husic, S.(1998). Structural and Mechanical Characterization of Compozites, Polimeri, 19(8-12), 183-189,0351-1871

Spanicek, D.; Smolcic, Z. &Husic, D.(1998). Odnos apsorpcije vode, udjela supljina i meduslojne cvrstoce polimernih kompozita, Polimeri, 19(8-12), 190-193, 0351-1871

Annals of DAAAM for 2006 & Proceedings of the 17th International DAAAM Symposium, ISSN 1726-9679
Editor B. Katalinic, Published by DAAAM International, Vienna, Austria 2006
150th Birthday of Nikola Tesla (1856-1943) Inventor of Mobile Robot Teleoperation

DETERMINATION OF CUTTING FORCES DIRECTION AND POINT OF ACTING

MATUSOVA, M.; KATALINIC,B. & HRUSKOVA, E.

Abstract: *Cutting force measurement is base on pressure sense actuating to toll. Basically this pressure is dedicated cutting force. Measurement cutting forces in milling are used special cutter blocks.*

Key words: *cutting force, centre pressure, force direction*

1. INTRODUCTION

Force which actuating cutting cline to chip term as cutting force F . Reaction to this cutting force is force of cutting F' resistance tooling material. Cutting force of one cutting cline we can decompose on three each other vertical components:

- force direction main cutting speed - F_c ,
- radial force direction to main – return component of force - F_p,
- axial – feed component of force - F_f .

The main cutting component of force is necessary to calculating torque and tension force.

Torque $M_k(N.m)$ which are necessary to tooling we defined from equation:

$$M_k = F_C \frac{d}{2} \qquad (1)$$

d - diameter tooling surface /mm/

Return component of force F_p is necessary for account of shape inaccuracy in tooling system, because they deform work piece and system. Feed component of force - F_f insure feed mechanism of machine. Reciprocal proportion of average value of this force components are:

$$F_c : F_p : F_f = 1 : 0{,}4 : 0{,}25$$

2. CUTTING FORCE AT MILLING AND DRILLING

Analyze of influence factors to cutting force is base on measurement forces pending milling at different conditions. Measurement cutting forces in milling are used special cutter blocks. In tooling process we indicate resulting cutting force and their components by empirical equation:

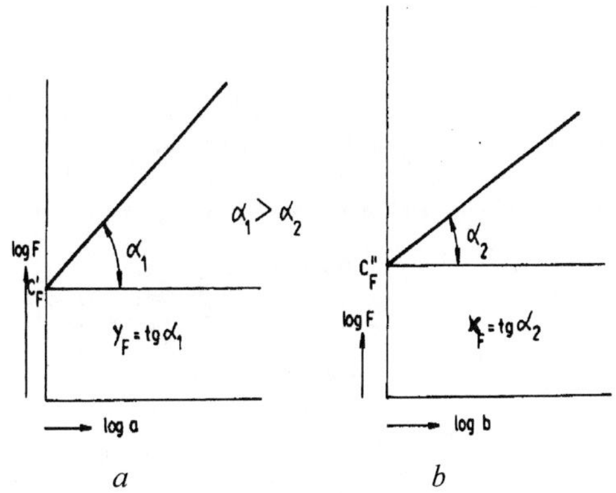

Fig. 1. Diagram of dependency cutting force on a- thickness of cutting layer , b- width of cutting layer

$$F = C_F S = C_F b^{x_F} F_{K_F} F_a^{y_F} \qquad (2)$$

C_F -material constant, depends on tooling material properties,

a -cutting thickness,

b -cutting width,

K_F -correction index, which respect tooling conditions, e. g. cutting speed influence, tooling and cutting materials, tool geometry and next actors to dimension of cutting force,

x_F, y_F- exponents, which are necessary determine.

Determination constants and exponents are performed follows:

At specific constant conditions we determine dependency $F = f(b)$ and $F = f(a)$ in logarithmic function (fig.1)

At specific value $a = a_1$

$$F = C'_F b^{x_F} \qquad (3)$$

In logarithmic function cutting force of dependency is determine as:

$$x_F = \tan \alpha_1$$

Constant C'_F is relevant ordinate F at $b = 1$.

At specific value $b = b_1$

$$F = C'_F a^{y_F} . \quad (4)$$

In logarithmic function of cutting force dependency is determine as:

$$y_F = \tan \alpha_2 . \quad (5)$$

Constant C'_F is relevant ordinate F at $a = 1$.
Apply:

$$F = C_F b^{x_F} a_1^{y_F} = C'_F b^{x_F} \quad (6)$$

Cutting force determining at milling:

At milling cutting chip are changed, which cause change of cutting force. Cutting force value depends on number of cutter teeth and their active surface. Resulting cutting resistance is count of resistances on individual cutter teeth.

Cutting force acting to cutting tooth is as two forces:

1. tangential tooth load F_t, which are tangent to circle trajectory,
2. radial force F_r ,which from centric this circle.

When cutter teeth is screwed F_o acting too F_1 in axial direction F . (fig.2)

F_H - horizontal component of cutting force
F_V - vertical component of cutting force.

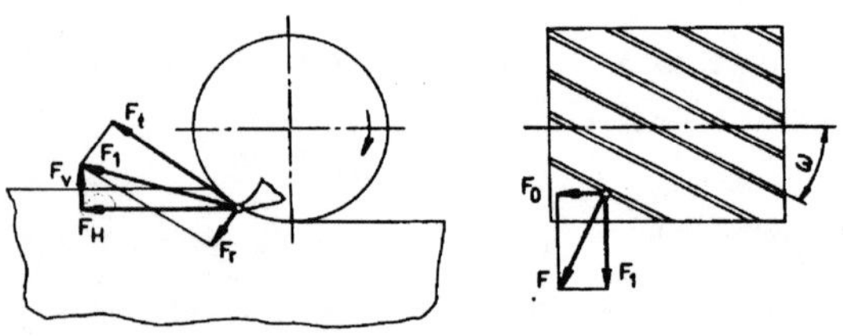

Fig. 2. Decomposition of cutting force at cylindrical milling an with screw cutter teeth

Proportion of tangential tooth load to diameter cutting surface determines specific cutting resistance. Diameter is changed therefore specific cutting resistance is changed too.

Value of specific cutting resistance is:

$$k_s = C_\omega C_1 a_x^k \quad (7)$$

C_1 -constant expressive general milling conditions,

C_ω -coefficient expressive influence angle of screw,

a_x -cutting chip thickness,

k - index - depends on materials, cutting liquid (-0,2 from to -0,4)

Cutting force determining at drilling:

At drilling cutting force are acting to cutting edge of drill.

Result forces we decompose to three axis x, y and z. (fig.3)

When cutting edge of drill is symmetrical radial forces F_y, F'_y is identical and their acting is removed. Force F_z acting at arm d/4 and cause torque M_k , which are transfer to spindle. Forces F_x, F_p are combined and created an axial force F_o , which negotiate in feed.
Torque is:

$$M_k = 2F_z \frac{d}{2} - 2F_y \frac{d_0}{2} (Nmm) \quad (8)$$

3. CONCLUSION

Value, direction and centre of cutting force are determined by value and centre of clamping forces. By tooling cutting force is raised from zero to specific maximum. Value of this cutting force depends on tooling material.

4. REFERENCES

Vasilko, K., Hruby, J.,Liptak,J.: Technologia obrabania a montaze., Alfa,Bratislava, 1991,

Buda,J., Soucek,J., Vasilko,K.: Teoria obrabania, Bratislava, 1988,

Chvala,B.,Votava,J.: Pripravky. Alfa,1988, Bratislava,

Velisek,K., Kostal,P., Matusova,M.:Analysing of clamping cases, In: Computer Integrated Manufacturing 2003, P.596, ISBN:83-204-2850-5

Kostal,P.,Katalinic,B.,Matusova,M.:Analysing of positioning and clamping cases, In:The 12th International DAAAM Symposium 2001, Vienna, Austria p.191 ISBN:1726-9679

Annals of DAAAM for 2006 & Proceedings of the 17th International DAAAM Symposium, ISSN 1726-9679
B. Katalinic, Published by DAAAM International, Vienna, Austria 2006
150th Birthday of Nikola Tesla (1856-1943) Inventor of Mobile Robot Teleoperation

ROBUST STABILITY ANALYSIS FOR POLYNOMIALS WITH NONLINEAR UNCERTAINTY STRUCTURE

MATUSU, R.; PROKOP, R. & VASEK, V.

Abstract: *Characteristic attribute of the systems with parametric uncertainty is that their structure is known but parameters can vary within given or supposed intervals. This type of uncertainty may occur in modelling and description of real systems as a consequence of inaccurate measurement, identification, due to ambient conditions, etc. This contribution is focused on description of systems with nonlinear (polynomial, general) parametric uncertainty and on demonstration of possible robust stability testing for this class of systems with special emphasis on utilization of The Polynomial Toolbox for MATLAB.*
Key words: *Robust Stability, Uncertainty, Value Set, Zero Exclusion Condition, Polynomial Toolbox.*

1. INTRODUCTION

The frequent effort of control engineers or researchers to create the mathematical model of controlled process as simple as possible almost always leads to the origin of uncertainty. Their emergence often consists in neglect of "less important properties", especially from the realms of fast dynamic effects, nonlinearities or time-variant behaviours of the system. Nevertheless, the presence of uncertainty can not be excluded even if the processes are in essence linear, because the physical parameters are never known exactly, possible they can vary according to operating conditions. Ergo, the principal problem is if the controller designed for nominal system will keep some properties of the feedback control loop also for system really controlled, which falls into certain neigbourhood, in other words, if the controller will keep these properties not only for one nominal system, but for the whole family of systems.

The uncertainty in constructed mathematical model and thus the size of neigbourhood which should the controller cope with can be taken into consideration and described in the two main ways – as parametric or nonparametric uncertainty. The former, nonparametric description of uncertainty lies in restriction of area of possible appearance of frequency characteristic. It is associated with unmodelled dynamics, truncation of high frequency modes, nonlinearities, randomness in the systems, etc. The latter, parametric approach then represents known structure but uncertain knowledge of actual physical parameters of the controlled system. Their possible values are usually bounded by intervals. This contribution deals with robust stability analysis for systems with nonlinear uncertainty structure. In such cases, the analytical methods are either quite complex or even do not exist. On that account, the graphical test is utilized in this work. The investigation of robust stability is accomplished via the value set concept and the zero exclusion condition, here practically with the assistance of The Polynomial Toolbox for MATLAB.

2. STRUCTURES OF UNCERTAINTY

From the control theory point of view, the common subject of investigation is the uncertain characteristic polynomial of the closed-loop control system. It can be described by:

$$p(s,q)=\sum_{i=0}^{n}\rho_i(q)s^i \quad (1)$$

In robustness problems, the vector of uncertain parameters q is often supposed to be confined by the uncertainty bounding set Q, which is usually given a priory, e.g. directly by user requirements. The most important problem consists in ensurance of stability and hence the control engineers are very often interested in a robust stability. It is familiarly known, that the polynomial $p(s)$ is stable if and only if all its roots are located in left complex half plane. The family of polynomials

$$P=\{p(\cdot,q):q\in Q\} \quad (2)$$

is robustly stable, if $p(\cdot,q)$ is stable for all $q\in Q$.

The uncertainty enters into the polynomial (2) through the coefficient functions $\rho_i(q)$. Nevertheless, the very significant is the way how the uncertain parameters enter into the coefficients of this polynomial. In accordance with this, several basic structures of uncertainty with increasing generality are distinguished:

- independent (interval) uncertainty structure
- affine linear uncertainty structure
- multilinear uncertainty structure
- nonlinear uncertainty structure (polynomial, general)

Moreover, the single parameter uncertainty is usually considered as a special case. This paper deals with nonlinear uncertainty structure and assumes Q to be a box.

3. TOOLS FOR ROBUST STABILITY ANALYSIS

Most of the analytical techniques are highly specialized and their applicability is restricted to particular type of uncertainty structure. The most known methods include Bialas eigenvalue criterion, the Kharitonov theorem, the edge theorem, the mapping theorem, etc. On the contrary, there is a very universal tool – the combination of the value set concept and the zero exclusion condition.

Assume a family of polynomials $P=\{p(\cdot,q):q\in Q\}$. The value set at frequency $\omega\in\mathbf{R}$ is given by:

$$p(j\omega,Q)=\{p(j\omega,q):q\in Q\} \quad (3)$$

In other words, $p(j\omega,Q)$ is the image of Q under $p(j\omega,\cdot)$ – e.g. substitute s for $j\omega$ in a family $P=\{p(s,q):q\in Q\}$, fix ω and let the vector of uncertain parameters q range over the set Q.

The zero exclusion condition for Hurwitz stability of family of continuous-time polynomials $P=\{p(\cdot,q):q\in Q\}$ says: Suppose invariant degree of polynomials in the family, pathwise connected uncertainty bounding set Q, continuous coefficient functions $a_i(q)$ for $i=0,1,2,\ldots,n$ and at least one stable member $p(s,q^0)$. Then the family P is robustly stable if and only if the complex plane origin is excluded from the value set $p(j\omega,Q)$ at all frequencies $\omega \geq 0$, i.e. P is robustly stable if and only if:

$$0 \notin p(j\omega,Q) \quad \forall \omega \geq 0 \tag{4}$$

Lots of generalizations, remarks and examples can be found e.g. in (Ackerman, *et al.*, 1993; Barmish, 1994; Bhattacharyya, *et al.*, 1995).

4. SIMULATIONS – THE POLYNOMIAL TOOLBOX

The visualization of the value set for systems with complicated uncertainty structure can be conveniently done in The Polynomial Toolbox for MATLAB (Šebek, *et al.*, 2000; www.polyx.com) through the commands "vset" and "vsetplot".

4.1 Example 1 – Polynomial uncertainty

Consider the family of polynomials given by:

$$\begin{aligned} p(s,q) = s^3 &+ \left(q_1^2 q_2 + 2\right)s^2 + \\ &+\left(2q_1^4 - q_2^{\,3} - 2q_1^2 q_2 + 2q_2 + 10\right)s + \\ &+\left(2q_1^4 + q_2^{\,3} + 2q_1^2 q_2 + 2q_2 + 10\right) \end{aligned} \tag{5}$$

and $q_1, q_2 \in \langle -1;1 \rangle$. The value sets for $\omega = \langle 0;5 \rangle$ and step size 0.25 are depicted in fig. 1. To see better what is happening in the neighbourhood of the origin of the complex plane the plot is zoomed in fig. 2. As can be seen, the zero point is excluded from the value sets, moreover, the family (5) has a stable member and thus it represents the robustly stable case.

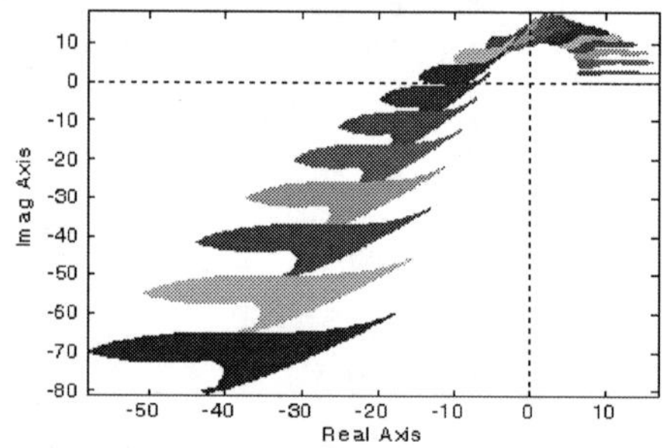

Fig. 1. The value sets of family (5) – full view

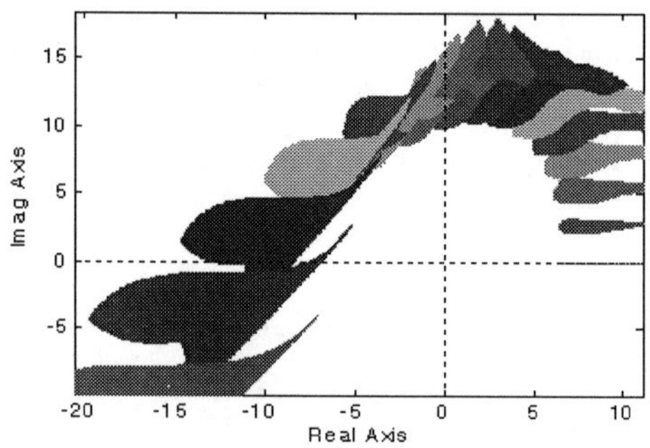

Fig. 2. The value sets of family (5) – detailed view

4.1 Example 2 – General uncertainty

Suppose the polynomial family:

$$\begin{aligned} p(s,q) = s^3 &+ \left[\cos(q_1 q_2)\right]s^2 + \\ &+\left[6\sqrt{|q_1|} - 5\sin q_2 - 2\cos(q_1 q_2) + 6\right]s + \\ &+\left[-3\sqrt{|q_1|} + 3\sin q_2 + \cos(q_1 q_2) + 0.1\right] \end{aligned} \tag{6}$$

again with $q_1, q_2 \in \langle -1;1 \rangle$. Fig. 3 shows value sets of (6) for frequencies $\omega = \langle 0;5 \rangle$ with step 0.25 while the zoomed version is in fig. 4. The origin of the complex plane is included in the value sets and consequently the polynomial (6) is not robustly stable.

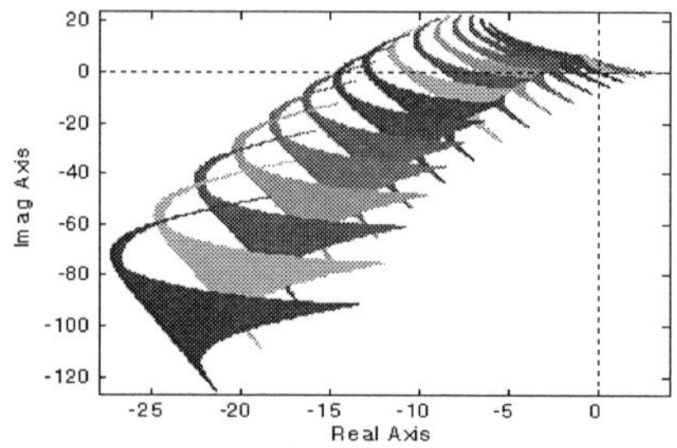

Fig. 3. The value sets of family (6) – full view

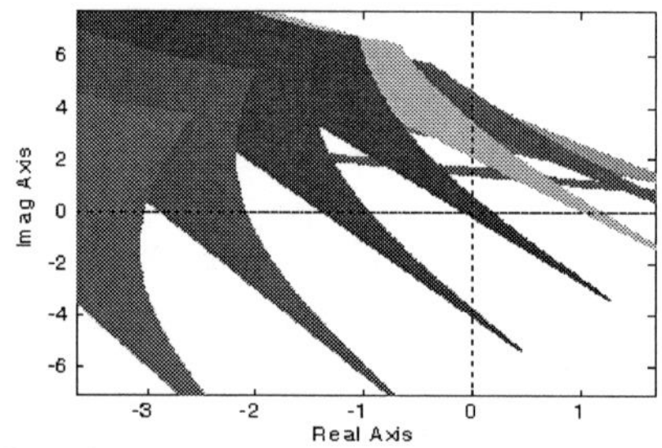

Fig. 4. The value sets of family (6) – detailed view

5. CONCLUSIONS

The contribution has been focused on problems connected with robust stability of systems with nonlinear uncertainty structure. Instead of very complex or even not existing analytic tools the highly universal value set concept and the zero exclusion condition have been applied. The practical tests of robust stability based on these key instruments can be handy done in The Polynomial Toolbox for MATLAB. Its possible utilization has been demonstrated on two examples for polynomials with polynomial and general uncertainty structure, respectively.

6. ACKNOWLEDGEMENT

This work was supported by the Ministry of Education, Youth and Sports of the Czech Republic under grant No. MSM 7088352102. This support is very gratefully acknowledged.

7. REFERENCES

Ackermann, J., *et al.* (1993). *Robust control – systems with uncertain physical parameters*, Springer-Verlag London, GB.

Barmish, B. R. (1994). *New Tools for Robustness of Linear Systems*, Macmillan, New York, USA.

Bhattacharyya, S. P.; Chapellat, H. & Keel, L. H. (1995). *Robust control: The parametric approach*, Prentice Hall, Englewood Cliffs, New Jersey, USA.

Kucera, V. (2001). Robustní regulátory. *Automa*, Vol. 7, No. 6, pp. 43-45. (In Czech).

Polyx: The Polynomial Toolbox. Available from WWW: http://www.polyx.com/.

Sebek, M.; Hromcik, M. & Jezek, J. (2000). Polynomial toolbox 2.5 and systems with parametric uncertainties. *Proc. of 3rd IFAC Symp. Robust Control Design*, Prague, Czech Republic.

Sebek, M. (2002). *Robustní řízení*, PDF slides for course "Robust systems", ČVUT Prague, 2002, Available from WWW: http://dce.felk.cvut.cz/ror/. (In Czech).

Annals of DAAAM for 2006 & Proceedings of the 17th International DAAAM Symposium, ISSN 1726-9679
B. Katalinic, Published by DAAAM International, Vienna, Austria 2006
150th Birthday of Nikola Tesla (1856-1943) Inventor of Mobile Robot Teleoperation

THE REMOTE MONITORING AND CONTROLLING VIA THE GSM NETWORK GPRS DATA TRANSFER

MATYSEK, M.; ADAMEK, M.; NEUMANN, P. & HAMRLICEK, J.

Abstract: *The aim of this work was to design and apply a remote access by means of a mobile phone running the Java based application routine. That Java mobile phone routine communicates with the controlled and monitored system through the GPRS data transfer by means of the internet TCP/IP protocol. The data are transferred in XML format.*
Key words: *Remote monitoring and controlling, GSM mobile phone, GPRS data transfer, JAVA language, XML data structures.*

1. INTRODUCTION

There are not many professional systems for the remote monitoring and controlling of simple devices via GSM GPRS data services in Czech Republic.
The mobile monitoring and controlling system for very simple systems embodying slow trends has been designed and is still in a development phase. The slow response of GSM GPRS processes is not the limiting factor in the application area then.
Some similar applications for mobile phones destined for free of charge SMS sending via the GSM operator gates were the inspiration for this work.

2. THE GSM-GPRS NETWORK STRUCTURE

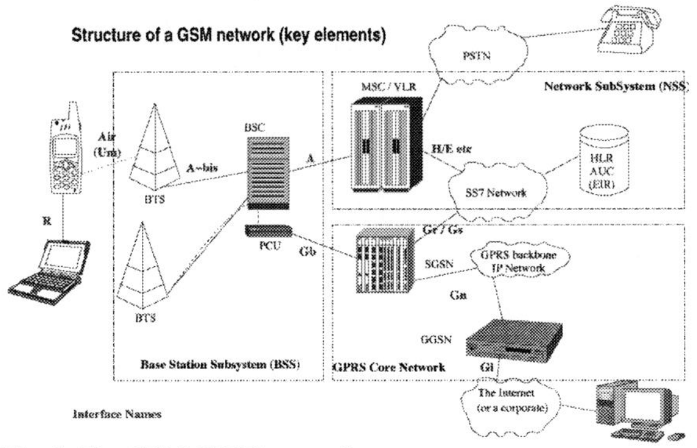

Fig. 1. The GSM-GPRS network structure.

The GPRS Core Network shown here is simplified to just have the SGSN (connected to the BSS by the Gb interface) and the GGSN. The two are connected together by a private IP network called the GPRS backbone shown as the Gn Reference Point.

2.1 GPRS

Unlike the CSD and HSCSD technologies, the GPRS is based on different principle. GPRS is based on the packet switching what is similar to LAN technologies. The GPRS technology creates something like a new network with packet switching overlaying the current one. In other words, the GPRS makes use of the network transfer capacity exclusively in case data sending or reception is needed. The transfer capacity is not permanently reserved like at CSD or HSCSD technology and the usage of the network is not blocked for another users. Thanks to the GPRS technology, more participants can use data transfer services than it is possible at the circuit switching of CSD and HSCSD.

2.2 EDGE

The EDGE (Enhanced Data rates for GSM Evolution) technology represents the last effort to increase the transfer rate of the existing GSM networks structure. The EDGE technology preserves the current frequency ranges and channels but it uses the different way of modulation. The original two-stage modulation used for radio signal transmission has been replaced by a more efficient 8-stage modulation. Due to that, it could reach theoretically the transfer rate up to 384 kb/s using 8 slots.

3. THE JAVA LANGUAGE TECHNOLOGY

The Java language is the most popular programming language nowadays. It is either used for Internet applications programming and for the programming of different hardware and software platforms applications.

3.1 Java 2 Micro Edition

This edition is called J2ME (Java 2 Micro Edition) or Java ME. J2ME and it represents the comprehensive collection of technologies and specifications that makes it possible to create and run Java applications on different devices and with limited resources. The resources are meant as processor performance and the accessible memory capacity. The devices fitting in that group are for instance pagers, mobile phones, set-top boxes and even car navigation devices.

3.1.1 Configuration CLDC Java 2 Micro Edition

That configuration is much less demanding in comparison to CDC and its goal is to create a lowest applicable common base fro J2ME devices. That configuration is well spread in the mobile phones area nowadays. The majority of mobile phones are supporting that configuration and only an unremarkable minority supports the CDC configuration. Analogous to CDC configuration, the CLDC configuration is passing the development phase and there is an up-to-date version CLDC 1.1 currently.

4. THE PRACTICAL IMPLEMENTATION

We need the necessary tools for Java application programming aimed at computers or mobiles. The development is more comfortable and quicker then. We need the JDK (Java Development Kit) installed as a base for Java programming. The JDK package contains the necessary tools for Java program development as was mentioned earlier. We manage only with runtime environment JRE (Java Runtime

Environment) packages in case we need only run Java applications. Those packages are to be downloaded from Sun Microsystem website. We need also a suitable development environment installed, so called IDE. There were two development environments suitable for our purposes, namely Eclipse and NetBeans. Both environments are supporting the standard application programming and there is an accessible support for both for development in J2ME. The internet download of both environments is free of charge what was also remarkable for us. We have chosen NetBeans after pondering advantages and disadvantages eventually. The main reason for our decision was the Mobility Pack existence. The Mobility Pack is a Net Beans supplement J2ME programming support. Furthermore, we had to add various devices emulators like mobile phones from different manufacturers.

4.1 The communication test program for PC

There was a PC test program created that sends the monitored and controlled systems simulated data to the mobile phone via internet and receives new data for them from the handy. This test-program is a standard Java application for a desktop PC. That program uses a simple graphic interface for displaying the program run information. The program was created in the NetBeans environment which has the graphic interface support inbuilt like many other modern development tools.

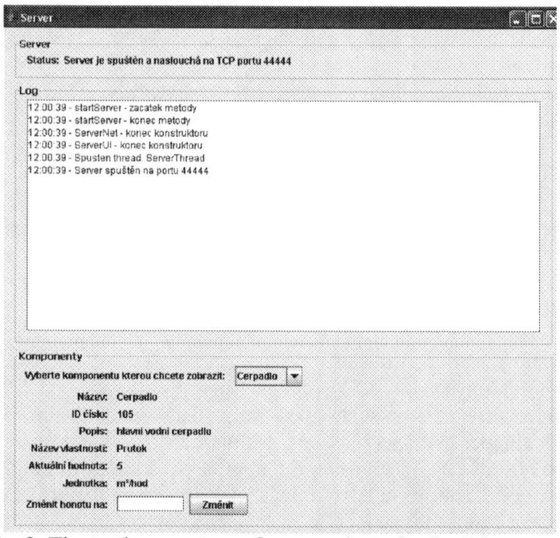

Fig. 2. The testing program of communicate for the PC.

You can see the graphic output from that program in Fig.2. The program graphic output is divided in three parts. There is only an information message on top informing on the fact that the server watching the incoming connection is running and the relevant port number. The state information on program activities is displayed in the middle area. That information was helpful during the program development and debugging. The main area is in the bottom. There are all simulated systems displayed and it is even possible to change set values there.
The PC test program expects the simple format commands from the mobile client. Those commands are similar to the XML format and PC returns the XML file containing demanded data to the mobile client.

4.2 The mobile phone program

The goal of the Java program for the mobile phone was to communicate with a remote monitoring and controlling system via Internet. It makes possible to monitor such system and change its parameters remote. According to the mobile phone mobility, we could monitor and control the respective system from anywhere. The program was created in the NetBeans development environment where the Mobility Pack and Nokia mobile phone emulator installed. Thanks to the Mobility Pack, it is possible to create applications for mobile phones the similar way the standard programs are created.
The mobile phone Nokia 6310i was employed for the communication testing and debugging. That mobile phone transfer data either in GPRS and EDGE structures in the Class 10 (4+1, 3+2 slots downstream/upstream) in both cases.
The main menu appears after starting the mobile phone program. See Fig.3.

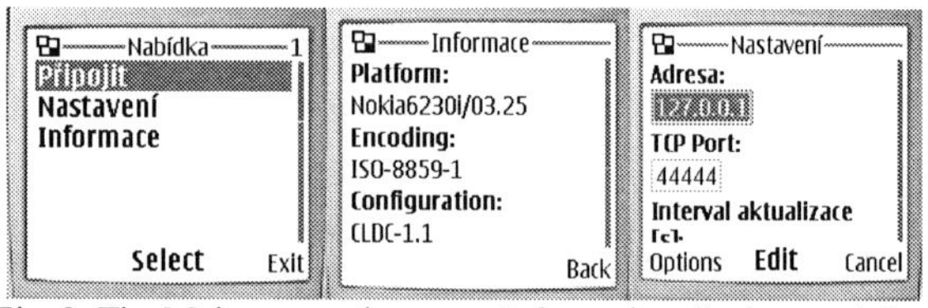

Fig. 3. The Main menu, the menu Information displays and the menu Set-Data.

As you can see from the first screen, there is a menu consisting from three items and also the possibility to stop application.
The menu Information displays the information on equipment the program is running on.
In the menu labeled Set-Data it is possible to set address and the server port for the client connection.
After the menu Connect has been chosen, the mobile phone connects to the server according to the Set-Data. It receives the data from the server and displays the list of simulated systems to be monitored and controlled with parameter changes.

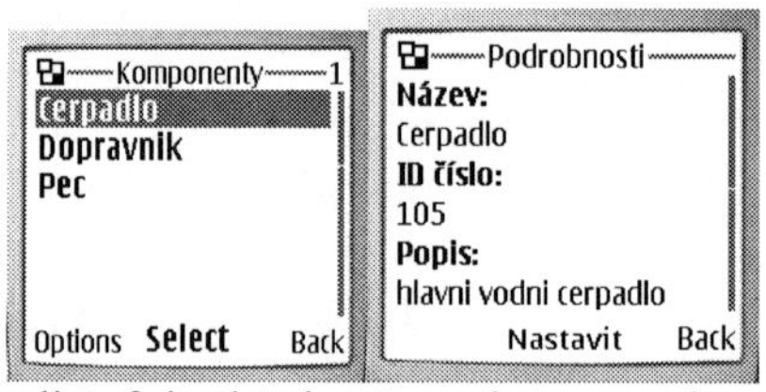

Fig. 4. The list of simulated systems, that are monitored and controlled and the further screen is displaying the information on the system.

After you choose one of displayed items, you get to the further screen displaying the information on the system selected and you can set the new parameter value to be sent to the server.

5. CONCLUSION

The communication between PC and mobile phone has been either tested with the Nokia 6310i emulator and with the real mobile phone Nokia 6310i. The system described above is fully functional. We would like to test the communication system parameters in nonstandard network situations like short mobile network drop-outs. The short signal drop outs because of short distance signal fluctuations (the handy in a moving car) are concerned. The short drop-out because of radio channel full slots occupation by other network users needs to be taken into account as well.

6. REFERENCES

Puzmanova, R. (2004). *Sirokopasmovy Internet – Prístupove a domácí sítě,* Computer Press, ISBN 80-251-0139-8, Brno.
Brůha, L. (2003). *Java Hotova řesení,* Computer Press, 2003. ISBN 80-251-0072-3, Brno.
Qusay, H, M. (2002). *Naucte se Java 2 Micro Edition,* Grada Publishing, ISBN 80-247-0444-7, Praha.

Annals of DAAAM for 2006 & Proceedings of the 17th International DAAAM Symposium, ISSN 1726-9679
B. Katalinic, Published by DAAAM International, Vienna, Austria 2006
150th Birthday of Nikola Tesla (1856-1943) Inventor of Mobile Robot Teleoperation

COMPUTER CONTROL SYSTEM FOR „ROBOFORMING" – A ROBOT BASED SHEET-METAL FORMING PROCESS AND FURTHER EXPERIMENTAL INVESTIGATIONS

MEIER, H.; ZHANG, J. & DEWALD, O.

Abstract: *"Roboforming" is a new robot-based incremental forming process for sheet metal. With two simple forming tools complex geometric parts can be produced incrementally. The potential use of the process is suitable for small-lot production and prototyping. The presented paper describes how a PC-based control system for Roboforming can be realized. Actual experimental investigations are shown in this paper.*

Key words: *robotic, forming, sheet-metal, control system*

1. INTRODUCTION

In the last decade, the incremental forming process has been widely discussed and investigated from academic researchers and industrial users. The aim of this research is a process to form sheet metal without complex and expensive dies. To produce axisymmetrical shapes Kitazawa (Kitazawa et al., 1999) has studied the incremental stretch expanding on a modified CNC lathe. The incremental forming process for asymmetric shapes has been distinguished by Jeswiet (Jeswiet 2005) with TPIF (Two Point Incremental Forming) and SPIF (Single Point Incremental Forming) by Duflou, Ambrogio and Jeswiet (Jeswiet 2002, Duflou, Ambrogio 2005). Incremental backward bulge forming has been introduced at first by Matasubara (Matasubara, 1994). In his TPIF process, a partial die or support post is positioned below the sheet. The forming tool moves around the fixed support on a pre-described path to pull down the sheet. Hirt used the TPIF process on an Amino machine (Hirt 2002). Beside the works on CNC machines, a single industrial robot was applied for the incremental sheet forming process by Lamminen (Lamminen 2005). Also Schaefer showed how the metal sheet could be formed by a hammer tool which is driven by an industrial robot (Schaefer et al., 2005). This paper describes a self-developed computer control system for a robot based incremental sheet metal forming process with two independently moved simple tools to produce asymmetric shapes. Focus of this abstract is the experimental investigation of the geometrical deviation.

2. ROBOFORMING

A new robot based incremental sheet-metal forming process – "Roboforming" – was developed at the Chair of Production Systems at the Ruhr-University Bochum. The sheet metal is formed on an equipment, which consists of two Industrial Manutec R15 robots and a test stand. The metal blank is carried by a clamping device which is fixed on the test stand. A simple forming tool and a supporting tool were mounted at the end of the robots' effectors. The path of the forming tool is generated directly from the CAD-files of the work piece. In order to get sharp edges on the bottom of the shape and to minimize the spring back of sheet, the supporting tool moves on the backside of the sheet synchronously to the forming tool. The particular description of the moving strategy can be found in previous articles (Meier et al., 2005). The robot based incremental sheet forming mechanism depends on the universal stretch expanding. The force between forming and supporting tool expands the sheet. The final shape of the sheet is formed by small incremental steps of the tools. There are two differences between the common incremental forming process and this new robot based forming process. One of the differences is the use of a moved supporting tool. With the usage of a supporting tool the flexibility of the incremental forming process can be increased. In the single point incremental forming process the backing plate must be pre-produced for various work pieces before the forming process. The production of the backing plate raises cost and time of the production. With Roboforming various work pieces can be produced with the same setup by using different moving strategies and paths. On the other hand, the supporting tool can also act in the role of the forming tool to produce a concave surface. The other difference is the use of two industrial robots. By installation of developed software on robots it is easy to set up the incremental forming process because of the large amount of degrees of freedom in movement.

3. EXPERIMENTAL SET-UP

The robot forming system consists of two robots, their control computers, the test stand and an operating computer. The metal blank is vertically clamped in the test stand between the robots, which have a maximum handling load of 15 kg. The robot control system rcs1 from Robolab ltd. Co. was installed on both of the robots´ control computers. The operating computer is connected with the two control computers by Ethernet in order to exchange the control information for the forming process. The forming control software is written in C++, so own functions can be implemented quite easily.

The robot control system rcs1 is an open system and the C++ libraries are available for the users to develop own software. In this way complex algorithms and extended calculation power can be combined with the robot control commands:

- Transfer the geometry information for the regular shape and realize the forming function for the free form.
- Offer different forming strategies for the regular geometry in order to achieve better experimental results.
- Automatic generation of the movement path of forming and supporting tool from the information of the shape.
- Supply a synchronization of both robots for forming process.
- Use different parameters for the forming process in order to investigate their influence.
- Measure the forming force in all directions.
- Take shape measurement with a tactile sensor that is mounted between the forming tool and the robots end effector.

The generation of the tool path is the most challenging part of the software. In the current experiments the robots work in 2½D. The regular geometrical shapes, which can be explained in mathematic expressions are taken apart to scatter plot by the pre-given incremental feed. For free form parts a STL file is used and the coordinate value of the points on its surface has been achieved by a slicing algorithm. By the radius correction and the sorting of the scatter plot the tool path is calculated.

A multi-thread function would be applied for the synchronization of two robots. The principle of threads programming is that multiple tasks run concurrently within the same program. They provide the way to divide the tasks of a program while sharing data. While the program is running on the operating computer two threads are created. In one thread the task of the forming tool is running while the other thread is carrying out the movement of the supporting tool. With the synchronization mechanism of the multi-threads the information between the two tools is exchanged. When a tool reaches the target position at first, it will let the other tool wait and will continue to move to the next position after the information about the other tool is given. Because of the different paths for the two tools the speed of two robots must be adjusted with the calculation of the length of the two paths.

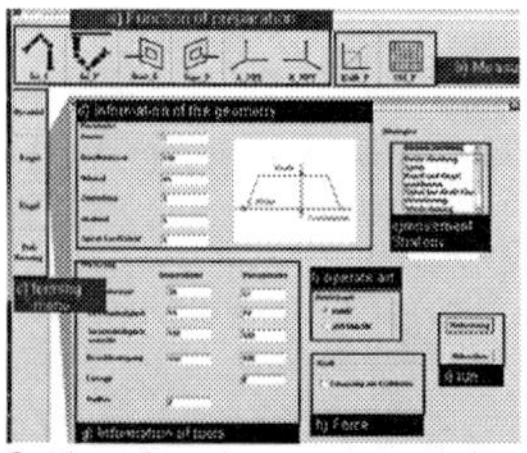

Fig.1. Interface of robot forming programming Software

Fig. 1 shows the interface of the developed robot control software. There are two menus for the preparation of the robots und forming functions. Fig.1a shows the functions of the preparation for two robots. Before the forming process the two robots must be initialized with the "Ini_S" and "Ini_P". The Tool Center Point (TCP) of robot is situated at the centre of the tool. The coordinate base of the work piece is situated in the middle of the metal blank. It is important that the coordinate system of the robot will be transformed from the robot coordinate system to the coordinate system of the work piece with the click the "R_NPT". The function of the "A_NPT" is to calibrate the situation precision between the two robots and the test stand because the test stand is not located at the precise place. The vertical menu offers the forming functions (fig. 1c). Besides the regular geometry – "frustum of pyramid and cone, spherical cap", the free form will be formed through the robot control with the control software. In the subform the information of the geometry (fig. 1d) and the form information of the two tools (fig. 1g) will be given. Also the velocity and the speed overrides can be set in the subform. Fig. 1e shows the different movement strategies for the forming process of regular geometry. To test the new movement strategy the "HAND" function is usually used. A force-torque-Sensor is amounted between the forming tool and the robot flange to sample the forces in all directions (fig. 1h). When all parameters are given, the forming process can be started (fig. 1i). After the forming process the geometrical data of the formed work piece can be measured with a tactile sensor which is driven by the robot (fig.1b).

4. EXPERIMENTAL RESULTS

The robot based forming process can be proved by several different samples. Because of the weak load of 15 kg, aluminum blank in thicknesses from 0.5 mm to 1.0 mm were used for the forming process. Fig. 2a shows the result of a 3D asymmetric geometry of a CAD file. The sheet material is Al 99.5 soft with a thickness of 0.5 mm. The infeed is 1.0 mm/circle. The initial forming / supporting tool speed is 10 m/min. While the forming tool moves in depth direction, the contour forming tool path is shorter than the supporting tool path and the forming tool speed is deceased in scale, so that the tools are running synchronously. The total forming time is 21 minutes. A complex product is shown in fig. 2b. It is a composition of a frustum of a cone (diameter: 110 mm, height: 30 mm, angle: 60°) and a negative spherical cap (Diameter: 50 mm, height: 15 mm) on the top surface. The components were produced by the exchange of the two tools: At first, the frustum of cone was formed in normal forming strategy. Then the forming tool and supporting tool moved to the starting point, the coordinate system of the work piece was resettled for the spherical cup and second forming process was started with a change of forming tool and supporting tool. The tool process speed was 10 m/min. The whole forming process was finished in 9 minutes.

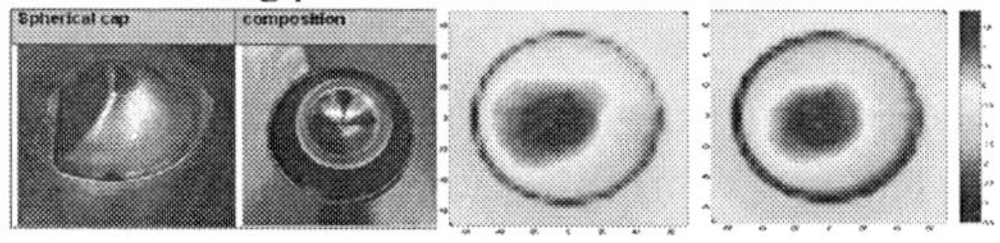

a) b) c) uni-directional d) bi-directional

Fig.2. Results of Roboforming processes

Fig. 2c/d shows the deviation of the frustum of cone in the Z direction. The dimension of the frustum of cone is 110 mm in diameter and 40 mm in height with a angle of 45°. The movement strategy of the forming tool is different. Fig. 2c used the uni-directional tool movement, and fig. 2d the forming tool moved bi-directionally. A maximum deviation of -3.3 mm occurred at the circle edge on the bottom surface. The space of 2.0 mm between forming tool and supporting tool causes the large error on the bottom surface. The deviation error on the top surface is 1.7 mm because of the spring back of the material. The cover face of the two parts has a good quality, the deviation is less than 0.5 mm, but the deviation error in left part was bigger than in the right part. The bi-directional movement strategy could not change this error trend.

5. SUMMARY

The paper presents research results regarding "Roboforming"- the new robot based incremental forming process and the self-developed computer control system to realize the forming process. The forming software has realized the automatic generation of two tool path from the CAD files. The produced spherical cup with a slanting face and the component part have proved the large capability and potential of the forming process for small-lot production and prototyping. Furthermore, the geometric accuracy of forming parts must be improved. Because of the tool paths depending deeply on the geometric accuracy the future research will focus on finding the properly manufacture technology for various formed geometries.

6. REFERENCES

Jeswiet. J. et al (2005). *Asymmetric Incremental sheet forming.* Annals of CIRP. 2005, 623-650.

Kitazawa, K., Nakajima, A. (1999). *Cylindrical incremental drawing of sheet metals by CNC incremental forming process*, Proceedings of 6th ICTP, Vol II, 1495-1500.

Matsubara, S. (1994). *Incremental backward bulge forming of a sheet metal with a hemispherical tool,* Journal of the JSTP, Vol.35/1, 1311-1316.

Jeswiet, J., Hagan, E., Szekeres, A. (2002). *Forming parameter for incremental forming of aluminium alloy sheet metal,* Journal of Engineering Manufacture, Vol 216, 1367-1371.

Schäfer, T.; Schraft, R.D. (2004). *Incremental foming of sheet metal by industrial robots.* Robotik 2004, 31-38

Laminen, L. (2005)*Incremental sheet forming with an industrial robot – Forming limits and their effect on component design.* SHEMET, 457-464

Ambrogio, G. et al (2005). *Sheet Thinning Prediction in Single Point Incremental Forming.* SHEMET , 479-486

Duflou, J.R. et al (2005). *Force Measurements for Single Point Incremental Study.* SHEMET, 441-448

Hirt, G. et al (2002). *Incremental Sheet Forming: Quality Evaluation and Process Simulation.* Proceedings of the 7th ICTP, Yokohama.

Meier, H.; Zhang, J.; Dewald, O. (2005). *Incremental Forming of Sheet Metal by Two Industrial Robots.* Proceedings of the 7th Int. Conference on AMST, 437-444

Meier, H.; Dewald, O.; Zhang, J. (2005) *A New Robot-Based Sheet Metal Forming Process.* SHEMET, 465-470

Annals of DAAAM for 2006 & Proceedings of the 17th International DAAAM Symposium, ISSN 1726-9679
B. Katalinic, Published by DAAAM International, Vienna, Austria 2006
150th Birthday of Nikola Tesla (1856-1943) Inventor of Mobile Robot Teleoperation

NEW TECHNOLOGY FOR MANUFACTURING GLOBOID WORM GEARS

MIHAI, M.; GYENGE, C.; CRISAN, N. & BOCA, V.

Abstract: After a short brief regarding the evolution of the globoid worm gear drives we present some advantages and disadvantages for different manufacturing technologies of these worm gear drives. After that we present a new technological solution for roughing and finishing operations of the globoid worm gear, with a high rate of universality.
Key words: *globoid worm gear, manufacturing technology*

1. INTRODUCTION

From the conceiving of the Globoid Worm Gear by Leonardo da Vinci, five hundred years ago, it has been experimented and obtained different technologies for processing the two elements of these worm gears. The theoretical basis of these globoid worm gears was build by Gohman in 1886. In 1992 Widhaber conceive the first globoid worm gear intended for the precision kinematics' chains. At short time after this, at Michigan Tool firm begins the industrial manufacturing of the globoid gears, started from the invention of one engineer named M Cone [Litvin, 1990]. Despite of the good mechanical and kinematical performances of these gearings, they hadn't been applied too much due to the geometry complexity and also to the technology.
For finishing the globoid worm it is trying different solutions. One of them is that of Dudas I [Dudas, 2000], who conceived a grinding machine with variable profile form of the abrasive tool. The precision processing of the worm gear is still a hard issue and a delicate problem especially due to complexity of manufacturing the necessary tool.
The authors conceived and tested a general device for manufacturing globoid worm gears that is going to be presented in this paper.

2. THE CONSTRUCTION OF THE UNIVERSAL TOOL USED FOR MANUFACTURING GLOBOID WORM GEAR

The tool conceived (fig. 1,2) can be adapted to the most gear cutting machines and it has the advantage of a good control possibility of manufacturing a large range of globoid worm gears.

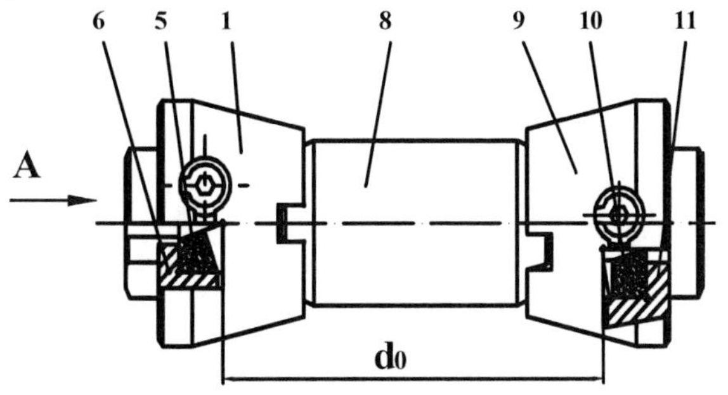

Fig.1. New tool for manufacturing globoid worm gears

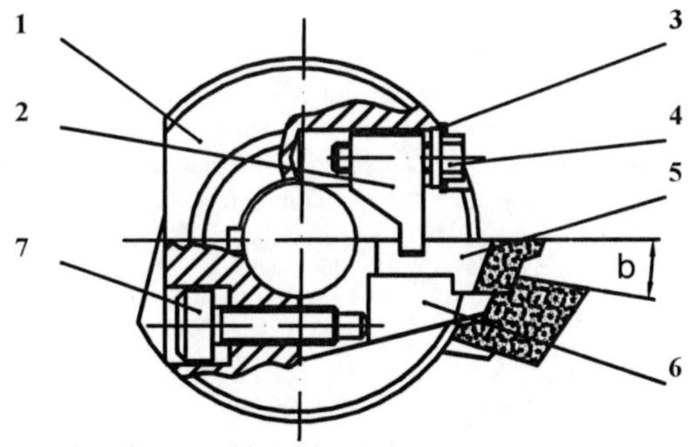

Fig. 2. View from A side of the tool

The tool is composed of two symmetric tool-holders 1 and 9, positioned between one and other by the help of a distance sleeve 8, which maintain the positioning of the tools 5 and 10 with the front cutting edges so that all the time they are tangent to the base circle of diameter d_0. (fig. 2)

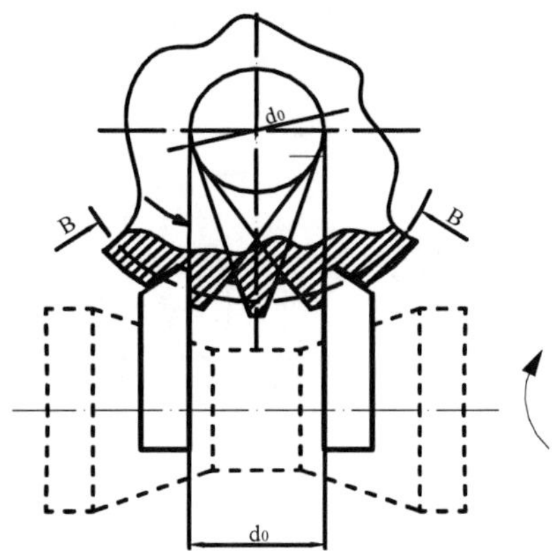

Fig.3. Relative position of the cutting edges of the tools during manufacturing globoid worm gear

The cutting tools are displaced with an angle β (see fig. 2) so that between their front cutting edges should be an odd number of half pitches *n*.
All the cutting tool assembly is clamped on the tool-holder spindle of the gear cutting machine. The cutting tools 5 and 10 are radial positioned through the special nuts 2 and the adjuster bolt.4. Radial position of the cutting tools is block up by the help of the retaining ring 3. The cutting tools are adjusted by the pair of wedges 6 and 11 (see fig.1) and by the screw 7.
The adjusting angle can be calculated by the following relation:

$$\beta = \frac{2}{m \cdot i}\left[\left(Z_n + 0{,}5 - \frac{Z_2}{9}\right)m \cdot \pi + a_p\right], \quad (1)$$

where:
m – module of the gear

z_n – the round number of pitches between edges of the tools
z_2 – the number of the worm hob teeth
a_p – the machining allowance on the flanks need for finishing operation, which is removed by the help of the compensating gear of the gear hobbing machine, that intercept the thinning of the flanks when the radial feed is used.

3. WORKING PRINCIPLE

For a good quality manufacturing of different type of globoidal worm gears it is necessary to accurately adjust the relative position of tools.
At the right hand toothing, for a positive value of the angle β (1), the cutting tool that manufactures the left flank get on the machining behind the cutting tool that manufactures the right flank, and for a negative value of the angle β, the cutting tool that manufactures the left flank get on the machining above the cutting tool that processes the right flank. (fig. 4)
At left hand toothing the entrance order for machining is reverted function of the β angle value. (fig. 5)

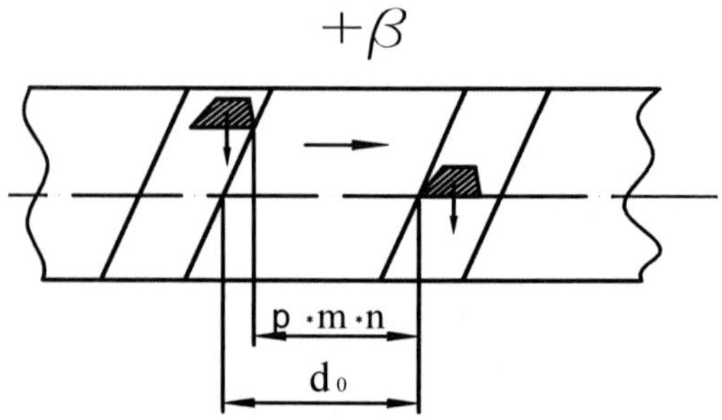

Fig.4. Entering of the cutting tools at the right handed toothing

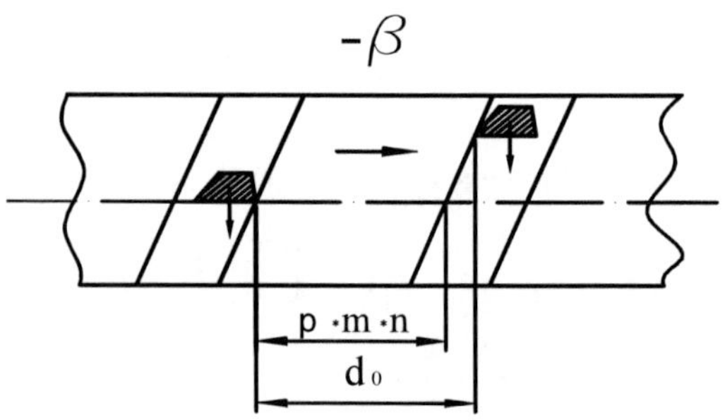

Fig.5. Entering of the cutting tools at the left handed toothing

The number of the cutting tools that can made the same flank is equal by the number of the beginnings *i* of the worm hob. The cutting tools are displaced between them with the value 2π/i and placed at the distance d_0 (see fig. 3)
By the help of the distance piece 8 it can be adapted to the cutting tool, a roughing cutting tool and after that the cutting tool 5 and 10 to do the finishing operation.
The main advantage of the new cutting tool in comparison with the globoid worm wheel hob is the one of the simple construction and the possibility of a precise execution. It is also eliminated the pitch errors that can appear at the manufacturing with the worm wheel hob. Concerning the productivity, it is not much smaller than the productivity with the worm wheel hob.
Adjusting the inter element position of the cutting tools can be done by the help of a gauge (fig. 6).
As a disadvantage that can be remarked, is the fact that it isn't possible the manufacturing of the globoid worm gears with different changes because it can't be obtain a good quality surface. In this case it is recommended to use a shaving worm hob (fig. 7).

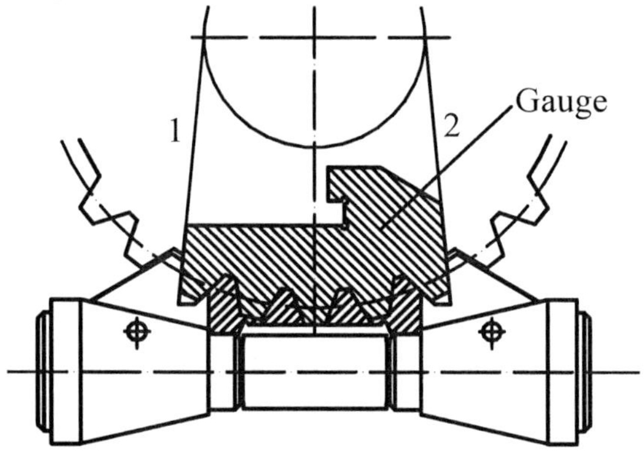

Fig.6. Adjusting the inter element position of the cutting tools

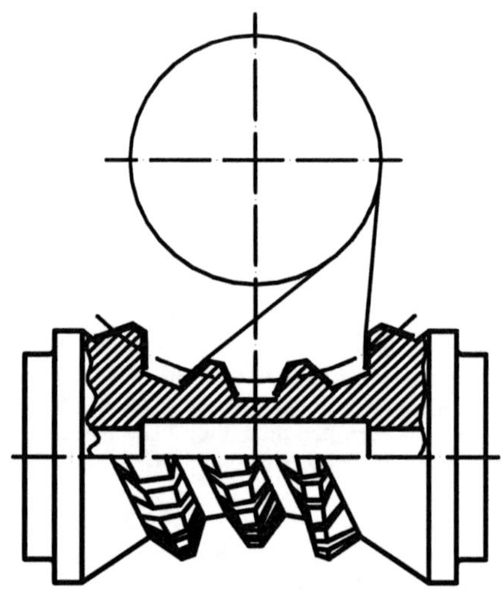
Fig.7. Shaving worm hob [Erney 1983]

The developed technology was experimentally tested at S.C. UNIO Satu Mare factory at some worm gears having
m = 4 …6 mm and number of teeth z = 30 … 60. The result was good.

4. CONCLUSIONS

The main advantages of the presented cutting tool are the following: a quite simple and compact construction, high general use, adapting possibility for different gear hobbing machines, useful for manufacturing gears for small production or prototypes. Symmetrical and geometrical precision of the manufactured flanks also eliminates the worm hob correction.

5. REFERENCES

Bocian I *Contributii la imbunatatirea angrenajelor melcate globoidale*, Teza de doctorat, Univ. Tehnica Cluj 1976
Drobni I, *Korrzaru csigahajtasok*, Miskolc, 2001
Dudas, I, *The Theory & Practice of Worm Gear Drives*, Penton Press, London 2000
Erney, Gy. *Fogaskerekek. Muszaki könyvkiado*. Budapest,1983
Litvin, F. I. *Theory of Gearing*, NASA Reference Publication 1990
Lebedev, I.A. *Technologia globoidnii predaci*, Oborongiz, 1957
Zak, P.S. *Globoidania Peredacia*, Moskova, Masghiq, 1962.

Annals of DAAAM for 2006 & Proceedings of the 17th International DAAAM Symposium, ISSN 1726-9679
Editor B. Katalinic, Published by DAAAM International, Vienna, Austria 2006
150th Birthday of Nikola Tesla (1856-1943) Inventor of Mobile Robot Teleoperation

THE DIAGNOSTIC OF SURGE PHENOMENA AT WINDINGS OF ELECTRIC MACHINES WITH USING PROGRAM EMTP-ATP

MISAK, S. & MACH, V.

Abstract: *Electric machines are important devices of each power industry. The technical diagnostic of these machines is necessary part for reliable and failure-free operation mentioned power industry. Using informatics technology is more and more popular in recent years. The modeling of surge phenomena at electric machines windings with using mathematical program EMTP-ATP -ATP is introduced within this manuscript.*
Key words: *The Technical Diagnostic, Program EMTP-ATP, The Surge Phenomena, Electric Machines Windings*

1. INTRODUCTION

For diagnostic of electric machines windings is possible to use so called comparative method with using surge generator PSG215A. The method of testing is follows: two capacitors with capacity 100 nF and voltage 15 kV are discharged to windings at the same time. Voltage's responses are measured at windings. Windings are in good order for case if responses are identical. If it be to the contrary, it is possible to assume on windings damage. Each fault of windings takes effect at deflection of measured response. The manner of deflection evaluation and accuracy of fault place identification is substantial for measuring system appliance. The obtaining of sufficient in numbers data, description of faults is very time-exacting. The using mathematical program is advantageous for time-exacting limiting. Results from modeling of surge phenomena at electric machines windings are introduced within manuscript. The program EMTP-ATP is used as a mathematical program. The verification of using EMTP-ATP for modeling surge phenomena was realized on physical model of mono-polar inductor.

2. THE DETERMINATION OF INDUCTOR'S PARAMETERS

For testing was projected mono-polar inductor with the Cu conductor (thickness: 0,65 mm, 700 turns) with cotton isolation. The Cu conductor is wound on tubular core (elastic asphalt). Turns are separated from second electrode by hydro-insolating sheet (thickness: 1 mm). Turnings of windings were disabused over 50 turns. Inductor's parameters were specified: inductance $L = 25{,}4$ mH, capacity to core $C = 7{,}2$ nF, resistance $R = 89\ \Omega$ and free frequency $f_r = 1{,}955$ MHz. The frequency analysis of program EMTP-ATP was used for determination of inter-phase's capacity K. Determination of inter-phase's capacity K parameter is very difficult by virtue of capacity coupling to inductor's core. The frequency analysis was applied on created model at graphic presentation ATPDraw (preprocessor EMTP-ATP). Parameter of inter-phase's capacity K was changed to the moment of coincidence simulated and measured free frequency f_r. The final scheme of inductor for determination of K parameter in ATPDraw is shown in Fig. 1, the graphical result from FFT analysis is shown in Fig. 2. From FFT analysis was defined value of inter-phase's capacity $K = 0{,}096$ pF. From Fig. 1 is evidently that inductor was modeled as circuit with distributed parameters according to theory. Serious resistors are used for suppression of mathematical oscillation. 700 turns are compressed to block over 50 turns by reason of transparent. For one of turns then reads: inductivity $L_1 =$ [illegible] $C_1 = 7/700 = 0{,}0$[illegible] $0{,}096 \cdot 700 = 67$ pF.

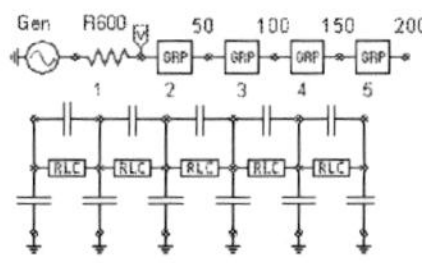

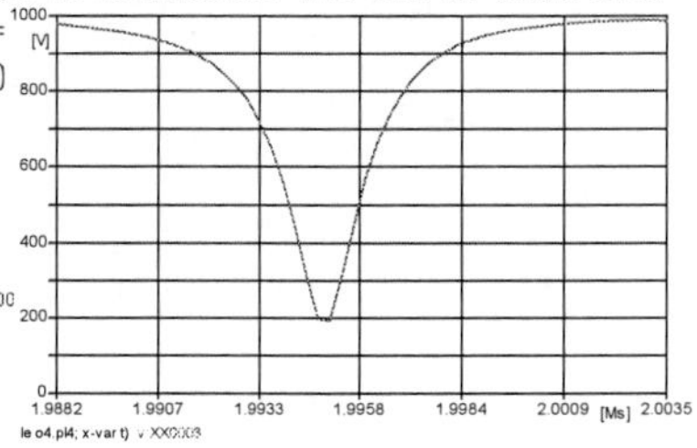

f_r (MHz)

Fig. 1. The scheme of inductor for FFT analysis in ATPDraw, detail of 5 turns

Fig. 2 FFT analysis of circuit in ATPDraw

Results from simulation in EMTP-ATP were compared with results from calculation according theory [1] and last but not least with results from measurement.

3. VERIFICATION OF INDUCTOR'S MODEL IN ATPDRAW WITH THEORY [1]

The program JEVY.EXE was used for definition of results, computing algorithm respects Wagner's theory [1]. According to Wagner's theory are turns stressed by voltage's unit impulse and voltage time distribution is calculated by program JEVY.EXE. Voltage time distribution is specified for inductor with insulated and earthed terminal of inductor too. Identical process was applied by EMTP-ATP; impact stress of turns is represented by DC source. The final course of voltage time distribution for inductor with insulated terminal (results from JEVY.EXE) is shown in Fig. 3a. According to theory is searched maximum of voltage and time of maximum. From graph in Fig. 3 is possible to find voltage's maximum at the moment of 16 µs with amplitude 218% of source's voltage at the terminal of inductor.

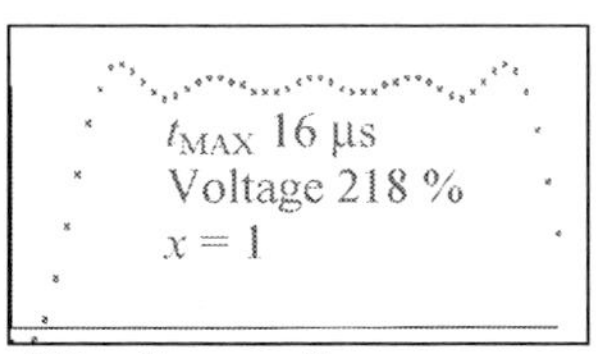

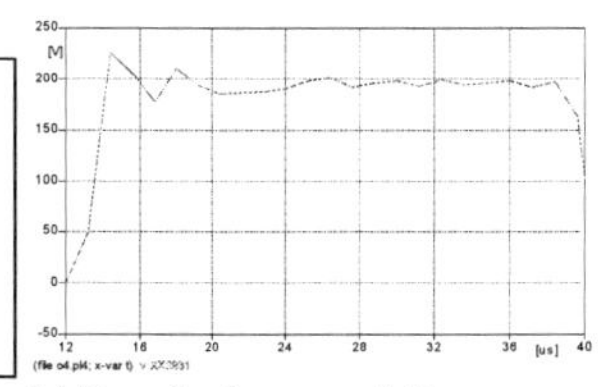

a) Results according to Wagner's theory (Program JEVY.EXE).

b) Results from modeling in EMTP-ATP.

Fig. 3. The voltage time distribution for inductor with insulated terminal

Graphical results from simulation voltage time distribution are shown in Fig. 3b for inductor with insulated terminal in EMTP-ATP. The initial and final voltage time distribution along inductor is introduced in Fig. 4. Influence of windings capacities is evidently for case of final voltage time distribution and influence of windings inductors is evidently for case of final voltage time distribution after suppressing of spherical oscillations. These graphical results are in compliance with theory [1]. The final scheme in ATPDraw for image generation is introduced in Fig. 5.

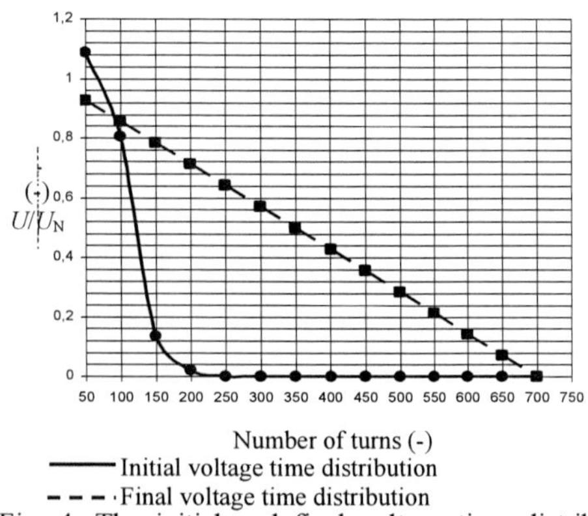

Fig. 4. The initial and final voltage time distribution along inductor

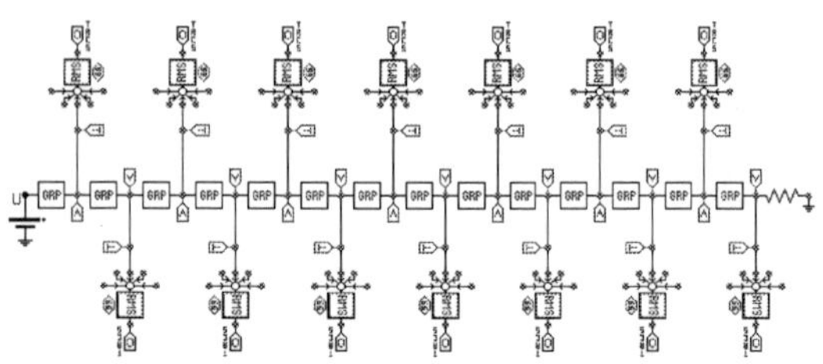

Fig. 5. The final scheme of inductor's model in ATPDraw for generation of initial and final voltage time distribution along inductor

4 MEASURING OF SURGE PHENOMENA BY SURGE GENERATOR PSG215A

The verification of inductor's model in EMTP-ATP was realized with using results from measuring surge phenomena by surge generator PSG215A too. Turns of inductor are not stressed by unit pulse for this case. Turns of inductor are stressed by voltage of capacitor discharging. The effect of internal impedance is applied. For this case, the model of inductor in ATPDraw must have been modified. The model was completed about source with impedance respecting surge generator. Voltage of surge generator was adjusted on value 100V.

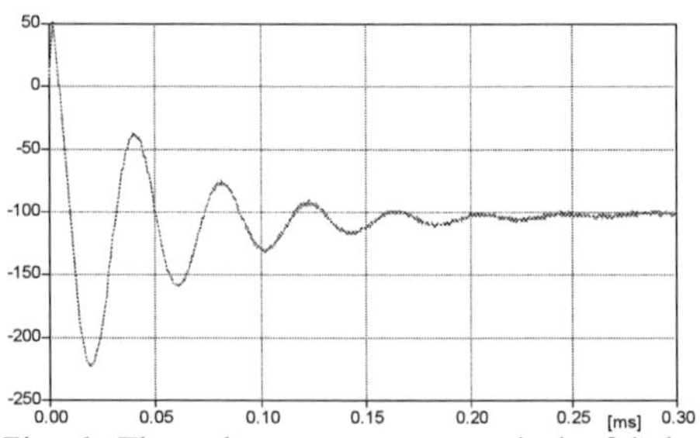

Fig. 6. The voltage course at terminal of inductor for case inductor with insulated terminal, measured course by PSG215A

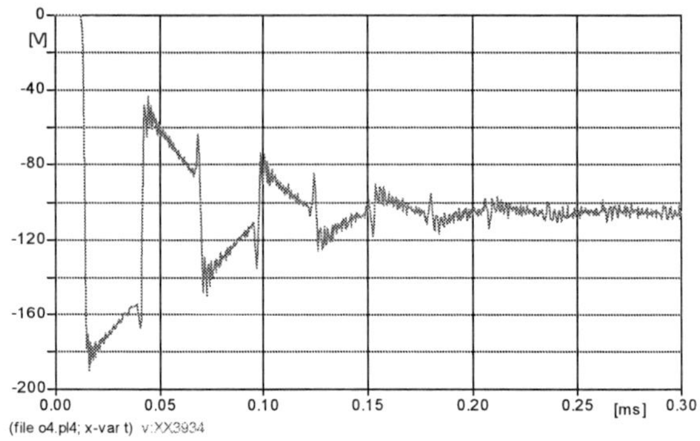

Fig. 7. The voltage course at terminal of inductor for case inductor with insulated terminal, simulated course by EMTP-ATP

5 CONCLUSION

The possibility of using program EMTP-ATP as a supporting program for diagnostic of electric machines was verified on physical model of inductor. Turns of windings of inductor were stressed by unit pulse at first. The final voltage time distribution was defined with using program JEVY.EXE for case inductor with insulated and grounded terminal. The model of inductor in EMTP-ATP was verified according to results from Wagner's theory. For case of inductor with insulated and grounded terminal were defined time (t_{max}), place (number of turns N) and amplitude of maximum voltage (U_{max}). Comparison of simulated (in EMTP-ATP) and calculated (in JEVY.EXE according to Wagner's theory) results is introduced in Table 1.

	Inductor with insulated terminal			Inductor with grounded terminal		
	t_{max} (μs)	U_{max} (%)	place (N)	t_{max} (μs)	U_{max} (%)	place (N)
JEVY.EXE	16	218	700	32,1	111	175
EMTP-ATP	15,2	222	700	31,2	126	175

Table 1 The comparison of results from programs JEVY.EXE and EMTP-ATP

The coincidence results from program JEVY.EXE and EMTP-ATP is evidently from Tab. 1 for both type simulations. From introduced graphical results is visible influence circuit's damping. According to theory [1] and mentioned graphical results is possible to calculate with voltage's amplitude 280% for case inductor with insulated terminal at end of terminal. For inductor with grounded terminal is maximum voltage's amplitude 150% of generator's voltage at fourth part of inductor.

Initial and final voltages time distribution along inductor with grounded terminal were simulated as a next, where final results are shown in Fig. 4 and the scheme for simulation in ATPDraw is introduced in Fig. 5.

Graphical results correspond to theory.

Finally, the model of inductor was verified with using results from measuring of surge phenomena by surge generator PSG215A. The voltage course at terminal of inductor for case inductor with insulated terminal, measured by PSG215A, is shown in Fig. 6 and simulated course by EMTP-ATP in Fig. 7.

The specification of inductor's model in EMTP-ATP is assumed for other type of winding's faults of electric machines.

6. REFERENCES

[1] Veverka, A.: *Technika vysokých napětí*, Praha, SNTL 1978

Annals of DAAAM for 2006 & Proceedings of the 17th International DAAAM Symposium, ISSN 1726-9679
Editor B. Katalinic, Published by DAAAM International, Vienna, Austria 2006
150th Birthday of Nikola Tesla (1856-1943) Inventor of Mobile Robot Teleoperation

THE ROLE OF THE ROTATION GROUP IN ROBOT KINEMATICS

MLADENOVA, C.

Abstract: *It is the well known that the rotations play an extremely decisive role in describing of the Euclidean motions. Nowadays, because of the modern and fast processors, the representation of the rotation group is still very important so that the hardware and the software to work efficiently together. The purpose of this work is to present the different parameterizations of the* **SO(3)** *group and to show their role in the efficient kinematics modeling. Here we accent on robot kinematics and trajectory planning, but naturally this reflects in dynamics modeling also and it is proved in our previous papers. The theoretical base is the refine knowledge of Lie groups and differential geometry in the problems of rigid body mechanics.*
Key words: **SO(3)** *parameterizations, Lie groups, robots, kinematics, dynamics, trajectory planning.*

1. INTRODUCTION

We consider the special orthogonal group **SO(3)** and its different representations for description of the motion of a rigid body with a fixed point. According to a theorem due to Euler a displacement in a general sense is equivalent to a rotation around an axis through the pole point (Corben & Stehle, 1960); (Bottema & Roth, 1979). There exist few analytical representations of a rotation (Ablamowicz & Sobczyk, 2004), (Bloch, 2003), i.e. a rotation is expressed by defining its action on :

a) vector x $\quad x' = Ax,\ (A \in \mathbf{SO(3)})$

b) quaternion x $\quad x' = qxq^{-1},\ (q \in \mathbf{Sp(1)})$

c) spinor ψ $\quad \psi' = U\psi,\ (U \in \mathbf{SU(2)})$

d) matrix X $\quad X' = UXU^{+}$,

where **Sp(1)** is the symplectic group of unit quaternions, **SU(2)** - the unimodular unitary group of 2×2 complex matrices, ψ - a one-index spinor and X is a complex, Hermitian, traceless 2×2 matrix.

2. VECTOR-PARAMETERIZATION OF SO(3) GROUP

Let us consider the special orthogonal group **SO(3)**

$$\mathbf{SO(3)} = \{O \in Mat(3,R);\ \det O = 1,\ OO^T = I\}, \qquad (1)$$

where $Mat(3,R)$ denotes the group of 3×3 real matrices together with its Lie algebra (infinitesimal generators) consisting of the real skew-symmetrical 3×3 matrices. In case that A belongs to the lie algebra of **SO(3)**, the Cayley transformation making the connection between the Lie algebra and the group explicit is given by the formulae

$$O = (I+A)(I-A)^{-1} = (2I-(I-A))(I-A)^{-1} = 2(I-A)^{-1} - I. \qquad (2)$$

As an exception in the three-dimensional space, there exists a map between vectors and skew-symmetric matrices, i.e., if $c \in \mathbf{R}^3$, then $c \to c^{\times}$, where $c^{\times}$ is the corresponding skew-symmetric matrix. Then we may write the **SO(3)** matrix in the form

$$O = O(c) = (I+c^{\times})(I-c^{\times})^{-1} = \frac{(1-c^2)I + 2c.c + 2c^{\times}}{1+c^2}, \qquad (3)$$

and consider it as a mapping from $\mathbf{R}^3$ to **SO(3)** for which the smooth inverse is

$$c^{\times} = \frac{[O-O^T]}{1+tr(O)}. \qquad (4)$$

Here I is the 3×3 identity matrix, $c.c$ means diadic, $tr(O)$ is the trace of the matrix O and "T" is the symbol for transposition of a matrix. The formula above provides us with an explicit parameterization of **SO(3)**. The vector c is called *vector-parameter*. It is parallel to the axis of rotation and its module $|c|$ is equal to $\tan(\alpha/2)$. The so defined vector-parameters form a Lie group with the following composition law

$$c' = \langle c_1, c_2 \rangle = \frac{c_1 + c_2 + c_1 \times c_2}{1 - c_1 c_2}. \qquad (5)$$

The symbol "$\times$" means cross product of vectors. Every component of c can take all values from $-\infty$ to $+\infty$ without any restrictions, which is a great advantage compared with the evident asymmetry in the Eulerian parameterization. Obviously, the vector $c \equiv 0$ corresponds to the identity matrix $O(0) \equiv I$ and $-c$ produces the inverse rotation $O(-c) \equiv O^{-1}(c)$.

Further, we shall show how the vector-parameter looks in the case when one uses other parameterizations of the **SO(3)** group, e.g. the case of a parameterization of the rotation group through Eulerian angles $(\psi, \vartheta, \varphi)$

$$c = (-\tan\frac{\vartheta}{2}\cos\frac{\psi-\varphi}{2}/\cos\frac{\psi+\varphi}{2},\ \tan\frac{\vartheta}{2}\sin\frac{\psi-\varphi}{2}/\cos\frac{\psi+\varphi}{2},\ \tan\frac{\varphi+\psi}{2}). \qquad (6)$$

In the case of Eulerian parameters (q_0, q_1, q_2, q_3), the vector-parameter is

$$c = (-\frac{q_3}{q_0}, -\frac{q_2}{q_0}, -\frac{q_1}{q_0}), \qquad (7)$$

while, for the Bryant angles $(\varphi_1, \varphi_2, \varphi_3)$, the components of the corresponding vector-parameter are expressed in the following way

$$c_1 = (\tan\frac{\varphi_1}{2}, 0, 0), \quad c_2 = (0, \tan\frac{\varphi_2}{2}, 0), \tag{8}$$

$$c_3 = (0, 0, \tan\frac{\varphi_3}{2}).$$

The Euclidean motions in $\mathbf{R}^3$, i.e. rotations about a fixed point and translations can be unified in the equation

$$r = O r_d + p, \tag{9}$$

where the matrix $O \in \mathbf{SO(3)}$, r_d and r are vectors in movable R_d and immovable R_o frames respectively and p is a vector of translation. If O and p are time dependent, the expression (9) gives a consequence of displacements. Introducing the skew-symmetric matrix of the angular velocity $\Omega = \dot{O}O^T$, it follows $\dot{\Omega} = \ddot{O}O^T + \dot{O}\dot{O}^T$. After differentiating the equation (9), we obtain the equation for the linear velocity vector $\dot{r} = \dot{O}r_d + \dot{p} = \Omega(r-p) + \dot{p}$, and the linear acceleration vector $\ddot{r} = [\dot{\Omega} + \Omega^2](r-p) + \ddot{p}$, where the last equation includes the tangential acceleration, parallel to the velocity and the normal acceleration. By using formula (3) and the relation $\Omega = \dot{O}O^T$ for the angular velocity vector, it can be verified that

$$\omega = \frac{2}{1+c^2}[c \times \dot{c} + \dot{c}], \tag{10}$$

where c is the vector-parameter defining the rotation of the movable frame R_d with respect to the fixed one R_o. Then the vector of the angular acceleration is

$$\varepsilon = \dot{\omega} = \frac{2}{1+c^2}[c \times \ddot{c} + \ddot{c} - \omega(c\dot{c})]. \tag{11}$$

The components of the vectors $\dot{c}$ and $\ddot{c}$ are respectively the first and second derivatives of the components of the vector-parameter c.

3. AN APPROACH FOR TRAJECTORY PLANNING

We have created also an approach for trajectory planning based on the vector-parameterization of the rotation group **SO(3)** (Mladenova, 2001a). In kinematical control the inverse kinematic problem is solved, and at the dynamical control - the direct dynamical problem. Our algorithm interpolates the rotational matrices. It is very important as for motion planning, so in computer animation, machine visualization and etc.

Problem statement:

Given an ordered set of $n+1$ rotational matrices $\{O_0, O_1, \ldots, O_n\}$, and $n+1$ scalars $t_0 < t_1 < \ldots < t_n$, two times differential curve $O(t)$ over the configurational group of the vector-parameters to be found, such that

$O(t_i) = O_i, \quad i = 0, 1, \ldots, n$.

Since $O_i = O(c_i)$, another definition of the solved problem is:

Given an ordered set of $n+1$ vector-parameters $\{c_0, c_1, \ldots, c_n\}$, and $n+1$ scalars $t_0 < t_1 < \ldots < t_n$, two times differential curve $c(t)$ over the configurational group of the vector-parameters to be found, such that

$c(t_i) = c_i$, $i = 0, 1, \ldots, n$.

On the basis of the above considerations, we may conclude that the suggested approach of the vector-parameterization is applicable with any choice of coordinate frames and with different generalized parameters. It is worth here to note that the composition laws of the other parameterizations are not used and that they have never been written in explicit form because of their complexity. We claim that the vector-parameterization reduces the elementary numerical operations in kinematic and dynamic modeling with about 40% (Mladenova, 1990); (Mladenova, 1999), (Ochier et al., 1995), (Mladenova & Müller, 1997), (Mladenova, 2001b). This is proved after a comparison of the number of the numerical operations (additions, multiplications and transcendental functions) between the matrix and the vector-parameter methods. In the matrix method the Eulerian angles, Bryant angles, Eulerian parameters and Denavit - Hartenberg parameters are used, and in the vector-parameter method – the corresponding vector-parameters expressed through them.

6. CONCLUSION

The present paper makes a short review of the rotation group and its representations and parameterizations . We stress on their calculation efficiency and possibility for application in real time control and computer visualization of complex mechanical systems such like attitude space systems, mobile robots, vehicle machines, human motion, and for any picture which has to be described through rotations. In the near future we plan to use the vector-parameter approach in modeling of manipulators with flexible links.

Acknowledgements: The paper is partially supported by Bulgarian NSF, Grant No. TH-1411/2004.

7. REFERENCES

Ablamowicz, R. & Sobczyk, G. (Eds.) (2004). *Lectures on Clifford (Geometric) Algebras and Applications*, Birkhäuser, ISBN 0-8176-3257-3, Boston

Bloch, A. M. (2003). *Nonholonomic Mechanics and Control,* Spinger-Verlag , ISBN 0-387-95535-6, New York

Bottema, O. & Roth, B. (1979). *Theoretical Kinematics*, North-Holland Publishing Co., ISBN 0-48666346-9, Amsterdam

Corben, H. C. & Stehle, P. (1960). *Classical Mechanics,* Dover Publications, INC., ISBN 0-486-68063-0, New-York

Mladenova, C. (1990). A contribution to the modelling and control of manipulators. *J. Intell.& Robotic Systems*, Vol.3 , pp. 349-363, ISSN 0921-0296

Mladenova, C. (1999). Applications of a Lie group theory to the modeling and control of multibody systems. *Multibody System Dynamics*, Vol. 4, No. 3., pp. 367-380 , ISSN 1384-5640

Mladenova, C. (2001a). Generation of rotation motions at trajectory planning, *Proc. 9th Nat. Cong. on Theor. and Appl. Mechanics*, Ivanov, Ya.; Baltov, A. & Manoach, E. (Eds.), pp. 363-367, ISBN 954-9526-10-0, Varna, September 19-22 , 2001, Bulgarian Academy of Sciences, Demetra Ltd, Sofia

Mladenova, C. (2001b). Modeling and control of multibody systems on a configurational space with a Lie group structure, *J. Theoretical and Applied Mechanics*, Vol. 31, No. 3, pp. 3-25, (invited paper), ISSN 0861-6663

Mladenova, C. & Müller, P.C. (1997). Dynamics and Control of Elastic Joint Manipulators. *J. Intell. & Robotic Systems*, Vol. 20, pp. 23-44, ISSN 0921-0296

Ochier Adongo, J.; Mladenova, C. & Müller, P. C. (1995). An approach to automatic generation of dynamic equations of elastic joint manipulators in symbolic language. *J. Intell. & Robotic Systems*, Vol. 14, pp. 199-218, ISSN 0921-0

Annals of DAAAM for 2006 & Proceedings of the 17th International DAAAM Symposium, ISSN 1726-9679
Editor B. Katalinic, Published by DAAAM International, Vienna, Austria 2006
150th Birthday of Nikola Tesla (1856-1943) Inventor of Mobile Robot Teleoperation

METHODOLOGY OF IMPLEMENTING INTERMODAL TRANSPORT IN THE CROATIAN TRAFFIC SYSTEM

MLINARIC, T.J. & ROGIC, K.

Abstract: *The intermodal transport chain deserves to be taken more seriously. As far as we can tell, after all, it is the only solution to the growing problem of transport infrastructure congestion. Because of the fact that road traffic threatens to come to a grinding halt, clearly other transport modes – especially railways, inland waterways and ocean shipping – will have to play a greater role if we are to avoid the chaos that looms ahead. To cope with the extra traffic prompted by economic growth, the different modes need to combine their services to create a sustainable and efficient transport system.*
Key words: *intermodal transport, transport modes, sustainable and efficient transport system*

1. INTRODUCTION

Cooperation of all the participants in intermodal transport in Croatia, in the Danube-Adriatic region and further afield should result in the shift of intermodal and especially container transport to the North Adriatic and Adriatic ports and to the organisation of transport with Central Europe(Figure 1.).

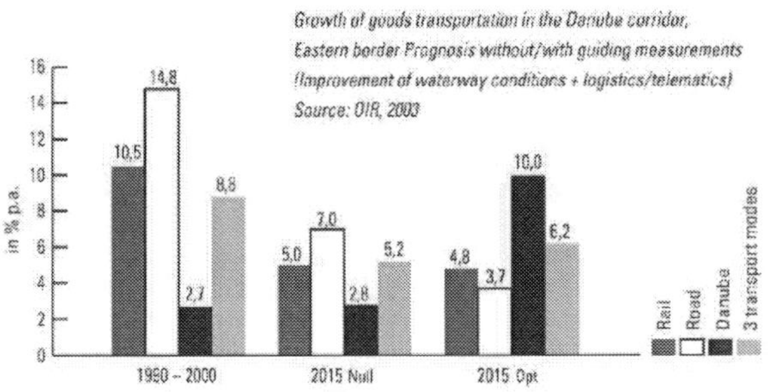

Fig.1. Predicted growth of transportation of goods in Danube region

Our vision is that: 1) the Adriatic ports should become "the Adriatic gateway to Europe"; 2) the infrastructure network should be developed with terminals for intermodal transport in the Danube-Adriatic region and with the elimination of bottlenecks, so that the network can be harmonised with and become an integral part of the European network; c) the Danube and Adriatic should be intermodally linked. This can only be achieved by providing the whole transport economy with an intermodal transport system (ocean shipping – railways – roads – inland waterways) using the already existing port capacities, free zones and transport terminals in the area.

2. ELEMENTS OF INTERMODAL TRANSPORT CHAIN

To analyse the logistics regions on a European scale, the scope of the region has to be determined. Based on present researches, it may be concluded that most existing definitions are found in terms of local, regional or national freight transport nodes and not related to elements like main industrial sectors and regional logistics policies.
Therefore, a logistics region could be defined by four main elements, as shown in Figure 2.

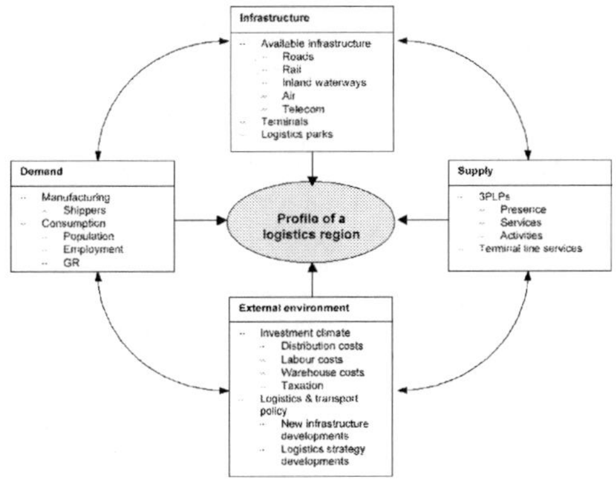

Source: PROTRANS partners. 2000

Fig. 2. Elements of logistics region

It is possible to identify five market segments with different historical developments and often with different stakeholders and customers, market structures, regulations, institutions, technologies and production systems. The segments are: short-sea feeder traffic consisting of the carriage of maritime containers between European regional ports and seaports that are the ports of call for intercontinental liner services; hinterland movement of maritime containers via inland waterways linking seaports with inland ports; hinterland traffic of maritime containers by rail linking seaports with inland terminals; continental short-sea shipping consisting of door–to–door movements originating in a bound for destinations in Europe, of which short-sea shipping forms an integral part. Most short-sea services are Roll On–Roll Off (Ro-Ro) services, in which cargo is shipped in semi-trailers or other road vehicle units and a few are Lift On–Lift Off (Lo-Lo) container services; continental intermodal rail traffic in which rail services are an integral part of continental door-to-door movements. Lo-Lo is the predominant technique, mainly using swap bodies, while in some parts of Europe Ro-Ro is more important, either in the form of trucks (Rolling Road) or semi-trailers (piggyback). Piggyback transport technology is a specific transport technology characterized by the horizontal and/or vertical loading, transport and unloading of road transport means, such as e.g. loaded, but also empty trucks with trailers, trailers and/or semi-trailers and loaded swap cases or tanks which are easily transported by road vehicles and all together at least along one section of the transport route on railway vehicles, i.e. transport of road vehicles and swap cases together with their load on railway cars, at least on one section of the transport route.

The technology combines and uses the advantages of both transport modes. In theory and practice several types of piggyback technologies may be distinguished, and these are A, B, C technologies and bimodal technology. The most important objectives of piggyback transport technology include: combining of road and railway transport in a very fast, safe and rational manner without re-loading from road vehicles to railway cars, and vice versa, from railway cars onto road vehicles; optimization of the effects of road and railway infrastructure and suprastructure; acceleration of cargo handling and carriage in combined road-railway traffic thus minimizing or completely eliminating the live work in the process of traffic service production; qualitative and quantitative maximization of technical, technological, organizational and economic effects of the traffic service production process; maximization of effects of the operation of creative and operative managers and other workers engaged in the piggyback transport system. If we follow this logic, another segment would be continental inland waterway traffic. However, this has not been included because, up to now, the segment has yet to attain significant proportions, especially in the case of the Republic of Croatia.

3. APPLICATION CRITERIA OF INTERMODAL TRANSPORT IN THE TRAFFIC SYSTEM

To cater to the growth in future Europe transport market intermodal transport will, however, have to cope with a number of serious constrains. These are most obvious for the infrastructure, which already faces problems of congestion and cannot be developed fast enough to keep pace with the growing transport demand in many European regions.

Rerouting is one of the strategies that should be adopted for solving these problems. The essence of methodology which is implemented for the realization of intermodal concept in the traffic system consists in iterative selection of the most justified combinations of traffic subsystems in the transport service realization, at the same time excluding those less justified ones. First, all the potential intermodal chains need to be isolated, the size of which meets the satisfactory condition at every en-route intermodal terminal (node). If at every en-route intermodal terminal the saving in hours per load unit due to the transit traffic flow of which it is part, is greater than the hours of collecting the load units, the required condition is met. Intermodal chains that meet the required condition are directly included in the optimal version of the train composing plan and are excluded from further considerations. Then, the routing graph is developed. The routing graph is formed on the basis of the step-wise chart of intermodal traffic flows, excluding previously those flows that satisfy the required condition and that are directly included in the formation plan, as well as on the basis of the calculation norms (t_u, c, m). It represents a scheme of all the possible routing of direct traffic flows that transit without handling at least one intermodal terminal. They are designed according to the principle of combining several traffic flows. This means, every traffic flow current is combined with other flows whose intermodal terminal of starting is the same or one of the previous ones, and intermodal terminal of routing is the same or one of the following ones in relation to the considered current. The routing graph is presented for every multi-current routing from one side intermodal traffic flows, then at every intermodal terminal – hours of saving per load unit as consequence of transiting without handling (nt_u), and on the other side total hours of saving per load unit on the entire travel path of the load unit subtracting the collecting hours, i.e. $\sum nt_u - cm$. Based on the formed graph we select the "input routing" which is included into the optimal version of the formation plan. Input routing shall satisfy the required condition at every en-route intermodal terminal ($nt_u \geq cm$) and insure the greatest saving of hours per load unit. If none of the input routings insures the required condition, the input routing selected is the one that insures maximal saving per load unit. After having selected the input routing, the routing graph is corrected taking into account the exclusion of the selected input routing in the optimal version of the formation plan. The re-selection of the new input routings and new graph corrections are being done until in the next routing graph there is none with the difference of total saved hours per load unit and hours of the collection of cargo units being positive.

4. CONCLUSION

Demand for freight transport in Europe has been rising steadily over the past decades. The increase, which has been most marked in road transport segment is expected to continue. Forecasts point to 50% growth in international road transport in the next ten years. All segments of intermodal transportation chain are expected to develop but it is likely that transport demand structures will change to a considerable degree. Container throughput in seaports will grow faster than other transport segments, boosting hinterland transport demand.

Logistics trends will also affect demand. The tendency towards unitisation of bulk and general cargo will foster road and intermodal transport solutions. The demand for transport of smaller consignments over longer distances will also increase, with more extensive consolidation, which again will favour road and intermodal transport solutions.

There is potential for further efficiency enhancement and quality improvement in more than one respect:

1) the players, whose role is vital in ensuring that the market functions properly;

2) the way resources are used in service production;

3) physical transport infrastructure and the efficiency of its use;

4) the technologies applied;

5) the standards of quality offered, including new logistics concepts;

6) production efficiency and the resulting costs;

7) changes in the legislation and polices to encourage freight transport innovation.

Integration in intermodal chains consists of different transport operators and/or terminals joining forces. This type of integration creates opportunities for economy of scale, for example by improving the use of larger vehicle fleets. Moreover, the advantages in terms of scale can arise in complex transport processes. For example, the skills developed in handling dangerous goods will pay off better in larger companies or systems. The implementation of the presented methodology allows finding of a formation plan for the traffic flows that are related to a great number of intermodal terminals, both for straight and for the branched traffic network polygons, and it is one of the more accurate methods of calculating the formation plan. The implementation of this methodology has not only justification because of the improved operative management of intermodal traffic currents, but also a significantly greater importance since its implementation directly influences the strategic thinking in the development of international intermodal traffic network which needs to include also the total traffic system of the Republic of Croatia.

5. REFERENCES

Vrenken, H.; Macharis, C.; Wolters, P. *Intermodal transport in Europe*, European Intermodal Association (EIA), 9090199136, Brussels, 2005

Wiegmans, B. W.; Masurel, E.; Nijkamp, P. *Intermodal Freight Terminals,* An Analysis of the Terminal Market, Transportation Planning and Technology, OPA, Vol. 23, 1999.

Mlinaric,T.J., Rogic,K., Safran, M. *The role of mobile management in the integration of Croatian economy into the European Union,* ECOMM 2000, pp. 298 – 301. Bregenz, Austria, May 2000.

Annals of DAAAM for 2006 & Proceedings of the 17th International DAAAM Symposium, ISSN 1726-9679
B. Katalinic, Published by DAAAM International, Vienna, Austria 2006
150th Birthday of Nikola Tesla (1856-1943) Inventor of Mobile Robot Teleoperation

EXPLORING THE POTENTIALS OF 5-AXIS MACHINING ROBOTS: CASE OF SCREW PROPELLER AND STAMPING DIE

MUNAR, A.C.; SINGH, M. & BOHEZ, E.L.J.

Abstract: *The ever-increasing demand for CNC machining of sculptured surfaces has highlighted the role of so called 5-axis machining robots in numerous industries including aerospace, die-mold making and shipbuilding. The efficiency of these robots inherently depends on the slowest link in the machine configuration.*
Key words: 5-axis, *propeller, stamping dies.*

1. INTRODUCTION

The most significant advantages of 5-axis machining robots are: reduced process time, reduced setup time and improved surface quality (Bohez, 2002; Lai et al., 2003). The ability of 5-axis machines to position the tool and the workpiece at any given relative point and angle allows them to produce the design part using several approaches (Bohez, 2002). Unlike 3-axis, 5-axis machines have considerable advantage in terms of efficiency and accessibility. For example, the effect of employing 5-axis machines in the manufacture of die molds has resulted in 10-20 times more than the efficiency set by its three-axis counterparts (Choi et al., 1998). Moreover, parts with irregular shapes such as turbo impellers and propellers can be machined using a single setup since areas previously inaccessible to 3-axis machines are made workable with added degrees of freedom although under certain constraints. There are several configurations of 5-axis machines that are classified according to their kinematic chain and rotary axes position. Theoretically speaking it is stated that there are 360 possible combinations yet only when 3 linear axes and 2 rotary axes are considered (Bohez, 2002). The Deckel Maho MH600E (Fig. 1) at CIM lab AIT, classified as a tilt-rotary table type (Warkentin et al., 2000), is a 5-axis horizontal machining robot consisting of XYZ linear slides and AB rotary tables. A-body holding the workpiece rotates at a full 360° about the horizontal X-axis while the B-body swivels at ±105° about the vertical Y-axis. The specifications are summarized in Table 1.

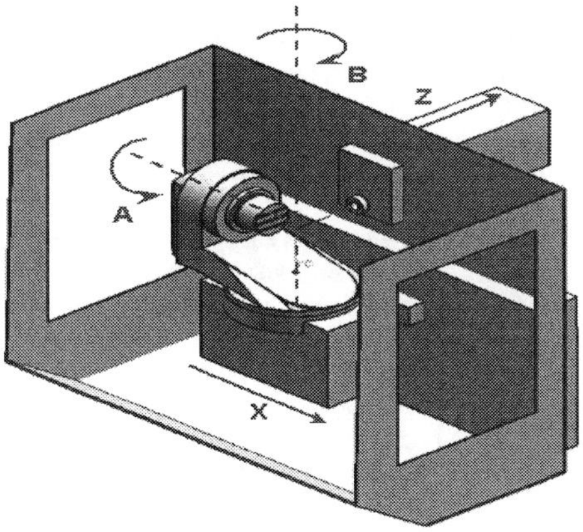

Fig. 1. Deckel Maho MH600E 5-axis milling robot

Table Size	320mm
Maximum Table Load	100 kg
Table Index Increments (A-Axis)	0.001deg
Table Swivel Range (B-Axis)	0± 105deg x 0.001deg
Traverses	X-Axis: 600mm
	Y-Axis: 450mm
	Z-Axis: 450mm
Spindle Speeds	20 - 5,000rpm
Maximum Feed Velocity (A-Axis)	235°/sec
Maximum Feed Velocity (B-Axis)	165°/sec
Feed Rates	1 – 4000mm/min
Feed Override	0-140%
Automatic Tool Changer (ATC)	30 stations
Maximum Tool Diameter	130mm
Maximum Tool Length	250mm
Machine Dimensions	3.2m x 1.83m x 2.54m

Table 1. Deckel Maho MH600E machine specifications.

2. 5-AXIS MACHINING OF SCREW PROPELLER

In order to maximize efficiency and productivity, all three machining operations i.e. roughing, semi-finishing, and finishing must be performed with one setup. However, in the manufacture of propeller master patterns for cavitation tunnel tests, it is still common practice that roughing of the cast preform would be done on a 2-1/2 copy-milling machine. This would require two setups viz. one for the front face and another for the rear. Thereafter, the workpiece is transferred to a 5-axis machine for semi-finishing and finishing. The present approach for propeller machining sets aside the redundant setups and fixturing by having the roughing sequence done straightaway on a 5-axis machine. Moreover, rather than starting off with a cast preform, the designated blank part is a cylindrical block. The aim of the roughing sequence is to first transform the workpiece into what is called the bounding-box preform geometry prior to the two remaining sequences at which the finished profile and surface of the propeller is achieved. The bounding-box is primarily an assemblage of hyperbolic paraboloids that crudely represent the geometry of the modeled propeller blades (Fig. 2). Among the benefits of 5-axis machines is their ability to mill ruled surfaces with the side of the cutter hence the major reason for developing the initial geometry.

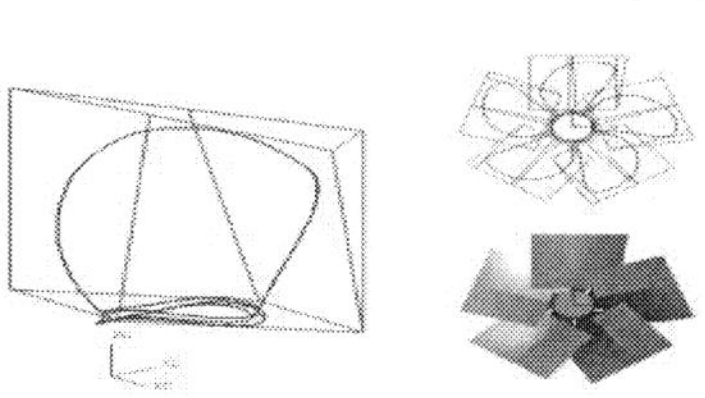

Fig. 2. Bounding box for single and five propeller blades

The toolpaths for roughing is generated on the front, side and back of the preform geometry. All three such features are cut via flank milling with a tilted flat end mill (Fig. 3). The estimated time for roughing with 300mmpm constant feedrate is 184.48min with at least 70% material removed from the workpiece. Following is the semi-finishing to machine the blade faces and to remove residual material at the root section between adjacent blades (Fig. 4).

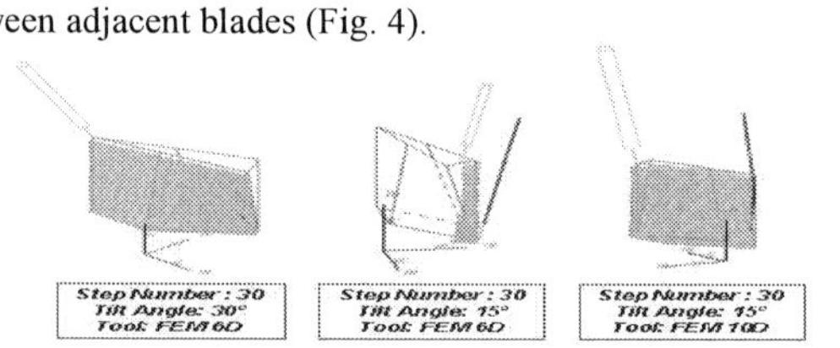

Fig. 3. Rouging toolpaths for the bounding box preform

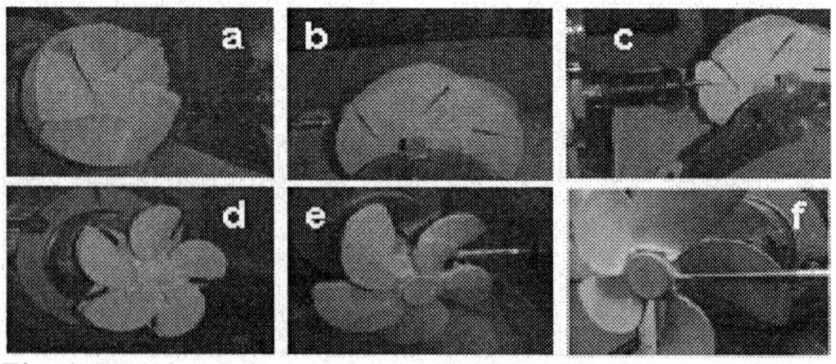

Fig. 4. Rouging (a,b,c) and semi-finishing (d,e,f) sequences

3. 5-AXIS MACHINING OF STAMPING DIE

Traditionally, stamping dies have been manufactured with 3-axis machines, which conventionally used ball nosed tools for machining the blank part to the finished die. The fact that the stamped product cannot have a surface quality superior to the die thus the requirement is of A-grade finish for the die surface. Machining with ball nosed tools does not give the required surface quality and the die would require further finishing processes viz. grinding, EDM and polishing to meet the aforesaid requirements. The designated blank part is a 345*125*105 block, of SKD11 (JIS) material. The freeform-feature for stamping dies can be defined as a sculptured surface region having a distinguishing, meaningful pattern, which may be thought of number sub-divided laminas/patches of zero thickness that are stitched together. (Fig. 5)

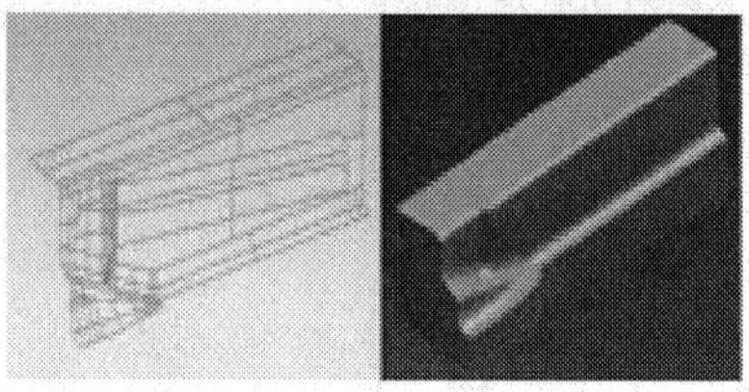

Fig. 5. Freeform Feature Geometry of Stamping Die

The roughing is performed to come up with a near net shape of the die. During the process, all the excess material is removed in order to begin with the 5-axis roughing. For 5-axis roughing, relative to drive option for generating tool path has been implemented with lead and tilt angles of 5 and 30 respectively. The semi-finishing and finishing operation were implemented with dual 4 axis relative to drive method, which gives the additional benefit to control the tilt angle as larger tilt angles gives poor surface finish (Lopez et. al. 2005). The tilt angle for semi- finishing and finishing is restricted to -3 for zig and 3 for zag and the lead angle for both the operations have been fixed at 10 degrees. Scallop height is reduced from .5 mm in roughing to 2 microns in finishing (Fig. 6). The total time for stamping die machining with 5-axis machine is about 22 hours as against 32 hours with a 3-axis machine.

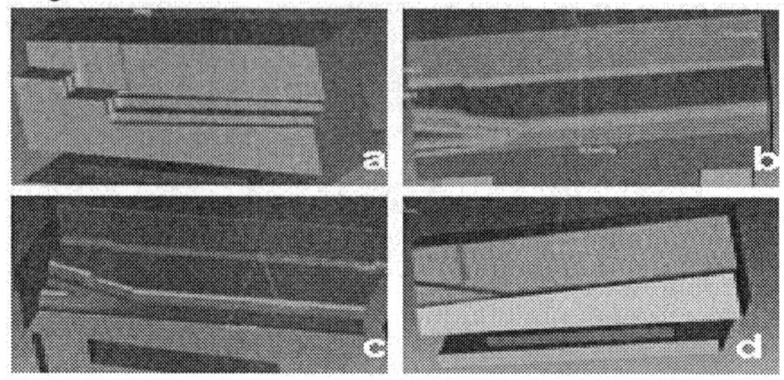

Fig. 6. Roughing (a,b); semi-finishing (c); finishing (d)

For the roughing operation flat end mill of diameter 16mm is used. Consideration for semi-finishing andd finishing operations is the smallest concave surface radii for the whole freeform geometry which is 13.2 mm. Conclusively, flat end mill of diameter 10mm is utilized for the last two operations. For the stamping die machining, MAHO 600E is not an ideal machine because of limited work space.

4. CONCLUSION

The possibilities are unlimited with the 5-axis machining robots. Exploring the full potential of the machines is highly beneficial which can prove to be a major cost and time saver for the manufacturing environment on the global scale. The gauging and collision avoidance is by large the biggest hindrance to avail the full advantage of these machines in terms of cost and time saving. The cost of the machines along with larger programming/machining time have been a barrier for large scale utilization.

5. ACKNOWLEDGEMENTS

The authors would like to acknowledge the contribution of AAPICO Hi-Tech Public Co. Ltd., Thailand for the collaborative effort for the research on stamping die inserts. The contribution of Asian Institute of Technology and people at CIM lab is also acknowledged for the patient and concerted efforts for the work.

6. REFERENCES

Bohez, E.L.J. (2002). "Five-axis Milling Machine Tool Kinematic Chain Design and Analysis." *International Journal of Machine Tools & Manufacture* Vol. 42, No. 4, 505-520, ISSN 0890-6955

Choi, B.K. & Jerard, R.B. (1998). *Sculptured Surface Machining Theory and Applications*. Kluwer Academic Publishers, ISBN 0 412 78020 8, Dordrecht

Lopez De Lacalle, L.N., Lamikiz, A., Munoa, J. & Sanchez, J.A. (2005). "The CAM as the Centre of Gravity of the Five-axis High Speed Milling of Complex Parts." *International Journal of Production Research* Vol. 43, No. 10, 1983-1999, ISSN 0020-7543

Lai, X.D., Zhou, Y.F., Zhou, J. Peng, F.Y. & Yan, S.J. (2003). "Geometrical Error Analysis and Control for 5-Axis Machining of Large Sculptured Surfaces." *International Journal of Advanced Manufacturing Technology* Vol. 21, No. 2, 110-118, ISSN 0268-3768

Warkentin, A., Hoskins, P., Ismail, F. & Bedi, S. (2000). "Computer Aided 5-Axis Machining." In *Systems Techniques and Computational Methods*, Chapter 3, Cornelius (EDT) Leondes, CRC Press LLC, ISBN 0 8493 0993 X, Boca Raton

Annals of DAAAM for 2006 & Proceedings of the 17th International DAAAM Symposium, ISSN 1726-9679
Editor B. Katalinic, Published by DAAAM International, Vienna, Austria 2006
150th Birthday of Nikola Tesla (1856-1943) Inventor of Mobile Robot Teleoperation

SUSTAINABLE DEVELOPMENT OF ISLAND TOURISM: CASE OF CROATIA

MUNITIC, A.; VIDUCIC, V. & VIDUCIC, LJ.

Abstract: *This paper deals with development of island tourism and with other environmental protection issues on the Croatian islands. In order to achieve a sustainable development of tourism, the management of resources (landscape, socio-cultural heritage, accommodation facilities and human resources) must satisfy manifold criteria, and a long-term development strategy must be defined. This paper argues strongly in favor of an local and a well-balanced development strategy that would emphasize the preservation of local community as well as the preservation of its natural environment, wildlife and cultural heritage.*
Key words: *sustainable development, island tourism, and ecotourism*

1. INTRODUCTION

Croatian islands make up 6% of the Croatian mainland, and, consisting of 718 islands, 389 rocks and 78 reefs, they are the second largest group of islands in the Mediterranean. Recently it has been widely accepted that a sustainable development of coastal tourism in Croatia is an optimal development trend, so it has been scientifically researched quite a lot lately. The development spiral of Croatian coastal region can be only triggered off by a sustainable development (Hitrec, 2002). Tourism carries many positive effects, but at the same time it carries the negative ones as well (Dovečar, 1999). The most outstanding negative effect of tourism is devastation of the authentic landscape. Sustainable development principles should be applied to all sorts of tourism activities (Kušen, 2002). A sustainable development of Croatian islands has to be strategically achieved by preventing and avoiding some negative effects (Klarić, 2002). Ecotourism has a positive effect on the environment, promoting an environmentally friendly approach. The island marinas should be also considered in the light of a sustainable development, in order not to jeopardize the present environment (Hitrec, 2002).
The paper further deals with the interdependence between tourism and ecology. Sustainable development strategy should be the basic development strategy of the islands. It should anticipate a future preservation of the preserved areas, some advantages and limitations of future development, and the role of the Government. Ecotourism should be encouraged and it should be involved in the structures of all types of tourism, having positive effects on the environment, economy and local population. Island development strategy should be tailored for each island separately, since unrestricted investments in tourism can be rather destructive for the environment and for the local population. The main development goal must be enhancement of the quality of life on the islands and prevention of their depopulation. Taking the interdependence between tourism and ecology as a starting point, the paper theoretically explores the specific interdependence on the case of Croatian islands, what is corroborated by empirical research on the island of Vis that is considered to be a jewel and paradise for tourists (being one of 10 best preserved islands in the world).

2. SUSTAINABLE DEVELOPMENT OF ISLAND TOURISM IN CROATIA

The Mediterranean, with its numerous historic and cultural monuments, is mostly visited tourism region in the world (Dovečar, 1999) Unfortunately, tourism carries not only positive effects, but also the negative ones. The most outstanding negative effect of tourism is devastation of the authentic landscape. The inflicted devastation, which has been esthetic, ecological, economic, historical, ethnical and architectural one, is simply irreparable (Klarić, 2002). However, devastation of natural and cultural environment in Croatia has not been only caused by development of coastal tourism, but it has also been the result of traffic congestion, motorization, urban development, forest fires etc. (Žuvela, 1996). Sustainable development comprises both an economic growth and social development, adjusted to the ecosystems in which it exists, and, as such, it is sustainable for a long term. (Jurković, 1991) The relationship between the natural environment and local population on Croatian islands is well preserved. The islands have a long tradition in tourism, and each one is a specific case. Sustainable development principles should be applied to all sorts of tourism activities. Ecotourism is merely a part of tourism that plays an important role in local community and that is very important for preservation of the island environment (Kušen, 2002). Nowadays, the attitude towards ecotourism has been clearly defined. It is believed that ecotourism actively promotes an environmentally friendly attitude and that it has a direct positive effect on the environment. After all, the prices are relatively high, and the number of tourists who ask for eco-services is relatively small (Klarić, 2002). The pattern of sustainable development of Croatian islands must be a strategic point of a long-term sustainable development. This is the only way to prevent a further devastation of environment and some negative effects of development. Sustainable development should be especially encouraged in island marinas, otherwise they could seriously jeopardize existing natural environment.
The development spiral of the Croatian coastal region can be only triggered off by sustainable development. However, the Government, with capital investing in infrastructure and with enactment of the appropriate legislative regulations, plays an important role in this case (Hitrec, 2002, Viducic, 2004).
In the light of sustainable development, the following issues should be analyzed:

- How to preserve the existing preserved areas,
- What the advantages of the development are,
- What the limitations of development are,
- What the role of the Government is,
- Whether the development should be based on cultural, historic and economic identifiability or whether to create some new ones,
- How to create a sustainable development that will not collapse by itself.

As regards the island tourism, two most important issues are organic farming and gastronomic offer. Fruit and vegetables and other food products, especially in the coastal area, are still traditionally produced, in an organic way (without use of pesticides) and they are an important economic and tourism brand-name. In addition to gastronomic offer, Croatian tourism must be recognized for its health tourism, cultural sites and environmental pecularities, such as well preserved wildlife, level and quality of urban development, especially in the coastal region. On one hand, the islands have become attractive destinations for tourists, and, on the other hand, the younger island population has been leaving islands, which can cause an imbalance in the existing environmental and socio-economic pattern. Development of tourism in general, and particularly on the islands, must be sustainable. The new generations of both islanders and tourists should be offered an opportunity to enjoy all the benefits of a relatively well-preserved island environment. However, it is indisputable that the quality of island tourism should be enhanced. By the year 2008, all accommodation facilities should be rehabilitated and privatized by strategic investors, who have already established brand name at some competitive destinations. At the same time, some new types of tourism, such as eco-tourism, health tourism etc., should be developed in order to revive the island interior areas. It is estimated that in 2010, the Republic of Croatia will be a full member of European Union, so the Croatian islands should offer an equal accommodation and catering service as the other most attractive island destinations of developed Mediterranean countries offer (France, Italy, Spain and Greece).

3. PREFERENCES OF THE TOURISTS VISITING ISLANDS – EXAMPLE OF THE ISLAND OF VIS

In order to support the theoretical assumption about the interdependence between tourism and sustainable development, which is especially emphasized on the islands, we have conducted a questionnaire examination of the tourists visiting the island of Vis. The island is geographically located in the middle of the Adriatic Sea. It has got numerous archeological sites and a rather ancient history, numerous coves with wooded hillsides, sandy and shingle beaches, rich and tasty gastronomic offer and very hospitable hosts. The questionnaire inquired their opinion about gastronomic offer, environmental preservation, accommodation, climate, and historical and cultural sights on the island of Vis. The total number of persons questioned was 212 (which makes 4,14 % of the registered visitors of the tourist population on 10th of August, 2005). They were asked to choose two responses to each of the nine items that covered some environmental and tourism issues.
The results have shown the preferences of the tourists visiting the island of Vis and the reasons why they have chosen Vis as their holiday destination, as well as the structure and specific features of tourist offer on the island of Vis. More than a half of the respondents stated that the main reason for their choice of this holiday destination was bathing and sunbathing, as well as possibility of having a quiet and peaceful holiday (62%), while their favorite holiday activity was walking by the sea and hiking (51%). A considerable number of respondents (45%) stated that the most prominent characteristics of the coast were clean beaches and crystal clear sea, which they preferred more than beach entertainment activities or skinny-dipping.
It is interesting to point out that the rural and traditional reasons for visiting the island, such as accommodation at rural farmsteads and eating of homemade food and drinks were rather highly ranked (82%). This speaks in favor of enhancement and promotion of both ecotourism and of traditional gastronomic offer. The villages of the island interior should be powerful development resources, especially in a developed information technology era. Traditional rural life, together with traditional food and drinks should be included in a Croatian tourist offer. It is evident that a well-preserved environment is one of the very important reasons for choosing the island of Vis as a holiday destination. This means that the government, both at the local and at the national level, should provide a long-term coordination between tourism and social environment. Therefore, all the negative effects of tourism should be minimized, while the positive ones should be encouraged: accommodation facilities should be built according to the available capacity, too exaggerated urban development should not be allowed, pollution of the land and sea should be avoided and the care should be taken not to cause an imbalance in the ecosystem. On the contrary, an environmentally friendly attitude should be encouraged, as well as the preservation of cultural sights and local traditional values, so that the life in local communities on the islands could be enhanced.

4. CONCLUSION

The balance between the environment and local population of the Croatian islands is well preserved. There has been a long tradition of tourism development on the islands, and every island is a specific case. Tourism development on the islands must be based on some local incentives that should encourage specific forms of island entrepreneurship and create specific partnerships, or some small-scale businesses, which should be subsidized by the Government. While making necessary strategic moves, a special care must be taken to include all the stakeholders on the islands (local population, tourism industry representatives, local politicians). This is of utmost important since the enhancement of tourism requires an acceptance of coexistence with traditional values and culture, respecting diversity of the tourist offer. As it has already been emphasized the main development goal must be enhancement of the quality of life on the islands and prevention of their depopulation. Too big investments in tourism can have a destructive effect both on the environment and on the local population. Mass tourism should not be encouraged since it would additionally devastate natural resources and it would have a negative effect on preservation of traditional local values, which are one of the main reasons for tourist visits to the islands. Existing accomodation facilities should be modernized according to the present demands of tourism market. In this way the tourists with higher purchasing power could be attracted, and accomodated in off-season as well.. New accomodation facilities, which are planned to be built on each island, should correspond to the number of tourists that can be accepted, and must be based on its sustainable development.

5. REFERENCES

Jurkovic, P. et.al., (1995). *Business Dictionary*, Masmedija, ISBN 953 157 – 036 - 1 Zagreb,

Dovečar, R., Vukčević, M. (1999), Ecological Component of Tourism in the Mediterranean and in Croatia, *Pomorski zbornik*, No. 1/1999.,pp. 115- 130, ISSN 0554 - 6397

Klarić, Z. (2002). Development of Ecotourism and of Rural Tourism Contributes to the Image of Croatia, *Okoliš,* No 111/ 2002., pp.11-12. ISSN 1330 - 6154

Kušen, E. (2002). Ecotourism and a Sustainable Development of Tourism *Okoliš*, NO 111/ 2002.., pp. ISSN1330 - 6154

Hitrec, T.(2002). Sustainable Tourism and Environment in the Focus of Interest of European Council, Turizam, No 2./ 2002.pp.4537- 459, ISSN 0494 – 2639

Žuvela, I. (1996). Ecological Aspects of Sustainable Development of Tourism on the Adriatic, *Pomorski zbornik*, No.1/1996., pp. 87 – 109, ISSN0554 - 6397

Annals of DAAAM for 2006 & Proceedings of the 17th International DAAAM Symposium, ISSN 1726-9679
Editor B. Katalinic, Published by DAAAM International, Vienna, Austria 2006
150th Birthday of Nikola Tesla (1856-1943) Inventor of Mobile Robot Teleoperation

CLEANING ROBOT: SELECTION AND ANALYSIS OF DRIVING MECHANISM

NAGLA, K.S.; UDDIN, M.; JHA, R.; ARORA, J.S.; KALRA, A.S. & KATYAL, A.

Abstract - This paper presents the selection and analysis of linear driving mechanism to support cleaning robot manipulator. The said robot has been designed to clean the surface below the Indian railway tracks at platforms. The existing cleaning process on the railway platform is manual, which is tedious particularly, when the frequency of the trains is very high. The end-effector of the robot manipulator is a nozzle to throw water as cleaning medium. The first part of the paper, deals with the selection and description of the driving mechanism of robot. The second part deals with it's analysis according to the schedule of dimensions and environmental conditions of the Indian railways.
Key words: Service robots, design of robot, cleaning robots

1. INTRODUCTION

The Indian railway runs thousands of trains everyday under unitary management system. The existing cleaning process on the tracks at Indian railway platform is manual, which is difficult particularly, when the frequency of the trains is very high. The manual work is not only costly and time consuming but also unsafe to the humans who work on the site. The proposed robotic system can be installed within the permissible area between the two railway tracks parallel to each other (Nagla, 2001). The end-effector of the proposed robot is a nozzle to throw water as cleaning medium. The cleaning schematic is shown in figure 1. The installation of the robot will not disturb the regular and irregular movements of mobile and stationary objects at railway stations. The number of robots required at particular platform will depend upon the length of the platform. The said robot will be on-line and off-line programmable system combined with various indicators and sirens. It can also be synchronized with the arrival and departure of the train. The system consists of two parts, stationary part i.e. manipulator and mobile part is a carriage of the robot. This paper deals with selection and analysis of the carriage actuating mechanism.

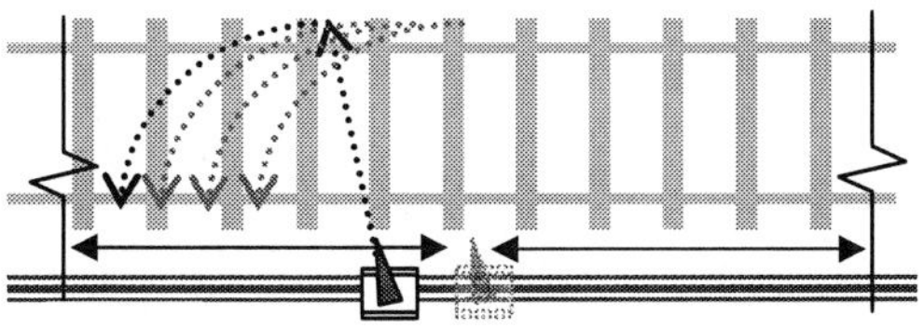

Fig. 1. Cleaning schematic of robot

2. SELECTION OF DRIVING MECHANISM

According to kinematics of the proposed robot (Nagla et al., 2005), it is required to move the robot on linear track. In order to accomplish such tasks many actuating systems (hydraulic, pneumatic, belt drives, gear drives, linear slides etc) have been proposed for the mobility of robots (Richard D. Klafter, 1989). Hydraulic and pneumatic actuators have brute strength, and excellent power-to-weight ratio but they tend to leak, and have lower reliability, and expensive (Mazumdar, 2004). Belt drives are not advisable for the said system because of the environmental constrains i.e. the varying temperature conditions may reduce the life of the belt. Chain drives get stretched with time and also require frequent lubrication. The linear slides may get clogged with dirt and dust with time which interrupts their smooth operation. The most advantageous drive is found out to be a gear drive in which the motion is transferred to pinion which provides linear motion moving on a rack.

3. DRIVING MECHANISM

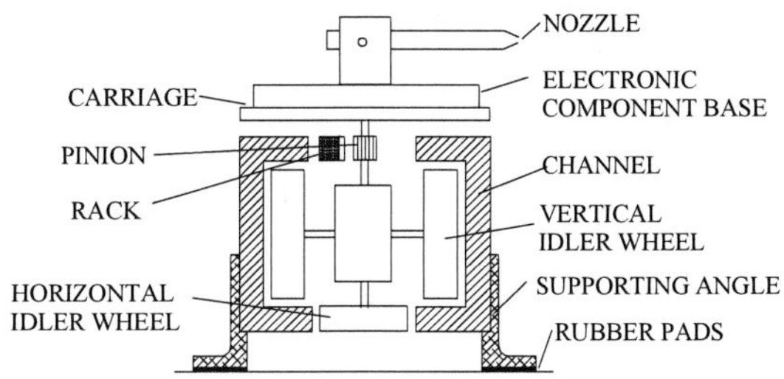

Fig 2. Mechanical design of driving mechanism (Section View)

The driving mechanism consists of a modular rack and pinion arrangement, which is enclosed inside the cast iron channels as shown in fig 2. The rack can be fastened to the inner section of the channel using nut and bolts at regular intervals. The lower section of the channel consists of vertical idler wheels which are mounted for providing support to the robot. In addition we mount two more horizontal idler wheels at bottom which provides support when the robot tilts to one side. The horizontal idler will help to minimize the reaction forces generated by fluid flowing through the nozzle, acting on the robot. At the top surface, we install a carriage over which manipulator and electric motor are installed. In order to avoid the complexities and number of mating parts (bevel gears for transfer of motion from motor to pinion) we have designed to install motor in vertical manner with the output shaft of motor connected to shaft of the pinion.

4. ANALYSIS FOR RACK AND PINION

The design procedure consists of selecting the proper material and conditions for the robot driving mechanism. As the only task of this robot is to clean the surface below the railway tracks an optimal linear velocity is required for efficient cleaning. From the preliminary experimental results it is presumed that the robot will clean the tracks effectively if it's linear velocity is in the range of 0.05 - 0.15 m/s. As the system is dynamic in nature, we used Buckingham's equations (1) & (2) for checking the maximum dynamic load, wear and feasibility of the proposed design. The service factors are also considered by incorporating overload factor (Sharma & Aggarwal 2004). The design contains comparison of results with certain threshold limits which has been examined. The different material combinations used for the analysis of rack and pinion are Steel with brinell hardness number (BHN) = 220, Young's modulus (E) = 210 x 10^9 N/m^2. Cast Iron (CI) with BHN = 160, Young's modulus (E) = 105x10^9 N/m^2. Plastic with BHN = 50, Young's modulus (E) = 5 x 10^9 N/m^2

5. EXPERIMENTAL ANALYSIS

5.1 Assumptions

For pinion: Teeth: 20° stub series. Diameter (D) = 0.05m; Face width (f) =0.05m; Number of teeth (z_p)=10; Module (m_o) = 0.005m; Lewis form factor (Y) = 0.285. Power transmitted (P) =1000 W (including overload factors for shock loading due to proximity of trains vibrations) has been assumed.

5.2 Design

Design transmitted load (Ft) = P/V = 1000/V

Using Buckingham's equation for dynamic load the maximum dynamic load (F_d) on the gear tooth is:

$$F_d = F_t + F_i$$

$$F_d = F_t + \frac{21 * V (c * f + F_t)}{21 * V + \sqrt{c * f + F_t}} \quad (1)$$

Where c is a constant. Using the values of F_t and f in equation (1), a graph is plotted between dynamic load(F_d) and pitch line velocity (V)

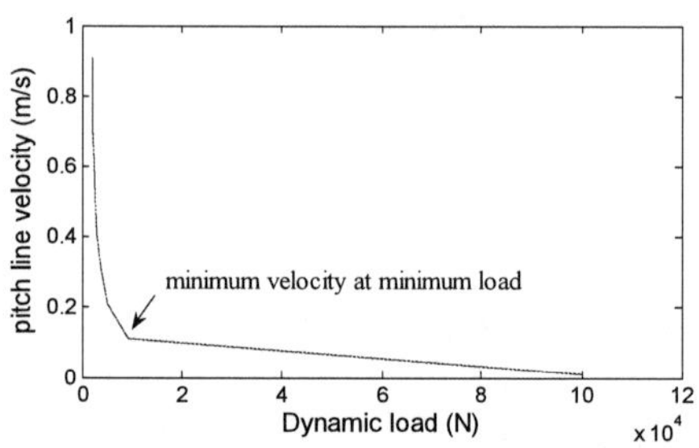

Fig. 3. Graph between dynamic load and pitch line velocity

From the above graph we concluded that when pitch line velocity (V) is 0.1 m/s the dynamic load (F_d) is approximately 10000 N which is the minimum load corresponding to the desired velocity range. To see whether the system is safe for wear or not we use Buckingham equation (2) for checking the limiting load of wear. According to this equation the limiting load for wear (F_w) is

$$F_W = D * f * K_w * Q \quad (2)$$

Where Q is ratio factor = 1.6

$$K_w = \text{load stress factor} = \frac{f_{es}^2 \sin\varphi [1/Ep + 1/Er]}{1.4} \quad (3)$$

f_{es} is surface endurance limit $= (2.8 * BH - 70) * 10^6$ N/m^2 (4)

Where BH is the average brinell hardness number of pinion and rack material. Using the values of BHN for different pairs in equation (4), we get different values of f_{es}. This value gives the numerical value of surface endurance limit for different materials of rack and pinion in mesh. Using this value and values of Young's modulus (Ep and Er) in equation (3) we get K_w which is further used to calculate F_w. from equation (2). To keep the system safe from wear $F_d \leq F_w$. This implies that if this condition is not fulfilled, the gears are not hard enough and they further need heat treatment to increase their hardness. Fig.4 shows that the combination of steel pinion - plastic rack and cast iron pinion - plastic rack has the F_w value greater than 10000 N, so this combination can be considered for their use in robot. If it happens to use any other combination of material, there hardness must be increased to increase their value of surface endurance limit f_{es}. Now comparing the wear load with endurance strength of gear tooth.

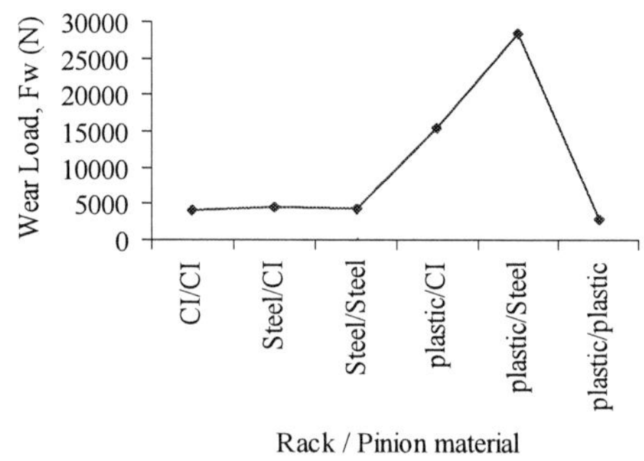

Fig. 4. Graph between limiting load for wear and material combination

The endurance strength (F_b) of the gear tooth is given as

$$F_b = f_{ef} * f * m_o * Y \quad (5)$$

here f_{ef} is flexural endurance limit. For steel f_{ef} is 385 x 10^6 N/ m^2 and for cast iron it is 84 x 10^6 N/ m^2. Now putting the values for steel and cast iron in equation (5) , we get endurance strength (F_b) for steel tooth as 27431.25 N and for cast iron tooth as 5985 N. For the system to sustain the shock load waves coming from the railway trains the required condition is $F_b \geq 1.5 F_d$, and $F_b \geq F_w$ Since the only the plastic rack and steel pinion combination is meeting this requirement, so only this combination is proposed for the robot.

6. CONCLUSIONS

A gear drive with rack and pinion arrangement has been found most suitable for the said system. The experimental analysis shows that the design parameters are within their respective threshold limits, the robotic system will clean the tracks effectively and sustain the vibrations coming from the nearby passing trains. Since only the plastic rack and steel pinion combination is meeting this requirement, hence only this combination is proposed for the robot.

7. REFERENCES

Nagla KS "*Robot for Railways (with case study of Indian railways)*" Proceedings of 12th DAAAM International symposium "Intelligent Manufacturing and Automation :Focused on Precision Engineering", Austria 2001. page 213-214

Nagla K.S., Saha S.K., Jha R., Akshay Mathur "*Robot for cleaning the surface below the Indian Railway Tracks at Platforms: Feasibility and Kinematics*" ACIAR-2005 (Asian conference on Industrial Automation & Robotics), 11th -13th May, 2005, Bangkok, Thailand. pp 55

Richard D. Klafter,Thomas A. Chmielewski.,Michael Negin., (1989)Robotic engineering and integrated approach, Prentice-Hall ,Inc ., New Jersey

Mazumdar, S.R.(2004). *Oil Hydraulics,* Tata McGraw-Hill, New Delhi

Sharma, P.C & Aggarwal, D.K.(2004). *Machine Design,* page numbers (261-270), S.K. Kataria Publication , New Delhi.

Annals of DAAAM for 2006 & Proceedings of the 17th International DAAAM Symposium, ISSN 1726-9679
B. Katalinic, Published by DAAAM International, Vienna, Austria 2006
150th Birthday of Nikola Tesla (1856-1943) Inventor of Mobile Robot Teleoperation

NUMERICAL VALIDATION OF ADHESIVE BONDED JOINTS MODELING

NEMES, O.; LACHAUD, F.; MOJTABI, A.; SOPORAN, V.; SURCIN, L.; DAN, V. & TATARU, O.

Abstract: *This work presents a numerical validation of a developed analytical model. For the assembly, the total force-displacement behavior is well defined. Thus the analytical model makes it possible to determine the rigidity of the assembly and to obtain a simple formulation very rapidly which gives the total behavior of the assembly*

Key words: *Adhesive joints, Finite element analysis, Stress transfer.*

1. INTRODUCTION

The mechanical performance of an adhesive bonded joint is related to the distribution of the stresses in the adhesive layer. Consequently it is essential to know this distribution, which, because of its complexity, makes prediction of fractures difficult. Many works was published on the determination of the stresses in multiple configurations of adhesive bonded joints. From the first works of Volkersen (1938) which give only a distribution of the shear stress in the adhesive joint to the more recent studies by finite elements, many formulations have made it possible to define the field of stresses in such assemblies better and better. Lubkin and Reissner (1956) present an analysis of stresses in tubular assemblies subjected to an axial loading and give a solution of the peeling stress distribution in the thickness of the adhesive. The tubes being supposed of small thickness, they use the theory of thin walls to build the stress field. Their analysis assumes that the work of shear and peel stresses in the two tubes is negligible relative to that of the same stresses in the adhesive. Alwar and Nagaraja (1976) made a study by finite elements of tubular joining subjected to an axial load taking into account the viscoelastic behavior of the adhesive. They also show that the viscoelastic behavior of the adhesive makes it possible to predict a considerable reduction in the maximum stresses at the ends of the joint. Shi and Cheng (1993) they built a first stress field using the equilibrium equations and the conditions of continuity of the stresses at the interfaces using an equation of compatibility. They then calculate the potential energy associated with this field and using the theorem of minimal complementary energy, they obtain a system of differential equations, the solutions of which are used to determine the optimal field. By minimizing the potential energy associated to the stress field using variational methods and some simplifying assumptions Nemes, Lachaud and Mojtabi (2004a, 2004b, 2005, 2006) has developed new analytical models to a fast pre dimensioning of the adhesive cylindrical bonded assemblies.

2. FINITE ELEMENTS MODELLING

In this work we consider an assembly of bonded tubes subjected to a tensile load whose geometrical definitions are given in fig. 1.

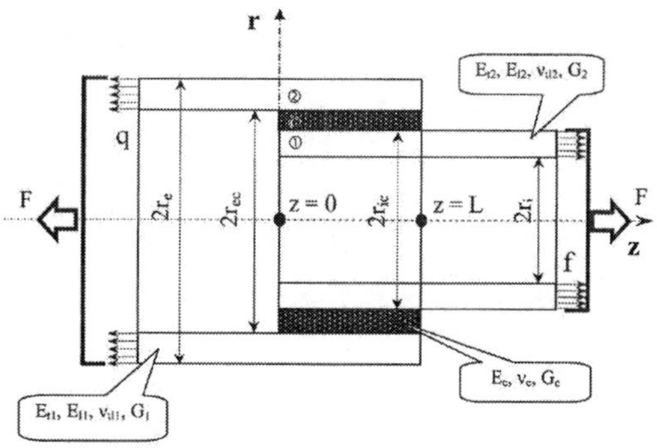

Fig. 1. Cylindrical assemblies

2.1 Meshing and boundary conditions

The objective of this study is to compare our analytical models of the adhesive-bonded joints with models made of finite elements. For the numerical analysis by finite elements of the adhesive-bonded joints, we used the computer code SAMCEF from SAMTECH®. The C.A.D. diagram, basis of the finite element model, is presented in figure 2. The diagram also describes the boundary conditions and the loading applied.

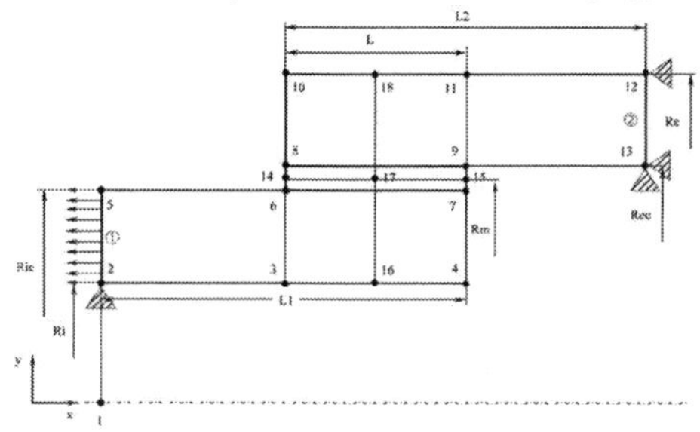

Fig. 2. CAD diagram of a cylindrical adhesive bonded joint

The cylindrical assembly is modeled by 2D quadrangles of degree 2 finite elements with the axi-symmetric assumption ($x \to z$, $y \to r$, $z \to \theta$). The displacements along x and y in face ② of the external tube and those along y in face ① of the internal tube are blocked. The load is applied as a pressure on face ① (Figure 13).

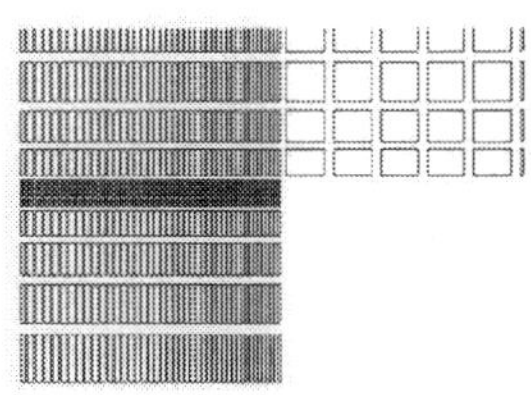

Fig.2. Numerical modeling of a cylindrical bonded joint with quadrangles elements

Figure 2 shows an example of the grid used in this study where all the finite elements are quadrangles. We imposed ten finite elements according to the adhesive.

2.2 Stress distributions

2.2.1 Load transfer

To compare the analytical model with the finite elements model we determine the load transfer in the middle of the bonded substrates, Figure 3, in a cylindrical metal-composite assembly. We can note some differences in the variation of the load given by the ideal model in the case of a metal-composite assembly, (Figure 3), while it still has a similar evolution.
The point of equivalence in stress in the two substrates is shifted (according to the length of the join) from approximately 15% in the finite element analysis. It should be noted that the position of this point varies according to the characteristics of the substrates: it is centered compared to the length of the join for substrates of equivalent total rigidities, and shifts on both sides as a function of the ratio of the rigidities of the bonded substrates.

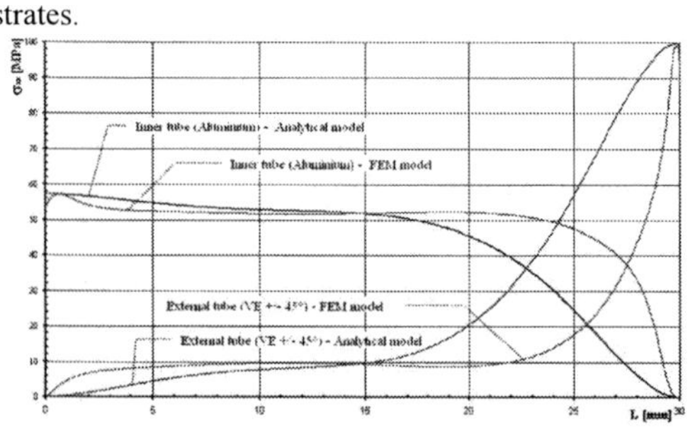

Fig. 3. Load transfer in an AU 2024 T3-AV 119-VE ±45° assembly

The ideal analytical model has the same aspect as the finite element model (FEM).

2.2.2 Stress analysis in the adhesive

The stresses in the adhesive make it possible to predict the failure of the adhesive-bonded joint. Their distribution is thus of primary importance on the prediction of this force.

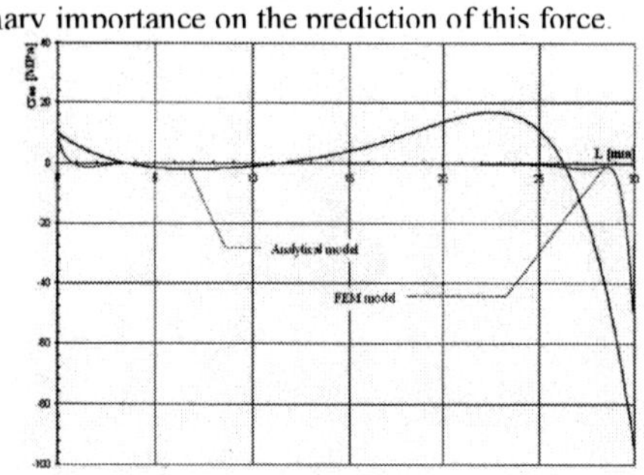

a)

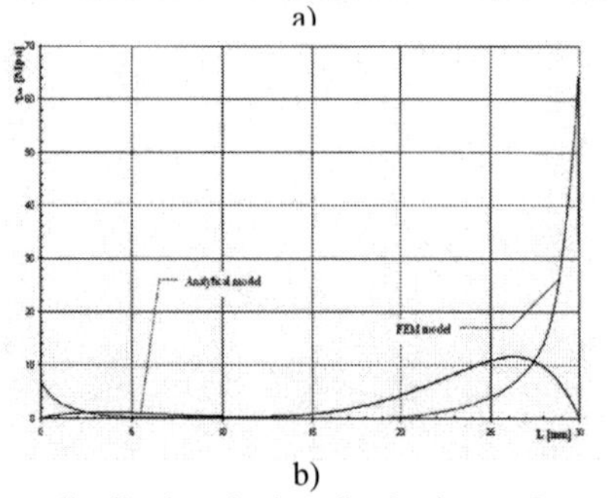

b)

Fig.4. Stress distributions in the adhesive layer of an cylindrical assembly AU 2024 T3-AV 119-VE ±45° for f = 100 MPa:
a) The circumferential stress ($\sigma_{\theta\theta}$); b) The shear stress (τ_{rz}).

Figure 4 present the distribution of the circumferential and shear stresses, according to the covering length. In this case, the stresses given by the ideal model are similar to those given by finite elements. The greatest differences (30%) are seen on the maximum amplitudes where the ideal model underestimates these values: the edge effect due to a local inflection of the substrates is not taken into account.

3. CONCLUSIONS

The objective of our study was to entirely develop analytical models for dimensioning adhesive-bonded joints. To this end we placed ourselves in the case of a cylindrical assembly.
The model validation is presented by comparison with finite elements models. For the assembly, the total force-displacement behavior is well defined. Thus the analytical model makes it possible to determine the rigidity of the assembly and to obtain a simple formulation very quickly, which gives the total behavior of the assembly. The comparison was also carried out on the distribution of stress in the substrates and the adhesive joint. We showed that the transfer of force by joining was well determined by the model. The distribution of stress in the adhesive remained very close to the solution given by finite elements. The analytical model underestimated the stresses in the adhesive leading to an over-estimate of the forces at rupture. However, this model is reliable and allows fast analysis of this type of assembly.

4. REFERENCES

Alwar, R.S. & Nagaraja, Y.R. (1976). Viscoelastic analysis of an adhesive tubular joint, *Journal of Adhesion*, Vol. 8, pp. 76-92.

Lubkin, L. & Reissner, E. (1956). Stress distribution and design data for adhesive lap joints between circular tubes, Trans. of ASME, *Journal of Applied Mechanics*, Vol. 78, pp. 1213-1221.

Nemes, O. (2004a), Contribution a l'étude des assemblages colles cylindriques et plans, *Ph.D. thesis*, INSA, Toulouse, France.

Nemes, O.; Lachaud, F. & Mojtabi, A. (2004b). Adhesive assemblies pre-dimensioning in aerospace applications, *Annals of DAAAM for 2004 & Proceedings of the 15th International DAAAM Symposium*, "Intelligent Manufacturing & Automation: Globalization - Technology - Men - Nature", 3 - 6th November, Vienna, Austria, ISBN 3-901509-42-9, ISSN 1726-9679, pp. 313-314.

Nemes, O., Lachaud, F., and Mojtabi, A. (2005), Adhesive assemblies optimization, Studia Univ. Babes-Bolyai, *Chemia*, XLX (2).

Nemes, O.; Lachaud, F. & Mojtabi, A. (2006), Contribution to the study of cylindrical adhesive joining, *International Journal of Adhesion & Adhesives*, DOI: 10.1016/j.ijadhadh.2005.07.09.

Nemeş, O., Lachaud, F., Mojtabi, A., Tătaru, O., (2006), Modelling and pre-dimensioning of bonded joints assemblies, *MITIP 2006*, Budapest, Hungary.

Shi, Y.P.; Cheng, S. (1993), Analysis of adhesive-bonded cylindrical lap joints subjected to axial load, *Journal of Engineering Mechanics*, Vol. 119, pp. 584-602.

Volkersen, O. (1938), Die Nietkraftverteilung in Zugbeanspruchten Nietverbindungen mit Konstanten Laschenquerschnitten, *Luftfahrtforschung*, Vol. 15, pp. 41-47.

Annals of DAAAM for 2006 & Proceedings of the 17th International DAAAM Symposium, ISSN 1726-9679
Editor B. Katalinic, Published by DAAAM International, Vienna, Austria 2006
150th Birthday of Nikola Tesla (1856-1943) Inventor of Mobile Robot Teleoperation

RESULTS RELATED TO THE OBTAINING, ON IRON-CARBON ALLOYS, OF COMBINED, CORROSION-RESISTANT LAYERS

NEMES, T.; BIBU, M.; DEAC, C.; PETRESCU, V. & DEAC, V.

Abstract: The researches targeted the obtaining of superficial layers for anticorrosive protection having determined thicknesses, without inner stresses, with controllable hardness and mechanical strength properties, easy and cheap to realise, applicable to carbon steels and grey cast iron used for the manufacturing of vessels and equipment for usage in the food processing industry. A combined treatment was adopted consisting in electroplating with chrome and thermochemical treatment for the chrome layer's enrichment with titanium and vanadium. Corrosion tests carried out in organic acids solutions present in vegetable and fruit juices, have confirmed that the corrosion resistance of parts treated as described is comparable with that of stainless steel types that are currently employed for manufacturing equipment for the food processing industry.
Keywords: *anticorrosive protection, electroplating, thermochemical treatment*

1. INTRODUCTION

One of the solutions to reduce the high costs implied by the usage of stainless steels in the manufacturing of equipment employed in the food processing industry and subjected to the corrosive action of various specific environments (humid heat, salty fog, organic acids etc.), is to generate corrosion-resistant superficial layers on common-use, low-cost types of iron-carbon alloys (Debenedetti, 1993; Zamfir, 1994).

For this, the authors of the current paper have used test samples made of low-alloyed steel 40Cr10 (0.4 % Cr and 0.9...1.1% Cr) and of an alloyed grey cast iron (Bibu, 2000; Deac, 2004), respectively, on whose surfaces first a chrome layer has been deposited initially, by means of electroplating, (galvanic coating) and which were then subjected to a thermochemical treatment for the enrichment of the achieved chrome layer with titanium and vanadium (Nemes & Petrescu, 2004).

As a result, on the tested materials, superficial layers were obtained that did not present internal stresses, that had well-determined hardness values and mechanical strength properties and which, when subjected to further experiments that were aimed at verifying the corrosive action of some organic acids environments on them, displayed significant increases of the corrosion resistance, thus making these materials ideal for the intended area of employment.

2. EXPERIMENTAL RESEARCHES

2.1. Materials used

The experiments detailed further on in this paper have been conducted on cylindrical test samples with a diameter of 20 mm diameter and a width of 30 mm, made of 40Cr10 steel (0,4%C and 0,9...1,1%Cr) and grey cast iron (2,8%C and 0,2...0,5%Cr), respectively, tempered and subjected, on the revolving surface, to a polishing process.

2.2. The chrome electroplating process

20 samples made of the mentioned steel and grey cast iron types have been processed with a covering with chrome process in a Watts-type galvanising bath, assuring the conformance of the following parameters:

- Electrolyte composition and characteristics:
  - 300 g/l chrome chloride, with the role of transporting the chrome ion from cathode to anode and easing the anodic dissolving;
  - 40 g/l boric acid, with the role of poising and pH blocking in the cathode film.
  - solution temperature: 50 ± 2 °C
  - electrolyte's pH: 3,2
  - cathode film's pH: 4,8
  - cathode density current: 22A/dm^2
- A purification of the electrolyte has been carried out periodically, by treating with oxygenated water (20%) and then by filtering, in order to remove the metallic and organic impurities.

As a result, chrome layers with a fine and homogenous structure have been obtained, with a hardness of 520 HB on the 40Cr10 steel surface and 500 HB for the cast iron surface, respectively. The thickness of the layers applied through the galvanising process described above was of 45...50 μm.

2.3. The thermochemical treatment

In order to achieve the enrichment of the electroplating layer with titanium and vanadium, after the electrochemical process the test samples have been thermochemically treated in sealed containers with liquid valves were used. The active environment with titanium and vanadium atoms was made of a powder mix obtained through the reduction of these metals' oxides with aluminium.

For the thermochemical treatment, the following process parameters were used:

- Temperature: 1100 ± 20 °C;
- Time: 4.5 h.

After the treatment, layers with a thickness of about 35...40 μm have been obtained on both tested materials, while the obtained hardness was 1200 HV for the steel sample and 750 HV for the grey cast iron sample.

2.4. The protective layers' structure

In order to analyse the structure of the layers obtained after the combined treatment - galvanising process followed by a thermochemical process, the treated test samples were studied by means of metallographical methods and analysed also by X-ray diffraction.

The main conclusions of this analysis can be summarised as follows:

- The protective layers show a uniform thickness, without cracks, blisters or inclusions;
- On the sample's surface there exists a superficial layer composed of a textured solid solution composed of simple or complex carbides: Cr_2C_6; Cr_7C_3, TiC, V_2C, but also of intermetallic compounds: Fe_2Ti, Cr_2Ti. Figure 1 shows the diffraction diagram obtained for this layer.

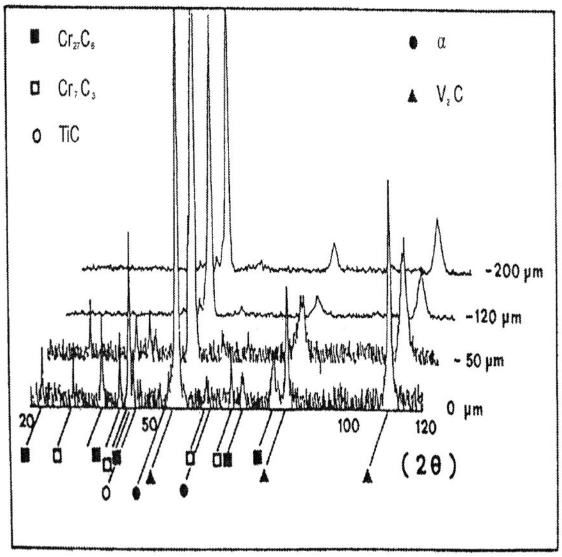

Fig. 1. The 40Cr10 steel diffraction diagram after the electroplating process and thermochemical treatment with titanium and vanadium

- The titanium content of this layer is approximately 15%, and the vanadium content is of approximately 12%, these values showing a significant decrease with the depth. Therefore, at 25 μm depth, no more titanium or vanadium traces can be found.
- Under this superficial layer, there is the electrolytically applied chrome layer;

The area of the alloyed solid solution containing chrome, titanium and vanadium, beneath the carbide layer, makes the two layers show similar values from the electrolytic point of view, assuring the material's protection against electrochemical corrosion for the case in which the superficial layer is deteriorated.

3. DETERMINING THE CORROSION RESISTANCE

The corrosion testing experiments on the test samples treated as described above, were conducted in the presence of specific food processing industry environments: humid heat, salty fog, organic acids (juices) etc.

The exposure times of the test samples to the corrosive action of the environments were of approximately 5, 10 and 20 days, respectively.

After every exposure period, the corrosion speed, expressed in $g/m^2 \cdot h$, was determined by comparing the results of the test samples' weighing before and after the corrosion process. Also, a visual analysis of the surface of the corroded surfaces has been carried out, and polarisation curves (I/U ratios) have been determined.

Some of the results obtained after the experimental determinations are presented in figure 2 and figure 3, respectively.

Figure 2 shows the polarisation curve of the 40Cr10 steel sample used in a NaCl solution with a concentration of 3%.

Figure 3 presents the polarisation curve of the grey cast iron sample used in an environment containing tomato juice, an environment typically found in the food processing industry.

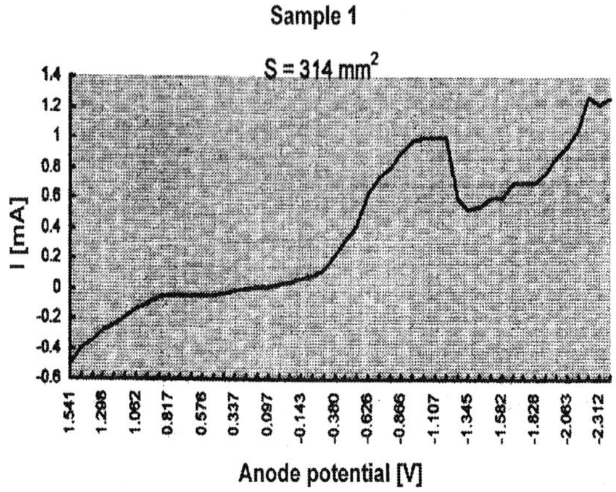

Fig. 2. The polarisation curve of 40Cr10 steel in a 3% NaCl solution

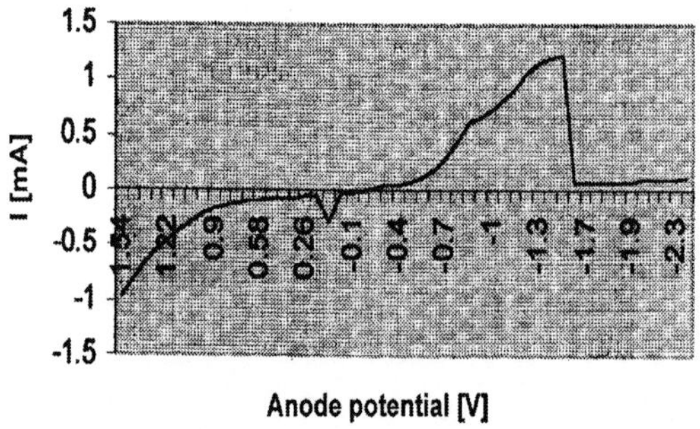

Fig. 3. Polarisation curve of the grey cast iron in a tomato juice environment.

4. CONCLUSIONS

As result of the researches and experiments conducted over some low alloyed carbide iron alloys but exposed to a combined treatment, which consists in an electrolytic covering process with chrome followed by a thermo chemical treatment of enrichment in titanium and vanadium, it was concluded that the formed superficial layers shows an self passivity process and a good corrosion resistance. In some cases the corrosion resistance is comparable with the stainless steel's corrosion resistance, which can be easy replaced as parts of some machines used in the food industry. According to the Romanian legislation, the treated samples were tested in sanitary and toxicology laboratories, but there was no evidence of metallic ions or other health damaging substances.

5. REFERENCES

Bibu, M. (2000) Metalografia aliajelor feroase şi neferoase, Editura Universitatii din Sibiu, Sibiu, Romania

Deac, C. (2004) *Studiul materialelor*, Editura Universitatii "Lucian Blaga" din Sibiu, Sibiu, Romania

Debenedetti, B. a.o. (1993) *Corrosion testing*, AIM International Conference on Innovation in Stainless Steel Products, Firenze, Italy

Nemes, T., Petrescu, V. (2004), *Tehnologia Materialelor*, Editura Universitatii "Lucian Blaga" din Sibiu, Sibiu, Romania

Zamfir, S., a.o. (1994). *Coroziunea materialelor metalice*, Editura Didactica si Pedagogica, Bucharest, Romania.

Annals of DAAAM for 2006 & Proceedings of the 17th International DAAAM Symposium, ISSN 1726-9679
B. Katalinic, Published by DAAAM International, Vienna, Austria 2006
150th Birthday of Nikola Tesla (1856-1943) Inventor of Mobile Robot Teleoperation

HOT-AIR TUNEL CONTROL USING MULTI-DIMENSIONAL PREDICTIVE CONTROLLER BASED ON NEURAL NETWORK MODEL

NEPEVNY, P. & PIVONKA, P.

Abstract: *This paper presents using of a multi-dimensional model predictive controller for hot-air tunnel control. Two quantities of hot-air tunnel are controlled – the air flow and the temperature. Mode predictive controller is a kind of optimal controller based on model. Model predicts future system output, which is used for finding an optimal control action. We used a feed-forward neural network model with back-propagation learning algorithm. Obtained controller is adaptive, because the neural network model is able to observe system changes and adapt itself. The algorithm was implemented in MATLAB-Simulink and tested on a physical model. Communication between PC and hot-air tunnel was provided by PLC (connected via Ethernet).*

Key words: *MPC, prediction, neural network, control, multi-dimensional.*

1. INTRODUCTION

The beginning of predictive regulators was in late 60s, when we can find terms as optimal system, optimal control. In the 70s, after the first optimal controllers, the branch of predictive controllers has been separated. Since that a lot of predictive algorithms were published, for example Generalized Predictive Control (Clarke et al., 1987). The most of methods belong in category Model Predictive Control (MPC). The methods have the same base, but difference is in used model of system and disturbance, cost function, optimization algorithm and using constrains.

The core of predictive controller is a model, which is used to predict a future value of system output. If we know predicted output, we can compute future error (if we know future reference trajectory) and this knowledge is used to find an optimal future action. Control action is optimal in accordance to the cost function, which is used. A number of predicted steps is known as prediction horizon. Prediction horizon is shifted in the time every sampling period (receding horizon).

2. MODEL PREDICTIVE CONTROL

2.1 One-dimensional MPC

MPC algorithm computes a future action increment vector by minimizing the cost function J (Camacho & Bordos, 1999).

$$J = \sum_{j=1+d}^{P+d} \left((\hat{y}(t+|t) - w(t+j\,|\,t)\right)^2 + \lambda \sum_{j=0}^{M-1} \left(\Delta u(t+j\,|\,t)\right)^2 \qquad (1)$$

where $\hat{y}(t+j\,|\,t)$ is predicted output of system in j-th prediction step in discrete time t, $w(t+j\,|\,t)$ is reference trajectory in coincidence points, $\Delta u(t+j\,|\,t)$ is j-th increment of control action, P is prediction horizon, M is control horizon, λ is cost constant and d is delay. Requirement $M \le P$ is always valid. The aim of predictive control algorithm is to compute a future control actions so that the system output tracks the reference trajectory with minimal error. The cost function also contains acost of control action increment (it limits an actuator damage).

Criterion (1) can be rewritten to a matrix form

$$J = \frac{1}{2}\mathbf{u}^{\mathrm{T}}\mathbf{H}\mathbf{u} + \mathbf{b}^{\mathrm{T}}\mathbf{u} + \mathbf{f}_0 \qquad (2)$$

where

$\mathbf{H} = 2\left(\mathbf{G}^{\mathrm{T}}\mathbf{G} + \lambda\mathbf{I}\right)$; $\mathbf{b}^{\mathrm{T}} = 2(\mathbf{f}-\mathbf{w})^{\mathrm{T}}\mathbf{G}$; $\mathbf{f}_0 = (\mathbf{f}-\mathbf{w})^{\mathrm{T}}(\mathbf{f}-\mathbf{w})$

$\mathbf{u}$... vector of action increments

$$\mathbf{u} = \left[\Delta u(t) \quad \Delta u(t+1) \quad \dots \quad \Delta u(t+P-1)\right]^{\mathrm{T}}$$

$\mathbf{w}$... vector of reference trajectory in coincidence points

$$\mathbf{w} = \left[w(t+1) \quad w(t+2) \quad \dots \quad w(t+P)\right]^{\mathrm{T}}$$

$\mathbf{f}$... free response of model

$$\mathbf{f} = \left[f(t+1) \quad f(t+2) \quad \dots \quad f(t+P)\right]^{\mathrm{T}}$$

$\mathbf{G}$... Jacobi matrix of model

$$\mathbf{G} = \begin{bmatrix} g_0 & 0 & \cdots & 0 \\ g_1 & g_0 & \cdots & 0 \\ \vdots & \vdots & \ddots & \vdots \\ g_{M-1} & g_{M-2} & \cdots & g_0 \\ \vdots & \vdots & \ddots & \vdots \\ g_{P-1} & g_{P-2} & \cdots & g_{P-M} \end{bmatrix}$$

where element g_j is j-th coefficient of model step response.

Minimum of the cost function (2) is obtained by making the gradient of J equal to zero. The result is equation (3), which is used for computation the future control action increment vector.

$$\mathbf{u} = -\mathbf{H}^{-1}\mathbf{b} = \left(\mathbf{G}^{\mathrm{T}}\mathbf{G} + \lambda\mathbf{I}\right)^{-1}\mathbf{G}^{\mathrm{T}}(\mathbf{w}-\mathbf{f}) = \mathbf{K}\,(\mathbf{w}-\mathbf{f}) \qquad (3)$$

Only the first element of vector $\mathbf{u}$ is used for control. In next sampling period the new control sequence is computed.

2.2 Multi-dimensional MPC

The most of the systems have more input and output quantities, which should be control. A suitable solution to control more outputs according to reference trajectories is to use multi-dimensional predictive controller, which is able to optimal control more quantities in the same time. We obtain solution of multi-dimensional MPC using equation (3) and its extending.

$$\mathbf{u}_m = \mathbf{K}_m\left(\mathbf{w}_m - \mathbf{f}_m\right) \qquad (4)$$

where

$$\mathbf{K}_m = \begin{bmatrix} \mathbf{K}_{11} & \mathbf{K}_{12} & \cdots \\ \mathbf{K}_{21} & \mathbf{K}_{22} & \cdots \\ \vdots & \vdots & \ddots \end{bmatrix} = \begin{bmatrix} \left(\mathbf{G}_{11}{}^{\mathrm{T}}\mathbf{G}_{11} + \lambda\mathbf{I}\right)^{-1}\mathbf{G}_{11}{}^{\mathrm{T}} & \left(\mathbf{G}_{12}{}^{\mathrm{T}}\mathbf{G}_{12} + \lambda\mathbf{I}\right)^{-1}\mathbf{G}_{12}{}^{\mathrm{T}} & \cdots \\ \left(\mathbf{G}_{21}{}^{\mathrm{T}}\mathbf{G}_{21} + \lambda\mathbf{I}\right)^{-1}\mathbf{G}_{21}{}^{\mathrm{T}} & \left(\mathbf{G}_{22}{}^{\mathrm{T}}\mathbf{G}_{22} + \lambda\mathbf{I}\right)^{-1}\mathbf{G}_{22}{}^{\mathrm{T}} & \cdots \\ \vdots & \vdots & \ddots \end{bmatrix}$$

$$\mathbf{u}_m = \begin{bmatrix} \mathbf{u}_1 \\ \mathbf{u}_2 \\ \vdots \end{bmatrix} \qquad \mathbf{w}_m = \begin{bmatrix} \mathbf{w}_1 \\ \mathbf{w}_2 \\ \vdots \end{bmatrix} \qquad \mathbf{f}_m = \begin{bmatrix} \mathbf{f}_1 \\ \mathbf{f}_2 \\ \vdots \end{bmatrix}$$

This solution is simply implementable in MATLAB using matrix algebra. The basic scheme of two-dimensional predictive control algorithm is in figure 1.

3. NEURAL NETWORK MODEL

Neural Network (NN) – feed-forward perceptron – is used for

realizing two-dimensional third order autoregressive model (fig. 2). Inputs of NN are process inputs and outputs. NN model dynamic is realized using step delays z^{-1} (Arahal et al., 1998).

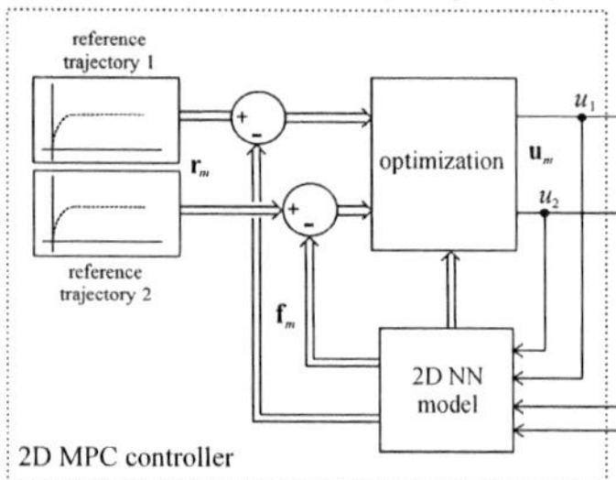

Fig. 1. Two-dimensional MPC control loop

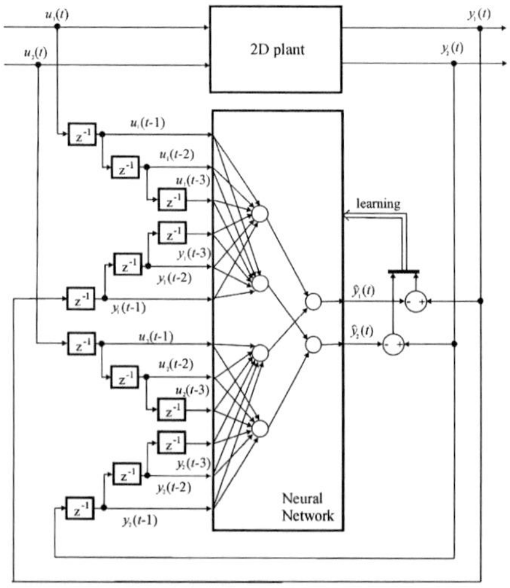

Fig. 2. Structure and learning of NN model

For learning is used back-propagation algorithm (Omatu et al., 1996; Cichocki & Unbehauen, 1993) with batch learning procedure with fixed-sized training set. Off-line measured data are used for NN weights initialization and NN model is learned in each sampling period (training set is updated). We obtain adaptive MPC, because model is adapted to system changes. For compute the equation (4) we need model Jacobi matrices $\mathbf{G}_{ij}$, which we obtain from process of NN model linearization.

4. APPLICATION

Application of multi-dimensional predictive controller is demonstrated on two-dimensional physical model – hot-air tunnel (fig. 3). It is non-linear thermal system. Two quantities of hot-air tunnel are controlled – the air flow (the actuator is a fan) and the temperature (the actuator is a light bulb). These quantities are controlled in feedback control loops, where air flow is measured by the vane with optical sensor of speed and temperature is measured by the thermistor. The aim is to control quantity of air and its temperature. Hot-air tunnel has its own actuator with output voltage from 0 to 10 V. The system was controlled in soft real-time realization (fig.4). The system control was realized via PLC (B&R), which was connected to the air-flow tunnel. For connection between PC and PLC the Ethernet was used. Sampling period for communication and control was 0.2 s. Control and identification algorithms were implemented in PC into S-function in MATLAB and control loop was created in MALAB-Simulink. Parameters of multi-dimensional predictive controller were set on $P = 50$, $M = 25$, $\lambda = 0.1$. For results look at the figure 5. Desired value of temperature and air flow is changing. The air flow has an effect on air temperature, but temperature don't have an effect on air flow.

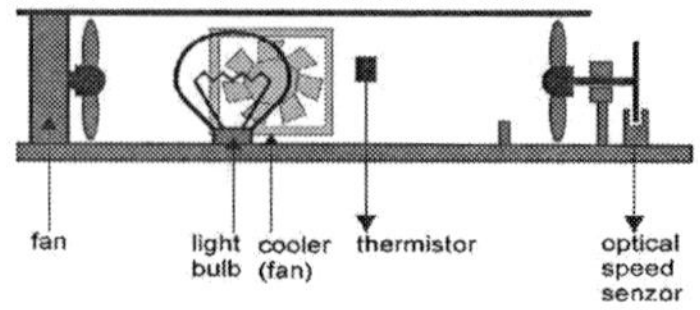

Fig. 3. Scheme of hot-air tunnel

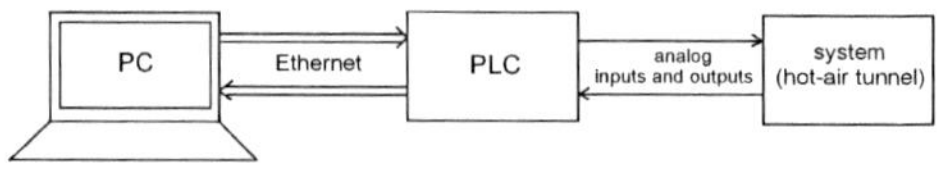

Fig. 4. Soft real-time realization of hot-air tunnel control

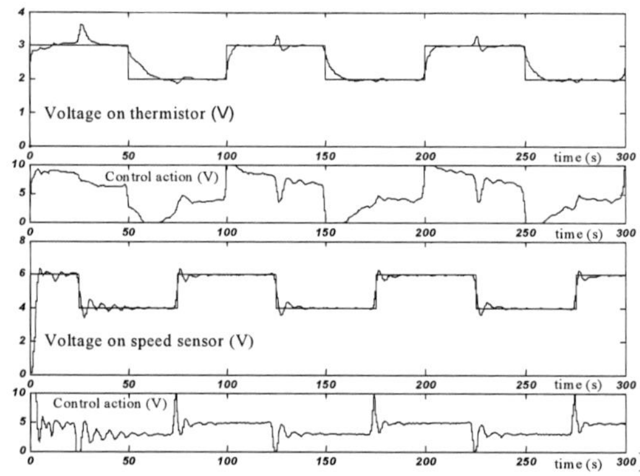

Fig. 5. Hot-air tunnel control

5. CONCLUSION

In this paper the application of model-based predictive controller with NN model is shown. Two-dimensional MPC controller based on NN model was used for control of hot-air tunnel. Adaptive predictive controller is able to track reference trajectory with minimal error and it reacts very fast to system changes and disturbances. There is shown how to convert simple one-dimensional predictive controller to multi-dimensional. If we extend dimension of controller, lengthen prediction horizon or shorten sampling period, the computational demands increase. There is also problem to guarantee stability and robustness in all situations within adaptive predictive control. In the next research we will focus on constraints implementation and optimization and identification algorithm robustness.

6. ACKNOWLEDGEMENTS

The paper has been prepared as a part of the solution of GAČR project TALENT No. 102/03/H116 and GAČR project Softcomputing Methods in Control No. 102/06/1132.

7. REFERENCES

Arahal, M. R., Berenguel, M. & Camacho E. F. (1998). Neural identification applied to predictive control of solar plant. *Control Engineering Practice*, Vol. 6, 333-344, ISSN 0967-0661

Camacho, E. F. & Bordos, C. (1999). *Model Predictive Control*, Springer, ISBN 3-540-76241-8, London

Cichocki, A. & Unbehauen, R. (1993) *Neural networks for optimization and signal processing*, Wiley, ISBN 0-471-93010-5, New York

Clarke, D.W., Mohtadi, C. & Tuffs, P. S. (1987). Generalized Predictive Control – Part I., The Basic Algorithm. *Automatica*, Vol. 23, No. 2, 137-148, ISSN 0005-1098

Omatu, S., Khalid, M. & Yusof, R. (1996). *Neuro-control and its applications*, Springer, ISBN 3-540-19965-9, London.

Annals of DAAAM for 2006 & Proceedings of the 17th International DAAAM Symposium, ISSN 1726-9679
B. Katalinic, Published by DAAAM International, Vienna, Austria 2006
150th Birthday of Nikola Tesla (1856-1943) Inventor of Mobile Robot Teleoperation

C.P.I. SCALES BETWEEN TRADITIONAL ETNOCENTRIC AND MODERN GEOCENTRIC CORPORATION

NICA, G.B.; AURITE, T.; HADAR, A.; CARUTASU, N.L. & PATRASCU, G.

Abstract: *In the new context of globalization, in terms of recruiting the human resources, is very important to understand the difference between traditional multinational corporation–ethnocentric and the modern multinational corporation-geocentric. Such conversion at the level of the enterprise will create truly global competitive advantages, because multinational corporations need to manage a complex structure of personality in human resources evaluation, in order to relieve the abilities of professional structure.*
Key words: *multinational enterprises, ethnocentric, geocentric, Achievement via Conformance, Achievement via Independence*

1. INTRODUCTION

Intensive competition forces multinational enterprises (MNE) to constantly scrutinize every subsidiary's performance. However, for the creation of truly global competitive advantages, MNEs need a growth and innovation strategy that effectively utilizes and channels the interplay between subsidiaries, divisions, corporate headquarters, and holdings. Such a strategy must be flexible and open enough for local excellence without loosing too much of its global reach and sustainability (Michael et al., 1985). The more productivity pressures force MNEs to disperse their activities throughout the world and the more competitive pressures impel close coordination of these operations, the more global growth strategies play a pivotal role in the success of the whole company.

2. VALUES IN GEOCENTRIC VERSUS ETHNOCENTRIC MULTINATIONAL CULTURES

Multinational Corporation - MNC is a firm having operations in more than one country, international sales, and a multinational mix of managers and owners. (e.g., Ford, Toyota, GE, IBM, Intel, Sony with headquarters both in N.Y. and Tokyo). A MNE ordinarily consists of a parent company and at least six subsidiaries typically with a high degree of strategic interaction among the units. Some MNC have more than a hundred subsidiaries and follow absolute and comparative advantages policies. MNC also have different number of foreign production sites, and thus, different numbers of international markets. This company earns profits in different markets and not only operates in different countries but, owns alliances in other countries and has no allegiance to a particular country. MNCs are based and owned in one country with manufacturing facilities in two or more countries in which profits are not invested (Dunning, 1993). The growing structural integration of the world economy is requiring firms to go outside their immediate boundaries to capture the complex realities of know-how trading and knowledge exchange in innovation, particularly where intangible assets are tacit and need to speedily adapt competitive enhancing strategies to structural change.

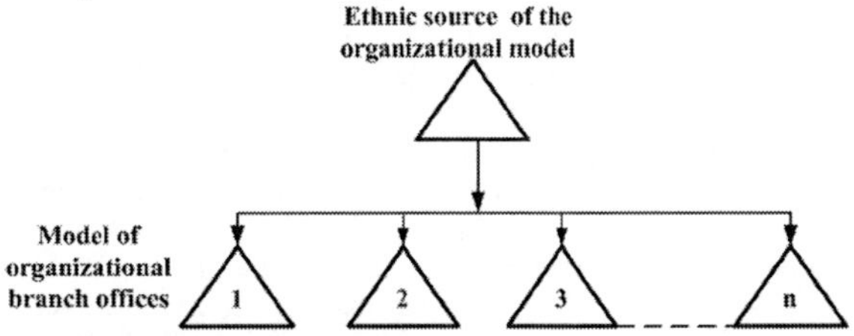

Fig. 1. Etnocentric Corporation

Alliances or network related advantages are those which prompt a voice rather than an 'exit' response to market failure; they also allow many of the advantages of internalization without the inflexibility, bureaucratic or risk-related costs associated with it. Such quasi-internalization is likely to be most successful in cultures in which trust, forbearance, reciprocity, and consensus politics are at a premium (Perlmutter, 1965). The first scenario analyzed is that of global corporations, also known as **ethnocentric corporations**, where there is a high degree of international cooperation in production, and a low degree of independence at the level of the branch companies (Figure 1). The label ethnocentric derives from the fact that the management is based on strictly one nation's set of criteria, that is, the country of origin of the corporation. Identity is mediated through the values, language and higher management of the same national derivation, found in the mother company and at the higher levels of the different branches of the corporation. It is an asymmetrical regime in which general company policy is planned from the centre, from whence the implementation of the different branches is monitored. The company's decision taking is thus controlled by the mother company and a group of managers of the same nationality (Nica & Aurite 2005). The geocentric companies is similarly a high degree of international coordination on production, but where there is a greater degree of independence observable across branch companies, and greater orientation towards satisfying the demand among local markets (Figure 2). The heads of the branch companies enjoy greater autonomy and decision-taking is more horizontal, tending to be carried

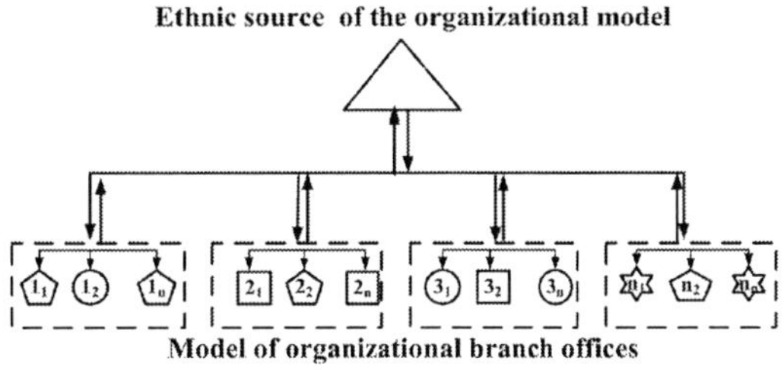

Fig. 2. Geocentric Corporation.

out among branch managers. Management by process is emphasized as well as the increased need for communication between the management of the different branches.

3. CALIFORNIA PSYCHOLOGICAL INVENTORY – C.P.I. CASE STUDY

Psychometric and conceptual analyses of the CPI instrument have identified three basic dimensions underlying scores on the folk and special purpose scales (*Gough & McAlliste 2004.)* The items on the inventory produce scores for 18 scales, which are divided into four classes: measures of poise, ascendancy, self-assurance and inter-personal adequacy; measures of socialization, responsibility, intra-personal values, and character; measures of achievement potential and intellectual efficiency; and measures of intellectual and interest modes. The scales of C.P.I. are structured in four dimensions.

The items on the inventory produce scores for 18 scales, which are divided into four classes: measures of poise, ascendancy, self-assurance and inter-personal adequacy; measures of socialization, responsibility, intra-personal values, and character; measures of achievement potential and intellectual efficiency; and measures of intellectual and interest modes. The first sector of the profile - DEALING WITH OTHERS - contains scales assessing interpersonal style and manner of dealing with others. The next sector of the profile - SELF-MANAGEMENT contains scales pertaining to the internalization and endorsement of normative conventions, including norms related to self presentation. The third sector of the profile - MOTIVATIONS AND THINKING STYLE - sheet contains three scales pertaining to cognitive/intellectual functioning and the need for achievement in either structured or open situations. The final sector of the profile - PERSONAL CHARACTERISTICS - sheet contains three scales that assess broadly stylistic or qualitative aspects of thinking and behaviour.

The purpose of the testing was to select the persons with the best psychological profile for specific requirements demanded characteristics of traditional corporation-etnocentric vs. modern corporation-geocentric. From the 18 scales of CPI, 4 were selected - **Ac, Ai, Ie, Py**.

By this test I want to propose a different evaluation for etnocentric and geocentric corporation. The high values for **Ac** scales, down values for **Ai** and moderate values for **Py** and **Ie** are specified for human resources whose are implied in the Ethnocentric Corporation (Figure 3).

Achievement via Conformance (Ac)	**Achievement via Independence (Ai)**	**Intellectual Efficiency (Ie)**	**Psychological Mindedness (Py)**
• is intelligent and resourceful, and functions well in situations having clearly specified rules and procedures • is ambitious and always tries to do well • is conscientious and efficient	• has a wide range of interests, independent ways of thinking, and a strong desire to do well in settings where ingenuity and innovation are favoured • is intelligent, clear-thinking, and imaginative	is an intelligent, resourceful, clear-thinking person • feels sure of self, equal to nearly any challenge • has a stable, optimistic view of the future	• has high personal aspirations and does well in most situations • is somewhat individualistic; may be seen by others as lacking in warmth or compassion

Table 1. CPI scales selected.

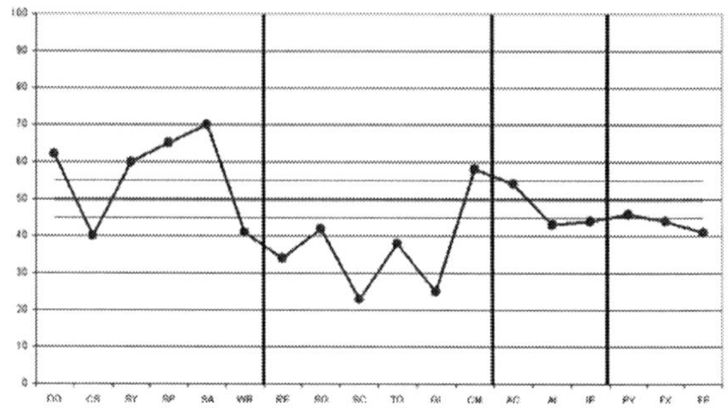

Fig. 3. CPI profiles for etnocentric corporation

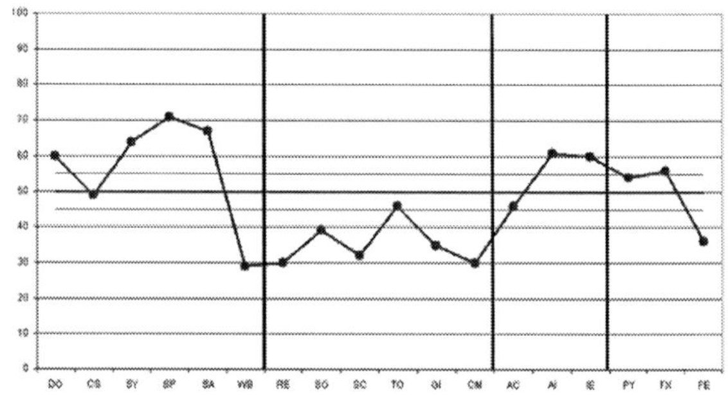

Fig. 4. CPI profiles for geocentric corporation

The high values for **Ac, Ie** and **Py** and the down values for **Ac** are specified for the human resources whose are implicated in the Geocentric Corporation (Figure 4).

C.P.I. was applied at a number of 46 students (24 female students and 22 male students).

4. CONCLUSION

The test results for the 4 relevant scales were related with the other 14 (considerate with a lower weight in the final appreciation) in order to allow a complex interpretation of the profiles. From 46 postulants only 1 has satisfying results according with the specific requirements implied by the specific tasks of multinational corporations: ethnocentric - traditional transition to. geocentric – modern enterprise culture.

5. REFERENCES

Dunning, J. H. (1993). *Multinational Enterprise and the World Economy*, Addison-Wesley Publishing Company.

Gough, H. & McAllister, L. (2004). Configural analysis report an interpretation of scale combinations, *Available from: http://www.acer.edu.au/articles/pdf Accessed:* 2006-06-20.

Michael F.; Baier, V.E. & Rhinesmith, S.H. (1985). *Before Takeoff: What Your Overseas People Need to Know*, Personnel, pp. 60-65.

Nica, B. & Aurite, T. (2005). *Applying Jung archetypes for virtual enterprises organizational culture*, Proceedings of the 8th International Conference on Management of Innovative Technologies MIT2005, Junkar, M. & Levy, P.R. (Ed.), pp. 85–88, ISBN 961–6238-96-5, Fiesa, Slovenia, September 2005, TAVO Slovene Society for Abrasive Water Jet Technology & LAT Laboratory for Alternative Technologies – University of Ljuljana, Ljuljana.

Perlmutter, H. V. The Tortuous Evolution of the Multinational Corporation, *Columbia Journal of World Business*, Jan-Feb 1969, pp. 9-18.

Annals of DAAAM for 2006 & Proceedings of the 17th International DAAAM Symposium, ISSN 1726-9679
Editor B. Katalinic, Published by DAAAM International, Vienna, Austria 2006
150th Birthday of Nikola Tesla (1856-1943) Inventor of Mobile Robot Teleoperation

TRANSFORMING CONSTRUCTION NOTIFICATIONS INTO BILL OF MATERIALS

OESTREICH, E. & TEICH, T.

Abstract: Construction notifications are documents that can be used to describe individual components in a very detailed way. By the use of these documents it is possible to optimize the procurement processes for individual parts within a company. In many instances there is also an additional demand for standard parts, needed for the final assembling of an individual component. In order to support the procurement processes of such parts too, it is necessary to transfer a construction notification directly into a bill of materials. Therefore this paper presents the basic requirements as well as an appropriate algorithm to accomplish the transformation process.

Key words: Mass Customization, product configuration, bill of materials, generic bill of materials, automobile industry, individualization

1. INTRODUCTION

One of the most important success factors for the introduction of a Mass Customization strategy is the ability to handle the high level of complexity of products and processes. Especially in the area of the individualization of cars it is often necessary to create additional documents that describe the individual items in a very detailed way instead of using unique item numbers that clearly identify the needed parts. The reason for this procedure is the almost unlimited diversity of variations that exists only for one individual component (e.g. an individual cover of a head rest (Oestreich & Teich, 2005)).
In addition to individual components, standard parts are needed for the assembly of the final package in many cases. The demand for these standard parts depends closely on the content of a certain construction notification. It is for this reason necessary to transform a given construction notification into a bill of materials in order to support the procurement processes for standard parts well.
As the generation of the construction notification can be carried out without nearly any user interaction, the transformation of construction notifications into bills of materials is not supported yet. Therefore this paper presents the basic requirements as well as an appropriate algorithm to perform this task.

2. INTIAL SITUATION

2.1 (Generic) bills of materials

The structure of a product describes the way, in which the product is build up from purchased parts and/or semi finished products. The relationship between two products in this structure is a so called gozinto-relationship (van Veen, 1992) that represents the fact that a product for the manufacturing or assembly of another product. Based on this assumption a bill of materials (BOM) can be considered as the set of gozinto-relationships of a certain product.
In order to cope with complexity in production environments which are coined by a high product variety an even more flexible product structure is required. While bills of materials are used to represent an order specific view of a product, generic bills of materials (GBOM) are used to model a complete product family with all of its variants. An order specific BOM can be derived from a GBOM. The decision whether a certain part of the GBOM is used in the BOM depends on the specification of the wished product. Therefore it is necessary to add constraints to the gonzito-relationships of the GBOM.

2.2 Definition of the problem

A Product configurator is "*software with logic capabilities to create, maintain and use electronic product models that allow complete definition of all possible product option and variation combinations ...*" (Bourke, 2000). One important function of configurators is the ability to transform a given configuration directly into a BOM (e.g. Lieberman & Leete, 2006). Prerequisite therefore is the existence of a complete product model. Thereby it is possible to represent one product family by the use of a generic bill of materials.
In connection with the relevance of this transformation function it seems to be necessary to enhance the construction notification configuration engine in order to support the transformation of construction notifications into bills of materials.

3. CONVERTING CONSTRUCTION NOTIFIACTIONS INTO BOMs

3.1 Extended configuration model

The basic model for the generation of construction notifications is independent from an ascertained product model. In order to convey a construction notification into a BOM it is first of all necessary to extend the existing model. Thereby two aspects are of high importance. On the one hand a GBOM must be assigned to every element of type "package". On the other hand it is necessary to determine those elements that specify the certain variant of the GBOM. Therefore a Boolean flag is used. Every specification element where the flag is set to true determines a certain variant of the GBOM. Further information about the extension of the configuration model can be found in (Oestreich & Teich, 2005).

3.2 Modeling (generic) bills of materials

A generic bill of materials *GBOM* can be expressed as a tuple $GBOM = (P, root, R_{\text{mandatory}}, R_{\text{optional}}, V, D, C)$. P is the set of all products or parts that are used within the GBOM. Based on the assumption that a GBOM is organised according to the low level code procedure (Günther & Tempelmeier, 1997), P can be divided into different subsets $P = \{P_0, \ldots P_{d-1}\}$ where d represents the number of low level codes. A subset $P_i = \{p_{i1} \ldots p_{ia}\}$, $i = 0 \ldots d\text{-}1$, $a = |P_i|$ contains all parts p with the same low level code i.

The root of a GBOM is represented by *root*. It applies $root \in P_0$, $|P_0| = 1$. *root* is the only part of the low level code of 0.

The presented model distinguishes between two types of gozinto-relationships. The first type $R_{mandatory} \subseteq P \times P$ describes all obligatory relationships between parts. Every edge $r_m = (p_x, p_y) \in R_{mandatory}\ p_x \in P_x, p_y \in P_y, x > y$ states that a part p_x has to be contained in the BOM obligatory in case that the BOM contains the product p_y. The demand that p_x is assigned to a higher low level code as p_y ensures that the structure is obligatory acyclic.

On the contrary $R_{optional} \subseteq P \times P \times C$ represents the set of optional edges $r_o = (p_x, p_y, c_a) \in R_{optional}, p_x \in P_x, p_y \in P_y, x > y, c \in C$. A product p_x is only required in a BOM if a condition c is fulfilled. The condition $r_o \in R_{optional} \Rightarrow r_m(p_x, p_y) \notin R_{mandatory}$ ensures that there is no obligatory and optional edge between two products p_x and p_y at the same time.

V describes the set of features which characterize the GBOM. It applies $V = \{V_1 \ldots V_n\}$ where n is the number of features. All possible specifications of the features are described in domains $D = \{D_1 \ldots D_n\}$. A tangible specification of a feature is presented as $d_{fg} \in D,\ f = 1 \ldots n,\ g = 1 \ldots |D_f|$.

C finally represents the set of constraints which are required for the modelling of the optional gozinto-relationships. The presentation of the constraints is carried out in conjunctive normal form as a AND relation of clauses K_z. It applies $C : \bigwedge_{z=1}^{n} K_z$.

Every clause K_z consists of a disjunction of an optional number of specifications of a feature V_z. It applies

$K_z : \bigvee d_{zr} \vee \bigvee \neg d_{zs}$ $r, s \in \{0 \ldots |D_z|\}$. If $r = s = 0$ then the clause is always fulfilled. Therefore the constraint is independent from the feature V_z.

An order specific BOM can be expressed as a tuple $BOM = (GBOM, P, root, R, OC)$. Basis for the creation of a BOM is the existence of a GBOM. The definition of P and $root$ is carried out according to the modelling of GBOM. It applies that the root of the BOM is equal to the root of the GBOM ($root = root^{GBOM}$).

In contrast to the GBOM the BOM contains only such gozinto-relationships R that are representing obligatory relationships between parts. Therefore the definition of R can be carried out according to $R^{GBOM}_{mandadory}$.

OC represents the values for each specification element d_{fg} in relation to a given order. By default, every specification d_{fg} is set to 0. If a specification d_{fg} is used in the given order, then d_{fg} is set to 1. It applies $OC: \forall d_{fg} \leftarrow 0/1$ $f = 1 \ldots n,\ g = 1 \ldots |D_f|$.

Furthermore there exists a function $SAT(c, OC) \rightarrow 0/1$, which checks a constraint $c \in C$ on its validity against the order code. If the constraint can be fulfilled then the function delivers 1 otherwise 0.

3.3 Algorithm

On the basis of the given definitions it is now possible to define the needed algorithm for the transformation of a package of a construction notification into a bill of materials (see figure 1). The presented algorithm requires three parameters: a part $p \in P$, the low level code of p and finally a package related order code list that contains all characteristic features of the package. The execution of the algorithm is initialized with $p = root^{GBOM}$ and the corresponding low level code of 0.

Algorithm 1 createBOM(product p, low level code llc, order code list oc)

(1) $\forall r_i(p, p_y) \in R^{GBOM}_{mandatory}$

  (a) $P_{llc+1} = P_{llc+1} \cup p_y$

  (b) createBOM(p_y, $llc + 1$, oc)

(2) $\forall r_i(p, p_z, C_c) \in R^{GBOM}_{optional}$

  (a) $if(SAT(C_c, oc))$

    (i) $P_{llc+1} = P_{llc+1} \cup p_z$

    (ii) createBOM(p_z, $llc + 1$, oc)

Fig. 1. Basic transformation algorithm

In the following the whole structure of the GBOM is processed by the use of a depth first search strategy. Thereby the algorithm adds each valid gozinto-relationship to the order specific BOM.

4. CONCLUSION AND FURTHER WORK

The algorithm, presented in this paper offers the possibility to transfer construction notifications into bills of materials. Therefore the algorithm can be considered as a bridge between existing ERP systems and the construction notification configuration system. The aim of further work on this concept is the definition and the development of a general interface between an arbitrary ERP system and the configuration system.

5. REFERENCES

Bourke, W. (2000) Product Configurators: Key Enablers for Mass Customization, *Available from:* http://www.bourkeconsulting.com/documents/Aug2000IntroCfgrs.pdf *Accessed:* 2006-05-12

Günther, H. & Tempelmeier, H. (1997) *Produktion und Logistik*, Springer Verlag

Lieberman, M. & Leete, B. (2006) Getting a Grip on Configurators – Comparing Configurators Part II, Available from: *http://www.plmic.com/news/plm-article-Getting-Grip-Configurators-part2.html#Business_Analysis_plm*, Accessed 2006-05-20

Oestreich, E. & Teich, T. (2005) Produktkonfiguration auf der Basis dynamischer Dokumentstrukturen. *Industrie-Management* Vol. 21 No. 3 pp. 39-42 ISSN 1434-1980

Olsen, K.; Sætre, P. & Thorstenson, A. (1997) A generic bill of materials based on a programming language notation, *Proceedings of the Norsk Informatikk Konferanse*

van Veen, E.A. (1992) *Modelling Product Structures by Generic Bills-of-Materials*, Elsevier Science Publishers B.V., ISBN – 0444896767, Amsterdam

Annals of DAAAM for 2006 & Proceedings of the 17th International DAAAM Symposium, ISSN 1726-9679
B. Katalinic, Published by DAAAM International, Vienna, Austria 2006
150th Birthday of Nikola Tesla (1856-1943) Inventor of Mobile Robot Teleoperation

RESEARCH OF THE GLOBAL DYNAMIC COMPLIANCE AND THE VISCOSE GLOBAL DYNAMIC DAMPER COEFFICIENT OF THE INDUSTRIAL ROBOT

OLARU, A. & OLARU, S.

Abstract: In the future, *the industrial robots have to be highly dynamic systems to sustain the required productivity, accuracy and reliability. Both the joints and robots bodies are necessary to be optimized for their usability performance to meet the productivity requirements of the tool center point (TCP). The global dynamic compliance (GDC) is one of the most important dynamic parameters of the dynamic behavior and with this parameter is possible to determine the viscose global dynamic damper coefficient (VGDDC) to obtain finally the desired dynamic behavior. The knowledge of the GDC, the global dynamic damper ratio (GDDR) and the transfer function (TF) between the displacement of the TCP and the applied force, are very important to chose, the rigidity in the joints, the optimal mechanical form of the bodies, to obtain finally one power vibration spectrum in concordance with the manufacturing application*

Key words: global dynamic compliance, viscose global dynamic damper coefficient, *assisted research, Fourier spectrum, LabVIEW instrumentation, dynamic behavior*

1. INTRODUCTION

The GDC is one of the most important parameter in the dynamic behavior of the industrial robot. In the manufacturing systems is necessary to know the vibration behavior of the robot, the VGDDC of his structure and how the dynamic variation of acceleration determines the damped mechanical vibrations, to ovoid the resonance frequency of the spectrum. In the paper is shown one assisted method with the virtual LabVIEW instrumentation of the GDC, VGDDC and GDDR of the industrial robots. This instrumentation was achieved in the LabVIEW soft 6.1 from National Instruments, USA, and simulates the real Fourier analyzer. This virtual apparatus is generally, it is possible to use in many others mechanical applications. Now, in the world, all determinations of GDC is made without the modern analyzer apparatus like show in this paper. The paper contains the assisted research of the GDC to know the transmission of the vibration between the floor to the TCP of one didactical arm type robot structure with U profile. The research was made by exciting the robot base modulus with the electromagnetic exciter and after that the acquisition data of the exciting force and the displacement of the TCP. For that on used one experimental setup presented below. Now, in the world must more and more will be used the LabVIEW instrumentation to assure the reduction of the time and of the research costs without reducing the research performances.

2. EXPERIMENTAL SETUP

The experimental stand contains the following components (fig.1): didactical arm type robot; the electromagnetic exciter type 11075 from RFT Germany; connector type CB- 68 LP from National Instruments USA; acquisition board type PCI 6024E from National Instruments USA; frequency generator type POF-1 from KABID Poland; generator amplifier type LV 102 from MMF Germany; personal computer from Taiwan; inductive displacement traducer type 16.1 IAUC Romania; Hottinger apparatus type KWS/T-5 from Germany.
With the virtual LabVIEW propre instrumentation have been determined the transmissibility between the robot base and the TCP, and finally the VGDDC, c and the GDDR, ξ.

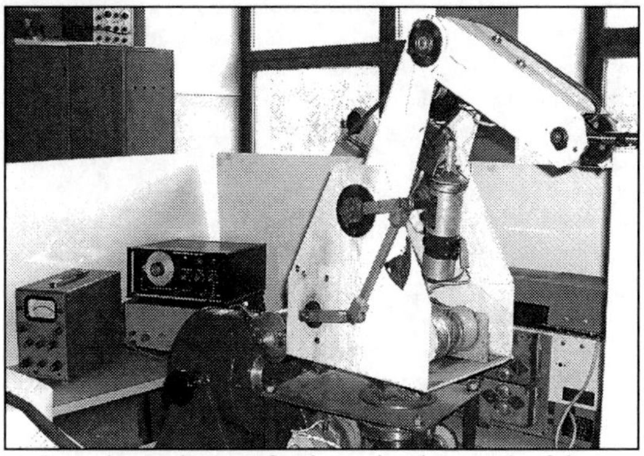

Fig.1. Experimental set up for the assisted research of the GDC and the VGDDC

3. MATHEMATICAL MODELING

Every mechanical robot structure consists of the multiple mechanical components witch are coupled together and it can be represented in a dynamic mechanical model as multiple mass, spring and damping elements. While the mass and the spring stiffness determine the natural frequency ν_n of the system, the damping element, represented by the GDDR, ξ, governs the resonance amplitude of the vibration and with it, the dynamic system stiffness. Modal testing theory has been successfully used for calculating the frequency spectrum of robots structure [1, 2]. A frequency spectrum includes the amplitude- frequency characteristic (dynamic compliance) and the phase- frequency characteristic of the mechanical robot structure. Frequency characteristic analysis is important to understand the dynamic performance of the bodies- joints robot system. If the dynamic impact with one periodical force on the base of the robot structure is *F(t)* and the displacement response of the TCP is *x(t)*, then the GDC of the robot structure is defined as:

$$\frac{1}{k(j\omega)} = \frac{\int_0^T x(t)e^{-j\omega t}dt}{\int_0^T F(t)e^{-j\omega t}dt} = \frac{FFT(x)}{FFT(F)} = \frac{E_x(j\omega)}{E_F(j\omega)} \quad (1)$$

where $E_F(j\omega)$ and $E_x(j\omega)$ are complex spectrum of energy of the input force on the base robot modulus and respectively output displacement of the TCP. After expansion with the complex conjugate, the GDC can be expressed as:

$$\frac{1}{k(j\omega)} = \frac{E_x(j\omega)E_F^*(j\omega)}{E_F(j\omega)E_F^*(j\omega)} = \mathrm{Re}(\omega) + j\,\mathrm{Im}(\omega) \qquad (2)$$

where E_F^* is the complex conjugate of E_F.
The viscose damping value c and damping ratio for each resonance frequencies ξ_i can be calculated with:

$$c_i = 2\xi_i \frac{k_i(\omega)}{\nu_{n_i}}; \; \xi_i = \frac{\nu_{i1} - \nu_{i2}}{2\nu_{iR}} \qquad (3)$$

where ν_{ni} is the natural frequency.

4. RESULTS AND DISCUSSION

The created virtual instrument with LabVIEW soft 6.1 is the propre virtual Fourier analyzer and used for the assisted determination of the GDC, GDDR and VGDDC. In comparison with the real analyzer, this one, assures the many real and frequency characteristics, some rapid changes of the input data and the presentation of the comparative results. The virtual analyzer assures the acquisition data on many input channels, the rapid change of the scale, or of the domain, the assisted comparison between the real and the frequency characteristics, the assisted calculus of the GDC, VGDDR and VGDDC for each resonance frequencies so necessary for the dynamic optimization of the robot structure. The complex virtual Fourier analyzer assures the complete analyze of the structure damping vibration. The knowledge of the TF [3] gives the possibility to choose the structure with the propre convenient frequency and with the small transmissibility of the vibration between the floor to the TCP. The virtual instrument assures also the change of the formula of the command function and the characteristics results of the TF, the GDC and the VGDDC. In all researched cases the robot's modulus was excited in a range between 5...70Hz. In all the cases have been designed the real and frequencies characteristics. In the figures 2, 3 and in the table 1 and 2 are presented some of these results about the GDC, VGDDC. After the comparative analyze of the characteristics, results the following remarks: the increase of the stimulus frequency determine the movement of the higher rigidity (the smaller compliance) to the small resonance frequencies; the VGDDC of the robot structure is bigger to the small resonance frequencies, when the force frequencies stimulus increase; the VGDDC decrease with the increase of the resonance frequencies for all stimulus frequencies. The curve of the maximum of the VGDDC is the same in all cases except the two small excitation frequencies. The 3D diagrams of the VGDDC are very important to choose the optimal values of the acceleration to avoid the bad resonance frequencies and to know the c values for all frequency domain. The determined values of the VGDDC could be used in many other mechanical applications and this general method can be used in many other research.

6. CONCLUSIONS

The virtual Fourier analyzer, the calculus of the GDC, VGDDC and VGDDR are very important for the assisted research of the vibration spectrum, of the dynamic behavior in all different mechanical applications with the final result, optimization of the industrial robot. The results, the comparative possibility of the analyze and the price of the research confirm that this method and the virtual LabVIEW instrumentation is the principal way in a vibration research.

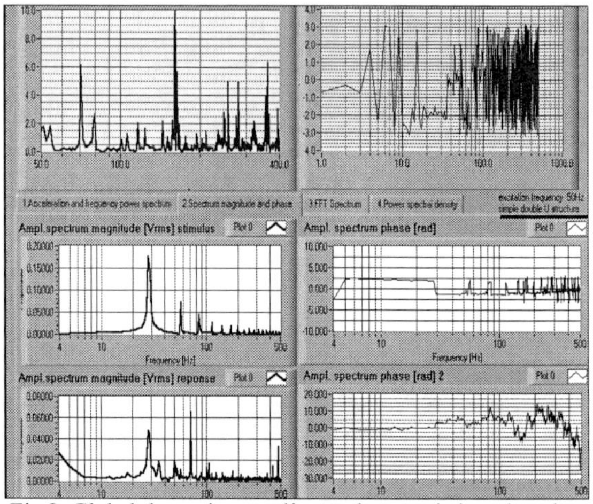

Fig.2. Global dynamic compliance characteristics when the base modulus was excited with 30Hz periodical force

	rezonance frequencies												
	50	65	70	80	90	100	110	120	130	160	170	180	200
csi	0.4	0.307692	0.285714	0.25	0.222222	0.2	0.181818	0.166667	0.153846	0.125	0.117647	0.111111	0.1
	0.6	0.6	0.6	0.01	0.5	0.01	0.01	1.2	0.1	0.5	0.8	0.5	0.01
	0.18	0.2	0.4	0.01	0.1	0.01	0.65	0.01	0.01	0.1	0.25	0.1	0.01
	0.2	0.2	0.6	0.25	0.01	0.1	0.1	0.2	0.18	0.01	0.01	0.01	1.2
1/k	0.01	0.01	0.7	0.01	0.1	0.1	0.1	0.01	0.11	0.1	0.2	0.2	0.05
	0.01	0.01	0.8	0.1	0.1	0.2	0.2	0.01	0.4	0.01	0.25	0.25	0.15
	0.01	0.01	0.25	0.01	0.05	0.1	0.1	0.01	0.1	0.2	0.01	0.2	0.45
	0.01	0.01	0.1	0.01	0.1	0.25	0.25	0.01	0.25	0.4	0.4	0.4	1.2

Table 1. The results of the global dynamic compliance

	0.026667	0.015779	0.013605	0.625	0.009877	0.4	0.330579	0.002315	0.023669	0.003125	0.00173	0.002469	0.1
	0.088889	0.047337	0.020408	0.625	0.049383	0.4	0.005086	0.277778	0.236686	0.015625	0.005536	0.012346	0.1
	0.08	0.047337	0.013605	0.025	0.493827	0.04	0.033058	0.013889	0.013149	0.15625	0.138408	0.123457	0.000833
c	1.6	0.946746	0.011662	0.625	0.049383	0.04	0.033058	0.277778	0.021517	0.015625	0.00692	0.006173	0.02
	1.6	0.946746	0.010204	0.0625	0.049383	0.02	0.016529	0.277778	0.005917	0.15625	0.005536	0.004938	0.006667
	1.6	0.946746	0.032653	0.625	0.098765	0.04	0.033058	0.277778	0.023669	0.007813	0.138408	0.006173	0.002222
	1.6	0.946746	0.081633	0.625	0.049383	0.016	0.013223	0.277778	0.009467	0.003906	0.00346	0.003086	0.000833

Table 2. The results of the viscose global dynamic damper coefficient

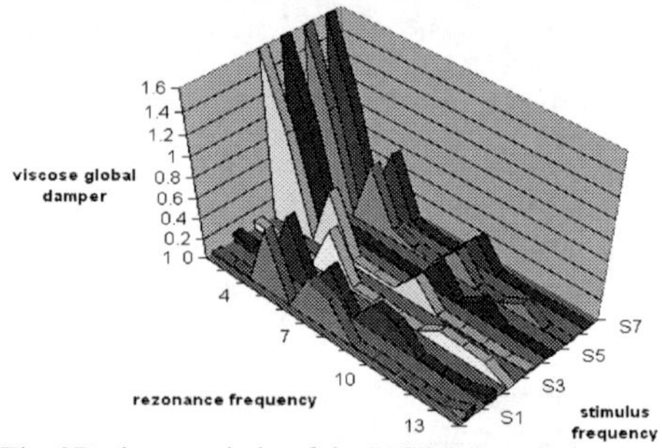

Fig.3. The 3D characteristic of the VGDDC variation

5. REFERENCES

Olaru, A. (2001). *Dynamic of industrial Robots,* Bren Edition 2001, Vol.II, ISBN 973-8143-65-9, Bucharest.

Olaru, A. & Mihai, N. (1999). *Dynamic of industrial robots,* Bren Edition 1999, Vol.I, ISBN 973-9493-35-1, Bucharest.

Olaru, A. (2001). Theoretical and experimental research of the cinematic and dynamic behaviour of industrial robots, *Proceedings of the 12th International DAAAM Symposium,* Katalinic, B., pp.333-334, ISBN 3-901509-19-4, Jena, DAAAM International, Vienna.

Annals of DAAAM for 2006 & Proceedings of the 17th International DAAAM Symposium, ISSN 1726-9679
B. Katalinic, Published by DAAAM International, Vienna, Austria 2006
150th Birthday of Nikola Tesla (1856-1943) Inventor of Mobile Robot Teleoperation

MODELING AND SIMULATION IN THE MINERAL PROCESSING INDUSTRY

OLEYNIK, A. & SKOROKHODOV, V.

Abstract: *The paper presents the modeling and simulation experience in field of research into the mineral row material separation. Features of the subject domain generate a need of different approaches using for modeling and simulation. Merits and demerits of different types of models are denoted. The general scheme of separation modeling process and some examples of developed simulators are presented.*
Key words: *modeling, simulation, separation of mineral resource, information support system.*

1. INTRODUCTION

As the raw materials source depletes, the requirements to processing efficiency of minerals mined increase. On-line control used in mineral processing and careful tuning of process parameters allow the technologies operated to be made maximum efficient. However, the only way to significantly improve the separation indices is to apply new separation technologies. A distinctive feature of new technologies is not only quantitative tuning of the parameters but, if necessary, enabling the separation process operated to be on-line redesigned.

The experts in mineral processing consider computer modeling systems to be the most effective approach in new technologies development. The modeling systems for mineral separation have been developed for some years by the researches of the Kola Science Centre of the Russian Academy of Sciences (Institute for Informatics and Mathematical Modeling and Mining Institute).

2. GENERAL SCEME OF SEPARATION MODELING

The investigations resulted in the structure of information technology synthesizing effective flowsheets and processing cycles (Gershenkop, A. at al. 2001). The data on properties and characteristics of mineral particles is of most significance here. Attention should be paid not only to the properties which the operating technologies are based on but also to the potentially "useful" ones. The apparatus and flowsheets used in separations have the following characteristics: separation characteristics $\varepsilon(\xi)$, acting forces **F** and control parameters **P**. Separation characteristics determine the extraction of mineral fractions to separation products. The parameters enable the separation characteristics and forces to be varied.

While choosing or synthesizing the appropriate flowsheet, the following steps are taken:

1. Mineral and fraction composition of mineral raw materials is analyzed to determine the properties ξ to be used in separation.
2. Possible prototypes of separation apparatus and flowsheets are chosen or synthesized for the determined set of properties ξ'.
3. Taking into account the limitations, the output parameters for the given flowsheets are calculated and required technological resources are estimated.

The algorithm of flowsheet selection (synthesis) can be used in iterative mode. In this case, middlings from the previous separation stage is used as a feed for the next iteration. A general flowsheet is synthesized as a structured association of several subflowsheets.

The third stage of flowsheet generation is the most complicated and important. This is the stage when the separation models should be used to determine the parameters which make separation more efficient. In the computer systems at the mineral-processing plants, the models should be developed beforehand and stored in the appropriate database. The research versions of the system include a special block for model development and/or tuning for mineral raw materials which have never been modeled before.

3. MODELS OF MINERAL RAW MATERIALS SEPARATION

Due to different features and standards of knowledge regarding the regularities in separation processes different formal models to describe and analyze separation processes should be used. An integrated information system applied in the field considered uses an approach allowing various models and modeling tools to be jointly used. The approach is based on the conceptual modeling technology (Emelianov, S.V., at al, 2004). A conceptual model provides an integrated formalization of experts' knowledge and notions about separation processes. An important advantage of the conceptual model is that it can be refined and developed without making essential alterations in the already generated part. The conceptual model is of a declarative nature. It functions as a coordinator between various computing models in the course of further simulation.

The choice of the type of a computer model simulating the separation process is determined by the standard of knowledge and the availability (or the possibilities to obtain) of experimental data concerning the process parameters, characteristics of the initial raw materials and separation products. The basic types of executors are modeling tools based on statistical, analytical or expert (logical) methods. Each of the modeling methods mentioned is characterized by its own merits and demerits.

For many flowsheets currently operated, a great deal of quantitative data of separation process has been accumulated. Using this data, it is possible to generate rather adequate statistical models. However, it is problematic to use statistical models based on the data characteristic of some separation processes for other separation processes. The analytical models describe actual physical and physical-and-chemical phenomena bringing mineral components to separation. Basing on the studies in the given area, an aggregated mathematical model of separation processes has been generated (Oleynik, A. &

Shalatonova, A. 2002). A set of properties used in separation is vector ξ. Then, the fractional composition is characterized by the function of mass fractional distribution γ (ξ, *x*, *y*, *z*, *t*). Mineral particles of any elementary fraction move with an averaged speed $\bar{v}$ (ξ,, *x*, *y*, *z*, *t*). An aggregated model of separation is presented by the following combined equations:

$$M_i = \begin{cases} \partial \gamma / \partial t = -\mathrm{div}(\gamma \vec{v}) \\ \sum F_i = 0 \end{cases} \tag{1}$$

The first equation of the model describes the interrelation between the fractional state function in some point of working volume of the separation apparatus γ and the function $\bar{v}$ that will determine the velocity field in the elementary fraction zone (the equation of this type is true under certain assumptions). The second equation is the equation of balance between statistically averaged forces exerted on elementary fraction particles. Substituting the exact functional expressions of the forces exerted in an apparatus into the model (1), we can generate a model of the apparatus. The main problem arising in analytical models development is the necessity to give a detailed study of the fractional structure of raw materials and separation products, to provide the analysis of various fractions properties and of the influence the separating and disturbing forces exert on them. Therefore, general equations reflecting the dependence between the separation products' characteristics and those of the separating processes (Tikhonov, O.N., 1984) sometimes do not give adequate results for industrial processes. Simulation has particular significance in actual industrial systems investigation. Although simulation does not allow to produce an optimal, in terms of mathematics, solution, it gives an opportunity to produce and compare quite adequate variants in real time. On-the-fly (within several minutes) repetition of the experiment with a computer model under various values of control parameters allows the model to be tuned by fitting of the interrelations' structure and/or their quantitative characteristics. The authors have developed some variants of simulators for separation at the processing plants of the Kola Peninsula. Many flowsheets contain middling. These products form a "circulating load" in the flowsheet. The values of separation products' characteristics are mainly determined by feedback implemented by circulating load. The simulators of these processes are system-dynamic models. Software from Powersim® was used to generate the simulators (Powersim Software AS, 2003). The models are based on integration of limited data obtained in laboratory experiments and on expert knowledge of the nature and regularities of these processes. The simulators of such type were developed for investigation of gravity separation of mica ore and of apatite-nepheline ore flotation (Biryukov, V. & Oleynik, A. 2004).

4. UNITS OF THE EXECUTIVE ENVIRONMENT

The integrated conceptual model provides a uniform formal description of the subject domain tasks and information-analytical resources ensuring these task solutions. The executers for particular tasks can be rather independently developed. The executive environment for a complicated task is formed by the executers integration according to the structure represented in the conceptual model. The research versions of the executive environment were generated for a local application mode. A full-scale implementation of the system requires multi-user distributed application. A tool has been developed to perform a regression analysis of the data stored in various formats (databases, spreadsheet, structured text files). It allowed to efficiently use the results obtained earlier by different researchers. A subsystem for dynamic generation of specialized databases has been developed to provide coordinated storage and flexible use of the data obtained in the course of the current experiments.

An application simulating separation processes has been developed. The application makes simulation control easier for users in an integrated environment. The integrated modeling environment includes Powersim-model and other components. To implement the system, software package Powersim Studio Software Development Kit (SDK) has been used. The components' interaction is based on the COM and ActiveX interfaces. With the application, it is not necessary for a user to install software package Powersim Studio. The significant advantage of Powersim Studio SDK is possibility to generate Web-simulations, which can greatly contribute to the development of distributed simulation systems.

5. CONCLUSIONS

The study has revealed and analyzed the peculiarities of application of the modeling methods and information technologies to solve the problems of mineral separations and increasing of separation processes effectiveness. The results obtained form the theoretical and methodical base to be used in complex information technologies development, which are to support all the stages of research work in this area. The tools developed allow to solve a number of important tasks concerning, for instance, organizing of structured storage of experimental data; to give a regression analysis of the data stored in various formats; to simulate some mineral separations operated at the mineral processing enterprises in the Kola Peninsula. Further investigations in the area considered are to be carried out in two basic directions. The first is a model base development. Within this direction, it is supposed to generate program modules libraries for analytical modeling of separation processes. It is also necessary to modify simulators on the basis of the data obtained in laboratory experiments carried out within the target studies of "problematic" situations. Another direction is to adapt laboratory results to the actual industrial processes. Here, it is necessary first to define the sets of process parameters which can be monitored in real conditions. It is these parameters that are the basis for possible real time control over the industrial separations dynamics, forecasting the separation results and the development of control techniques increasing separation processes effectiveness.

6. MAIN REFERENCES

Biryukov, V. & Oleynik, A. (2004) Information technology of decision making support in mineral processing control *Transactions of Institute of System Analysis, RAS.* Vol.8. (2004) pp. 186-193, M.: Editorial URSS.

Emelianov, S.V., at al (2004) *Information technologies for regional management.* M.: Editorial URSS, ISBN 5-354-00799-2, Moscow.

Gershenkop, A. at al (2001) Mathematical modeling of stationary systems in process of mineral complex separation. *Obogashchenie Rud,* No 6, (2001) pp.35-39, ISSN 0202-3776

Powersim Software AS (2003) Powersim Studio 2003: Academic Documentation. *Available from: http://www.powersim.com*

Oleynik, A., & Shalatonova, A. (2002) An aggregated mathematical model of mineral components separation, In: *Information technologies in regional development: conceptual aspects and models.* Putilov, V (Ed.), pp. 71-74, the KSC RAS Publ.H, Apatity

Tikhonov, O.N. (1984) *The regularities of effective mineral separation in mineral separation processes.* M: Nedra, Moscow

Annals of DAAAM for 2006 & Proceedings of the 17th International DAAAM Symposium, ISSN 1726-9679
B. Katalinic, Published by DAAAM International, Vienna, Austria 2006
150th Birthday of Nikola Tesla (1856-1943) Inventor of Mobile Robot Teleoperation

AN INTELLIGENT SYSTEM FOR MEASURING SOIL ELECTRICAL CONDUCTIVITY

OLTEANU, C.; ZAMFIRA, S.; OLTEAN, G. & OLTEANU, F.

Abstract: Romania's integration within the European Community structures imposes the alignment of the Romanian agriculture with the quality, economic efficiency and environment protection-oriented standards required in the developed countries. The reference parameters used in the European Union countries may be reached by increasing the resources monitoring and control upon all factors involved in the management of production agriculture system. In the paper we present the results obtained in the designing of the experimental model for soil conductivity measurement. The final result will be an integrated system able to measure, to acquire and to map the soil conductivity chart. This will give the opportunity to agriculture engineers to plan and obtain enhanced quality and quantity yields per hectares.
Key words: soil conductivity, agricultural mechatronics, terminal method, measuring.

1. INTRODUCTION

The Department of Fine Mechanics and Mechatronics from *Transilvania* University Library of Brasov in collaboration with the Research and Development Institute for Potatoes and Beetroots of Brasov - I.C.D.C.S.Z Brasov - has had intense preoccupations in the field of mechatronics applied in agriculture (Olteanu et al., 2004):

- computer-aided installation for measuring the starch and dry substance content in potatoes tubers;
- installation for automated dosage of chemical liquids used as fertilizers;
- research and experiment of a mobile laboratory and of the automated system of acquisition and processing of phytoclimate data in the potatoes growing.

The precision agriculture, which comprises the agriculture as an application of mechatronics, enhances a new methodology (that aims at a new agricultural system) that could be the key to many current problems (Oltean et al., 2005).

The opportunities that have favored the development of precision agriculture are (Oltean et al., 2004):

- capacity to understand the complexity of agricultural systems – systemic and holistic approach;
- capacity to monitor the phenomena and processes – computer-controlled data acquisition;
- achievements in computing techniques – hardware, software, fineware and data bases;
- improvement of interpretation and computing methods – statistics, modeling, simulation, decision support systems – DSS;
- development of geographic informational systems – GIS;
- occurrence and development of spatial analysis and statistics – Geostatistics;
- progresses of spatial technique – teledetection, GPS;
- technical achievements in automating and improving agricultural machines – agricultural mechatronics.

2. SYSTEM FOR MEASURING, ACQUIRING AND TRACING MAPS WITH REGARD TO SOIL ELECTRICAL CONDUCTIVITY

At the beginning of XX-th century Conrad Slumberger developed the plotting technique of geological sub-layers by inducing electric currents in the ground and by noting the distribution of the potential field resulted. Later on, the measurement of soil conductivity was applied in many areas of research: determining the rocks lithology, plotting the aggregates and clay deposits, plotting the subterranean water, locating the geo-thermal areas, plotting the archeological sites.
The conductivity is a measurement of a material's ability to transmit an electric charge. It stands for an intrinsic property of the material, similar to other properties such as density and porosity. The usefulness of soil conductivity stems from the fact that sands have low conductivity, silts have medium conductivity and clays have high conductivity.
Soil electrical conductivity (EC) is a property of soil that is determined by standardized measures of soil conductance (resistance^{-1}) by the distance and cross sectional area through which a current flows.
The movement of electrons through bulk soil is complex. Electrons may travel through soil water in macropores, along the surfaces of soil minerals (i.e. exchangeable ions), and through alternating layers of particles and solution. Therefore, multiple factors contribute to soil EC variability, including factors that affect the amount and connectivity of soil water (e.g. bulk density, structure, water potential, precipitation, timing of measurement), soil aggregation (e.g. cementing agents such as clay and organic matter, soil structure), electrolytes in soil water (e.g. salinity, exchangeable ions, soil water content, soil temperature), and the conductivity of the mineral phase (e.g. types and quantity of minerals, degree of isomorphic substitution, exchangeable ions).The electrical conductivity is influenced by a series of soil properties:

Porosity – the greater the porosity, the greater the electrical conductivity (the clayey soils have greater porosity as compared to sandy soils);

Water content – the electrical conductivity gets greater as the soil gets more humid;

Salinity level - the electrical conductivity is strongly influenced by the solution concentration in electrolytes, being directly proportional with it;

Cations exchange capacity influences the soil electrical conductivity through its content in positive ions of Ca, Mg, Na, NH_4 or H, which are held especially in soils with increased clay humus and minerals level;

The temperature influences the soil electrical conductivity, being in relation of direct proportionality.

The block diagram for tracing maps on soil conductivity by using the systems of global positioning, acquisition and data processing is presented in figure 1.

The experiments were carried out in the areas belonging to I.C.D.C.S.Z Brasov on chernozemic cambic soil, characteristic of about 5% of the agricultural fields in the county and 15% of the agricultural fields in the depression of Barsa.

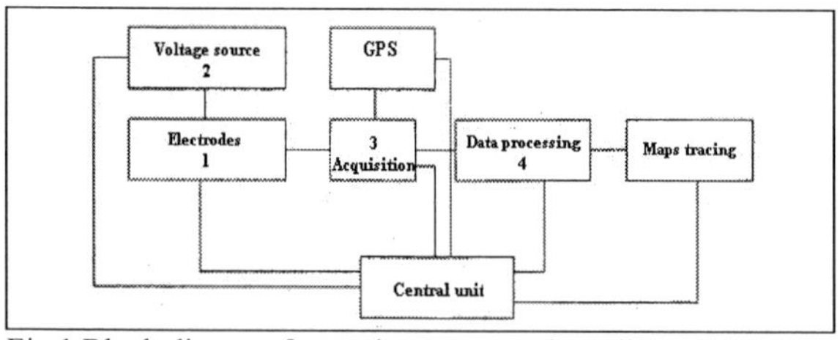

Fig.1 Block diagram for tracing maps on the soil conductivity

The depression of Barsa, where the Institute of Brasov is located, belongs to the intra-mountainous plain naturally drained, situated in Tara Barsei and formed through stratified and carbonated deposit.
In the area of I.C.D.C.S.Z Brasov there are soils belonging to molic, hydromorphic and weakly developed classes. The first class is made up of chernozemic, rendzina, cambic-gley and underground soils from the class of weakly developed soils. The chernozemic, rendzina, gley soils are predominant.

3. EXPERIMENTAL SETUP FOR SOIL CONDUCTIVITY MEASUREMENT

In order to measure soil conductivity we used four- terminal method (figure 2 –block diagram for measurements). The measuring principle consists in injecting a current with known value trough two external terminals and in measuring the voltage between the inner terminals. Taking into account the mechanical parameters, h the depth of terminals in the soil, and the distance between electrodes we can calculate soil conductivity (Olteanu et al., 2006).

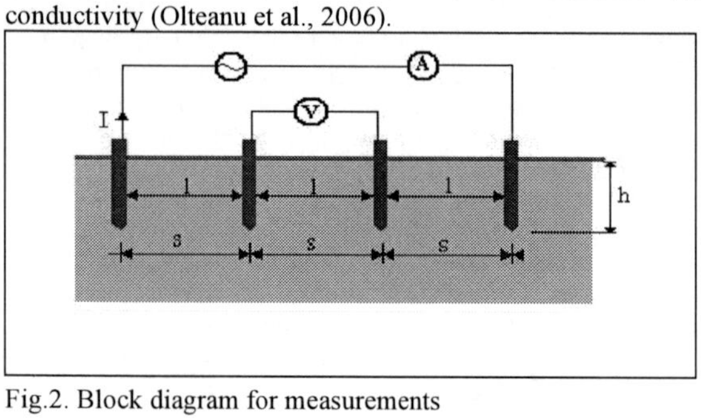

Fig.2. Block diagram for measurements

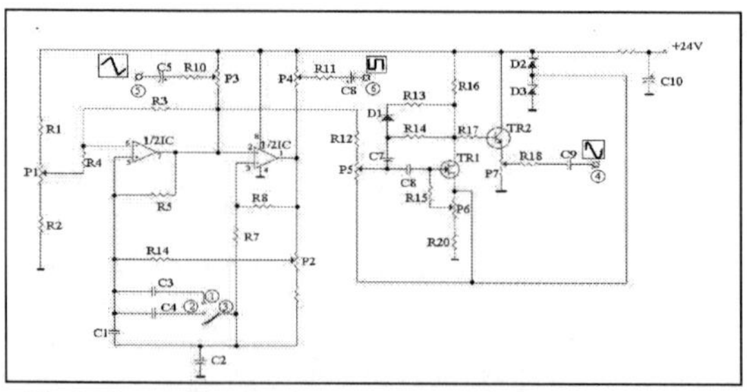

Fig.3 Electronic schemes for the signal generator

We designed and created a signal generator with various types of signals. Signal generator can supply three types of signals: sinusoidal, triangular and rectangle forms in three decades from 25 to 25000 Hz. The signals are amplified in a low distortion power amplifier. The TL082 integrated circuit has two amplifier stages. The first amplifier is used as an oscillator whose frequency is adjustable by P2 potentiometer. The frequency domain is established by S1 switch and C3 capacitor. P1 trimmer adjusts the signal symmetry. The second amplifier acts as an integrator and it is used to turn the triangular signal into a rectangular one. P3 and P4 potentiometers tune the amplitudes of the triangular and rectangular signals. TR1 and TR2 transistors are used to turn the triangular signal into a sinusoidal signal. To obtain minimum distortions we can adjust the sinusoidal signal by P5 and P6 trimmers. The amplitude of the sinusoidal signal is adjusted by P7 trimmer.

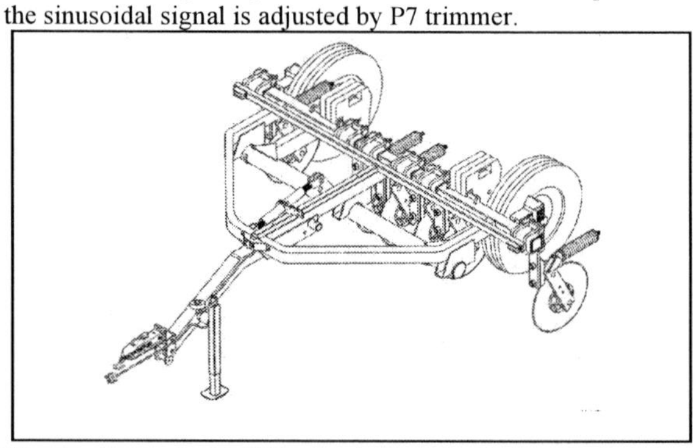

Fig. 4. Continuous soil conductivity measurement device

In figure 4 there is presented the continuous soil conductivity measurement device that is attached to the vehicle especially designed for this purpose. Measurement electrodes, disk-shaped, assure the continuous measurement in soil. The disks are rolling at the displacement of the device. The measurement device and the computer which routes the soil conductivity map, according to the block diagram in figure 1, are placed on the vehicle.

4. REFERENCES

Oltean, Gh.; Cramariuc, R.; Oltean M.I.; & Turcu, C, (2004). *Preliminary Results Regarding the Creation of a Device for Measuring Electrical Conductivity of Soil.* (Ed. AcademicPres,) ISSN: 1454-2382, 430 Lucrarile simpozionului ştiinţific: "Perspectivele agriculturii în mileniul IIP", , 20-23 octombrie 2004, Cluj-Napoca .

Olteanu, C.; Zamfira, S.; Oltean, Gh.;, Olteanu, F.; & Turcu,.C (2004). *Mechatronics in Agriculture - One of the Research Directions within the Precision Mechanics and Mecachatronics Department with Transilvania University of Brasov.*(Ed. MondoR), pg. 72-76, ISSN 1583-7653, Revista Mecatronica nr. 2, Bucureşti, 2004.

Oltean, Gh.;, Cramariuc, R.; Oltean, M.I.; Ianosi, M.; Turcu C.; Pamfi, Gh..; & Aldea, C. (2005). *Using Soil Electrical Conductivity Maps in Improving the Potato Crop Technology in Precision Farming.* Abstracts of papers and posters, 16th Triennial Conference of the EAPR 2005, July 17-22, Bilbao, BI-1.688-05, Spain, 554-546.

Olteanu C.; Zamfira, S.; Oltean Gh.; Turcu, C.; & Olteanu, F. (2006). Designing and Implementing an Intelligent System for Measuring, Acquiring and Tracing Maps with Regard to Soil Electrical Conductivity Experimental Model. Acta Technica Napocensis, series: Applied Mathematics and Mechanics 45 vol III 601-606 ISSN 1221-5872, Technical University of Cluj-Napoca, 7-9 July 2006.

Annals of DAAAM for 2006 & Proceedings of the 17th International DAAAM Symposium, ISSN 1726-9679
B. Katalinic, Published by DAAAM International, Vienna, Austria 2006
150th Birthday of Nikola Tesla (1856-1943) Inventor of Mobile Robot Teleoperation

MONITORING OF TECHNOLOGICAL COMPETENCES

OTTO, T. & RIIVES, J.

Abstract: *Online resource databases allow production enterprises to cooperate effectively. The elaborated system concept supports the strategic planning of technology transfer, it also could be used as a basis for the industrial enterprises in elaborating co-operation networks and developing towards extended enterprises. Description and evaluation of innovation capacity through human resources development is a novel solution taken into consideration in the building of the model. It enables to improve factor conditions of regional productivity, enhancing Porter's Competitive Diamond Model. Results at this phase are used to develop the technological and human resources database test version.*

Key words: *Network monitoring, e-community, resource allocation, databases.*

1. INTRODUCTION

A survey of e-Work applications (Nof, 2005) has shown, that e-X is not the same as X. E-Work and e-Production/e-Business are not the same as work, production, and business. A traditional operation cannot be copied as-is to an e-Activity, some changes must be made. For example, comparing a search for information by three generations of specialists (a. Search through manual records; b. Search of a database by a query; c. Search of the Web by a browser) one can realise that they all are different.

In modern systems the most erroneous part of technological systems – human resource – should be made measurable, improving also factor conditions of Porter's Competitive Diamond Model (Fig. 1). The four attributes are self-reinforcing and catalyse the process of continuous innovations. The model focuses upon the conditions that support firm competitiveness at the national scale. In a study of Estonia in transition Porter shows that the quality and efficiency of Estonian workers was regarded by many as ahead of Finland, known for its hard-working culture. However, Estonia was experiencing a shortage of highly qualified labor (Porter, 2006). At a cluster level, knowing the real-time functionality expectations and evaluating the experience on speed performance and limits of data interaction amount of commercial solutions drive the cluster to build up a new system (Viharos, 2003).

To solve the Porter's Diamond Model task an online system of technological and human resources is proposed. In similar manufacturing situations and with similar equipment results can be different depending on competence of workforce. The most erroneous part of technological systems – human resource – is analysed on the basis of competence, thus making it possible to elaborate a methodology for evaluation and development of the database model for adding innovation capacity of labour force and entrepreneurs of the metal engineering, machinery and apparatus sector. As to the human resource evaluation only most widely known engineering professions of the machinery sector have been studied.

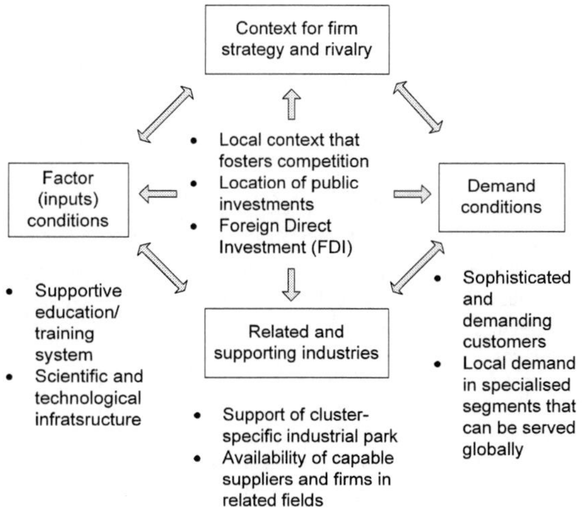

Fig.1. Porter's Competitive Diamond Model: determinants of regional productivity (Porter, 1990)

2. COMPETENCE MANAGEMENT

Skills/knowledge to be evaluated are grouped according to professional standards. A professional standard is a paper, defining requirements of professional qualification for knowledge, skills, experience, values and personal qualities.

In practice prescription of professional requirements is substantial. Professional requirements are divided into four groups beginning with more general skills and ending with specific personal qualities essential for working in a profession. The general principles of professional requirements are as follows:

general skills – requirements for general skills and knowledge originating from economic affairs;

basic skills – special professional requirements for skills and knowledge;

extra skills – special professional requirements for skills and knowledge, characterised in narrow specialisation and/or necessity for executing additional assignments at current position;

personal qualities – expected personal identities and abilities required in a certain profession (Riives, 2006).

The definition of specific skills/knowledge depends on the field of activity of an enterprise. The definition process should be started in every particular case from job descriptions of a current enterprise.

Evaluation of skills/knowledge is the next step towards human resources development at an enterprise. For assessment a scale of grades [0…5] is used. The lowest grade is "1" and the highest "5". Grade "0" is used if the respective skill is not relevant. Both needed (NL) and existing (EL) levels are to be evaluated.

3. FACTOR CONDITIONS

Companies try to find skilled employees to fulfil their strategy. From the other side the study programmes reflect the competency of particular educational institution. Missing are coherent activities between industry and the academic world.

The real situation should be as follows.

- In companies one should do all defined by the list of skills related to a particular job on the needed qualification level.
- In educational institutions a student after passing a course "should know" the subject (passive skills) and "should be able to do" related to the subject (active skills).
- The sum of active skills of the programme should match the lowest set of skills by the related qualification standard.

In the system there are separate processes which are tightly connected with each other:

- Determination of the Human Resources (HR) competence and the training necessities in the company, taking into consideration the strategy of the company and operating needs.
- Matching the training necessities with the possibilities and carrying out the real courses through the system.
- Fixing the needs for vocational examinations and developing a national professional award system in the field of machine building and apparatus industry.

In the first stage of the process training needs are determined according to competency charts of the company. Competency charts are filled personally for each employee or vocation (e.g. CAD engineer, CNC operator, welder, etc.). The charts can be filled in through the Internet in every enterprise, with the sensitive information of the enterprise remaining undisclosed. The analysis is based on average indicators of vocations, industrial fields (toolmaking, machine-building, etc.) or on regional basis. For general professions special standard templates have been created, e.g. the results of queries concerning CAD designers in 18 Estonian enterprises are depicted in Fig. 2.

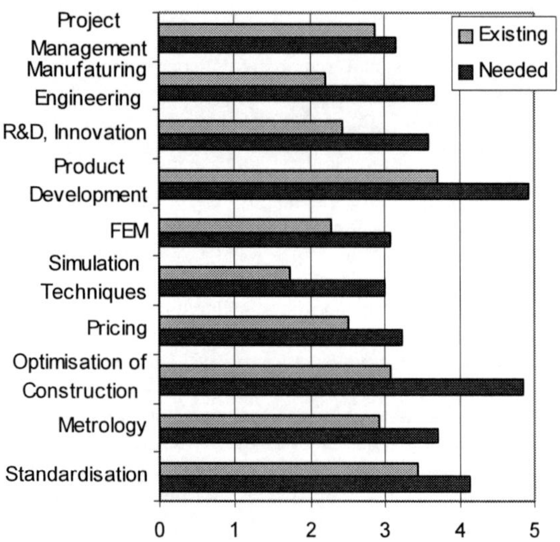

Fig.2. Existing (EL) and needed (NL) competences of a CAD Engineer in Estonian enterprises

The second stage includes arrangement of training courses according to the needs mapped. The input is obtained from an analysis of all Estonian educational organisations of the sector. The training activities are organised based on unified training calendar and the corresponding documentation.

On the information of Fig. 2 educational institutions can draw conclusions and offer corresponding training courses, e.g. a course of CAD systems could be advantageous, at the same time a course of technical drawing appears to have less potential.

Before the system was implemented vocational training courses were organised independently by taylor-made course plans. The new monitoring system makes it possible to support organisational work by adding value through different queries, statistical calculations and prediction mechanisms

4. DISCUSSION

The system results can support the strategic planning of technology transfer, it also could be used as a basis for the industrial enterprises in elaborating co-operation networks and developing towards extended enterprises.

The proposed model is capable of monitoring the quality and quantity of technological resources in machine-building enterprises of the network. Influence of human resources can be evaluated successfully when using proper taxonomy and expert estimations. The described methodology of human resources development has been thoroughly tested in five institutions: in enterprises of diverse type (specialising in machinery, tools engineering, metal engineering and road engineering), and Tallinn University of Technology. Testing results turned to be successful and the system elaborated has proved its place as a carrier of competence development.

Results at this phase are used to revise and develop a test version of the human resource database. However, in a long perspective when a critical mass of companies are involved in the system the results could be used as a basis for educational institutions to elaborate complementary study and training programmes and modify the existing ones.

The current solution is focused on the sector of metalworking, machinery and apparatus engineering. The proposed model can be transferred also to other industrial sectors (wood processing, chemical industry, construction materials industry, etc).

The developed database solutions were tested in a restricted area, having common cultural and language basis. Development at a multicultural level needs additional research of common understanding – development of ontology for building automated libraries assuring similar knowledge-based input from different languages.

Acknowledgements The Estonian Science Foundation (Grant No 6795) is acknowledged for supporting the research.

6. REFERENCES

Nof, S.Y. (2005). "Collaborative E-Work: Research Progress and Challenges", (Plenary), *Proceedings of ICPR–18*, Salerno, Italy, August.

Porter, M. (1990). The Competitive Advantage of Nations, London: Macmillan.

Porter, M., Solvell, O. (2006). "Estonia in Transition", Harvard Business School Case, 702–436.

Riives, J., Otto, T. (2006). Integrated human resources development and monitoring system. *Proc. 5th Int Conf DAAAM Baltic*, Tallinn, 219-224.

Viharos, Zs. J.; Monostori, L.; Csongrádi, Z. (2003). Realizing the Digital Factory: Monitoring of Complex Production Systems, *Intelligent Manufacturing Systems*, 2003, Elsevier, 29–34.

Annals of DAAAM for 2006 & Proceedings of the 17th International DAAAM Symposium, ISSN 1726-9679
Editor B. Katalinic, Published by DAAAM International, Vienna, Austria 2006
150th Birthday of Nikola Tesla (1856-1943) Inventor of Mobile Robot Teleoperation

VALUE MANAGEMENT AND NEW PRODUCT DEVELOPMENT

PALCIC, I.; POLAJNAR, A.; LEBER, M. & PIZMOHT, P.

***Abstract**: This paper deals with the concept of value and customer satisfaction that is directly related to the value that a product or a service brings to customer. The concept and importance of value management is also briefly discussed. The final part of the paper presents a value management methodology used for new product development.*
***Key words**: value, value management, new product development, project*

1. INTRODUCTION

It is recognised to be successful in business, we must satisfy our customers. High customer satisfaction is thought to build loyalty and retain customers while also attracting new customers. Consequently, improving customer satisfaction is viewed to be a means to grow business (Dumond, E. J., 2000). It is also well recognised fact that it is extremely hard to measure customer satisfaction. One way is to relate customer satisfaction to the concept of value. The customer will be satisfied if the product or service provides them with value. There are many definitions of value concept, common themes are through these definitions are:

- Customer value is linked to the use of a product or service, thereby removing it from personal values;
- Customer value is perceived by the customers rather than objectively determined by the seller;
- Customer value typically involves a trade-off between what the customer receives (e.g. quality, benefits, and worth) and what he or she gives up to acquire and use a product or service (e.g. price, sacrifices).

Kelly and Male (1993) define value as a measure expressed in currency, effort, exchange, or on a comparative scale which reflects the desire to obtain or retain an item, service or ideal. To manage value, one must recognise the distinction between "price" and "worth". Price is assigned to goods and services at a level to attract customers and profit from their sale, whereas worth reflects the buyer's view of the price as it relates to the perceived benefits, or functions and attributes, of the product or service. When worth equals or exceeds price, the customer is likely to buy the product or service. Value management methodology with its unique techniques and tools is often used as a help for new product development. Developing new product or a new service can be described as a project. Value management and project management are interrelated in many ways. We will present authors' methodology to develop new products using value management.

2. VALUE MANAGEMENT

Value management is an organised effort directed at analysing the functions of goods and services to achieve those necessary functions and essential characteristics in the most profitable manner. Organised effort means that value management is a well structured process – a job plan. Value management makes a deliberate effort to identify what the market needs, as opposed to the supplier's perceptions of wants. The process interfaces engineering and marketing to define the priority requirements from the point of view of the customer and includes the target price and life cycle cost. In addition to achieving the basic functions of the product or service, the product or service must also satisfy other requirements and attributes such as quality, time to market, safety, maintainability etc. Beside that value management methods determine cost generation and evaluate a range of alternatives including new concepts, reconfiguration, eliminating or combining items, and process or procedure changes. These elements interface marketing, engineering and manufacturing (Kaufman, 1998). Value Management is a style of management particularly dedicated to motivating people, developing skills and promoting synergies and innovation, with the aim of maximizing the overall performance of a company or organization. The value management approach involves three root principles:

- a continuous awareness of value for the organization, establishing measures or estimates of value, monitoring and controlling them;
- a focus on the objectives and targets before seeking solutions;
- a focus on function, providing the key to maximize innovative and practical outcomes (The Institute of Value Management, 2006).

At the heart of value management tool is focus on analysis of functions and requirements. Done properly, this seeks to achieve the most cost effective alternative, or best way to obtain all performance and capabilities at optimum price. This does not necessarily mean cost reduction. Typical application is during planning, to develop conceptual alternatives, and during design or construction for cost reduction efforts. Recently, value management applications have begun to include work processes, systems, and programs as part of any activity to achieve results needed by fixed funding or reengineering activities (Department of Energy, 1997).

3. VALUE OF THE PRODUCT

The value of a product is a measure expressed in units of currency, which reflects the desire to obtain or retain the product, and is equal to the cost of the product and a subjective marginal value. Interrelation of cost and value could be expressed as follows:

$$\text{Value} = \text{Cost} \pm \text{marginal value} \qquad (1)$$

where marginal value is the subjective part of the value and depends on the owner(s)'/buyer(s)' value system.

Based on the above definition, the value of a product is the cost of the product plus or minus a marginal value. The cost of the product is the total price paid for the product. The marginal value of the product, which depends on the owner(s)'/buyer(s)' desire to obtain or retain the product is a subjective value and would be referred to as the owner(s)'/buyer(s)' value system. A criteria weighing method can be used in obtaining the marginal value. The methodology is illustrated in Figure 1. The first column represents the steps of the method, the second column represents market input and third column represents the owner(s)'/buyer(s)' confirmations (Shevket Neap, Celik, 1999).

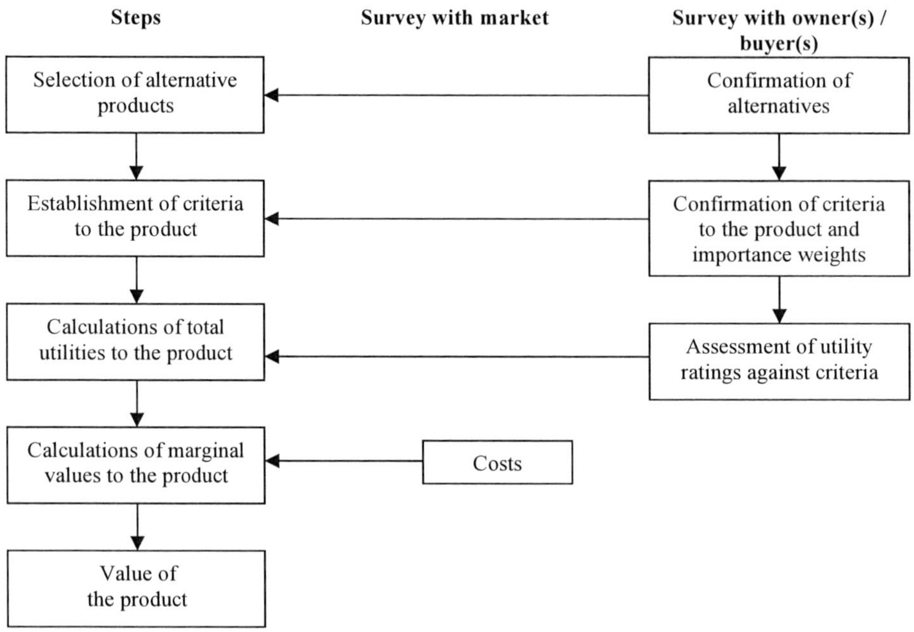

Fig 1. Determination of the value of the product

4. NEW PRODUCT DEVELOPMENT PROJECT AND VALUE CONCEPT

For the purpose of new product development the authors of this paper have defined the methodology that bases on the use of value management methods and techniques. It consists of four phases:

1. preparation for value management,
2. value planning,
3. value engineering,
4. final value management analysis and report.

The preparation for value management phase deals mostly with selecting the members of the project team that will be involved in the new product development. As in any team it is vital that it consists of experts from different company areas and posses different expert background. This phase also deals with securing the right conditions for the team to work and with the formation of detailed job plan with basic goals and time schedule.

The second phase consists of three sub-phases:

- Gathering data – the first step in value management is the definition of the problem. Why do we have to develop a new product or improve the existing one? What are the goals of the company because of this? The second step is actual data gathering. We need data about our existing products, market potentials, out competitors etc. We must gather data from different areas in the company: sales, marketing, purchase, engineering, manufacturing, quality assurance, finance, cost accounting, general management etc. We can use different methods, e.g. SWOT analysis, benchmarking, stakeholders analysis. The last step is "info" analysis, where we actually analyse all received data and select the one we need for our future work.
- Function analysis – cornerstone of value management, where function is defined as intent or purpose of a system, product or process operating in its normally prescribed manner. We assign basic and secondary functions of the product, rank them according to their importance, assign costs to them and look for their solutions with different creative thinking methods.
- Criteria selection – we have to determine and rank few criteria that we will use later to assess the relevance of proposed solutions of a product. This sub-phase is a start of a method called Combinex that reflects in the third phase of our methodology, called value engineering.

Value engineering deals with problems or opportunities involving the physical sciences as the principle discipline in its resolution (product oriented). It mostly deals with finding the (technical) final solution of the product, based on all the data and methods used in previous phases of value management. It concludes with final product report.

Final value management analysis and report summarises the whole project and gives recommendations for future projects.

4. CONCLUSION

This paper describes the basic characteristics of value management, with the focus on the importance of value for the customer satisfaction. Designing a methodology to systematically develop product that will assure appropriate value to the customer was the focal point of this paper and this methodology is used also in practise in companies that are involved in a new product development.

5. REFERENCES

Department of Energy, Office of Field Management, Office of Project and Fixed Asset Management, (1997). *Good practice guide – Value Management*

Dumond, E. J. (2000). Value management: an underlying background. *International Journal of Production and Operations Management*, Vol. 20, No. 9, pp. 1062-1077

Kaufman, J. J. (1998). *Value management – creating competitive advantage*, Crisp Publications, ISBN 1-56052-484-7, London

Kelly, J. & S. Male (1993). *Value Management in Design and Construction*, E & FN. Spon, London

Shevket Neap, H. & Celik, T. (1999). Value of a Product: A Definition, *International Journal of Value-Based Management*, Vol. 12, pp. 181-191

The Institute of Value Management. (2006). What is value management? *Available from*: http://www.ivm.org.uk/vm_whatis.htm#intoduction *Accessed:* 2006-09-02

Annals of DAAAM for 2006 & Proceedings of the 17th International DAAAM Symposium, ISSN 1726-9679
B. Katalinic, Published by DAAAM International, Vienna, Austria 2006
150th Birthday of Nikola Tesla (1856-1943) Inventor of Mobile Robot Teleoperation

COMPUTER AIDED DESIGN OF NUMERICALLY CONTROLLED MACHINE TOOLS DRIVES

PANDILOV, Z. & DUKOVSKI, V.

Abstract: *Characteristics of main spindle and feed drives highly depend upon skillfulness of composing variable speed motors (AC or DC) and mechanical transmission elements. In order to enable interactive computer aided design of numerically controlled machine tool drives original computer programs in C-language have been developed. Our empirical and theoretical knowledge is implemented in the created software. During our investigation in order to get a higher acceleration of the machine slide, a new optimization procedure for transmission ratio and feed screw lead has been developed.*

Key words: *CAD, main spindle drives, feed drives, numerically controlled machine tools*

1. INTRODUCTION

The object of this paper is to present our investigation of principles for analysis and design of NC machine drives. Both, main spindle and feed drive systems analysis and design are very seldom presented in the literature. We can mentioned only the investigations of (Gross et al., 2001) and (Push, 1986) which give only the basic equations for selection of the main spindle and feed drives.

Contemporary development of machine tools highly depends on improvements in drive systems.

A special feature of numerically controlled machine tool drives is use of variable speed (AC or DC) motors which provide continuous changing of cutting speed and feed rate.

The use of variable speed motors creates a question of their appropriate composing with mechanical transmission elements in order to get better output characteristics and to satisfy some conditions as a system.

Results of our long-term empirical and theoretical investigation are implemented in an originally created software for interactive CAD of main spindle and feed drives. During our research we have developed an original optimization procedure for the mechanical transmission elements in order to obtain a higher acceleration of the machine slide.

2. CAD OF MAIN SPINDLE DRIVES

Dominantly, main spindle drives are built with the use of variable speed motors combined with a gearbox group. The gearbox number of speeds Z_E should be in accordance with the variable speed motor range ratio to avoid overlapping of the stepless speeds of the drive. We can calculate the necessary number of speeds of the gearbox Z_E:

$$Z_E = \log Rmsp / \log Rmp \qquad (1)$$

where: Rmsp-range of regulation of main spindle speeds at constant power and Rmp-range of regulation of variable speed motor with constant power. The equation (1) is also recommended by (Push,1986).

Because Z_E is usually a decimal number, it is round to the nearest full number Z.

If $Z > Z_E$ we get characteristic with overlapping speeds.

In case $Z < Z_E$ we get characteristic with step decrease of the power ΔP.

Percentual decrease of the power ΔP in relation with the nominal power Pm of the motor, when $Z < Z_E$ can be calculated with the equation (2),

$$\Delta P / Pm = \left(1 - Rmp \cdot \sqrt[z-1]{Rmp / Rmsp}\right) \cdot 100\ [\%] \qquad (2)$$

According us and (Push,1986) $\Delta P/Pm$ should not be greater than 30%.

Selected performances of the variable speed motor and stepped gearbox directly influent power/torque-rpm characteristics of the main spindle drive and overall NC machine energy consumption.

Based on the empirical investigation we have developed basic schemes for building efficient motor-gearbox group in main spindle drives. A appropriate procedure for kinematics analysis of variable speed motor-gearbox is developed, which serves as a theoretical base for our computer based design of NC machine main spindle drives.

The computer program is written for PC in C-language and can perform the following functions:

-accept input data (power and speed requirements),

-selection of appropriate AC/DC motor from a list of different motor producers,

-graphical presentation of power/torque-rpm characteristics of the chosen motor,

-calculation and selection of appropriate gearbox for the chosen motor and input data,

-graphical presentation of power/torque-rpm characteristics and design concept for the chosen motor-gearbox group,

-speed chart for the selected design,

-main spindle characteristics for the selected motor-gearbox group.

3. CAD OF FEED DRIVES

Although feed drives have very simple kinematics structure their optimal design is problem which consists of selection of servo motor and mechanical transmission elements which must satisfy some requirements as a system.

During this process the drive angular nominal frequency ω_{od} and nominal angular frequency of the mechanical transmission elements ω_{omech} is calculated.

In order not to affect the properties of the highly dynamic AC or DC servo motor, the nominal angular frequency of the mechanical transmission elements ω_{omech} must be higher than the drive nominal angular frequency ω_{od}.

According (Gross et al., 2001) end (Weck, 2001)

$$\omega_{omech} / \omega_{od} \geq 2 \qquad (3)$$

is recommended.

Particular attention has to be paid to the selection of feed drive servomotors. Necessary for the servo motor selection is calculation of the total load torque Mtot, necessary motor speed ne for a rapid feed rate and total moment of inertia Jtot. Details about calculation of the Mtot, ne and Jtot are given by (Gross et al., 2001). For the selected servo motor an analysis of dynamic behavior must be performed (calculation of the acceleration time to rapid feed rate for loaded motor ta, nominal angular frequency of the drive ωod and position loop gain Kv). The acceleration time to maximal speed for loaded motor ta can be calculated as:

$$ta = \frac{Jtot \cdot n_m}{9.55 \cdot Ma} \cdot 10^3 \quad [ms] \qquad (4)$$

where: n_m is maximal motor speed $[min^{-1}]$ and Ma is acceleration torque [Nm].

According (Gross et al., 2001) nominal angular frequency of the drive ωod can be approximately calculated as

$$\omega od \approx 1/Teld \cdot [1 + Teld / (2 \cdot Tmechd)] \quad [s^{-1}] \qquad (5)$$

where: Teld is a drive electrical time constant [s] and Tmechd is a drive mechanical time constant [s].

Another important element which can be approximately calculated is the position loop gain Kv. The position loop gain Kv is a ratio of nominal speed vn [m/min] and difference between nominal and actual position Δx [mm].

The analysis by (Gross et al., 2001) shows that in ideal condition the optimal value of Kv must lie in the range of:

$$0.2 \cdot \omega od \leq Kv \leq 0.3 \cdot \omega od \qquad (6)$$

For real conditions it is recommended:

$$Kv < (0.2\text{-}0.3) \cdot \omega od \qquad (7)$$

The calculated values for Kv from equations (6) and (7) are approximate. The exact value can be obtained experimentally during the fine tuning procedure of the drives (Kakino et al.., 1994, 1995, 1997) and (Pandilov & Dukovski, 2004).

One of the most important requirements for good dynamic behavior of the feed drive is high acceleration of the NC machine tool slide due to the demand for minimal mechanical time constant (Gross et al., 2001) end (Weck, 2001). For the already selected type of servo motor and mechanical transmission elements using the appropriate optimization procedure for transmission ratio i and feed screw lead h we may obtain a higher acceleration a of the machine slide.

The acceleration of the machine slide for the constructive variant with a ball feed screw and nut can be calculated with the following equation:

$$a = \frac{Ma}{Jtot} \cdot \frac{h \cdot i}{2\pi} \quad [m/s^2] \qquad (8)$$

where: h is a feed screw lead [m] and i is a transmission ratio. The optimization of a transmission ratio i and feed screw lead h will be done by using the following procedure:

1. For every standard value of the feed screw lead h we can calculate the transmission ratio range i1≤i≤<i2, to satisfy the following conditions:

-the calculated necessary motor speed ne for the desired rapid traverse feed rate must be smaller or equal to the maximum motor speed nm (ne≤nm),

-the total load torque Mtot must be smaller or equal to the nominal motor torque Mn (Mtot≤Mn).

2. In the range [i1,i2] we will find the maximum of the function of acceleration a=f(i) at constant h or:

$$\max\{a(i)\} = \max\{a(i1),a(i2),a(e1),.....,a(ej)\} \qquad (9)$$

where e1,.....,ej are extremes in the range [i1,i2].

The extremes can be found by the equation

$$\frac{da(i)}{di} = 0 \qquad (10)$$

The function of the acceleration a=f(i) in the range [i1,i2] may have one, more or no extremes.

When the function of acceleration a=f(i) gets a maximal value for the constant feed screw lead h the transmission ratio obtains the relative optimal value iop.

3. To get the absolute optimum of the transmission ratio iopt and optimal values of the feed screw lead hopt we should repeat n times procedures described above in 1 and 2 for all standard values for h . In that way we will get n relative optimal transmission ratios iopi for the appropriate n different standard values for hi where i=1,...,n. The pair (iopi,hi) that gives the maximal value for the acceleration function, will provide the absolute optimum for the transmission ratio iopt, and, the optimal value for the feed screw lead hopt.

It means

$$\max\{a(iop,h)\} = \max\{a(iop1,h1),...,a(iopn,hn)\} \qquad (11)$$

Using equation (11) we obtain the pair (iopt,hopt) which provides a maximal value for the function a=f(i).

This optimization procedure is different from procedures shown by (Gross et al., 2001) and (Motika & Ciglar, 1986) where the relative optimal transmission ratio iop is calculated using equation (10) without taking in consideration that ne≤nm and Mtot≤Mn. The theoretical assumptions treated in the text above, are implemented in the computer program, written for PC in C language.

4. CONCLUSION

The created computer programs enable an efficient interactive and optimal design and analysis of different variants of NC machine main spindle and feed drives. The presented software also reduces the design time and modernizes the design process.

5. REFERENCES

Gross H., Hamann J., Wiegaertner G. (2001). *Electrical Feed Drives in Automation: Basic, Computation, Dimensioning,* MCD Corporate Publishing, Erlangen and Munich.

Kakino, Y., Matsubara, A., Li, Z., Ueda, D., Nakagawa, H., Takeshita, T., Maruyama, H. (1994). A study of the Total Tuning of the Feed Drive System in NC Machine Tools (1st Report)-Modeling of Feed Drive Mechanism and Identification of the Parameters. *Journal of the Japan Society of Precision Engineering,* Vol.60, No.8, pp. 1097-1101.

Kakino Y., Matsubara A., Li Z., Ueda D., Nakagawa H., Takeshita T. & Maruyama H. (1995). A study of the Total Tuning of the Feed Drive System in NC Machine Tools (2nd Report)-Single Axis Servo Parameter Tuning, *Journal of the Japan Society of Precision Engineering,* Vol.61, No.2, 1995, pp.268-272.

Kakino Y., Matsubara A., Li Z., Ueda D., Nakagawa H., Takeshita T., Maruyama H. (1997). A study of the Total Tuning of the Feed Drive System in NC Machine Tools (4th Report)-Multiple Axes Tuning, *Journal of the Japan Society of Precision Engineering,* Vol.63, No.3, 1997, pp.368-372.

Motika R. & Ciglar D. (1986). Optimization of feed drives for NC machine tools, *Strojarstvo* Vol. 28, No.4, 1986, pp.229-233, (In Croatian).

Pandilov Z. & Dukovski V. (2004). Analytical calculation of the NC machine tool position loop gain, *Proceedings of the 21st International Manufacturing Conference, IMC-21, "Servicing Manufacturing",* pp.408-413, ISBN:1-8746537-7-1, Limerick, Ireland, 1st-3rd September, 2004.

Push V. E. (1986). *Machine tools,* Machinostroenie, Moscow, (In Russian).

Weck M. (2001). *Werkzeugmaschinen und Fertigungssysteme, Band 3, Mechatronisch Systeme: Vorschubantriebe and Processdiagnose,* Springer.

Annals of DAAAM for 2006 & Proceedings of the 17th International DAAAM Symposium, ISSN 1726-9679
B. Katalinic, Published by DAAAM International, Vienna, Austria 2006
150th Birthday of Nikola Tesla (1856-1943) Inventor of Mobile Robot Teleoperation

ON LINE MONITORING AND CONTROL OF MOBILE ROBOTS USING JAVA

PARIDA, S.K.; UMESH-BABU, S. & MADHAB, G.B.

Abstract: *This paper describes the development of a system for controlling the motion and monitoring of robots through any browser-rendered virtual environment, irrespective of the platform. A unified method for communication and control of robots is discussed. The unification comes from the use of web technologies such as browsers, Java language, and various socket communications. Several experiments have been conducted using LEGO mobile robots. We have successfully controlled robots, guiding them easily along a corridor, through the use of the interface developed. The communication is established via http protocol from the remote client's browser to the application server by streaming data between applet and servlet and through infrared tower from server to mobile robot and between the robots for collaborative action.*
Key words: *Mobile Robotics, LEGO Robots, JAVA, RCX, lejJOS.*

1. INTRODUCTION

With the advent of Java's "write once, run any where" technology, a fundamental change in the application development and deployment has been taken place. So Java based web application has been accepted as a medium for development of collaborative, remote control of devices over internet (Barnes,2002), (Monteiro et. al., 1997). There are few methods and frameworks for building web-based systems in the field of automation e.g. webCADET (Caldwell & Rodgers, 1998) for distributed design support and CyberCut (Smith & Wright, 1996) for web-based rapid prototyping. For remote monitoring, GE FANUC Automation allows clients to view operation in progress in their factory. HITACHI and MAZAK also provide similar kind of simulation, but all of them require specific hardware and software for smooth running of the system. Our attempt to control a mobile robot through any standard browser over internet wirelessly over a small area has its own advantages, although it has real-time constraints (Caldwell & Rodgers, 1998). The aim of this paper is to develop a robust distributed system using J2EE technology and its design pattern which is very well tested and adopted in the fields like online banking. This approach can be used to control various house hold appliances in a smart home. As a prototypical model we have chosen LEGO mobile robot and its wireless communication facility for establishing a wireless link between application server and the RCX. A behavior based robotics is also given importance while following J2EE specification.

2. ARCHITECHTURE

The overall system architecture in which the way to control mobile robot(s) over internet is shown in Fig.1. A client who wants to control the robots has to connect to the server through the login page and then chose the type of control he want to do between the two options, (i) manual control by pressing appropriate button on the webpage, and (ii) the other is co-ordinate control and viewing its actual motion on the graphs drawn on the applet area.

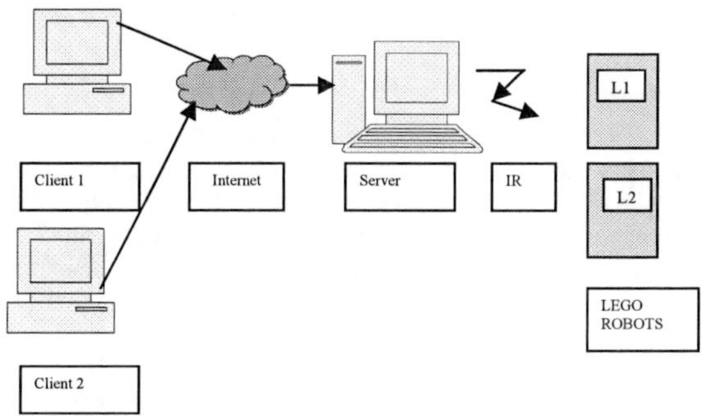

Fig. 1. Schematic of overall system architecture

The server program manages more than two requests from different users by method synchronizing mechanism and allows shift of control from one to the other with permission of the former for collaborative operation, but all can view the status who logged in to the page. This system can be used for running experiments with group of people who want to monitor (or even interact) with the experiments. Examples are mining students doing some experiment inside the mine want to see the status of the experiment while sitting in the office, people studying the behaviour of the wild animals with touch sensors and lots of other applications. The next sections briefly describe different APIs used in the system.

2.1 The Communications API

leJOS allows communication on many levels (Barnes, 2006), (Solorzano, 2006): IR tower to RCX, RCX to RCX, and IR tower to IR tower as shown in Fig.2. The communication classes use the streams for sending data both to the RCX and to the IR tower of computer.

View Controller Lego Hardware
Client side browser Remote application server

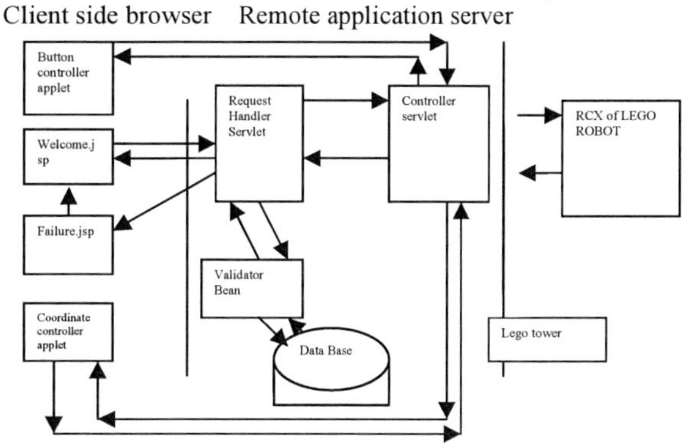

Fig. 2. Data flow in the communications API

Fig. 3. Client-Server webpage

A model-view-controller design pattern is used as per the J2EE specification for future modification and modularity. The components here are jsp, servlets, applets, Java bean. The control communication is done by of streaming of data between the applet and servlet and with the special API RCXCOMM of leJOS which manages dataflow from application server to RCX via IR wireless communication tower attached to the USB port of the host computer. User's name and password is forwarded from the login page as shown in Fig. 3 to the request handler servlet after being initially checked for empty text field and data format using Java script at the client side. If the authentication failed the request shows the failure message to the client and asks for login again through the link provided. If authentication succeeded the request is forwarded to the main controller servlet which has two options to choose the type of operation to be performed.

First one is the manual controller which is actually an applet as shown in Fig. 4. It has a set of buttons, text fields etc. on it to set speed, send control command and view the feedback from the RCX of the mobile robot. On pressing the button an event is listened by the appropriate listener class and performs an action of opening a connection to the main controller servlet to pass the control data to it through object streams. On receipt of the object the controller servlet desterilize it and insatiate the class to establish the connection between the application server and the RCX through the wireless IR tower connected to it and send the character stream of the control command embedded in the desterilized object came from the Applet.

2.2 Co-ordinate controller

In case of the co-ordinate controller the link between the client and server is taken place by applet-servlet communication. The user's mouse position on the controller area is captured as shown in Fig. 5 and then sends to the server. The sequences of mouse click the client made on the applet control area is send to the servlet and then to the RCX. The program in the RCX calculates the distance it should move through that sequence of points clicked by the user for which robot's working area and initial position is fixed. The size of the area always depends upon the size of the wheel used and the gear ratio of the drive. The positional accuracy is maintained by a proportional controller. The program always takes input from the rotation sensor attached to the motors in the loop and check with the reference command input coming from the coordinate controller applet at client page through applet-servlet communication.

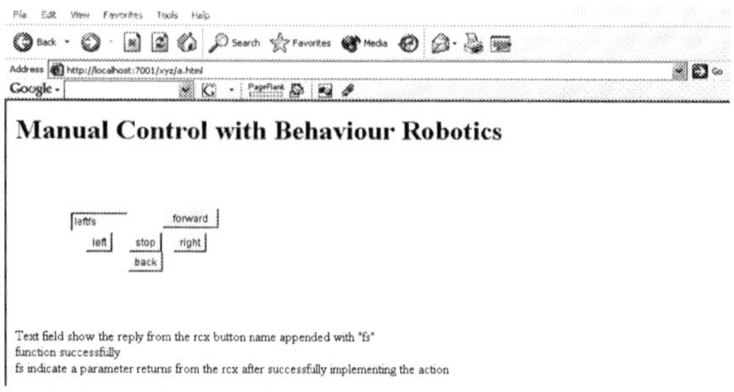

Fig. 4. Manual controller

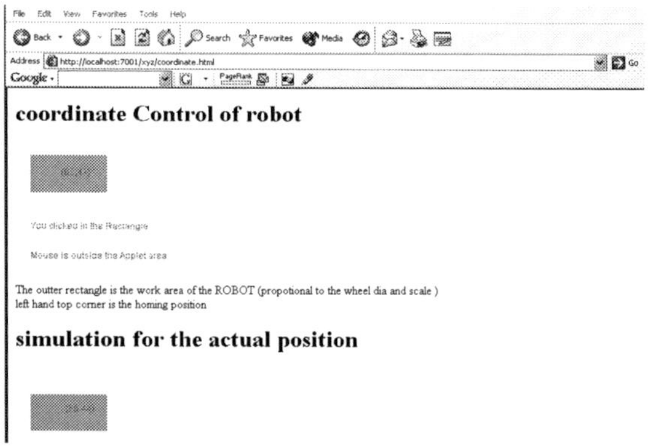

Fig. 5. Co-ordinate controller

3. CONCLUSIONS AND FUTURE WORK

This paper describes an experience using the LEGO Mindstorms platform to support the learning of mobile robotics topics. This work can also be extended to distance learning and remote web-based education, where students able to gain experience using the actual hardware about the establishment of high-speed and reliable internet technologies. In the future, we intend to expand the application of this platform to support the learning of other areas of knowledge, such as electronics, computation, pneumatics and artificial intelligence. There are still many problems to solve until the remote control of mobile robots through the web becomes reality in the complex environment.

4. REFERENCES

Barnes, D. J. (2002). Teaching introductory java through LEGO MINDSTORMS models. *Proceedings of SIGCSE'02*, pp. 147-151, Kentucky, USA, Feb27-Mar03, 2002.

Barnes, David J. An API for the leJOS platform intended for introductory Java instruction. *Available from:* http://www.es.ukc.ac.uk/people/staff/dib/rcx/ *Accessed:* 2006-03-25.

Caldwell, N. H. M. & Rodgers, P. A. (1998). WebCADET: facilitating distributed design support. *IEE Colloquium on web-based knowledge servers*, Digest No. 1998/307,pp. 9/1-9/4.

Monteiro, F., Rocha, P., Menezes, P., Silva, A. & Dias, J. (1997). Teleoperating a mobile robot: a solution based on JAVA language. *Proceedings of ISIE'97*, pp.SS263-SS267, Guimariies, Portugal.

Smith, C. S. & Wright, P. K. (1996). CyberCut: a world wide web based design to fabrication tool. *Journal of Manufacturing Systems*, pp. 432-442, vol. 15, no. 6.

Solorzano, J. leJOS: Java based OS for Lego RCX. *Available from:* http://lejos.sourceforge.net/ *Accessed:* 2006-03-25.

Annals of DAAAM for 2006 & Proceedings of the 17th International DAAAM Symposium, ISSN 1726-9679
B. Katalinic, Published by DAAAM International, Vienna, Austria 2006
150th Birthday of Nikola Tesla (1856-1943) Inventor of Mobile Robot Teleoperation

INTEGRATION OF PROCESS CHAIN: "CAD-CAPP-CAM" FOR MACHINING MOLD DIE WITH 5-AXIS MACHINE

PARK, H.S.; LE, V.S.; KIM, J.M. & LEE, G.B.

Abstract: *This paper demonstrates integration of process chain: "CAD-CAPP-CAM" for machining mold die with 5-axis machine. The integrated system helps to select the planning operations and define their sequence in a way providing the optimum manufacture of the product with respect to the optimal cutting parameters, cost and machining time. The system also combines the rule, strategy and manufacturing knowledge from empirical database and manufacturing handbook.*

Key words: *Integration, Process Chain, Knowledge DB, Machining Strategy, CNC, Converter.*

1. INTRODUCTION

The mold die design is a complex and highly specialized process which usually uses 5-axis CNC to machine the complex surfaces. Because of the complexity of the machining process, its optimization as well as its optimal control is difficult to perform. Optimization of the cutting process needs efficient means for determining the optimal cutting parameters in order to minimize production time and simultaneously preserve the quality of the machining surfaces. Manufacturing process system with the use of empirical and academic knowledge database should be treated as solving the inheritance problems in company. The designer can use knowledge of previous design processes and apply to new design process. Also, the knowledge gained by the die design experts after long years of experience is often available to others within the same company. It helps the freshmen can fill in vacuum whenever the expert retires or leaves the company.

In addition, the integrated system proposed the complete process chain: "CAD-CAPP-CAM" without requesting the auxiliary activities for each stage of integrated process, such as paper drawings, some technological attributes…etc. In this case, the bottle-neck between CAD and CAM will be solved by a good bridge. This bridge is known as process planning activities. Whole system is integrated into unique environment and used feature technology, and STEP as the technology and data exchange standard. In some of recent work, feature technology and STEP standards are excellent candidates to use for the system integration. So, integration difficulty lying in both information interfaces before and after CAPP connected with the commercial CAD/CAM software will be solved.

2. INFORMATION

The integration of the process chain and added functions are programmed in VBA and VBS language, using automatic objects of CATIA V5. They can be used to extract and control all features and their geometric information for making automatic information interface. We will use powerful functions of each CATIA's Workbench and then embed them into unique environment. There are two possible system architectures for integrating commercial CAD, CAPP and CAM domains: Internal and External. Here, we used internal approach with CATIA V5. The process chain is developed and integrated in CATIA environment. A command button added in a new toolbar will control whole activities of integrated environment (Fig. 1).

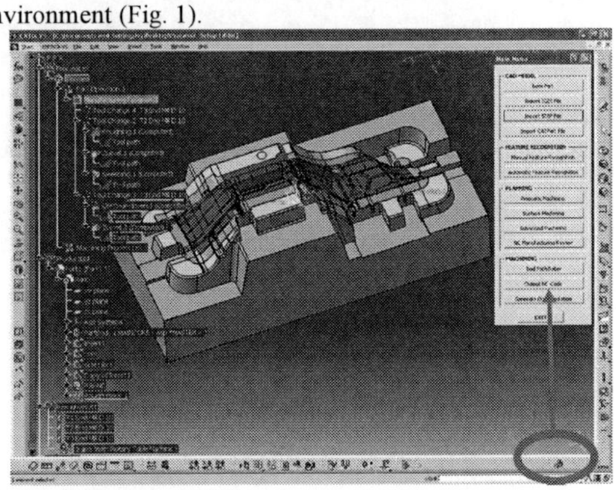

Fig. 1. Integration interface

The integration of process chain includes two stages: CAD/CAPP and CAPP/CAM (Fig. 2). The main functions of CAD/CAPP integration include three parts: feature data interpreter, feature data recognition and machining transformation. On the view of feature technology, CAD model is divided into two kinds: Feature based model and Non-feature based model, compatible with some exchange formats such as STEP (& their APs), IGES, DXF, PDES…etc. Integrated CAD/CAPP environment must interpret them. The feature based recognition technologies and their algorithms are presented in many materials as boundary models, volume decomposition, CSG Models, and hybrid method. There are two application methods: automatic feature recognition and human-assisted (or interactive) feature recognition. Each method has its advantage and disadvantage.

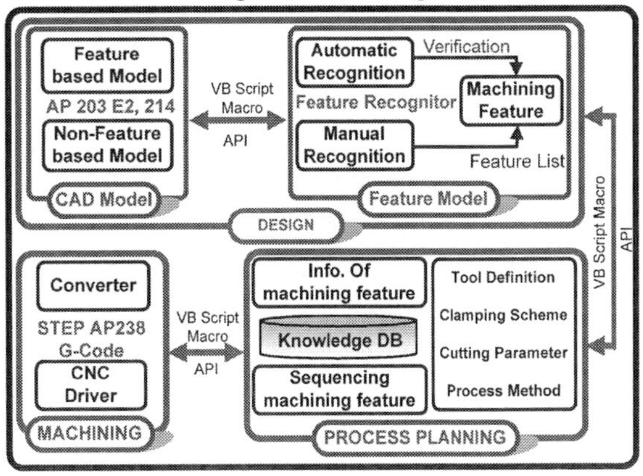

Fig. 2. Integration of process chain CAD-CAPP-CAM

The main functions of CAPP/CAM integration are to receive geometric extracted information from CAD/CAPP integration and perform other works as setup, operation sequencing, tool definition, and machining parameters (cutting parameter, clamping scheme…etc). The second task has a little particularity and can be time consuming even for expert process planning. By using the process template and organizing machining process in catalog will reduce process time and variability. In addition, the partial automatic process planning also reduces significantly process time by using custom macro.

3. THE PARTIAL AUTOMATION OF CAPP

Process planning can be time consuming even for expert process planners. Process plans also vary from process planer to process planer. By partly automating process planning, machining feature can be automatically recognized by the process planner, thus reducing process time and variability.
The partial automation of CAPP is performed by the automatic process selection and sequencing algorithm. They comprise three sub-algorithms, namely, the Transformation algorithm, the Refinement algorithm, and the Linearization algorithm respectively. Those algorithms are presented in many materials. The transformation algorithm is basically a backward planning procedure. In the refinement mechanism, a preliminary operation precedence generated by the transformation algorithm will be refined if necessary. In the present implementation of the refinement algorithm, preliminary operation precedences involving drilling operations are required to be refined. The linearization algorithm aims at generating an operation sequence by grouping machining operations using the same machine tools and/or cutting tools together. A sequencing knowledge-base is provided in the system, numerous process selections and sequencing knowledge are resided in this database. In order to generate the operation sequence for a machined part, it is necessary for all three sub-algorithms to infer from the knowledge in the sequencing knowledge-base. Also, since design and machining knowledge database is extracted by custom macro in CAD/CAPP and CAPP/CAM stage, less experiment process planners will be able to produce reasonable and consistent process plans with minimal effort.

4. KNOWLEDGE DB OF THE INTEGRATED SYSTEM

Figure 4 shows the system architecture of knowledge database established in the integrated system. This is called the sequence knowledge database as knowledge concerning process selection and sequencing is recorded. The sequence knowledge database provides the basic knowledge of automatic process selection and sequencing.

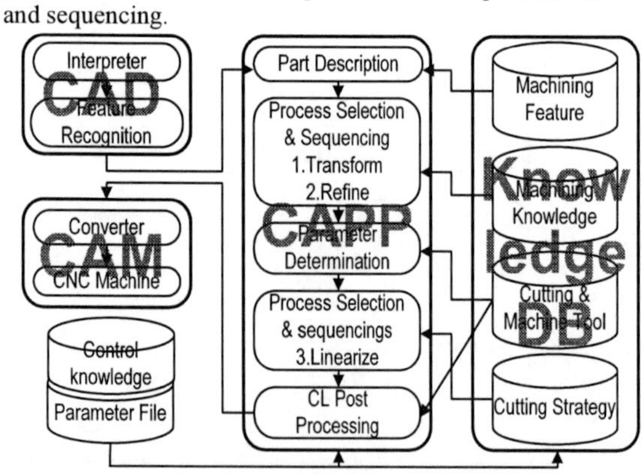

Fig. 4. The system architecture of Knowledge Database

The system architecture of knowledge in the sequencing knowledge database includes: control knowledge database and planning knowledge database. Control knowledge database determines the set of planning operations adequate to the current planning context, so it is used to determine the possible sequence of planning operations. The role of the control knowledge is supervisory - it controls the rule-based knowledge database. The acquisition of the data was supported with the results of continuously conducted investigation in the structure of manufacturing processes of typical machine parts and the model of behaviour of the processing engineer during process planning. Planning knowledge database, represented as production rules, specifies the conditions which must be satisfied to execute elementary process planning activities. It includes the sets of knowledge database with respect to the process planning activities. Users can access automatically data through macro, or can list, append, or remove a piece of knowledge database in each set.

5. OPTIMAL CUTTING PARAMETERS

Normally, the design of mold die requests many complex surfaces and the quality of manufactured surface. Therefore, optimization of the cutting process should be performed. In the system, a cutting force is generated by material flow against the tool. Non linear dependency of the cutting force on the cutting velocity can cause chaotic vibrations of the cutting tool which influence the quality of a manufactured surface. They are very complex due to the variety of nonlinear phenomena involved. Therefore, it can not be analytically by linear models. To model the machining process, various empirical methods have been used. These methods are mainly based on a large amount of experiment data. In order to avoid storing a large database, we model the cost function empirically by applying an artificial self-organizing neural network. By a self-organization process, the relation between the quality of the cutting process and the cutting parameters is implicitly represented by the prototypes. With the helps of the implicit empirical model of the cost function, the optimal value of cutting depth is estimated using conditional average estimator.

6. CONCLUSION

The integrated process chain has orientated the goals: reduction of manufacturing process time, reduction of process planning time, integration of CAD, CAPP, and CAM modules into a simple process chain. This process chain combined both empirical and academic knowledge in aspect of optimizing cutting process and guaranteeing the quality of manufactured surface.

7. REFERENCES

A.Nassehi, S.T.Newman, R.D. Allen. The application of multi-agent for STEP-NC computer aided process planning of prismatic components. *International Journal of Machine Tools & Manufacture*, 2005, pp 1-16.

David Odendahl and Martin Hardwick. Recommend Practices for AP 283 CC1. 2005

Jami J.Shah & Martti Mäntylä. *Parametric and Feature-Based CAD/CAM: concept, techniques, and applications*, John Wiley & Sons, Inc. 0-471-00214-3, USA

Hong Seok.Park, VanSy.Le. Integration of End-to-End process: "Design-Planning-Manufacturing" Based on Feature Technology and STEP. *IJCC Workshop 2006 on Digital Engineering*, 2006.

Miao, H.K., Sridharan, N., Shah, J. Integration of Commercial CAD with Commercial NC Programming Packages. *Proceedings of ASME Design Engineering Conference*. Baltimore, MD, 2000.

Annals of DAAAM for 2006 & Proceedings of the 17th International DAAAM Symposium, ISSN 1726-9679
B. Katalinic, Published by DAAAM International, Vienna, Austria 2006
150th Birthday of Nikola Tesla (1856-1943) Inventor of Mobile Robot Teleoperation

THE PROFILE OF THE HYPERBOLIC ROLLS FOR THE VRM TUBE STRAIGHTENING MACHINE

PARPALA, R.C. & CIOBANU, L.F.

Abstract: *This paper present a method of approximated the hyperbolic profile of the rolls used for the VRM straightening machine. The hyperbolic profile needed for the VRM tools is usually replaced by a simple curve. This conduct to rapid tools wear. By using a mathematical and graphical method one is able to shorten the design and set-up time for the VRM tubes straightening machine. The results obtained from the mathematical algorithm are check with the 3D parametrical model.*

Key words: *Straightening rolls, hyperbolic profile, VRM-90, parametrical model.*

1. STRAIGHTENING EQUIPMENTS

The straighteners are composed of three, four or five pairs of driven rolls mounted skewed, with opposite skewing for each pair of rolls (Fig.1); the arrangement of the roll axis combined with a hyperbolic profile of rolls allow the tube to rotate and to advance contemporary.

Like all machinery, the various components and systems join together to form the working straightener, however the heart of a tube straightener, is the set of work rolls, specially contoured to enable the machine to straighten a range of different tube diameters.

2.CALCULATING THE ROLL PROFILE

In order to determinate the roll profile it is necessary to define the O'P vector from the B-B section (Fig.2). From the O'NP triangle we get the following relation:

$$R_z^2 = X_c^2 + Y_c^2 \Rightarrow R_z = \sqrt{X_c^2 + Y_c^2} \quad (1)$$

The section of the object that has to be straightening is en ellipse having the center in O'' and semi axes r/cos β and r

The equation of the ellipse is:

$$\frac{(x - z\cdot tg\beta)^2}{r^2}\cdot\cos^2\beta + \frac{(y-A)^2}{r^2} - 1 = 0 \quad (2)$$

or:

$$tg\beta = \frac{\sin\beta}{\cos\beta} \Rightarrow F(x,y) = (x\cdot\cos\beta - z\cdot\sin\beta)^2 + (y-A)^2 - r^2 = 0 \quad (3)$$

Fig. 1. VRM machine straightening method

The coordinates of the contact point P are determinate by finding the minimum of the following function:

$$f(x,y) = \sqrt{x^2 + y^2} \quad (4)$$

$$D(f,F) = \frac{\partial f}{\partial x}\cdot\frac{\partial F}{\partial y} - \frac{\partial f}{\partial y}\cdot\frac{\partial F}{\partial x} = 0 \quad (5)$$

$$\frac{\partial f}{\partial x} = \frac{x}{\sqrt{x^2+y^2}} \quad (6)$$

$$\frac{\partial f}{\partial y} = \frac{y}{\sqrt{x^2+y^2}} \quad (7)$$

$$\frac{\partial F}{\partial y} = 2\cdot(x\cdot\cos\beta - z\sin\beta)\cdot\cos\beta \quad (8)$$

$$\frac{\partial F}{\partial y} = 2\cdot(y - A) \quad (9)$$

$$x\cdot(y-A) - y\cdot(x\cdot\cos\beta - z\cdot\sin\beta)\cdot\cos\beta = 0 \quad (10)$$

In order to solve this equation system we must compare the value of y from both expression (3) and (10).

After solving this system we obtain the following equation:

$$X_c = [1 - \frac{\sqrt{r^2 - (x_c\cos\beta - z\sin\beta)^2}}{A}]\cdot(x_c\sin\beta + z\cos\beta)\cdot\sin\beta \quad (11)$$

This IV gr. equation is very difficult to solve this is the reason why it's easier to approximate the result.

X_C can be calculated as the limit of the convergent expression (11) starting with a random value between 0 and ztgβ.

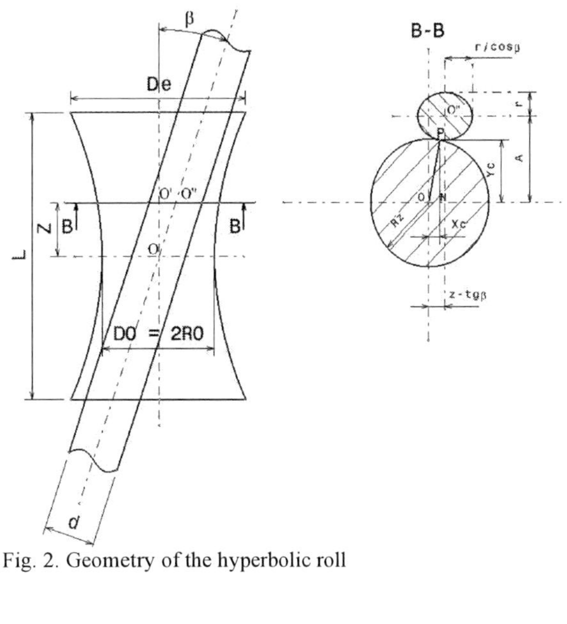

Fig. 2. Geometry of the hyperbolic roll

The appropriated result is obtained when the difference between two consecutive values are the same for a specified number of decimals.

Y_c and Z_c are then calculated:

$$Y_c = A - \sqrt{r^2 - (x_c \cos\beta - z\sin\beta)^2} \qquad (12)$$

$$Z_c = \sqrt{x_c^2 + y_c^2} \qquad (13)$$

This way for every value of the β parameter and for each Z between 0 and L/2 a value for the hyperbolic profile (R_z) is obtain.

If the value for the exterior diameter it's not in the range of needed diameters a new calculation must be done using a different value for β.

This procedure has to be repeated many times before getting the best set of parameters. This was possible by using a Visual Basic program.

3. PARAMETRICAL MODEL

The parametrical model must contain all the parameters needed to completely define the part (Fig. 3). From this set of parameters only a few are entered by the users, the others are constant or the program calculates them.

Using different sets of parameters it is possible to visually compare different hyperbolic profiles and chose the correct set-up for the VRM machine (Fig.4).

The user is also able to check the numerical algorithm by visualising the virtual contact line between the two hyperbolic rolls and the pipe (Fig. 5).

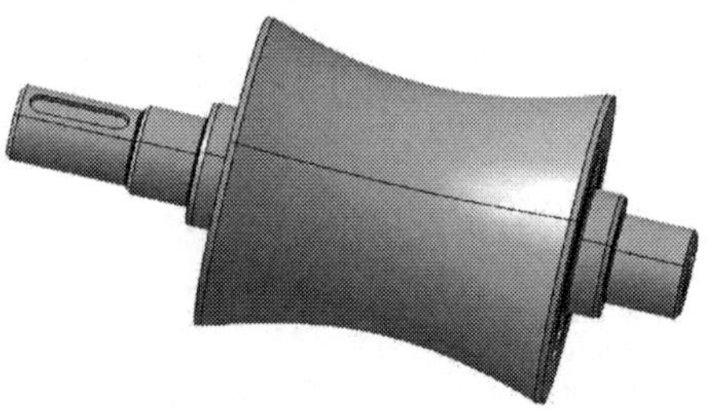

Fig. 5 Parametrical model of the hyperbolic roll

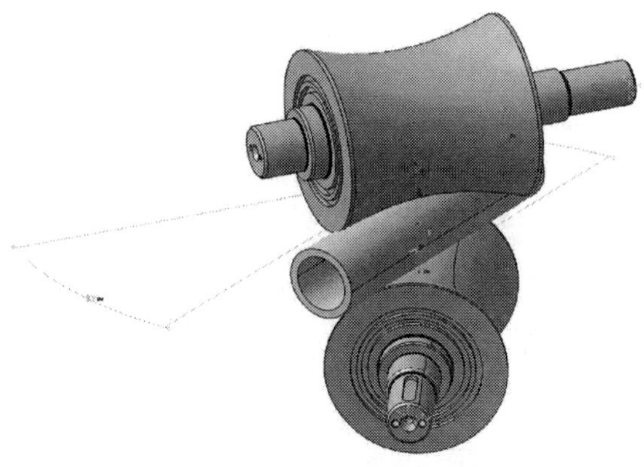

Fig. 3 Parametrical model of the VRM rolls assembly

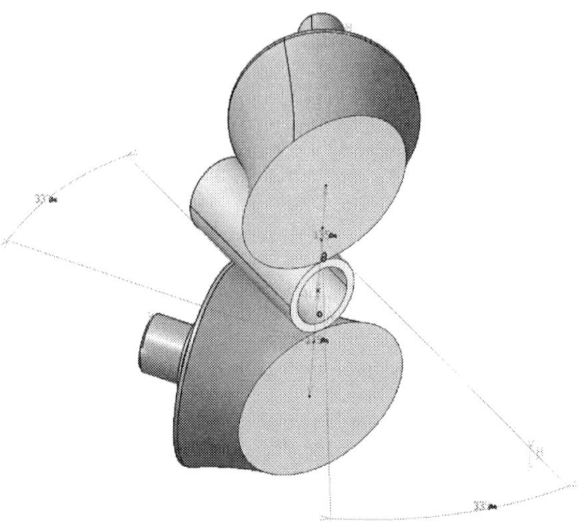

Fig. 4 Section through the VRM rolls 3D model

4. CONCLUSION

Because of the wide range of tube's diameters that are loaded on the same machine the hyperbolic profiles of the tools are very difficult to calculate using classical methods.

Thanks to programming it is possible to customize a generic system according to ones specific needs and therefore make it more efficient. Specifically it is possible to automate the repetitive phases of the design process through the development of ad-hoc procedures, which are able, for example, to generate the models according to functional specifications; generate customized parts lists; obtain from the model the geometric information necessary to perform the required simulations and verifications in order to identify any criticality in the first steps of the process.

Using the correct profile could reduce the wear of the tools by 60-80 %.

By using a parametrical 3D model various values for the β parameter can be visualized and tested in a virtual environment, this provide a better set-up of the real machine.

5. REFERENCES

Randall S., Prieur H. (1967) *Tubular Production in the Cold Pilger Machine*, Iron and Steel Engineer - August 1967

Stroud R. (2005) Tooling, the key for mill production, Available from: http://www.thefabricator.com/articles /tube_and_pipe_article?ID=189, Accessed: 2005.07.10

VRM-90 technical documentation

Parpala, R.C. (2005) Automatic mandrel design for cold pilger mill using CATIA V5 CAD/CAM/CAE Integrated Software, *Proceedings of 16th DAAAM International Symposium DAAAM Intelligent Manufacturing & Automation:*, pp. 287-288, ISBN 3-901509-46-1, , Katalinic, B. (Ed.), October 2005, Opatija, Croatia

Parpala, R.C., Ciobanu, L.F., Predincea, N. (2006) Collaborative design in CATIA V5 using Visual Basic for application and IBM-DB2 databases, *Proceedings of 6th International Conference on Integrated Design and Manufacturing in Mechanical Engineering – IDMME*, Tichkiewitch, S., Tollenaere, M. (Ed), ISBN 2-9523979-1-0, May 2006, Alpexpo - Grenoble, France

Annals of DAAAM for 2006 & Proceedings of the 17th International DAAAM Symposium, ISSN 1726-9679
B. Katalinic, Published by DAAAM International, Vienna, Austria 2006
150th Birthday of Nikola Tesla (1856-1943) Inventor of Mobile Robot Teleoperation

REGRESSION ANALYSIS OF RELATIONSHIP BETWEEN DEGREE OF WOOD PROCESSING AND BUSINESS RESULT

PASIC, M.; SUNJE, A. & KARIC, E.

Abstract: *In this paper results of the research of influence of degree of wood processing on business result are presented. The analysis is done using regression method. In developed mathematical model business result is defined as independent variable, while degree of wood processing is defined as independent variable. The research sample consists of organizations from Bosnia and Herzegovina from wood processing industry that was observed as the reference industrial branch for this research.*
Key words: *Business Result, Degree of Wood Processing, Regression, Variables.*

1. INTRODUCTION

This research was based on posture that related relationship exists between business result, as a dependent variable, and degree of wood processing as independent variable. As such, the construction of statistical models like linear regression can serve as a tool to verify or otherwise disprove the presence of relationships between interacting variables. The technical approach adopted here was constructed that represents anticipated relationship between business result and related independent variable of degree of wood processing. This model was tested by using statistical tools such ANOVA analysis (analysis of variance) and coefficient of determination, R^2.
Studies in this areas show relationship between dependent and independent variables. Study (Gautam et al. 2003) shows that, in some circumstances, adopting the effectiveness of business processes as a dependent variable may be more appropriate than adopting overall firm performance as a dependent variable. Results are consistent with resource-based expectations, and they show that distinctive advantages observable at the process level are not necessarily reflected in firm level performance. Creation (Yamaguchi et al. 2001) of a new business-model is considered to be one of the important factors of market growth. The creation of a new business-model means an appearance of new demand. Study (Evans 1987) on some aspects of firm dynamics finds that firm growth, the variability of firm growth, and the probability that a firm will fail decrease with firm age. It also finds that firm growth decreases at a diminishing rate with firm size even after controlling for the exit of slow-growing firms from the sample. Organizational characteristics, especially number of employees and amount of capital invested, and organizational strategies, especially businesses aiming at a national market, are the most important determinants of business survival (Bruderl et al. 1992). Research (Davidsson et al. 2002) shows that business age, beginning size, ownership form, industrial sector, and legal form are the most important factors related to growth. Although business growth differs among industrial sectors, youth, ownership independence, and small size are major factors that underlie growth across all industries. Empirical research (Kang & Sørensen 1999) based on the approach that shareholders are homogenous and that their influence on firm performance is directly proportional to the percentage of equity they hold has failed to produce definitive evidence. Influence of ownership structure, variables of external and organizational discipline on financial and economic performance suggests a non-linear relationship between ownership structure and performance (Séverin 2001).
Data used in mathematical analysis, for this research, were obtained through empirical methodology of polling organizations using carefully designated Questionnaire. Sample data consists of 48 organizations from wood processing industry in Bosnia and Herzegovina.

2. PROBLEM APPROACH AND MATHEMATICAL MODEL

Regression is perhaps the most widely used data analysis tool in statistics. Many fields used it as away to analyze simple or very complex models. The base of research is representing by dependent variable of business result, and degree of wood processing as independent variable. The specific definitions and units of measurement of these variables were defined as given below.

Variable	Symbol	Description
Business result	BR	Is continuous variable that represents the annual profit or loss of a company at the end of the financial year, measured in BAM (1€=1,9558BAM)
Degree of wood processing	DWP	Is categorical variable that represents degree of wood processing as shown below: 1.primary degree of wood processing; 2.final degree of wood processing; 3.primar and final degrees of wood processing

Table 1. Variable definitions and units of measurement

Linear regression model will be constructed and statistical tools such ANOVA analysis, coefficient of determination, R^2, F-test, t-test, and standardized coefficients will be applied to prove existence of relationship, and to research the influence of degree of wood processing influence on business result. Simple linear regression model used in this research is given by below equation:

$$BR=b_0+b_1DWP \qquad (1)$$

Where b_0, and b_1 represents regression coefficients: b_0 is the intercept and b_1 is slope. The slope coefficient b_1 represents the change in BR (business result) for a unit change in DWP (degree of wood processing).

3. RESULTS AND INTERPRETATIONS

Tables numbered as 2. and 3. show statistical results of research. It is based on two-tailed test. Below tables are given detailed interpretations of obtained results after which is given final equation. This equation determines relationship between business result and degree of wood processing.

	df	Sum of Squares	Mean Square	F	Sig. F	R^2
Regression	1	7,7561E+10	7,76E+10	7,997	0,006866	0,1454
Residual	47	4,5586E+11	9,7E+9			
Total	48	5,3342E+11				

Table 2. ANOVA table for linear regression

The ANOVA analysis shown in Table 2. shows that a computed value for the F-ratio is 7.997. The corresponding table value for F-ratio is 4.052 at 0.05 level of significance, where degrees of freedom are $df_1=1$ and $df_2=47$. F-ratio value shows that the linear regression was significant and valid. The R^2 value is 0.1454, indicating that regression model could explain 14,54 percent of the variation of business result by variable degree of wood processing.

	Coefficients	Standard Error	t Stat	P-value
Intercept (b_0)	-24989,67	33732,05	-0,741	0,462483
X Variable (b_1)	48828,62	17267,12	2,828	0,006866

Table 3. Coefficients of simple linear regression, t test and p-value

Table 3. shows that regression coefficients b_0, b_1 are 48828,62 and –24989,67 respectively. Results of the t-test indicate that regression coefficients b_0 and b_1 are statistically significant and not equal to zero at 0.05 level of significance (t-value for sample size greater than 30 is 1.96 at 0.05 level of significance). So, the simple linear regression equation of business result and degree of wood processing can be given by:

$$BR = 48828{,}62\ DW - 24989{,}67 \quad (2)$$

Next figure shows graphical relationship of variables with the simple linear equation between them.

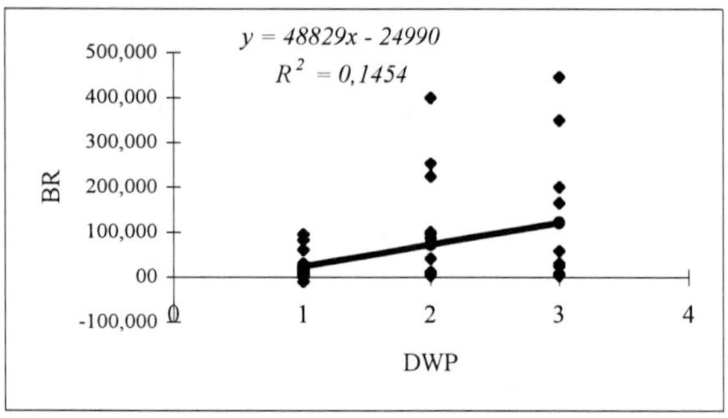

Fig.1. Scatter Plot of Business Result versus Degree of Wood Processing

The p-value shows the level of significance for the test. It is based on two-tailed test and shows the probability of finding a sample with an estimate slope given from the model when the real value is zero. Since the t value is sufficiently large (t=2,828) and p value is sufficiently small (p=0,006866), we can conclude that there is a linear relationship between the two variables. The residuals are an important part of regression. Residual analysis shows the fit of the model and violations in assumptions. We want a random pattern with residuals that are within ±3 standard deviations. Figure 2. proves this for the model developed in this paper.

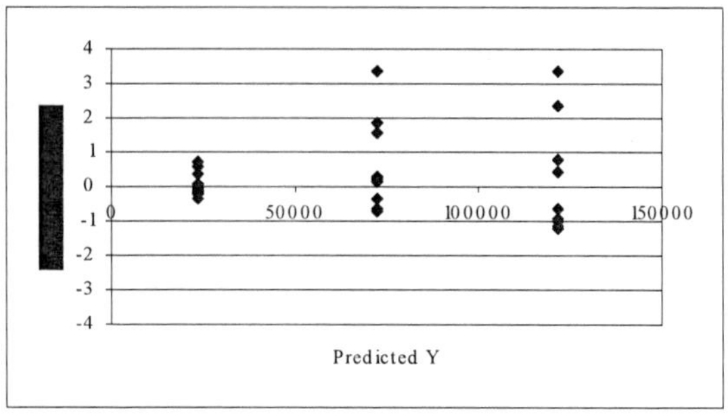

Fig.2. Scatter Plot of Standard Residuals vs. Predicted Y

4. CONCLUSION

This research proves that there is a relationship between business result and degree of wood processing. Results of simple linear regression show that significant relationship exists between business result and degree of wood processing. Positive sign related to degree of wood processing, or the slope in the equation (2), means that if independent variable (degree of wood processing) increases, dependent variable business result will also increase.

So, best business result can be expected if degree of wood processing is on high level, including primary and final wood processing. In our case, as presented in Table 1, degree of wood processing variable is set as categorical variable, with the categories as follows: 1. Primary degree of wood processing, 2. Final degree of wood processing and finally 3. Primary and final degree of wood processing. In our model it is represented by category number three. So, from the results it is obvious that the business result corresponds with the degree of wood processing. The higher degree of wood processing the better business result.

Future research should be focused on multivariate analysis with the aim to determine the most important variable that influences business results of a company.

5. REFERENCES

Bruderl J., Preisendorfer P., Ziegler R. (1992). Survival Chances of Newly Founded Business Organizations, *American Sociological Review*, Vol. 57, No. 2, April 1992. pp. 227-242.

Davidsson P., Kirchhoff B., Hatemi–J A. (2002). Gustavsson H., Empirical Analysis of Business Growth Factors Using Swedish Data, *Journal of Small Business Management*, Vol. 40, No. 4, October 2002, pp. 332 – 344.

Evans S. D. (1987). The Relationship Between Firm Growth, Size, and Age: Estimates for 100 Manufacturing Industries,
Journal of Industrial Economics, Vol. 35, No. 4, June 1987, pp. 567-581.

Gautam R., Barney J. & Waleed M. (2003). Capabilities, business processes, and competitive advantage: choosing the dependent variable in empirical tests of the resource-based view, *John Wiley & Sons, Ltd*, OH, USA.

Kang D. L., Sørensen A. B. (1999). Ownership Organization And Firm Performance, *Annual Review of Sociology*, Vol. 25, August 1999, pp. 121-144.

Séverin É. (2001). Ownership structure and the performance of firms: Evidence from France, *European Journal of Economic and Social Systems*, Vol. 15, No 2, pp. 85-107.

Yamaguchi T., Gemba K. & Kodama F. (2001). Quantitative analysis of business-model, Portland Internat. Conference on Management of Engineering and Technology, Vol. 1, pp 34, ISBN: 1-890843-06-7 Portland, OR, USA.

Annals of DAAAM for 2006 & Proceedings of the 17th International DAAAM Symposium, ISSN 1726-9679
B. Katalinic, Published by DAAAM International, Vienna, Austria 2006
150th Birthday of Nikola Tesla (1856-1943) Inventor of Mobile Robot Teleoperation

COMPARISON OF RANDOM SURFACES 2D AND 3D PARAMETERS

PAWLUS, P. & REIZER, R.

Abstract: *The paper presents comparison of various 2D and 3D parameters of random surface topography. The slope, parameters connected with peaks (summits) as well as amplitude parameters were analysed. The modelled and measured isotropic and anisotropic surfaces were the objects of investigations.*

Key words: *surface topography, parameters, random surfaces*

1. INTRODUCTION

All surfaces are rough. Surface roughness exists in 3 dimensions. The stylus instrument is widely used technique for measuring surface topography. In industry 2D profiles are mainly measured. 3D measurement is time consuming and expensive. No standard exists about 3D surface topography measurement. However some information about 3D surface topography can be obtained based on 2D profile measurement. Some differences between 3D and 2D measurements were analysed in the past. However it is now possible to check correctness of the theoretical presumptions.

2. THEORETICAL INVESTIGATIONS

Longuet-Higgins analysed some properties of random surfaces (Longuet-Higgins,1957). Nayak developed formulae for these surfaces (Nayak, 1971). Rough surfaces were treated as two-dimensional isotropic random processes. Nayak developed the following equation for 3D slope of isotropic surface of normal ordinate distribution determination:

$$S_{\Delta a} = \left(\frac{\pi m_2}{2}\right)^2 = \frac{\pi}{2}\Delta_a \qquad (1)$$

where Δ_a is profile average slope.

Summit density D_s should be:

$$D_s = \frac{1}{6\pi\sqrt{3}}\left(\frac{m_4}{m_2}\right) \approx 1.2(D_p)^2 \qquad (2)$$

where D_p – profile peak density, m_2, m_4 – second and fourth profile spectral moment. Bush and others (Bush et al., 1976) derived the following formula for the summit curvature determination:

$$S_{sc} = \frac{8 m_4^{\frac{1}{2}}}{3\pi^{\frac{1}{2}}} \qquad (3)$$

Sayles and Thomas believed that for two-dimensional surface topography (like ground surface) spectral moments of second and fourth order should be equal to square root of product from moments obtained in two perpendicular directions (Sayles & Thomas, 1979). Substituted average slope should be:

$$\Delta_a = \sqrt{\Delta_{ax}\Delta_{ay}} \qquad (4)$$

and 3D slope should be equal to that value magnified by $\pi/2$.

Nayak found that generally summit curvature should be bigger than peak curvature, summit height bigger than peak height, but summit density smaller than peak density. In 3 dimensions various methods of summit identification are used. The point can be identified as a summit, when its ordinate is higher than ordinates of 4 or 8 neighbouring points. Sayles and Thomas (Sayles & Thomas, 1979) found that when the summit was identified based on 8 neighbouring points, the summit density was in accordance with Nayak presumptions.

The average height parameters of 3D surface topography (like Sa or Sq) are similar to corresponding parameters of profiles (Pa, Pq, Ra, Rq) of random isotropic surfaces. However the maximum height parameters of random 3-D surface topography (like St) are considerably bigger than Pt parameters of 2D surface profiles. The range of height parameters depends on the number of measuring points. When the number of points is bigger - the maximum height is also bigger. In 3D measurements the number of measuring points is considerably bigger than the number of profile points (when the sampling interval is the same – see Figure 1a). Figure 1b presents the similar dependence for surface after 2 processes.

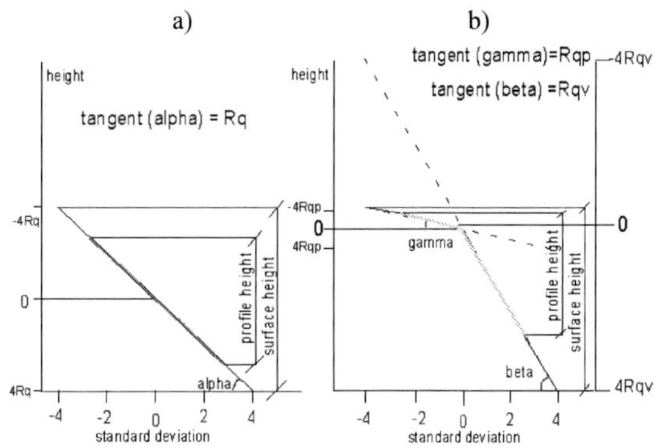

Fig. 1. Maximum height of 2D profiles and 3D random surface topography for surface after 1 (a) and 2-processes (b)

3. THE RESULTS AND THEIR ANALYSIS

Iso- and anisotropic random surfaces of Gaussian ordinate distribution surfaces of exponential shape of auto correlation function were generated. The special procedure was used in order to obtain surfaces after 2 processes.

In addition measured one-process surfaces after vapour blasting, shot peening, barrel finishing and other processes were analysed. The multi-process cylinder surfaces after 'zero-wear' and plateau honing were also studied.

For simulated isotropic surfaces of random ordinate distribution we found that presented above equation in which average surface slope was equal to average profile slope magnified by $\pi/2$ was confirmed. The differences in some rare cases were not bigger than 6%.

During investigation of computer created two-dimensional surfaces dependence that substituted average slope should be geometric mean from average slopes from perpendicular directions was confirmed. Errors usually were smaller than 4%

only for strong anisotropy they were equal to 10%. In this case the differences were smaller when substitute slope was equal to arithmetic mean of average slopes in perpendicular directions:

$$\Delta_a = \frac{\Delta_{ax} + \Delta_{ay}}{2} \tag{5}$$

However for very strong anisotropy the errors were very big, when substitute slope was equal to the geometric average of slopes in perpendicular directions (about 40%). When substitute slope was equal to arithmetic mean the errors were smaller (about 12%) but they were also great. So we found that if deviation of the average slope values in 2 perpendicular directions was bigger than 3.5 the smaller error was obtained when the slope of 3D surface was equal to the bigger 2D slope. This dependence was obtained for surfaces of normal ordinate distribution.

The present authors tried to find out if the described dependencies between slopes took place also for 2-process surfaces. When substitute average profile slope was equal to geometric mean of slopes in two perpendicular directions, average errors were in the range 3.5-4.1%. For the case of anisotropic surfaces change of geometric mean into arithmetic mean caused decrease of errors to 1.8-2.9%. These results were achieved when slopes in perpendicular directions were obtained from whole surfaces. When surface slopes were calculated based on profile slopes, errors were bigger (6-7%).

For some random surfaces of various shape of ordinate distributions we tested if the 3D surface slope is equal to 2D profile slope magnified by $\pi/2$. For nearly isotropic surfaces the average errors were smaller than 1%. For honed and plateau honed cylinders the average error using geometrical mean was 3.3%, arithmetical mean 1.4%. For some surfaces after running-in the mean error of surface average slope calculation using geometrical mean was 4.5%, arithmetic mean 2.7%. In these cases the relations between average profile slopes in perpendicular directions were smaller than 3. When ratio of slopes of strongly anisotropic surfaces is bigger than 3.5 it is better to use the bigger value of slope in 2 directions.

Summit can be recognised based on 4 and 8 neighbouring points. Scott proposed method based on 6 points (Scott, 2000). In order to confirm the presented above presumptions of Nayak (Nayak, 1971) and Bush (Bush et al., 1976), the analysis of computer generated random surface topographies was carried out. It was found that predicted summit density was closest to summit density obtained using 8-point method - the average error was 8%. This result was also confirmed based on the analysis of real surfaces. The predicted summit curvature was also the most similar to the results obtained using method based on 8 points (the average error was 5%). The method based on 6 points gave the values of average summit height, curvature and density between values received based on 4- and 8-point methods.

The method of spectral moments calculation for anisotropic surfaces with small degree of anisotropy (Sayles & Thomas, 1979) was tested. The prediction was confirmed for computer generated random surfaces. The average errors of obtaining parameters connected with summits were smaller than 11%. These results were also confirmed with regard to measured surfaces.

It was found that information of profile peaks disturbed information of surface summits. The summit height was bigger than peak height, the difference was smaller, when ordinates of points were more correlated. It was confirmed that summit curvature was bigger than height curvature. The theoretical dependencies between summit and peaks curvatures and density were confirmed. However the observed tendencies were confirmed for real surfaces not necessary isotropic, or not having Gaussian ordinate distributions. The difference between summit and peak density was noticed. The average height of summit was bigger than height of peak. The peak curvature was bigger than summit curvature.

The analysis of height 3D and 2D parameters was also done. The 3D surface average height parameters of cylinder surfaces were very similar to mean values of axial profiles Pq or Pa parameters (a little bigger than circumferential parameters). Difference between mean value of Pa parameter of axial profiles and 3D surfaces was smaller than 1% for all analysed surfaces. However maximum height parameter St was about 1.6 times bigger than Pt parameter mean values of axial and circumferential profiles.

3D surface average height parameters of one-directional surfaces were very similar to mean values of Pq or Pa parameters in direction perpendicular to lay. Difference between mean value of Pa parameter of axial profiles and Sa of 3D surfaces was smaller than 0.5-7%. However maximum height parameter St was about 1.4-2 times bigger than Pt parameter mean value.

The average value of Sa and Sq parameters of isotropic surfaces were a little bigger than of corresponding 2D profile parameters, like Pa and Pq (the differences were smaller than 2-10%). However the maximum parameter St was about 1.4-1.6 times bigger than mean value of Pt parameters in perpendicular directions.

4. CONCLUSIONS

It is possible to predict 3D surface parameters based on 2D parameters of random surfaces. The prediction of 3D surface slope based on 2D profiles slopes in perpendicular directions was confirmed with regard to one-process and two-process surfaces. Modification of slope calculation was recommended. The parameters connected with 3D summits like summit density and curvature can be also calculated based on profile measurement of random surfaces of normal ordinate distribution. The theoretical predictions of dependencies between peak and summit parameters were also confirmed.

The average 2D height parameters of profiles of isotropic surfaces are similar or a little smaller than corresponding 3D parameters. These 3D parameters of one-directional surfaces are similar to 2D profile parameters in direction perpendicular to lay. However maximum 3D height parameters are bigger than maximum 2D profile parameters.

5. REFERENCES

Bush, A. W.; Gibson, R. D. & Keogh, G. P. (1976). The limit of elastic deformation in the contact of rough surfaces. *Mechanics Research Communications*, Vol. 3, pp. 169-174, ISSN: 0093-6413

Longuet-Higgins, M. S. (1957). The statistical analysis of a random moving surface. *Philosophical Transactions of the Royal Society*, Vol. 249, Series A, pp. 321-387, ISSN: 0080-4614

Nayak, P. R. (1971). Random process model of rough surfaces. Transactions of the ASME. *Journal of Lubrication Technology*, Vol. 93, pp. 398-407

Sayles, R. S. & Thomas T. R. (1979). Measurements of the statistical properties of engineering surfaces. Transactions of the ASME. *Journal of Lubrication Technology*, Vol. 101, pp. 409-417

Scott P. J. (2000). An algorithm to extract critical points from lattice height data, *Proceedings of Metrology and Properties of Engineering Surfaces*, pp. 26-28, United Kingdom, April 2000, Huddersfield

Annals of DAAAM for 2006 & Proceedings of the 17th International DAAAM Symposium, ISSN 1726-9679
B. Katalinic, Published by DAAAM International, Vienna, Austria 2006
150th Birthday of Nikola Tesla (1856-1943) Inventor of Mobile Robot Teleoperation

SIMULATION OF RANDOM PROFILES USING AR METHOD

PAWLUS, P. & REIZER, R.

Abstract: *The paper presents the review of various methods of isotropic and anisotropic random surfaces of Gaussian ordinate distribution modeling. The procedure of digital simulation of one-process profile using AR method of various degrees is presented. The examples of using this method for the simulation of machined profiles are given.*
Key words: *simulation, random profile, surface topography*

1. INTRODUCTION

Most engineering surfaces have height distributions which are approximately Gaussian. Simulation techniques can be used to create reference data. The tribological behaviour of engineering surfaces can be predicted numerically. The numerical solution of these problems involves input of surface data. It can be obtained either from digital input from profilometer or from numerical simulation of the rough surfaces. Randomly generation of surface roughness by numerical means is simple and offers some advantaged. The hardware and software requirements can be eliminated. The numerical simulation is also more flexible. However the simulated surface should be similar to real surface.

2. THE EXISTING METHODS OF RANDOM SURFACE SIMULATION

There are 2 groups of methods of random surface modeling. First of them is based on modeling of machining or wear processes. In this case it is possible to simulate changes in machined processed, for example vibration, change of depth of abrasion, feed etc.
The methods from second group depend on similarity between modeled and measured surface. There are criteria of matching accuracy of the model to measurement results, based on the autocorrelation function, power spectral density function, spectral moments and set of parameters describing the surface topography. The matching accuracy is usually better than that of the first group of methods. Sometimes these 2 methods can be included in the simulation model.
Several methods of 3D surface topography modeling were suggested by researchers. They were mainly concentrated on random surfaces of Gaussian ordinate distribution. A growing need was developed to computer generate surfaces with desired properties.
The numerical characterisation of stylus-measured data is based on the recording of 2D profiles. The time series model of rough surface (Staufert, 1979, deVries, 1979) was applied to one-dimensional profile generation by the authors of paper (Watson et al., 1979). The Johnson transformation system was introduced into the simulation of non-Gaussian distribution profile with the given skewness and kurtosis (Watson & Spedding, 1979). The need to measure surface microgeometry in 3 dimensions is very substantial.

A numerical procedure for randomly generating any general three-dimensional surface topography with prescribed statistical properties was described in Reference (Patir, 1977).
Auto – Regressive and Moving Average (ARMA) method is popular in generating surfaces. The simulation of 3D random surfaces was done by Whitehouse (Whitehouse, 1983). AR time series model was used. Gu and Huang (Gu & Huand, 1990) used ARMA model to create 3D surfaces of desired autocorrelation function. A modeled surface which agrees with the actual surface can be obtained through measurement of some profiles along different directions. Hong and Ehmann (Hong & Ehmann, 1995) attempted to synthesise engineered surfaces. The proposed method was based on 2-dimensional difference equations and two-dimensional linear autoregressive models. In new paper (Uichidate et al., 2004) a generation procedure of surface texture using the non-casual 2-D AR model was suggested and its efficiency was examined.
Fast Fourier Transform is also popular in generating surfaces. Newland, in 1984 used it (Newland, 1984). Wu (Wu, 2000) developed a numerical procedure of three-dimensional surfaces. The method, based on FFT can simulate surfaces with given spectral density or autocorrelation function. Hu and Tonder (Hu & Tonder, 1992) used finite impulse response filter.
In majority of presented above papers the surfaces of Gaussian normal ordinate distribution were generated. Interesting procedure was proposed for generating non-Gaussian surfaces (Wu, 2004). This method based on FFT approach preserves the skewness, kurtosis together with autocorrelation function or spectral density of surfaces.
In paper (Bakolas, 2003) a numerical method for the generation of arbitrarily oriented three-dimensional surfaces was presented. The procedure extends the method of the linear transformation matrix, first proposed by Patir (Patir, 1977).

3. THE AIM AND SCOPE OF INVESTIGATIONS

The majority of current time series models can consider few orders system. It means that only the features near the origin are well simulated. Therefore the aim of the investigations is to compare the various orders AR models of profile of normal ordinate distribution modeling. The scope of investigations includes:

- development of random profile modeling procedures,
- modelling of machined profiles using AR models of various orders,
- the comparison of the results of modeling,
- presentation of examples of simulated profiles.

A model definition assumed a normally distributed profile and normally distributed residuals a_x. It must be accompanied by a value for the residual variance σ_a^2. It is required that the heights produced by the model ought to have a particular variance σ_a^2 to represent the actual profile, so the relation between σ_a^2 and σ_y^2 is needed. For a general AR(p) model:

$$\sigma_a^2 = (1 - \phi_1\rho_1 \ldots - \phi_p\rho_p)\sigma_y^2 \quad (1)$$

where ρ_k is the autocorrelation at lag k defined as:

$$\frac{1}{\sigma_y^2} E\{(y_x - \mu_y)(y_{x+k} - \mu_y)\} \quad (2)$$

E denotes expectation. The procedures were carried out by computer program (programmed in C++)).

4. THE RESULTS OF THE MODEL

Various measured profiles were the objects of investigations. The modeled autocorrelation function was close to measured autocorrelation function when the correlation length was rather small. The AR(3) model approximates autocorrelation function well for not more than 6 points. In this case in majority of results the autocorrelation function was better approximated than when AR(6) and AR(9) models were used. For bigger number of measuring points AR(6) or AR(9) models were better. However the AR(9) model (and AR(6)) well approximated autocorrelation function for bigger number of measuring points. However AR(6) model was good not only for 6 but also for about 10 and sometimes more points of autocorrelation function. When the correlation length was bigger than 15 points, only first points were approximated well. So the resulting correlation length after using 3 methods was smaller than the real correlation length.
In addition, the parameters of real and simulated profiles were compared. Standard deviation of height Pq, rms. slope, spacing parameter PSm and peak density were selected. The tendency was found that Pq parameter was the closest to measured surface profiles when order of model was bigger. However it is possible to obtaining correct value of Pq by changing the magnification. Generally when correlation length was smaller than 12 points the spacing parameters and slope of simulated profiles were similar to those of measured profiles. The maximum differences were smaller than 10%. However sometimes the spacing parameters (like PSm) of measured profiles were bigger than those of simulated profiles, even when the shapes of autocorrelation functions were similar. The differences can be caused for example by mechanical filtration by the tip probe. Not big differences between results obtained using various models were noticed. When correlation length corresponds to bigger number of points, the AR(6) model seemed to be better than AR(3) with regard to spatial properties similarity. Generally tendency was found that the model which allowed to obtain the better shape of autocorrelation function was also the best with regard to spacing parameters similarity. Figure 1 presents the measured profile with its autocorrelation function and the modeled profile with autocorrelation function.

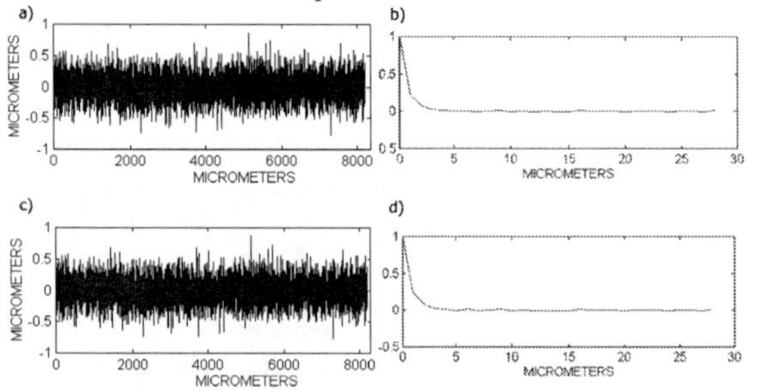

Fig. 1. a) Original profile (Pq = 0,198 μm, PSm = 4,751 μm, PΔq = 3,025), b) its autocorrelation function, c) modeled profile AR(6) (Pq = 0,197 μm, PSm = 4,633 μm, PΔq = 3,022), d) its autocorrelation function

5. CONCLUSIONS

The AR models can be used to simulate profiles of normal ordinate distribution. The autocorrelation function can be well simulated for the number of points bigger than order of AR model.
The AR(3) model was the best for modelling the initial points of the autocorrelation function. AR(6) model is better for modelling the further points. In order to precise model the shape of the autocorrelation function the bigger number of order required the bigger number of profile points. When the autocorrelation function was modelled precisely, the parameters of the modelled profile were closed to original profile. It can be the result of mechanical filtration of the real profile. When the number of autocorrelation function corresponding to correlation length is bigger than the order of the model, the spatial parameters of modelled profile is smaller than those of measured profile.

6. REFERENCES

Bakolas, V. (2003). Numerical generation of arbitrarily oriented non-Gaussian three-dimensional rough surfaces. *Wear,* Vol. 254, pp. 546-554, ISSN: 0043-1648

DeVries, W. R. & Colwell L. V. (1979). Autoregressive time series models for surface profile characterization. *CIRP Annals*, No 28/1, pp. 437-440, ISSN: 1726-0604

Gu, X. & Huand Y. (1990). The modelling and simulation of rough surface. *Wear,* Vol. 137, pp. 275-285, ISSN: 0043-1648

Hong, M. S. & Ehmann K. F. (1995). Three-dimensional surface characterization by two-dimensional autoregressive models. *Journal of Tribology,* Vol. 117, pp. 385-393

Hu, Y. Z. & Tonder K. (1992). Simulation of 3-D random surface by 2-D digital filter and Fourier analysis. *International Journal of Machine Tools and Manufacture,* Vol. 32, pp. 82-90, ISSN: 0890-6955

Newland D. E. (1984). *An introduction to random vibration and spectral analysis*, Longman, New York

Patir, N. (1977). A numerical procedure for random generation of rough surfaces. *Wear,* Vol. 45, pp. 263-277, ISSN: 0043-1648

Staufert, G. (1979). Characterization of random roughness profiles - a comparison of AR model technique and profile description by means of commonly used parameters. *CIRP Annals*, No 28/1, pp. 431-435, ISSN: 1726-0604

Uichidate, M.; Shimizu T.; Iwabuchi A. & Yanagi K. (2004). Generation of reference data of 3-D surface texture using the non-casual 2-D AR model, *Proceedings of 9th International Conference on Metrology and Properties of Engineering Surfaces*, pp. 67-76, Sweden, Halmstad

Watson, W.; King T. G.; Spedding T. A. & Stout K. J. (1979). The machined surface – time series modelling. *Wear,* Vol. 57, pp. 195-205, ISSN: 0043-1648

Watson, W. & Spedding T. A. (1982). The time series modelling of non-gaussian engineering processes. *Wear,* Vol. 83, pp. 215-231, ISSN: 0043-1648

Whitehouse, D. J. (1983). The generation of two dimensional random surfaces having a specified function. *CIRP Annals*, No 32/1, pp. 495-498, ISSN: 1726-0604

Wu, J. J. (2004). Simulation of non-Gaussian surfaces with FFT. *Tribology International,* Vol. 37, pp. 339-346, ISSN: 0301-679X

Wu, J. J. (2000). Simulation of rough surfaces with FFT. *Tribology International,* Vol. 33, pp. 47-58, ISSN: 0301-679X

Annals of DAAAM for 2006 & Proceedings of the 17th International DAAAM Symposium, ISSN 1726-9679
B. Katalinic, Published by DAAAM International, Vienna, Austria 2006
150th Birthday of Nikola Tesla (1856-1943) Inventor of Mobile Robot Teleoperation

EVALUATION OF IN PROCESS STORAGE AND WAREHOUSE CAPACITIES IN AN MDF PRODUCTION PLANT

PENKER, A.; BARBU, M.C. & GRONALT, M.

Abstract: *Material flow analysis by means of discrete event simulation proved to be a useful tool for decision support by several studies. This paper presents a bottleneck analysis for an Austrian MDF production plant. The developed model was linked to actual production data to generate results as a useful input for the real system.*

Key words: *medium density fibreboard (MDF), material flow analysis, simulation of automated processes, discrete event simulation, warehouse and in-process storage capacity.*

1. INTRODUCTION

The European Wood based panel industry uses production lines with a serious degree of automation (Gronalt et al., 2005). However especially in MDF-production the high variety of products and the volume of the boards may cause logistical problems in handling and storage of finished goods.

Simulation models are widely used to describe the technical performance of the material (Wu and Xiong, 2001). Material flow analysis as a tool for decision support in logistical issues was used before in other industries like steel manufacturing (Schmid, 2003). This paper presents a bottleneck analysis for an Austrian MDF production plant using discrete event simulation. A simulation model for the discrete part of MDF board production was developed and linked with actual production data. Different scenarios were analysed to acquire information for tactical and strategic planning.

2. METHODOLOGY AND DATABASIS

2.1 Application Area

The material flow considered starts with the hot pressing of the boards in a continuous press, proceeds with cooling, conditioning, sanding, eventually cutting to size, and ends up with the storage of MDF and loading of batched customer orders on trucks and railway carriages.

2.2 Problem Description - Scenarios

Two main issues were analysed in detail: The first one deals with the alteration of the finished products warehouse and its effect on the utilization of the warehouse and the transporters used for batching loads and loading waiting trucks. This rebuild includes both an increase of storage area and the construction of an extended railway track into the warehouse in order to speed up the loading of customer orders by reducing the average loading distance. In connection with that, new warehouse allocation principles were evaluated and proposed.

The second issue was to work out the effects of changes in the production program on in-process and finished products inventory and on transporters and process utilization as well. More precisely the effect of a shift in the production program to a higher percentage of thin boards on one hand and to an increased proportion of boards passing the cut-to-size saw on the other hand was analysed. The two different warehouse situations and the three production programs resulted in six distinctive scenarios to analyse. The scenario "existing warehouse situation and actual production program" was used as a reference scenario.

2.3 Research Method and factory data analysis

For the simulation the software *Arena 8.01* was used. It allows a very flexible design of the model using a variety of modules. Modelling in *Arena* is based on discrete event simulation and stochastic components (Banks et al., 2000).

As input for the numerical experiments factory data covering one year was used. Among others following data was taken into account: material flow, process intersections, product specifications, process parameters depending on the product and season, characteristics of storage, order picking and shipping. Events of failures like repairs, accidental stops and maintenance with the frequencies and durations according to the records for each process have also been implemented.

By means of regression analysis and statistical distributions it was possible to describe the processes of production, transport and storage with sufficient accuracy. As an example Figure 1 shows the calculation of the continuous press throughput depending on board type, thickness and size.

2.4 The Simulation Model: Logic and Animation

The main function of the in-process storage is the conditioning of the MDF boards to improve their properties. In this area two automated transporters (AGV's) are used. In the simulation model the exact distances in the transport network, the speed of the units and their loading and off-loading times were considered. In the finished product warehouse five stackers of varying capacity are used for transportation. Two were used for in-feed of MDF boards and three for batching customer orders and loading on trucks and railway carriages. Additional transport in the system occurs in form of conveyors. Breakdowns and maintenance for all the transporters were taken into account to capture the system with sufficiency.

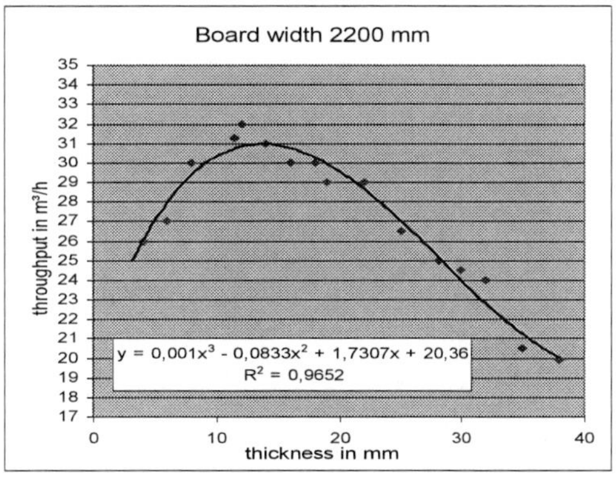

Fig. 1. Calculation of hot-press throughput in the model

The main challenge in the modelling phase was the creation of the logic for the two inventories. The time for conditioning of the MDF boards in the in-process inventory is dependent on their type and thickness. A batch of standard 18 mm boards e.g. has to mature for at least 48 hours to proceed to the sander. In addition the sequence of jobs at the sander has to be such that the number of sanding belt replacements is minimized.
The logic of the finished products warehouse is of high complexity. First the company deals with more than 1200 products with varying dimensions and qualities at this stage of the plant and the feed-in of the product has to be in a way that a minimum of storage volume is occupied.
Up to this point the plant works 24 hours every day except for defined periods for maintenance which were considered in the model. Batching of orders and loading on trucks and railway carriages is performed only on workdays in two shifts.
Some logic was constructed to capture this discontinuity, the process of batching orders for customers and the loading of goods. Seasonal influences on production, sales and weekday dependent fluctuations of shipping were also integrated.
The model was then animated, proving a useful tool for the evaluation of the logic, the communication of the model and understanding of the system. A section of the animated simulation model is shown in Figure 2.
As a next step a warm-up run was carried out to "fill up" the inventory and verify the model. In the warm up information like throughput, utilisation and failure time of processes and transporters, were gained. The comparison of these findings to real production data from the same time span showed good alignment. According to Carson (2005) this is an indication of the ability of the model to picture the real system accurately. The simulated time span for the scenarios described before was three month in each case. As initial inventory the data generated in the warm-up run was used.

3. RESULTS OF THE SIMULATION EXPERIMENTS

As expected the planned extension of the warehouse showed no influence on the utilisation of continuous press, sander and cut-to-size saw. There was also no significant effect on the adjacent transport systems and the in-process inventory. But the average occupancy rate of the finished products warehouse was found to be reduced by 10 % at common production and by 7 and 5 % respectively at a production program with increased amounts of thin boards or higher portions of sawn boards.
Alteration of the production program showed minor influence on the performance of the continuous press. The utilisation of the sander increased by 5 % at a raised amount of thin boards. A higher proportion of goods passing the saw did not affect the sander but the saw itself and the adjacent transporters.

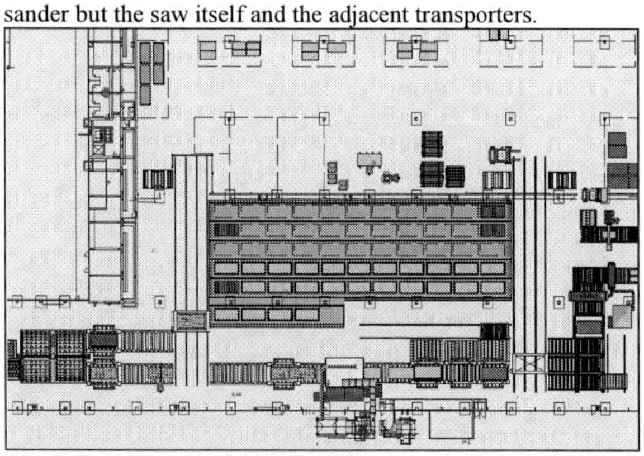

Fig. 2. Partly image of the animation

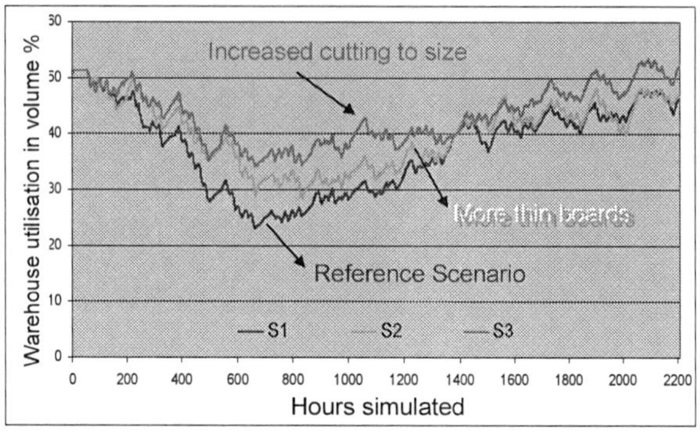

Fig. 3. Behaviour of warehouse utilisation in different scenarios

The occupancy rate of the finished goods warehouse was found to be increased especially in the scenarios with higher amounts of boards passing the saw but also to some extent in those with thinner boards (Figure 3). A very interesting result of the simulation was an increase of waiting time and loading duration for trucks in all scenarios compared to the reference scenario. It is comprehensible that higher amounts of thin and small boards lead to longer waiting and loading times. The negative influence of the extended warehouse is possibly described by the reduced number of loading stations and by the fact that in the new warehouse the loading space for railway carriages is positioned behind the truck loading stations. Due to this a carriage can only be loaded when no truck blocks the space. The shorter distance of the railway carriage loading space to the stock does not take effect because of this blockade.

4. CONCLUSION

By means of discrete event simulation it was found out that the extension of the warehouse has a positive effect on the occupancy rate but also that the laying of the railway track into the warehouse leads to increased waiting and loading times.
In addition the simulation provided useful information about the capability of the production, transport and storage system and its performance at altered production programs and maximum load. The Case Study showed that especially automated processes are an excellent scope for simulation experiments. Consequently it is an applicable tool for the analysis and assessment of advanced and complex manufacturing, transport and storage systems in the wood based boards industry.

5. REFERENCES

Banks, J.; Carson, J.S. & Nelson, B.L.(2000) *Discrete Event System Simulation*, Prentice-Hall, ISBN: 0131446797 NJ

Carson, J.S. (2005) Introduction to Modelling and Simulation; In: *Proceedings of the 2005 Winter Simulation Conference*, p.16 - 23; ISBN: 0-7803-9519-0; Omnipress Madison

Gronalt, M.; Häuslmayer, H. & Teischinger, A. (2005) Mass Customization in the Wood-Working Industry - Concepts for customized wooden floorings; In: *Proceedings of International Mass Customization Meeting 2005*, p. 163-174; ISBN 3-936771-73-1. GITO-Verlag Berlin

Schmid V. (2003) *Materialflussanalyse bei der logistischen Beurteilung einer Ersatzinvestition,* Diplomarbeit Universität Wien

Wu Q. Xiong M. (2001). A Simulation of Moisture Diffusion Process in Furniture - Grade Medium Density Fibreboard, *Available from:* http://www.rnr.lsu.edu/lfpdc/publication /present/woodscience/pdf/MDFmoisturepaperjapannew.pdf *Accessed:* 2006-07-30

Annals of DAAAM for 2006 & Proceedings of the 17th International DAAAM Symposium, ISSN 1726-9679
Editor B. Katalinic, Published by DAAAM International, Vienna, Austria 2006
150th Birthday of Nikola Tesla (1856-1943) Inventor of Mobile Robot Teleoperation

THE IMPACT OF INTRODUCING E-LEARNING SYSTEM IN THE TEACHING PROCESS AT THE FACULTY OF TRAFFIC AND TRANSPORT SCIENCES

PERAKOVIC, D.; KAVRAN, Z. & REMENAR, V.

Abstract: *The development of technology has brought to additional improvement in the quality of education by using the e-learning technology. Following the advanced trends of improving the teaching processes the Faculty of Traffic and Transport Sciences started several years ago the development of own e-Learning system modules. The paper presents the phases of the system realization. Special attention has been paid to the analysis of time-independent learning by the students and the work of the teaching staff members using the system. The obtained results indicate an increase in the number of users and their satisfaction in using the system thus saving time and allowing easier and more efficient access to information.*

Keywords: *Internet, web, study, e-Learning*

1. INTRODUCTION

Continuous development of information and communication technologies represents the necessary assumption for the realization of project called *e-Learning*, along with the reduction in prices, both of the necessary equipment and of the price for accessing the Internet. Good will, knowledge and volunteer work of a group of students under the mentorship of the head of the pilot project have brought about the start of the project regarding the development of the e-Learning system at the Faculty of Traffic and Transport Sciences through several phases. The entire system consists of several modules that have been designed and implemented into the study system of the Faculty of Traffic and Transport Sciences over the last three years. Generally, the own developed systems consist of the following modules: system for content management (*Content Management System – CMS),* system for managing processes and documents (*Document Management System – DMS*), and system for relationships with students (*Customer Relationship Management – CRM*).

2. DEVELOPMENT AND IMPLEMENTATION PHASES

The first and basic phase of introducing the e-Learning system at the Faculty of Traffic and Transport Sciences was the development and implementation of the system of control and authorization (*Sustav autorizacije i nadzora* - SAN) for the control and authorization of the users of PCs in computer classrooms of the faculty. This created the foundation for further development and implementation of the e-Learning system.
The second phase of development and implementation of the e-Learning system saw the design and development of the web system called *e-Student* which allows students to send either from their own or a public PC seminar papers, seminar assignments and to solve entire ranges of pre-tests called *e-Test*, *e-Blic* etc. During this phase a control system of the e-Student system was created and implemented, called *DMSFpz*, which allows the teaching staff to quickly and simply check and mark the topics for seminar papers, seminar assignments, various tests and to organize, control and carry out practical drills in the faculty computer classrooms. Moreover, this facilitates the publication of the teaching materials, thus making these more available to students.
The current phase of the development and expansion of the functionality of the e-Learning system is directed to the increase of the care for the users i.e. students, such as the development of the CRM system based on the short textual message services (*Short Message Service*, abbreviated *SMS*) and applications for mobile devices. The CRM system is based on the *SMSCentar* application which is used to send and receive SMS messages by means of mobile phones, *smsCRM* application which is used for handling students based on the SMS messages and *FPZmobile* application developed for mobile terminal devices.
All the systems are based on communication by means of the SQL database, thus having achieved modularity, i.e. efficient upgrade and modification of the system, which makes it possible to adapt the system for operating at other institutions.

3. E-LEARNING SYSTEM INFORMATION AND COMMUNICATION INFRASTRUCTURE

For the design, development and implementation of the e-Learning system the following information technologies have been used:

- operation systems: Microsoft Windows 2000, 2003 and XP,
- database: Microsoft SQL Server 2000.

For the development of *desktop* applications, the following software tools have been used:

- Microsoft Visual Basic, version 6,
- Borland Delphi, version 7.

For the development of *web* applications the following technologies and applications have been used:

- IIS (Internet Information Services), version 6,
- ASP (Active Server Pages),
- VBScript and JavaScript,
- Macromedia Dreamweaver, version 8,
- Adobe Photoshop, version 8.

All *desktop* applications have been developed for working on the Microsoft Windows platform and therefore IBM PC compatible computers are used to work with them.
The communication infrastructure is based on the CARNet access network and on the intranet (local) network. The Faculty of Traffic and Transport Sciences has its own official premises and lecture-rooms at several locations throughout the city of Zagreb which are interconnected by the CARNet network or by the leased T-Com operator cable.

4. EFFECTS OF INTRODUCING E-LEARNING SYSTEM AT THE FACULTY OF TRAFFIC AND TRANSPORT SCIENCES

E-Learning system was experimentally introduced two years ago in the subject of Information Systems in Postal and Telecommunication Traffic, when the system was used by approximately some 120 students and one teacher for practical drills and the work on and receiving of seminar papers. Today, the system is used in 29 subjects, by about 3400 students who have accessed the e-Learning system about 57000 times. The system is used by a total of 127 employees of the Faculty of Traffic and Transport Sciences, whereas it has been administrated by 25 teachers who have accessed the system in a total of approximately 5500 times. The analysis of the system operation has considered the data for the academic years 2004/2005, in the time period from 1 October 2004 to 1 October 2005 and 2005/2006 in the period from 1 October 2005 until 15 April 2006. The statistical data for the academic year 2005/2006 need to be considered only as frame data since these are not complete, i.e. the academic year has not been completed yet. With the analysis of the current trend according to forecasts, a significant increase in the usage rate of the system is expected. However, even these data are sufficient proof of the usage rate and popularity of the e-Learning system at the Faculty of Traffic and Transport Sciences. The introduction of the *Bologna System* in the study at the Faculty of Traffic and Transport Sciences has resulted in additional growth of the number of users and the number of visits to the e-Learning system. While in the academic year 2004/2005 a total of 8843 accesses to the e-Learning system were achieved, only in January 2006 there were 16052 accesses (presented in Graph 1) out of a total of 47002 accesses in the academic year 2005/2006 until 15 April 2006. The analysis of statistical data about the students' papers in the academic year 2004/2005 shows that there was a total of 438 topics submitted for seminar papers, and a total of 378 seminar papers were sent, whereas the teaching staff members published 43 files of teaching materials. The academic year 2005/2006 marks 2200 submitted topics for seminar papers, 1210 sent seminar papers and 753 solved seminar assignments. The number of published *OnLine* pre-tests (e-Test) amounted to 9 whereas a total of 3860 questions were answered. The teaching staff members have published a total of 596 files of teaching materials. Since the system has been designed so as to "break through" the time and space accessibility barriers, thus allowing the students and professors to work even outside the "working hours", special analysis was done on the accesses to the system during the non-working days and hours. Graph 2 shows the number of accesses of students to the e-Learning system of the Faculty of Traffic and Transport Sciences per days in a week. It may be observed that the system was accessed many times on Saturday and Sunday, or more concretely, out of 47000 visits, as many as 7647 visits, which is 16 percent, were realized on Saturday and Sunday. Graph 3 shows the number of visits per hours in a day, and it may be observed that during the faculty non-working hours, i.e. from 8 p.m. to 8 a.m. there was a total of 11008 visits, i.e. 23 percent of the total number of visits to the e-Learning system. It should be noted that there were as many as 1834 visits in the period from midnight to 8 a.m.

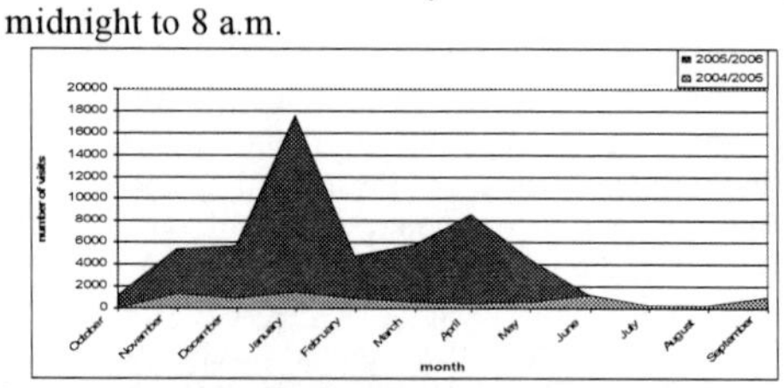

Graph 1. Access rate to e-Learning server per months during the observed academic year

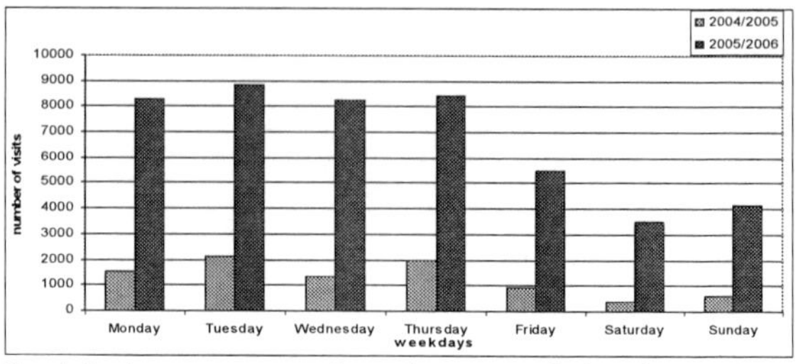

Graph 2. Access rate to e-Learning server during the day in the week of the observed academic year

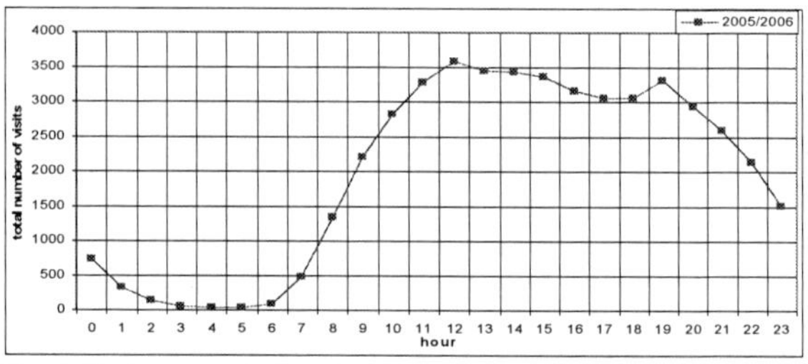

Graph 3. Access rate to e-Learning server during the day in the observed academic year

5. CONCLUSION

The application rate of the own developed e-Learning system implemented at the Faculty of Traffic and Transport Sciences is constantly growing over time. Every month new syllabuses are added thus increasing also the number of accesses to the system, number of students who use the system and the total volume of data that are transferred on the faculty network. The volume of teaching materials is also increasing. The growth of the system has resulted in relocation and replacement of the server. The relocation and the purchase of a new, *more powerful*, server have solved the drawbacks of the limited data transfer speeds.
The presented analysis of data about the operation of the e-Learning system shows that the system has justified its introduction, improved the quality of studying, especially of the teaching process (through easier access to the teaching materials, on-line knowledge testing, and submission of seminar papers as well as numerous other functions) according to the operation model 24/7/365, i.e. non-stop. Continuous monitoring of the work, carrying out surveys among users and analyses of their requirements, will provide the necessary further improvements, upgrades and modifications of the current system.

6. LITERATURE

Baranovic, M., et al. (2003). *Information System of Higher Education Institutionsn*, MZT, 953-6755-08-4, Zagreb.

Fallon, C., Brown, S. (2002). *E-Learning Standards: A Guide to Purchasing, Developing and Deploying Standards-Conformant E-Learning*, St. Lucie Press, 1-57444-345-3, Boca Raton, USA.

Meic, K., Remenar, V., Sasek, Z. (2005). *E-Learning System Implementation Possibilities at Faculty of Transport and Traffic Sciences*, FPZ, Zagreb.

Perakovic, D. et al. (2003): *Project of SAN Implementation*, FPZ, Zagreb.

Perakovic, D., Saric, S., Kavran, Z. (2004). Web-Based Applications for Improving the Quality of Studies, Conference, *Proceedings of 15th Conference on Information and Intelligent Systems*, pp. 27-33, 953-6071-23-1, Varaždin, September 2004, FOI, Varaždin.

Annals of DAAAM for 2006 & Proceedings of the 17th International DAAAM Symposium, ISSN 1726-9679
Editor B. Katalinic, Published by DAAAM International, Vienna, Austria 2006
150th Birthday of Nikola Tesla (1856-1943) Inventor of Mobile Robot Teleoperation

DECENTRALIZED REDUCED-ORDER SWITCHING CONTROL WITH PRE-IDENTIFICATION: THEORETICAL PRELIMINARIES

PERUTKA, K.

Abstract: *The paper deals with decentralized control. Firstly an introduction part is given. After this part there is given the theoretical background, which includes the description of multivariable controlled systems with disturbance and one way of decentralized control, namely the decentralized control of "P" structure of two variable system. This is followed by the main part of the paper, the formulation of the theoretical preliminaries of new modification of decentralized control including the order reduction, switching algorithm and pre-identification.*

Key words: *Decentralized control, identification algorithms, switching algorithms.*

1. INTRODUCTION

Multivariable controlled circuits (MIMO) are the controlled systems where several quantities are necessary to control simultaneously nevertheless there are not independent one-dimensional circuits but interconnected circuits, which is given in Balate (Balate, 2003). According to this book, the change of one control quantity influences using a set of interactions among the circuits the remaining control quantities, so called interactions between inputs and outputs. The relations between inputs and outputs are described by matrixes of thransfer funtions.

If the decentralized approach is used, the task of control is split into subtasks and their corresponding simpler controllers. They give together the whole result of the original task of control. The main two advantages of decentralization are: the more complex task is divided into the simpler ones and the resulting controller is more flexible. Sequential control using relay, see Palmor et al. (Palmor *et al.*, 1995), might be a good example.

The area of usage can be enlarged when using continuous-time identification. Nice toolbox for MATLAB that includes a lot of methods for continuous-time identification was done by Mensler (Mensler, 1999).

2. THEORETICAL BACKGROUND

Multivariable deterministic controlled system can be schematically shown as in fig. 1 (Balate, 2003)

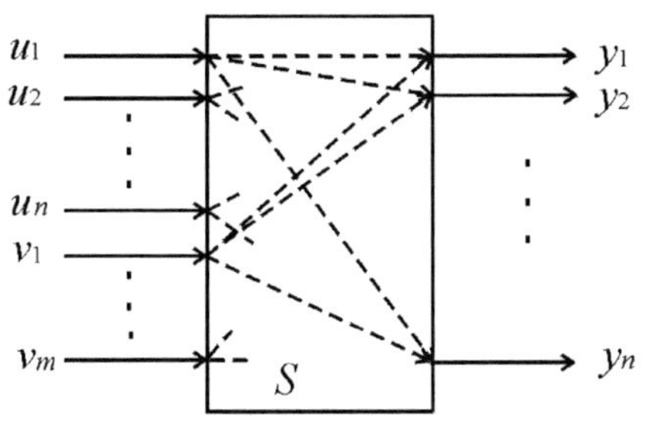

Fig. 1 Multivariable controlled system *S*

Input quantities are classified into action quantities u_i (i =1, 2, ..., n) and disturbance quantities v_j (j =1, 2, ..., m). Action quantities are used to control the output, controlled quantities y_i (i =1, 2, ..., n). The disturbance quantities cannot be affected but measured only. Each of input quantities influences all the output ones.

System S is generally nonlinear, the parameters are time dependent. The complete description of system with distributed parameters goes to the system of partial differential equations. If the system is described by the set of ordinary differential equations, the system is so called the system with concentrated parameters.

3. PROBLEM SOLUTION - PROPOSED ALGORITHM

Decentralized controllability *Suppose the existence of stable minimal-phase linear time-invariant continuous-time multivariable n x n system* $\mathbf{S}(t)$ *and its Laplace transformation* $\mathbf{S}(s):\mathbf{S}(t)\rightarrow\mathbf{S}(s)$, *denoted as the transformed system, in the form*

$$\mathbf{S}(s)=\begin{pmatrix} S_{11}(s) & S_{12}(s) & \cdots & S_{1N}(s) \\ S_{21}(s) & S_{22}(s) & \cdots & S_{2N}(s) \\ \vdots & \vdots & \ddots & \vdots \\ S_{N1}(s) & S_{N2}(s) & \cdots & S_{NN}(s) \end{pmatrix} \quad (1)$$

where $S_{ij}(s)$, $i=1,2,\ldots N$, $j=1,2,\ldots,N$ *is the Laplace transformation of the ijth subsystem* $S_{ij}(t)$ *of the transformed system* $\mathbf{S}(s)$. *The transformed system* $\mathbf{S}(s)$ *is decentralized controllable if it is main diagonal dominant.*

Model of system *Assuming Equation 1 there exists model of the system* $\mathbf{S}(t)$ *for the purpose of decentralized control, denoted as* $\mathbf{M}(t)$, *and its Laplace transformation* $\mathbf{M}(s)$, *which is stable minimal-phase linear time-invariant continuous-time multivariable main diagonal, in the form*

$$\mathbf{M}(s)=\begin{pmatrix} M_1(s) & 0 & \cdots & 0 \\ 0 & M_2(s) & \cdots & 0 \\ \vdots & \vdots & \ddots & \vdots \\ 0 & 0 & \cdots & M_N(s) \end{pmatrix} \quad (2)$$

where $M_i(s)$, $i=1,2,\ldots N$, *is the Laplace transformation of the ith sub-model* $M_i(t)$ *of the model* $\mathbf{M}(t)$ *of the transformed system* $\mathbf{S}(s)$. *Moreover, let as assume that the matrix of reference signals* $\mathbf{R}(s)$ *is always in the form*

$$\mathbf{R}(s)=\begin{pmatrix} R_1 & 0 & \cdots & 0 \\ 0 & R_2 & \cdots & 0 \\ \vdots & \vdots & \ddots & \vdots \\ 0 & 0 & \cdots & R_N \end{pmatrix} \quad (3)$$

where $R_i(s)$, $i=1,2,\dots N$, *is ith reference signal of the matrix of reference signals* $\mathbf{R}(s)$ *and has the form*

$$R_i(s)=\frac{h_i}{s} \quad (4)$$

where $h_i \in R$, $i=1,2,\dots N$, *is ith piecewise constant function, i.e., reference signal that is only a combination of step changes of the signal of its different values*

$$h_i=\begin{cases} j_{i1} & \text{for } t\in\langle 0,t_{i1}) \\ j_{i2} & \text{for } t\in\langle t_{i1},t_{i2}) \\ \vdots & \\ j_{ip} & \text{for } t\in\langle t_{ip-1},t_{ip}\rangle \end{cases} \quad (5)$$

$$0<t_{i1}<t_{i2}<\dots<t_{ip-1}<t_{ip},$$

where $j_{il}\in R$, $i=1,2,\dots N$, $l=1,2,\dots p$, *is lth constant function of the ith piecewise constant function* h_i, *t is time and* $t_{il}\in R^+$, $i=1,2,\dots N$, $l=1,2,\dots p$, *is lth time instant of the ith piecewise constant function* h_i. *Than* $M_i(s)$, $i=1,2,\dots N$, *the Laplace transformation of the ith sub-model* $M_i(t)$, *expands as follows*

$$M_i(s)=\begin{cases} M_{i1}(s) & \text{for } t\in\langle 0,t_1) \\ M_{i2}(s) & \text{for } t\in\langle t_1,t_2) \\ \vdots & \\ M_{ir}(s) & \text{for } t\in\langle t_{r-1},t_r\rangle \end{cases}$$
$$0<t_1<t_2<\dots<t_{r-1}<t_r \quad (6)$$

where $M_{ir}\in R$, $i=1,2,\dots N$, $k=1,2,\dots r$, *is kth element of Laplace transformation of the ith sub-model* $M_i(t)$, *t is time and* $t_k\in R^+$, $k=1,2,\dots r$, *is kth time instant of Laplace transformation of the ith sub-model* $M_i(t)$, *which is same for all Laplace tranformed sub-models* $M_i(s)$, $i=1,2,\dots N$.

The some of industrial processes can be described in the Laplace form using the second order without time-delay, we only uses the second-order of the Laplace transformation of ith sub-model $M_i(t)$. *If the model of the real subsystem is of higher order, we can call this control as a reduced-order control.* In our case, we only assume the I/O system description, i.e., no state-space model is used.

Decentralized continuous-time controller *Assuming the previous equations, a decentralized feedback continuous-time controller* $\mathbf{C}(s)$ *of the transformed system* $\mathbf{S}(s)$ *exists in the form*

$$\mathbf{C}(s)=\begin{pmatrix} C_1(s) & 0 & \cdots & 0 \\ 0 & C_2(s) & \cdots & 0 \\ \vdots & \vdots & \ddots & \vdots \\ 0 & 0 & \cdots & C_N(s) \end{pmatrix} \quad (7)$$

where $C_i(s)$, $i=1,2,\dots N$, *is the ith continuous-time sub-controller of the decentralized feedback continuous-time controller* $\mathbf{C}(s)$.

Supervisory switching control. *Combining the previous equations we obtain the fact that each part of control of such a system, where all reference signals have the constant value, has its own sub-models and therefore its own sub-controllers. Therefore the sub-controllers are switched during the control using the supervisor, which can be viewed as a supervisory switching control.*

Overall stability. *If all the sub-controllers are strongly stable and equation 1 holds, the overall system is stable. The strong stability of all sub-controllers is guaranteed from their definition. Moreover, it is possible to test the stability using some of the well-known methods such as the Hurwith criterion of stability.*

Continuous-time multi-model pre-identification *Assuming the previous equations, the system* $\mathbf{S}(s)$ *is viewed as "a black box" model and it is identified by direct and/or indirect continuous-time algorithms. Than continuous-time identification of this model is realized as follows:*

1. *The values of matrix of reference signals* $\mathbf{R}(s)$ *are send to the input of the system* $\mathbf{S}(t)$. *They are same and for the same time as those one which are going to be used during the control.*
2. *Each time interval of history of control of the system* $\mathbf{S}(t)$, *where all reference signals have the constant value, is identified separately, in the so called identification elements.*
3. *Every identification element is identified several times, every time with different identification algorithm, and the obtained model is verified with the measured data. The obtained model which gives the best agreement with the measured data is used for control. This can be viewed as a multi-model identification.*

The system $\mathbf{S}(s)$ is supposing to be completely identified before the control task starts, i.e., there is no identification during the control therefore it is denoted as a pre-identification.

Minimal overshooting of output signal caused by interactions Assuming the previous paragraphs we can set the sub-controllers using the suitable method of control such that the system $\mathbf{S}(t)$ has a minimal overshooting of output signal caused by interactions during control or at least smaller overshooting than during the "classical decentralized control" when the values of parameters of the sub-models are the same for all the control time.

4. RESULTS

To formulate the preliminaries of new algorithm of decentralized control was the main aim of this contribution. They will be used to verify them by simulation and control in real-time using MATLAB.

5. REFERENCES

Balate, J. (2003). *Automatic Control.* BEN – Technical literature, Prague, ISBN 80-7300-020-2. (in Czech)

Mensler, M. (1999). *Analyse et étude comparative de méthodes d'identification des systèmes à représentation continue : Dévelopment d'une boîte à outils logicielle.* PhD. thesis, CRAN, Université Henri Poincaré, Nancy, France. (in French)

Palmor, Z.J., Halevi, Y., Krasney, N. (1995). Automatic Tuning of Decentralized PID controllers for TITO processes. *Automatica, 31,* pp. 1001-1010.

Annals of DAAAM for 2006 & Proceedings of the 17th International DAAAM Symposium, ISSN 1726-9679
Editor B. Katalinic, Published by DAAAM International, Vienna, Austria 2006
150th Birthday of Nikola Tesla (1856-1943) Inventor of Mobile Robot Teleoperation

COMPUTER AIDED EDUCATION OF PROGRAMMING WITH AID OF COMPLEX GAME IN MATLAB

PERUTKA, K. & HECZKO, M.

Abstract: *The paper deals with the computer aided education of programming with the aid of complex game in MATLAB. This approach has many positive features primarily aimed at the education of programming in MATLAB at our faculty. It attracts student attention in the way students see MATLAB used in a different way not only for automation and in industry. The source code of game is easily readable and includes many areas used not only in basic but also in advanced programming, for instance the usage of its own database which can be easily editable in Notepad of MS Windows without harming the game core. The game is described in detail in the paper together with the sample screenshots. The game was successfully used during the lessons of programming at our faculty.*
Key words: *Computer Aided Education, MATLAB, programming.*

1. INTRODUCTION

By the reason that the game was programmed in MATLAB, the introduction part of the paper is all about this software for those who are not familiar with it.
MATLAB is a complex system that is primarily used by technicians and researchers in industry, mainly in Control Engineering. The software is one of those "happy ones" that survived for many years, it has celebrated 20 years of its usage when including all its versions. This fact proves that the software became the standard in its area of usage.
It has its core and its extensions that are named as toolboxes. Toolboxes solve special area of problems and were mainly programmed by group of researchers covered with the world-known scientist in the special area, System Identification Toolbox or Statistics Toolbox may be two examples of it. The most widely used toolbox is denoted as Simulink. Simulink is nowadays almost an independent system that has also its extensions, also denoted as toolboxes. As its name indicates, Simulink is primarily used for simulation of behaviour of arbitrary system from the chosen area of usage, SimHydraulics and SimPowerSystems might be two illustrative examples. Besides the standard toolboxes that are in the installation of MATLAB and Simulink, there are tens of toolboxes that are usually free of charge for research and they are available on the Internet. This fact proves that the software is popular all around the world. You probably can find example of toolbox that was created at your university. For instance, there was created such one at the university we are coming from (http://www.utb.cz/stctool). This toolbox was created in purpose the ease the teaching of Self-Tuning Controllers in Simulink.
The popularity of MATLAB evoked the wave of publication about programming in it. Nice introduction to this software was written by Sigmon (Sigmon, 1993). Interesting book about programming in MATLAB for electrical and computer engineers with a lot of practical examples was written by Mannasah (Mannasah, 2001). These books are in English because the most of users come from USA. But MATLAB is also used by technicians in other countries. In the country, we are coming from, the book written by Perutka (Perutka, 2005) might be a sample book including basic description of the program, basics of programming, sample source code, original examples, reference of commands etc.
The paper is organised as follows. After the problem formulation, where the reasons for such a game are formulated, the problem solution is given. In this part, there is the description of the game together with its screenshot is provided.

2. PROBLEM FORMULATION

Students of arbitrary subject should be motivated as much as possible. If they are motivated, their attention is oriented in the right way, in their study results. Finally, if students see good example, they know from it the software, they are programming in, is the right tool for them. It is generally know that programming a game is one of the most complicated tasks the programmer can realize. Therefore the necessity of some complex game occurred. The solution is now going to be described.

3. PROBLEM SOLUTION

The game was named as "Riskuj" and was programmed in MATLAB 6.5. All dialogs are in Czech language because the game was primarily intended for the usage at our university. The code of the game is fully available for the students. The game itself is a modification of a popular game that has run on the NOVA TV, television channel in the Czech Republic. The initial game settings can be seen in figure 1, the number of players is chosen in pop-up menu. Depending on this fact, the number of edit boxes for setting the names of players appears. Although this dialog seems simple, it has "hidden inside" several important bases of programming in MATLAB for students. It was created by GUIDE, the tool for creating GUIs. There is solved the communication between code generated by GUIDE and code we had to add. There is hidden also the multiple condition meaning practicing this part of programming together with storing the data, names of players, in global variables. All these parts of programming are demonstrated there.

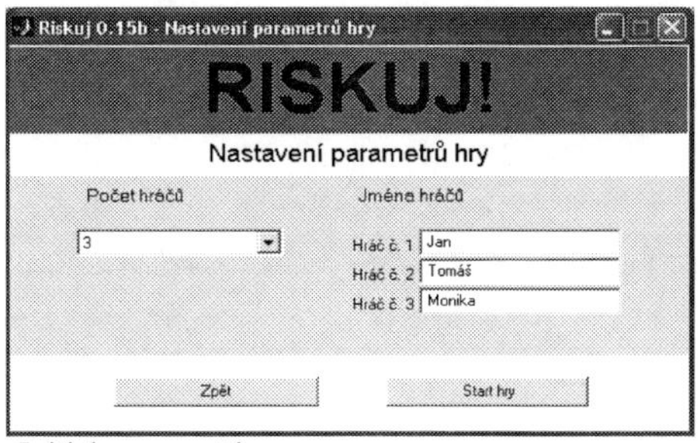

Fig. 1. Initial game settings

The dialog is followed by the dialog of round one of the game, see figure 2 – 4. All these dialogs are divided into four parts. Top left has the column, where is written topic for questions. Rows include questions from the topic hidden under buttons 1000 – 5000.

Under this part of the dialog, there are two bars changing according to the time, grey bar gives the total remaining time of the round one while blue bar give the total remaining time for answering the question by enterer player.

Bottom left part of the dialog shows the questions chosen by the button with or the name of the player who entered the question or the answer with edit box where the answer has to be written. Also right answers are shown there.

The right part of the dialog includes the name of players, their score and the key they have to use if they want to answer the chosen question. The game has three rounds, round two play players without the worst player in the previous round and now they do not choose who is going to answer but they take turns.

Third round plays only one player, the one who has the best score after round two. In this round, there is going to be decided how much money/many points she or he is going to win.

What is practiced here? Almost all what students should know after the course of bases of programming plus some extras, such as locking and unlocking keys from keyboard, multitasking (two timers and state bar). The questions and answers are stored in files that are easily editable outside the program in arbitrary editor. This part demonstrates working with the given rules of new structure of new database. More areas that were practiced will be presented in lecture.

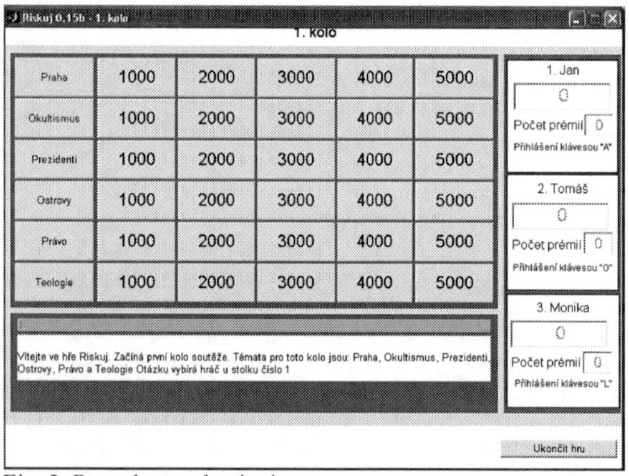

Fig. 2. Round one – beginning

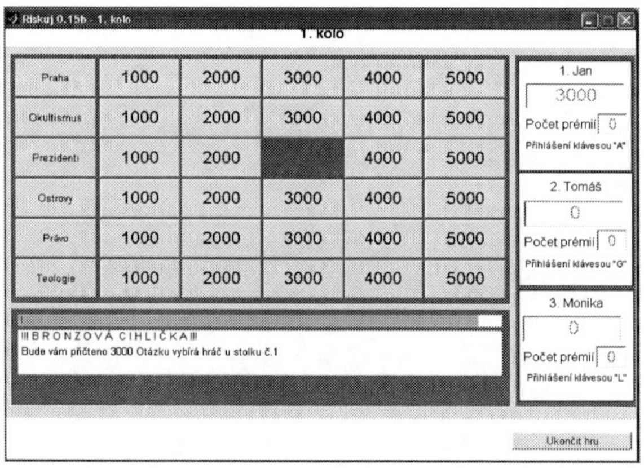

Fig. 3. Round one – bronze brick

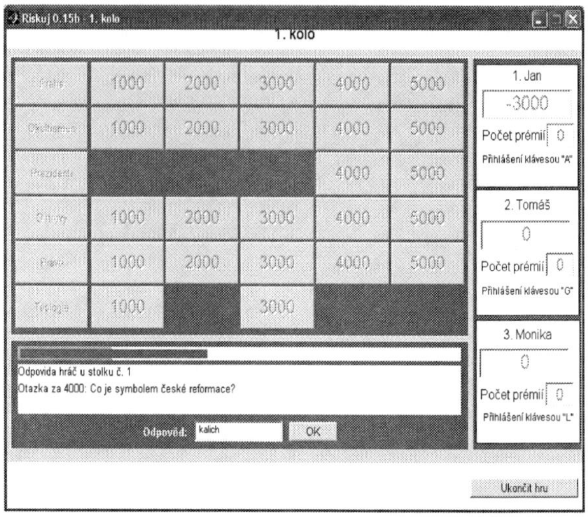

Fig. 4. Round one – asking question

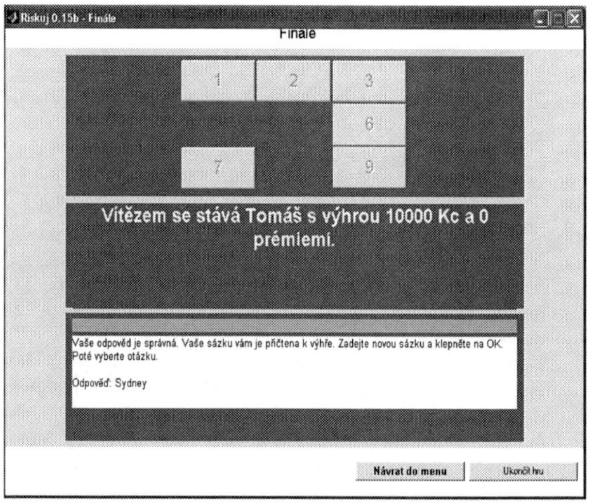

Fig. 5. Round three – end of game

RESULTS

There was outlined the solution of complex game in MATLAB, which was proven useful during the lessons of programming in MATLAB at our faculty. Students have a nice example and a complex program which demonstrates all information they should know after the course of bases of programming. The source code is available to them to, so they can see how it is implemented and they use it in a similar way.

REFERENCES

Manassah, J.T. (2001). *Elementary mathematical and computational tools for electrical and computer engineers using MATLAB*. CRC Press LLC, Boca Raton, Florida, USA.

Perutka, K. (2005). *MATLAB – Bases for students of automation and information technologies*. Tomas Bata University in Zlin, Zlin, Czech Republic, EU

Sigmon, K. (1993). *MATLAB Primer*. 3rd edition, Department of Mathematics, University of Florida, Gainesville, Florida, USA.

Annals of DAAAM for 2006 & Proceedings of the 17th International DAAAM Symposium, ISSN 1726-9679
B. Katalinic, Published by DAAAM International, Vienna, Austria 2006
150th Birthday of Nikola Tesla (1856-1943) Inventor of Mobile Robot Teleoperation

LASER-BASED STRUCTURAL VIBRATION ANALYSIS OF A CAR BODY

PIETRUSEWICZ, K.; OKULIK, T.; HOFFMAN, M. & PARUS, A.

Abstract: *The paper presents results from the non-contact vibration analysis of a passenger car body. The non-contact measurements ensures us that the shape of the structural vibrations does not change during the measurement – for instance by the number of the mounted accelerometers.*
Key words: *laser, vibrometry, Doppler Effect, non-contact measurement.*

1. INTRODUCTION

In the paper there is an analysis of the structural vibrations which was made with the use of the non-contact method. The main concept of the used measurement instrumentation is based on the Doppler Effect.
The paper presents the results from the real-world test of the vibrations of a passenger car body.
As an alternative of a presented here method the accelerometers could be used. But the tests with the use of accelerometers are usually time-consuming procedures. It is obvious that the accelerometers have to be carefully mounted on the measured surface. For high spatial resolution a lot of accelerometers have to be used. Sometimes it is not possible due to limited number of channels of the used Data Acquisition system. But the main drawback of the contact methods is: the change in mass of a car body, introduced by the mounted accelerometers.
So the non-contact methods could be helpful alternative. The contribution of this paper is the use of laser vibrometer for the real-world analysis of the structural vibrations of a car body, excited with the use of a modal hammer.

2. LASER DOPPLER EFFECT VIBROMETRY

Laser vibrometry uses the well-known Doppler Effect to measure vibrations. If light is scattered from the moving (vibrating) object its frequency is changed slightly. Within a used vibrometer a high precision interferometer detects the minute frequency shifts of the back-scattered laser light. To achieve this, the interferometer splits the light into two parts: the so-called reference beam is pointed directly to the foto-detector, while the measurement beam is incident on the test object, where the light is scattered by the moving (vibrating) surface. Depending on the velocity and displacement the back-scattered light is changed in frequency and phase. It is obvious, that the characteristics of the motion (vibrations) are completely contained in the back-scattered light.

3. THE PSV-3D FOR CAR BODY STRUCTURAL DYNAMICS TESTING

Performing a typical measurement test with the use of PSV-3D laser vibrometer involves these simple steps:

- Free positioning of the sensor heads in front of the object to be measured,
- Train the space coordinate system,
- Define the sample points on the object,
- Set the parameters for analog and digital acquisition of measurement values,
- Start scan,
- Evaluate and/or export the data (in the best way: as a tables of values, as a graphic files, or as an animations).

Laser vibrometer consists of two parts (Figure 1):

- Three laser scanning heads (Figure 1a),
- Control and data acquisition module (Figure 1b).

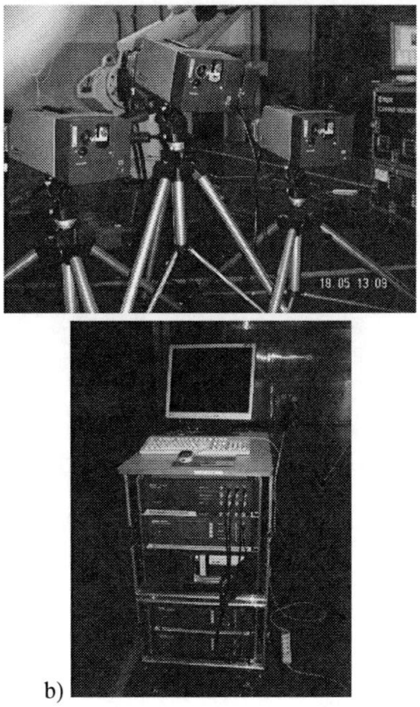

a) b)

Fig. 1. Laser PSV-3D vibrometer used in tests

3.1 Structural dynamics tests setup

A Skoda Fabia Sedan has been used for the test. The test object has been excited with the use of a modal hammer. The car was staying at the hard surface. During this simplified test, we performed measurements on the selected sections of a car:

- Front left door part,
- Rear left fender.

In-between the measurements of the individual segments, the scanning heads have been manually moved to their new positions to scan the next segment.
Overall the 80 scan-points on each segment have been measured (averaging 3 measurements for each point), each one with its x, y, z vibration components, hence a total of 480 degrees of freedom. This is largely sufficient for experimental

modal tests on the selected surfaces and for the frequency range of interest (up to 200Hz).

3.2 Workflow

In order to provide an insight on how a measurement with the PSV-3D is performed, a simplified workflow can be outlined as follows:

- Physically setup the heads on their tripods to cover the desired scan area,
- 2-D alignment to establish the connection between position on the video image and scan angles,
- 3-D alignment – definition of the coordinate system by pointing the sensor head with the distance sensor to 3 known points on the car,
- Measurement grid definition,
- Geometry scan to get the 3D coordinates of the measurement points,
- Focus setting,
- A/D setting of the measurement system,
- Perform measurement – points are automatically scanned,
- Data visualization in the PSV software to check deflection shapes,
- Data export into customer's modal analysis software package.

4. RESULTS

Here are the results obtained during the tests performed:

- Displacement at the different frequencies,
- Spectral analysis of the structural vibrations of a car body.

4.1 Front left door part

There are results from the spectral analysis presented on the Figure 2.

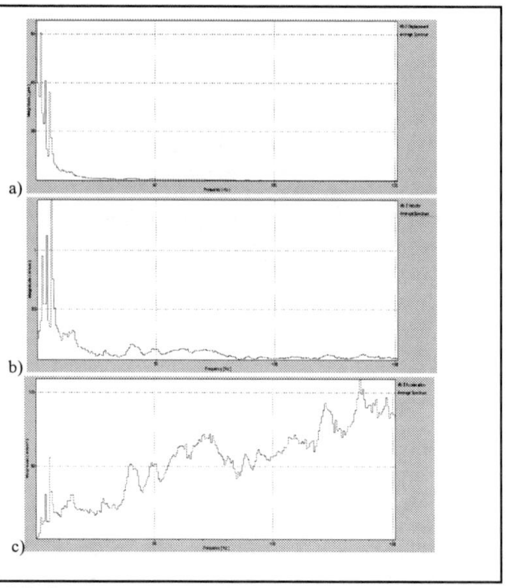

Fig. 2. Displacement (a), velocity (b) and acceleration spectrum (c) in the front left door test case

Figure 3 shows the siplacements in z-axis at difeerent frequencies.

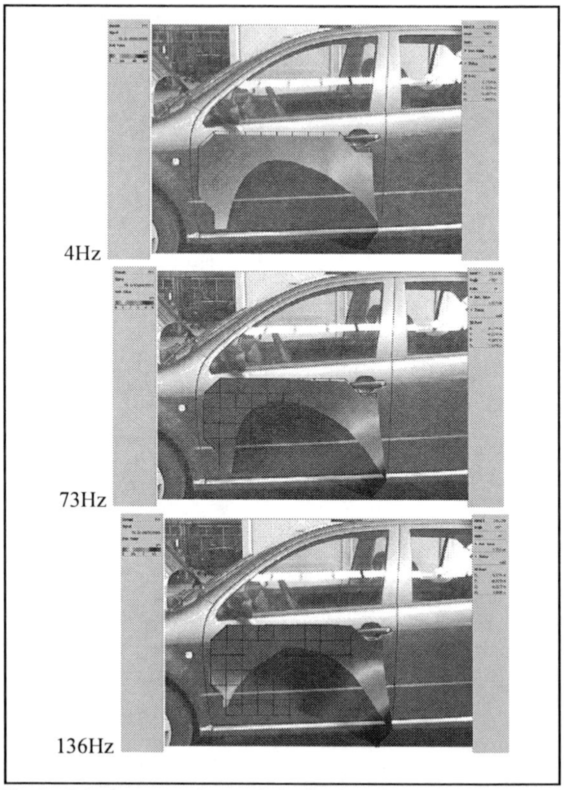

Fig. 3. Displacement in z-axis in the front left door test case

4.2 Experiment time

The required time for the entire test was up to 6 hours. It is composed out of the time for the initial setup and the setup and measurement time for each segment of a car body.

The total time depends on the selected measurement segment. If the surface is well accessible, has 'easy' geometric shape and good backscattering properties, the total setup time will be shorter than for hidden, complicated surfaces with bad backscattering properties.

The time saving as compared to accelerometers is considerable. It gets even more important as soon as the number of simultaneously measurable accelerometers is exceeded, as in the case accelerometers would have to be displaced during the measurement process.

For higher density grids, the time saving highly increases, as for the PSV-3D only measurement time increases, the setup time remains nearly constant.

5. CONCLUSIONS

The laser vibrometry proves to be very well suited for data acquisition for structural dynamics test on car body. The main advantages as compared to conventional accelerometers are: significant decrease of test time, higher point density (test resolution) allows more precise results, ease of setup, no error checking of cabling, and the most important - no alteration of structure (no change of mass and stiffness),

Presented results shows the displacements in z-axis of a front door part as expected. Also the frequency range for the vibrations of a body of car used in tests was evaluated as less than 200Hz.

Annals of DAAAM for 2006 & Proceedings of the 17th International DAAAM Symposium, ISSN 1726-9679
B. Katalinic, Published by DAAAM International, Vienna, Austria 2006
150th Birthday of Nikola Tesla (1856-1943) Inventor of Mobile Robot Teleoperation

SOME ASPECTS OF INFORMATION SYSTEM INTEGRATION/MIGRATION

POGARCIC, I.; PAP, K. & ZILJAK-VUJIC, J.

Abstract: *This research deals with the significance of the information system integration phase and the detection of some critical factors and circumstances that may significantly slow down or stop the process of modification/integration. Special emphasis is put on data migration that had been used in previous systems and their consolidation into a new common base. The results are based on the study of cause and consequence of such activities in the mentioned phase such as, for instance, the need to apply new information technologies, necessary alterations in the information system and its environment, alterations in the business system. The need for strategic planning of this phase is also considered in the range of the information system development strategic planning.*

Key words: *system integration, data migration, information system modification, information system life cycle, information system strategic development*

1. INTRODUCTION

With strategic planning of the information system development or with planning of its modification it is necessary to foresee the possibilities of its integration with other information systems. Here it is assumed that there are risky activities and processes that may be generated in such circumstances. In the strategic plan,for instance, the data collection phase requires careful preparation and planning. This is often a phase with a high risk level and unpredictable influence on the required time period and executors, and therefore on the budget necessary for realization.

Creating the so-called service information in the database(most often relational) that usually covers static data - various masters and similar is common for all parts of the information system. For example, data coverage during the modification process as a generic activity was not a problem in the „physical sense" as much as it was its adaptation to new conditions.

When an information system begins to use concrete application solutions, alterations of such solutions are possible, sometimes even frequent, having a corrective character and also the quality of being permanent, i.e. they last as long as there is a need for the information system. On the other hand, each business system will tend to keep its competitive position in the market and will try to keep its technical importance. Indirectly this means that the purchase of new equipment and new technological solutions will be reflected onto both business and information processes as well.

2. MIGRATION OR INTEGRATION?

The question arises as to what may provoke the information system migration/integration,when and why.Let us first define integration as: „a combination of two or more information systems that allow joint operation and joint information generation". Of course, this definition is reduced to the information systems level and it may be applied to business systems if we exchange the term information for the term meaning business. We are stressing this point because, of course, an information system all by itself is pointless.

During its life cycle a business system may alter its technical and technological settings due to different reasons. Two reasons are the most prominent ones:

- Alterations in business operation
- Alterations in hardware and/or software (Morris, 2006)

Alterations in business operations may be provoked by internal reorganization, like for instance by significant extension of the range of products, its altering and alterations in carrying out business operations. Business operation alterations may also appear as a consequence of integration with other business systems, sale/purchase of a business system (Barton, 2003).

The integration process is started by, such systems that have existed until then in different circumstances. Figure 1 shows two most frequently evidenced possibilities where systems exist that might be candidates for integration. Theoretically there is a border possibility where an information (sub) system exists completely on its own.

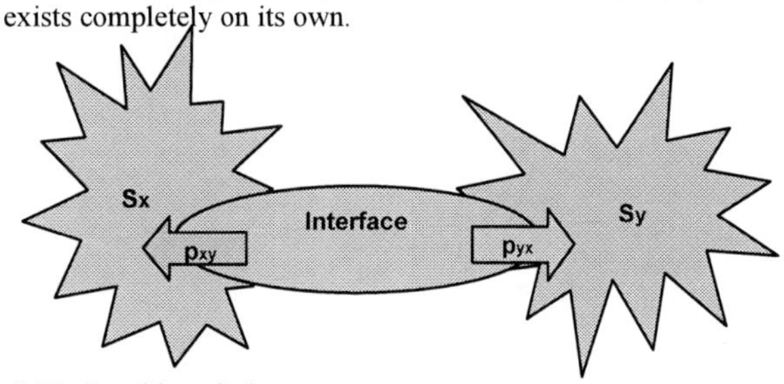

a) Mediated interfacing

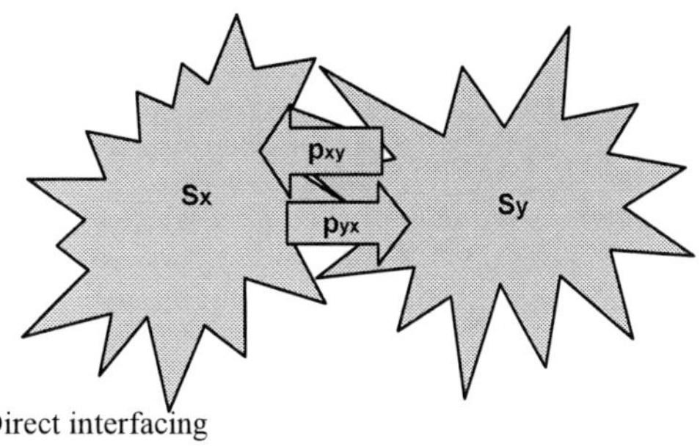

b) Direct interfacing

Fig. 1. A possible relation between two systems

Taking into consideration the given circumstances a business system is faced with planning requirements for all levels beginning from the operation level and continuing up to the strategical level. In integration/migration practice there have been system combinations as shown in Table 1. For the purpose of explanation, let us determine the information system as a set of information processes {pxi} that service processes in a certain system Sx. Of course, the given process set is in accordance with minimalist conditions and the holistic approach supplementary.

System/relationship	Sx	Sy	Integration (result)	Note
Independent and incomparable	{pxi}	{pyi}	{pxi}U{pyi}	Senseless
Independent but supplementary	{pxi}	{pyi}	{pxi}U{pyi}	Business function supplement
Independent with partly the same processes	{pxi}	{pyi}	({pxi}U{pyi})/ ({pxi}∩{pyi})	Business function supplement with reduction of repetition

Table. 1. Integration of two systems

During integration, in times of accelerated ICT advance as well as rapid market requirement alterations, there is the question of the possibility/necessity for strategic planning.

Namely, the emphasis in planning is shifted towards tactical planning and opens the necessity for transactional planning or the planning of each business/use case respectively. Business system status change following the sale and intergration to a larger consortium leads to the situation where both migration and integration took place at the same time. Software alteration was necessary due to the business system integration and the merging of software solutions on the consortium level. Thereby the informatics process modification and data migration together with their adaptation took place in several phases. The final goal is the intergration of the information system into the consortium information system.

3. PROBLEMS AND CHALLANGES

Working for twenty years on information system design and the integration/migration through three significant modifications the same amount of time was needed for Data Take Over phase, and for instance, for software testing. However, the problems were regularly linked with the user's readiness and capabilities to get involved.. When changes do take place, the overall business system segments change as well. The most frequent reason for introducing new applicative solutions or for the modification/upgrading of old applications is the increase of the requirement for delivering new functions necessary for business operation improvement.

An especially risky integration effect is the situation where the informatics system modification causes significant business process modifications, after which the business system functions may be disturbed to such extent that functioning is threatened. In such circumstances it is necessary to carefully plan activities such as:

- Engaging additional lifeware resources such as independent external consultants and experts for various fields
- Checking and understanding systems that are integrated, requiring a whole group of certain specialized skills
- Controlling whether the documentation is complete and authentic; some important documents may be missing
- Finding solutions for some possibly incomplete segments in processes
- Eliminating data redundancy
- Avoiding and eliminating all unprofessional activities (unorganized approach or intuitive approach)
- Avoiding all actions that may provoke delay in the integration realization

4. SOME EXPERIENCE BASED FACTS

The user and especially the management need to be aware of the fact that the existing informatics system is no longer adequate (due to any cause whatsoever), and that a system was in the preparation stage, most often in the testing phase (Chatterjee, 2002). The necessity to transfer data into a qualitatively new environment assumes the possibility to comprise the correct data in the right place and the right time. Impossibility to transfer data objectively, i.e. matching or compressing are necessary on one hand or decomposition and separation on the other hand. There is need for very complex programmer preparation and engaging and for creating custom programs-interfaces and scripts that translate data automatically. High quality preparation is necessary for all this, of course (Morris, 2006).

During the integrated system stabilization process it is necessary to perform the following:

- Perform testing on the new system
- Provide parallel operation during the new system stabilization period,
- Solve the problem of historical data,
- Solve the problem of informing in circumstance of double system data storage,
- Predict the time period for the system interfacing.

5. CONCLUSION

The integration migration from one technological application into another technological application system is not by all means a harmless situation in the life cycle of a business informatics system. As it is caused by different impulses and circumstances, it may have far-reaching consequences for the informatics business system functioning if it is not carefully executed. The most frequent negative effects are breaking of deadlines and increase of expenses, as well as indirect crossing of budget means limits. Recently there have been many business system activities excluded from the business system and used as outsourcing services. For instance: Application Outsourcing, Business Process Outsourcing, Credit Services, Customer Relationship Management, Enterprise Solutions, Hosting Services, IT Infrastructure Outsourcing , Knowledge Management etc. .From the user's point of view it would be ideal if the user's standards could be defined in advance and if the users could be warned as to possible alterations. However, it is evident that on the market as a part of the possible solutions there are quite specific use cases behind general methodologies, making this problem area specific and very sensitive. Therefore, there is a very frequent dilemma in the integration/migration area: to buy or to do it yourself. Hence the decisions on the manner of executing the described integration/migration activities should be brought with great care.

6. REFERENCES

Barton, R. (2003). *Global It Management - a Practical Approach*, John Wiley and Sons Ltd

Morris, J. (2006). *Practical Data Migration*, British Computer Society

Chatterjee, P. (2002). *Legacy Data: A Structured Methodology for Device Migration in DSM Technology* , Kluwer Academic Publishers

Annals of DAAAM for 2006 & Proceedings of the 17th International DAAAM Symposium, ISSN 1726-9679
B. Katalinic, Published by DAAAM International, Vienna, Austria 2006
150th Birthday of Nikola Tesla (1856-1943) Inventor of Mobile Robot Teleoperation

TESTING APPLICATION FOR THE SPEED OF ALGORITHMS

POKORNY, P. & SYSEL, M.

Abstract: *Development of the computer technology makes to use their always wider possibilities. With this, it creates a strap and hardware sophisticated programs. The reason is, to get a most satisfaction of the single computer users. It concerns especially complicated applications. These applications are working with big quantity of data, for example graphics or database applications. Most user computers have not a modern state7 of the art configuration today. This is a reason to speed optimization of algorithms. These programs can be work on older slower configurations of computers. This paper presents a description of programme for the simple testing time heftiness algorithms.*
Key words: *optimalization, programming, application, software.*

1. INTRODUCTION

Developers use algorithms and data structures every day of their working lives. Having a good understanding of these algorithms and knowledge of when to apply them is essential to producing software that not only works correctly, but also performs efficiently.

It is often tempting to try to make our code as fast as possible right from the start. Characteristic optimization and performance is that the bottlenecks are almost never where we expect them, nor of the nature we first suspected. Preempting where these hot-spots might be is a costly exercise. It is much better to get the design of our code right and leave performance improvement as a separate task requiring separate skills.

A good design is much easier to profile and optimize than the many fibers that results from "clever" code. In fact our experience is that simple designs actually result in code that performs well with little optimization required. (Harris & Ross, 2005) Videliced, dealing with complexity is the essence of software engineering and we need to build a system, with good manageable.

2. OPTIMIZATIONS

Optimization is an important part of software development. It is recommended, to take time to accumulate an awareness of the types of issues that affect performance in our particular programming environment, and keep them in mind as our code. However, it is not simple answer to the question, when we want know, how we need cause the change of design algorithms. This is called premature optimization, but it is recommended to resist the temptation to build an implementation that is faster but harder to understand.

The practical experience is that a clean and simple design is much easier to optimize than one that the original developers thoughtfully optimized while writing it. It is also very important to choose the right algorithm initially. We should always be aware of the performance profile of your chosen implementation before trying to optimize it. Experience shows that the first cut of a program is very unlikely to be the best performing. Unfortunately, experience also shows that it is unlikely that you can guess the exact reason why performance is suffering in any nontrivial application. When first writing a given program, we should make it work before making it fast. (Harris & Ross, 2005)

Optimalization depends on many other factors too. For example, they are selection of programming language or possibilities of used programming libraries and templates. Optimization has a big importance at processing of a great data quantity, for example at work with databases or near sophisticated graphics and multimedia applications. The practical example is optimization of graphics library Direct3D, a part of Microsoft DirectX. Direct3D supports more techniques that can help us optimize our applications. The 3D developers at Microsoft have found to be essential for producing the most efficient applications possible. The more of these suggestions we use, the faster your programs will be. Here are selected techniques: The Direct3D framework (the CD3DFramework9 class), conversing texture memory (make optimal use of memory when working with Direct3D textures.), enhancing our 3D effect (multitexturing, triple buffering, true color rendering), optimizing geometry handing (computersphere visibility, vertex buffers, floating-point precision and multihread setcooperativelevel flags, etc.), and minimizing resources (changes number of polygons, texture changes, visibility grid, etc.). (Kovach, 2000)

Likewise, optimization has a graphics platform OpenGL - it supports many optimized functions and these functions can be used in user algorithms. It supports 3D transformations, work with a colors, shading a lighting, functions for the processing of Digital images, texture mapping, rasterization and fragment processing, Windows platform integration, multiple rendering passes, antialiasing, compositing, blending, transparency, etc. (McReynolds & Blythe, 2005)

3. PROGRAMME COMMANDS

Any algorithms needs to be support with programming language commands and functions – here must be a two way link. Any programming languages and operating systems have a different function and programming structures. We are created a simple time testing application in language C (C++), therefore we present a functions here, which support work with the optimalization. We have specialized to the Windows platform. Therefore we use Win32 API functions.

An operating system is defined by its API. An API encompasses all the function calls that an application program can make of an operating system, as well as definitions of associated data types and structures. In Windows, the API also implies a particular program architecture that we'll explore in the chapters ahead. Generally, the Windows API has remained quite consistent since Windows 1.0. A Windows programmer with experience in Windows XP would find the source code for a Windows 1.0 program very familiar. One way the API has changed has been in enhancements. Windows 1.0 supported fewer than 450 function calls; today there are thousands. Using C and the native APIs is not the only way to write programs for Windows 98. However, this approach offers you the best performance, the most power, and the greatest versatility in exploiting the features of Windows. Executables are relatively small and don't require external libraries to run (except for the Windows DLLs themselves, of course). Most importantly, becoming familiar with the API provides you with a deeper understanding of Windows internals, regardless of how you eventually write applications for Windows. (Petzold, 1999)

When we want to optimize algorithms, we need to know, how time it is running. And therefore we use a timer functions. The Microsoft Windows timer is an input device that periodically notifies an application when a specified interval of time has elapsed. Our program tells Windows the interval, in effect saying, number of milliseconds that has elapsed since Windows was started or get number of increment timer of delivery system in one second. We are interesting in the function, which returns number of milliseconds, that has elapsed since Windows was started. The name of this function is GetTickCount() and its returned value is type of DWORD (4 bytes). It means that this value is overflow after 49 days, when Windows are continuous running. When we call this function before algorithm and after this algorithm, the difference between returned values gives us time duration of algorithm. When is the algorithm very quick, we can give here a loop and we can this algorithms number of time repeat.

4. APPLICATION AND TESTING

In the previous chapter we describe a theory about its duration. Now we demonstrate a simple example. We have following equation:

$$H[\%] = 100 * \frac{AR}{AR + DR} * 2 * \frac{Alvl}{Alvl + Dlvl}. \qquad (1)$$

This equation uses a very popular role playing game Diablo II. It calculates chance to enemy hit [%]. (Gustafsson, 2004) Here are: *AR* - attack rating of the attacker, *DR* – defense rating of the defender, *Alvl* – level of the attacker and *Dlvl* – level of the defender.

We want know a calculation time. We were created a program, which calculates this time. Unfortunately is this program too long and we can not to write here complete source code written in C++ language. Here is a main part with comments:

```
#include<windows.h>
#include<iostream>

using namespace std;

int main(void)
{
DWORD First, Last;
double Duration;
int H, i;

First=GetTickCount();
// tested source code
for(i=0;i<10000000;i++)
  // here is calculated H - see equation (1)
Last=GetTickCount();
Duration=double(Last-First)/1000;
cout<<endl<<lfDuraton;
cout<<endl<<endl;
}
```

At first we need include standard libraries. It is a Windows application and we use programming interface Win32 API. Also, we include a header library windows.h. Here is implemented used function GetTickCount(). Next, we include a header library iostrem. This library gives us access to the data streams. One of these streams we use for the output numbers and texts on the screen (cout). We must to define using namespace, where these streams are. This makes the command using namespace std. In the next part we have a main part. Here begins the program. Inside a main loop we defined the needed variables. Types of variables DWORD are First and Last – into these variables we save a acquired numbers, which returns the function GetTickCount(). The third variable is Duration. It contains a value of difference between end time and begin time. Also, the calculation of the equation is inside it. The function GetTickCount() returns value in milliseconds. And the value Duration is in seconds. For this, the difference must be divided with number 1000. The source code contains a cycle. We have the cycle 10 000 000. Also, the value of attack rating is calculated 10 million times. The reason is, that algorithm is very quick on modern computers and we want to know a reasonable value. On our computer gives cycle 10 million times a calculated value Duration 2,034 seconds. This application can be modified with the compare of different algorithms very simple. It is true, calculation time is not long, but, when the Diablo is played, this equation must be multiple calculated (a hero can to have a helpmate and there are many enemies). Beside this, equation (1) is only one part of section calculation hit. It must be calculated damage with many modifications (reduced damage, resistances...), artificial intelligence, graphics and animation engine, sounds and music, controlling game, etc. Each of these parts is to be optimize. Only so we can have the optimized application, which needs minimal hardware requirements.

5. CONCLUSION

In this paper we present a simple application for the detection of the time difficulty algorithms. This application was created in the C (C++) language and it uses functions of the programming interface Win32API. The basis of the application is the function GetTickCount(). It returns the number of milliseconds that has elapsed since Windows was started. When we call this function before algorithm and after this algorithm, the difference between returned values gives us time duration of algorithm. When is the algorithm very quick, we can give here a loop and we can this algorithms number of time repeat.

6. REFERENCES

Gustafsson, T. (2004). To Hit Calculations, *Available from:* http://users.tkk.fi/~tgustafs/tohit.html *Accessed:* 2006-07-14

Harris, S. & Ross, J. (2005). *Beginning Algorithms,* Wiley Publishing, ISBN: 0-7645-9674-8, Indianapolis, Indiana, United States of America

Kovach, P. (2000). *Inside Direct3D,* Microsoft Press, ISBN: 0-7356-0613-7, Redmond, Washington, United States of America

Liberty, J. (2004). *Teach yourself C++ in 21 days,* Sams publishing, ISBN: 0-672-3271-12, Indianapolis, Indiana, United States of America

McReynolds, T. & Blythe, D. (2005). *Advanced graphics programming using OpenGL,* Morgan kaufmann publisher, ISBN: 1-55860-659-9, San Francisco, California, United States of America

Petzold, Ch. (1999). *Programming in Windows,* Computer Press, ISBN: 80-7226-206-8, Prague, Czech Republic

Simon, J., Gouker, M. & Barnes B.C. (1997). *Win32 API – user guide,* Unis Publishing, ISBN: 80-86097-06-4, Brno, Czech Republic.

Annals of DAAAM for 2006 & Proceedings of the 17th International DAAAM Symposium, ISSN 1726-9679
B. Katalinic, Published by DAAAM International, Vienna, Austria 2006
150th Birthday of Nikola Tesla (1856-1943) Inventor of Mobile Robot Teleoperation

A COOPERATIVE SYSTEM DESIGN BASED ON AGENT TECHNOLOGY

POPA, C.L. & HADAR, A.

Abstract: *A cooperative system represents a system where different users or agents work together at the same project from different locations. The cooperative systems came from the distributed applications family and they are distinguish because the actors from the system work together to realize the same purpose. Also it is necessary an interaction between all the partners involved in the project: they have common access to information, they can exchange information and verify every partner situation and progress in the project. This paper presents the main steps and the main elements needed for designing a cooperative system. Also the paper present some type of agents used in cooperative system architecture.*

Key words: *cooperative system, agent technology, virtual enterprises, communication.*

1. INTRODUCTION

In today economy context, concurrency, specialization and product diversification enterprises must respond promptly to the market demands to sustain their activities. In these conditions the enterprises consider collaboration with others to realize the product development in short time, with a lower cost, a short delivery time and a product that satisfy the client demand. Actual tend is to realize multidisciplinary cooperative systems for collaboration in real time between all professions involved in product development at the enterprise level. This collaborative environment actually constitutes a virtual enterprise. (Dragoi, 2003). Distributed and collaborative systems represent a new paradigm, and constitute a new interdisciplinary field between economic, computer, cybernetics, psychological and management field.

From the activity cooperative time perspective we have:
- real time communication
- asynchronous time communication

From the users place perspective communication can be from:
- single locations (face to face) – user are in the same place
- different locations (distributed) – users cooperates from different geographic locations

Agents that work in a cooperative system have access to common files, can exchange information and verify the project evolution. (Farley, 1998)

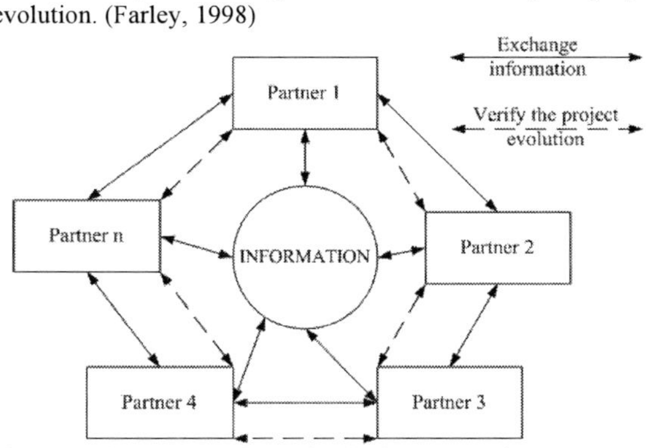

Fig. 1. Exchange the information in the cooperative system

2. COOPERATIVE SYSTEM DESIGN

The design of a cooperative system should start from:
- activities type
- the number of the partners involved in the project
- a better knowledge about de technologies needed in the system and their implementation
- the infrastructure and the relation with the outside the system partners
- what techniques and methods will be used in the system
- the cost of design, realization, maintenance and development of the system
- the lifecycle time of the product

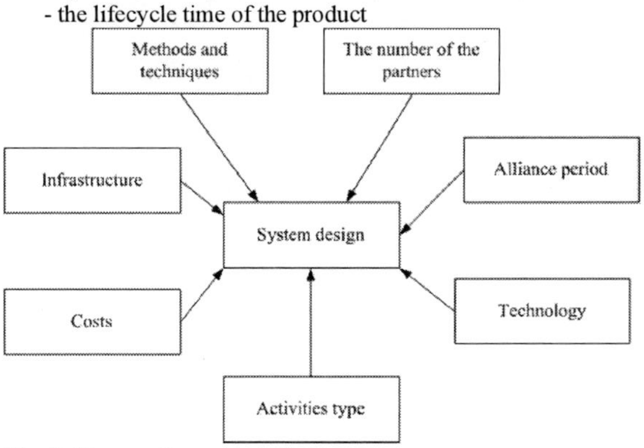

Fig. 2. Cooperative system design

Next elements must be accomplished for the technologies and product integration:
- to realize a good communication and synchronization
- opening to new technologies and methods
- previous verification of potential risks
- to be able to realize a lower cost and to reduce time
- focused to client's demands
- methods, information and instruments integration
- systems able to support collaborative environments
- education and personal training have an important role

Some questions must be put before a company will join to a virtual enterprise based on cooperative system:

Why? – a final purpose must be specified, a final product focused to client's demands

Fig. 3. Essential question put by organizations before they join to a virtual enterprise

What? – what products and technologies should be involved in the process

How? – the organization strategies, the methods used for the final purpose realization

Who? – from the beginning all the future partners and their responsibilities in the project should be known

When? –when the alliance and main activities start and end

Where? - the location of each actor from the project and the interaction methods with the group

All the partners from the virtual enterprise should answer to the following questions:

- what methods should we use to realize new and innovative product?
- how the workflow can be accelerated?
- how can we make a better product with a lower cost?
- how can be sure we will be first on the market with this product?
- how can we prevent the delays in fabrication?

3. AGENTS TYPES USED IN COOPERATIVE SYSTEMS

When we decide to use agent technology for a cooperative system implementation first we must study the main types of agents that are necessary in the system. Next we will present some types of agents that we can use in a common distributed system.

The user agent main task is to guide the user to use the system at his best performances. For example a costumer make a request in the system, this agent will help the user to make all necessary steps to accomplish the request.

Interface agents main task is to adapt the information from the application in a format that can be understood by the user or his personal agent. They are autonomous and utilize learning to perform tasks for their users. This class of agent is also used to implement assistants, guides, memory aids, and filters; perform matchmaking and referrals; or buy and sell on behalf of the user.

In a design system agents must assure the communication, cooperation and to improve coordination for the actors involved in the system. They have to allow to all actors to change information and because in such systems are involved actors from different domains, they need to be able to take decisions, to cooperate and to negotiate the relations between these actors. An agent that will realize a software version check is useful in the system. He must be able to offer some solutions to prevent the appearance of errors, losing information and he must keep records about all the activities from the system.

For example agent A ask to process a part on M1 machine, if our agent will have all the information he can tell to agent A that the part will be process faster and chipper on M2 machine. Or an error example: agent A asks to acquire a tool-machine, but the task is impossible to realize because the price is too high for this department budged and our agent have to prevent that. To prevent the errors is an important task because these errors can stop all the process and will delay the delivery time.

In the system we must have agents for machine and robots control. These agents have all machines or robots information, the types of parts that can be machined, the process time and cost, the material and tools needed. Also they will know a tool is damaged and need to be replaced. For example when a problem needs to be solved the agent will follow a totem-pole. First the agent will check the general condition of machine. He will check if the agent which controls the energy system had cut the energy to our equipment and if is true will ask the reason. He also will check the tool if is damaged or worn and need to be replaced. In this case he will send a note to the agent that manages the store department to deliver a new one. Also he will gather information about the causes of damage. An agent that manages the delivery system is needed. He will have all the information about the products, delivery times and costumers. He must be able to check on the Internet: the roads condition, the weather, traffic and he will decide which way is the best for delivery. Also an agent that will manage all the agents that have responsibilities in the manufacturing department is useful. For example he can decide to replace equipment E1 with a new one E2 to continue the production in good condition. Some problems can appear. First he will check if the new equipment can be placed in the department (maybe is too big), he can decide to put it between E3 and E4 for the best process optimization etc.

Many other types of agents can be involved in the system. For example: a security agent that manages the inputs and the outputs of other agents in the system, coordination agents, auxiliary agents etc. (Popa, 2005)

4. THE ADVANTAGES OF COOPERATIVE SYSTEMS

The advantages of using cooperative systems are:

- a good intellectual property protection, because the organization gives to the partners just the information needed
- a lower cost for infrastructure adaptation to project and costumers demands
- a good flexibility
- better relation inside the organization
- access to the date base is made by security levels according to the partners importance, his implication in the project and the type of information demanded
- gives a workflow automation level proper to the company demands
- gives access to all web services, and also for the teams that work in a distributed environment
- must be able to keep the standards for operation systems and for information security
- allow real-time interaction and visualization for the partners that realize a lower cost and reduce time for the product development
- gives the same interface to all actors involved that allow communication, organization and collaboration

5. CONCLUSIONS

Agent technology represents a new way of analyzing, designing, and implementing complex software systems. In the future we intend to realize a cooperative system architecture supporting the exchange of information and first we will focus on agents for manufacturing department (agents for machine and robots control).

6. REFERENCES

Caglayan, A.; Harrison, C. (2002) *Agent Sourcebook: Commercial Agent Software*, Wiley Computer Publishing

Dragoi, G. (2003). *Intreprinderea integrata: metode, modele, tehnici si instrumente de dezvoltare si realizare a produselor*, Editura Politehnica Press, ISBN 973-8449-25-1, Bucuresti

Farley, J. (1998). *Java™ Distributed Computing*, ISBN 1-56592-206-9E

Orlikowski, W.; Yates, J. (2002). Structuring Interaction through Communicative Norms. *Journal of Business Communication*, pag. 13-35

Popa, C.L. (2005). An Agent Classification From the Environment Perspective, *Proceedings of 16th International DAAAM Symposium*, Katalinic, B. (Ed.), pp. 309-310, ISSN 1726-9679, Opatija-Croatia, Published by DAAAM International, Vienna

Annals of DAAAM for 2006 & Proceedings of the 17th International DAAAM Symposium, ISSN 1726-9679
Editor B. Katalinic, Published by DAAAM International, Vienna, Austria 2006
150th Birthday of Nikola Tesla (1856-1943) Inventor of Mobile Robot Teleoperation

FUSED DEPOSITION MODELING PROCESS FOR BUILDING CELLULAR SOLIDS WITH CLOSED CELLS

POPESCU, D.P.

***Abstract**. The manufacturing of solid cellular structures using Rapid Prototyping (RP) processes offers the possibility to precisely control the shape, distribution, dimensions and interconnection of cells, thus controlling their micro and macro-structural properties. Nevertheless, not all RP processes can build cellular solids with closed cells, the present paper studying the capability of Fused Deposition Modeling (FDM) process to manufacture cellular solids with completely enclosed spherical and cubic cells. An important application of this type of closed-cell structures is their use as test parts for detecting voids in injection moulded parts.*
***Keywords**: Rapid Prototyping, Fused Deposition Modeling, cellular structures, closed voids*

1. RAPID PROTOTYPING PROCESSES FOR MANUFACTURING CELLULAR SOLIDS

The study of the literature in field of using Rapid Prototyping (RP) processes for manufacturing cellular solids showed the following main research directions:

- Analysis of the capability of different RP processes to build cellular structures. In (Stampfl. et al., 2004) are mentioned as selection criteria for RP processes: resolution and possibility to remove support structures, the comparison being made between processes such as stereolithography (SLA), FDM, Selective Laser Sintering (SLS) and MoldelMaker (MM). In (Williams et al., 2005) four RP processes used for manufacturing metallic prototypes (SLS, Direct Metal Laser Sintering (DMLS), Selective Laser Melting (SLM) and Electron Beam Melting (EBM) are rigorously evaluated from the perspective of manufacturing cellular solids with open cells. The comparison is made based on a requirements list: geometry, material, quality, cost and recycling.
- Building RP cellular solids with open cells of different geometries. In (Manjubala et al., 2005) and (Stampfl et al., 2005) are presented direct and indirect modalities to build cellular solids with open cylindrical, spherical, hexagonal or cubic cells.
- Experimental investigations regarding the mechanical properties of cellular solids with open cells manufactured via RP processes (Avalle et al., 2005), (Stampfl et al., 2004). Properties such as stiffness, strength, density, impact resistance for different open cells dimensions and shapes are studied.
- Development of libraries of geometrical structures which can be used for modelling cellular solids with open cells (Chua et al., 2003).

As can be noticed, the researches in the field are focused on manufacturing cellular solids with open cells using different RP processes, the main reason for that being the difficulty to eliminate the supports, which are blocked inside the cellular structure (Popescu, 2003). A solution for this problem is the use of liquid or powder support structures or which dissolves in water, but that is only applicable for cellular structures with open cells.

RP processes can build objects with very complicate geometry, but the layer-by-layer manner of manufacturing imposes limitations in obtaining objects with completely enclosed voids, usually a small hole for eliminating the support being necessary. Nevertheless this is not the case of the FDM process because it has the particularity that support structures are built only when the drafting angles of the object's surfaces exceed a certain value. Thus, for certain dimensions of the geometrical features (in this case, cells), is sufficient the support provided to the material layer by the previous layer, with which bonds.

In this context, the purpose of the present paper is to prove that FDM process can be used for manufacturing cellular solids with closed cells, but only for certain cells dimensions. In order to show that, a test part containing cubic and spherical closed voids of different dimensions was built.

2. MANUFACTURING CELLULAR SOLIDS WITH CLOSED CELLS USING FDM PROCESS

For analyzing the capacity of FDM process to manufacture objects with completely enclosed cells, a 3D CAD model of a cubic part (24x24x24mm) containing voids with spherical and cubic shapes (for other geometries, the same pattern can be applied) was built. The cubic cells dimensions vary between 1 and 1.9mm, and the radius of the sphere varies between 1 and 1.4mm. The common axis of the spherical voids is positioned at 8mm from two sides of the cube, and the cubic voids are placed in a plan parallel at 8mm distance from the axis of the spheres and at 8mm from the base of the cube. Using Catalyst 4.3 software (for FDM Prodigy machine) the deposition paths for part material and support material were generated, in order to find out (without actually building the physical object) for which cubic and spherical cells dimensions support structures are required. The results are presented in table 1. Figure 1 shows a section in the test part at z=8.1788mm height (measured from the base) with the deposition paths for part and support structures (the layer width is 0.254mm). In figure 2 is shown an image from Catalyst software presenting the voids for which support structures are built.

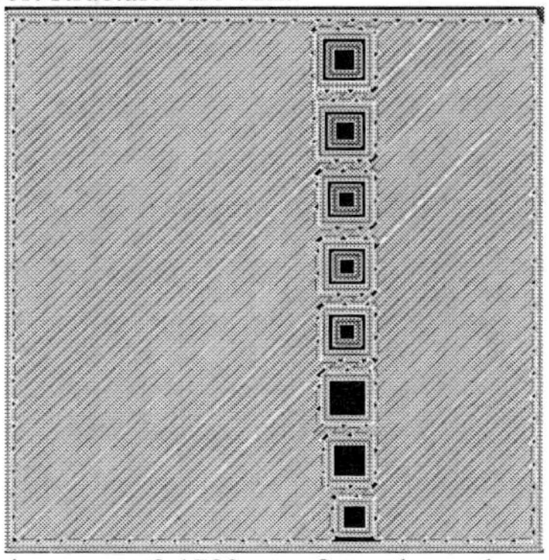

Fig.1. Section at z=8.1788mm from base, layer width of 0.254mm (Catalyst software)

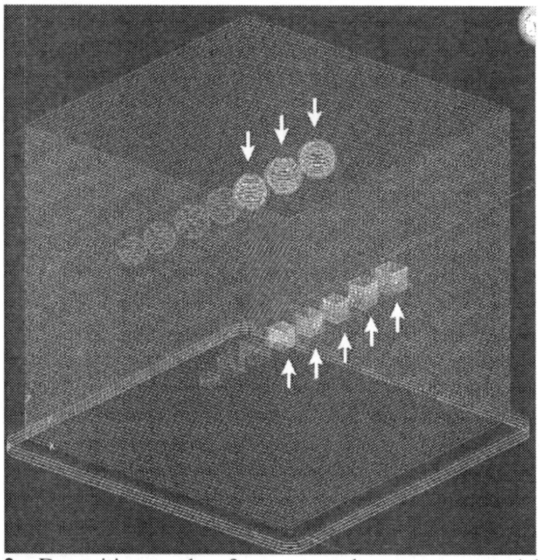

Fig.2. Deposition paths for part and support material at 0.254mm layer width (Catalyst software)

Layer width	Characteristics of cellular structure with completely enclosed cubic and spherical voids								
0.178 [mm]	Cube side [mm]	1	1.4	1.6	1.62	1.65	1.7	1.8	1.9
	Support	N	N	N	Y	Y	Y	Y	Y
	Sphere Radius [mm]	1	1.1	1.2	1.25	1.3	1.35	1.4	
	Support	N	N	N	N	Y	Y	Y	
0.254 [mm]	Cube side [mm]	1	1.4	1.6	1.62	1.65	1.7	1.8	1.9
	Support	N	N	N	Y	Y	Y	Y	Y
	Sphere Radius [mm]	1	1.1	1.2	1.25	1.3	1.35	1.4	
	Support	N	N	N	N	Y	Y	Y	
0.330 [mm]	Cube side [mm]	1	1.4	1.6	1.62	1.65	1.7	1.8	1.9
	Support	N	N	N	N	N	Y	Y	Y
	Sphere Radius [mm]	1	1.1	1.2	1.25	1.3	1.35	1.4	
	Support	N	N	Y	Y	Y	Y	Y	Y

Table 1. The need for support structures for voids at different layer widths (Y-yes, N-no)

The results obtained using Catalyst software were confirmed by the physical prototype, manufactured on a FDM Prodigy machine.

Fig.3. Physical prototype with support structures in some of the interior voids

After the manufacture the prototype was cut in order to show the presence of interior support structures (Fig.3).

3. CONCLUSION

The purpose of the research presented in this paper was to determine whether or not the FDM process can be used for building objects with completely enclosed voids. First, using Catalyst software, the manufacturing of the object was simulated for different voids dimensions and different layer widths, in order to see for which range of voids dimensions support structures are required.
It was shown that FDM, unlike other RP processes, can build objects with closed voids, but only for some dimensions.
Thus:

- For layer widths of 0.178mm and 0.254mm, cubic closed cells can be built without support structures if the cube side is smaller that 1.6mm;
- For a layer width of 0.330mm, cubic closed cells can be built without support structures if the cube side is smaller than 1.7mm;
- Spherical voids cannot be built without support if their radius is bigger than 1.3mm, whatever layer width;
- For a 0.330mm layer width spherical voids can be built without support for radius smaller than 1.1mm;
- For 0.178mm and 0.254mm layer width, spherical voids do not require support if their radius is smaller than 1.25mm.

These results obtained with the Catalyst software were confirmed by the physical prototype.

4. FURTHER RESEARCH

Further research will be focused on manufacturing FDM cellular solids with closed voids, in order to used them as test parts in experiments for determining their mechanical properties. Also, an analysis of the manner in which shape, distribution, dimension and number of cells are influencing the mechanical properties of ABS FDM closed-cell cellular solids, will be performed.
Another important research direction will be to use this type of objects with closed voids for investigating different modalities to detect voids in injection moulding plastic parts.

5. REFERENCES

Avalle, M., et al., Mechanical models of cellular solids: parameters identification from experimental tests, *Transactions on Engineering Sciences, Impact Loading of Lightweight Structures*, 2005, p.75-87

Chua, C.K., Leong, K.F, et al., Development of a Tissue Engineering Scaffold Structure Library for Rapid Prototyping. Part 1: Investigation and Classification, *International Journal of Advanced Manufacturing Technologies,* vol. 21, n.4., 2003, p.291-301

Stampfl, J., et al., Regular, low density cellular structures - rapid prototyping, numerical simulation, mechanical testing, Mat. Res. Soc. Symposium Proc., vol.823, 2004, p.W8.8.1-W8.8.6

Stampfl, J., Fouad, H., Seidler, S., Fabrication and moulding of cellular materials by rapid prototyping, *International Journal of Materials and Product Technology*, vol.21, no.4, 2004, p.285-296

Manjubala, I., et al., Biomimetic Mineral-Organic Composite Scaffolds with Controlled Internal Architecture, *Journal of Materials Science: Materials in Medicine*, vol.16, 2005, p.1111-1119

Popescu, D., *Procedee de fabricare rapidă a prototipurilor*, AIUS Press, Craiova, 2003, p.45-113, ISBN 973-9490-83-2

Annals of DAAAM for 2006 & Proceedings of the 17th International DAAAM Symposium, ISSN 1726-9679
B. Katalinic, Published by DAAAM International, Vienna, Austria 2006
150th Birthday of Nikola Tesla (1856-1943) Inventor of Mobile Robot Teleoperation

NEURAL NETWORKS AS THE TOOL OF KNOWLEDGE EXTRACTION FOR FIXTURE DESIGN

PORTJANSKI, L. & NEKRASSOV, G.

Abstract: *The aim is to develop an integrated design space for CAD of technological equipments for SMEs. This report shows the possibility of creating the wizard master, which is invariant to designing objects. It is offered to apply a method of neural networks to realization procedure of forecasting of decisions. The considered algorithms of knowledge extraction are realized by development of invariant nucleus design space for fixture design. The neural network is a black box to which can ask questions and to receive from it answers and explanations.*
Key words: *CAD, integrated design environment, neural networks, Knowledge Discovery*

1. INTRODUCTION

There are two ways of products' engineering:

- to realize engineering only using CAD system resources. Such of designing is implemented for original designs, required much time and great qualification of a designer;
- to realize engineering using the resources of preliminarily created environment of engineering for this CAD system. In this case the high qualification of a designer is required at the stage of designing the object-oriented environment, when is defined structure of a product, its geometrical characteristics and assembling rules. It takes less time and energy to execute the designing of specific parts of this group. It may be used methodology when creating environment of designing.

In the design space development the problem arises how to decompose the design process so that individual design tasks are communicating only over the supervisory subsystem (Kuttner et al., 2003).

Wizard is the main element of environment, supporting to design objects of this group.

This report shows us the possibility of creating the wizard master, which is invariant to designing objects. The scheme of wizard master's work is corresponded with CBR methodology - description of a problem, 4-RE cycle performance (retrieve, reuse, revise, retain) (Aamodt & Plaza 1994), with the need of the problem's decomposition with its specification performance and etc. Creating environment of designing with using wizard master is implemented on supervisor- level.

Tooling wizard master to the given group of tools is made with the help of worked out language of high level as well as CAD system means. In the latter case a designer puts together goals from parts, pointing out conditions of impurities or a dialogue's call for choosing the properly element of a design. In that way the script of a designing process is created. This script is and/or a graph. There are procedures at the top of a graph which is necessary for designing of this type of parts; there are terms of procedures' choosing on arches. It is determined, that the least developed is the problem of construction of procedure of forecasting of decisions invariant to objects of designing. It is offered to apply a method of neural networks to its realization.

2. NEURAL NETWORKS

Neural networks is the group of analytical methods, based on (hypothetical) principles of teaching intellectual beings and brain functioning and premising to forecast values of some variables in new observations according to other data of observations after passing the teaching stage according to existing data (Wasserman, 1989).

Neural networks are one of methods of so-called extraction of the data.

Last years huge databases in which the information of scientific, economic, business and political character is stored are created. For job with similar bases the computer technologies are developed, allowing keeping, sort and visualizing the data, to provide fast access to them, to carry out their statistical processing. Considerably achievements in development of methods and the programs, capable to find out in the data important, but the latent information are smaller, however. Similar *not trivial* extraction *implicit, before the unknown person and potentially the helpful information* from the big databases also is called as Data Mining or Knowledge Discovery. We shall use further for the description of this area of computer science term - knowledge extraction.

Knowledge extraction uses the concepts developed in such areas as Machine Learning, Database Technology, statistics and others. Job with very big databases demands efficiency of algorithms, and discrepancy and, frequently, incompleteness of the data generate additional problems for extraction of knowledge. Filling of misses in databases is one of the problems solved by neural networks. However, the main claim to neural networks always was absence of an explanation. Extraction of rules from neural networks means their preliminary training. This procedure demands a lot of time for the big databases.

Let's consider one of methods of extraction of rules from the neural networks, trained to the decision of a problem of classification (LU, Setico and Liu, 1995). This method refers to **NeuroRule.**

The problem will consist in classification of some data set with the multilayered perseptron and the subsequent analysis of the received network with the purpose of a finding of the classifying rules describing each class.

3. ALGORITHM OF KNOWLEDGE EXTRACTION

Let A designates a set from N properties

$A_1, A_2, \ldots, A_n$, and $\{a_i\}$

- set of possible values which can accept property A_1.

We shall designate through C set of classes $c_1, c_2, \ldots, c_m$.

For training sample the associated pairs vectors of entrance and target values $(a_1, \ldots, a_m, ci)$, where $c_i \in C$.

The algorithm of extraction of classifying rules includes three stages:

- **Training of a neural network.**

On this first step two-layer perseptron trains on a training set down to reception of sufficient accuracy of classification. We shall assume that the training data set is necessary for classifying on two classes A and B. In this case the network should contain N the entrance and 2 days off neurons. To each of classes there will correspond the following of activity of the days off neurons (1,0) and (0,1). The suitable quantity neurons in an intermediate layer, generally speaking, cannot be defined beforehand - their too big number conducts to conversion training while small does not provide sufficient accuracy of training. All methods of adaptive search of number neurons in an intermediate layer share on two classes, according to that, from small or big number intermediate neurons will start algorithm. In the first case, in process of training in a network are added additional neurons, in opposite - after training there is a destruction excessive neurons and connections. NeuroRule uses last approach so the number intermediate neurons get out big enough. We shall notice that NeuroRule destroys as well superfluous entrance neurons which influence on classification is not enough. As function of activation intermediate neurons the hyperbolic tangent so their conditions change in an interval [-1, 1] is used. At the same time, function of activation of the days off neurons is Fermi's function (conditions in an interval [0, 1]).

- **Filtration of a neural network.**

The trained neural network contains all possible connections between entrance neurons and neurons on the latent layer, and also between last and target neurons. The full number of these connections is usually so great, that from the analysis of their values it is impossible to take foreseeable for the user classifying rules. Filtration consists at a distance excessive connections and neurons, not resulting to increase in a mistake of classification by a network. The resulting network usually contains a little neurons and connections between them and its functioning gives in to research.

- **Extraction of rules.**

Even if the parameters describing attributes of classified objects represent continuous sizes, for their representation it is possible to use binary neurons and a principle of coding such as the thermometer. At such way of coding the area of change of parameter shares on final number of M of intervals and for representation of all values lying in *m* interval the following condition of M binary neurons is used (Fig. 1):

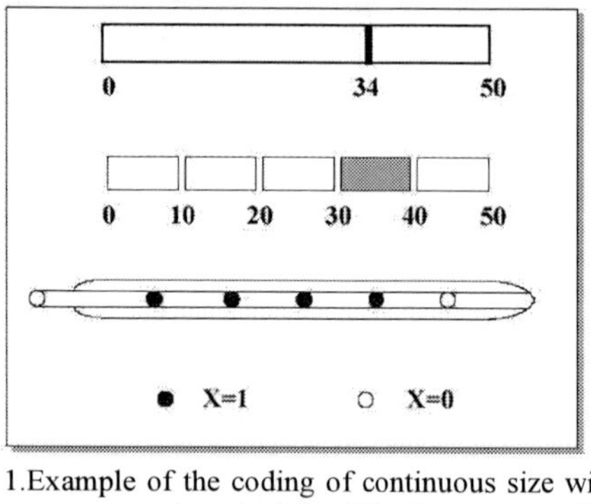

Fig. 1.Example of the coding of continuous size with the help binary neurons and a principle of the thermometer. The interval (0, 50) is broken into 5 equal parts. Value 34.0 gets in 4-th interval. Thus of a condition of the first 4 from 5 coding binary neurons 5-th is equal to unit, and - to zero.

At this stage from a neural network the rules are taken having the form

if $\{a_1\Theta q_1\}$ *and u* $(a_2\Theta q_2)$ *and ... and* $(a_n\Theta q_n)$ *then* $c_{j\cdot}$.,

where $q_1,\dots,q_n$ constants,

Θ - the operator of the relation (=> <,> <)

It is supposed, that these rules are obvious enough at check and are easily applied to the big databases.

Let's mention even one procedure which is carried out at extraction of knowledge from neural networks - correction. Similar operation has been offered Weigend and colleagues and as a matter of fact is used in parallel with training (Weigend, Zimmermann, and Neuneier 1996). Hybrid use of training and correction of the data carries name **CLEARNING** (CLEARING+LEARNING). The given procedure includes ascending process of training at which the data change connections in a neural network and descending process in which the neural network changes the data on which training is made. Craven and Shavlik (Craven and Shavlik, 1996) have developed algorithm TREPAN generating a tree of decisions, approximating behavior of the trained neural network. The important advantage of algorithm is that he does not show any requirements to architecture of a network, to number of its elements and connections (we shall recollect as it was important to simplify structure of a network at use of NeuroRule). It is quite enough of that the neural network is a black box to which can ask questions for it and to receive from it answers. Accuracy of the prediction, given by the generated tree of decisions, is close to accuracy of a neural network prediction.

4. CONCLUSION

In the development of design space for fixture design the greatest difficulties arise at a formulation of conditions of a choice of decisions. The information received from qualified experts, should be written down directly in programs. In this case it is static and does not take into account accepted decisions at operation design space for fixture design. Application a neural network allows not only to predict the accepted decision, but also to take knowledge from databases.

The offered technique allows to form „if then" rules dynamically. The considered algorithms of knowledge extraction are realized by development of invariant nucleus design space for fixture design. The neural network is a black box to which can ask questions and to receive from it answers and explanations.

Acknowledgements. Hereby we would like to than the Estonian Science Foundation, for the grant G6183 "Methodology of Case-Based Reasoning (CBR) for development design environment of technological equipment." enabling us to carry out this work.

5. REFERENCES

Aamodt A. & Plaza E. (1994). Case-based reasoning: Foundational issues, methodological variations, and system approaches. AI Com-Artificial Intelligence Communications, 7(1), March 1994, pp. 39–59.

Craven, M.,W., & Shavlik, J.,W.(1996) "Extracting tree-structured representations of trained networks". In Touretzky, D., Mozer, M. and Hasselmo, M., eds. Advances in Neural Information Processing Systems (volume 8). MIT Press, Cambridge MA/

Kuttner,R.; Nekrassov,G.; & Sutt, A. (2003) Design Space for Collaborative Concurrent CAD of Technological Equipment. DAAAM International Scientific Book 2003, ed. Branko Katalinic, pp. 347-358. Vienna 2003

Lu Hongjun, Setiono, R. and Liu Huan (1995). "NeuroRule: A connectionist approach to Data Mining". Proc.ofthe 21st VLDB Conference, Zurich, Swizerland

Wasserman, P. D. (1989). Neural Computing: Theory & Practice. Van Nostrand Reinhold: New York. ISBN 0-442-20743-3

Weigend, A.,S., Zimmermann, H.,G., and Neuneier, R. (1996) "Clearning. In Neural Networks in Financial Engineering", World Scientific, Singapore

Annals of DAAAM for 2006 & Proceedings of the 17th International DAAAM Symposium, ISSN 1726-9679
Editor B. Katalinic, Published by DAAAM International, Vienna, Austria 2006
150th Birthday of Nikola Tesla (1856-1943) Inventor of Mobile Robot Teleoperation

MICROWAVE HEATING OF NATURAL AND BUTADIENSTYREN RUBBER

POSPISIL, R.; LOUCKOVA, P.; DVORAK, Z. & MANAS, M.

Abstract: *The aim this work is the defining of preheating dependence by microwave power source on energy absorption of mixing mechanism. As the power source was used mircowave technology. The experiment is realized on the specified elastomer compound.*
Key words: *microwave heating, rubber coumpound, undercooling rubber, natural and butadien rubber.*

1. INTRODUCTION

The heating by microwave energy is essentially an interaction between the molecules of the material to be heated and the electric field component of microwave energy. In an energy-efficient heating system transfer of energy from generator to workload (input of the device for applying microwave energy to a product) must clearly be achieved with minimum loss.
Elastomers pose typically very low thermal conductance. Conventional heating requires external heating of surface first, creating of thermal gradient enabling the heat transfer into the material volume that can possibly deteriorate its properties. On the other side dielectric heating enables immediate affecting into the depth of material. Various natural and synthetic rubbers are using different fillers posing variable dielectric properties reacting in effect very diversely in high frequency electromagnetic field. Using the radio frequency heating was discovered in 1935 while the first industrial installements applied in 60s. The pioneers of the technology were well-known companies as Elliot, Dunlop and Goodyear.

2. EXPERIMENT

2.1 Microwave rubber heating natural (PK SIR 20SED) and butadienstyren rubber (Kralex) at standard temperature

The measurement plan at rubber heating with microwave technology
Individual samples in the cube form of natural rubber (PK SIR 20SED) and butadienstyren rubber (Kralex) in the size cca 100 x 100 x 100 mm and average wage m = 400 g were heated in microwave PROFESSOR MTE-235. Each of samples were heated in the selectec time t = 2, 5, 7, 10 min at power P = 20, 40, 60, 80, 100 %. The samples were heated three times and each of sample was quenched (after heating) at inlet temperature T_1 (°C). After a period of selected heating time with microwaves was determined temperature T_2 (°C) at three parts and it was averaged.
Measurement conditions:

- average inlet temperature of samples (rubber temperature before heating) T_1 = 18 °C
- rubber samples homogenization, i. e. put sample in microwave and for a period t_h = 1 min is heated at power 100% (900W)

All measured and calculated values of natural rubber (PK SIR 20SED) and butadienstyren rubber (Kralex) are displayed in Fig. 1. and Fig. 2.

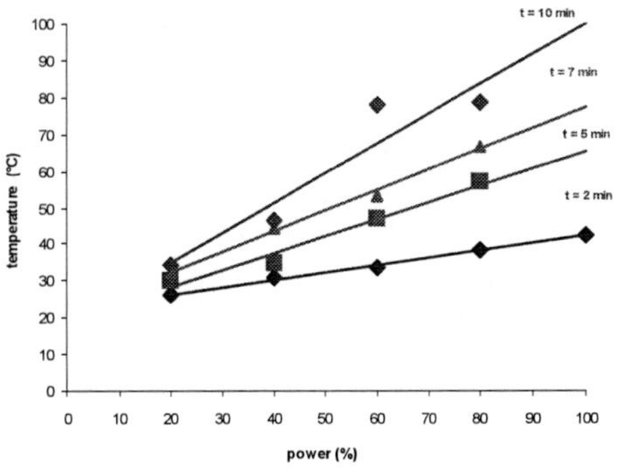

Fig. 1. Microwave heating of natural rubber – SIR 20SED

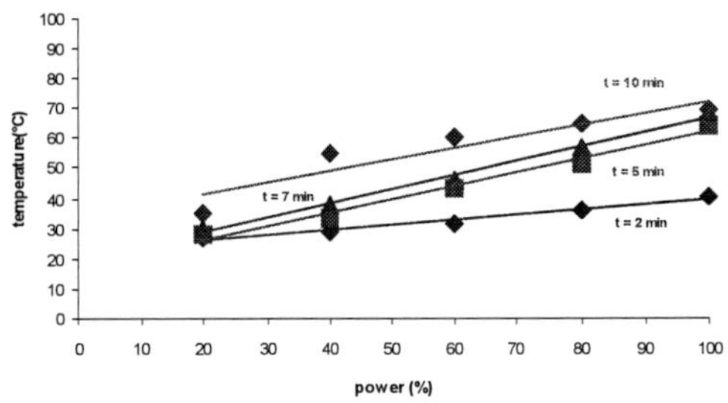

Fig. 2. Microwave heating of synthetic rubber – Kralex

2.1 Microwave heating of "frozen" rubber natural rubber (PK SIR 20SED) and butadienstyren rubber (Kralex)

Experiments of rubber undercooling at temperature T_1 = -18 °C should simulated storage time period and storage conditions. Fast undercooling or longterm storage of rubber caused changing rubber internal structure. There was confirmed with experiment that at fast undercooling of rubber reach the crystallization as ahow Fig. 3 (colour change). The undercooled synthetic rubber SBR Kralex doesn´t crystalize, remained unchanged as show Fig. 4.

Fig. 3. Rubber comparing PK SIR 20SED – crystallization effect

Fig. 4 Rubber comparing SBR Kralex –doesn´t crystallize

The measurement plan at "frozen" rubber heating with microwave technology

Rubber samples were for selected time put into the freezing box and frozen at average temperature T_1 = -18 °C . Then samples were heated for the time t = 10, 20, 30, 40, 50 min in microwave at maximum power 100% (900W). At each of sample was measured temperature T_2 (°C) in three sample parts and averaged. All measured and calculated values are displayed in Table 1, 2. Run of de-freezing at individual samples is shown in Fig. 3,4.

Samples indication	Power (W)	time (min)	inlet temperature T_2 (°C)			average
			A	B	C	
5		0	-19	-23	-17	-19,67
6	900	10	7	10	11	9,33
11	900	20	35	38	37	36,67
7	900	30	52	59	56	55,67
23	900	40	75	68	71	71,33
13	900	50	88	90	87	88,33
average			51,4	53	52,4	52,27

Table 1. De – freezing of natural rubber PK SIR 20SED

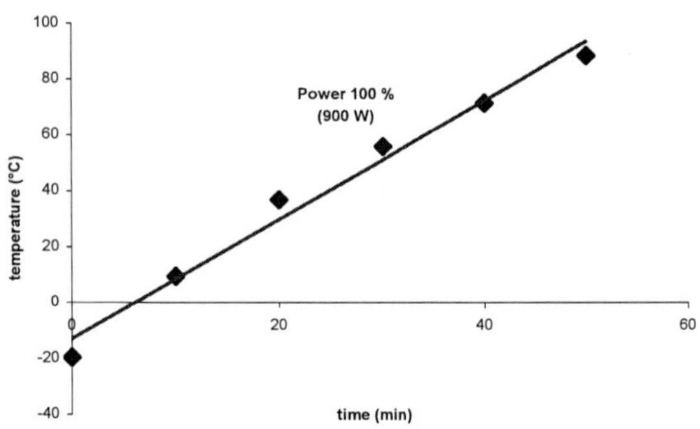

Fig. 5. De-freezing of natural rubber PK SIR 20SED

Samples indication	Power (W)	time (min)	inlet temperature T_2 (°C)			average
			A	B	C	
12		0	-25	-27	-23	-25,00
21	900	10	5	7	9	7,00
7	900	20	28	26	30	28,00
25	900	30	50	48	46	48,00
24	900	40	70	68	73	70,33
27	900	50	80	82	84	82,00
average			46,6	46,2	48,,4	47,07

Table 2. De – freezing of butadienstyren rubber (Kralex)

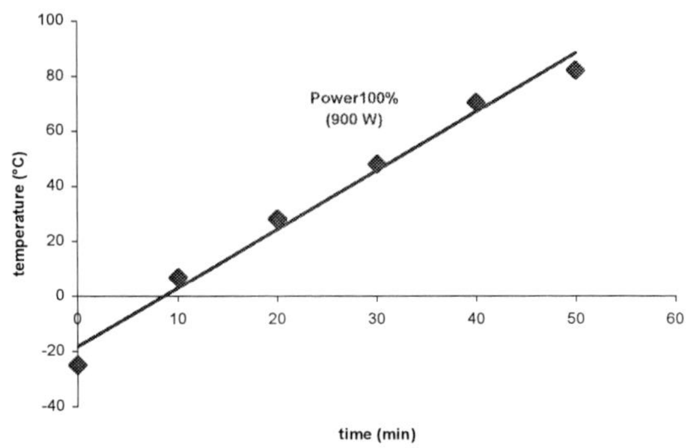

Fig. 6. De-freezing of butadienstyren rubber (Kralex)

3. CONCLUSION

There was proved, that non-polar material (rubber) is question of heating influence with microwave technology. The natural rubber PK SIR 20SED and synthetic rubber Kralex de-freezing run linearly. As like as in previous experiment we can find on the basic of evaluation that the outlet temperature of monitored samples are increasing in dependence on time as we expected. The running of rubber de-freezing is basically same.

4. FUTURE WORK

For influence evaluating on rubber properties in mikrowaves will be used experiment of repeated heating and cooling of each sample. After heating will be sample cooled at average temperature T = -10°C and again heated. From this samples will be cut specimen for viscosity determinating. To solve influence of microwave heating on rubber compounds properties. To assess using of microwave heating with reference to energy savings at rubber compounds mixing. To determine economic size of using microwave technology at rubber compound process. The experiment implement in rubber sphere with mikrowave technology, from temperature under the freezing-point to temperatures where is real to heat rubber. Individual parts realize in laboratory and in production.

5. ACKNOWLEDGEMENT

This work was supported by the Ministry of Youth, Education and Sports of the Czech Republic, under grant MSM 7088352102. The support is very gratefully acknowledged.

6. REFERENCES

Louckova, P. (2005). *Microwave heating rubber,* Diploma thesis

Adamski, W.; Kitliński, M. (2001). *On measurement applied in scietific researches of microwave heating processes,* Measurement science review,Vol. 1, No. 1,

Romill-Krmivarske stroje. *Industrial microwave heating (online) Available from* :http://www.romill.cz

Dvorak, Z.; Svoboda, I. (2003) *Vývoj spotřeby energií na přípravě gumárenských směsí,* Nove smery v gumarenstvi, ISBN 80-7318-149-5

Mleziva, J.;Snuparek, J. (2000) *Polymery.výroba, struktura, vlastnosti a použití.* Sobotáles, 2000.537 s. ISBN 80-85920-72-7

Annals of DAAAM for 2006 & Proceedings of the 17th International DAAAM Symposium, ISSN 1726-9679
B. Katalinic, Published by DAAAM International, Vienna, Austria 2006
150th Birthday of Nikola Tesla (1856-1943) Inventor of Mobile Robot Teleoperation

THEORETICAL AND EXPERIMENTAL RESEARCH REGARDING LONG AND VERY LONG LEAD SCREWS

PRODAN, D. & GHEORGHIU, H.

Abstract: *The paper presents a number of solutions to some specific problems arising from using long and very long ball screws, as a final mechanism. Ball screws usually have maximum lengths of 6-8 m. Longer lengths require special systems to compensate length variations due to temperature and to absorb the slack. These systems and their evaluation method are the subject of this paper. Here are presented a number of solutions and calculi that can be used in the projection and fabrication of mechanisms which include screws whose length exceed 20 m.*

Key words: *ball screw, machine tool, hydraulic system, slack, clearance.*

1. INTRODUCTION

The bolt-and-nut mechanism is one of the most frequently used ultimate mechanism in feed linkages. It can be used in two different ways:

- the screw rotates and the nut shifts, moving the slide;
- the screw is fixed and the nut rotates and shifts together with the slide.

Ball screws are also preferred, nowadays, for the refabrication of machine tools, instead of the gear rack mechanisms (Prodan & Constantin, 2004), (Prodan & Marinescu, 2005).

Regardless of the chosen alternative, in the case of horizontal driving, the screw propped at both ends has a slack which, for large lengths, affects the linkage's good functioning and might induce permanent strains.

Temperature variations can lead to considerable length variations in screws longer than 8-10 m, regardless of the mode of action: horizontal or vertical.

In order to reduce such phenomena, hidraulic systems of propping and stretching are recommended, as presented in the following.

2. SLACK REDUCTION

2.1. Installation

For the slack's reduction, it is recommended to use supplementary supports, hydro driven (Prodan & Marinescu, 2004). The skeleton diagram with two supplementary props is shown in fig. 1. The authors used a maximum number of five prop elements.

In fig. 1, we used the following notations: 1- bedframe, 2- lead screw, 3- supporting, 4 – revolving nut, 5 – limiters confirming the rolls' position; 6 – system of pre-tension and absorbtion of temperature strain; 7- rolls controlling limiters; 8 – motor which drives the feed linkage on the Z direction; 9 – reductor.

While the machine is out of work, the screw shall be propped on the rolls. When the slide-reductor unit shifts, the rolls descend and allow the latter's passage, then ascend again. Consequently the screw has in this case two supplimentary abutments, symmetrically placed, besides the two abutments in its ends.

The two hydro driven rolls provide the line of the two end abutments. They are driven by a hydraulic installation, whose diagram is shown in fig. 2.

The system includes an accumulator and an anticipatory control. The accumulator's loading is provided by the two pressure relays, 21.1. and 21.2. The sense valve separates the accumulator circuit from the source of pressure: the pump and the valve 19. The rolls are endowed with a brake system at stroke end and interlocking systems, which are not shown in the diagram.

2.2 Backing rolls' evaluation

Following the ball screw's calculus, we determined the force and the stroke providing a zero slack and a correct support. Rolls with a piston diameter of 50 mm were chosen, to be used at a pressure of approximately 30-40daN/cm^2.

The equations describing the functioning of such roll while rising are (Prodan, 2004):

$$M\frac{d^2x}{dt^2} + bv + Ff = p_1S_1 - (p_2S_2) \qquad (1)$$

$$Q = S_1\frac{dx}{dt} + \frac{V_m}{E}\frac{dp_1}{dt} \qquad (2)$$

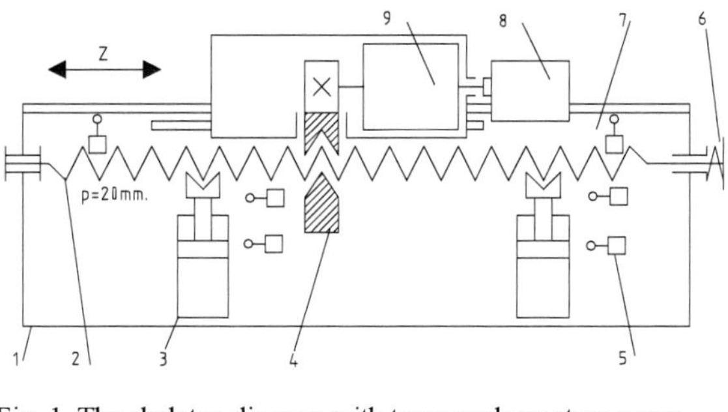

Fig. 1. The skeleton diagram with two supplementary props

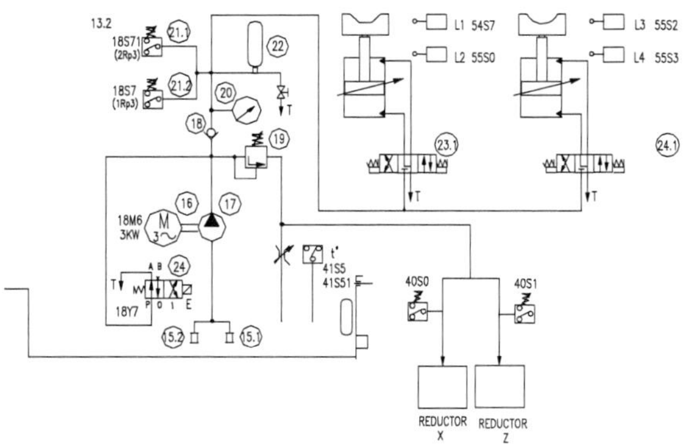

Fig. 2. Hydraulic installation diagram

In the equation above, we used the following notations: m- reduced mass, x- excursion, t- time, b- linear coefficient of force losses, in proportion to speed, v- speed, F_f- friction force, p – pressure on the piston's large area, S_1- piston's large area, p_2- pressure on the piston's small area, S_2- piston's small area, Q – used flow rate, V_m- medium volume of the liquid in roll's big chamber, E – coefficient of elasticity of the used oil.

The term (p2S2) appears solely in the braking stage and we consider it to be constant. If equation (1) is differentiated and $\frac{dp_1}{dt}$ in equation (2) is replaced, the resulting equation is:

$$M\frac{d^3x}{dt^3}+b\frac{d^2x}{dt^2}+S_1{}^2\frac{E}{V_m}\frac{dx}{dt}=QS_1\frac{E}{V_m} \qquad (3)$$

By resolving the differential equation, it is possible to calculate the speed and stroke's development in time.

The differential equation was resolved through simulation.

The speed characteristic is shown in fig. 3, and the stroke characteristic, in fig. 4. Following the calculus, we determined that the running time for a maximum stroke, corresponding to the screw's correct support, is about 1-2 s.

For the dimensioning of rolls' control cams, it was considered that the lifting time is 6 s. Under these circumstances, for a high speed rate of travel of 0,1 m/s, the resulting response to the derivative is 0,6m. In order to avoid accidents, the descending command is double, both from the cam and from the position transducer on the Z axis. The rolls' positions are confirmed by inductive limiters. In cases of accidental falls of potential, the hydraulic locks block the rolls. In order to avoid oil's heating and to shorten the driving times, a gas-loaded accumulator was provided. The latter maintains the pressure in the system, after cancelling the anticipatory control.

The backing rolls were projected and fabricated to include special seals of raylon. Braking at stroke end is performed by special sockets, which are also made of raylon. The hydraulic equipment is DN6 and DN10. One of the backing rolls is shown in fig. 5.

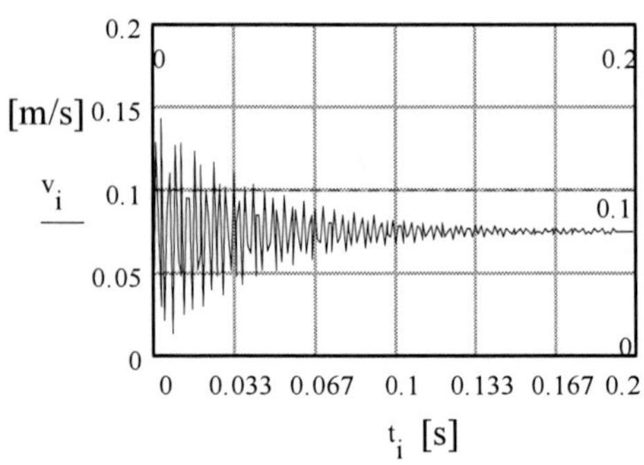

Fig. 3. The speed characteristic

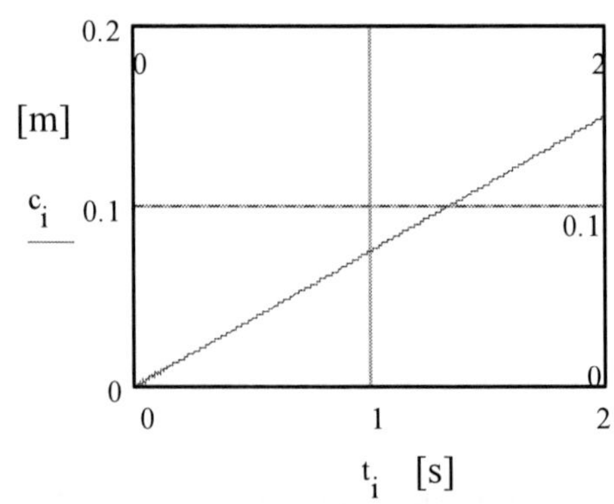

Fig. 4. The stroke characteristic

Fig. 5. One of the backing rolls

3. CONCLUSIONS

Ball screws can be used for strokes that are longer than 8-10 m, provided that strains due to gravity and temperature variations are prevented.

It is recommended to use supplementary abutments, with a maximum length of support of 6-8 m. These supports are mobile; they allow the passage of the slide, which is driven by the linkage. Depending on the slide's position, the supports' control is performed by limiters and cams and/or their position as read by the transducer.

The backing rolls' driving speed is chosen depending on the slide's speed and the length of support. The rolls must be endowed with a brake system at stroke end.

The system's hydraulic installation is sized according to the above.

Each abutment element is adjusted during mounting, in order to effectively absorb the slack. The entire system must include safety elements able to prevent malfunctions and to exclude any possibility of dabbing the slide.

5. REFERENCES

Prodan, D. & Constantin, G. (2004). Refabrication of Heavy Machine Tools, *Procedeeings of The 15th INTERNATIONAL DAAAM SYMPOSIUM "Intelligent Manufacturing & Automation: Globalisation – Technology – Men - Nature"*, Katalinic B. (Ed.), pp. 375-376, ISSN 1726-9679 ISBN 3-901509-42-9, Vienna, Austria, November 2004, DAAAM International Vienna.

Prodan, D. & Marinescu, S. (2005). *Refabricarea masinilor-unelte sisteme hidraulice*, Editura Tehnica (Ed.), Bucharest.

Prodan, D. & Marinescu, D. (2004). Sistem hidraulic pentru preluarea sagetii la suruburile conducatoare din lanturile cinematice de avans de la strungurile paralele grele, in *Automatizari si instrumentatie,* No. 5/2004, Asociatia pentru Automatizari si Instrumentatie din Romania (Ed.), pp. 28-29, ISSN 1582-3334.

Prodan, D. (2004). *Hidraulica masinilor-unelte*, Editura Printech (Ed.), Bucharest.

Prodan, D. & Marinescu, D. (2005). Sistem hidrostatic pentru pretensionarea suruburilor cu bile, in *Automatizari si instrumentatie*, no. 2/2005, Asociatia pentru Automatizari si Instrumentatie din Romania (Ed.), pp.16-18, ISSN 1582-3334.

Buzdugan, Gh. (1974). *Rezistenta materialelor*, Editura Tehnica (Ed.), Bucharest.

Annals of DAAAM for 2006 & Proceedings of the 17th International DAAAM Symposium, ISSN 1726-9679
Editor B. Katalinic, Published by DAAAM International, Vienna, Austria 2006
150th Birthday of Nikola Tesla (1856-1943) Inventor of Mobile Robot Teleoperation

CLEARANCE ABSORBTION IN LEADING LINKAGES OF HEAVY MACHINE TOOLS WITH NUMERICAL COMMAND

PRODAN, D. & GHEORGHIU, H.

Abstract: ***Most leading and/or positioning linkages used in machine tools with numerical command consist in: an a frequency control alternomotor, a mechanical reductor and a mechanism for the conversion of motion. The following are used as mechanisms for the conversion of motion: ball screws for strokes of a maximum length of 6-8 m and a gear rack, for longer strokes. There is also the possibility to use longer screws nowadays, provided that constructive changes are made. If such changes are impossible to make, the only solution remains the gear rack. However, the clearance absorbtion is essential for achieving the required accuracy. The paper includes a theoretical and experimental study of clearance absorbtion systems.***

Key words: *clearance ,accuracy, machine tool.*

1. INTRODUCTION

Electromotors are endowed with special command equipment and work in a double feedback loop. The final position is read on linear transducers. The rotating transducer on the electromotor's axle performs the second control loop. Given the above, one might infer that the eventual clearances which exist in the mechanical system bear no influence on the processing precision, as linear transducers "correct" any possible error. This is true only in the case of a single axis processing. Simultaneous processing on at least two axes, i.e. interpolation, raises special problems. The reversal clearance is ideally zero (Prodan et al., 1999). This clearance is in fact the effect of the clearances resulting from mechanical transmissions: gears, cog belt transmissions, etc., as well as those resulting in the final mechanism for conversion of motion: screw, short screw – long half-nut or gear rack. The paper further presents a real case, regarding a two-axes processing. It involves processing on the "X"-"Y" axes in a Gantry-type milling machine, as shown in figure 1. On the two bedframes with guides 1, the racks 2 are fixed. Reductors 3 shift the slide 4 along the guides. The guides 5 are attached to the slide. The working head 6 is moved along by the gearmotor 7 and the ball screw 8.

We consider that the final mechanism on the "Y" axis is a ball screw-nut with zero reversal clearance. On the "X" axis, the two reductors with gear rack final mechanisms perform the shifting.

We consider that, at the working head, a V-shaped trajectory must be obtained.

In order to illustrate the influence exerted by clearance on processing, we consider two cases:

- A clearance-free processing, i.e. processing in ideal conditions ;
- A real conditions processing, with reversal clearances on the "X" axis.

The ideal trajectory for a clearance-free processing is shown in figure 2a.

In the presence of clearances, the real trajectory is the one shown in figure2b. If senses in figure 1 are taken into account, one can notice that the shift –X +Y, up to the peak of the letter

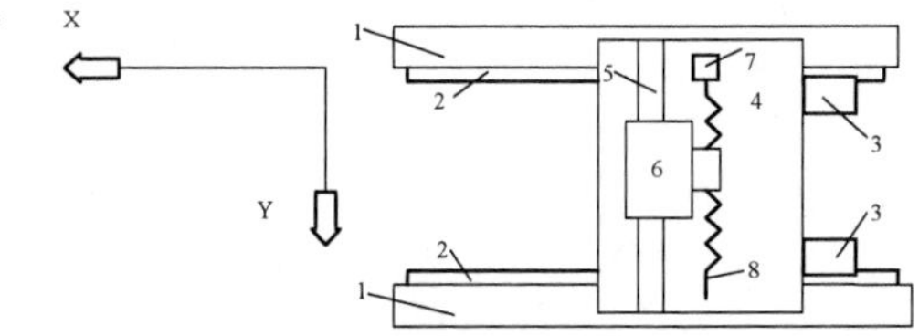

Fig. 1. A Gantry-type milling machine

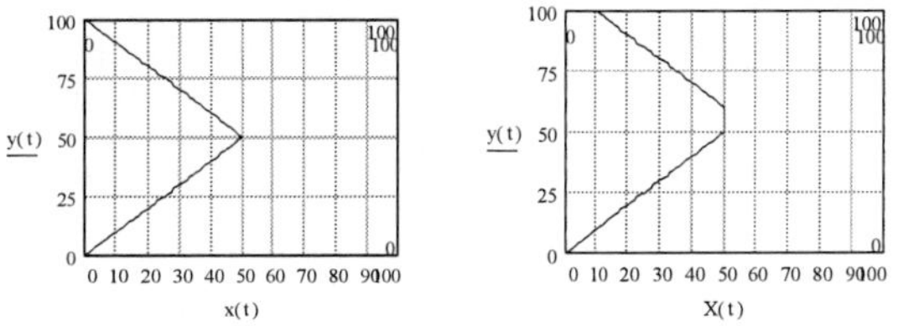

Fig. 2. A comparison between trajectories: *a*-ideal trajectory, b-real trajectory

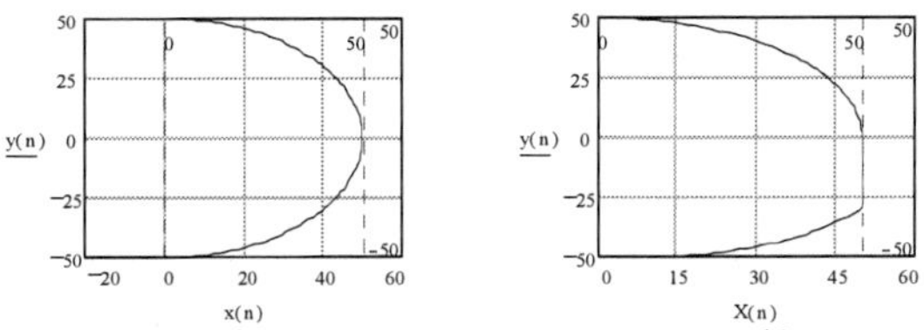

Fig. 3. Trajectories results for processing: *a* – ideal processing, *b* – real processing

V, is followed by a shift +X+Y. Therefore the shift on the Y axis is continuous, while the shift on the X axis interrupts and switches the sense. The sense change is delayed by the existence of clearances.

A comparison between trajectories in figure 2, *a* and *b* shows that, when the shifting sense on the "X" direction changes, due to the continuous shift on the "Y" axis, the trajectory is deformed. If the processing is performed along a curve, the following result:

- for an ideal processing, the trajectory in figure 3a;
- for a real processing, the trajectory in figure 3b.

In a circular processing, the piece's processing can fail, too, due to the reversal clearance on the "X" axis.

2. HYDRAULIC SYSTEMS FOR CLEARANCE ABSORBTION (PRODAN ET AL., 2004)

A solution, used in reaming and milling machines, is shown in figure 4. Electromotor 1 transmits the motion to gear 2 and then to gears 3. These gears have inclined toothing, as shown in the picture. Gears 4 transmit the motion to pinions 6.

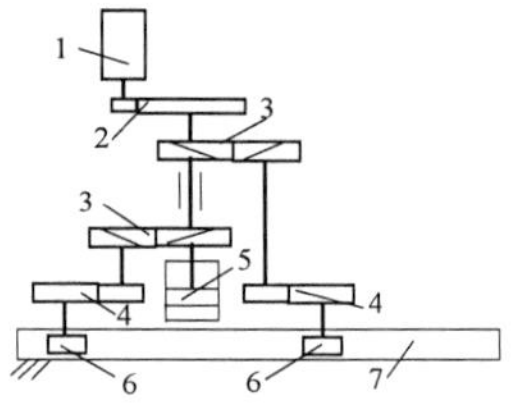

Fig. 4. A solution used in reaming and milling machines.

Hydraulic cylinder 5 absorb the clearance in the pinions due to the different senses of toothing's inclination in gears 3. In the presence of pressure, clearance on gear rack 7 is zero.

3. MATHEMATICAL SIMULATION OF CLEARANCE ABSORBTION HYDRAULIC SYSTEMS

The hydraulic cylinder assures permanent straining of the two pinions on the opposite flanks of the gear rack. In these circumstances, the following equations apply:

$$F = pS \tag{1}$$

$$dp = E\frac{dx}{x} \tag{2}$$

In equations (1) and (2), we noted: F- the straining force; p – the pressure provided by the hydraulic system; S – the active surface of the clearance absorbtion system. If the gear rack mechanism had been perfect, the pressure would have been constant. In reality, due to pitch errors and other causes, the "F" force varies, leading to other pressure variations as well. We note the pressure variations by "dp". We consider that the liquid's coefficient of elasticity is "E" and occupies a volume defined by a surface "S" and a length "x". The pump provides a constant flow rate, and the pressure is regulated at a certain value by the piloted pressure valve. The hydraulic motors used in the clearance absorbtion, especially the oscillant ones, register large flow rate losses. In the following, we present the results of a simulation regarding the functioning of system. We shall consider that the input is the pitch error in figure 5. On x, time in [s], on y pitch error [m]. The outputs are: the flow rate lost through motor (figure 6,a) and the pressure in the system (figure 6,b). On x, the time [s] and on y, the flow rate in m^3/s. On x, the time [s] and on y, the pressure in N/m^2.
It is worth noting the existence of pressure variations, around the state point. As a rule, these oscillations take place on the downward characteristic of the pump-pressure valve subsystem.

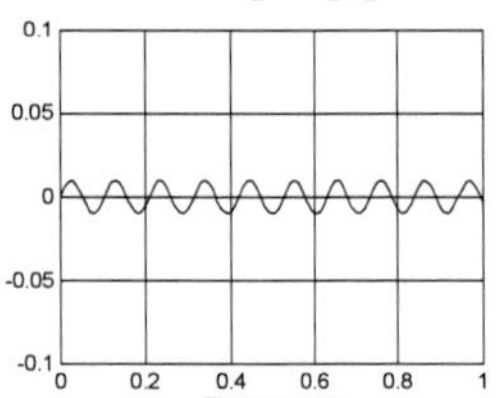

Fig. 5. The pitch error

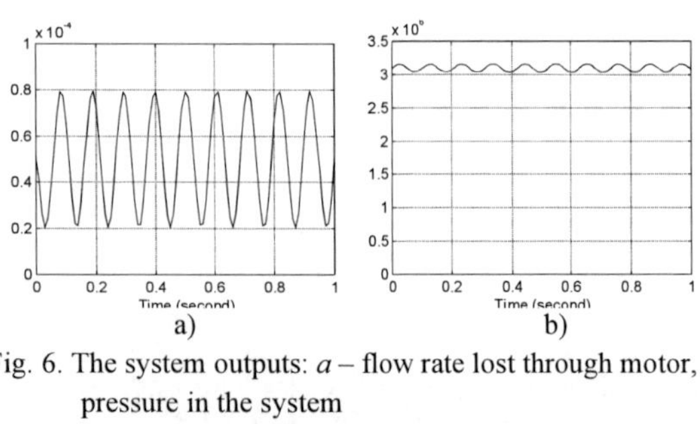

a) b)

Fig. 6. The system outputs: *a* – flow rate lost through motor, *b* – pressure in the system

Fig. 7. The driving reductor

4. EXPERIMENTAL RESEARCH

Experimental measurements were performed on a GMX30 type machine. Picture 7 presents the driving reductor. The current's value was measured in ten points, with and without a clearance absorbtion system.

5. CONCLUSIONS

Clearance absorbtion systems are useful, especially in "CNC" type machines. There are several solutions for clearance absorbtion in mechanisms of the gear rack type, depending on the machine's type and dimensions.
The optimum alternative is chosen by taking into account the machine's type, the stroke's length, the maximal speeds and the required precision. In the case of refabrication, the option of using ball screws can be analysed, provided that they can be assemblied and driven (Prodan & Constantin 2004), (Prodan & Marinescu 2005).
In the case of hydraulic driving systems, the optimum functioning parameters shall be determined, especially the pressure. The latter's reduced value leads to the emergence of a reversal clearance. High pressures stress the leading linkage and can produce its excessive wear, especially of the gear rack's.

6. REFERENCES

Prodan, D., Predincea, N. & Pupaza, C. (1999). Preluarea jocului la inversarea sensului miscarii in lanturile cinematice de avans/pozitionare cu mecanisme pinion-cremaliera, *Proceedings of TehnomuX edition X*, Tehnologii si Produse Noi in Constructia de Masini, TehnomusX 1999, pp.166-169, ISBN-973-9408-27-3, Suceava, mai 1999.

Prodan, D., Catrina, D., Constantin, G. & Bucuresteanu, A. (2004). *Preluarea jocului din lanturile cinematice de avans/pozitionare la masini-unelte grele, tehnici si tehnologii moderne utilizate in realizarea masinilor si echipamentelor pentru agricultura, sesiune de comunicari stiintifice*, Piatra Neamt, 9-10-Octombrie 2004, Publisher Bren, Bucharest, ISBN 973-648-334-4, pag.121-127, 2004

Prodan, D. (2004). *Hidraulica masinilor-unelte*, Publisher Editura Printech, ISBN 973-718-109-3, Bucharest.

Prodan, D., Pupaza, C. & Predincea, N. (1999). Metodologii de determinare a jocului la inversarea miscarii de avans/pozitionare, *Proceedings of the X-th TEHNOMUX*, 28-20 mai 1999, Tehnologii si Produse Noi in Constructia de Masini, TehnomusX Suceava, 1999, ISBN-973-9408-27.

Prodan, D. & Constantin, G. (2004). Refabrication of Heavy Machine Tools, *Proceedings of The 15th INTERNATIONAL DAAAM SYMPOSIUM "Intelligent Manufacturing & Automation: Globalisation – Technology – Men - Nature"*, pp.375-376, ISSN 1726-9679 ISBN 3-901509-42-9, Vienna, November 2004, DAAAM International Vienna.

Prodan, D. & Marinescu, S. (2005). *Refabricarea masinilor-unelte sisteme hidraulice*, Publisher Editura Tehnica, ISBN 973-31-2255-6, 2005, Bucharest.

Annals of DAAAM for 2006 & Proceedings of the 17th International DAAAM Symposium, ISSN 1726-9679
B. Katalinic, Published by DAAAM International, Vienna, Austria 2006
150th Birthday of Nikola Tesla (1856-1943) Inventor of Mobile Robot Teleoperation

MODEL FOR GEAR CONTACT SIMULATION

PUPAZA, C.; KALOUDIS, A. & ATHANASIADIS, G.

***Abstract**: The paper deals with modeling solutions regarding Hertz contact simulation. The mesh preparation procedure allows reliable contact simulation results for cylindrical gears and accurate values for stress calculation. The steps are the followings: generation of gear geometry using ProENGINEER; mesh preparation and refinement using ANSA CAE Translator; FEA using ABAQUS and complete model postprocessing with μETA System. The procedure is simple, assures full mesh control and can be further developed by including friction, thermal effects and film lubrication, as well as shafts and bearing effects in the model. It can also be used for helical gear design where the mesh has to follow the path of contact.*
Key words: *Modeling, contact, mesh, stress, gears.*

1. INTRODUCTION

For gears design evaluation, FEA is used to simulate the product's behavior and thereby predict its performance. Using CAD packages and FEA solvers, the companies offer their own solutions. However, because the accuracy of the results depends on the mesh quality, model preparation is still a preprocessing problem. On the other hand, FEA on 3D models can improve gear design, providing useful and reliable results for industry. These remarks point out the importance of the present topic in mechanical engineering transmission research.
Studies on FEA of cylindrical gears have been reported in recent literature. Heavily loaded spur gears with effects on transmission shafts and gear blanks (Tian et al., 2003) were studied using the substructuring method. An analytical approach to calculate the tooth load sharing and contact stress was also developed (Mohanty, 2003). The calculation is based on some assumptions for both tooth deflection and contact stress. Integra Engineering Inc. performs mechanical and structural engineering simulation and analysis. Using 3D CAD packages, like Solid Edge and solvers like ALGOR's Accupack for nonlinear transient dynamics, the company offers sophisticated design solutions with mechanical event simulation (Orgeron, 2002). FZG (Institute for Machine Elements-Gear Design Center) developed an integrated design environment for calculating gears. The procedure is known as RIKOR (FZG, 2004). For tooth engagement the simulation is performed with Nastran MARC and the pre- and postprocessing are done through PATRAN. The model is solved on IBM parallel computers. The whole gear design environment presumes several data transfers. The present paper presents a mesh preparation procedure that allows obtaining accurate contact simulation for cylindrical gears and reliable results for stress calculation. The procedure is simple and assures full mesh control, can be further developed by including friction, thermal effects and film lubrication, as well as shafts and bearing effects in the model.

2. CONTACT MODELING

The general objective of the contact analysis is to determine contacting areas and stress transmitted. Contact is a severely discontinuous form of geometric nonlinearity.
When performing a contact analysis, the concept regarding the local support, where nodes and elements communicate with the nearest neighboring is no longer sufficient. Points and surfaces on one body have to recognize points and surfaces on other bodies in order to avoid penetration. The contact analysis in ABAQUS Standard is based on the Master/Slave approach (Hibbitt et. Al., 2002).
The advantages when using the Master/Slave approach are as follows: accurate results can be obtained and the simulation can be done in a very realistic way; a complex model can be processed with 2D solid elements; the solver uses automatic contact path detection and optimization procedures in order to decrease the wavefront; the contact elements are automatically created by the solver.

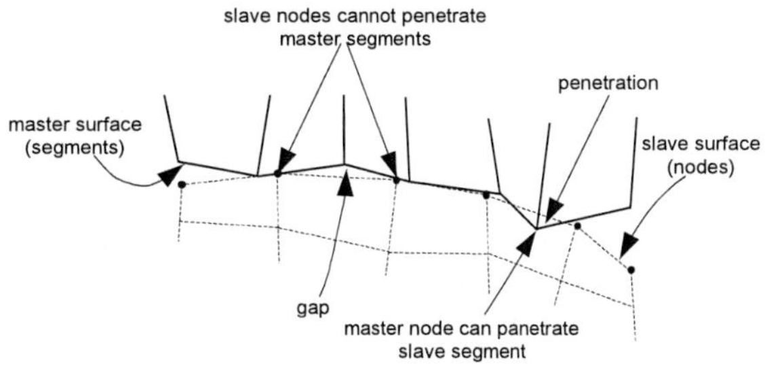

Fig. 1. Penetrations due to coarse discretization

The overall accuracy of the solution depends on mesh refinement. Penetrations may occur (Fig. 1.) (ABAQUS, 2004) when slave surfaces are coarsely discretized. In this case the master surface nodes can grossly penetrate the slave surface.
Nonuniform contact pressure distributions are likely to occur when very different mesh densities are used on the two deformable surfaces making up a contact pair. Unless perfectly matching meshes are performed, local oscillations in the contact stress may still be observed, even in refined models.

3. MODEL PREPARATION

3.1 Mesh preparation

The geometry of the FZG-A gears was used for a 3D model (Pupaza & Athanasiadis, 2004). The geometry was created in Pro/ENGINEER and exported to ANSA Translator. The thickness of the model was 20 mm. A 90 degree model for ABAQUS Standard was than obtained. Only five pairs of teeth were kept from the whole geometry and the gear faces were meshed using CPE4R2D solid elements. The preparation strategy used additional cuts, offsets and curves creation in order to avoid triangle elements in the contact zone and to assure a sufficiently fine mesh for the computational algorithm. Fig. 2 and 3 show some plots to illustrate the model preparation strategy. The mesh was refined on the flanks coming in contact during the simulation. The element length for the surfaces coming in contact was less than 0.06 mm and a very tender pass from the model boundary to the inside regions was controlled. The mesh of master and slave surfaces were refined with the same density, avoiding oscillating contact pressures.

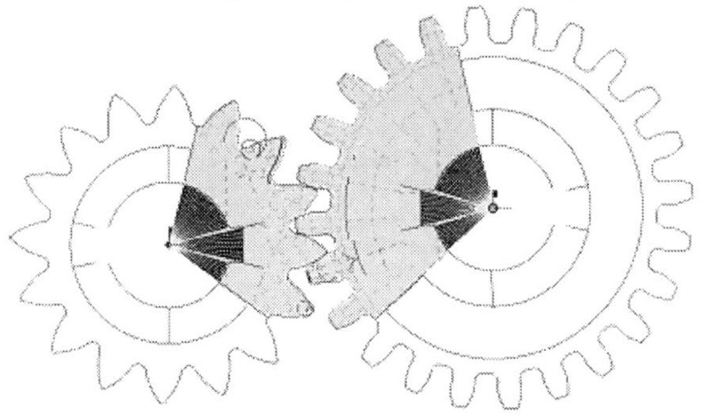

Fig. 2. Model preparation. Assembly and linked faces

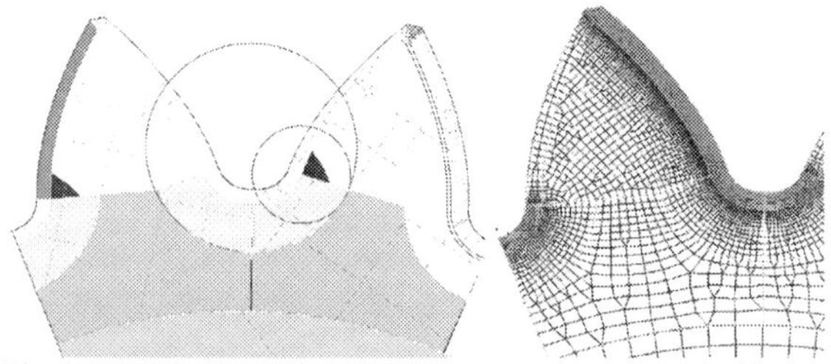

Fig. 3. Model preparation. Additional curves, cuts and offsets

3.2 Contact definition

The type of contact chosen was finite sliding of deformable bodies against each other, which is cinematically the most general. Special elements were introduced in the center of the pinion to initiate the contact between the gears and multiple point constraints were used to support the gears and to input the torque in the center of the assembly, which is the most accurate input in gear calculation. The contact definition used 15 sets containing nodes, faces and axes (Fig. 4).

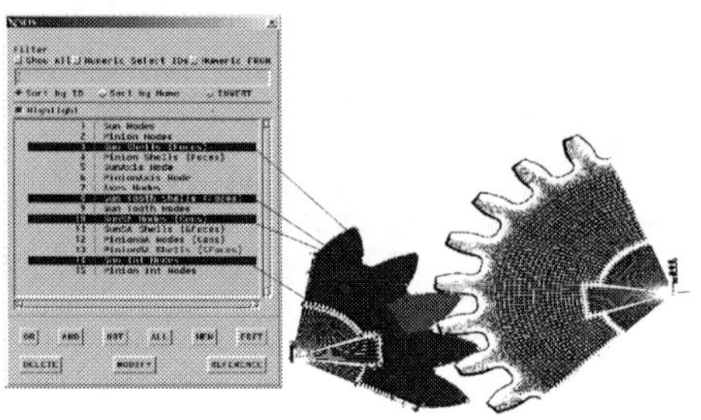

Fig. 4. Contact definition

3.3 Load

In order to prevent rigid body motion in contact simulation two load steps were defined. A first load step gave a small initial overclosure to the contact pair. The boundary conditions were removed after the body was adequately constrained through the contact with other components. In this first load case a constant torque of -67700 Nmm was applied as a concentrated moment on the driven gear Z axis. A second load step was used to run the simulation, where only boundary conditions were imposed. The maximum number of increments declared in the load steps were 100 and 300, respectively.

4. RESULTS

The contact analysis was performed on a Pentium 4 PC under Linux Operating System. The computational time was six hours and the results file was 7 GB.

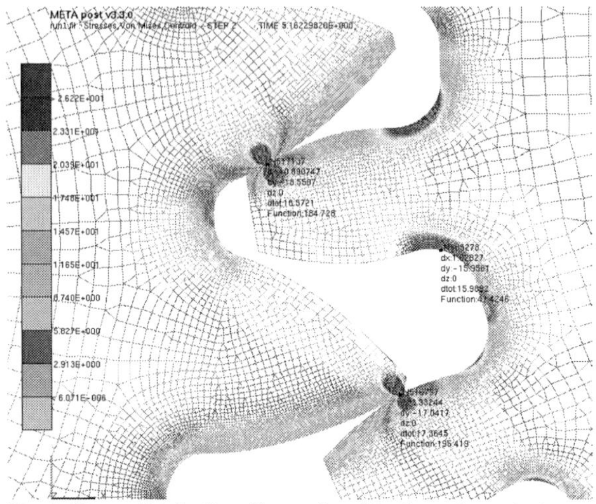

Fig. 5. Stress distribution. Two pairs of teeth are in contact.

Figure 5 contains a plot when two pairs of teeth are in contact. When processing the results, no bending effects were put in evidence on the deformed shape. The reason was to clearly monitor the contact evolution, free from other effects. The maximum Von Misses stress is 380 MPa and the maximum bending stress is about 80 MPa.

5. CONCLUSION

No suitable ideas regarding mesh generation for this special type of complex geometry and severe form of nonlinearity were found in the literature. At present, a good solution for gear Hertz contact simulation is the Master/Slave approach, which gives good accuracy if the slave surface is sufficiently refined. The modeling procedure has the following advantages: it is simple and fast; no assumptions have to be done regarding the load input and no empirical factors need to be used; the 3D simulation is based on a 2D model; even the model is complex, no parallel computing is necessary; the analyst has full control on the mesh and on the simulation input; the model can be developed by including spindles, bearings and other elements, as well as prestressed forces and lubrication conditions.

6. REFERENCES

FZG (2004). Institute for Machine Elements Homepage. Gear Research Center. *Available from:* www.fzg.mw.tum.de/forschung/edv_en. php4. *Accessed:* 2006-04-02

Hibbitt, Karlsson & Sorensen (2002). Contact in ABAQUS/Standard. Lecture Notes. Hibbitt, Karlsson & Sorensen Inc., 2002

Mohanty, S.C. (2003). Tooth load Sharing and Contact Stress Analysis of High Contact Ratio Spur Gears in Mesh. In *IE(I) Journal-MC*, July 2003, p. 66-70

Orgeron, K.J. (2002). Standardized Calculation Methods Complement Finite Element Analysis to Hasten Gear Design. *Design News*. Calculate, Integrate, Evaluate, May 20th, 2002. *Available from:* http://www.designnews.com *Accessed:* 2004-02-01

Pupaza, C.; Athanasiadis, G. (2004). Hertz Contact Simulation for Spur Gears. Romanian Academy, *Proceedings of the ICMaS'2004 Conference*, p.599-602, ISBN 973-27-1102-7

Tian, Y.; Li, C.; Tong, W. & Wu, C. (2003). A Finite-Element-Based Study of the Load Distribution of Heavily Loaded Spur Gear System With Effects of Transmission Shafts and Gear Blanks. *Trans. of the ASME. Journal of Mechanical Design*, Vol. 125, September 2003, p. 625-631.

Annals of DAAAM for 2006 & Proceedings of the 17th International DAAAM Symposium, ISSN 1726-9679
Editor B. Katalinic, Published by DAAAM International, Vienna, Austria 2006
150th Birthday of Nikola Tesla (1856-1943) Inventor of Mobile Robot Teleoperation

THE RELIABILITY OF THE METAL CUTTING TOOLS ON THE BASES OF TECHNOLOGICAL CRITERIA OF CONSUMPTION METHOD

QEHAJA, N.; BUNJAKU, A.; SALIHU, A. & ZEQIRI, H.

Abstract: *The estimation of the reliability of the metal cutting tools represents a complex issue which depends on a number of factors: the process of cutting, geometric complexity, dynamic and thermal tensions.*
The estimation of reliability of metal cutting tools in volume productions is done on the bases of monitoring failure rates of metal cutting tools in a long period of time in order to obtain a large number of samples possible. However, in a medium quantities, where the number of samples is smaller then 50 the method of medial rank is preferable.
Key words: *cutting, tool, reliability, failure, consumption.*

1. INTRODUCTION

The reliability of the metal cutting tools is a relatively new scientific field. This field includes study, analysis and development of the cutting edge characteristics in the definite conditions and time interval of exploitation, that will not change the used parameters of the allowed limits.
The aim of the paper is to research, identify and analysis the factors which bring to the failure of the instrument during the cutting process.
If we are limited on the investigation of the reliability of the metal cutting tools, within the processing system one can conclude that its depend the several factors and presents very complex phenomena because the tool could fail during the work. These failures mostly happen cause of consumption, crack and fracture of the cutting instrument. The cutting process characterized by; material of working piece, the material of instrument and the conditions of the realized processing (cutting regime, geometry of instrument, cooling equipment and lubricators and the dynamic state of system: machine–instrument–equipment-working piece).
The current researches indicate that the probability functions of the cutting instruments failure will relied on the Weibulls disperse:

$$F(t)=1-\exp[-(t/\eta_o)^{\beta o}] \quad (1)$$

The reliability is the compliment of the probability;

$$R(t)\ 1- F(t)=\exp[-(t/\eta_o)^{\beta o} \quad (2)$$

Frequency of failures is defined as follows:

$$f(t)=dR(t)/dt \quad (3)$$

The intensity of failures:

$$\lambda(t)=f(t)/dt \quad (4)$$

Investigations have been realized in theoretical and experimental ways, hereby to get approximately data about failure occurrence of the instrument while metal cutting process from the aspect of consumption, crack and fracture.
The problem that has been considered in this paper was the determination of reliability of the metal cutting tool at small and medium series by applying the medium rank method.

2. EXPERIMENTAL CONDITIONS

The reference material for work piece is rolled steel CM45 (according to ISO). The initial work piece diameter is 20,2 mm and the length is 360 mm.

The research has been realized on processing conditions of the piston of absorber (Tab. 2) Factory for Absorbers in Prishtina, where the data for the technologic consumption, crack and fracture of the instrument failure are gathered.

Data is gathered:

- verifying the number of processed pieces between two instrument substitution,
- the good pieces are identified measuring control dimensions (in every 5 pieces),
- the processing process is interrupted after identification of the first piece over the allowed tolerances after what the instrument is replaced,
- the work of the instrument is monitored visually estimating the reasons for replacement of the cutting plate; normal consumption of the edge, cracks or instrument fractures,

The criteria for defining the effective duty of the instruments:

- the exactness of dimensions in the work part (it is realized through special control metering) serves as a principal criteria,
- monitoring the edge (it is done in a visual way) and serves as preventive criteria,
- the form of the chip (monitoring in a visual way) and serves as preventive criteria,
- the conditions for monitoring the instruments are the same for all instruments,
- the examinations are done through regular technological process in conditions of exploitation.

The change of instruments:

- the change of instruments is done one by one, in the moment of failure of any crack, fracture or consumption according to the criteria mentioned above,
- the sharpening is done in the tool room, for the cases of normal consumption or if the instrument is cracked or fractured,
- if the edge is broken the entire instrument should be changed, when there should be done the repair of the base handler,
- the machine paused during the sharpening of the instrument.

The plan for executing the instrument:

- the monitoring of duty of instrument is realized according to the existing state and the existing technological process,
- the lowest number of the monitored samples between two cuttings is 36,
- the data of monitoring of the work of instruments are done in (Tab. 1).

The experiments have been performed by using the similar processing parameters in volume productions of the piston of shock absorber at the shock Absorber Factory in Pristina, Kosovo. During the experiment there have been used the hard metal plates TNMGS04 10 FR according the ISO standard No.1832 the product from "Sintal"-Zagreb, of the quality SCM-105 (covered by three layers TiC-Al2O3-TiN on the hard metal base P25) (Fig.1). The plate is attached mechanically to the holder PSBNR2020K12 with the fastening system PROMAX-C. The testing has been realized on automatic 6 axis machine Wickman .

3. EXPERIMENTAL RESULTS

The recovered data from the sustained average lifetime of the cutting edge between two changed instruments, from the monitored samples according to the experimental conditions, ordered in 9 value groups, medial rank and the reliability limit of exactness $\alpha_z = 0.05$ (5% and 95% respectively), are shown in (Tab.1). The graphical presentation of Weibull functions according the results based of the grapho-analitical calculations are shown in (Fig. 2).

4. CONCLUSION

According to the theoretical researches and experiments carried earlier and the analysis based on diagram at (fig. 2), can conclude that the plates from the hard metal P25 covered by three layers TiC-Al_2O_3-TiN during the exploiting process offered a high reliability (95%), in the period of time of 2800-3200 s.

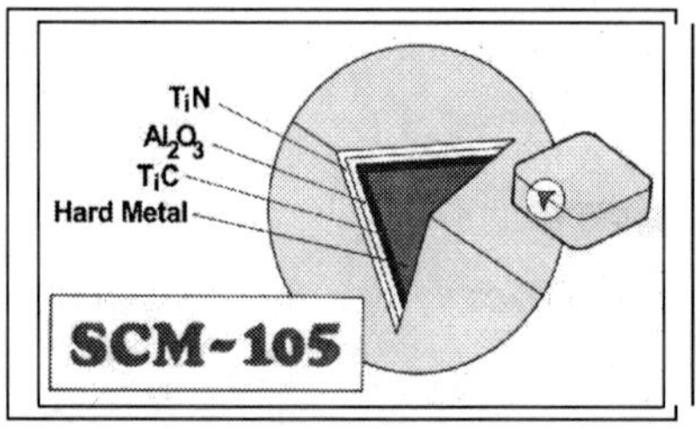

Fig.1. Cutting plate "Sintal"

i	T_{mi} [s]	5 %	RM %	95 %
1	2800	0.60	7.40	28.30
2	2850	4.10	18.00	42.90
3	2900	9.80	28.60	55.00
4	2950	16.90	39.30	65.50
5	3000	25.10	50.00	74.90
6	3050	34.50	60.70	83.10
7	3100	45.00	71.40	90.20
8	3150	57.10	82.00	95.90
9	3200	71.70	92.60	99.40

Tab.1. Resistance of cutting plate

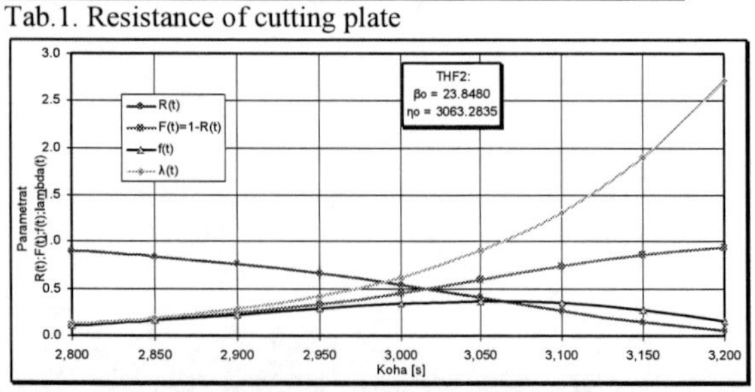

Fig.2. Graphical presentation of Weibull's functions

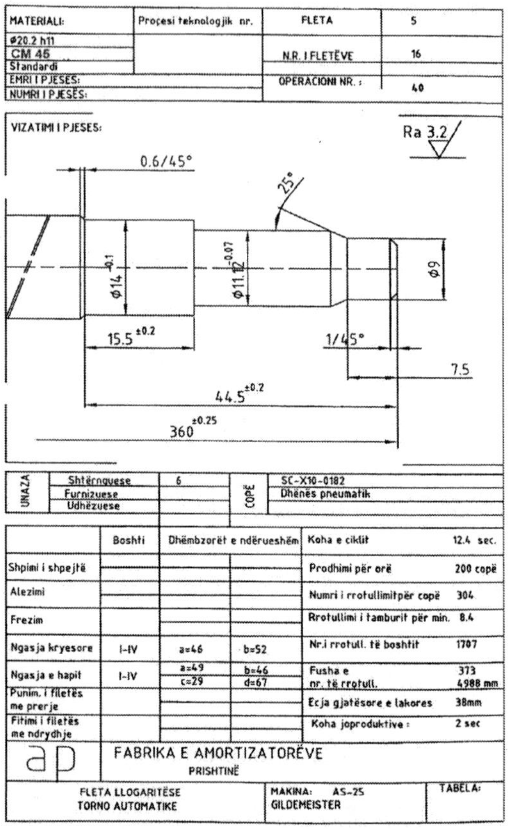

Tab. 2 Technological process of the piston of absorber

5. REFERENCES

Qehaja N. (2006). *Hulumtimi i besueshmerise te instrumenteve metalpreres gjate perpunimit me tornim te pistonetave te amortizatoreve,* punim i doktoratures, Fakulteti i Inxhinierise Mekanike, Prishtine.

Qehaja, N. (1989). Zavistnost reznih alata i produktivnosti rada u procesu eksploatacije, magistarski rad, FSB Zagreb.

Qehaja N. *"The reliability of the metal cutting tools on the bases on method of consumation"*, The 13th International DAAAM Simposium, Vienna, 2002.

Qehaja, N. (1997). *Besueshmëria e instrumenteve metal prerëse* Krasniqi, Z. Bunjaku, A, & Qehaja, N. (2001). *Shpenzimet e përpunimit gjatë regjimeve optimale të prerjes*, Revistë shkencore për teori dhe praktikë në makineri, Fakulteti i Makinerisë, Prishtinë Vol.III N_0.1 Prishtinë.

Griffiths, B.J.; (1987). *Mechanisms of White Layer Generation with Reference to Machining Processes and Deformation Processes, ASME Transactions, Journal of Tribology*, Vol. (109) pp. (525-530), http://hardingeus.com

Sekulic, S. (1983). *Odredjivanje pouzdanosti reznog alata skracanim postupkom,* XVIII. Savjetovanje proizvodnog masinstva Jugoslavije, Zbornikradova, Budva.

Sekulic, S. (1985). *Graficki postupak za odredjivanje pouzdanosti reznog alata zasnovan na broju obradaka obradjeni pojedinim alatima,*XIX. savetovanje proizvodnog masinstva Jugoslavije, Zbornig radova, Kragujevac.

Dowson,G., (1993). *Introduction to Powder Metallurgy- The Process and its Products*, European Powder Metallurgy Ass., Shrewsbury, England,. www.epma.com

Annals of DAAAM for 2006 & Proceedings of the 17th International DAAAM Symposium, ISSN 1726-9679
Editor B. Katalinic, Published by DAAAM International, Vienna, Austria 2006
150th Birthday of Nikola Tesla (1856-1943) Inventor of Mobile Robot Teleoperation

DESIGN OF ROTATIONAL PARTS IN CAPP SYSTEMS USING OBJECT ORIENTED METHOD

RADASLIC, K.; RAHIMIC, S. & MAKSUMIC, I.

Abstract: *Using functional requirements,geometrical as well as technological, as the starting point, this paper analyzes possible applications of advanced CAD-CAM systems in production of rotational parts and transfer from conventional to CNC technology. An example is given of the application of advanced CAD-CAM system for parametric design of rotational parts as the basis for CNC manufacturing technology.*
Key words: CAD-CAM, CAPP, object oriented, rotational parts, methodology

1. INTRODUCTION

The manufacturing industry has been pushed to adopt more effective and efficient production strategies to meet the challenge of shorter life cycle, higher quality, lower cost, wider variety of customer demands. This increased emphasis on achieving highly adaptive manufacturing to reduce manufacturing costs and to better utilize manufacturing capacity has led to a critical focus on agile manufacturing as a strategy to achieve these goals. In manufacturing, process planning is the task that transforms the design information into the manufacturing processes and determines the operation sequence. CAPP is considered a crucial link between Computer-Aided Design (CAD) and Computer-Aided Manufacturing (CAM). Research of over 30 years in CAPP has resulted in a wealth of knowledge on CAPP and many experimental and commercial CAPP systems have been developed as a result.

2. CAD/CAM SYSTEMS

Modern CAD/CAM systems represent an association between different system components (for example CAD and CAM module). [3]
Rotational parts, as a rule, share significant geometrical and technological likeness which makes them the ideal candidates for parametric design technique.

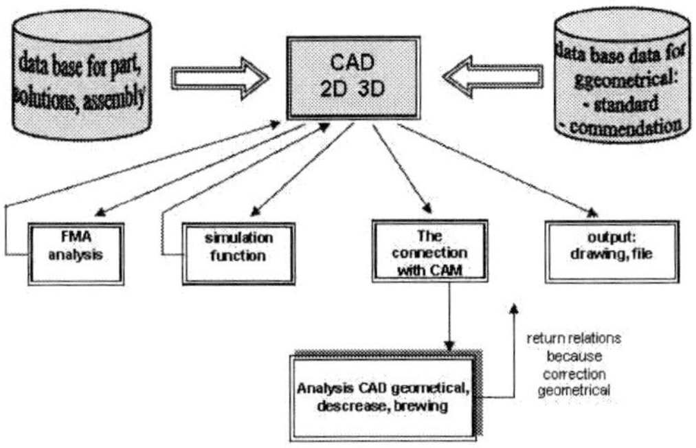

Fig. 1. Product analysis and design

This paper gives an example of using parameteric design technique for rotational parts in a CAD-CAM system. The CAD model was made using SolidWorks, and CAM model using proprietary application developed using MS VBA. Schematics for this system are given in the Figure 1. above.

2.1. CAD/CAM interactive programming

Geometrical information is memorized in a CAD system's database, so it can be called upon and used in programming – a CAM system interface is shown in figure 3.
Interactive programming takes two forms:

- CAD graphical workstation's functionality is extended to CNC programming using software tools.
- CNC programming is done on a workstation running afore mentioned application which transforms design data obtained from CAD-CAM database to G code. G code thus obtained is ready to be used on the appropriate CNC machine.

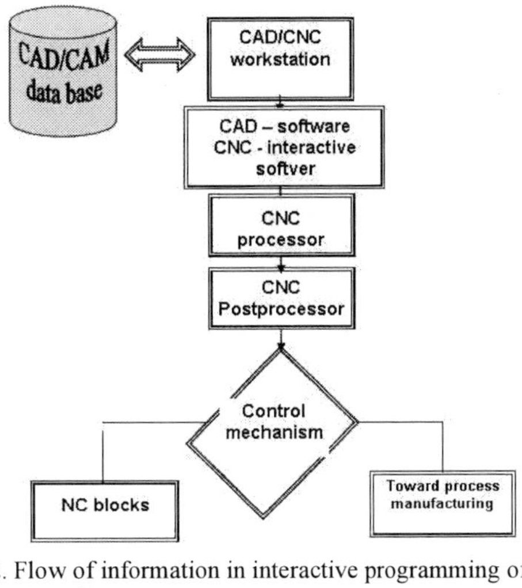

Fig. 2. Flow of information in interactive programming of CAD/CAM systems

3. PARAMETRIC DESIGN IN CAD/CAM SYSTEMS

Modern CAD/CAM systems offer extended range of different approaches for product design and technology; these alternative approaches include: explicit, parameteric, variant and modelling based on typical forms ("Features"). Today's standard approach in CAD/CAM systems is explicit modelling whose basic shortfall is lack of flexibility when products share basic geometrical-technological likeness and there's a possibility of taking advantage of product family likeness in forming of parts. The parametric modelling is rooted in CAD-

3D model – 2D drawing associativity (where changes in 3D model cause changes of 2D drawing). Parametric design gives control over model; parametric definition of "lead" dimensions enables realization of geometrical and other limits and relations (parallelism, normality etc.) Parametric information fully defines the model, on all system levels and modern CAD-CAM systems have bi-directional associativity (change of parameters on any model level causes all other model levels to change). Figure 4 shows a typical example of rotational part suitable for parametric modelling and automatic CAD to G code conversion using object oriented methodology.

4. MINIMUM COST PER PIECE

Correct G code, which can be used for CNC machining without any modification, represents the desired output of a CAD-CAM integrated system. Ideally, G code encompasses information which makes machining process economically viable. An attempt to give economical significance to a technological parameter (cutting speed) is described below.
The average cost to produce a work piece consists of the following costs: nonproductive cost per piece, machining time cost per piece, tool changing cost per piece and tooling cost per piece.
Mathematically, this can be expressed as:

$$c_1 = c_o t_1 + c_o t_c + c_o t_a \left(\frac{t_{ac}}{T} \right) + c_a \left(\frac{t_{ac}}{T} \right) \quad (1)$$

Upon partially differentiating c1 with respect to v, equating the result to zero (2), and solving for v, we obtain the cutting speed (v_{min}) (3) which corresponds to minimum unit cost as follows

$$\frac{\partial c_1}{\partial v} = 0 \quad (2)$$

$$v_{min} = \sum_{j=1}^{i_p} \frac{C}{\left[\left(\frac{1}{n} - 1 \right) \left(\frac{c_o \cdot t_{aj} + c_a}{c_o} \right) \right]^n} \quad (3)$$

Graphical representation of the above mentioned equation is given in Figure 3. This approach enables us to calculate the value of cutting speed using financial parameters thus giving a new dimension to the proposed system.

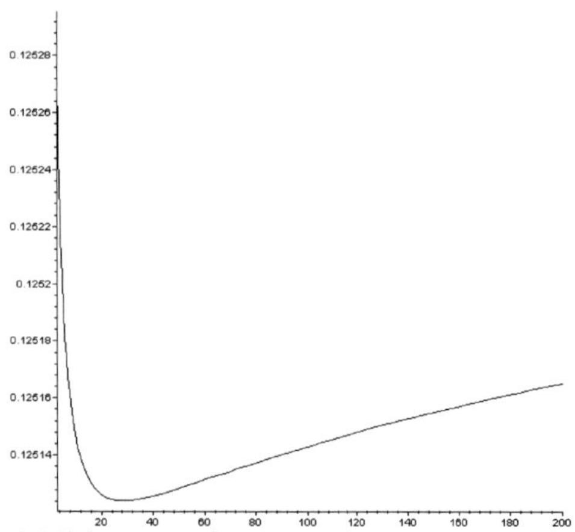

Fig. 3. Minimum cost function (c_1) by v_{copt}

In majority of cases, an automatic process planner should be able to reason about manufacturing features that may be machined in a single opration and/or with a single tool and/or on a single machine.

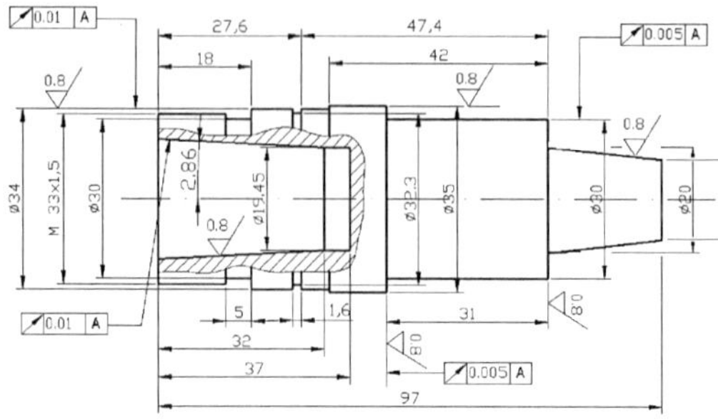

Fig. 4. 2D drawing of the part

The set of frames is usually obtained through the interface between process planning and design. (Višekruna, V.; Petković, D.; & Rahimić, S.) One of the forms which make such an interface is shown in Figure 5.

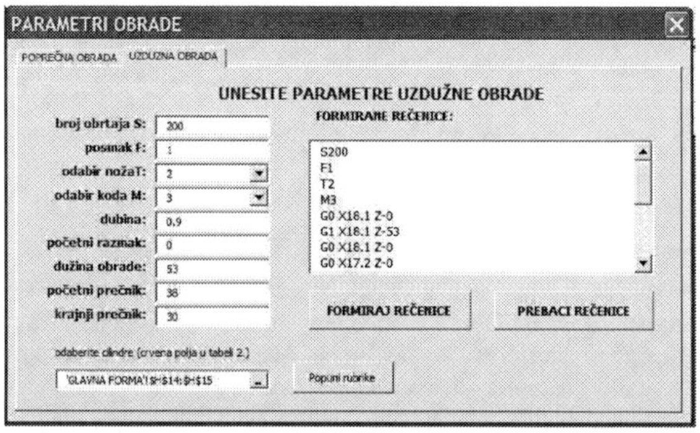

Fig.5. Interface form for alternative process planning

5. CONCLUSION

This paper proposes a methodology and architecture suitable for generating the process plan for machining of parts. The proposal takes into the account the peripheral processes needed and provides the bases for development of alternative process plans. The introduction of modern CAD/CAM systems represents an important step in an effort to devise an alternative to the production systems available on the market. Parameter modelling and integration realized using modern CAD/CAM systems, represents an example of rotational parts modelling and enables integration of CAD and CAM modules. The goal is to link CAD and CAM systems in order to realize flow of product information from its design to its manufacture.

6. REFERENCES

Ben-Arieh, D.; Gutin, G.; Penn, M.; & Zverovitch, Y.; *Process planning for rotational parts using the generalized travelling salesman problem*, int. j. prod. res., 2003, vol. 41, no. 11, 2581–2596

Gonzalez, F.; & Rosado, P.; *General and flexible methodology and architecture for CAPP: GF-CAPP system*, int. j. prod. res., 2003, vol. 41, no. 12, 2643–2662

Hang-Wai, L.; & Hon-Yuen, T.; *Object-oriented analysis and design of computer aided process planning system,* Int. Journal CIM – vol 13, 2000.g.,

Kusiak, A.; *Decomposition in Data Minning: An Industrial Case Study,* IEEE Transactions on electronics packaging manufacturing, 2000

Visekruna, V.; Petkovic, D.; & Rahimic, S. (2004). *A contribution of the generative gt-system in the filed of forging industry,* Proceedings of 4th International Symposium on Intelligent manufacturing systems – IMS Sakarya, Turkey

Annals of DAAAM for 2006 & Proceedings of the 17th International DAAAM Symposium, ISSN 1726-9679
Editor B. Katalinic, Published by DAAAM International, Vienna, Austria 2006
150th Birthday of Nikola Tesla (1856-1943) Inventor of Mobile Robot Teleoperation

INTERMODAL APPROACH TO PASSENGER HANDLING

RADOS, B.; RADOS, J. & KOLAR, V.

Abstract: *Intermodality is a technical expression which specifies the type of travel which tends to provide the passenger with maximally simple connections between transport means. As such it can contribute to integrated and efficient transport system that will establish a network of interconnected nodes, where the transfer from one mode to another is simple and smooth and provides the passenger with multi-criteria choice.*
Key words: *combined transport, quality criteria, Intermodality*

1. INTRODUCTION

The notion of intermodal transport appeared for the first time in the USA and has become today the international term for global logistics and transport. The notion "combined transport" is a European term which originates from the German, i.e. French expression "transports combines" although this term is not an adequate equivalent for intermodality.

D'Este definition is: Technical, legal, commercial and management framework for door-to-door transporting of goods using more than one mode of transport. The definition emphasises that this is more a service than a technology. Therefore, more emphasis is placed on the "soft" aspects of the service, which facilitate the delivery, rather than on the multi-modal transport technology.

D'Este definition is mostly used since it includes the physical side of transport, service which is related to intermodal transport systems and the organization of such systems. However, intermodal transport should not be confused with express and parallel transport systems.

Transport systems are characterised by successive movement of goods between the points of production and distribution, which are defined as nodes (terminals) in which all the activities are performed, such as consolidation, sorting, warehousing and transhipment, transport between vehicles as well as the shift between traffic branches. For every transport task every node can be defined as the origin, end point or place of transloading. The throughput of goods is always delayed at the node. Connections which represent transport and connect by their activities the nodes, determine the network. In this way the physical transport network is defined, unlike the abstract networks that can be defined as the connection i.e. shortcut between the physical network sections. Following the terminology proposed by Churchman (1968), the nodes (terminals) and links are designated by the term *components*, while physical parts of the operative technology are designated by the term *resources*.

In studying the transport networks two to four modes of the goods flow are of special interest. The flow of goods which is directed one-way from the consigner to the consignee and the flow of resources which is two-way and ideally overlaps the other flow in the opposite direction. In reality the mobile resources are often redirected along the network in order to realize their turnover. UTI and railway cars move more freely than the road vehicles that are dependent on their headquarters. Only a small part of the entire transport of goods is performed by direct transport using a single transport means. The majority of shipments have to be collected, stored and transloaded between the transport means. By collecting and arranging cargoes on the principle of collective transport regarding their characteristics into standardised transport units, economic profit is realized.

In studying the transport chains two general principles of development can be isolated, which are described as the technological openness or commercial openness. The technological openness in general terms can be generalized as the level of restriction in technical handling of different transport units, wagons, trucks and in some aspects also the transloading machinery. The system with the lowest technical openness allows receiving and usage of only specialised non-standard transport units, wagons, trucks and transloading machinery. The operation freedom of these systems is reduced, but the narrowly specialised subsystems can function in this way. The terminals have to decide on their own about their transport development policy, and whether they will become an integral part of the general systems of the flow of goods or offer their customers door-to-door service through their specialised and isolated system. Analogue to the technological openness we can define also the commercial openness as the level of restrictions in commercial handling of different clients. The system with the lowest commercial openness will accept only one client, whether a shipper or an intermediary. These services can be officially restrictive, they may evolve as services for only one client or they can be accompanied by adequate tariff policy as regulator. Technological openness is a matter that is dealt with by the system designers, whereas commercial openness is traditionally strongly subjected to government development strategies. In realising the uniformity of transport chains it is essential to create the one-sided regulations which will achieve uniformity through various measures and thus also a quick and high-quality throughput of goods. These measures result from the legally adopted standards and development guidelines and they are based on the external effects such as emission of air pollution, noise, greater number of traffic accidents and working conditions.

In order to plan and construct a compatible infrastructure a decision on standard dimensions of vehicles has to be brought. The allowed dimensions vary from one transport model to another, which is illustrated by the lack of uniformity in railway infrastructure.

External effects are extremely important in designing the transport system. In order to satisfy the regulation it is important to realize accurate assumptions of the costs for every transport mode. The tariff policy makes it possible to achieve the ban on pollution, noise or transport of harmful materials; it is precisely here that the possibility opens up for the development of new, safer and cleaner technologies such as intermodal transport. In intermodal transport chains the most important is the standardisation since it results in optimal flow of goods.

2. DISCUSSION

Intermodal transport chain is defined by four functions:

a. Composition. The process of arranging and consolidating the cargo at the terminal that offers intermodal interface between the local and the regional system of distribution. It would be ideal if the cargo groups arrived from various consigners already grouped at the distribution centres so that they could be dispatched by high capacity models, such as rail and maritime traffic. The transport is further operated by road which offers flexibility of door-to-door delivery. The activities such as packing and warehousing are included in the compositioning, which is closely connected to the functionality of production.

b. Connecting. It influences a determined flow model, such as cargo train or container ship or even truck convoy, between at least two terminals, operated in internal or international traffic. The success of such connections depends greatly on the economic measures such as double cargo stacking.

c. Replacement. The majority of intermodal functions is carried out at terminals whose function is to insure efficient continuity in the transport chain. These terminals are the dominant places of domestic or international cargo distribution system.

d. Decomposition. At the moment when the cargo reaches the terminal which is closest to the given destination, it should be transloaded and transferred to local, regional cargo distribution system. This function, which is related to the function of compositioning is usually performed within the limits of bigger towns and usually causes problems in the distribution which are called urban logistic problems.

e. Intermodality has been implemented in several EU documents. "Transport Policy White Paper (2001)" identifies integrated issuing of transport documents, and luggage handling. The dokument related to this "White Paper DG TREN" places emphasis on the cargo transport sector.
The objective of implementing intermodality in transport of both cargo and passengers is to achieve door-to-door transport chains, and from this aspect the last phase of travelling will be analysed, the so-called urban kilometres.

3. CONCLUSION

Journeys over longer distances (>100 km) have low market share, only 1-2 per cent out of the total journeys, but they account for about 20 per cent of passengers per kilometre. This percentage has high ecological impact and continues to grow. The traffic forecast given in TEN-STAC study gives a journey increase over longer distances of 28 per cent for the period from 2000 to 2020 in the current EU member states and 74% in the new member countries. Growth is especially expected in air traffic from 88 to 133 per cent.

Great many journeys on longer relations are for private reasons, 31% holidays, 47% other private journeys, whereas about 22% of passengers travel for business.

In journeys over longer distances, the use of cars is prevailing (65%), aircraft (14%), trains (12%), coaches (6%) and the rest 3%. There is strong asymmetry both in the number of passengers and in the intensity of travelling.

Intermodal travelling of passengers includes several different modes of travelling.
Co-operation is essential for the optimal development of long distance passenger transport. There is generally no specific national or regional intermodal strategy. Similarly, at this moment no central independent institution or structure is responsible for the coordination role in long-distance intermodality - which is a barrier to the development of a platform for data exchange between stakeholders, an integrated timetable and ticketing system and other structures of importance for passenger intermodality. The national legal and regulatory frameworks in place are incapable to impede the possible negative effects of the liberalized market. A co-operation and competition is possible in certain market environments if a win-win situation can be created. Although co-operation between competing companies seems difficult to achieve, it is unanimously considered a prerequisite to a fully integrated transport network.
In order to implement intermodality in the transport of passengers, it has to satisfy the following requirements:

1. Networks and terminals – Infrastructure networks have to be flexible and interoperable. Additionally passenger terminals need to be developed and modernised as connections between two or more modes (especially the air-rail traffic connections, rail-urban traffic connection). Transport services of various public and private operators have to be coordinated in order to satisfy the passengers' needs for more flexible, adequate and fast transport system, especially if one considers the competition in the use of passenger cars it is necessary to develop systems oriented towards passengers' needs. Mode shift results in reduced comfort and/or the shift includes additional costs, and the terminals appear as the key points which can contribute to intermodality.

2. Door-to-door information. Integrated door-to-door systems that function in real time are the key to the development of functional and attractive European intermodal transport of passengers over longer relations. The technology has crucial influence on the development of intermodality. Information that is necessary and crucial for the passengers' continuation of the journey is very often inaccessible or impossible to obtain just for a certain travelling module. The development has been achieved at the regional urban level, but has no equivalent intermodal structure which would promote integrated information at the national or EU level.

3. Tariffs and fares. Organisational technical aspects are strongly interwoven. Technical solutions that are today integrated into the tariff and fares systems (including booking and charging) are already available and in use and will continue to develop. The development in smart card technology has not been adequately used, so that standardisation in this field is of highest priority. The greatest barrier is the lack of cooperation among all the transport stockholders, especially on longer journeys and those journeys that include border crossing.

4. REFERENCES

Toward Passenger Intermodality in the EU, Report 3 (Final Version) , Dortmund, December 2004

Analysis of the Key Issues for Passenger Intermodality, Report 1 (Final Version) for the European commission unit G3, Dortmund, July 2004.

Analysis of the National Inventories on Passenger Intermodality
European Commission unit G3, Motorways of the sea And intermodality, Dortmund, October 2004

Annals of DAAAM for 2006 & Proceedings of the 17th International DAAAM Symposium, ISSN 1726-9679
Editor B. Katalinic, Published by DAAAM International, Vienna, Austria 2006
150th Birthday of Nikola Tesla (1856-1943) Inventor of Mobile Robot Teleoperation

MODEL OF OPTIMIZING THE OPERATION OF SMALL AIRPORTS

RADOS, B.; RADOS, J. & KOLAR, V.

Abstract: *Air traffic certainly has great significance due to the extremely strong transit area for air routes towards all the continents. Because of its geotraffic position and territorial distribution and tourist orientation, Croatia is inevitably oriented to further development of air traffic.*
Key words: *optimizing small airports, operation*

1. INTRODUCTION

Seven international airports have been constructed until now in Croatia:
(Zagreb, Osijek, Pula, Rijeka, Zadar, Split, Dubrovnik)

and eleven sport-tourist ones:
(Brač, Lošinj, Osijek-Čepin, Vrsar, Lučko, Varaždin, Čakovec, Borovo, Slavonski Brod, Grobnik and Sinj).

For further development of air traffic the construction of minor airports is planned, that would accommodate short take-off and landing aircraft for the tourist needs:

(Karlovac, Rab, Pag, Dugi otok, Hvar, Korčula, Vis, Lastovo, Ploče, Krapina, Koprivnica and Udbina).

These airports would significantly contribute to meeting the tourist demands, and those on the islands would provide better connections between the island and the continental part of Croatia.

Based on the geo-traffic position, Croatia has an unavoidable transit function in the air traffic of Europe. In order to bypass the Croatian space, agreements on air traffic control with all the neighbouring countries need to be signed, which are in the interest of both parties and in the interest of the international community.

Today, a part of north-south transit traffic has mainly returned, i.e. on the lines from Central Europe to Italy and North Africa.
The Croatian air traffic control is already technically and regarding personnel ready to take over the majority of the pre-war traffic.
In order to realize the optimal operation of such a diverse system, it is necessary to apply one of the methods which correspond to the given system.
For centuries the solving of certain problems had been considered in a classical analytic way, and there was a deterministic way of looking at the world, the problems had been solved in isolation from the environment and inductive research prevailed.
The development of science and technology at the beginning of the 20th century brought also the beginning of systemic thinking and systemic approach in theory and practice.

The basic guideline of the theory of system is systemic thinking, which is in fact the way of thinking in which every problem, relation, phenomenon or real phenomenon is considered as a set of elements connected into a whole.

2. DISCUSSION

Changes in transport logistics or just decisions in general affecting logistics can be manifold. The corporate manager decides to open or close down a production facility thinking in a different way than the truck driver deciding which way to drive his lorry.

These different decision levels are used by McKinnon (1998) to propose an analytical framework differentiating between levels of decision making within the field of logistics
Holism is an evolutionist theory according to which organism as a whole is something completely different and more than a mere sum of activities of all its parts.
For example, if an increase in transportation causes congestion, accidents, pollution, etc. it may lead to government intervention giving new conditions for transport and thereby inducing new logistic procedures.
The basis of holistic problem solving is the systemic approach based on systemic thinking according to which every problem can be defined as a system, i.e. a whole, and where every system is at the same time also a subsystem of a system of a higher level, and vice versa as well.

The notion of system in modern theory and practice is defined in various ways, i.e.:

- A system is a set of interconnected elements.
- A system is a set of interconnected elements (parts, processes, subsystems) which function according to certain rules, directed towards a certain objective, and form a relatively independent whole.
- A system has a structure, realizes a function, and transfers information or compares them.
- A system is a set of objects united by rules of mutual interaction.

According to the given items, a system can be defined also as a set of mutually, purposefully connected and interdependent subsystems and elements that by means of transport infrastructure, transport suprastrukture and intellectual capital allow production of transport services, i.e. transport, transfer and relocation of transport items from one place to another, overcoming spatial and time dimensions.
Or, it may be defined more narrowly as a set of mutually, purposefully connected and interdependent subsystems and elements that by means of traffic infra- and supra-structure and the intellectual capital allow production of traffic services.
The most complete and the most demanding approach applicable in analysing the work of transport systems is by all means the Churchman's systemic approach, which is defined through the systemic approach and analysis of the environmental factors, components (stockholders) and resources.
Churchman defines the basic aspects of the system: objective, environment (stockholders who influence the system but cannot be influenced within the same system), resources (stockholders

who influence the system but they can be influenced within the very system), and the very resource components and the system management.
Technical standards guide the manufacturers and system designers. Standards for equipment for transportation generally define the interfaces between system resources in terms of dimensions and fastening points but they also stipulate the required construction strength.

Most significant for development of intermodal transport systems are standards stipulating the snize of unit loads. These standards are narrowly linked to regulations for use of infrastructure. The purpose is that vehicles loaded with suitable unit load combinations shall benefit from the maximum vehicle weights and dimensions.

Technical standards are obviously stipulated in order to simplify the development of complex systems, but it also implies restrictions for the system's designer. The European intermodal system has not been hampered by too many standards but rather the opposite. Unit load standards have been established after discussions over many years meaning that some standards have been obsolete from the beginning. Swap bodies have constituted a special problem since they have long lacked a standard. When the standard was finally approved it included the lengths 7.15, 7.45 and 7.82 m. Three standards within 70 cm is regarded as almost a non-standard.
A transport system designer is advised to follow standards as far as possible. Especially the choice of unit loads should only diverge from standards when there are very good reasons for doing so.

Even when firm standards are missing, technological development can be hampered by the presence of dominant technologies, so called de facto standards. This does not only include technical resources but also whole procedures and principal solutions. One example of de facto standards of transport procedures is the over-night traffic solution in European intermodal transport. Trains stand at the terminals all day and travel between terminals over night. This has long been the prevailing way of doing things, but demand for more advanced logistics services, separate tracks for high speed passenger trains and new technologies for short stops at terminals has opened up for increased frequencies during daylight.
One reason for choosing another technology rather than the prevailing one is obviously that new and better technology is available.

WOXENIUS MODEL

Synthesised reference model was developed by Woxenius (1998), and is based on the following three perspectives: 1. classical / technical perspective, 2. networks, and 3. channels / chains.

His model was influenced by Jensen (1990) with his model of division to administrative and physical strata, Churchman (1979) with the described model, and network approach which is based on the work of Hakansson and Johansson,(1992, Uppsala) and in which the division is done to stockholders, activities and resources.

In 1998 Woxenius gives his reference model, based on the work of the mentioned authors, as a synthesis of the models that are based on the classical / technical, network and channel / transport chain perspective.
Based on the Churchman's systemic approach as the base model, other theories have evolved, one of the more significant ones apart from the mentioned Synthesised reference model developed by Woxenius (1998), the theories of OO Model are being developed.

THE OOM THEORY

The OOM Theory i.e. object-oriented model is used in the development of a wide range of computer software models.

The characteristics of OOM, such as the possibilities of generating quality, modularity, reusability, data encapsulation, occur also in the major systems of the authors, such as: Wache (1998), Zhang (1999), Kovacs (1999) who used OOM in the production processes and McCormac and Roberts (1996), Bullinger (1997), Foster and Zhao (1998) in transport.

In 1996 D' Este introduced also a perspective which was based on an event, which is similar to the OOM model in modelling intermodal transport system. Arnas (2002) developed several application examples of OOM model in transport systems.

In selecting a model, OOM visualisation provides strong comparative advantage since it allows faster and easier understanding of the system, saving into bases is one of the reasons for easy use of this model since the in-advance input and/or used data can be used, data encapsulation means that the data are used but all are not displayed for easier monitoring.

This methodology has developed in order to reduce the complexity of conceptual models of intermodal roads-railways of the transport systems.

THE THEORY OF NETWORKS

The theory of networks applied in transport has been developed by Woxenius (1998), Waidringer (1999), Hultkrantz (1999), Acharjee (2000) and Rohani (2000), Department of Logistic and Transportation.

Transport network consists of consignees and consigners with their cargo transport requirements, various transport operations that need to be performed and different traffic and terminal models.

In case of so many stockholders the number of combinations in the network grows and becomes complex, which is also influenced by the requirement of the consigner and consignee for the maximally efficient and good transport.

3. CONCLUSION

This paper presents a ways to model the logistical way to operative functioning of a small airport.
The different types of indicators (quantitative and qualitative) combined with the model describe the interaction between logistics and transport. The theory is relevant in this context as - according to the theory - modifications of the observed transport of goods will be either due to changes in infrastructure or changes in logistic principles concerning the dominating companies (and/or a combination of the two). By using this definition it is difficult to determine whether the changes in logistics are due to infrastructure or other conditions concerning the logistic principles.

4. REFERENCES

Woxenius, J; *Development of small-scale intermodal freight Transportation in system contenty*, 1998,Chalmers University of Technology, Göteborg, Sweden
Stigller, H; *The theory of price.,ucla NY editioning, 1991*

Annals of DAAAM for 2006 & Proceedings of the 17th International DAAAM Symposium, ISSN 1726-9679
Editor B. Katalinic, Published by DAAAM International, Vienna, Austria 2006
150th Birthday of Nikola Tesla (1856-1943) Inventor of Mobile Robot Teleoperation

AIR TRAFFIC DEVELOPMENT GUIDELINES

RADOS, B.; RADOS, J. & LOVRIC, I.

Abstract: *In this work the criteria of the optimal development of air traffic in Croatia have been set by using the SWOT analysis. The main factor of success of the project is the capacity of involved institutions and organizations. SWOT analysis is the tools of estimating the possibilities of the institutions / organizations.*

Key words: *Air Traffic, external factors, organization*

Strengths – positive internal attributes of an organization;
Weaknesses – negative internal attributes of an organization;
Opportunities – external factors that can have positive influence on the project or organization;
Threats – external factors that can have negative influence i.e. reduce the opportunities of an organization.

1. ANALYSIS OF THE ORGANIZATION CAPACITIES

PE (E) ST analysis

The second part of the SWOT analysis was used to perform a good analysis of the opportunities and external threats that jeopardize the organization – from its surroundings, environment. In this way the strategy has been developed which will be based on the objective evaluation of the environment within which the organization is operating. The analysis was performed per groups. The name PE(E)ST is derived from the initial letters of the main groups: political, economic, (ecological), social, technological.

Political factors

Passenger air traffic represents the strategic orientation for the future development of the total traffic in this region, primarily because it represents the shortest connection between the Adriatic Sea and the Mediterranean and the countries of Central and Eastern Europe (Austria, Slovakia, Hungary, Serbia and Montenegro, Bulgaria and Rumania).

Political factors

Passenger air traffic represents the strategic orientation for the future development of the total traffic in this region, primarily because it represents the shortest connection between the Adriatic Sea and the Mediterranean and the countries of Central and Eastern Europe (Austria, Slovakia, Hungary, Serbia and Montenegro, Bulgaria and Rumania).

Economic factors

The Republic of Croatia will join the implementation of the "Marco Polo" program. The program supports the commercial actions on the market of services for cargo transport, and the promotion projects of modal transfer in all the segments of cargo market, and not only in combined transport.

Ecological factors: optional element

The ISPA program of EU plans financial support for the investments in the field of environment protection and transport improvement in the 50:50 ratio, so that a part of these means needs to be invested into the construction of traffic infrastructure.

SWOT analysis

	Strengths – strong points		**Weaknesses – weak points**
OORGANIZATION	- safe transport method of a large number of passengers; - all the existing terminals are located at intersections of large corridors; - environmentally friendly and sustainable transport method		- undeveloped terminal network; - only one operator; - insufficient usage of the existing terminals; - failure to accept
	Opportunities	**P EST**	**Threats**
ENVIRONMENT	- closeness of strategic markets; - stabilization of political and economic conditions in the environment and demand growth; - transit traffic potential; - raising of ecological awareness in Europe and redirecting of passenger traffic from road to air		- lack of investments; - inadequate condition of the air infrastructure; - poorly developed infrastructure; - lack of clearly defined development plan.

SWOT analysis of transport needs to be invested into the construction of traffic infrastructure.

Social factors

The national producers of equipment and transport capacities intended for air transport need to be additionally supported and adequately assisted at all levels, especially regarding issues of granting and using favourable credit lines with the aim of achieving overall competitiveness on the national and international market.

Technological factors

The Republic of Croatia as the Mediterranean, Danube and transit country has a favourable geo-strategic position, representing an important factor for its integration into the European traffic system.

Quality criteria

For the assessment and obtaining of correlation indicators the method "weight criterion" was used, which has been found useful for evaluating the alternative solutions in case there are several input criteria (parameters). This method has proven best in the majority of criteria that cannot be simply ranked by

Criteria

	K1	K2	K3	K4	K5	Corr	Pi	K i (p i/Σ p i)
K1	1	2	2	1	0	1	7	0,27
	K2	-2	1	0	1	3	3	0,11
		K3	-3	2	0	5	4	0,15
			K4	-3	2	7	6	0,23
				K5	-3	9	6	0,23
						Σ :	26	1,00

numerical values but are added to the value obtained by comparison according to in-advance set variables. The method was originally developed by Bjarenemo in 1982 for the evaluation of alternative solutions in designing machines, but it is used more for the evaluation of intermodal transport technologies.

The method requires:

1. defining of conditions to evaluate the situation;
2. list of requirements and criteria;
3. list of alternative solutions;
4. multi-criteria estimate of parameters.

Criteria given for evaluation:

K1	Intermodality of air traffic
K2	compatibility with the currently dominant types of ITU, vehicles and ships
K3	allows simple, flexible, fast, reliable transloading from ship to ship
K4	logistic support, unnecessary coordination of vehicles and terminals
K5	possibility of system expansion

Analysis of the given criteria:

where the following values are used:

If:	Value amounts to:
K 1 is more important than K2 (to K n)	2
If the values are equal	1
If K2 is more important than K1 (to K n)	0

IMT quality criteria, designed by the author

The analysis in the matrix yields numerically expressed values of correlation of the given criteria. Column P i is added as a column which sums the value in the rows.

Weight factor K i is calculated according to: Ki = Pi / Σ Pi . If the sum of all Ki amounts to 1, then the matrix is correctly completed

2. CONCLUSION

Evaluation of the variants according to the set criteria: In order to evaluate the variants that are assumed as the possible quality criteria, we place them in the matrix which weights the obtained results in the scale. The criteria are weighed according to the given options:

Options:

The strengths of the variants are given numerically, and can be compared.

3	meets the criteria well
2	likely to meet the criteria
1	unlikely to meet the criteria
0	impossible to meet the criteria

		V 1		V 2		V 3	
Cri.	Ki	Opt.	fKi	Opt.	fKi	Opt.	fKi
K1	0,27	3	0,81	1	0,27	2	0,54
K2	0,11	2	0,23	3	0,69	1	0,12
K3	0,15	2	0,31	3	0,69	1	0,15
K4	0,23	3	0,69	3	0,69	2	0,46
K5	0,23	1	0,23	1	0,23	2	0,46
Σ :	1,00	11	2,27	11	2,58	7	1,73

Evaluation of the development variants, designed by the author

At the national level, little attention is given so far to the study of this market and its potential. Nor is much information available on the possible impacts of air traffic and services through cost-benefit analyses or impact assessment studies. Nevertheless, it is becoming a greater topic and consequently viewed with greater importance.

REFERENCES :

Woxenius, J; *Development of small-scale intermodal freight Transportation in system contenty,* 1998, Chalmers University of Technology, Göteborg, Sweden

Annals of DAAAM for 2006 & Proceedings of the 17th International DAAAM Symposium, ISSN 1726-9679
Editor B. Katalinic, Published by DAAAM International, Vienna, Austria 2006
150th Birthday of Nikola Tesla (1856-1943) Inventor of Mobile Robot Teleoperation

AIR TRAFFIC DEVELOPMENT STRATEGY

RADOS, B.; RADOS, J. & LUBURIC, G.

Abstract: *The traffic forecasts are the basis for determining the development of the subsystem in air traffic. Depending on the expected number of passengers, cargo volume, number of aircraft operations, aircraft are purchased, airports constructed or reconstructed, air traffic control developed. The majority of service users in air traffic in Croatia are tourists (about 60 to 65%).*
Key words: *traffic, strategy, quality criteria*

1. INTRODUCTION

Scheduled traffic shows a more stable growth for which it is possible to determine the trends of further development with more reliability. The return of tourists who arrive by air to the forecast level is expected in the period from 2003 and 2005. According to the carried out studies, it is estimated that the increase in tourist demand, expressed in the number of realised tourist overnight stays in Croatia, in the first five years of observation, will be accelerated until reaching the pre-war traffic. This will be followed by slower growth, which would mean a realization of about 84 million tourist overnight stays in 2010, or about 22 per cent more than in the most successful pre-war year (1986).

EC regulation 2320/202 establishes common rules in the field of civil aviation security such as control of access of sensitive areas of airport and aircraft, control and monitoring of passengers and cargo, training and physical security measures. The basis for a faster growth of tourist demand in the next five years can be found in the fact that the capacities that allow such growth already exist. This means not only the accommodation capacities, but also the human resources, traffic and other tourist infrastructure, renewal of previous business connections and improvement of the offer. Another important basis for such a forecast lies in the fact that a new, big, emission tourist market has been opened, of the so-called New Europe. The tourists from these countries have already clearly changed the usual structure of tourists in Croatia, as well as in other receptive European countries. It would be realistic to expect a constant growth of the tourist demand from these countries in the years to come. In order to be able to absorb the growth of tourist demand in the years following the return to pre-war figures, it is necessary not only to make best use of the existing tourist capacities but also to construct new ones, and especially to improve the entire tourist product of Croatia, which has been planned by the Development Strategy of the Croatian Tourism. This document emphasises the development of quality in tourism, which means improvement of the tourist product which should result in faster increase of the total revenue from tourism with a somewhat slower growth of physical indicators. Therefore, the growth dynamics of tourist demand in the period following the return to pre-war figures is more moderate and declines towards the end of the considered period. Because of the moderate demographic potential of Croatia, development of tourism relies to a great extent on foreign tourists so that the improvement of traffic accessibility to these markets is one of the major factors which represent the basis of the constant growth in the demand for tourist destinations in Croatia. It is realistic to expect a greater inclusion of continental destinations into the tourist flows, which means greater number of arrivals and realised tourist overnight stays.

The aforementioned may lead to the conclusion that there are real foundations for the projection of the development of tourist demand in Croatia. Modern development and its constant acceleration enabled by economic processes have affected the changes in the relations between humans, their civilization achievements and natural-geographic basis of life. Since space is in fact the main resource, and the quality of living the main goal, there are warnings about the space not actually being only the resource, but in a certain sense also the limiting factor of development. Based on this, the ecological movements are constantly getting stronger and from the European environment they contribute to activating the ecological awareness in the Croatian environment as well.

The aims of environmental protection, when speaking of traffic, can be realized at the same time by adequate planning and construction, as part of exploitation through suitable measures and technical regulations, even bans. Some general orientations, such as stimulating greater use of public transport limiting individual traffic, and in a wider sense the general development policy, regarding distribution of the economic activities can contribute to the reduction of general needs for communication and reduction of traffic loads. The Republic of Croatia is a country of valuable and preserved natural richness, and her extremely high interest in environment protection is understandable, and her interest in the development of advanced traffic technologies that will contribute to the preservation of the environment in the conditions of intensive internal traffic and transit. Croatia is a Central European and Mediterranean and Danube country, and as such, it occupies an important geo-traffic position in Europe (including intensive traffic). Therefore, it has to develop advanced technologies in transport which are to meet these requirements. The ecological awareness is getting stronger in Europe, and the issues generated by road traffic are increasingly coming to the fore. Roads are and will remain for quite a long time yet the main land routes of transport, but the advantages of air transport are getting greater all the time. The energy consumption per unit of cargo transport for trucks is over four times greater than rail and water transport, and over sixty times greater is the energy consumption in air cargo transport compared to railway or ship.

Environmental protection is there one of the strategic orientations in the development strategies. The development of environmental awareness in the European countries has contributed to the fact that the air traffic is considered with much more understanding than in the recent past. The external costs of negative impact on the environment caused by single branches of traffic which had not been quantified until now and had been included in the general cost of the community are increasingly taken into consideration in the planning of traffic infrastructure. In relation to its biggest competitor – road, railway has least negative impact on the environment. Compared to other branches, it shows a number of advantages. Railway generates least emission of harmful gases energy consumption per unit of transport is minimal at railway and in water transport, about three times less land occupancy by the construction of new railway lines compared to road construction, greater transport safety, lower impact of noise, etc. With regulation of the permitted aircraft noise level, and limiting of harmful substances in exhaust gases by aircraft engines, air traffic is complying with the world trends in environmental protection. Additional restrictions are brought by civil aviation and airport authorities by night curfews, not permitting landings of older, noisier aircraft, etc. In order to achieve efficient effects in the area of environmental protection, apart from declarative orientations and their strict observation, it is necessary to count also with the costs that are becoming more and more obviously present, and in greater amounts, so that the total costs of the society in the area of environmental protection would be transferred to a possible extent to the generators of this phenomenon. Such approximations in the field of traffic are measurable, and in the development criteria and in adequate target decisions they need to be taken into account.

The preservation of natural and civilisation values of the Croatian space is an important strategic orientation and general objective of the development of the Republic of Croatia, including also separately the traffic development. In achieving the given objectives the legislative framework is important, and it is given in the proposal of the strategy of regional development of the Republic of Croatia (proposal of the Strategy of Regional Development of RH) which represents a means for establishing and implementation of a coordinated approach to sustainable economic development of all the parts of the country.

The overall objectives of the strategy are as follows:

1. Existence of functional policy of regional development in the Republic of Croatia which contributes to the sustainable development and national competitiveness until 2013;

2. The objective refers to the establishment of a coherent legal framework and construction of the necessary institutional structures and capacities required for the management of regional development policy.

2. DISCUSSION

Hirschman and Frey have provided two development strategies, especially related to the development of the developing countries, and they are:

1. The strategy of development using excessive infrastructure capacities, and

2. The strategy of development using the lack of infrastructure capacities.

The former strategy assumes, as a rule, that a state insures and/or stimulates investments into infrastructure systems and thus creates attractive conditions (assumptions), which motivate individual entrepreneurs to invest into directly manufacturing capacities expecting high profits through the internalization of external effects insured by the already constructed infrastructure.

When somebody wants to control or manage a system he will undoubtedly have to make certain decisions. The decision is normally based on information about the system status. Most logistics systems can be regarded as systems with high complexity, i.e. the systems have a large variety. This large variety puts high requirements on the availability of information in the systems.
Woxenius et al. (2001) targeted the trend towards a higher degree of standardisation in transport networks. The increasing degree of standardisation suggests utilisation of existing transport networks to produce new services. This would also be a way to obtain synergy effects in the organisation of the production, and in that the same production system could be used for several transport services.
The latter strategy assumes that, due to the lack of infrastructure capacities in a certain area the direct manufacturing activities (operating in the region) and the population will exert force to eliminate the lack in infrastructure (which limits efficient development) by constructing the lacking infrastructure capacities. This strategy, of course, assumes that there is always a certain minimum in the infrastructure capacities in a certain area so that the directly manufacturing processes could function, and so that the population (households) could satisfy their minimal infrastructure service requirements.

3. CONCLUSION

Less developed countries are recommended to apply the "strategy of development with the lacking infrastructure" since poor investment capital does not allow simultaneous harmonised development of infrastructure and directly manufacturing capacities. Besides, the investments into infrastructure are high and their return takes much longer (long acquisition period due to long time of construction, long time of depreciation – capital coefficients are not measurable) from the temporally much faster return on investments into directly manufacturing capacities. Of course, this does not mean that in many developing countries there is no spatial and sector excess and insufficiently used infrastructure capacities on the one side and/or that one should not advocate, require and realize the construction and permanent improvement of certain infrastructure facilities or systems on the other side. On the contrary, in the long run it is necessary to tend in operation towards optimal relation between the infrastructure and the directly manufacturing capacities from the sub regional, regional, macro regional and territorially integral national aspect.

A permanent development policy is necessary; in the initial phase regarding construction of infrastructure priority should be given to supplementing the drawbacks of the capacities, in the second phase a real and consistent plan of new possibilities needs to be developed and implemented, whereas in the third phase (higher development phase) a balanced development has to be insured. Therefore, it is first of all necessary to make maximal use of the existing infrastructure, and only then (when the existing one has been fully used) to construct new capacities. However, the growth of the GDP and the infrastructure capacities have the same changes, so that it has been empirically determined that 1% of growth in infrastructure capacities is accompanied by 1% of GDP growth, in all the countries. This means that the growth in production (economic growth) requires the infrastructure to adapt to the changes in demand.

3. REFERENCES:

Bruton, M.J: *Introduction to Transportation planning HT Education, London 1990*

Samanovic, J: *Logisticki I distribucijski sustavi,* Ekonomski fakultet, Split, 1999.

Annals of DAAAM for 2006 & Proceedings of the 17th International DAAAM Symposium, ISSN 1726-9679
Editor B. Katalinic, Published by DAAAM International, Vienna, Austria 2006
150th Birthday of Nikola Tesla (1856-1943) Inventor of Mobile Robot Teleoperation

INTERMODAL TRANSPORT MODEL IN CROATIA

RADOS,B.; RADOS, J. & PERAKOVIC, D.

Abstract: *The main elements of the production of traffic services in all traffic modes, including also intermodal transport consist in the transport of various items, cargo and passengers by means of infra- and supra-structure facilities, and clear development guidelines*

Key words: Intermodal*, transport, combined transport*

1. INTRODUCTION

The EU Commission Directives for the next ten years include refocusing of the European transport policy according to the needs and requirements of its citizens. EU Commission has adopted a document – "White Paper: Time of Decision", in which the EU Commission has, for the first time, placed the users' needs into the central position and has proposed a strategy and 60 measures in order to achieve them.

The forecasts of transport in EU indicate an increase in the transport of goods by +38% until 2010 (EU 15) and in passenger transport by +24% until 2010 (EU 15), which leads to a conclusion that if nothing is undertaken, cargo transport will increase by + 60% until 2013 (EU 15).

Transport is the generator of industry development, goods exchange, way of living and the future European integration. The economic significance of the sector is that transport generates over 10% of European Union GDP and employs over 10 billion people. The main priorities in the development are directed to the re-distribution of the balance between the traffic branches, on the roads – improve the standards, at railways – cargo cabotage, safety standards, integration of maritime and river transport by using maritime highways, intermodality – programs Marco Polo, Freight Integration.

The tendency in air pollution CO_2 indicates that gas emissions will increase by more than 50% and if such tendency continues, transport will be the main factor in failing to meet the obligations according to the Kyoto Agreement of 8% until 2010. In the meantime heavy industry has continued to reduce the pollutant emissions. The problems that occur in relation with transport is the fact of existing road congestion, higher costs and the increase in the traffic volume will not adequately follow the increase in the need for new infrastructure, local development, lack of sufficiently educated drivers, and the need for fossil fuels. Intermodal transport model in the Republic of Croatia should be based on the following guidelines:

1. Maximal usage of the maritime orientation of Croatia through the project of "maritime highways";
2. Construction and equipping of the port of Rijeka and the port of Ploče by introducing the concept of (IPSI) terminals;
3. Construction of connected cargo and logistic centres and free zones – basis of the development of intermodal transport and economy;
4. Standardization of transport units (EILU), harmonization of tunnel profiles;
5. Integration of railways into the existing programs " Euroorient" cargo trains;
6. Estimate of the necessary requirements in the infrastructure, in the short-term period shifting of passenger transport from the railway lines that are necessary for Corridors V.b and X, and equipping of these with new ETCS systems of railway traffic control;
7. Investing into the development of river infrastructure through achieving better navigability, and by opening of the river ports in Sisak, Slavonski Brod, Karlovac and Osijek;
8. Networking of the members of the logistic intermodal chain into an computer network;
9 Improvement of cooperation between various transport modes and enabling the use of their systemic advantages, alleviating the burden on roads and elimination of "bottlenecks"

2. DISCUSSION AND CONCLUSION

1. Maximal usage of the Croatian maritime orientation of Croatia through the project"maritime highway». As part of the project "Maritime Highways" the Adriatic ports should become the "Adriatic gate to Europe", and thus realise the infrastructure network with the intermodal traffic terminals in the Danube – Adriatic region. This can be achieved only if the traffic system is considered as a whole through intermodal transport system: sea – rail – road – inland waterways, using the already existing port capacities. Cargo transport is growing every year in EU so that the existing road infrastructure already shows difficulties in supporting such burden. An increase of 70% until 2020 is foreseen. Maritime highways represent the first line of intermodal transport offering more frequent, more precise, more punctual and more efficient transport service, recognising the environmental requirements, information exchange, door-to-door delivery, with their capacities can compete with road traffic. Maritime highways will be the carriers of intermodal transport which implies quality which is based on precise, efficient, transport service, information exchange and updated administration.

2.Construction and equipping of the port of Rijeka and the port of Ploče introducing the concept (IPSI) terminals Introduction of IPSI (Improved Port Ship Interface) technology is a concept for intermodal transport which combines the advantages of Ro-Ro methods of transport, i.e. based on the Ro-Ro operations with containers and semi-trailers. The studies show that the terminals equipped in this way can have maximum annual operation of 1.75 mill. 20́ containers which would satisfy the needs for the operation of the port of Rijeka. Modernization of the Adriatic ports The Croatian seaports have a significant role and great potential significance based on the favourable geographic position. (5 ports open to public transport of special (international) economic interest for the Republic of Croatia: Rijeka, Zadar, Split, Ploče and Dubrovnik). Rijeka and Ploče are primarily cargo ports, whereas Zadar, Split and Dubrovnik are primarily intended for handling of passenger ships. Recognising the significance of ports for the overall development of both traffic and economy, the modernization processes have started in the ports.

3.Construction of connecting cargo and logistic centres and free zones – basis of the development of intermodal transport and economy Currently, there are 55 free zones in Europe, zones which are a significant instrument of international economic policy whose aim is the development of free international commerce. The founding of free zones in the Republic of Croatia is one of the ways to improve the current economic situation in Croatia. Modern logistic centre represents a complex system which consists of such as terminals, cargo-transport centres, cargo-distributive centres and lower subsystem units, such as warehouses, customs warehouses, free customs zones, etc. For a logistic centre to have adequate significance, it has to be based on market needs and the role in transport chain.

4. Standardization of transport units (EILU), harmonization of tunnel profiles. Several different types of containers and pallets are used currently for transport. It is necessary to harmonize and standardize the transport units in order to reduce the additional costs. By introducing IPSI terminals and by using Ro-Ro ships the usage of EILU (European intermodal loadings units) will result in the reduction of costs, better utilisation of all the transport capacities, because EILU are compatible with road, rail and maritime transport. EILU are intermodal cargo units specially designed for the needs of interoperability given by COM (2003) 155 Article 3. In this sense the capacities should be gradually adapted and preparations should be undertaken for the transfer to the use of new transport units. Harmonization of tunnel profiles: Intermodality means fast and continuous transport of cargo. In Croatia the corridors have been defined and modifications to railway infrastructure should be performed and it should be harmonized with the existing condition and requirements along entire corridors (Vb V c and X). The height of Euro-tunnels varies from 5.60 to 4.16m and it is necessary to check the condition along the Croatian corridors.

5. Integration of railways into the existing programs "Euro-orient", cargo trains. The initiative to start the project Euroorient – "railway on the sea" is the creation of intermodal path, line, for the transport of containers from the Near East to Central Europe using railways in the starting and final section, and ships for the transport of containers in the central leg of the journey, has originated from the IPC-intermodal promotion centre Danube-Adriatic, Zagreb and EIA-European Intermodal Association, B-Brussels, and the company leading the project is OPTIMA Tours GmbH from Munich. The company's founders of the consortium that will participate in the project come from Austria, Slovenia, Turkey, Switzerland, Bulgaria and Croatia.

6.Estimate of the necessary requirements of the infrastructure in the short period shifting of passenger transport from railway lines which are necessary for Corridors Vb and X, and equipping them with new ETCS systems of railway traffic guidance The volume of railway traffic on the most important pan-European traffic corridor which passes through Croatia, and that is Corridor X which connects West and Central Europe with its south-eastern part, was through the entire last decade of the last century only symbolic. Apart from overhauling the railway lines which are the Croatian sections of the pan-European corridors, intensive works are performed on the modernization of the passenger and cargo wagon and traction vehicles. Until the end of the mid-term period from 2006 to 2010 cargo transport expects further increase of transport on pan-European corridor X. The solution for the transition period to full functionality of the mentioned routes is also part shifting of passenger transport which would free the line capacities until opening to full traffic operation.

7.Investments in the development of river infrastructure through achievement of better navigability, and opening of the river ports in Sisak, Slavonski Brod, Karlovac and Osijek The Danube basin and the Adriatic need to be connected through several phases and through many investment and other interventions for strengthening the connections between sea and river ports and connecting of computer systems, and finally the development of the "Short Sea Shipping (SSS)" system and the river traffic. With the construction of the Danube-Sava canal, the river traffic in Croatia and beyond will be several times improved and will allow further development of the Port of Vukovar and the usage of river traffic on the Danube and the Sava and allow the economic development especially of industry and energetic. This project includes a greater number of various teams and institutions and it is necessary to form a single joint coordination body with the task of coordinating all the activities related to this project. Regarding its geo-traffic position the port of Vukovar is in direct relation to the important international railway, road and river corridors (corridors Vc, VII and X), which provides it with great chances of becoming quite soon the leading transport hub in this part of Europe. In order to realize the comparative advantages of the port of Vukovar it is necessary to modernize and increase the port capacities (to practically construct a new port), construct the secondary facilities for the port requirements (e.g. various warehouses), equip the port with modern port facilities for the transhipment of all types of goods, improve the railway connections to Zagreb, Rijeka and especially to the port of Ploče. It is also necessary to construct adequate road infrastructure which will connect Vukovar with other parts of the Republic of Croatia and beyond.

8.Networking of the members of logistic intermodal chain in the computer network Intermodality is a service, rather than a technology, with the aim of internationalizing and solving the operative and organizational problems which occur due to the use of multiplied methods and technologies that occur in the logistic chain. In order to overcome this, a good logistic network is needed, whose organisation (Scheme 7) realises the "door-to-door" service. All the subjects and capacities need to be networked into a unique network which will accompany the supply and demand thus controlling the speed of throughput of goods and eliminate the phenomenon of bottlenecks in the traffic flow. The year 2005 has seen the implementation of the integrated e-solution "Multilink" which will serve for the exchange of disposition with the environment, and will be implemented in the container terminal the Adriatic gate and will allow automation of the business processes related to container cargo. The system is a complete integration of all the business subjects necessary for the carriage of containers in the region of the port of Rijeka. It is necessary to support the introduction of such systems both in the port of Ploče and in the Croatian river ports.

9.Improvement of cooperation between different transport modes and allowing the usage of their systemic advantages, reduction of burden on roads and elimination of "traffic bottlenecks" The Republic of Croatia has started the realisation of the "Marco Polo" program and in the year 2006 will become a full user. The main aim of the "Marco Polo" program (2003-2006) is to reduce the load on road traffic and improvement of the impact on the environment of the entire transport system in the way of transferring cargo from road to short sea shipping, railway and inland waterways. The program supports the commercial actions on the market for cargo transport service, and promotion projects of modal transfer in all the cargo market segments, and not only in combined transport. Elimination of bottlenecks in traffic infrastructure (restructuring of Croatian Railways, modernization of railway lines, connection of ports with the main traffic corridors) is set as the quality requirement in this project.

3. REFERENCES:

Communication from the commission to the European Parliament and the council: *Intermodality and intermodal freight transport in the EU*

Strategija prometnog razvitka RH, Ministarstvo pomorstva Prometa i veza RH, Zagreb, Studeni 1999.

Real cost reduction of door-to-door intermodal transport DG TREN

http://europa.en.int./comm/energy-transport

Annals of DAAAM for 2006 & Proceedings of the 17th International DAAAM Symposium, ISSN 1726-9679
B. Katalinic, Published by DAAAM International, Vienna, Austria 2006
150th Birthday of Nikola Tesla (1856-1943) Inventor of Mobile Robot Teleoperation

COST ANALISYS OF ABRASIVE WATERJET CUTTING

RADOVANOVIC, M.

Abstract: *Cutting with waterjet costs less than traditional machining or other cutting methods. Waterjet is now the most cost effective method to cut steels. The cost ranges provide a basis for an cost estimation prior to plan of buying order of waterjet machines. The cost analysis of abrasive watejet cutting is presented in this paper.*

Key words: *Economic, cost, cutting, abrasive waterjet*

1. INTRODUCTION

Waterjet cutting is an accelerated erosion process. Waterjet can cut or erode through virtually any material known, making it one of the most versatile machines available. Application of waterjet cutting machines made possible good quality of products, flexibility of production and enlargement of economy. For many potential users of waterjet cutting machines, it was strategically important to invest in future technology and, if finances allowed, just to accept the financial risk. Waterjet is now the most cost effective method to cut steels. As one of the fastest growing machine tool industries, waterjet cutting has proven to save time and money on countless applications. Generally, cutting with waterjet costs less than traditional machining or other cutting methods. The high investment and operating costs for waterjet cutting machines, compared to conventional cutting machines, must be justified according to economic criteria with corresponding financial advantages. The question whether waterjet cutting technology has economic advantages compared to equivalent conventional cutting technologies cannot always be immediately answered. Often, only a detailed analysis of the technology solution together with an exact estimate of the piece costs expected during production provides an acceptable basis for an investment decision. Knowledge of a waterjet system's investment and operating costs should form the basis for evaluating its profitability.

2. ABRASIVE WATERJET CUTTING

Abrasive waterjet stream is the cutting tool. The cutting process is like grinding, except that abrasives are moved through the material by water rather than by a solid wheel. Most abrasive waterjet cutting theories explain abrasive waterjet cutting as a form of micro erosion. By abrasive waterjet cutting the high pressure pump produces the required pressure up to 400 MPa. A high pressure supply line direct the pressurized water from the pump to the cutting head. The high pressure supply line is terminated by a water orifice at which the pressure is relieved. Water orifice has a diameter of approx. 0.08 to 0.50 mm. The water pressure is relieved in the orifice and a waterjet is created. The water is pressed out of the orifice at a speed of approx. 900 m/s – nearly three times the speed of sound. The result is a very thin, extremely high velocity waterjet. Solid abrasive particles are added and mixed with the waterjet and subsequently focused by an abrasive nozzle. Abrasive nozzle is with diameter from 0.8 to 1.2 mm and length from 50 to 100 mm. In figure 1 is shown abrasive waterjet cutting.

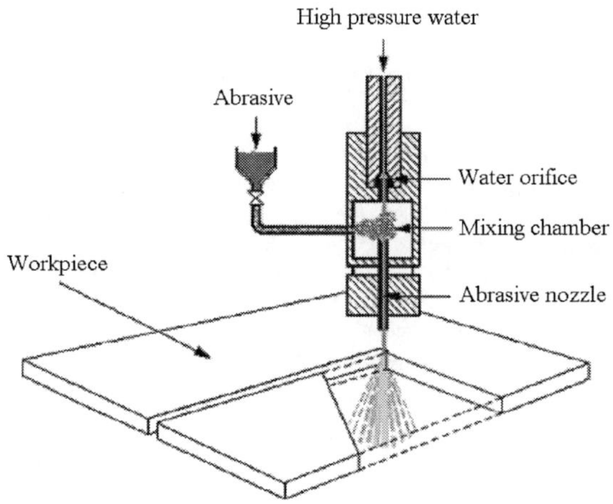

Fig.1. Abrasive waterjet cutting

The high speed of the waterjet creates a partial vacuum in the mixing chamber so that abrasive material and air are sucked in and flushed away by the waterjet. This is known as the injector principle. The abrasive nozzle has two functions: it accelerates the abrasive particles and it focuses the jet. The cutting head makes a circular or pendular motion over the piercing point. The abrasive waterjet penetrates the material and progressively cuts the material guided by the cutting head along programmed contour.

Abrasive waterjet cutting machines designed for a specific applications are just as varied as the field of application. Concepts range from systems with fixed cutting head, machines equipped with an XY guided cutting head or robot guided cutting head. Machines designed in the form of a portal are used primarily in sheet metal processing. The basic components of abrasive waterjet cutting machine are: high pressure pump, high pressure delivery system, cutting head, storing, delivering and matering abrasive material, mechanical machine for movements, workpiece rest, catcher, disposal unit and control unit. The high pressure pump is the heart of a waterjet machine. It creates the required operation pressure and transports the water through the system. The pressure pump for waterjet cutting machine is high performance pump. The back and forth motion of the pressure intensifier causes pressure fluctuations in the piston's end of a travel. For that reason, pumps for this pressure area are usually equipped with a pulsation damper to smooth out pressure fluctuations. The high pressure delivery system safely conducts the water under pressure from the pump to the cutting head. It consists of thick walled high pressure pipes made of strain hardened austenite steel. The cutting head is the most important component of a waterjet cutting machine. It consists from water orifice, mixing chamber and abrasive nozzle. Water orifice is made from a sapphire, ruby or diamond. Abrasive nozzle is made from hard metal. Container for the abrasive material is located next to the machine. The abrasive material is transported from the container by means of a vacuum system to either the cutting head or to a metering device located on the cutting head. During processing, the workpiece lies on a rest. It is normally located directly on the machine frame. Depending on the machine concept, the workpiece is situated either above or below the water surface during processing. The catcher is filled with balls. It moves along with the cutting head and absorbs the reamining energy after the cutting process. Sludge (cutting chips and abrasive material residue) and waste water are separated in the disposal unit. Modern machines can hold tolerances of ±0.1 mm or better. Automating the process with programmable systems in conjunction with the CNC control provides high flexibility and a wide variety of parts.

3. COST OF ABRASIVE WATERJET CUTTING

Three categories make costs of water jet cutting: investment costs, operating costs, and labor costs. Investment costs are associated with the equipment purchase. It is the initial price of the equipment amortized over a specific amount of time. Since investment costs are fixed, they occur whether the machine is working or idle. The investment cost of a machine is a function of the power of the pump, the size of the workpiece and the number of axes required to manipulate the workpiece around while performing the operation. While the investment cost of a waterjet machines is high, improved productivity and product quality, often make a system economical, especially for companies that produce a wide range of parts. Operating costs are associated with operating the process, including electrical energy consumption, water consumption, abrasive consumption, wearing parts consumption, and maintenance and repair. These costs occur only when the machine is operating. The pump is the determining factor in the operating costs. The key parameter that controls cutting speed is the water orifice diameter or the cutting water flow rate. The other two parameters that greatly influence the speed are the system pressure and the abrasive flow rate. The optimum abrasive flow rate can be related to the cutting water flow rate. Labor costs are associated with running the machine, including the time to handle raw material, finished parts and remnants. It must know the hourly cost of an operator, the percentage of time allotted for machine setup, and the percentage of time an operator actually attends the machine. Costs of abrasive waterjet cutting can be calculated in exactly the same way as for conventional machining, i.e. based on investment, operating and labor costs. Cost per hour of abrasive waterjet cutting can be calculated by equation:

$$C = C_i + C_o + C_p \quad (1)$$

Where C (EUR/h) – waterjet cutting cost per hour, C_i (EUR/h) – investment cost, C_o (EUR/h) – operating cost, C_p (EUR/h) – labor cost.
Or by equation in form:

$$C = \frac{I}{D \cdot L_a} + c_e \cdot E + c_w \cdot Q_w + c_a \cdot Q_a + \frac{c_n}{L_n} + \frac{c_m}{L_m} + \frac{M}{L_a} + C_p \quad (2)$$

Where I (EUR) – capital investment cost of abrasive waterjet machine, D (year) – depreciation period, L_a (h/year) – machine utilization, c_e (EUR/kWh) – unit cost of electrical energy, E (kW) – electrical power consumption, c_w (EUR/m^3) – unit cost of water, Q_w (m^3/h) – water consumption, c_a (EUR/kg) – unit abrasive cost, Q_a (kg/h) – abrasive consumption, c_n (EUR/part) – water orifice price, L_n (h/part) – water orifice lifetime, c_m (EUR/part) – abrasive nozzle price, L_m (h/part) – abrasive nozzle lifetime, M (EUR/year) – maintenance costs per year.

Capital investment	I=120 000 EUR
Depreciation period	D=5 years
Machine utilization	L_a=4000 h/year
Investment cost	C_d=6 EUR/h
Electrical power consumption	E=45 kW
Unit cost of electrical energy	c_e=0.105 EUR/kWh
Cost of electrical energy	C_e=4.725 EUR/h
Cutting water consumption	Q_{cw}=0.24 m^3/h
Cooling water consumption	Q_{qw}=0.90 m^3/h
Unit cost of water	c_w=0.5 EUR/m^3
Cost of water	C_w=0.57 EUR/h
Abrasive consumption	Q_a=21.6 kg/h
Unit cost of abrasive	c_a=0.7EUR/kg
Cost of abrasive	C_a=15.12 EUR/h
Water orifice wear cost per hour	C_n= 0.078 EUR/h
Abrasive nozzle wear cost per hour	C_m= 0.78 EUR/h
Cost of wearing parts per hour	C_r= 0.858 EUR/h
Maintenance cost per year	M=5000 EUR/year
Maintenance cost	C_m=1.25 EUR/h
Labor cost	5 EUR/h
Cost per hour of cutting	33.523EUR/h

Table 1. Cost calculation example

The cost per meter of abrasive waterjet cutting is established as:

$$C' = \frac{C}{v} \quad (3)$$

Where C' (EUR/m) – cost per meter of waterjet cutting, v (m/h) – cutting speed.
Table 1 show cost calculation example of cutting by abrasive waterjet machine "PTV Hostivice", with two axes numerical control. The technical characteristics are: high pressures pump 37 kW, 3.73 l/min, 413 MPa, water orifice diameter 0.33 mm, abrasive nozzle diameter 0.76 mm, table size 2000 x 4000 mm.

4. CONCLUSION

The costs of abrasive waterjet cutting must be justified according to economic criteria with corresponding cutting methods. Cost per hour of abrasive waterjet cutting is calculated based on investment, operating and labor costs.

5. REFERENCES

Kramar D., Junkar M. (2003). The Development of a Software Tool for the Selection of Contour-Cutting Processes, *Journal of Mechanical Engineering,* 49(2003)6, 346-365, ISSN 0039-2480

Kramar D., Junkar M. (2000). Laser and Abrasive Waterjet Cutting Economics, *International Sheet Metal Review*, 2: 38-41

Olsen J. (2002). Buying an abrasive jet machine, *The Fabricator*, April 15(2002), http://www.thefabricator.com

Radovanovic M. (2005). Abrasive Water Jet Machining, *Proceedings of Seventh Scientific Conference "Smolyan-2005"*, pp. 229-234, ISBN 954-91073-9-6, University of Plovdiv, Technical College-Smolyan, Smolyan, Bulgaria

Radovanovic M. (2004). Precision cutting by abrasive waterjet, *Proceedings of Scientific Conference "Manufacturing and management in 21st Century"*, pp. 94-99, ISBN 9989-2325-0-4, Faculty of Mechanical Engineering, Ohrid, Macedonia

Annals of DAAAM for 2006 & Proceedings of the 17th International DAAAM Symposium, ISSN 1726-9679
Editor B. Katalinic, Published by DAAAM International, Vienna, Austria 2006
150th Birthday of Nikola Tesla (1856-1943) Inventor of Mobile Robot Teleoperation

THE METHODS AND TECHNIQUES OF CEMENT PLANT ROTATING MACHINERY VIBRATION MEASUREMENT AND ANALYSIS

RAHMANOVIC, S. & DOLECEK, V.

Abstract: The methods of acquisition and analysis of vibration signals produced in cement plant rotating machinery are presented in this paper. The main causes of vibration and forcing frequencies for specific cement plant rotating machines are also discussed. An overall vibration measurement and the main aspect of time and frequency domain vibration analysis are further described. It is also explained how the imbalance, eccentricity, misalignment, looseness and problems in rolling bearing and gearboxes can be detected using appropriate methods of vibration analysis. Finally a common configuration of the equipment for vibration monitoring and analysis has been presented.
Key words: vibration signal, frequency analysis, cement plant, rotating machinery vibration, forcing frequencies.

1. INTRODUCTION

The main rotating machines in a cement plant are rotary kiln, ball mills for raw material and clinker grinding, vertical roller mill for coal and raw material grinding, dynamic separator and process fans. Besides these there are a lot of common machines in a cement plant such as compressors, blowers, pumps etc. All specified machines produce a certain vibration during operation. The first sign that appears after a fault starts to develop inside of a machine is a change in vibration. The other signs of the incipient fault such as wear particle in oil, strange noise and heat generation can be detected later on. This is the main reason why vibration monitoring and analysis has become the most powerful tool for diagnosing faults in cement plant rotating machines long before failure occurs. Condition monitoring through vibration combined with oil analysis, root cause analysis, and some other techniques is the latest technique for achieving higher availability of the cement plant rotating machinery.

2. A CEMENT PLANT MACHINERY VIBRATION

2.1. Causes of cement plant rotating machinery vibration

The main reasons causing vibration in cement plant rotating machines are: imbalance, misalignment, increased clearences, mechanical looseness, wear of fan blades, wear, spalls, cracks and other defects in journal- and rolling bearings and gears, eccentricity of rotating components, permanent deformation such as bent shafts, or bent cilindrical shell of a ball mill and other reasons. Very often, due to depositing of the dust from the flue gases an increased imbalance arises on the rotor of a rotary kiln exhaust fan while the other fans in cement plant are exposed to a significant wear. All these causes of vibration leave their trases in the vibration signature of machine making it possible to detect them using vibration monitoring.

2.2. Cement plant rotating machinery forcing frequencies

All cement plant rotating machines have some set of forcing frequencies of which the most important is the RPM of the shaft or fundamental frequency, 1X, because it is usully conneted to other forcing frequencies. All rotating machines have a certain amount of residual imbalance which imparts a force on the bearings, causing the structure to vibrate at 1X.

The rolling bearings have four specific forcing frequencies or four bearing tones. They are: fundamental train frequency (FTF), ball spin frequency (BS), outer race frequency (OR) and inner race frequency (IR).

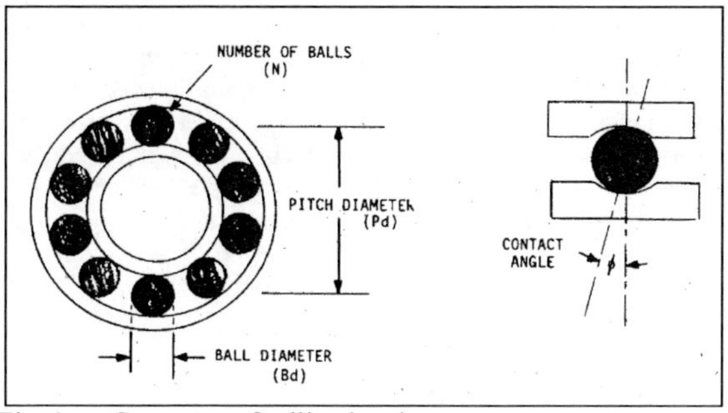

Fig. 1. Geometry of rolling bearing

The bearing forcing frequencies are calculated as follows:

$$FTF = \frac{rps}{2}\left(1-\frac{B_d}{P_d}\cos\phi\right),\quad BS = \frac{rps}{2}\cdot\frac{P_d}{B_d}\left[1-\left(\frac{B_d}{P_d}\right)^2\cos^2\phi\right]$$

$$OR = N\cdot\frac{rps}{2}\left(1-\frac{B_d}{P_d}\cos\phi\right),\quad IR = N\cdot\frac{rps}{2}\left(1+\frac{B_d}{P_d}\cos\phi\right), \quad (1)$$

where: rps – rev. per second of inner race, B_d – ball diameter, P_d – bearing pitch diameter, N – number of rolling elements, ϕ – contact angle

When a defect such as crack or spall appears on the inner or outer race or if a rolling element is deformed a strong peak will be produced in the vibration signal at some of bearing forcing frequency depending on the location of defect (Bodre, 2005). Gear mesh frequency is equal to the number of teeth on the gear multiplied by the gear RPM. In the vibration spectrum of a gearbox the hunting tooth frequency, (HTF) may also be present. Hunting tooth frequency HTF for a given gear pair can be calculated using the following equation:

$$HTF = \frac{f_{GM}\cdot N_A}{T_G\cdot T_{PN}}, \quad (2)$$

where: f_{GM} – gear mesh frequency (pinion speed x gear teeth)
T_G – number of gear teeth , T_{PN} – number of pinion teeth
N_A – number of unique assembly phases

Hunting tooth frequency is important for low speed planetary gear box in a drive of a roller mill used for coal grinding in a cement plant. It can be calculated acc. (Klein,1999). Fan blade pass frequency is equal to the number of blades in a fan times the RPM of fan rotor. Rotor bar pass frequency of an electric motor and of a dynamic separator used in coal, clinker and raw meal grinding plants are calculated in a similar way.

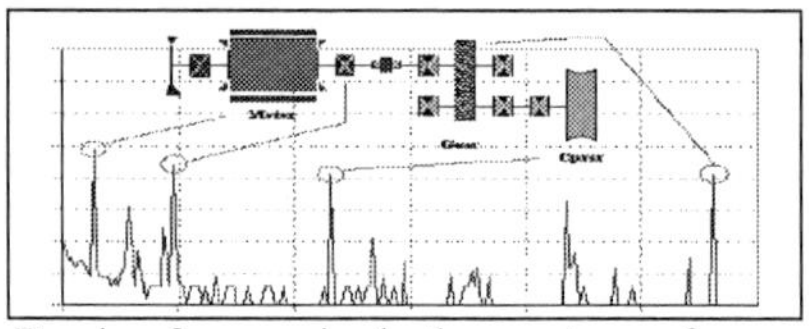

Fig. 2. Forcing frequencies in the spectrum of a compressor

3. OVERALL VIBRATION MEASUREMENT

Measurement of overall vibration level over a broad band of frequencies of rotating machinery is the most basic vibration monitoring technique used in the cement plants. Data are collected using transducers that can measure displacement, velocity or acceleration. Several simple metrics are used for overall vibration measurements or overall readings. They are peak amplitude Pk, peak-to-peak amplitude Pk-Pk and the root mean square amplitude (RMS). By combining these metrics the crest factor is defined as the peak-to-RMS-ratio and the kurtosis as amplitude^4/RMS^4. Both the crest factor and the kurtosis indicate the spikiness of a signal by giving more weight to amplitudes greater than the RMS. Measured vibration amplitude level can be trended against time in order to indicate deteriorating machine condition. These levels can also be compared to the published vibration criteria to estimate severity of vibration. Trending of overall vibration may indicate deteriorating condition of a machine but it cannot detect the faults. The overall vibration measurement can be easily made by using a velocity transducer or an integrating accelerometer and a RMS meter such as Integrating vibration meters — Types 2513 and 2516 both products of Brüel&Kjär.

4. TIME AND FREQUENCY DOMAIN ANALYSIS

Vibration signal in the time domain is displayed as amplitude versus time and in the frequency domain as amplitude versus frequency. Any periodic waveform can be transformed into a series of discrete sine waves each of them having its own frequency and amplitude. This transformation is done by discrete Fourier transform (DFT), which nowadays is performed by spectrum analyzer using fast Fourier transform algorithm (FFT). Using waveform in the time domain it is possible to detect changes in the vibration caused by faults but the diagnosis of faults is very difficult (Forrester,1996). It is much easier to find out the reason of change in vibration that occurs at a specific frequency in the frequency domain. An incipient fault such as a small defect in rolling bearing can also be detected easier in the frequency domain. The time waveform can be used in low speed applications where vibration signal has a defect at such a low frequency that cannot be seen in a spectrum (Wowk,1991). A rotary kiln-or ball mill low speed gearbox that has one worn tooth on a gear is an example of this. This defect will be seen in a waveform.

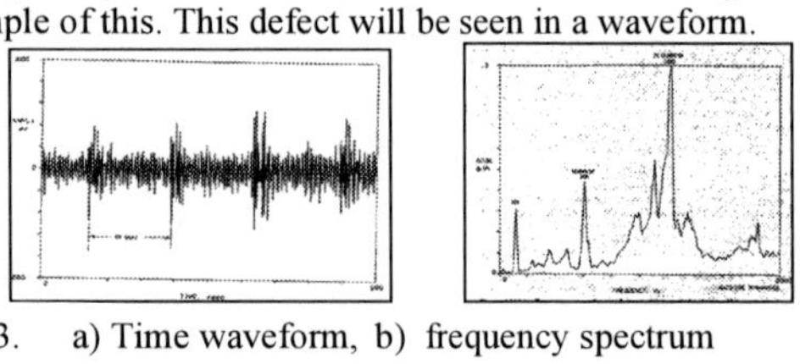

Fig. 3. a) Time waveform, b) frequency spectrum

5. EVALUATING VIBRATION SPECTRA

Each forcing frequency will create a peak in the vibration spectrum. A change in the amplitude of that peak indicates its severity. Other frequencies such as harmonics and sidebands are often present in the spectrum of a vibration signal. Harmonics are multiples of 1X while the sidebands are always result of a frequency or amplitude modulation. The sidebands are placed in spectrum on either sides of a modulated frequency at distances equal to the modulating frequency. Usually the higher frequency is modulated by a lower one such as in a case of modulation of gear mesh frequency by the gear shaft rotating frequency. In that case a defective gear produces sidebands at either sides of the gear mesh frequency at the distances equal to the gear shaft rotating frequency.

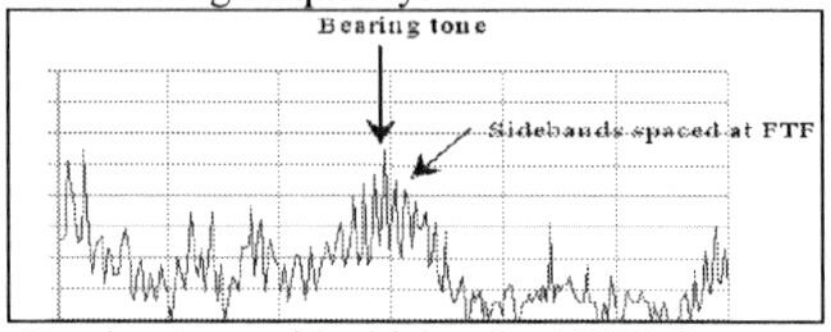

Fig. 4. Bearing tone with sidebands at FTF frequency

Gear boxes produce very complicated spectrum the analysis of which can be simplify by using cepstrum analysis that shows frequencies which repeat themselves in the frequency spectrum (Barron, 1996). An increase in 1X at a fan bearing indicates an increased imbalance. Harmonics of 1X indicate damage such as increased clearance in bearing and/or looseness of the structure.

6. VIBRATION MONITORING IN CEMENT PLANT

Integrating vibration meters type 2513 or 2516 can be used for the simple overall readings. The next instrument that can be used is Vibrotest 60 a hand held device capable to perform FFT of a spectrum, bearing condition analysis, BCU, field balancing in 1 and 2 planes, cepstrum analysis and other diagnostic task. This instrument can be equipped with an accelerometer transducer for manual offline data collection but it can also be connected to the permanently installed monitoring units such as Vibrocontrol 1100, that are parts of an on-line vibration monitoring system as shown on Fig. 5b. That system as a minimum has settings for alarm and shutdown in case of excessive vibration. The more developed system includes a dedicated computer with an appropriate software for vibration analysis for examle VIBROEXPERTCM-400, (Brüel-Kjär).

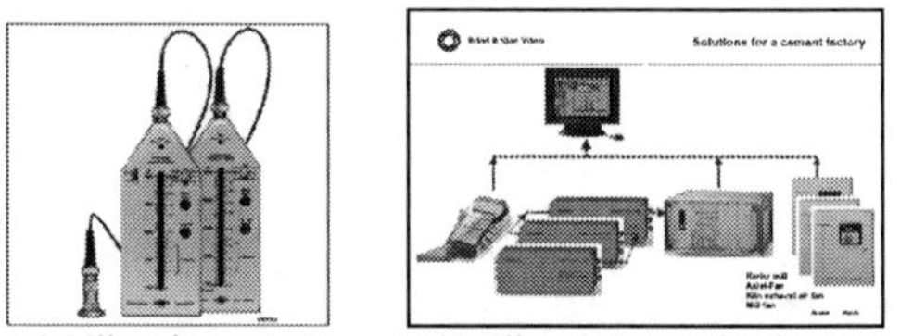

Fig. 5.a) Vibration meter 2513, b)Vibration monitoring system

7. CONCLUSION

Using appropriate methods and techniques of vibration measurement and analysis makes it possible to detect faults in the cement plant rotating machinery at the early stage of their appearance. It is important to know the specific forcing frequency produced in those machines as well as their harmonics and sidebands that can be found in the spectrum of the recorded vibration signal. To enhance results obtained by frequency analysis the time waveform analysis and overall vibration measurement can be performed as well.

8. REFERENCES

Barron R: Engineering Condition Monitoring, Practise, methods and applications, Addison Wesley Longman, New York, 1996.

Bodre R: Introduction to Machine Vibration, DLI Engineering Corporation, Bainbridge Island WA 98110, 2005

Forrester, D.: Advanced Vibration Analysis Techniques for Fault Detection and Diagnosis in Geared Tranmission Systems. Dissertation TU Swinburne 1996.

Klein, U.: Schwingungsdiagnostische Beurteilung von Maschinen und Anlagen. Düsseldorf: Verlag Stahleisen, 1999.

Wowk, V.: Machinery Vibration: Measurement and Analysis, McGraw Hill inc.,USA, 1991.

Annals of DAAAM for 2006 & Proceedings of the 17th International DAAAM Symposium, ISSN 1726-9679
Editor B. Katalinic, Published by DAAAM International, Vienna, Austria 2006
150th Birthday of Nikola Tesla (1856-1943) Inventor of Mobile Robot Teleoperation

THE HEAT TREATMENT OF AXIS AND SHAFTS REPAIRED WITH WELDING IN THE "PALAJ"-KASTRIOT

RAMAJ, V.; BYTYCI, B.; OSMANI, H. & KACANI, J.

Abstract: *The heat treatment of axis and shafts, repaired with welding, has the purpose to improve the mechanical, physical and technological attributes by changing the internal construction of these details in rigid state. These attribute improvements will be fulfilled in three main phases: warming up until reaching the temperature of head treatment, continuing with the heat treatment temperature and cooling.*
In our case, we will handle the thermal processing of the active-tumbler's shaft after finishing the cutting and weld-processing - surfacing.
Key words: *heat treatment, surfacing, annealing, hardening,*

1. INTRODUCTION

Welding and surfacing is used a lot not only for producing new details, but also for their reparation after breakage or wearing out.
Reparation and surfacing of axis, shafts and other elements of the machines is very successfully used in Kosovo's Energy Corporation – KEK.
Shafts and axes which are used in KEK, operate in very difficult exploiting conditions. Considering the load, these machine elements belong to the group of elements that work in most difficult conditions, and we use these factors to calculate them and set the dimensions.
Since we are talking about repairing shafts and axes, and this problem is not particularly treated in literature, we should consider the following:

- The technical justification of repairing,
- The economical justification of repairing.

During analysis we determined that surfacing is reasonable by all means, because the total price does not amount 20% of new shafts or axes.

2. REPAIRING OF SHAFTS AND AXES AND PREPARING THEM FOR SURFACING

Following, there are images of shafts which are damaged in many places, and the procedure of surfacing the shafts.
The preparation of the details for welding implicates:

- Preparation
- Formation of elements
- Tube and profile forming,
- Detail fixation and concentration before the surfacing,
- Preparation of the work area, etc.

The preparation for axes and shafts reparation implicates:

- Placement and foundation,
- Griping,
- Groove preparation,
- Preheating.

3. ANALYSIS OF BASIC MATERIALS

To choose the right filling material, we should consider: preheating – which matches the shaft's material and also the behavior of annealing after surfacing – the chemical analysis is performed and we gained these results:
0,51 % C; 0,60 % Mn; 0,29 % Si; 0,17 % Cr; 0,04 % Ni; 0,04 % Mo; 0,025 % S; 0,027 % P.
The steel of this chemical composition leads us to a conclusion that this steel is for improving Ck 45.
The chemical composition and mechanical properties of steel Ck 45 are given in the following table:

Fig. 1. Repairing of shafts.

Material	Chemical composition, %				
DIN	C	Si	Mn	Pmax	Smax
Ck 45	0,42-0,50	0,15-0,35	0,50-0,80	0,035	0,035

Table 1. Chemical composition

Material	Rv, N/mm^2	Rm, N/mm^2	A, %	Z, %
Ck 45	360	600-720	18	45

Table 2. Mechanical properties

4. FILLER MATERIAL

While selecting the filling material, we considered first the quality and state of heat treatment of shaft.
The electrode with these properties is selected as follows:

Electrode	Chemical composition, %				
	C	Cr	Mn	Si	Mo
AWS/ASTM; E 9018-G	0,18	1,0	0,9	0,3	0,2

Table 3. Chemical composition

Electrode	Rv, N/mm^2	Rm, N/mm^2
AWS/ASTM; E 9018-G	80-85	85-90

Table 4. Mechanical properties

5. SELECTING THE PROCEDURE OF SURFACING

While selecting the procedure of surfacing we kept notes of the following:

- the quality of material from which the shaft is composed
- dimensions of shafts
- operating options for the welder
- the required measures for achieving good quality for welding joining point.
- preparing the performer to whom the repairing is credited.
- controlling the works in the training phase and performing the surfacing.

Analyzing the mentioned factors we conclude that for the repairing works we should apply the GMAW (CO_2) method.

6. AFTER SURFACING CONTROL

The ultrasound control after surfacing should conclude if joining is carried out without errors. If we respect all the measures, having these errors is difficult

7. HEAT TREATMENT AND EXAMINATION

In our case we have done the heat treatment of an axis of 480 mm length and Φ440mm diameter.
After these two operations the measurement of the axis hardness is needed, which will serve for control. This service will then gives the permission for the materials hardness, whether it agrees with the base material conditions or not.
The heat treatment of the axis is processed this way:
The axis is heated from 0°C up to 450°C and remains in this temperature for one hour. Afterwards it is heated from 450°C up to 840°C. It stays at the temperature of 840°C for one hour. After that we take the axis out of the furnace and dipped in hardening oil (UTO20) and we do the hardening.
By controlling the joining we conclude that it is homogeny, therefore, we can go on to the heat treatment. To attain full success, an improvement is made to the whole axis. In this case the extension, joining and the axis material gain uniform structure.

The results of hardness measurement, of the axis of acting tumbler are these:

- Before surfacing: 26 HRC
- After surfacing: 37 HRC
- After the heat treatment: 45 HRC.

Unit MIC 10, Fig. 3, after three consecutive measures we calculated the average value of the hardness.

Fig 2. ARC-MET ; Optical Emission analyser

Fig .3. Hardness tester MIC-10

8. CONCLUSION

The main problem during detail welding, is not knowing the chemical content of the material and its mechanical virtues. This information is the bases for choosing the right technology, welding or regenerating method, the base material and also the heat treatment.
Detail reparation is done with the GMAW-CO_2 method, choosing the technology of material's preparation, the controlling method and the heat treating method.
The results are positive: details are looser (up to 80%), the realization time is shorter, sufficient quality, longtime serving, and the reparation is carried out at the KEK department by the workers.
The reparation of these parts, even if we are talking about sizes from few millimeters up to a few meters, is adopted and successfully carried out at FVPX – KEK – Kosova, with the experts of Mechanical Engineering Faculty.

9. REFERENCES

Bytyçi, B. & Osmani, H. (2006). *Saldimi I*, University of Prishtina, Prishtinë.

Osmani, H. (1997). *Optimalizimi i paramvetrave të regjimit të saldimit përmes izotermave në zonën e ndikimit të nxehtësisë*, doctoral dissertacion, Faculty of mechanical Engineering, Prishtinë

Katalogje të FPX – Palaj - Kastriot

Annals of DAAAM for 2006 & Proceedings of the 17th International DAAAM Symposium, ISSN 1726-9679
Editor B. Katalinic, Published by DAAAM International, Vienna, Austria 2006
150th Birthday of Nikola Tesla (1856-1943) Inventor of Mobile Robot Teleoperation

A TOPOLOGICAL APPROACH TO PRESENT INPUT DATA TO PERCEPTRON NEURAL NETWORK FOR PASSING IN MIDDLE SIZE ROBOT SOCCER

RASTI, J.; MOVAHHEDINIA, N. & JAMSHIDI, K.

Abstract: *One of the problems in middle size robot soccer, which also can be applied to scanner robots routing in unpredictable environments, is leading passing among robots that comprises choosing the best team-mate to receive the ball without any need to explicit communication among robots. In this paper we have developed an algorithm based on Perceptron neural network for this problem which determines the best passing angle based on the topological data of the play field (i.e. position of robots). With a modification to Perceptron structure and proper data presentation approach a considerable improvement in solution performance has been achieved.*
Key Words: *Topological Data, Perceptron Neural Network, Robot Soccer, Scanner Robots.*

1. INTRODUCTION

Various ideas in robot intelligence, specially enabling a robot to make independent decisions, have been presented in recent works. One of applications is robots routing in dynamic environments. For instance, a scanner robot which is supposed to go through an unpredictable environment with varying obstacles, should select the best route for passage with the least impact to impediments. A similar scenario is passing in robots soccer where the opponent robots and the outlines of the play field are obstacles. In this case, to lessen the possibility of pass termination by opponents, the robot who handles the ball, after considering the position of other robots, should select the best team-mate to pass. Tambe, 1997, and Yokota, et al. 1999 have used explicit communication and message interchange among robots to set up a passing tactics. But the more appealing idea is an independent passing strategy without any necessity to inter-communication between robots or even a central controller (Matsubara, et al. 1996; Pagello, et al. 1999). The problem is so complicated due to the considerable number of parameters for correct passing such as number and direction of visible team-mates and opponents and the distance among them, distance between team-mates and the out lines and the motion angle of the robots. In addition to that, some of the parameters are contradictory to each other. For example, while existence of a team-mate in front of the ball-handler is an advantage, however it may change to an obstacle if it comes too close. Burkhard 2000, Munandar & Siagian 2001, Vecht & Lima 2004 and Wang, et al. 2004 have proposed heuristic search based algorithms which use some of passing parameters (extracted from the image received by robot's camera) to assign positive or negative scores to the passing routes and finally select the best one. Although these algorithms are fast, but are not so successful because of the contradictory nature of some parameters and lack of harmony among positive and negative scores. For such kind of problems that numerous and antagonistic parameters reject mathematical characterization, neural networks seems to be proper solutions. Using Perceptron neural network is a sample presented by Matsubara, 1996. In this approach, plenty of images recorded by robots' cameras in real soccer matches are used to form a training set composed of extracted passing parameters. Due to lack of particular basis to determine the weight and role of each parameter, this algorithm does not respond well in many cases. In this paper, a topological approach to present input data is examined that instead of extracting passing parameters and assigning weights to them by designer, position of observed robots form the network input to let it extract necessary parameters and their weights by itself. This method considerably improves the number of true passes. Section 2 describes this algorithm. A modification to network structure and its results will be expounded in section 3. Section 4 discusses a general method for scanner robots routing in dynamic environments. In the final section, the conclusions of the paper are presented.

2. PERCEPTRON NEURAL NETWORK FOR PASSING IN ROBOT SOCCER

2.1. Input Data

The image of robot's camera is the basis of all its activities. The ball-handler robot should make a topological plan composed of the position of other observable robots and select the best team-mate for passing. Fig.1 illustrates some plans which team-mates are shown by light gray color, opponents are shown by black color and the invisible part is shown by dark gray color. This plan forms the input of the decision maker neural network. To do this, we divide the cited plan to some equal square sections. The size of each section is almost twice the size of the ball to warrant ball passage through sections. Each section will be scored based on the position of robots and then its value will be exerted as an input node to the neural network. Our general idea is to use a neural network to evaluate the eligibility of each team-mate to receive the ball on the basis of total value of sections located between it and the ball-handler. When the ball-handler robot stands exactly in front of its goal, it has the widest view of the play field. In this case, the visible part of the plan is divided to 278 sections (Fig.2). Therefore the neural network has 278 input nodes. Fig.3 shows samples of section evaluating procedure. The evaluation rules are listed below:

I) Empty sections are scored by 1.1 (Fig.3.a).

II) Sections that include even a small part of the opponent robot will be assumed completely under the possession of the opponent and scored by -2.2 (Fig.3.b and Fig.3.c). This is done to exaggerate opponents and reduce number of risky passes.

III) Value of sections that include a team-mate is 1.1 plus the team-mate's presence percent (Fig.3.d).

IV) Invisible parts of sections are not included in value calculation (Fig.3.e).

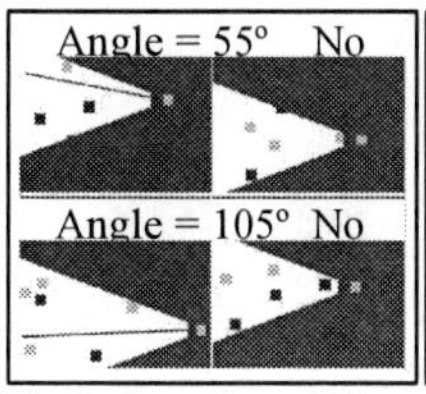

Fig.1- Samples of Plans

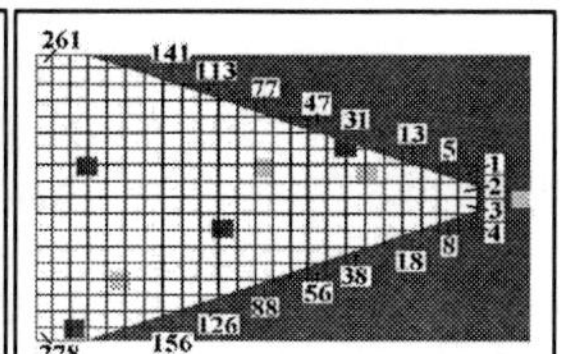

Fig.2- Sections Numbering

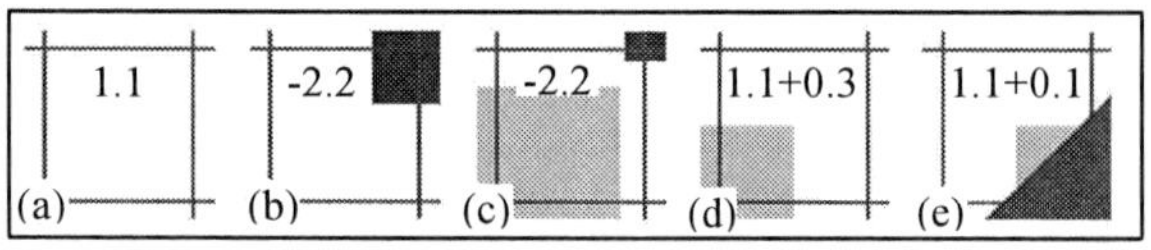

Fig.3- Values of Different Sections

V) As the ball-handler robot goes towards opponent goal, some of 278 sections will exclude because of settling behind opponent goal. Value of these invisible invalid sections is zero.

To train the neural network, 1000 random samples of cited plans were made. Their relative 278-elements vectors were taken as inputs and the best passing angle for each case was considered as desired output.

2.2. Perceptron Neural Network Structure and Training Method

Since there are 18 sections at the top of the plan (end-points of passing routes), it seems that neural network should extract 18 main pieces of information out of 278 input pieces. So our Perceptron network has two 18-nodes hidden layers. The output layer includes 6 nodes that each one determines a range of passing angles and one of them will be triggered at instance. In fact, the output layer summarizes the neural computations in terms of angle ranges. All layers were initially fully connected to each other. After training course with the set cited in 2.1, whenever a robot tends to pass, it converts the play field plan to a vector and exerts it to the neural network to obtain the passing angle range. For simplicity, we assume that when a robot catches the ball, all its team-mates go towards opponent goal. A set of 100 test vectors extracted from real robot soccer races was presented to the network. The results were 52 true passes, 17 faults and 31 pass refusal despite of a good situation..

3. MODIFICATION TO NETWORK STRUCTURE

To improve network performance, the interconnections between two first layers were modified based on the input data presentation approach. In modified structure (Fig.4), only input nodes located in the i-th passing route (out of 18 routes), are connected to the i-th node of the next layer. For example, only input nodes are connected to the first node of the next layer that are located in the first passing route (ending in node no. 261). Exerting previous test set to modified network results in 79 correct passes, 5 faults and 16 pass refusal. In addition to improved performance, the modified network offers faster convergence. Fig.5 compares the convergence speed of the fully connected Perceptron and modified network.

4. SCANNER ROBOTS ROUTING USING TOPOLOGICAL DATA

This section describes how to use mentioned algorithm for the scanner robots routing in dynamic environments.

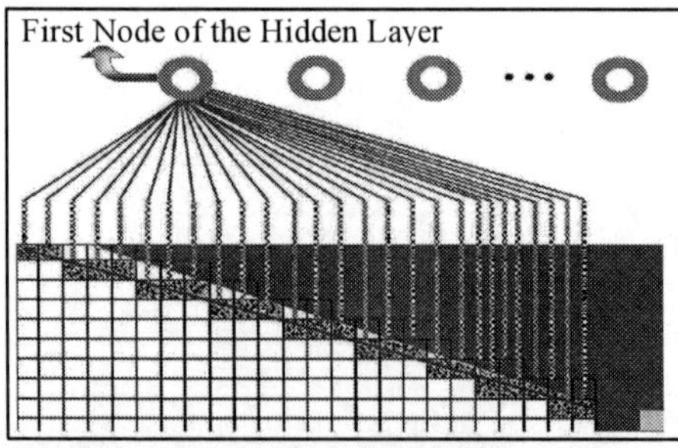

Fig.4- Modified Perceptron Structure
(Connections for the 1st passing route are shown)

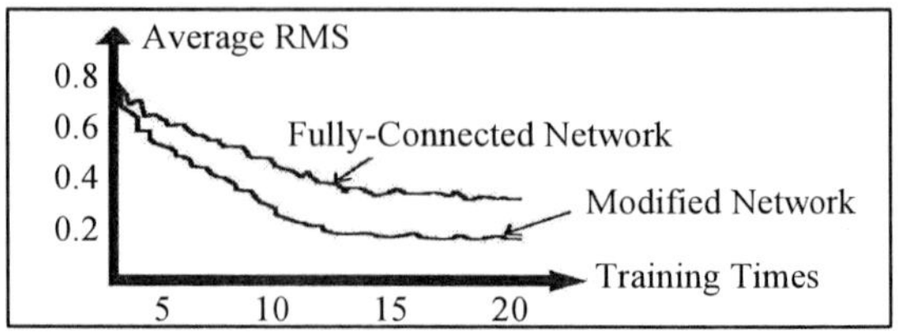

Fig.5 – Comparison of Convergence Diagrams

I) The image received from robot's camera is mapped to a plan of obstacles. This plan is divided to equal sections (e.g. N sections) that each one is a little bigger than the robot to assure confident passage among impediments. Each section should be assigned a value. It is advisable to score proper parts of the plan for passage by a positive value and the improper parts by a negative value to avoid large total scores.
II) The input layer of the Perceptron neural network includes N nodes that each one corresponds to the value of a section. Number of passage routes (number of sections at the top of the plan) specifies the number of nodes in hidden layers. To achieve better performance, only the input nodes that are located in the i-th passage route should be connected to the corresponding i-th node in the next layer.
III) In order to train the network, numerous images of the robot's camera from real situations are taken as inputs to the network (after conversion to N-elements vectors) and the best passage angle is considered as desired output for each case.
IV) In the usage course, when the robot tends to select a passage route, it exerts the plan of the image to the network (as an N-elements vector) and obtains the best passage angle.

5. CONCLUSIONS

In this paper, a solution based on Perceptron neural network for passing problem in middle size robot soccer was developed. Necessary input data for correct passing is fed as topological data to the network. The network extracts the essential parameters by itself and determines the best passing angle. A modification to network structure based on input presentation approach results in additional true responses as well as faster convergence. An approach is suggested to use this algorithm for scanner robots routing in dynamic environments.

6. REFERENCES

Burkhard, D. (2000). Learning of Kick in Artificial soccer, *RoboCup Symposium*, Melbourne, Australia.

Matsubara, H.; Noda, I. & Hiraki, K. (1996). Learning of Cooperative Actions in Multi-Agent Systems: a Case Study of Pass in Soccer., *Proceedings of the AAAI-96 Spring Symposium*, California, USA.

Munandar, E.& Siagian, C. (2001). Passing Simulation in robot soccer, *Proceedings of BOOM symposium*, India.

Pagello, E.; D'Angelo, A. & Montsello, F. (1999). Cooperative Behaviors in Multi-Robot Systems Through Implicit Communication, *Proceedings of Robotics and Autonomous Systems, Vol 29.*

Tambe, M. (1997). Towards Flexible Teamwork, *Journal of Artificial Intelligence Research, Vol 7.*

Vecht, B. & Lima, P. (2004). Formulation and implementation of relational behaviors for multi-robot cooperative systems., *Proceedings of RoboCup Symposium* , Lisbon , Portugal.

Wang, C.; Chen, X. & Zhao, X. (2004). Design and Implementation of a General Decision-making Model in RoboCup Simulation , *International Journal of Advanced Robotic Systems, Volume 1, Number 3*, ISSN 1729-8806.

Yokota, K.; Ozaki, K. & Watanabe, N. (1999). Cooperative Team Play Based on Communication., *Proceedings of RoboCup 1998 Symposium*, Paris, France.

Annals of DAAAM for 2006 & Proceedings of the 17th International DAAAM Symposium, ISSN 1726-9679
B. Katalinic, Published by DAAAM International, Vienna, Austria 2006
150th Birthday of Nikola Tesla (1856-1943) Inventor of Mobile Robot Teleoperation

SYSTEM RELIABILITY ASSESSMENT - FAULT TREE ANALYSIS (FTA)

ROMAN, S.M.; SIMON, A.E.; DUMITRASCU, D. & LEPADATESCU, B.

Abstract: *In this paper we present a fault tree analysis of a pressure tank using the probabilistic evaluation based on the function of structure. We derive and present a dynamic fault tree model that can be used to assess the reliability of the pressure tank.*
Key words: *Fault Trees Analysis, pressure tank.*

1. INTRODUCTION

Fault Tree Analysis (FTA) is a top-down approach to failure analysis starting with an undesirable event called a top event, such as a failure or malfunction and then determining all the ways it can happen.
Fault Tree Analysis is used to investigate the reliability and safety of complex and large systems for diagnostic applications.

2. FAULT TREE ANALYSIS (FTA)

The analysis starts by determining how these top events can be caused by individual or combined lower level failures or events. If there is a critical failure mode, then all possible ways that mode could occur, must be discovered. Fault trees combine graphical and mathematical representation of events to capture the failure properties of the whole system. The graphical part shows how events are related; the mathematical part provides the failure rates.

2.1 Drawing fault trees: gates and events

A fault tree is constructed by relating the sequences of events, which individually or in combination, could lead to the Top Event. Fault trees are built using gates and events (blocks). The two most commonly used gates in a fault tree are the AND and OR gates. As an example, consider two events (or blocks) comprising a Top Event (or a system). If occurrence of either event causes the top event to occur, then these events (blocks) are connected using an OR gate. Alternatively, if both events need to occur to cause the top event to occur, they are connected by an AND gate.

2.2 Quantification of FTA – Probabilistic evaluation based on the function of structure

A fault tree provides a framework for both quantitative and qualitative analysis of the system reliability. Let it be considered a fault tree that contains a logic operator OR. Relationship between top event reliability, top event unreliability and failure probability are:

$$P_F = 1\text{-}R_T = 1\text{-}\ R_A R_B = F_T, \tag{1}$$

$$P_F = 1\text{-}\ [(1\text{-}P_A)(1\text{-}P_B)] = P_A + P_B - P_A P_B. \tag{2}$$

Let it be considered P_i failure probability of events i in the fault tree, R_i reliability of events i, F_i unreliability of events i.
The failure probability of events i is shown as follows:

$$P_F = 1 - \prod_{i=1}^{n} (1\text{-}P_i). \tag{3}$$

Let it be considered a fault tree that contains a logic operator AND. Relationship between top event reliability, top event unreliability and failure probability are:

$$R_T = R_A + R_B - R_A R_B, \tag{4}$$

$$P_F = 1\text{-}R_T = 1\text{-}\ (R_A + R_B - R_A R_B) = F_T, \tag{5}$$

$$P_F = 1\text{-}\ [(1 - P_A) + (1 - P_B) - (1 - P_A)(1 - P_A)], \tag{6}$$

$$P_F = P_A P_B. \tag{7}$$

Let it be considered P_i failure probability of events i in the fault tree. The failure probability of events i are shown as follows:

$$P_F = \prod_{i=1}^{n} P_i. \tag{8}$$

3. CASE STUDY - FTA APPLIED TO A PRESSURE TANK

We present a fault tree analysis of a pressure tank where the top event it is chosen as "Pressure tank does not work at the designed parameters values", for a hydraulic pressure test.
From the beginning it is considered that the tank is adequate for the conditions and parameters of work requested by design specification. In this case it follows to develop the secondary faults of the tank meaning those that were not qualified as a burst due to some hidden material defects or faulty manufacturing. In order to simplify the analysis it may neglect the failures that might occur into the fittings and pipes separate of the system, as well as the failures due to disturbance of discharge valve that are fitted on the compression pump. The main interest is to find out the risk of bursting the pressure tank when the pump starts for fluid head test.
Some statements were shown into rhombus frames as not being enlarged base events. They might be excluded from the failure tree.
The events of the fault tree for the system analyzed are (fig. 3.):
E – Top event – Pressure tank does not work at the designed parameters - hydraulic pressure test;
E1 – The tank can burst;
A – The tank could be of a faulty design;
A1 – Faulty calculation when design;
A11 – No knowledge/ neglect of design matter;
A12 – Not following the initial design parameters;
A13 – Not following the technical specification;
A2 – Choosing inadequate base materials;
A21 – Wrong choice due to miscalculation;
A22 – Not following the quality specifications for materials;
A3 – Details of damaged welded joints;
B – There is a material defect;
B1 – Hidden defects of base material (construction failure);
B11 – Not following the heat treatment;
B12 – Not following the casting and forging;
B13 – Not following the after-casting and after-forging parameters;
B14 – Material testing (metrology for plates according to ISCIR specifications);

B2 – Input material defects;
B21 – Not compatible with the base material;
B21-c – If the technician is incompetent;
B211 – Choosing the inadequate input materials;
B22 – Stale or improper use;
B221 – No heat treatment technology for welding wires;
B222 – Incompetent welder;
B223 – Improper storage or out of life materials;
B3 – Inadequate welding procedure or unaccomplished;
B4 – Heat treating--Suggested and failed;
C – The hydraulic pump is over pressurized;
C1 – Gauges do not work;
C11 – Deteriorated gauges – not metrological checked;
C12 – Inadequate gauges (different accuracy rating);
C2 – Missing or malfunction of pressure relief valve of the pump; C3 – Inattentive operator;
D – Filling modulus--Filling speed/ ventilation;
E2 – Tank can fracture; E3 – Tank can crack;
E4 – Testing fluid loss.

The failure probabilities of the primary failure events are shown in table 1.

The functions of structure corresponding to each event are:

$$P(B22)=1-(1-P(B221))\cdot(1-P(B222))\cdot(1-P(B223)), \quad (9)$$

$$B21=B211\cdot B21\text{-}c, \quad (10)$$

$$P(B1)=1-(1-P(B11))\cdot(1-P(B12))\cdot(1-P(B13))\cdot(1-P(B14)), \quad (11)$$

$$P(C1)=1-(1-P(C11))\cdot(1-P(C12)), \quad (12)$$

$$P(B2)=1-(1-P(B21))\cdot(1-P(B22)), \quad (13)$$

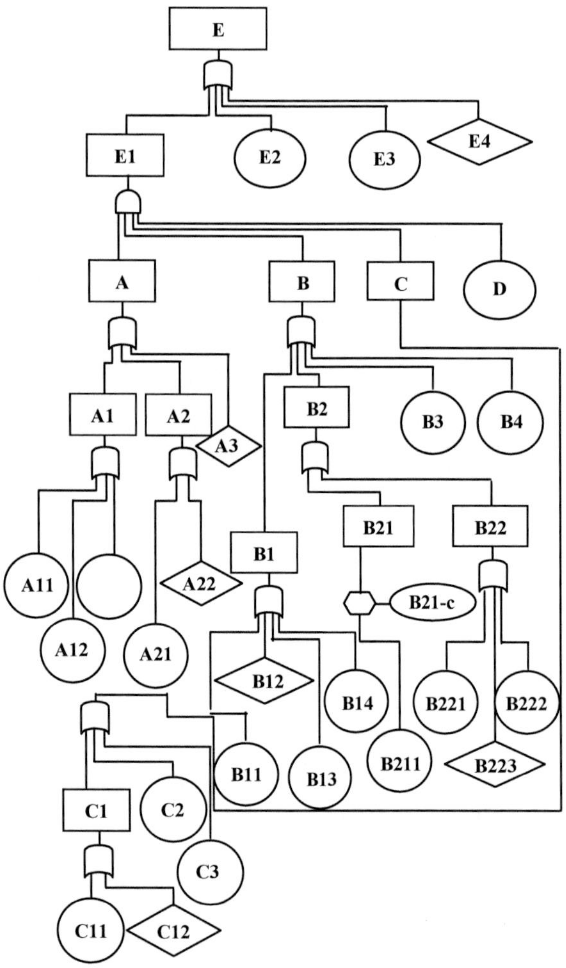

Fig. 3. Fault tree scheme for the pressure tank

Symbol	failure probabilities	Symbol	failure probabilities
E2	$2\cdot10^{-7}$	A22	$6\cdot10^{-7}$
E3	$5\cdot10^{-7}$	B11	$2\cdot10^{-7}$
E4	$5.5\cdot10^{-6}$	B12	$2\cdot10^{-7}$
D	$5.5\cdot10^{-5}$	B13	$2\cdot10^{-7}$
A3	$6.5\cdot10^{-5}$	B14	$2\cdot10^{-7}$
B3	$1.25\cdot10^{-4}$	B21-c	$9\cdot10^{-4}$
B4	$3.5\cdot10^{-5}$	B211	$8\cdot10^{-7}$
C2	$8\cdot10^{-7}$	C11	$6.2\cdot10^{-9}$
C3	$6\cdot10^{-3}$	C12	$4.5\cdot10^{-7}$
A11	$3\cdot10^{-7}$	B221	$1.4\cdot10^{-5}$
A12	$2\cdot10^{-7}$	B222	$2.5\cdot10^{-4}$
A13	$3\cdot10^{-5}$	B223	$1.8\cdot10^{-5}$
A21	$6\cdot10^{-7}$		

Table 1. Failure probabilities

$$P(A2)=1-(1-P(A21))\cdot(1-P(A22)), \quad (14)$$

$$P(A1)=1-(1-P(A11))\cdot(1-P(A12))\cdot(1-P(A13)), \quad (15)$$

$$P(C)=1-(1-P(C1))\cdot(1-P(C2))\cdot(1-P(C3)), \quad (16)$$

$$P(B)=1-(1-P(B1))\cdot(1-P(B2))\cdot(1-P(B3))\cdot(1-P(B4)), \quad (17)$$

$$P(A)=1-(1-P(A1))\cdot(1-P(A2))\cdot(1-P(A3)), \quad (18)$$

$$P(E1)=P(A)\cdot P(B)\cdot P(C)\cdot P(D), \quad (19)$$

$$P(E)=1-(1-P(E1))\cdot(1-P(E2))\cdot(1-P(E3))\cdot(1-P(E4)). \quad (20)$$

The probability of the pressure tank to burst is:

$$P_T = 1.417\times10^{-14}. \quad (21)$$

The failure probability of the top event "Pressure tank does not work at the designed parameters values" for the pressure tank analyzed is $P_T = 6.2\times10^{-6}$.

4. CONCLUSIONS

- For large fault trees, the calculating of the occurrence probability of the top event can be difficult because of their complexity. In this case, more efficient approaches need to be used for assessing the reliability of a system, such as the minimal cut set approach;
- The top event probability is equivalent to system unreliability;
- It is used as one of the best methods for systematically identifying and graphically displaying the many ways something can go wrong. This is a graphical technique that provides a systematic description of the combinations of possible occurrences in a system, which can result in;
- FTAs are used in safety critical systems especially where human life is involved;
- FTAs are used to evaluate other potentially damaging events during test, build or operation;
- FTAs complement FMEAs keying in on the worst identified failure modes.

5. REFERENCES

Pavel, A., (1987). *Work safety ness of petrol-chemical equipment*, Editura Tehnica, Bucureşti.

Bilal, A. & Popescu, C. (1998). - Web-based System Reliability Assessment (WSTAR): Fault Tree Analysis (FTA), *Technical Report for Contracts DACA31-96-D-0063*, Army Corps of Engineers Washington, D.C. 20314-1000, U.S.

Martinescu, I. & Popescu, I. (2002). - *Analiza fiabilităţii şi securităţii sistemelor tehnologice*. Editura Universităţii „Transilvania" Braşov, 2002.

Baron, T. (1988). *Quality and Reliability – Practice Instructions*, vol.1 and 2, Editura Tehnica, Bucureşti.

Renert, M. & Tănase, G. (1987). *Calculation and construction of chemical equipment*. Vol.1, Ed. Didactica si Pedagogica.

Annals of DAAAM for 2006 & Proceedings of the 17th International DAAAM Symposium, ISSN 1726-9679
Editor B. Katalinic, Published by DAAAM International, Vienna, Austria 2006
150th Birthday of Nikola Tesla (1856-1943) Inventor of Mobile Robot Teleoperation

PALLETIZATION AND CONTAINERIZATION AS MODERN TECHNOLOGIES OF LOAD TRANSPORT

RONKO, D.

Abstract: *Nowadays modern technologies of load transport are implemented in combination on the part of operators of international multi-modal transport. Investments into modern technologies of load transport result in significant advantages when compared with classical transport. To be able to choose the best and the most cost saving aspect of contemporary transport of goods it is necessary to know advantages and disadvantages, as well as the elements of modern technologies of load transport – among them – palletization and containerization.*
Key words: *palletization, pallets, containerization, containers*

1. INTRODUCTION

Economical rewards of palletization and containerization are undisputable; their effectiveness in faster freight handling, when compared with old technologies of load transport, being incontestable. In road transport it enables "door to door" forwarding of goods in pallets and containers, reducing thereby road transport costs for 35%, on rails for 50% and in the air for 20%. We are going to explain some basic properties and features, indicating advantages and disadvantages of these contemporary transport technologies, for which necessary preconditions required for growth and development already exist. Likewise, we are also going to deal with equipment required for implementation of these modern transport technologies. Palletization and containerization are present as technologies in the Republic of Croatia, but their application has not reached the expected level of development, which has already and for longer time been achieved in countries of the European Union.

2. PALLETIZATION

Pallet is a specially manufactured packing platform on which load items are being placed according to certain rules to form larger standardized freight patterns, which can be easily, safely and quickly handled.
Palletization is a set of interconnected means and technological procedures for handling and transportation of pieces of load, which procedure, after completion of packing, forms a higher level of contemporary load transport unit, which exists for almost 100 years. Due to possibility of combination of palletization with other transport technologies, such as Huckepack, RO/RO, future of palletization is indisputable. .
The biggest advantage of palletization is simple and cheap way of packing, leading to reduction of weight up to 75%, and price up to 25%; or total elimination, or reduction of costs of handling; leading to speeding up of freight handling (loading, over- and unloading) and forwarding - it's possible to achieve acceleration of freight handling up to 400%); savings in the storage areas (up to 50%) and reduction of investment expenditures for construction of storage space, decrease of damages and losses, injuries at work; greater degree of exploitation of means and equipment and many other betterments.
Main disadvantages are reflected in maintenance and repair of pallets, in addition to loss, damage and taking away of pallets.
All this had an effect that 90% of load units and circa 75% of wagon deliveries out of total road and rail transport volume in developed European Community countries has been brought into effect by palletization. In overseas transport this amounts to circa 50%, while in air transport it equals 30 to 40%, showing that uni- modal transport of goods is unimaginable without palletization.
Basic working means of palletization are pallets, fork lifters, diverse pallet shape formers, frames, movable launch pads, automatic machines for palletization and de-palletization, transport vehicles, etc.
Nowadays several sorts of pallets are in operation. Most of them are standardized, especially those which are used in international freight forwarding.
Exactly standardized pallets enable multi-modal transport, which includes the application of several means of transport in international freight transport, easier handling of pallets due to identical port or ship loading and unloading launching pads and facilities, and reduced costs due to standardization of pallets being accepted as a rule.
Flat wood pallets appear in several variants, but basic one has got one or two weight supporting surfaces. They are mostly made from wood, but can also be manufactured of plastics and aluminum.
Box pallets are wooden or metal made, closed or opened, made mainly from metal sheets, with grid sides so as to reduce their weight.

3. CONTAINERIZATION

Containers are portable boxes, special devices, transport crates and vessels which can be completely or partially closed, with at least one door, designed so as to ensure safe and fast loading and emptying, and to speed up freight delivery with one or more transport vehicles without overloading, equipped with devices enabling fast and safe handling, sufficiently strong and made of solid material, resistant to weather conditions and suitable for multiple usage, having volume of at least one cubic meter.
Fundamental purpose of container is to be used as a loading and handling unit in freight forwarding, and as "door to door" storage and packing unit.
Containerization is organizationally interconnected set of working equipment and technological procedures designed for automatic handling and transport of enlarged units of load – containers, from storage place to consumers.
Containerization is, after packaging and palletization, the first modern transport technology which has found its broad application in all European countries, and so in Republic of Croatia, in the last fifty years.

The most important goal of containerization is joining of load units packed in boxes, cartons, bales, barrels and similar, into standardized, handling and transport units of load, optimization of effects of traffic infra and super structure, qualitative and quantitative maximization of technical, technological and economic effects of production process, traffic service, and safe, simple, fast and economic handling and transport of load.

Basic equipments for work with this type of load transport technology are containers and container ships. Containers may be divided according to their application into: universal containers (for transport of goods packed in factory packaging designed for wide consumers), which makes 75% of total container fund; containers for general application (water and dust proof), special purpose containers, opened and closed containers with ventilation, container platforms with opened sides, then special containers with isothermal features, and container tanks for transport of liquid and gaseous goods.

According to their size containers may be categorized into small sized (1-3 cubic meters), middle sized (3-10 cubic meters) and large containers (volume over 10 cubic meters and external length greater then 6 meters).

At the present time three basic types of containers, standardized according to ISO standardization have found wide application. ISO has defined technical conditions for manufacture of containers, which have been accepted by the UIC (International Railway Union).

Present day standardized ISO containers correspond by their form to one part of ship storage area in overseas and river transport, half or one third of open railway freight wagons, and partly to freight cell (storage area) of large transportation planes.

The advantage of standardized containers lies in their exchangeability on overseas, railroad and road transport vehicles, which gives Republic of Croatia possibility to link overseas transport with inland transport and to develop in this way modern multimodal transport network which can compete with European traffic links. Nonexistence of standardization would slow down handling and moving of containers in the process of giving of transportation services, and complicate further inland delivery of containers after completion of overseas transport and vice versa.

Container ships as fundamental equipment in this transport technology must at least in one of their sections be equipped for carrying of large ISO containers.

Container ships may be categorized into four groups, such as: ships for exclusive container transport, then ships of the first, second, third, fifth and sixth generation, having nowadays capacity of 5,500 TEU, achieving speed of 30 knots. There are already projects and plans for construction of ships of the seventh generation, with capacity up to 6,500 TEU, which would be modernly equipped and automated with minimal exploitation costs.

The second groups of container ships are RO/RO ships, which are characterized by horizontal loading and unloading, often with on land loaded wheel transport vehicles on special, so called RO/RO ships. The third groups of container ships make LO/LO ships characterized by vertical loading and unloading of containers by application of port or ship mechanization.

The fourth groups of container ships are FO/FO ships characterized by horizontal or vertical overloading of containers. Other production means with this transport technology depend upon traffic branch applied in multimodal transport, such as diverse wagons and block trains in rail transport, platform vehicles deep bed loader, half loaders and special trucks for transport of loaders in road transport, and freight and passenger planes and helicopters, special transporters and transport conveyors in air freight transport. All these tell us about rapid development of this transport technology which has its economic justification. In Republic of Croatia container transport and containerization started already in 107', when Rijeka based "Croatia Line" introduced its five classical liners into container transport from Adriatic Sea to USA (East Coast), and the same ship owner opened in 1973 its first container ship line on the same route. The most important advantages of containerization are reduction of goods packaging costs, transport of goods in containers excludes overloading, unification of technical and technological solutions, automation of the entire transport chain, reduction of trade and administrative procedures, solid protection of goods followed by reduction of storage costs and increased handling speed. The most important disadvantages being observed are the necessary initial investment capital into production means and equipment, specialization and standardization of infra and superstructure, professional personnel requirements and operational management and high quality employees, and highly synchronized coordination of work with other participants in the process of containerization.

4. CONCLUSION

Palletization and containerization are very efficient and cost saving modern technologies of goods transport. Although palletization has been used for circa hundred years, and containerization for circa fifty years, they are nowadays irreplaceable means in transport of goods.

Users of transport in modern load transport technologies are in advantage because such system guarantees lower price of freight handling and transport when compared with classical transport. Likewise, these systems guarantee many other advantages such as minimization or total elimination of manpower with accompanying avoidance of injures at work, reduction of trade, customs and other administrative procedures, reduction of storage costs and many other improvements. On average more favorable handling and transport of loads in containers in relation to classical, conventional way of handling and load transport of not containerized goods results to circa 25% savings. Coordination of work of all participants in the chain represents industrialization of the transport service itself. This means that the entire preparation and investments into the system of palletization and containerization must be gradually coordinated. Only then container transport grows into integrated system of containerization, including palletization itself, with all other subsystems, and all their features. Only then can we enjoy to the full extent quantitative and qualitative, technical, technological, economical and legal benefits of containerization. Subsequent research showed that the entire handling transportation and storage costs in Europe up to store yards in USA have been reduced for 52,32 USD each ton in the "door to door" system of transport in relation to conventional method of transport "port to port".

5. REFERENCES

Bozicevic, J. (1985). *Infrastructure of railway traffic*, Faculty of Traffic Sciences, Zagreb

Nikolic, G. (2003). *Multimodalni transport u funkciji uključivanja Hrvatske u europski prometni sustav*, Faculty of Economics, Rijeka

Zelenika, R. (2001). *Traffic systems*, Faculty of Economics, Rijeka

Zelenika, R. (1994). *Interdisciplinary technologies of railway transport, «Railway in theory and practice»*, HŽ – Office for informative and publishing activities, 1-2/XVIII, Zagreb

Zupanovic, I. (1994). *Technology of road traffic*, Faculty of Traffic Sciences, Zagreb

Zmegac, D. – Kauzljar, D (1997). *Transport of goods with railway – types of goods wagons*, Croatan railway, Zagreb

Annals of DAAAM for 2006 & Proceedings of the 17th International DAAAM Symposium, ISSN 1726-9679
B. Katalinic, Published by DAAAM International, Vienna, Austria 2006
150th Birthday of Nikola Tesla (1856-1943) Inventor of Mobile Robot Teleoperation

KNOWLEDGE MANAGEMENT IN THE CESICED PLATFORM TO IMPROVE COLLABORATIVE PRODUCT DEVELOPMENT

ROSU, S.M.; DRAGOI G.; GURAN M. & COTET, C.E.

Abstract: *The knowledge management (KM) has received attention from designers responsible for the product development process because many of the design activities require a creative thought and is highly dependent on individual's knowledge. Furthermore, the product development process deals with a large amount of knowledge which makes the process more critical. Therefore, this work analyses the state of the art of the knowledge management and it proposes a methodological model used in the CESICED platform, based on the occurrence of types of conversions of the knowledge to be used during the product development process.*

Key words: knowledge management, collaborative systems, design, virtual enterprise

1. INTRODUCTION

Today, one of the main objectives of the virtual enterprise (VE) is to create a collaborative integrated design platform allowing the different members of the network to participate either in synchronous or asynchronous mode in collaborative design projects. Each member will bring in the knowledge related to his own expertise as part of a larger whole. Therefore, each member has to be connected to a common data base and has to be able to understand in detail the part of the content, which he/she needs to use and also the scope of the knowledge which can be delivered by other partners involved in the network. The sharing of information in the right context needs a transformation of information into knowledge, in order to disseminate the same meaning to the different actors. In this paper the workers (each professional actor) construct different models of an agent used in their applications owing to their different aims. An agent ought to be modeled as consisting of the following basic parts: knowledge base (containing the data and domain knowledge necessary for the agent to carry out all its activities), problem solver (carrying out independently learning, planning, reasoning, decision making to execute corresponding activities to accomplish task) and coordination unit or knowledge management (controlling interactions with other agents including communication, negotiation, coordination and cooperation).

2. KNOWLEDGE MANAGEMENT OVERVIEW

Knowledge has become the more important economical factor for competitiveness (Radcliffe, 1996). In this new era of information, the fundamental sources of wealth are knowledge and communication, and not natural resources or labor work. During the first decades of the computer science, the emphasis was data management. In order to transform data into information it is required tools. However, in order to transform information into knowledge it is needed time.

Knowledge is to use information (and as a consequence data) to generate new ideas or solutions. Also, today, are differentiating these three classes of elements as (Pugh, 1993): data (a discreet and objective group of facts of a certain event), information (a message containing an originator and a receiver and whose meaning involves a new interpretation based on a group of data); and knowledge (a mixture of experiences, values, contextual information and intuition, forming a framework in a person's mind that enables him/her to evaluate and to obtain new experiences and information). The larger is individual's knowledge the best will his/hers appreciation and analysis of the data and information available. As a consequence, the better is the quality of the decisions taken within the product development process. Today, the main studies are identified two important distinctions of types of knowledge that has been used (see figure 1):

- The tacit knowledge (it is the knowledge that the people possess but it is not described in any place. It is just residing in your heads);
- The explicit knowledge (it is the knowledge that is registered in some ways and therefore it is available for the other people)

Many studies on knowledge management are based on the successive passages from tacit knowledge to explicit knowledge and vice-versa. Today, these studies have also suggested four basic conversion patterns for the knowledge creation in an organization. These four basic standards are presented in figure 1 in such a way that the creation of the organizational knowledge is based on a continuous and dynamic interaction between the tacit knowledge and the explicit knowledge:

- From tacit knowledge to tacit knowledge: it is a process of sharing experiences and, therefore, the creation of tacit knowledge and the base input for the acquisition of this knowledge type is experience.
- From tacit knowledge to explicit knowledge (it is a process of articulation of the individual's tacit knowledge in explicit concepts. This conceptual knowledge happens through: symbolic representation of the tacit knowledge).
- From explicit knowledge to explicit knowledge (it is a conversion process of some type of explicit knowledge generated by an individual to add up to the explicit knowledge of an organization). Usually this systemic knowledge happens by grouping and processing different explicit knowledge that could generate into a new knowledge.
- From explicit knowledge to tacit knowledge (it is the process of incorporating explicit knowledge from the organization into individual's tacit knowledge.

Two problems were already present today (Tichkiewitch et al., 2006) and are always waiting for an answer:

- What does one capitalize?
- How does one capitalize?

The organizational knowledge is not only found in documents, databases and information systems. They can also be found in the business processes, group practices and in the accumulated experience of individuals.

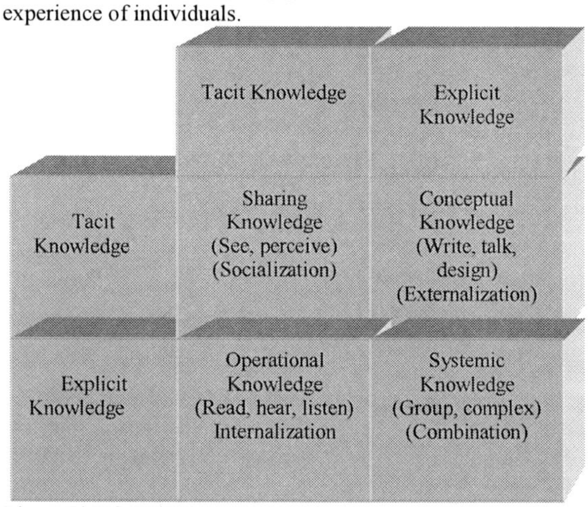

Fig.1. The four basic standards for the knowledge creation or ways of converting knowledge

The knowledge is transmitted from people to people through means such as videos, books, documents and Internet. Furthermore, individuals can gather knowledge from those who already have it by interpersonal learning and sharing experiences and ideas (Rosu et al., 2005).

3. A METHODOLOGICAL VIEW FOR KM

A methodological model of knowledge management for the product design process in the CESICED platform is presented in figure 2. We presents the company's knowledge and we shows a simplified manner to present how organizations can generate, retrieve and dispose its knowledge to strategically create adding value for their products. Besides the four conversions of the knowledge, many other steps were added in order to increase the knowledge of the company and, consequently, to improve the product development process. Also, the additional steps are: obtain and use, learn and contribute, evaluate and sustain, discharge and support. The knowledge of the organization is composed by the sharing knowledge of each individual. Based on the steps described in the figure 2 an efficient and effective management of the intellectual capital of the company is obtained. The steps obtain and uses are well known within organizations. People always seek information and use them later to solve their problems, to take decisions or to create new products. Therefore, new technologies (e.g. intranet/internet/ extranet) allow that the large amount of information that flows within organizations can be correctly managed. The steps learn and contribute are relatively new for organizations. It has been difficult to convince employees to contribute to the organization's knowledge base. New technologies have helped companies easily organize, send and transfer certain types of information. However, this facility has been seen by the employee as a threat for his/hers own job security. The most difficult task is to convince individuals that their contribution will give return to their organization as well as to themselves. The step evaluate indicates that the organization should define its own necessary knowledge for its mission and classify its own currently intellectual capital. The knowledge manager does more than organize the content in system on-line, he/she should understand and foresee the community's needs. The step sustain or maintain should assure that the future intellectual capital will maintain the organization viable and competitive. Organizations tend to build their own intellectual capital through their relationships with customers, employees, suppliers etc.

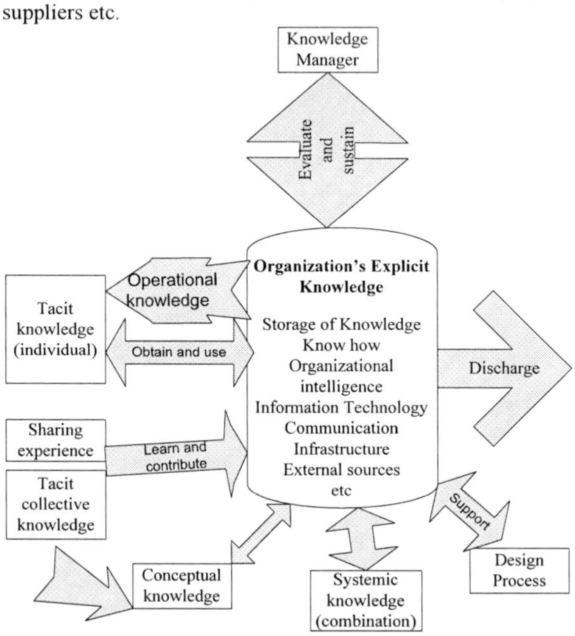

Fig. 2. Organization knowledge management model

The knowledge manager should also be responsible for the maintenance of the organizations knowledge base. The step discharge excludes any useless knowledge from the organizations knowledge base. However, some knowledge can be more valuable if it can be transferred to outside of the organization. The step support can be used for the continuous improvement of the product design process.

4. CONCLUSION

In this paper, a methodological model for knowledge management within the product development process was developed based on an extensive theoretical review on knowledge management, organizational learning and product development process management issues. The validation of this methodology will be carried out based on a practical application in a CESICED project and Romanian SME. The aim of the CESICED project (university-SME's partnership) is to determine the new organization type for integrating in the virtual enterprise medium and to outsourcing shared resources service PREMINV center for industrial partners.

5. REFERENCES

Pugh, S. (1993). *Total design,* Addison-Wesley, Reading, Massachusetts.

Radcliffe, D. (1996). Concurrency of Actions, Ideas and Knowledge Displays within a Design Team, in *Analysing Design Activity*, N. Cross et al., pp. 343-364, John Wiley & Sons. Ltd: West Sussex, England.

Tichkiewitch, S.; Radulescu, B. & Dragoi, G. (2006). Knowledge management for a cooperative design system, in *Advances in Design,* ElMaraghy & Hoda A. Eds., pp. 97-110, Springer Verlag, ISBN 1-84628-004-4.

Rosu, S.M.; Drăgoi, G.; Guran, M.; Cotet, C.E. & Rosu, L. (2005). Knowledge management support to product design process within CESICED platform, *Proceedings of The Second International Conference of Integrated Engineering*, pp. 189-190, with CD-ROM support for the full papers, ISBN 973-625-259-0, .16th – 18th October 2005, Timisoara, Romania.

Annals of DAAAM for 2006 & Proceedings of the 17th International DAAAM Symposium, ISSN 1726-9679
B. Katalinic, Published by DAAAM International, Vienna, Austria 2006
150th Birthday of Nikola Tesla (1856-1943) Inventor of Mobile Robot Teleoperation

THE INFLUENCE OF MATERIAL COMBINATION ON CONTACT STRESSES IN CYLINDER ROLLING BEARING

RUBESA, J. & SIMINIATI, D.

Abstract: *The influence of material combination on contact stresses in cylinder rolling bearings has been analysed in this paper. Both concave-convex and concave-concave contact of inner and outer races with the cylinder was simplified as cylinder-planar surface contact. A short review of commonly used material combinations is given, followed by FEM model description, and obtained contact stress analysis results. The four different values of load have been applied at cylinder-plate contact model.*
Key words: *contact stress, mesh, bearing, FEM*

1. INTRODUCTION

Considering the wide range of the practical applications of bearings in machinery, extensive experimental analyses of contact stresses have been done by various researchers in order to determine changes in behaviour for different exploitation conditions as well as influences of bearing construction.
Nowadays, development of the computers and FEM software, enables fast and accurate numerical calculations of various exploitation parameters for different material combinations.

2. THE MATERIALS OF ROLLER BEARINGS

2.1 Steel-steel combination

The conventional steel rolling bearings are still most applicable in practical use. The most widely used and most suitable materials for rolling bearings are high carbon steels such as AISI 52100 which means that it has one percent chromium and one percent carbon. For applications requiring high corrosion resistance, stainless steel AISI 440C is used. Bearing materials must have a low friction coefficient, to be weightless and be able to withstand contact stresses. However, depending on the application, stainless steel is most commonly used in combination with steel rolling elements.

2.2 Steel-plastic combination

Plastics such as acetal, nylon, PTFE and other non-metallic materials are increasingly used as self-lubricating bearings. Their composition is refined over many years to obtain favourable bearing characteristics included low friction, corrosion resistance, ability to function over wide temperature ranges and significant load-carrying capability. Although temperature ranges, dimensional stability and load quantity of plastic bearings are in general smaller than for metallic bearings, plastic bearings are significantly economical. The bearing consists of one material plastic and the other steel, are an excellent solution for corrosive operating environments.

2.3. Steel-ceramic combination

Ceramic materials such as ZrO_2, SiC especially Si_3N_4 have the advantages such as low density, high strength, low friction factor and wear resistance (Zhao et al., 1998). The best choice for rolling elements of the hybrid constructions is ceramics with metallic inner and outer race.

3. NUMERICAL MODEL OF CYLINDER ROLLING BEARING

Numerical model of cylinder roller bearing is simplified as 2D contact between cylinder and planar surface.
The contact stresses were calculated using MSC.Marc 2005 r2 nonlinear analysis (Bathe, 1996) for an elastic cylinder and elastic plate (Wriggers, 2002). Both cylinder and plate were modelled using 2D Quad4 elements with two translations on every node (Msc. Corp, 2004). It has been found that meshing with 2414 elements for cylinder and 3520 elements for plate, gives the results that are very closed to analytical. Material is considered elastic, following Hooke's law with appropriate modulus of elasticity E and Poisson's ratios ν .
The plate was fully constrained at the bottom and side edges of the model and the load per unit length was divided by cylinder diameter and then applied at the top of the cylinder half (Fig. 1). The four different values of radial load (2500, 4375, 6250, 10000 N/mm) were applied on cylinder model.

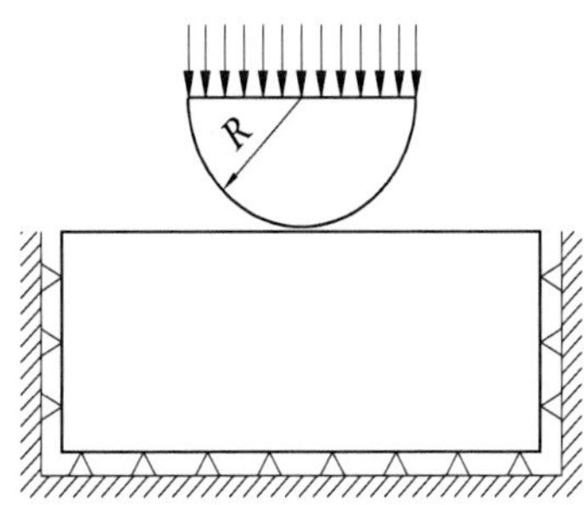

Fig.1. Boundary conditions and loading

Three different material combinations for cylinder and plate are considered as shown in Table 1.

Cylinder material	Plate material
Steel $E = 210000$ N/mm^2 , $\nu = 0,3$	Steel $E = 210000$ N/mm^2 , $\nu = 0,3$
Steel $E = 210000$ N/mm^2 , $\nu = 0,3$	Acetal $E = 2800$ N/mm^2 , $\nu = 0,35$
Si_3N_4 $E = 314000$ N/mm^2 , $\nu = 0,46$	Steel $E = 210000$ N/mm^2 , $\nu = 0,3$

Table 1. Material combinations

For this problem an exact solution for the contact pressure was obtained analytically (Boresi & Lynn, 1974), hence the results obtained by finite element method were compared directly with analytical results. According to Von Mises theory of equivalent strength in dependence of compressive yield strength and Poisson's ratios of considered materials (Brnic, Turkalj 2006; Slocum, 1992), the value of maximum pressure p_{max} for each material combination are given, too.

4. RESULTS

Maximum normal stress σ_{yy} and maximum shear stress τ_{max} were calculated for the above described finite element mesh density, materials and load cases. The given results are obtained by both numerical and analytical analysis.

Fig. 2 shows obtained normal and shear stresses with limited pressure that material can last before plastic deformation for steel-steel combination, Fig. 3 for steel-acetal combination and Fig.4 for Si_3N_4-steel combination of material.

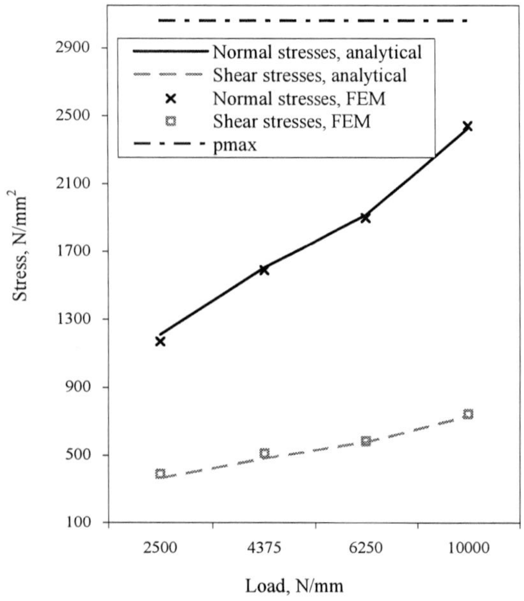

Fig. 2. Analysis result, steel-steel combination

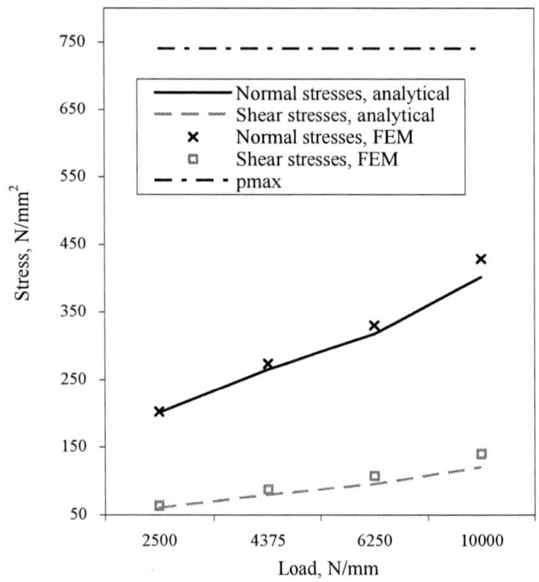

Fig. 3. Analysis result, steel-acetal combination

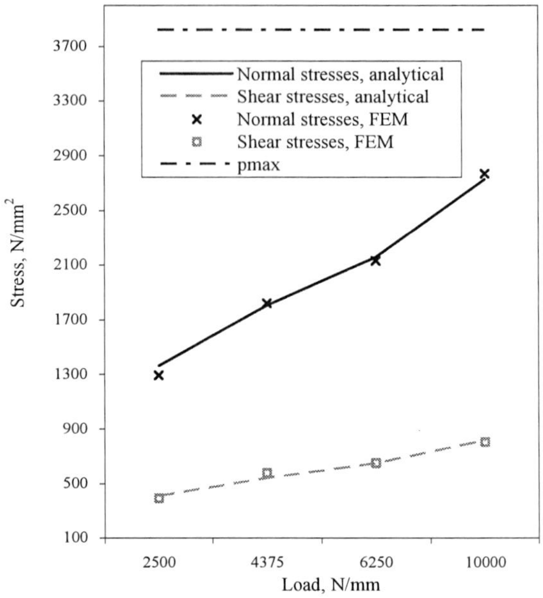

Fig. 4. Analysis result, Si_3N_4-steel combination

5. CONCLUSION

The numerical analysis results, with optimal chosen mesh density, confirm analytically obtained results of contact stresses. The simulation results of contact stresses shows that the principal stresses are largest at the contact surface and the maximum shear stresses are largest just beneath the contact surface.

A deviation of 9% to 30% for steel-acetal material combination for greater load has been noticed. The reason for that phenomenon we can find in various Hooke's law for metal and non-metal materials. Beside that, the deformations of non-metal materials are of the different rank that for metallic. So, we can talk about some kind of non-Hertzian contact problem.

Generally, the analysis shows that all materials combinations ensure applied loads do not produce pressure greater than is allowed, which enables safe operation and duration of bearings of described construction.

6. REFERENCES

Bathe, K. J. (1996) *Finite Element Procedures*, Prentice Hall, ISBN 0-13-301458-4, New Jersey

Boresi, A. P; Schmidt, R. J. & Sidebottom, O. M. (1992). *Advanced Mechanics of Materials,* John Wiley and sons, ISBN 0-471-55157-0, USA

Brnic, J. & Turkalj, G. (2006). *Nauka o cvrstoci II,* Zigo, ISBN 953-7142-13-2, Rijeka

Msc. Software Corporation (2004). *Msc. Marc User's Guide v.2005, Los Angeles.*

Slocum, A. H. (1992) *Precision Machine Design*, Society of Manufacturing Engineers, ISBN 0-872-63492-2, Dearborn, Michigan

Wriggers, P. (2002). *Computational Contact Mechanics,* John Wiley and sons, ISBN 0-471-49680-4, West Sussex, England

Zhao, X. Z.; Liu, J. J. & Fischer, T. E. (1998). Effects of lubricant rheology and additives chemistry in the wear of Si_3N_4 sliding on steel, *Wear,* Vol. 223, No. 1, (December 1988), pp. 37-43, ISSN: 004

Annals of DAAAM for 2006 & Proceedings of the 17th International DAAAM Symposium, ISSN 1726-9679
B. Katalinic, Published by DAAAM International, Vienna, Austria 2006
150th Birthday of Nikola Tesla (1856-1943) Inventor of Mobile Robot Teleoperation

TARGETS FOR DEVELOPING COST EFFICIENCY IN ELECTRIC UTILITIES

SAARISTO, H.; TAKALA, J. & HAKOLA, T.

Abstract: The purpose of this paper is to demonstrate activity based accounting projects in six electric utilities, give a general picture of the cost structure and give some improvement ideas to the case companies. The research data was gathered by using mass tailorized activity based costing. The results seemed to be supporting each other in all case companies and the standard deviation.seemed to be inside reasonable limits.
Key words: *ABC, electric utilities, cost structure, cost efficiency.*

1. PROBLEM STATEMENT

The government regulator authorities in Finland demand electric utilities to develop their cost efficiency and limit the profit the companies may get from the invested capital. The aim of this activity is to guide electric utilities to improve their efficiency without cutting the quality. This causes that utilities has to improve their efficiency even they are operating monopoly business.
Utilities are using a lot of resources to maintain their grid but the resources are generally not focused optimally because of the lack of cost information that supports asset management properly. Case utilities wanted to know how efficient their processes are compared to the other utilities in the research.

2. APPLICATION AREA

Case companies are electric utilities that are owners of electricity distribution networks. They are taking care of the electricity consumers service in a local level and distribute the electricity from the regional 110kV grid.

3. THEORETICAL BACKGROUND

Activity Based Costing (ABC) is an accounting technique that allows organizations to determine their actual costs associated with each product and service produced by the organization. The basic idea of activity based costing is that activities use resources and the products use activities and materials. The goal in ABC is to recognize activities, define their costs and outputs. Compared with traditional cost accounting, which arbitrarily allocates overhead to the cost object, ABC can be considered to be a more accurate cost management methodology. ABC converts indirect expenses into direct expenses by means of allocating each expense category to the particular cost object. ABC is especially suitable in cases where overhead costs are high and products are diverse (complexity, volume, amount of direct labor). Representative steps in ABC are 1) identify activities, 2) determine cost for each activity, 3) determine cost drivers, 4) collect activity data, and 5) calculate cost (Brimson 1992.)

4. DATA ACQUISITION

Cost data was acquired by activity-based accounting. The basic idea to mass tailor the analysis came from ABC2000TM-concept developed by ABB Quality Management Services. (Wingren 2005). The flow chart of the processes in electric utility was made only once in the first case company. For the creation of flown chart all the managers from power distribution business were summoned together. In the meeting different kind of activities were presented to the participants with a question: "Does anyone in your organization do this activity?" After finding out all activities, they were organized to the form of a flow chart. The meeting lasted 7 hours and after that the process chart was ready. The same process chart was used in all analysis in six case companies of this research. The suitableness of the chart was tested in each company by presentating the chart to the executive body of the company. The ABC-analysis was carried out by a chain of meetings described in next sub-paragraphs. (Saaristo 2005)

4.1 Start up meeting

First meeting in a chain is the presentation meeting, where we lock the ready planned project plan to the deadlines and decide the wanted focus areas and targets. The process chart is also presented to the executive body witch accepts it for the project use and in some cases demands more details to the companys focus areas. It's also very sensible to agree the wanted level of debriefing and the costs of the project in this meeting. As a result from the start up meeting you should have an agreed project plan and schedule. Typically the start up meeting will take about 2-4 hours. The preparations from the customer side will take about 1-2 hours including the target planning, selection of persons, convening and organizing the meeting place.

4.2 Data from the financial management

This meeting must be held before next meetings. The participants will be defined by the project manager. Usually the financial director or the chief accountant with the researcher is enough. However project managers often want to participate. The purpose of this meeting is to define what cost-information and in which form needs to be delivered to the researcher. This meeting usually takes 1-3 hours and the data acquisition completed by the customer representative takes about 2-6 hours depending of the programs the company uses.

4.3 Finding out the use of working hours

In most cases with electric utilities this meeting is the place which has the biggest effect to the results. An efficient way to find out the use of working hours have been a directed meeting where the typical process chart of a utility is presented and the personnel divides their working hours to the processes. A form with processes on X-axis and the personnel names on Y-axis is filled by the foremen so that every employee in the business unit will be processed.

Name	Preventive maintenance	Corrective maintenance
Smith, John	30%	70%
Doe, Joe	80%	20%

Table 1. Example: Form to find out the working hours.

This meeting takes typically 4-6 hours and the preparations will take 2-3 hours working time in customers organization. In some cases the surveillance of working hours in customer companys own system can be used as an aid. However there is often differences between companies and because of the wanted possibility to benchmark companies the use of working hours must be found out in same form from every case company.
The purpose of this meeting is to direct the working hours to the activities. This information can be used to target most of the personnel dependent costs to the activities.

4.4 Defining the use of non-working hour dependent resources
This meeting is pretty similar as the previous one, but the aim is to find out how does the non-working hours dependent costs fall to the activities. Depreciation, the use of expensive tools like computer programs and cars are often things that will be found out in this meeting.
Persons who know the use of resources handled needs to participate in this meeting. Participants will fill a form similar to table 2.

Name	Preventive maintenance	Corrective maintenance
Jeep	50%	50%
Program 1.	80%	20%
Deprecation from buildings	60%	40%

Table 2. Example: Form to find out the non-working hour related costs.

After getting the information presented in table 2 the next part of the project is mainly researcher own work. Building up a regular Excel-based model takes usually 2 working days.

4.5 Presenting the results
In this meeting the results from the analysis will be presented to the executive body of the company. The validity of the results will be ascertained by the executive body and comparisons to the other utilities. In some cases executive body wants changes or corrections to the results but usually this is the closing meeting for the project.

5. RESULTS

Key figures from all six case companies seemed to be supporting each other but they were very responsive to the density of the customers in the distribution area. Standard deviations seemed to be in reasonable limits, so ABC seems to be adequate tool to analyze electric utilities.
In average the amount of maintenance costs was about 20 percent of turnover in power distribution business and 48 percent of all operational costs. Metering and customer service was the second biggest process when measured by resource consumption. The costs of supporting processes are divided to the main processes according to the amount of usage.

	Average	Standard Deviation
Maintenance costs / all costs	48%	14%
Maintenance costs / turnover	20%	11%
Energy metering and customer service / all costs	27%	6%
Energy metering and customer service / turnover	10%	1%
Operation costs / all costs	7%	4%
Operation costs / turnover	2%	2%

Table 3. Figures describing the average cost structure of the analyzed electrical utilities (6 cases)

Maintenance activities seem to take most of resources, but the differences in efficiency were bigger in operation activities. Most of these costs were personnel costs when the depreciation from the grid, distribution losses and payments to the transmission system owner were excluded.
In area of maintenance the optimum balance between corrective and preventive maintenance can be approached for an example by the means of reliability centered maintenance. Also the optimum focus of investments should decrease the amount of maintenance cost. Finding optimum points demands information, so the need of well managed IT-infrastructure is strong. Well considered outsourcing may make the process also more efficient; especially in activities that are rarely done but the support cost still exists all the time.
In the area of energy metering the new wireless meter-reading technology may offer the solution. Synergy possibilities between other infrastructures should also be considered, for an example water and district heating consumption meters could be read all together. Pipes for optical fiber could be rented for the teleoperators. Real-time metering would provide benefits also to other processes. Finding faults from the infrastructure would be much faster if every customer would have real-time meter.
Customer service could be improved with better standardization of products. Also the use of internet could be in a higher level. (Tuominen 2000)

6. REFERENCES

Antila, Pertti & Wingren Tom. 2000. ABC2000 –concept. ABB Quality Management Services, Vaasa. Finland. Unpublished in Finish.

Tuominen, Juha Mikael. The potential for rationalization of electricity distribution companies in Norway. Lappeenranta. Finland. 2000.

Brimson, J., “Toimintolaskenta Activity-Based Accounting”. Reports and publications series, University of Jyväskylä, 1992.

Saaristo, Hannu. “Sähköverkkoyhtiön ylläpitoprosessin tehostaminen”. Master thesis, University of Vaasa, 2005. Published only in Finnish.

Wingren, Tom. Essays in Activity-Based Costing. Mass-Tailorization, Implementation and New Applications. Ph.D. Thesis. University of Vaasa. 2005.

Annals of DAAAM for 2006 & Proceedings of the 17th International DAAAM Symposium, ISSN 1726-9679
Editor B. Katalinic, Published by DAAAM International, Vienna, Austria 2006
150th Birthday of Nikola Tesla (1856-1943) Inventor of Mobile Robot Teleoperation

INVESTIGATION OF BEARING RATIO OF THE MACHINED SURFACE PROFILE AS A FUNCTION OF MACHINING CONDITIONS

SALIHU, A.; BUNJAKU, A.; QEHAJA, N. & ZEQIRI, H.

***Abstract**: In this work are presented results on the research of bearing ratio during turning. The turning is performed on the machined steel C45, applying the methods of planning of poly-factorial experiments, using as cutting tool the ceramic inserts MC2 of the company Hertell. The measurements are carried out by means of profile graphmeter model Talasurf 6 of company Taylor Hobson.*
Key words: *roughness, bearing ratio, turning, cutting.*

1. INTRODUCTION

The result of the many researchers show the necessity of presentation of geometric structure of the surface of the plane lay, with the height parameters, horizontal, plane metric and functional, in order to get more completed data to define the roughness of the surface .
In functional parameters of the roughness is included as well the bearing ratio. From the size of the bearing ratio are depended the exploited characteristics of contact details. Therefore the knowledge of changes of bearing ratio in functions of cutting regimes is important. By using the theory of experimental planification with many factors are obtained the mathematic models of the bearing ratio . The obtained results are presented in this work.

2. CONDITIONS FOR EXPERIMENT REALIZATION

Machine: The experiments for measurement of roughness parameters of the surface process are realized by numeric lathe model MD 5S GILDEMEISTER P = 1.85 – 25 kW with rotating nr.fields n=100–4000 rot/min and feed 0.001–39.99 mm/rot
Metal cutting instrument: The cutting inserts: SNGN 120708 -120712-120720 from mixed ceramic (Al_2O_3 + TiC) of HERTELL company are used.
Also, the tool holder is used CSRNR 25x25 M12H3 that the cutting edge obtains these values:

$\chi = 75^{\circ}$, $\chi_1 = 15^{o}$, $\gamma = -6^{\circ}$,

$\alpha = 6^{\circ}, \lambda = -6^{\circ}, r_{\varepsilon} = 0.8 - 1.2 - 2.0\,mm$

$\gamma_f = -20^{\circ}, b_f = 0.2\,mm$.

Measure device: Measurement of parameters of the roughness on the processed surface is done with the computer measurement equipment Talasurf 6 of Taylor Hobson company.

Research material: The rings are made of machined steel C45 (pursuant to DIN) in normal state with strength in limits of 185-200 HB with dimensions Ǿ 170 x 80 x 25 mm.

Processing parameters: The research process is realised with the change of *v, s, a* and *r* presented in Tab. 2.1 using the plan with many factors of the first row ($2^4 + 4$).

CHARACTERISTICS OF INDIPENDENT DIFFERENT SIZES					
Nr	Note	Level Code	Maximal 1	average 0	Minimal -1
1	v (m/min)	X1	700,000	458,258	300,000
2	s (mm/rot)	X2	0.400	0,283	0,200
3	a (mm)	X3	1,600	0,894	0,500
4	r (mm)	X4	2,000	1,265	0,800

Table 2.1 Characteristics of the factors

3. ANALYSE OF THE RESEARCH RESULTS

Chosen plan and results obtained are shown in tab. 3.1. Basing in the obtained results, and the processing of data in computer there are presented the curves of bearing ratio for the depth p=0.2Rt and p=0.5Rt [1]. After processing the data there are obtained the mathematic models 3.1 and 3.2 . Graphic interpretation is shown in fig. 3.1. In depth p= 0.2Rt the cutting speed *v and the round radius* of cutting insert r influence in reduction of the size t_{p20t}, although the feed *s* and the cutting depth a influence in the increase of the size t_{p20t}. The situation for p=0.5Rt changes. Here the cutting speed *v*, rounding radius *r* and the cutting depth *a* have proportional influence, although the feed has no proportional influence.

	REAL PLAN OF MATRIX				REZULTS	
Nr rend	v (m/min)	s (mm/rot)	a (mm)	r (mm)	tp20 (%)	tp50 (%)
1	300.000	0.200	0.500	0.800	7.100	25.866
2	700.000	0.200	0.500	0.800	4.700	27.300
3	300.000	0.400	0.500	0.800	8.460	29.330
4	700.000	0.400	0.500	0.800	8.360	29.600
5	300.000	0.200	1.600	0.800	7.870	28.330
6	700.000	0.200	1.600	0.800	4.030	25.500
7	300.000	0.400	1.600	0.800	8.000	27.460
8	700.000	0.400	1.600	0.800	13.600	39.530
9	300.000	0.200	0.500	2.000	4.800	37.530
10	700.000	0.200	0.500	2.000	6.670	55.830
11	300.000	0.400	0.500	2.000	8.500	31.800
12	700.000	0.400	0.500	2.000	7.330	31.300
13	300.000	0.200	1.600	2.000	8.360	43.233
14	700.000	0.200	1.600	2.000	6.060	46.330
15	300.000	0.400	1.600	2.000	7.260	28.900
16	700.000	0.400	1.600	2.000	9.930	36.700
17	458.258	0.283	0.894	1.265	4.800	23.300
18	458,258	0.283	0.894	1.265	8.400	27.800
19	458.258	0.283	0.894	1.265	4.100	21.000
20	458.258	0.283	0.894	1.265	4.800	30.600

Table 3.1

The increase of cutting speed during turning causes vibrations. The increase of the cutting feed *s* changes the character of the curve of bearing ratio. For $s = 0.2\,(mm\ /\ rot)$ the curve of bearing ratio has digressive –regressive character, although with the increase of the cutting feed $s = 0.4\,(mm\ /\ rot)$

the curve of the bearing ratio has typical digressive character [1].

$$tp_{20t} = 19.529\ v^{-0.0576}\ s^{0.541}\ a^{0.105}\ r^{-0.0151} \quad 3.1$$

$$tp_{50t} = 9.883\ v^{0.149}\ s^{-0.138}\ a^{0.0284}\ r^{0.3029} \quad 3.2$$

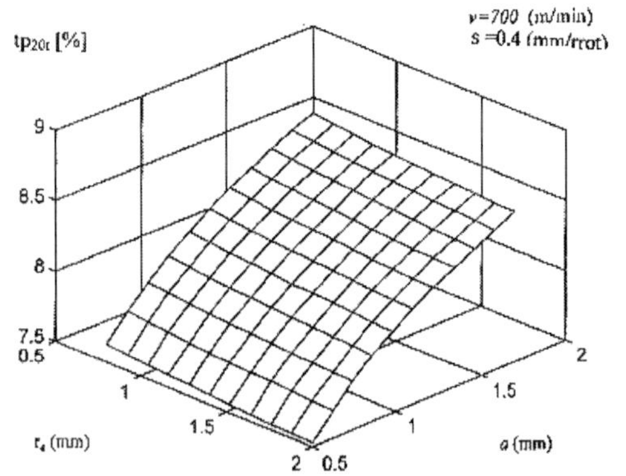

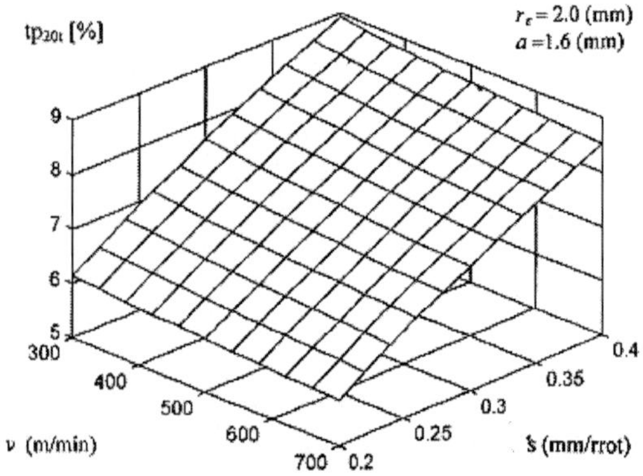

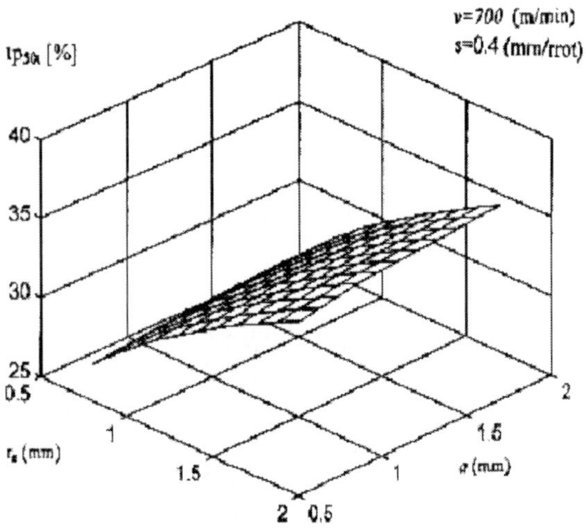

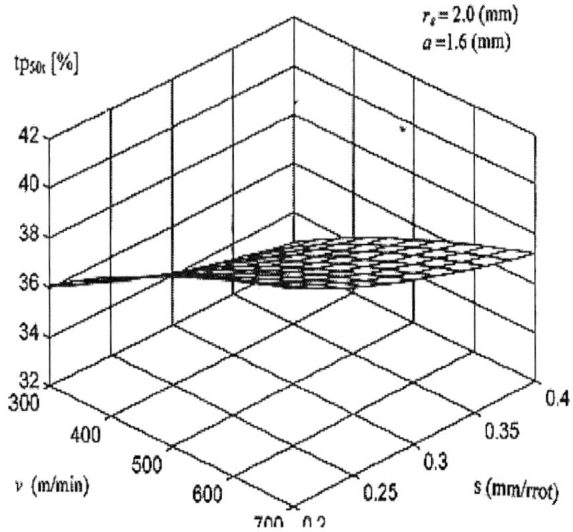

Fig.3.1 Graphical interpretation of the mathematical models3.1and 3.2

4. CONCLUSION

The analyses of the mathematic models obtained makes us possible to conclude:

-The change of the bearing ratio in the function of cutting parameters can be represented in step function.

-Larger influence in the bearing ratio has the feed then the radius *r*.

-The influence of the speed is direct.

The results of the research are of the interest to show the exploited characteristics in the surface lay by choosing the cutting parameters.

5. REFERENCES

Salihu, A.(2001) *Hulumtimi i perpunueshmerise se materialit me prerje me shpejtesi te rritura,* punim i doktoratës, Fakulteti i Makinerisë Prishtinë.

Bushati, A. (1975). *Teknologjia mekanike*, dispenca I, Fakulteti i inxhinjerisë Mekanike, Tiranë, 1975.

Salihu, A. (1995). *Korrelacioni i përpunueshmërisë së materialit me prerje përmes rezistencave të prerjes për çeliqet Č.4571 dhe Č.1530,* punim magjistrature, F M, Prishtinë

Stankov, J.(1982), *Osnove merne tehnike, metode planiranja eksperimenata*, FTN, Novi Sad..

Pantelic,I (1984), *Primena statistickih metoda i istrazivanjima i procesima proizvodnje*, FTN, Novi Sad.

Joko, S. (1989). *Teorija obrade metala* I, M F. Beograd,

Savar, S. (1977), *Obrada odvajanjem čestica, I deo,* M.F Zagreb.

Annals of DAAAM for 2006 & Proceedings of the 17th International DAAAM Symposium, ISSN 1726-9679
B. Katalinic, Published by DAAAM International, Vienna, Austria 2006
150th Birthday of Nikola Tesla (1856-1943) Inventor of Mobile Robot Teleoperation

ADAPTIVE PREDICTION USING ADALINE

SAMEK, D. & MANAS, M.

Abstract: ***The*** *goal of this paper is to introduce ADALINE as an adaptive predictor. Special attention is paid for applicability in nonlinear model predictive control. This ability is demonstrated by two simulations in the software MATLAB and Simulink. The first simulation presents prediction of changing harmonic signal. The second simulation describes prediction of nonlinear SISO system output also with parameter changes.*
Key words: *prediction, adaptation, ADALINE.*

1. INTRODUCTION

The term prediction is mostly understood as the future value estimation. In the system control theory it usually means the prognosis of state or output of the system based on current state, history, system characteristics and its eventual mathematical description. As far as signal prediction is concerned we aim at the concrete system output prediction.
The classic approach in prediction is based on a set-upping of mathematical model of system. The deterministic model is usually concerned, but in case of stochastic system it is necessary to use the stochastic model. Many predictive control techniques are established on this approach (Camacho and Bordons, 1995). They are mainly based on the assumption that the system to be controlled can be regarded as linear and that its model is available a priori. The disadvantage of this approach is the necessity to know system, its parameters and to describe them. Moreover, difficulties rise during the prediction of strongly nonlinear systems and the eventual control performance is not satisfactory.
Artificial neural networks (ANNs) offer interesting possibility for modelling and predicting any nonlinear process output without a priori knowledge. ANNs can be regarded as non-linear black-box models. The most used type of ANN for predictive control is multilayer feed forward neural network (Kanjilal 1995, Najim at al. 1997). Nevertheless, there are many other ANNs that are suitable for nonlinear system prediction, for example adaptive linear neuron (ADALINE).

2. ADALINE

ADALINE was described by B. Widrow and M.E. Hoff as an adaptive threshold logic element in 1960 (Widrow & Hoff, 1960). Though, the original version of ADALINE had only simple two-state threshold transfer function with the range of function {-1;+1}, nowadays ADALINE is also used with linear transfer function (Arbib, 2002; Demuth & Beale, 2002). Although this structure has limited skills, it is possible to connect more of ADALINEs together to obtain MADALINE (Multiple ADALINE).
Despite the fact that ADALINEs are able to solve only linearly separable problems, it has been shown in practice that they can approximate nonlinear functions with the sufficient accuracy while using enough number of neurons (Freeman & Skapura, 1991). Because of their main advantage that is the very fast learning, they have many practical applications, e.g. noise reduction, signal processing and signal prediction in control and communication systems. The learning procedure is based on an iterative search process, where performance feedback is used to guide the search process. In other words, a designer "trains" the system by "showing" it examples of inputs and the respective outputs. In this way, the system competence is directly and quantitatively related to the amount of experience the system was given.
The most popular learning method is simple LMS (Least Mean Square) algorithm (Widrow & Hoff, 1960), often called the Widrow-Hoff Delta Rule, which is adopted in this paper. This method is based on the minimization of Mean Square Error (MSE) which is for j-th ADALINE defined:

$$E_j = \frac{1}{n}\sum_{i=1}^{n}\left(\tau(i) - y_{out}(i)\right)^2 \qquad (1)$$

where $\tau(i)$ is i-th target neuron output, $y_{out}(i)$ is output from i-th ADALINE, n is number of training data. For Multiple ADALINE (MADALINE) with m neurons is necessary to compute global MSE:

$$E = \frac{1}{m}\sum_{j=1}^{m} E_j \qquad (2)$$

New weighting matrix $\boldsymbol{W}(k+1)$ and bias vector $\boldsymbol{b}(k+1)$ in the step $k+1$ equal to (for ADALINE $\boldsymbol{W}$ has only one row and $\boldsymbol{b}$ is scalar):

$$\boldsymbol{W}(k+1) = \boldsymbol{W}(k) + \alpha \cdot \varepsilon(k) \cdot \boldsymbol{u}_{in}(k) \qquad (3)$$

$$\boldsymbol{b}(k+1) = \boldsymbol{b}(k) + \alpha \cdot \varepsilon(k) \qquad (4)$$

where $\boldsymbol{W}(k)$ is previous weighting matrix, $\boldsymbol{b}(k)$ is previous bias vector, α is learning rate from the interval $<0; 1>$, $\boldsymbol{u}_{in}(k)$ is the vector of input data and $\varepsilon(k)$ is output error in the step k:

$$\boldsymbol{\varepsilon}(k) = \boldsymbol{\tau}(k) - \boldsymbol{y}_{out}(k) \qquad (5)$$

where $\boldsymbol{y}_{out}(k)$ is actual output from MADALINE:

$$\boldsymbol{y}_{out}(k) = \boldsymbol{b}(k) + \boldsymbol{W}(k) \cdot \boldsymbol{u}_{in}(k) \qquad (6)$$

Faster learning can be reached by higher learning rate α, however too high learning rate could lead to instability and errors. For stable learning process the learning rate should be lower than reciprocal value of highest eigen value of correlation matrix $\boldsymbol{u}(k)^T \boldsymbol{u}(k)$ of the input vector.

3. SIMULATIONS

Presented method was tested on two nonlinear prediction problems. The first was sine-based signal with changing parameters. The second was nonlinear system of two interconnected cylindrical tanks. In both cases the predictors used five recent values for prediction (in case of the tanks input and output values were used).

3.1 Sine-based signal

In order to test adaptivity of ADALINE, the sine-based signal with two steep changes of signal parameters was used. In the time of 10s the magnitude is changed and in the time of 20s the magnitude and frequency is changed. The sampling period of the signal is 0.025s.

$$f(t) = \beta \cdot \sin(\pi \cdot \gamma \cdot t) \tag{7}$$

where t is time and β and γ are signal parameters defined as:

$$\begin{aligned} \beta = 1; \gamma = 1 \quad &\text{pro} \quad t \in \langle 0;10 \rangle \\ \beta = 2; \gamma = 1 \quad &\text{pro} \quad t \in (10;20\rangle \\ \beta = 1; \gamma = 2 \quad &\text{pro} \quad t \in (20;30\rangle \end{aligned} \tag{8}$$

In the figure 1 there can be seen that ADALINE was adapting very quickly to the predicted signal and after few seconds it is only improving the prediction accuracy. Changes of signal parameters in time of 10s and 20s have only very small influence to the prediction error because adaptation works well. In the figure 2 are depicted weights and bias of the ADALINE during the simulation.

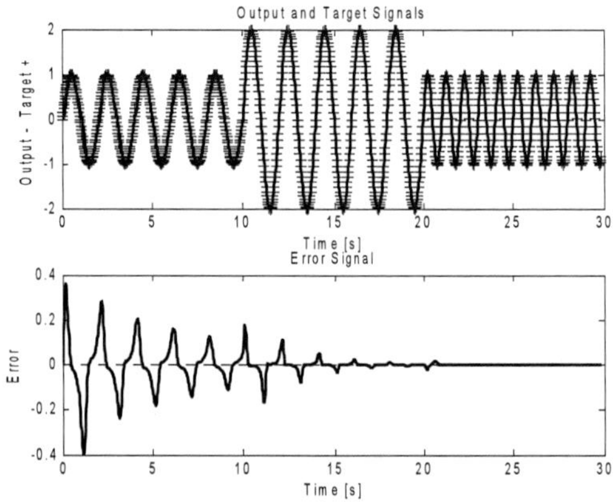

Fig. 1. Adaptive signal prediction by ADALINE

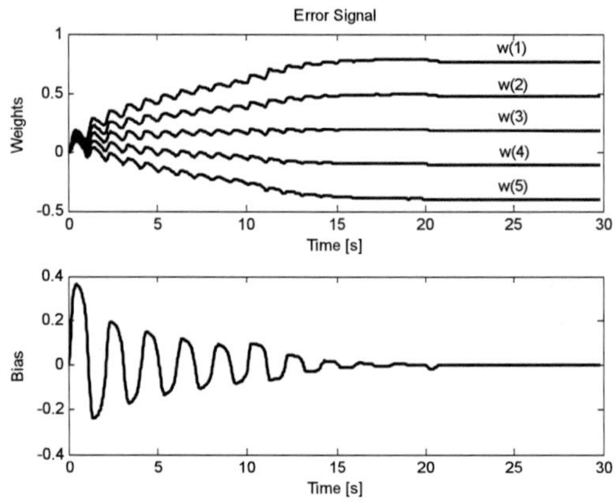

Fig. 2. Adaptation of the ADALINE

3.2 Nonlinear system

Simulink model of nonlinear system of two interconnected ball tanks was used. The model can be written (assuming usual simplifications):

$$\pi h_1(d_1 - h_1)\frac{dh_1}{dt} + q_1 = q_{1v} \tag{9}$$

$$\pi h_2(d_2 - h_2)\frac{dh_2}{dt} - q_1 + q_2 = q_{2v} \tag{10}$$

$$q_1 = k_1\sqrt{|h_1 - h_2|} \text{ if } h_1 - h_2 < 0 \text{ then } q_1 = -q_1 \tag{11}$$

$$q_2 = k_2\sqrt{h_2} \tag{12}$$

where d_j are tanks diameters, h_j are levels of liquid, q_j are output flow rates, q_{jv} are input flow rates and k_j are parameters representing pipeline properties. Initial conditions of equations (9) and (10) are $h_1(0)=1.5\text{m}^3/\text{min}$ and $h_2(0)=1.3\text{m}^3/\text{min}$. Parameters of the system model were $d_1 = 2$ m, $d_2 = 2$ m, $k_1 = 0.85$ m$^{2.5}$/min, $k_2 = 0.5$m $^{2.5}$/min, $q_{1v}(0) = 0.38$ m^3/min and $q_{2v} = 0.19$ m^3/min = constant. However, to test adaptivity of ADALINE the parameters d_1 and d_2 changed to 20m. The system was fed by sine-based signal (q_1). Simulation results are presented in the figure 3.

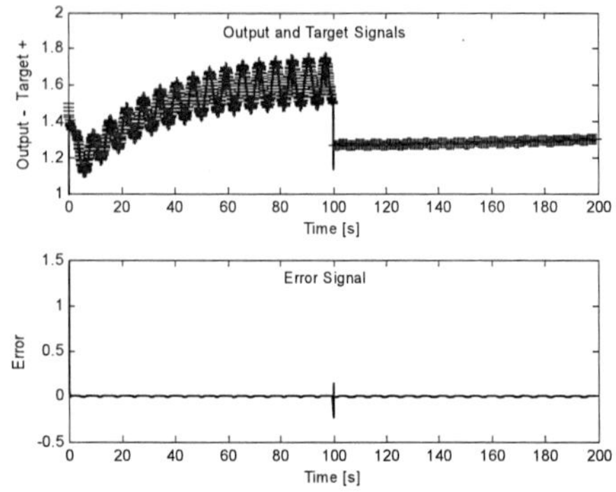

Fig. 3. Adaptive prediction of the nonlinear system

4. CONCLUSION

The simulations proved ability of ADALINE to predict signals with changing parameters as well as nonlinear system output with changing parameters. This predictive method should be considered as adaptive predictor in predictive control methods. Nowadays, most of predictive control methods with artificial neural network as a predictor use only offline identified predictor. It is caused by the fact that usage of artificial neural networks in control algorithms significantly increases computational demands of the controller as a whole. What is more, the online identification of the predictor computational cost increases even more. However the usage of the offline identified predictor has many disadvantages, such as high demands for the predictor accuracy and uselessness of offline identification for t-variant processes.

5. ACKNOWLEDGEMENT

This work was supported by the Ministry of Education, Youth and Sports of the Czech Republic under grant MSM 7088352102. This support is very gratefully acknowledged.

6. REFERENCES

Arbib, M. A. (2002). *The Handbook of Brain Theory and Neural Networks*, The MIT Press, ISBN 0–262–01197–2, Cambridge, Massachusetts

Camacho, E. F. & Bordons, C. (1995). *Model Predictive Control in the Process Industry*, Springer-Verlag, ISBN 3-540-19924, Berlin

Demuth, H. & Beale, M. (2002). *Neural Network Toolbox for Use with MATLAB*, The Mathworks, Inc., Natick

Freeman, J. A. & Skapura, D. M. (1991). *Neural Networks Algorithms, Applications, and Programming Techniques*, Addison Wesley Publishing Company, ISBN 0-201-51376-5, Reading, MA.

Kanjilal, P.P. (1995). *Adaptive prediction and predictive control*, Peter Peregrinus Ltd., ISBN 0863411932, London

Najim, K.; Rusnak, A.; Meszaros, A. & Fikar, M. (1997). Constrained long range predictive control based on artificial neural networks. *International Journal of Systems Science*, Vol. 28, No. 12, p. 1211-1226, ISSN 0020-7721

Widrow, B. & Hoff, M. E., Jr. (1960). Adaptive switching circuits. *IRE WESCON Convention Record*, Vol. 1960, No. 4, p. 96–104, ISSN 0097-6636

Annals of DAAAM for 2006 & Proceedings of the 17th International DAAAM Symposium, ISSN 1726-9679
Editor B. Katalinic, Published by DAAAM International, Vienna, Austria 2006
150th Birthday of Nikola Tesla (1856-1943) Inventor of Mobile Robot Teleoperation

TOOTH FILLET GEOMETRY IN GEARS MANUFACTURING WITH HOBBING MACHINE

SANDU, C.; GHIONEA, A. & SANDU, CO.

Abstract: *The fatigue at teeth root is an important cause for the reduction of the gears life span. Multiple factors are influencing the fatigue, one of them being the shape of the tool cutting edge. In this paper, we present different cases for the edge shape, where filleting, chamfering or protuberance modifies the hob profile. We present graphical and numerical results of some x cylindrical gears machining with the virtual hobbing machine, such as: fillet and the most influencing geometrical factors between those of the tool and those of the gear.*

Key words: *Generating profile, Virtual hobbing machine, Generated tooth profile, Addendum modification, Tooth fillet.*

1. INTRODUCTION

Cylindrical gears represent an important part in many machines and equipments. Gear machining with a hobbing machine is the most productive and cost-efficient production type. Variable strains appear in gears functioning (Henriot, 1999), resulting in fatigue, noise and wear. Geometrical discontinuities of the tooth flanks, which are determined by the tool profile geometry (Minciu, 1996), influence the fatigue phenomenon. This paper is distinguished by others (Litvin, 1987) because it uses the concept of "Virtual hobbing machine" (Sandu & others, 2004) and it presents in which measure the tool cutting edge geometry is influencing the fillet. The generated surface is the real one, resulted from the interaction between the hob and the hobbing machine.

2. MODELLING OF THE PROCESS

At the base of this study is a new mathematical theory (Sandu 1987), (Sandu & Sandu, 1996), (Sandu & others, 2000) resulting in the creation of a virtual hobbing machine (Sandu & others, 2004).
On machining with the virtual hobbing machine we utilize tools (hob) with different profile geometries (fig. 1).

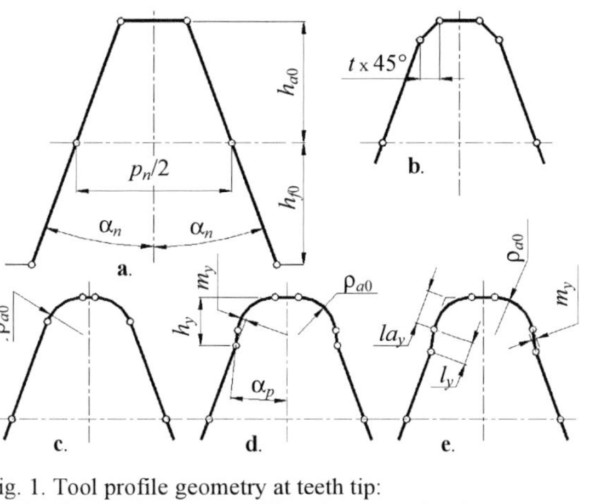

Fig. 1. Tool profile geometry at teeth tip:
a - simple; *b* - chamfer; *c* – fillet; *d*, *e* - complex form.

Notations description for fig. 1 is:
h_{a0} = 1.25 × m - Tool addendum; h_{f0} = 1.25 × m_n - Tool dedendum; α_n = 20° - Normal pressure angle; p_n - Normal pitch; ρ_{a0} = 0.38 × m_n or 0.3 × m_n (fig. 1,*e*) - Tip radius of the tool; α_p - 6° ; h_y - 0.5 × m_n ; la_y - 0.37 × m_n ; l_y - 0.23 × m_n; m_y - 0.05 × m_n or 0.02 × m_n (fig. 1,*e*);
To apply the concept of virtual hobbing de machine at machining we follow the steps:

- Extracting the parameters from the gear technical documentation (number of teeth, pitch diameter, pressure angle, Addendum modification coefficient, reference rack);
- Setting the virtual hobbing machine;
- Hob selection, with one of the cutting edge geometries;
- Teeth flanks generation.

3. SIMULATION RESULT AND THEIR INTERPRETATION

The window (fig.2) of a virtual hobbing machine is presented with all settings and ready to start the machining. We used for our example a x cylindrical gear with 15° reference helix angle and reference rack 20°-1-0.25. We considered the machining of two x cylindrical gear pairs with the tooth count of 77 and 100 mm distance between the axes. First x gear pair has: z_1 = 13 and z_2 = 64 tooth, addendum modification coefficient x_1 = 0.5 mm. Second x gear pair has: z_1 = 26 and z_2 = 51 tooth, addendum modification coefficient x_1 = 0.3 mm. Also, two cylindrical gears were machined, with z_1=40 teeth and, respectively, z_2=80 teeth, and addendum modification coefficient x_1=-0.5; -0.25; 0.0; 0.25; 0.5 mm.
Frontal section (Fig. 3.) of the x cylindrical gear pair with z_1=13 teeth z_2=64 teeth machined on virtual hobbing machine shows the influence of the hob geometry over the teeth profile.
The detail (fig. 4) is made for the x gear with z_1=13 teeth. The letters used in the figure shows the profile geometry utilized in the figure 1 representation. We can see the notable differences in the filleted area at the base of the tooth. Also we represented the theoretical profile of the tooth without the fillet.

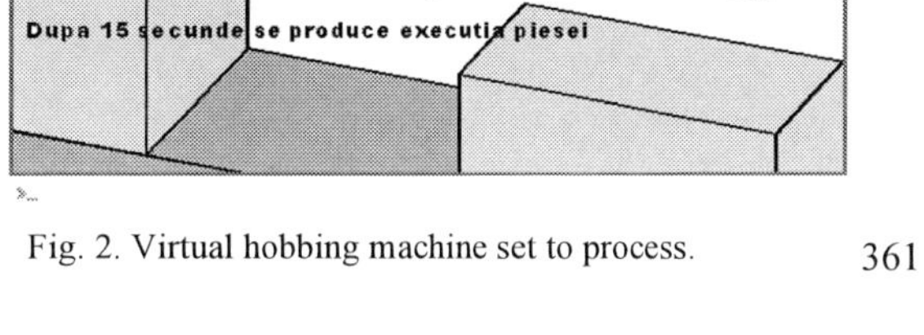

Fig. 2. Virtual hobbing machine set to process.

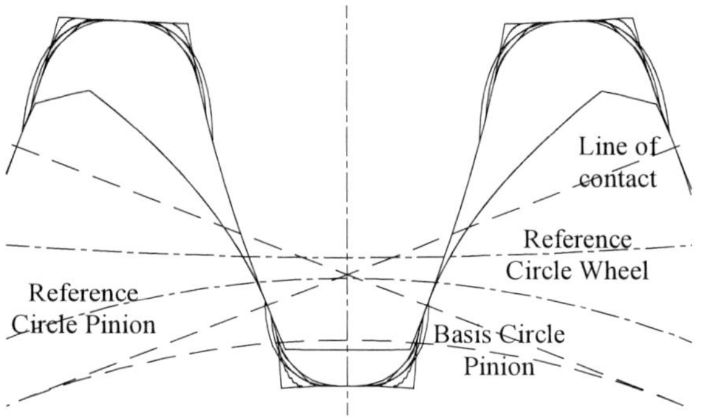

Fig. 3. X cylindrical gear pair with: x_1= 13; x_2= 64 teeth.

We can see that the simple profile (fig. 1,*a*) or the one with the chamfer (Fig. 1,*b*) generate a geometry that increases the risk of breaking from fatigue. The round part of the hob edge can be approximated with 6-8 straight segments. At gears with less than 16 teeth with standard profile, the gear tooth makes contact with the pinion tooth fillet, with unwanted consequences in functioning.

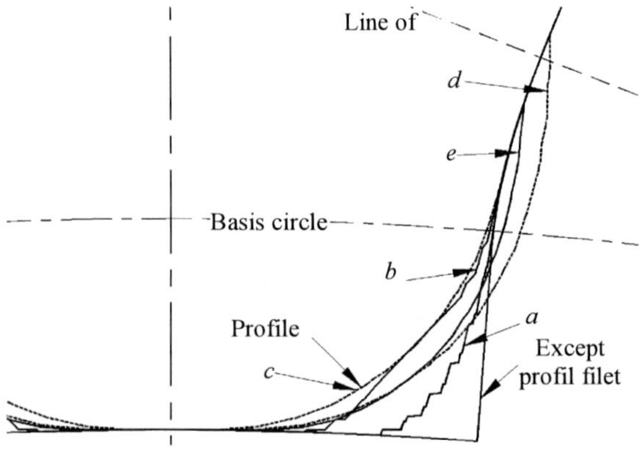

Fig. 4. Tooth fillet geometries for z_1=13 teeth.

For the 13 teeth pinion, curvature radius varies from 0.90 to 1.01 mm, with the bigger value towards the base, and for 64 teeth gear, curvature radius varies from 1.01 to 5.56 mm, with the lower value at base. Clearance dimensions (width/length) generated with complex profile (fig. 1, *d*, *e*.) are: for z_1=13 0.112/1.001 mm, respectively 0.046/0.625 mm, and for z_2=64 are 0.123/1.638 mm or 0.051/1.026 mm. Therefor, we can't obtain after finishing by grinding meshing to the tip of the other gear. So, at the 64 teeth gear we don't have contact at the tip for 0.15 mm or 0.45 mm of the flank, and, for 13 teeth, for 0.21 mm or 0.30 mm. This modifies the meshing degree of coverage for the gears.

In the same way we determined and analyzed the results for the

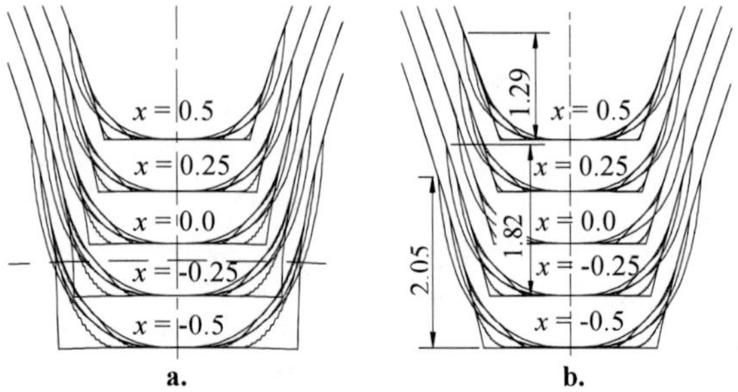

Fig. 5. Fillet form for varied addendum modification coefficient: *a* - gear with 40 teeth; *b* - gear with 80 teeth.

X cylindrical gear pair x_1= 26; x_2= 51 teeth.

We can see from fig. 5,*a* and fig. 5,*b* where we show the frontal sections for a gear with z =40 teeth, respectively z =80 teeth with different addendum modification coefficient. It shows that the curvature radius decreases from la 1.57 to 0.99 mm for the first gear and from 1.27 to 0.99 mm for the second one when the addendum modification coefficient is increased. The height of the undercut, ordered by the magnitude of the addendum modification coefficient, has the values: for z =40 teeth is 2.50/2.18/1.87/1.59/1.33 mm (profile 4) or 2.12/1.84/1.45/ 1.216/1.09 mm (profile 5) and respectively for z = 80 teeth is 2.05/1.82/1.50/1.43/1.29 mm (profile 4) or 1.57/1.37/1.24/ 1.16/1.09 mm (profile 5).

4. CONCLUSION

Applying the concept of virtual hobbing machine tool, in particular for cylindrical gears, allowed us a fast determination, graphically and numerically, of different basic aspects of the gear tooth profile. We showed graphically different aspects: the geometrical elements of the meshing gears, the visualization of the gears and the shape of the fillet at the tooth root. This way, we verified the correctitude of the geometrical parameters up to a high zoom where we can see the surface roughness. The software is modular and it has a great flexibility. The cutting edges were mathematically defined for the considered profiles and they are used in practice too. By choosing the correct movements the virtual hobbing machine can be used for worm gears. With other software modules we can study the variance of the transmission ratio or different aspects of the flanks contact. If in the files of cinematic programming we are introducing harmonic functions, we can study as well how the machining vibrations influence the profile geometry for gears (Sandu, 1987).

5. REFERENCES

Henriot, G. (1999). *Engrenages. Conception. Fabrication. Mise en oeuvre,* 7th édition, Industries Techniques, Dunod, ISBN 2 10 003903 2, Paris.

Litvin, F. (1994). *Gear Geometry and Applied Theory,* Prentice-Hall, Inc., ISBN 0-13-211095-4, Englewood Cliffs.

Minciu, C., et al. (1996). *Scule aşchietoare,* Vol. 2, Editura Tehnică, ISBN 973-31-0790-5, Bucharest,.

Sandu, C., (1987). *Contribuţii privind îmbunătăţirea calităţilor dinamice ale lanţului cinematic de rulare la maşina de prelucrat roţi dinţate, Ph. D. Thesis,* Institutul Politehnic, Bucharest.

Sandu, C. & Sandu, Co. (1996). Determinarea profilului sculelor aşchietoare folosite pentru prelucrarea unor suprafeţe elicoidale, *TCMM,* 15, pp.120-126, ISBN 973-31-0766-2, Editura Tehnică, Bucharest.

Sandu, C.; Dobre, G. & Sandu, Co. (2000). On contact of conjugated flanks generated in the system machine tools - tools. *ICMaS 2000, TCMM,* 41, pp. 263-268, ISBN 973-31-1492-8, Editura Tehnică, Bucharest.

Sandu, C., Sandu, Co. & Dobre, G. (2000). New Principles for Surface Generation, *Scientific Bulletin of North University of Baia Mare,* Volume XIV, pp. 233-238, ISSN 1224-3264.

Sandu, C.; Dobre, G.; Mirică, R. F. & Sandu, Co. (2004). Virtual machine tool for generation of the helical gears, *Proceedings of the International Conference on Manufacturing Systems ICMaS 2004,* pp. 133-136, ISBN 973-27-1102-7, Editura Academiei Române, Bucharest.

Wächter, K., (1987), *Konstruktionslehre für Maschinen-ingenieure,* Verlag Technik, ISBN 3-341-00045-3, Berlin.

Annals of DAAAM for 2006 & Proceedings of the 17th International DAAAM Symposium, ISSN 1726-9679
B. Katalinic, Published by DAAAM International, Vienna, Austria 2006
150th Birthday of Nikola Tesla (1856-1943) Inventor of Mobile Robot Teleoperation

THE EXPERIMENTAL STUDY OF STRUCTURAL VISCOSITY WHICH CHARACTERIZED THE NON-NEWTONIAN FOOD MIXTURES AND PROCESS OPTIMATIZATION USING FINITE ELEMENT METHOD

SAVIN, C.; ZICHIL, V. & JUDELE, A.

Abstract: *In the food industry we call non-newtonian food mixtures, those mixtures that have a flow resistance which is changing with change of application speed of shearing force, generating the irregular flow of them. This paper content the study of correlation between the torsion moment produced by mixing device which is in rotary motion with the value of structural viscosity created into the non-newtonian mixture, propriety (the structural viscosity) which characterized the degree of homogenization and power consumption needed to result the mixture. This paper content the study of correlation between the torsion moment produced by mixing device which is in rotary motion with the value of structural viscosity created into the non-newtonian mixture, propriety (the structural viscosity) which characterized the degree of homogenization and power consumption needed to result the mixture. Reducing the weight of the moving parts in the machine building technology it's a big goal, this low weight designs has to resist on the same loads and using them has to have as result reducing the inertia momentum, economy on the materials and other mechanical components like bearings.The target of this paper is to compare the stress and strain values of a cylindrical shaft and a tubular shaft used in the same application and loading conditions.*

Key words: *cylindrical shaft, tubular shaft, experiment, finite element method*

1. INTRODUCTION

To validate the experimental method it was used also analytical relations that allows to determinate the torsion moment using known parameters.

The torsion moment of mixer shaft can be determinate by the relation (1):

$$Mt = \frac{\eta_a \cdot z \cdot k^2 \cdot n \cdot d}{4} \quad (1)$$

where:

M_t represent the torsion moment of mixer shaft, (Nm)

η_a represent the apparent viscosity, kgf s/m^2

K =2,5 is the coefficient for working condition

n the revolution of the mixing device, (rot/s)

d the diameter of mixer shaft, (m)

z the number of mixer arm

2. EXPERIMENTAL STAND

For getting the experimental researches, it was designed and made an experimental stand which allows the creation of the non-newtonian mixtures in laboratory conditions and also allows experimental measuring of torsion moment of the shaft.

The experimental stand (figure no.1.) with discontinuous motion is formed by the mixing tank 12, put on the supporting plate 1, the movement of the mixing device 11 it is made from the asynchronous engine 13, by a frequenced static converter.

On the stand it was put a measure installation which can put together data about technological parameters of mixing process.

The mixer shaft has a full cylindrical section and it is built by a stainless steel semi-fabricated.

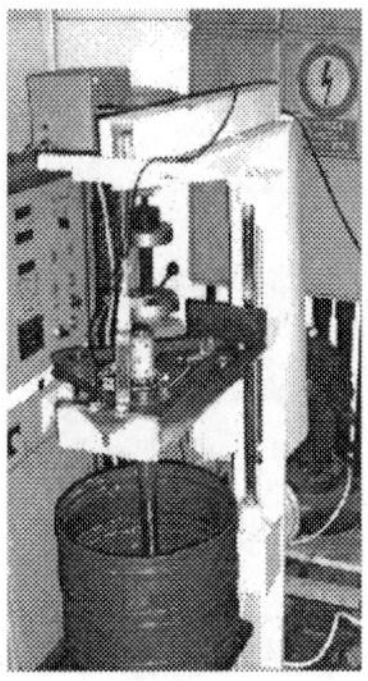

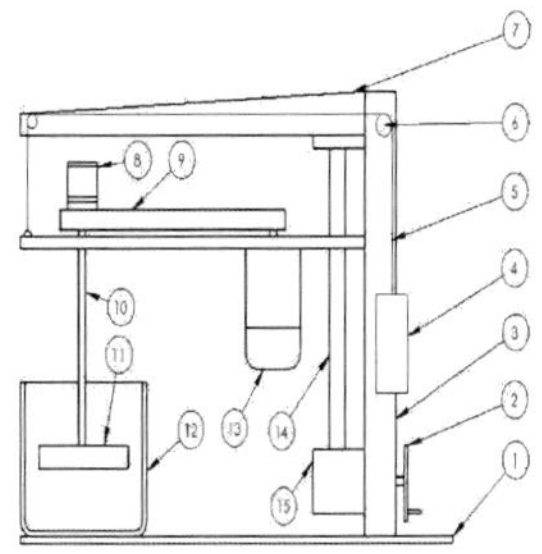

Fig. 1. The experimental stand Fig. 2. The components element of stand

3. THE MEASUREMENT METHOD

To measure the torsion moment transmitted by the mixer shaft (10) it was used a full bridge of electro resistive traductors (TER) applied into a shaft section, at quadrants, arranged at 45°.

The electro resistive traductors are 1- XK11E – 3/380 types and are designed to measure the deformations produced in the shaft while the torsion occurs.

Fig. 3. The mixer shaft with electro

To be able to have real-time signal from the traductors while the shaft is turning it was used a mobile contact by shape of a collector ring with many brushes (with holders designed to allow the take off from contact during the time that will be no measuring). The collector ring with brushes is put on the free extremity of the shaft which presents an internal opening,

through which are passing the connection cables from electro resistive traductors to mobile contacts.
To process the electric signal it is used the external digital module Spider 8 that convert the signal from analogic to digital and allows digital processing of the deformations measured by the electro resistive traductors.

4. RESULTS

Becouse the mix was an non-newtonian one, with variable viscousity, the torsion moment of the shaft was proccesed as a graph representing variation of torsion in time (4).

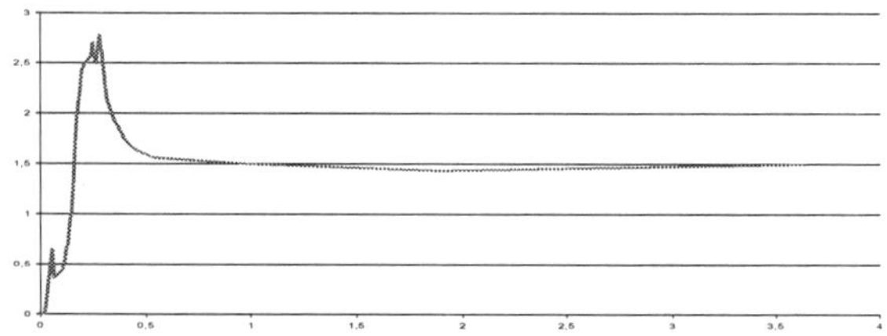

Fig. 4. The variation of the torsion moment (Mt) for a non-newtonian mixtures with average consistence depending on mixing time (t), at the constant revolution - experimental

The same graph was processed by analytical methods (5) to be able to compare the results

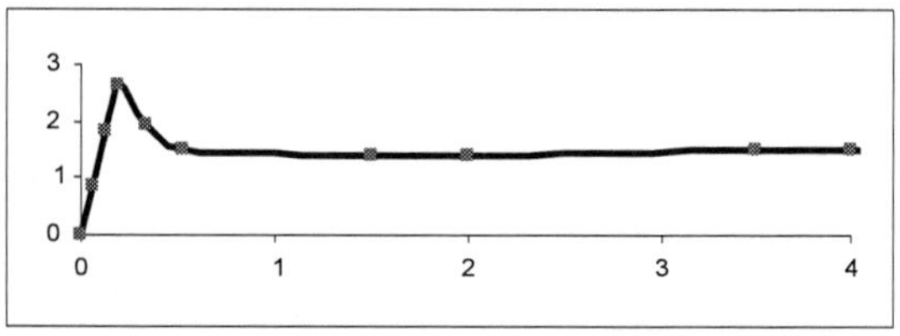

Fig. 5. The variation of the torsion moment (Mt) for a non-newtonian mixtures with average consistence depending on mixing time (t), at the constant revolution - analytic

The experimental values and the values analytical obtained are similar so the experimental method is valid.

5. FEM ANALYSIS

The target of this paper is to compare the stress and strain values of a cylindrical shaft and a tubular shaft used in the same application and loading conditions.
For this first it was made an experiment to have the experimental dates. The experiment was carried out using the cylindrical shaft, then the experiment was validated by the analytical results.

Using the ANSYS software it was possible to run analysis for the cylindrical shaft and tubular shaft. Comparing the experimental data, analytical data and the results of the FEM analysis for the cylindrical shaft we see that the finite element model used is valid, and we can assume that the results of the finite element analysis for the cylindrical shaft and tubular shaft could be compared.
The FEM simulation was done for cylindrical shaft and the tubular shaft, using the same loading conditions and the same material.
The obtained results shows a lower value for the shearing stress in the case of the tubular shaft, also the level of the global stress (by von Misses) is comparable for the tubular and cylindrical one, therefore, the tubular shaft can be successfully used.

Fig. 6. Mesh of the mode l

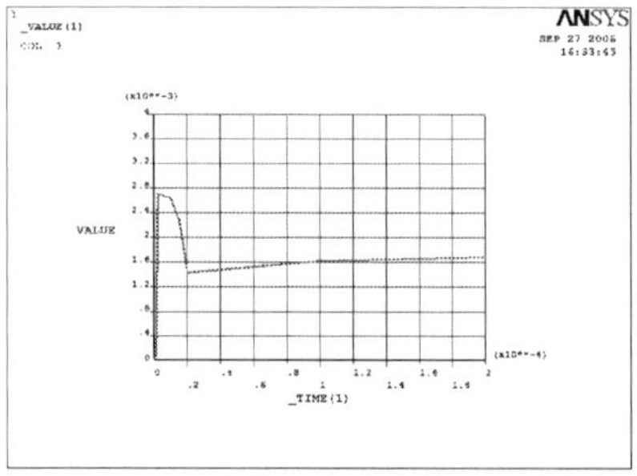

Fig. 7. Loading curve for the analysis

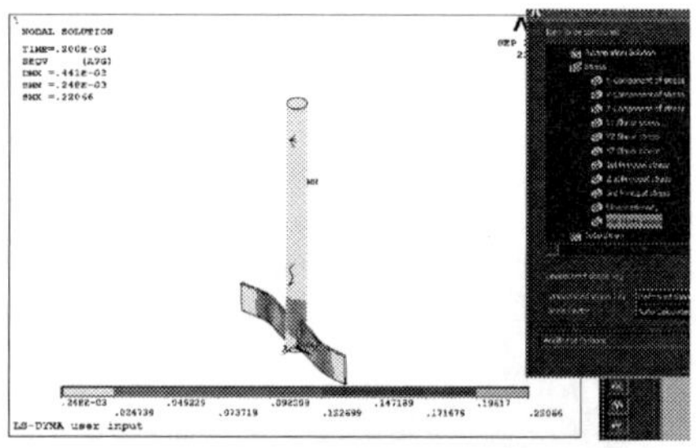

Fig.8. Shearing stress for the cylindrical shaft

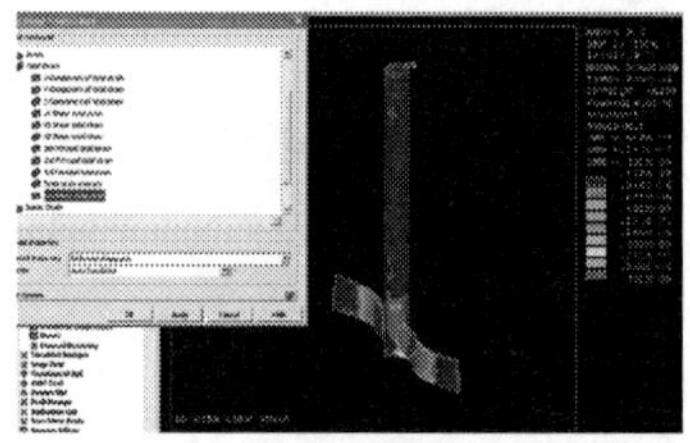

Fig. 9. Shearing stress for the tubular shaft

6. CONCLUSIONS

Comparing the results of the FEM analysis of the cylindrical and tubular shaft we can say that using the tubular shaft could be a good solution. Economy of the material and reducing the loads on the bearings are only a few reasons to do that.
The values of viscosity determinates by the torsion moment are different variation in function of mixing technological process time.We can consider that the maximum homogenization degree is corresponding with time which the torsion moment constantUsing analytic methods it was deducted the stress and strains during the load cycle to be compared with the ones obtained by finite element method analysis.

7. REFERENCES

[1] Savin C. (2004) – *Referatul 3, Rezultate experimentale*, Teza de dodctorat – Optimizarea procesului de amestecare a produselor agro-alimentare, Gheorghe Asachi Iasi

Annals of DAAAM for 2006 & Proceedings of the 17th International DAAAM Symposium, ISSN 1726-9679
B. Katalinic, Published by DAAAM International, Vienna, Austria 2006
150th Birthday of Nikola Tesla (1856-1943) Inventor of Mobile Robot Teleoperation

QUALITY IMPROVEMENT FOR ELASTIC ADHESIVE BONDING OF LARGE COMPONENTS WITH HIGH TOLERANCES

SCHLICKENRIEDER, K. & ZAEH, M.F.

Abstract: *The paper presents a method for the enhancement of quality and flexibility of the elastic adhesive bonding of large components with high tolerances. The main improvement is based on an innovative methodology for the alignment of two parts, e. g. a windshield to a truck body. This is realized by a deformation of one piece immediately before the joining process, based primarily on geometric information and on emerging stresses due to the deformation.*
Key words: *Actoric & Sensoric, Control Strategies & Algorithms, Industrial Automation.*

1. INTRODUCTION AND FUNDAMENTALS

For the vehicle production are especially the customer requirements for more safety and improved equipment contrary to the necessity to reduce fuel consumption. An essential approach is the reduction of weight through lightweight construction strategies (Friedrich & Meschut 2003). Thus synthetic materials, aluminum, titan as well as magnesium are established as construction materials (Schumann & Friedrich 1998, Zäh et al. 2004). Unfortunately, with traditional joining technologies it is often difficult to realize the resulting combinations of the materials (Bauer 2003). For that is adhesive bonding a promising alternative (Büdgam et al. 2004), which is for example used for the joining of windows and secondary structure in railcars (Koch et al. 2004). Nowadays windshields are integrated as load bearing structures. This procedure enhances the static and dynamic stability of the bonded joints of around 30 %. Furthermore the impermeability of the joint is improved and the driving-noise as well as the weight is reduced (Habenicht 2002). The assembly process can be separated in several steps. In the first step the windshield is fixed on racks to provide a precise gripping position. Then the windshield is manipulated by an industrial robot for the application of the adhesive bead. The next step is the joining of windshield and car body. At last detaches the gripper from the windshield.

2. STATE OF THE ART

Windshields for commercial or railway vehicles are produced with a gravity-based bending process, which is less costly but also less precise. As a result the windshields feature large tolerances of up to 5 mm. In addition tolerances up to 10 mm and more can occur in truck or railcar bodies. Therefore three main problems arise. First, the windshield can contact the steel body partially, whereas the adhesive will be pressed out of the joint. Second, the adhesive bead has no contact to the body and therefore a gap appears. Third, strains can arise in the windshield during the preparatory fixing process, if the windshield is formed according to the ideal joint geometry by force. During the handling process the geometry of the strained windshield is held by the manipulator. Yet after the separation of manipulator and windshield, glass and adhesive bead can disconnect due to relaxation processes. A presently established method for the improvement of the joining process is the use of sensor technology for the optimized alignment of windshield and joint surface. This can be achieved with triangulation sensors, which measure contactless the position and the orientation of windshield and body in relation to each other before the joining process. Based on the collected data a zero shift of the robot coordinate system in up to six degrees of freedom is calculated and a correction of the working coordinate system is accomplished (Sakai & Yasumatzu 1993). Even though the joining process can be optimized by sensor systems it is not always possible to fit the windshield perfectly into the body due to the following reasons:

- A comprehensive equalization of shape tolerances of windshield and body is not always possible
- Instead only an adaptation of position and orientation of the windshield as a whole is feasible
- Local deviations can therefore not be compensated

Consequently the necessity to improve the existing assembly process for large work pieces with high tolerances, e. g. windshields for trucks, through innovative sensor technology and control strategy is evident. The compensation of the work piece tolerances during the joining process is of utmost importance. Furthermore it is necessary to provide possibilities for the adaptation of the joining process according to the work pieces by increasing the flexibility of the actual joining process. In addition it is important to minimize the stresses induced into the work piece as a result of the handling and joining processes. In order to achieve the best results it is necessary to keep the state of the art sensor technology for the optimization of position and orientation of the work piece. However further improvement in the quality of elastic adhesive bonding of large components with high tolerances can be gained through partial shape adaptation of one or both work pieces during the joining process.

3. METHODOLOGY AND REALIZATION

Therefore an extension of the given joining process with a control algorithm for partial work piece deformation during the joining process was evident. Said control algorithm is based on precise alignment data of both components and individually addressable actuators. With knowledge of the alignment, the necessary work piece deformation can be derived, which is then realized by the actuators during the joining process (see Fig. 1).

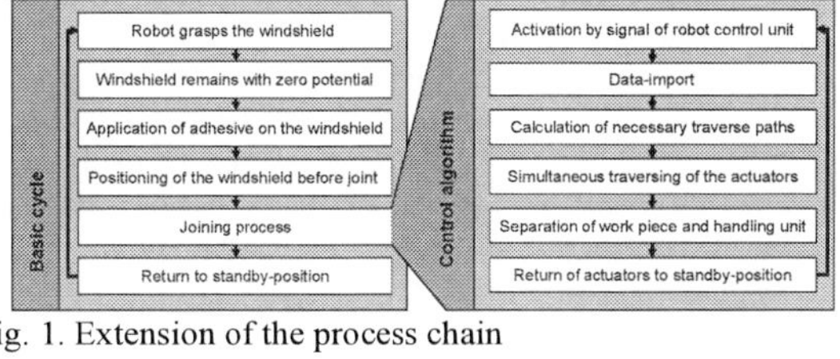

Fig. 1. Extension of the process chain

Varied data are suitable for the calculation of the deformation. These can be the resulting forces, which appear during the joining process, or the resulting torque, emerging from the pressure on the adhesive bead, or the distances between the components. A benchmark in terms of technical feasibility, costs, efficiency and process reliability shows, that the utilization of distance measurement as an input into the control algorithm is the most efficient approach.

For the implementation several technological requirements for the sensors have to be considered:

- Measuring range at least between thickness of windshield plus thickness of adhesive after joining and thickness of windshield plus thickness of adhesive before joining
- Accuracy of measurement of less than 0,2 mm
- Output signal proportional to distance (analog output)
- Contactless and direct measurement
- Measurement through windshield and adhesive bead

Mechanical distance transducers and optical distance measurement sensors are not relevant, because the windshield and the adhesive bead obstruct a direct measurement. However, inductive sensors are suited for the direct measurement of the distance through solid materials like glass, ceramic coating of the glass and adhesive bead. Special consideration was given to the calculation of the traverse paths. The first step is dedicated to the appointment of traverse paths based on the actual distances between the work pieces. This is done by allocating the distance measured by each sensor to a specified value for the distance of the joint through a linear equation. The greatest distance between both components will be assigned to the upper limit and the smallest distance to the lower limit of the specifications. All other distances in between will be mirrored accordingly.

Subsequently the determination of the gradient of inclination between three adjacent points and of the maximal allowable relative displacement of two adjacent actuators follows. For that, distances, strains, materials of the work piece and capabilities of the adhesive have to be considered, whereby the implied strains are of highest relevance. A promising approach for the determination of implied strains due to deformation is the utilization of a finite-element model, which calculates the emerging strains due to a given local deformation of the windshield done by the actuators.

The second step contains the definition of the final traverse path for each actuator. For this, a verification whether all gradients of inclination between three adjacent points reside inside the maximum allowable relative displacement has to be done. This is necessary for the minimization of the strain load in the windshield during the deformation process, since large relative deformations lead to high strains, which can lead to a partial separation of component and adhesive due to a relaxation process. In case of a transgression the traverse paths have to be adjusted as can be seen in figure 1.

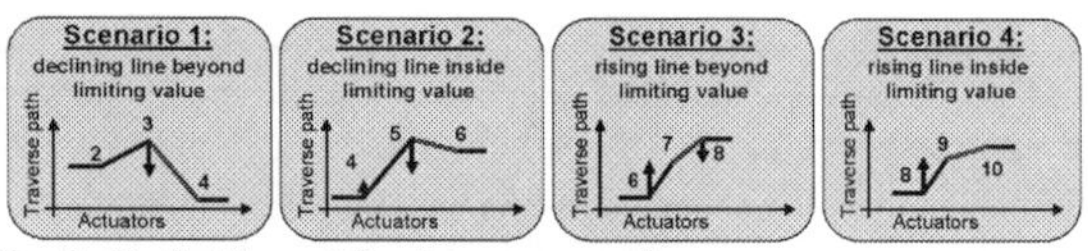

Fig. 2. Definition of final traverse paths

The transgressions can be divided into four different scenarios. In scenario 1 the gradient of inclination between point 2 and 3 is falling and exceeds the determined limit. In order to homologize the gradient, the calculated traverse path of actuator 2 has to be decreased. In case of a declining gradient of inclination, which remains inside the limit as can be seen in the line between actuator 5 and 6 in fig. 1, the traverse path of actuator 4 should be increased and that of actuator 5 decreased. Scenario 3 and 4 can be treated accordingly. The adjustment of the traverse path has to be done iteratively in order to achieve the best possible solution. As final step the joining process has to be carried out by the actuators. The above mentioned methodology was invented and verified through examinations on the joining process of a large windshield for heavy duty trucks. In the beginning, the robot handles the windshield for adhesive application and positioning in front of the truck body. Then the actual distances between the windshield and truck body are measured by inductive sensors and the required traverse path for each actuator is calculated according to the methodology. For the actuators, precise stepping motors have been chosen. In the next step, the stepping motors join the windshield to the truck body. The robot is used only as a primary positioning system, whereas the actual joining process is carried out precisely by the actuators.

4. CONCLUSIONS

Several advantages can be achieved by the utilization of the presented approach. The process stability and reliability can be improved with the precise control of the joining process. Trough the association of the extended joining algorithm with a standard positioning systems the process quality can be increased essentially. Existing tolerances of the work pieces can be compensated inline. Therefore leaky spots and the affiliated manual refinishing processes can be reduced. In general, a customization of the joining process according to the work pieces is possible. Thus the strains induced by handling processes can be reduced significantly. The above mentioned methodology was invented and verified through examinations on the joining process of a large windshield for trucks and the transfer of method to other large components will be promoted.

5. REFERENCES

Bauer, F. E. (2003): Zugfest und korrosionsfrei. Automobilwoche 1 (2003) 23, S. 16.

Büdgam, S.; Freitag, V.; Hahn, O.; Ruther, M. (2004): Fügesystemoptimierung für Mischbauweisen im Karosseriebau. ATZ Automobiltechnische Zeitschrift 106 (2004) 12, S. 1132-1141.

Friedrich, H. E.; Meschut, G. (2003): Autovision: Zukünftige Werkstoffe und Fügekonzepte im Automobilbau. In: Gemeinsame Forschung in der Klebtechnik; Düsseldorf 18. / 19.3.2003. Düsseldorf: DVS-Verlag 2003, S. 5-10. (DVS-Berichte 222).

Habenicht, G. (2002): Kleben Grundlagen, Technologie, Anwendungen. 4. Auflage Aufl. Berlin: Springer 2002.

Koch, S.; Starlinger, A.; Wang, X. (2004): Advanced Mass Transportation with Elastic Bonding of Sandwich Components. 7th European Adhesion Conference EURADH 2004. Freiburg im Breisgau 5. - 9.9.2004, S. 300-305.

Sakai, H.; Yasumatzu, S. (1993): Development of Body Auto Fitting System. Proceedings of the 26th International Symposium on Automotive Technology and Automation. Aachen 13.-17.9.1993.

Schumann, S.; Friedrich, F. (1998): The Use of Magnesium in Cars - Today and in Future. In: Mordike, B. L. et al. (Hrsg.): Proceedings of the International Conference on Magnesium Alloys and Their Applications; Wolfsburg 28.-30.04.1998. Frankfurt: MATINFO Werkstoffinformationsgesellschaft 1998, S. 3-13.

Zäh, M. F.; Eireiner, D.; Mosandl, T.; Roeren, S.; Schlickenrieder, K.; Trautmann, A.; Ulrich, C. (2004): Aktuelle Forschungstrends in der Leichtbau-Fügetechnik. In: Zäh, M. F. (Hrsg.): Fügetechnik im Leichtbau; Stuttgart 16./17.03.2004. Stuttgart: Hanser 2004, S. 1-20.

Annals of DAAAM for 2006 & Proceedings of the 17th International DAAAM Symposium, ISSN 1726-9679
B. Katalinic, Published by DAAAM International, Vienna, Austria 2006
150th Birthday of Nikola Tesla (1856-1943) Inventor of Mobile Robot Teleoperation

LEVEL CURVE BASED GEOMETRY REPRESENTATION FOR CUTTING FORCE PREDICTION

SCHMIDT, C.

Abstract: *This paper gives an insight into a new approach for the representation of geometries for simulation based cutting force prediction in end milling processes. The shown calculation of tool engagement is based on level curves representing the geometries of the workpiece and NC code driven sweep volume. Derived from the tool engagement cutting forces are predicted using a parametric force model. Experiments are conducted and the measured forces are compared to the results of the simulation to verify the proposed method.*
Key words: *cutting simulation, end milling, polygon clipping*

1. INTRODUCTION

The efficiency of semi-analytical process models for the cutting force simulation has already been shown in several applications (Denkena et al., 2005, and Weinert et al. 2004). In their area of validity, limited to experimentally gained parameter fields, they are adequately precise in predicting cutting forces. Only the analytical geometry representation assuming constant contact conditions proves to be disadvantageous, however. For the analysis of processes with arbitrary workpiece geometries and changing contact conditions, a process model is required which combines the advantages of the analytical approach (Denkena et al., 2006) with a new and flexible geometry representation. This article gives an insight into the fundamental procedure for the prediction of cutting forces on the basis of a process progress depending flexible geometry model. The results presented here have been developed in the CRC653 which is funded by the Deutsche Forschungsgemeinschaft (DFG).

2. GEOMETRY REPRESENTATION BY LEVEL CURVES

The representation of workpiece geometries is carried out effectively two-dimensionally. For this purpose, the three-dimensional workpiece geometry is discretized in axial tool direction before simulation. According to a predefined approximation resolution, a number of l surface vertices V^i define a workpiece contour $\Omega\Pi^i$ on the level curve i.

$$\mathrm{WP}^i = \left\{V_1^i, V_2^i, \ldots, V_l^i\right\}, i = 1 \ldots n, \qquad (1)$$

Therefore, n of these workpiece contours shape the three-dimensional form of the workpiece (Fig.1 a). Like the representation of the workpiece, the representation of the tool, or more exactly of its sweep volume, is carried out in form of level curves, too. The representation of the sweep volume is based on the NC code driven movement of the milling cutter edges resulting in an edge pattern $E\Pi^i$

$$\mathrm{EP}^i = \left\{\mathrm{T}_1^i, \mathrm{T}_2^i, \ldots, \mathrm{T}_z^i\right\}, \qquad (2)$$

forming a self-intersecting polygon through z trochoid paths T^i.

$$\mathrm{T}^i = \left\{C^i_{\varphi_{min}}, C^i_{\varphi_{min}+\Delta\varphi}, \ldots, C^i_{\varphi_{max}}\right\} \qquad (3)$$

Again, every trochoid is an ordered list of vertices C^i which if connected to each other leads to the cutting edge path from the start angle φ_{min} in parameterizable increments $\Delta\varphi$ to the final angle φ_{max} (fig. 1 b).

By connecting the outer contour vertices C^i of the edge pattern, a closed concave polygon is composed. The vertices of the polygon describe a level curve ($E\Pi'$), which leads in an n-fold occurrence to the sought-after volumetric representation of the sweep volume. From the intersection between workpiece contour and sweep volume the geometry of the material removal IZ can be determined (Schneider, 2003).

$$\mathrm{EP}'^i \cap \mathrm{WP}^i = \mathrm{IZ}^i. \qquad (4)$$

Using this kind of geometry representation, the precision in z-direction is determined by the number of level curves n to which a processing zone is subdivided.

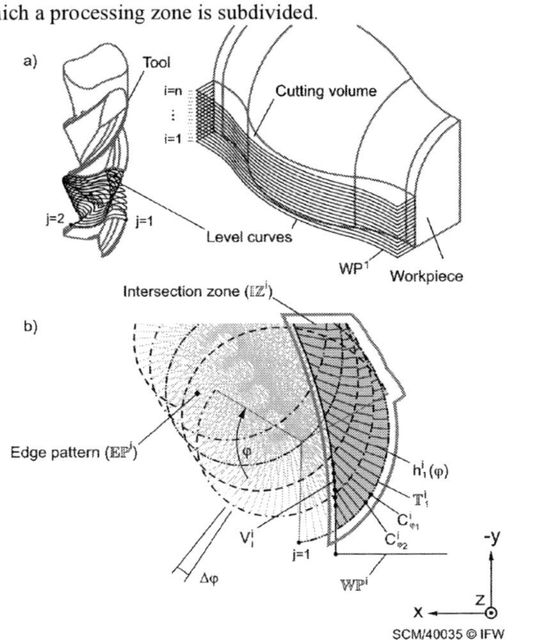

Fig. 1. Geometry representation by level curves and calculation of the intersection zone

3. CUTTING FORCE PREDICTION BASED ON SPECIFIC FORCE COEFFICIENTS

In addition to the calculation of the material removal for the representation of the resulting workpiece geometry, cross section surfaces A^i required for the prediction of cutting forces are determined. The necessary uncut chip heights h^i result from edges between vertices and intersection points within the intersection zone. Through the integration of the surface which is enclosed by two uncut chip heights h^i und h^{i+1}, the surface area A^i, which is crucial for the calculation of process forces, can be determined.

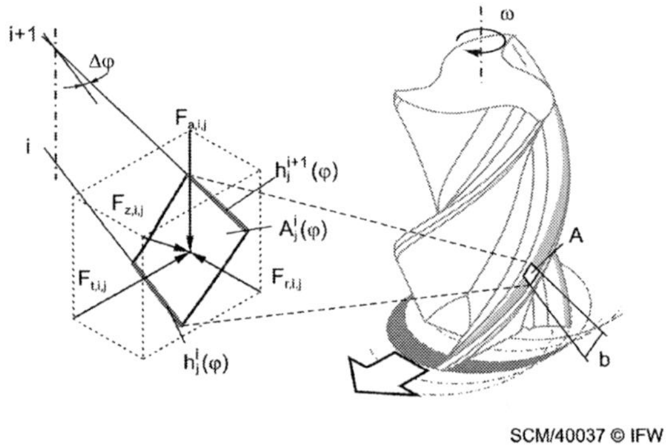

Fig. 2. Geometry and cutting forces in end milling process

By analogy with the fragmentation of the cutting force into infinitesimal force rates (Altintas, 2000) in tangential, radial and axial direction (Fig. 2), the forces $F_{tra}{}^i$ are related to the surface A^i and to the arc length b^i of one cutting edge segment. Thereby, the proportionality of the surface is adjusted by the shear force coefficient $K_{tra,c}$ and that of the arc length by the cutting coefficient $K_{tra,e}$.

$$\{F_{tra}\}_j^i = \{K_{tra,c}\} A_j^i + \{K_{tra,e}\} b_j^i . \qquad (5)$$

Related to the workpiece coordinate system, the transformation of forces takes place in Cartesian coordinates

$$\{F_{xyz}\}_j^i = [T]\{F_{tra}\}_j^i . \qquad (6)$$

By summation of the infinitesimal force portions along the operative cutting edges z from angle of entrance φ_{en} to angle of exit φ_{ex} over the number of cutting edges z, the forces can be calculated as follows under consideration of the milling cutter angularity of φ.

$$\{F_{xyz}(\varphi)\} = \sum_{j=1}^{z}\sum_{i=1}^{n}\{F_{xyz}\}_j^i . \qquad (7)$$

For predicting the cutting forces the knowledge of the milling cutter geometry and the boundary conditions resulting from the cutting process it is necessary to determine the remaining shear force and cutting coefficients (Denkena et al., 2006).

4. COMPARSION OF SIMULATION AND EXPERIMENTAL RESULTS

The comparison of the simulation results and the measured cutting forces is carried out by means of experiments. In one experiment a three-flute uncoated carbide end mill with a radius R = 5 mm and a helix angle γ = 45° was used. The feed per tooth rate f_z = 0.28 mm, the effective cutting width a_e = 5 mm and the cutting depth a_p = 10 mm at the cutting speed of v_c = 157.08 m/min leads to the measurement and simulation results shown in Fig. 3. For the validation of the simulation results, the measurement signal was nearly freed from interfering signals. The unfiltered and filtered measured forces are represented by grey and grey-dotted lines respectively. Despite the signal filtering, low-frequency signals resulting from the periodic stimulation by tool deflection can be recognized well. Nevertheless, the simulation values represented in black reproduce the real force progression adequately exact.

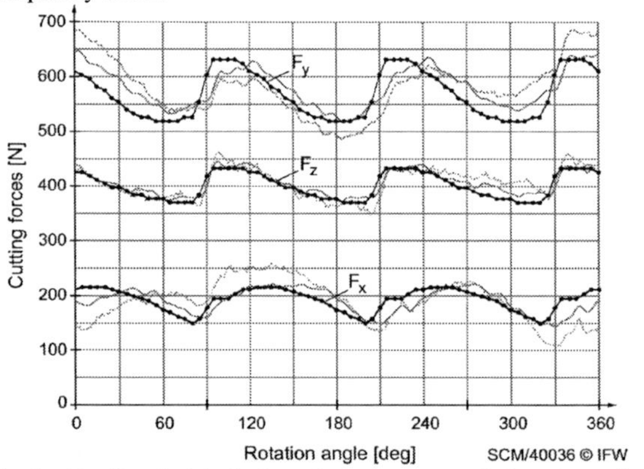

Fig. 3. Predicted (black dotted) and measured (grey) cutting forces in up-cut end milling. Workpiece: AlCuMgPb

5. CONCLUSION

In this article, a new approach for the prediction of cutting forces based on a level curve geometry representation has been proposed. The method is based on a two-dimensional polygon clipping algorithm to calculate infinitesimal edge segments describing the intersection zone of material engagement. It could be shown that the calculation of cutting forces leads to results of high accuracy by the use of experimentally determined cutting force coefficients. Thereby this approach is applicable to almost every type of workpiece geometry and can be used for an NC-code driven process simulation pre-calculating desired force characteristics.

6. REFERENCES

Altintas, Y. (2000). Manufacturing Automation (Metal Cutting Mechanics, Machine Tool Vibrations, and CNC Design) Cambridge University Press, Cambridge

Denkena, B.; Tracht, K.; Schmidt, C. (2006). A flexible force model for predicting cutting forces in end milling, Handed in: Production Engineering WGP

Denkena, B.; Tracht, K.; Clausen, C. (2005). Predictability of Milling Forces based on Specific Cutting Forces, 8th CIRP International Workshop Modeling of Machining Operations, pp. 259-266

Schneider,P. J.; Eberly, D. H. (2003). Geometric Tools for Computer Graphics, Morgan Kaufmann Publishers, ISBN: 1-55860-594-0, USA

Weinert, K.; Gradisek, J.; Kalveram, M. (2004). Mechanistic Identification of Specific Cutting and Edge Force Coefficients for a General End Mill, Production Engineering, Vol. XI/2, pp. 33-38

Annals of DAAAM for 2006 & Proceedings of the 17th International DAAAM Symposium, ISSN 1726-9679
B. Katalinic, Published by DAAAM International, Vienna, Austria 2006
150th Birthday of Nikola Tesla (1856-1943) Inventor of Mobile Robot Teleoperation

NEURAL NETWORK-BASED SEMI-INVERSE CONTROLLER

SCHMIDT, M. & PIVONKA, P.

Abstract: *The paper presents a neural network-based semi-inverse controller. While an inverse controller uses an inverse model of the controlled plant, the semi-inverse controller is based on a forward model instead. This avoids some of the problems of inversion. The control algorithm can work with a short sampling period and is suitable for real-time implementation in a PLC.*
Key words: *Neural network, semi-inverse controller.*

1. INTRODUCTION

A popular neural network-using control algorithm design is a neural controller with an inverse model. It takes advantage of neural networks' ability to approximate the inverse transfer function of the controlled system. Serial connection of the inverse model with the plant then theoretically suffices to respond well to the desired value (Cichocki & Unbehauen, 1994; Widrow & Walach, 1996; Omatu et al., 1996).
Unfortunately the inversion approach has several problems. The inverse model can be hard or even impossible to obtain. Other problems arise from the fact that the inverse controller is a dead-beat controller and as such it is highly sensitive to the precise learning of the model. When working with short sampling periods, it also produces extreme control actions. These problems are emphasised in real systems where noise is present. Noise can be mitigated by filtration, but every filter worsens the resulting dynamics of the system.
The semi-inverse controller attempts to avoid the problem of learning of the inverse model by being based on a forward model instead (Krupanský, 2003).

2. THE SEMI-INVERSE CONTROLLER

2.1 The inverse controller in a closed loop

Suppose the plant is approximated by an ARX model:

$$F_{\mathrm{M}}(z)=\frac{b_0+b_1z^{-1}+b_2z^{-2}+\ldots+b_mz^{-m}}{1+a_1z^{-1}+a_2z^{-2}+\ldots+a_nz^{-n}} \tag{1}$$

where: m, n ... orders of the model
b_i, a_i... num./denom. polynomial coefficients
Fig. 1 shows the structure of the model.

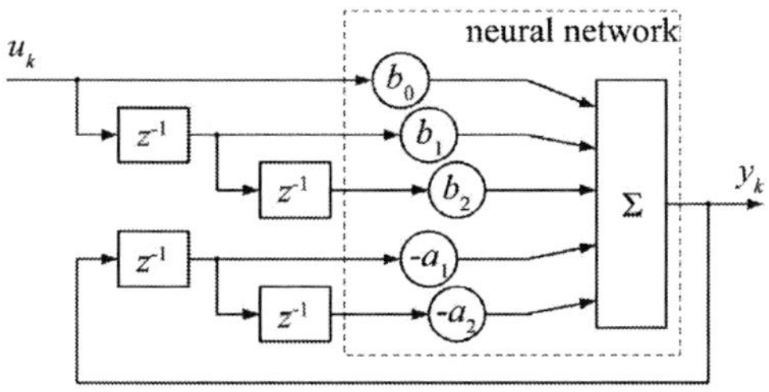

Fig. 1. The model: linear ARX or a neural network model

An idealized inverse controller would have a transfer function:

$$F_{\mathrm{C}}(z)=\frac{1+a_1z^{-1}+a_2z^{-2}+\ldots+a_nz^{-n}}{b_0+b_1z^{-1}+b_2z^{-2}+\ldots+b_mz^{-m}} \tag{2}$$

The inverse controller is designed to operate in an open loop, without a real feedback. That is undesirable for disturbance elimination purposes. If used in a closed control loop, it would compute its action as:

$$u_k=\frac{1}{b_0}(e_k+a_1e_{k-1}+a_2e_{k-2}+\ldots+a_ne_{k-n}-\\ -b_1u_{k-1}-b_2u_{k-2}-\ldots-b_mu_{k-m}) \tag{3}$$

where: k .... discrete time step
u_k....... control action in step k
e_k....... control error in step k

The big sensitivity of the controller on the model accuracy comes mainly from the parameter b_0, especially when its value is close to zero.
After the finish of the transient response, the control action would be:

$$u_{\mathrm{S}}=\frac{1}{b_0}\left[e_{\mathrm{S}}(1+a_1+a_2+\ldots+a_n)-u_{\mathrm{S}}(b_1+b_2+\ldots+b_m)\right]=\\ =e_{\mathrm{S}}\frac{1+a_1+\ldots+a_n}{b_0+b_1+\ldots+b_m} \tag{4}$$

It's clear that in the steady state (indicated here by the index S) the control error e_{S} would not be completely eliminated. Let's introduce a modified control error:

$$e_{\mathrm{M},k}=w_k+e_k=2w_k-y_k \tag{5}$$

where: $e_{\mathrm{M},k}$.... modified control error in step k
w_k...... desired value in step k
y_k....... system output in step k

This can be implemented by a filter $F_k(z)$ operating on the desired value. If the controller works with the modified control error, then in the steady state this modified error will remain non-zero, but the real control error will be able to reach zero. The modified error approach works as long as the gain of the open loop equals 1.

2.2 Modification for a semi-inverse controller

Let's start with the inverse controller operating on the modified error. In a steady state, its action will be:

$$u_{\mathrm{S}}=\frac{1}{b_0}\left[e_{\mathrm{M,S}}(1+a_1+a_2+\ldots+a_n)-u_{\mathrm{S}}(b_1+b_2+\ldots+b_m)\right] \tag{6}$$

In order to simplify the situation, let's assume that the gain of the plant model is A_{M}=1. This assumption will be generalized away in the next section. Let's look at the steady state when the desired value is w_k=1. Under these simplifying conditions it holds:

$$u_{\mathrm{S}}=y_{\mathrm{S}}=e_{\mathrm{M,S}}=w_{\mathrm{S}}=1 \tag{7}$$

Then some of the members in the equation (6) can be cancelled, resulting in:

$$1=\frac{(1+a_1+a_2+\ldots+a_n)-(b_1+b_2+\ldots+b_m)}{b_0} \tag{8}$$

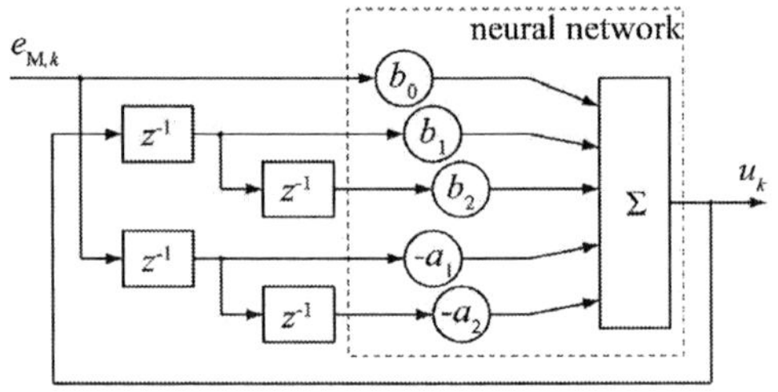

Fig. 2. The semi-inverse controller is based on the model. Its inputs are reconfigured.

Thus it holds that:

$$1 = b_0 - a_1 - a_2 - \ldots - a_n + b_1 + b_2 + \ldots + b_m \tag{9}$$

At this point we'll introduce the semi-inversion modification of the inverse controller and rewrite the action of the semi-inverse controller in the steady state as:

$$\begin{aligned} 1 = u_S &= e_{M,S}(b_0 - a_1 - a_2 - \ldots - a_n) + u_S(b_1 + b_2 + \ldots + b_m) = \\ &= 1 \cdot (b_0 - a_1 - a_2 - \ldots - a_n) + 1 \cdot (b_1 + b_2 + \ldots + b_m) \end{aligned} \tag{10}$$

The idea behind this modification is that the parameter b_0 is transferred instead of the constant 1 in the inverse controller. In the non-steady case the action is computed as:

$$\begin{aligned} u_k = b_0 e_{M,k} - a_1 e_{M,k-1} - a_2 e_{M,k-2} - \ldots - a_n e_{M,k-n} + \\ + b_1 u_{k-1} + b_2 u_{k-2} + \ldots + b_m u_{k-m} \end{aligned} \tag{11}$$

The corresponding transfer function of the semi-inverse controller is:

$$F_{SI}(z) = \frac{b_0 - a_1 z^{-1} - \ldots - a_n z^{-n}}{1 - b_1 z^{-1} - \ldots - b_m z^{-m}} \tag{12}$$

The core of the semi-inverse controller's structure is the same as the model's. The same trained neural network is used to compute the control action. The difference is in the input signals of the neural network, as seen in Fig. 2.

2.3 Gain normalization

The total gain of the serial connection of the semi-inverse controller (with the gain A_{SI}) and the plant can be calculated as:

$$A = A_{SI} A_M = \frac{b_0 - a_1 - a_2 - \ldots - a_n}{1 - b_1 - b_2 - \ldots - b_m} \cdot \frac{b_0 + b_1 + b_2 + \ldots + b_m}{1 + a_1 + a_2 + \ldots + a_n} \tag{13}$$

To fulfil the requirement of the unit gain in the open loop, it's necessary to add a multiplication by $1/A$ to the loop. The complete control loop is shown in Fig. 3.

3. TESTING THE CONTROLLER

The semi-inverse controller was implemented and tested in simulations and on laboratory physical models. To facilitate the implementation into a PLC, a portable approach based on an abstract controller programming interface was used (Pivoňka & Schmidt, 2005).

A neural network was used in the controller for on-line identification. The controller was adaptive and worked well even when we changed the time constants of the physical model. It showed better resistance to noise than an otherwise comparably operating non-adaptive PID controller.

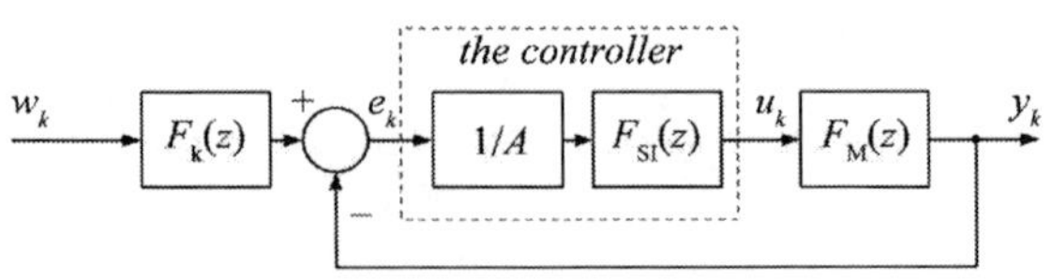

Fig. 3. The semi-inverse controller in a closed loop

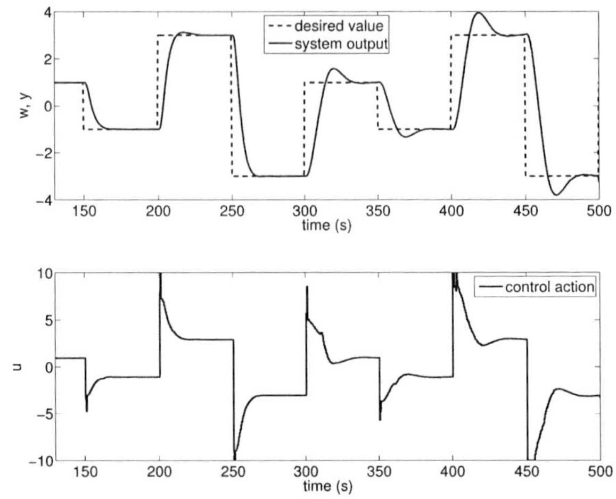

Fig. 4. Adaptive semi-inverse control of a physical model

A result of testing on a physical model is shown in Fig. 4. The controlled system had a transfer function approximately:

$$F_{S,1}(s) \approx \frac{1}{(10s+1)(s+1)^2} \tag{14}$$

The dynamics of the model were changed at time $t = 280\text{s}$ to approximate a different transfer function:

$$F_{S,2}(s) \approx \frac{1}{(7s+1)^2(s+1)} \tag{15}$$

The controller operated with a sampling period T=0.5 s.

4. CONCLUSION

The semi-inverse controller is able to adapt to a wide range of controlled plants. It works with short sampling periods, which is an advantage for disturbance cancellation. It performed well in testing on simulations and on physical models. Thanks to the on-line training of the neural network, it was able to adapt to changes in the time constants of the plant. It also showed good resistance to noise.

The algorithm is computationally simple. By porting the semi-inverse controller to the PLC, its usability for real-time control was demonstrated.

Acknowledgement: *The paper has been prepared as a part of the solution of GAČR project Softcomputing Methods in Control No. 102/06/1132.*

5. REFERENCES

Cichocki, A. & Unbehauen R. (1994). *Neural Networks for Optimization and Signal Processing*, John Wiley & Sons, ISBN 0-471-93010-5, Chichester

Krupanský, P. (2003). *The Possibilities of Using Neural Networks in Control* (in Czech), Ph.D. thesis, ÚAMT, FEKT VUT, Brno, Czech Republic

Omatu, S.; Khalid, M. & Yusof, R. (1996). *Neuro-Control and its Applications*, Springer-Verlag, ISBN 3540199659, London

Pivoňka, P. & Schmidt, M. (2005). Rapid Development and Implementation of Intelligent Control Algorithms, *Proceedings of the 12th East-West Fuzzy Colloquium*. Hampel, R. (Ed.), pp. 162 - 167, ISBN 3-9808089-6-3, Zittau, Germany, September 2005, Hochschule Zittau/Görlitz

Widrow, B. & Walach, E. (1996). *Adaptive Inverse Control*, Prentice Hall, ISBN 0-13-005968-4, Englewood Cliffs, NJ

Annals of DAAAM for 2006 & Proceedings of the 17th International DAAAM Symposium, ISSN 1726-9679
Editor B. Katalinic, Published by DAAAM International, Vienna, Austria 2006
150th Birthday of Nikola Tesla (1856-1943) Inventor of Mobile Robot Teleoperation

COMPARISON OF THE PROCESSING OF SIRO-SPUN YARNS AND CONVENTIONAL PLY YARNS INTO THE FABRIC

SCHWARZ, I.; BRNADA, S. & KOVACEVIC, S.

Abstract: *Advantages and disadvantages of the Siro-spun yarn compared to the conventional yarn will be analyzed. The advantage of Siro spinning process is elimination of processing stages, like electronic cleaning phase which results in higher yarn breakages, a more difficult passage over the machine elements, and eventually in a bad fabric quality. Conventional ply yarn has a more compact structure resulting in a higher resistance to mechanical loads during weaving. Due to an opener structure Siro-spun yarn cannot withstand frictions and stresses resulting in a greater number of process neps during weaving. The result is a bad fabric appearance.*
Key words: *Siro-spun yarn, conventional yarn, neps*

1. INTRODUCTION

This paper compares the processing of two yarn kinds that are spun after the different spinning procedures and are produced to a woven fabric. It concerns the conventional ply yarn and Siro-spun yarn. In the conventional ring spinning process, the single yarn is electronically cleaned and wound from a cop onto a cross-wound bobbin, and it is doubled and twisted. In the case of the Siro spinning process twisting is effected continuously on the same machine. With the Siro process two rovings, each separated by its condenser, are led in parallel through the drawing system and they are combined at the exit from the drawing system. The result is a two-fold ply yarn, which is possible to make in a continuous process, making it very cost-effective because cleaning, doubling and twisting stages have been completely eliminated (www. swicofil.com). The Siro-spun yarn has satisfactory mechanical properties, but just the absence of the electronic cleaning stage of yarn with the purpose of removing thick and thin places as well as neps, causes a more difficult passage over yarn guides, dents, healds and thread stop motions. In contrast to the Siro-spun yarn the conventional ply yarn is much more compact and withstands mechanical stresses during weaving (Najar et al., 2006). The structure of the Siro-spun yarn cannot withstand frictions and stresses, so in the weaving process a higher amount of process neps are formed and the fabric appearance is unacceptable. The presence of process neps in the fabric manifests itself as a problem only in cases when they are visible (Frydrych & Matusiak, 2002). In the case of thinner, light-coloured fabrics or the fabrics with contrast systems of warp and weft yarns, process neps are very easily recognizable, unlike in thicker, single-coloured, darker or mottled articles. For articles with a larger number of heald frames (higher mechanical stresses), the use of the Siro-spun yarn may result in the stretched threads that are formed during higher stresses.

2. THEORY

Neps in any form or size create a fault in the finished fabric. Figure 2a displays a top containing neps, which cannot be opened up, but they should be removed. Neps are formed at all stages following the combing process, where fibres may be broken. After spinning the nep structure in the yarn is loose and almost separated from the yarn (Figure 2b). In the course of electronic cleaning thin and thick places and neps are removed from the yarn. By passing over yarn guides yarns are rubbed off so that short fibres are piled up and detained in the friction point. As soon as the accumulated fibres reach the critical mass, they fall down on the yarn and form a process nep, which is compressed along the yarn structure, but it is not contained within the structure (Figure 2c). In the downstream processes, such as doubling and twisting, the yarn passes over various machine elements, so the further formation of process neps is inevitable. The process neps formed up to the twisting stage, become a part of the yarn structure itself during twisting (Figure 2d). In the course of weaving, yarns are exposed to high loads and friction, which make the surface of the fibrous thread layer to be degraded. Thus, protruding or whole fibres are physically separated from the structure and accumulated on the machine elements, forming fibre accumulations which fall on the thread and get woven into the fabric, which are later visible as a slub (Figure 2e) (Ethridge & Simonton, 2004). Because of its smaller compactness the Siro yarn is more prone to neps.

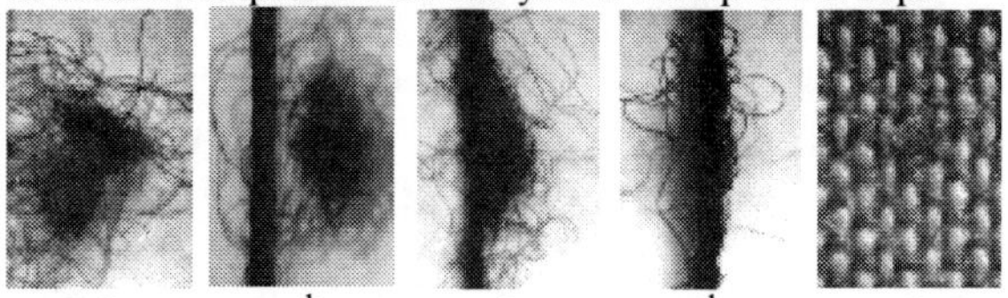
a b c d e

Figure 2. Neps in the spinning process

A fabric fault known as a stretched thread includes local appearances in thread sections with visual qualities different from the acceptable part. In single-coloured fabrics they are visible as brighter places in the fabric, whereas in the case of contrast effects of the thread system it is observed as a more emphasized effect of one system. This problem is exclusively connected with the weaving process. They are formed as a consequence of creeping within the yarn structure. The reasons may include inadequately stringent criteria of requirements for mechanical yarn properties or inadequate adjustments of the tension or speed of the weaving machine. In the case of the fabrics woven from the Siro-spun warp yarn the occurrence of stretched threads is very frequent, in particular when weaving with a number of heald frames higher than 8, where the differences in the tensions of threads drawn in different heald frames are much greater. The structure of the conventional ply yarn is much more compact. Observed as the whole its twist level is higher, so that slippage within its structure is much lower. As such, it is much more adequate for the range of fabrics more prone to this kind of faults (Internal scripts).

3. EXPERIMENTAL

Experimental tests of mechanical properties were conducted on samples of the Siro-spun and conventional ply yarn, as well as

on samples of fabrics woven from the Siro-spun and conventional ply yarn, woven at the textile mill Varteks, Varaždin, on the Sulzer G 6200 weaving machine.
The first part of investigations analyzed the Siro-spun and conventional ply yarns, whereby the yarn count was determined in accordance with the standard HRN ISO 2060:2003, and the twist level of the yarn in accordance with the standard HRN ISO 17202:2003, as well as tensile strength and elongation at break tested in accordance with the HRN ISO 2062:2003 standard. The results obtained are shown in Table 1.

	Siro-spun yarn (SSY)	Conventional ply yarn (CPY)
Count (tex)	25	25
Twist level (twists/m)	580	511
Tensile force (cN)	770,4	734,4
Elongation at break (%)	32,28	24,83

Table 1. Physical mechanical yarn properties

The same yarns were used for weaving fabrics in twill weave 2/2 with the same density of warp and weft threads. The investigations of construction and physical mechanical properties were conducted. This implies the determination of mass per unit area in accordance with HRN ISO 3801:2003 standard, the investigation of tensile force and elongation at break in accordance with the HRN F.S2.017 standard (Table 2).

	SSY fabric	CPY fabric
Warp density (threads/cm)	19	19
Weft density (threads/cm)	18	18
Warp tensile force (daN)	95,4	78,1
Weft tensile force (daN)	64,1	93
Warp elongation (%)	45,5	33,7
Weft elongation (%)	39,3	39,3
Mass per unit area (g/m^2)	200	198

Table 2. Construction parameters and physical mechanical properties of the fabrics woven from the tested yarn

Additionally, three different fabric samples were tested. Each was woven from the Siro-spun yarn and the conventional ply yarn of the same count, with the same density of warp and weft threads within one fabric sample. All the fabrics were woven in the same twill weave 2/2. The same tests were conducted in accordance with the same standards as for the fabrics aforementioned and tested. The results are shown in Table 3.

	Sample 1		Sample 2		Sample 3	
	SSY fabric	CPY fabric	SSY fabric	CPY fabric	SSY fabric	CPY fabric
Warp density (threads/cm)	20	20	27	27	24	24
Weft density (threads/cm)	19	19	22	22	22	22
Warp tensile force (daN)	78,9	75,5	106,5	103,2	52,7	45,8
Weft tensile force (daN)	67	68,8	75	82,8	37,1	38,7
Warp elongation (%)	36,8	34,3	34,5	32	31,5	23,8
Weft elongation (%)	36,3	30,3	34,5	30,5	45	39,4
Mass per unit area (g/m^2)	206	207	255	267	186	183

Table 3. Construction parameters and physical mechanical properties of the other fabrics

4. DISCUSION

By analysis of two yarn kinds of the same count, it can be concluded that the Siro-spun yarn has a higher tensile force and elongation at break, than the conventional ply yarn (Table 1). Twists make a contribution to the strength and compactness of yarn. Although the Siro-spun yarn has more twists than the conventional ply yarn, one cannot speak about a greater compactness since the twist level of the single components of the Siro-spun yarn is several times lower than in the conventional ply yarn. Therefore, it can be concluded that both yarn kinds with the same twist level in twisting possess an incomparably different compactness. The total twist is higher in the conventional ply yarn than in the Siro-spun yarn. Under the condition of the same weave and density of warp and weft yarns, the fabrics woven from these yarns show differences in mechanical properties. The tensile force in warp direction of the fabric woven from the conventional ply yarn is lower compared to the fabric woven from the Siro-spun yarn, while the situation regarding the tensile force in weft direction is reverse (Table 2). The elongation at break in warp direction of the fabric woven from the Siro-spun yarn is higher compared to the fabric woven from the conventional ply yarn. By contrast, the elongation at break in weft direction is equal in both fabrics. The mass per unit area is similar in both fabrics. In the second part, three groups of fabric samples were analyzed. The same weave, thread densities and weaving conditions were used (Table 3). According to the analyses obtained, it can be concluded that dynamometric investigations of the fabrics are very similar and the tensile force of the fabric in warp direction is always higher in the case of the Siro-spun yarn, and in the case of the conventional ply yarn in weft direction. However, the elongation at break is always higher in the fabric woven from the Siro-spun yarn. That means that the Siro-spun yarn can be elongated and deformed more easily in the weaving process than the conventional ply yarn, and that the Siro-spun yarn is "opener" and more prone to forming neps when rubbing against healds and reed. Fabric masses per unit area within one sample differ substantially, what has been expected since the fabric does not differ in construction parameters and raw material.

5. CONCLUSION

The Siro procedure has been developed to accelerate the process and to reduce production costs. This was achieved by uniting spinning, doubling and twisting into one continuous process. The Siro-spun yarn has ousted the competition by its acceptable price, so many fabric manufacturers replaced the more expensive ply yarn with the Siro-spun yarn in the warp. But, the disadvantages of the Siro-spun yarn were recognized which manifest themselves as faults in the finished fabric, which arise during the weaving process. Since the mechanical parameters are satisfactory, it is assumed that the disadvantage of its processing lies in structure compactness. Altogether the Siro-spun yarn has a lower twist level than the ply yarn. It may be assumed that the fibres under the influence of tensile force change their position more easily in the structure than it is the case with the conventional yarn. The Siro-spun yarn is more and more available on the market and has an acceptable price, and therefore it is desirable to integrate it into the fabric. Its application is not recommendable for product ranges in which contrast yarn systems are present, for brighter shades of fabrics, for weaves for which a big number of heald frames is used - large repeats, and on high-speed weaving machines.

6. REFERENCES

Ethridge, M.D. & Simonton, J. (2004). Renewed focus on neps. *Textile topics*, No. 2, Spring 2004, 2-7

Frydrych, I. & Matusiak, M. (2002). Predicting the nep number in cotton yarn-determining the critical nep size. *Textile Research Journal*, Vol. 72, No. 10, Oct 2002, 715-723

Najar, S.; Khan, Z. A. & Wang, X. G. (2006). The new Solo-Siro spun process for worsted yarns. *Journal of the Textile Institute*, Vol. 97, No. 3, Mar 2006, 205 - 210, 0040-5000

Internal scripts of textile industry Varteks d.d.

SiroSPUN™ – Spinning Process, *Available from:* http://www.swicofil.com, *Accessed:* 2006-02-07

Annals of DAAAM for 2006 & Proceedings of the 17th International DAAAM Symposium, ISSN 1726-9679
B. Katalinic, Published by DAAAM International, Vienna, Austria 2006
150th Birthday of Nikola Tesla (1856-1943) Inventor of Mobile Robot Teleoperation

THE INFLUENCE OF THE HYDRAULIC DRIVING SYSTEM ON THE DYNAMIC BEHAVIOUR OF THE CLAMPING DEVICES

SEGHEDIN, N.

Abstract: *The study of the dynamic behaviour of the clamping devices has a great importance because the clamping device, as a sub-system of the technological system, is most times the weak element. The paper presents the factors which influence the dynamic behaviour of the clamping devices, the influence of the hydraulic driving system being emphasized. The frequencies spectrum which characterizes the dynamic behaviour of two types of multiple hydraulic-driven clamping devices is being presented. It is shown the fact that the hydraulic driving system is a source of vibrations which can dynamically influence the clamping device in the same degree as some other elements of the machine-tool do.*

Key words: *fixtures, clamping, hydraulics, dynamic behavior, machine-tools.*

1. INTRODUCTION

The clamping devices are used for the aligning-positioning and clamping of the workpieces, tools or measuring units within the technological systems for mechanical processing. The clamping devices greatly influence the total processing accuracy by their drafting, designing, manufacturing, running, etc. The well-known specialized literature contains many references to the the accuracy of the aligning-positioning and clamping of the workpieces in devices. To simplify the computations, the determination of the devices accuracy are often done by taking into account the hypothesis that their loading is of a static type. Because of the fact that the cutting process and the entire running of the machine-tool have a well-marked dynamic character, the problem of the devices behaviour under dynamic operation regime has often been studied. Thus, the dynamic behaviour of the devices during the milling operations have been studied in the papers (Zgură & Ispas, 1971). The researches accomplished on the modal analysis of the ensemble cutting tool – tool clamping device are presented in the paper (Lacerda & Lima, 2004). The researches concerning the dynamic behaviour of the multiple clamping devices during a shock excitation have been carried out in the paper (Seghedin, 2005). A methodology and an equipment to investigate the interaction workpiece – device under the action of the cutting forces are presented in the paper (Nee et al., 2006). On taking into account the rule of "the six points", the dynamic development of the reaction forces of the device in six contact points with the workpiece is studied in the paper. In the paper (Fang et al., 2002), a dynamic model that considers the vibrations of both the workpiece and fixture elements in order to accurately predict the friction damping over a wide range of clamping forces has been developed.

This paper approaches the hydraulic driving system of the devices as a source of vibrations in the elastic system machine-tool – device – workpiece – tool – device – machine-tool. The influence of the hydraulic driving system on the dynamic behaviour of the devices as compared with the influence of some other vibration sources within the processing system (kinematic chains, toothed wheels, etc.) is being estimated.

2. HYDRAULIC DRIVING OF THE CLAMPING DEVICES

The clamping devices can be driven in two ways, namely, hand-driven or power-driven. The power-driven one is of several types, namely, pneumatic, hydraulic, pneumo-hydraulic, mechanical, electromechanical, magnetic, electromagnetic, vacuum, etc. The hydraulic driving is characterized by a high output and allows the application of some high clamping forces by means of reduced-size hydraulic cylinders. The hydraulic driving devices are often employed in industrial practice, more typical driving diagrams being known. These diagrams include elements such as hydraulic pumps, filters, valves, pressure controllers, flow controllers, hydraulic cylinders, etc.

3. SUGGESTED METHODOLOGY

To accomplish a research on the dynamic behavior of the clamping devices, we suggest a methodology which includes the following stages and activities: E1 – establishing of the type of device to be studied; E2 – establishing of the type of dynamic study; E3 – establishing of the type of the machine-tool on which the device is fixed; E4 – establishing of the tool type; E5 – establishing of the equivalent dynamic model for the analysed system; E6 – establishing of the cutting operation parameters; E7 – establishing of the parameters to be measured; E8 – establishing of the research equipments; E9 – establishing of the programming language for the data control, analysis and displaying; E10 – choosing (designing) of the virtual instruments for the data control, analysis and displaying; E11 – establishing of the variable parameters of the virtual instruments ; E12 – establishing of the variable parameters of the clamping devices; E13 – working out of the experiment strategy; E14 – accomplishing of the experiments; E15 – interpretation of the results. This methodology can be completed with other stages and activities depending on the specific research conditions.

4. EXPERIMENTAL RESULTS

For a preliminary research of the dynamic behaviour of the hydraulic clamping devices, we have chosen the establishing of the global level of the chatters to which the device is subjected to, under normal running conditions and characteristic working environment, the magnitude of the excitating forces remaining unknown (Ispas et al., 1998). There have been experimentally analysed the hydraulic devices for series or parallel multiple clamping devices with elastic-type clamping (directly from the hydraulic motor) from figs. 1 and 2.

A FUS 250 type milling machine and a face end mill-type tool with changeable plates have been chosen to carry out the researches. To measure the global level of vibrations we have used a 333B52 type PCB Piezotronics accelerometer, a 480C02 type PCB Piezotronics signal conditioner, a PCI-6023E type National Instrument data acquisition card, a CB-68LP type National Instrument connector block and a Pentium III computer. A virtual Spectrum Analyzer from the LabVIEW 6.1 program has been used for the data control, analysis and displaying. A hydraulic installation with a maximum 400 bars driving force has been used to drive the clamping device. Fig. 3 presents the frequency spectrum corresponding to the series clamping device, only the hydraulic driving installation with the RS-04 gear pump being switched on. One can notice the frequencies of 220 Hz, 440 Hz, due to the pump of the hydraulic installation.

Fig. 1. Hydraulically-driven series multiple clamping device

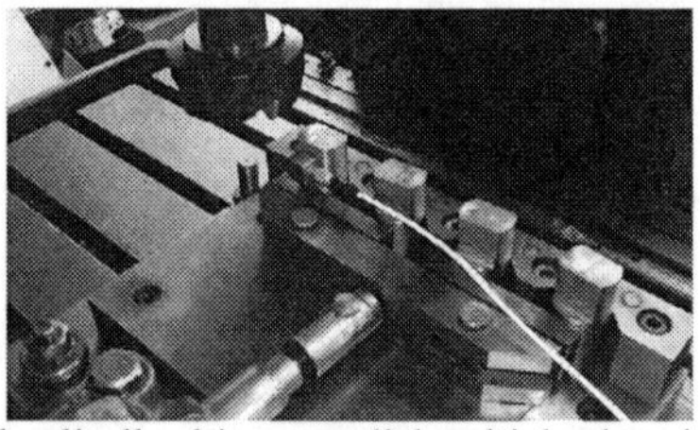

Fig. 2. Hydraulically-driven parallel multiple clamping device

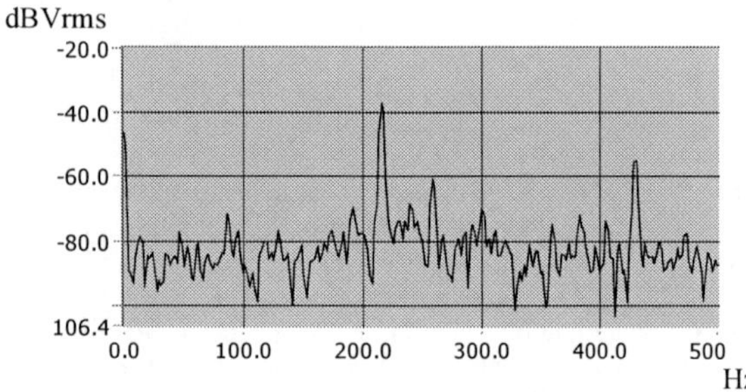

Fig. 3. The frequency spectrum for the series clamping device at the switching on of the driving hydraulic installation

Fig. 4 presents the frequency spectrum corresponding to the parallel clamping device, only the hydraulic driving installation being switched on.

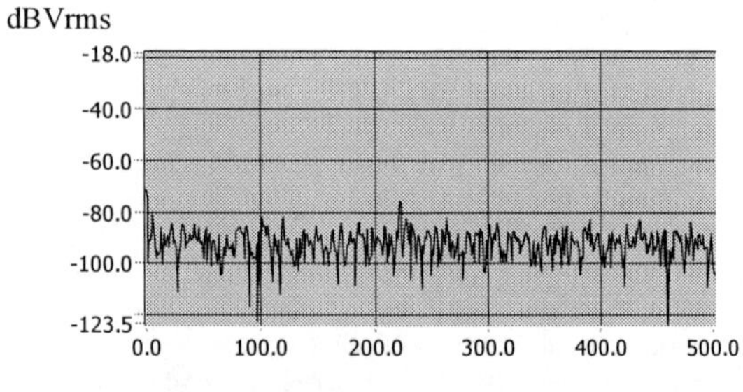

Fig. 4. The frequency spectrum for the parallel clamping device at the switching on of the driving hydraulic installation

As in the previous case, one can notice the frequencies of 220 Hz, 440 Hz, due to the pump of the hydraulic installation, but the amplitudes of these frequencies are not so well-marked.
Fig. 5 presents the frequency spectrum corresponding to the series clamping device, the main engine of the machine-tool, the gear box and the hydraulic installation being switched on. One can notice the spectral components at 220 Hz, 440 Hz, due to the hydraulic installation and at 360 Hz, due to the main engine and gear box of the machine tool.

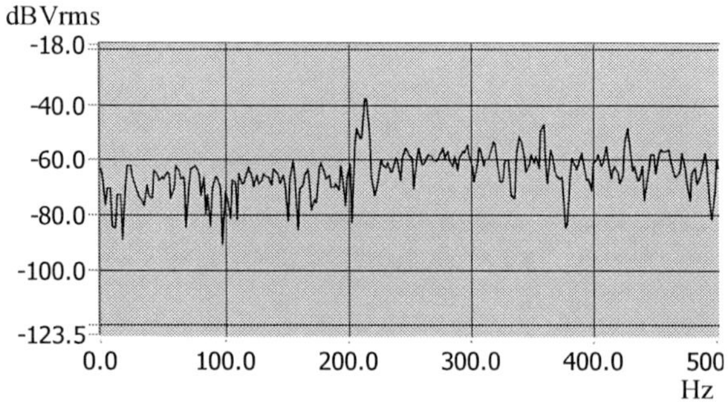

Fig. 5. The frequency spectrum for the series clamping device at the switching on of the main engine of the machine-tool, of the gear box and of the hydraulic installation

5. CONCLUSIONS

As a result of the experiments, the considerable influence of the hydraulic driving installation in the case of the series clamping devices has been rendered evident. The vibrations due to the hydraulic installation are comparable to those transmitted to the workpiece by the main engine and the main kinematic chain. Not the same thing happens at the parallel clamping devices in which the workpieces are more stable owing to their locating on individual supports. These researches can be carried on by studying the influence of the pump type on the dynamic behaviour of the device.

6. REFERENCES

Fang, B.; DeVor, R.E. & Kapoor, S.G. (2002). Influence of Friction Damping on Workpiece-Fixture System Dynamics and Machining Stability. *Journal of Manufacturing Science and Engineering. Transactions of the ASME* , Volume 124, Issue 2, pp. 226-233.

Ispas, C.; Predincea, N.; Zapciu, M.; Mohora, C. & Boboc, D. (1998). *Maşini-unelte. Încercare şi recepţie*. Ed. tehnică, ISBN 973-31-1183-X, Bucureşti.

Lacerda, H. B.& Lima, V. T. (2004). Evaluation of cutting forces and prediction of chatter vibrations in milling. *Journal of the Brazilian Society of Mechanical Sciences and Engineering*, 26, *1*, Rio de Janeiro, Jan./ Mar., ISSN 1678-5878.

Nee, A. Y. C.; Senthil Kumar, A. & Tao, Z. J. (2004). *An Advanced Treatise On Fixture Design And Planning*. World Scientific Publishing, ISBN 981-256-059-9, Singapore.

Seghedin, N. (2005). Considerations On The Dynamic Behaviour Of The Multiple Clamping Devices Of The Machine-Tools. *Buletinul Institutului Politehnic Iaşi (published by the Technical University „Gh. Asachi" Iaşi)*, tomul LI (LV), fasc. 5, pp. 217-222, ISSN 1582-6392.

Zgură, Gh. & Ispas, C. (1971). Unele consideraţii privind influenţa dispozitivului asupra comportării dinamice a sistemului tehnologic. *Construcţia de maşini*, XXIII, *12*, decembrie, Bucureşti, pp. 731-734.

Annals of DAAAM for 2006 & Proceedings of the 17th International DAAAM Symposium, ISSN 1726-9679
Editor B. Katalinic, Published by DAAAM International, Vienna, Austria 2006
150th Birthday of Nikola Tesla (1856-1943) Inventor of Mobile Robot Teleoperation

STUDY OF PULSED ELECTRO JET DRILLING PROCESS

SEN, M.

Abstract: *Electro jet drilling (EJD) is an electrochemical machining based fine hole drilling process which is primarily used for machining macro and micro holes in difficult-to-machine materials. This paper presents the effect of pulsed current on the quality of micro holes produced by the EJD process. It has been found that pulsed-EJD offers significant improvements in dimensional accuracy in comparison with direct current-EJD.*
Key words: *electro jet drilling, micro hole, pulsed current, super-alloy.*

1. INTRODUCTION

The success of the electrochemical jet machining processes in electronics and computer industries has paved the way for the renewed research interest in the electro jet drilling (EJD) process (Dutta et.al. 1989). The ever increasing demand of micro-sized machine components has created a need for small and micro size holes in difficult-to-machine materials. The complexity of shapes and degree of precision required on the components used in these industries need such holes to be straight, accurate and exactly positioned and also be produced at a faster rate. EJD is one such non-traditional machining process, which possesses all the requisite capabilities in meeting the modern day demands of drilling small and micro holes (Sen & Shan 2005a).
In EJD, a negatively charged stream of acid electrolyte is impinged on the workpiece to form a hole. The acid electrolyte (10-25% concentration) is passed under pressure (0.3-1.0 MPa) through a finely drawn glass tube nozzle. The electrolyte jet gets charged when a platinum wire, inserted into the glass tube is connected to the negative terminal of DC power supply. The workpiece acts as anode. With the power supply switched on, the material removal takes place through electrolytic dissolution as the charged electrolyte stream strikes the workpiece. The metal ions thus removed from the work surface are carried away with the flow of the electrolyte.

2. EXPERIMENTAL SETUP AND RESULTS

An indigenous experimental setup was developed which consists of the following major elements: (i) Work feed assembly, (ii) Glass nozzle holding arrangement, (iii) Electrolyte flow system, and (iv) DC power supply. In the designed setup as shown in Fig.1 the feed is imparted to the workpiece against the stationary glass tube nozzle. Borosilicate glass material was used for making the glass nozzles with the capillary end. A DC power supply (direct as well as pulsed) with working voltage range 0-650V and current 0-5A was used for conducting the experiments. The cathode wire material used was platinum while work material selected was SUPERNI 263A. A stainless steel screw pump with tubings made of corrosion resistant material was used to pressurize acid electrolyte solutions at different concentration levels. A Simulink based fuzzy logic controller was developed to control the gap between the workpiece and the glass capillary tip (Sen & Shan 2005b). The hardware implementation of the gap control system was carried out with a PCI bus based multi-function data acquisition card (PCI-1711 Advantech Taiwan).

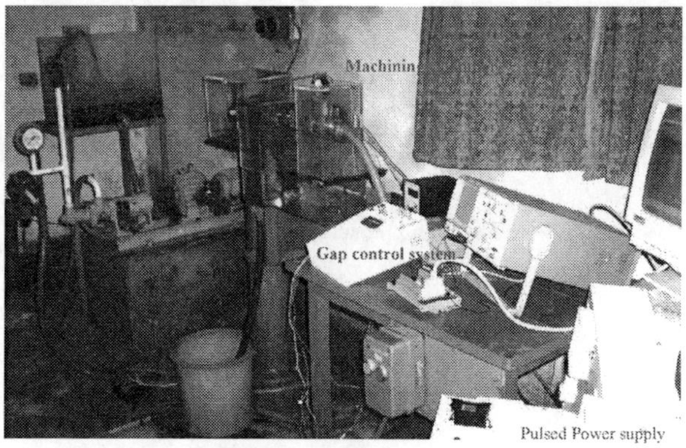

Fig.1 Pulsed EJD experimental setup

The successful application of the electrochemical micromachining in drilling fine holes in metallic foils by using pulsed current has prompted the author to study the effect of the pulsed current on the quality of the small holes produced by the EJD process in the nickel base super alloy and to compare them with those obtained from using the direct current. For this purpose, an experimental design based on RSM was selected for performing the experiments on SUPERNI 263A workpieces. A second order central composite rotatable design (CCD) with 20 runs was selected. The levels of the selected process variables in CCD are given in Table 1. Pulsating current has three parameters: pulse on-time, pulse-off time and peak current density, which can be varied independently in order to achieve the desired machining rate. In the present case 70μs of pulse-on time and 20μs of pulse-off time were taken as the constant parameters for conducting the experiments.

Process parameter	Level/code				
	-1.682	-1	0	+1	+1.682
Applied voltage	100	190	325	460	550
Electrolyte conc.	10	13	17.5	22	25
Feed rate	0.0	0.2	0.5	0.8	1.0

Table 1 Process parameters and their levels

The three independent process variables selected were applied voltage, electrolyte concentration and feed rate. The response parameters identified for this purpose were material removal rate (MRR), radial overcut (R_{oc}) and hole taper (T_a).
The SEM images have been used to compare the results of direct current-EJD with those of pulsed current-EJD. From the Fig. 2 it is amply clear that the use of pulsed current in EJD has yielded better-shaped holes not only at the hole entrances for

the different combinations of the operating parameters but even at hole exits. The use of pulsed DC currents has significantly reduced the radial overcut and hole taper. The response characteristics i.e. MRR, radial overcut and hole taper were quantitatively compared with those obtained with direct current.

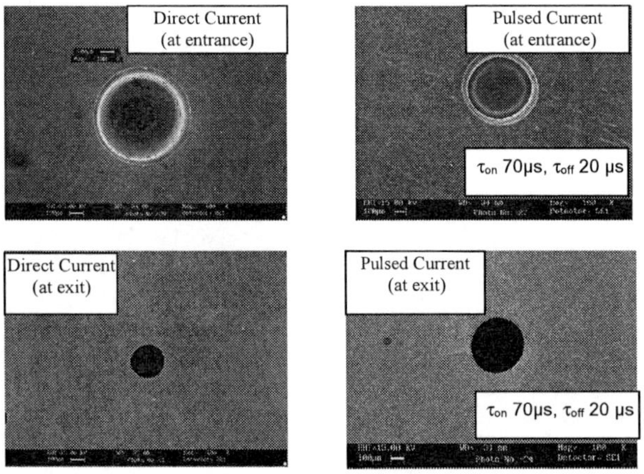

Fig.2 Comparison of the SEM images at the hole entrance and at exit with direct and pulsed current.

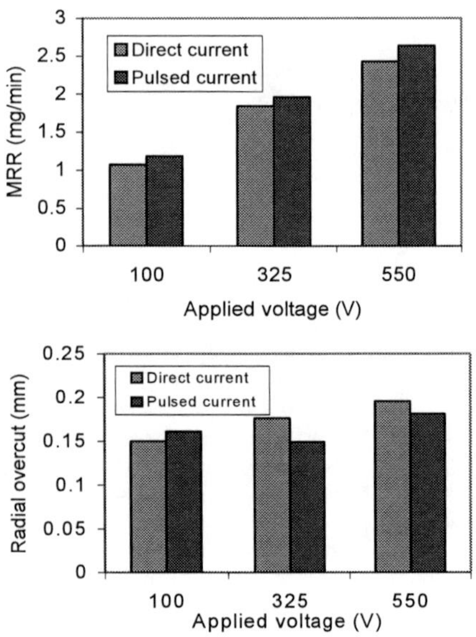

Fig.3 Comparison of overcut with direct and pulsed current EJD

Figs. 3 and 4 reveal that irrespective of the nature and type of the process variables, the response characteristics obtained with pulsed current have found to have shown marked improvement in the responses for the mid-range to higher-range values of the selected process parameters. However the direct current results particularly those obtained at the lowest level of the electrolyte concentration and zero feed rate have outperformed the pulsed current results. The reason for this can be attributed to the low machining current during the pulse on-time may not be sufficient enough to overcome the resistance of the long electrolyte flow-path of the EJD process.

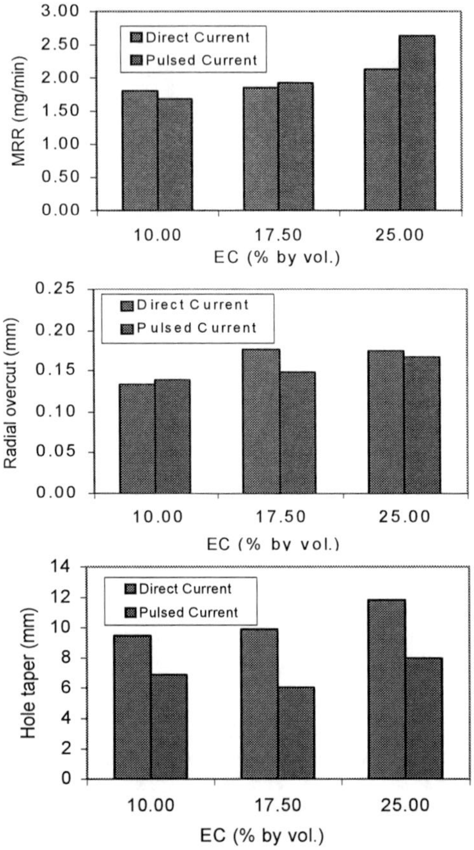

Fig.4 Comparison of taper with direct and pulsed current EJD

3. CONCLUSIONS

Encouraging results have been obtained with pulsed-EJD process during machining of micro-holes in nickel base super alloy. The application of micro-second pulse durations and fuzzy logic based gap control system assures the better quality of micro holes.

4. REFERENCES

Datta, M.; Romankiw, L.T.; Vigliotti, D.R. & Von Gutfeld, R.J. (1989) Jet and laser-jet electrochemical micromachining of nickel and steel. *Journal of Electrochemical Society*, Vol.136, No.8, pp. 2251-2256, ISSN 0013-4651.

Sen, M. & Shan, H. S. (2005a). A review of electrochemical macro-to-micro hole drilling processes. *International Journal of Machine Tools and Manufacture*, Vol.45, No.2, pp. 137-152, ISSN 0890-6955.

Sen, M. & Shan, H.S. (2005b) Design of a fuzzy logic based gap control system for the EJD process. *Proceedings of the 16th International DAAAM symposium*, pp.341-342, Vienna, Austria, ISSN 1726-9679.

Annals of DAAAM for 2006 & Proceedings of the 17th International DAAAM Symposium, ISSN 1726-9679
B. Katalinic, Published by DAAAM International, Vienna, Austria 2006
150th Birthday of Nikola Tesla (1856-1943) Inventor of Mobile Robot Teleoperation

THE MODEL FOR DECENTRALIZED ACCESS CONTROL IN P2P ENVIRONMENS

SHISHAEV, M. & SHEMJAKIN, A.

Abstract: *The paper touches upon the problem of access control in distributed peer-to-peer environments. The model for distributed access control based on levels of authority and decentralized trust management techniques is presented.*
Key words: *peer-to-peer; access control; trust management; computational model.*

1. INTRODUCTION

One of the most perspective modern architecture of distributed information systems (DIS) is a peer-to-peer architecture and corresponding model of nodes' interactions. The essence of peer-to-peer architecture is organizing of equal in rights, symmetric interactions between all nodes of the distributed system. It means that in different situations nodes may carry out both the role of the client and the server. Like any other DIS, peer-to-peer systems face with the problem of effective resource access control. In equal in rights systems of nodes the well-known centralized techniques of access control can not be applied. To decide whether to provide access to a resource in a peer-to-peer system specialized mechanisms of distributed access control are required. In this paper the model for distributed access control based on authority levels and decentralized trust management techniques are offered. These techniques are by no means revolutionary, they serve exceptionally to implement one of the possible models for distributed access control and encapsulate a great number of similar ideas stated in the studies of other authors. In the first part of the article the main problem is stated and positioned among similar problems. The second part concerns the formalism of the model. In the conclusion brief description of simulation experiments and further directions of model development are given.

2. PROBLEM STATEMENT. RELATED STUDIES

The most effective and flexible mechanisms of access control are based on access matrixes and their most common implementation is access lists (Sandhu & Samarati, 1994). The access matrix is a table, lines of which correspond to subjects, and columns that is to objects of access provision. In the cells of the table static or dynamic (depending on the function) lists of the subject's access rights for the certain object are stored. Flexibility of this mechanism requires storing of great volumes of information used to make decision as to access provision. This problem is especially critical for distributed environments as it increases the requirements to storage and communications capacities.

In the view of requirements to system resources, access control based on authority levels are opposite to full-fledged access matrixes. The principle of the control mechanism is as follows: resource r is assigned the minimum level of the authority required to provide access to it (access level) – $A(r)$. Access to a resource is provided, if the subject gets the level of authority P, not smaller than $A(r)$. Such approach minimizes the information content to store on nodes and to transfer on a network in order to provide or deny the access. However the subjects of access are not differentiated in this case, that is the access level for some object is the same for all subjects. The optimal solution here is to use levels of authority combined with the role mechanism of access control. In this case there is some set of user roles $U = \{u_1, \ldots, u_L\}$ in the system and each subject is associated with one or more role. L access levels is assigned to resource r – $A(u_i, r)$ – for each $u_i \in U$. Access to a resource is given, if the subject associated with role u_i, has level of authority P, not smaller than $A(u_i, r)$.

The problem of development of the access control model in distributed environment may be divided into three sub problems (Aberer & Despotovic, 2001):

1. Developing global access model determining the criterion of decision making as to access provision.
2. Developing computational model determining algorithms used to calculate access levels.
3. Problem of data and communication management: the ways of such algorithms implementation assuming that the data is decentralized.

This paper covers the first two subproblems. To solve the third task there exist various effective mechanisms, based on indexing or broadcasting (Adar & Huberman, 2000; Aberer, 2001).

In case of centralized access control authority level of the subject is determined directly by the system administrator. There is no common administrator if peer distributed nodes are equal. In this case the quantitative estimation of the authority level for a node is defined as its' trustworthiness toward the nodes included in distributed community.

One of the first well-known studies concerning formalization of degree of trust and development of computational models to perform its quantitative estimation was Marsh's (Marsh , 1994). In Marsh's models sociological and psychological factors of trust relations are used. Quantitative estimation of access subject's trust (in Marsh's study – agent) is defined as a real number in a range [-1; 1]. The trust, addressed from one agent to another, is calculated on the basis of general trust of trusting agent, importance and potential utility of current situation using linear formula.

Such approach to formalization of trust is quite comprehensive and universal, but is too complex to be implemented on the computer (in distributed environment in particular). For this reason simplified, adapted to implementation in distributed computer environments models of trust were developed by various authors (Aberer & Despotovic, 2001; Yu & Singh, 2000).

As a whole the models considered assume fully automatic definition of trust values in distributed multiagent environment. This makes computational models of trust and techniques of their distributed implementation rather complicated. Initialization of the distributed access control system, that is initial values setting for the parameters used in subsequent calculations is also quite a problem here. Besides numerical values used in models are rather subjective as a rule. Thus we came to a conclusion that only partly automated techniques may be actually implemented to provide effective distributed access. These techniques involve active human participation to define basic authority values.

3. DISTRIBUTED ACCESS CONTROL MODEL

Presented approach to define authority level reveals social mechanisms of trust relations between subjects. In this respect proposed global model of trust reproduces, with some modifications, the ideas stated in the study (Yu & Singh, 2000). The authority level is defined as a scalar continuous value of the agent reputation, representing the subject in distributed computer environment. Each agent may delegate a part of the authority (reputation) to any other, having transferred the corresponding mandate of trust to it. The current agent reputation is determined by quantitative and time parameters of the mandates received by it from other agents.

The following requirements were considered:

1. The agent cannot delegate the authority exceeding its own reputation.
2. The reputation of the idle agent decreases in time.
3. The delegated level of authority can be reduced or increased at any time.
4. Mandates of trust should provide positive, neutral and negative levels of trust. Neutral mandate implies that the trusting agent-source is acquainted with the trusted agent-recipient.
5. The agent reputation cannot grow indefinitely as it is limited by some limiting value.
6. The quantity of the positive and negative mandates asserted by the agent should be considered when defining its reputation.
7. Escalation of the reputation by recursive delegation of authority between a pair or a number of agents should be excluded.

Further we present computational model of reputation, satisfying the given requirements.

The mandate of trust addressed from agent i to agent j, M_{ij} represents the triplet:

(P_{ij}^0, t^0, R_i^0), where t^0 – the time of mandate generation; P_{ij}^0 - authority level delegated by agent i to agent j; R_i^0 - authority level of reputation of the delegating agent for a moment t^0. In case the number of mandates with similar second indexes exceeds the latest one is used for calculation of agent's reputation.

The reputation level of the agent is a function of the sum of the authorities, delegated to it by other agents: $R_j=f(p)$. Graph of function f is submitted on fig.1.

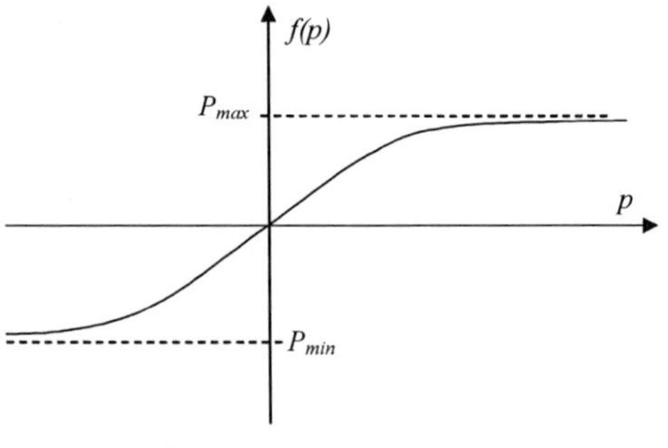

Fig.1. Function $f(p)$ graph

Function's argument p is calculated by the formula:

$$p = g(N^+)\sum_{i=1}^{N^+} P_{ij}^+ R_i^0 + g(N^-)\sum_{i=1}^{N^-} P_{ij}^- R_i^0 \qquad (1)$$

Here: N^+, N^- – numbers of positive and negative mandates addressed to agent j;

P_{ij}^+, P_{ij}^- – authority levels contained in positive and negative mandates, addressed from i to agent j;

$g(x)$ – function determining a degree of influence of mandates' quantity on agent's reputation; $g(0) = 0$ and $g(x) \in [1, g^{max}]$ for $x \in [1, +\infty[$; $g^{max} = const$.

In its turn, the level of authority once delegated by one agent to another can not be treated as a constant, but there is a decreasing function of time:

$$P_{ij}(t) = P_{ij}^0 - h(\Delta t), \qquad (2)$$

where $h(\Delta t)$ – increasing function, $\Delta t = t - t^0$, and $h(\Delta t) \xrightarrow[\Delta t \to \infty]{} P_{ij}^0$.

4. CONCLUSION

The developed computational model for reputation of agents on the basis of distributed initial data has been approved in course of simulation. Specially developed software tool was used in simulation. This tool emulates distributed computing environment with various numbers of nodes and various time schedules of mandates' exchanges. As functions involved in formulas, the following were used:

$$f(p) = p \cdot th(a \cdot p); \qquad (3)$$

$$g(x) = g^{\max} \cdot th(x-1) + 1; \qquad (4)$$

$$h(\Delta t) = \frac{1 - th(a \cdot \Delta t - b)}{2}, \qquad (5)$$

where th - hyperbolic tangent; $a=const$; $b=const$.

Experiments have shown that the developed model meets the requirements formulated above.

Ways of the model implementation in distributed environment can be the subject of the further studies. The most important issues here are maintenance of agent's authenticity, maintenance common system time, searching and gathering of positive and negative mandates among distributed nodes.

5. REFERENCES

Aberer K. (2001). P-Grid:A self-organizing access structure for P2P information systems *Proc. of the Ninth International Conference on Cooperative Information Systems.* (Trento, Italy), 2001.

Aberer K. and Despotovic Z. (2001). Managing trust in a peer-2-peer information system. *Proc. of 10th International Conference on Information and Knowledge Management,* pp. 310-317, 2001.

Adar E. and Huberman B. A. (2000). *Free riding on Gnutella Technical report*, Xerox PARC, 10 Aug. 2000.

Bin Yu and Munindar P. Singh (2000). A Social Mechanism of Reputation Management in Electronic Communities *Proc. of the 4th International Workshop on Cooperative Information Agents*, Matthias Klusch, Larry Kerschberg (Eds.), Lecture Notes in Computer Science, Vol. 1860, Springer, 2000.

Marsh S. (1994). *Formalising Trust as a Computational Concept. Ph. D. Thesis*, University of Stirling, 1994.

Sandhu R.S., Samarati P. (1994). Access control: principles and practice. *IEEE Communications magazine*, Sept 1994. P.40-48.

Annals of DAAAM for 2006 & Proceedings of the 17th International DAAAM Symposium, ISSN 1726-9679
B. Katalinic, Published by DAAAM International, Vienna, Austria 2006
150th Birthday of Nikola Tesla (1856-1943) Inventor of Mobile Robot Teleoperation

CHAOS TECHNOLOGY: A NEW METHOD FOR DESIGNING AN AGILE MANUFACTURING LINE

SIBILIA, A.; WULLSCHLEGER, C. & MALISA, V.

Abstract: Using a new concept called chaos technology, products handle their own movements and arrive right where they are needed! Production series are becoming ever smaller, product diversity ever greater and the requirements products have to meet ever more complex. It is increasingly common for product specifications to be changed at short notice or even during production. Modern production operations have to respond quickly to even the most specialized customer requirements. This is a standard expectation. Future production will demand greater flexibility of conveyor systems. Logistics is faced with increasingly complex tasks that have to be performed in ever shorter periods. As a result of these technical advances, the requirements placed on conveyor systems are growing steadily. Montrac conveyors utilize a control system based on so-called chaos technology. The Montrac System allows any logical connection desired to be established between individual processing stations. This in turn allows chaotic production and complex logistical paths to be handled with ease and efficiency.

Key words: *agile manufacturing, intelligent and flexible conveyor, chaotic manufacturing, mobile robot.*

1. THE IDEA OF CHAOS TECHNOLOGY

Chaos technology designates a method by which the distance covered by a Montrac conveyor system (Fig. 1) is determined independently of the coordinating control system.
The monorail (Trac) of the Montrac System is made of clear anodized aluminium extrusion. The bus bars are on one side, and the T-slots accept flexible control elements on the other.
The switch serves to divide and merge traffic into/away from multiple paths. The TracSwitch operates electrically.
The shuttles are intelligent self-propelled vehicles. Their maximum speed is 30 m/min. They are available with a single drive axle or two drive axles. A single axle shuttle can carry up to 12kg, while a dual axle shuttle can carry 24kg. Each drive axle has a maintenance-free, 24V brushless DC motor and sensors to avoid collisions with other shuttles or obstacles.
A Trac Station is used if a shuttle has to be positioned exactly and/or if the plate requires additional support while being processed.
The coordinating control system is relieved of the task of activating actuators along the conveyor path at the right time and for the right function. This approach leaves the coordinating control level free to concentrate on processes at processing stations.
Processing itself is what improves, refines and finishes products. Transport should play a subordinate role at most.
The Montrac System is designed with this principle in mind. The components of the travel path have intentionally been kept simple and consistently modular. Consequently, complete systems can be created from a handful of basic modules for handling anything from the simple circulation of materials to complex logistical tasks.

This is a crucial aspect nowadays because product changes are ever more frequent and often even occur during a product's life cycle. Furthermore, today's products are tailored ever more specifically to customer needs and are produced on a just-in-time basis. As a result, production batches can be as small as just one unit in extreme cases.

2. UNDERLYING PRINCIPLES

All processing stations connected to the components of the Montrac travel path to create a logistics system are, for example, numbered consecutively (Fig. 2).
Stations for feeding unprocessed materials and those for removing finished pieces are also considered as processing stations.
The idea behind chaos technology is that after the work process in one station is finished, the shuttle is given the next target address. All actuators configured for chaos technology respond autonomously to the target address, which allows the shuttle to reach the next station of its own accord.

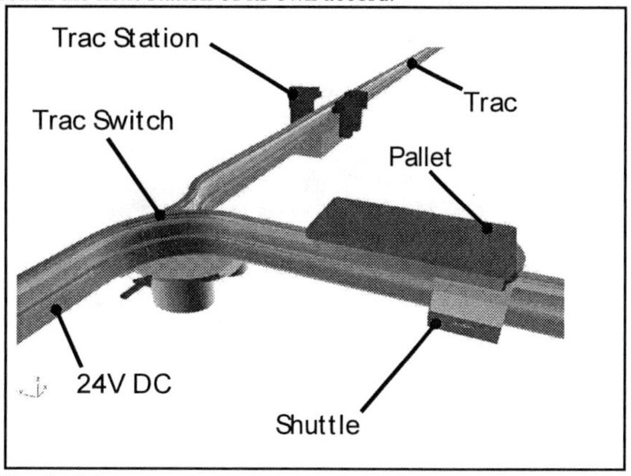

Fig. 1. Montrac conveyor system

The basic principle is similar to that of a multi-master bus system. The processing stations are the masters. At the very start, they transmit a unit of information to a shuttle in the form of an address. The shuttle moves along the path represented by the bus until it arrives at the target destination specified by the address. There the procedure repeats itself.
The address is nothing more than an ID number saved in the shuttle and changeable as desired.

3. EXECUTION

The travel path – the Montrac System – logically connects the processing stations. The path is put together according to rules that support the process and ensure an optimum supply of goods to the stations.

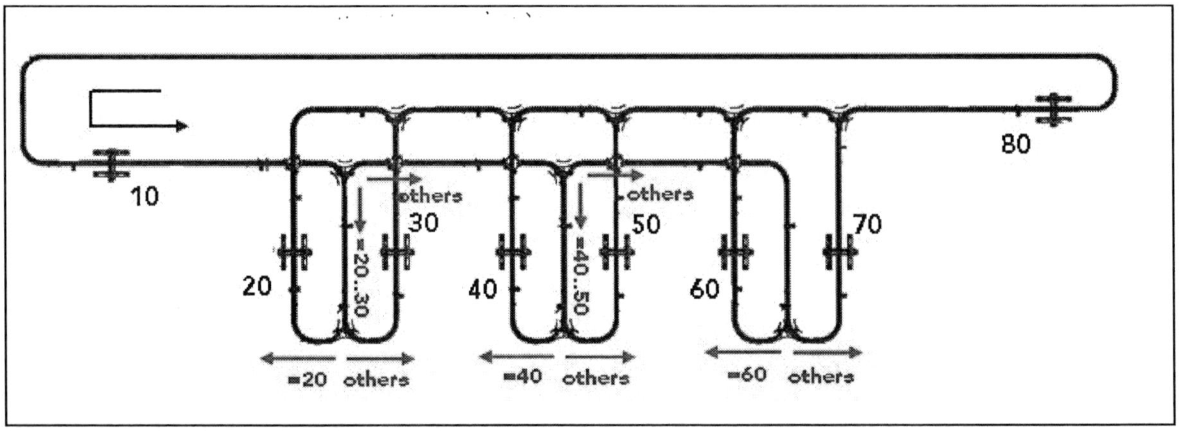

Fig. 2. Montrac travel path based on chaos technology

An ISM (communications module in the shuttle) signals the presence of a shuttle by transmitting a specific ID number, which the IRM (communications module along the path, the track module) acknowledges with "detect".
The IRM starts the shuttle by sending the start command and the ID number to the ISM.
The track module is notified of a shuttle's ID number at each stop position and the ID can be linked thereto a decision.
This link ensures that a shuttle with a certain ID number turns a certain way at a junction and that any shuttle with a different address does not (Fig. 3).
A distribution switch can be operated autonomously under these conditions. The only information that has to be deposited there is a station address in the form of a certain ID number for comparison purposes.
That means that when a shuttle arrives at a processing station it is assigned an ID number related to the state of the product, with which the next processing step is simultaneously defined. The path travelled is controlled with this same ID number.
To put it more briefly, the product controls the route taken.

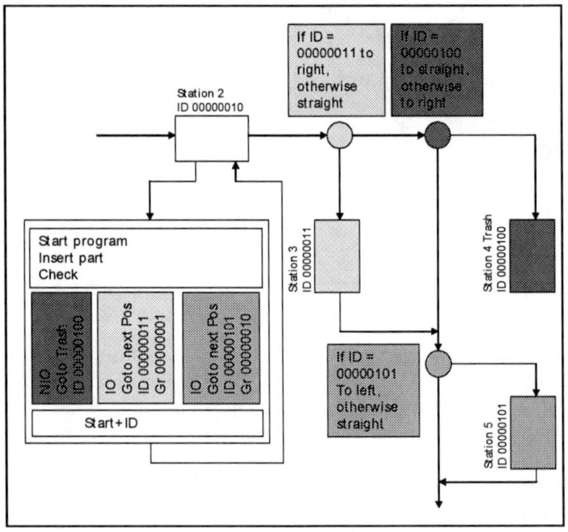

Fig. 3. Diagram of logic for addressing based on chaos technology

The advantages of chaos technology are not confined to being a simpler way of achieving outputs down to a batch size of 1. Other advantages also deserve mention:

- Programming work is substantially reduced and limited to the individual stations.
- Wiring work is virtually eliminated.
- The logistics path can be configured, changed and reconfigured with ease.

4. CONCLUSION

Intelligent production requires flexible conveyor systems. The Montrac System based on principles of chaos technology ensures full freedom with regard to processing stations. Shuttles for carrying products are assigned addresses, which enables the process to be controlled autonomously. The number of shuttles can be increased or decreased to alter capacity without any need for writing or changing a control code.
Today, the shuttle and the travel path communicate via an infrared interface. If products were fitted with RFID chips, all information about the manufacturing process could be deposited directly in a product.
Communication modules (ISM, IRM) with implemented RFID-technology would provide chaos technology and product data interchange.
Using the new generation of RFID technology, it will be possible to design a handling system that enables the product to guide its own production.

5. REFERENCES

Heinz Pfeifer, Gerald Kabisch, Hans Lautner (1995): *Fördertechnik: Konstruktion und Berechnung.* Braunschweig Wiesbaden: Vieweg, 1995

Hesse, S. (2006): Grundlagen der Handhabungstechnik, published in German by Hanser, 2006, ISBN 3-446-40473-2

Malisa, V. & Katalinic, B. (2004). Next Generation of Production Systems: Original Concept of Self-organizing Production Systems, Eighth International Conference on Manufacturing & Management Proceedings, pp. 1-14, 0-9578296-1-2, Queensland, Australia; October, 2004

Mehrabi, M. G., Ulsoy, A. G., Koren, Y., Heytler, P. (2004), Trends and perspectives in flexible and reconfigurable manufacturing systems, Journal of Intelligent Manufacturing, 13, 2, pages 135-146, 12 2004, DOI 10.1023/A:1014536330551

BüyüKözkan, Gülçİn, Derelİ, Türkay, Baykasoğlu, Adİl (2005), A survey on the methods and tools of concurrent new product development and agile manufacturing, Journal of Intelligent Manufacturing, 15, 6, pages 731-751, 1 2005, DOI 10.1023/B:JIMS.0000042660.77571.81

Annals of DAAAM for 2006 & Proceedings of the 17th International DAAAM Symposium, ISSN 1726-9679
Editor B. Katalinic, Published by DAAAM International, Vienna, Austria 2006
150th Birthday of Nikola Tesla (1856-1943) Inventor of Mobile Robot Teleoperation

ATTITUDES OF CROATIAN MANAGERS TO THE STAKEHOLDER APPROACH

SISEK, B.; KOLAKOVIC, M. & LAZIBAT, T.

***Abstract**: This paper focuses on the implementation of a new organisational model in Croatian companies. The model is known as the «stakeholder approach» or the «stakeholder perspective». Analysis of privatization in Croatia has shown that managers, as the key stakeholders, have full control in companies with no majority owner. However, in companies with a majority owner, the owners have also left key decisions to managers. Therefore, the main goal of this paper is to analyse the attitudes of Croatian managers to the stakeholder approach, because their views will have a great impact on the introduction of the stakeholder approach in Croatian companies.*

***Key words**: stakeholder, stakeholding, stakeholder management, business performance*

1. INTRODUCTION

The stakeholder approach has been designed to assert the general principles of equality, unity, and solidarity. The key value of the stakeholder approach is inclusion, which is given precedence over equality and individual freedoms. The partisans of this approach emphasize the desire to form a just society compatible with private property, profit-making, and decentralized decisions on the market. The concept is a reaction to negative consequences of market mechanisms in contemporary conditions, and it is based on a major challenge: how to harmonize demands for economic efficiency with those for social justice. The term «stakeholder» refers to the interest-and-influence group within and around a company. The «stake» refers to every interest or claim which a group or an individual has in the outcome of corporate policies, procedures, and actions towards others. Interest-and-influence groups differ not only over different priorities and motives, but also over different levels of involvement in a company, and consequently over the degree of risk they take. Therefore, stakeholders could be divided into different groups: external and internal, primary and secondary, key and strategic stakeholders. This division proves that not all stakeholders have the same importance for a company. Depending on whether they have one or more characteristics (power, legitimacy, urgency of reaction), stakeholders' relevance differs over the extent to which managers give priorities to competitive claims of stakeholders. The stakeholder approach points out that a corporation's management must act in the interest of various interest-and-influence groups. It must safeguard profits and economic growth (for investors and shareholders), meet expectations of customers and employees, demands of both immediate and broader environment, and ethical and ecological goals. Thus, it is obvious that the harmonization of stakeholders' interests within an organization can be complicated due to differences in knowledge, goals, competence, various compensation systems, and different business policies.

In contrast to the "shareholder approach", which primarily focuses on financial relationships, the stakeholder approach threats every corporation as a social organisation with emphasis on ethical provisions. "Ethical reasoning in the stakeholder analysis means asking: what is equitable, just fair, and good for those who affect and are affected by business decisions?" (Weiss, 2003, p. 48).

2. STAKEHOLDER MANAGEMENT

It is necessary to harmonize general aspirations of a company with interests and values of its individual stakeholders in order for such a conglomerate of wishes to be able to put all its human, organizational, and intellectual potentials into the service of new value creation.

Thus, managers are required to bear in mind possible conflict between their legal obligation to act in the interest of all stakeholders, and their own role as a corporate stakeholder. They form a special interest-and-influence group that pays attention to job security, the degree of discretion over making use of the company's resources, and the level of their compensation (Thomsen, 2005). According to the stakeholder approach, the task of management in the context of global competence is to increase efficiency and wealth, and to handle conflicts between various demands of stakeholder groups. These conditions have led to the formulation of the principles of stakeholder management the task of which is to reconcile those frequently opposed goals. Stakeholder management is trying to achieve an efficient combination of contributions, risk, and benefits which takes into account roles and interests of interest-and-influence groups. There are strong reasons to believe that the introduction of stakeholder management will contribute to the long-term success of a company. This type of corporate management opens up the possibility of including stakeholders into the process of corporate decision-making, and thereby it can increase their support for the corporation's activities. This is important since there is a reasonable possibility of management making a move towards a reduction in expenses and a boost in profits, which could jeopardize the interests of other interest-and-influence groups, and lead to instability in the corporation (Sachs and Rühli, 2005).

Such a complex position of management in the stakeholder model has provoked criticism of both: the approach and the model. In addition to stating that the character of the approach is too normative, the critics emphasise that the stakeholder approach serves as a covert justification for bringing a company under unlimited control of its managers, whereby assumes the position of an independent and final arbiter in case of conflicts of interests between various stakeholders. According to some critics, the stakeholder model and approach have reduced the importance of ownership and capital in the formation of stakeholder company.

3. RESULTS OF THE SURVEY

In this paper we would like to identify the extent to which Croatian managers are acquainted with the stakeholder approach, and their attitudes to its impact on business

performance. The data have been obtained by survey method at the end of 2003 and 2005 respectively (e-mail or direct polling). The sample comprised 100 businesses from all economic activities and sectors, and the sample proportion was adequate to the percentage of large, medium, and small businesses in the total number of Croatian companies. We started from the assumption that only a small number of managers was acquainted with the approach, its fundamental components, and its application in other countries, since the model was formulated in the early 1980s. In addition to that, an extremely small number of Croatian managers used the term «stakeholder» in media interviews.
The first question referred to the extent to which they were acquainted with the stakeholder approach: «Are you familiar with that approach»?

	The year of survey	
	2003.	**2005.**
a) Yes	67	63
b) No	33	37
Total:	100.00	100.00

Table 1. Managers' information level about stakeholder approach

The high percentage of positive answers, from 63% (2005) to 67% (2003), may be the result of a preliminary definition of the concept offered to the respondents: Stakeholders are the individuals and groups who participate, intentionally or consequentially, in the process of the corporation's wealth creation. The stakeholder approach points out that are corporation's management must act in the interest of various stakeholder groups. Interestingly, in the USA it is reckoned that 75% of those working in corporations now recognize the meaning of the term stakeholder
The second question was related to the implementation level of the stakeholder approach in the surveyed businesses.

	The year of survey	
	2003.	**2005.**
a) not at all	13.4	31
b) little	21.6	17
c) medium	35.1	27
d) strongly	20.6	23
e) extremely	9.3	2
Total:	100.00	100.00

Table 2. Implementation level of the stakeholder approach in surveyed firms

Only 30% (2003) i.e. 25% (2005) of the respondents defined the implementation level as high (the answers «strongly» and «extremely»).
The third question was related to the impact the implementation of the stakeholder approach had on business performance.

	The year of survey	
	2003.	**2005.**
a) not at all	12.40	29.59
b) little	17.50	15.31
c) medium	35.10	23.47
d) strongly	26.80	29.59
e) extremely	8.20	2.04
Total:	100.00	100.00

Table 3. The impact of the stakeholder approach on business performance

Answers to this question affirm the usefulness of the approach more than answers from the previous questions. If we take into account only the answers «strongly» and «extremely», then the percentage of unambiguously positive answers is 35% (2003) i.e. 31.6% (2005).
The last question of the survey was more concrete: «To which extent could the implementation of the stakeholder approach influence a particular segment of your company's activity?
The answers are ranked according to the percentage of positive answers («strongly» and «extremely») given by the surveyed companies for particular fields: (2003) 1) Increased productivity (49.5%), 2) Improvement of the quality of corporate governance (49.3%) 3) Reduced turnover (48.4%), 4) Improved compensation system (44.3%), 5) Increased net profit (40.2%), 6) Increased stock value (39.2%)
The rank list for 2005 reads as following: 1) Improving the quality of corporate governance (61.18%), 2) Improved compensation system (55.3%), 3) Reduced turnover (55.3%), 4) Increased net profit (55.29%), 5) Increased productivity (54.11%), 6) Increased stock value (43.53%).
Only two fields of the companies' activities have the same place on the rank-list in both surveys: reduced turnover (the third place), and increased stock value (the sixth place).

4. CONCLUSION

In this paper we would like to identify the extent to which Croatian managers are acquainted with the stakeholder approach, and their attitudes to its impact on business performance.
In spite of only 25% (2005) i.e. 30% (2003) respondents claiming a high implementation level of the stakeholder approach, such data are surprising if we take into account the outcome and the model of privatization in Croatia, dominated by the «shareholder approach» i.e. the maximization of shareholder value. The most surprising attitudes are those regarding the potential usefulness of the implementation of the stakeholder approach, that is, its impact on profit, productivity, governance quality, shareholder value, etc. They show that the respondents are aware of the potentials of the stakeholder approach. Against the backdrop of an affirmation of small and medium businesses in Croatia's economic development, such findings are encouraging, because we can expect a further affirmation of the model.
The stakeholder model and practice have not been sufficiently investigated, and more research in the field is needed. Therefore, it is necessary to identify the attitudes of other stakeholders to the problems which were the subject of this survey.

5. REFERENCES

Hutton, W. (1999) *The Stakeholding Society,* Polity Press, Cambridge.

Kelly, G. et. all (1999) *Stakeholder Capitalism,* Macmillan, London.

Nutti, M. (1998.) *Stocks and Stakes: the case for protecting stakeholders' interests*, Economics Analysis, No1

Sashs, S. & Rühli, E. (2005.) *Changing managers' values towards a broader stakeholder orientation,* Corporate Governance, VOL.5, NO.2

Sisek, B. (2003.) *Implementation Level of the Stakeholder Approach in Croatian Companies,* Annals of DAAAM for 2003 & Proceedings of the 14th International DAAAM Symposium, 22-25th October 2003, Sarajevo, Bosnia and Herzegovina

Thomsen, S. (2005.) *Corporate governance as a determinant of corporate values,* Corporate Governance, VOL.5, NO.4

Weiss, J.W. (2003.) *Business Ethics: A Stakeholder and Issues Management Approach,* Thompson South Western, Canada

Annals of DAAAM for 2006 & Proceedings of the 17th International DAAAM Symposium, ISSN 1726-9679
B. Katalinic, Published by DAAAM International, Vienna, Austria 2006
150th Birthday of Nikola Tesla (1856-1943) Inventor of Mobile Robot Teleoperation

DEVELOPMENT OF EXCITER INTENDED TO TESTING RAILWAY BRIDGES DYNAMIC PERFORMANCES

SKOCZYNSKI, W.; SZYMKOWSKI, J.; ROSZKOWSKI, A. & MAREK, B.

Abstract: *Determination of dynamic behaviours of railway bridges needs to develop some testing methods. The aim of this study was to adopt existing eccentric mass exciter to use it in railway bridges dynamic tests. A special frame transferring the excitations to the rails was designed and produced. Then it was tested together with the exciter. Some results of laboratory tests were presented. It can conclude the exciter together with a new construction of the supporting frame fulfill all the functional assumptions and allows to obtain an excitation force with a stable frequency in the range of 5–22Hz.*
Key words: *inertial exciter, dynamic properties, testing*

1. INTRODUCTION

The dynamic behaviours of railway bridges under train loads is one of the fundamental problems to be solved. These problems have more and more importance with development high-speed railways being built in the world. The train running with high speed induces dynamic impact on the bridge structures, influencing their working state and service life. So it makes develop some methods to test dynamic characteristics of railway bridges structures. This study is concentrated on excitation method called measured – input test. Measuring the input forcing function is required for determining the modal characteristics (resonant frequencies, mode shapes and modal damping ratios) of testing structures.

The aim of this study was to adopt existing eccentric mass exciter to use it in railway bridges dynamic tests. It needs to design and produce a special frame with dynamometers transferring the excitations to the rails. Concerning the new construction the following assumption were made. It should be possible to permanently measure the excitation force components during the railway bridges inspections. It should enable to locate the exciter in various positions in relation to the rails. The temporary fastening of the road wheels for the transport purposes from its unloading place to the investigation plant should be provided.

2. A NEW CONSTRUCTION OF SUPPORTING FRAME

Figure 1 presents the inertial exciter with designed frame and dynamometers fixed to the rails. The frame consists of two placed lengthwise channel irons connected with cross-bars. The spacing between channels corresponds to the exciter's base width. Threaded holes made at the top part of each channel iron enable fastening of exciter to the frame with screws in three different placements: symmetrically according to the rails, and above each of them in such a way, that the resultant excitation force lies at the symmetry center of the rail. The cross-bars made of flat iron and connecting two placed lengthwise channel irons at their bottom serve as a base to fasten dynamometers. Two transversal channel irons placed bellow the frame play dual role: they support four screws allowing to raise and lower the frame with the exciter and allow to temporarily fasten driving system.

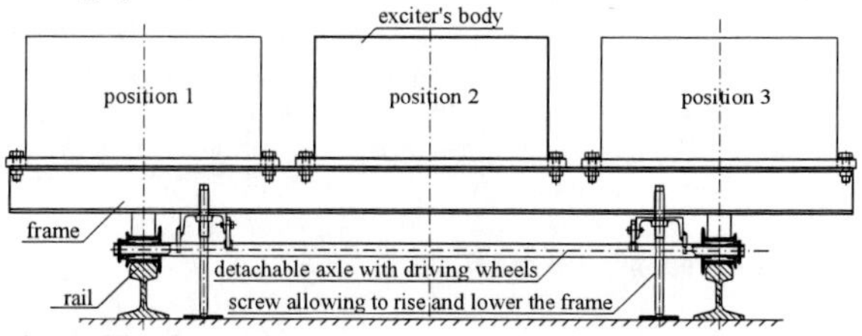

Fig. 1. The frame structure with sliding support wheel system, allowing to fasten the exciter in three different positions

To measure the components of the excitation forces three tensometric bow shaped dynamometers (Fig. 2) were designed and produced. These dynamometers can be detached from the frame with the exciter for the transport purposes. On the railway bridge the frame is raised with help of four screws, what allows to detach the wheel system and fasten all dynamometers. After lowering the frame with the exciter the dynamometers rest on the rails: two of the dynamometers on the one of them and the third on the other. Thus the whole construction is supported at three points, and bullets applied at the dynamometer – rail contact points allow for stabile support of the exciter's frame. To attach the dynamometers to the rails dedicated screw clamping rings were constructed (Fig. 2). They prevent the frame with the exciter from moving during the tests (falling off them due to vibrations).

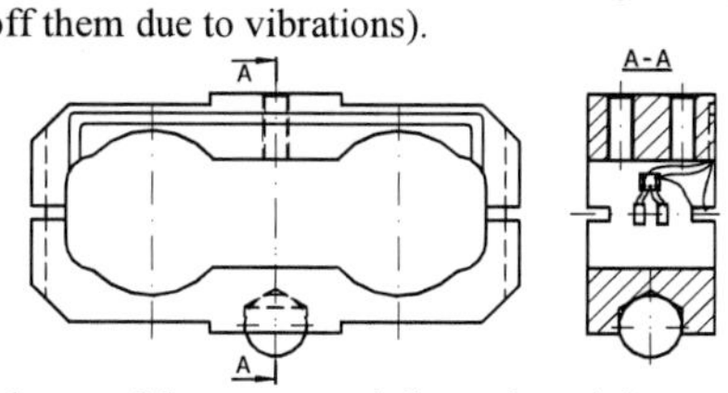

Fig. 2. Scheme of the tensometric bow shaped dynamometer

The driving system of the exciter's supporting frame contains two detachable axles, which consist of a tube with wheels mounted on it with sliding bearings. Both the axles with wheels were supported and mounted with screws to the channel irons, which were welded to the exciter's supporting frame.

3. LABORATORY TESTS OF THE EXCITER'S WITH SUPPORTING FRAME

A series of tests of the exciter mounted on the supporting frame were conducted at the laboratory. The whole structure together with dynamometers was mounted to the rails, which were screwed with clamps to the base plate made of cast iron (see Fig. 1). The tests were performed for three different exciter placements on the supporting frame: in its middle, and at both its edges. At the utmost positions of the exciter the resultant excitation force acted in the symmetry axis of the rail. The aim of the laboratory tests was to determine components of the excitation force at the support points of the frame while applying excitations of frequencies in the range 1–22Hz, as well as to check whether at any of these frequencies there occur whole system resonance with any of the excitation force component having opposite phase to the others, what can be observed when the system is swinging.

First of all the excitation force components at each dynamometer for different exciter placements and increasing frequencies were analyzed. As actual force values their RMS values on period of 20s were taken. When the exciter was mounted at the edge of the supporting frame above the single dynamometer (Fig. 3), the force acting on that dynamometer (no. 3) dominates the others. Regarding the characteristics of excitation force components dependence on the excitation frequency, which shows no influence of the resonant vibration of the supporting frame with the exciter on the force components values, the placement of the exciter over the single dynamometer seems to be the best. Moreover it displays additional advantage i.e. during railway bridges dynamical characteristics measurements it is enough to record only the single force component acting directly below the exciter, since the other components could be neglected.

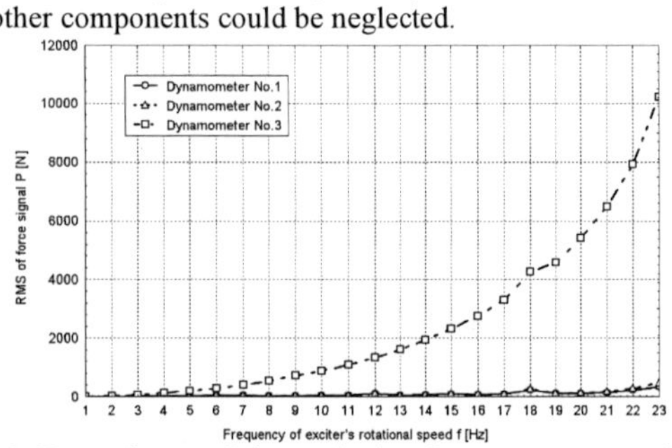

Fig. 3. Dependence of the excitation force components (RMS values) on the frequency for the exciter mounted at the edge of the supporting frame above the dynamometers no. 3

Estimating the value of the excitation force, which relies on calculating its RMS value, does not provide additional information about the force variation nature. Regarding the efficiency of bridge excitations it is important to assure that the exciter force the bridge vibration with a frequency set by the operating personnel. The amount of harmonic components in the excitation force should be as low as possible. Keeping in mind this claim all the force components for different excitation frequencies were analyzed. As expected at the lowest frequencies the force signal contains significant amount of harmonic components relatively to its base component (Fig. 4). While increasing the excitation force frequency the force signal approaches sinusoid, with almost no harmonic components. It is connected with force value increases. The usefulness of the exciter to excite vibration in bridge constructions results from both, the excitation force value as well as “clarity” of its shape. Due to this, one can consider application of the analyzed exciter to determine dynamical characteristics of bridge constructions for frequencies larger than 5Hz.

The supporting frame was also exposed to impact excitations in three directions, while the free vibration of the exciter placed in three positions on the frame were measured. First of all the lowest natural frequencies were analyzed. They were observed for the horizontal system excitations across the frame and varied from 18.75Hz to 6.6Hz (Fig.5). It was noticed the appropriate mode shape is expressed in frame swinging on the flexible dynamometers. Although the exciter's supporting system natural frequencies lied within the range of possible excitation frequencies its swinging did not influence the resultant excitation force for the exciter placement on edge of the frame, above the single dynamometer. The resonant vibration appearing at this position resulting from the system swinging do not influence this dynamometer signals due to its one-point supporting and the domination of the force component measured by this sensor.

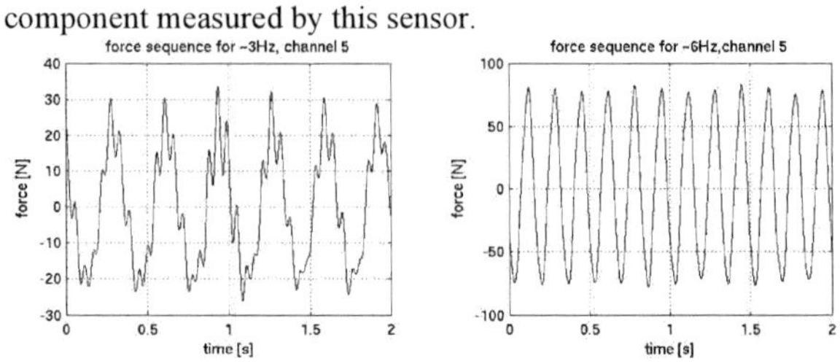

Fig. 4. Plots of the excitation force for the exciter mounted at the edge of the supporting frame over the dynamometer no. 3 at the frequencies of 3Hz, 6Hz, 10Hz and 14Hz

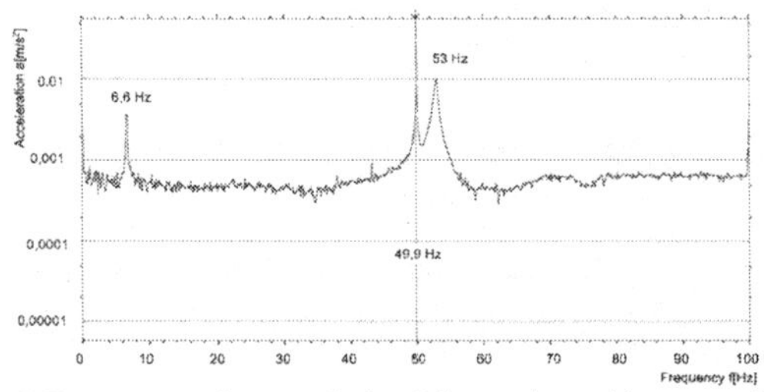

Fig. 5. Frequency characteristic of the exciter with supporting frame for the horizontal impact excitations across the frame

The next step of analysis lead to evaluate of a frequency distortion and a stabilization of exciting force amplitude while an actual exciter's parameters were adjusted. A special algorithm was prepared. No frequency deviations were observed using resolution of 0.125Hz. So the real exciting frequency was stable for each adjusted frequencies. The relative deviations of excitation amplitudes did not exceed 6% for 3Hz and they decreased with frequency rising.

4. CONCLUSION

The exciter together with a new construction of the supporting frame, adopted to its fastening to rails, fulfill all the functional assumptions. A set of three tensometric dynamometers allows for permanent measurement of the excitation force components during the railway bridges inspections. Placement of the exciter above the single dynamometer seems to be the best solution.

The exciter with the supporting system allows to obtain an excitation force with a stable frequency and the amplitude deviations no larger then 5% of nominal value. The deviations get smaller when the excitation frequency and resulting excitation force increase. The excitation contains the base harmonic signal with frequency set by an operator as well as the harmonic frequencies and measurement noise. Usually the amplitudes of the latter components were significantly smaller than the base component and thus their influence on the resultant force could be neglected. Application of the exciter to determine dynamical characteristics of railway bridge constructions is considerable for frequencies larger than 5Hz.

5. ACKNOWLEDGEMENTS

The work was co-funded by the European Commission within the Sixth Framework Programme Project No. TIP3-CT-2003-001653 entitled *Sustainable Bridges - Assessment for Future Traffic Demands and Longer Lives*

Annals of DAAAM for 2006 & Proceedings of the 17th International DAAAM Symposium, ISSN 1726-9679
Editor B. Katalinic, Published by DAAAM International, Vienna, Austria 2006
150th Birthday of Nikola Tesla (1856-1943) Inventor of Mobile Robot Teleoperation

THE DIAGNOSTIC OF INDUCTION MACHINES WITH USING METHOD OF MEASURING MAGNETIC FIELD

SOKANSKY, K.; CHMELIK, K.; MISAK, S. & FOLDYNA, J.

Abstract: *The technical diagnostic is process for determination of functional state of diagnostic's devices. The functional state of these devices is necessary to determine with respect veracity and accuracy. Vibrating diagnostic and diagnostic using harmonic analysis of windings currents are methods for determination and identification faults of induction machines. Each of those methods is useful for a specific diagnostic area. With using vibrating diagnostic it is possible to diagnose faults of mechanical character as an overbalance, bearing faults, dislocation of rotor and stator axis and so on. For determination of faults of mechanical and electrical characters is applicable diagnostic method harmonic analysis of windings currents. Interference of magnetic field in the air gap of induction machines is common feature of mechanical and electrical faults. Field incidence between stator and rotor creates mentioned changes of magnetic field in the air gap. Incurred exciting field evokes response at vibrating spectrum. The change of magnetic field evokes deformation of stator current too. Static and dynamic eccentricity, interruption of rotor cage, short-circuit of stator's winding, short circuit or released armature stampings belongs to faults evocatory change of magnetic field in the air gap of induction machines.*
Key words: *The Technical Diagnostic, The Magnetic Field, The Surge Generator, The Squirrel Cage Induction Machine*

1. INTRODUCTION

Three balanced alternating voltages applied to the stator cause balanced stator currents to flow. These currents produce a rotating mmf that can be represented as a rotating magnetic field. This rotating magnetic field induces voltages in the rotor windings, by Faraday's law. These induced voltages, in turn, cause balanced currents to flow in the short-circuited rotor. These rotor currents then produce a rotor mmf, which can also be represented as a rotating magnetic field. The interaction of these two rotating magnetic fields produces an electromagnetic torque T_e, which is used to turn a mechanical load T_m. At steady state, when the motor losses are neglected, T_m and T_e are equal.
The speed of rotation of the rotor magnetic field, when viewed from a stationary position on the stator, is equal to the speed of rotation of the stator magnetic field, which is the synchronous speed. However, the rotor speed is different from the synchronous speed. If the rotor speed were equal to synchronous speed, there would be no variation of flux linkage in the rotor, or there would be no net flux cutting; hence, no voltage would be induced in the rotor. Therefore, rotor speed has to be lower than synchronous speed. In the ideal case, the course of flux density is sinusoidal in space and time. In fact, the magnetic field in the air gap is deformed by over-saturation of sprockets as well as final number of stator and rotor slots. These deformations are sources of all spectrum harmonics, so these are visible as a modulation of stator current. Current harmonics induce additional losses.
The saddle form of speed-torque characteristic is caused by current harmonics too.
It stands to reason, that the magnetic field in the air gap does not have an ideal form, although parameters of motor are corresponding to the relevant norms.
Diagnostic method measuring magnetic field of induction motor is based exactly on principle of evaluation of the form of the mentioned magnetic field. The results from this diagnostic method are introduced in this manuscript. The results of the mentioned diagnostic method are compared with the results from diagnostic method by using surge wave.
These diagnostic methods are applied to determinate faults of the squirrel cage induction motors. The fault of stator windings (such as short circuit winding) was assumed.

2. THE MAGNETIC FIELD OF INDUCTION MACHINE

The magnetic field excited by windings of induction machine is induced outside of motor's cage too. This magnetic field is called scattered and its amplitude is depending on:
The form of the motor's cage (form of motor's fin, form of motor's cap, etc.), • the number of poles, • running state of the machine, • used constructional material (material of cage, material of ventilator case, material of motor terminal board, etc.), • the symmetry of magnetic and electric circuit form (separated magnetic circuit, etc.). It is possible to use sensors using variety of physical principles for measuring of the magnetic field. To choose the optimal principal of measurements it is necessary to reflect some aspects, such as: • the range of the measured field, • the velocity of the fluctuating of the measured field, • the direction of the measured field. For measuring it is also possible to use:• magnetic-optical probe for measuring magnetic field in range (1-10)T, field of superconducting magnet, and pulsed inductor, • Hall-effect sensor for measuring magnetic field in range mT-T, magnetic field at conventional magnetic circuits, • ferromagnetic probe or resonance method for measuring of magnetic field in range pT-mT (magnetic filed of Earth (50 μT), its variation and anomaly (tens of nT) as well as bio-magnetic field (pT), • etc.
The diagnostic method of measuring scattered field was verified on squirrel cage induction machines 1,1 kW with the suspicion of fault of stator's windings. The measurement was realized according to the scheme in Fig. 1. Within this manuscript there are introduced methods for determination and localization of faults of stator windings by measuring scattered field of induction motor. The 3D H/E field-meter ESM100 is used for measuring of scattered field. The 3D H/E field-meter ESM-100 is a patented, unique hand held measuring instrument. It can easily measure electric and magnetic alternating fields simultaneously, independent of the direction of the field. With this equipment it is possible to make quick, professional and error-free measurements for anybody right from the beginning. The scattered field of stator's windings was measured at horizontal and vertical direction. Absolute TRUE RMS value of flux density was determined with using field-

meter ESM100 such as vector sum three dimensional values (See equation 1).

$$H_{\text{FIELD}} = \sqrt{H_x^2 + H_y^2 + H_z^2} \qquad (1)$$

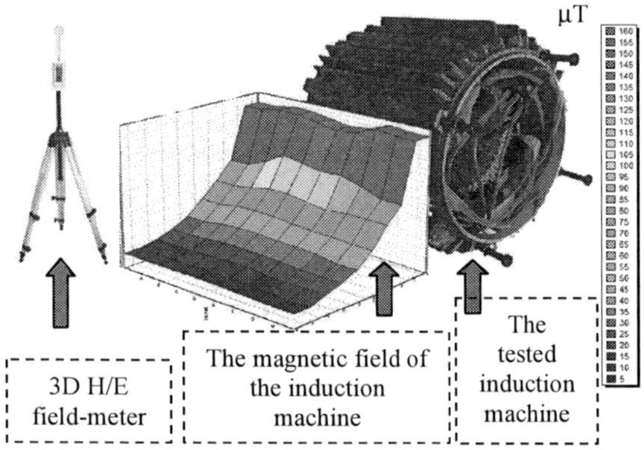

Fig. 1 The measuring of magnetic field of squirrel-cage induction machine

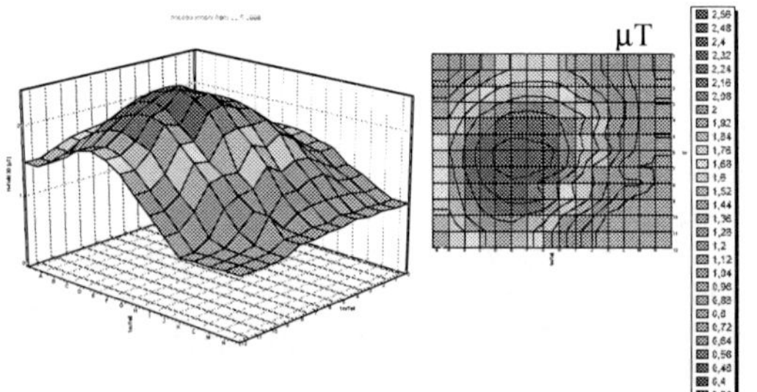

Fig. 2 The 3D and 2D Magnetic field of stator's windings without faults

Magnetic field of squirrel cage induction machine without faults is shown in Fig. 2. Magnetic field was measured while the stator was placed in horizontal position; probe of field-meter ESM100 defined flux density on surface of cage. The size of basic grid was selected 25x25 mm. Amplitude of flux density was in range 1,2-2,56 μT, form of equipotential lines is corresponding with magnetic circuit of induction machines. Therefore the maximal value of flux density is in the area of coil group and the field is decreased with the increasing distance from this coil group (See 2D magnetic field in Fig. 2).

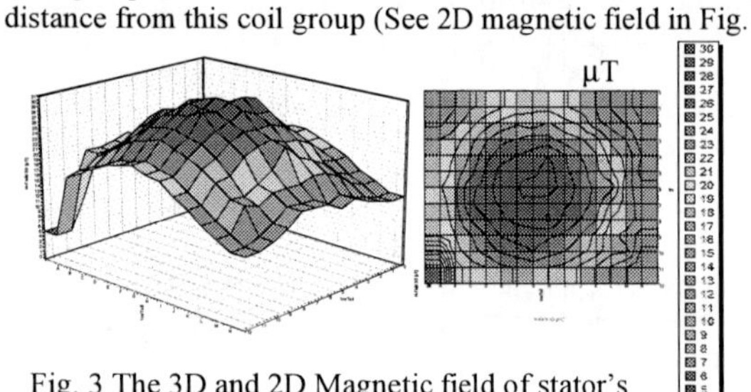

Fig. 3 The 3D and 2D Magnetic field of stator's windings with fault

In case of damaged stator windings the situation is different (See Fig. 3). In this case is typical not only increasing value of flux density (from 2,56 μT without fault to 30 μT with fault) but the form of equipotent lines is deformed too. The expressive flexure in the area of coil group (25-30 μT) is evident from 2D measured magnetic filed in Fig. 3. The insulated area with constant high flux density (c. 30 μT) is visible in the 2D measured magnetic field in Fig. 3 and this is typical for fault of stator windings, such as short-circuited turn of stator windings. This assumption was verified after demounting and analyzing of the stator.

3. THE TEST OF STATOR WINDINGS WITH USING SURGE GENERATOR PSG 215A

The principle of the method is following: two inductors of stator windings are connected to source of voltage pulses for a very short time (μs). Impulsive tension between coils is created by voltage gradient too. Each of the inductors has its own explicit character of response on voltage pulse. The character of response is dependent on insulating state of winding against earth or insulation state of winding between phase windings. This diagnostic method belongs among D.C. methods of testing. The amplitude of tested voltage is determined as a double of nominal voltage plus one kilovolt.

As mentioned above in the introduction of manuscript, the use of surge generator was verified for determination of fault of stator windings squirrel cage of induction motors.

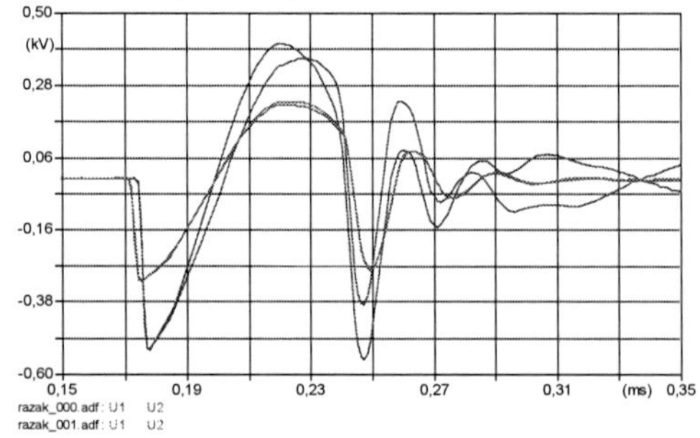

Fig. 4 The response on voltage pulse,
U_1, U_2 - windings without fault, U_1, U_2 - damaged windings.

The response on voltage pulse at inductor for windings with and without fault is introduced in Fig. 4. Courses of responses are identical for undamaged windings, (see U_1, U_2), in case of the tested squirrel cage induction 1,1 kW motor is situation different. From courses of responses (see U_1, U_2) there are evident differences, these are typical for fault of stator windings such as short-circuited turn. Therefore, results from this method confirmed deduction of previous analyze.

4. CONCLUSION

The use of more diagnostic methods and comparisons of results from these methods is needed for determination of functional state of every device. Obtained results are authentic in case of their correspondence, which is evident from the results introduced in this paper. Faults of stator windings, in the concrete short-circuited turn, were determined by method of measuring magnetic field as well as by method using surge generator. Each of these methods has its pros and cons. The method of measuring magnetic field is suitable for technical diagnostic as an on-line method. A determination of concrete type of fault is very difficult. Therefore, this method is applicable as a comparative method. On the other hand, determination of concrete type of fault by the method using surge generator is much greater. However, this method is impossible to use as an on-line method. The development of diagnostic method based on principle of measuring magnetic field of induction motors is realized at present.

5. REFERENCES

Chmelík, K.; Misak, S. & Foldyna, J. (2006). *Diagnostika asynchronních strojů*, Slovenská Poľnohospodárská Univerzita v Nitře, ISBN 80-8069-707-8, Nitra, Slovakia

Annals of DAAAM for 2006 & Proceedings of the 17th International DAAAM Symposium, ISSN 1726-9679
B. Katalinic, Published by DAAAM International, Vienna, Austria 2006
150th Birthday of Nikola Tesla (1856-1943) Inventor of Mobile Robot Teleoperation

ANTI-SLIP CONTROL SYSTEM OF A LOCOMOTIVE

SPIRYAGIN, M.; VIVDENKO, Y.; SPIRYAGIN, V. & KLIPAKOV, M.

Abstract: *This article describes the anti-slip control system of a railway vehicle. The proposed system has two control stages. This system works based on adhesive force as a main criterion, which provides a robust working locomotive with effective adhesion between wheels and rails.*
Key words: *control system, railway vehicle, adhesive force, torque, slip.*

1. INTRODUCTION

Improvement of tractive and braking efforts of railway rolling stock is a very difficult problem. For the solution of this problem it's necessary to provide a composite of design and technological efforts, which will require the investment of funds. What's more, the realization that tractive and braking efforts is connected with questions of resource savings and train driving safety.

The analysis of worldwide experiences confirms that there is a real possibility of solving this problem with application of microprocessor control systems to produce a new generation of tractive railway rolling stock.

One of the important parts of a microprocessor control system for a railway vehicle is the control method for their accurate and reliable work. A very good review and analysis in the field of control systems for railway vehicles was made by Frylmark & Johnsson in 2003. From this ,we can see, that control methods are not absolutely perfect for the attainment of the most effective utilization of adhesion between wheels and rails (Gadjar et al., 1997; Park et al., 2001; Watanabe &Yamashita 2002) Moreover almost all of control algorithms have the same input signals – velocity of a railway vehicle and the adhesion. The difference is how they work with these input signals.¶

2.PROPOSED CONTROL SYSTEM

Based upon what is written above, the authors propose a variant for anti-slip control system of a locomotive. The proposed system has two control stages.

The first stage is built upon a determination of a value of adhesion force. The proposed control system is shown in Figure 1.

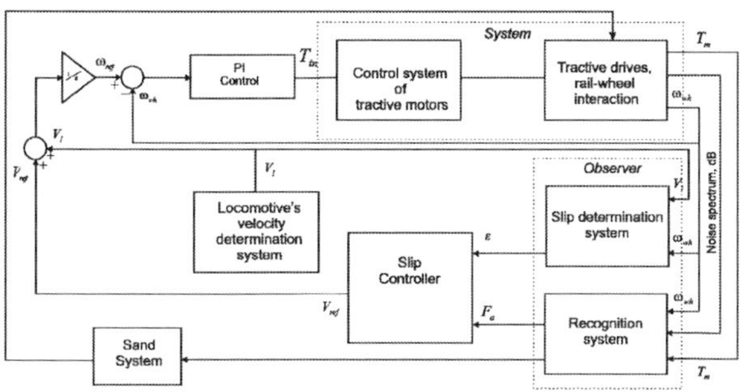

Fig. 1. Block diagram of the proposed anti-slip control system.

The variables and parameters presented in Figure 1:

F_a – adhesive force; T_m – real motor torque; T_{in} – input motor torque; ω_{wh} – real angular velocity of a wheel; ω_{ref} – reference angular velocity of a wheel; ε – relative slip; V_l – real locomotive velocity; V_{ref}–reference locomotive velocity.

The main of the proposed anti-slip control system is the regulator of angular velocity for each wheel pair. The assigned value of angular velocity makes it possible to shift a characteristic of the asynchronous three-phase motor; therefore it's possible to set a needed point on any part of a tractive characteristic. To reach the most effective adhesion between wheels and rails, the needed point should be set on a maximum of the tractive characteristic. What's more, it's necessary to say, when the velocity of a railway vehicle is changed, we have a shift of a tractive characteristic. So, it is necessary that a characteristic of tractive drive should automatically follow to the shift for setting a work point to maximum of a tractive characteristic.

Therefore, we can see that it's important for a slip control to set a rational speed depending on a value of adhesive forces in wheel-rail contacts.

The determination of adhesive force F_a is possible with the help of getting a value of relative slip and real characteristics of wheel-rail interaction. We can use the following mathematical relation:

$$\vec{F}_a = P\frac{\vec{S}}{|\vec{S}|}/\begin{pmatrix} A_1/\exp(\varepsilon\cdot B)+A_2\ln(\varepsilon\cdot B)+ \\ +A_3/(\varepsilon\cdot B)+A_4(\varepsilon\cdot B)+A_5 \end{pmatrix} T_2T_4T_9/T_7T_8 \quad (1)$$

where A_1, A_2, A_3, A_4, A_5 – coefficients of relation; $\vec{S}$ - vector of wheel slip; ε - relative slip, [%]; $B=T_1T_3T_5T_6$; T_1, T_2 – coefficients, which are depended on friction conditions and noises in rail-wheel contact; T_3, T_4 – vertical load coefficients; T_5 – velocity coefficient depended on a velocity of a locomotive; T_6, T_7 – coefficients, which are depended on cross motion of wheel relatively rail; T_8 – coefficient of angle of attack of the wheel; T_9 – braking mode coefficient.

This relation is confirmed by extensive experimental research. This research was made at Railway Department of Dal's East Ukrainian National University (Golubenko et al., 2000).

This relation allows computation of maximum adhesion forces for tractive and braking mode, which makes it possible to effectively use the anti-slip control system for these two modes of motion for a locomotive.

Coefficients T_1 and T_2 could obtained from data received from tractive drive as a feed-back control signal. As distinct from (Patent UA65187 A B60T8/58), we can correct these coefficients, the correction data depends upon noises in rail-wheel contacts. To acquire a noise spectrum, we should mount directional microphones to scan for noises.

To determine the values of V_{ref} and T_{in} we should use a method, based on that of a peak of adhesive curve is reached on the assumption of equaling to null for a coefficient of tilting for slip curve k :

$$k = \frac{\partial \mu_a}{\partial \varepsilon} = \frac{1}{N}\frac{\partial F_a}{\partial \varepsilon}, \quad (2)$$

where $\mu_a = \frac{1}{N} F_a$ – adhesion coefficient,

N – normal force.

Normal force N can be calculated:

$$N = m_a g, \quad (3)$$

where m_a – adhesive mass of bogie,

g – gravity acceleration.

Write our relation (1) in the following form:

$$\vec{F}_a = P\frac{\vec{S}}{|\vec{S}|} / \xi(\varepsilon) T_2 T_4 T_9 / T_7 T_8, \quad (4)$$

where

$\xi(\varepsilon) = A_1 / \exp(\varepsilon \cdot B) + A_2 \ln(\varepsilon \cdot B) + A_3 /(\varepsilon \cdot B) + A_4(\varepsilon \cdot B) + A_5$.

We should use scalar F_a, which is equalled to module $\vec{F}_a$:

$$F_a = P\tau / \xi(\varepsilon) T_2 T_4 T_9 / T_7 T_8, \quad (5)$$

where τ – constant, equaled to 1 or -1. The value is equaled to1 in case of if vector's direction $\vec{S}$ is the same with a direction of movement. In the other case the value of τ is equaled to -1.

After that, we should calculate $\frac{\partial F_a}{\partial \varepsilon}$ as:

$$\frac{\partial F_a}{\partial \varepsilon} = P\tau\varphi(\varepsilon) / \xi^2(\varepsilon) T_2 T_4 T_9 / T_7 T_8, \quad (6)$$

Where

$$\varphi(\varepsilon) = -A_1 B / \exp(\varepsilon \cdot B) + A_2 / \varepsilon - A_3 /(\varepsilon^2 \cdot B) + A_4 B.$$

In this case:

$$\frac{\partial \mu_a}{\partial \varepsilon} = \frac{1}{N}\frac{\partial F_a}{\partial \varepsilon}. \quad (7)$$

From (1),

$$V_{s,ref}(t+1) = V_{s,ref}(t) + \alpha k, \quad (8)$$

where α – variable depended on the value of k.

After that, we can calculate:

$$V_{ref} = \omega_{wh} r - V_{s,ref}(t+1). \quad (9)$$

As result:

$$\omega_{ref} = \frac{V_{ref}}{R}. \quad (10)$$

After getting of ω_{ref}, we can calculate the value of reference motor torque by means of using of a PI-controller:

$$T_{ref} = K_p(\omega_{wh} - \omega_{ref}), \quad (11)$$

where K_p - proportional gain of a PI-controller. The calculated moment we should use as the input motor torque T_{in}.

This ensured that the assigned tractive force has a maximum value. It means that a tractive drives of locomotive work in normal modes.

We can see from Fig.1 that the proposed system can limit a tractive torque by a locomotive's driver. It guaranteed, that for good adhesive condition tractive or braking efforts couldn't exceed the assigned force inputted by a locomotive's driver. In this mode the regulator of rotational speed works as a rotation limiter.

In the case, when it's impossible to reach good characteristics of adhesion in rail-wheel contact, we can use our second control stage. The second stage is based on the use of fuzzy logic.

The set of logical rules are presented in Table 1.

R1	IF	$T_m \leq 0{,}3..0{,}5T_{in}$ for a tractive mode	THEN	sand system starts working
R2	IF	$T_m \leq 0{,}2..0{,}3T_{in}$ for a braking mode	THEN	sand system starts working

Table 1. The set of rules for the slip controller.

The proposed conception of using the sand system of a locomotive allows for provision of a compromise between the improvement of adhesive conditions and the consumption of sand. We can see that in the braking mode we have the lowest relation, which gives a possibility to more accurately apply the brake even in bad adhesive conditions. In tractive mode, we can save sand, it also allows for clean working surfaces of wheels during motion with a peak adhesion.

For accurate sand feeding we recommend to use a special nozzle of send system of a locomotive (Patent UA38096 A B61C15/10).

The evolution of the proposed system was performed on a mathematical model of a locomotive. We used some blocks, which imitated rail conditions and some sensors. As a result, we achieved a satisfactory control system.

3. CONCLUSION

The proposed system is able to provide the best performance control of adhesion in the tractive and braking modes. It gives the possibility of using tractive drivers effectively, reducing the wear of wheels and rails.

Finally, we should say – our proposed anti-slip control system of a locomotive has room for further improvements.

4. REFERENCES

Frylmark, D. & Johnsson S. *Automatic slip control for railway vehicles*, M.Sc. Thesis, Linköpings Universitet, Vehicular System, Dept. of Electrical Engineering, 581 83 Linköping, Sweden, 2003.

Gadjar T., Rudas I. & Suda Y.. *Neural networks based estimation of friction coefficient of wheel and rail//* 1997 International Conference on Intelligent Engineering System. – September, 1997 – p.315-318.

Golubenko A., Kashura A., Gorbunov N., Spiryagin V., Spiryagin M. *Results of experimental research of friction properties for a wheel-rail interaction with a friction machine//* Omsk railway state university, Russia. - 2000. - p. 274-277.

Park D-Y., Kim M-S., Hwang D-H., Kim Y.-J. & Lee J-H.. *Hybrid re-adhesion control method for traction system of high-speed railway//* IEEE, Proceedings of the Fifth International Conference on Electrical Machines and Systems. - August, 2001. - 2. -p.739-742.

Watanabe T. &Yamashita M.. *Basic study of anti-slip control without speed sensor for multiple motor drive of electric railway vehicles//* IEEE, Proceedings of the Power Conversion Conference. – April, 2002. - 3. -p.1026-1032.

Annals of DAAAM for 2006 & Proceedings of the 17th International DAAAM Symposium, ISSN 1726-9679
B. Katalinic, Published by DAAAM International, Vienna, Austria 2006
150th Birthday of Nikola Tesla (1856-1943) Inventor of Mobile Robot Teleoperation

METHODOLOGY FOR APPRAISAL OF THE PARTNER ENTERPRISES FINANCIAL STABILITY WITHIN THE ENTERPRISES NETWORK

STANCIK, P. & KLEINOVA, J.

Abstract: *Enterprise is having more network character and the original meaning of the network enterprise is getting new dimension. We are talking about the relations not only among the enterprises but also among other institutions with the purpose of collective effect. Measurement is the key problem for cluster to ensure the present state and for other cluster members to compare members among them and with competitors and to measure cluster suppliers and customers.*
Key words: *Performance, Cluster, Financial Analysis, Methodology*

1. INTRODUCTION

Currently we have to quantify other, till now non-specified parameters, which are necessary for information about the development of cluster and of cluster members as well. The cluster and its management need the creation of methodology for finding present and supposed future state with the help of public accessible data about the individual cluster members and accessible data of cluster as well.
Creation of methodology for measurement of potential cluster members from the view of cluster preferability requires careful setting of individual indicators and parameters. Therefore the analysis of financial analysis methods was carried out. Because we are working on the field of complex measurement of the economic subjects, the analysis of systems for performance measurement of companies was carried out as well.

2. INITIAL THEORETICAL CONSIDERATIONS CONCERNING THE DEVELOPMENT OF THE EVALUATION METHODOLOGY OF THE CLUSTER MEMBERS PERFORMANCE

Measurement of the enterprise performance or (if you like) the enterprise processes, is also known as the Performance Measurement (PM). PM represents a style of the organization management based on the management of targets, both financial and non-financial. These targets are referred to as the key performance indicators (KPI – Key Performance Indicators). The Performance Measurement links various specific pieces of information in relation to the performance indicators. The Key Performance Indicators are such indicators, which have to be met in order to meet enterprise targets.
At present, there are many different systems of the Performance Measurement type. In this article, we shall be dealing only with the most significant of them. They are:

- Balance Scorecard;
- European Foundation for Quality Management;
- National Award of Malcolm Baldrige;
- Value Based Management; and
- Throughput Accounting.

A basic comparison of individual methods is given in the table below:

	Place of origin	Date of origin	Target	The method of realization
Balanced Scorecard	USA	1992	Target performance	Coordination of long-term and operative system targets
EFQM	Europe	1989	The best evaluation	Audit
National Award of Malcolm Baldrige	USA	1987	The best evaluation	Audit
Value Based Management	USA	The nineties	Maximum market value	Coordination of long-term and operative system targets
Throughput Accounting	USA	The eighties	Maximum profit	Maximum flow

Fig. 1 Comparison of the selected Performance Measurement systems

3. THEORETICAL CONSIDERATIONS CONCERNING DEVELOPMENT OF A METHODOLOGY FOR EVALUATION OF CLUSTER MEMBERS PERFORMANCE

The development of an evaluation methodology of potential cluster members, which would be beneficial to the cluster concerned, requires careful defining of individual indicators and parameters. This is the reason to undertake an analysis of methods of the financial analysis at the beginning. Because we are working within the area of a complex evaluation of the economic subjects, we should also undertake an analysis of systems for performance measurement. The Balanced Scorecard (BSC) principle will be used in the development of the methodology. Methodology is formulated on the BSC principle for these reasons:

- With the aspect of both internal and external benchmarking.
- With the aspect of simplicity for cluster management, that will measure benchmarking comparison of cluster members and benchmarking of individual clusters.
- With the aspect of simplicity and lucidity for company management (company = cluster member).
- With the aspect of simplicity and lucidity for company employees (company = cluster member).
- Input information is easily accessible in every company.
- With analogical BSC approach from the reason of cluster member measurement and cluster measurement in the four perspectives – it is not only the financial measurement but it is a complex view on company.
- With the big emphasis on symmetry between financial and non-financial indicators.

- For possibility of measurement of any type of company without the aspect of company size and legal form of a measured company.
- With the aspect of measurement of members that will create the added value for cluster and with the aspect of measurement of members that will deliver for cluster.
- With the aspect of individual perspectives measurement with the help of gravimetric coefficient for priority determination for individual indicators.
- For measurement of more types of cluster members and with this aspect weights for individual indicators will be adjusted.
- For possibility of quick result analyzing and searching causes in individual perspectives (advantage over against the systems of timely caution).
- The result measurement of cluster members is carried out by the order method.
- For the software processing and for quick measurement and easy utilization by the common user.

The goal is to make a proposal for such an evaluation methodology, which enables relatively fast identification of status and management of relevant enterprise on the base of indicators system, which would cover individual areas of management of the economic subjects. The methodology proposed is depicted in Fig. 2.

Financial area	*Area of supplier-customer relations*
Portion of the own capital *Performance* *Profitability of total capital* *Period of extinction of a debt (in years)* *Available liquidity*	*Debts turnaround time* *Liabilities due date to the suppliers* *Share of claims* *Turnaround time of finished products reserves*
Area of internal enterprise processes	***Personnel area***
Turnaround time of production reserves *Material intensity* *Waste rate* *Process performance* *Share of new products revenues on the total revenues* *Investment in research and development*	*Personnel intensity* *Productivity of labour* *Extensive utilization of working time* *Added value per employee*

Fig. 2 The methodology proposed, proportional indicators applied

Financial area

Portion of the own capital

Portion of the own capital = Own capital/ whole capital

Performance

*Performance = (Cash Flow/Revenues)*100*

Profitability of the total capital

Profitability of the Total Capital = Profit + Interest Cost / Total Capital

Period of extinction of a debt in years

Period of extinction of a debt = (Loan Capital – available means)/Net Profit + Depreciation

Available liquidity

Available Liquidity = (Short-term assets – Reserves) / Current liabilities

Area of Supplier - Customer Relations

Debts turnaround time

Debts turnaround time = Average State of Debts x 360 /Revenues

Liabilities turnaround time to suppliers

Liabilities turnaround time to suppliers = Current Liabilities from Business Relations x 360 / Purchases

Share of claims

*Share of claims = Number of Claims / Total Number of Pieces Supplied * 100*

Turnaround time of finished product reserves

Turnaround time of finished products reserves = Averages State of Reserves x 360 /Revenues

Area of internal enterprise processes

Turnaround time of production reserves

Turnaround time of Production Reserves = Average State of Reserves x 360 / Total Production Costs

Material intensity of production

Material intensity of production = Material costs /Production volume

Waste rate

*Waste rate = (Costs for Wasters / Production Volume) * 100*

Production process performance

Pass time = Process Time + Checkout Time + Transfer Time + Storage Time

Performance = Process Time / Pass Time

Share of new products revenues on the total revenues

*Share of new products revenues on the total revenues = (New products revenues / Total Revenues) * 100*

Investments in research and development

*Investments in Research and Development = (Investments in Research / Revenues)*100*

Personnel area

Personnel intensity

Personnel Intensity =Personnel Costs / Volume of Production

Productivity of labour

Productivity of labour per hour = Volume of Production / Number of Delivered Working Hours

Productivity of Labour per 1 Labourer = Volume of Production / Number of Labourers

Productivity of Work per 1 Worker = Volume of Production / Number of Workers

Extensive utilization of working time

Extensive Utilization of Working Time = (Working Time Delivered / Available Time Fund)

Added value

Added Value per 1 Labourer = Added Value / Number of Labourers

Added Value per 1 Worker = Added Value / Number of Workers

4. CONCLUSION

The evaluation methodology can not be used as a tool which replaces experience of the management. However, it can serve the management as a support when making decisions as well as a tool for the management of cluster and individual cluster members. Methodology itself does not make decision, but it is supposed to support decision making of the manager. The evaluation system is set for each of four areas to include about five indicators, because the usage of more indicators would make the whole situation worse. This number of indicators is recommended also by authors of BSC. The developed methodology has been processed into a Software and has been experimentally applied to one cluster in the Czech Republic.

5. ACKNOWLEDGEMENT

The paper has been prepared with the support of the research project AV Czech Republic: 1ET 201450508.

6. REFERENCES

Brigham E. F., Houston J. F. (2004), *Fundamentals of Financial Management,* Thomson, ISBN 0-324-25872-0, South-Western

Kaplan R., Norton D. (2001), *Balanced Scorecard - Strategický systém měření výkonnosti podniku,* Management Press Praha, ISBN 80-7261-124-0, Praha

Annals of DAAAM for 2006 & Proceedings of the 17th International DAAAM Symposium, ISSN 1726-9679
Editor B. Katalinic, Published by DAAAM International, Vienna, Austria 2006
150th Birthday of Nikola Tesla (1856-1943) Inventor of Mobile Robot Teleoperation

CARBON FIBER ALUMINUM MATRIX COMPOSITE FOR ADVANCED AEROSPACE STRUCTURES

STANCIU, V.; DOBRE, T.; STOICA, A.; TACHE, F.; TOMA, A.C. & CHICIUDEAN, T.G.

Abstract: *Preliminary results of the design and manufacturing tehnique of a new metal-matrix composite are presented. Having aluminum matrix and carbon fibers as main constituents, the new CFRAl material (Carbon Fiber Reinforced Aluminum) will make use of its constituents' properties and enhance them to a yet unattained level. Also, the envisaged installation for producing this new material is briefly described.*
Key words: *carbon fiber, aluminum matrix composite*

1. INTRODUCTION

Carbon fiber composites are attractive structural materials in aerospace structures. A large amount of work has been done in connection with the design, construction and testing of these materials. There are two main types of composite structures which are employed: *monolithic stiffened* or *sandwich* (Mahdi et al., 2002). Mechanical properties of metal-matrix composites are conditioned by bulk properties of the constituents, but also by the mechanical behavior of the matrix – reinforcement interfaces. For taking advantage of the excellent mechanical properties of carbon fiber in composite materials, an optimal interfacial adhesion between fiber and matrix is required (Urena et al., 2005).

2. ALLYING THEORY

Aluminum alloys are reinforced to increase their mechanical properties while keeping their low density and good electrical and thermal conductivities. Carbon fibers are one of the most appealing reinforcements, reducing the friction coefficient of aluminum and improving its wear and tear resistance. The downside is that manufacturing problems have limited the industrial application of carbon fiber reinforced aluminum composites. Under the typical temperatures used for casting aluminum composites (*700–800°C*), the wettability of all carbon fiber by molten aluminum is limited, even for long time exposures, causing the segregation of the reinforcement from the matrix at the solidification and resulting in highly heterogeneous composites. Temperatures above 900–1000°C increase the wettability by molten aluminum due to the interfacial reaction between aluminum from the matrix and carbon fibers, but the reaction product is brittle aluminum carbide, which expands in humid environments, stressing and breaking the composite surface (Rams et al., 2006).
Therefore, short and continuous fibers are coated prior to allying with aluminum. One solution is to use metal coatings of refractory materials, because of their inertness towards both the molten metal and the carbon reinforcement, but these types of coatings are very expensive. The use of pyrolytic carbon around the fibers has also been suggested, because it improves the toughness of the interfaces. Finally, metallic coatings of nickel, copper or silver are used to improve the wettability of fibers by the molten aluminum alloys (Torralba et al., 2003).

In engineering designs, many structural components applied in the field of aerospace and transport demand for high specific stiffness, high specific strength and excellent damping capacity in order to stabilize the system. Therefore, of great interest for scientists is to develop new materials simultaneously exhibiting both good mechanical properties and high damping. Particular attentions have been paid to the unidirectional carbon fibers reinforced metal matrix composites (MMC), due to their high mechanical properties, low thermal expansion coefficient in the fiber direction, good electrical and thermal conductivity and potential of high damping capacity (Gu et al., 2004). Among these fiber composites, the fiber coating is commonly implemented for several reasons: *(1)* to change the overall composite properties, such as elastic properties, toughness, thermal expansion and thermal conductivity; *(2)* to affect the stress distribution around the fiber–matrix interface; *(3)* to improve the bonding quality between the fiber and the matrix; and *(4)* to form barriers against the interfacial reaction and diffusion.

3. STRUCTURAL PROPERTIES OF POROUS BODIES

A porous environment is considered to be a solid containing pores. A geometrical definition of the pore concept is quite difficult. Usually, pores are considered hollow spaces distributed inside the solid that can communicate or not with each other, being closed or open between faces. The communicating portion of a pore is commonly refered to as *effective pore space* or simply *effective porosity*. This term is largely used in filtering technology where porous bodies can have a high total porosity related to a low effective porosity. Total porosity is always superior to the effective one.
General (volume) porosity is defined as the ratio between pores volume and form volume:

$$\varepsilon = \frac{V_{pore}}{V_f}. \tag{1}$$

Together with *volume porosity* there is the concept of *surface porosity*, defined as the ratio between the pores ending area existing onto a sectioned surface of the porous form and the area of the sectioned surface:

$$\varepsilon_s = \frac{A_{pore}}{A_f}. \tag{2}$$

In a first approximation, the equality $\varepsilon = \varepsilon_s$ is acceptable.

The simplest model of the porous body is a system of short radius and equal forms particles (*monodisperse granular system*). Any *polidisperse granular system* can be fragmented into monodisperse granular systems by real or hypothetical granulometric analysis.

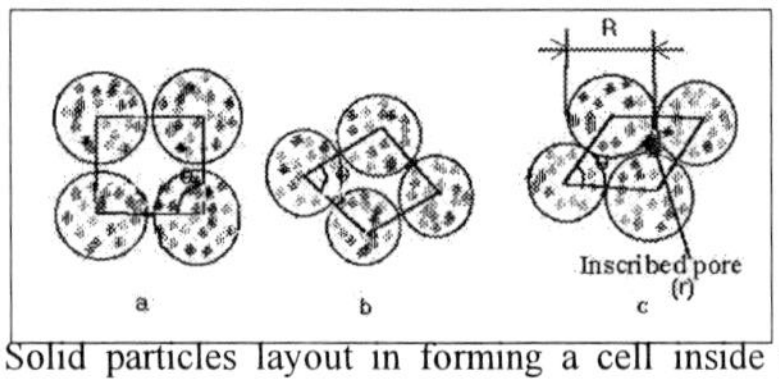

Fig. 1. Solid particles layout in forming a cell inside the porous layer (a – cubic layout $\theta = 90^o$; b – general layout; c – hexagonal layout $\theta = 60^o$)

Figure 1 shows that spherical particles can be arranged in different ways. The limits are a rare layout (*a*, cubic) and a dense layout (*c*, hexagonal). Between these two limits there are all real layouts.

4. MATERIAL POROSITY

Hollow spaces inside a porous material can be interior (closed) pores, exterior (open at both ends) pores and intergranular pores or unoccupied spaces. Monofilament fabrics and granular layers of non-porous particles generally have intergranular spaces. In operations involving flows through porous structures (filtering, capillary pumping etc.), the porosity corresponding to intergranular spaces and sometimes to open pores presents interest. Pore size distribution is also important, though pore "size" or "diameter" is a simplified notion, because pores are far from having cylindrical form. Most commonly, they have irregular forms with contractions and ramifications. This is why a porous material (layer) is preferably characterized by the notion of permeability. But there are cases, especially regarding membranes, that consider as important the size of the pores.

5. EXPERIMENTAL RESULTS

Preliminary samples of CFRAl have already been produced, using the sinterization process which has the advantage of allowing to modify the aluminum percentage within one composite section and also to modify the aluminum alloy composition used from one layer of the composite to another. This method can be used to obtain parts having, for instance, soft core and tough outer skin that would withstand corrosion, sablation and other hostile external actions.

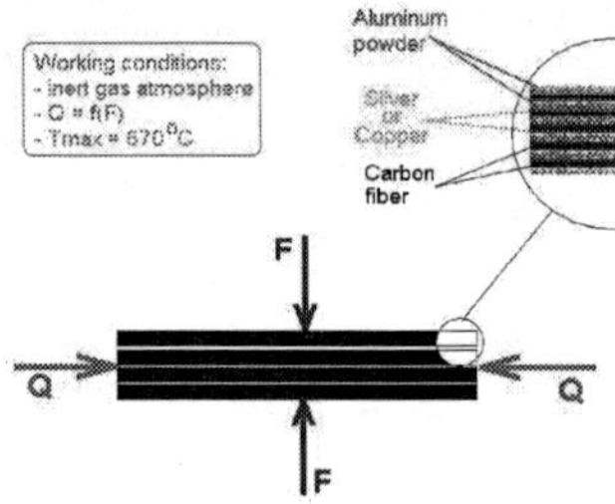

Fig. 2. Sinterization base method for obtaining CFRAl

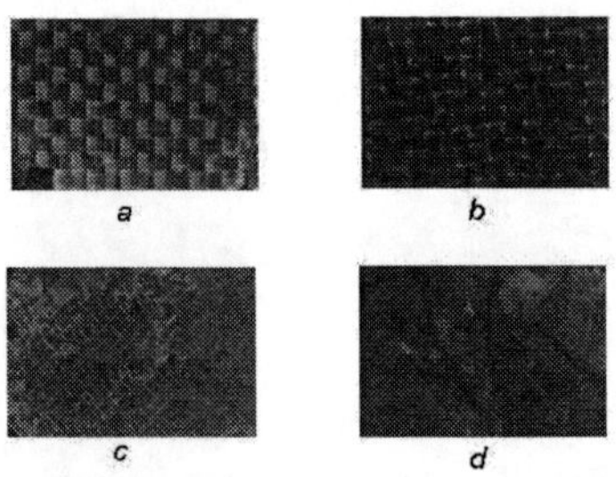

Fig. 3. CFRAl preliminary samples
a – 0/90° Carbon fibers fabric with high Al content;
b – 0/90° Carbon fibers fabric with low Al content;
c – +45°/-45° Carbon fibers fabric with medium Al content;
d – Aluminum reinforced with chopped Carbon fibers.

6. CFRAl PRODUCTION

The envisaged installation for continuously producing CFRAl as a laminated product is presented below.

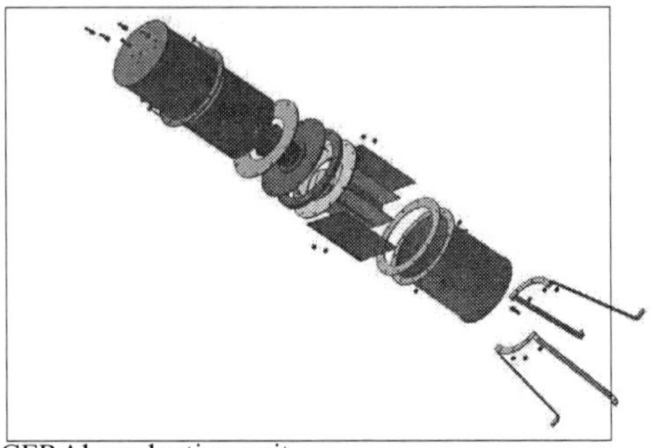

Fig. 4. CFRAl production unit

The actual manufacturing process of CFRAl using this unit is presented in Figure 5. The great advantage of this process is the continuous production of CFRAl in a relatively small facility. Limitations are only dictated by the length of the carbon fibers themselves and by the need to continuously supply aluminum in the upper melting unit.

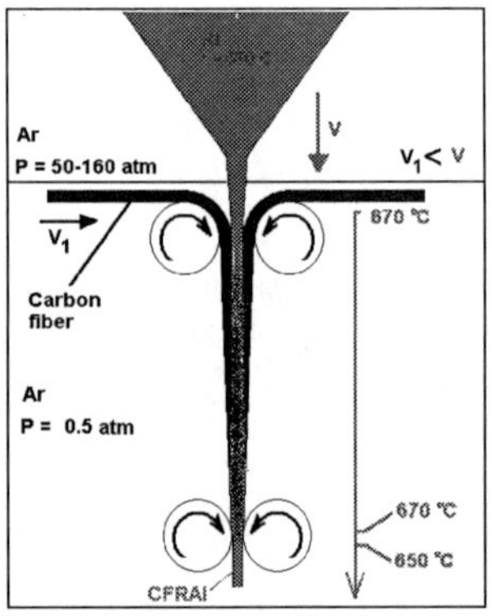

Fig. 5. CFRAl manufacturing technique

7. CONCLUSIONS

CFRAl is being developed by University 'POLITEHNICA' of Bucharest and the Romanian Space Agency, having applications that range from reinforcement of structures exposed to small-dimension high-speed bodies, such as cosmic dust, small-caliber bullets, to reinforcement of closed compartment walls in order for the construction to withstand the explosion of a bomb inside it.

This paper briefly presents the results of a Romanian national research and development grant on security issues regarding new materials for high risk applications that prove to have promising use in aerospace technology.

8. REFERENCES

Dobre, T. (2000). *Diffusion through porous materials (in Romanian)*, Electronic book, University 'POLITEHNICA' of Bucharest, 2000

Gu, J., Zhang, X., Gu, M., Zheny, L., Zhang, G. (2004) *Materials Letters 58,*3170-3171

Mahdi, S., Kinloch, A. J., Matthews, F. L., Crisfield, M. A. (2002) *Composites:Part B* 33, 343-354

Rams, J.,Urena, A., Escalera, M. D., Sanchez, M. (2006) *Composites*: *Part A* in press

Torralba, J. M., da Costa, C. E., Velasco, F. (2003), *J. Material Proc. Technol.* 133, 203-206

Urena, A., Rams, J., Escalera, M. D., Sanchez, M. (2005) *Comp. Sci. Technol* 65, 2025-2038.

Annals of DAAAM for 2006 & Proceedings of the 17th International DAAAM Symposium, ISSN 1726-9679
B. Katalinic, Published by DAAAM International, Vienna, Austria 2006
150th Birthday of Nikola Tesla (1856-1943) Inventor of Mobile Robot Teleoperation

CHIP-CHUNK RESISTANCE OF TIRE TREADS

STANEK, M.; MANAS, M.; DRGA, T. & MANAS, D.

Abstract: *Rubber (rubber mixture) is an indispensable material as an element for automobile because of its original characteristics such as big elongation (transformation) and ability of restoration which is unattainable by other materials. Rubber is the main raw material for production of automotive components such as tires, bushes, insulators and seals. All these products have to be safe and wear resistant. Good wear resistance is very important for tires. The article describes the testing possibilities of tires used in very hard traffic conditions (off-road tires).*
Key words: *rubber, rubber mixture, wear.*

1. INTRODUCTION

Wear of tire treads at road surfaces is measured as abrasion resistance. Off-road behaviour of tyre treads on surfaces with sharp stones is not well characterised by abrasion resistance as the mechanism of rubber damage is here rather different. The sharp edges of stones can cut a rubber tread surfaces and gradually tear-off bigger pieces of rubber (chips or chunks).

2. EXPERIMENTAL

The test for cutting and chipping of rubber compounds which correlates with service behaviour and gives test results with reasonable speed and accuracy was described by J. R. Beatty and B. J. Miksch in RCHT, vol. 55, p.1531. In accordance with their description we constructed rather modified apparatus. The enhanced laboratory apparatus where conditions of testing can be widely changed makes possible to measure different characteristics of chip-chunk processes.

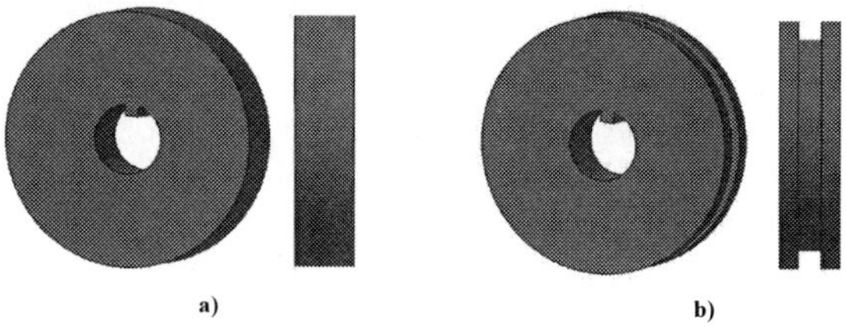

Fig. 1. Sample for chip-chunk test: a) before the test, b) after the test

Tests were carried out on cylindrical samples for Lüpke resilience test with diameter 55 mm and thickness 13 mm. The rotating vulcanised rubber cylinders were abraded by sharp edge mounted on beam lifted and dropped on the rubber sample perimeter by pneumatic cylinder. The samples were weighed before and after test. The samples were made from compounds designated for production of motocross, motoenduro fork lifter tires and off-road tires. The evaluation of wear progress during the testing period was tested as well.

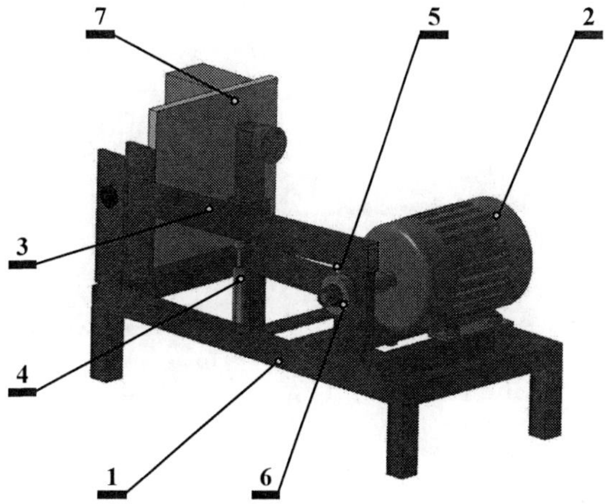

Fig. 2. Design of cutting and chipping tester
1 – frame, 2 –electric motor, 3 – beam, 4 – pneumatic cylinder, 5 – ceramic tool, 6 –rotating sample, 7 – control panel

Fig. 3. Samples wearing during the test period

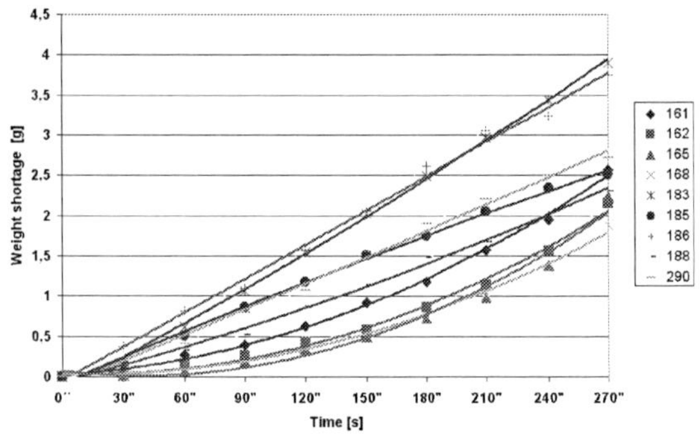

Fig. 4. Wear of samples during the test period

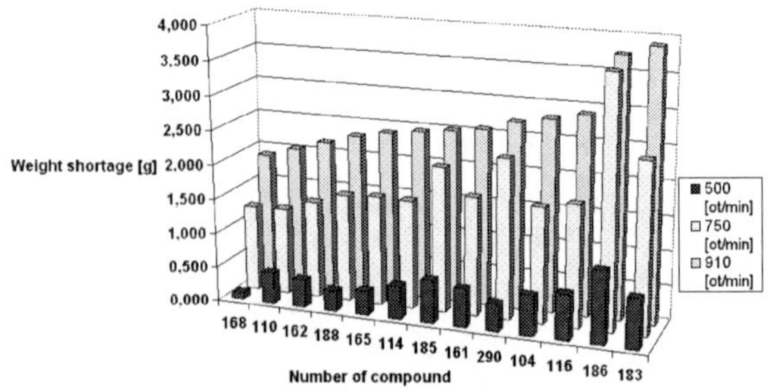

Fig. 5. Wear of samples by various conditions

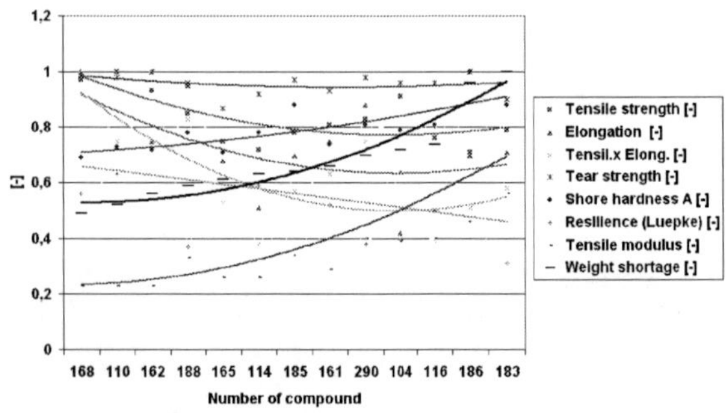

Fig. 6. Comparison of all measured properties

The measured dates were statically calculated and processed. For evaluation of measured dates the multiple linear regression was used. We used the linear statistical model of Chip – Chunk resistance in the form.

$$Y = \beta 0 + \beta 1\ X1 + \beta 2\ X2 + \beta 3\ X1\ X2 + \beta 4\ X3 + \beta 5\ X4 + \beta 6\ X5 + \beta 7\ X6 + \varepsilon \tag{1}$$

Where:
Y ... Weight Shortage [g]
Xi ... measured value of properties of rubber compounds, i.e.
βi ... material constants
X1 ... Tensile Strength [MPa]
X2 ... Elongation [%]
X3 ... Tear Strength [N/mm]
X4 ... Shore Hardness [ShA]
X5 ... Resilience [%]
X6 ... DMA (Tensile Modulus) [MPa]
βi ... regression parameters
ε ... error

By the least – square method we obtained the estimates of unknown parameters βi . The answered regression function takes the form:

$$Y=2{,}619273 + 0{,}034052\ X1 - 0{,}001868\ X2 + 0{,}000125\ X1\ X2 - 0{,}129815\ X3 + 0{,}116571\ X4 + 0{,}012328\ X5 + 0{,}000121\ X6 \tag{2}$$

(suitability of the model describes the determinacy index: R2 = 0,850893).

Correlation matrix	X_1	X_2	X_3
X_1	1,000000	0,352717	0,489553
X_2	0,352717	1,000000	0,273991
X_3	0,489553	0,273991	1,000000
X_4	-0,573941	-0,223167	0,136835
X_5	0,355879	-0,274672	0,121560
X_6	-0,522965	-0,131786	0,203524
Y	-0,595387	-0,320170	-0,207062

X_4	X_5	X_6	Y
-0,573941	0,355879	-0,522965	-0,595387
-0,223167	-0,274672	-0,131786	-0,320170
0,136835	0,121560	0,203524	-0,207062
1,000000	-0,622169	0,967055	0,832489
-0,622169	1,000000	-0,600251	-0,486060
0,967055	-0,600251	1,000000	0,782667
0,832489	-0,486060	0,782667	1,000000

Table 1. Correlation matrix.

3. CONCLUSION

Presented testing method shows the possibility of evaluation of wear (chip – chunk) resistance of tire treads on the small samples. This method makes possible to compare various types of compound with a standard and follow the progress of wear during the test period. The wear of sample during the test period depends on the properties of rubber compounds and testing conditions. Significant correlation of chip-chunk resistance with usual rubber properties was found. According statistical calculation all measured data are statistical significant.

4. ACKNOWLEDGEMENT

This article is financially supported by the Czech Ministry of Education, Youth and Sports in the R&D project under the title 'Modelling and Control of Processing Procedures of Natural and Synthetic Polymers', No. MSM 7088352102.

5. REFERENCES

Manas, D. (2005). *Rubber Workability and Wear of Rubber Parts*, VUT FSI Brno, ISBN 80-214-3026-5, Brno

Batty, J.R. & Miksh, B.J. (1989). *A laboratory cutting and chipping tester for evaluation off-the-duty tires treads*, Rubber chemistry and technology, vol.55

Manas, D.; Stanek, M.; Manas, M. & Dvorak, Z. (2005). *Off - Road Tires Behavior*, IRC 2005, Yokohama, Japan

Johnson S.J (2001). *Rubber processing*, Hanser Publishers, ISBN 3-446-21578-6, Munich

Dick, J.S. (2001). *Rubber technology. Compounding and testing for performance*, Hanser Publishers, ISBN 3-446-19186-0, Munich

Annals of DAAAM for 2006 & Proceedings of the 17th International DAAAM Symposium, ISSN 1726-9679
B. Katalinic, Published by DAAAM International, Vienna, Austria 2006
150th Birthday of Nikola Tesla (1856-1943) Inventor of Mobile Robot Teleoperation

POLYMER FLUIDITY TESTING

STANEK, M.; MANAS, M.; DRGA, T. & MANAS, D.

Abstract: *Injection molding is one of the most extended plastic processing technologies. Delivery of polymer melts into the mold cavity is the most important stage of the injection molding process. In the mold cavity the melt freezes and the product gain the final shape and form.*
Key words: *Injection mold, injection molding, polymer, fluidity.*

1. INTRODUCTION

Injection molding is one of the most extended plastic processing technologies. Delivery of polymer melt into mold cavity is the main stage of injection molding. The melt freezes in the mold cavity and the product gains the final form and dimensions. The injection molding process and solidification of polymer melt runs under non-isothermal conditions affected by complex of rheological properties of polymer.

2. INJECTION MOLDING

2.1 Injection molding process

Injection molding represents such a way of polymer processing that requires injection of polymer melt into the mold cavity with very high injection rate. This technology makes the production of high quality and precise products possible. The injection molding cycle time is relatively short and the process can be widely automated. In case of cold runner mold the sprues and runners can be reused.
The flow front therefore is like a fountain in which the most of center material flows out and deposits itself on the cavity wall. Plastic material does not slide along the mold surface but is instead rolled over much like rolling out a carpet. This type of laminar flow is usually described as a "fountain flow" (Fig.1).
It is necessary to fill the mold cavity fully during the injection molding process. The ability of cavity filling could be affected by the polymer properties (esp. melt temperature) and the properties of cavity walls especially mold temperature and surface quality of cavity or runners.

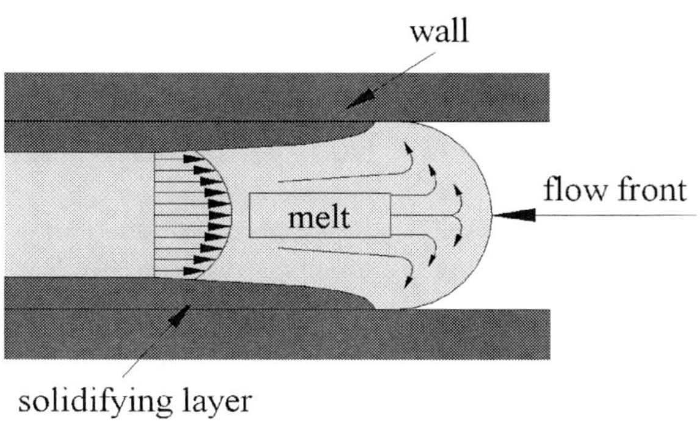

Fig. 1. Flow of polymer melt into mold cavity

3. EXPERIMENTAL

3.1 Injection mold

To be able to study the influence of surface quality of runners and cavities on the flow of polymer melts, a special injection mold has been designed and machined.
The mold is equipped by pressure and temperature transducers to be able to measure the really value of pressure and temperature in the mold cavity during the injection molding process. Length of testing samples has been measured by measuring jig. The experimental dates have been statistical processed and evaluated.
The design of this mold makes the exchange of the gates kind and size possible. The various types of gates come to the spiral channel (Fig.2) whose bottom is formed by exchangeable plates. The width of spiral is 6 mm, depth 1 mm (Fig.3) and length is 2000 mm.

Fig. 2. Spiral sample

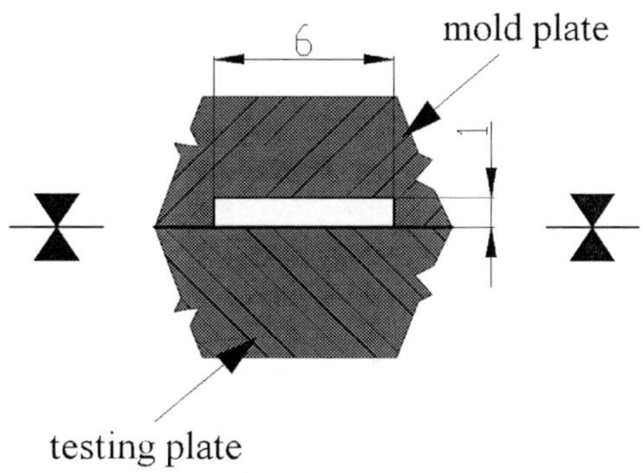

Fig. 3. Channel cross-section

The surface of the plates (bottom of spiral) is machined using various types of working technologies: milling, grinding, polishing and electro-spark cutting. The quantitative expression of surface quality of plates is given in the Table 1.

Tooling technology	Ra [µm] (contact method)	Ra [µm] (non-contact method)
Polishing	0,1020	0,4208
Grinding	0,1720	0,4530
Electro-erosion machining („fine“)	4,0550	4,3638
Electro-erosion machining („rough“)	9,5660	12,7434
Milling	4,4990	4,9996

Table 1. Tooled surfaces of testing plates.

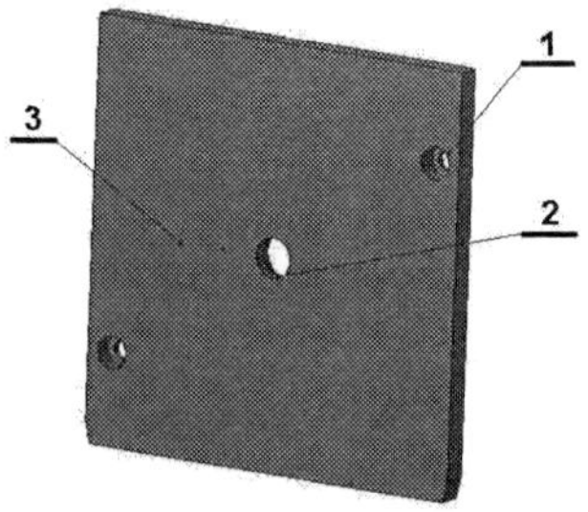

Fig. 4. Testing plate
1 – hole for sprue bushing, 2 – hole for screws, 3 – hole for temperature and pressure sensor

3.2 Injected polymers

- polypropylen (PP) Mosten GB 003, MFI = 3,3
- polyethylen (LDPE) Bralen VA 20-60, MFI = 20
- termoplastic polyurethan (TPU) Ellastolan C 78 A, MFI = 6,1
- akrylonitril-butadien-styren (ABS) Polylac PA 757, MFI = 2,4
- polypropylen filled by 20% of chalk Taboren PH 89 T20, MFI = 14,4
- polypropylen filled by 10% of chalk Keltan TP 7603, MFI = 16,9

4. EXPERIMENTAL

The major results are presented in the Figure 5-6. Better cavity (spiral) filling by materials with better flow properties was expected. Also the results obtained during injection moulding at various injection rates or under different pressures of polymer melts were expected. Unusual (not expected) behaviour of polymer flow using various surface qualities of spiral cavity walls was observed. Better quality of wall surface worsened the flow conditions (the length of injected sample was shorter).

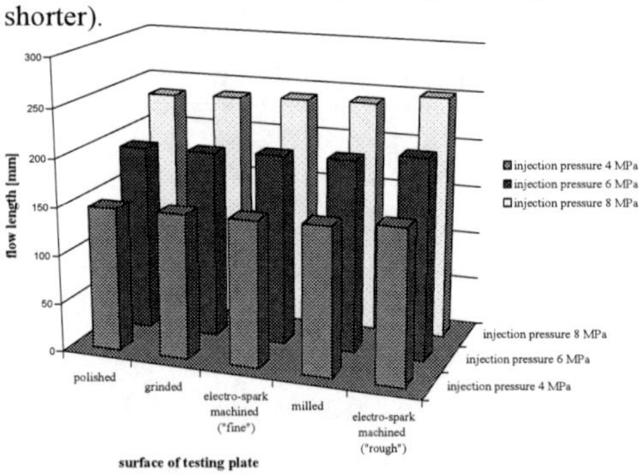

Fig. 5. Dependence of flow length on surface quality and injection pressure (Keltan, gate 6 mm, injection rate 30 mm.s^{-1})

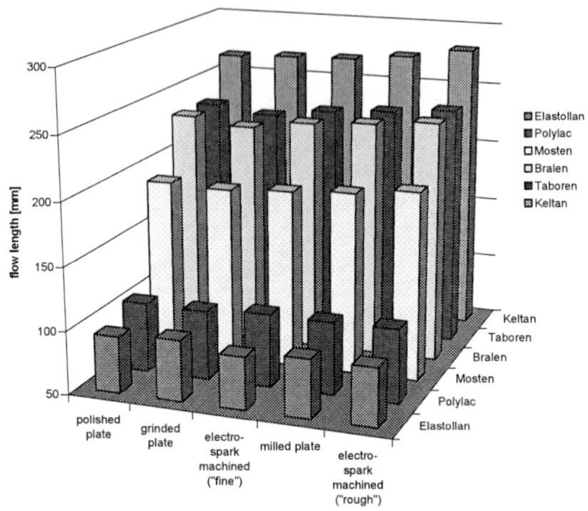

Fig. 6. Dependence of flow length on surface quality and type of polymer (injection pressure 8 MPa).

5. CONCLUSION

Measurement shows that surface quality has substantial influence on the length of flow. Samples which were injected into the spiral (cavity) with the worst surface quality have the longest length of flow.

For verification of these results further experiments have to be carried out using different polymers and various testing conditions.

6. ACKNOWLEDGEMENT

This article is financially supported by the Czech Ministry of Education, Youth and Sports in the R&D project under the title 'Modelling and Control of Processing Procedures of Natural and Synthetic Polymers', No. MSM 7088352102.

7. REFERENCES

Stanek, M. (2005). *Modelling of polymers forming process inc. tools design,* VUT FSI Brno, ISBN 80-214-3025-7, Brno

Agassant, J.-F.; Avenas, P.; Sergent, J.-Ph. & Careau, P. J. (1991). *Polymer processing,* Hanser Publishers, ISBN 3-446-14584-2, Munich

Beaumont, J. P. (2004). *Runner and gating design handbook,* Hanser Publishers, ISBN 3-446-22672-9, Munich

Lindner, E. & Unger, O. (2002). *Injection mold,.* Hanser Publishers, ISBN 3-446-21448-8, Munich

Manas, M., & Vlcek, J. (2001). *Applied rheology,* UTB Zlin, ISBN 80-7318-039-1, Zlin

Rees, H. (2002). *Mold engineering,* Hanser Publishers, ISBN 3-446-21659-6, Munich

Sykorova, L. & Samek, D. (2004). *Results evaluation of the plastics laser micro-cutting by artificial neural networks,* Nove smery vo vyrobnych technologiach, ISBN 80-8073-136-5, Presov

Annals of DAAAM for 2006 & Proceedings of the 17th International DAAAM Symposium, ISSN 1726-9679
B. Katalinic, Published by DAAAM International, Vienna, Austria 2006
150th Birthday of Nikola Tesla (1856-1943) Inventor of Mobile Robot Teleoperation

TESTING INJECTION MOLDS FOR POLYMER FLUIDITY EVALUATION

STANEK, M.; MANAS, M.; DRGA, T. & MANAS, D.

Abstract: *An injection mold is generally assembled from a number of single components. The forming parts, the cavity in connection with core, provide configuration and surface texture. It stands to reason that these parts demand particular attention to handling. Details about the molded part should provide information concerning the plastic material to be employed (e.g. reinforced or unreinforced, tendency to decompose, etc.). They determine minimum cavity dimensions, wear of the mold under production conditions, and the quality demands on the molding with respect to dimensions and surface appearance. This paper describes testing injection molds and their main parts. This injection mold was developed and tested in our department where it is also used for measuring of polymer melt fluidity.*
Key words: *injection mold, design, surface roughness, gate, cavity.*

1. INTRODUCTION

Injection molding is a method of producing parts with a heat-meltable polymer material. This is done by the use of an injection molding machine. The shape which is produced is controlled by what is called a "mold." The injection molding machine has two basic parts; the injection unit, which melts the plastic and then injects or moves it into the mold, and the clamping unit, which holds the mold. The unit clamps the mold in a closed position during injection, opens the mold after cooling, and ejects the finished part.

2. INJECTION MOLDING

2.1 Injection molding machine

The main units of a typical injection machine are the clamping and the injection unit (Fig. 1.). The clamping unit holds the injection mold. It is capable closing, clamping and opening the mold. Its main components are the fixed and moving platens, the tie bars and mechanism for opening, closing and clamping. The injection unit melts the polymer and injects it into cavity of the mold. Injection molding machines are often classified by the maximum clamp force that the machine can generate, capacity of injection unit and maximum mold dimension.

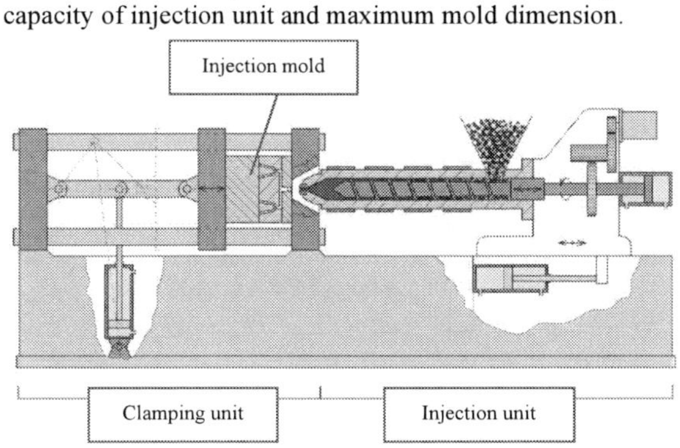

Fig. 1. Injection molding machine

2.2 Injection mold

Injection mold is a arrangement, in the one assembly, of one (or number of) hollow cavity spaces built to the shape of the desired product, with the purpose of producing (usually large numbers of) plastics parts or products. The cavity space is generated by the female mold part, called cavity, and a male mold part, called the core. It is necessary to fill the mold cavity fully during the injection molding process. The ability of cavity filling could be affected by the polymer properties and the properties of cavity walls.

3. TESTING INJECTION MOLDS

To be able to study the influence of surface quality of runners and cavities on the flow of polymer melts, two special injection molds have been designed and machined.

3.1 Injection mold for thermoplastics

Injection mold for thermoplastic materials is equipped by pressure and temperature transducers to be able to measure the really value of pressure and temperature in the mold cavity during the injection molding process. Length of testing samples is measured by measuring jig. The design of this mold makes the exchange of the gates kind and size possible. The various types of gates come to the spiral channel (Fig.2) whose bottom is formed by exchangeable plates. The width of spiral is 6 mm, depth 1 mm and length is 2000 mm.

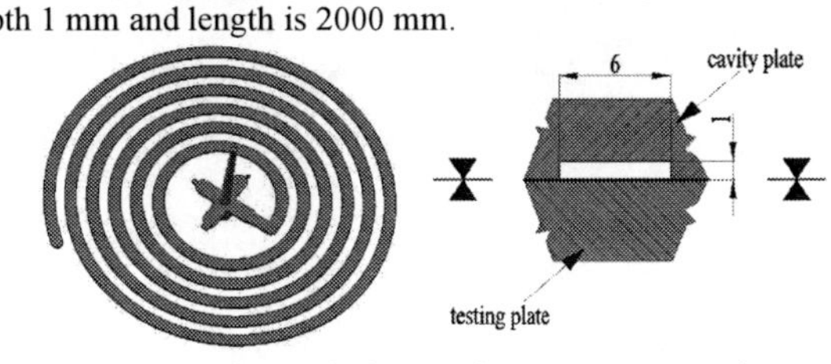

Fig. 2. Thermoplastics spiral sample and cross-section of channel

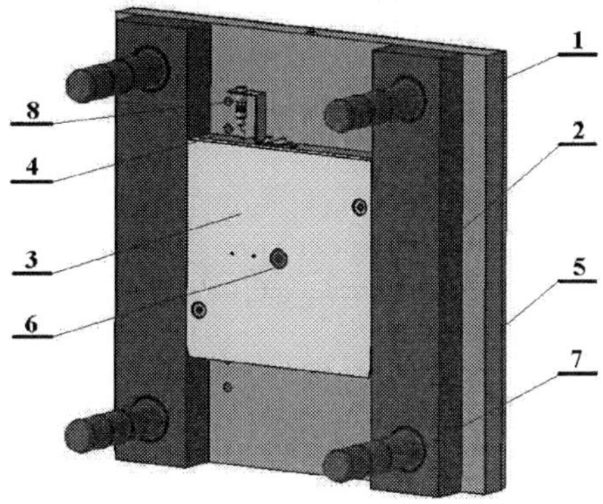

Fig. 3. Injection mold (sprue side)
1– clamping plate, 2– spacer plate, 3– testing plate, 4– plate, 5– insulating plate, 6– sprue bushing, 7– guide pillar, 8– connector of pressure and temperature sensor

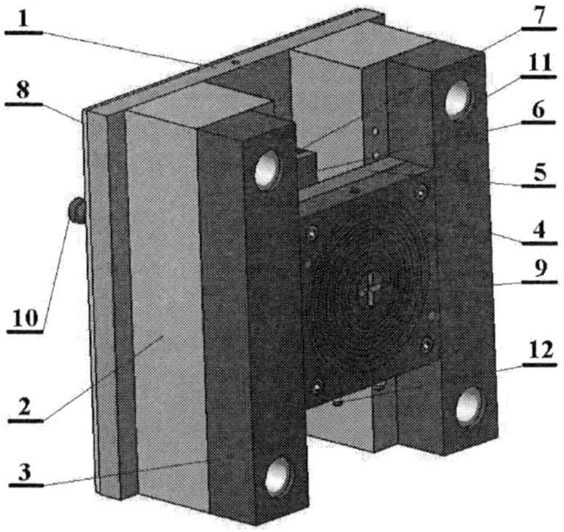

Fig. 4. Injection mold (ejection side)
1– clamping plate, 2,3– spacer plate, 4– cavity plate, 5– plate, 6,7– ejector plates, 8– insulating plate, 9– sprue puller insert, 10– ejector rod, 11– guide bush, 12– hose nipple

3.1 Injection mold for elastomers

Injection mold for elastomer materials is similar to injection mold for thermoplastic. The main difference between these injection molds is heating of cavity plates (elastomers) against cooling of cavity plates (thermoplastics). For exact measurement of spiral length there is line scale on cavity plate surface.

The design of this mold makes the exchange of the gates kind possible. The width of spiral is 5,6 mm, depth 2 mm and length is 2000 mm.

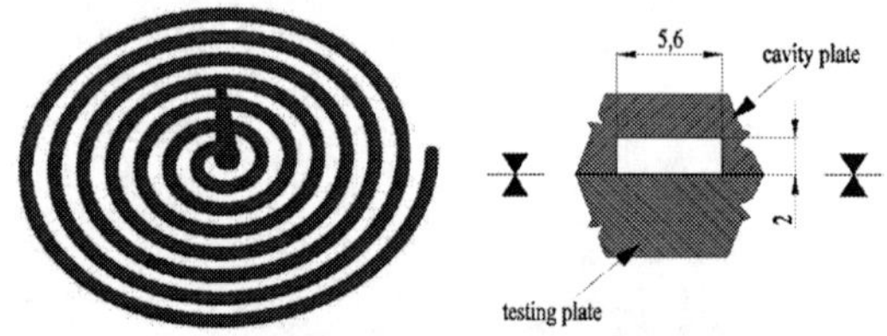

Fig. 5. Elastomers spiral sample and cross-section of channel

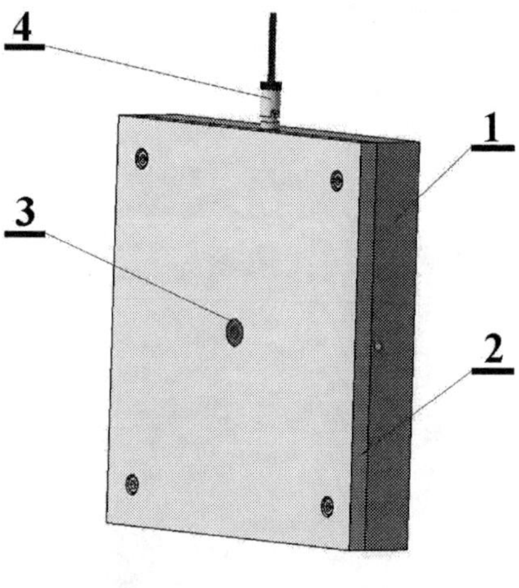

Fig. 6. Injection mold (sprue side)
1– clamping plate, 2– testing plate, 3– sprue bushing, 4– temperature sensor

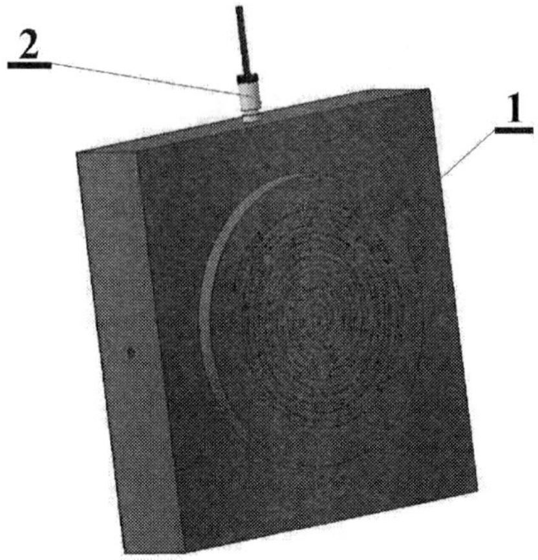

Fig. 7. Injection mold (ejection side)
1– cavity plate, 2– temperature sensor

Thermoplastic				
Polished plate Ra[μm]	Ground plate Ra[μm]	Electroerosion machined plate ("fine") Ra[μm]	Milled plate Ra[μm]	Electroerosion machined plate ("fine") Ra[μm]
	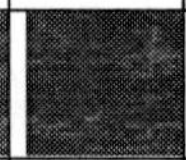	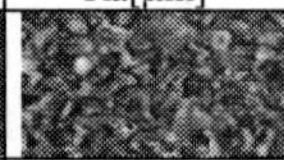		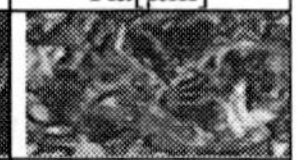
0,102	0,172	4,055	4,499	9,566
Elastomer				
	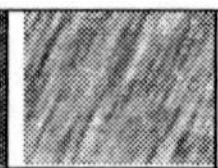	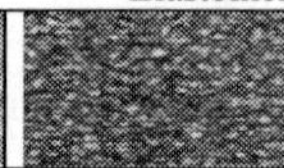	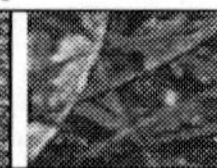	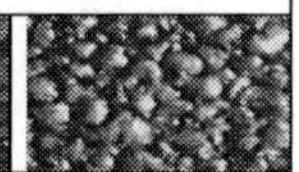
0,029	0,369	3,52	9,368	17,393

Table 1. Surface roughnes of testing plates

4. CONCLUSION

Two injection molds were designed and made. These injection molds can be used for almost all type of injection polymers for evaluation of material fluidity. There is also possible to change the type and size of gate.

5. ACKNOWLEDGEMENT

This article is financially supported by the Czech Ministry of Education, Youth and Sports in the R&D project under the title 'Modelling and Control of Processing Procedures of Natural and Synthetic Polymers', No. MSM 7088352102.

6. REFERENCES

Stanek, M. (2005). *Modelling of polymers forming process inc. tools design,* VUT FSI Brno, ISBN 80-214-3025-7, Brno

Agassant, J.-F.; Avenas, P.; Sergent, J.-Ph. & Careau, P. J. (1991). *Polymer processing,* Hanser Publishers, ISBN 3-446-14584-2, Munich

Beaumont, J. P. (2004). *Runner and gating design handbook,* Hanser Publishers, ISBN 3-446-22672-9, Munich

Lindner, E. & Unger, O. (2002). *Injection mold,.* Hanser Publishers, ISBN 3-446-21448-8, Munich

Manas, M., & Vlcek, J. (2001). *Applied rheology,* UTB Zlin, ISBN 80-7318-039-1, Zlin

Rees, H. (2002). *Mold engineering,* Hanser Publishers, ISBN 3-446-21659-6, Munich

Annals of DAAAM for 2006 & Proceedings of the 17th International DAAAM Symposium, ISSN 1726-9679
Editor B. Katalinic, Published by DAAAM International, Vienna, Austria 2006
150th Birthday of Nikola Tesla (1856-1943) Inventor of Mobile Robot Teleoperation

ANALYSIS OF THE EROSION ON 45 MW STEAM TURBINE MOVING BLADES

STANISA, B. & POMENIC, L.

Abstract: *During the first regular opening and repair, the severe erosion damage was detected at the last stage of moving blades in the 45 MW condensing steam turbine with controlled steam extraction. Moving blades which are damaged due to erosion reduce the usefulness and reliability of condensing steam turbines. On the basis of long-time measurements and erosion monitoring the approximate law of erosion process was obtained at the last stage of moving steam turbine blades.*

Key words: steam turbines, moving blades, wet steam, erosion

1. INTRODUCTION

In the condensing steam turbines of intermediate power with controlled steam extraction, the considerable erosion damage of the last stage moving blades in some cases can occur due to effects of wet steam stream (Stanisa, 1979; Krzyzanowski, 1991). Erosion damage reduces the usefulness and reliability of condensing steam turbines and durability of moving blades (Stanisa, et al., 1987; Stanisa, 1987). Because of severe erosion damage in a few steam turbines, moving blades have to be replaced with new blades earlier than expected. So far the law of erosion process in real intermediates power steam turbines in exploitation has not been enough researched. There are no reliable data on erosion process law on moving steam turbine blades in exploitation as the mathematical model basis for erosion rate estimation and eroded moving blades durability. Durability of eroded blades can be estimated on the basis of this model and new blades can be ordered on time. In this paper, as a contribution to this problem, great effort has been invested in the research of an erosion process law of the last stage moving blades of condensing 45 MW steam turbine with controlled steam extraction which was built in Osijek Heat and Thermal Power Plant. During each steam turbine opening and repair the erosion of last stage moving blades was measured and analysed. According to the long-term measurements and erosion monitoring the approximate law of erosion process of the last stage moving blades was obtained. This law of erosion process is performed as a ratio between size of erosion damage on last stage moving blades and turbine operating hours.

2. THE BASIC TECHNICAL DATA OF STEAM TURBINE AND EROSION MEASUREMENT

The 45MW condensing steam turbine consists of one controlled steam extraction and three uncontrolled steam extractions (Stanisa & Pomenic, 2004). This turbine was put into operation in 1979. The purpose of this turbine is the production of electrical energy and heat. On the turbine inlet the steam pressure is 8.7 MPa with steam temperature of 510 °C. The amount of steam on the turbine inlet is 69.44 $kg{\cdot}s^{-1}$. Controlled steam extraction is performed at pressure of 0.14 MPa. Steam amount of controlled steam extraction is 38 $kg{\cdot}s^{-1}$. The pressure in condenser is 8 kPa. Steam humidity degree on the turbine output is 10.2%. The rotation velocity of the steam turbine is $50s^{-1}$. This steam turbine consists of 22 action stages. The last five stages operate in wet steam due to erosion of moving blades. In Figure 1 the longitudinal cross-section through the last three turbine stages with chambers for water separation from the steam for erosion protection was presented. Water separation from the wet steam decreases moving blades erosion. The length of the last stage moving blades is 1100mm whereas the diameter on the blades root is 530mm. Blade material is X20Cr13 whose hardness amounts from 212 to 230 HV_{10}.

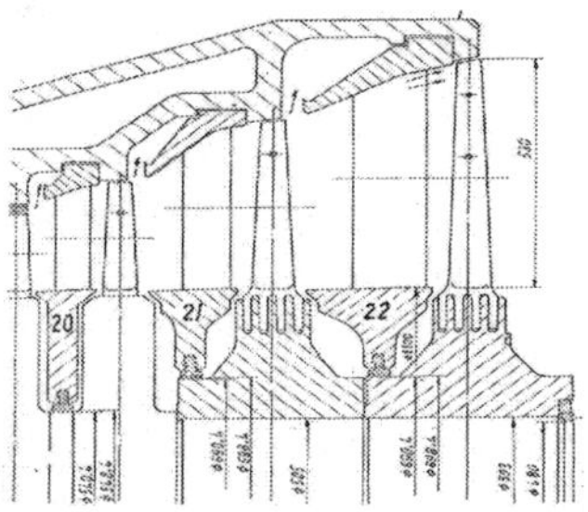

Fig.1. Longitudinal cross-section of the last stages of the 45MW condensing steam turbine

Monitoring and measurements of erosion damage were performed on the chosen five last stage steam turbine moving blades on disc rim, which were the most endangered. For the each measured size the average value of erosion was calculated. Both the size of damage of the blade chord profile and the length of damaged chord from the top to the blade root were measured. Also, the width of the pitting erosion trace, on the back part of the profiles inlet edge, and the length of the pitting trace from the top to the blades root were measured. Damage on the chord of blades profile was measured with benchmark where the error of the method can be 0.1mm. The length of the damage from the top to the blade root was measured with steel metre where the error of the method can be 1mm. These measured values were performed as a ratio between size of erosion damage on moving blades and turbine operating hours. Characteristics of erosion damage on moving blades were photographed with added metre where approximate size of erosion can be seen. The change of erosion damage can be monitored from the catalogue of these photographs. The influence of turbine operation exerted on erosion process was analysed at each measurement of erosion on steam turbine moving blades.

3. RESULTS OF THE MEASUREMENTS AND ANALYSIS OF EROSION

During the last inspection of the 45MW steam turbine, in year 2006 after 82208 lifetime hours of operating turbine, the very low pitting erosion of the inlet edges on top of the moving blades of the third stage, counted from the last stage, was detected. On the moving blades, from next to the last stage, the relative low pitting and crater like erosion occurred. All the inlet edges on the last stage moving blades were periphery similarly damaged by severe erosion. Figure 2 shows moving blades from the last three 45MW turbine stages after 82208 hours of turbine operating where only the chord of inlet edges on the top of moving blades from the last stage were damaged by erosion.

Fig. 2. Moving blades from the last three stages of 45MW steam turbine after 82208 hours of turbine operating

Figure 3 shows an example of the chord inlet edge erosion damage on moving blade number 86 from the last stage after 822008 hours of turbine operating. This is one of the five last moving blade stages on which the erosion process was monitored where the size of erosion damage on blade inlet edge can be seen with an aid of a metre. Erosion was damaged chord of inlet edge profile on the top of blade Δb - 13.4mm. Pitting erosion damage from the top to the blade root is 245mm. Each five monitored blades were similar damaged as a blade number 86.

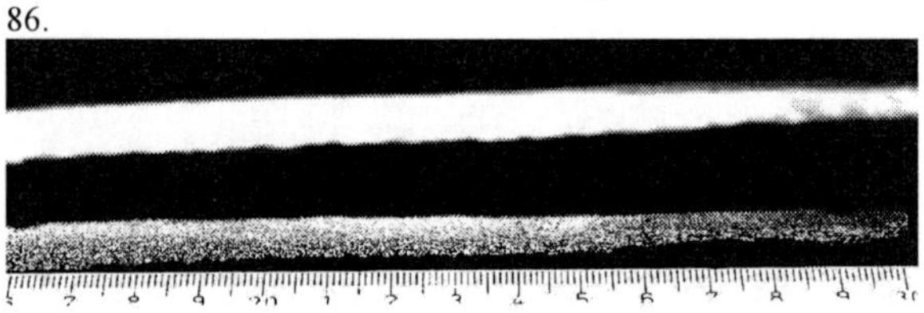

Fig. 3. An example of erosion damaged inlet edge on moving blades periphery from last stage 45MW steam turbine after 82208 hours

Year	Turbine operating hours	Average erosion damage on chord of inlet edge blade profile, mm
1991	11463	5.2
1997	40200	11.3
2002	63843	12.6
2004	82208	13.3

Table 1. Results of long-time erosion monitoring on five moving blades of the last stage of 45MW steam turbine

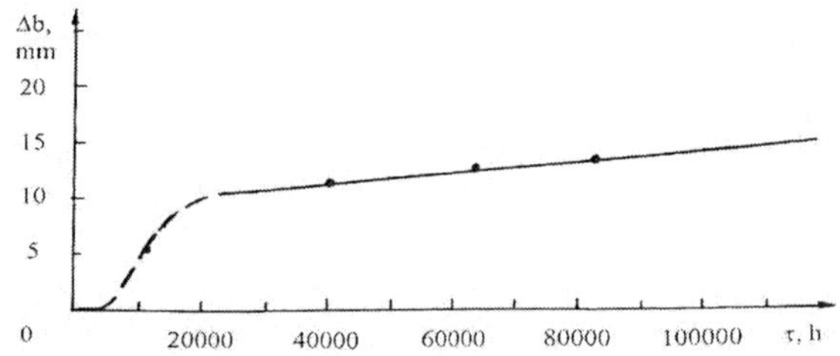

Fig. 4. Dependence between average erosion damage on chord of inlet edge blade profile (Δb) on the top of last stage moving blades 45 MW steam turbine and turbine operating hours (τ)

Figure 4 shows the first, second and third stadium of erosion process detected on moving blades inlet edge, based on Table 1. During the first stadium of incubation - there is no erosion whereas the second stadium – the stadium of accelerated erosion is shown as a broken line because it was approximately valuated. There are not enough measurements data for these two stadiums and these data were taken from the measurements performed on other similar steam turbines (Stanisa, 1989). The third one - a slow down stadium of erosion process is shown with solid line. Researches of erosion process on steam turbines in exploitation have shown that the third stadium is the most important for erosion of moving blades because they operate in these conditions most of their life time. Erosion propagation on moving blades inlet edge of the last stage 45MW steam turbine can be approximately estimated on the basis of erosion law in the third slow down erosion process. From the period between 2002 and 2006 the average erosion damage on chord profile on the top of the last stage moving blades was propagated only 0.7mm. This erosion was occurred because of impingement attack of water droplets in wet steam stream and shown in the Figure 5, where profiles of last stage turbine stator and moving blades are showed with triangles of steam velocity and water droplet phase. Because of their inertia, water droplets have lower inlet velocity from the stator blades (c'_1) in relation to steam phase of wet steam (c_1). Because of that water droplets under different angle (β'_1) and velocity (w'_1) enter moving blades from angle (β_1) and velocity (c_1) of steam phase and attack the back part of moving blades inlet edge and causes their erosion. Severe erosion damage, occurred on the top of the last stage 45MW steam turbine moving blades, can be explained by water film flow from the rim of the distribution wall and his break in large droplets, which are most dangerous to cause erosion. Also, because of centrifugal force the largest droplets are on the stage periphery.

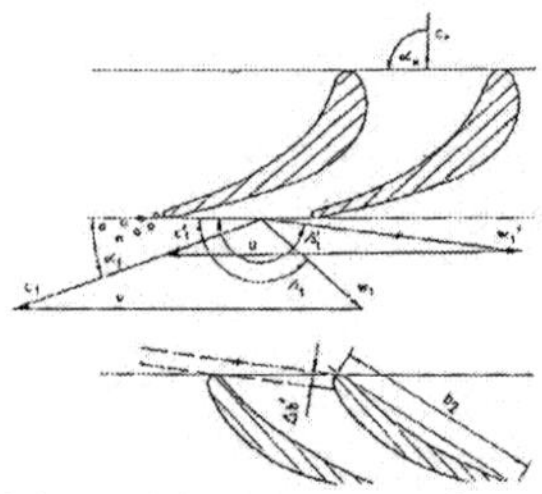

Fig.5. Scheme of the periphery cross-section profiles of the last stage stator and moving blades of 45MW steam turbine with triangle of steam and water droplet phase velocity

4. CONCLUSION

On the basis of long time measurements and erosion monitoring the law of erosion process for the third slow down stadium was obtained which is approximately linear.

5. REFERENCES

Krzyzanowski, J. (1991). Erozija lopatek turbin patowyech, *Zaklad Narodowy im,* Ossolnickich Wydawnietwo, Wroclaw

Stanisa, B. (1979). Utjecaj strujanja vlažne pare na eroziju turbinskih lopatica, *Strojarstvo,* Vol. 21, No. 3/4, pp. 161-187, Zagreb

Stanisa, B.; Dicko, M. & Puklavac, K. (1987). Erosion Process On the Stage Rotor Blade of Turbine in Service, *Proc. of ELSI VII Cambridge,* 6-10 September, pp.16-17, Cambridge

Stanisa, B. (1987). Procjena vijeka trajanja erodiranih rotorskih lopatica parnih turbina većih snaga, *Elektroprivreda,* Vol.40, No. 9/10, pp. 357-364, Zagreb

Stanisa, B. (1989). Erozija rotorskih lopatica toplifikacijskih parnih turbina i metode zaštite, *Strojarstvo,* Vol. 31, No. 5/6, pp. 259-265, Zagreb

Stanisa, B. & Pomenic, L. (2004). The analysis of corrosion damage in a 45 MW steam turbine, *3rd DAAAM Conference ATDC'04,* June 23-26, pp. 181-186, Split

Annals of DAAAM for 2006 & Proceedings of the 17th International DAAAM Symposium, ISSN 1726-9679
Editor B. Katalinic, Published by DAAAM International, Vienna, Austria 2006
150th Birthday of Nikola Tesla (1856-1943) Inventor of Mobile Robot Teleoperation

RECENT GENERATION ROBOT TEACH PENDANT UNIT

STOPPER, M. & WANG, J.

Abstract: *An essential requirement in making robots easy to use is to adapt programming units, control panels and other operator tools to diverse needs and different knowledge. In general robots are delivered as all-purpose machines with a common user interface, designed to work with the same set of controls, buttons and screens independent of the task to solve. However, different applications require different operator use, and thus varied operating units, which are adapted to the specific tasks. But also for a single application there are diverse needs during the life cycle of the installation. Former generations of robot teach pendant units do not fulfil these requirements of flexibility and adaptability to application needs. This paper presents the recent generation robot teach pendant unit and especially the new software architecture to change it to the - now fully customizable - user interface.*

Key words: *robot teach pendant unit usability, robot application builder, robot embedded systems, advanced robot programming, compact framework technology*

1. INTRODUCTION

In general, a customized solution is superior to a standard solution. This is especially true when it comes to operator interfaces on the factory floor. For instance, a robot installation will span over several phases, like installation and set up, configuration and calibration, robot programming, test and debug, production supervision and production changes. Each of these phases will have its own specific demands, and to make operation easy and safe each phase should use an optimized interface. A key advantage of customized user interfaces is, that the system becomes 'easier to use'. This helps to significantly reduce downtime because of operator error and also reduces the need for training. To realize the benefits of customized interfaces a short common retrospect to some key points of robot TPU generations, without claim for completeness, should be argued in the next section.

2. TEACH PENDANT UNIT GENERATIONS

An industrial robot is officially defined by ISO as an automatically controlled, reprogrammable, multipurpose manipulator programmable in three or more axes. The field of industrial robotics may be more practically defined as the study, design and use of robot systems for manufacturing, a top-level definition relying on the prior definition of robot (Wikipedia, 2006). A robot system typically consists of a manipulator, a robot controller and a robot teach pendant unit (TPU). The robot TPU is a handheld device which is used to online program and/or control the motion of a robot. The TPU is often disconnected after programming and the robot then runs on the program that has been installed in its controller. Another way for online setup or programming of motions and sequences for an industrial robot is typically by linking the robot controller via network or serial interface of a laptop computer (the computer is installed with corresponding interface software), which leads to a fundamental different approach. Further, a TPU should come with safety features including an emergency stop and a live-man switch (TrueForce, 2006). The first generation of robot TPU's has been very rudimentary and identified itself primarily by a variety of fixed function buttons for robot programming. The robot movement in manual mode was also realized by function buttons. In general there were no customizable function buttons. The display consisted of just a few digits showing e.g. instruction numbers, instruction types or error codes. Commonly the second generation of robot TPU's has introduced an emergency stop and a live-man switch. A significant step was the establishment of a joystick with A/D-converter for robot motion control replacing the corresponding function buttons of the first generation. Its deflection is converted to digital signals which control the robot axes. A keyboard is used for entering program instructions and other input data. Additionally the display grew up, to at least two lines, showing e.g. robot programs, function menus or system and error messages. The designer of TPU's then began more and more to have a squint to computer user interfaces. This led to the next generation of TPU's, which implemented a big display consisting of multiple lines able to show menus and windows in text mode, much better keyboard features and organisation as well as a more computer-oriented operating philosophy. Within this generation the possibilities of character based user interfaces were fully exhausted. It is worth to mention that up to this generation the TPU's software technology basement was fully proprietary and the user interface only less customizable by fabricator's special tools. Reaching the limits of character based user interaction the next logical step must be the transition to graphically based TPU's with open system architecture.

3. PROBLEM STATEMENT & CHALLENGES

Normally robots are delivered as general machines with a general user-interface, designed to work with the same set of controls, buttons and screens independent of the task or application. On the other hand, different applications or processes require different functionality to be truly efficient and effective. A well-designed custom operator interface will present the appropriate information and functionality at the appropriate time. Thus reducing time and minimizing the potential for costly errors. In addition, a single application has different needs during the life cycle of the installation. However, the problem has been that customized user interfaces have been expensive and can be very time-consuming to create. Additionally the fabricator's proprietary TPU software system architecture has been an immense brake. In an open system, components do not depend on a single administrative domain and are not known at system design time. The objective of TPU system software architecture must be a domain specific open dependable platform. To overcome the limits of the former TPU software architectures the following technical issues have to be solved:

- dependable platform architecture
- support for the development of extensions
- certification of platform extensions
- predictable assembly of the platform and extensions

Dependable platform architecture is a fundamental account. The questions are: What platform interfaces should be created? How can platform dependability be assured? Which software component model for adding third party software has to be chosen? The right software development kit (SDK) and tools for development extensions as well as the definition of criteria and implementation of tools for certification are necessary. Especially a closer look is mandatory of how the properties of extensions can be described and measured and how the verification of the assembly properties will take place.

As a result of the unsatisfactory situation as well as the unsolved technical challenges for the time being, most companies have relied on standard solutions and have therefore sub-optimized their investment in automation.

4. RECENT GENERATION ROBOT TPU

The described situation has changed now with the introduction of the recent generation of robot TPU coming with its new open software architecture. Now it is possible for all companies to realize the benefits of customer operator interfaces by making development quick, easy and less expensive. Detailed and application oriented considerations of recent generation robot TPU should be outlined with a robot company representative in the following. Due to the outstanding features as well as the uncompromising realization of recent generation robot TPU requirements, including the implementation with latest system software technology, the so called 'FlexPendant' TPU from ABB Company is chosen (ABB, 2005a). The 'FlexPendant' TPU is on the one hand a part of the ABB IRC5 controller unit, but on the other hand an independent computer running Windows CE .NET as operating system. Two exceptional characteristics are the excellent three way joystick for intuitive and direct robot motion control and the touch screen feature. The TPU possesses only eight fixed buttons and an emergency stop. Any further inputs can be done by a virtual keyboard appearing at the touch screen on demand. This represents a real step up against the former complex, keyboard based TPU's. At TPU design time usability stayed always on top. The use of a one-finger touch screen is fast and grades in a natural ease of use. The user interface is very similar to Microsoft Windows operating system user interface with international understandable symbols. That fact minimizes the training effort since most users are familiar with the Microsoft environment. Built in multiple language support (including Japanese and Chinese) and the ability to turn the display for lefty's usage indicate the focus for usability during TPU development. The development of the FlexPendant TPU started end of 2001. After defining technical requirements, the operating system choice was easy since a new generation of Microsoft embedded operating system was already on the market which fulfilled the specified needs. However, the choice for the appropriate programming and component model was not easy. Aiming to be ready on the market within a few years and to provide a user friendly SDK to enable third-party companies to develop customized applications, the main problem was the complexity of the available technology (COM/ATL component model with MFC user interface library). At the same time Microsoft Company started the development of a new framework for embedded systems for devices, the .NET Compact Framework (.NET CF), which has been expected to be a better, less fault-prone and less time consuming programming model. The attendance at the adequate Microsoft Early Adopters Program (EAP) for Windows CE .NET yielded in the FlexPendant SDK and further optimizations of the TPU base system. Since 2004 the FlexPendant SDK is part of ABB Robot Application Builder (RAB) software hosted by Microsoft Visual Studio .NET environment, which enables customers and partners to create own robot applications for the FlexPendant TPU. RAB is fully based on the Microsoft .NET Framework technology (including CF) and uses therefore the corresponding programming and component model. An exceptional feature of the Microsoft .NET environment is the independence from the programming language. That means that RAB developers may choose their own preferred programming language (supported by Visual Studio .NET) for developing applications for the FlexPendant TPU. A further important characteristic of the .NET environment is that program code runs on any supported software platform. This yielded into the development of a virtual FlexPendant. Using Virtual IRC5 controller – based on Virtual Robot Technology (Stopper, 2005) – and the virtual FlexPendant leads to the possibility of writing and testing customer specific TPU applications on the desktop. In addition the debugging becomes much easier since the programmer just has to set a break point and step through the code during program execution. A further key component is the Controller Application Program Interface (CAPI). The CAPI is the link between the commands and the controls appearing on the screens of the FlexPendant, and the internal functions of the robot, which have to be controlled. This CAPI includes a large number of useful commands reaching from read/write of any data to the control of the input/output interface or robots operating mode. The CAPI is more than a common interface for further diverse products communicating with the controller (ABB, 2005b). As a product, the FlexPendant SDK is a tool which will offer a lot of functionality to enable the programmer to create an application oriented design. The output is a package, which is ready to be loaded into the controller and appears on the FlexPendant touch screen as active buttons, fields, text boxes etc. - whatever is wanted for an efficient operator control.

5. CONCLUSION AND FURTHER RESEARCH

A customized operator interface on factory floor is the key for an 'easy-to-use' robotic system. A well-designed custom operator interface will present the accurate amount of information at the right time in the correct format to the operating user, so that the training time and the downtime (due to the operating errors) are reduced to a minimum. The recent generation of robot TPU provides this thesis and overcomes the limitations of all former character based TPU user interfaces. Additionally to the now graphic's interface with mostly touch screen operation the open system architecture can be appraised as a significant key for technical and economical success. On the market can be observed that almost any of the big robotics companies have already solutions or at least can show prototypes of recent generation TPU's as described in this paper.

Currently there are two common ways to create operator interfaces on recent generation TPU: A low end way using mostly robot programming language extensions with very limited flexibility and a high end way using .NET programming based on TPU software development kits for boundless flexibility. It is expected that further research yields into middle-ground-solutions as a third way.

6. REFERENCES

ABB (2005a). *Robot Application Builder – FlexPendant SDK Users Guide*, ABB Automation Technologies AB, Robotics, 3HAC 024914-00, RobotWare v5.07, Mölndal, Sweden

ABB (2005b). *Robot Application Builder – PC SDK Users Guide*, ABB Automation Technologies AB, Robotics, 3HAC 024913-001, RobotWare v5.07, Mölndal, Sweden

Stopper, M. (2005). *Virtual Engineering for Industrial Robotic Work Cells*, Final Program and Abstracts of the 4th Asian Conference on Industrial Automation and Robotics (ACIAR'05), Paper ID: F-35, ISBN 974-8208-58-3, Bangkok, Thailand

TrueForce (2006). *Encyclopaedia Robotica – Teach Pendant, Available from*: http://trueforce.com *Accessed:* 2006-09-06

Wikipedia (2006). *Encyclopaedia – Industrial Robot, Available from*: http://en.wikipedia.org *Accessed:* 2006-09-06

Annals of DAAAM for 2006 & Proceedings of the 17th International DAAAM Symposium, ISSN 1726-9679
B. Katalinic, Published by DAAAM International, Vienna, Austria 2006
150th Birthday of Nikola Tesla (1856-1943) Inventor of Mobile Robot Teleoperation

A REAL-TIME FRAMEWORK FOR PERFORMING COMPUTER VISION ALGORITHMS ON EMBEDDED PLATFORMS

SULZBACHNER, C.; KUBINGER, W. & HEMETSBERGER, H.

Abstract: *For embedded systems the usage of reliable software frameworks is essential. These frameworks guarantee stable and correct system behaviour over time. In this paper development and implementation details of such a framework for computer vision processing are presented. This includes a description of the concept, the hardware used, the implemented software, and a final demonstration.*

Key words: *embedded system, embedded vision processing, real-time system, digital signal processing*

1. INTRODUCTION

An infrastructure for embedded vision processing consists of an embedded system and peripheral components. A processing framework is an application which is executed on an embedded system. It allows the execution of computer vision algorithms.
This paper presents the results of the integration of computer vision algorithms in an embedded platform. The framework is responsible for capturing, processing images, executing a computer vision algorithm and showing results. The requirements for this framework are various. The main aspects are real-time behaviour, a high data throughput and a low latency period from the input of an image to the output of results computed by an algorithm. These are opposed characteristics, hence a balance between these conditions has to be found.
The focus of this paper was on a new concept for processing computer vision algorithms considering these requirements. In the following section, the realization of the framework will be described. Finally, in the last section, the workflow of the tasks of the system will be presented. Applications for such an embedded computer vision real-time framework can be driver assistance systems for intelligent vehicles (Behringer et al., 2005), for example.

2. REALIZATION OF THE FRAMEWORK

2.1 Concept

The infrastructure, shown in Fig. 1, consists of the embedded system and peripheral components like a camera and/or a display. The application, which is executed on the embedded system, represents the real-time processing framework. The application consists of different types of functions that are characterized in three groups: (i) configuration functions, which are responsible for communication with external devices, (ii) system functions, which keep a continues processing frame rate and observe the system concerning stability and (iii) image processing functions, which capture images from an input peripheral component, process the images with a computer vision algorithm and show the results on a screen and/or control output peripheral components to perform actions.
The required components for the infrastructure are the embedded system and at least one capturing device, which can be a camera or a memory card with images stored. Optional components are a computer, which allows a dynamic configuration of the system, a display, which shows the results of image processing and finally a communication device for further processing of results.

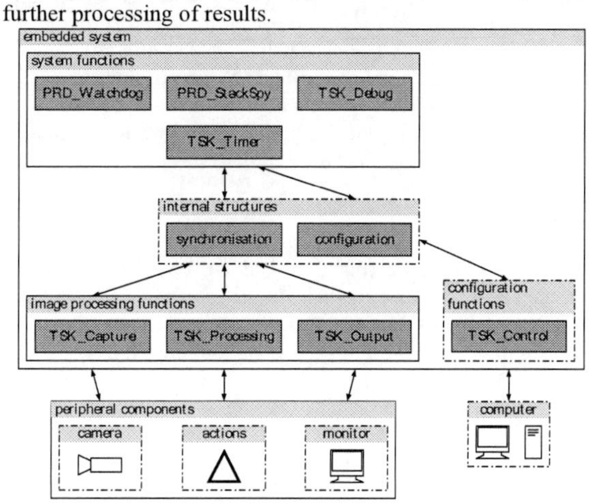

Fig. 1. Assembly of the infrastructure

2.2 Hardware

The embedded system consists of a basis board and peripheral equipment that can be assembled considering the requirements of the infrastructure. The core of the embedded system is a TMS320C6414 fixed-point *digital signal processor* (DSP) from Texas Instruments. Furthermore, the DSP has 1024 kByte internal memory that can be mapped as cache and/or internal memory. Data, which is stored in the external memory, can be copied via *direct memory access* (DMA) transfers to the internal memory, where much faster data processing is possible, which speeds up the performance significantly. Afterwards, the processed data is copied back to the external memory (Zinner & Kubinger 2006). The system is equipped with a memory module and *field programmable gate arrays* (FPGA), which are connected to the *external memory interfaces* (EMIF) of the DSP.

2.3 Software

The processing framework that is executed on the embedded system uses the real-time kernel DSP/BIOS from Texas Instruments as operation system (Texas Instruments, 2000).
The application uses tasks and periodic functions which are a special form of software interrupts. Both are supported by the kernel. The kernel uses a pre-emptive scheduler, which synchronizes the tasks in reference to their priority (Texas Instruments, 2004). The tasks, shown in Fig. 1, are now described in more detail. At the start up of the system an *initialization task* initializes the internal structures and configures all peripheral devices. After execution, this task

terminates. The *control task* communicates with an application executed on a computer. Thereby, the system can be dynamically configured during image processing.
The *timer task* is responsible to ensure a continuous processing speed. There are two mechanisms that assure the stability for the application. These are a *run-time check function* and a *memory check function*. Furthermore, there is a *debug task* that is responsible for data transfers for debug purposes. This task has the lowest priority of the system. Thus, this will be only executed if the more important image processing has finished.
The image processing block consists of a *capturing task* which grabs an image i_0 from an input peripheral component at the execution-time t_0. At the same execution-time t_0 an image i_{-1}, which has been grabbed at the system-time t_{-1}, is processed by the *processing task*. The results of the image processing are displayed with an *outputting task* which operates on an image i_{-2}. Because of the fact that tasks operate on autonomous memory regions $i_{0...-2}$, they can be executed at the same time t_0.
Both, the capturing and the outputting task use a double buffering concept. Up to this, the capturing task provides a static buffer for the image processing and an alterable buffer for the image that will be captured. The outputting task provides a static buffer for the image update on the screen and an alterable buffer for the changing image from the image processing.

2.4 Image Handling

The data formats of the input and output equipment are frequently different. Reasons therefore are special data formats from the vendors or optimizations of the processing framework. Thus, a *pre-processing function* allows converting the data format from the source equipment to the required format for the algorithm, see Fig. 2.

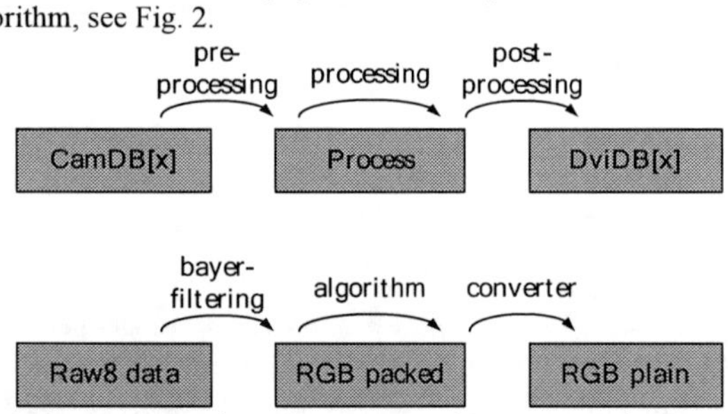

Fig. 2. Image processing; above: pre- and post-processing, below: colour image processing

Consequently, the *processing function* processes the images in an autonomous format. The destination equipment may also have another data format. Thus, a *post-processing function* allows converting those different data formats.
Examples for pre- and post-processing are colour processing algorithms, see Fig. 2. Input equipment, e.g. a camera that is configured in raw mode captures the images. Raw means a pre-stage format of an image. Hence, no filters are automatically processed in the input device. The pre-processing function converts the raw data to a colour image using a Bayer filtering function (Brainard, 1994). Thus, the colour processing algorithm can be executed. In that case, the output equipment, e.g. a graphic card, needs another colour format. Hence, a post-processing function converts the different data formats.

3. WORKFLOW OF THE SYSTEM

The framework consists of various tasks and periodic functions. Thus, the interactions of the parallel operating tasks have to be synchronized. For that, semaphores and mailboxes are used. The context switches of the tasks are handled by the pre-emptive scheduler provided by the kernel. Depending on the status (ready, running, blocked, terminated) and the priorities of the tasks, the scheduler handles the corresponding system state.

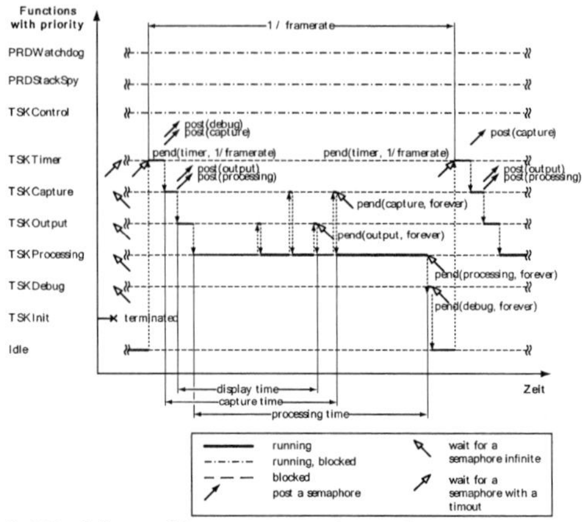

Fig. 3. Workflow of image processing tasks

The workflow of the image processing tasks is shown in Fig. 3. The *timer task* triggers a new image processing cycle at a constant frame-rate by starting the capturing task. Before triggering, the task verifies that the last processing cycle has finished successfully. The corresponding data structures are modified and a new image grab is started. Then the *outputting task* and the *processing task* are launched. The *outputting task* has a higher priority and starts displaying an image. Both, the capturing and the outputting task use DMA transfers. Thus, these transfers are handled automatically at the same time by a controller without using the processor core. Hence, the processing core of the DSP is available for the *processing task* for the whole processing cycle. During the execution of the *processing task*, the scheduler of the kernel switches to the other tasks for stability purposes.

4. CONCLUSION

This paper deals with the realization of a real-time computer vision framework running on an embedded real-time system. Both, the realization of the processing framework and the workflow of the system were discussed. The processing framework, which is executed on the embedded system, distinguishes with real-time behaviour and versatility. Finally the workflow, the synchronization and the parallelism of the processing tasks were shown.

5. REFERENCES

Behringer, R.; Travis, W. & Daily, R. & Bevly D. & Kubinger, W. & Herzner, W. & Fehlberg, V. (2005). RASCAL – An Autonomous Ground Vehicle for Desert Driving in the DARPA Grand Challenge 2005, Proceedings of IEEE ITSC 2005.

Brainard, D. (1994). Bayesian Method for Reconstructing Color Images from Trichromatic Samples, Proceedings of the IS&T 47th Annual Meeting

Texas Instruments (2000). DSP/BIOS User's Guide, Literature Number: spru303b.

Texas Instruments (2004). Synchronizing DSP/BIOS Threads, Literature Number: spraa25.

Zinner, C.; Kubinger, W. (2006). ROS–DMA: A DMA Double Buffering Method for Embedded Image Processing with Resource Optimized Slicing, Proceedings of the IEEE RTAS 2006.

Annals of DAAAM for 2006 & Proceedings of the 17th International DAAAM Symposium, ISSN 1726-9679
B. Katalinic, Published by DAAAM International, Vienna, Austria 2006
150th Birthday of Nikola Tesla (1856-1943) Inventor of Mobile Robot Teleoperation

CUTTING DEPTH DETERMINATION WITHIN CO2 LASER MICROMACHINING

SYKOROVA, L.

Abstract: *The results of CO_2 laser micro-machining research are presented in this paper. There was researched laser beam and its using on different types of polymeric materials in details. The quality of microstructures is depended on the laser parameters and the type of polymeric material. Optimal combination of laser cutting parameters (power and feed) was tried to find for PA6, ABS, POM, PS and PMMA. The experimental results were evaluated and depicted into the graphs. Machinability of used polymers was investigated.*
Key words: *CO_2 laser, polymeric material, correlation coefficients, slot depth .*

1. INTRODUCTION

There are several types of plastics that have been processed by laser. These include acrylics, polytetrafluoroethylene, polymethyleneoxide, polyethylene, polypropylene, polycarbonate, and polyvinyl chloride. They absorb electromagnetic energy at 10.6 micrometers – the wavelength of light emitted by CO_2 lasers. Results of the laser micro-machining – surface quality of product and his utility in specific application – depend on the laser parameters and the polymer material type.

2. EXPERIMENT

2.1 Preparation of specimen

From the obtained scientific knowledge it follows, that the character of surface machined by laser depends among others on the thermal conductivity of polymer. Therefore it was decided to test more kinds of polymers - PMMA, PP, PA, ABS, PS, POM. These tests are presented in given study.Commercial CO_2 laser Mercury L-30 by firm LaserPro, USA was used for cutting of specimens. It is possible to change power and feed rate of laser system. Ray of laser could be focused on mark diameter d = 185 μm. The maximum value of density of energy flow is q = 1,1 GWm^{-2}. Laser is cutting with the software help of Corel Draw.

2.2 Machining of specimens on the CO_2 Laser

Desired symbol was created in program Corel Draw, the width of slot was 2 mm in all cases. Cutting parameters (output power and feed) were set and these were changed gradually. Values of power and feed are presented as percents from maximum power (P = 30 W) and maximum feed (fmax = 1066 mms^{-1}) in charts of parameters combination and graphs.

2.3 Dimension and profiles measuring of machined slots

Dependence of slot depth on the laser parameters – output power and feed - was measured. The dimension and profiles measuring of machined slots were realized on the optical microscope ZEISS 2772. Suitable microscope optics which ensured 117x enlargement was set first of all.

2.4 Processing measurement data

Slot depth (d in μm) in dependence on laser parameters combination is presented in following results tables. Values of power and feed are presented as percents from maximum power (P = 30 W) and maximum feed (f max = 1066 mms^{-1}) in charts of parameters combination and graphs. Description 20/70 means 20% value from power 30W and 70% value from maximal feed 1066 mms^{-1}.It was not possible to measure the slot depth for specimen number 1 (PP) for the reason of short resolution capability of using microscope optics. We can assume that slot depth will be increased in connection with increasing values of power.

f [%]/ P [%]	10	20	30	40	50	60	70	80
50	10	18	31,2	53,3	86	112	129	145
70	-	-	-	12	27	40	80	93
100	-	-	-	-	8	15	25	34

Table 1. Slot depth in μm for PA6

f [%]/ P [%]	10	20	30	40	50	60	70	80
50	10	18	31,2	53,3	86	112	129	145
70	-	-	-	12	27	40	80	93
100	-	-	-	-	8	15	25	34

Table 2. Slot depth in μm for POM

f [%]/ P [%]	10	20	30	40	50	60	70	80
50	-	10	75	90	105	118	162	206
70	-	-	10	54	68	75	102	127
100	-	-	-	20	31	65	75	101

Table 3. Slot depth in μm for ABS

f [%]/P [%]	10	20	30	40	50	60	70	80
50	8	15	42	70	105	142	210	238
70	-	10	13	31	55	98	138	197
100	-	-	18	32	51	63	66	76

Table 4. Slot depth in μm for PS

f [%]/P [%]	10	20	30	40	50	60	70	80
50	24	38	102	126	178	200	246	248
70	-	3	15	39	51	116	146	160
100	-	-	8	25	31	37	47	63

Table 5. Slot depth in μm for PMMA

Especially at short power slot values were very short and it was not possible to measure. These experimental results were evaluated and depicted into the graphs. Approximating straight-line and their correlation coefficients were created for these dependencies.

Graphic representation of experimental results

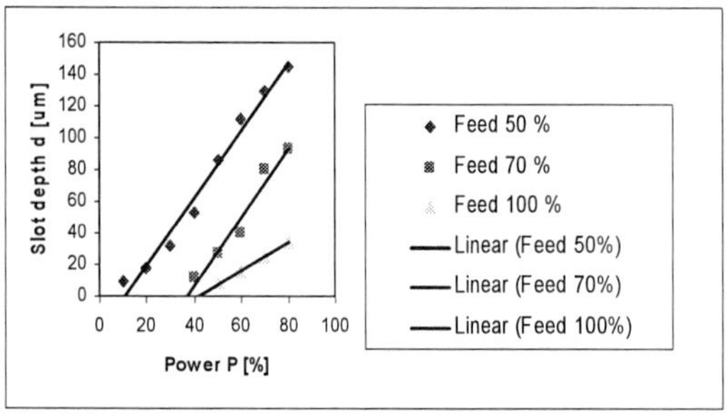

Fig.1. Slot depth linear dependence PA6

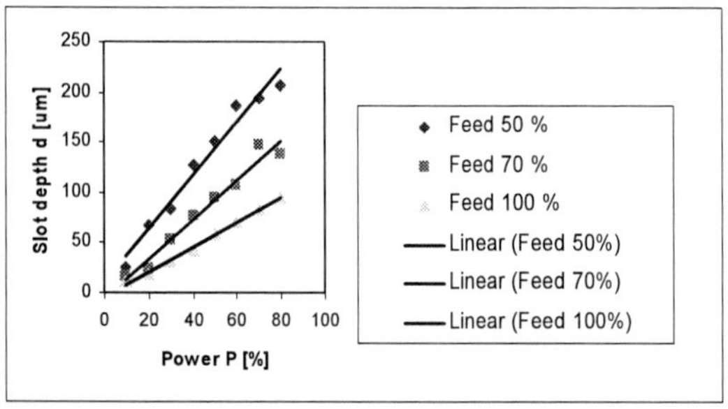

Fig.2. Slot depth linear dependence for POM

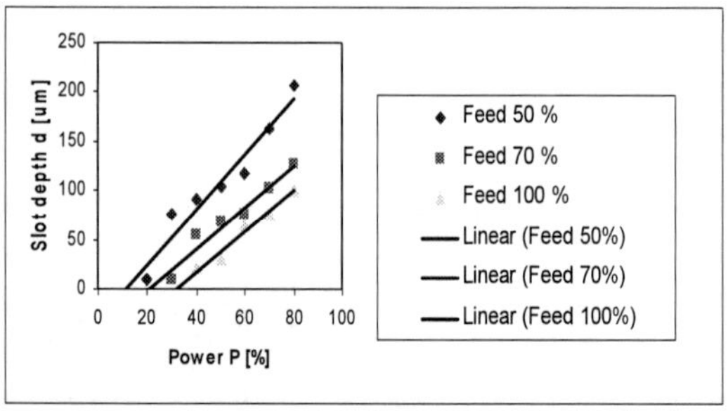

Fig. 3. Slot depth linear dependence for ABS

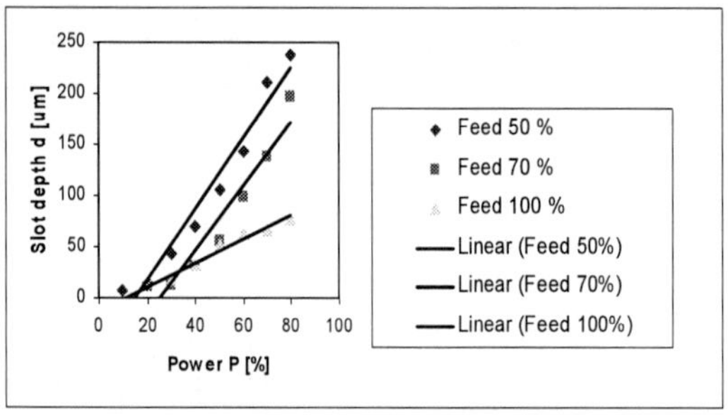

Fig.4. Slot depth linear dependence for PS

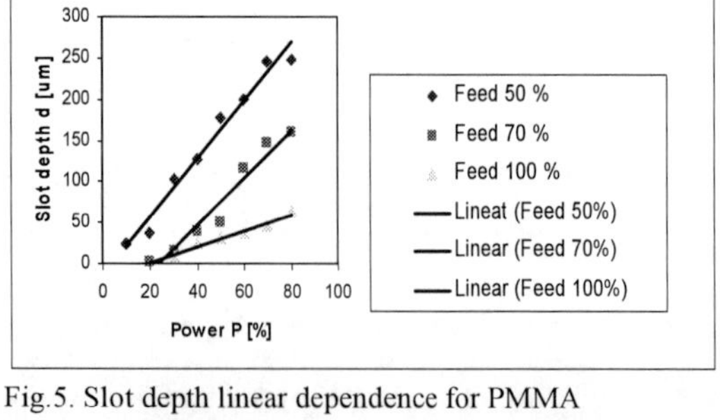

Fig.5. Slot depth linear dependence for PMMA

Material	f [%]	equation	correlation coefficient R^2
	50	y = 2,1132x	0,98
	70	y = 2,1500x	0,9588
PA6	100	y = 0,8800x	0,9954
	50	y = 2,6619x	0,9705
	70	y = 1,9429x	0,9654
POM	100	y = 1,2536x	0,9952
	50	y = 2,8214x	0,9432
	70	y = 2,1029x	0,9536
ABS	100	y = 2,0600x	0,9735
	50	y = 3,4762x	0,9617
	70	y = 3,1429x	0,9257
PS	100	y = 1,1543x	0,954
	50	y = 3,5167x	0,9759
	70	y = 2,8929x	0,9525
PMMA	100	y = 0,9914x	0,9681

Table 6. Regression equation of slot depth dependencies on laser parameters combination designated in program EXEL

3. CONTRIBUTION

From this experiment it is evident the slot depth grows linear with increasing values of power at concrete adjusted value of feed. Biggest depth is at low values of feed. This phenomenon was observed at all materials. Biggest slot dept vas measured at PS and PMMA materials. Gradient of lines and correlation coefficients offer information about dependence character of slot depth on adjusted power and feed at graphic presenting results. Value of gradient (k) is used for exact control of stock removal (slot depth). Gradient of line gives information about sloth depth at increasing of power about definite number of percents. Correlation coefficient gives approximate quality of linear dependence (compare with $R^2 = 1$).

Resulting structures can be very exact and with high quality of surface in dependence on laser parameters and on type of machining materials. The biggest problem is the transformation of unremoved material again into solid state. This phenomenon causes deterioration of both accuracy of dimension and quality of surface.

4. REFERENCES

Lukovics, I. & Sykorova, L. (1997). *Study of the effect of concentrated energy on plastics and metals.* WORKSHOP 97, pp. 1433-1436, Janu. 20-22, Czech tech. un. in Prague

Varga, V. & Mankova, I.(1995). *Aplikácia CO_2 lasera pri rezaní oxidickej keramiky.* Technologia 95 STU Bratislava

Halaska, P. & Manas, M.(2002). *Laser Cutting Optimization of the Polymeric Plates and Films.* 42nd Sc. Week Laser Sc. and App., pp. 80-84, 2.-4. 11. 2002, Un. Aleppo, Syria

Stanek, M.; Manas, M. & Drga, T.(2005). *Influence of Surface Quality of Injection Molds on Fluidity of Polymer Melt.* ICPM 2005, pp. 287-293, ISBN 3901888314, 18.-19.10.2005 Vienna University of Technology, Austria

Halaska, P. ; Manas, M. ; Stanek, M. & Pop Mircea T. (2003).*Modeling of Polymeric Plates Laser Cutting.*, Editura Universitatii Din Oradea, pp. 128-132, Annuals of the Oradea University, Vol. II, 2003, ISSN 1583-0691

ACKNOWLEDGEMENT

This work was supported by the Ministry of Education and Youth of the Czech Republic under grant MSM 7088352102. This support is very gratefully acknowledged.

Annals of DAAAM for 2006 & Proceedings of the 17th International DAAAM Symposium, ISSN 1726-9679
Editor B. Katalinic, Published by DAAAM International, Vienna, Austria 2006
150th Birthday of Nikola Tesla (1856-1943) Inventor of Mobile Robot Teleoperation

PREVIEW OF TOOLBOX GSM

SYSEL, M.; PINOS, R. & POKORNY, P.

***Abstract**: Toolbox GSM is the name for recently developed complex system of goods supplies and take-off. This system allows fully automate demands on supplies of new goods, consumer getting 24 hourly access to goods, that have had to order before. The Toolbox GSM consist of the case that contains goods, barcode scanner, GSM modulus for communication and program with WWW interface that caters entire logic. The basic pillar of the system is automatic record of take-off of tools and goods. Flow of goods and person identification is done by barcode.*
***Keywords**: Trading system, work automatization, WWW*

1. PREVIOUS STATE OF ORDERS A SUPPLIES

Each productive engineering company has in his shop floor big consumption of tools, for example tools for turning operations, milling, cutting operation and making screw-threads. The system of purchase of this material should be very sophistry; flow of tools in the bigger company often required specialized labor force.

Previous state of production of orders and verification of supplies of goods requires a lot of manual work. When computer age coming to us, It is getting better. But it's innovation only from spectacle of company leaders, because they could obtain better statistic information (computer could offer this). However, when you ask common worker, many of them say that his work is more complicated and he has to think about. The computer is something mystical for common workers and he approach to computer with big distrust. He should pursue a lot of operations (that previously does not make) for statistical output. Important element: "think about work" is insuperable obstacle for many workers. In system of supplies, orders, consumption and subsequent invoicing approach other factors, which can influence faultless functioning of whole company. It is impossible order tools at the moment of the requirement, because at that point can come to suspension of production and subsequent financial losses. The responsible stuff should be appointed.

All of these problems are possible pass by the help of developed Toolbox GSM, which simplify above described operations and shifts some duties to supplier.

2. TOOLBOX GSM

2.1 Benefits

Toolbox GSM is dispenser of necessary tools, which is placed right on the production shop floor. Box contains all necessary tools. Goods contained in the toolbox is property of supply organization, it is a big benefit for company. When workers need a new tool, it is possible take out the tool from the box and system of Toolbox GSM consecutively generates invoice for all taken tools at the end of the settled time period. Production company (customer) can save fairly big amount, because they do not need stock and have instant access to necessary tools. Supplier should keep constant quantity of tools. Developed system provides evidence of issued instruments, notice supplier on necessity of refilling of box and itself produce required statistic.

2.2 Principle of function

Toolbox GSM consists of case that can have arbitrary proportions, everything depend only on consumer requirement. Typical proportions are between 700x700x200 mm and 1400x1800x200 mm. All of cases have transparent front door, so naked eye can check contents of the case (figure 1). The case contents small Liquid Crystal Display which report status information and provide easy user guide.

Fig. 1. ToolBox GSM

GSM module (Siemens TC45) with implementation of programming language JAVA is one of the components of the case. This GSM module (figure 2) provides internet access using GPRS technology. Advantage of GPRS technology is permanent on-line internet connection with high speed (Siemens, s.r.o., 2006). The GSM module connects directly to central server in the Prosperity System company and sends security and service information there. This server supplies entire logic of developed system. Centralized logic has big advantage in the case of possible problems. As the server store very important data, the data are backed up each 30 minutes.

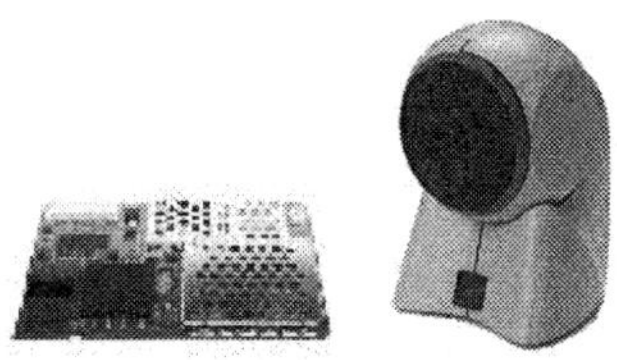

Fig. 2. GSM module and barcode scanner

The information about goods flow are obtained by the help of omni directional laser barcode scanner Metrologic MS7100

Orbit (figure 2), which is based on holographic scan technology (GABEN, spol. s.r.o. 2001). Barcode reading in visual angle of sensor is done fully automatic without activating any button. Service life of light source (laser diode) is possible extend by sleep mode. The laser is switched off in the sleep mode; this mode is automatically finished by barcode reading.
Each worker should be identified by access card, which is based on barcode (Figure 3).

Fig. 3. Access card

There are the name of the worker and his identification barcode on the identification card. The information about Toolbox (Identification and location) is written on the right down. Access rights are possible define via WWW interface.

2.3 Taking tools away from the Toolbox

The procedure of taking tools from the Toolbox is very simple and it is possible describe this procedure in a few steps:

- Identification of worker by access card
- Display writes progress information's out
- Scanner read barcode of detracted tool.
- Display writes confirmation out and twice shortly beeps, than Toolbox expect next operation.

The owner of the all goods in the case is supplier (just till to the take away). If the tool is taken away by mistake, it is possible return the tool by the help of special return card. However it is possible return only unused package.

2.4. Business system accessible via WWW interface

Each consumer obtains access to the business system by the installation of Toolbox GSM. Whole system consists of several modules. Administrator is able to set up access rights (including devolution of setting up access rights). The user can setup up just only access rights which are subset of his access rights. The system support several language mutation, but in this time just Czech language is fully supported. The English translation will be finished soon. System with all modules is displayed on figure 4.

3. CONCLUSIONS

Toolbox GSM is the name for recently developed complex system of goods supplies and take-off. This system allows fully automate demands on supplies of new goods, consumer getting 24 hourly access to goods, that have had to order before. Box contains all necessary tools required by customer. Goods contained in the toolbox is property of supply organization, it is a big benefit for production (customer) company. Production company can save fairly big amount, because they do not need stock and have instant access to necessary tools. Developed system provides evidence of issued instruments, notice supplier on necessity of refilling of box and itself produce required statistic.

4. ACKNOWLEDGEMENTS

This work was supported by the GAČR under grant No. 102/05/0271, and by the Ministry of Education of the Czech Republic under grant No. MSM 7088352101.

5. REFERENCES

GABEN, spol. s.r.o. (2001). MS7100 Orbit. Available from http://www.gaben.cz/Download/Datasheet/orbit.pdf.

Gilmore, W. J. (2004). *PHP5 a MySQL*. ZONER press, ISBN 80-86815-20-X, Praha.

Prokop, M. (2003). *CSS - kaskadové styly pro webdesignéry*. Mobil média a.s., ISBN 80-86593-35-5, Praha.

Prosperity System s.r.o. (2005). Toolbox GSM. Available from http://www.toolboxgsm.cz/

Schlossnagle, G. (2004). *Pokročilé programování v PHP5*. ZONER press, ISBN 80-86815-15-5, Praha.

Siemens, s.r.o. 2006. Datové moduly GSM – TC45 Available from http://www.siemens.cz/siemjet/cz/home/ic/wirelessModules/gsm/Main/13566.jet

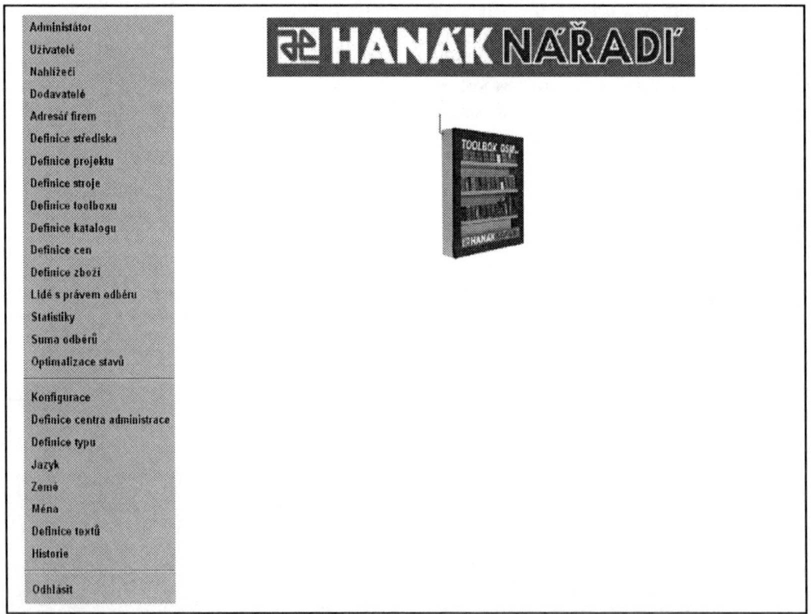

Fig. 4. Business system of company HANÁK NÁŘADÍ with all modules

Annals of DAAAM for 2006 & Proceedings of the 17th International DAAAM Symposium, ISSN 1726-9679
Editor B. Katalinic, Published by DAAAM International, Vienna, Austria 2006
150th Birthday of Nikola Tesla (1856-1943) Inventor of Mobile Robot Teleoperation

EXPERIMENTAL DEMONSTRATOR OF THE DRAUGHT DRIVER FOR INFRA-TURBULENCE AERODYNAMICS

TACHE, F.; RUGESCU, R.D.; SLAVU, B.; CHICIUDEAN, T.G.; TOMA, A.C. & GALAN, V.

Abstract: *The results of new experimental measurements and numerical simulations with the draught tower driver for an aero-acoustic wind tunnel are presented. Besides the small-scale model, a full-scale concept is described, which includes a high efficiency turbine that can be used to extract energy from the air flow passing through the 80-meter tall tower.*
Key words: *Draught tower, infra-turbulence wind tunnel, high-efficiency turbine*

1. INTRODUCTION

The destination of the facility here described is to allow the development of numerical simulation activities and laboratory experiments in the new and promising field of aero-acoustics, considered as fruitful for the emerging technologies, mainly related to aerodynamic noise suppression. The potential use of the new thermal up-draft mechanism for accelerating air through a infra-turbulence, noise-free wind tunnel is investigated. In addition to the aero-acoustic application of this installation, the effects of a high efficiency turbine are predicted by analytical calculations and numerical simulations.

2. GRAVITY DRAUGHT FOR AIR ACCELERATION

The source of the gravitational draught is the ascending effect of lower density volumes of warm gases in contrast with the higher density surrounding air as a result of their different weights, quantitatively described by the famous Archimedes' law. As far as we keep standing in the frames of this effect of gravity, the problem remains entirely immobile. The accompanying buoyant force supplies the maximal estimate for the draught effect. However, when the upward motion of the rarefied gas from the inside is considered, the more elaborated dynamic equilibrium must be approached by the rules of gasdynamics, although its roots remain in the realm of crude Archimedes' principle.

A vertical stack is considered, circulated by the heated air that defuses into the atmosphere after exiting its top opening. It is surrounded by higher density cold air and the effect of gravity must be accounted for in computing the different inner and outer pressures that act on the stack's walls. Depending on the construction of the heat exchanger, the drag largely varies. For simple, tubular channels the pressure loss due to frictions stands negligible (Chiciudean et al., 2005) and the reduced mass flow rate (RMF) results from the simple equation

$$R^2 \equiv \frac{\dot{m}^2}{2g\ell\rho_0{}^2 A^2} = \frac{\gamma\cdot(1-\gamma)}{\gamma(2-\Gamma)+\Gamma}, \tag{1}$$

where $\Gamma \equiv \frac{\kappa-1}{\kappa}$ and γ is the air heating intensity:

$$\gamma = \frac{\rho_0-\rho_2}{\rho_0} = 1-\beta, \tag{2}$$

an alternative to the previous solution of Unger (Unger, 1988)

$$R^2 = \frac{\gamma\cdot(1-\gamma)}{\gamma+1} \tag{3}$$

or to the more recent one (Rugescu et al., 2005)

$$R^2 = \frac{\gamma(1-\gamma)}{1+\gamma-\Gamma} \tag{4}$$

and gives optimistic values in the region of smaller levels of heating (Fig. 1).

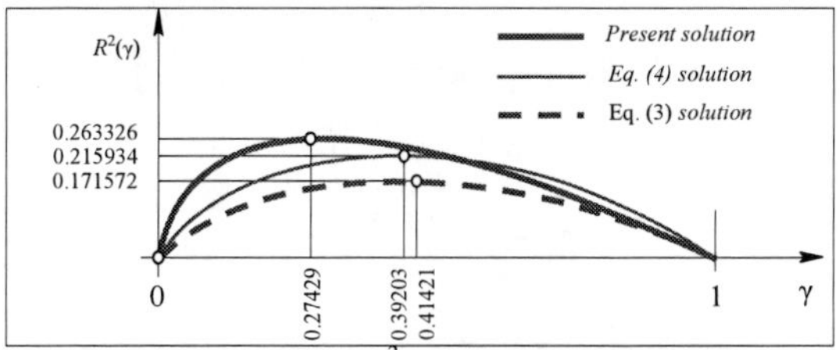

Fig. 1. Stack discharge R^2 versus air heating intensity γ

3. MEASUREMENTS AND CFD SIMULATIONS

With the existing small-scale model at UPB, several tests have been conducted, leading to the values for air velocity in the tube as given in the diagram below. The average values, measured at *1.7 m* distance from the inlet section in the tube, were *2.115 m/s* air speed with the contracted area effect and *6.216 m/s* without that contraction. Air temperature at the exit section was recorded to *195°C* and *123°C*, respectively.

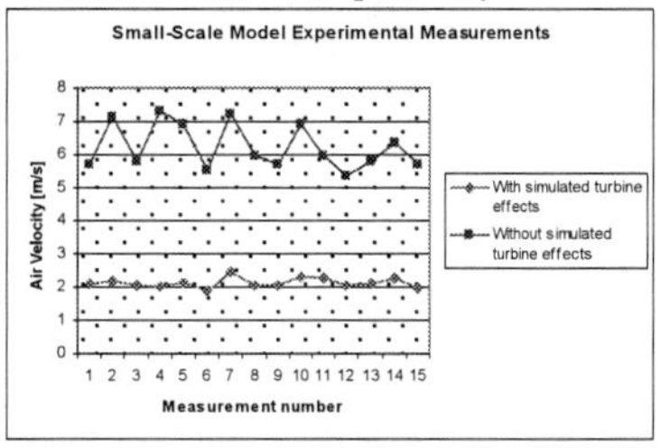

Fig. 2. Experimental measurements on the small-scale model

Fig. 3. Small-scale model of the draught tower driver
(overall view, ¼ contraction area, hot resistors, exit temperature)

The air flow through the experimental rig was further investigated by CFD simulations. Both cases, with and without the contraction area for simulating the turbine effects were analyzed and the results are briefly presented below. The very good concordance between the experimental measurements and the numerical solutions is noticeable.

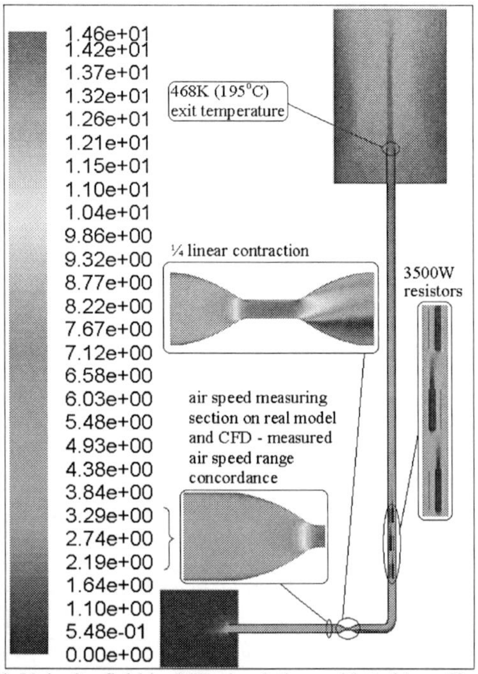

Fig. 4. Velocity field in CFD simulations with turbine effects

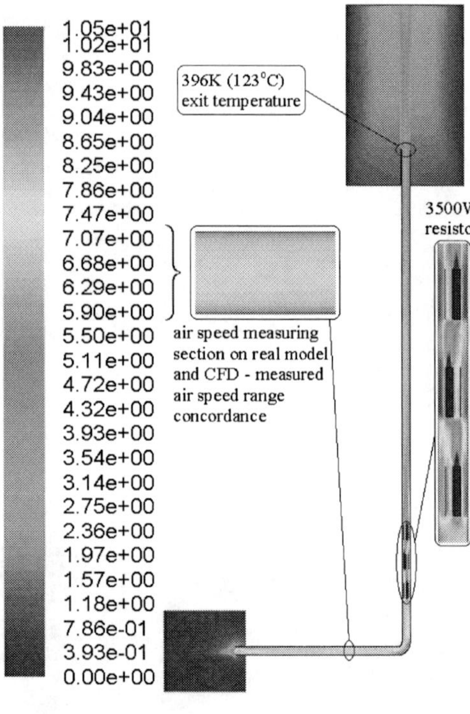

Fig. 5.Velocity field in CFD simulations without turbine effects

All recorded experimental values and the ones obtained from numerical simulations are listed in Table 1. The differences between these values are small, with greater values (*~29.55%*) when accounting the turbine effects and much smaller values (*~4.33%*) in the absence of a contraction zone.

No.	Measured Air Velocity [m/s]			Simulated Air Velocity [m/s]		
	With Contraction	Without Contraction	Speed Ratio	With Contraction	Without Contraction	Speed Ratio
1	2.101	5.703	2.714	Vmin = 2.19	Vmin = 5.90	SRVmin = 2.37
2	2.190	7.110	3.247			
3	2.051	5.767	2.811			
4	1.996	7.310	3.662			
5	2.127	6.920	3.253			
6	1.867	5.521	2.957			
7	2.414	7.208	2.986			
8	2.027	5.966	2.943	Vmax = 3.29	Vmax = 7.07	SRVmax = 2.15
9	2.051	5.703	2.780			
10	2.307	6.920	3.000			
11	2.276	5.966	2.621			
12	2.027	5.351	2.639			
13	2.076	5.767	2.778			
14	2.276	6.329	2.780			
15	1.937	5.703	2.945			
Mean values	**2.115**	**6.216**	**2.941**	**2.740**	**6.485**	**2.260**

Table 1. Experimental and simulated air velocity values

4. FULL-SCALE MODEL OF THE DRAUGHT TOWER

The principle described herein can be used in real-life applications such as aero-acoustic wind tunnels, ideal for noise investigations, and energy generation using solar heating and a high-efficiency turbine. The huge progress is the complete removal of all moving parts and, consequently, of all driving noise sources.

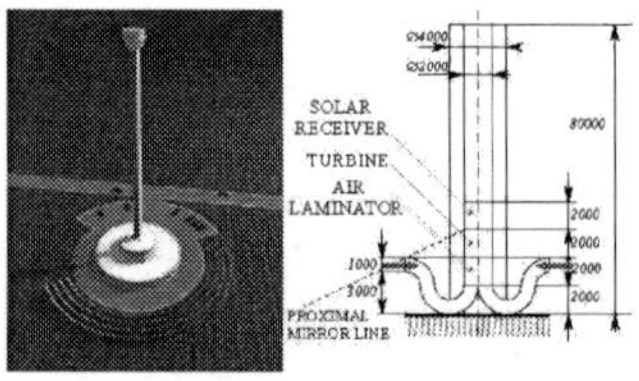

Fig. 6. Full-scale model general overview and dimensions

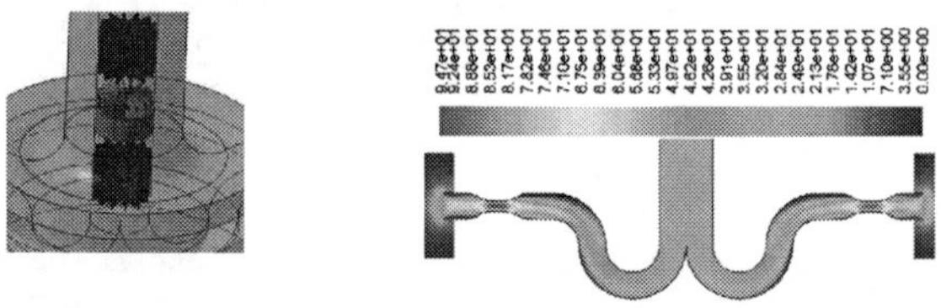

Fig. 7. 3D model of the tall tower and predicted velocity range

5. CONCLUSIONS

Named SEATTLER from ***S**olar **E**nergy **A**ctuator for **T**all **T**ower **L**ow-cost **E**lectricity **R**esearch*, the 80m tall tower creates a considerable air acceleration and allows the air noise to be definitely perceived. Numerical simulations conducted on this architecture reveal astonishing air velocities close to 100 m/s in the test chamber area.

6. REFERENCES

Bejan, A. (1984). *Convection Heat Transfer*, New York, Wiley and Sons, 1984

Raiss, W. (1970). *Heiz und Klimatechnik*, Springer, Berlin, vol. 1, pp. 180-188, 1970

Rugescu, R. D.; Chiciudean, T. G.; Toma, A. C. & Tache, F. (2005, a). *Thermal Draught Driver Concept and Theory as a Tool for Advanced Infra-Turbulence Aerodynamics*, Chapter 45 in DAAAM International Scientific Book 2005, Katalinic, B. (Ed.), published by DAAAM International, ISBN: 3-901509-43-7, ISSN 1726-9687, Vienna

Schlaich, J.; Bergermann, R. & Partners. (2005). *EuroDish System Description*

Unger, J. (1988). *Konvektionsströmungen*, B. G. Teubner, ISBN 3-519-03033-0, Stuttgart.

Weinrebe, G. (2004), *Das Aufwindkraftwerk-Wasserkraftwerk der Wüste*, Nova Acta Leopoldina NF91, Nr.339, pp. 117-141.

Chiciudean T. G., Rugescu R. D., Tache F., Toma C. A. (2005), *Draught Tower Driver for Infra-Turbulence Aerodynamics*, Proceedings of the 16th DAAAM Symposium, 19-22nd October, Opatija, Croatia, pp.065-066.

Annals of DAAAM for 2006 & Proceedings of the 17th International DAAAM Symposium, ISSN 1726-9679
B. Katalinic, Published by DAAAM International, Vienna, Austria 2006
150th Birthday of Nikola Tesla (1856-1943) Inventor of Mobile Robot Teleoperation

LOGISTICS AND INVENTORY SYSTEMS OF JUST-IN-TIME MANUFACTURING

TAKAKUWA, S.

Abstract: *The characteristics of logistics and inventory systems for the realization of just-in-time manufacturing are described from the standpoint of management and information systems. First, the methods of implementing just-in-time manufacturing in a factory producing personal computers are detailed. Following this, the inventory control systems adopted in JIT manufacturing are clarified. Finally the methods to supply parts to a company from domestic and international suppliers are highlighted in meeting just-in-time manufacturing in the factory.*
Key words: *Just-in-time manufacturing, Kanban system, Cell production system, Logistics.*

1. INTRODUCTION

The developers of the just-in-time (JIT) concept often utilize the Mizusumashi system as well as the Kanban system. The Kanban system has been applied globally (Monden, 1983). The Mizusumashi (fixed-course pick-up) system remains one of the most popular means for maintaining the industrial philosophy of JIT manufacturing used to produce or retrieve the required amount of necessary items at the correct time. Workers regularly move between assembly lines and parts-storages and bring the required amount of necessary types of parts onto the line. Such workers are called Mizusumashi in the context of popular Japanese production system; that is, the Toyota production system (TPS), also known as the JIT manufacturing system. The cell production system remains one of the major means in realizing the philosophy of just-in-time (JIT) manufacturing, along with the Kanban system and Mizusumashi system. The cell production system is mainly applied to the assembly lines of manufacturers in Japan.

In this paper, the characteristics of logistics and inventory systems of just-in-time manufacturing are described from the standpoints of management and information systems, including the parts-suppliers and customers. The methods to supply parts to the company from the domestic and international suppliers are introduced to meet the need of just-in-time manufacturing.

2. A TYPICAL JUST-IN-TIME MANUFACTURING SYSTEM

A typical just-in-time manufacturing system comprises cell production system, a series of Mizusumashi, and Kanbans processes. The layout of a typical just-in-time manufacturing is illustrated in Fig. 1.

2.1 Cell Production System

The cell production system has been widely introduced in Japan as a way to improve just-in-time manufacturing especially in assembly line work since 1990s. The system comprises one or a series of multiple workers performing direct operations on an assembly line. The cell production system consists of a production system in which one or several workers completely produce one specific product. There are some advantages in introducing a cell production system. A relay-type of cell system can be introduced easily from a traditional conveyor line. After removing the conveyor, the entire line is divided into some moderate size cells with workers are assigned to each cell. The relay-type of cell is adopted when many itemized parts are used in production and when precise operations are required.

2.2 Mizusumashi (The fixed-course Pick-up) System

Mizusumashi carriers are the workers who supply parts or goods to assembly lines by hand or human-powered carts (Takakuwa & Nomura, 2003). They typically work in a manufacturing system that has been realized just-in-time manufacturing. Today, product life cycle is becoming increasingly shorter, and a more flexible manufacturing system is required to meet the requirements of demand. Therefore, the Mizusumashi system plays an important role for manufacturing partially because it provides more flexible characteristics than the automated parts supply system.

2.3 Kanban System

Two major kinds of Kanbans systems exist: the production Kanban (P-Kanban) and the conveyance Kanban (C-Kanban) that has direct workstations to produce or convey items. In the typical case, a P-Kanban is used as a tool for ordering production to the designated assembly line (Nomura & Takakuwa, 2004). The production manager prepares the P-Kanban at the appropriate occasion. When the demand or a C-Kanban arrives from the succeeding workstation, the C-Kanban is exchanged for the P-Kanban. In addition, for making withdrawals from a vendor or a parts-supplier, a supplier Kanban is used. The supplier Kanban containes instructions that request the supplier to deliver specific parts.

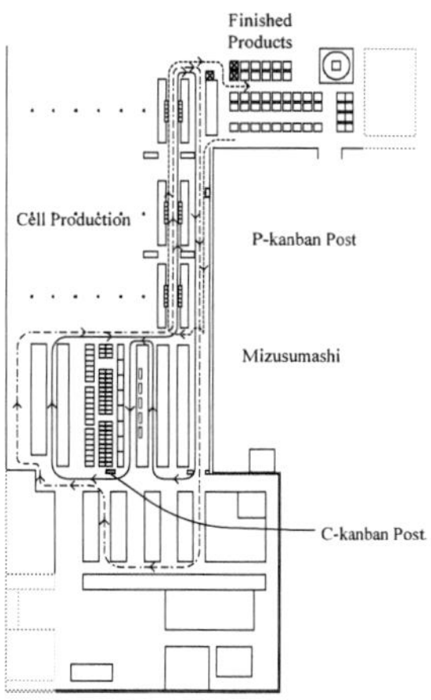

Fig. 1. Layout of a just-in-time manufacturing

3. INVENTORY SYSTEMS

Inventory plays a role in the buffering of material flow. Parts are put in an individual container which is served at the assembly line of a cell production. Furthermore, another type of container is used in the Kanban system

3.1 Cell Production and Mizusumashi

In this system, a predetermined number (for example, three) of workers perform the operation elements to assemble, inspect, and package notebook computers in a manufacturer on a line. The standard operational elements have been assigned to each worker. However, they cooperate with each other in performing their operations, as mentioned in the previous section. Mizusumashi regularly moves between assembly lines and parts-storages and brings the required amount of necessary types of parts onto the line. A transition of the amount of inventory together with corresponding containers is illustrated in Fig. 2. This is a type of fixed-order quantity system. In this case, the expected value of the number of containers (M_{ij}) for a specific item is given by:

$$M_{ij}^{E} = \left[\frac{2E(L_i)}{E(d_{ij})N} \right] + 1 \qquad (1)$$

where d_{ij}: production speed (sec./pc),
i: Mizusumashi No. ,
j: visiting sequence No. of Mizusumashi i,
L_i: leadtime of Mizusumashi i and sequence No. j, and
N: capacity of container (pcs.).
[A] is a Gaussian symbol, denoting a largest integer equal to or less tan A.

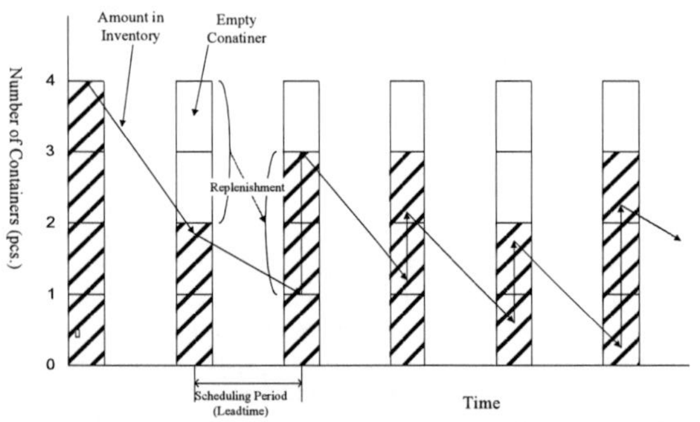

Fig. 2. Amount in inventory in cell production line

3.2 Mizusumashi and Kanban System

When a Mizushumashi picks out the parts from a completely full container, the associated Kanban is removed, and brought to the assigned Kanban post. Following this step, an associated part supplier or a person in charge will supply as many numbers indicated in a Kanban as the needed parts. The workers periodically review to check amount in inventory, for example, in one hour intervals. A transition of the amount of inventory together with the corresponding containers is illustrated in Fig. 3. Again, this is a type of fixed-order quantity system. In addition, the replenishment quantity is determined in terms of the capacity of the container.

4. LOGISTICS FOR IMPLIMANTING JIT MANUFACTURING SYSTEM

The logistics including just-in-time manufacturing system is shown in Fig. 4. The domestic and international suppliers provide the associated parts in order to maintain the just-in-time manufacturing in the factory.

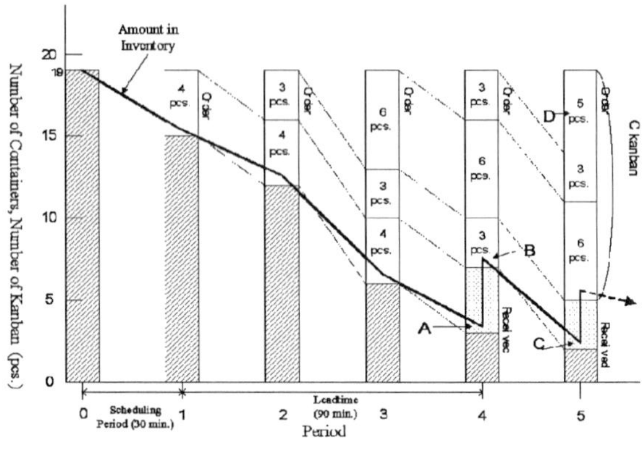

Fig. 3. Amount in inventory in parts-storage

4.1 Domestic Suppliers

The domestic suppliers are using a supplier Kanban, and transport parts, by using the joint distribution system. In the joint distribution, multiple companies deliver their items jointly in the same truck, where they used to deliver them independently.

4.2 International Suppliers

For the international suppliers, VMI (Vendor Management Inventory) warehouses and Bounded JIT warehouses are typically used. In the VMI warehouse, a supplier performs inventory control of parts in the warehouse which is provided by the JIT manufacturer. In the bounded JIT warehouse, the manufacturer pays import duties on as many parts as used in the factory.

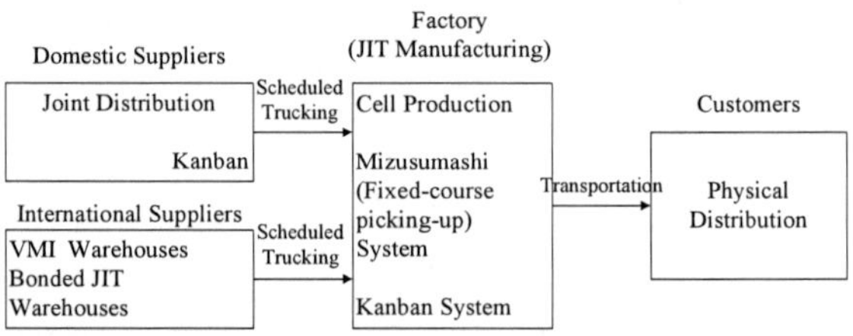

Fig. 4. Logistics for JIT system

5. CONCLUSION

The characteristics of logistics and inventory systems of just-in-time manufacturing on a PC manufacturer were described from the standpoint of management. In addition, the methods of implementing just-in-time manufacturing in supply chain were explained, taking both domestic and international suppliers into consideration.

6. REFERENCES

Monden, Y. (1983). *Toyota Production System*, Industrial Engineering and Management Press, ISBN 0-89806-034-6, Atlanta.

Nomura, J. & Takakuwa, S. (2004). Module-based modeling of flow-type multistage manufacturing systems adopting dual-card Kanban system, *Proceedings of 2004 Winter Simulation Conference*, (eds) M. D. Rossetti, et al., pp.1065-1072, Piscataway, ISBN 0-7803-8786-4, New Jersey: Institute of Electrical and Electronics Engineers

Takakuwa, S. & Nomura, J. (2003). Optimization analysis of Mizusumashi system as just-in-time manufacturing, *Annals of DAAAM for 2005 & Proceedings*, Katalinic, B. (Ed.), pp. 461-462, ISSN 1726-9679.

Annals of DAAAM for 2006 & Proceedings of the 17th International DAAAM Symposium, ISSN 1726-9679
B. Katalinic, Published by DAAAM International, Vienna, Austria 2006
150th Birthday of Nikola Tesla (1856-1943) Inventor of Mobile Robot Teleoperation

THE FUNCTIONAL MODEL OF A ROBOT SYSTEM

TARCA, R.; TARCA, I.; TRIPE VIDICAN, A.; DANUT TOCUT, P. & TRIPE VIDICAN, C.

Abstract: *This paper presents some achievements regarding the link components models of a robot system, which presents a sensory control. The information given by a visual system of the position of a target, materialized by a moving light source, is used to realize the robot control.*
Key words: *visual servoing, CCD camera, MATLAB Simulink, real time driving.*

1. INTRODUCTION

The functional model of a robot system which presents a visual servoing control is presented in this paper. In fact, the robot system control is a position based visual servoing, the system using for control loop two CCD cameras. The robot system cinematic scheme for the system realized by authors is presented in fig. 1.

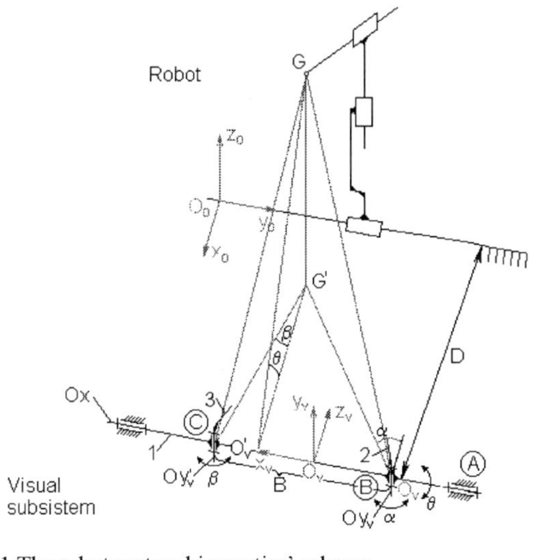

Fig. 1 The robot system kinematics' scheme

The robot system contains a Cartesian robot (TTT structures) respectively a visual subsystem materialized through two CCD cameras.
The visual subsystem has to follows the robot end-effector (point G in fig. 1). The CCD cameras present two DOF (pan and tilt).
The TTT robot joints are actuated by DC actuators, and the visual device are actuated by three step by step actuators.

2. THE SYSTEM COMPONENTS FUNCTIONAL MODEL

2.1 Functional model

The functional model of the robot was realized for one axes targeting system – fig. 2. In order to study the dynamic behavior of a command system with visual sensor the functional model of the system for a robot axes was realized.

2.2 Functional model of the system

"Commanded position" block generates a step signal which represents the programmed position of the axles.
At the receiving time of the signal "Chopper" block generates the signal that represents the necessary supply voltage for the d.c. motor. "MCC" block has as output the signal that represents the relative position of the modeled c.c. elements.
This position is seized by the "visual sensor" and evaluated by the "computing system", being after compared with the programmed value.
The comparison result is the position error and represents the new command level. As the value that characterizes the realized position increases, the difference between the programmed values and the real value decreases, and as a fact the command voltage of the d.c. motor decreases, in the end the d.c. motor stops in the programmed position.

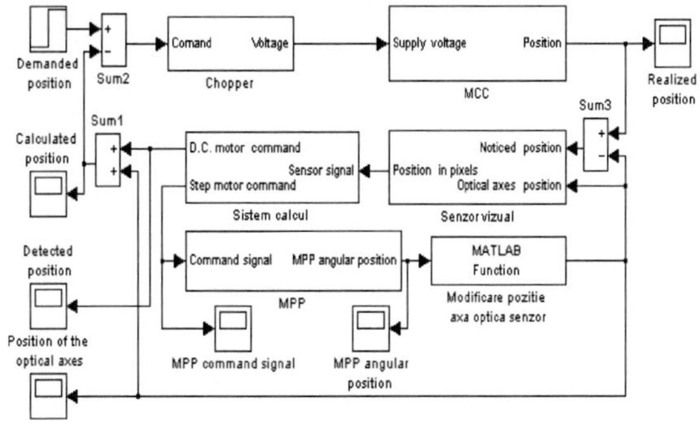

Fig. 2. Functional model of the robot sensorial driving system (for one cinematic couple of the robot)

Because the visual field of the sensor doesn't fit the entire moving domain of the modeled axle this is moving in such a way that the characteristic point should remain in the visual field of the sensor.
The displacement is realized using a step motor commanded by the computing system in such a manner that it executes a step each time the characteristic point approaches to a certain distance from the sensor visual field limit.
To evaluate the behavior of the modeled system in the block schema, graphical displaying elements were introduced for different signals, as functions of time ("Real position", "Optic axes position", etc.)

2.3 Simulation results on the experimental model for one cinematic couple drive

The results of the robot system simulation for one cinematic couple that have relative movements along O_0x_0 axes were achieved for a step signal. This signal represents the prescribed position of the lightning target along O_0x_0 axes. The command signal was generated with a delay of 0.01 [s].
The visual sensor response is presented in figure 3, and is collected from the "Sensor signal" block output.

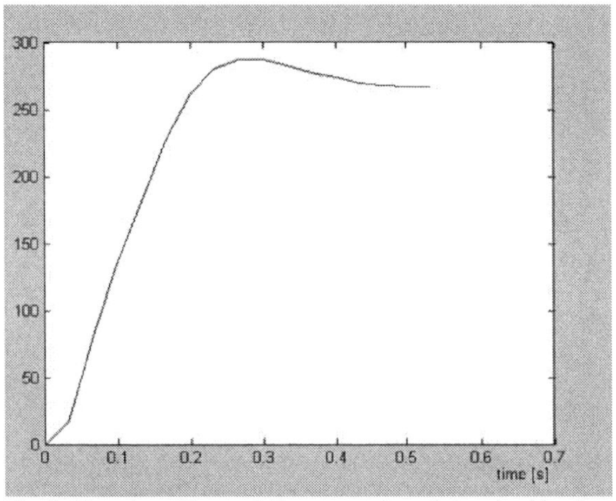

Fig. 3. Target position variation versus time determined by the visual sensor

During target displacement "Computing system" block concluded the necessity of two command signal for the step motor so that the target image should be placed inside the interest rectangle from the image plane.

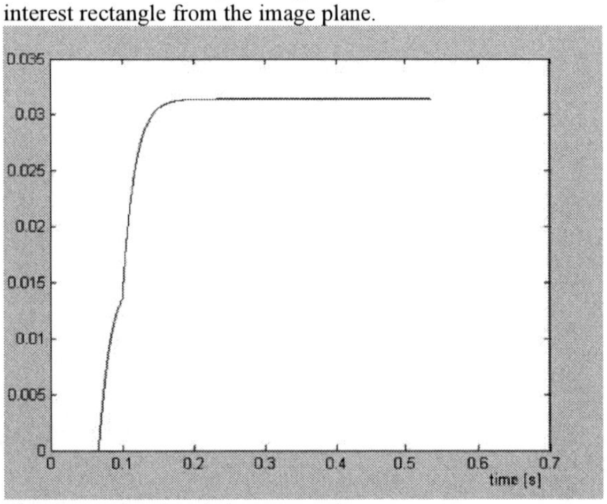

Fig. 4. x_{opt} signal variation versus time

The step motor output signal is transformed in "Sensor optical axes position modification" block into a signal corresponding to x_{opt} , having the evolution with time presented in fig. 4.

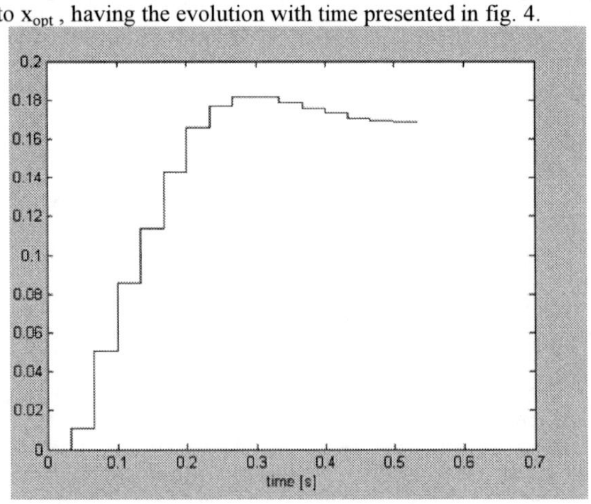

Fig.5 Instant position of the characteristic point x_r variation respectively with the optical axes of the visual sensor versus time

Over this signal the signal presented in fig. 5 is superimposed, thus the position detected by the mean of the visual sensor (real position) being obtained. Fig. 6 presents this signal.

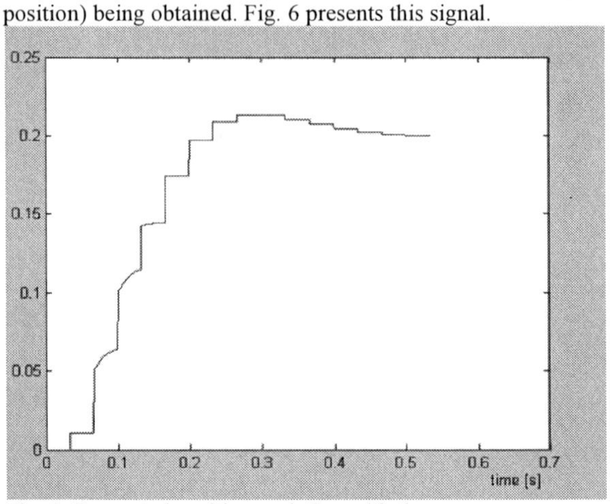

Fig. 6. Computed position variation with time

3. CONCLUSIONS

The robot system response time to a step signal input necessary to achieve the prescribed position with 1% error is 0.53 [s], time in which the adjustment loop of the system was run for 16 times.

Consequently, the average time for one transition of the loop is 0.033 [s].

The time for the generation of the d.c. and step motors commands is under 0.01 [s]. The corresponding time for the image acquisition results from the Matrox® board acquisition frequency, which is 30 frames per second. Consequently, the limitation regarding the execution time of the loop came from the visual sensor.

Table 1 presents the time achieved at the robot system functioning simulation.

Simulated process	**Times obtained during simulation**
D.C. motor signal generating	0,01
Image acquisition	0,03
Compute in „Computing system" block	0,01
Step motor signal generating	0,01

Table 1. Times obtained during simulation [s]

Concluding, considering the times obtained in the simulation process, one can say that the system of the sensorial robot driving which uses information from a visual tracking system functions in real time.

5. REFERENCES

Bogdanov, I., (1989) *Microprocesorul în comanda acționărilor* electrice, Editura Facla, Timişoara.

Kovács, F.,V., Rădulescu, C., (1992), *Roboți Industriali, vol.I-II*, Lito. Universitatea "Politehnica" Timişoara

Țarcă, R.C. (2001) *„Utilizarea informațiilor cu privire la situarea efectorului final achiziționate prin senzori în vederea conducerii în timp real a roboților"*, PhD Thesis, Timişoara.

Corke, P.I. (1994) *High-Performance Visual Closed-Loop Robot Control*, PhD Thesis, University of Melbourne.

Hashimoto, K. (1991) *Model-Based Control of Manipulator with Image-Based Dynamic Visual Servo*, PhD Thesis, Osaka University.

Annals of DAAAM for 2006 & Proceedings of the 17th International DAAAM Symposium, ISSN 1726-9679
Editor B. Katalinic, Published by DAAAM International, Vienna, Austria 2006
150th Birthday of Nikola Tesla (1856-1943) Inventor of Mobile Robot Teleoperation

AUTOMATED SYSTEM FOR MANUFACTURING OF ACCOMODATIVE CUSTOM FOOT ORTHOTICS

UDILJAK, T.; OBROVAC, K.; VUKOVIC-OBROVAC, J. & CIGLAR, D.

Abstract: *Manufacturing of foot orthotics is typical example of custom made production. As it possess a characteristic of mass production (it is evident that same concept could be applied in shoe production industry), manufacturing of foot orthotics could be considered as an example of mass customization concept. This work describes prototype of automated system for manufacturing of accommodative custom foot orthotics, and it is developed on Faculty of Mechanical Engineering and Naval Architecture University of Zagreb. System encompasses component for foot scanning, CNC machine tool with subsystem for chip suction, subsystem for workpiece and insole handling, and software. Foot scanning is done by laser and digital camera; machining is done with 4-axis milling machine tool with two coaxial, opposite milling spindles, each one machining different surface; workpiece handling is done by pneumatic cylinders; software is developed using visual studio package. System is automated and autonomous when applied for accommodative insoles but it could be also applied for other type of insoles and other products.*
Key words: *custom made manufacturing, foot orthoses, automation, body part scanning, CAD/CAM*

1. INTRODUCTION

Through the centuries man has wear footwear as a protection, and to help foot to perform the three main functions: weight bearing support, adaptation to ground and propulsion. Through the evolution, except primary protection of the foot, footwear has also evolved in negative sense, mostly because of fashion (Scherer & Morris 1996). The insoles are feet orthoses, and are applied in case of an ill, injured or deformed foot, in order to improve its function or prevent further deformation (Miller et al., 2000). However, more and more, they are also applied with healthy feet (for example with athletes who use them to deal with larger strains, etc.).
The prototype of automated system for manufacturing of orthopedic insoles-PASIO is primarily concerned with insoles that should be applied with generally healthy feet. These types of insoles are most commonly described as *comfortable insoles*. Equipped with adequate software for insole design and manufacturing, (Lange, 1996), system could be very efficiently used for the manufacturing of insole prescribed by the clinician (Udiljak et al., 2002). No adjustment is necessary except postprocessor, what is necessary for any other CNC machine tool for insole machining.

2. RECENT WORK

Past decade is characterized with intensive development and application of new technologies in production of orthotics. Application of CAD/CAM started in the mid-nineties, and approximately at the same time digitizing found its place in the field of orthotics, (Obrovac et al., 2005). Today, both technologies are in standard use. The most of the developed systems applies one-side machining while the other side of insole is pre-formed (templates) or it is adjusted manually. The

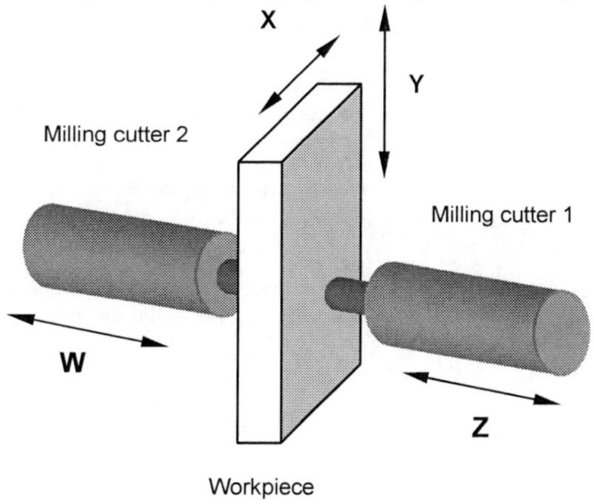

Fig. 1. The principle of two coaxial opposite milling spindles

PASIO system uses two coaxial, opposite milling spindles and machines both insole surfaces simultaneously, figure 1. Both spindles run up to 24000 rpm, enabling high speed machining.

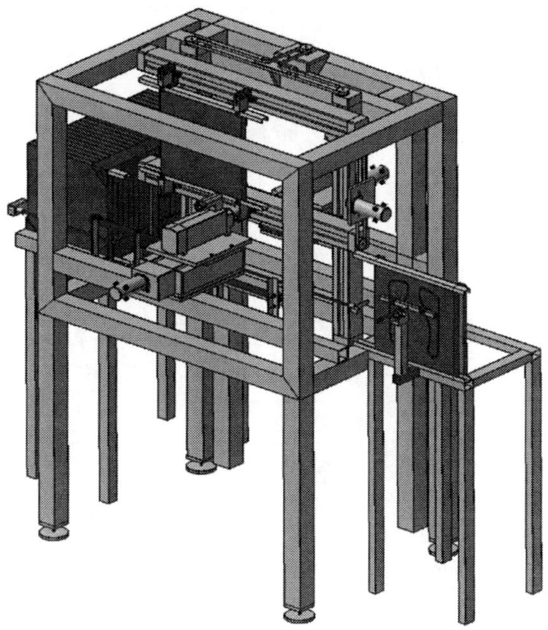

Fig. 2. Schematic of machining and workpiece handling system, PASIO

Since the system should be autonomous, the digitizing

(Rocchini et al., 2001) of foot is very important. Among many possible methods (Taylor & Kleeman 2004), we selected the method based on laser and single digital camera, figure 3.

3. GENERAL SYSTEM REQUIREMENTS

Design of machining module is based on several following requirements that must be satisfied:

- Machining time < 10 minutes,
- Possibility to machine 10 insole pairs without human intervention,
- Basic system price < 20.000,00 EUR,
- User friendliness.

To fulfill the requirements the feedrate velocity should be in the range of 5m/min. To obtain acceptable surface roughness (less then 0.1 mm) radial depth for ball-nose end mill of 12 mm in dia, should be 1.5 mm or less, (Udiljak et al., 2002). With all other times such as clamping time and loading-unloading time it is very important to have digitalization time as short as possible. Having in mind the price of the components and customer satisfaction, the project digitalization time was 1 minute.

4. DIGITAZING METHODOLOGY

As it is already stressed, the importance of digitalization is one of the crucial requests, necessary for the system autonomy.

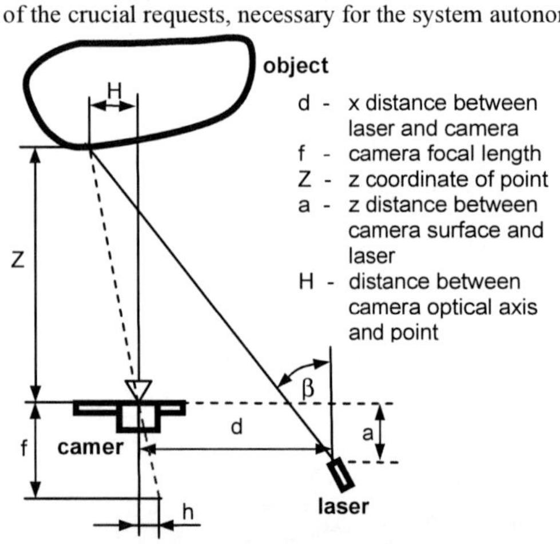

Fig. 3. Digitizing with laser and single digital camera

PASIO system uses digitizing methodology based on the laser and single digital camera, (Askold, 2003). Fig. 3 is schematic presentation of depth calculation of single point for single plane, while the figure 4 is photo of realized digitizer. The

Fig. 4. The photo of the digitizer

values of *a and* β are defined during setting, while the value of f *is* obtained during calibration procedure. The Z coordinate (depth) of point could be calculated with simple trigonometric relations.

$$\frac{h}{H}=\frac{f}{Z} \quad and \quad \tan\beta=\frac{d+H}{a+Z} \qquad (1)$$

$$Z=\frac{a\tan\beta-d}{h/f-\tan\beta} \qquad (2)$$

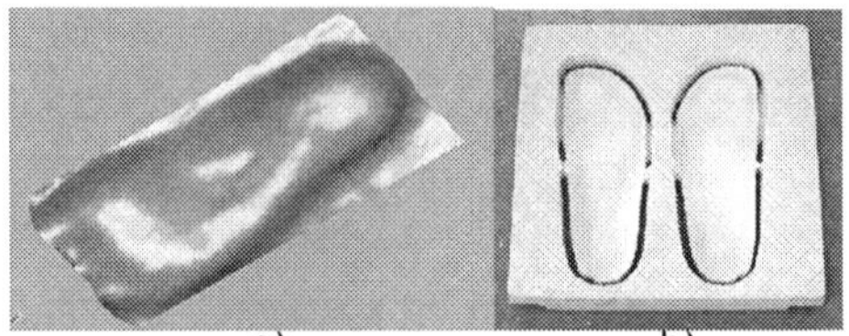

a) b)

Fig. 5. The result of foot digitalization (a) and photo of insole after machining (b)

5. CONCLUSIONS

PASIO system can manufacture pair of insoles for economical price in any store or clinic fully automatic. Basic advantage of the system is manufacturing of quality insoles with high level of autonomy, i.e. without need for operator. All necessary actions customer can do on himself. Main disadvantage of the system is low level of flexibility when system works autonomously. In autonomous mode, system is working with only one type of material and can be used only for manufacturing of insoles. However, equipped with appropriate software system could be used in other applications. Further work will be focused on research of other digitizing methodologies with intension to reduce digitizing time, and without significant influence on system price. Considerable efforts will be given to research of software development in order to extend PASIO application area.

6. REFERENCES

Scherer, P.R, Morris, J.L. (1996). *The classification of human foot types, abnormal foot function, and pathology*, In: Valmassy RL, Clinical biomechanics of the lower extremities, Mosby, St.Lois

Miller, J.E, et all. (2000). *"Influence of Foot, Leg and Shoe Characteristics on Subjective Comfort"*, Foot and Ankle International/Vol.21, No.9, pp. 759-767

Lange, LR. (1996). *The Future of CAD/CAM in O&P, Biomechanics*, Vol.III No.2

Udiljak, T, Granberry, W.M, Obrovac, K. (2002). *New Approach to Foot Orthosis Design and Manufacturing*, Proceedings of World Manufacturing Congress, Rochester, New York

Obrovac, K., Udiljak, T., Mihoci, K. (2005). Foot Orthosis Design and Manufacture, FAMENA, 29(2005)1, ISSN 1333-1124, pp. 49-60

C. Rocchini, C., Cignoni, P., Montani, C., Pingi, P., &Scopigno, R. (2001). *A low cost 3D scanner based on structured light,* EUROGRAPHICS 2001, *Volume 20 (2001), Number 3,* ISSN 1017-4656

G. Taylor, G., & Kleeman, L. (2004). "Stereoscopic Light Stripe Scanning: Interference Rejection, Error Minimization and Calibration", International Journal Robotics Research, Vol 23 No 12, pp 1141-1156

Udiljak, T, Obrovac, K, Mihoci, K. (2002). *Design and Technology Guidelines of CAD/CAM System for Foot Orthoses*, The 13th International DAAAM Symposium "Intelligent Manufacturing & Automation: Learning from the Nature", Wienna

Askold V. Strat, Manuel M. Oliveira, (2003). "A Point-and-Shoot Color 3D Camera," Proceedings of the 4th International Conference on 3-D Digital Imaging and Modeling (3DIM), Banff, Canada, pp. 483-491.

Annals of DAAAM for 2006 & Proceedings of the 17th International DAAAM Symposium, ISSN 1726-9679
B. Katalinic, Published by DAAAM International, Vienna, Austria 2006
150th Birthday of Nikola Tesla (1856-1943) Inventor of Mobile Robot Teleoperation

INVESTIGATION AND ANALYSIS OF BIPEDAL HUMAN WALKING TO CONTROL THE MOTION OF KNEE PROSTHESES

UYAR, E.; BASER, O.; CETIN, L. & OFLAZ, H.

Abstract: *In this paper a general motion analysis of normal human bipedal-walking under various circumstances is carried out and the optimal control of knee joint for an upper knee amputee is investigated. Real measurements of both hip and knee angles of healthy human are obtained on a walking platform. After exposing the motions of hip and knee angles, a theoretical analysis of both measured stochastic signals are made using auto and cross correlation functions. These results are then used to improve a computer based control of the knee prosthesis motion of an amputee without knee joint with respect to the motion of hip (Baser, 2003).Various measurements are made to investigate the real parameter variations. In this way, it is aimed to reach an optimum theoretical model of the human gait dynamics to manufacture upper knee amputee prosthesis according to this model.*

Keywords: *Hip, knee joint control, upper-knee amputee, human gait analysis*

1.INTRODUCTION

The purpose of the study is to improve a prosthesis, which would help to the amputees to walk as good as possible without using any aid equipments like a walking stick. In the first section of this study, the gait analysis of a bipedal walker is made by using a convenient measuring device which is designed in the physical therapy laboratories of the medical Faculty of the University of D.E.U. For this reason a jointed measuring prosthesis is constructed for the investigation of hip and knee motion. This prosthesis is equipped with well-situated potentiometers on the relevant joints.

The analogue outputs of these potentiometers as hip and knee motion are then transferred digitally to a PC. These data are investigated to introduce the relations between the joints of a leg in the standard walking.

2. EXPERIMENTAL ANALYSIS

For the walking experiments, from four different people, sufficient number of data are taken as voltage signals and captured by computer and a statistical evaluation is done to find out the average values, which can be taken in consideration as significant characteristics of normal walking.

Real measurements of both hip and knee angles of a healthy human being during bipedal walking under various circumstances are registered to make a certain decision of normal walking conditions. In these experiments, it is seen that, these hip and knee angles are quasi-periodic with random amplitude and phase which make these signals stochastic (Fig 1). After using a software filter hip angle versus time and knee angle versus time graphs are employed in Fig. 2(a) and 2(b).

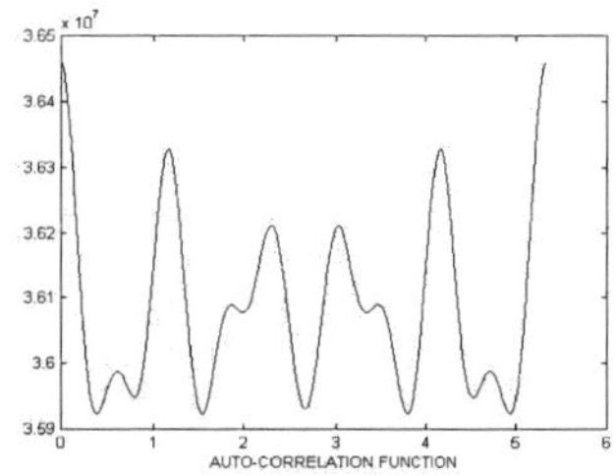

Fig. 1 Auto-correlation function of one of the data of hip angle in a 4 period of time.

3. THEORETICAL SIGNAL BASICS

It is known that walking is not a deterministic motion, since the walking of different people or even the same individual changes with the changing conditions (Baser, 2003). For this reason, walking can be modeled as a stochastic process. As a new approach, a linearized statistical analysis based on finding out the relationship between hip and knee joints of the same walker is investigated by using estimated values of stochastic changes. Through the correlation analysis of hip and knee signals, the stationary and ergodic behaviour of both motions are investigated.

Correlation is a mathematical expression that closely resembles convolution. It is known that convolution can be used to calculate response of a system to arbitrary inputs by using the impulse response of a system. It can be defined as:

$$y(t) = h(t) * x(t) = \int_{-\infty}^{+\infty} f(\tau) \cdot h(t-\tau) d\tau \qquad (1)$$

where $h(t)$ is the response of the system to a unit impulse. In contrast to convolution, the goal in the computation of the correlation between two discrete signals is to measure the degree to which the two signals are similar and thus to extract some information about the nature of the signals and relationship between them (Larmouth et al., 1992).

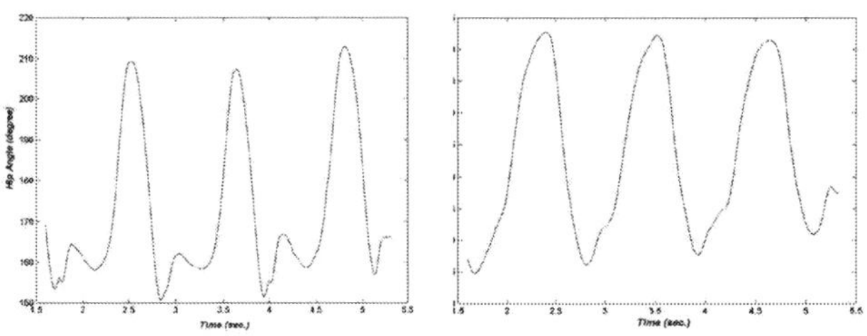

Fig. 2 (a) hip angle vs. time graph (b) knee angle vs. time graph

From the ergodic hypothesis, it is known that the ensemble average which is the *auto-correlation function* $R_{xx}(\tau)$ for $x_1 = x(t)$ and $x_2 = x(t+\tau)$ is equal to the time average:

$$R_{xx}(\tau) = \lim_{T\to\infty} \frac{1}{2T} \int_{-T}^{T} x(t) \cdot x(t+\tau) \cdot dt \tag{2}$$

When we consider two different random signals $x(t)$ and $y(t)$, we have *cross-correlation function* $R_{xy}(\tau)$ and according the the ergodic hypothesis:

$$R_{xy}(\tau) = \lim_{T\to\infty} \frac{1}{2T} \int_{-T}^{T} x(t) \cdot y(t+\tau) \cdot dt \tag{3}$$

If we compare convolution and cross-correlation terms, we can say that they look similar. In fact, we can perform cross-correlation as if we are performing the convolution of signal $x(t)$ obtained from the angular changes at hip joint and the folded signal $y(-t)$ obtained from the angular displacement of knee joint. This means:

$$R_{yx}(\tau) = y(\tau) * x(-\tau) = h(\tau) * [x(\tau) * x(-\tau)] \tag{4}$$

The term in parenthesis gives the auto-correlation function. Hence, Eq. (4) yields:

$$R_{yx}(\tau) = h(\tau) * R_{xx}(\tau) \tag{5}$$

The operation here is convolution; so to make the analyses easier, we have to define correlations in frequency domain in terms of energy density spectra. Taking the Fourier Transform of the Eq.(5) gives the relation between correlation functions, thus between hip and knee angles. Here, $S_{xx}(w)$ denotes the energy density spectrum and $S_{yx}(w)$ is the cross-energy density spectrum.

$$H(w) = \frac{S_{yx}(w)}{S_{xx}(w)} \tag{6}$$

4. CONTROL APPROACH

After the above mentioned theoretical analysis a control model and approach is applied. For this purpose a control circuit as given in Fig.3 is carried out. The measured hip angle signal with hip potentiometer is used as the input data of a microcontroller which produce from hip angle also the hip deviation, so that the instant hip position can be investigated correctly. By means of above mentioned transformation the relevant reference knee signal is then evaluated as a look up table. Comparing the reference signal with the instant knee feedback signal, the microcontroller produces the relevant signal for knee motion which will be executed over a geared DC motor.

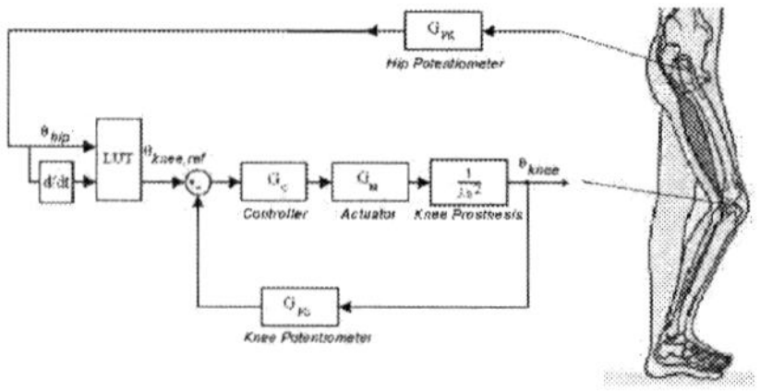

Fig. 3 Block diagram of the control system

5. CONCLUSION

In this paper, analysis of an upper knee leg plant and its control applications are studied. As a statistical approach, the agreement between maxima or minima of angles measured in walking cycles of four participants is searched. From the results it is seen that the general shapes of gait cycles of different people are not exactly the same. These differences are mostly due to different anatomical properties of people. These properties affect the walking characteristics of walkers.

Another analysis, which helps us to understand the walking dynamics, is the investigation of walking process of the participants individually. The results of the variance analyses for this investigation show us that after the dynamic balance, there is no significant difference between the gait cycles of the walker

So we can conclude that when the walking dynamics of an upper-knee amputee is investigated in steady state, we might have an opinion about his/her walking behaviour. The correlation between hip and knee angles has to be worked out for the control of prosthetic knee according to the angular changes at the hip joint. By using correlation functions we can control the knee angle according to the given hip angle sequence as input to the system which is represented by the transformation function obtained. The results are given in Figure 4 and it is seen from the figure that the knee can follow the hip with a negligible error by using this control algorithm.

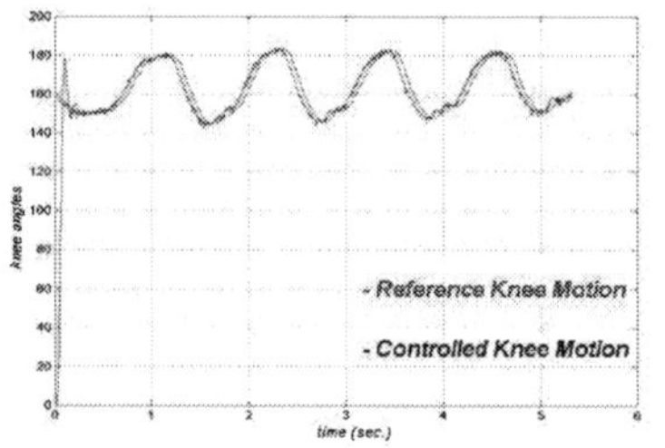

Fig. 4 Comparison of reference knee and controlled knee motion

6. REFERENCES

S. Deutsch, A "Understanding Nervous System", IEEE Press, Newyork, 1992.

O.Baser "Analysis and Control of bipedal human locomotion", M.S. Thesis, Dept. Mechanical Eng., Dokuz Eylul University, Izmir, Turkey, 2003.

A.Torige, "Analysis of eonsumption of energy on biped robots", *Syrocco 2000 Conf.*, Austria-Wien, pp. 139-145

E.Uyar, O.Baser, E.Ozcivici and R.Baci, "Investigation of bipedal human gait dynamics and knee motion control", MDM2002 Conf., Cappodocia, Turkiye, 2002.

J.Larmouth, F.N.Nagy, "Personaş assistant robots for the 21th century (intelligent bipeds and multipeds)", IFAC International Workshop on Automatic Control for Quality and productivity, 3-5 June 1992, Istanbul, Turkiye, pp.1-8

Annals of DAAAM for 2006 & Proceedings of the 17th International DAAAM Symposium, ISSN 1726-9679
Editor B. Katalinic, Published by DAAAM International, Vienna, Austria 2006
150th Birthday of Nikola Tesla (1856-1943) Inventor of Mobile Robot Teleoperation

INVESTIGATION OF OPTIMAL CONTROL PARAMETERS OF A BRIDGE CRANE AND ITS MODEL IMPLEMENTATION

UYAR, E.; CETIN, L. & ORAL, O.

Abstract: *In this work optimal control of a bridge crane like system is investigated with a theoretical and implemented model. After the investigation of stability conditions, the LQR principle is applied to the theoretical model for evaluation of optimal design parameters.*
Keywords:*Optimal control, System Modeling, Computer Controlled Systems*

1. INTRODUCTION

Bridge cranes are used in various industrial establishments like steel works, factories, shipyards, construction and building businesses, etc. to carry heavy loads. (Abdel-Rahman, 2003), (Corriga, 1998).
In the following work a theoretical model of such a crane is given and the relevant motion equations in form of state variables are represented. A closed loop control by using LQR principle is applied to achieve a stable system and a desired transient behavior (Bugaric et al., 2002), (Giua et al., 1999).
Relevant theoretical results are derived then compared with a built up implementation model.

2. THEORETICAL MODEL AND DERIVATION OF MOTION EQUATIONS

The system which will be mathematically modeled is shown in Figure 1 below. It is important to notice that the hanging load-crane model can represent the inverted pendulum model as well (Abdel-Rahman, 2003), (Ogata, 2002), (Dorf et al., 2005). The hanging load can be thought as a simple pendulum.
To obtain the first equation of motion, the summation of total forces applied to the whole system along x direction is obtained (Ogata, 2002), (Dorf et al., 2005). From the free body diagram of the cart-and-crane system, it is found (1).

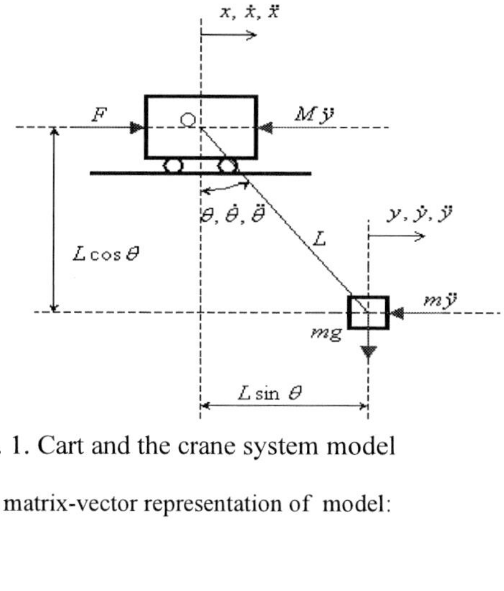

Fig. 1. Cart and the crane system model

The matrix-vector representation of model:

$$\begin{bmatrix}\dot{x}_1\\ \dot{x}_2\end{bmatrix}=\begin{bmatrix}0 & 1\\ -\dfrac{g(m+M)}{ML} & -\dfrac{C(m+M)}{mML}\end{bmatrix}\cdot\begin{bmatrix}x_1\\ x_2\end{bmatrix}+\begin{bmatrix}0\\ -\dfrac{(m+M)}{ML}\end{bmatrix}\cdot \qquad (1)$$

3. LQR DESIGN PROBLEM AND SPECIFICATION OF OPTIMAL PARAMETERS

Given any state variable model of the form (1), it is assumed that all the states are measurable and the desired closed-loop properties are to be determined such that the cost function, or performance index is minimized (Corriga, 1998), (Ogata, 2002)

$$J=\frac{1}{2}\int_0^{\infty}\left(\mathrm{x}^T\mathrm{Qx}+\mathrm{u}^T\mathrm{Ru}\right)dt \qquad (2)$$

The objective in optimal design is to select the feedback matrix **K** that minimizes the performance index J. The two matrices **Q** and **R** are selected by the designer and **Q** is n x n symmetric and positive semi-definite; whereas **R** is m x m symmetric positive definite, (Dorf et al., 2005). The problem of determining the **K** to minimize J is called the Linear Quadratic Regulator (LQR).
To obtain the matrix **P**, algebraic Riccati equation is to be solved

$$\mathrm{A}^T\mathrm{P}+\mathrm{PA}-\mathrm{PBR}^{-1}\mathrm{B}^T\mathrm{P}+\mathrm{Q}=0 \qquad (3)$$

Here, the system that is analyzed is of order two, hence Q, R, P and K matrices will be

$$\mathrm{K}=\mathrm{R}^{-1}\mathrm{B}^T\mathrm{P}=-\frac{B_{21}}{r}\begin{bmatrix}p_{12} & p_{22}\end{bmatrix} \qquad (4)$$

p_{12} and p_{22} was calculated from the solution of Riccati Equation as

$$p_{12}=\frac{A_{21}}{B_{21}^{\;2}}r+\frac{1}{B_{21}^{\;2}}\sqrt{(A_{21}r)^2+B_{21}^{\;2}q_1r} \qquad (5)$$

$$p_{22}=\frac{A_{22}+\sqrt{(A_{22}r)^2+(p_{12}+q_2)B_{21}^{\;2}r}}{B_{21}^{\;2}} \qquad (6)$$

the characteristic equation of the controlled system is:

$$s^2+\left(\frac{p_{22}}{r}-A_{22}\right)s+\left(\frac{p_{12}}{r}-A_{21}\right)=0 \qquad (7)$$

If equation (7) is set to be equal to the general form of characteristic equation of second order systems, damping ratio and natural frequency of the system can be found by means of **P** and **R**. Proceeding with placing equations (5) and (6) into (7), direct relation between system parameters and **Q** and **R** as

$$w_n=\sqrt{\frac{A_{21}}{B_{21}^{\;2}}-A_{21}+\frac{1}{B_{21}^{\;2}}\sqrt{A_{21}^{\;2}+\frac{B_{21}^{\;2}q_1}{r}}} \qquad (8)$$

$$\xi = \frac{1}{2w_n}\left(\frac{A_{22}}{B_{21}^{\ 2}} - A_{22} + \right.$$

$$\left. \frac{1}{B_{21}^{\ 2}}\sqrt{A_{22}^{\ 2} + \frac{q_2}{r}B_{21}^{\ 2} + 2A_{21} + 2\sqrt{A_{21}^{\ 2} + B_{21}^{\ 2}\frac{q_1}{r}}}\right) \quad (9)$$

4. CONTROL OF THE MODEL

In the control design of this experiment we shall have one eminent objective in mind. The purpose is to minimize the angular deviation of the pendulum (minimal swing).

A .Analysis of the Implementation Model

We are farther aware that the following parameters compose the inverted pendulum system.

Fig. 2. Implementation Setup

Physical Variables	Values
M	0.65 (kg)
m	0.405 (kg)
L	0.28 (m)
g	9.81 (m/s^2)

Table 1.Values Of The Physical Variables

After using the numerical values, ω_n and ξ becomes

$$\omega_n = \sqrt{55.18 + \sqrt{2.86 + 0.03\frac{q_1}{r}}} \quad (10)$$

$$\xi = \frac{1}{2\omega_n}\left[7.267 + \sqrt{-0.05 + 0.03\frac{q_2}{r} + \sqrt{0.01 + 1.1x10^{-4}\frac{q_1}{r}}}\right] \quad (11)$$

Proper **Q** and **R** values are calculated for requested time domain design specifications.These equation pairs give only the ratio between r and q values so control action is to be set before any further calculation of **Q** (Ogata, 2002), (Dorf et al., 2005).

B. Experiments

Since we are using LQR to control the system, we must adjust the value of the **Q** and **R**, in order to meet the target of minimal oscillations. In this stage of operation, to visualize the results, parameters are defined using time domain specifications that are overshoot value and steady state time.

As regards, implementation layout turns out as this:

1. Specify overshoot value and steady state time as constraints.
2. Calculate damping ratio from overshoot value and and natural frequency from steady state time.
3. Solve (24) for q_1/r and (27) for q_2/r
4. Set r related to control power and obtain q_1 and q_2
5. Evaluate (15) (16) and obtain K feedback matrix.

5. RESULTS AND CONCLUSION

This paper presents a novel selection routine for optimal control parameters. Specific contribution to the problem is to expand the formulation through time domain specifications and formulate a mathematical connection to optimal control parameters. Formulation seems more practical approach in LQR when considered trial and do method. An example of calculations are made for an experimental crane model for control of oscillations and results are verified the specifications. The unit impulse response of the optimal controlled system output, and the corresponding control input as percentage of the maximum control signal are given below in Figure 3 and Figure 4:

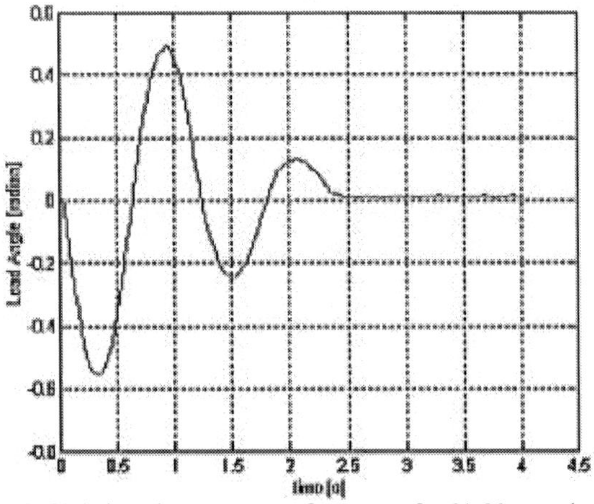

Fig. 3. Unit impulse response of system for % 30 overshoot and 2.5 s steady state time

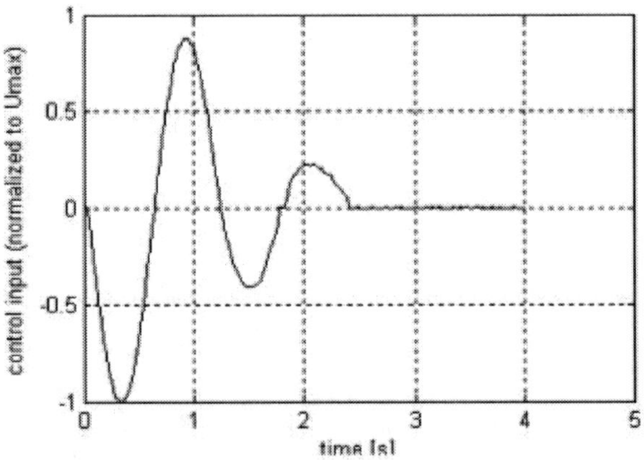

Fig. 4. Control signal magnitude of the system for % 30 overshoot and 2.5 s steady state time

6. REFERENCES

E. M. Abdel-Rahman, "Dynamics and control of cranes: A review," *Journal of Vibration and Control*, vol. 9, pp. 863-908, 2003.

G. Corriga, A. Giua, G. Usai, "An implicit gain-scheduling controller for cranes," *IEEE Transactions on Control Systems Technology*, Vol. 6, pp. 15-20, January 1998.

K. Ogata, Modern Control Engineering, 4th ed., Prentice Hall, Inc., New Jersey, 2002, pp. 86-90, 897-910.

R. C. Dorf, R. H. Bishop, Modern Control Systems, 10th ed., Prentice Hall, Inc., New Jersey, 2005, pp. 151-152, 689-693.

U. Bugaric, J. Vukovic, "Optimal control of motion of the system based on mathematical pendulum with constant length," FME Transactions, Vol. 30, pp. 1-10, 2002.

A. Giua, C. Seatzu, G. Usai, "Observer controller design for cranes via Lyapunov equivalence," Automatica, Vol. 35, pp. 669-678, 1999.

Annals of DAAAM for 2006 & Proceedings of the 17th International DAAAM Symposium, ISSN 1726-9679
B. Katalinic, Published by DAAAM International, Vienna, Austria 2006
150th Birthday of Nikola Tesla (1856-1943) Inventor of Mobile Robot Teleoperation

AC INDUCTION MACHINE SPEED ESTIMATION BASED ON ROTOR SLOT HARMONICS ANALYSIS

VACLAVEK, P. & BLAHA, P.

Abstract: *Control of drives based on AC induction motors is a quite complex task. Provided the vector control algorithm is used we need to know not only the rotor speed but also the position of the magnetic flux inside the motor during the control process. Flux position is usually estimated, but there are also applications in which even speed sensors should be omitted. In the current paper we present a method of AC induction machine speed estimation based on rotor slot harmonics analysis.*

Key words: *AC induction motor, sensorless control, rotor slot harmonics, motion control*

1. INTRODUCTION

Induction motors become more and more popular due to their reliable construction. If we intend to use AC induction motor in a low cost application (e.g. mass produced washing machine) it is necessary to optimize production costs. In many applications precise speed control is required. The speed sensor is a quite expensive device comparing to the other parts of the drive. That is why we are trying to develop a reliable control system which estimates the rotor speed from electrical quantities instead of using a speed sensor.

2. MODEL BASED SPEED ESTIMATION

The most straightforward idea of AC induction machine speed estimation is based on the AC induction machine model

$$\frac{d\mathbf{\Psi}_s}{dt} = \mathbf{u}_s - R_s \mathbf{i}_s \quad (1)$$

$$\frac{d\mathbf{\Psi}_r}{dt} = j\, z_p\, \omega \mathbf{\Psi}_r - R_r \mathbf{i}_r \quad (2)$$

$$\frac{d\omega}{dt} = \frac{1}{J}\left(T - T_{load}\right) \quad (3)$$

$$\mathbf{i}_s = \frac{\mathbf{\Psi}_s}{L_M} - \frac{\mathbf{\Psi}_r - \mathbf{\Psi}_s}{L_L} \quad (4)$$

$$\mathbf{i}_r = \frac{\mathbf{\Psi}_r - \mathbf{\Psi}_s}{L_L} \quad (5)$$

$$T = 1.5\, z_p\, \Im\left(\overline{\mathbf{\Psi}}_s \mathbf{i}_s\right) \quad (6)$$

where $\mathbf{\Psi}_s = \Psi_{s\alpha} + j\Psi_{s\beta}, \mathbf{\Psi}_r = \Psi_{r\alpha} + j\Psi_{r\beta}$ are stator and rotor fluxes in $\alpha\beta$ coordinates, $\mathbf{u}_s = u_{s\alpha} + ju_{s\beta}$ is stator voltage, $\mathbf{i}_s = i_{s\alpha} + ji_{s\beta}, \mathbf{i}_r = i_{r\alpha} + ji_{r\beta}$ are stator and rotor currents, ω is rotor angular velocity, T, T_{load} are driving and load torques, R_s, R_r are stator and rotor resistances, z_p is the number of pole-pairs, J is the rotor inertia and L_M , L_L are magnetizing and leakage inductances.

Model based methods includes several approaches - Extended Kalman Filter applications (Holtz, 2001), non-linear state observers (Vaclavek & Blaha, 2003), MRAS structures (Vas, 1998). A common problem of all model based methods is their dependency on actual machine parameters. Some of them can vary rapidly during the drive operation (especially resistances) resulting in speed estimation errors. It is clear that it would be great to find other speed estimation method which is machine parameters independent.

3. ROTOR SLOT HARMONICS ANALYSIS

Three phase induction machine is normally supplied by a system of symmetrical three phase harmonic voltages. The magnetic flux will contain not only fundamental but also higher harmonics. Some of them are caused by nonlinearities in magnetizing characteristics and by asymmetry in the machine parameters or geometry. But there are also harmonics produced by the reluctance variation caused by stator slots and moving rotor slots. Frequency of these harmonics is determined be the rotor rotational speed.

It can be shown (Nandi at al., 2003), that the fundamental rotor slot harmonic can be expressed as

$$f_{sh} = \frac{N_r}{z_p} f_r - f_s \quad (7)$$

where f_{sh} is fundamental rotor slot harmonic frequency, N_r is number of rotor slots, z_p is number of pole pairs, $f_r = \omega/(2\pi)$ is rotor rotational frequency and f_s is synchronous stator frequency.

The flux variations caused by rotor slots generate rotor slot harmonics in the stator current. Unfortunately it can be quite difficult to detect these harmonics in the stator current due to their very low amplitude comparing to the fundamental stator current harmonic. That is why we use another method. It is possible to monitor voltage u_s between the neutral point *0* of the stator windings and the neutral point *0'* constructed by the three phase resistive network as shown on Fig. 2. It is possible to show, that the amplitude of the harmonic at stator frequency f_s and also its multiplies Nf_s is significantly reduced in the voltage u_s (it is not zero because of machine parameters and geometry asymmetry).

Actual rotor speed can be computed by following algorithm:

1. Signal u_s is filtered by low pass filter to remove high frequency high-voltage components introduced by PWM.
2. A signal spectrum is computed using FFT algorithm.
3. Stator frequency f_s is determined.
4. Frequency range $\left\langle \frac{N_r}{z_p}\frac{\omega_{\min}}{2\pi} - f_s ; \frac{N_r}{z_p}\frac{\omega_{\max}}{2\pi} - f_s \right\rangle$ is searched for the harmonic f_{sh} with the highest amplitude. The harmonics with frequency Nf_s where $N = 1,2,3,\ldots$ are excluded from the search. Drive is

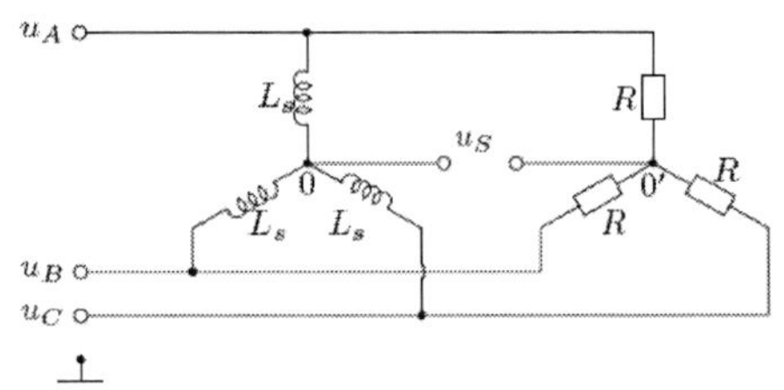

Fig. 1. Signal measurement for rotor slot harm. analysis

supposed to work at the rotor speed in the interval $\langle \omega_{\min}; \omega_{\max} \rangle$. The speed bounds are determined from stator frequency and maximal possible slip value.

5. Rotor speed is computed using

$$\omega = (f_{sh} + f_s)\frac{2\pi z_p}{N_r} \quad (8)$$

Some of the rotor slot harmonics can be attenuated because of the machine geometrical construction. If it is possible, the machine construction should be selected to provide easy rotor slot harmonics detection at reasonable frequencies [10]. In any case we do not know the rotor slots number N_r. Then we have to make experiment when the rotor speed is being measured by a speed sensor and the rotor slots number is computed from (7) as

$$N_r = \frac{f_{sh} + f_s}{f_r} z_p \quad (9)$$

4. ALGORITHM IMPLEMENTATION

The algorithm has been tested on a real small 250 W induction machine with 2 pole pairs and 36 rotor slots. The machine has been supplied by inverter driven by Freescale DSP56F805 Hybrid Controller. At this time the electronics used in the experiment does not provide functionality needed for capturing of the signal needed for rotor slot harmonics analysis. That is why the data for the experiment has been captured using HP 89410A vector signal analyzer and then processed in Matlab environment. We plan to make slight modifications in the electronics to be able to process rotor slot harmonics analysis directly on Freescale Hybrid Controller.

The process of searching for the rotor slot harmonics can be described using fig. 2. The machine was running with stator frequency $f_s = 47\text{Hz}$ and rotor mechanical speed $\omega = 1000\text{rpm}$. There are several significant spectral lines. The first one at point A is at the frequency of the stator voltage fundamental harmonic. In the ideal case there will be no harmonic at this frequency and this harmonic component occurs only due to machine stator windings parameters asymmetry. The spectral lines at points C and E stand for the third and ninth stator voltage harmonics and are excluded from the search. As we know the breakdown slip of our machine to be $s_b = 50\%$ we can compute the minimal possible rotor speed

$$\omega_{\min} = f_s(1 - s_b)\frac{2\pi}{z_p} = 73.8\,\text{rad.s}^{-1} \quad (10)$$

and minimal possible rotor slot harmonics frequency

$$f_{sh_{\min}} = \frac{N_r}{z_p}\frac{\omega_{\min}}{2\pi} - f_s = 164.4\text{Hz} \quad (11)$$

Because of this minimal possible rotor slot harmonic frequency also the spectral line B is excluded. If we search for the harmonic component with the highest amplitude omitting excluded spectral lines we find the spectral line D at frequency $f_{sh} = 254\text{Hz}$ and we can compute rotor speed using (8) as

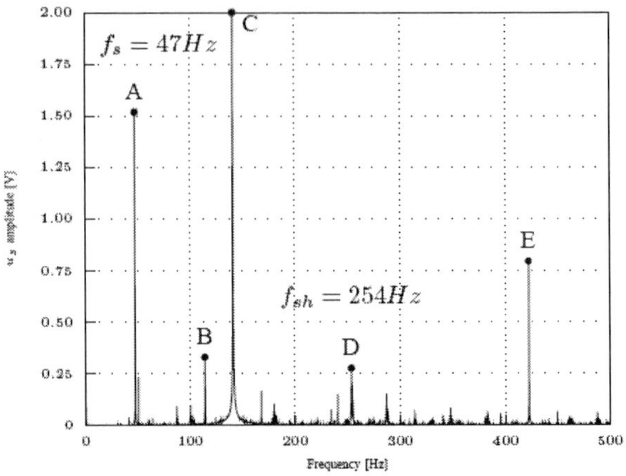

Fig. 2. Voltage spectra for rotor speed 1000 RPM

$$\omega = (f_{sh} + f_s)\frac{2\pi z_p}{N_r} = 1003\text{rpm} \quad (12)$$

It can be seen that the speed estimation error is as low as 0.3%. The algorithm testing proved that such a high accuracy can be achieved in wide range of rotor speed. The only problem is poor dynamical response in low speed range when the algorithm can be compared to the performance of a low resolution encoder.

5. CONCLUSION

The presented algorithm has been proven in testing on real AC induction machine. The algorithm seems to be efficient and provides machine parameters independent speed estimate. Parameter independency is a very promising feature as it will allow development of AC induction machine sensorless control scheme with machine parameters adaptation ability.

Future research will concentrate on the hardware and software improvement to be able to perform rotor slot harmonics analysis in real-time.

6. REFERENCES

Holtz, J. (2001). Sensorless Speed and Position Control of Induction Motors. In: Proceedings of the 27th Annual Conference of the IEEE Industrial Electronics Society, Nov. 29 - Dec. 2, 2001, IECON, Denver/Co.

Nandi, S.; Toliyat, S. Ahmed, H. & Bharadwaj, R. (2003) Selection criteria of induction machines for speed-sensorless drive applications. *IEEE Transactions on Industry Applications*, vol. 39, no. 3, 2003, pp. 704–712.

Vaclavek, P. & Blaha, P. (2003). Lyapunov function based flux and speed observer for AC induction motor sensorless control implementation on motorola DSP. In: Proceedings of the IEEE Conference on Industrial Technology 2003, pp. 536-541, Dec. 10 - Dec. 12, 2003, Maribor.

Vaclavek, P. & Blaha, P. (2005). Lyapunov-function-based flux and speed observer for AC induction motor sensorless control and parameters estimation. *IEEE Transactions on Industrial Electronics*, vol. 53, no. 1, 2006, pp. 138–145.

Vas, P. (1998). Sensorless Vector and Direct Torque Control. Oxford University Press, ISBN 0-19-856465-1, New York.

Acknowledgment: This work was supported by Czech Science Foundation under the project GA 102/06/0949, the Ministry of Education of the Czech Republic under Project 1M0567 and by Freescale Semiconductor Czech Republic (former division of Motorola).

Annals of DAAAM for 2006 & Proceedings of the 17th International DAAAM Symposium, ISSN 1726-9679
Editor B. Katalinic, Published by DAAAM International, Vienna, Austria 2006
150th Birthday of Nikola Tesla (1856-1943) Inventor of Mobile Robot Teleoperation

INTERNET BASED DEVELOPMENT OF AN OPEN SOURCE LOW-COST CNC

VANOLI, R.; LICARI, R. & LO VALVO, E.

Abstract: *In this paper we show how to realize a low-cost numerical system, based on an "open source" software called EMC (Enhanced Machine Control). First, we assembled a low cost CNC workstation with a remote control by means of an online computer with a webcam, as an example of the software's potentiality. Moreover, we translated a graphical user interface (AXIS), in co-operation with the international development community, in order to improve its user-friendliness.*

Key words: *CNC, Open Source, EMC, Low Cost.*

1. INTRODUCTION

The aim of the present work is to implement a low-cost numerical system, based on an "open source" software called EMC (Enhanced Machine Control).

At first, in co-operation with the international development community, we translated the user graphical interface, in order to both improve its user-friendliness by the nationalisation of one of its graphical interfaces and to show how efficient the updating system is. Then, we created a workstation remote control by means of an online computer with a webcam, as an example of the software's potentialities. Eventually, we performed a number of demonstrative projects in order to check the system's functionalities.

Fig. 1. The CNC mini milling machine used

2. SYSTEM DESCRIPTION (OVERVIEW)

EMC is a Linux CNC-executable control program, freely downloadable on the Internet, able to control servo-motors, stepper motors, robots and other automated devices. It allows us to manage a numeric control machine from the parallel port of an even low-performing PC (as an example, for the present work we used an embedded PC produced by FabiaTech).

It provides motion planning with cutter radius compensation, tool length offsets, and constant velocity control.

Also have support for non-Cartesian motion systems. This includes hexapods (Stewart platforms and similar concepts) and systems with rotary joints to provide motion.

EMC runs on Linux using real time extensions. Support currently exists for version 2.4 and 2.6 Linux kernels with real time extensions applied by RT-Linux or RTAI patches

The operative system we have chosen for the controlling PC is Linux-Ubuntu. It is a free, community-supported, complete distribution, based on Debian. Ubuntu has proved to be a very good (stable, easy and powerful) distribution: we have not found any problem with either peripheral recognition or mass memories.

2.1 Configuration

After connecting the machine with the PC, we had to set the INI file parameters up, in order to adapt the new controller to our mini milling machine features. This operation, being the "core" of the whole connection procedure, has implied the modification of a few lines of the EMC standard INI files. As a matter of fact, there is a range of modified INI files, ready to be used in a number of different systems. We took the <stepper_mm.ini> INI file as a reference, which is set to run with such stepper engines as ours, with TkEmc (GUI) graphical interface. It was enough to modify a few lines of [Axis_0], [Axis_1] and [Axis_2] sections - equal respectively to axis X, Y, and Z - in order to obtain a correct functioning of the machine. Starting from this modified, adapted file, we have created two INI files, different from each other only for their Graphical User Interface (GUI). This way, for each start of EMC, thanks to the INI files opening menu at our disposal, it has been possible, from time to time, to immediately select our desired GUI between the most functional two (TkEmc and AXIS).

2.2 AXIS Graphical User Interface

While using EMC, we focused our attention on AXIS, one of the graphical user interfaces at our disposal. AXIS (developed by Epler & Radek) is written in the Pyton programming language and released under GNU-GPL open-source licence. Not only it has a very good preview tool, but also backplot functionality.

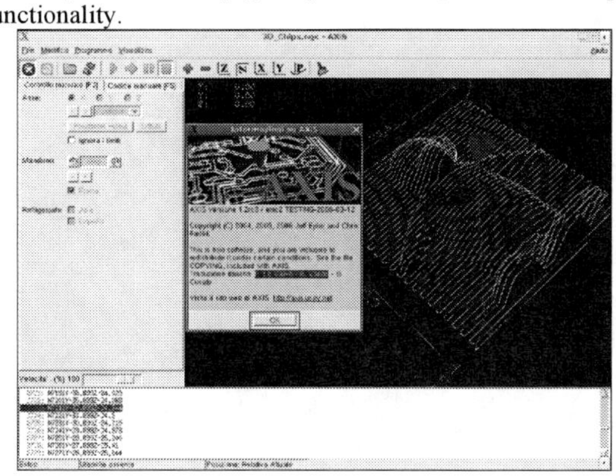

Fig. 2. A view of AXIS graphical user interface

When NGC program is open, these two tools allow us to see the 3D model of the tool path, to follow "live" the manufacturing. It is also possible to easily associate one specific command line to the real tool movement: one window shows the ISO code lines of the program which is suitable to the piece realisation. When a line of the model is selected, the preview shows the corresponding command line of the part program; similarly, the other way round, when a line of the part program is selected, the preview shows the path. This way it is easier for the user to find errors in the code of the program he is either writing or modifying. It is possible to select one of the two control modes cat our disposal. Manual control allows us to both reset the axes to zero and start the milling machine; the MDI (*manual data input)* mode allows us to insert single ISO code lines. The backplot shows the different movement modes by different colours.

3. SYSTEM DEVELOPMENT

EMC more interesting quality is the result of the relentless maintenance work done by both the Internet community of users and developers (www.linuxcnc.org) and the EMC mailing-list, hosted by the SourceForge.net (www.sourceforge.net) site, which keeps in contact all those who have joined it.

3.1 Internationalization project

We engaged in one of the mailing-list activities: the Axis internationalization project. One particular file with extension <.po> of the AXIS source code has been modified and sent to the coordinator. Thanks to this procedure, we have had an opportunity not only to easily reach our objective of internationalising our working environment, but also to test and verify the effectiveness of the development and updating method which makes the basis for such a free software as EMC.

3.2 Remote Control

In order to show how powerful and flexible EMC is, we realised a computer remote control of the milling station: thanks to a webcam pointed on the working table, the remote computer shows the mini milling machine movements. SSH client and server are installed -or it is possible to install them- not only on a great number of such UNIX versions as Linux and Mac OS X, but also on Windows. So, first we made a connection on a host computer running Ubuntu, then we made the same connection with a Windows computer.

In the first case - a very easy procedure thanks to the Ubuntu package manager - we have first installed on both machines an SSH program, and then, through the ssh command from the host computer shell, realised a connection between the two PC.

Fig. 3. Remote control system overview

In the second case, using a Windows client, the procedure is substantially the same. Again, we have taken advantage of one of the many free SSH shell programs for Windows systems available on the web. The only difference is that we have installed on the Windows XP host computer an X-server emulator allowing us to run, in a Windows environment, applications which were created to run, in X-Windows mode, on UNIX-like operative systems. Among all the programs available in Internet, we have selected WinaXe. The remote programs can be seen on separated windows: this way we can open multiple windows at the same time, one with EMC and the other with the webcam images visualisation program. In this aim, we have selected a program which is able to send images to a server by an ordinary Internet browser.

This has proven to be a more than satisfactory solution as far as image quality and fluency are concerned. Moreover, it gives us an opportunity to visualise webcam images on different computers at the same time, if they are in network. EMC remote control could result to be an interesting tool, as an example in all those case when, for safety reasons, the operator cannot closely follow either the operations or the manufacturing; such an idea is made realer when we think about the chance to command robots by EMC.

4. CONCLUSIONS

In conclusion, this numerical control system based on EMC software is a very good alternative to the expensive systems we had up to now, for the following reasons:

- it is a low-cost, flexible system;
- it is open-source and always updated;
- its working environment can be nationalised;
- it has user-friendly graphical interfaces;
- it performs the working station remote control.

5. ACKNOWLEDGEMENTS

This work was performed with the financial support of M.I.U.R. of Italy.

6. REFERENCES

AXIS official home page,
Available from: http://axis.unpy.net

EMC official home page,
Available from: http://www.linuxcnc.org

Olszta, M. W. & Klemm, K., *Interactive Qualifying Project Report: The Transfer of Technology to Small and Medium Sized Manufacturers*, Worcester Polytechnic Institute,
Available from: http://www.linuxcnc.org/Dropbox/iqp.PDF

Proctor, F.; Damazo, B.; Yang, C. & Frechette, S., *Open architectures for machine control*, National Institute of Standards and Technology Gaithersburg, Maryland 20899,
Available from:
http://www.isd.mel.nist.gov/documents/proctor/sim/sim.html

Proctor, F.; Shackleford, W. & Yang, C., *Simulation and Implementation of an Open Architecture Controller*, National Institute of Standards and Technology, Gaithersburg, MD 20899,
Available from:
http://www.isd.mel.nist.gov/documents/proctor/315openarchit.pdf

Yang M.-Y., Hong W.-P., *A PC–NC milling machine with new simultaneous 3-axis control algorithm*, International Journal of Machine Tools & Manufacture 41 (2001) 555–566

EMC SourceForge.net official home page,
Available from: http://sourceforge.net/projects/emc

Annals of DAAAM for 2006 & Proceedings of the 17th International DAAAM Symposium, ISSN 1726-9679
Editor B. Katalinic, Published by DAAAM International, Vienna, Austria 2006
150th Birthday of Nikola Tesla (1856-1943) Inventor of Mobile Robot Teleoperation

OPTIMIZATION OF PATH OF MOBILE ROBOT BY GENETIC ALGORITHMS

VAUPOTIC, B.; BREZOCNIK, M.; FICKO, M. & BALIC, J.

Abstract: *The paper deals with the optimization of the path of the mobile robot by genetic algorithms. The concept imitates the natural selection of living organisms, where in the struggle for natural resources the successful individuals gradually become more and more dominant, and adaptable to the environment in which they live, whereas the less successful ones are present in the next generations rarely. The responsibility of the mobile robot is to cover the shortest possible path from the starting point to the goal. The robot is self-learning and gathers the information from the environment by means of sensors. It processes the obtained information and then utilizes it for adopting its decisions. Throughout evolution it becomes more and more intelligent so that it manages to perform the set task.*
Key words: *Evolutionary method, Genetic algorithms, Mobile robot, Path optimization*

1. INTRODUCTION

Optimization of navigation and looking for the path can be ranked into the group of difficult problems. The problem appears in different areas: looking for the path by the mobile robot, transport systems in company, ship navigation, planning of visits of sights considering the wishes of the participants etc. By changing the environment the problem becomes dynamic. Due to highly specific characteristics of such problems each of them needs its heuristic algorithm which becomes complex and, consequently, difficult for additional requirements in order to satisfy all the needs. The problem can be successfully solved by the genetic algorithm. So far, similar problems have been addressed by many authors (Mitsuo 1997; Kovačič & Balič 2003; Brezočnik 2000). This paper proposes improved genetic operations.

2. LOOKING FOR THE PATH

Figure 1 presents a model of the production area. Gray rectangles represent the barriers. The aim of the robot is to travel from start (point S) to end (point E) without colliding with barriers. Different paths from initial to final point represent the population of solutions. Each path consists of the path sections which may cross the barrier or not. In case one of the path sections crosses the barrier an incorrect path is in question.

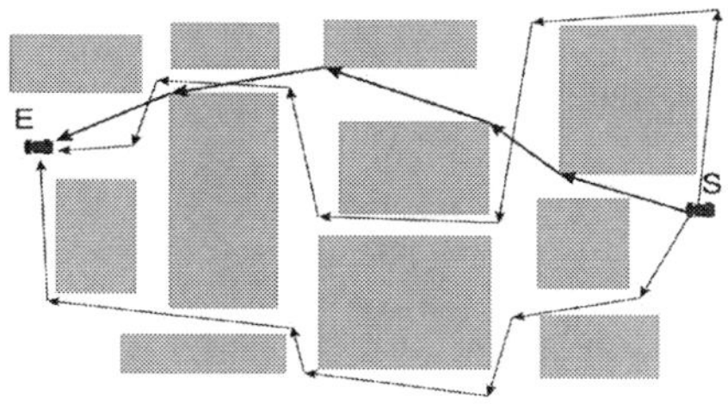

Fig. 1. Three different successful solutions

Various improvements in searching have led to a significant number of different operators which can be divided into simple ones causing a small change is the path (mutation) and more exacting ones creating a descendant with a combination of two or more paths (crossover). Each path in the evolutionary process is evaluated by means of the fitness function.

2.1 Initial population

The initial population can be created randomly or by means of the existing knowledge about the problem (Michalewicz 1996). In the random creation many incorrect sections are obtained (Figure 2a).

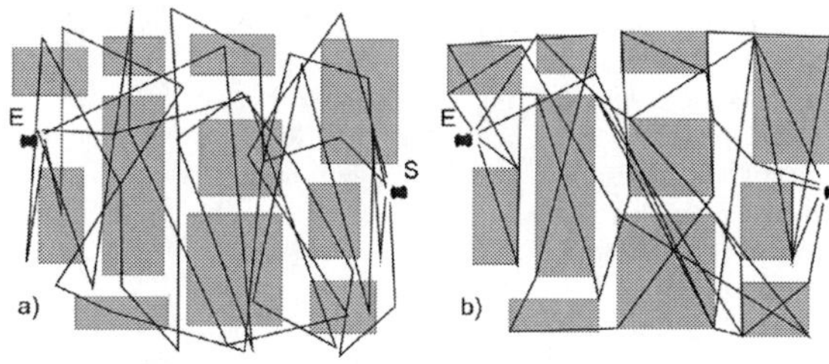

Fig. 2. Initial population: a) with randomly selected paths and b) with location of points in corner points of barriers

To reduce the number of incorrect paths it is possible, when creating the paths, to use the information about the position of barriers and to locate the control points in the corner points of barriers. Figure 2b shows locating of points in the corner points of barriers.

2.2 Presentation of path

The path consists of sections. The individual section is described with two points. The first point represents the beginning and the last point the end of the path. All other points describe the beginning of the path of the new section and the end of the path of the previous section. As the number of sections is not known, the so - called presentation with variable length is in question.
The points are connected into a list of points representing the path (Figure 3). In addition to coordinates each point contains also the information about correctness of the previous section. The section is correct if it does not intersect the barrier. The information is used for evaluation of the path.

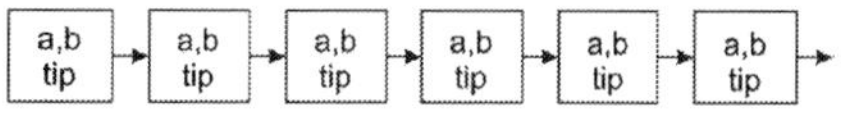

Fig. 3. Presentation of path with continuous list

2.3 Evaluation

Fitness of the solution is influenced by three factors, namely the path length, the number of sections and the number of incorrect sections. For evaluation of the path x the linear combination of those factors can be used. The fitness function can be described with the equation:

$$eval(x)=k_1\times length(x)+k_2\times sections(x)+k_3\times incorrect(x) \qquad (1)$$

The constants k_1, k_2 *and* k_3 in equation 1 represent the weights for the path length, number of sections and number of incorrect sections. By means of weights the importance of the individual properties of the path x is adjusted.

2.4 Operator

The following nine operators, into which our knowledge about the problem is incorporated (Figure 4), are introduced for successful solution of the problem.

Crossover combines the paths of two parents. In each point the crossover point is to be found and, thus, the paths are divided into the head and tail. The crossover brings about new paths.

Mutation 1 randomly selects a point on the path and randomly moves it into the near vicinity. The operator proves to be excellent for optimization of the path already found.

Mutation 2 similarly as the operator in mutation 1 it randomly selects the point, but it moves it into a random position in the entire space. This results in great changes of the path which, however, prove to be useful in the beginning of the evolutionary process.

Adding is an operator which divides the random section into two sections by means of a new point.

Deleting acts oppositely to the adding operator. It randomly selects a point in the list and removes it.

Changing takes care of changing the order of two consecutive points. The operator is used to eliminate two consecutive sharp angles.

Change of corner points is an operator which moves the point to another corner point. The corner point may be located on any barrier. The operator uses the knowledge about barriers.

Correct 1 is an operator which moves the point with coordinates inside the barrier (incorrect position) into the corner point of the barrier.

Correct 2 is more complex than the operator correct 1. It changes the section intersecting the barrier so that it changes the section into several smaller sections avoiding the barrier.

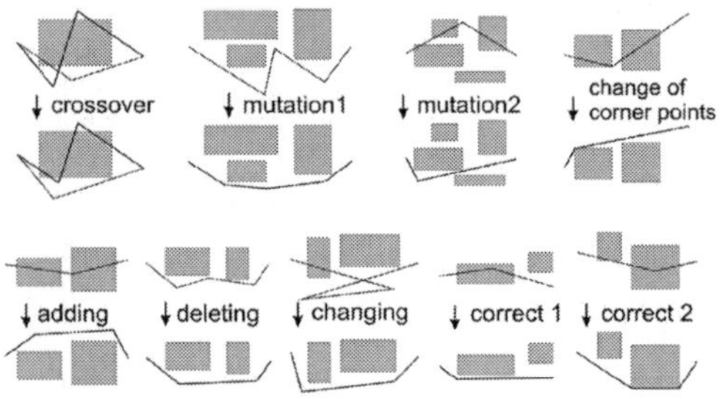

Fig. 4. Path navigation operators

2.5 Different approaches

The basic algorithm for searching for the path uses the random initial population and randomly selects the operators. Such algorithm uses the basic evolutionary principle for searching for the path between barriers (Mitchell 1997). Due to relatively slow progressing various derivatives of the algorithm are used.

Corner point algorithm: The optimum path frequently runs in the corner point of the barrier. This property is utilized for conceiving of the algorithm whose path points represent the corner points of the barriers. The mutation operator replaces the corner point by a random corner point. The crossover operator uses single point crossover Due to smaller space (only corner points are examined) the algorithm is faster and more efficient. The principal disadvantage is that algorithm does not know how to find the optimum path, when it does not run through corner points.

Algorithm with direct path: The main characteristic of the approach is the description of the path of the initial population by means of straight line from the initial to final point. During the evolutionary process the straight line is divided into several sections by means of operators of changing (adding of points, mutation etc.) This algorithm is slower than the corner point algorithm, but more efficient than the random algorithm.

Hybrid algorithm: The hybrid approach selects a part of the initial population randomly, a part by means of corner points and a part by direct paths. In such an approach a wide set of operators can be used. Due to the combination of properties of all the above algorithms this algorithm is fit and robust.

3. EXAMPLE

The figure 5 and 6 show the program of searching for the path. The search space is the production area of square form accommodating nine machines which represent the barriers. The transport robot must transport all finished products located on the collecting place at the end of the production line to storage area.

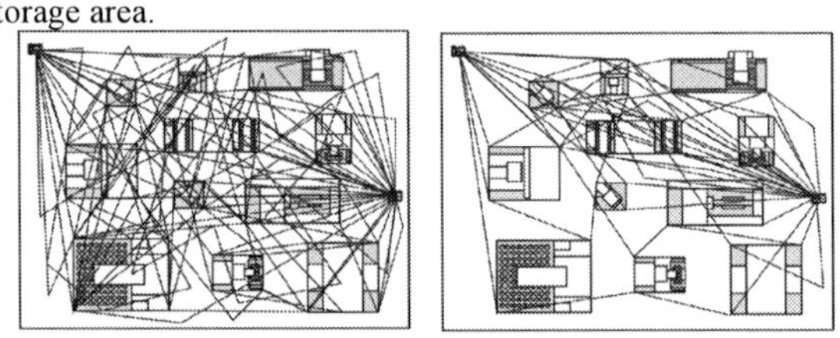

Fig. 5. Evolution (initial generations and generations 20)

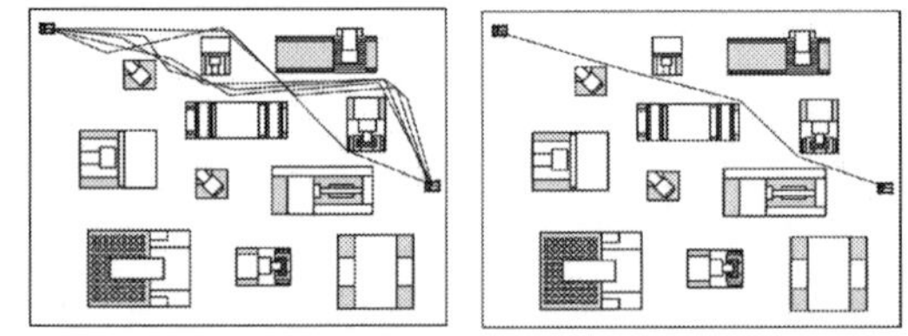

Fig. 6. Evolution (generations 70 and generations 100)

The system for optimization of the path of the mobile robot was run 100 times. For all runs the identical parameters of evolution were used, namely the population size 1000 and the maximum number of generations 30.

4. CONCLUSION

The paper discusses the approach to searching for the shortest path of the robot y means of the evolutionary method. Some improved genetic operations were used which have proved to be adequate. By means of them the robot become more and more intelligent throughout evolution and performed the set task. In future, it would be appropriate to test the improved operations also in the room with moving objects and to enable the robot to have full autonomy.

5. REFERENCES

Mitsuo, G. (1997). *Genetic algorithms and engineering design*, A Wiley-Internascience Poblicatin, ISBN 0-471-12741-8, New York

Kovačič, M. & Balič, J. (2003). *Evolutionary programming of a CNC cutting machine*. International journal of advanced manufacturing technology, Vol. 22, No. 1/2, pp. 118-124. ISSN 0268-3768, Maribor

Brezočnik, M, (2000). *Uporaba genetskega programiranja v inteligentnih proizvodnih sistemih*, Faculty for mechanical engineering, ISBN 86-435-0306-1, Maribor

Michalewicz Z. (1996). *Genetic algorithms + data structures = evolution programs* Springer-Verlag, ISBN 3540606769, New York

Mitchell, T. M. (1997). *Machine Learning*, The McGrawHill Companies, Inc., ISBN 0-07-042807-7, New York

Annals of DAAAM for 2006 & Proceedings of the 17th International DAAAM Symposium, ISSN 1726-9679
B. Katalinic, Published by DAAAM International, Vienna, Austria 2006
150th Birthday of Nikola Tesla (1856-1943) Inventor of Mobile Robot Teleoperation

FAST INITIALIZATION AND RECONFIGURATION OF ASSEMBLY LINE CONTROL USING VISUAL PARAMETERIZATION

VERYHA, Y.; MECHLER, G.; JALOWICKI, S. & NIDAMARTHI, S.

Abstract: *A static control system architecture for assembly lines in automotive industry can hardly meet the growing demand for continually introducing changes in the manufacturing systems while maintaining high production efficiency. A reconfigurable assembly line control framework with fast and easy configuration is needed to be able to accommodate any required control changes as fast as possible to reduce production stops. In this paper, we describe a reconfigurable control framework for assembly line conveyor. We present our exemplary testbed application built with the use of OPC DA, Microsoft Visio and Excel (Ms V & E) that enables fast initialization and reconfiguration of the conveyor control to adapt to any possible conveyor layout changes using simple visual parameterization of the control program. Finally, an application of such framework for the automotive business of Power Train Assembly is discussed.*

1. INTRODUCTION

Assembly manufacturing systems for automotive business are usually built to continually repeat a specified set of sequences of operations. Because the capital investment into an automatic manufacturing system is very high, most of assembly systems have to operate non-stop and, thus, if an assembly system can be reconfigured to continue production (even if it will be in a degraded mode) after a need has occurred, the overall benefit of the system will significantly increase. One of the important requirements for manufacturing systems is to have simple and easy-to-use configuration tools to be able to create the initial system layout and to introduce fast changes like, change of station positions, bypass for stations being repaired, etc.

The problem of a formal definition of the control system layout and functions that a control system must perform has been a challenging problem for years. A number of interesting proposals should be mentioned, e.g., automatic logic generation and validation for cell-based manufacturing systems (Almeida & Tilbury, 2004), agent-based approach to reconfiguration of real-time distributed control systems (Brennan et al., 2002), common visualization of PLC (Programmable Logic Controller) programs using XML (Bani & Frey, 2004), generative programming (Cote et al., 2005) or modular finite state machines for logic control (Endsley & Tilbury, 2004). However, the main trend which became widely popular in the last 5-10 years was the definition of standardized models for both system functions and controller code that can be adaptable in real scenarios according to the fast varying customer demands (Baresi et al., 2000). This trend lead to modularization and standardization of control components of manufacturing systems, which allowed reducing costs in the design, implementation, testing, installation and maintenance of the overall control system, and most importantly gaining real flexibility, reconfiguration and reuse of control solutions. One of the proposed approaches from the methodological and a practical point of view is the separation of system functions from code, which allows the semi-automatic generation of the control code and clear correspondence between design and implementation models.

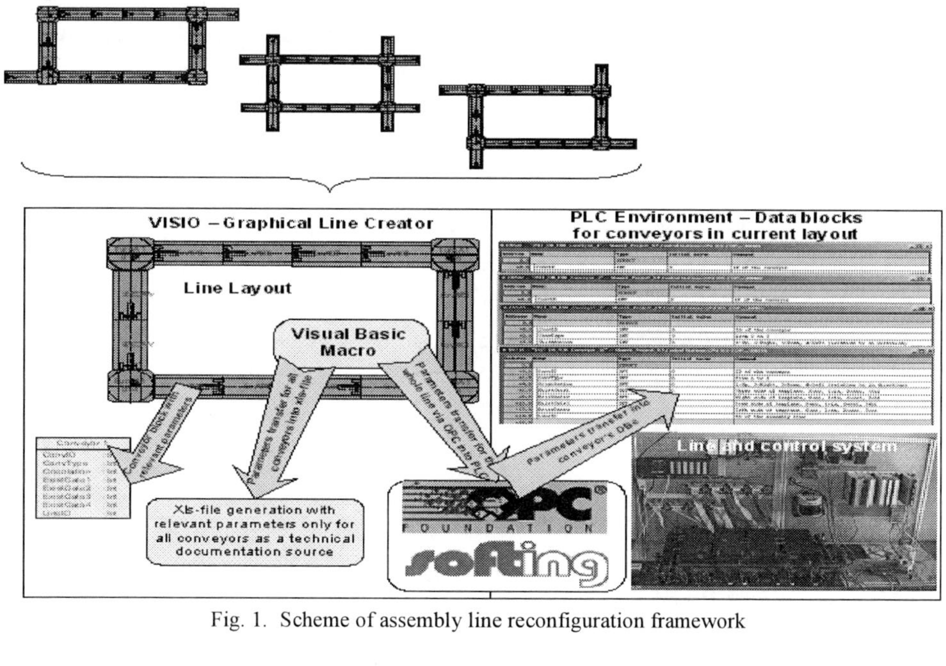

Fig. 1. Scheme of assembly line reconfiguration framework

As regards reconfigurable manufacturing systems, relatively small attention was paid to the usage of common office software packages like Ms V & E for visual line control parameterization. These tools are widely accepted by customers. Our developed testbed for PTA (Power Train Assembly) enables easy initialization and further reconfiguration of the conveyor control to adapt to any possible conveyor layout changes using simple parameterization of conveyor control program from Ms V & E.

2. PARAMETERIZATION OF CONVEYOR CONTROL

One of the visual engineering tools used within PTA on the early assembly line design is Microsoft Visio. It allows quickly drawing an assembly line by instantiating base elements from pre-defined templates. Basic elements, which define the transport topology of the assembly line, are conveyor modules, turn-table modules and connections between modules. Once all conveyor modules are placed at their position in the Microsoft Visio layout page, a real transport line can be constructed. The control parameters for the conveyor modules are mapped to an OPC server items, which are in their turn mapped to the variables in the given PLC. The scheme of the assembly line configuration framework is shown in Fig. 1. The same framework is used for fast and easy dynamic reconfiguration of line control parameters.

In a conventional approach, the static parameters of the assembly line are documented on CAD drawings and/or additional control documentation. Without our line configuration framework, line parameters were set by manual entries on the PLC operator panel. Our line configuration framework allows setting and loading line control parameters for the given line using visual line presentation model in Microsoft Visio. The Macro from Microsoft Visio connects our model to OPC server, automatically identifies OPC item groups, which correspond to control parameter groups of Visio model, and loads configuration data to the PLC.

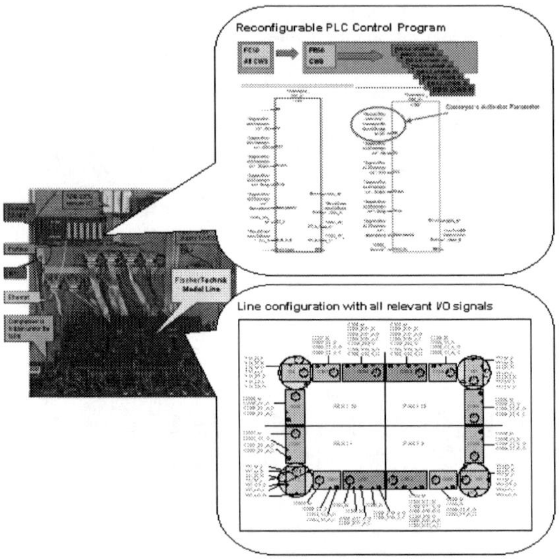

Fig. 2. Control of reconfigurable assembly line testbed

The line is constructed in a modular way, which means that it is divided into four parts built in exactly the same way. Each of those parts includes the same modules. Modules are conveyors as a mechanical construction with electrical elements like motor, switches, stoppers and pneumatic elements, which are externally controlled by valves. Fig. 2 shows the mechanical and electrical model of assembly line on the left with the controller program and I/O layout in callouts. An exemplary application area in which fast and easy parameterization of a line control is of high importance is PTA, in which assembly lines for transmissions, engines and axles of cars are produced for automotive customers. Most of line configuration data (e.g., positions of modules, recipes, decision points, etc.) can be parameterized in similar way as in our testbed from well-known PC office tools, like Ms V & E, and with the use of OPC DA uploaded to PLCs and robot controllers. We developed a special PLC program capable of being parameterized from outside. This means the control loops in PLCs access special parameterization blocks (values of which are initialized from parameterization tools on the PC through OPC DA server) defined in the PLC memory to obtain actual configuration data for line control.

3. CORRESPONDING ADDRESS

ABB Corporate Research Centre, Wallstadter Str. 59, 68526, Ladenburg, Germany. e-mail: yauheni.veryha@web.de

4. CONCLUSION

We have presented the concept and testbed application for fast initialization and reconfiguration of assembly line control using parameterization from office software tools like Ms V & E. We have shown that the static control system architecture for assembly lines in automotive industry can hardly meet the growing demand for continually introducing changes in the manufacturing systems while maintaining high production efficiency. A reconfigurable assembly line control framework is needed to be able to accommodate any required control changes as fast as possible to reduce production stops. We described a reconfigurable control framework for assembly line conveyor that we have developed at the ABB Corporate Research Centre as a mock-up for future implementation on the Power Train Assembly lines to enable a fast initialization and reconfiguration of line control and to adapt to any possible layout changes using simple visual parameterization of line control program.

5. REFERENCES

Almeida, E. & Tilbury, D. (2004). Automatic logic generation for cell-based manufacturing systems. *Proceedings of IFAC Workshop on Discrete Event Systems*, pp. 31-36, Reims, France

Bani, M. & Frey, G. (2004). Visualization of PLC programs using XML. *Proceedings of American Control Conference*, pp. 3082-3087, Boston, USA

Baresi, L.; Mauri, M.; Monti, A. & Pezze, M. (2000). PLC tools: Design, formal validation and code generation for programmable controllers. *Proceedings of IEEE Conference on Systems, Man and Cybernetics*, pp. 2437-2442, Nashville, USA

Brennan, R.; Fletcher, M. & Norrie, D. (2002). An agent-based approach to reconfiguration of real-time distributed control systems. *IEEE Trans. on Robotics & Autom.*, Vol. 18, No. 4, pp. 444 - 451.

Cote, D.; St-Denis, R. & Kerjean S. (2005). Generative programming for programmable logic controllers. *Proceedings of IEEE Conference on Emerging Technologies and Factory Automation*, pp. 741-748, Catania, Italy

Endsley, E. & Tilbury, D. (2004). Modular finite state machines for logic control. *Proceedings of IFAC Workshop on Discrete Event Systems*, pp. 403-408, Reims, France

Annals of DAAAM for 2006 & Proceedings of the 17th International DAAAM Symposium, ISSN 1726-9679
B. Katalinic, Published by DAAAM International, Vienna, Austria 2006
150th Birthday of Nikola Tesla (1856-1943) Inventor of Mobile Robot Teleoperation

IDENTIFICATION OF POTENTIAL INDUSTRIAL CLUSTER IN THE REGION

VEZA, I.; DULCIC, Z. & GJELDUM, N.

Abstract: *This paper deals with potentials identification model building up analysis for a cluster development and clustering model development by using the well-known theoretical and practical developed countries experience where clusters have become a basic stronghold for contemporary industrial and overall economic development. Moreover, it offers a simplified model for key competent authorities identification in the regional economy based on selected criteria. Based on defined competent authorities we can identify potential cluster development. The identification model described in this paper deals with the principles and other supporting measures needed for cluster development.*
Key words: *cluster, regional development, core business*

1. INTRODUCTION

In a globalize world, regional economy and companies have to be adjusted to new requirements, which can lead to an increased investment risk. At the same time there is an increased opportunity that a country or a region participates in a joint product. It is obvious that the competitiveness of a regional company can be stimulated by different measures and, consequently, the living standard of the population in the region will be enhanced.
Here are some of the global tendencies (Porter, 1990):

- Companies will focus on their core business.
- In a global market, there will be the companies focused on design and research of a specific product, the manufacturing companies will be concerned with its production and the commercial companies will be concerned with selling of the product.
- In order to guarantee the quality, the safety and the traceability, companies will establish strict manufacturing rules for their subcontractors.
- The sharing of the works among several companies will require networking models at a global level.
- Specialization and networking skills will be the key to SME success.
- Networking requires a permanent connection between the automatic centralized managing systems of the enterprises.

It is certain that innovations have an impact on the increase of productivity within a company. By increasing its competitiveness, a company can increase its economic effect, and, regarded from a global point of view, the employment opportunities and the standard of living will be also increased. Only innovations can ensure new jobs and, as such, they are basic prerequisite for regional and overall national economic development.

2. IDENTIFICATION OF POTENTIAL INDUSTRIAL CLUSTER IN THE SPLIT-DALMATIAN COUNTY

Split-Dalmatian County economy is still burdened by previous economic system inherited anomalies and some transitional problems. Here are some of the most evident ones: low productivity additionally burdened by a great number of employees and obsolete technology; a predominant lack of products and services demanded by developed markets; Government subsidizes lame ducks; enormous foreign debt; inflationary economic policy; Government meddling with economy; neglected agriculture; uneven development of the urban and rural areas; gray economy etc. However, public property/companies privatization, which is a sine qua non prerequisite for sound market basis establishment and prospective growth, has not been successfully implemented. Most public companies completely disappeared in the privatization process, and those that managed to survive, have undergone numerous recovery programs or have gone into liquidation. In these conditions, small and medium-sized enterprises development could not have been supported by big industrial systems; therefore economic development has been mostly turned to the service sector. Generally speaking, during the transition process there was no development whatsoever, particularly there were no new technologies or new products introduced. Cooperation between economy and science was rather weak, and the accompanying infrastructure required to support technological development and innovations was developing rather sluggishly. Taking all this into consideration, a basic prerequisite for making a turnaround to a successful economic development in the region is to restructure its economy.
In the lack of a unique and commonly agreed economic strategy at the national level, the Split-Dalmatian County has decided to define its own goals and priorities and to develop its own potentials. As a result of this decision there is a recently brought Regional operational program (Petrić et al., 2005) that defines the County development guidelines in the next five-year period.

One of the primary strategic goals, defined in ROP, is to develop a competitive, diversified, technologically advanced and environmentally sustainable economy that will be oriented to enhance the living standard of the local population. Consequently, such clearly defined goals require a radical change of the existing settings, in which an inadequately competitive economy still prevails. Moreover, it is additionally burdened with obsolete technologies, insufficiently educated and unskilled personnel, particularly in the marketing and management field. On the other hand, in the places where there are some potentials, economy is not adequately integrated, the products have a rather low VAT, while the export products, excluding rather inefficient shipbuilding industry, are mainly concerned with service industries (tourism and transportation). In order to set things in motion, the following priorities have been defined:

- To strengthen cooperation between research institutions and entrepreneurship that will enable new technologies implementation and technologically innovative products production.

- To restructure organization in traditional manufacturing sectors, agriculture, fishing industry and tourism, in order to achieve bigger competitiveness.
- To support export-oriented, specialized production of products and services with higher VAT.
- To encourage regional and inter-regional integration processes (including transnational ones) and to support cluster organizations in order to strengthen overall synergetic impact in the sectors and between them.
- To ensure business, entrepreneurial and managing training,
- To increase employment opportunities.

Pointing out the need for regional and inter-regional networking, integration and County economy clusterization, is the first step that has to be taken in order to start more intensive research that will help in identification, organization and development of the above-mentioned networks. Bearing this in mind, this paper presents model for identification of basic technologies and industrial clusters development potentials in the County. The following parameters have been adopted in order to identify basic technologies and industrial clusters development potentials of in the Split-Dalmatian County, taking into account some objective impediments in availability of relevant statistical data:

- Revenue structure, the each industry share in the total revenue of County economy, shown in terms of percentage,
- Employment structure, each industry employed number as oppose to the County's total economy employment number, shown in terms of percentage,
- Cluster index CI (Vieregge, 2004). This index shows how high the concentration of some activity is, at the County level as oppose to an average value at the national level.

The basic activities are defined by analyzing industrial activity position in quadrant quarter shown in Figure 1. A measure adopted for the size of an activity in the County, is "share of the revenue structure of an activity compared with total revenue in economy" on the abscissa, and a measure adopted for activity expansion degree is "number of entrepreneurs".

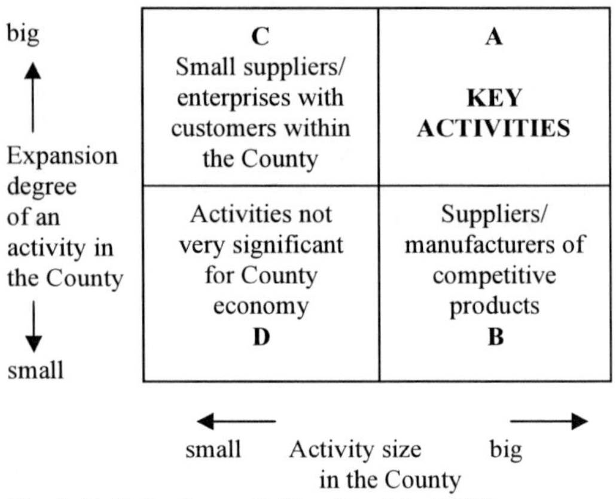

Fig. 1. Defining key activities (Roethig, 2002)

Based on activity position in Figure 1, two basic industrial activities have been defined (quarter A), which are considered to be the most significant for Split-Dalmatian County development at the moment. They are:

1. Other vehicles production
2. Food and drinks production

The quarter B includes high revenue structure percentage activities/enterprises in the County economy, but there are very few of them. They deal with motor vehicles, caravans, trailers & semi trailers production, and other nonmetallic mineral products production. These enterprises goal should be increased number of customers at the home and foreign markets. On the other side, the quarter C includes the activities/enterprises with rather low revenue structure percentage in the County economy, but they are rather spread. These enterprises could easily increase their revenue percentage by networking in clusters. This way they will be able to grow and develop. They include:

- Publishing and printing activity,
- Metal products production, machines & equipment not included,
- Furniture production,
- Rubber and plastic products production.

All other activities (17 out of total 25 industrial activities) are in the quarter D at the moment, and they have a very small significance for Split-Dalmatian County economy.

3. CONCLUSION

In this paper we have tried to emphasize the need for actual development potential identification in the regions where economy has been affected by transition and restructuring processes. We have used theoretical and practical developed countries experience, in which networking and clustering model is based on innovative potential and on knowledge-based economy, which is a basic stronghold for both contemporary industrial development and overall economic development. Regional development strategic goal is to provide conditions for innovations based on regional identity. However, in order to achieve this goal it is necessary to do the following:

- To ensure transparency with regard to the existing technological competence,
- To link regional resources and competent institutions (universities, institutes, enterprises, chambers of commerce etc.),
- To identify the existing clusters development potentials in attractive regional business lines development context,
- To establish new regional exchange cultures of knowledge and experience.

Therefore, it is necessary to make a transition from the regional strategy based on predominantly industrial production to the knowledge based economic development strategy. Economy, scientific institutions and regional management linkage can be achieved using a triple helix model (Etzkowitz, 2002), which provides required prerequisites for the society towards knowledge transition. In order to redefine the interrelationship between institutional knowledge, economy and regional management, it is of utmost importance to enhance the local conditions for innovative processes development by linking together research activities with others. Our aim is to continue our preliminary research and to develop such linkage models.

4. REFERENCES

Roethig, H. (2004). Standortentwicklung und Wirtschaftsfoerdeerung, *Tagungsband zur Fachtagung «Strukturwandel in Wirtschaftsstandorten»*, ISBN 3-937487-01-8, Chemnitz

Etzkowitz, H. (2002). The Triple Helix of University-Industry-Government: Implications for Policy and Evolution, *Working paper 2002-11*, http://www.sister.nu/

D. Vieregge (2004). *Clusteranalyse and Umsezungsstrategien*, Stadt Neumuenster, 2004

Petrić, L.(2005). Regionalni operativni program Splitsko-dalmatinske županije (Regional Operational Program of The Split-Dalmatian County), *The Split-Dalmatian County and Economic Faculty of Split*, ISBN 953-6121-79-9, Split

Porter, M. (1990). The Competitive Advantage of Nations, Free Press. ISBN 0-684-84147-9, New York

Annals of DAAAM for 2006 & Proceedings of the 17th International DAAAM Symposium, ISSN 1726-9679
B. Katalinic, Published by DAAAM International, Vienna, Austria 2006
150th Birthday of Nikola Tesla (1856-1943) Inventor of Mobile Robot Teleoperation

BANK FINANCE OF SME SECTOR: THE CASE OF CROATIA

VIDUCIC, L.

Abstract: *This paper deals with finance of the SME sector in Croatia. After review of the literature, bank lending to the SME sector in Croatia is investigated. Attitudes and experience of managers of the SME in the Split-Dalmatian County are analyzed. Finally, conclusion brings some consideration regarding bank lending improvement in Croatia.*
Key words: *SME, bank finance, access to finance, transition.*

1. INTRODUCTION

EU SMEs participate in the total number of all business activities with 99.8 percent, and employ 66.5 percent of the total workforce. Croatian SMEs have almost the same importance for the national economy participating in the total number of all business activities with 99.7 percent and employing 82.8 percent of the workforce. However, administrative and lending procedures and overall climate may be severe obstacles for establishment and operation of SMEs. To spur the private sector in transition economies calls for proper legislation and law enforcement, provision of a friendly environment in the administrative and financial sector, as well as an increase of financial strength for the institutions designated to foster SMEs.
It is widely argued in the literature that SMEs are disadvantaged in securing access to finance in appropriate amount and/or competitive terms (Stiglitz, Weis, 1981; Deakins, 1996, Viducic, 2000, Pisarides, 2001, Klapper, Allende, Sulla, 2002). This is especially the case for certain SME groups. Examples include start-ups, deprived areas, technology based SMEs and those in countries in transition, deprived areas or owned by ethnic minorities and females.
The characteristics of the SME owner-manager profile may represent further obstacles for optimal SME financing. These obstacles embrace reluctance to share control by raising external equity, reluctance to raise debt due to risk aversion, as well as fear of bureaucratic procedures and unfriendly attitude of officers/local politicians. Furthermore, impediments to qualitative financing may stem from lack of an adequate experience and education, including financial illiteracy, and insufficient corporate finance knowledge.
Access to appropriate finance with adequate terms in transition economies, is further aggravated due to uncompleted financial institutional structure - absence of (well funded) equity investors, a shallow and illiquid securities market, and bank reluctance to fund SMEs.

2. CHARACTERISTICS OF BANK FINANCE OF SMEs

Banks represent major sources of external finance for SMEs, even in developed market economies. Banks participate in SME finance with their own programs designed for this sector; programs that are realized in cooperation with the state development bank (CBRD) as well as with common credit programs. The results of the research, undertaken for SMEs reveals finance as the major limitation in the process of establishment (Vidučić, 2006).
Bank credits have become more accessible for the SME sector following privatization by foreign banks, which now participate with more than 90 percent in total bank assets for the country. Interest rates have decreased and special programs have been developed in the process of broadening bank products. Undoubtedly, this activity needs to be stronger. An increase in transparency for both banks and the business sector will support this aim.
Entrepreneurs in Croatia use unconditioned credits for investment; furthermore, they use credit cards for business purposes (bank drafts, revolving credits), so business sector finance is presented more favorably. However, commercial banks seem to be too cautious in lending to SMEs, and too profit oriented. For example, when participating in special programs in SME lending, they prefer borrowers that take unsubsidized direct credits, and sometimes combine subsidies lending for fixed assets with expensive lending for working capital, knowing the borrower is under time pressure.
Market prejudice, foreign ownership pseudo monopoly, and the situation in the banking market are detrimental for SME access to commercial bank lending. Although foreign banks had a positive impact on the domestic banking market in terms of lower interest rates, increased reputation, bank product innovations, they tended to prefer lending to LSE. Furthermore, market prejudice is responsible for excluding from the commercial bank lending SMEs that do not meet collateral requirements. Also, commercial banks seem to use a pseudo monopoly situation in the financial sector. As a consequence, banks skim deals from the market that have appropriate collateral, excluding SMEs that do not have it, which is, unfortunately, sound business and more profitable to the bank. Since banks are not compelled to earn a risk adjusted return on capital they may earn a high return while taking only low risk. They also tend to be slow in loan decision making and too conservative in assessing risk ruling out higher risk and a medium rate of return on investments. A combination of proper credit analysis, some collateral, and a risk pricing orientation will provide access to the commercial debt market for many profitable and growing SMEs. Besides, from these loan activities there is an informal credit market operating through usurious loans of certain saving and credit cooperatives, as well as through some limited liability companies, which offer usurious loans backed by real estate or other properties. In spite of the positive changes (introduction of new programs, establishment of new funds), and tendencies (decrease of interest rates) owners/ managers of SMEs still have problems in providing financial sources under suitable conditions and at favourable terms, which indicates still unfavourable environment. The research on the SME segment in the Split-Dalmatian County may be a good illustration of SME business performance in Croatia. The results of the analysis reveal the similar problems which are faced by SMEs in Europe, and particularly SMEs in advanced transition countries, which have

recently become the members of EU or are due to become EU members shortly. The provision of the funds presents the major limiting factor for SME in the establishment phase, followed by administrative barriers (Table 1).

Barrier	Average estimate
Finance	3.2
Administrative barriers	3.0
Space	2.4
Entrepreneurship knowledge	1.9

Table1. Start - up barriers

Administrative barriers also appear as the limiting factor in other performance aspects, which suggests the inevitable and urgent need to implement the reforms in the administrative sector as well, which is anyhow set as one of the priorities for the accession to EU. Access to external sources, among which the most important ones are still bank loans, is rather difficult, therefore the financing of the investments, consequently the development, is mostly provided from the internal sources: amortisation and retained earnings. It is followed by long term and short term bank loans, while the loans granted by other companies are rarely used, and the funds approved by the subsidized programs of the state (loans of the respective Ministries and HBOR – Croatian Bank for Reconstruction & Development). In evaluating the access to bank loans, dissatisfaction with the loan approval terms & conditions is the most common attitude among the managers / financial managers of small and medium enterprises (Table 2). Bank loans do not present the problem, considering the possibility of obtaining the loan and the conditions for its use, for only one quarter of the companies. Most of the companies (almost 2/3) face problems, either due to high interest rates, loan collateral, or long and complicated procedure of loan application evaluation required prior to the approval. The main characteristic of credit financing is the excessive number of documents and high costs of financing (Table 3). 3/4 of the companies think that too many documents are required to approve the loan, while almost 2/3 think that the requested collateral is too high. It is widely thought that the interest rates, as well as bank loan fees are high (2/3 of managers). For 2/5 of managers, the loans are hardly available and the procedure of the evaluation of the loan request application is too long. For less than 1/5 of managers, the system of supporting the entrepreneurship is rather problematic and so is the low level of information on the possibilities of incurring debts. Similar attitudes on the limitations in obtaining bank loans, and particularly short term loans, are present also with managers/ owners of SME in the transition countries of Central and Eastern Europe (Nagarayan, Meyer, 1999). Namely, private SMEs as a rule have the access to the short term loans in a rather limited extent, and the owners identify the problem of high interest and high collateral requests (properties are preferred , which is not the rare case in Croatia either). High interest rates have emerged as the problem mostly in Poland and Republic of Slovakia, followed by collateral requests (mostly in Hungary), high bank fees and complicated procedure of loan application evaluation, although frequency of the occurrence of these limitations is lower than in Croatia.

Attitude	Rating by firms
Access to finance at unfavorable terms (interest, collateral, time consuming and complex procedure)	1.
No access to bank loans	2. i 3.
Access to bank loans	2. i 3.

Table 2. Access to bank loans

Characteristic	Spread of attitude(%)
Hard accessibility	40
High interest rates	62
High fees	62
Numerous loan documentation	75
Time consuming credit analysis	42
Poor information accessibility on loans	15
Too demanding collateral requirements	57
Modest system of support	15

Table 3. Characteristic of SME bank finance

Although only 15 % of managers think that the system of support to entrepreneurship is not developed, and in spite of the extension of the system of supporting programmes and institutions, the state activities, tax policy, administration attitude and quality of the business environment are still given low grade. It suggests the inevitable need to undertake further improvements in terms of general attitude in the society, and education of the administration staff.

3. CONCLUSION

One of the crucial factors hampering growth of small entrepreneur sectors in countries in transition and less developed countries is the lack of a complex infrastructure. In addition, lots of SMEs, especially those new-born and fast-growing, are not able to provide financial resources in appropriate form under the given circumstances. A large number of unfavorable loans approved to the SME sector, extreme precaution by foreign venture capital funds and commercial banks, lack of individual and networks of business angels, lack of clusters and rather slow progress for creation of business associations and networking practices at local levels conrifm the above statements.

According to experiences of developed countries (e.g., Germany, Italy, Austria, Denmark, Japan etc.), it could be expected that primary credit sources for SMEs would be local and large banks as well as development banks/funds. Access to finance needs to be eased in terms of conditions as well as the overall period of time, and better documentation is needed to secure proper financing as well. Furthermore, lengthy procedures for loan approval need to be shortened as well as the related entry and exit procedures. The Project one-stop-shop currently started by the government through the FINA agency is a good step in the right direction, while a more friendly state and local administration staff, as well as the entire business environment need to be secured if private sectors are to flourish and prosper.

4. REFERENCES

Deakins, D. (1996). *Entrepreneurship and Small Firms*, McGraw Hill, Berkshire, ISBN

Klapper, L., F; Sarria-Allende, V. & Sulla, V.(2002). Small and Medium-Size Enterprise Financing in Eastern Europe, *The World Bank Policy Research Paper*, No. 2933.pp, 1-36.

Nagarajan, G; Meyer, R. (2001): Financial and Advisory Services to SMEs in Transition Counitreis, Office of development policy studies, Bureeau for development studies, United nationes development programme, pp- 1-26

Pisarides, F.,(2001): Financial Structure to Promote Private Sector Development in South: Eastern Europe, EBRD Working paper, No 64., pp.1-22.

Stiglitz, J. and A. Weiss (1981). Credit Rationing in Markets with Imperfect Information. *American Economic Review*, No 71, pp. 393-410., ISSN

Viducic, Lj., (2006), *Financijski menadzment (Financial Management)*, RRiF, ISBN 953-6121-49-2, Zagreb

Annals of DAAAM for 2006 & Proceedings of the 17th International DAAAM Symposium, ISSN 1726-9679
Editor B. Katalinic, Published by DAAAM International, Vienna, Austria 2006
150th Birthday of Nikola Tesla (1856-1943) Inventor of Mobile Robot Teleoperation

THE PRINCIPLES OF TREATMENT OF RAILWAYS RAIL DEFECTS

VITEZ, I.; KLADARIC, I. & ARACIC, S.

Abstract: *Increases in railway traffic, heavier axleloads, higher speeds and introduction of new generations of rolling stock have increased the loads on rails in curves especially. Most railways today opt as standard equipment for their track steel grades 260 (R0900A) and to a very limited degree 260Mn (R0900B). Over last decade railway infrastructure has increasingly suffered from rolling contact fatigue (RCF) damage. That are lateral and vertical wear problems arise later on heads of railway rails. This paper elaborate the principles of monitoring and eliminating of railway rails defects caused by operation according to new regulations (Catalogue of Rail defects UIC 712 R, Recommendations for use of rail grades UIC 712 R and Treatment on rail defects UIC 725 R)*

Key words: *railway rails defects, new regulations, principles of treatment*

1. INTRODUCTION

With the continually increasing demands on the rail system, methods of increasing system capacity are continually being sought. Further increase of railway productivity and thus the system capacity is one of the main challenges the railway society is facing. To reach this target, a qualitative and quantitative improvement of wheel / rail system is essential. When looking at today's capacity – and productivity restraining factors it can be noticed that:

- Maintenance intervals / lifetime of railways components are shortened due to increasing demands (heavier axleloads, higher speeds, traffic intensity),
- Accessibility of track is becoming more and more difficult which feeds the demand for low maintenance components,
- Rolling contact fatigue (RCF) damage is increasing all over European rail, increasing safety risk and reducing availability,
- Noise levels are too high to allow the necessary growth in capacity (Hiensch et.al., 2001).

In the project two application technologies for processing two – material rail are studied: rolling technology for application to new rail during the production process (Corus process), and laser cladding technology for application to the existing infrastructure or in the production process of new rail (Hiensch et.al., 2003 a). Rolling contact fatigue (RCF) is one of the major current limitations of railway infrastructure productivity. Squats, Tache Ovals, Shelling and Head Checks are all forms of RCF. These surface–initiated cracks can ultimately lead to complete failure of the rail. In addition to RCF, high noise emission (up to 100–110 dB) caused by stick–slip at the wheel–rail interface is one of the main environmental problems in Europe (Hiensch et.al., 2003 b).

2. RAIL DEFECTS

Handbook of rail defects includes definitions, recommendations, location and general coding system of defects. For broken, cracked and damaged rails are given a code that may comprise up to four digits. The general coding system for rail defects and the classification of the different types of defects are given in the table 1 (Code UIC 712R, 2002).

1st digit	2nd digit	3rd digit	4th digit
		Pattern, nature	
Situation 1. Rail ends 2. Zone away from rail ends	Location 0. Full section 1. Rail head 3. Web 5. Foot	1. Transverse (111) 2. Horizontal (112) 3. Longitudinal vertical (113) 4. Corrosion (134) 5. Passing through a hole 9. Lap (139)	
	2. Surface of rail head	0. Wear (220) 1. Surface defects (221) 2. Shelling (222) 3. Crushing (223) 4. Local batter etc.	Additional characteristics and diferentiations
3. Defects caused by damage to rail	Location 0. Full section	Origin, cause 1. Bruising 2. Faulty machining 3. Permanent deformation	(no 4th digital)
4. Welding and resurfacing defects	Welding method and defects 1. Electric flash-butt welding 2. Thermit welding 3. Electric arc welding 4. Oxyacetylene (autogenous) welding etc.	Pattern nature 1. Transverse 2. Horizontal or shelling	Additional characteristics and diferentiations

Table 1. Rail defect coding system (shortened)

A few typical damages to railway rails caused by exploitation on the head of rails are damages No:

1. Defects in rail ends on head (11 /12) and web (13)

111 Progressive transverse cracking (kidney – shaped)
112 Horizontal cracking
113 Longitudinal vertical cracking
121 Surface defects
122 Shelling of running surface
123 Crushing
124 Local batter of running surface
125 Wheel burns.

2. Defects away from rail ends on head (21/22) and on web (23)

211 Progressive transverse cracking (kidneyshaped fatigue)
212 Horizontal cracking
213 Longitudinal vertical cracking
220 Wear
221 Surface defects
222 Shelling
 2221 Shelling of the running surface
 2222 Shelling of the gauge corner
 2223 Head checking / Fissuring / Scaling
223 Crushing
224 Local batter of the running surface
225 Wheel burn
227 Squat / Cracking and local depression of the surface.

Defect type is defined by the potential detecting methods, growth rate and risk of breakage in relation to size.
Rail defects can therefore be divided into 6 categories:
1 - Transversal cracks (111, 211, 411, 421, 431, 471),
2 - Longitudinal cracks (112, 212, 239, 432, 1321, 472 etc.),
3 - Longitudinal vertical cracks (113, 153, 233, 133, 213, 253),
4 - Head checking (UIC code 2223)

Head checking is a rolling contact fatigue (RCF) phenomenon and it occurs mainly on the high rail in curves with high shear stresses and relatively low wear. It is a growing concern for infrastructure menagers and difficult to detect in the early stages.,

5 - Squats (UIC code 227)

Squats are rolling contact fatigue phenomenon which occurs mainly on straight lines with high shear stresses, especially zones where accelerations and breaking occurs.,

6 - Oblique cracks (cracks around fish bolt holes 135 and 136)

Oblique cracks in fishplate chambers usually start at bolt holes due to lacking or inadequate chamfering. It is not possible to inspect this area visually without dismantling the joint (Draft UIC 725R, 2005).

3. TREATMENT OF RAIL DEFECTS

A crack will have certain detectable size (P – potential). This depends on the detection technique used. From this size propagation of the crack can be followed until it reaches critical size where a rail break can be expected (F – failure). The time or traffic load between these sizes can be defined as the P – F interval.

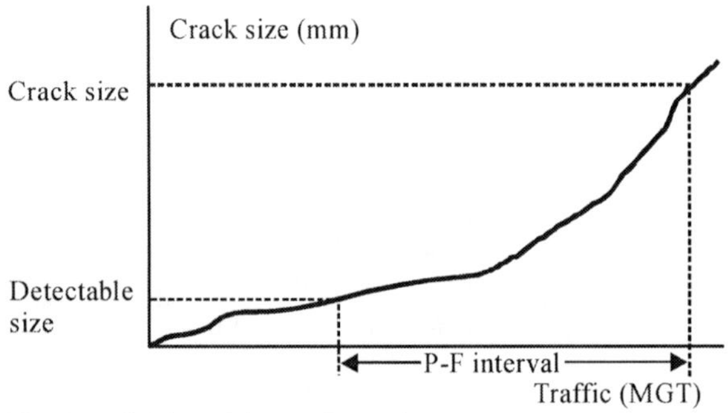

Fig. 1. Definition of the P-F interval

Irrespective or defect type, crack growth rates can vary considerably. Simple crack growth models can be produced for transversal defect in the rail head area.

3.1. Key parameters for planning of inspection cycles

Minimum requirements:

- Track class: national rules, normally based on traffic load and / or speed
- Traffic load: may be real traffic load or an equivalent traffic load, for example based on the UIC categorisation

3.2. Recommended requirements

In addition to the minimum requirements, the following conditions are to be met:

- Line criticity,
- Special areas,
- Rails susceptible to production related defects,
- Environmental consequences,
- Track which are susceptible to RCF defects,
- Machined rail components.

3.3. Use of the state of the art technology

The evaluation of the data gathered from previous inspections will indicate tendencies for the number and growth of rail defects. The inspection interval can be adjusted according to whether there is an increase or decrease in the defect rate.

3.4. Rail inspection cycles

Rail inspection is a vital component in rail maintenance. It is a way of ensuring an acceptable security level for traffic and a viable maintenance level for rail and the underlying track.

3.5. Planning of inspection cycles

The planning of rail ultrasound inspections is a complex issue on account of the high number of variables and consequences.

3.6. Classification of defects and minimum measures to be taken

The minimum measures to be taken for each type of defect (transversal cracks, horizontal longitudinal cracks, vertical longitudinal cracks, head checking, squats and oblique cracks around bolt holes) and according to the size of the crack are subdivided into 4 categories characterised by type of action to be taken and the deadline that should not be exceeded:

- Category 0: Prohibition of traffic and immediate removal of the rail,
- Category I: Immediate removal of the rail,
- Category II: Removal of the rail,
- Category III: Keep rail under inspection

(Draft UIC 725R, 2005 and UIC 712R, 2003).

4. CONCLUSIONS

Each railway track maintenance always involve the removal of a number of rails because of breakages or damage resulting from manufacturing defects and/or traffic loads.

Monitoring the behaviour of rails in the track must be closely with computer processing, so that users and producers can focus their efforts to improve the quality of the rails and the conditions for their use.

Efficient rail defect management is dependent on the defects being classified into six categories (chapter 2.-from transversal cracks to oblique cracks.

Each defect type is allocated to four a category of measures depending on the size of the defect in question (see 3.6.). The handling of rail defects is often referred to as rail defect management.

The main goal of the European research project "InfraStar" is to prevent rolling contact fatigue (RCF) and to reduce squeal noise in curves by applying an additional surface layer material on the top of the railhead, resulting in a two – material rail. Different models can be used to find the optimum test interval based on the local track and traffic conditions.

Acknowledgment

This work was supported by the Program TEST – Technological Research and Development Projects of Ministry for Science, Education and Sports Republic of Croatia.

5. REFERENCES

Hiensch, M.; Kapoor, A.; Josefson, B.L.; Ringsberg, J.W.; Nielsen, J.C. & Franklin, F.J. (2001). Two-material rail development to prevent rolling contact fatigue and reduce noise emissions in curved rail track, *Proceedings of the World Congress on Raiway Research*, 16 pp, Germany, November 2001, Cologne

Hiensch, M. & Lefranc, B. (2003). Development of a two-materijal rail allows engineering squeal noise behaviour, *Proceedings of EuroNoise*, paper ID:339/p.1 to 6, Italy, 2003, Naples

Hiensch, M. (2003). Two-material rail combats rolling contact fatigue, *Railway Gazette International* (September 2003), 3 pp

*** Code UIC 712R (2002). *Catalogue of Rail Defects*, 4th edition

*** Draft UIC 725R (2005). *Treatment of rail defects*

*** Draft UIC 721R (2003). *Recommendations for use of rail steel grades*

Annals of DAAAM for 2006 & Proceedings of the 17th International DAAAM Symposium, ISSN 1726-9679
B. Katalinic, Published by DAAAM International, Vienna, Austria 2006
150th Birthday of Nikola Tesla (1856-1943) Inventor of Mobile Robot Teleoperation

FEASIBILITY ISSUES IN ROBOTIC MILLING

VOSNIAKOS, G.-C. & MATSAS, E.

Abstract: *Robotic milling is investigated from the point of view of designing a first-principles impedance controller for an industrial robot. The performance of the controller is assessed using a simulation model. The exercise takes into account the milling force which varies periodically with time and demonstrates feasibility of robotic milling using standard industrial robots for soft materials and/or 'gentle' cutting conditions.*

Key words: *Robot, milling, impedance control, simulation*

1. INTRODUCTION

Robotic milling is much more flexible than CNC milling, equipment is less expensive, but accuracy is still low and rigidity is achieved only by special robotic structures. Milling applications have been reported mainly in the domain of rapid prototyping (Vergeest & Tangelder, 1996), prototyping (Chen & Hu, 1999) and human bone milling (Cohan, 2001). However, applications emerge continuously and new possibilities are worth exploring through modeling and simulation.

In general, research literature focuses either on dynamics of end-effector tool - workpiece contact or on system kinematics or, even, on controller design. For instance, (Cetinkunt and Tsai, 1990) proposed a model for compensating position errors in robotic milling with a flat end mill based on milling dynamics models and parallel position-force control. (Anderson & Johansson, 2001) present a control method for an industrial robot equipped with a force/torque sensor emphasising machining parameters and forces in wood milling.

In this paper, a study is presented on the feasibility of robotic milling from the position accuracy point of view. An impedance control scheme is used to prove via model-based simulation that robotic milling performance can be adequate.

2. IMPEDANCE CONTROL

Position control tries to achieve real trajectory of each joint to match as close as possible the desired trajectory. In the current work position control is not sufficient because the end-effector (tool) interacts with the machined workpiece through cutting forces, so that small position errors may result in large control signals, which may have catastrophic results for the robot and the workpiece. Therefore interaction control is appropriate, so much so that force variation is available through milling models. Note that if focus is on real-time control, rather than on feasibility of milling for a particular robot, sensors may measure forces and feed measured values to the controller.

Interaction control strategies may be categorized as indirect and direct force control. The former controls force through position without clearly closing the force feedback loop, whereas the latter uses force feedback. Impedance control falls in the first category, where position error is related to force through mechanical impedance with variable parameters.

The robot is described as a dynamic system with mass, stiffness and some damping, the interaction force being an input variable. The impedance in any direction of the functional space is usually non-linear and coupled. Use of a force/torque sensor enables use of force measurements in the control law, which would lead to linear and de-coupled impedance. A characteristic of impedance control is the balance between position errors and external force that is achieved.

In the problem under consideration, the end-effector trajectory is known, therefore desired position, speed and acceleration of the end effector and (through inverse kinematics) of the joints are known. Contact force variation is also known. Therefore an elementary off-line impedence control scheme can be drawn as follows.

Robot dynamics is represented by the following equation

$$B(\vec{q})\cdot\ddot{\vec{q}}+C(\vec{q},\dot{\vec{q}})\cdot\dot{\vec{q}}+F_u\cdot\dot{\vec{q}}+G(\vec{q})=\vec{\tau}-J^T(\vec{q})\cdot\vec{h} \qquad (1)$$

where q represent joint coordinates, B the inertia matrix, C the Coriolis and centrifugal force matrix, F_s the viscous friction force, G the weight, τ motor torques and J^Th represents the tool-workpiece interaction moments developed to counterbalance the torque emerging at the joints due to contact forces. Non-linear control inverse dynamics law on the functional space may be applied, see (Sciavicco & Siciliano, 2000) as

$$\tau=B(q)\cdot y+C(q,\dot{q})\cdot\dot{q}+F\cdot q+G(q) \qquad (2)$$

Combining equations (1) and (2) yields

$$\ddot{q}=y-B^{-1}(q)\cdot J^T(q)\cdot h \qquad (3)$$

where y is the new control variable.

By choosing the control law as follows (4)

$$y=J_A^{-1}(q)\cdot M_d\cdot\left(M_d\cdot\ddot{x}_d+K_D\cdot(\dot{x}_d-\dot{x})+K_P\cdot(x_d-x)-M_d\cdot\dot{J}_A(q,\dot{q})\cdot\dot{q}\right)$$

where x is the current position and x_d the desired position and M_d, K_D, K_P are positively defined diagonal matrices and by considering the known kinematics equations:

$$x=kine(q)\Rightarrow\dot{x}=J_A(q)\cdot\dot{q}\Rightarrow\ddot{x}=\dot{J}_A(q)\cdot\dot{q}+J_A(q)\cdot\ddot{q} \qquad (5)$$

equation (3) becomes:

$$M_d\cdot(\ddot{x}_d-\ddot{x})+K_D\cdot(\dot{x}_d-\dot{x})+K_P\cdot(x_d-x)=M_d\cdot B_A^{-1}(q)\cdot h \qquad (6)$$

where:

$$B_A(q)=J_A^{-T}(q)\cdot B(q)\cdot J_A^{-1}(q) \qquad (7)$$

is the equivalent of the inertia matrix in functional space and is positively defined if the Jacobian matrix J_A has no singularity.

Equation (6) relates the acting forces vector (right hand side) with the displacement vector x_d-x in functional space, through a generic impedance, expressed by a mass matrix, M_d, a damping matrix K_D and an elasticity matrix, K_P.

2. SIMULATION ENVIRONENT

The robot used in simulation modeling was the PUMA 560, see Fig. 1, not because of its particular suitability for milling, but because all its data were available. The robot's D-H parameters are shown in Table 1. The cutting tool which is fixed as end-effector was a flat ended four flute end mill with diameter 19 mm, helix angle 30°, constant feed 0.1 mm/tooth/rev, rotating at 500 rpm, axial depth of cut 0.5 mm and radial depth of cut 10 mm. The workpiece material was assumed to be Ti6Al4V.

According to this data and following the procedure outlined in (Altintas, 2000) the variation of milling forces with time along a 33.2 mm straight line toolpath (10 sec cutting time) was calculated using Matlab™, see Fig. 3.

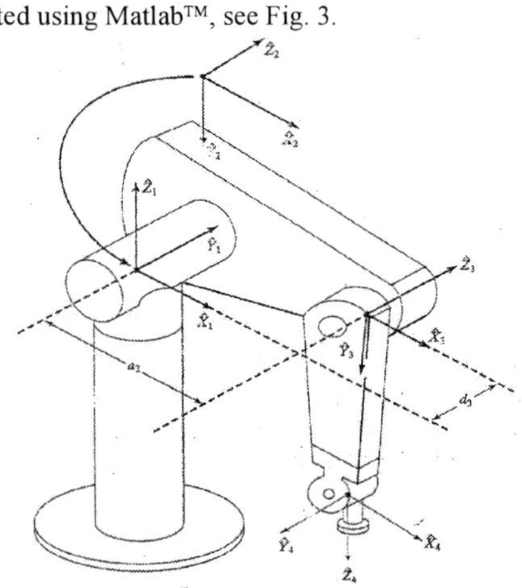

Fig.1 The PUMA 560 robot used in the case study

Joint i	α_{i-1}	a_{i-1}	d_i	θ_i
1	0	0	0	θ_1
2	-90°	0	0	θ_2
3	0	a_2	d_3	θ_3
4	-90°	a_3	d_4	θ_4
5	90°	0	0	θ_5
6	-90°	0	0	θ_6

Table 1. D-H parameters of the robot

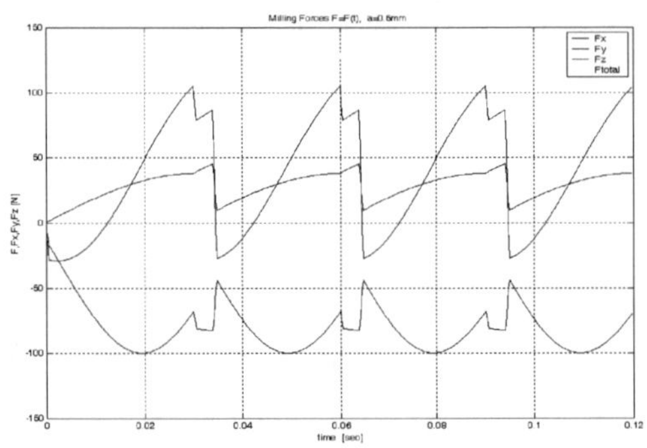

Fig. 2 Variation pattern of cutting forces at the end-effector

The simulation model of the robot was setup in Matlab™ Robotics Toolbox (Corke, 1996), which possesses all necessary functions, e.g. for forward and inverse kinematics including Jacobian manipulation, forward and inverse dynamics etc. It was assumed that contact between tool and workpiece was at a single point and that the tool axis was vertical, thereby keeping the last three joints of the robot at a constant angle throughout.

3. IMPEDANCE CONTROL IMPLEMENTATION

The code created in Matlab™ for implementing the impedance control scheme consists of the following parts.

First, the cutting force vector is defined as well as the final coordinates of joints, the cutting tool initial and final cartesain coordinates and the initial configuration in joint space. The path and velocities in joint space are calculated.

Then, the Jacobian is calculated at every intermediate position of interest. The orientation of cutting tool (roll, pitch and yaw angles) is calculated for the final position, so that position, velocity and acceleration are defined there.

The gains of inertia, elasticity and damping matrices are defined. These values were 'intuitively' determined in a trial-and-error mode formally known as 'gain tuning'.

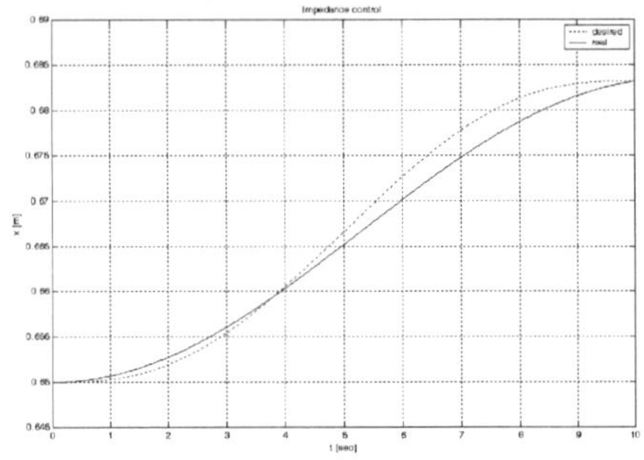

Fig. 3. Simulated position error of the tool in Cartesian space

The derivative of the Jacobian at every moment of interest is calculated according to a finite difference philosophy. This enables cutting tool velocity calculation at all points of interest according to differential kinematics. Then, the control variable y is calculated according to equation (4) as well as B_A in functional space and the rest of equation (3). Joint acceleration is calculated according to equation (3) . This facilitates numerical calculation of velocity and joint position, yielding x coordinate variation according to the control scheme.

4. RESULTS AND CONCLUSIONS

Fig. 3 depicts the position error in cartesian space for the straight line motion of the tool. The actual trajectory is shown as a continuous line and the desired one is shown as a dotted line. The difference is deemed small enough to prove that millling is possible from the point of view of accuracy, i.e. at the feasibility study stage.

A different control scheme needs to be used, if actual robotic milling had to be implemented, probably using real-time force measurements. This is the subject of on-going work.

5. ACKNOWLEDGMENT

This work is supported by General Secretariat of Research and Technology partly under project number 01EΔ131 and partly under project number 05ΔΣΒΕΠΡΟ-99.

6. REFERENCES

Altintas, Y. (2000). *Manufacturing Automotation: Metal Cutting Mechanics, Machine Tool Vibrations and CNC Design*, Cambridge University Press, London.

Andersson, J.-E. & Johansson, G. (2001). Robot control for wood carving operations, *Mechatronics* , 11, 4, 457-490.

Cetinkunt, S. & Tsai, R.-L. (1990) Position error compensation of robotic contour end-milling, *International Journal of Machine Tools & Manufacture*, 30, 4, 613-527.

Chen, Y.H. & Hu, Y.N. (1999) Implementation of a robot system for sculptured surface cutting. Part 1: Rough Machining, *International Journal of Advanced Manufacturing Technology*, 15, 9, 624-629.

Cohan S. (2001). ROBODOC achieves pinless registration, *Industrial Robot*, 28, 5, 381-386.

Corke, I.P. (1996). A Robotics Toolbox for Matlab, *IEEE Journal of Robotics and Automation*, 3, 1, 24-32.

Sciavicco, L. & Siciliano, B. (2000). *Modelling and Control of Robot Manipulators*, 2nd edition, Chapter 6, Springer-Verlag, London.

Vergeest, J.S.M. & Tangelder, J.W.H. (1996) Robot machines rapid prototype, *Industrial Robot*, 23, 5, 17-20.

Annals of DAAAM for 2006 & Proceedings of the 17th International DAAAM Symposium, ISSN 1726-9679
B. Katalinic, Published by DAAAM International, Vienna, Austria 2006
150th Birthday of Nikola Tesla (1856-1943) Inventor of Mobile Robot Teleoperation

FLOWSHOP SCHEDULING USING GENETIC ALGORITHMS

VOSNIAKOS, G.-C. & MILLAS, V.

Abstract: *This paper reports on the use of a simple genetic algorithm as a means to schedule manufacturing systems of the flowshop type. The parameters used in the algorithm were selected according to the Design of Experiments approach and the total makespan is shown to be minimised fast in a 15 machine 30 job problem used as an example.*
Key words: *Flowshop, Scheduling, Genetic algorithms.*

1. INTRODUCTION

A flowshop, as considered in this work, consists of a number of machines of different type each and a number of jobs that have to be ordered in an optimum way. Each job needs a number of operations (machines) up to a certain maximum. All jobs are available to start at time point 0 and each job follows the same 'flow path' from machine to machine. Each machine executes one operation of a job completely and jobs never return to a machine for performing a second operation. Machine buffer capacity is endless. Setup time for an operation is included in the operation's processing time. Machines are available all the time without downtime of any kind. All technological constraints for each job are known beforehand and remain constant.
The main target in flowshop scheduling is usually total makespan. In addition, criteria such as mean machine utilisation, mean job tardiness, number of tardy jobs may also be taken into account.
Heuristics ere the first approach to be used extensively for flowshop problems due to combinatorial explosion.
Heuristics for flowshop scheduling are classified into solution generation and solution improvement heuristics. The two-machine flowshop problem with total makespan as objective function was solved by Johnson [12] with what, then, became a classic heuristic, effectively putting jobs with lowest operation times on one of the machines first in its queue, and at the same time putting jobs with lowest operation times on the other machine last in its queue.
Classic flowshop scheduling heuristics are presented in (Pinedo, 2001). Recent examples (Ho & Chang, 1991) introduce the notion of gap between operations and give generally better solutions than the other known heuristics.
One of the early attempts to exploit genetic algorithms (GAs) in flowshop scheduling (Gen et al, 1994) employed a chromosome indicating job order. Due to the nature of the problem this coding does not leave any real alternative representations, therefore it is widely adopted. However, there have been attempts to combine GAs with other optimisation techniques. For instance, in (Reeves, 1995) a GA is used as solution generator and simulated annealing as solution improver. (Chen et al, 1995) employ the CDS heuristic as solution generator and a GA as solution improver. In (Zegrodi et al, 1995) simulated annealing is combined with a knowledge base assigning job priority indices to guide search.
In this work, the classic GA formulation for flowshop scheduling is employed with improved genetic operators and a Design of Experiments (DoE) methodology to select their values.

2. GENETIC ALGORITHMS

These are optimisation algorithms that rely on a 'survival-of-the-fittest' philosophy and natural evolution mechanics (Goldberg, 1989). They divide the solution space into subsets and then evaluate them in order to find the best solutions. This process is repeated and good solutions from each subset are kept until the optimum is reached.
The variables of the problem have to be coded into chromosomes (usually binary strings). A number of chromosomes, called population, represent solution space, searched by changing the chromosomes of the population using operators. Selection operator picks chromosomes for reproduction. Crossover generates children by combining parents genes. Mutation is used to change genes in a chromosome. An objective function, related to a fitness function, assesses quality of chromosomes and ranks them.

3. FLOWSHOP GENETIC ALGORITHM

The flowshop problem tackled in this work is essentially a permutation problem of a number of jobs / operations. As an example, a static-deterministic flowshop was studied comprising 15 machines and 30 jobs, which ensured job bottlenecks, thus uncovering strengths and weaknesses of the approach proposed. The characteristics of jobs were operation times on each machine, see Table 1, with values randomly assigned to each job in the interval [0,100] time units, making for a pure flowshop. The chromosome defined represented a natural coding of the problem, comprising alternative forms of job permutations, for instance : Chromo=[4 5 22 14 8 9 ...... 17 2 6] represents job order : 4-5-22-14-8-....-2-6.
This representation ensures that jobs are not constrained to each other (feasibility), that no unacceptable solutions can be arrived at (legality) and that there is a one-to-one correspondence between chromosome genes and job-order (uniqueness of mapping). For this type of chromosome the partially matched crossover (PMX) operator has been suggested (Goldberg, 1989), working as follows. Suppose two parents with the following chromosomes:

P1=[9 8 4 | 5 6 7 | 1 3 2 10]
P2=[8 7 1 | 2 3 10 | 9 5 4 6]

Two positions are chosen randomly, e.g. 3rd and 5th. These define a matching section for each parent, the genes of which are simply swapped between the parents, i.e.

P1'=[9 8 4 | 2 3 10 | 1 3 2 10]
P2'=[8 7 1 | 5 6 7 | 9 5 4 6]

At this intermediate stage genes 2,3 and 10 appear twice in parent P1', whilst genes 5, 6 και 7 are missing. To maintain chromosome legality, matching is applied outside the matching sections, too, replacing 2 by 5, 3 by 6 and 10 by 7. The offspring become :

O1=[9 8 4 | 2 3 10 | 1 6 5 7]
O2=[8 10 1 | 5 6 7 | 9 2 4 3]

The genetic algorithm was implemented in C++.

M# / J#	1	2	3	4	5	6	7	8	9	10	11	12	13	14	15
1	74	72	54	57	52	60	4	8	40	8	85	45	74	67	48
2	99	77	58	50	31	67	19	96	93	29	27	6	85	22	48
3	15	10	85	2	92	53	60	63	11	94	44	71	19	99	94
4	63	18	44	30	80	94	63	28	50	55	78	83	8	68	65
5	94	42	20	92	73	62	45	86	76	11	80	53	29	3	70
6	16	15	97	30	7	31	82	10	28	13	63	55	24	26	49
7	51	57	19	87	81	17	8	27	93	72	1	19	30	80	86
8	18	46	17	12	54	54	90	52	69	82	47	96	90	14	12
9	77	29	52	40	83	53	44	49	87	60	2	88	26	18	30
10	19	95	57	94	93	19	36	14	82	74	94	7	90	40	39
11	25	28	72	89	5	87	15	87	14	20	24	91	93	41	36
12	92	98	56	35	64	15	95	22	67	61	12	98	73	6	10
13	47	80	88	77	77	60	63	66	8	10	63	74	90	1	56
14	38	84	99	21	13	73	4	26	68	99	9	72	42	43	27
15	77	65	38	88	95	9	13	13	42	55	51	36	27	78	12
16	22	81	30	45	63	94	44	78	98	57	26	85	61	82	71
17	6	3	43	96	51	39	79	28	50	27	49	30	73	53	85
18	32	28	61	62	18	76	92	50	5	26	1	53	33	79	17
19	28	97	87	10	79	35	55	72	98	32	2	58	87	86	12
20	16	73	45	90	76	86	75	38	58	49	61	90	23	92	24
21	47	8	30	52	53	96	26	76	18	65	32	18	35	71	36
22	65	90	73	28	34	58	20	10	46	94	38	34	1	17	72
23	6	62	42	42	95	26	90	83	82	54	47	44	80	22	2
24	39	12	28	15	77	22	46	17	19	18	83	5	28	55	96
25	4	75	57	1	24	46	99	43	17	22	36	22	12	22	38
26	14	97	19	81	92	44	66	92	71	66	43	61	11	42	82
27	38	55	10	76	22	61	40	59	46	39	33	2	92	81	99
28	82	38	7	12	2	28	88	89	2	93	88	60	60	53	82
29	68	55	77	51	34	25	35	45	38	77	96	3	27	23	43
30	86	4	95	81	63	51	73	80	1	10	11	47	87	80	54

Table 1. Operation times for jobs (J#) on machines (M#)

4. RESULTS, DISCUSSION AND CONCLUSIONS

The best solution corresponds to GA parameters ps=0.50, pc=0.75 and pm=0.10 and the following job order : 3-23-6-27-8-24-21-17-19-1-28-9-11-22-7-16-30-13-4-10-14-5-29-26-20-2-25-18-15-12. The GA 'convergence' path is shown in Fig. 1. Convergence takes less than 4 min on a Pentium IV with 512 MB RAM. The optimisation result is shown in Fig. 2 as a Gantt chart for each machine on a common time frame. The GA achieved a decrease in makespan from 2876 to 2463, i.e. about 15% in 500 generations. Fig. 1 shows good performance, since the mean solution quality improves steadily from the first generations. However, in order to overcome local minima between generations 70 and 370 more than 50 generations are required. Waiting times are quite restricted. Long waiting times that emerge between jobs 3 and 23 are advantageous to makespan, i.e. in all machines, except 3 and 5, the current operation of job 23 is executed without delay. The final schedule conforms to Johnson's rule, because the initial operations on machine 1 as well as the final operations on machine 15 exhibit low completion times, as in the two-machine flowshop.

Ii is relatively straightforward to include constraints in this approach, such as tardiness with respect to due dates, setup time variation according to job order or even machine breakdown patterns. This is the subject of on-going work.

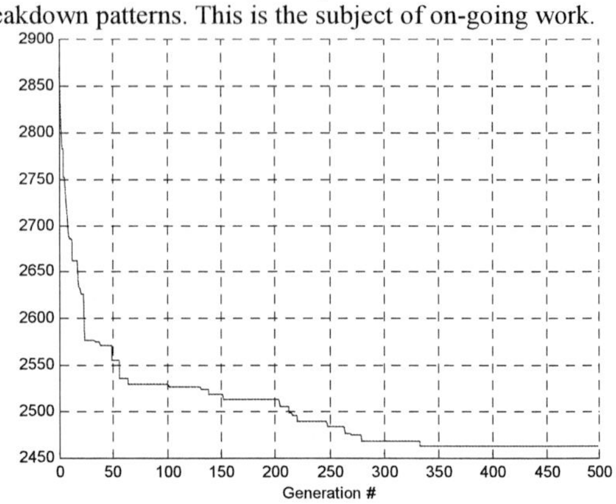

Fig 1. GA convergence in terms of total makespan

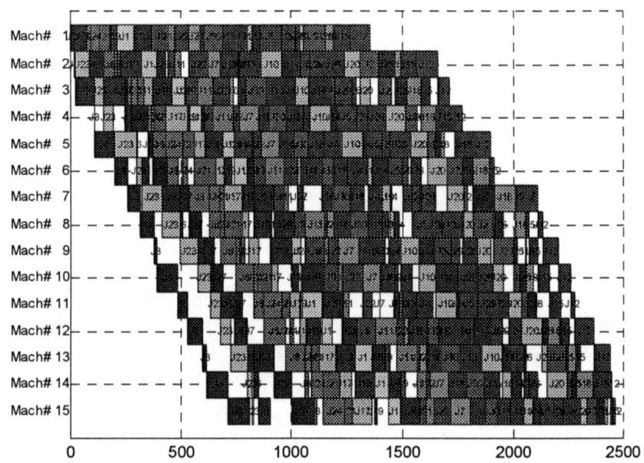

Fig 2. The optimised schedule

5. REFERENCES

Chen, C.L., V.S. Vempati, & Aljaber, N. (1995). An application of genetic algorithms for flow-shop problems, *European Journal of Operational Research*, 80,2, 389-396.

Gen, M.; Tsujimura, Y. & Kubota, E. (1994). Solving Job-Shop scheduling problem using genetic algorithms, *Proceedings of the 16th International Conference on Computers and Industrial Engineering*, pp.576-579, Ashikaga (Japan), March 1994, IEEE.

Goldberg, G.E. (1989). *Genetic Algorithms in Search, Optimization and Machine Learning*, Addison-Wiley, 0201157675, Reading, Massachussets.

Ho, J.C. & Chang, Y.L. (1991). A New Heuristic for the n-job, m-machine Flowshop Problem, *European Journal of Operations Research*, 52, 194-202.

Johnson S. (1954). Optimal Two- and Three-Stage Production Schedules with Setup Times Included, *Naval Logistics Quaterly*, 1, 61-68.

Pinedo M. (1995). *Scheduling Theory, Algorithms and Systems*, 2nd edition, Prentice-Hall, ISBN: 0130281387, Amherst.

Reeves C.-R. (1995). A genetic algorithm for flowshop sequencing, *Computers and Operations Research archive*, 22, 1, 5-13.

Ross, P. (1996). *Taguchi Techniques for Quality Engineering*, Mc Graw Hill, ISBN 0070539588, New York.

Zegordi, S.H.; Itoh, K. & Enkawa, T. (1995). Minimizing makespan for flow-shop scheduling by combining simulated annealing with sequencing knowledge, *European Journal of Operational Research*, 85, 3, 515-531.

Annals of DAAAM for 2006 & Proceedings of the 17th International DAAAM Symposium, ISSN 1726-9679
B. Katalinic, Published by DAAAM International, Vienna, Austria 2006
150th Birthday of Nikola Tesla (1856-1943) Inventor of Mobile Robot Teleoperation

ADVANTAGES AND DISADVANTAGES OF PURCHASE AND RENT WITHIN THE REAL ESTATE MANAGEMENT OF A RETAILER

VOUK, R.; MARKOVIC, M. & KOVACEVIC, Z.

***Abstract**: Whether he is considering entering a new market or restructuring the existing portfolio of stores, a retailer has to decide either to purchase or rent his store premises. The advantages of purchase are reflected in capital gains, full freedom of disposal of the existing premises and long-term planning. On the other hand, the rent ensures financial flexibility and higher liquidity, to begin with. To prepare an adequate information base for the above decisions, and to ensure the adequate implementation thereof, retailers tend to organise all these activities within a single department: Corporate Real Estate Management Department.*
***Key words**: retailer, real estate, purchase, rent, CREM.*

1. INTRODUCTION

Unsurprisingly, in the literature we can often find a statement saying that the retail success depends on the following three things: location, location, and of course, location. The reasons for the above statement are more than justified taking into account the specific requirements that should be met while choosing the location. Normally, a large number of buyers will choose a specific store only because they find it convenient due to its proximity, car park or a large number of other stores around. On the other hand, prior to making his decision on the choice of the right location, a retailer must solve a number of problems. Basically, choosing the right location is one of the key issues for a retailer because it ensures the creation of a relatively long-term sustainable competitive advantage. But the real ''sweet spot'' locations have become increasingly difficult to find because most of them are already occupied, and if there are some left in the market, their prices are very high. This paper is aimed at providing an insight, through a descriptive analysis, into the problems of strategic decision-making related to real estate management within the store portfolio of a retail business.

2. THE ANALYSIS OF THE RETAILER'S DECISION TO PURCHASE REAL ESTATE

Once he has made his decision as to the specific location and specific form of the store, a retailer only has to decide between the two basic strategies: Whether to purchase the premises or make a lease agreement. Whatever the decision, one thing is for sure: as a rule, once the decision is made, it will be quite difficult and costly to change it. That making the above decision a very serious task, especially if we consider the fact that as much as 30-50% of the capital owned by retailers who decided to purchase the premises is bound by the real estate ownership (Levy&Weitz, 2004). One of the most serious financial implications of the real estate purchase is the issue of the so-called *sunk costs*. These costs may be defined as unrecoverable costs of investment in a specific kind of property. In principle, a retailer may face three kinds of sunk costs. First of all, there are *set-up sunk costs* which refer to unrecoverable costs that were, for some reason, incurred at the initial purchase of the store. Secondly, the term *accumulated sunk costs* refers to the costs incurred during the useful life of the store and that may not be recovered by means of a market valuation of the store. As a result of dynamic and competitive development of retail forms, and the consequential need for frequent changes of locations in order to maintain the strategic advantage, while selling the property there are additional unrecoverable costs which in this case may be referred to as *exit sunk cost* (Guy, 1999). The extent to which these costs may significantly affect a purchase decision may be best determined by the analysis of potential problems related to the selling of the respective store. The fact is that, as a rule, the success indicators are worse for the stores located in the poor areas, with a decreasing purchasing power, not many passers-by, and with negative demographic indicators. Since, quite obviously, the above described stores must be eliminated from the portfolio if the shareholder value is to be increased, it is quite clear that the short-term negative off-set will certainly not be irrelevant.

3. SPECIFICS OF THE DECISION TO RENT A STORE

As a result of the globalisation process and the increasingly fierce competition in the retail markets, retail businesses now attempt to take a more flexible position with regard to challenges they have to face in order to create conditions for a timely change of their strategy. One step in this direction, important for the subject of the paper, is to gradually abandon the ownership of the entire store portfolio and to rent a certain number of stores in order to achieve a dynamic balance between the capital frozen in the real estate and finances required for current operations. Specifics of this type of store management first of all depend on the choice of the lease type with respect to interest rates – fixed or variable (generally, price-index related). In addition, there are specific clauses and provisions of contractual relations, such as:

- *Prohibited Use Clause* – this may refer to the companies whose clients make no contribution to the retail business turnover, but they may occupy a significant portion of the parking area (dentist's offices, etc.).
- *Exclusive Use Clause* – based on this clause a retailer tries to impose restrictions upon the owner so that neighbouring premises may not be rented out to competition.
- *Escape Clause* – this is a clause that is very difficult to include in the contract due to a potentially adverse affect that it may have for the lesser. Basically, it stipulates that the retailer/lessee may unilaterally terminate the contract unless the agreed turnover is realised within the specified term.

Favourable trends for retailers – renters started with the radical change in the real estate market in the late 1980s (Krumm 2001). As a result of the recession in the market of real estate for rent, retailers suddenly gained some negotiating power and managed to win certain concessions which mainly consisted in the following:

- Decrease in the lease period from the standard 25 years to 10-15 years,
- Revision of interest rates every five years,
- Capital maintenance is made part of owner's obligations.

The overall conclusion is that the above changes resulted in a larger flexibility on part of retailers, and in the opening of the real estate market to a large number of retailers who until then had a very limited negotiating power and signed contracts based on the *take-it-or-leave-it* principle.

4. CORPORATE REAL ESTATE MANAGEMENT WITHIN THE EFFICIENT PORTFOLIO MANAGEMENT

The expansion into new markets through the process of globalisation led to the increasing value of the real estate portfolio. This put a strong pressure on a retailer involved in the management of the processes related to the acquisition, maintenance and sale of the real estate. Taking into account the fact that there are no two identical properties, and the fact that there are medium-to-long-term frozen resources, will add to a better understanding of the importance of this process. Due to the above facts, in addition to the need for a local modification of the appearance of the store and its assortment and marketing in different countries, it seems that centralised management of the real estate portfolio is almost impossible. For the reasons mentioned above and earlier in the article, companies that diversify their business, find a solution in the setting up of special departments to deal only with the Corporate Real Estate Management. It is true that these departments are no news; however, the scope of their work has significantly changed so that today they have a lot more responsibilities and authorities. In order to enable them to cope with the new situation, it is necessary to set the guidelines for the implementation of their strategic tasks for the purpose of efficient management. The basic steps in the process of strategic management of the delegated tasks include as follows:

- Creation of partner-based relationship with the local professional real estate agencies. As a rule, finding of such agencies results in significant savings because of their excellent knowledge of specific local circumstances and the time needed for negotiations and finalisation of the rental/purchase agreement. Because of the fact that retailers will work with several agencies, it is important that they share their mission, vision, and objectives, and work methodology in order to minimise the mismatch between what is demanded and what is supplied.
- Creation of the complete portfolio data base of all real estate in the possession of the respective company. Information is power. The advantages of the creation of the above base are reflected in the complete and fast accessibility of the entire portfolio in order to obtain the latest information on the specifics of real estate around the world. The information from such base provides an excellent *benchmark* while looking for the information about the new markets and potential acquisitions for the portfolio. Information may provide us with an insight into possible costs, acquisitions, maintenance, i.e. operating management, sale and so on. With such bases, decision makers will have more certainty in their evaluation of the advantages and disadvantages of the decision on the purchase or rent.
- Development of the quality-oriented real estate management. To ensure that the retail real estate management may function at the satisfactory quality level in the described conditions, it must be process oriented. Furthermore, the process itself must include the quality principles as the basic guidelines of the process. The four quality principles of this process would be as follows:
  1. Conformance to Requirements. What are the main objectives of the retailer?
  2. Prevention. The system must be designed in such a way as to prevent potential problems after the purchase/rent.
  3. Zero Defects. There must be no mismatch between the strategy of the retailer and the results of the real estate management process.
  4. Price of Non-conformance. It is necessary to develop the awareness of the price of mistakes on part of the CREM department staff and to point, by means of quantification, to the significance they have for the entire system.
- Ensure the corporate support for the activities of the CREM department. It is an illusion to expect that all the managers and lower ranked staff will accept, without reserve, its principles and work methods. The CREM will surely assume certain activities that used to be within the competence of other departments. Within this context it is advisable to ensure not just the support but also the active participation of the senior management with lover ranked staff regarding the measures taken by the CREM, in order to minimise their resistance towards the activities of the department (Zappile 2004). The central corporate place for any region will ensure not just the concentration of know-how and experience in the field of real estate management, but it will also result in the significant cost savings for the company as a whole.

5. CONCLUSION

Ad – hoc approaches and uncritical evaluation of the important factors regarding purchase or rent may not only cause a direct financial loss for the retailer but may also affect the retailer's capacity to adjust to the constantly changing market. The observing and understanding of the current market trends and development assumptions provides a contextual framework for the respective decision-making. The forming of special, largely decentralised departments to deal with the issues of purchase, rent, maintenance and selling of real estate may result in a direct competitive advantage that will help the retailer to cope with specifics of individual markets and regions. Since the specifics of the purchase-rent decision are largely based on the historical data, this department may, based on its experience and relevant information obtained from such historical data, bring the key advantage while entering the new markets. The constant pointing to the growing relevance of the above department and its processes for the successful implementation of the retailer's strategy undoubtedly indicates a need to know and implement the state-of-the-art methods and techniques of the corporate real estate management.

6. REFERENCES

Guy, C. (1999) Exit strategies and sunk costs: the implications for multiple retailers. *International Journal of Retail & Distribution Management,* Vol. 27, No. 6, (1999) pp. 237-245, ISSN 0959-0552

Levy, M.; Weitz, B. A. (2004). *Retailing management*, fifth ed., McGraw Hill, ISBN 0-07-121481-X, New York

Roper, O. K.; (2001) Aligning corporate real estate with the business of your company. *Journal of Corporate Real Estate,* Vol. 3, No. 3, (2001) pp. 222-231, ISSN 1463-001X

Krum P. I. M. M. (2001) History of real estate from a corporate perspective. *Facilities,* Vol. 19, No. 7/8, (2001) pp. 276-286, ISSN 0263-2772

Zappile, R. P. (2004). A quality process approach to corporate real estate management. *Journal of Corporate Real Estate*, Vol 6. No. 3, (2004) pp. 215-226, ISSN 1463-001X

Annals of DAAAM for 2006 & Proceedings of the 17th International DAAAM Symposium, ISSN 1726-9679
Editor B. Katalinic, Published by DAAAM International, Vienna, Austria 2006
150th Birthday of Nikola Tesla (1856-1943) Inventor of Mobile Robot Teleoperation

INTERNATIONALIZATION IN CROATIA

VRANESEVIC, T.; BOGUNOVIC, B. & MANDIC, M.

Abstract: *Global economy can be judged through various indicators, e.g. size of supply and demand, or indicators of changing social environment under which economic processes are taking place and the similar. Thus, internationalization embraces not only the economy, but also the conditions relevant for life and work of the people. Croatia is one of the European transitional economies where SMEs like in many other former socialist countries are relatively recent phenomena. Most SMEs have been established in the early and middle 1990s. The process of internationalization of enterprises that started in many former transition countries has taken place in basically larger companies. In fact, relatively small number of medium and small number of enterprises have been embraced by internationalization. The case of Croatia is not much different than in most east European countries.*
Keywords: *Croatia, SME, Internationalization, FDI.*

1. INTRODUCTION

The process of internationalization of enterprises that started in many former transition countries has taken place in basically larger companies. Privatization of larger Croatian enterprises has left its trace on the current condition of Croatian SME. SME's are usually leaned on the larger companies that are the basis of each economy. For an investigation of the internationalization process of an enterprise, one needs to define the meaning of "internationalization" and set its measure. The process of internationalization of companies can be defined in various manners. To our mind the best way to judge the degree of internationalization of companies is to measure the degree of foreign ownership, import-export exposure to foreign market, inflow/outflow of investments. One of the important dimensions of this problem lies in the over all division structured as small, medium and large size companies. A number of sources define internationalization. One approach to internationalization is from quantitative point of view (Westhead et al., 2002) and (Calof & Beamish 1995), proposing FDI and level of exports as its measures. The other approach is look at some 'soft' indicators such as market knowledge, strategy, and success (Chetty & Campbell-Hunt 2003). The later approach, which to us seems better suited to our investigation, requires primary data collection. Unfortunately, data on indicators on firm's success in Croatia is not easily available, since there is no systematic auditing done by authorized institution. For applying a macro-statistical approach, a set of secondary data can be used. Such data is available from published sources e.g. The Croatian National Bank, and the Croatian Bureau of Statistics. It must, however, be stated that different data sources provide very different information for the same indicator for the same period. Sometimes, the data from various sources varies even by 30%. Accordingly, the data sources need to be quoted. Another problem that one is confronted with in case of Croatia is that consistent statistics on these indicators exist only for the late 1990s. Any attempt to compare the early period with the later needs every caution for the data is either nonexistent or deficient. Further, no institution in the country uses specific FDI statistics according to the size of the firms. Also, there is no systematic approach to statistical coverage of internationalization of enterprises. For these very reasons, investigation needs to be based on the FDI statistics and structure of enterprises. The inward FDI is considered as the most significant of the internationalization. Since, parallel to the growth and development of SMEs, Croatia was going through a war and a transition of ownership process, internationalization of enterprises had a specific path. Transformation of economy and effects that followed must be carefully interpreted to understand and explain the changes in structure of enterprises and FDI. Explaining macroeconomic and political conditions at certain point of time are therefore one of the cornerstones of data interpretation in this paper. Our secondary analyses in particular include analysis of FDI in Croatia.

2. RESULTS

Process of internationalization in economy is, in relative sense, stronger in financial than in the real sector. It is because the conditions of internationalization of financial sector are defined by global institutions like IMF, Word Bank and World Trade Organization and are not dependent on the state of information systems, while the real economy and competitive market in real sector need developed information system. Process of change in ownership in Croatian financial sector came after process of reconstruction of damaged monetary and banking system over short and long term. Changes took place in personal expenditure and state budget. Increased private consumption expenditure resulted in increased fiscal revenues, investments and foreign debt. Internationalization in Croatia took place in financial sector only while most of the real sector remained untouched except a few large enterprises. Process of internationalization in Croatian economic structure is mostly directed by change in ownership, but not based on evolution and expansion of companies. This can be confirmed by the decrease in the volume of business turnover, employment and finally the GDP. Change in ownership in real sector in favour of international owners is visible e.g. in telecommunications, energy (oil and gas), metal industry, hotel industry and especially in trade (large shopping centres). FDI statistics is strongly connected to the privatization process. According to Table 1, highest FDI in Croatia were in 2001 when both the Croatian Telecom and Croatian Oil Company among others, were privatized causing high monetary inflows in Croatian economy. Considering the fact that still each year brings another government managed privatization, it is difficult to observe FDI as a measure of internationalization or free market activity. FDI outflows are more significant parameter, and according to data in Table Appendix, Croatian companies show significant progress in internationalization (88 million EUR in 1998. to 273 million EUR in 2004.). There is a constant rising trend if overall FDI inflow in Croatia is observed (Table 1, Croatian national bank, 2005). Somewhat higher FDI in years 1999, 2001 and 2003 is the consequence of privatization of Croatian banks and Telecom in those years when the same were sold to foreign companies, and that inflow is being considered as FDI. What concerns us is the decrease in 2004 in relation to 2000. There was no major privatization in those years so the data is comparable and show that Croatia is becoming less attractive to foreign investors. The official statistics on inflow of funds is shown in Table 1.

Year	FDI (mln EUR)
1993	120,3
1994	117,0
1995	114,2
1996	510,8
1997	538,0
1998	833
1999	1.472,0
2000	1.086,5
2001	1.746,8
2002	1.126,1
2003	2.041,6
2004	871

Table 1. FDI inflows in Croatia (1993 – 2004)

Structure of FDIs in Croatia in last twelve years confirms the privatization of state-owned companies as the main source of FDI (Table 2).

Industry	in %
Other monetary intermediation	21,42
Telecom	15,88
Manufact. of pharma. Preparations	10,52
Manufact. of refined petroleum products	7,79
Extraction of crude petroleum and natural gas	3,78
Hotels and motels, with restaurant	3,29
Manufacture of cement	3,27
Other retail sale in non-specialized stores	3,12
Manufacture of beer	2,08
Business and management consultancy activities	1,02
Others	27,83

Table 2. Foreign direct equity investments in Croatia 1993 - 2004 (6.562.5 million USD)

Highest shares are in monetary intermediation (banks), telecommunication (Croatian Telecom), oil and petrol industry (30% of domestic petroleum company sold to Hungarian MOL). Even the beer industry (Heineken and Interbrew bought two biggest Croatian breweries) can be spotted on the bottom list. In that sense, it is hard to take FDI as a measure of internationalization because it is parameter that better describes selling out process than internationalization, at least in inflow direction. However if the structure of FDI inflow according to industry is linked with the analysis of SMEs, relative high share of FDI in hotel industry can be observed. We have already noted that the hotel industry and trade are Croatian industries that have high potential of internationalization. Hotel industry is ranked sixth in FDI inflow structure, but we have to bear in mind that first five industries according to FDI inflow volume are the industries where the privatization of largest Croatian enterprises took place (telecommunications, banks, oil and gas industry, etc.). FDI in hotel industry also comes from privatization, but as against large companies, new hotel owners are in most cases influencing on company strategy and influencing on management structures and thus contributing to internationalization of such companies. Unfortunately, there are only summarized data on FDI outflows available, with no further division which could provide us with a base for deeper FDI outflow analyses of SMEs in Croatia.

3. CONCLUSION

Like most transitional countries, Croatia has experienced large increase in FDI inflows in last 10 years. In order to determine extent of the influence of FDI in Croatia on internationalization of all enterprises and specially SMEs, changes in enterprise structure, trend in SME enterprise segment and FDI analysis on the basis of secondary data were made. Results show large increase in overall number of enterprises in Croatia (but with only one third of all with any economic activity) in last 10 years. Similarly, number of SMEs is also growing, but with decreasing average number of employees. It means that an average SME is becoming smaller. Considering that an FDI is more likely to be placed by a larger company that has negative effect on ability of Croatian SME to place an FDI abroad. Structure and trends of FDI inflow show large dependence of FDI inflow with privatization of large companies in Croatia. If the years when the largest companies were privatized are not considered in trend analyses, FDI inflow in Croatia is stagnating since 2000. Encouraging fact is the high percentage of FDI in hotel industry, such that hotels are mostly SMEs directed to internationalization because of export nature of hotel industry. The highest rate of Croatian inward FDI in last twelve years was in monetary intermediation, telecommunication and oil/petrol industry which corresponds with privatization of largest companies in Croatia that happened in that period. The state of Croatian economy suggests that in order to improve conditions for internationalization of SMEs, overall structure of enterprises and Croatian economy should be enhanced.

4. REFERENCES

Bogunovic, A. (2001): *Economic integrations and regional policy*, Faculty of Economics Zagreb, Faculty of Economics and Tourism "Dr. Mijo Mirković" Pula and Mikrorad d.o.o., Zagreb, 2001.

Calof, J. and Beamish, P. (1995) Adapting to foreign markets: explaining internationalisation, *International Business Review*, Vol. 4 No.2, pp. 115-31

Chetty S. and Campbell-Hunt C. (2003) Paths to internationalisation among small to medium-sized firms, *European Journal of Marketing*, Vol. 37 No. 5/6, pp. 796-820

Croatian Accounting and Finance Union web site (2006), http://rif.hr

Croatian national bank (2006), www.hnb.hr

Griffiths, A. & Wall, S. (1991): *"Applied Economics"*, Longman, London, 1991, p. 68.

Lu, Jane W. and Beamish, Paul W. (2001): The internationalization and performance of SMEs, *Strategic Management Journal*, Vol. 22, pp. 565-586

OECD Economic Survey (1999), *Enterprise policy – SME*, 1999. and Eurostat, Product 3-1003., - EN – BP, Luxembourg, 2000.

Ostwirtschaftsreport [1998] cited in *International Herald Tribune*, Wednesday, June 24, 1998, p. 12.

Republic Croatia – Central Bureau of Statistics (2006), www.dzs.hr

Wei, H.C. and Christodoulou (1997) An examination of strategic foreign direct investment decision processes: the case of Taiwanese manufacturing SMEs, *Management Decision*, Vol. 35 No. 8, pp 619-630

Welch, L.S. and Luostarinen, R. (1988) Internationalisation: evolution of a concept, *Journal of General Management*, Vol. 14 No. 2, pp. 34-55

Westhead P., Binks M., Ucbarasan D. and Wright M. (2002) Internationalisation of SMEs: a research note, *Journal of Small Business and Enterprise Development*, Vol. 9 No. 1, pp. 38-48

Annals of DAAAM for 2006 & Proceedings of the 17th International DAAAM Symposium, ISSN 1726-9679
Editor B. Katalinic, Published by DAAAM International, Vienna, Austria 2006
150th Birthday of Nikola Tesla (1856-1943) Inventor of Mobile Robot Teleoperation

MOTIVATION AND DECISION MAKING AS A KEY FACTORS OF DOMINANT LEADERSHIP STYLE IN HOSPITALITY INDUSTRY

VRDOLJAK RAGUZ, I. & PAVLIC, I.

***Abstract**: Hospitality industry has always attracted the attention of theoretic and scientist because of often chopping in the surroundings, tourist demand, competitiveness at the market and complexity of hotel operations. Because of numerous of specifics and diversity of hospitality industry from the other economic fields, it was expected that the tested variables of leadership styles will be much different from the other industries. The paper is the result of empirical research conducted in the summer 2004 at the sample of hospitality enterprises in Dubrovnik-neretvian county in Republic Croatia.*

Key words: hospitality industry, motivation, decision making, leadership styles

1. INTRODUCTION

This paper analysis the motivation and decision-making as a key variables of dominant leadership style in Dubrovnik-neretvian county. Most of the hotels in the County are privatised so it can be said that the foreign proprietories have engaged the international hotel expertise for the improvement of the moment state of hotel supply, quality of services and implementation of new management concepts in their doing business.

The business internationalization has enabled the differentiation of hospitality industry in Dubrovnik-neretvian county from other Counties in Republic Croatia. Leadership and variables of leadership have dramatically changed in these new terms in hospitality industry. The changes have been caused with the development of information technologies, new organisational structure and the reconstruction of hotel enterprises. The paper also applies different methods such as analysis, description and comparison of the data, in order to describe the variables of motivation an decision-making with the main purpose to determinate dominant leadership style in the mentioned County.

2. THE DIAGNOSTICAL MODEL FOR THE DETERMINATION OF THE LEADERSHIP VARIABLES

Starting with an assumption that in transition economies the leadership styles are more rigid, surveyed research of leadership styles of hotel managers in Dubrovnik-neretvian County have been conducted established on the Likerts leadership styles. Likert has defined four dominant leadership styles, known as a ″participate-democratic system of efficient work groups″ (Buble, 2000., Koontz, Weihrich, 1998.)

Based on previous researches Likert has defined four dominant styles: exploitive-authorative system-system I, benevolent-authoritative system-system II, consultative system-system III and participative-group system-system IV.

In the paper the diagnostic model is based on Likert's model. Managers have given the answers to the six variables that enactive the leadership styles and are included in the survey:

1. leadership as a first variable of leadership style is measured with: reliability of leader to the subordinates, freedom of subordinates in communication with leader and;
2. motivation is measured with flow of information, types and formats of motives, responsibility for achieving goals and team work;
3. communication in the enterprise is measured with these variables: flow of information, acceptance of information from employees, accuracy of information;
4. decision making is measured with variables like levels of decision making, participation of employees in decision making and motivation of employees;
5. goals are measured with the way of setting the goals and with resistance on settled goals;
6. control is measured with the holder of control function, resistance of roles and purpose of control data.

3. RESEARCH METHODOLOGY

The research of leadership styles was conducted in the summer 2004. on a sample of 32 hotels in Dubrovnik-neretvian county. The researched population was consisted of 59 hotels. The share in the population result from the mentioned was 54,24%. The size of representative sample was 58,75%.The survey consisted of 46 questions formulated on Likert's scale. Two management levels were incorporated in the research, top and middle management (hotel managers and sector managers).

4. THE RESULT OF EMPIRICAL RESEARCH OF THE VARIABLES OF LEADERSHIP STYLE

In the variables of leadership, communication, goals and control the obtained averages means of leadership variables are on the level of system III., that replies to the consultative leadership style while the variables of motivation and decision-making are much more different from the mentioned variables. Motivation is closely connected with the leadership.

The survey has comprehended four questions connected with the motivation that in hotel enterprises with the contemporary organisation plays an important role in doing the successful business.

Variables	Leadership styles (%)				Average means of leadership variables
	Styles 1	Styles 2	Styles 3	Styles 4	
Types of motives	2,13	0,00	44,68	53,19	3,49

Formats of motives	2,13	19,15	42,55	36,17	3,13
Responsibility for achieving goals	38,30	29,79	29,79	2,13	1,96
Team work and cooperation	6,38	27,66	42,55	23,40	2,83

Table 1 Share of variables of motivation in leadership styles and average means of leadership variables

The research has shown that managers often use rewards and participation as the main type of motives, 97,87%. Therefore averages means of the leadership styles is among styles 3 and styles 4. Motives formats vary from the economic and socially, 19,15%, through economic and social rewards and through the rewards and laurels, 42, 55%. The data assigned by research demonstrates that the managers of hospitality industry in Dubrovnik–neretvian county have not got development of methodological motivation policy yet. Motivation is the reflex of their personally persuasion versus the employees results. Most of managers consider that the top management is responsible for achieving the goals, even 38,30% managers that are inquired said so, while 29,79% of managers consider that for the achievement of goals are equally important middle management and all the employees of the firm as well as the top management.

To define leadership styles in some firms or in the group of firms it is very important to learn process of decision mode as well as the process of decision making.

Variables	Leadership styles (%)				Average means of leadership variables
	Styles 1	Styles 2	Styles 3	Styles 4	
Levels of decision making	14,89	42,55	36,17	6,38	2,34
Participation of employees in decision making	2,13	59,57	34,04	4,26	2,40
Decentralization of decision making	6,38	14,89	17,02	61,70	3,34

Table 2 Share of decision making variables in leadership styles and average means of leadership variables

As it seems the concentration of decision making is on the top, where only routine decisions are delegated, 42,55%. Research results show that the hotels managers of the Dubrovnik–neretvian county are not hard autocrats when it is about the level of decision making. Even 36,17% managers consider that the most important decisions are brought at the top with the delegation of all the others at the lower levels of management. That argument is confirmed with the average means of leadership styles that is 2,34. When it is the question about the subordinate participation in the decision making the managers are liable to occasionally consulted with them even them 59,57%, while 34,04% of them usually consult the subordinate levels. As it is well known that the decision making except on the top, is attendance also at the lower management levels, it is logically to expect that the decentralization in the hotel industry of the Dubrovnik–neretvian county will be enough pronounced. That is confirmed with the questionnaire. Namely 61,70% of managers that are included in this research have confirmed that the decentralization is present occasionally in the hotels that they manage.

5. CONCLUSION

The research of leadership styles has shown that firms behaviour and leadership praxis of the researched hotels managers of the Dubrovnik–neretvian county is at the level of the consultative leadership style (style 3). Leadership style of the hotel managers has shown different manifestations. It is closest to leadership style 4. That style is deemed as the most efficient leadership style in the three cases when it is about the motive formats, acceptance of information from employees, the way of setting the goals and with resistance on settled goals. The managers that have been included in the research process are most deemed to the leadership style 1 when it is about responsibility for achieving the goals. Most of them have considered that for achieving the goals in the firms is primer responsibility on the top management and on the middle management. The same situation is with the level of the decision making where the most managers considered for the two options from the questionnaire. That is about making decisions mostly on the top or generally on the top level with the delegation of the routine decisions. Managers have shown that the responsibility for the rest of answers is on their leadership styles, between style 2 and style 3. But it is closely to style 3 then to style 2. It is obvious that managers should timely react also on the changes that are happened in the surroundings like they react on the internal changes. That will be improved by the knowledge and skills that are undertaken from the competition. Of course if that mode is contributed in their successfully business and to their better positioning at the tourism market.

6. REFERENCES

Buble, M., (2000.) *Management,* Ekonomski fakultet Split, Split

Dujanić, M., (1993), *Vođenje) u funkciji managementa,* Zbornik radova Ekonomskog fakulteta Rijeka, Rijeka

Hooijberg, R., Hunt, J., G., Dodge, G. E., (1997), *Leadership Complexity ND Development of the Leaderplex Model,* Journal of Management, Vol. 23., No. 3., JAI Press Inc.

Koontz, H., Weihrich, H., (1994), *Menedžment,* deseto izdanje, Mate d.o.o., Zagreb

Tipuric, D., (ured.)., (1999), Konkurentska sposobnost poduzeca, SinergijA, Zagreb

Annals of DAAAM for 2006 & Proceedings of the 17th International DAAAM Symposium, ISSN 1726-9679
B. Katalinic, Published by DAAAM International, Vienna, Austria 2006
150th Birthday of Nikola Tesla (1856-1943) Inventor of Mobile Robot Teleoperation

ULTRASONIC WELDING - NEW METHOD IN SAILMAKING

VUJASINOVIC, E.; JANKOVIC, Z. & DRAGCEVIC, Z.

Abstract: *Although contemporary racing sails are made of high-performance fibres, such that constitute the top of the technology in global material science, sail making itself is still limited to conventional sewing of individually cut parts of sails. Having in mind recent advances in the field of textile technology, primarily in joining the cut parts of textile fabrics, the authors have aimed the investigations described in this paper at employing one of the contemporary techniques of joining (ultrasound, thermal, high-frequency) in sail making.*
Key words: *sails, sailcloth, ultrasonic welding, objective measurement.*

1. INTRODUCTION

Historically speaking, textile fabrics of various constructions and raw material content have been used in sail making. In the past, the selection of the fabric was dictated by the available raw materials and technology. However, the advances in material science and general technology offer today a wide range of various materials to be used in sail making, designed with the aim to meet all of the requirements imposed on modern sails. Contemporary racing and cruising sails are made of high-performance fibres, such as Dyneema®, Spectra®, Kevlar®, Twaron®, Vectran®, Zylon®, etc, in the form of woven fabrics or fibre composites, which often constitute the very top of the technology in the field of engineering materials globally (Vujasinovic & Cunko, 2002). However, it is strange to see that 3D sails from so sophisticated 2D materials are today made employing the conventional sewing of individually cut sail parts, using sewing threads in the process. Considerable advances in the field of textile technology, primarily in the techniques of joining the cut parts, have inspired the authors to try to implement one of the contemporary techniques of joining (ultrasound / thermal / high-frequency) in sail making. The possibilities of joining sail parts using the apparatus for ultrasound welding is investigated, the joining (seam) is evaluated and described and compared to the conventional seam used in sail making.

2. EXPERIMENTAL PART

2.1 Methodology

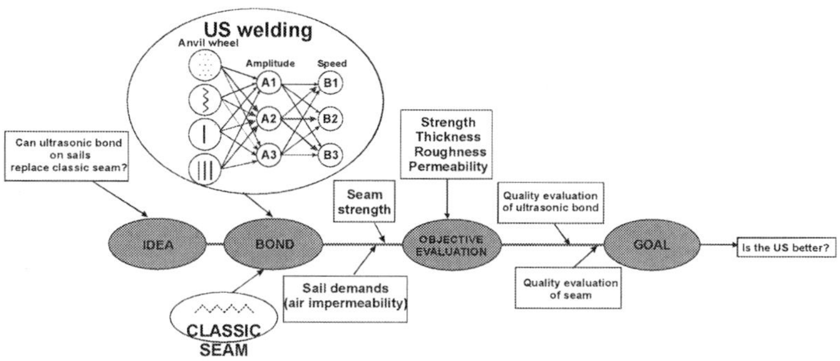

Fig. 1. Experiment plan

Considering the aim and the purpose of the investigations proposed, and having in mind the experiment plan (Fig. 1), sails were joined employing conventional sewing and the contemporary ultrasound method. The parameters of joining were varied (amplitude and speed) for individual seams, obtained using the four geometrically different feed rollers, i.e. anvil wheels.
Elements necessary for the characterisation of the seam quality were defined, e.g. for its objective evaluation, so that the individual seams could be compared, regardless of the method and type of joining them.

2.2 Procedures and equipment used in joining

Fabric samples used in sail making made of Dacron® 3,7 us oz, Polyant, the USA, were joined employing the following equipment:

- ultrasound machine for welding Seamsonic 8310 Pfaff. To achieve various seam geometries, four different anvil wheels were used: a feed rolled with point engraving, a zigzag engraving, a single-line engraving and a feed rolled with three-line engraving. Welding parameters were varied for each of the anvil wheels and the samples were joined employing different welding speeds - 6, 18 and 30 dmmin^{-1}, and the amplitudes of 50, 76 and 100 %. The sonotrode is at the constant distance 0.18 mm from the anvil wheel and the pressure is constant as well - 384 N.
- special sewing machine Minerva-Boskovice, of the class 525-101, designated ID ZZ 525 TD, specially adapted to sail making (longer headstock, so called long-arm). The seam was made using a triple zigzag stitch, the needle used was of the gauge 110 and the sewing thread selected was DABOND 2000 VVR Polyester thread Heminway & Bartlett; V 92 Shade White.

To evaluate the joining objectively, standardised (Bona 1994) and specially adapted methods were used, together with the adequate testing equipment (roughness and sealing were determined microscopically, employing the method of measuring the intensity of the light passing through).

3. RESULTS

To evaluate objectively seams on a sail, it is necessary, apart from tensile strength, which is of utmost importance for any sail as sails are exposed to considerable tensile stresses in use (Doyle, 1997), to further determine thickness, roughness and sealing (impermeability), which determine the level of fulfilling the aerodynamic requirements imposed on a sail (Vujasinovic, 2004).

Tables 1 and 2 show the results of measuring breaking strength (σ_B) and thickness (T) of a conventional seam on a sail, and an ultrasound joining on sailcloth, obtained using various anvil wheels.

v [dmmin^{-1}]	Amplitude 50%		Amplitude 76%		Amplitude 100%	
	$\sigma_{B\,a}$ [Nmm^{-2}]	s [Nmm^{-2}]	$\sigma_{B\,a}$ [Nmm^{-2}]	s [Nmm^{-2}]	$\sigma_{B\,a}$ [Nmm^{-2}]	s [Nmm^{-2}]
Anvil wheel with point engraving						
6	30.13	0.37	28.07	0.84	24.19	0.75
18	14.37	12.18	31.77	0.60	28.97	0.64
30	0.44	0.40	30.85	0.71	29.71	1.3
Anvil wheel with zigzag engraving						
6	9.84	0.53	4.13	0.55	2.21	0.30
18	13.69	10.32	9.49	1.19	7.23	1.63
30	3.88	2.51	13.84	1.98	5.74	1.06
Anvil wheel with single-line engraving						
6	6.47	0.71	3.18	0.35	2.99	0.11
18	5.62	0.57	0.71	0.17	2.86	0.62
30	1.39	0.35	2.43	1.14	3.88	0.35
Anvil wheel with three-line engraving						
6	5.10	4.15	0.89	0.04	0.73	0.13
18	8.00	3.67	5.35	1.45	2.78	0.55
30	0.52	0.52	17.34	2.58	2.72	0.73
Conventional seam						
$\sigma_{B\,a}$ [Nmm^{-2}] = 13.66			s [Nmm^{-2}] = 0.12			

Table 1. Results of measuring sail seam/joining strength

v [dmmin^{-1}]	Amplitude 50%		Amplitude 76%		Amplitude 100%	
	T_a [mm]	s [mm]	T_a [mm]	s [mm]	T_a [mm]	s [mm]
Anvil wheel with point engraving						
6	0.4267	0.0047	0.4366	0.0125	0.5400	0.0589
18	0.4167	0.0047	0.4467	0.0189	0.4533	0.0047
30	0.3867	0.0094	0.4133	0.0047	0.4200	0.0082
Anvil wheel with zigzag engraving						
6	0.9100	0.0735	1.3433	0.1543	1.8133	0.1109
18	0.4967	0.0047	1.0200	0.0698	1.1933	0.1347
30	0.4600	0.0163	0.7700	0.0829	1.0667	0.5302
Anvil wheel with single-line engraving						
6	0.9067	0.0818	1.1033	0.2098	0.9767	0.1725
18	1.1333	0.1034	1.8967	0.3256	1.1833	0.2037
30	0.4700	0.0245	1.4867	0.1914	0.9833	0.0579
Anvil wheel with three-line engraving						
6	1.3367	0.1915	1.6433	0.1759	1.8267	0.2156
18	0.6167	0.0826	1.6733	0.3108	1.6300	0.3844
30	0.4967	0.0309	0.8900	0.0424	2.1000	0.3876
Conventional seam						
T_a [mm] = 1.1667			s [mm] = 0.0094			

Table 2. Results of measuring sail seam/joining thickness

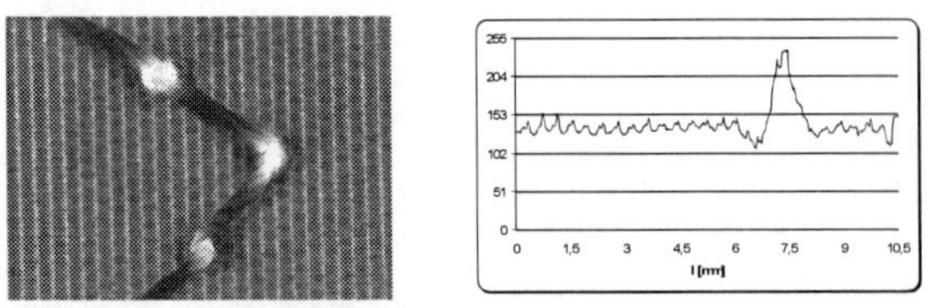

Fig. 2. Microscopic picture and the changes of the intensity of the light passed through a conventional seam

Figures 2 and 3 show roughness (as the change in the intensity of the light passed through), as well as the permeability/impremeability of the ultrasound seams/joinings (optimised as regarding their strength and thickness) and conventional seam for a comparison.

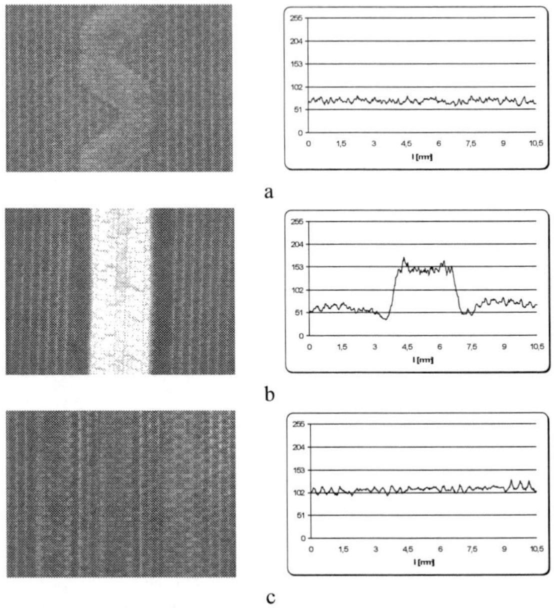

Fig. 3. Microscopic appearance and the changes in the intensity of the light passed through the ultrasound seam: a) anvil wheels with a zigzag engraving, v=18 [dmmin^{-1}], A=50%; b) anvil wheel with a single-line engraving, v=6 [dmmin^{-1}], A=50% and c) anvil wheel with a three-line engraving, v=30 [dmmin^{-1}], A=76%

4. CONCLUSION

The results presented above indicate that the ultrasound joining, provided optimal parameters are selected, can successfully substitute conventional sailcloth sewing. Ultrasound seams are generally stronger that the conventional sailcloth seams, which contributes to the safety in sailing. Initial modulus is also higher, which results in an improved aerodynamic efficiency of the designed sail shape, additionally aided by considerably lower elongation at break.

The ultrasound seam/joining on a sail is by 2/3 thinner than the conventional seam, less rough and, provided optimally selected welding conditions, completely sealing (impermeably).

5. REFERENCES

Bona, M. (1994). *Textile Quality – Physical methods of product and process control*, Nuova Oflito Torino, Italy

Doyle, B.E. (1997). Strong Fabrics for Fast Sails, *Scientific American*, Vol. 277, No. 7, 46-53, ISSN 0036-8733

Vujasinovic, E. & Cunko, R. (2002). Sailcloth – past, present and future, *Proceedings of the 1st ITC&DC*, Z. Dragcevic (Ed.)pp. 130-135, ISBN 953-96408-8-1, Dubrovnik, October 2002, Croatia

Vujasinovic, E. et al (2004). Micro & Macro Construction Features of Technical Textiles for Sailmaking, *DAAAM International Scientific Book 2004*, B. Katalinic (Ed.), pp. 645-670, Pub. DAAAM International, ISBN 3-901509-38-0, Vienna, Austria

Annals of DAAAM for 2006 & Proceedings of the 17th International DAAAM Symposium, ISSN 1726-9679
Editor B. Katalinic, Published by DAAAM International, Vienna, Austria 2006
150th Birthday of Nikola Tesla (1856-1943) Inventor of Mobile Robot Teleoperation

AJAX CLIENT FOR OPC DA XML SERVER

VUKMIROVIC, S.; ERDELJAN, A. & LENDAK, I.

Abstract: *During the previous years almost all computer based techologies were gradually moving towards the Internet. This trend is not so evident in industrial systems, but based on client demands it is expected that SCADA users will eventually need to access their data over the Internet. To our knowledge there are no such commercially available solutions. In this paper we propose a combination of existing techologies that could be used to create web based SCADA software. The heart of this proposed solution are OPC servers and an AJAX web client. OPC servers are used to access process data in a standard way, and AJAX is a new techology that allows asynchronous data access over the Internet.*

Key words: *OLE for Process Control (OPC), Exstensible Markup Language (XML), Asynchronous JavaScript and XML (AJAX)*

1. INTRODUCTION

The concept of component-based SCADA software development has been widely accepted in automation software system development. However, before the appearance of OPC specifications, these vendor-defined interfaces were different from each other. These components based on proprietary interfaces could not be reused by others (Hong & Jianhua 2005).

OPC (OLE for Process Control) foundation is a non profit organisation which aims to standardize communications in SCADA systems. The majority of specifications introduced by this foundation are based on Microsoft DCOM technology. The specification for Data Access (DA) server was introduced first, and it is most widely used. Later the OPC foundation introduced specifications for Alarm and Event (OPC 2002), Historical Data Access, Batch and Data eXchange. The fact that all of them are based on Microsoft DCOM technology disables them from being used on computers that do not use Microsoft Windows. Communication over the Internet is not reliabe due to the fact that DCOM technology does not function through most available firewalls.

The OPC XML-DA specification describes functions of a web service that transforms data from DCOM to XML format. The Communications protocol used in OPC XML-DA is Simple Object Access Protocol (SOAP) which is based on HTTP.

2. THE PROBLEM

Internet, the global network, became an irreplaceable part of the most of computer based systems. So, the users of SCADA system logically want to view what is happening in their system over the Internet. An internet browser (Internet Explorer, Mozilla FireFox, Opera) is the place where user expects to see online data from the supervised system.

The first problem is that the communications between browsers and web servers is done by HyperText Transfer Protocol (HTTP) which is based on a strict request/response architecture. This means that the whole page is submitted to the server every time the client needs to refresh data from the server. This can be very slow, and it is redundant in a scenario where only small amount of data is changed since the last client request.

The second problem is that only Internet Explorer supports Microsoft DCOM technology. In order to be sure that a web client would work in every browser, the client has to be written with standard HTML/XML features.

Currently users of SCADA software are used to Windows like software, since most commercial SCADA software are written using the windows forms user interface. Windows programs are richer and more user friendly then programs written for web browsers. They have drag-and-drop functionality, more flexible controls (datagrid, treeview, list view etc.), and most importantly they can load data asynchronously. This means that new data can be loaded into the form in windows program at the same time while the user interacts with the user interface.

The main issue discussed in this paper is how to create a windows-like SCADA client that works in web browsers. This solution uses standard specifications in order to avoid different software interfaces and problems discussed in the introduction of this paper.

3. SYSTEM ARCHITECTURE

The main advantage of OPC Data Access server is that it supports asynchronous data transfer. This means that a client can send to the server a list of items it is interested in. The server will notify the client when items values are changed. The client can also specify deadband (minimum change of item value) and update rate (minimum time between two updates). This reduces network trafic and that is essential in systems with huge numbers of items (OPC 1998).

The OPC XML-DA server brings data from SCADA systems to the Internet. This can be achieved with XML-DA server directly accessing values from the field (OPC 2003). However, XML-DA specification recommends an architecture in which XML-DA server accesses data via DA server. Figure 1. shows this architecture and the use of an XML-DA server. Internet clients access data via the XML-DA server, and it reads data from DA server. DA server accesses items from devices (PLC and RTU).

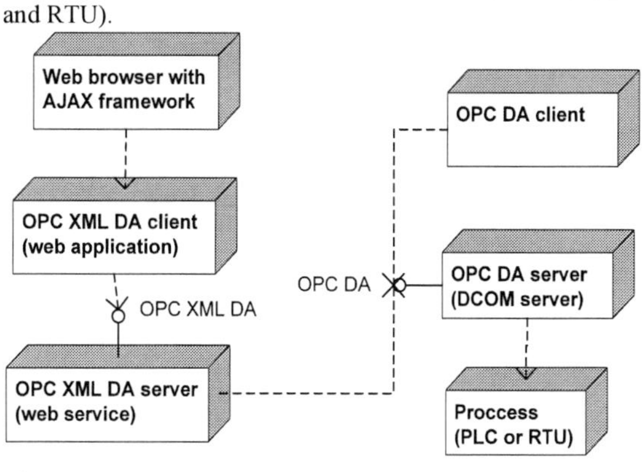

Fig. 1. OPC servers interaction

In order to preserve the idea of asynchronous data transfer the XML DA specification introduces the Subscription mechanism. With the Subscribe method the client sends to the server a list of items it is interested in. Later on, the client periodically calls the function SubscriptionPolledRefresh which returns latest data value changes.
The XML DA client is created as a web application that will dynamically create final output displayed in the web browser. Later on, web browser client will reload process data asynchronously with the use of AJAX technology.

4. AJAX CLIENT

Ajax, a shorthand for Asynchronous JavaScript and XML, is a Web development technique for creating interactive web applications. The intent is to make web pages feel more responsive by exchanging small amounts of data with the server behind the scenes, so that the entire web page does not have to be reloaded each time the user makes a change.
Ajax isn't a single technology. Rather, it's a collection of four technologies that complement one another: JavaScript, Cascading Style Sheets (CSS), Document Object Model (DOM) and XMLHttpRequest object (Crane & Pascarello 2006). Basically, on some user action (or on timer event) new XMLHttpRequest object is created and sent to the web server (with data about which data are needed). After the web server proceesses the request, results are returned to the web page in a function defined during request submission. If response is complex data it is useful to encode it in an XML document.
This is just what was required for the process of data transfer from OPC DA server to the web client. Now clients can view data from SCADA, or even change some values while fresh data is downloaded from the XML DA server. This solution can leverages advantages of the Subscription mechanism.

```
function RefreshData(serverHandle)
{ var url = 'GetOPCData.aspx';
  var strParams = 'h='+ serverHandle;
  var loader1 = new net.ContentLoader
        (url,UpdateNodes,null, "POST",strParams); }
```

Listing 1. RefreshData function

Listing 1. shows the RefreshData JavaScipt function that submits a request to the web application ('GetOPCData.aspx' page). A ServerHandle that uniquely identifies the client is sent with the request. Submission is done with the ContentLoader class that is part of the AJAX library. When data processing is finished, the UpdateNodes function is called and it updates user interface control with the new values.

5. FINAL OUTPUT

Final output of oru work is a web client application that shows data from a SCADA system. The data transfer from the process to the web client is truly asynchronous in every phase of the transfer. The web client can watch item values changing without delay or dead time (time while web page is reloaded). This gives the user feeling of having more control over the system. Also, the client can change some values in the same time while others are refreshing from the server.
A small drawback of AJAX is that the "Back" button in web browsers will not work as expected. It will transfer the user to the previous page, and will not show the previous item values since those values are replaced by new ones in th UpdateNodes function. This drawback can be solved by buffering values into local arrays and creating a new button that will load data from a previous timestamp to the user interface.

Security of the data transer can be achieved by using the HTTPS that will add an additional encryption/authentication layer between HTTP and TCP. Another security issue that has to be considered here is the fact that AJAX is based on JavaScript and web browser security settings can block usage of JavaScript functions. This could make the AJAX client unusable, unless the user marks the web site hosting the SCADA web application as trusted site.
An example for usage of AJAX clients as web user interface for SCADA systems is given in Figure 2. It represents the part of water distribution system of Novi Sad (Serbia). Process data is loaded from "Matrikon.OPC.Simulation" (free OPC DA server). OPC XML DA server used was developed by the authors of this paper. The web SCADA application is tested in Internet Explorer 6.0 on Mozilla FireFox 1.5.

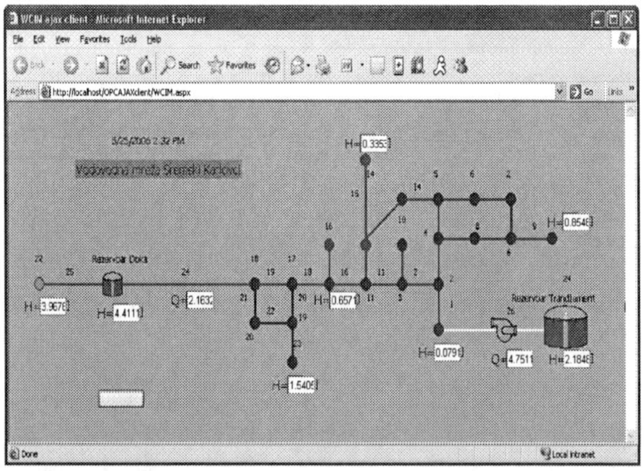

Fig. 2. Ajax web client for water distibution system

6. CONCLUSION

The solution proposed in this paper has a real life benefit, as it transfers data from industrial processes to Internet clients with maximum exploitation of benefits of the asynchronous data transfer. The proposed solution is based on OPC DA and XML DA specifications which give it modularity and reusability since this specifications are accepted by the leading SCADA vendors.
Refreshing data in web browsers is done asynchronously with the help of AJAX technology. It refreshes data from the XML DA server without submission of the current web page. This gives user a feeling of smooth data transfer.
Finally, the concept was proved by creating a web SCADA client for a water distribution system and results were tested in various web browsers.
The authors would like to express their thanks to the Ministry of Science of Serbia, which allowed them to conduct their research within its National Project "Information system of water systems in large settlements".

7. REFERENCES

Dave Crane & Eric Pascarello *"Ajax in Action", Manning, Greenwich, 2006*

OPC Foundation, *"OPC Data Access Custom Interface Specification 2.0"*, 1998

OPC Foundation, *"OPC Alarms and Events Custom Interface Specification 1.10"*, 2002

OPC Foundation, *"OPC XML-DA Specification Version 1.0"*, 2003

Xu Hong & Wang Jianhua *"Using standard components in automation industry: A study on OPC Specification"*, 2005

Annals of DAAAM for 2006 & Proceedings of the 17th International DAAAM Symposium, ISSN 1726-9679
B. Katalinic, Published by DAAAM International, Vienna, Austria 2006
150th Birthday of Nikola Tesla (1856-1943) Inventor of Mobile Robot Teleoperation

SYSTEM TO REALIZE A CONVEYING BELT SYNCHRONOUS ASSEMBLY

WERNER, J.; ZAEH, M.F. & LANGE, F.

Abstract: *This paper presents a methodology to realize a fully automated assembly process which can be synchronized to the conveying system. The system is build up by using standardized components like industrial robots which are equipped with a camera and a force-torque sensor. Advantages and requirements of the synchronized assembly in the final assembly of vehicles are pointed out.*

1. INTRODUCTION

During the last years a steady increase of the level of automation was observed within the industrial branches. About 44 % of the industrial robots were delivered to the automobile industry. Even though many assembly and handling processes can be automated with today's technology most enterprises rather use manual flow assembly lines. The main reason for this development is the need for clocked lines while using automated systems. Large buffers are needed as a result of process automation in flow assembly lines which lead to higher costs and therefore often ineffectiveness.
Within this research project an automated assembly system is to be implemented which is able to synchronize the movement of the conveying system and the robot in all degrees of freedom. This enables an assembling process of parts e.g. on a moving car body. The system is composed of standardized components in order to realize an economic and efficient assembly process.

2. STATE OF THE ART

The multiple advantages of an automated and synchronized assembly have been known for over 30 years. Even though different approaches to the problem were evaluated there is no economic system to date, which is capable of realizing the accuracy of positioning which is required by most assembly processes investigated. These approaches can be differentiated by the principle of synchronization between the conveyer belt and the handling system. As shown in figure 1 there are approaches using mechanical, guided or controlled synchronizing principles.

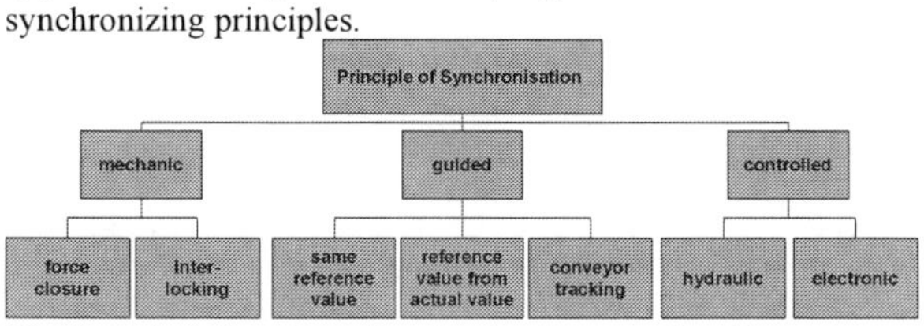

Fig. 1. Principles of synchronization between handling system and conveyor system

The most promising principles are the guided and the controlled synchronization. The guided conveyor tracking is today implemented in most robot control systems (ABB 1998, Dirndorfer 1993, KUKA Roboter 1998). Here the speed of the conveyer system is measured and the difference to the Tool-Center-Point (TCP) is calculated within the interpolation clock. Then the movement of the robot is corrected by the robot control. According to Dirndorfer (Dirndorfer 1993) and Walther (Walther 1985) this system cannot achieve a maximum accuracy in synchronization of less than one millimeter. This results from vibrations and the following error of the robot.
In controlled synchronization the relative differences between the robot and the conveyor belt are constantly measured and provided to the robot control by a closed loop control system. These systems may achieve the required accuracy but are composed of specialized, non-standard equipment and therefore not economic.

3. REQUIREMENTS

The reason for the difficulties to implementing a conveying belt synchronous assembly is the tolerance of the conveyer belt regarding the reference speed and the stiffness. Conveying belts used for heavy weight show a very unstable performance. Studies have detected that non reproducible variations of the line speed can occur which lead to tolerances up to 10 % within a frequency of 2-3 Hz. In addition most conveyer belts do not only transport items in flow direction (x-axis) but also allow a change in heights (z-axis) which results from ergonomically designed manual workstations. The movement to the side (y-axis) is only known for power&free conveyers and is a result of their reduced stiffness. Table 1 shows the different directions, the respective tolerances and the anticipated frequencies, which a synchronized automated assembly system has to cope with.

Direction	Speed	Tolerance	Frequency
x-axis	8 m/min	+/- 2 m/min	3 Hz
y-axis	0 m/min	+/- 70 mm	0.5 Hz
z-axis	0 m/min	+/- 150 mm	0.5 Hz

Table 1. Tolerances of an average power&free conveying system

4. METHOD & DESIGN

Within this project a methodology to realize a synchronized assembly is to be designed which can be adapted to different assembly processes and can be combined with different conveying systems. To show the functionality of the general algorithm a wheel assembling process has to be implemented.
The system consists of a six-axis industrial robot on a linear unit and a power&free conveyer. The robot is equipped with a force-torque sensor and a camera system. This camera is used to detect the moving object in all 6 degrees of freedom. It is needed for the approximated positioning of the gripper in front of the car body. The compliant force torque sensor (Grewe et. al. 2001) which is developed by the DLR is able to detect and

compensate tolerances in all 6 degrees of freedom. This sensor is used to balance the deviations of the movement of the conveyed object after a force lock has been made.
The applied guidance concept is divided into the non tactile guiding of the robot by the camera system and the tactile control by the force-torque sensor. The use of this combined approach allows a fast positioning of the robot guided by the camera systems. As soon as the robot contacts the conveyed object the force-torque sensor will start guiding the handling system. Therefore an algorithm has been developed that uses all seven axis (six on the robot and the linear unit) to guide the robot to do the assemble process. Besides movements with high frequencies such as the acceleration in x-axis are partially equalized by the flexibility of the gripper system (see figure 2).
The reference position and orientation after the connection between the handling system and the car body was made are measured by the force-torque sensor from the current position of the TCP (Lange & Hirzinger 1996a).

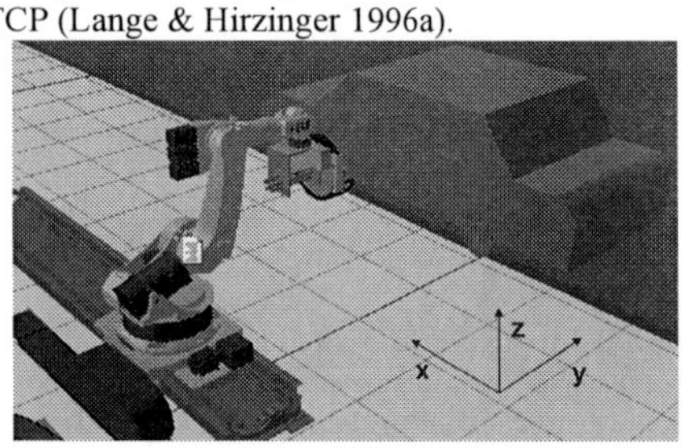

Fig. 2. Simulation of the assembling process

By using multiple scans the reference values are calculated in advance. Useful for this calculation are the speed of the conveyor or the pendular like movement of the car body around the fixed point of the power&free conveyer. Before the connection is made both the pendular movement and the conveyer speed are measured by the mounted camera system and are used for tracking.
The guiding of the robot is performed similar to the predictive guidance (Clarke et. al. 1987): the prediction of the robot movement is compared to the programmed reference movement. For that an algorithm described by Lange (Lange 2006) is used which also allows a smoothing of the values and considers the maximum acceleration of the robot. Predicted differences are compensated by an adaptive feedforward control according to Lange&Hirzinger (Lange&Hirzinger 1996a). This is possible by using the standardized feedback functionality of the robot control which is already implemented. Problems of this guidance concept are the resiliences of the robot because they influence both the positioning control and the sensors which can lead to an over steering of the control systems. Further tests have to show whether this scenario is relevant and which implications this would have.

5. REALIZATION

The prototype consists of an industrial robot which is mounted on a linear unit and can carry up to 180 kg. The car body is conveyed by a power&free system with a speed of 8m/min. The conveying system has been adapted so that a movement of 0,5 m in height can also be realized. The assembly process to show the functionality of the system is the mounting of the wheels as shown as a simulation in figure 2. The entire process from detecting the car, to gripping the wheels, synchronizing the robot and finally assembling the wheel takes less than 60 seconds. The construction of the gripper is always depending on the focused assembly process. It is necessary to provide a wide angle of sight for the camera. Figure 3 shows that in the wheel mounting scenario the camera is located in the middle of the gripper so that the relevant parts of the car body, here the axis and the brake disc, can be seen through the wheel. The force torque sensor will be mounted between the robot flange and the gripper in order to achieve the best result.

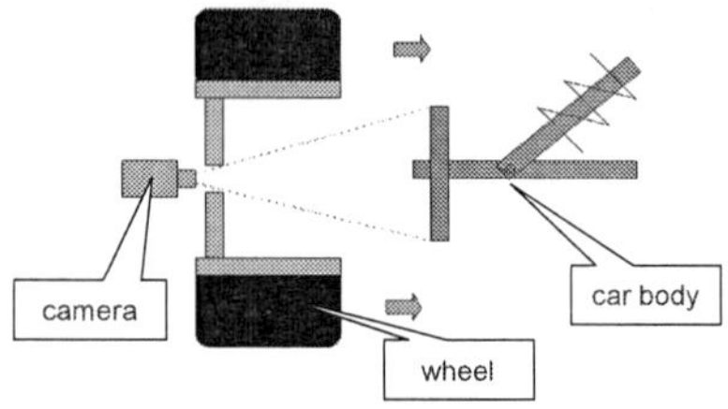

Fig. 3. Angles of sight of the camera while located within the gripper

All necessary tasks needed for the automated assembly of wheels on a moving car body are included in the experimental system. The tasks are detecting the car, delivering the wheel, gripping of the wheel, synchronizing the robot to the car body and assembling the wheel.
Partner of the *iwb* are the German Aerospace Center (DLR e.V.), BMW AG, KUKA Roboter GmbH, Schunk GmbH&Co. KG, Quiss GmbH and Sturm Maschinenbau GmbH. The project is funded by the Bavarian Research Foundation.

6. CONCLUSION

Using the described method of synchronizing handling and conveying systems enables an economic automation of final assembly lines in the car industry. Major advantages are the loss of buffer prior and after automated assembly cells, the avoidance of special conveyor systems for automation and of ergonomically critical manual workstations. Possible scenarios for the conveying belt synchronous assembly are all assembly processes where large and heavy items are mounted like the cockpit, the seats or the retractable roof.

7. REFERENCES

ABB Flexible Automation GmbH (1998): Produktspezifikation Conveyor Tracking. Väesteras: 1998.

Clarke, D. W.; Mohtadi, C.; Tuff, P. S. (1987): Generalized predictive control - part I. the basic algorithm. Automatica, 23 (2):137-148, 1987.

Dirndorfer, A.: Robotersysteme zur förderbandsynchronen Montage. Berlin: Springer 1993. (iwb Forschungsberichte 63).

Grewe, J.; Hirzinger, G.; Koeppe, R.; Strobl, C.; Willberg, B. (2001): Compliant-Force-Torque-Sensor. In: 10th International Workshop on Robotics in Alpe-Adria-Danube Region (RAAD2001), Vienna 2001.

KUKA Roboter GmbH (1998): Standart-Conveyor mit KR C1. Augsburg: November 1998.

Lange, F.; Hirzinger, G. (1996a): Learning Force Control with Position Controlled Robots,Proc. IEEE Int. Conference on Robotics and Automation, Minneapolis, Minnesota, April 1996;pp 2282-2288

Lange, F.; Frommberger, M.; Hirzinger, G. (2006): Impedance-Based Smoothing for Visual Servoing along Edges, Proc. International Symposium on Robotics (ISR 2006), Munich, Germany, May 2006

Lange, F.; Hirzinger, G. (1996b): Learning of a Controller for Non-Recurring Fast Movements, Advanced Robotics 10 (2): 229-244, 1996

Walther, J. (1985): Montage großvolumiger Produkte mit Industrierobotern. Berlin: Springer 1985. (IPA-IAO Forschung und Praxis 88).

Annals of DAAAM for 2006 & Proceedings of the 17th International DAAAM Symposium, ISSN 1726-9679
B. Katalinic, Published by DAAAM International, Vienna, Austria 2006
150th Birthday of Nikola Tesla (1856-1943) Inventor of Mobile Robot Teleoperation

A ROBOTIC FOOT MECHANISM WITH MULTI-PLATFORMS USING A PARALLEL MECHANISM

YOON, J.

Abstract: *This paper proposes a new humanoid foot device that can allow humanoid robot to walk naturally like normal human. The device can generate rotations at two toe joints and two rotations at ankle. The new mechanism consists of front-left, front-right, and rear platforms, and four limbs. Base of three platforms is located at a humanoid shin. A new 4-dof mechanism with three platforms can generate pitch and roll motions at each platform. Based on this mechanism, the foot device was developed for natural walking of humanoid robot.*
Key words: *Humanoid, Foot device, Multi platforms.*

1. INTRODUCTION

There are a lot of researches trying to achieve creature-like artificial motilities on robots. However, until now, human-like skillful mobility has not been achieved, because the robotic feet are far from adaptation to keep stable contact on the ground. If the toe and foot mechanism could enable the robots to make efficient body motions on various terrains, skillful walking of robots will be available.

To improve the walking abilities of humanoid-type robots, toe mechanisms with 1-dof (Ahn et. al, 2003 and Takahashi et. al, 2002) have been suggested. For walking in the straight direction, the 1-dof toe mechanism can allow faster robot walking. In addition, relative rotation at the metatarsal joint of the foot increases the naturalness of robot walking and helps reduce the load on the knee joints, where high force and speed are required to achieve robot locomotion (Nishiwaki and Kagami, 2002). However, the foot device with 1-dof toe mechanism cannot adapt the turning walking (Takao et. al, 2004) and may not satisfy safe and natural walking under uneven terrains. Takao et.al (2004) concluded that the robotic foot mechanisms with multi degree-of-motions can contribute smooth and effective body motion in the stance phase through gait experiments. Therefore, it is necessary to develop a novel toe and foot mechanism with multi-dof motions for humanoid-type robots, satisfying natural walking motions on various terrain situations.

This paper presents a new 4-dof serial-parallel mechanism with three platforms. The suggested mechanism can generate pitch and roll motions at each platform, which motions can be applied to ankle and two toe joints of humanoid robot. Then, by using the suggested 4-dof serial-parallel mechanism, humanoid foot was developed for natural walking of humanoid robot.

2. MECHANISM DESCRIPTION

In areas such as clinical medicine, robots with more than two platforms are necessary to support multi-body objects such as the human foot. Research (Hunt et. al, 2001) in biomechanics areas showed the importance of inter-segments foot motions based on the extent of forefoot motion during stance phase, and foot could be simply modeled as two segments with one revolute joint to represent more closely natural foot motion during gait (Gilchrist et. al, 2006). For humanoid robot, this motion will allow more diverse motions which cannot be achieved with the single rigid platform. For more natural foot motions, a 4-dof parallel mechanism with three platforms has been developed.

The proposed 4-dof mechanisms with three platforms consist of front-left, front-right, and rear platforms and four limbs, as shown in Figure 1. In the figure, the letters *P*, *R*, *U*, and, *S* represent prismatic, revolute, Universal, and spherical joints, respectively. An underlined letter represents an actuated joint. Two outer limbs with 6-dof serial joints (S-$\underline{P}$-U) are attached to each front platform, while the rear limb with 3-dof serial joints ($\underline{P_e}$-$\underline{R_e}$-R) and the four-bar limb (S-S) are attached to a rear platform. The rear limb is driven by the 2-dof driving mechanism that is equivalent active serial prismatic and revolute joints ($\underline{P_e}$-$\underline{R_e}$) with two base-fixed prismatic actuators. The four-bar limb will allow the rear platform to generate pitch motions according to active prismatic joint motions of a 2-dof driving mechanism. In result, the rear limb can generate two rotation of the rear platform with equivalent 3-dof serial joint ($\underline{R_e}$-$\underline{R_e}$-R). Since front platforms have one limb with one 6-dof serial joint (S-$\underline{P}$-U) and rear platform has 3-dof serial joint ($\underline{R_e}$-$\underline{R_e}$-R), the final output displacement of each platform is dependent only on that of the rear limb with its 3-dof serial joint ($\underline{R_e}$-$\underline{R_e}$-R), which is the intersection of the special Plücker of two limbs and equals each platform. Therefore, the suggested mechanisms have four degrees of freedom in total. For the proposed mechanism shown in Figure 1, each platform has 2-dof, which is described by two rotations about the x- and y-directions. Therefore, the mechanism with three platforms can generate pitch and roll motions at each platform, which motions can be applied to ankle and two toe joints of humanoid robot.

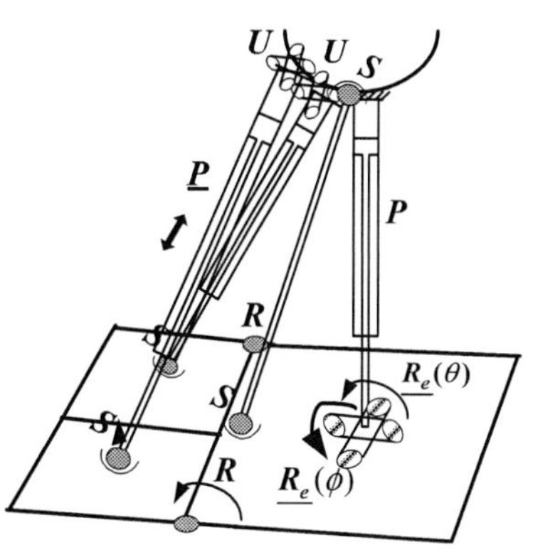

Fig. 1. Novel Foot mechanism with 4-dof motions

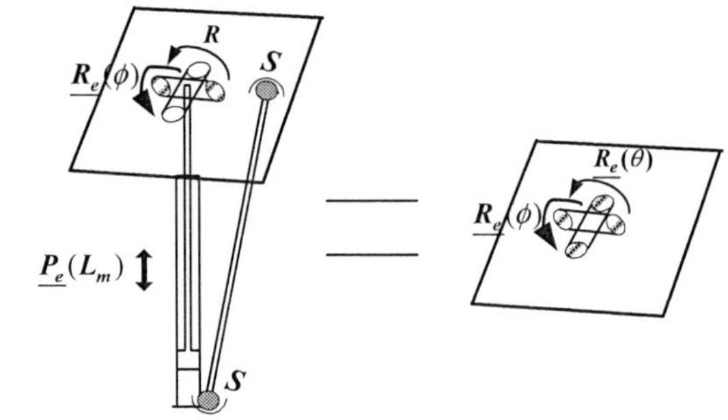

Fig. 2. 2-dof driving mechanisms with R-R type

Figure 2 shows that R-R-type mechanism can be generated by attaching an additional rotation bar from T-R type mechanism, which will change translation motion to rotation motion.

Figure 3 shows the link-pair relationship diagram for the suggested mechanism. The white boxes represent passive joints and the hatched boxes represent active joints. Lines between letters represent links. It is possible to consider each driving mechanism as equivalent to the two actuated joints ($\underline{P_e}$-$\underline{R_e}$ *or* $\underline{R_e}$-R_e) in terms of the number and type of degrees of motion. The motions of the given mechanism can be verified using Grübler's formula as:

$$M = d(n-g-1)+\sum_{i=1}^{g} f_i \tag{1}$$

where *d*, *n*, *g*, and f_i represent the dimension of the feasible motion space of all the joints of a given mechanism, the number of links including the ground, the number of constraints of the joint, and the degree of freedom at the *i*-th joint, respectively.

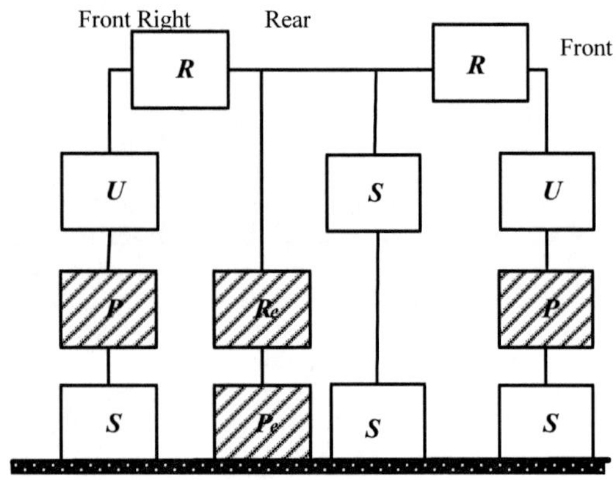

Fig. 3 The link-pair relationship diagram of the 4-dof mechanisms with three platforms

If the R-R type driving mechanisms are considered equivalent to serial ($\underline{R_e}$-$\underline{R_e}$) joints, the mobility of the new 4-dof mechanisms with three platforms is:

$$M_{4dof} = 6(10 - 12 - 1) + 1\times6 + 2\times2 + 3\times4 = 4$$

3. DEVELOPMENT OF THE PROTOTYPE

A humanoid foot attached to the shin of humanoid robot should be designed to generate natural walking trajectories. In addition, the footpad device should have sufficient workspace to simulate natural walking on various surfaces, and should be able to support humanoid robot weight. Based on normal gait requirements (Yoon et. al, 2003), a humanoid foot was developed based on the suggested mechanism.

Figure 4 shows the 3D modeling and the workspace envelope of the designed rear plate of the suggested mechanism based on a discrete workspace analysis using the inverse kinematics of the mechanism and actuator length constraints. With this design, the maximum pitch angle of the rear plate exceeds ±85°. The maximum roll angle of the platforms exceeds ±40°.

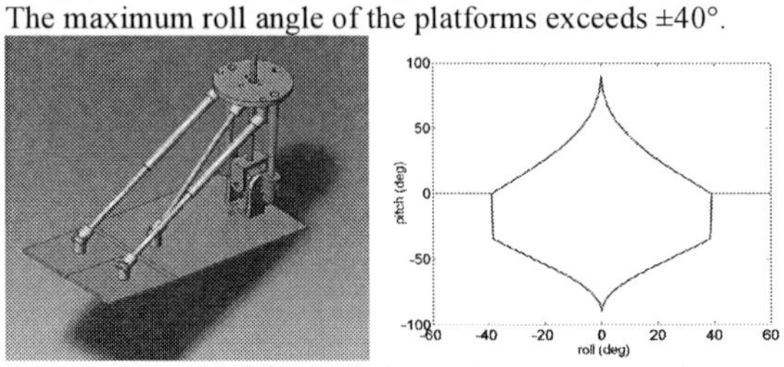

(a) 3D modeling (b) Workspace of rear plate

Fig. 4 Design of the robotic foot

4. CORRESPONDING ADDRESSES

This paper presents a new humanoid foot mechanism with a 4-dof–serial-parallel mechanism with three platforms. The proposed 4-dof robot is composed of two front platforms, rear platform, and four limbs driven by four prismatic actuators. This new design can generate toe joint motions between the fore and rear foot and two rotations about the ankle joint. These motions are adequate for natural foot and ankle motions of a humanoid robot. The developed foot device with the suggested mechanism will allow humanoid robots to walk more stable and natural ways. For future works, the gait simulations with the combination of the suggested novel foot with multiple platforms and a humanoid robot will be performed at various situations such as turning and stairs walking.

5. ACKNOWLEDGEMENT

This work was supported by the Korea Research Foundation Grant funded by the Korean Government (MOEHRD) (KRF-2005-005-J09902).

6. REFERENCES

Ahn, C.; Lee, M. & Go, S. (2003). Development of a biped robot with toes to improve gait pattern, *Proc. Of the IEEE/ASME Int. Conf. on Advanced Intelligent Mechatronics (AIM2003)*, pp.729-734.

Takahashi, T. & Kawamura, A. (2002). Posture control using foot toe and sole for biped walking robot Ken, *AMC*, pp.437-442, Maribor, Slovenica, 2002

Nishiwaki, K. & Kagami, S. (2002), Toe joint that enhances bipedal and full body motion, *IEEE Int. Conf. on Rob. and Auto.(ICRA 2002)*, pp. 3105-3110, 2002.

Takao, S. ; Ohta, H. ; Yokokohji, Y. & Yoshikawa, T. (2004), Function analysis of human-like mechanical foot, using mechanically constrained shoes, *IEEE/RSJ Int. Conf. on Intelligent Robots and System*, pp.3847-3852, 2004.

Hunt, A. E.; Smith, R. M.; Torode, M. & Keenan, A. M. (2001), Inter-segment foot motion and ground reaction forces over the stance phase of walking, *Clinical biomechanics*, Vol. 16, No. 7, pp. 592-600, 2001.

Gilchrist L. A. & Winter, D. A. (1996), A two-part, viscoelastic foot model for use in gait simulation, *Journal of Biomechanical Engineering*, Vol. 29, No. 6, pp. 795-798, 1996.

Yoon, J.; Ryu, J. & Burdea, G. (2003), Design and Analysis of a Novel Virtual Walking Machine, *IEEE Haptic Symposium 2003*, pp. 374-381, Los Angles, USA, 2003.

Annals of DAAAM for 2006 & Proceedings of the 17th International DAAAM Symposium, ISSN 1726-9679
B. Katalinic, Published by DAAAM International, Vienna, Austria 2006
150th Birthday of Nikola Tesla (1856-1943) Inventor of Mobile Robot Teleoperation

DECOMPOSITION OF A MACHINED SURFACE IMAGE FOR THE ESTIMATION OF SURFACE ROUGHNESS PARAMETERS

ZAWADA-TOMKIEWICZ, A. & STORCH, B.

Abstract: *Image of a machined surface is a perspective projection of a magnified surface on a converter plane. Change in image is distinctly noticeable from the energetic point of view. For this purpose there was singular value decomposition (SVD) method applied as a factorization technique which effectively reduces machined surface image into a smaller portion of data. In this paper the analysis of the eigenvalues of a machined surface image was applied in the estimation of surface roughness parameters.*

Key words: *machined surface, digital image, eigenvalue decomposition.*

1. INTRODUCTION

Increased emphasis upon product quality standards connected with the introduction of flexible manufacturing techniques, has tended to highlight the potential benefits of vision based quality control within the component production environment. Surface roughness is a widely used index of product quality. Achieving the desired surface quality is of great importance for the functional behaviour of a part. But the nature of the surface roughness formation mechanism along with uncontrollable factors, make almost impossible a straightforward prediction of the surface parameters.

In this paper it is presented research focused on the application of vision system in a quantitative surface inspection. The vision system consisted of a computer with a frame–grabber card, a digital camera, lenses, a stand for a camera with movable worktable and a lighting system. There was applied a cold light source for the illumination of the surface. An image magnified through an optical system was changed into a digital form. Obtained image was described by image features and then the surface roughness parameters were calculated with the use of neural network. So far a machined surface image was described by statistical methods, image texture description methods (Zawada-Tomkiewicz at al.,2004-1), and wavelets (Zawada-Tomkiewicz at al.,2004-2). In this paper for the estimation there were chosen features of the singular value decomposition (SVD) of the image.

2. DESCRIPTION OF A MACHINED SURFACE IMAGE

The machined surface formed in turning operation reflects all the changes and interactions that co-exist during the formation of the surface (Storch, 1989). With the use of vision system it is a machined surface image created by sampling the quantity $z(x,y)$ at x and y spatial locations, leading to a matrix of data describing the intensity of colour in each pixel, As there is a great correlation between the elements of the machined surface image, matrix Z can be decomposed as a product

$$Z = U \cdot S \cdot V^T \tag{1}$$

U and V are respectively N x M and M x M the unitary matrices $U^T \cdot U = V^T \cdot V = I$ and S is an M x M diagonal matrix that consists of the eigenvalues of the matrix Z. The eigenvalues - weighs S_k can be positive or equal to zero values. They are sorted in descending order such that $S_1 \geq S_2 \geq ... \geq S_{K-1} \geq S_K$.

Decomposition of the exemplary surface image is presented in Table 1. There can be seen a few components of the image. The first structure is the single pattern that represents the most variance in the data. In the following vectors of the matrix U there can be distinguished the averaged changes in a profile at the angle parallel to the lay direction in the image.

Image	U_k	S_k	V_k
	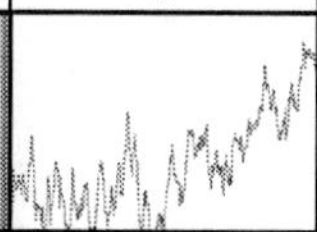	(k=1) 395	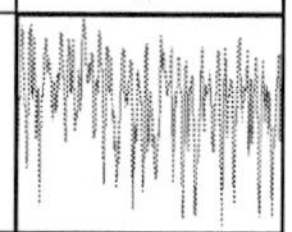
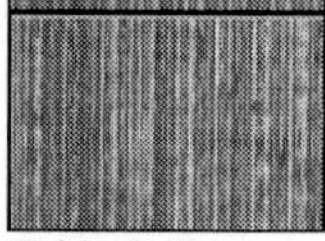	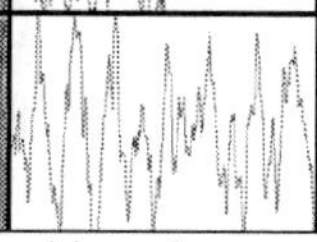	(k=20) 7	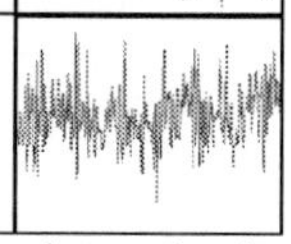

Table 1. Decomposition of an exemplary surface image for the first and twentieth components

The second structure V is the pattern that describes the second largest amount of variance, calculated in the same way as the first structure. In the following vectors of the matrix there was shown the averaged profile of a machined surface digital image at the angle perpendicular to the lay direction in the image, in pitch surface generator. A very important property of the second structure is that it is completely uncorrelated with the first structure. All the structures are mutually uncorrelated.

3. PARAMETERS OF A MACHINED SURFACE IMAGE

Singular values are strongly ordered and the largest one exceeds the smallest one by a few orders of magnitude. This allows the signal to be reconstructed with the use of only the most significant biorthogonal components. The ability of SVD to eliminate a large portion of the data was a primary reason for its use in surface roughness estimation. Comparisons between original image data and truncated one reveal a great similarity. Original images contain more details, more high frequencies but the general character of the data remains the same. SVD is in this way similar to wavelet decomposition. As was shown in (Zawada-Tomkiewicz at al., 2006) most information is contained only in the first component.

A single image which was acquired with the use of a vision system could be described by almost 40 thousands of twenty-four bit numbers. The input data were of great redundancy which was reduced by the conversion of an RGB image into a grey scale one. Accordingly to the previous research (Zawada-Tomkiewicz, 2003) where there had been checked various portions of data for their usability in surface roughness estimation, there was applied only a part of data from the central part of the image. But this subimage was chosen large enough to preserve sufficient information about the whole image. To fulfil this assumption there ware conducted experiments and was chosen subimage equal to 64 x 64 pixels. Such an image contained enough energy and was treated as sufficiently representative.

4. SURFACE ROUGHNESS ESTIMATION

Accordingly to the experiment there were machined surface digital images acquired. There was acquired more then one thousand images. These images constituted the base for a monitoring system. As was stated earlier most of the information of the machined surface was contained in singular values of SVD decomposition. These first 20 singular values calculated with the use of data from the central part of the image were chosen as features describing machined surface image. These values were applied as the input of the neural network estimator. The problem of feature selection in case of applying the neural network became reduced to the problem of optimization of the input layer i.e. selection of such inputs, which changed, produce the biggest output error during training and eliminate those inputs, which produce small change of output error. For this purpose there have been applied the Optimal Brain Surgeon (OBS) Method (Hassibi at al., 1992). There was used feed-forward neural network composed of three layers. It was trained with the use of Lavenberg-Marquard algorithm. The neurons in the hidden layer had hyperbolic tangent activation function; the single neuron in the output layer had a linear activation function. The data were divided into training and testing sets. The number of examples in a training set was two times bigger then that in a testing set. The sets came from two independent experiments. During the optimization procedure the neural network was trained several times and after the process of learning the input layer was pruned. Each time the pruned weights were counted again. At the end of the optimization procedure there were a rank of each input neuron determined. Neurons from input layer, which weights were frequently pruned, were rejected (Fig.1).

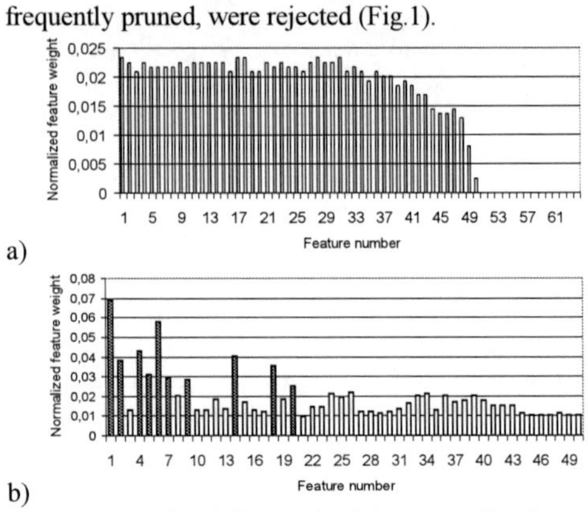

Fig. 1. Normalized feature weight (normalized sum of neuron connections which were not pruned)

At the beginning, the network was composed of sixty four neurons in an input layer, three neurons in a hidden layer and one neuron in an output layer (64–3–1). The network was trained and after training process the weights were pruned. The results of these experiments are presented in Fig. 1a. Each bar represents the normalized sum of connections which were not pruned during fifty experiments. The higher the bar the more important feature was for the surface roughness estimation. The neurons which output signals were frequently eliminated by pruning were removed. The network was reduced to the 50–3–1 structure and then to 10-3-1 (Fig.1b).

Limiting Absolute Error was applied as a measure of the estimator structure quality. During the optimization procedure when the number of input features was reduced the error for the training set increased and for the testing set consequently declined. Further reduction of the network structure caused significant increase of the error. This reasoning made it possible to come to the conclusion that ten features had the greatest influence on the output error.

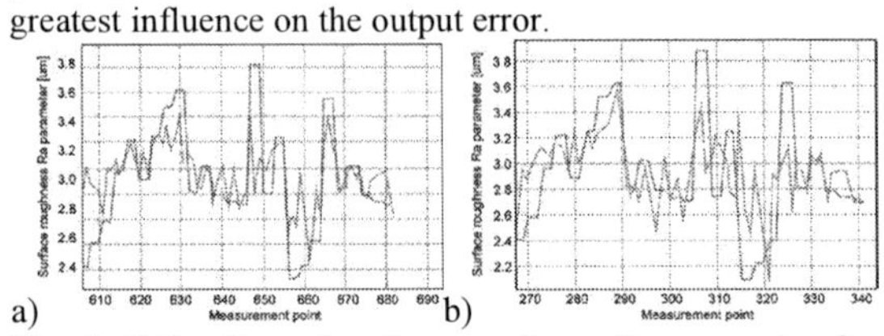

Fig. 2. Estimation of surface roughness Ra parameter for the part of training (a) and testing sets (b)

Neural network was then applied as an estimator of a surface roughness Ra parameter. In Fig.2 there is presented training and testing of neural network. The error of estimation for different cutting parts (uncoated sintered carbides, coated TiN and TiAlN) was less then $0.4\,\mu m$ (16%). The results of the estimation authorize one to draw a conclusion that digital images of a machined surfaces contain enough information of a surface Ra parameter to apply them in monitoring.

9. CONCLUSION

An innovative methodology for a quantitative surface inspection has been presented. There was used optical method for the acquisition of a machined surface digital image. This image was decomposed with the use of singular value decomposition and the eigenvectors were used as features describing the image. For the feature selection and surface Ra parameter estimation there have been applied neural networks. The results of estimation confirm the usability of a SVD in a digital machined surface image description.

10. REFERENCES

Hassibi B., Stork D. (1992). Second Order Derivatives for Network Pruning: Optimal Brain Surgeon, Neural Information, ISBN:1-55860-274-7, Processing Systems–92,

Storch B. (1989). Influence between the Corner of the Edge and Machined Material (in Polish), ISBN 0324-9352, Monograph 8, Technical University of Wroclaw

Zawada-Tomkiewicz A. (2003). Neural Networks in Surface Roughness Estimation, pp. 173-179, ISBN 83-911675-7-7, 3rd Conference on Computer Recognition Systems, KOSYR 2003

Zawada-Tomkiewicz A., Tomkiewicz D. (2004). The Application of Optimal Brain Surgeon Method for optimization of tool wear estimator structure, pp. 218-225, ISBN 83-204-3006-2, Polyoptimization and CAD, Vol. III, WNT, Warszawa

Zawada-Tomkiewicz A., Storch B. (2004). Introduction to the Wavelet Analysis of a Machined Surface Profile, pp. 91-100, ISSN 0137-4478, Advances in Manufacturing Science and Technology, Vol. 28, No 2

Zawada-Tomkiewicz A., Storch B. (2006). SVD Analysis of a Machined Surface Image for the Tool Wear Estimation, pp. 171-176, ISBN 9985-894-92-8, DAAAM Baltic Industrial Engineering Tallin

Annals of DAAAM for 2006 & Proceedings of the 17th International DAAAM Symposium, ISSN 1726-9679
Editor B. Katalinic, Published by DAAAM International, Vienna, Austria 2006
150th Birthday of Nikola Tesla (1856-1943) Inventor of Mobile Robot Teleoperation

INTERNET IN FUNCTION OF MANAGING MONETARY TRANSPARENCY OF TRANSITIONAL COUNTRIES

ZIVKO, I.

Abstracts: *Transparency means something clear and distinctly, and on that way monetary policy with its objectives, activities and results must be presented to public. Precisely for that reason it's important that central bank makes an information which must satisfy a few criteria like as: easy access (information available to all), relevance, quality and reliability (information must be fair, timely, complete and presented in clear and simple terms and be available). But more important things is that central banks transfer transparency from theory to practice. This paper will analyze monetary transparency in central banks communication of select transitional countries through Internet in practice (Croatia, Bosnia and Herzegovina, Macedonia, Bulgaria, and Serbia and Montenegro).*
Key word: *communication, transparency, monetary authority, ex-ante and ex-post test of transparency*

1. INTRODUCTION

In every there are several functions of central banks: issue national currency, acts as banker to both government and private banks "lender of last resort", and supervise the financial system (Mishkin, 2004). The most important functions of central bank are to formulate, to adopt, and to control national monetary policy in the purpose to implement macroeconomic policies. To realise all of these functions, transparency is very important thing for central bank.

Transparency of monetary policy means that the central bank provides important information on its goals to it's public, strategy and policy decisions as well as its procedures in an open, clear and timely manner (Geraats, 2001). Transparency helps the public to understand the monetary policy which one country has been realized. Better public understanding makes the policy more credible and effective.

This paper will analyse monetary transparency of central banks in transitional countries through the Internet as media in communication (Hanke, Sekerke, 2000). Also, we will direct attention on ex-ante and ex-post testing monetary transparency through Internet at central banks of select transitional countries in practice (Croatia, Bosnia and Herzegovina, Macedonia, Bulgaria, Serbia and Montenegro).

Our hypothesis in this paper is that central banks of transitional countries, in spite of all problems, works very hard to insure good monetary transparency. The aim of this work is to analyze timeless and frequency in using Internet for managing monetary transparency in transitional countries. On the basis of those results we will give recommendation to improve that situation, especially when we talk about accessibility of information through the Internet.

2. MONETARY TRANSPARENCY AND INTERNET

In terms of monetary policy, the most important is to make the right decisions. The right decision is the central issue, but not the starting-point for monetary authorities. First of all, you have to properly prepare the public, and the financial markets.

Monetary policy must be seen as a progression from communication to action, from announcement to decision.

There are two forms of central banks communication: internal (inside of central bank) and vertical communication. The communication dimension of monetary policy is reflected through vertical communication - communication between central banks and public. In this form of communication central banks are faced with a problem of managing monetary transparency.

Monetary authorities must made transparency communications with public and financial market players. So central banks must be able to present their information in communication on clearly and precisely terms. On the other side, public must be able to form those information to interpret and understand effects of monetary policy strategy (figure 1.).

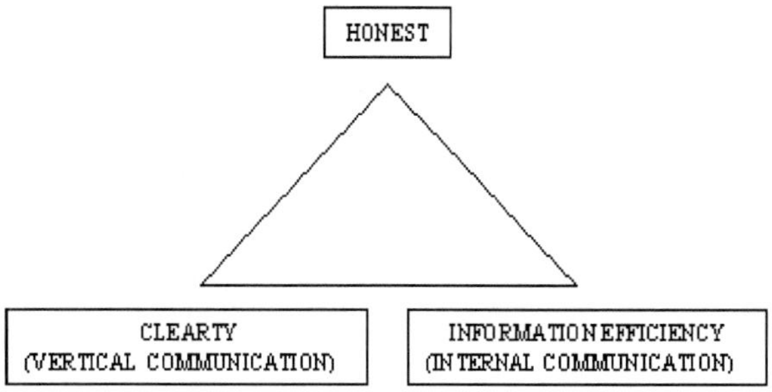

Fig. 1. The transparency triangle of monetary policy strategy

Central banks in their choices of monetary policy strategy are located in different points of this triangle. Triangle shows trade off between principles of monetary transparency. Honesty in monetary policy present degree to which representation of information employed in external communication correspondents to the actual structuring of information adopted internally. To uphold the closest possible correspondence between the internal efficiency to the internal communication and effectiveness and clarity in external communication central banks are faced with trade offs (Winkler, 2000).

There are, also some other advantages of adoption transparency in working of central banks:

1. transparency reduces "information asymmetries" between central bank and private sector
2. transparency could alter the incentives that central banks face as they attempt to overcome information asymmetries and influence private sector perceptions
3. transparency also include greater flexibility to stabilise economic disturbances
4. reduces output volatility
5. reduces inflation.

In communication central banks may use different channels of communication. Usually, we can see this information in newspapers, television or theirs own publication. Today this information about monetary strategy central banks may make available on theirs web site or Internet. On that way central bank may used all benefits which offers Internet network. Development of monetary authorities communication and public and manage monetary transparency through Internet in transitional countries is determinate with introducing IT (for all selected countries that is 1995.) and establishing Low of Central Banks those countries which obligate central banks on transparency communication with public.

Country	Number of computer per 1000 people	Number of Internet users per 1000 people
Bosnia and Herzegovina	45	58
Bulgaria	59	283
Croatia	190	293
Macedonia	65	78
Serbia and Montenegro	48	147
EMU	421	443

Table. 1. Number of computer and Internet users per 1000 people in 2004, for selected countries.

But more important is that central banks transfer transparency from theory to practice, for example, with annuconing time and schedule of meeting decision-making body, publishing central bank an inflation or monetary report of some kind and etc on Internet.

3. EX-ANTE TESTING MONETARY TRANSPARENCY

To test ex-ante transparency in working of central banks in transitional countries we shall analyse estimation of Code of Good Practices on Transparency in Monetary and Financial Policies in those countries. The Code has developed Interim Committee and Bank for International Settlements 1998. The Code of Good Practices on Transparency in Monetary and Financial Policies identifies desirable transparency practices for central banks in their conduct of monetary policy and for central banks and other financial agencies in their conduct of financial policies. The Code focuses on:

1. clarity of roles, responsibilities and objectives of central banks and financial agencies
2. the processes for formulating and reporting of monetary policy decisions by the central bank and of financial policies by financial agencies
3. public availability of information on monetary and financial policies and
4. accountability and assurances of integrity by the central bank and financial agencies.

Results of analyses availability of these key elements of Code on central banks' web sites are present in table 2.

Criteria	Fill up
clarity of roles, responsibilities and objectives of central banks	Yes
the processes for formulating and reporting of monetary policy decisions by the central bank	Yes
public availability of information on monetary policy	Yes
accountability and assurances of integrity by the central bank	Yes

Table 2. Results of analyses availability of elements of Code of Good Practices on Transparency in Monetary and Financial Policies on central banks' web sites selected transitional countries.

The objective and related responsibilities of central banks is clear, simple and simplified on web-site. This objective and related responsibilities of the central banks are also disclosed and explained in theirs annual report, bulletins, newsletters and legislation published in the State Gazette, as well as on theirs web site in languages of nations and English.

4. EX-POST TESTING MONETARY TRANSPARENCY

In order to test ex-post transparency of central banks we can apply a simple test which use Hanke and Sekerke in them analyse. If central bank give current information on in English about balance sheet, off-balance sheet transactions and it's enabling laws on web site it's transparent. For as, is important frequency and timeliness of presentation balance sheet. Frequency is defined as the time interval at which dates are reported, and timeliness as the lag between a reporting period ended and when date are published. Results of testing frequency and timeliness are present in tables 3 and 4.

Frequency	Result
weekly or more often	-
monthly or more often	yes
less often than monthly	-

Table 3. Frequency of balance sheet publication by central banks of transitional countries.

Timeliness	Result
Lease the one month elapsed	yes
Lease the two month elapsed	-
Lease the three month elapsed	yes
Lease the six month elapsed	-
Lease the one year elapsed	-

Table 4. Timelines of balance sheet publication by central banks of transitional countries.

5. CONCLUSION

Valuation of transparency of central banks in transitional countries we can summarized in a few following sentence. Central banks has got a web site and written publications (annual report, bulletins, and newsletters). All publications which central banks publishing are available on central banks web site on English and mother languages. Publication has frequency from one to three months depend on the type of publication. It's clear that Low of Central bank incorporate articles about transparency in monetary policy of transitional countries. Central banks must work very hard on all aspects of monetary transparency.

6. REFERENCES

Internation Monetary Fund (IMF), *Code of Good Practices on Transparency in Monetary and Financial Policies: Declaration of Principles, Available from:* http://imf.org *Accessed:* 2006-05-05

Geraats, P. M. (2001) *Why adopte transparency? The publication of Central Bank Forecasts*, ECB, Working paper No. 41, ISSN 1561-0810

Mishkin, F. (2004). *The Economics of Money, Banking and Financial Markets*, Addison Wesley, ISBN 0-321-20463-8, New York

Hanke, S., Sekerke, M. (2002), *A shroud of secrecy, The Central banking Publications LTD.*, Ouarterly Journal, Vo. XIII, No. 1,

Winkler, B. (2000), *Which kind of transparency? On the need for clarity in monetary policy-making*, ECB, Working paper No. 26, ISSN 1561-0810

Annals of DAAAM for 2006 & Proceedings of the 17th International DAAAM Symposium, ISSN 1726-9679
Editor B. Katalinic, Published by DAAAM International, Vienna, Austria 2006
150th Birthday of Nikola Tesla (1856-1943) Inventor of Mobile Robot Teleoperation

THE INFLUENCE OF THE ICC PROFILE ON THE GRAPHIC REPRODUCTION QUALITY

ZJAKIC, I.; BOLANCA, S. & BARISIC, M.

Abstract: *Colour management is a system which transfers the information on colour from one device to another The information transfer happens so that the system reproduces the colours we see on monitor onto the print with minimal information loss about the colour. The basic aim of colour management in printing is to achieve the least possible difference or almost none between the original presented on the monitor and the obtained print. ICC profiles are the basic component of such system. They enable the correct reproduction of colours during their conversion from one system into another. As the independent colour space ICC has chosen CIE L*a*b* colour space, because it is possible to reproduce in it the colour range of any device or of any technology taking into consideration that such a device has a specific ICC profile. ICC colour profiles define the characteristics for colour reproduction on each device and in each technological process. ICC profiles that are completely open colorimetric and operationally independent systems for colour management are the components of the ICC size.*

Key words: *ICC Profiles, Colorimetry, Printing, Quality.*

1. INTRODUCTION

1.1 International Color Consortium

The device profiles are used to transform the information about colours created on another device into the original form of the colour space of another device. The acceptance of that size by the producers of the operational systems allows the end users to move profiles and images with certain profiles transparently among the different operation systems. Such a system allows great flexibility to users and producers.

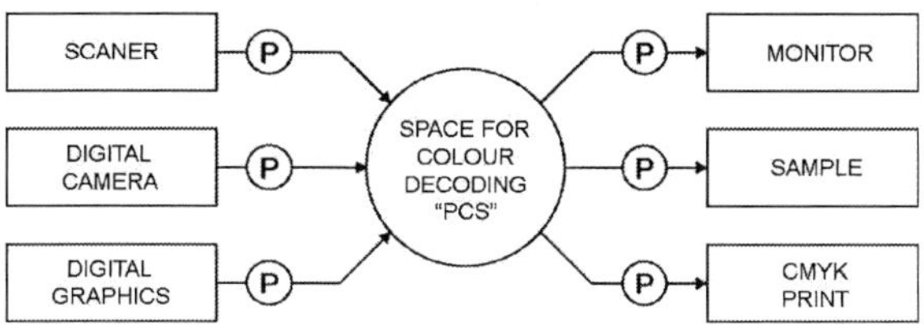

Fig. 1. Schema of colour transformation among the different devices

1.2 Components of Color Management System

Colour management is the activity that in open technological systems for digital communication takes care of optimal colour reproduction. The optimal colour reproduction is done by means of the CMS (Colour Management System). The colour management system CMS can be understood as the simultaneous transformer of colour feelings among the different devices and technologies, that is, among the different colour spaces. The transformation system uses independent colour specification that can be interpreted by each technology. These are most often reference colour models CIE-XYZ or CIE L*a*b*. In principle the question is about the prescription adaptation so that the colour perception would always be the same.

The basic function of CMS is to ensure the clear colorimetric transformation among the primary colour systems included into the reproduction data flow. In order to achieve the objectivity, the following tools are necessary:

1. Application software to generating the input and output profiles and the corresponding "colour management" technology.
2. Profiles which describe the reproduction characteristics of the input and the output devices colorimetrical.
3. Working station (CMM = Color Management Module) that performs the colour data transformation itself from one colour space into another by means of the colour profile.
4. Application that can use different functions of the working station that would start the colour conversion process.

The key components of the functional CMS system are the colour profiles for the system of the input and the output units and the CMS modules for colour management. ICC profile size specification defines the standard structure for the input and the output units that is independent on the platform. It describes very precisely and correctly how the profile should look like and which information from the header of the data file to the concrete colour data had to be included in it.

The movable ICC modul for colour management on the level of the operating systems offers great advantages" all the programs if they are compatible with the ICC sizes, use the same mechanism for colour control and give the identical results independently on the used working station.

2. EXPERIMENTAL

The testing in the paper was the testing of the quality perception with 32 examinees. It was performed so that in visual evaluation the examinees had given points from 1 to 5 to great number of samples with different ICC profiles and they determined the profile nearest to the original presented on the monitor. The samples were marked at the back and this fact was unknown to the examinees.

The testing was performed by the method of binocular simultaneous synchronizing under the D50 illumination with the observing distance of the original of 1 meter on monitor. Before the testing the calibration of the CRT monitor was performed which had the defined white dot at 6500K with the gamma of 1,8.

22 samples for comparison were printed in the offset printing technique on the Heidelberg machine Speedmaster 102F. On two printed sheets 8 samples were printed among which the white areas of 1,5 cm exist. The mutually differ on artificially modified ICC profiles for particular colours.

For each colour on the print (CMYK) at 50% halftone value of the optimal reproduction, ICC profile is modified for 2%, 4%, 6% and 8% as shown in figure 2. 6 samples were printed on one sheet with the white areas of 1.5 cm in which the ICC profile is modifies for equal values in the areas of brightness. Standardized motive according to ISO suggestions was used as the sample.

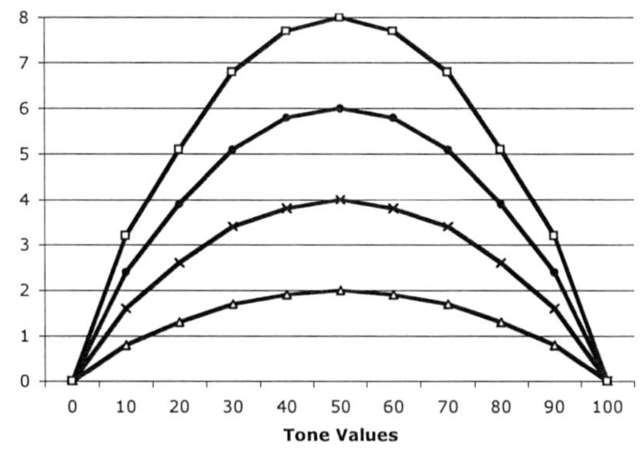

Fig. 2. Presenationf of the values of the artificially modified ICC profiles

3. RESULTS

After the performed calculations the following results were obtained:

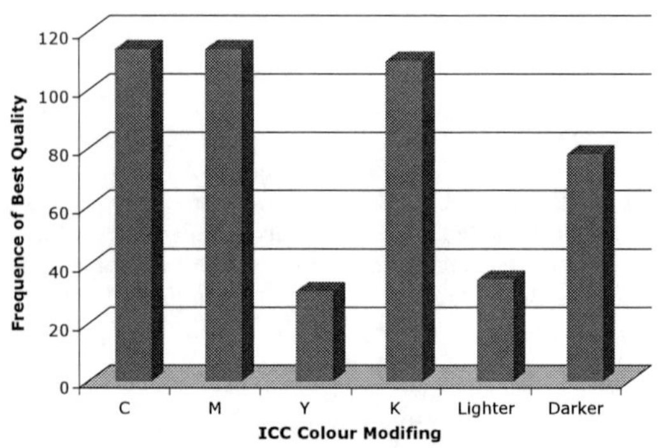

Fig. 3. Frequency of appearance of the modified ICC profiles in perception of the printed product quality

4. DISCUSSION

The perception of the graphic reproduction depends on the reproduction quality of the original. The reproduction of the original depends on the devices that are used in the production chain. The testing was performed with the aim to find out the connection between the influence of the device and the influence of a human on the reproduction of the original. As the majority of the examinees scored the samples almost equally, the results were analyzed by observing the samples of the examinees. The choice of the sample which would be the nearest to the original, depends on the subjective perception of the examinees. A man has a specific visual system and the fact is that the permeability of light up to his eye depends on various wave lengths of the incoming energy. The lens of the eye absorbs mostly the wave lengths of about 400 nm. Samples with the dominant yellow tone got the smallest number of points, although the yellow colour has the strongest stimulus frequency, that is, a man notices it most quickly.
It is visible from the figure 3 that the greatest majority of the examinees have given the greatest number of points to the samples on which cyan, magenta and black tones were dominant. From all the diagrams can be seen that each of the examinees has given points at least once to samples with earlier mentioned dominant tones. Repetition and scoring the samples with cyan, magenta and black dominant tones has been continuously recreated with all the examinees. All of them, who scored the yellow sample, scored it only once, most often with the smallest number of points. In scoring, one should take into consideration that the halftone values are changed in dependence on the used printing substrate and ink. It is possible that they increase or decrease. The decrease will result in the decrease of inking density, saturation and print chromaticity, while the increase will result in the increase of the inking density, saturation and print chromaticity. As the values of the artificially modified ICC profile increase, the aberration of the samples from the original presented on monitor will also increase.

5. CONCLUSION

The aim of the graphic production is to obtain the most qualitative reproduction of some original. Technological operations in the production are mainly lead by densitometric and spectrophotometric as well as with visual methods. The most reliable of them is the method of binocular simultaneous synchronization. Introduction of the digitalization has brought new solutions and the new problems which appeared by appearance of greater and greater number of the producers, technique and technology which becomes more accessible on the market, but with the different quality. With such system opening the prepress becomes multidisciplinary.
Colour which we see depends on different devices such as digital apparatuses, scanners, monitors, printers etc. Each of them has its own colour range. In regard to that the aberrations will appear in colour reproduction in greater or smaller degree. It was the reason that caused the development of the colour management system CMS as well as the development of ICC profiles. ICC profiles are the basic components of such system and they serve for colour information transfer from one colour space into another from the input to the output device with the minimal colour information loss. In the open system the technological processes are digitalized and managed colorimetrically with the known input and output parameters.
This paper is based on the fact that without the colour management in the open technological systems it is difficult to obtain the optimal colour reproduction. It is found out that ICC profiles change the perception of the graphic product quality so, that the human eye notices mostly the fall of the product quality by the change of ICC profile in cyan, magenta and black, and by the change of ICC profiles by which the product is experienced as darker and then lighter. The smallest product quality fall was found at the change of ICC profiles that define yellow colour.
Under the optimal reproduction it is understood that regardless the used devices and technologies the perception of colours in final reproduction is always the same. On the other hand, it is impossible to conclude that the perception of colour will be the same because the people observing some reproduction cannot experience it in the same way because of their own visual system, which proves the testing performed in this paper.

6. REFERENCES

Thompson E.. (1995). *Colour Vision*, Routlege, 0-415-07717-6, USA

Birren F., (1998). *Color Human response*, John Wiley & Sons, 0-471-28864-0, USA

Wyszecki G. & Stiles W. S. (2000). *Color Science: Concepts and Methods, Quantitative Data and Formulae*, Wiley Classic Library, 0-471-02106-7, USA

Fehrman K. R.; (2004). *Color, the Secret Influence*, Pearson Education, 0-13-035859-2, USA

Berns, R. S. (2000). *Principles of Color Technology*, John Wiley&Sons, 0-471-19459-X, USA

Annals of DAAAM for 2006 & Proceedings of the 17th International DAAAM Symposium, ISSN 1726-9679
Editor B. Katalinic, Published by DAAAM International, Vienna, Austria 2006
150th Birthday of Nikola Tesla (1856-1943) Inventor of Mobile Robot Teleoperation

DRIVERS' AGGRESSIVENESS AND ROAD TRAFFIC SAFETY

ZUBER, N.; BOCKAL, Z. & KAVRAN, Z.

Abstract: *Road traffic safety is a complex, interdisciplinary issue. The driver, as one of the factors, has significant influence on the functioning of traffic and the quality of relations in the safety system. There is a certain level of aggressiveness in traffic. The carried out survey analyzed the drivers' behaviour and the level of aggressiveness in road traffic flows. The obtained results indicate that the tendency to aggressive behaviour is inversely proportionate to age and drivers' experience. A series of activities need to be undertaken in order to exert positive influence on the traffic safety, primarily the education and information of media about the adverse consequences of aggressive behaviour.*
Key words*: traffic safety, drivers' aggressiveness, survey*

1. INTRODUCTION

Reducing the casualties in traffic and the number of traffic accidents in general to an acceptable level is a very difficult and demanding task. The basis for the calculation of traffic accident costs lies in the economic and social benefit which results from the avoided accident, that is, when nobody is killed, there are no seriously or lightly injured persons, and the material damage is avoided. Because of all this, the protection of traffic participants has nowadays grown into universal responsibility and exceeds the borders of individual countries, mainly due to mobility which is, undoubtedly a universal category. Road traffic safety is a complex, interdisciplinary issue in which various factors play an important role and interact significantly with one another. The traditional classification of safety measures on roads into three groups (infrastructure, technical characteristics of vehicles, and participants' behaviour in traffic) indicates an approach which does not consider sufficiently the influence of measures on the system as a whole and the possible side effects that can reduce efficiency. The reduction of the number of victims would mean improvement of this system in such a way that the system errors occur less frequently and/or that they may be compensated within the system, i.e. to create conditions which would reduce the consequences when the accident occurs. One of the segments of the complex safety system is, no doubt, a good and confident driver. The driver is the key factor of road traffic safety system, which directly influences it by behaviour, experience, psychophysical capabilities, knowledge about traffic and skills in managing the vehicle. The driver affects significantly also the traffic functioning and the quality of relations in the safety system. The society needs to protect the traffic participants from those who endanger the health and lives of others, and must invest into the systems which will provide the necessary conditions for this, and not only into measures to reduce the fatalities in traffic.

2. PROBLEM OF AGGRESSION IN TRAFFIC

Car driving, under the influence of a number of factors, can become a stressful activity. Hurry, impatience, jams, competition with other drivers, the desire to show off, supplementing the feeling of inferiority, and immaturity, wrong actions of other drivers are just some of the factors which cause frustration and stress in traffic, which in turn result in hostile, destructive behaviour and aggression. Misperceptions and wrong estimates and misbehaviour, as well as sudden, wrong or insufficiently fast decisions in the condition of stress may be fatal. Aggression can be characterized by hostile feeling or hostile behaviour towards other traffic participants. Aggressive drivers impose their behaviour upon others, not paying attention to their rights in traffic and their desires. The reaction to such behaviour is often countering with similar behaviour as a necessity to survive on the roads. Retaliation for aggressive behaviour of other drivers additionally makes the already dangerous situation even worse and more complex. It is very questionable where the limits of socially acceptable behaviour are. Special problem lies in the fact that insufficient attention had been paid until now to the issues of driver rage and aggression in traffic, and that this is precisely the reason for extremely high potential danger. A special group of young motorists commits bigger mistakes in driving due to their behaviour than due to ignorance or driving techniques! They are generally more exposed to risk, not only because of their poorer driving experience but also because they are less capable of controlling their feelings. Anyway, aggressive behaviour is not specific either of a certain age, group or gender. The cause is when the motorists exceed their limits, and even when this happens, the hostility toward others as a rule is not personal, but rather a consequence of the previous frustration and stress. However, the basic question is whether aggressive tendencies in road traffic can be completely suppressed? Aggression, as a form, should not become part of the drivers' behaviour, especially of the young ones. The rage and aggression in drivers should be fought against with all the known and available means and methods, but also new means and methods should be found in order to eliminate this problem. For this purpose studies and research should be planned and implemented in order to record the actual condition among the widest population of motorists. According to results obtained in this manner, the fight against aggression and wild behaviour of traffic participants should be conceived and organized.

3. PURPOSE AND OBJECTIVES OF THE SURVEY

It is a fact that traffic accidents do not occur only because of ignorance or insufficient skills, but rather precisely because of the behaviour that is unacceptable in traffic. We are witnessing every day the lack of culture in traffic, extreme carelessness and absence of tolerance towards other traffic participants, needless nervousness and unacceptable reactions of drivers and other participants. Conditionally speaking, namely, traffic accident as an unexpected event caused by misperception, inattention or inexperience might be tolerated, but it would be quite unacceptable to tolerate traffic accidents caused by unacceptable behaviour of traffic participants. The intention of the survey is to use a representative sample of over three thousand surveyed drivers of all age groups and genders, to determine the level of their socially acceptable or unacceptable behaviour. This survey, and the obtained and presented results were to assist in considering one segment of safe traffic flow – the drivers' behaviour in road traffic. The objective was not to test general aggressiveness, but rather the level of aggressiveness as socially unacceptable behaviour in traffic. There was a total of 32 questions. For each question three possible answers were offered and an adequate number of points. The total number of points and characteristics that determine the behaviour and level of aggressiveness in traffic are presented in Table 1. There were 3500 correctly filled ones for processing.

Points	Behaviour
0	acceptable behaviour
1 to 13	acceptably aggressive behaviour
14 to 39	aggressive behaviour
40 to 69	extremely hostile behaviour
70 and more	warlike behaviour

Table 1. Distribution of points

4. RESEARCH RESULTS

Figure 1 presents the share of participants in the survey procedure according to age, with the share of surveyed female drivers being 17 percent. Out of the total number of the respondents according to the criterion of driver's experience: 563 respondents or 16.09% of respondents have one year of driver experience, 911 respondents or 26.03% have between two and four years of driver's experience, and 2026 respondents or 57.88% have over five years of drivers' experience.

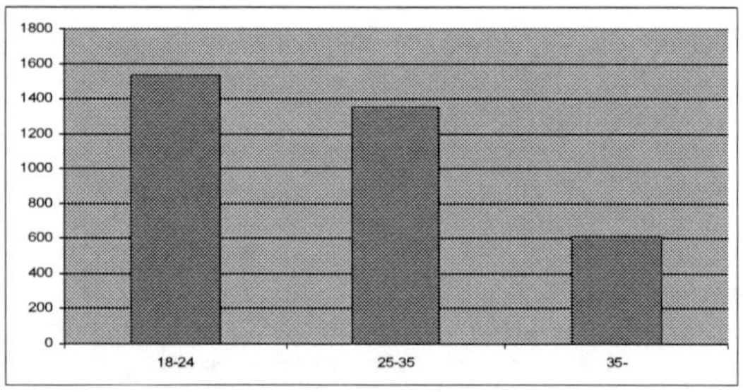

Fig. 1. Share of drivers according to age

Points	Number of respondents	Age		
		18 - 24	25 - 35	over 35
0	55	21	16	18
1 - 13	619	208	229	182
14 - 39	1653	654	708	291
40 - 69	638	343	249	46
70 -	535	308	152	75

Table 2. Survey results –age variable

Points	Number of respondents	Drivers' experience		
		1	2 - 4	5 -
0	55	4	16	35
1 - 13	619	116	125	378
14 - 39	1653	183	1015	455
40 - 69	638	117	182	339
70 -	535	143	133	259

Table 3. Survey results– drivers' experience variable

4.1 Results of female drivers

The results of female drivers show a tendency to less aggressive behaviour. The study of relation between the age of female drivers and the tendency to aggressive behaviour gives results that indicate that there is little negative dependence between the age of female drivers and the tendency to aggressive behaviour, and negative dependence between the drivers' experience and aggressive behaviour in traffic.

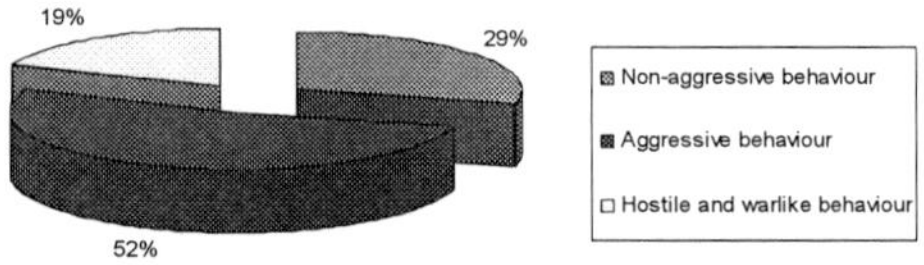

Fig. 2. Level of aggressiveness among surveyed female drivers

4.1 Results male drivers

The results of male drivers tend towards more aggressive behaviour. The study of the relation between the age of male drivers and the tendency to aggressive behaviour gives results that indicate that there is little negative dependence between the age of the drivers and the tendency to aggressive behaviour, and negative dependence between the drivers' experience and aggressive behaviour in traffic.

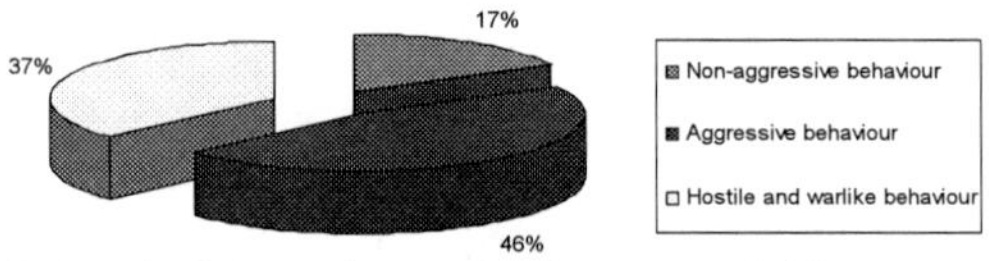

Fig. 3. Level of aggressiveness of the surveyed drivers

6. CONCLUSION

Based on the obtained and processed survey results, the tendency to aggressive behaviour is inversely proportionate to age and drivers' experience, whereas male persons are more prone to aggressive behaviour in traffic than female drivers. Obviously, more time and more attention should be paid to aggressive behaviour and aggressive drivers if the traffic safety level is to be increased. A wide range of measures, activities and methods that have to be undertaken and that could positively influence the traffic safety, range from education and information of the media about adverse consequences of aggressive behaviour, usage of cameras at points of expected aggression, development of the system of assistance and self-assistance for aggressive drivers, modifications of curricula in training the driver candidates and their training for safe participation in traffic, even to the final legal definition of aggressive behaviour as extremely dangerous behaviour and introduction of severe penalties for those who behave like that on the road.

7. REFERENCES

Asbridge M, Smart RG, Mann RE (2003), The 'homogamy' of road rage: understanding the relationship between victimization and offending among aggressive and violent motorists, *Violence Vict,* Vol. 18, No. 5, October 2003, 517-531, ISSN 0886-6708

Parker D., Lajunen T. & Stradling S. (1998). Attitudinal predictors of interpersonally aggressive offences on the road, *Transportation Research Part F,* Vol. 1, No. 1, August 1998, 11-24, ISSN 1369-8478

Shinar D. (1998). Aggressive driving: the contribution of the drivers and the situation, *Transportation Research Part F,* Vol. 1, No. 2, December 1998, 137-160, ISSN 1369-8478

Smart R, Mann R, Goldbloom D, (2005). Road Rage, Are Our Patients Driving Angry? *Psychiatric Times*, Vol. 22, No. 4, April 2005, ArticleID 162100161, ISSN 0893-2905

Zvonarević, M. (1989). *Social Psychology*, Školska knjiga ISBN 86-03-9958-7, Zagreb

Annals of DAAAM for 2006 & Proceedings of the 17th International DAAAM Symposium, ISSN 1726-9679
B. Katalinic, Published by DAAAM International, Vienna, Austria 2006
150th Birthday of Nikola Tesla (1856-1943) Inventor of Mobile Robot Teleoperation

DIGITAL ADAPTATION OF CUTTING PARAMETERS DURING HIGH SPEED MILLING

ZUPERL, U.; CUS, F. & KIKER, E.

Abstract: *A combination of off-line feedrate optimization and on-line adaptive force control is used to maintain a reference peak force during end milling for safe, accurate, and efficient machining. An adaptive control system is built which controlling the cutting force and maintaining constant roughness of the surface being milled by digital adaptation of cutting parameters. In this way it compensates all disturbances during the cutting process: tool wear, non-homogeneity of the workpiece material, vibrations, chatter etc. The basic control principle is based on the control scheme (UNKS) consisting of two neural identificators of the process dynamics and primary artificial controller.*

Key words: *machining, force control, adaptive control with optimization, cutting conditions.*

1. INTRODUCTION

In this paper, neural adaptive controller is developed and some experiments with the neural control strategy are carried out. The results demonstrate the ability of the proposed system to effectively regulate peak forces for cutting conditions commonly encountered in end milling operations. Force control algorithms have been developed and evaluated by numerous researchers. Among the most common is the fixed gain proportional integral (PI) controller originally proposed for milling by (Liu & Wang, 1999). (Lee & Lin, 2000) proposed an adjustable gain PI controller where the gain of the controller is adjusted in response to variations in cutting conditions. The purely adaptive model reference adaptive controller (MRAC) approach was originally investigated by (Liu & Wang, 1999). These controllers were evaluated and implemented by (Balic, 2000). Both studies found that adaptive control alone cannot effectively control cutting forces. There is no controller that can respond quickly enough to sudden changes in the cut geometry to eliminate large spikes in cutting forces. Therefore, we implement on-line adaptive control in conjunction with off-line optimization.

2. STRUCTURE OF THE SYSTEM FOR OPTIMIZATION AND ADAPTIVE CONTROL

The basic idea of this approach is to merge the off-line cutting condition optimization algorithm and adaptive force control (Figure 1). The objective of the developed combined control system is keeping the metal removal rate (MRR) as high as possible and maintaining cutting force as close as possible to a given reference value. Combined control system is automatically adjusted to instant cutting conditions by adaptation of feedrate. When spindle loads are low, the system increases feeds above and beyond pre-programmed values, resulting in considerable reductions in machining time and production costs. When spindle loads are high the feed rates are lowered, safeguarding cutting tool from damage and breakage. Sequence of steps for on-line optimization of the milling process is presented below.

1. The recommended cutting conditions are determined by ANfis (adaptive neuro-fuzzy inference system) models, which are basic elements of the software for selecting the recommended cutting conditions (Cus & Balic, 2000).
2. The pre-programmed feed rates determined by off-line optimization algorithm are sent to CNC controller of the milling machine
3. The measured cutting forces are sent to neural control scheme,
4. Neural control scheme adjusts the optimal feedrates and sends it back to the machine,
5. Steps 1 to 3 are repeated until termination of machining.

The neural adaptive force controller adjusts the feedrate by assigning a feedrate override percentage to the CNC controller on a 4-axis Heller, based on a measured peak force.

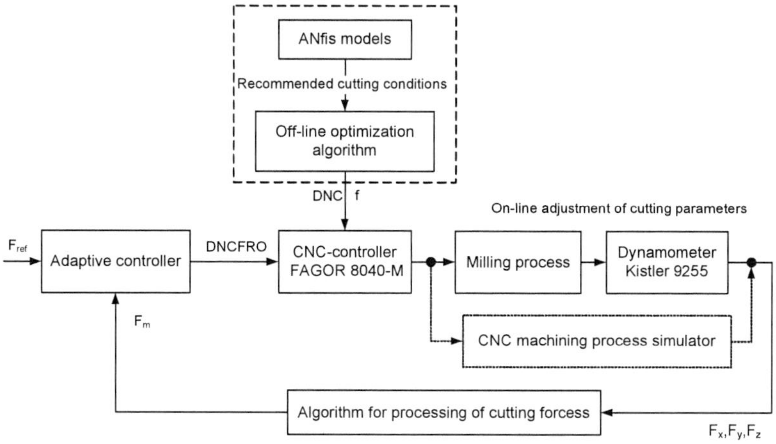

Fig. 1. Structure of combined system for off-line optimization and adaptive adjustment of cutting parameters.

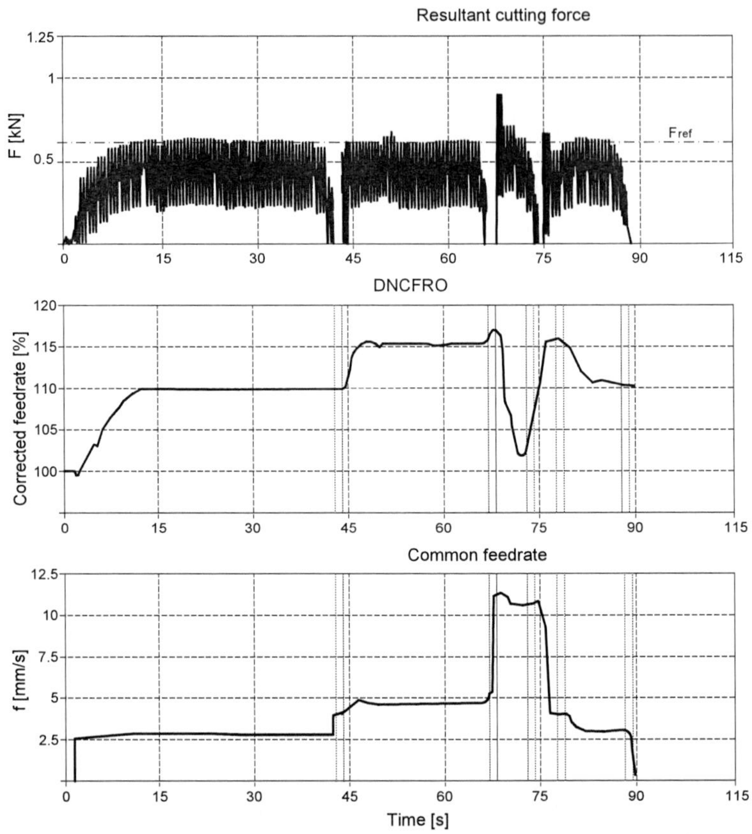

Fig. 2. Machining of irregular profile by off-line optimizing of cutting conditions and adaptive adjusting of feedrate.

The basic control principle is based on the control scheme (UNKS) consisting of three parts. The first part is the loop known as external feedback (conventional control loop). The feedback control is based on the error between the measured (F) and desired (Fref) cutting force. The second part is the loop connected with neural network 1 (NM-1), which is internal model of process dynamics. It acts as the process dynamics identifier. The third part of the system is neural network 2 (NM-2). The NM-2 learns the process inverse dynamics.

A CNC milling simulator is used to evaluate the controller design before conducting experimental tests. The CNC milling simulator tests the system stability and tunes the control scheme parameters. The simulator consists of a neural force model, a feed drive model and model of elasticity. The neural model predicts cutting forces based on cutting conditions and cut geometry. The feed drive model simulates the machine response to changes in desired feedrate. The elasticity model represents the deflection between the tool and the workpiece.

3. TESTING OF ADAPTIVE CONTROL SYSTEM

To examine the stability and robustness of the proposed control strategy, the system is first analysed by simulations using LabVIEW´s simulation package Simulink. Then the system is verified by experiments on a CNC milling machine for Ck 45 steel workpieces with variation of cutting depth (workpiece with irregular profile). The experiment with small and large step changes is run to test system stability over a range of cutting conditions (Cus & Balic, 2003). The system remains stable in all experiments, with little degradation in performance (Figure 2). In presented experiment, the UNKS increases the feedrates to obtain peak forces close to 650N. The slower response of the neural control scheme is noticeable at the beginning of cut one and three. The experimental results show that the milling process with the designed controller has a high robustness, stability, and also higher machining efficiency than standard controllers.

4. CONCLUSION

On the basis of the cutting process modeling, off-line optimization and feed-forward neural control scheme (UNKS) the combined system for off-line optimization and adaptive adjustment of cutting parameters is built. Applicability of methodology of adaptive adjustment of cutting parameters is experimentally demonstrated and tested on a 4-axis CNC milling machine. Experiments have confirmed efficiency of the adaptive control system, which is reflected in improved surface quality and decreased tool wear.

5. REFERENCES

Balic, J. (2000). A new NC machine tool controller for step-by-step milling. *Int. j. adv. manuf. technol.*, Vol. 2, pp. 399-403.

Cus, F. & Balic, J. (2000). Selection of cutting conditions and tool flow in flexible manufacturing system. *The international journal for manufacturing science & technology*, Vol. 2, pp. 101-106.

Cus, F. & Balic, J. (2003). Optimization of cutting process by GA approach. *Robot. comput. integr. manuf.*, Vol. 19, pp. 113-121.

Lee, T.S. & Lin, Y.J. (2000). A 3D Predictive Cutting-Force Model for End Milling of Parts Having Sculptured Surfaces, *International Journal of Advanced Manufacturing Technology*, Vol. 16, pp. 773-783.

Liu, Y. & Wang, C. (1999). Neural network based adaptive control and optimisation in the milling process, *International Journal of Adv. Manufacturing Technology*

Annals of DAAAM for 2006 & Proceedings of the 17th International DAAAM Symposium, ISSN 1726-9679
Editor B. Katalinic, Published by DAAAM International, Vienna, Austria 2006
150th Birthday of Nikola Tesla (1856-1943) Inventor of Mobile Robot Teleoperation

AUTHOR INDEX

A

B

C

D

E

F

G

H

I

J

K

L

M

N

O

P

Q

R

S

T

U

V

W

Y

Z

Annals of DAAAM for 2006 & Proceedings of the 17th International DAAAM Symposium, ISSN 1726-9679
Editor B. Katalinic, Published by DAAAM International, Vienna, Austria 2006
150th Birthday of Nikola Tesla (1856-1943) Inventor of Mobile Robot Teleoperation

SUBJECT INDEX

F

G

H

I

J

K

L

M

N

O

P

Q

R

S

T

U

V

W

X

RESEARCH IS WHAT I'M DOING WHEN I DON'T KNOW WHAT I'M DOING.

WERNER VON BRAUN

- Acceptance without proof is the fundamental characteristic of Western religion, Rejection without proof is the fundamental characteristic of Western science. Gary Zukav, "The Dancing Wu Li Masters"
- Any sufficiently advanced technology is indistinguishable from magic. Arthur C. Clarke
- Basic research is like shooting an arrow into the air and, where it lands, painting a target. Homer Adkins
- Every revolutionary idea - in science, politics, art, or whatever - evokes three stages of reaction in a hearer:
  1. It is completely impossible - don't waste my time.
  2. It is possible, but it is not worth doing.
  3. I said it was a good idea all along.
- Every successful person has had failures but repeated failure is no guarantee of eventual success.
- I do not believe that it is possible to create living art out of anything but the direct visual experience of nature, combined with the heritage of a Tradition, unless it be by the practice of magic ritual. Michael Ayrton 1921-1975
- In theory, there is no difference between theory and practice. In practice, however, there is.
- Murphy's Law of Research: Enough research will tend to support your theory.
- Never worry about theory as long as the machinery does what it's supposed to do. R. A. Heinlein
- No committee could ever come up with anything as revolutionary as a camel anything as practical and as perfectly designed to perform effectively under such difficult conditions. Laurence J. Peter
- Not until a program has been in production for at least six months will the most harmful error then be discovered.
- Practical sciences proceed by building up; theoretical sciences by resolving into components. [Saint] Thomas Aquinas ca 1225-1274
- Putt's Law: Technology is dominated by two types of people: Those who understand what they do not manage. Those who manage what they do not understand.
- Rule of Creative Research: (1) Never draw what you can copy. (2) Never copy what you can trace. (3) Never trace what you can cut out and paste down.
- Science is the meeting place of two kinds of poetry: the poetry of thought and the poetry of action. George Agostinho da Silva
- Science is what happens when preconception meets verification.
- So, He was a great teacher. But what did he publish? From an Israeli academic
- The goal of Computer Science is to build something that will last at least until we've finished building it.
- The goal of science is to build better mousetraps. The goal of nature is to build better mice.
- The habit of apprehending a technology in its completeness-, this is the essence of technological humanism, and this is .what we should expect ed ucation in higher technology to achieve. I believe it could be achieved by making specialist studies the core around which are grouped liberal studies which are relevant to these specialist studies. But they must be relevant; the path to culture should be through a man's specialism, not by-passing it.... A student who can weave his technology into the fabric of society can claim to have a liberal education; a student who cannot weave bis technology into the fabric of society cannot claim even to be a good technologist [Lord] Eric Ashby 1904-
- Theft from a single author is plagiarism. Theft from two is comparative study. Theft from three or more is research.
- To get to know, to discover, to publish-this is the destiny of a scientist. Frangois Arago 1786-1853
- To know / To know what to do / To do it / To make it known. Savoir / Savoir faire / Faire Faire savoir. Department of Physics, Trinity College, Dublin
- Watson's Law: The reliability of machinery is inversely proportional to the number and significance of any persons watching it.
- What is a magician but a practicing theorist? Obi-Wan Kenobi
- Whenever anyone says, "theoretically," they really mean, "not really". Dave Parnas
- There is no national science just as there is no national multiplication tables; what is national is no longer science. Anton Chekhov (1860-1904)

AFTER ALL IS SAID AND DONE, A HELL LOT OF A LOT MORE IS SAID THAN DONE